Medium/Heavy Duty Truck Engines, Fuel and Computerized Management Systems

2nd Edition

Medium/Heavy Duty Truck Engines, Fuel and Computerized Management Systems

2nd Edition

Sean Bennett

Centennial College
Toronto, CANADA

THOMSON ™

DELMAR LEARNING

Australia   Canada   Mexico   Singapore   Spain   United Kingdom   United States

THOMSON

DELMAR LEARNING

Medium/Heavy Duty Truck Engines, Fuel and Computerized Management Systems 2nd Edition
Sean Bennett

Vice President, Technology and Trades SBU:
Alar Elken

Editorial Director:
Sandy Clark

Acquisitions Editor:
David Boelio

Development:
Dawn Jacobson

Marketing Director:
Cyndi Eichelman

Channel Manager:
Fair Huntoon

Production Director:
Mary Ellen Black

Production Coordinator:
Dawn Jacobson

Project Editor:
Dawn Jacobson

Art/Design Specialist:
Cheri Plasse

Editorial Assistant:
Kevin Rivenburg

Library of Congress Cataloging-in-Publication Data
Bennett, Sean.
   Medium/heavy duty truck engines, fuel and computerized management systems / Sean Bennett.—2nd ed.
      p. cm.
Includes index.
ISBN-13: 978-1-4018-1499-1
ISBN-10: 1-4018-1499-9
   1. Trucks—Motors (Diesel)
2. Trucks—Fuel systems.
3. Trucks—Motors—Computer control systems.   I. Title.
TL210.B4143 2004
629.25'06—dc22
                                          2003066343

NOTICE TO THE READER

Contents

Photo Sequences

Preface

While the primary focus of the second edition of *Medium/Heavy Duty Truck Engines, Fuel and Computerized Management Systems* continues to be current truck diesel engines, its spectrum has been expanded somewhat to cover the smaller diesel engines used to power light-duty trucks, automobiles and off-highway equipment. The reason for doing this has been the demand for a textbook offering up-to-date coverage of the full range of mobile diesel engines for those college programs that do not distinguish between truck and general diesel programs.

This edition has been enlarged to cover the rapidly changing engine technology used in trucks today. New chapters on three recently introduced electronic common rail systems, multiplexing electronics, emissions testing, and alternate fuel systems have been added. To make way for these new technologies, coverage of some of the hydromechanical diesel fuel systems that are becoming obsolete has been reduced, but not eliminated. While some would argue that there is little evidence of pressure-time/common rail and mechanical unit injector fuel systems on our roads today, they continue to be addressed by NATEF standards and ASE task lists, and remain a component of college curricula. These factors were considered sufficient justification to retain at least some coverage on these older technologies in this edition.

The rate of innovation in truck technology generally and diesel engines in particular has made this one of the most exciting fields to study. This rate of change is a real challenge to the learner, and programs of study must attempt to keep pace with technological innovation if they are to have relevance. A good program of study should not only be as up-to-date as possible but should emphasize performance learning outcomes: there is little point in merely understanding concepts if they cannot be put into practice. Today's truck technicians must become lifelong learners if they are going to be successful. This textbook is designed to suit the learning requirements not just of those entering the field but also those certified and active in it.

A large textbook such as this is actually the product of the many students, technicians, and teachers, located coast to coast throughout the United States and Canada, who liaise with me on a regular basis. Their input can be as brief as pointing out a simple typo or as complex as submitting detailed suggestions for improvements. In this respect, I would like to thank all those friends and colleagues who make the effort to contact me with feedback, but I am especially indebted to Darrin Bruneau of Freightliner LLC Training Department and Bernie Andringa of Skagit Valley College, Mount Vernon, Washington.

In the preface of the first edition, I said that the truck technician today was required to be smarter, better trained, computer literate, work more accurately and to higher ethical standards than those of previous generations. None of this has changed in five years, but a trend toward specialization is becoming more noticeable. Because of the training and equipment required to work on the complex electronics networked off a data bus in a modern truck, a greater proportion of diesel engine diagnostics and repair is being undertaken by the OEM dealer service facilities. This means that truck technicians employed by fleets and independent garages tend to be responsible for first level diagnosis only. Extended warranty has also played a role in directing repair work to OEM dealers, though some of the larger fleets often strike purchase agreements that permit them to undertake warranty operations. There is no doubt that the role of technicians is changing and is strongly influenced by their employers: the challenge is to remain on top of technological change even when not directly exposed to it. In addition to its more obvious use as an entry-level, diesel program textbook, this book attempts to help technicians maintain technical currency once they are in the field.

FEATURES OF THE TEXT

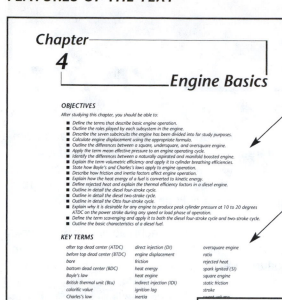

Chapter

4

━━━━ Engine Basics

OBJECTIVES

After studying this chapter, you should be able to:

- Define the terms that describe basic engine operation.
- Outline the roles played by each subsystem in the engine.
- Describe the seven subcircuits the engine has been divided into for study purposes.
- Calculate engine displacement using the appropriate formula.
- Outline the differences between a square, undersquare, and oversquare engine.
- Apply the term mean effective pressure to an engine operating cycle.
- Identify the differences between a naturally aspirated and manifold boosted engine.
- Explain the term volumetric efficiency and apply it to cylinder breathing efficiencies.
- State how Boyle's and Charles's laws apply to engine operation.
- Describe how friction and inertia factors affect engine operation.
- Explain how the heat energy of a fuel is converted to kinetic energy.
- Define rejected heat and explain the thermal efficiency factors in a diesel engine.
- Outline in detail the diesel four-stroke cycle.
- Outline in detail the diesel two-stroke cycle.
- Outline in detail the Otto four-stroke cycle.
- Explain why it is desirable for any engine to produce peak cylinder pressure at 10 to 20 degrees ATDC on the power stroke during any speed or load phase of operation.
- Define the term scavenging and apply it to both the diesel four-stroke cycle and two-stroke cycle.
- Outline the basic characteristics of a diesel fuel.

KEY TERMS

after top dead center (ATDC)	direct injection (DI)	oversquare engine
before top dead center (BTDC)	engine displacement	ratio
bore	friction	rejected heat
bottom dead center (BDC)	heat energy	spark ignited (SI)
Boyle's law	heat engine	square engine
British thermal unit (Btu)	indirect injection (IDI)	static friction
calorific value	ignition lag	stroke
Charles's law	inertia	swept volume
clearance volume	joule	
combustion pressure	kinetic energy	
compression ignition (CI)	manifold boost	
compression ratio	mean effective pressure (MEP)	
cylinder volume	naturally aspirated (NA)	
diesel cycle	Otto cycle	

Objectives

Objectives at the beginning of each chapter introduce the main topics of the chapter and define what should have been learned by studying the material.

Key Terms

A list of the terminology that will be introduced in the chapter appears after the chapter Objectives. Definitions for these terms can be found when they first appear in the chapter and also in the glossary.

---

10   Chapter 1

FIGURE 1–5 ADS TechCert Logo. (Courtesy of ADS)

support the automotive equipment service technician. Membership to the SAE and STS is inexpensive and opens access to detailed information on just about every technical aspect of any vehicle. The address for the SAE and STS is:

Society of Automotive Engineers
400 Commonwealth Drive
Warrendale, PA 15096
Telephone: 1–800–STS–9596
http://www.sae.org

The **Truck Maintenance Council (TMC)** division of the **American Trucking Association (ATA)** sets standards and practices in the industry. Their *Recommended Practices Manual* should be available in every truck shop; most of the procedure and practice is consensually agreed to by OEM and member experts. The TMC's address is:

Truck Maintenance Council
American Trucking Associations, Inc.
2200 Mill Road
Alexandria, VA 22314
Telephone: 1–703–838–1763
http://www.trucking.org

PROPRIETARY ASSOCIATIONS

Many OEMs have in-house professional associations whose objective is educating technicians, keeping them up-to-date, and maintaining high professional standards. The oldest of these is probably the Detroit Diesel Corporation (DDC) Guild. Membership is available for technicians working in DDC service dealerships. Its testing is designed to challenge the technician to keep

FIGURE 1–4 Logo used by an ASE certified training program. (Courtesy of ASE)

WARNING: ASE certification examinations are intended to test your performance skills. Performance skills cannot be learned from just textbook study. Before challenging any ASE certification, make sure that you have some hands-on knowledge or you will be wasting money.

 PROFESSIONAL ASSOCIATIONS

The Society of Automotive Engineers (SAE) has governed most of the standards in the automotive industry, whether it be for a standard screw thread, the viscosity of a lubricant, or the electronic protocols required to integrate two separate computer control systems on a vehicle. The SAE regulates the industry from both an engineering and technical perspective and produces a vast and comprehensive amount of literature on every subject relevant to a vehicle. Most of these are accessible on hardcopy, through the Internet and on CD-ROM. More recently the SAE has formed the **Service Technicians Society (STS)** with a specific mandate to

Warnings

Warnings appear in the text to advise the reader of things that can go wrong if instructions are not properly followed or if a part or tool is misused.

---

of force to separate ... and combination ...siderable force and pullers with the ap... usually required. In ...ers fail to move the ...lding bead using a ...electrode (such as ...e immediate appli...rk. Whenever such ..., it should be rec...stroy the cylinder ...e welding is per...ny engine compo...down technique to

...gs and threaded oil ...from the cylinder

...pection is covered ...t that an engine is ...s should get in the ...art as it is removed and tagging it when it is important that it is reinstalled in the same location.

all particles and sludge produced by the block boiling are removed.

3. Visually inspect the cylinder block, checking all the coolant passages and ensuring they are clean and unobstructed. Ream or drill out if necessary to dislodge any deposits.
4. Check to see that there are no casting fins or residues that might obstruct coolant flow by removing any casting irregularities with a pry bar.
5. Run a cylindrical wire brush in all the oil passages to ensure they are unobstructed.
6. Flush the oil passages with air and solvent.

MAGNETIC FLUX TEST

It makes sense to **magnetic flux test** engine cylinder blocks, crankshafts, and all connecting rods at every major engine overhaul. These processes are neither expensive nor time consuming and as the consequences of a single warranted engine failure out of 20 overhauls will demolish the profits of the other 19, it is shortsighted to overlook it. Remember, this expense is passed on to the customer, and if it is to be passed over, it is the customer who should make the decision: get a refusal in writing. The usual drawback is not having the equipment on site, but in most cases, the equipment is accessible and machine shops usually pick up and deliver.

 CLEANING AND INSPECTING COMPONENTS

The cleaning and inspection of an engine is a critical stage in the engine reconditioning procedure. If the reason for the engine rebuild is a failure, ensure that the cause is identified before any attempt at reassembly is made.

1. Use a gasket scraper to remove all the gasket material and heavy dirt from the cylinder block. Install the cylinder block into a preferably heated, **soak tank** using a heavy-duty alkaline soak cleaner for a period of one to two hours. One OEM reports that ⅛″ of coolant scale has the insulating properties of 4″ of cast iron. It is therefore important not to skip this procedure.

CAUTION: Use extreme care and wear protective clothing when working with alkaline solutions.

2. Remove the cylinder block from the cleaning or soak tank. Thoroughly flush the cylinder block using a shop **high-pressure washer**, ensuring that

 **ENGINE REASSEMBLY GUIDELINES**

Although engines may be disassembled with little or no reference to a workshop manual (not recommended practice), they should be assembled precisely according to the sequencing in the OEM workshop manual. Maybe the only excuse for not observing this practice could be made by the technician who continually overhauls the same engine series and who keeps abreast of every OEM technical service bulletin (TSB). Even for the experienced engine overhaul technician, using the manual as a guide is simply good practice. For the novice engine overhauler, the workshop manual should guide every move: it is important that short cuts are never experimented with. As every experienced diesel technician will acknowledge, there are a few short cuts to some procedures, but these must involve a zero risk factor, and that requires an intimate knowledge of the engine that only comes with experience. To provide a general guide to an engine assembly procedure would simply reverse the general guidelines for disassembly (Photo Sequence #1 shows the disassembly/reassembly of an engine), and the point is engines cannot be successfully rebuilt according to general guidelines but

Cautions

Cautions throughout the text alert the reader to potentially hazardous materials or unsafe conditions.

Tech Tip

This feature provides a technical tip or special procedure when appropriate. These tips are generally those things commonly done by experienced technicians or recommendations from manufacturers.

---

142   Chapter 9

Cylinder Head Installation

On in-line multicylinder heads, it is usually required that separate heads be aligned with a straightedge across the intake manifold faces before torquing. Failure to observe torquing increments and sequencing can result in cracked cylinder heads, failed head gaskets, and fire rings that will not seal. Because of the large number of fasteners involved, a click type torque wrench should be used. Some OEMs require that a **template torque** method be used: this requires setting a torque value first and then turning a set number of degrees beyond that value using a template or protractor. Cylinder head bolts should be installed lightly oiled. Excessive quantities of oil should be avoided because the excess can drain into the bolt hole and cause a hydraulic lock.

TECH TIP: Installing a cylinder head onto an engine that uses nonintegral fire rings in the cylinder head gasket can be made easier by using four guide studs inserted into cylinder bolt holes. This reduces the chances of a fire ring misalignment occurring during head installation.

INTAKE AND EXHAUST MANIFOLDS

In a CI engine, the **intake manifold** is required to deliver air only to the engine cylinders and is bolted to the cylinder heads enclosing the intake tracts. Because of

the almost universal use of turbochargers on diesels, intake manifold design is less complex than that for naturally aspirated SI engines. This usually means that the runners that extend from the intake plenum can be of unequal lengths and this does not compromise engine breathing due to turbocharging. A tuned intake manifold is one in which the shape and length of each runner is similar and designed to establish optimum gas dynamics for engine breathing. A single box manifold fed off the turbocharger compressor pipe can often meet a boosted engine's breathing requirements. Intake manifolds can either be wet (coolant ports) or dry; the latter is generally used in truck and bus applications. Materials used are aluminum alloy or cast iron, but some OEMs are experimenting with plastics and carbon-based fibers. The gaskets used are usually fiber based and must be able to accommodate component creep and the peak boost pressures. When a single section cast aluminum manifold is bolted to a multicylinder head configured engine, it is critical that the cylinder heads be aligned with a straightedge before they are torqued down. Figure 9–14 shows a sectional view of an engine: note the location of the engine housing components.

EXHAUST MANIFOLD

The function of the **exhaust manifold** is to collect cylinder end gases and deliver them to the turbocharger or in the case of a naturally aspirated engine, directly to

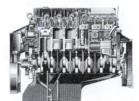

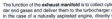
M8-906: note the location of the engine housing components. (Courtesy of

---

248   Chapter 15

PHOTO SEQUENCE 2
Chapter 15 Chassis Dynamometer Installation and Test Procedure

PS2–01   The chassis dynamometer to be used for this test sequence is a double roller, water brake model. Note that it is housed in a bay isolated from the main truck shop. The bay is also soundproofed.

PS2–02   The dynamometer display console is designed to slide on a rail, so it can be positioned to give the operator optimum view of the readouts from the cab during the test.

PS2–03   The truck should be backed onto the dynamometer test bed on to the double rollers: the bogie axle should be measured and the roller spread adjusted prior to backing the truck over the rollers.

PS2–04   When backing the truck up over the dynamometer rollers, the roller brake must be actuated.

PS2–05   Next, the truck chassis must be chained to the deck from both front and rear. Secure the rear chain with a moderate tensile load.

PS2–06   The front axle is secured to the dynamometer test bed by means of a chain assembly that holds the front tires on either side.

PS2–07   The dynamometer test sequence should be performed as per the OEM test procedure and the safety guidelines outlined in this chapter.

PS2–08   Following the test sequence, the engine should be permitted a cool-down period of at least 5 minutes prior to removal from the dynamometer test bed.

PS2–09   Releasing the rear holddown chain prior to removing the truck from the dynamometer test bay.

Photo Sequences

Photo sequences in the text illustrate the steps of several practical procedures. The detailed photographs help the student visualize the steps in some of the diagnostic and repair procedures presented in the chapter.

---

266   Chapter 16

duration of *rapid combustion* are closely associated with the length of the delay period. Generally, as ignition delay is prolonged (for whatever reasons), the rate and resultant pressure rise increase in the second phase. In modern diesel engines with electronically managed variable timing, ignition timing can be controlled so that the period of rapid burning produces peak cylinder pressure at an ideal crank angle under all operating conditions from idle speed/load up to rated speed/load. However, in many older engines with static injection timing (fuel delivery timing could neither be advanced or retarded), this period of high cylinder pressure could occur out of phase with the mechanical dynamics of the crank throw angle.

THIRD PHASE OF COMBUSTION

The third phase of the combustion cycle begins at the moment of peak cylinder pressure (wherever that happens to occur) and ends when combustion is measurably complete; that is, the available fuel has been oxidized. Under conditions of an extended fuel delivery pulse (high engine loads), some portion of the fuel will be injected into the cylinder during this third phase, so the burn rate will be influenced by the rate of injection as well as the mixing rate. Generally, engines are designed with cylinder gas dynamics to enable rapid mixing of fuel and air during the third phase so that the combustion process is completed as early as possible.

AFTERBURN PHASE

Afterburn in the diesel combustion cycle is a period in which any unburned fuel in the cylinder may find oxygen and burn. Most modern engines incorporate this phase in the third phase of combustion because with improvements in cylinder gas dynamics and computer-controlled injection timing, it should not be a significant factor in the present day diesel engine.

DETONATION

Detonation describes the phenomenon that the diesel technician describes as "**diesel knock**" and a driver of any car knows as "ping." In the diesel engine, it is identified by sound caused by intense pressure rise that vibrates the cylinder walls.

Following the ignition of the fuel charge as the flame **propagates** (spreads) through the combustion chamber, the portion of the charge farthest from the primary flame front is subject to both radiated heat and com-

pression caused by flame front. The heat burned portions of the of the primary flame b dition causes an ab and the resultant pre diesel knock. When gine cylinder, the pr the same; however, scribe an oxidation speeds. Detonation combustion occurrin late generally to the nation is seldom a si electronically mana American fuels. It wa engines fueled by h condition can be observed when starting a cold engine during winter conditions.

SUMMARY

- An element is any one of more than a hundred substances that cannot be chemically resolved into simpler substances.
- Elements consist of minute particles known as atoms.
- A mixture is composed of two or more elements and/or compounds, all of which retain their own characteristics and identity.
- A compound is composed of two or more elements combined in definite proportions and held together by chemical force.
- A molecule is the smallest particle of a compound that can exist in a free state and take part in a chemical reaction.
- Electrons carry a negative charge and orbit in shells around the atom's nucleus.
- Protons carry a positive charge and are located in the atom's nucleus.
- Neutrons are electrically neutral and are located in the atom's nucleus.
- Matter can be classified into three states: solid, liquid, or gas.
- Water is the only substance that is familiar in all three states: ice, water, and steam.
- Hydrogen is the simplest of the chemical elements and one of the most reactive.
- Carbon exists in a number of forms and combines to form compounds more readily than any other element.

Summaries

Each chapter concludes with summary statements of the important topics of the chapter. These are designed to help the reader review and study the contents.

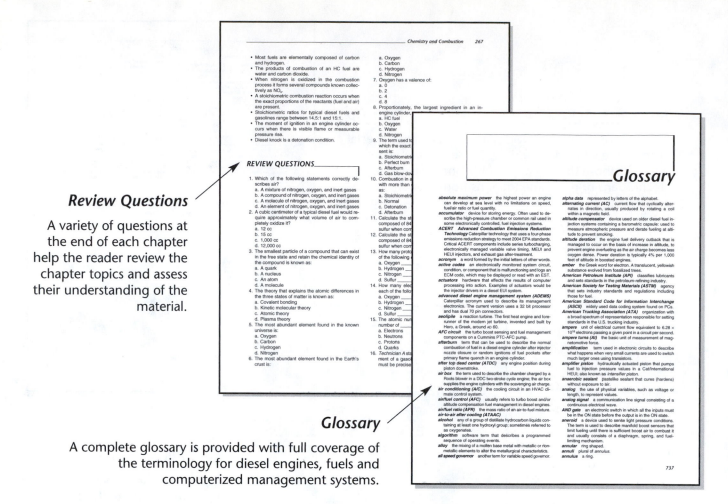

Review Questions

A variety of questions at the end of each chapter help the reader review the chapter topics and assess their understanding of the material.

Glossary

A complete glossary is provided with full coverage of the terminology for diesel engines, fuels and computerized management systems.

NEW TO THE 2ND EDITION

This edition has been substantially updated in all areas but some features that are entirely new to this edition are:

Diesel Engine additions

- Specing power: how much do you need?
- Cracked conn rod technology
- Timing concept gears on Cummins ISX
- Cooled-exhaust gas recirculation (C-EGR)
- CI-4 and PC-10 (CJ-4) HD lubricating oils
- Multistage turbo wastegates
- Urea-injection catalytic converters
- Intebrake engine retarders
- MB constant throttle valve (CTV) retarders

Fuel System additions

- Electrohydraulic injectors for common rail (CR) systems
- Rotary distributor pumps: inlet metering, opposed plunger (Stanadyne and Delphi)
- Rotary distributor pumps: sleeve metering (Bosch VE)
- Biodiesel fuels including B100 and B20

- Hybrid diesel electric powertrains
- Fuel cell operating principles
- Proton exchange membrane (PEM) fuel cells
- Hydrogen fuel
- Regenerative braking

Engine Management additions

- Electronic circuit testing procedure
- Wireless vehicle to base communications
- GPS trilateration
- Multiplexing CAN 2.0 and J-1939
- Data bus troubleshooting
- Ladder switches and FETs
- Electronic rotary distributor pumps
- Electronic common rail (CR) systems (Bosch)
- DDEC Series 60 14 liter engine
- Volvo VECTRO EUI systems
- Caterpillar ACERT technology
- HEUI developments including Caterpillar ACERT and Navistar International 6.0
- Cummins HPI-TP on ISX engines
- Cummins CAPS on ISC and ISL engines
- Latest emission control developments
- Opacity testing using SAE J-1667 procedure

Acknowledgments

INDIVIDUALS

Ray Amlung, Cummins Engineering, Columbus, Indiana

Bernie Andringa, Skagit Valley College, Mount Vernon, Washington

Jim Bardeau, Centennial College, Toronto

Bev Blaine, Cummins Ontario, Toronto

Darrin Bruneau, Freightliner Training, Toronto

George Clark, Centennial College, Toronto

Dave Coffey, Caterpillar Engines, Peoria, Illinois

George Czata, Harper Detroit Diesel, Toronto

Enzo DiPietroantonio, Cummins Ontario, Toronto

Owen Duffy, Centennial College, Toronto

Donald Felterolf, Mack Trucks Incorporated, Allentown, Pennsylvania

Scott Furr, Cummins Atlantic, Charlotte, North Carolina

Terry Harkness, Toromont Caterpillar, Toronto

Alan Hertzog, Mack Trucks Incorporated, Allentown, Pennsylvania

Bob Huzij, Cambrian College, Sudbury, Ontario

Serge Joncas, Volvo Training, Toronto

Gord King, Fanshawe College, London, Ontario

Marty Kubiak, Detroit Diesel Corporation, Detroit, Michigan

Patrick Leitner, Freightliner NE Training Center, New Jersey

George Liidemann, Centennial College, Toronto

Sam Lightowler, Toronto Transit Commission, Toronto

Ben Macaro, Cummins Ontario, Toronto

Kerry Matthews, Cummins Atlantic, Charlotte, North Carolina

Timothy Meyer, Cummins Fuels Division, Columbus, Indiana

Don Mitchell, Case Corporation, Moline, I

Roger Morin, Volvo Training Center, Toronto

John Murphy, GM Sales, Toronto

Stephen Pang, Mack Trucks Training, Toronto

George Parsons, Sault College, Sault Ste Marie, Ontario

Mike Perreira, Cummins Ontario, Toronto

Martin Restoule, Algonquin College, Ottawa, Ontario

Wayne Scott, Freightliner LLC, Portland, Oregon

Craig Smith, Nova Freightliner, Truro, Nova Scotia

Dan Sullivan, Sullivan Solutions, Durham, North Caroline

Gary Van Nederynan, Orion Bus, Oriskany, New York

ORGANIZATIONS AND CORPORATIONS

Association of Diesel Specialists, Kansas City, Missouri

Robert Bosch Corporation, Chicago, Illinois

Caterpillar Engines, Peoria, Illinois

Cummins Engine Company, Columbus, Indiana

Cummins Ontario, Toronto, Ontario

Davco Manufacturing, Saline, Michigan

Detroit Diesel Corporation, Detroit, Michigan

Freightliner LLC, Portland, Oregon

Harper Detroit Diesel, Toronto

Imperial Oil, Toronto, Ontario

Institution of Mechanical Engineers, London, England

Jacobs Manufacturing, Bloomfield, Connecticut

Kent Moore, Warren, Michigan

Mack Trucks Incorporated, Allentown, Pennsylvania

Mack Trucks Canada, Toronto

Mercedes-Benz, Montvale, New Jersey

Mid-Ontario Freightliner, Toronto

International Trucks, Chicago, Illinois

Service Technicians Society, Warrendale, Pennsylvania

Society of Automotive Engineers, Warrendale, Pennsylvania

Stanadyne, Windsor, Connecticut

Sullivan Training Systems, Raleigh, North Carolina

Superflow Corporation, Colorado Springs, Colorado

Toromont Caterpillar, Toronto

Toronto Transit Commission, Toronto

Vibratech TVD, Springfield, New York

Volvo, North America

Williams Controls Incorporated, Portland, Oregon

REVIEWERS

David Bailey
Walla Walla Community College
Walla Walla, WA

David Biegel
Madison Area Technical College
Madison, WI

Jeff Callen
Universal Technical Institute
Phoenix, AZ

Dennis Chapin
Rogue Community College
Grants Pass, OR

Alan Clark
Lane Community College
Pleasant Hill, OR

Danny Esch
Southwest Mississippi Community College
Summit, MS

Terry Harkness
Caterpillar
Utopia, Ontario

Jerry Johnson
Nashville Diesel College
Nashville, TN

Jocelyn Lepage
International Truck and Engine Corp.
Fort Wayne, IN

Terryl Lindsey
Southern Community College
Fort Gibson, OK

Steve Musser
Wyoming Technical Institute
Laramie, WY

Ralph Papidero
New River Community College
Dublin, VA

Leslie Pike
Elizabethtown Technical College
Hodgeville, KY

David Reynolds
Lincoln Technical Institute
Indianapolis, IN

Dennis Vickerman
Southeast Technical Institute
Harrisburg, SD

Section

1

Diesel Engine Fundamentals

Section 1 begins with an introduction to the trucking industry and its technology. This chapter is followed by chapters on tools and safety; study of these is recommended before progressing to actual shop procedures. However, the section is mainly devoted to introducing the diesel engine, beginning with its operating fundamentals and historical development and then examining it on a system-by-system basis. Chapter 14 provides some tips on diesel engine disassembly and reassembly procedures, and the final chapter in the section presents engine run-in and dynamometer testing.

1. Introduction
2. Hand and Shop Tools, Precision Tools, and Units of Measurement
3. Personal and Safety Awareness
4. Engine Basics
5. History of the Heat Engine
6. Power
7. Engine Powertrain Components
8. Engine Feedback Assembly
9. Engine Housing Components
10. Engine Lubrication Systems
11. Engine Cooling Systems
12. Engine Breathing
13. Engine Retarders
14. Engine Removal, Disassembly, Cleaning, Inspection, and Reassembly Guidelines
15. Engine Run-In and Performance Testing

Chapter 1

Introduction

OBJECTIVES

After studying this chapter, you should be able to:

- Describe the overall objectives of this textbook.
- Define the role of the trucking industry in North America.
- Describe some of the recent technological advances that have changed trucks in the past decade.
- Outline the role of the contemporary truck technician.
- Understand the role that the truck technician is expected to play in the delivery of customer service.
- Outline popular customer service trends in the truck OEM industry.
- Describe the qualifications required to practice as a truck or bus technician.
- List some of the professional associations to which truck technicians may belong and identify some of the benefits of each.

KEY TERMS

American Trucking Association (ATA)

Association of Diesel Specialists (ADS)

chief executive officer (CEO)

electronic engine management

Environmental Protection Agency (EPA)

hydromechanical engine management

kaizan

(National Institute for) Automotive Service Excellence (ASE)

original equipment manufacturer (OEM)

Service Technicians Society (STS)

Society of Automotive Engineers (SAE)

total quality management (TQM)

Truck Maintenance Council (TMC)

INTRODUCTION

The primary objective of this textbook is to provide a basic understanding of the truck diesel engine and its fuel management circuits to the extent required by the truck technician. Much of the book deals with electronically managed engines (Figure 1–1). These systems have been handled with a focus on principles of operation, the objective being to help the technician understand the systems rather than address the detail of the procedure required to repair them. Most engine **original equipment manufacturers (OEMs)** mandate that those seeking to repair their computerized engine management systems attend proprietary courses; these courses deliver the information and procedures required to diagnose and repair the systems. Because of the delivery time constraints, these courses seldom touch on the principles of operation. This textbook is very concerned with why and how components and systems function and less concerned with reproducing procedure and specifications that are best obtained from OEM literature and training courses.

The focus of this book is the truck/bus engine and its hydromechanical and electronic fuel management systems. Because of the rapid increase in statutory noxious emission controls required of the commercial highway diesel engine, it has evolved rather differently from its off-highway and marine counterparts over the past decade. Within a 10-year period from 1987 to 1997, truck engine management systems sold in North America evolved from almost 100% **hydromechanical management** to almost 100% management by electronics. Today, all new medium and heavy duty trucks are managed by computers. At the onset of the electronic age there was widespread skepticism, but this has dissipated to almost complete acceptance of computerized management systems by the industry. The facts speak for themselves, and as most of these have something to do with the pocketbook, computer-controlled engines and many other vehicle systems won the day. **Electronically managed engines** cost fractionally more than their hydromechanical counterparts, but last much longer, make vehicles much easier to drive, require less maintenance, and produce much better fuel mileage. **Environmental Protection Agency (EPA)** noxious emission requirements, which triggered the introduction of computerized engine controllers and have since driven their progress, have brought additional benefits of fuel economy and much greater engine longevity.

This textbook attempts to handle engine and fuel management technology from the perspective of the general technician in a modern truck service or fleet operation. The emphasis today is on diagnosis followed by a remove/replace procedure. Diagnosis is a culminating skill. The individual responsible for diagnosis in a truck shop was traditionally widely experienced and probably had many years of field practice. While today's technician is not required to perform that much disassembly/reconditioning, a high level of diagnostic skills is required. Today's technician is expected to be computer literate. Computers are used to diagnose engine malfunctions, program customer and proprietary data to engine control modules (ECMs), and track every aspect of working life in a business. A heavy emphasis is placed on computer literacy in this book, and again, the focus is on principles of operation.

ROLE OF THE TRUCKING INDUSTRY AND THE TRUCK TECHNICIAN

The trucking industry in North America grows by the year. In many ways, new truck sales provide a barometer of how the economy is doing in any given year. The saying "if you got it, a truck brought it" is true for most of the consumer items we purchase. Even if that item was transported by train, boat, or plane for a portion of its journey, trucks would have played a role in pickup and delivery at stages of the journey.

Millions of people are directly and indirectly employed by the trucking industry. Trucks must be designed, built in factories, marketed, and then operated and maintained. The role of the technician in this industry is a small but crucial one. If all the truck technicians in the United States withheld their labor for 2

FIGURE 1–1   *Caterpillar C-12 truck engine. (Courtesy of Caterpillar)*

weeks, a large percentage of the economy would shut down. The trucking industry, besides employing mechanical technicians to maintain and repair equipment, employs drivers, dispatchers, warehouse personnel, people to market services, and people to manage operations.

For some truck technicians, practicing as a technician will be just one of the career roles they play in the industry. A recent **chief executive officer (CEO)** of a major truck OEM described his years with the company, beginning when he started as a janitor, apprenticed as a mechanic, practiced as a technician, and progressed with the company as a service manager, sales manager, production operations manager through upper management, to the CEO position. While this is unlikely to be the objective of most aspiring truck technicians, a recent study indicated that within 5 years of achieving their certification status, 50% of truck technicians no longer worked as hands-on technicians. Most were employed in other positions in the trucking industry. There are many opportunities in management, sales, training, field service operations, and information systems that the qualified truck technician can explore.

Many truck technicians have no desire to do anything but repair trucks. They will, over time, become the experts the industry needs to maintain the ever more sophisticated equipment on the highways. There are engine specialists in the truck garages of North America who know and understand a given engine model better than the engineers who designed it. Over a period of time they learn to diagnose every rattle and burp the engine can produce.

THE TRUCK TECHNICIAN OF THE TWENTY-FIRST CENTURY

An objective of this textbook is to provide an understanding of the operating principles behind the engine and fuel technology of the modern transport truck. For a technician to effectively work on modern electronically managed engines, a large amount of general experience, knowledge of mechanical principles, and product-specific training are required. Persons choosing a career as a truck technician in North America usually begin by attending a one- or two-year college program and then undergoing a period of hands-on training, which culminates in professional examinations conducted by the **National Institute for Automotive Service Excellence (ASE),** the **Association of Diesel Specialists (ADS),** or other licensing agencies. In Canada, most provinces mandate both the structure and duration of apprenticeship and then license techni-

cians by their ability to pass an examination. A proportion of the professional examinations that the technician must pass to practice focuses on the engine, fuel system, and management electronics.

The role of the truck technician has probably changed more in the past 15 years than it did in the previous 40 years. First, there has been a rapid shift from hydromechanically managed systems to electronically managed systems. Next, many labor intensive procedures (such as the out-of-chassis overhaul of major components) have been moved out of the truck garage to a remote remanufacturing center. This remanufacturing center is often located in a jurisdiction where labor rates are low and the remanufacturing processes can be subdivided, allowing lesser skilled but specialist workers to perform most of the labor. Finally, the repair industry has become highly cost conscious and every aspect of the repair procedure is analyzed for efficiency and cost-effectiveness. The truck engine technician of a generation ago who may have diagnosed, disassembled, reconditioned, and tested an engine, is probably responsible today only for the diagnosis, followed by somebody of lesser experience who will remove and replace the engine on a rebuilt/exchange basis. Another factor that has powerfully influenced change in the trucking industry is the improvement in communications technology. Satellite telecommunications are used by many fleet operators to track vehicle location and communicate with onboard electronic systems.

The changes that have occurred during the past 15 years have resulted in a different type of truck and a different kind of truck technician. (Figure 1–2 shows some of the many computer-controlled subsystems on a modern truck.) The modern technician is required to be computer literate, have excellent comprehension of the written word, and because a fundamental role is that of diagnostician, possess a thorough understanding of the various vehicle systems and components. There is no doubt that the technician of today is required to know much more than that of a generation ago and is also required to perform to higher levels of efficiency, competency, and ethics.

The truck technician has always been expected to possess an array of skills. Some of those skills are mechanical and incorporate many abilities that some might regard as specialties in their own right: machinist, plumber, electrician, welder, and truck driver, to name a few. Additionally, the technician is probably expected to have sound computer skills to work with vehicle and office data management systems, play a role in maintaining good customer relations, and have the ability to interpret written technical data with absolute accuracy (Figure 1–3).

Ideally, the truck technician should have a broad base of training that accommodates at some point

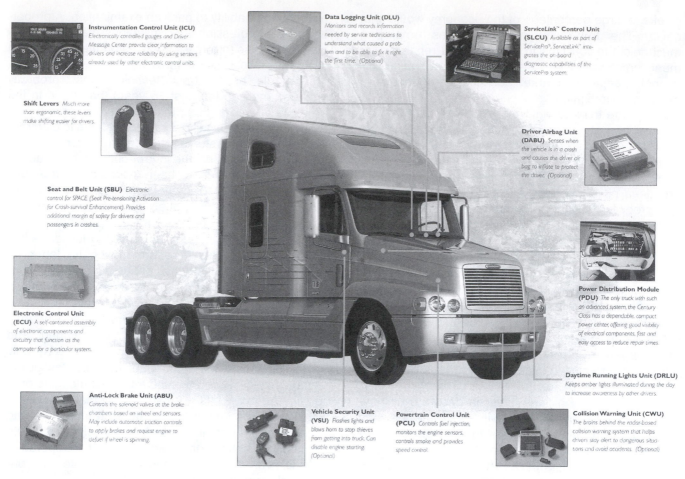

Instrumentation Control Unit (ICU) *Electronically controlled gauges and Driver Message Center provide clear information to drivers and increase reliability by using sensors already used by other electronic control units.*

Data Logging Unit (DLU) *Monitors and records information needed by service technicians to understand what caused a problem and to be able to fix it right the first time. (Optional)*

ServiceLink™ Control Unit (SLCU) *Available as part of ServicePro™, ServiceLink™ integrates the on-board diagnostic capabilities of the ServicePro system.*

Shift Levers *Much more than ergonomic, these levers make shifting easier for drivers.*

Driver Airbag Unit (DABU) *Senses when the vehicle is in a crash and causes the driver air bag to inflate to protect the driver. (Optional)*

Seat and Belt Unit (SBU) *Electronic control for SPACE (Seat Pre-tensioning Activation for Crash-survival Enhancement). Provides additional margin of safety for drivers and passengers in crashes.*

Electronic Control Unit (ECU) *A self-contained assembly of electronic components and circuitry that function as the computer for a particular system.*

Power Distribution Module (PDU) *The only truck with such an advanced system, the Century Class has a dependable, compact power center, offering good visibility of electrical components, fast and easy access to reduce repair times.*

Anti-Lock Brake Unit (ABU) *Controls the solenoid valves at the brake chambers based on wheel end sensors. May include automatic traction controls to apply brakes and request engine to defuel if wheel is spinning.*

Vehicle Security Unit (VSU) *Flashes lights and blows horn to stop thieves from getting into truck. Can disable engine starting. (Optional)*

Powertrain Control Unit (PCU) *Controls fuel injection, monitors the engine sensors, controls smoke and provides speed control.*

Daytime Running Lights Unit (DRLU) *Keeps amber lights illuminated during the day to increase awareness by other drivers.*

Collision Warning Unit (CWU) *The brains behind the radar-based collision warning system that helps drivers stay alert to dangerous situations and avoid accidents. (Optional)*

FIGURE 1–2 *Chassis computer control systems used on a modern truck. (Courtesy of Frieghtliner)*

every vehicle system and component. It is a poor strategy to specialize in one vehicle system and to learn it at the exclusion of all others; every system on a vehicle enables it to function and therefore is to some extent in-

FIGURE 1–3 *Technician using software to conduct an ECM reprogramming procedure.*

terdependent. This interdependence has increased with the introduction of a data bus to optimize the performance of multiple chassis computer-controlled systems. Specialization is desirable only when the "specialist" has a broad base of experience and knowledge in all vehicle systems and then develops expertise in the subject of the specialty.

The nature of the work in truck repair facilities is changing. The fuel injection pump shop specializing in the overhaul and testing of hydromechanical pumps is fast becoming an anachronism as far as the trucking industry is concerned. Truck and engine OEMs offer extended warranties of such length that fleets may employ technicians to perform little other than running repairs such as lighting repairs and vehicle servicing. This means that proportionally, more of the work is being performed in the OEM dealerships. This probably makes sense. The OEM dealership is more likely to have properly trained personnel and specialized equipment to perform the repairs. The per hour labor rates at such a facility are likely to be high but then the financial investment in training and specialized tools is significant.

TRAINING AND THE TRUCK TECHNICIAN

Some of the training a truck technician requires will take place in college. However, much of the most valuable training is less formal. Often the best teachers are not found in classrooms, but are expert practitioners who take pride in helping others by sharing their expertise. Most people are born not knowing much about mechanical technology and, therefore, must learn. The apprentice technician will depend on those with greater experience and knowledge in order to learn. In most cases, knowledge cannot be hoarded or controlled. The student who wishes to learn will usually find a means and will come to resent the individual who tries to obstruct that goal. It is often a measure of the health of a company to observe how it handles the learning and training of its employees. Sound organizations will nurture an atmosphere in which every person understands the benefits of a trained, knowledgeable colleague. Technicians who are disinclined to teach others what they have been taught themselves are usually motivated by insecurity. This type of insecurity is more often a fault of the employer rather than the individual, and it most often occurs in operations in which every person is tiered according to their perceived ability and in which advancement is discouraged.

Many apprentice truck technicians are first employed because they are young and have strong backs. There are many heavy, dirty jobs to be performed in a truck garage, and there is no doubt that a 20-year-old can remove and reinstall a transmission in better average times than a 50-year-old. However, it is in the interest of any employer to develop the potential of apprentice technicians, from the employer point of view, to prepare for the future, and from the learning technician's point of view, to give that person a sense that he or she is gradually acquiring the skills that will result in expertise.

A generation ago, the certified technician may have been able to consider himself or herself a finished product and exempt from further in-school training. This is not so today; the modern technician requires regular upgrading as the technology changes. One major truck OEM uses the Japanese word *kaizan* meaning "continuous improvement" to describe its corporate philosophy, and technicians would do well to adopt kaizan as their motto. The learning process for anyone working with today's technology is a lifelong one. Learning for a truck technician is seldom a process confined to classrooms, although this is an important element of it. The training process for a technician really consists of three stages: formal, self-development, and experience.

FORMAL

Formal training involves some study in classrooms and is structured to produce a result such as knowledge of a practice or procedure, backed by certification or a diploma. Most technicians are required to successfully complete some formal training before they are certified to practice. Additionally, depending on the specialties of their employer's operation, technicians should attend formal upgrading on a regular basis. Formal upgrading courses are usually designed to address some very specific subject matter or procedures and usually last from 1 to 5 days; again they often culminate in testing and the awarding of a certificate.

SELF-DEVELOPMENT

This type of training is self-motivated, so many overlook it. Smart technicians read. It is a good idea to make reading a habit. Reading trade magazines is an excellent means of keeping up-to-date; most truck shops make these available to their technicians but few take advantage of them. There are many excellent publications, some of which cover very specific subject matter, while others cover the industry in general. Some recommended publications include:

Automotive Engineering

Published monthly by the **Society of Automotive Engineers (SAE)** and sent to all of its members, it covers the whole automotive spectrum from an engineering/technological perspective focusing on innovation. Most innovations in the truck, bus, and heavy equipment industries are covered.

Automotive News

Covers the whole automotive industry (including trucks) from the sales perspective. It tracks the monthly and yearly sales data for all vehicles, including each category of truck by class size.

Commercial Carrier Journal (CCJ)

Published monthly and covers the trucking industry from the service and maintenance perspective. It is highly recommended as a means of keeping up-to-date with the changes in truck technology as they happen.

Diesel Progress

Covers the diesel engine industry from the smallest to the largest engines. It is a great magazine for the engine and fuels specialist. Most new diesel engine technologies are profiled in *Diesel Progress* ahead of production.

OEM Service Literature and Electronic Information/Education

OEM workshop manuals are written with the objective of sequencing a procedure and seldom make reading entertaining. However, many OEMs have their own in-house magazines to keep their employees up-to-date, and the technical service bulletins (TSBs) of today are usually presented in a more readable format than they were ten years ago because OEMs realize their importance as an information tool. Many OEMs also provide learning packages deliverable on CD-ROM or Internet/Intranet format; these are often interactive and may result in in-house certification at their conclusion. Technicians should never forget that self-development tends to be rewarding and has a way of enhancing interest while those who do not bother, stagnate.

EXPERIENCE

For the technician, there is no substitute for the hands-on experience of tackling a procedure. Employers generally try to ensure that after investing in sending a technician on a one-week course training for a specific procedure that the technician practice the procedure at the first opportunity. If six months slip by between the training course and the hands-on application that the techniques target, much of the trained content will have been forgotten. Apprentice technicians should make a practice of using their eyes and ears a lot. They should watch successful technicians and learn from them: how they organize their work area, how they strategize each task they undertake, how they test the results of a repair procedure, and how they approach and deal with customers. Novice truck technicians have a steep learning curve to adjust to and during the process it will be important to learn how to think. There is little room for impulse, guesswork, or carelessness in working with today's technology and the consequences of a mistake can be very costly.

CUSTOMER SERVICE AND PROFESSIONALISM

A decade ago, few truck diesel technicians had to concern themselves with what is today known as customer service. Most larger operations had personnel on their payroll to deal with customer service issues and, in fact, many technicians worked with the expectation that the customer adapt to their way of working and certainly never the other way around. The catchphrase was take it or leave it. Many service facilities had signs directed at customers expressly preventing them ac-cess to the working area of service garages usually citing insurance regulations. Things have changed. The pioneer of modern customer service philosophy was an American, Edward Demmings. Demmings's ideas on customer service initially met with little acceptance in North America, but he had an easier time convincing the Japanese. In his lifetime, Edward Demmings acquired almost heroic status in Japan. Japanese companies adopted his teachings and used them to step into the world arena and establish Japan as a major economic force, second only to the United States.

Edward Demmings is credited with coining the phrase **total quality management (TQM).** TQM is used by many companies and often in a sense that has little to do with Demmings' original concepts. It is best described as a whole corporate philosophy that views customer service as being the very reason for that company being in business and includes every employee as a stakeholder. In other words, every person within a corporation is regarded as a member of a team, and every member of the team contributes significantly to the image that the organization projects. The notion that every person in a company had an essential role to play was alien to many North American and European organizations in the 1970s and 1980s, which maintained strictly hierarchical structures in which those at the base of the pyramid tended to be little valued. The Japanese exploited this and essentially, among their many imports to North America, re-imported the teachings of Edward Demmings. Buying a Japanese product was often derided during the 1960s with accusations of low quality; however, this reversed during the 1970s and today, we, more often than not, associate Japanese manufactured products with good quality backed by excellent customer service support.

It should be noted that more than half of the main players in the truck manufacturing industry claim to practice some form of TQM, including the company with the largest market share, which has set many of the trends of the industry over the past two decades, forcing the rest to either keep up or lose market share. TQM has been dismissed as a trend by some and indeed many companies have treated Demmings's teachings as a sort of smorgasbord, picking and choosing those items that they feel might result in an increase in business profitability, and perhaps omitting those items that they feel might cost them money. This is a case of missing the point. But TQM has influenced the way in which all companies do business. Many facilities are offering extended service hours—often at 24-hour, 7 days a week facilities. Service operations are recognizing the crucial role played by the service technician in delivering customer service and many are training their personnel with seminars with titles such as "How to Communicate with Customers" and "Handling the Difficult Customer." Operations that used to discourage customers from even entering the service garage are now welcoming the customer, encouraging them to talk to the technician performing the work, and providing them with waiting facilities more in keeping with a hotel lobby than with what used to be associated with a truck service garage.

This makes for a working environment that can be difficult for the novice technician. A truck owner/operator with 20 years in the business is likely to have learned a thing or two about the technology he operates and can easily intimidate a first-year apprentice truck technician. It is essential to underline the fact that one of the core premises of TQM is honesty: technicians will find that in dealing with customers, the relationship will always be made easier when the approach is open and sincere. Until recently, two of the worst traits of the automotive garage were the fact that many customers felt that either they were being talked down to or that their repairs were misrepresented either to inflate a bill or mask a lack of competence in diagnosing a problem. Using an honest approach will probably not reduce the number of policy adjusted (write-off time) work orders, but it will have the effect of increasing the number of repeat customers. As the competitive environment in which the truck technician must work becomes more so, every person working in it must cultivate customer service techniques and a sense of teamwork. It is simply not enough to be a competent technician anymore.

All technicians should develop a sense of professionalism. One seldom witnesses a doctor or dentist criticizing the handiwork of a colleague who previously worked on the body or mouth in question although there are surely cases where there is justification to do just that. Unfortunately, it is all too easy to hear a technician deriding the abilities of whomever previously worked on a piece of equipment. This is simply unprofessional and reflects as badly on the person doing the criticizing as the person supposedly at fault. It reflects badly on the whole industry. Mistakes happen. Remember that what goes around, comes around. Accept that and the fact that the very best technicians can have bad days, commit errors, and have lapses in concentration. Technicians should make a practice of learning from their own mistakes and those of their colleagues and actively work to improve the level of professionalism in the industry.

QUALIFICATIONS AND CAREERS FOR THE TRUCK TECHNICIAN

Certification for truck technicians in the United States is managed by the National Institute for Automotive Ser-

vice Excellence (NIASE), usually abbreviated ASE. ASE tests for truck technicians are as follows:

T1 Gasoline Engines	60 questions
T2 Diesel Engines	65 questions
T3 Drivetrains	50 questions
T4 Brakes	60 questions
T5 Suspension Steering	50 questions
T6 Electrical System	50 questions
T7 Heating, Ventilation, and Air Conditioning	40 questions
T8 Preventive Maintenance	50 questions
L2 Electronic Diesel Engine Diagnosis Specialist	45 questions

Technicians who succeed in passing tests T2, T3, T4, T5, T6, T7, and T8 qualify for master technician status.

The Association of Diesel Specialists (ADS) also offers voluntary certification and their testing is managed by the ASE, permitting their tests to be held in a large number of centers across the United States and Canada. ADS testing is directed at the fuel pump specialist technician. Truck technicians should be aware that ADS testing addresses hydromechanical systems that are obsolete in current trucks. The following lists current ADS certification test areas:

TC1 Diesel Engine Theory and Operation	70 questions
TC2 Distributor Fuel Injection	60 questions
TC3 In-line Fuel Injection	60 questions
TC4 Rail Fuel Injection	60 questions
TC5 Turbochargers and Blowers	60 questions
TC6 Injectors/Unit Injectors	60 questions

In Canada, technicians must be licensed to practice in most provinces; licensing requires undertaking a five-year apprenticeship followed by a provincially administered examination. Figure 1–4 shows the insignia used by a ASE certified program.

The addresses are:

National Institute for Automotive Service Excellence
101 Blue Seal Dr., SE, Suite 101
Leesburg, VA 20175
(703) 669-6600
http://www.asecert.org

Association of Diesel Specialists
P.O. Box 26487
Overland Park, KS 66225-6487
(913) 851-9840
http://www.diesel.org

FIGURE 1–4   Logo used by an ASE certified training program. (Courtesy of ASE)

WARNING: ASE certification examinations are intended to test your performance skills. Performance skills cannot be learned from just textbook study. Before challenging any ASE certification, make sure that you have some hands-on knowledge or you will be wasting money.

 ## PROFESSIONAL ASSOCIATIONS

The Society of Automotive Engineers (SAE) has governed most of the standards in the automotive industry, whether it be for a standard screw thread, the viscosity of a lubricant, or the electronic protocols required to integrate two separate computer control systems on a vehicle. The SAE regulates the industry from both an engineering and technical perspective and produces a vast and comprehensive amount of literature on every subject relevant to a vehicle. Most of these are accessible on hardcopy, through the Internet and on CD-ROM. More recently the SAE has formed the **Service Technicians Society (STS)** with a specific mandate to

FIGURE 1–5   ADS TechCert Logo. (Courtesy of ADS)

support the automotive equipment service technician. Membership to the SAE and STS is inexpensive and opens access to detailed information on just about every technical aspect of any vehicle. The address for the SAE and STS is:

Society of Automotive Engineers
400 Commonwealth Drive
Warrendale, PA 15096
Telephone: 1–800–STS–9596
http://www.sae.org

The **Truck Maintenance Council (TMC)** division of the **American Trucking Association (ATA)** sets standards and practices in the industry. Their *Recommended Practices Manual* should be available in every truck shop; most of the procedure and practice is consensually agreed to by OEM and member experts. The TMC's address is:

Truck Maintenance Council
American Trucking Associations, Inc.
2200 Mill Road
Alexandria, VA 22314
Telephone: 1–703–838–1763
http://www.trucking.org

PROPRIETARY ASSOCIATIONS

Many OEMs have in-house professional associations whose objective is educating technicians, keeping them up-to-date, and maintaining high professional standards. The oldest of these is probably the Detroit Diesel Corporation (DDC) Guild. Membership is available for technicians working in DDC service dealerships. Its testing is designed to challenge the technician to keep up-to-date with DDC technology, and it offers a number of benefits as a reward for membership.

SUMMARY

- The objective of this textbook is to provide the truck technician with an understanding of the operating principles of modern heavy- and medium-duty engines and their fuel management systems.
- There is a heavy emphasis on the electronically controlled engines that power most contemporary highway trucks.
- It is important for the truck technician to have a good understanding of customer service trends prevalent in the industry.
- Truck technicians become qualified by passing ASE examinations in the United States and provincial certification in Canada.
- Because of the rapid changes in technology, the truck technician should regard education, both self-driven and formal, as an ongoing process.
- Practicing truck technicians should consider membership in both professional and proprietary organizations because they provide an easy means of remaining up-to-date with the trends of the industry.
- For many, becoming a truck technician is just the first step of a career in the trucking industry, which grows at an accelerated rate even during times of recession.
- The truck technician of today must come to terms

with electronic vehicle management systems, shop data tracking systems, and communications technology.

REVIEW EXERCISES

1. Log on to the Internet and search http://www. asecert.org, http://www.diesel.net, and http://www. dieselnet.com. Explore the Web pages of some of the major truck engine OEMs. Select a company that has an e-mail query page and solicit information on one of their products.
2. Contact the ASE online or by mail and request electronic or hard-copy information on their testing criteria and locations.
3. Investigate the role of the trucking industry in your community. Contact a local trucking association and request data on the business activity of its members.
4. Obtain copies of *Diesel Progress,* the *CCJ,* and other trade publications. Remember that many publications make electronic versions available and accessible via the Internet.
5. Contact some OEMs in the automotive and truck manufacturing fields and ask them about their customer service philosophy. A lack of response can sometimes be as informative as a response.
6. Check out the ATA Web site at http://www.trucking.org. Bookmark this and http://www.dieselnet.com into your favorites file and consult weekly.

Chapter 2

Hand and Shop Tools, Precision Tools, and Units of Measurement

OBJECTIVES

After studying this chapter, you should be able to:

- Identify the hand tools commonly used by truck technicians and describe their function.
- Categorize the various types of wrenches used in shop practice.
- Describe the precision measuring tools used by the engine and fuel system technician.
- Outline the operating principles of a standard micrometer and name the components.
- Identify different types of torque wrench.
- Calculate torque specification compensation when a linear extension is used.
- Read a standard micrometer.
- Outline the operating principles of a metric micrometer and name the components.
- Read a metric micrometer.
- Understand how a dial indicator is read.
- Define TIR and how it is determined.
- Understand how a dial bore gauge operates.
- Outline the procedure for setting up a dial bore gauge.
- Describe some typical shop hoisting equipment and its application.

KEY TERMS

bar	outside diameter (od)
calipers	outside micrometer
chain hoist	scissor jack
dial bore gauge	spreader bar
dial indicator	tensile strength
dividers	total indicated runout (TIR)
electronic digital calipers	units of atmosphere (atms)
inside diameter (id)	yield strength
inside micrometer	

INTRODUCTION

This chapter is intended to provide a guide to tools for novice truck technicians. The tools are loosely divided into the categories of hand tools, precision measuring, and shop tools (Figure 2–1). There is also a guide to the contents to a truck technician's toolbox; however, the rookie technician should invest in a minimum of tools before obtaining employment and then develop a tool collection with the job requirements in mind. Some guidance in standard to metric conversion units is also provided. The modern truck technician is usually expected to work on metric engineered engines using specifications generally presented in standard or English values, so it helps to be familiar with both systems.

HAND TOOLS

The technician will need to possess a basic set of hand tools and while some basic guidelines are provided later in this chapter, the tools selected and their quality will be largely determined by the nature of the work. Hand tools vary considerably in price and before spending large sums of money, the technician should determine whether the expenditure is justified by the amount of use they will be put to. Most better quality hand tools carry lifetime warranties; however, this may not cover a tool that wears out, so it is well to question the extent of the warranty offered. Diesel technicians seldom wear out tools and the main problem is usually loss. Because of the high price of hand tools, most technicians learn to check the contents of their toolboxes carefully after completing each job. Most loss of tools is usually the result of carelessness on the part of the technician. Thousands of wrenches are lost every week because they are left on a truck, bus, or car chassis.

OPEN-END WRENCHES

Open-end wrenches have open jaws on either side of the wrench, usually with different sizes at either end and slightly offset (Figure 2–2A). The wrench should be of sufficient quality that the jaws do not spread when force is applied and the jaws made not too bulky to restrict access to difficult-to-get-at fasteners. They may damage softer fasteners (such as brass pipe nuts) because their design permits them to impart force to only two of the six flats of a hex nut.

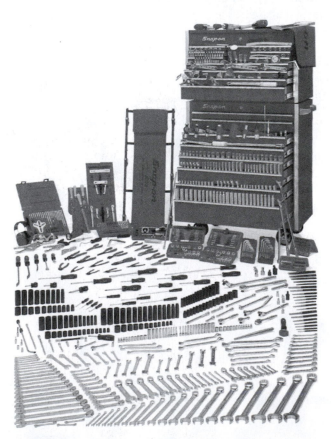

FIGURE 2–1   *A complete tool kit. (Courtesy of Snap-on Tools Corporation)*

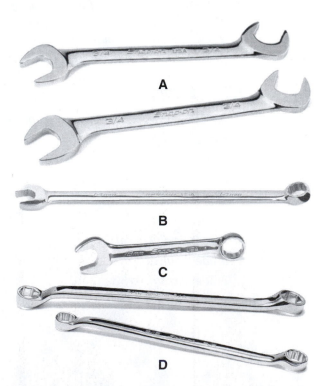

FIGURE 2–2   *Wrenches: A. open-end; B. and C. combination; D. box-end. (Courtesy of Snap-on Tools Corporation)*

COMBINATION WRENCHES

Most technicians would usually own a couple of sets of combination wrenches through the common sizes. A combination wrench is manufactured with a box end and an open end, both of the same nominal size (Figure 2–2B and C). Most truck diesel technicians should only consider better quality combination wrenches that are guaranteed for life. There exists a wide range of prices with lower quality wrenches that tend to be heavy and clumsy (but often as strong as the best quality) and the best quality, which are light and smooth to handle. It is useful to own a set of top quality combination wrenches in sizes up to ¾″ (19 mm) because these are able to access fasteners on engines that heavier, clumsier wrenches do not access. Less expensive but fully warrantied wrenches can be considered for use on sizes larger than ¾″ (19 mm). There are many cheaper, poor quality wrenches on the market, many of them imported. These are seldom guaranteed. Poor quality wrenches are dangerous and should be used by no technician, shadetree or professional.

BOX-END WRENCHES

A box-end wrench surrounds the fastener and may be of hexagonal or double-hexagonal design (Figure 2–2D). As most bolts and nuts are of a hexagonal design, the hex box-end wrench will grip more securely; however, it would be less versatile where access is restricted as it can only fit on the fastener in six radial positions through a rotation rather than the twelve radial positions of the double hex, box-end wrench.

ADJUSTABLE WRENCHES

The adjustable wrench consists of a rigid jaw integral with the handle and an adjustable jaw moved by a worm adjuster screw. The truck technician should probably own a couple of these and then resolve to use them as little as possible. Their advantage is versatility and their ability to sometimes grip a worn fastener. Their disadvantage is that they cause wear because the adjustable jaw never fits tightly to the flats on a hex fastener and it tends to round them out. Never apply excessive force to an adjustable wrench.

LINE WRENCHES

A line wrench is designed to grip to a pipe or line hex nut and act on four of the six flats of its hex. It has the appearance of a box-end wrench with a small section removed so that it fits through the pipe to enclose the pipe nut. The line wrench should be used in place of the open-end wrench to avoid damaging pipe nuts.

SOCKET WRENCHES

Diesel technicians will require complete socket sets in ¼″, ⅜″, and ½″ drive sizes and may consider ¾″ drive. What constitutes a complete set of sockets varies a little by manufacturer but typically ¼″ drive sets are provided with sockets up to ½″ (12 mm), ⅜″ drive up to ¾″ (19 mm), and ½″ drive up to 1¼″ (30 mm). Sockets may be of the hex or double-hex design and enclose the fastener. The socket may be hand rotated by a ratchet or flex bar and power rotated by a pneumatically powered wrench or impact wrench. Impact sockets are manufactured out of softer alloys than those designed to be driven manually, to prevent fracture. Technicians should purchase good quality sockets because the consequence of a failed socket is personal injury. Deep sockets permit access to a nut in which a greater length of the bolt or stud is exposed. A crowsfoot socket is essentially an open-end wrench that can be turned by a ratchet; it grips two of the six flats of a nut and is probably mostly used for final torquing a difficult-to-access nut. A line socket is the socket counterpart to the line wrench. It grips four of the six flats of a nut and its main use is to deliver final torque to a pipe nut.

RATCHETS AND BREAKER/FLEX BARS

Reversible ratchets used in conjunction with sockets are tools much used by any technician. They are used to rapidly turn fasteners by hand and should be of good quality because the consequence of failure is personal injury. However, they are not designed to accommodate high torque loads. The ratchet spur wheel is locked to one direction of rotation by a single or double cog. The spur and cog cannot be observed because they are enclosed in the ratchet head, but this determines the ultimate strength of the tool. A breaker bar (also known as a flex bar, power bar, and Johnson bar) has a grip bar and pivoting drive square to engage with a socket in the same way a ratchet does so they are available in ¼″, ⅜″, ½″, ¾″, and 1″ sizes. A breaker/flex bar can be used to release fasteners that require considerably more force than could be safely applied to a ratchet; however, the use of "helpers" such as a pipe over the handle should be avoided.

TORQUE WRENCHES

Torque wrenches measure resistance to turning effort (Figure 2–3). The objective of torquing fasteners is to ensure that a specified clamping force between two components is achieved. However, an estimated 90% of the applied force to a torque wrench is required to overcome the friction of the fastener thread surface area with about 10% contributing to clamping force (Figure 2–4). Most technicians possess at least

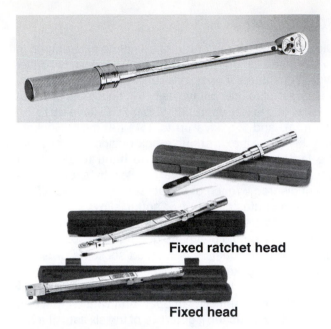

Fixed ratchet head

Fixed head

FIGURE 2–3   *Types of torque wrenches. (Courtesy of Snap-on Tools Corporation)*

one torque wrench and probably do not use it as much as they should. In assembling engine and fuel system components, every fastener should be torqued to specification. Studies indicate that when technicians fail to use torque wrenches, they overtorque fasteners to values 50% to 100% over the specification. This action can damage fasteners and distort components including cylinder blocks and heads. The commonly used torque wrench is the sensory or click type: when the selected torque value is attained, the wrench produces an audible click. Diesel technicians are often required to torque large numbers of fasteners to the same specification such as when torquing cylinder heads to a cylinder block; it makes sense to use the click-type torque wrench when performing this type of procedure. Click-type torque wrenches should always be backed off to a

zero reading after use and their calibration should be routinely checked. Dial-type torque wrenches have circular dial scales with a needle that indicates the applied torque value. These tend to be more useful when torquing sequences of fasteners at different torque values: dial-type torque wrenches are usually higher priced than click-type torque wrenches. Beam-type torque wrenches use a flexible, middle alloy steel shaft (beam) that deflects when torque is applied: a needle pointer indicates the applied torque effort. These tend to be the most economical but have good accuracy and seldom require calibrating. They should be stored carefully as the needle pointer is vulnerable.

Technicians should probably own a ⅜″ and ½″ drive torque wrench of the sensory or click type. The torque wrench calibration should be checked annually with a torque wrench tester. Torque specifications are provided by OEMs in order to obtain reasonably consistent clamping pressures between mated components. Torque values for fuel injection and engine component fasteners are usually factored for lubricated threads.

TECH TIP: Sensory or click-type torque wrenches are set to a specified torque value by setting internal spring tension to a specified torque value either by rotating the handle or by a dial and latch. Always relieve the spring pressure after use; in other words, store the torque wrench with the torque specification set at zero.

Torque wrenches are mainly used with sockets, line sockets, or crows-foot jaw wrenches. Torque readings are a product of the length of the handle measured in feet or inches and the force applied to the handle. For example, a 2-foot (ft.) long torque wrench when applied with 10 pounds (lb) of force would produce 20 lb.-ft. of torque. To express any torque value specified in lb.-ft. in lb.-in. simply multiply by 12. So 20 lb.-ft. is equivalent to 240 lb.-in. When a standard extension (one that is at right angles to the plane of the wrench) is used with a

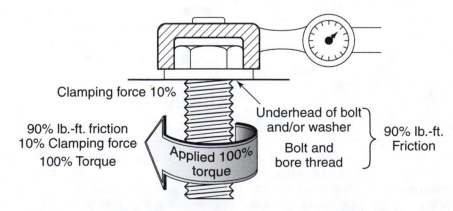

Clamping force 10%

90% lb.-ft. friction
10% Clamping force
100% Torque

Applied 100% torque

Underhead of bolt
and/or washer

Bolt and
bore thread

90% lb.-ft.
Friction

FIGURE 2–4   *Torque to overcome friction. (Courtesy of Navistar International Corp., Designer and manufacturer of International Brand diesel engines)*

torque wrench, no torque reading adjustment is required. If a linear extension (one on the same linear plane as the wrench) is used, the mechanical advantage is increased, and the torque reading should be adjusted using the following formula:

$$TS = \frac{T \times L^1}{\left(L^1 + L^2\right)}$$

T = torque in lb.-ft.

TS = torque scale in lb

L^1 = torque wrench frame length in inches

L^2 = torque wrench extension length in inches

So if a torque specification is 50 lb.-ft. and an 18″ torque wrench is to be used in conjunction with a 12″ linear extension, the following calculation would be used to determine what the correct reading on the torque wrench scale should be:

$$TS = \frac{T \times L^1}{\left(L^1 + L^2\right)} = \frac{50 \times 18}{(18 + 12)}$$

$$TS = \frac{900}{30}$$

TS = 30 lb.-ft.

Torque wrenches may be calibrated in the English system or metric system of Newton-meters.

1 lb.-ft. = 1.356 Newton-meters

1 Newton-meter = 0.7375 lb.-ft.

HAMMERS

Mechanical technicians mostly use ball peen hammers in various different weights. The specified weight of a hammer is the head weight, which starts at ½ lb and should go up to about 4 lb in weight. While there is no place for the carpenter's claw hammer in the technician's toolbox, a 4 lb cross peen hammer can be a useful addition, and the engine specialist should also own a 5 lb rubber mallet and a couple of soft-faced or fiberglass hammers. Safety glasses should be worn whenever using any striking tool. The impact faces of hammers should be inspected regularly and discarded when the face becomes damaged. Properly heat-treated hammers should possess highest hardness at the contact face and be relatively soft behind it to buffer the shock loads. Hammer handles are also important, and a hammer should not be used when its handle is damaged. The handle may be made of hickory, in which case it is susceptible to damage, or steel and integral with the head with a rubber-cushioned grip. Some examples are shown in Figure 2–5 A and B.

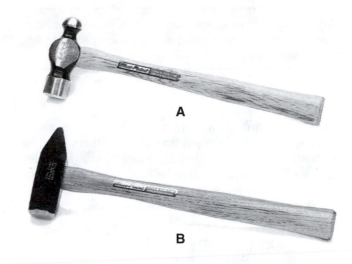

A

B

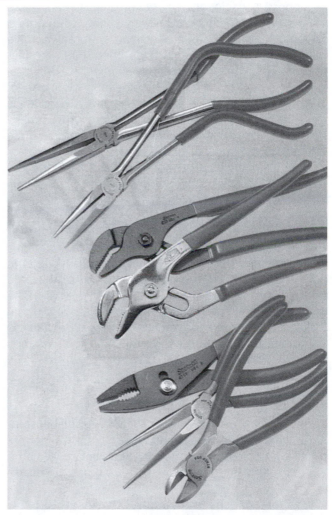

C

FIGURE 2–5 *Types of hammers and pliers: A. Ball peen hammer; B. Cross peen hammer; C. Selection of different types of pliers. (Courtesy of Snap-on Tools Corporation)*

WARNING: Never strike a hammer with another hammer. The hardened impact surfaces can shatter and cause serious injury.

PLIERS

Most technicians will require a large selection of pliers. They are used for gripping and cutting. Pliers used for working on electrical circuits should have insulated handles. Each type of pliers is named, and some examples are needle nose, slip joint, lineman, and side-cutter. Figure 2–5C shows some examples.

SCREW EXTRACTORS

Fasteners occasionally fail when the fastener head sheers. When methods such as welding a nut onto a fastener that has had its hex head sheered off have failed, a screw extractor must be used. First, the fastener must be drilled as centrally as possible to 75% of its depth, then the screw extractor must be inserted. Two types exist: the taper square screw extractor is designed to bite into and grab the bore of the drilled hole; it can be progressively driven into the hole if the edges round out. The left-hand twist screw extractor works by tapping its way into the drilled hole in the fastener as it is turned counterclockwise.

STUD EXTRACTORS

Stud extractors may be used to extract a fastener only when not sheered flush. Two types are used: the collet-type fits over the exposed length of the stud and locks to it as it is rotated counterclockwise. The wedge-type consists of a splined circular wedge that locks to the stud as it is rotated. Wedge-type stud extractors should only be used with hand tools as they are driven eccentrically.

TAPS AND DIES

Taps and dies are designed to cut threads in both standard and metric specifications. Taps cut internal threads in bores; quite often the technician will use a tap to repair damaged threads. Three types are used:

1. Taper tap: used to cut threads to a virgin bore
2. Plug tap: used to finish cut or repair threads
3. Bottom tap: used to cut the final threads in a blind hole

Dies cut external threads. Most are designed with graduated teeth and a taper, enabling them to cut threads to a shaft. It is important to use an appropriate cutting medium. Again, technicians are more likely to

use these to repair damaged threads. In fact, most toolbox quality taps and dies should not be used to cut virgin threads in hardened steels especially if the fastener is critical.

THREAD CHASERS

A thread chaser is a die designed for the sole purpose of repairing minor damage to an existing thread—it cuts in much the same way a die does, but it is not designed to cut new threads.

THREAD PITCH GAUGES

Thread pitch gauges are designed to measure the thread pitch (angle) and number of threads per inch (tpi) of any thread pattern. They are used to determine which tap, die, or fastener is appropriate.

REAMERS

Reamers are rotary cutting tools that enlarge an existing hole to an exact dimension. They are used where a greater degree of accuracy is required than a drill is capable of delivering. Reamers are available in several types: adjustable reamers can be set to cut holes through a range of dimensions, and spiral fluted, taper reamers are designed to be driven by air or electric power tools and enlarge a hole accurately to a specific size.

DRILL BITS

Drill bits are driven by an electric or pneumatic power tool. Twist drills must be machined to the correct cutting pitch (angle) normally by grinding on a fine abrasive wheel. The point angle of a drill bit is the combined angle of the dressed cutting edges, usually about 120 degrees for drilling most steels and cast irons. This would produce a cutting pitch of about 30 degrees. The point angle is increased when harder steels must be drilled. The two cutting edges of the twist drill must be angled identically from the center point of the drill and have the same radial dimension; a slight difference will result in a larger hole than the drill shank specification. Drills sharpened by hand seldom produce exactly sized holes. Drill speeds should be adjusted for the material being cut and the size of the hole.

HACKSAWS

Most technicians will be required to use a hacksaw from time to time. A hacksaw is designed specifically to cut metals. A hacksaw should have a rigid frame and the blade selected should have the appropriate number of teeth per inch (tpi) for the metal to be cut; generally,

TABLE 2-1 *Recommended Cutting Fluid Processes*

METAL	DRILLING	REAMING	TAPPING
Mild steel	Soluble oil/water	Lard	Lard
Middle alloy	Soluble oil/water	Lard	Lard
Cast iron	Dry	Dry	Lard
Brass	Dry	Water	Dry
Copper	Soluble oil/water	Soluble oil/water	Soluble oil/water
Aluminum	Kerosene	Kerosene	Soluble oil/water

the harder the metal, the more tpi required. Better quality hacksaw blades tend to be cost effective as they last until they wear out, while cheaper blades tend to break. Inspect the blades and replace them when dulled or missing teeth are evident. A hacksaw should be used with a light but firm grip. An even horizontal stroke with no rocking will produce the highest cutting rates. A relaxed, calm approach to the cutting task has a way of producing fast results, whereas those who attempt to power their way through end up breaking blades and losing their tempers.

CUTTING FLUIDS

Soluble oil is machinist's oil mixed with water according to the manufacturer's recommendations. Remember that engine oil is not soluble and cannot be used as a substitute. See Table 2–1.

PRECISION MEASURING TOOLS

The following are some examples of precision measuring instruments used in a typical truck service garage (Figure 2–6 and Figure 2–7). Precision measuring tools tend to be high-cost items. They are sometimes provided by employers but technicians frequently using a specific tool may wish to purchase their own.

ELECTRONIC DIGITAL CALIPERS

Electronic digital calipers (EDCs) are a great addition to the toolbox. EDCs are an invaluable tool for the diesel technician who works extensively in the engines and fuels area. These dimensional measuring tools will perform inside, outside, and depth measurements to one ten thousandth of an inch (0.0001″) accuracy. EDCs perform with at least the accuracy of a micrometer and have the advantage of being easier to read.

Additionally, they will perform metric to standard linear conversions at the push of a button. Figure 2–6 shows some of the uses of an EDC.

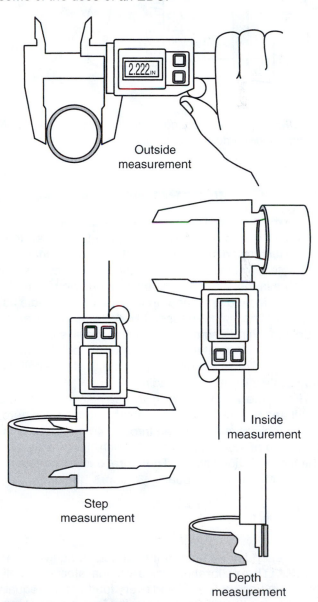

Outside measurement

Step measurement

Inside measurement

Depth measurement

FIGURE 2–6 *Various uses of a set of digital calipers.*

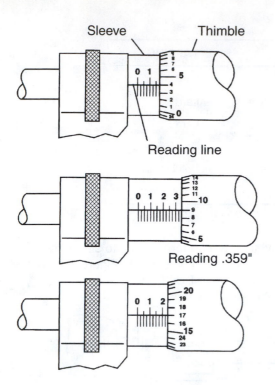

Sleeve   Thimble

Reading line

Reading .359"

FIGURE 2–7   *How to read a micrometer. (Courtesy of Navistar International Corp., Designer and manufacturer of International Brand diesel engines)*

EXAMPLE

The 1 line on sleeve is visible, representing   .100"

There are three additional lines visible, each representing .025   3 × .025" = .075"

Line 3 on the thimble matches up with the reading line on the sleeve, each line representing .001"   4 × .001" = .004"

The micrometer reading is   .179"

ADDITION METHOD

.300"
.050"
+ .009"
_____
.359"

.____
.____
+.____
_____
?

STANDARD MICROMETERS

Since most of the dimensional specifications on a metric engine are still recorded using the English system, every technician must know how to read a standard micrometer. A standard **outside micrometer** consists of a frame in the shape of the letter G: on one side of the frame is a fixed anvil and on the other is a spindle assembly. The spindle assembly consists of an accurately machined screw that rotates in the spindle nut. The spindle is rotated by a thimble: the point at which the thimble registers on the micrometer sleeve calibration scale indicates the reading.

From the point at which the spindle contacts the anvil and the micrometer dimensional readout is zero, the thimble must be rotated through 40 complete revolutions to produce a reading of 1 inch. This means that the inch is divided by 40. Therefore, each complete rotation of the thimble equals:

$$\frac{1.00}{40} \text{ inch} = 0.025''$$

So each full rotation of the thimble is equivalent to 0.025". The calibration scale on the sleeve is calibrated in units of 0.025" and every fourth stroke equals 0.100" usually indicated by single digits 1 through 10. The leading edge of the thimble is divided into 25 cal-

ibration strokes, each representing 0.001". Understanding exactly how the micrometer is calibrated enables the technician to read it easily. All standard micrometers measure dimensions through 1 inch and are read in the same manner. Changing the frame and anvil assembly permits the micrometer to be used for measuring dimensions larger than 1 inch but the readings are still confined to the 1-inch window of spindle travel. An outside micrometer is used to measure the outside dimensions of a component such as a shaft or set of shims whereas an **inside micrometer** is used to measure internally such as a bore dimension (Figure 2–7).

METRIC MICROMETERS

All metric micrometers have a scale that reads 25 mm and from the zero measurement when the spindle contacts the anvil, must be turned through 50 rotations. Therefore, each revolution of the thimble equals:

$$\frac{25.0 \text{ mm}}{50} = 0.5 \text{ mm}$$

The thimble on a metric micrometer is divided into 50 graduations so:

$$\frac{0.5 \text{ mm}}{50} = 0.01 \text{ mm}$$

As all metric micrometers measure dimensions from zero through 25 mm once again, changing the frame and anvil assembly permits the micrometer to be used for measuring dimensions larger than 25 mm, in which the readings will be confined to the 25 mm window of spindle travel. Metric micrometers are as easy to read as standard micrometers especially when the technician understands how they are calibrated.

DIAL INDICATORS

Dial indicators are used to measure travel or movement in values of thousandths to one hundred thousandths of an inch. Metric dial indicators are calibrated to read in tenths to thousandths of a millimeter. Dial indicators are used for many general and job-specific functions in the diesel engine shop. Most have a total travel range of 1″ but job-specific indicators may have a total travel range exceeding that and some much less. Dial indicators may have a balanced dial in which case, in an indicator in which one revolution of the indicator needle represents 0.100″, the dial would be calibrated in units of 0.001″ to 0.050″ through 180 degrees of the dial and then graduate back down to 0.001″ through the other side of the dial; the 0 at the center of the dial face would be marked with a + on one side and a – on the other. This type of dial indicator is useful in determining the **total indicated runout (TIR)** of a rotating component. Figure 2–8 demonstrates dial indicator terminology.

For instance, when measuring flywheel concentricity, a magnetic base dial indicator would be placed on the engine crankshaft and the engine rotated manually through 360 degrees. Using chalk on the flywheel housing face, the indicator dial would be set to zero at the start point and as the engine is rotated through 360 degrees, readings both above and below zero would be added to indicate the TIR. A secondary dial gauge is

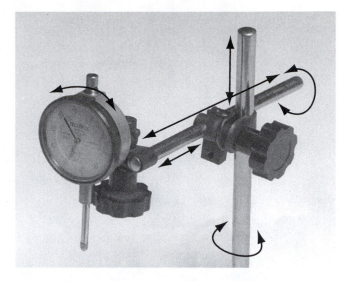

FIGURE 2–9   *Adjustment features of a dial indicator.*

used to indicate complete revolutions of the primary face gauge.

Dial indicators should always be mounted so that the plunger is at right angles to the component being measured. Figure 2–9 shows the various positions of a universal dial indicator.

TECH TIP: The TIR is always the sum of the highest positive and the highest negative reading.

DIAL BORE GAUGES

Dial bore gauges are used to measure bore dimensions such as **inside diameter (id)**, taper, and out-of-round. They are calibrated in 0.001″ or 0.0001″ and are a fast way of rapidly assessing cylinder bore dimensions. The typical dial bore gauge consists of a shaft on top of which is the dial indicator; at the base of the shaft is a measuring sled consisting of guides and an actuating plunger. One of the three guides is located diametrically opposite to the actuating plunger. The actuating plunger is responsible for producing the indicator readings.

The guide located diametrically opposite to the actuating plunger is both removable (to permit the dial bore gauge to measure different bore dimensions) and adjustable. The dial bore gauge should be mounted in a soft-jawed vice wrapped in a rag to adjust it. Using the adjustable guide shaft, the dial bore gauge should then be set close to the minimum service specification for the bore to be measured, then zeroed at precisely the minimum specification. If this method is used, every dimension measured should read on the positive side of the zero on the dial indicator. For instance, if the required specification tolerance on an installed liner bore

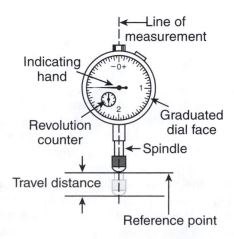

FIGURE 2–8   *Dial indicator terminology.*

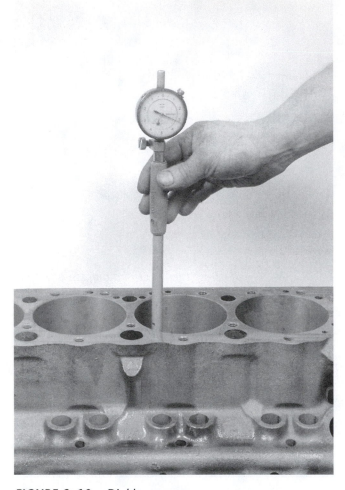

FIGURE 2–10   *Dial bore gauge.*

same way as regular micrometers. The dial-type depth gauge is simply a dial indicator mounted on a block. It is read in an exactly similar manner to a dial indicator.

COMBINATION SQUARE

The combination square consists of a right-angle square, a protractor, and a center gauge assembled on a steel ruler. A good quality, precision combination square may be used as a square, protractor, center gauge, depth gauge, height gauge, level, straight edge, and ruler. It is a valuable addition to the technician's toolbox.

TELESCOPING GAUGES (SNAP GAUGES)

Telescoping gauges are used to measure internal dimensions. They have no integral calibration and must be used in conjunction with a standard outside micrometer. In other words, they are a comparison measuring instrument. A set of telescoping gauges is usually capable of measuring dimensions from ½″ up to 6″. The gauge has the appearance of a T: the T bar is equipped with a spring-loaded plunger, which, when released by the locking handle, expands to the dimension to be measured. The gauge may then be locked by the locking handle and removed from the bore; an outside micrometer is then used to measure the T bar dimension.

SMALL-HOLE GAUGES

Small-hole gauges, like telescoping gauges, are comparison measuring instruments used to measure small cylindrical bores in conjunction with an outside micrometer. In the typical small-hole gauge, a tapered spindle is rotated by a handle to spread split ball halves, moving them outward to contact the bore walls being measured. The gauge is adjusted for minimal drag in the bore, then measured with an outside micrometer.

(Mack E7) is 4.875″ to 4.8770″, zero the dial bore gauge so that the actuating plunger is loaded to produce at least one full rotation of the indicator gauge and measures exactly 4.870″. Readings of up to +0.005″ will be within specification, while readings above 0.005″ will be outside the required specification. Figure 2–10 shows a dial bore gauge in a setting fixture.

DEPTH GAUGES

Depth gauges may be of the micrometer or dial gauge type. In each case, the instrument consists of a block to which either a micrometer assembly or dial indicator mechanism is attached. The micrometer depth gauge is read opposite from the standard micrometer: when the spindle is flush with the block (that is, it is in its most retracted position), the micrometer reads zero. As the thimble is rotated, the plunger extends beyond the flush position on the block to produce readings on the sleeve calibration scale. The calibration scales on both standard and metric depth micrometers are read in the

PLASTIGAGE

Plastigage™ is used to check friction-bearing clearances. It consists of a cylindrical plastic thread enclosed in an envelope calibrated in the dimensions that the plastigage is designed to measure. To measure bearing-to-shaft clearance, a small strip should be cut and placed across the width of the bearing shell. Next, the cap should be torqued to specification, which results in flattening the plastigage. The cap is then removed and the width that the plastigage has been flattened to measured against the calibration scale on the envelope. The wider the flattened dimension of the plastigage, the less the bearing clearance. Plastigage is used to measure rod and main bearing clearance on engines. When crankshaft main journal clearance is measured, the engine must be inverted so that the weight of the crankshaft is fully supported by the cylinder block and not by the main caps. Plastigage is available in three size ranges:

Green: clearance range −0.001″ to 0.003″
Red: clearance range −0.002″ to 0.006″
Blue: clearance range −0.004″ to 0.009″

• Carefully remove plastigage from the shaft when the measurement is complete.
• A plastigage test strip that is flattened irregularly indicates journal taper.

DIVIDERS AND CALIPERS

Dividers are used for measuring dimensions between lines or points and scribing reference points and arcs. **Calipers** are designed with internally or externally arced legs to perform internal and external measurements. Dividers and calipers are comparison measuring instruments that require the use of a calibrated measuring instrument such as a micrometer or ruler to produce a specific dimension.

PRECISION STRAIGHTEDGE

A precision straightedge is manufactured from a middle alloy carbon steel. It should be encased in a protective wood or plastic cover and hung vertically when stored. A precision straightedge is used for such tasks as measuring cylinder block wear and warpage in conjunction with a set of thickness gauges.

THICKNESS GAUGES (FEELER GAUGES)

Thickness gauges are precisely machined blades of tool steel usually packaged in sets. They are available in standard and metric dimensions and tend to be one of the most used items in the technician's toolbox. Thickness gauges are used for adjusting valve lash, checking connecting rod endplay, checking backlash on gear sets, checking ring end gap and, used with a precision straightedge, checking cylinder head and cylinder block warpage and wear.

The technician should be aware that thickness gauges wear with frequent use and should be measured from time to time with a micrometer. Individual blades can be easily replaced in a thickness gauge set.

TRUCK TECHNICIAN'S TOOLBOX

Not every technician's toolbox will be equipped as shown in Figure 2–1, but the following information might help as a guideline.

¼″ drive ratchet
⅜″ drive ratchet
½″ drive ratchet   ½″ drive flex/breaker bar
¾″ drive ratchet   ¾″ drive flex/breaker bar

Sockets: Purchasing in sets is usually more economical. See Table 2–2.

Allen sockets—an assortment of sizes

½″ drive torque wrench—click or dial type to 250 ft.-lb (swivel/flex head useful for engine work in tight locations)

⅜″ drive torque wrench—preferably dial type to 120 in.-lb

¾″ and 1″ drive torque wrenches—usually provided by shops

4 × 4 torque multiplier—usually provided by shops

Combination wrenches—standard: ⁵/₃₂″–1¼″; Metric: 4 mm–30 mm

TABLE 2–2   *Ratchet Drives and Socket Sizes*

	DOUBLE HEX	HEX	HEX DEEP	SOFT IMPACT
¼″ drive		4 mm–12 mm ⁵/₁₆″–½″	4 mm–12 mm ⁵/₁₆″–½″	
⅜″ drive	10 mm–19 mm ⅜″–¾″	10 mm–19 mm ⅜″–¾″	10 mm–19 mm ⅜″–¾″	
½″ drive	½″–¹⁵/₁₆″ 12 mm–19 mm	½″–1¼″ 12 mm–24 mm	½″–⁵/₁₆″ 12 mm–19 mm	½″–1¼″ 12 mm–24 mm
¾″ drive		⅞″–1½″ 20 mm–30 mm		

Open and box-end wrenches—standard: $\frac{5}{32}''$–$1\frac{1}{4}''$; Metric: 4 mm–30 mm

Note that purchasing wrenches in sets, rather than individually, is usually significantly more economical.

Allen keys—standard and metric sets

Screwdrivers—purchase in sets: slotted, Phillips, torx (to #30)

Digital multimeter (DDM)—2½- or 3½-digit resolution ammeter not essential

Circuit test light

Circuit testing clips and cables

Breakout Ts

Breakout boxes—specialty diagnostic breakout boxes are usually provided by shops.

Reader/programmer (ProLink) head—usually provided by shops

ProLink Software cartridges and MPC cards—provided by shops

Coolant hydrometer (not a recommended instrument)

Refractomer (for coolant and battery electrolyte)—usually provided by shops. Ensure that it is calibrated for electrolyte, ethylene glycol (EG), and propylene glycol (PG).

4 lb cross peen hammer (optional)

1½ lb ball peen hammer (wood handle)

1½ lb nylon/rubber head

Hacksaw—selecting a good quality frame with high rigidity and using the best blades can pay off in saved frustration and sweat.

Prybar set—to 18″ size (length)

Cold chisel set—to 1″

Coldpunch set—to 1″

Brass drift—1″ × 8″

Mild steel drift—1″ × 12″

Stud extractor set + wheel stud extractor collar

Nut splitter

Bolt cutter

Seal and bearing drivers—normally provided by shops

Lineman pliers 8″ and 12″

Terminal crimpers—bent nose electronic pliers

Speciality terminal crimpers/connector disassembly tools—usually provided by shops

Wire strippers—needle nose pliers

Sidecutters

Tin snips—straight cut

Slip joint/waterpump pliers—12″, 24″, 36″

Vise grips (Never let anyone kid you that these are the tools of an amateur. There are thousands of valid uses for them in the truck shop.)

10″ Pipe jaw/10″ straight jaw/8″ needle nose vise grips

18″ pipe wrench

Adjustable wrenches—6″, 8″, 12″

½″ chuck pneumatic drill

HS drill bits to ⅜″ size

⅜″ drive air ratchet

½″ drive impact gun (see Figure 2–12)

Pneumatic chisel/hammer

Truck tire air chuck

Tire gauge to 150 psi–1 MPa

Air blower nozzle

Air hose—often provided by shops (see Figure 2–12)

Hand primer pump + #8 hydraulic hose and couplers

Variable focus flashlight

Soldering gun

Telescoping mirror

Telescoping magnet

0–1″ micrometer

0–1″ depth micrometer

0–25 mm micrometer

Micrometer sets exceeding 1″ are provided by shops.

6″–150 mm vernier caliper—digital preferred

12″ tape measure

Dial bore gauge—telescoping gauges (usually provided by shops)

Stethoscope (usually provided by shops)

Infrared thermometer (usually provided by shops)

Trouble light—often provided by shops

Heavy-duty hand tools—Wrenches sized over 1¼″ or 30 mm, ¾″ and 1″ drive sockets, and other heavy-duty specialty tools are normally provided by shops.

Safety glasses

Hearing protection muffs

Roller cabinet—many shallow drawers preferable to fewer deep ones

Top box—also consider a side cabinet for shop manuals and fluid containers

Creeper—often provided by shops

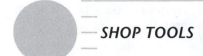

SHOP TOOLS

Shop tools are those tools generally provided by the employer. Tools that are too large to fit into a toolbox, high in cost, or highly specialized to a specific procedure should be provided by the service garage.

SLEDGEHAMMERS

Sledgehammers are designed so that the weight of the head (and perhaps the length of the arms holding the hammer) defines the force imparted. Sledgehammers have a variety of functions in the truck shop and while they are unlikely to be called on in engine reconditioning, they may be required in some of the procedures required to remove components from a chassis. They are usually manufactured in 8, 12, and 16 lb weights. Before using a sledgehammer, the handle should be inspected for damage and the head to handle securement checked. When swinging a sledgehammer, the hand grip should be firm but relaxed and the weight of the head allowed to define the amount of force delivered. On no account should the operator attempt to amplify this force with muscle power as it usually results in missing the target. If the force is insufficient to achieve the objective, select a heavier hammer.

TECH TIP: The neck of a sledgehammer handle is vulnerable when its operator misses the target. Help protect the neck of the sledgehammer handle against accidental damage by binding it with a split section of appropriately sized rubber coolant hose.

PRESSES

Most truck and bus garages will have at least one power press. Extreme caution is required when operating a power press: components should be properly supported, and mandrels/drivers should be used when required. Arbor presses are hand actuated. Whenever pressing components using any kind of power press, always consider both the consequences of component slippage and where separated components will fall. Personal safety and that of those working in the vicinity must always be considered. Figure 2–11 shows a press driving a bearing onto a shaft.

SCISSOR JACKS

Scissor jacks are designed to quickly raise one end of a truck to heights of up to 8 feet above the shop floor.

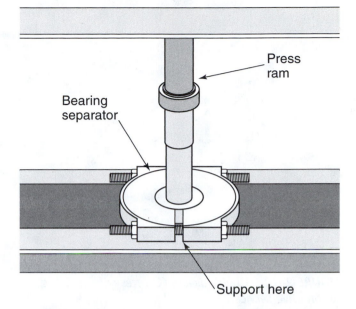

FIGURE 2–11   *Press and bearing puller used to install bearing on shaft.*

Clevises on the jack fit under each frame rail and the truck is raised by an air-actuated piston. Scissor jacks are an invaluable shop tool but they must be safely used. Ensure that the lift clevises are positioned at a safe location on the frame rails and allowance is made for the relative movement between the truck and jack during raising and lowering. When the truck has been raised, engage the mechanical stops and make sure that the weight is supported on them and not the power piston. Also check the weight of the vehicle to be lifted and the load capacity of the scissor jack. Chock the wheels on the end of the chassis not being raised.

"A" FRAME HOISTS

"A" frame hoists are often large enough to pass over the top of a truck and use a block and tackle (chain falls) lift mechanism. They can be used to lift components such as an engine or a cab from the chassis. They may also be capable of hoisting one end of the chassis, but extreme caution should be exercised because of the tendency of anything lifted by chains to swing. The hoist mechanism should be inspected annually whether or not local regulations require the inspection.

CHERRY PICKERS

These are portable, hydraulically actuated, one-arm boom hoists that have many uses in the truck and bus garage. They are available in a variety of sizes and load-carrying ratings. The pickup arm is usually

FIGURE 2–12   *Cherry picker. (Courtesy of Tim Gilles, Automotive Service 2e)*

adjustable in length, and the longer the adjustment arm setting, the less the load lift potential. If the load-carrying ability is exceeded, they will topple. Figure 2–12 shows a cherry picker.

TRANSMISSION AND CLUTCH JACKS

Transmission and clutch jacks are usually hydraulically actuated and designed to fit under the truck frame and support the transmission/clutch. It is important to ensure that transmissions are chained securely to the jack especially when the mass of the unit is top heavy such as in triple countershaft units. When it is expected that the jack must support the transmission stationary such as when removing an engine from a chassis, ensure that the transmission jack load is supported mechanically. Clutch jacks should be used when installing heavy-duty clutches: a 15½ in. clutch pack weighs somewhere around 175 lb (80 kilograms [Kg.]) and should never be handled without some kind of assistance.

SPREADER BARS

A **spreader bar** is a rigid bar, usually adjustable in length used for lifting engines out of the chassis. The spreader bar should be adjusted to the length of the engine; the spreader bar is normally attached to the engine to be lifted on three or four points. Chains should be installed so that the chain length is at a minimum with the spreader bar clear of the engine. The chains should be attached to the engine by means of hooks to lifting eyes located on either the cylinder head or cylinder block assembly. Never fit lifting eyes to the rocker housing fasteners. Some engines have just two

permanently fitted lifting eyes. It is usually safe to lift these engines using the spreader bar on a two-point lift.

LOAD ROTOR

The load rotor is a less preferred method of hoisting an engine than the spreader bar. Load rotors consist of a ratcheting chain block with a single chain equipped with hooks at either end. The hooks fit to the engine lift eyes and the load rotor chain block locks the chain, permitting a different length of chain on either side of the block. This feature often permits the chains to be fitted to the engine and clear the upper engine components.

CHAINS

Chains are rated by working load limit—a value that is normally equivalent to about 25% of the tensile strength of the chain material. Many jurisdictions require that chains be inspected annually. Additionally, technicians should visually inspect all chains before using them. The saying that every chain is only as strong as its weakest link bears true. Truck diesel engines can weigh more than a ton and while relying on a chain to support this kind of weight, the technician should try to avoid working under the load. Apart from inspecting the **chain hoists** and checking their load rating, the technician should check the hooks, lifting eyes, lifting eye fasteners, and connecting links.

SLINGS

Slings may have to be used to hoist engines from the chassis in some applications. These are normally manufactured from synthetic fibers. Again, the load capacity must be checked and the sling integrity inspected. Avoid using steel cable slings unless also using the engine lift eyes.

AIR TOOLS

Air tools are extensively used in any truck service location. Some of the air tools are owned by technicians, others are provided by the shop. Technicians working around pneumatic equipment sometimes forget that it can be dangerous. It makes sense to wear eye protection and be aware that dusts driven into the air by pneumatic tools can cause breathing problems. Most shops provide heavy-duty pneumatic tools such as 1″ drive air guns. A ½″ drive impact wrench such as that shown in Figure 2–13 will be one of the most frequently used tools in the truck technician's cabinet. Buying a good quality air gun and properly maintaining it by keeping it moisture-free and oiled (look at the oiler and filter in Figure 2–13) will help ensure that it functions well for a number of years.

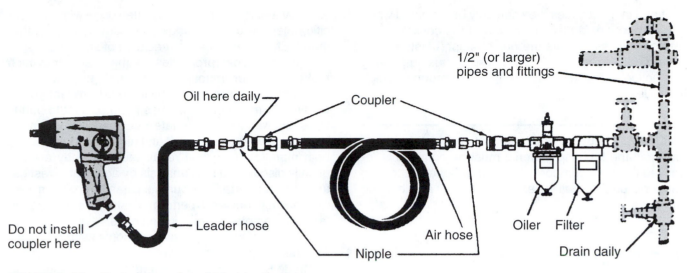

FIGURE 2–13   *Typical setup for a ½" drive impact gun.*

TECH TIP: Purchase a good quality ½" drive air impact wrench and with a little care, it should last for many years. Low-cost impact guns are generally a poor investment for a truck technician who uses one daily.

OXYACETYLENE EQUIPMENT

Truck and bus technicians use oxyacetylene for heating and cutting probably on a daily basis. Less commonly, this equipment is used for braising and welding. Technicians using this equipment require some basic instruction in the safety requirements and handling of this equipment. An explanation of oxyacetylene handling and safety is provided in Chapter 3. Oxyacetylene equipment must be used by every technician, and seldom is the novice technician provided with adequate instruction. At least consult this section before using this equipment.

STEAM AND HIGH-PRESSURE WASHERS

Hot water, high-pressure washers have generally replaced the steam cleaners more commonly used a decade ago. Hot water, high-pressure washers are safer and usually require less maintenance than steam washers. The technician should be aware of the potential for damage when using any type of high-temperature, high-pressure washers; eye protection and gloves should be worn when operating this equipment. Environmental regulations in most jurisdictions mandate that the runoff from this type of cleaning operation not be permitted to enter sewage systems. Power wash runoff should be filtered through a water separator system, and the separator tanks should be pumped out regularly. Heavy fines may be imposed when sewage is contaminated with oil and road dirt washed off trucks, buses, and their engines.

PULLERS

Shops usually have a selection of general use and specialty pullers that can be power- or mechanically actuated. Ensure that puller jaws and legs are capable of handling the force to which they will be subject. Safety glasses should always be worn when operating pullers. A bearing puller was shown in Figure 2–11.

BUSHING DRIVERS

Bushing drivers consist of a mandrel, which should fit tightly in the bushing bore with the shoulder having an identical **outside diameter (od)** to the bore that the bushing is pressed into. This should enable bushings to be removed and installed without damaging either the bushing or the bore to which it is fitted. Bushings may be installed using direct mechanical force such as a hammer or slide hammer, or by using hydraulic or pneumatic power drivers. Always wear safety glasses when using bushing drivers.

GLASS BEAD BLASTERS/SANDBLASTERS

Most shops rebuilding engines are equipped with a glass bead blaster or sandblaster. Usually these are encased in an enclosed housing and powered pneumatically. They are an ideal method of cleaning up components, especially when the components are coated with high-tech adhesives or gasket remains that can be difficult to remove. In enclosed housing glass bead blasters, protective gloves are integral with the unit and in most cases they can only be actuated when the component is placed inside and the cover sealed. Armored glass permits the object being blasted to be observed. Technicians should consider using hearing protection as these units are capable of producing high noise levels.

TECH TIP: Following any bead or sandblasting procedure, all the beading material must be completely removed from the components treated. Special attention must be paid to bolt holes, oil galleries, and bearing surfaces.

TACHOMETERS

Tachometers measure rotational speed. Mechanical tachometers consist of a pickup button that directly contacts the rotating component (it should be held close to its axis) and produces a direct rpm reading. Electronic tachometers use a sensor that reads a magnetic strip and reports the rpm digitally. Electronic tachometers should be capable of producing mean (average) readings when reading rotational speed on components whose rpm is fluctuating.

SAE FASTENER GRADES AND TORQUES

Diesel engine technicians should be able to identify SAE capscrew (bolt) and nut identification grades (Figure 2–14 and 2–15). However, most engine OEMs use a large quantity of special fasteners that the OEM clas-

sifies by a part number. This is often done with the specific objective of discouraging crossover to any other than OEM fasteners either because of some very specific metallurgical properties of the fastener, which make crossover impossible, or simply to prevent the use of substandard quality fasteners. In recent years, the North American market has been subject to bogus fasteners with the appropriate SAE markings that have found their way onto aircraft and highway equipment at both manufacturing and repair facilities; they are not usually discovered until analysis by accident investigators. It is important to purchase fasteners from reputable suppliers and perform destructive testing on fastener samples at random intervals.

The SAE fastener grades commonly used on trucks and their engines are:

Grade 5 category—manufactured from medium carbon steels and heat treated to provide:

	Proof Loads	Tensile Strength
to ¾″	85,000 psi	120,000 psi
¾″–1″	78,000 psi	115,000 psi
1″–1½″	74,000 psi	105,000 psi

Grade 8 category—manufactured from medium carbon alloyed steels and heat treated and roll threaded to provide:

	Proof Loads	Tensile Strength
up to 1½″	120,000 psi	150,000 psi

SAE Grade 5 fasteners are identified by three radial strokes.

SAE Grade 8 fasteners are identified by six radial strokes.

TECH TIP: Replacing an SAE Grade 5 fastener with one of Grade 8 may be inviting problems in certain fastener applications. Grade 8 fasteners have lower elasticity (flexibility) and can fail or cause the components they are fastening to fail. Always use the OEM-recommended fastener.

CLAMPING FORCE

Fasteners are designed to provide clamping force. When the fasteners responsible for clamping a cylinder head to a cylinder block are torqued to a common torque value, the objective is to ensure a consistent clamping force throughout the assembly. Fasteners such as Huck™ fasteners provide a more reliable clamping force value because they are set by defining the clamping force and eliminate the variables encountered when using a torque value to set clamping force.

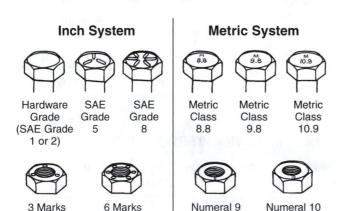

FIGURE 2–14   Fastener grade identification. (Courtesy of Mack Trucks)

Grade Marking	Specification	Material	Nominal Size, Dia. In.	Proof Load PSI (MPs)	Tensile Strength Min. PSI (MPa)	Bolt Rockwell Hardness		Nut Rockwell Hardness	
						Min.	Max.	Min.	Max.
	ASTM A307 Grade A SAE J429 Grade 1	Low carbon steel	¼ thru 1½	33,000	60,000	B70	B100	—	—
	SAE J429 Grade 2	Low carbon steel	¼ thru ¾ over ¾ to 1½	55.000 33,000	74,000 60,000	B80	B100	—	C32
 5.8	ISO SAE J1199 Property Class 5.8	Low or medium carbon steel	M5 thru M24	55,100 (380)	75,400 (520)	B82	B95	—	C32
	ASTM A449 Type 1 SAE J429 Grade 5	Medium carbon steel, quenched and tempered	¼ thru 1 over 1 to 1½	85,000 74,000	120,000 105,000	C25 C19	C34 C30	— —	C32 C32
 8.8	ISO/DIN SAE J1199 Property Class 8.8	Medium carbon steel, quenched and tempered	M3 thru M16 M17 thru M36	84,100 (580) 87,000 (600)	116,000 (800) 120,350 (830)	C20 C23	C30 C34	—	C32
	SAE J429 Grade 5.1 (SEMS)	Low or medium carbon steel, quenched and tempered with assembled washer	No. 6 thru 5/8	85,000	120,000	C25	C40	—	—
 9.8	ISO SAE J1199 Property Class 9.6	Medium carbon steel, quenched and tempered	M1.6 thru M16	94,250 (650)	130,500 (900)	C27	C36	—	C32
	SAE J429 Grade 7	Medium carbon allow steel, quenched and tempered, roll threaded after heat treatment	¼ thru 1½	105,000	133,000	C28	C34	—	—
	ASTM A354 Grade BD Bowma-Torq®/Grade 8 SAE J429 Grade 8	Medium carbon alloy steel, quenched and tempered	¼ thru 1½	120,000	150,000	C33	C39	C24	C36
	SAE J429 Grade 8.2	Low carbon boron martensite steel, fully killed, fine grain, quenched and tempered	¼ thru 1	120,000	150,000	C35	C42	—	—
 10.9	ISO SAE J1199 Property Class 10.9	Medium carbon alloy steel, quenched and tempered	M6 thru M36	120,350 (830)	150,800 (1040)	C33	C39	C26	C36
 12.9	ISO Property Class 12.9	Medium carbon alloy steel, quenched and tempered	M1.6 thru M36	140,650 (970)	176,900 (1220)	C38	C44	C26	C36
	Bowmalloy®	Proprietary medium carbon alloy steel, quenched and tempered	¼ thru 1½	156,000	180,000 Min. 2000,000 Max.	C38	C42	C26	C36

˙Manufacturer's identification symbols are required per ASTM, ISO, or SAE

FIGURE 2–15   *Bolt head identification chart. (Courtesy of Navistar International Corp., Designer and manufacturer of International Brand diesel engines)*

TENSILE STRENGTH AND YIELD STRENGTH

Fasteners are rated by **tensile strength**, the amount of force that must be exacted on a round bar of 1″ sectional area to make it fracture. The **yield strength** is the amount of force that must be exerted using the same test to cause the bar to permanently deform; typically this force is about 10% lower than the tensile strength value in steels. The difference between the tensile strength and yield strength is sometimes used to denote the elasticity of a material.

SHEAR STRENGTH

Shear strength is a measure of a fastener's ability to withstand force applied at a 90-degree angle to the axis of the bolt. Some bolts are made to withstand shear forces such as body bound bolts, manufactured with an interference fit shank.

THE METRIC SYSTEM AND ENGLISH/METRIC CONVERSION

The metric system has been authorized by an act of Congress in the United States and by federal legislation in Canada. The metric system is used by most countries in the world, and has been generally adopted by industry for use in the United States as a replacement for the stndard system. For example, every truck diesel engine engineered in the United States since 1980 has been a metric engine; however, the specifications reproduced in every service manual are usually presented in both standard and metric systems with the consequence that most mechanical technicians are forced to have an understanding of both. It should be noted that the British replaced the "English" system of weights and measures with the metric system more than 25 years ago.

The metric system is a decimal system, the meter being the basis of all measures, whether of length, surface, capacity, volume, or weight. The meter measures 39.37 inches and is theoretically one ten-millionth of the distance from the equator to the pole. The unit of weight is the gram (15.432 grains), and is the weight of a cubic centimeter of water at its greatest density at about 39°F.

Multiples of the units are expressed by the Greek prefixes "deca," "hecto," "kilo," "mega", and "giga," indicating respectively tens, hundreds, thousands, millions, and billions. Decimal parts of the units are indicated by the Latin prefixes "deci," "centi," "milli," "micro," "nano," and "pico," meaning respectively tenth, hundredth, thousandth, millionth, thousand millionth, and billionth.

METRIC WEIGHTS AND MEASURES

Listings of the commonly used metric weights and measures as well as a conversion table for common metric measurements to standard units follow.

Metric Weights

Milligram	(1/1000 gm)	= 0.0154 gr.
Centigram	(1/100 gm)	= 0.1543 gr.
Decigram	(1/10 gm)	= 1.5432 gr.
Gram		= 15,432 gr.
Decagram	(10 gm)	= 0.3527 oz.
Hectogram	(100 gm)	= 3.5274 oz.
Kilogram	(1000 gm)	= 2.2046 lb
Myriagram	(10,000 gm)	= 22.046 lb

Metric Dry Measures

Milliliter	(1/1000 L)	= 0.061 cu. in.
Centiliter	(1/100 L)	= 0.6102 cu. in.
Deciliter	(1/10 L)	= 6.1022 cu. in.
Liter		= 0.908 qt.
Decaliter	(10 L)	= 9.08 qt.
Hectoliter	(100 L)	= 2.838 bu.
Kiloliter	(1000 L)	= 1.308 cu. yd.

Metric Liquid Measures

Milliliter	(1/1000 L)	= 0.0338 fl. oz.
Centiliter	(1/100 L)	= 0.338 fl. oz.
Deciliter	(1/10 L)	= 0.845 gill
Liter		= 1.0567 qt.
Decaliter	(10 L)	= 2.6418 gal.
Hectoliter	(100 L)	= 26.417 gal.
Kiloliter	(1000 L)	= 264.18 gal.

Metric Measures of Length

Millimeter	(1/1000 m)	= 0.0394 in.
Centimeter	(1/100 m)	= 0.3937 in.
Decimeter	(1/10 m)	= 3.937 in.
Meter		= 39.37 in.
Decameter	(10 m)	= 393.7 in.
Hectometer	(100 m)	= 328.1 ft.
Kilometer	(1000 m)	= 0.62137 mi.
	(1 mile = 1.6093 km)	
Myriameter	(10,000 m)	= 6.2137 mi.

Metric Surface Measures

Centare	(1 sq. m)	= 1,550 sq. in.
Are	(100 sq. m)	= 119.6 sq. yd.
Hectare	(10,000 sq. m)	= 2.471 acre

METRIC TO STANDARD CONVERSION FORMULAE

The following formulae can be used to translate metric values to standard and vice versa.

Linear Measurements

Centimeters × 0.3937 = in.

Meters = 39.37 in.

Kilometers = 0.621 mi.

Kilometers × 3280.89 = ft.

Square centimeters × 0.155 = sq. in.

Cubic centimeters × 0.06102 = cu. ins.

Cubic meters × 35.3144 = cu. ft.

Liters × 0.2642 = gal. (231 cu. in.)

Kilograms × 2.2046 = lb

Kilograms per square millimeter × 1422.3 = lb per sq. in.

Kilograms per square centimeter × 14,223 = lb per sq. in.

Torque Conversion

1 lb.-ft. = 1.355 Newton-meters (Nm)

1 Nm = 0.738 lb.-ft.

Temperature Conversion

$$\text{Degrees Fahrenheit} = \frac{9 \times C}{5} + 32$$

$$\text{Degrees Celsius} = \frac{5 \times (F - 32)}{9}$$

Pressure Conversions

Fuel injection test instruments are often calibrated in **units of atmosphere (atms)**. Technicians should become familiar with the process of converting units of pressure into the metric and standard systems. This is an easy way:

Atmospheric pressure @ sea level = 14.7 psi = 101.3 kPa = 1 unit of atmosphere or 1 atms

Remembering the following is not as mathematically accurate as the previous equivalent, but is a fast method of converting pressure values, accurate enough for quick conversions:

15 psi = 100 kPa = 1 atms

So, to convert 45 psi to kPa:

$$\frac{45}{15} = 3 \text{ units of atmosphere or 3 atms}$$

3 atms × 100 kPa = 300 kPa

In Europe, a unit of pressure measurement known as **bar** is used. You should be familiar with bar because Bosch uses this unit for specifications in the many diesel fuel systems it manufactures. One bar is equivalent to 105 newtons per square meter, not precisely equivalent to one atmosphere. Although units of bar and atms are often used as if they were exactly equivalent, this is not so, and, where precise values are required, they should not be confused. Some fuel injection comparator equipment is calibrated in bar.

1 atms = 14.7 psi = 101.3 kPa = 1.033 bar = 29.92″ Hg = 407.19″ H_2O

Power Conversion

1 hp = 550 lb.-ft. per second = 0.746 kW = 42.4 Btu per minute

SUMMARY

- The truck technician must get used to working with both standard and metric systems of weights and measures because both are widely used in the industry.
- The actual contents of a truck technician's toolbox will be determined by the type of work performed. However, 80% of the contents are probably common among all truck technicians.
- Cheaper tools are often more bulky and more prone to breakage.
- The personal safety of the user is always on the line when hand tools are being used, so it makes sense for the professional to invest in reliable tools.
- The apprentice technician should acquire a mastery of precision measuring tools before using them; this is best done using the instruments and measuring actual engine components.
- Reading both standard and metric micrometers becomes much easier when the technician understands exactly how they are constructed and calibrated.
- A standard micrometer must be rotated through 40 complete revolutions from the point at which the spindle contacts the anvil producing a zero reading to the point at which it reads 1″. Each complete revolution of the thimble therefore represents 0.025″.
- A metric micrometer must be rotated through 50 complete revolutions from the point at which the

spindle contacts the anvil producing a zero reading to the point at which it reads 25 mm. Each complete revolution of the thimble therefore represents 0.05 mm.

- Shop hoisting apparatus should be routinely inspected by qualified personnel and by the technician before using it. This may be a legal requirement in some jurisdictions.
- Shop power equipment must be checked out by the technician before each use.
- The technician should know how to identify SAE fastener grades and understand the importance of selecting the correct grade for the job being performed.
- The technician should get used to standard and metric systems and be prepared to work in both. Formulae need not be remembered but the technician should get used to rapidly converting values from each system.

REVIEW QUESTIONS

1. When the spindle contacts the anvil on a standard 0–1″ micrometer, it should read:
   a. zero
   b. one thousandth of an inch
   c. 0.025″
   d. one inch
2. How many complete rotations must the thimble of a standard micrometer be turned to travel through a reading of zero to a reading of 1 inch?
   a. 25
   b. 40
   c. 50
   d. 100
3. How many complete rotations must the thimble of a standard metric micrometer be turned to travel through a reading of zero to a reading of 25 mm?
   a. 25
   b. 40
   c. 50
   d. 100
4. When the thimble of a metric micrometer is turned through one full revolution the dimension between the anvil and the spindle has changed by:
   a. 0.1 mm
   b. 0.5 mm
   c. 2.5 mm
   d. 5.0 mm
5. When using a dial indicator to check the concentricity of a flywheel housing, during a single rotation of the flywheel the reading on the positive side of the zero on the dial peaks at .003″ while the read-

ing on the negative side peaks at 0 .006″. What is the TIR?
   a. 0.003″
   b. 0.006″
   c. 0.009″
   d. 0.018″
6. Which of the following precision measuring instruments would be required to measure a valve guide bore?
   a. Dial indicator
   b. Inside micrometer
   c. Split ball gauge and micrometer
   d. Dial bore gauge
7. If 300 kPa is converted to pounds per square inch, the result would be closest to which of the following values?
   a. 15 psi
   b. 30 psi
   c. 45 psi
   d. 300 psi
8. To which of the following values would 1.5 mm be closest?
   a. 0.0625″
   b. 0.125″
   c. 0.250″
   d. 1.50″
9. Convert 550°Fahrenheit into Celsius.
   a. 240
   b. 288
   c. 385
   d. 550
10. A wrench with box and open ends at either end, both of the same nominal dimension is known as a:
   a. torque wrench
   b. combination wrench
   c. box-end wrench
   d. adjustable wrench
11. Which of the following is used to identify an SAE grade 8 bolt?
   a. 3 radial strokes on the capscrew head
   b. 5 radial strokes on the capscrew head
   c. 6 radial strokes on the capscrew head
   d. 8 radial strokes on the capscrew head
12. If a 12″ linear extension is used on a torque wrench with a 24″ bar, calculate the reading on the torque wrench scale required to produce an actual torque value of 250 lb.-ft.
   a. 36 lb.-ft.
   b. 120 lb.-ft.
   c. 167 lb.-ft.
   d. 323 lb.-ft.
13. Convert 250 lb.-ft. to Newton-meters and select the closest value from the answers below.
   a. 167 Nm
   b. 340 Nm
   c. 410 Nm

d. 500 Nm

14. Convert 600 hp into kW and select the closest value from the answers below.
    a. 350 kW
    b. 450 kW
    c. 550 kW
    d. 650 kW

15. The cutting fluid recommended for use when cutting threads in mild steel is:
    a. soluble oil and water
    b. lard
    c. dry
    d. kerosene

16. When drilling into cast iron, the correct method calls for the procedure to be performed:
    a. with kerosene
    b. with lard
    c. preheated
    d. dry

17. A bolt designed so that the shoulder has a small interference fit with the bore it is to be fitted to is called a(an):
    a. SAE #5
    b. SAE #8
    c. body bound bolt

d. Huck fastener

18. The working load of a chain is normally what percentage of the rated tensile strength of the chain material?
    a. 10%
    b. 25%
    c. 50%
    d. 75%

19. Which Plastigage™ color code should be selected to measure a main bearing clearance that the manufacturer specifies must be between 0.0023″ and 0.0038″?
    a. Red
    b. Green
    c. Blue

20. After Plastigage™ checking a main bearing, using red plastigage, the measuring strip has not been deformed. *Technician A* states that the bearing clearance must be greater than 0.006″. *Technician B* states that a strip of green coded plastigage must be used to perform the measurement. Who is right?
    a. A only
    b. B only
    c. Both A and B
    d. Neither A nor B

Chapter
3

Personal and
Safety Awareness

PREREQUISITE

Chapter 2

OBJECTIVES

After studying this chapter, you should be able to:

■ *Identify the basic personal safety equipment required in a truck service garage.*
■ *Outline the importance of wearing the appropriate clothing and footwear on the shop floor.*
■ *Explain the importance of using eye protection in the shop environment.*
■ *Describe two methods used to protect hearing.*
■ *Understand how to lift heavy objects in the safest manner and the importance of using power lift equipment whenever possible.*
■ *Explain the function of OSHA.*
■ *Identify the four categories of fire and the fire extinguishers required to put them out.*
■ *Explain the legislation pertaining to an employee's Right to Know.*
■ *Interpret the acronyms WHMIS and MSDS.*
■ *Interpret the emergency and first-aid policies used in a service garage.*
■ *Understand the importance of basic training in first aid and fire suppression techniques.*
■ *Explain the safety requirements of handling oxyacetylene gases and heating, cutting, and welding processes.*
■ *Describe the safety devices used on oxygen and acetylene cylinders.*
■ *List the federal agencies responsible for administering hazardous waste disposal and shop and personal safety in the United States and Canada.*

KEY TERMS

backfire

compressed air

corrosive

flammable

flashback

Hazard Communication Legislation

material safety data sheets (MSDS)

Occupational Safety and Health Administration (OSHA)

oxyacetylene

radioactive

reactive

Resource Conservation and Recovery Act (RCRA)

Right-to-Know legislation

toxic

Workplace Hazardous Materials Information Systems (WHMIS)

PERSONAL SAFETY EQUIPMENT

Ownership of personal safety apparel is important but not as important as developing the habit of using it properly (Figure 3–1). The following are some tips.

EYE PROTECTION

In some shops, wearing safety glasses is mandatory. Although most technicians probably own at least one pair of safety glasses, they tend not to be worn nearly enough. For the person who does not regularly use eye glasses, wearing safety glasses is only an irritation for the first few days, after which they will not be noticed. Part of the problem is that many technicians purchase or are provided with safety glasses of the poorest quality that tend to be both uncomfortable and actually impair vision. A technician who will spend $50 on a wrench that might be used four times a year balks at spending the same amount on a good quality pair of safety glasses worn every day. Buy a good quality pair of safety glasses and get used to wearing them all the time.

HEARING PROTECTION

Two types of hearing protection are available: internally worn plugs made of a sponge or wax fiber and the external type that have the appearance of ear muffs. They tend to be uncomfortable and can be dangerous if they work too effectively. The noise levels in truck service facilities vary considerably, but in most it is not consistently at levels that will result in hearing damage. The technician should own hearing protection devices and use them on an as-necessary basis. Some machining procedures require the use of hearing protection. When operating a chassis dynamometer or entering an engine test cell, hearing protection must be worn; in fact, using both internal and external hearing protection is recommended.

SAFETY FOOTWEAR

Legal requirements aside, anyone working in an automotive, truck, or heavy equipment service facility should wear safety footwear, preferably boots so that the ankle is properly protected. These boots should have a steel toe and heel. Poor quality safety shoes are extremely uncomfortable and may in themselves end up damaging the feet they are supposedly protecting. Purchase footwear that breathes and is capable of adapting to the shape of the wearer's foot; this usually means leather. Synthetic materials may suffice for the shadetree mechanic who wears them once a week but not the professional who wears them daily.

CLOTHING

Generally, technicians are to some extent exposed to **oxyacetylene** and various electric welding processes. Anyone exposed to working around heat and flame should be aware of the dangers of many synthetic fibers when ignited: when they burn, they melt and fuse to the skin. It is good practice to wear cotton clothing and coveralls. Synthetic fibers treated with fire retardant tend not to breathe and can be uncomfortable in hot weather.

GLOVES AND BARRIER CREAMS

Generally, gloves are not recommended when performing engine work although it is probably good practice to wear heavy leather gloves when performing heavy-duty suspension work. Many service shops make a variety of gloves available for their technicians. Most technicians find it difficult to work effectively wearing any of the gloves currently on the market designed for their purposes. Those gloves that least compromise the sense of touch also seem to be the ones that rip the easiest and do not breathe. Considering that the truck technician comes into contact with many potentially harmful fluids, the wearing of gloves is probably justified but is currently not practical.

Barrier creams are wax-based hand creams that provide some protection; however, most of these dissolve when in contact with solvents.

Barrier creams work to help clean up hands rather than protect them from harmful liquids. Perhaps the technician is best advised to be aware of the potential harm in the fluids and materials in the workplace and practice good personal hygiene. For instance, because diesel fuel has been identified as a carcinogen, it is good practice, after coming into contact with it, to wash thoroughly with soap and water. The same applies to used engine oil and all types of antifreeze solution.

BACK PROTECTION

Most technicians start young and seldom have to think much of the physical abuse they will subject their bodies to. While truck technicians usually have a comprehensive array of lifting apparatus available to them, many do not use them enough. Lifting a 15½" clutch assembly would be heavy if the feat had to be performed in a weight room; it lacks the convenient hand grips provided with barbells and the mass is concentrated in the center. Performing this feat in the cramped conditions under a truck is an invitation to back problems. One major OEM performed an in-house survey in

Ⓐ **GAC1495B Protective Eyeshields.** Will fit over prescription glasses or can be worn alone. Wrap around side shields, impact resistant, antiglare guard.

Ⓑ **Eyelights.** Lights on each side of glasses allow hands free while you conduct your work. Lightweight design. Unfold glasses and light goes on; fold and light goes off.
    EYELIGHT1. Black.
    EYELIGHT10. High visibility orange.

Ⓒ **Safety Glasses.** Single panoramic lens with molded side shield. 100% polycarbonate lens for clear vision and impact resistance. Patented lens with 3 position tilt, as well as patented adjustable temple length for personalized comfort. 4C+™ antifog coating absorbs and emits moisture. 4C+™ coated lenses also absorb 99.9% of harmful ultraviolet light, reduce dust and particle attraction, and resist scratching. Meets ANSI Spec. Z87.1-1989 and CSA Z94.3-1992.
    GLASS1R. Red frame, clear lens.
    GLASS1BK. Black frame, clear lens.
    GLASS1BL. Blue frame, clear lens.
    GLASS2R. Red frame, mirror UD lens.
Safety Glasses. Contemporary style with low profile, four length adjustable slide temple. Stylish black frames made of durable nylon. Six-base lens curve and folded back lens design provides increased peripheral vision and protection. ANSI Spec. Z87.1-1989.
    GLASS4. Black frame, clear antifog lens.
    GLASS5. Black frame, gray lens.
    GLASS6. Black frame, blue mirror lens.

Ⓓ **Safety Glasses.** Scratch-resistant polycarbonate lenses offer heavy-duty protection from flying chips and particles. Feature the classic look of dual-lens eyewear, although actually an exclusive single-lens design. Lenses are easily replaceable for different lighting and work application. Adjustable temple length. Complies with ANSI Spec. Z87.1-1989.
    GLASS3BR. Burgundy frame with clear lens.
    GLASS3GR. Gray frame with clear lens.
    GLASS3GRG. Gray frame with gray UD lens.

Ⓔ **YA346 Safety Glasses.** Large lenses offer increased visibility. Protective lens coating resists fogging, scratching, static, and ultraviolet rays. Frames adjust both horizontally and vertically for comfort. Side impact protection. Complies with ANSI Spec. Z87.1-1989.

Ⓕ **GAC1575B Safety Goggles.** Plastic lenses conform to American National Safety Standards and OSHA requirements for strength and piercing. Soft, form fitting plastic frame is vented. Meets ANSI Spec. Z87.1-1989.
GAC1580 Safety Goggles. Similar to **GAC1575B**, but instead of straight perforations for air vents, these goggles feature six protected vents—debris would have to make a 90° turn to enter. Meets ANSI Spec. Z87.1-1979.
GOGGLE1 Protective Goggles. Features durable molded frame with a soft and flexible elastomer that seals comfortably for a custom fit around nose, cheeks, and forehead. Quick adjust headband clips permit fast and easy fitting. Indirect ventilation system channels and directs air flow over the lens. This minimizes fogging and improves vision potential while still providing superior splash and impact protection. Has uvex 4C+™ lens coating that provides antifog, antiscratch, and anti-UV protection. Meets ANSI Spec. Z87.1-1989.

Ⓖ **GA3000 Ear Protectors.** ABS plastic domes are fitted with PVC covered ear cushions. The cushions contain a liquid/foam nontoxic glycerine filled inner bladder layered with polyurethane foam to provide a close, comfortable fit. Headband is also cushioned. Adjust at each ear piece for proper alignment. Meets ANSI Spec. S3.19-1974 (R1990). Noise reduction rating of 27.

Ⓗ **YA160 Ear Protectors.** Worn under the chin, the YA160 has a noise reduction rating of 17 dB. Swiveling ear plugs adapt to the angle of your ear canals. Weighs only 10 g. Meets ANSI Spec. S3.19-1974 (R1990).
EEND500A Digital Sound Meter. Designed to meet the needs of safety engineers and quality control professionals in measuring noise levels in factories, school offices, airports, and other environments. Features MAX level hold, over and under range indication, LCD display (3½ digits), analog signal output for data recording, and fast /slow dynamic response settings to check peak/average noise levels.

Ⓙ **GA224A Face Shield.** Consists of spark guard, adjustable headpiece, and visor. Room for prescription glasses. 8" x 12" x .040" visor. Cushioned headpiece.
GA224A5 Clear Replacement Shield. 9" x 15½".
GA224A6 Dark Green Replacement Shield. 9" x 15½".

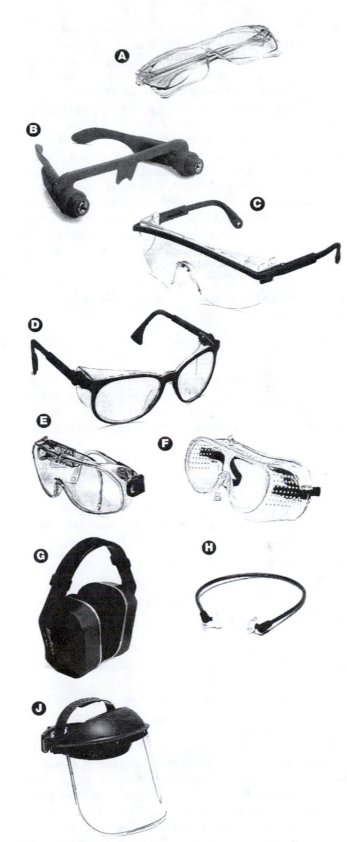

FIGURE 3–1 *Personal safety equipment. (Courtesy of Snap-on Tools Corporation)*

which it determined that 50% of its service personnel had taken some time off work due to a back-related problem before they had reached the age of 30 years. Back problems are a fact of many technicians' lives. The best strategy is avoidance—bad lifting habits are developed young and reinforced by the fact that it takes some years before a problem develops. Lift smart! Lift with the legs rather than the back and do not lift heavy weights in confined spaces. There is always a jack or hoist that can make the job easier. One national chain of hardware stores provides its employees with flexible back braces. These employees wear the braces as part of their uniform. A back brace spreads the focal point of the load being lifted over a larger area of the back, so it makes a whole bunch of sense for anyone exposed to routine heavy lifting to be wearing one. If there were a single factor that could threaten the technician's ability to work through a career, it would be problems related to the back.

BREATHING PROTECTION

The truck engine technician should be aware of the atmosphere in the shop and wear a protective mask when necessary. The environment of the truck shop is not generally unhealthy but is mostly based on the type of work being performed. Exhaust fume extraction pipes should always be fitted to vehicles when they have to run on the shop floor and engines should be warm when moving vehicles in and out of the garage. It goes without saying that extra precautions are required when painting, sand blasting, brake servicing, and other operations.

WORKPLACE HAZARDS AND SAFETY REGULATIONS

Again, it should be stressed that the environment of the typical truck service garage is not usually an unhealthy one. The environmental hazards associated with a trucking operation are generally determined by the specific nature of the business conducted. However, every truck technician should make safe practice part of his or her working routine. Repairing trucks requires the use of an extensive array of equipment. Apprentice technicians should make it their business to learn how to safely operate shop equipment. This often involves asking questions of those who do know how. If an apprentice technician pretends a knowledge in operating a piece of equipment, the result can be fatal. Note the American National Standards Institute (ANSI) symbols in Figure 3–2: these are mostly common sense, but make sure you know them.

DRIVING TRUCKS

Licensing requirements vary by state and province in North America but in the majority of jurisdictions, the licensed automobile driver can drive an uncoupled tractor and unloaded, straight trucks. Trucks brought in for repair must be moved from the yard into the shop in any case, so apprentice technicians usually get their first experience of driving a truck as a *yard jockey*. There is nothing especially difficult about driving a truck but it is not a car and a lesson or two from a driver trainer can help develop good driving habits. Certain vehicles should be handled with some caution. An example would be a garbage packer with bucket fork hydraulics: whenever the hydraulics must be actuated to access components on a chassis, ensure that someone familiar with the system provides some instruction in the procedure.

When road testing trucks, it is obviously important to ensure that the licensing requirements of the jurisdiction are met. Technicians should also be aware of the fact that the handling characteristics of a tractor uncoupled from a trailer are quite different from those of the tractor/trailer combination. The weight distribution of an uncoupled tractor is uneven, being focused over the steering axle; this results in comparatively little weight over the driven rear axle(s). When bobtailing, that is, driving an uncoupled tractor on a highway, special care should be exercised when the road surface is either wet, icy, or snow covered.

HOISTS, CHAINS, SLINGS, AND JACKS

Any lifting apparatus should be inspected by a qualified person on an annual basis; this may be statutorily mandated in some jurisdictions. Never take risks with visibly damaged lifting apparatus. A minor leak in a jack lift ram may be the first step in a seal blowout. Tag and report any defective equipment.

FIRES

Some basic fire suppression training will equip an individual with the skills required first to assess the extent of a fire and next, to handle a fire extinguisher. Fire departments and fire fighting equipment suppliers will provide training at the most basic level in how to assess the seriousness of a fire and how to extinguish small fires safely. At least a percentage of employees in a service facility should be trained in basic fire safety. In shops with a health and safety committee, one of the functions of such a committee is to identify potential fire hazards and rectify them. The objective of training in basic firefighting techniques is not to take over the role of the fire departments but rather to do everything possible in a safe manner to control a fire until the arrival of fully trained firefighters.

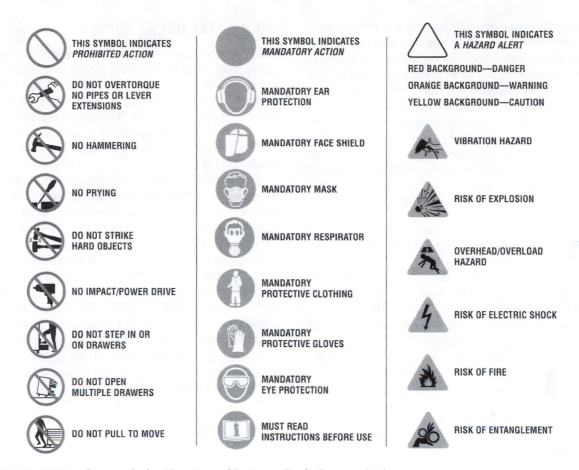

FIGURE 3–2   *ANSI safety symbols. (Courtesy of Snap-on Tools Corporation)*

All shops must be equipped with a variety of fire extinguishers. A fire extinguisher is categorized by the type of fire it is capable of extinguishing. There are four categories of fire:

Class A—Combustible materials such as wood, paper, textiles. Extinguished by cooling, quenching, and oxygen deprivation.

Class B—Flammable liquids, oils, grease, fuels, and paints. Extinguished by smothering (oxygen deprivation).

Class C—Fires that occur in the vicinity of electrical equipment perhaps caused by a current overload. Extinguished by shutting down power switches; smothering with a nonconducting liquid or gas.

Class D—Combustible metals such as magnesium and sodium. Extinguished by smothering with an inert chemical powder.

The corresponding types of fire extinguishers are:

Soda-acid—Consists of bicarbonate of soda and sulfuric acid. Used for Type A fires only. Not a suitable fire extinguisher for a garage.

Water—Consists of pressurized water. Used for Type A fires only. Not a suitable fire extinguisher for a garage.

Carbon dioxide—Consists of compressed carbon dioxide. Used for Type B and C fires; not so effective for Type A fires. This type of fire extinguisher has uses in the shop beyond putting fires out; it provides a safe means of killing a runaway engine without damaging it.

Dry chemical—Consists of mostly sodium bicarbonate. Used for Types A, B, C, and D fires. Most shops should be equipped with dry chemical fire extinguishers. They are best used by directing the stream at the base of the fire and then upward.

CAUTION: Fire suppression in any shop or industrial facility is a job for a trained expert. A technician should always tackle any fire fighting with a safety-first approach.

EMERGENCIES

Emergencies are going to arise from time to time in the workplace, and it makes sense to ensure that every employee has a clear idea of how to react. If the nature of the emergency is medical, then whatever first-aid policy is in effect should kick in, and within this policy is at what point outside aid is sought. First-aid kits and eyewash stations should be clearly identified, and a shop procedure for periodically checking the contents should be in place. Lists of emergency telephone numbers that include fire, medical emergency, and police should be posted. All employees should be aware of the fire drill procedure both in written form and by occasional test drills.

First-Aid Training

Many employers offer training courses in basic first-aid procedures; these courses are usually of short duration but it makes a lot of sense to have a workforce trained in how to react to a medical emergency. The role of the individual administering the first aid is primarily one of assessment. Most people do not require a first-aid course to teach them how to apply an adhesive strip on a cut, but knowledge of how to assess the extent of an injury and how to sequence the steps required to enlist external or expert assistance can save lives.

USE OF FLAMMABLE SOLVENTS FOR CLEANING FLOORS

This generally illegal practice is actually widespread. The potential for environmental damage is great even when the shop floor drain-off is passed through separator tanks before discharge to public sewage systems. Pouring solvent onto an oil-soaked concrete floor does effectively lift oil and grease, which when power washed into drain systems floats on top. It only takes a spark from a torch to ignite this mixture. The resultant danger in a truck shop is amplified by the range of trucks that may be present. When floors are cleaned, ensure that the cleaning agent is both environmentally and workplace safe.

GENERAL SHOP CLEANLINESS

Apart from the fact that it appears to the outsider to be totally unprofessional, dirty and cluttered shop floors are dangerous. Larger shops usually employ cleaning personnel so cleanliness tends to be less of a problem. In smaller service garages, cleanliness should be the responsibility of each technician. It is common sense to clean up after the completion of a job; or better said, cleanup is part of the job. Apart from the obvious danger of a cluttered shop floor, customers are inclined to see a dirty shop facility as a backyard operation, even if the output quality of the work is satisfactory.

MAINS ELECTRICAL SUPPLY

Most service garages use mains electrical supply at three pressures: 110–120 V, 220 V, and 550 V. In most cases, truck technicians are not expected to service this equipment and in some cases may be prohibited from doing so. It is not an objective of this book to cover any mains pressure electrical systems, but merely to underline the danger of working around high-voltage equipment.

Dynamometers, welding equipment, hot tank heaters, and machine shop equipment often use 550 V, three-phase circuits. Welding equipment, machine shop equipment, drill presses, high-intensity lighting, and so on, use single- or three-phase feeds. When a problem occurs, allow a properly qualified person to repair the problem and avoid taking personal risks.

The truck technician should also understand the potential dangers of 110 V systems. Extension cords should never be used in place of permanent outlets and these and all shop electrical tools and trouble lights should be properly grounded. Take care not to run extension cords through puddles of water, and routinely inspect for indications of insulation failure.

VENTILATION

In most jurisdictions, it is mandatory to have a shop exhaust system, a network of flexible exhaust pipes connected to an air pump whose function it is to expel exhaust gas from the building. Uncombusted and incompletely combusted diesel fuels are known carcinogens (cancer-causing agents), and the inhalation of fumes produced by a running engine should be avoided. Today's truck engines burn fairly clean once they are at operating temperature, so it is good practice to at least warm up the engine of a truck for 5 minutes before driving it from the yard into the shop. Whenever an engine has to be run inside the shop, always connect the shop exhaust pipe(s) even when starting and warming the engine before removing the truck.

OSHA

All workplaces in the United States and Canada are protected by **Right-to-Know legislation**, and most companies have an implicit role in ensuring that their employees are fully aware of any harmful substances, hazardous chemicals, or potentially dangerous practices. In the United States, Right-to-Know legislation is covered by the federal **Hazard Communication Regulation** and administered by OSHA. **OSHA** or the **Occupational Safety and Health Administration** is also the federal organization that establishes rules for safe work practices. All workplaces are required to observe OSHA regulations, and technicians should also be somewhat familiar with them; they are posted in most places of work.

In both the United States and Canada, employers have an obligation to ensure that their employees properly understand **Workplace Hazardous Materials Information Systems (WHMIS)** and specifically the **material safety data sheets (MSDS)** that accompany any potentially hazardous substance. As the burden is on each employer to prove that this training has been provided, in most cases they offer WHMIS training on a regular basis and track it by testing and awarding certificates.

CATEGORIES OF HAZARDOUS SUBSTANCES

The technician should understand the terms that describe dangerous substances. The same terms are used on trailer and tanker safety placards and they may determine whether a vehicle can be safely brought into a shop to be worked on.

- **Flammable:** identifies any materials that can be combusted. Includes substances that may be slightly flammable to explosives.
- **Inflammable:** means capable of inflammation so it is synonymous with flammable.
- **Corrosive:** materials of high acidity or alkalinity that may dissolve other substances and destroy animal tissue.
- **Toxic:** materials that may cause death or illness if consumed, inhaled, or absorbed through the skin.
- **Reactive:** materials that may become chemically reactive if they come into contact with other materials resulting in toxic fumes, combustion, or explosion.
- **Radioactive:** any substance that emits measurable levels of radiation. When containers of highly radioactive substances have to be brought into a shop environment, they should be tested by qualified personnel with the appropriate equipment.

HAZARDOUS WASTE DISPOSAL

The disposal of hazardous waste is covered by federal legislation in the United States under the **Resource Conservation and Recovery Act (RCRA)** and reinforced by fines heavy enough to put small operations out of business if they fail to comply; jail sentences are also used to reinforce this legislation. It is administered by a national response center, the phone number of which appears at the end of this chapter.

COMPRESSED AIR

Compressed air is used in truck shops to power equipment and clean components. Using compressed air presents a potential for danger and the rookie trainee is seldom instructed in how to use it. The typical truck technician usually possesses an assortment of air-powered tools, which are coupled by means of pneumatic couplers to a shop air supply. Certain basic rules should be observed and great care should be exercised when using compressed air to clean and air dry components. Eye protection should always be worn when using any air-driven tools and equipment. Special care should be exercised when using compressed air to hydrostatically test components.

CAUTION: Compressed air is potentially dangerous and must be treated with respect. Hydrostatic testing must always be performed in exact accordance with the manufacturer's testing guidelines.

OXYACETYLENE SAFETY

Truck and bus technicians use oxyacetylene for heating and cutting probably on a daily basis; less commonly this equipment is used for braising and welding. Technicians using this equipment require some basic instruction in its safety requirements and handling. The following information should be known by every person using oxyacetylene equipment.

ACETYLENE

Acetylene is an unstable gas produced by immersing calcium carbide in water. It is stored in a compressed state, dissolved in acetone at pressures of approximately 250 psi. Acetylene cylinders are fabricated in sections and seam welded; next a paste of cement, lime silica, and asbestos is baked within the cylinder, forming a honeycomb structure. The cylinder is then charged with liquid acetone, which floods the honeycomb structure and is itself capable of absorbing acetylene. The base of the acetylene cylinder is concave and has two or more fusible plugs threaded into two apertures. The fusible plugs are made of a lead base alloy and are designed to melt at around 100°C (212°F), so if the cylinder is exposed to heat, the fusible plugs are designed to melt and permit the acetylene to escape and avoid exploding the cylinder.

Acetylene regulators and hose couplings use a left hand thread. The regulator gauge working pressure should *never* be set at a value exceeding 15 psi (101 kPa); acetylene becomes extremely unstable at pressures higher than 15 psi. The acetylene cylinder should always be used in the upright position; using the

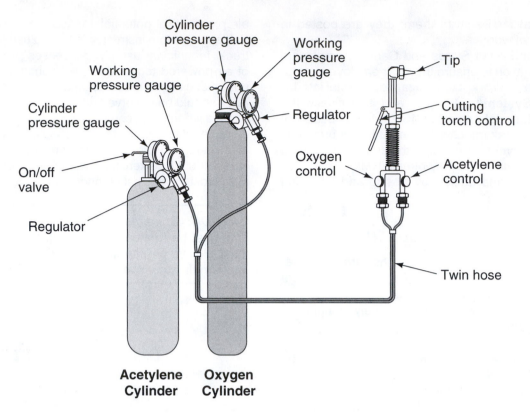

FIGURE 3–3   *Oxyacetylene station setup with a cutting torch*

cylinder in a horizontal position will result in the acetone draining into the hoses. Figure 3–3 shows an oxyacetylene station set up for a cutting torch.

OXYGEN

Oxygen cylinders are forged in a single piece, no part of which is less than ¼″ thick; the steel used is armor plate quality, high-carbon steel suitable for pressure vessels. Oxygen is contained in the cylinder at a pressure of one ton per square inch so the design is consistent with those for high-pressure vessels with radial corners; they are periodically hydrostatic tested at 3,300 psi. The safety device on an oxygen cylinder is a rupture disc designed to burst if cylinder pressure exceeds its normal value such as when exposed to fire. It should be noted that oxygen cylinders tend to pose more problems than acetylene when exposed to fire. They should be stored in a designated place when not in use (this should be identified to the fire department during an inspection) and not left randomly on the shop floor. Oxygen regulator and hose fittings use a right-hand thread.

Oxygen is stored in the cylinders at a pressure of 2,200 psi (15 MPa). The hand wheel-actuated valve forward-seats to close the flow from the cylinder and back-seats when the cylinder is opened; it is important

to ensure, therefore, that the valve is fully opened when in use. The consequence of not fully opening the valve is leakage past the valve threads.

REGULATORS, GAUGES, HOSES, AND FITTINGS

A regulator is a device used to reduce the pressure at which gas is delivered: it sets the working pressure of the oxygen or fuel. Both oxygen and fuel regulators function similarly in that they increase the working pressure when turned clockwise. They close off the pressure when backed out counterclockwise.

Pressure regulators are usually equipped with two gauges. The cylinder pressure gauge indicates the pressure in the cylinder. The working pressure gauge indicates the working pressure and this should be trimmed to the required value while under flow.

The hoses used with oxyacetylene equipment are usually color coded: green is used to identify the oxygen hose and red identifies the fuel hose. The hose connects the regulator assembly with the torch. Hoses may be single or paired (Siamese). Hoses should be routinely inspected and replaced when defective.

Fittings couple the hoses to the regulators and the torch. Each fitting consists of a nut and gland. Oxygen fittings use a right-hand thread and fuel fittings use a left-hand thread. The fittings are machined of brass, which has a self-lubricating characteristic. Never lubricate the threads on oxyacetylene fittings.

TECH TIP: Acetylene fuel fittings use a left-hand thread so they cannot be connected to oxygen fittings.

CAUTION: Lubricating the brass fittings used on oxyacetylene equipment can cause an explosion.

BACKFIRE

Backfire is a condition where the fuel ignites within the nozzle of the torch producing a popping or squealing noise; it often occurs when the torch nozzle overheats. Extinguish the torch and clean the nozzle with tip cleaners. Torches may be cooled by immersing in water briefly with the oxygen valve open.

FLASHBACK

Flashback is a much more severe condition than backfire. It takes place when the flame travels backward into the torch to the gas-mixing chamber and upstream. Causes of flashback are inappropriate pressure settings (especially low pressure settings) and leaking hoses/fittings. When a backfire or flashback condition is suspected, close the cylinder valves immediately, beginning with the fuel valve. Flashback arresters are usually fitted to the torch and limit the extent of damage when a flashback occurs.

TORCHES AND TIPS

Torches should be ignited by first setting the working pressure setting under flow for both gases, then opening the fuel valve only and igniting the torch using a flint spark lighter. Set the acetylene flame to a clean burn (no soot), then open the oxygen valve to set the appropriate flame. When setting a cutting torch, set the cutting oxygen last. When extinguishing the torch, close the fuel valve first, then the oxygen; the cylinders should be shut down finally and the hoses purged.

Welding, cutting, and heating tips may be used with oxyacetylene equipment. Consult a welder's manual to determine the appropriate working pressures for the tip/process to be used. There is a tendency to set gas working pressure high. Even when using a large heating tip often described as a rosebud, the working pressure of both the acetylene and the oxygen should be set at no more than 7 psi (50 kPa).

EYE PROTECTION

Safety requires that a #4 to #6 grade filter be used whenever using an oxyacetylene torch. The flame radiates ultraviolet light, which can damage eyesight.

 SAFETY HOTLINES

Chemical Emergency Preparedness Hotline
CERCLA (SARA Title III)   1-800-535-0202
Chemical Transportation Emergency Center
(CHEMTREC) 24-Hour   1-800-424-9300
CMA Chemical Referral Center   1-800-CMA-8200
EPA RCRA, Superfund, Hazardous
Waste Hotline—Office of Solid Waste and
Emergency Response   1-800-424-9346
EPA, Small Business Hotline   1-800-368-5888
National Response Center   1-800-424-8802
(Report chemical releases, radiological incidents.)
National Safety Council   312-527-4800
NIOSH (National Institute of
Occupational Safety and Health)   1-800-356-4674
OSHA, Health Standards   202-523-7075
Safe Drinking Water Hotline   1-800-426-4791
Substance Identification   1-800-848-6538
United States Department of Transportation
Hotline   202-366-4488

SUMMARY

- Every technician should own and use personal safety attire.
- Because of the high risk of eye injury in the typical service garage, technicians should develop the habit of wearing safety glasses, even when it is not mandatory.
- Novice technicians should be aware that their chances of suffering a back injury during their working life are greater than any other type of injury and should work to develop safe lifting techniques at the beginning of their careers.
- A clean, well-organized shop floor will always produce lower accident rates than cluttered dirty facilities.
- The objectives of first-aid and basic fire suppression training is to teach employees how to react until expert intervention is available.

- Every technician should be fully aware of the danger potential of oxyacetylene equipment and be instructed in how to use it safely.

REVIEW EXERCISES

1. Using an actual service garage, perform a fire safety report noting the location of fire extinguishers, where oxyacetylene gases are stored, and potential fire hazards.
2. Access OSHA on the Internet and note the contents of their information package.
3. Make a list of all emergency services in your local area and how to access them.
4. Using an actual service garage, perform a plant safety inspection noting the condition of hoisting apparatus, electrical wiring and breakers, and any potential dangers.
5. List ten clothing and equipment items that play a role in personal safety on the shop floor.
6. Explain why the cylinder valve on a compressed oxygen cylinder should either be in the fully open or fully closed position. State the maximum setting pressure on an acetylene regulator and explain why it is important to observe this.
7. Outline the internal construction of an acetylene cylinder.
8. Describe the operation of the cylinder safety devices on both oxygen and acetylene cylinders.

REVIEW QUESTIONS

1. Materials that have either high acidity or alkalinity are described as:
   a. Flammable
   b. Toxic
   c. Radioactive
   d. Corrosive
2. Materials that may emit toxic fumes or explode when brought into contact with other materials are described as:
   a. Reactive
   b. Toxic
   c. Radioactive
   d. Corrosive
3. Which of the following is required to have a left-hand thread?
   a. Oxygen cylinder fitting
   b. Acetylene cylinder fitting
   c. Cutting nozzle fitting
4. Which of the following would be the more serious condition when operating oxyacetylene cutting equipment?
   a. Flashback
   b. Backfire
   c. Popping
5. Which of the following grades of eye protection filter would be recommended when performing oxyacetylene cutting?
   a. 2
   b. 5
   c. 10
   d. 13
6. Which type of fire extinguisher would effectively snub a runaway engine without itself causing any damage?
   a. Foam
   b. Carbon dioxide
   c. Dry chemical powder
   d. Water
7. The reason an exhaust pipe of an engine run inside the shop must always be connected to the shop exhaust system is to protect:
   a. The employees
   b. Paintwork on the trucks
   c. Paintwork within the building
8. What is a carcinogen?
   a. Cancer causing agent
   b. Respiratory illness
   c. Fire hazard
   d. Skin irritant

Chapter 4

Engine Basics

OBJECTIVES

After studying this chapter, you should be able to:

- Define the terms that describe basic engine operation.
- Outline the roles played by each subsystem in the engine.
- Describe the seven subcircuits the engine has been divided into for study purposes.
- Calculate engine displacement using the appropriate formula.
- Outline the differences between a square, undersquare, and oversquare engine.
- Apply the term mean effective pressure to an engine operating cycle.
- Identify the differences between a naturally aspirated and manifold boosted engine.
- Explain the term volumetric efficiency and apply it to cylinder breathing efficiencies.
- State how Boyle's and Charles's laws apply to engine operation.
- Describe how friction and inertia factors affect engine operation.
- Explain how the heat energy of a fuel is converted to kinetic energy.
- Define rejected heat and explain the thermal efficiency factors in a diesel engine.
- Outline in detail the diesel four-stroke cycle.
- Outline in detail the diesel two-stroke cycle.
- Outline in detail the Otto four-stroke cycle.
- Explain why it is desirable for any engine to produce peak cylinder pressure at 10 to 20 degrees ATDC on the power stroke during any speed or load phase of operation.
- Define the term scavenging and apply it to both the diesel four-stroke cycle and two-stroke cycle.
- Outline the basic characteristics of a diesel fuel.

KEY TERMS

after top dead center (ATDC)	direct injection (DI)	oversquare engine
before top dead center (BTDC)	engine displacement	ratio
bore	friction	rejected heat
bottom dead center (BDC)	heat energy	spark ignited (SI)
Boyle's law	heat engine	square engine
British thermal unit (Btu)	indirect injection (IDI)	static friction
calorific value	ignition lag	stroke
Charles's law	inertia	swept volume
clearance volume	joule	thermal efficiency
combustion pressure	kinetic energy	top dead center (TDC)
compression ignition (CI)	manifold boost	undersquare engine
compression ratio	mean effective pressure (MEP)	volumetric efficiency
cylinder volume	naturally aspirated (NA)	
diesel cycle	Otto cycle	

INTRODUCTION

This chapter begins by introducing some basic engine terminology with some fairly comprehensive definitions. It then uses this terminology in describing the diesel cycle, the two-stroke diesel cycle, and Otto cycle. It is essential that these terms are understood at least to the extent defined here before progressing to later sections in this textbook. Many of the terms briefly explained here are discussed again and expanded on in later chapters.

ENGINE TERMINOLOGY

The first step to understanding how an internal combustion engine operates is to understand the language used to describe it (Figure 4–1). The terms explained here are part of the vocabulary required by the technician. Every effort has been made here to use the correct term. In some cases the terminology used on the shop floor differs, sometimes considerably. All language is concerned with communicating. So the terms and expressions common on the shop floor express the messages and ideas that are appropriate for that setting. In learning to understand technology, the technician must, for the sake of accuracy, become familiar with technically correct terminology. In many cases, this requires the technician to develop two parallel sets of vocabulary, one for communicating effectively on the shop floor and another for learning and accurately interpreting the technology.

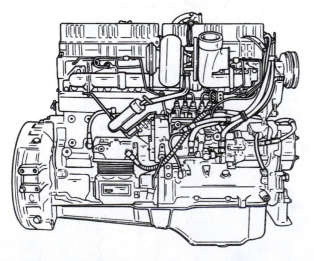

FIGURE 4–1   *Typical in-line, six-cylinder diesel engine.* (*Courtesy of Mack Trucks*)

The following text simply interprets many of the words that are later used to describe the various engine cycles and key events within those cycles.

HEAT ENGINE

Heat engine describes any engine that converts the potential **heat energy** of a combustible fuel into mechanical work.

KINETIC ENERGY

Kinetic energy is the energy of motion. The term is used to describe that portion of the potential heat energy of a fuel that is converted into mechanical work.

TDC

Top dead center. The uppermost point of the piston travel in the engine cylinder.

BDC

Bottom dead center. The lowest point of piston travel in the engine cylinder.

BTDC

Before top dead center. A point of piston travel through its upstroke.

ATDC

After top dead center. A point of piston travel through its downstroke.

BORE

Bore is cylinder diameter. It defines how the sectional area over which cylinder pressures developed in the engine will act.

STROKE

Stroke is the distance through which the piston travels from BDC to TDC. Stroke is established by the crank throw offset; that is, the distance from the crankshaft centerline to the throw centerline multiplied by 2 equals the stroke dimension.

SWEPT VOLUME

Swept volume is the volume displaced by the piston in the cylinder as it moves from BDC to TDC. It can be calculated if both stroke and bore are known.

CLEARANCE VOLUME

Clearance volume is the volume in an engine cylinder when the piston is at the top of its travel or TDC. Clearance volume influences actual compression temperatures and cylinder breathing efficiencies.

CYLINDER VOLUME

Cylinder volume is the total volume in the cylinder when the piston is at BDC: swept volume plus clearance volume. The clearance volume on older indirect injection (IDI) diesel engines was considerable.

ENGINE DISPLACEMENT

Engine displacement is the swept volume of all of the engine cylinders expressed in cubic inches or cubic centimeters/liters.

Engine displacement = bore × bore × stroke × 0.7854 × number of cylinders

Examples:
Detroit Diesel Series 60

Engine displacement calculation data: 6-cylinder engine, bore 130 mm, stroke 160 mm

Engine displacement = 130 × 130 × 160 × 0.7854 × 6

= 12,742,329.6 cubic millimeters

= 12.7 liters

Caterpillar C-15

Engine displacement calculation data: 6-cylinder engine, bore 5.4″, stroke 6.5″

Engine displacement = 5.4 × 5.4 × 6.5 × 0.7854 × 6

= 893 cubic inches

TECH TIP: To convert liters to cubic inches or cubic inches to liters, use the following simple formulae in which 61 is either multiplied or divided into the value to be converted:

12.7 liters × 61 = 774.7 cubic inches or rounded to 775 cubic inches

893 cubic inches ÷ 61 = 14.639 liters or 14.6 liters

SQUARE ENGINE

Square engine is the term used to describe an engine in which the cylinder bore diameter is exactly equal to the piston stroke dimension. When bore and stroke values are expressed, bore always appears before stroke.

OVERSQUARE ENGINE

Oversquare engine is the term used to describe an engine in which the cylinder bore diameter is larger than the stroke dimension.

UNDERSQUARE ENGINE

Undersquare engine is the term used to describe an engine in which the cylinder bore diameter is smaller than stroke dimension. Most truck and bus diesel engines are undersquare as are most high compression engines.

CI

Compression ignition. CI is the acronym commonly used to describe any diesel engine or one in which the cylinder fuel air charge is ignited by the heat of compression.

SI

Spark ignited. SI is the acronym used to describe any engine in which the fuel air charge is ignited by a timed electrical spark.

DI

Direct injection. A CI or SI engine in which a liquid fuel charge is injected directly into the engine cylinder rather than to a precombustion chamber or part of the intake manifold.

IDI

Indirect injection. A CI or SI engine in which the fuel charge is introduced outside of the engine cylinder to a precombustion chamber, cylinder head intake tract, or intake manifold.

RATIO

Ratio is the quantitative relationship between two values expressed by the number of times one contains the other. The term is commonly used in automotive technology to describe the drive/driven relationships of enmeshed gears, the mechanical advantage of levers, and cylinder compression ratio.

COMPRESSION RATIO

Compression ratio is a measure of the cylinder volume when the piston is at BDC versus cylinder volume when the piston is at TDC. Compression ratios in diesel engines fall between 14:1 and 24:1. Current, high-speed, turbocharged truck and bus diesel engines have compression ratios typically around 16:1.

COMPRESSION PRESSURE

Compression pressure is the actual cylinder pressure developed on the compression stroke. Actual compression pressures developed range from 2.41 MPa (350 psi) to 4.82 MPa (700 psi) in CI engines. The higher the compression pressure, the more heat developed in the cylinder. However, the work done by the piston on its compression stroke must be subtracted from the work it receives through the power stroke. Truck diesel engines typically produce compression pressures of ± 600 psi.

COMBUSTION PRESSURE

Combustion pressure is the peak pressure developed during the expansion or power stroke. In CI engines, this typically approximately doubles the compression pressure but in electronically controlled truck diesel engines, combustion pressures may peak at up to four times the compression pressure.

MEAN EFFECTIVE PRESSURE (MEP)

Mean effective pressure is the expansion pressure (pressure developed in the engine cylinder during the power stroke) minus the compression pressure. In a four-stroke cycle CI engine it is assumed that cylinder pressure through the intake and exhaust strokes is zero for purposes of calculating MEP, because through both there is no significant amount of pressure. The MEP describes the relationship between the work performed by the piston (in compressing the air charge) to the work received by it (through its downstroke on the power stroke). It, therefore, expresses the net gain in terms of work.

CONSTANT PRESSURE CYCLE

The theoretical diesel cycle engine presumes that the fuel supplied to the cylinder during the expansion stroke will be at a rate permitting combustion pressure to remain constant through a large portion of the stroke. The theoretical diesel cycle is neither practical nor desirable (see the following section).

EXPANSION STROKE-VECTORS-THROW LEVERAGE

The objective of the modern CI engine is to transfer the power developed in the engine cylinders as smoothly and evenly as possible to the power take-off mechanism, usually a flywheel. The relationship between the crankshaft throw and the crankshaft centerline is that of a lever. The extent of leverage or mechanical advantage depends on the rotational position of the throw, which ranges from no leverage when the throw is positioned at TDC to maximum leverage when the throw is positioned at 90 degrees before or after TDC. This makes the relationship between cylinder pressure (gas pressure acting on the piston) and throw leverage (the position of the piston) critical in meeting the objective of smooth/even transfer of power from engine cylinders to drivetrain. When the piston is at TDC beginning a power stroke, it is desirable to have minimum cylinder pressure, because in this position throw leverage is zero; therefore, no power transfer is possible. A properly set up fuel system attempts to manage cylinder pressure so that in any given cycle it peaks somewhere between 10 degrees and 20 degrees ATDC when there is a small amount of throw leverage. As the piston is forced down through the power stroke, gas pressure acting on the piston diminishes but as it does so, throw leverage increases. Optimumly, this would result in consistent torque delivery from a cylinder through the power stroke to 90 degrees ATDC.

NATURALLY ASPIRATED

Naturally aspirated (NA) describes an engine whose only means of inducing air (or air/fuel mixture) into its cylinders is the low cylinder pressure created by the downstroke of the piston.

MANIFOLD BOOST

Manifold boost describes the extent of charge pressure above atmospheric delivered to the cylinders in a turbocharged engine. Most current truck and bus diesel engines are boosted; that is, they are turbocharged.

VOLUMETRIC EFFICIENCY

Volumetric efficiency is a measure of an engine's breathing efficiency. Often defined as the ratio between the volume of the induced cylinder charge versus the cylinder swept volume, but correctly expressed, it is a ratio of *masses*, not volumes. The term can be generally used to describe pump aspiration (breathing) efficiencies. If an example of a liquid filling a pump cylinder is used, it is the amount of actual liquid drawn into the cylinder in a cycle versus the maximum quantity of the liquid the cylinder could contain. In a diesel engine the fluid being induced to the cylinder is air, which happens to be compressible, so in a turbocharged engine volumetric efficiency can often exceed 100%. Therefore, another way of defining the term would be that it is the amount of air charged to the engine cylinder in actual cycle versus the amount it would contain if it were at atmospheric pressure. Volumetric efficiency is usually expressed in percentage terms.

BOYLE'S LAW

(Robert Boyle, U.K., 1627–1691)

Boyle's law states that the absolute pressure that a given quantity of gas at constant temperature exerts against the walls of a container is universally proportional to the volume occupied. In other words, assuming a constant temperature, the pressure of a specific quantity of gas depends on the volume of the vessel it is contained in. So to use an example of this law as it applies to a diesel engine, it means that a constant temperature mass of gas (air) in a cylinder as its volume is reduced by moving the piston will exert pressure on the cylinder walls because the number of molecules will remain the same, but they will have less room to move in, thereby causing the pressure rise.

CHARLES'S LAW

(Jacques Charles, France, 1746–1823)

Charles's law states that the increase in temperature in gases produces the same increase in volume if the pressure remains constant. In other words, heating a gas must result in an increase in volume if the pressure is to remain unchanged. Using the Celsius scale, it can be proved that the volume of a gas increases by 0.003663 of its volume at zero Celsius for every one degree of temperature rise. On the Fahrenheit scale, the volume increases by 0.002174 for every one degree of temperature rise above 32°. By graphing this equation negatively, a point would be reached at which the gas would have no volume; the vibration of the molecules would cease, the gas would contain no heat energy and would cease to exist as a substance. This would occur at absolute zero or – 273°C (– 460°F.) To conclude, if the volume of a gas is changed by increasing its temperature while keeping its pressure constant, then its volume will increase proportionally with temperature rise.

LAWS OF THERMODYNAMICS

First law: states that heat energy and mechanical energy are naturally convertible. This is predicated by the law of conservation of energy, which states that energy can be neither created or destroyed. This means that the total energy available remains constant. However, energy can change its form. Heat energy can be changed into mechanical energy (the operating principle of any internal combustion engine) or vice versa.

Similarly, heat energy can be changed into electrical energy and vice versa.

Second law: states that heat will not flow from a cool body to a warmer body without some kind of assistance but that it will flow from a warm body to a cooler body. If the objective is to force heat from a cool body to a hot body such as in an air-conditioning system, some form of external assistance must be applied.

FRICTION

Force is required to move an object over the surface of another. **Friction** is the resistance to motion between two objects in contact with each other. Friction is factored by both load and surface condition. Smooth surfaces produce less friction than rough surfaces, and if a lubricant such as water or oil is added, friction diminishes. Lubricants coat and separate two surfaces from each other and reduce friction, but the lubricant itself provides some resistance to movement, which is known as viscous (fluid) friction. A friction bearing such as a crankshaft main bearing provides a sliding friction dynamic, while ball bearings provide a rolling friction dynamic that usually offers less resistance to motion.

STATIC FRICTION

Static friction describes the characteristic of an object at rest to attempt to stay that way. For example, the engine piston at its travel limit stops momentarily before the crankshaft and connecting rod reverse its movement. When the piston is momentarily stopped, the crankshaft must overcome the static friction of the stationary piston, which places both the connecting rod and a portion of the crankshaft under tension. This tensile loading of the connecting rod and crankshaft is amplified as rotational speed increases.

INERTIA

Inertia describes the tendency of an object in motion to stay in motion or conversely, an object at rest, to remain that way. Kinetic inertia describes the characteristic of an object in motion to stay in motion. For example, an engine piston moving in one direction must be stopped at its travel limit and its kinetic inertia must be absorbed by the crankshaft and connecting rod. The inertia principle is used by the engine harmonic balancer and the flywheel—the inertial mass represented by the flywheel would have to be greatest in a single cylinder, four-stroke cycle engine. As the number of cylinders increases, the inertial mass represented by the flywheel can be diminished due to the greater mass of rotating components and the higher frequency of power strokes.

JOULE'S HEAT EQUIVALENCY

(James Prescott Joule, U.K., 1818–1889)

Joule established the relationship between units of heat and work that could be done. This relationship is used to describe the potential energy of a fuel and is known as Joule's mechanical heat equivalent.

1 Btu of potential heat energy = 778 ft.-lb of mechanical energy

1 J (joule) of potential heat energy = 1 J of mechanical energy

1 J = 1 Newton/meter (Nm) = 0.7374 ft.-lb

CALORIFIC VALUE

Calorific value is the potential heat energy of a fuel. A heat engine attempts to convert the potential heat energy of a fuel into kinetic energy: the thermal efficiency of the engine is a measure of how successful this is.

The calorific value of fuels is measured in Btu (English system) or joules and calories (metric system).

BTU

One **British thermal unit (Btu)** is the amount of heat required to raise the temperature of 1 pound of water 1 degree Fahrenheit.

THERMAL EFFICIENCY

Thermal efficiency is a measure of the combustion efficiency of an engine calculated by comparing the heat energy potential of the fuel (calorific value) with amount of work produced. Electronically controlled CI engines can have thermal efficiency values exceeding 40%.

REJECTED HEAT

Rejected heat is that percentage of the heat potential of the fuel that is not converted into useful work by an engine. If a CI engine operating at optimum efficiency can be said to have a thermal efficiency of 40%, then 60% of the calorific value of the fuel can be described as rejected heat. Half of the rejected heat is typically transferred to the engine hardware to be dissipated to the atmosphere by the engine cooling system, and the other half exits in the exhaust gas.

THE DIESEL CYCLE

A cycle is a recurring sequence of events. The **diesel cycle** is usually described by the four strokes of the pistons made as an engine is turned through two revolutions. A complete cycle of a diesel engine requires two full rotations and this translates into 720 crankshaft degrees. Each of the four strokes that comprise the cycle involve moving a piston either from the top of its travel to its lowest point of travel or vice versa; each stroke of the cycle therefore translates into 180 crankshaft degrees. The four strokes that comprise the four-stroke cycle are: intake, compression, power, and exhaust (Figure 4–2). The diesel cycle by definition is a four-

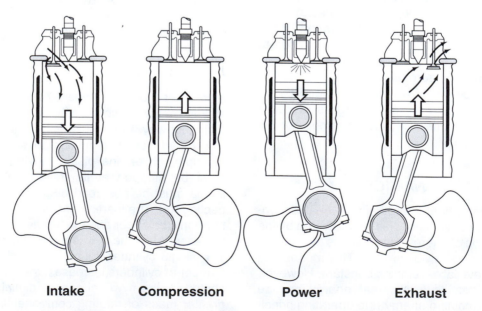

Intake   **Compression**   **Power**   **Exhaust**

FIGURE 4–2   *The four-stroke diesel cycle.*

stroke cycle. Two-stroke cycle compression ignition engines exist but correctly, these should be qualified as two-stroke cycle diesel engines.

DESCRIPTION OF THE FOUR-STROKE CYCLE IN A DIRECT IGNITION, COMPRESSION IGNITION ENGINE

1. Intake Stroke

The piston is drawn from top dead center (TDC) to bottom dead center (BDC) with the cylinder head intake valve(s) held open. The downstroke of the piston creates lower-than-atmospheric pressure in the cylinder and in a naturally aspirated engine, this induces a charge of fresh, filtered air into the cylinder. Because most current truck and bus engines are turbocharged, the cylinder will actually be filled with charged air (that is, at a pressure above atmospheric) when the intake valve(s) open; the degree of cylinder charge will depend on the manifold boost value. Air is a mixture of gaseous elements: approximately four-fifths nitrogen and one-fifth oxygen. The oxygen is required to combust the fuel that will be introduced to the cylinder later in the cycle. By pressurizing the air charge using a turbocharger, more oxygen can be forced into each engine cylinder. All diesel engines are designed for lean burn operation; that is, the cylinder will be charged with much more air than that required to combust the fuel. Volumetric efficiency in most phases of engine operation will usually exceed 100% in turbocharged engines.

2. Compression Stroke

The piston is now driven from BDC to TDC with the intake and exhaust valves closed, compressing the charge of air in the cylinder and in doing so, heating it. Compression pressure in diesel engines varies from 400 psi (27 atms/2750 kPa) to 700 psi (48 atms/4822 kPa). The actual amount of heat generated from these compression pressures varies, but it usually substantially exceeds the minimum ignition temperature values of the fuel. Compression ratios used to achieve the compression pressure required of diesel engines generally vary from a low of 14:1 to a high of 25:1. However, in modern turbocharged, highway diesel engines, compression ratios are typically around 16:1.

3. Expansion or Power Stroke

Shortly before the completion of the compression stroke, atomized fuel is introduced directly into the engine cylinder by a multi-orifii (multiple hole) nozzle assembly. The fuel exits the nozzle orifii in the liquid state in droplets appropriately sized for combustion in a DI engine. Once exposed to the heated air charge in the cylinder these liquid droplets are first vaporized, then ignited. The ignition point is designed to usually occur just before the piston is positioned at TDC with the objective of peaking the gas pressure acting on the piston during the expansion stroke at 10 to 20 degrees ATDC. However, because this is the objective at all engine speeds and loads, it is difficult to uniformly achieve it with hydromechanical fuel systems. Noxious emissions requirements of newer electronically managed engines often mean that the power stroke may not be managed to produce optimum mechanical efficiency because of the requirement to remain within legal emission specifications.

In managing the power stroke it is desirable to have little pressure acting on the piston at TDC. Cylinder gas pressure should peak at 15 to 20 degrees ATDC when throw leverage exists but is close to minimum. As gas pressure acts on the piston and forces it through its stroke, cylinder pressure will decrease, but as it does so, throw leverage increases maximizing when the angle between the connecting rod and crank throw is at 90 degrees. This relationship between pressure and throw leverage helps to transmit the energy produced in the engine cylinder as smoothly as possible to the flywheel (Figure 4–3).

4. Exhaust Stroke

Somewhere after 90 degrees ATDC during the expansion stroke, most of the heat energy that can be converted to kinetic energy has been converted and the

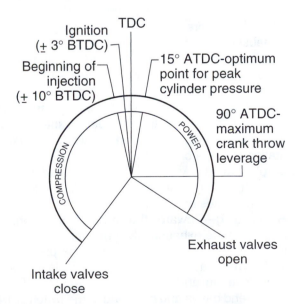

FIGURE 4–3 *Events of the compression and power strokes.*

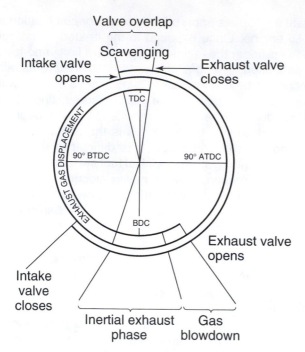

FIGURE 4–4   *Events of the intake and exhaust strokes.*

exhaust valve(s) open. The products of cylinder combustion are known as end gas. The exhausting of combustion end gases occurs in four distinct phases and the process begins during the latter portion of the power stroke (Figure 4–4):

1. Pressure differential—At the moment the exhaust valves open during the latter portion of the power stroke, pressure is higher in the cylinder than in the exhaust manifold. High pressure end gas in the cylinder will therefore flow to the lower pressure in the exhaust manifold.
2. Inertial—Next the piston comes to a standstill at BDC at the completion of the power stroke. However, gas inertia established during the pressure differential phase will result in the end gases continuing to flow from the cylinder to the exhaust manifold while the piston is in a stationary and near stationary state of motion.
3. Displacement—As the piston is forced upward through its stroke, it positively displaces combustion end gases above it.
4. Scavenging—Toward the end of the exhaust stroke, as the exhaust valve(s) begin to close, the intake valve(s) begin to open with the piston near TDC. The scavenging phase takes place during valve overlap and can be highly effective in expelling end gases and providing some piston crown cooling. The efficiency of the scavenging process is greatest with turbocharged engines.

THE TWO-STROKE CYCLE DIESEL ENGINE

The intake and exhaust strokes are eliminated, meaning that in 360 degrees of rotation, each engine cylinder has fired. Theoretically, the two-stroke cycle diesel engine should develop twice as much power as a four-stroke cycle engine of the same displacement, but in reality this is not achieved mainly due to reduced cylinder breathing efficiency. The two-stroke cycle diesel engine principle is widely used but the only examples in the truck and bus industry are manufactured by Detroit Diesel Corporation. Currently, these engines are not able to meet EPA noxious emissions criteria when fueled with diesel fuel. However, for more than a generation they have been the bus/coach engine of choice. These engines have also been adapted for using fuels other than diesel fuel oil with some success. Transit corporations recondition chassis and powertrain components to a greater extent than is general in the trucking industry, so these engines will be around for many years yet. Many of the two-stroke cycle engines in existence are managed electronically; retrofitting electronic controls on some Detroit Diesel hydromechanical engines can make them more fuel efficient.

The physical characteristics of the two-stroke cycle engine differ from those of the four-stroke cycle diesel engine mainly because cylinder breathing must take place in less than one-fifth of the time. All the valves in the cylinder head, usually four per cylinder, are exhaust valves. After combustion, the cylinder end gases must be expelled and to enable this, air must be pumped through the cylinder from an air box charged by a Roots blower, sometimes aided by a turbocharger. In truck and bus applications, because of widely variable speed and load variations, engine breathing requires the use of a Roots blower. A Roots blower is a positive displacement pump that works efficiently at all rotational speeds. Its disadvantage is that it is gear driven and leeches engine power. A turbocharger may be used in applications such as a genset in which engine loading would result in adequate exhaust gas heat to drive the turbine with sufficient efficiency for it to act as the air box pump. The cylinder liners are machined with ports designed to be exposed when the piston is in the lower portion of its downstroke; when these ports are exposed to the air box by the downward travelling piston, the cylinder is charged with air for scavenging and breathing. The ports are usually canted (angled) to encourage a vortex (cyclonic) air flow dynamic.

The two-stroke cycle sequence begins with the piston at BDC when what is termed cylinder *scavenging* takes place. At this moment, the piston has fully exposed the canted intake ports and the exhaust valves are fully opened. Air from the air box rushes into the cylinder and displaces the combustion end gases, spiralling them upward to exit through the exhaust valves. Air from the air box continues to charge the cylinder until the piston reverses and its upward travel closes off the intake ports; the exhaust valves close almost simultaneously. Every upward stroke of the piston is therefore a compression stroke. If all the end gases were effectively expelled, only air is compressed; however, scavenging efficiency is a problem with these engines and any end gases remaining in the cylinder after the exhaust valves close will dilute the incoming air change. Shortly before TDC, the fueling of the cylinder begins directly into the engine cylinder. After a short delay, ignition occurs and expanding combustion gases act on the piston and drive it downward through the power stroke. Every downward stroke of the piston is a power stroke. Shortly before the liner intake ports are exposed by the piston, the exhaust valves open beginning the exhaust process, which must take place quickly and in two stages: pressure differential and scavenging (Figures 4–5 and 4–6).

THE OTTO CYCLE

Every technician should have a clear understanding of the **Otto cycle**, the engine cycle used in most SI (spark-ignited) gasoline-fueled engines. The first stroke of the cycle is appropriately called the induction stroke. A charge of air/fuel mixture is induced into the engine cylinder by the low pressure created by the downstroke of the piston in the engine cylinder as it moves from TDC to BDC. The air/fuel charge is mixed outside of the engine cylinder, either by a carburetor located upstream from the intake plenum or by injectors located at the intake tract of the cylinder head. When the piston

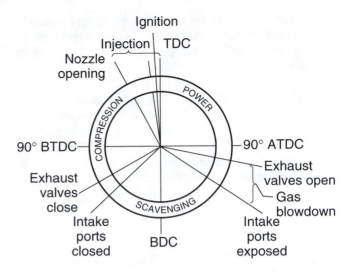

FIGURE 4–6   _Two-stroke diesel cycle events._

stops at BDC and the intake valve(s) close, the cylinder contains all of the ingredients, that is, the fuel and air required for the subsequent power stroke. The mixture of fuel and air is compressed on the second stroke of the cycle, the compression stroke. Compression ratios must be somewhat lower on SI engines because compression ignition of the more volatile fuel used (gasoline) must be avoided: 8:1 is typical and this produces compression pressures around 150 psi (1 MPa). The cylinder charge is ignited just before the completion of the compression stroke.

The power stroke can produce peak combustion pressures approximately five times the compression pressure. The gasoline fuels used are chemically more complex than diesel fuels and must be formulated with a mixture of fractions (hydrocarbon compounds derived from crude petroleum) that ignite in a sequential chain reaction, expanding to apply pressure to the piston evenly through the power stroke. As with the diesel engine, cylinder pressure and the crank throw angle must

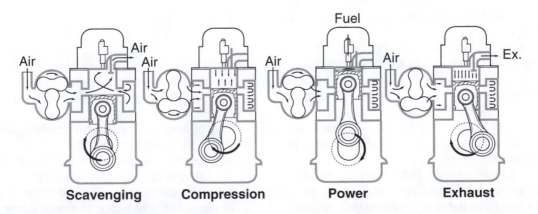

FIGURE 4–5   _The two-stroke diesel cycle._

be phased to transfer the power smoothly and evenly to the flywheel. At some point after the piston has passed 90 degrees ATDC on the power stroke, the exhaust valves open and begin a four-phase exhaust process more or less identical to that of the diesel engine, with the exception of the scavenging phase, which must be held short to prevent the discharge of unburned air/fuel mixture (the incoming charge) into the exhaust system. Thermal efficiencies tend to be lower in Otto cycle engines than in diesel engines while weight per horsepower values are also lower, making the diesel engine a less attractive option in an automobile.

ENGINE SYSTEMS AND CIRCUITS

Many of the components in the diesel (compression ignition or CI) engine are identical to those in the Otto (spark ignition or SI) cycle engine. For study purposes, the engine components are divided as follows:

1. Engine housing components—the cylinder block, cylinder head(s), oil sump, rocker covers, timing gear covers, and flywheel housing.
2. Engine powertrain—the components directly responsible for power delivery including the piston assemblies, connecting rods, crankshaft assembly, vibration damper, and flywheel.
3. Engine feedback assembly—the engine's self-management components also known as the valve train assembly. This term is used to describe the diesel engine timing gear train, camshaft, valve trains, valves, fueling apparatus, and accessory drive components.
4. Engine lubrication circuit—the oil pump, relief valve, lubrication circuitry, full flow filter(s), bypass filters, and heat exchangers.
5. Engine cooling circuit—the coolant pump, thermostat(s), water jacket, coolant manifold, filter, shutters, fan assembly, radiator, and other heat exchangers.
6. Engine breathing system—the engine intake and exhaust system components including precleaners, air cleaners, ducting, turbocharger, Roots blower, charge air heat exchangers, tip turbine assemblies, intake and exhaust manifolds, pyrometer, exhaust piping, engine silencer, catalytic converter, and other external noxious emission control apparatus.
7. Engine Fuel Management System—the fuel storage, pumping, metering, and quantity control apparatus including a management computer, sensors and actuators, hydraulic injectors, mechanical

unit injectors (MUIs), electronic unit injectors, (EUIs) hydraulically actuated, electronic unit injectors (HEUIs), electronic unit pumps (EUPs), common rail injection, hydromechanical injection pumps, fuel tanks, filters, and transfer pumps.

DIESEL FUEL

Diesel fuel is covered in some detail in later sections of this book, so this is just an introduction. The fuel used in modern diesel engines on North American highways is composed of roughly 85% carbon and 12 to 15% hydrogen, much the same as the chemical composition of gasolines. Unlike gasolines, diesel fuel does not vaporize readily at ambient temperatures, so it is less likely to form combustible mixtures of fuel and air. The heat required to ignite the fuel oil is defined by the most volatile fractions of the fuel. This temperature is usually around 250°C; it would be higher, around 290°C (550°F), using fuel with the poorest ignition quality that can be sold legally for use on North American highways.

Using diesel fuel oils with a fire point of around 250°C, at least this temperature must be achieved in the engine cylinder during the compression stroke of the piston if the fuel is to be ignited. In fact, actual cylinder temperatures generated on the compression stroke tend to be considerably higher than the minimum required to ignite the fuel. The greater the difference between these two temperature values, the shorter the ignition lag. **Ignition lag** is the time between the entry of the first droplets of fuel to the engine cylinder and the moment of ignition.

TECH TIP: The information contained in this chapter is theoretical but contains the building blocks required to properly understand engine operation from the repair technician's perspective. Make sure you understand these concepts!

SUMMARY

- Most diesel engines are rated by their ability to produce power and torque. The tendency is to rate gasoline-fueled auto engines by their total displacement.
- Diesel engines have high compression ratios so they tend to be undersquare.
- MEP is the average pressure acting on the piston through the four strokes of the cycle. Usually the intake and exhaust strokes are discounted, so MEP is equal to the average pressure acting on the piston through the compression stroke subtracted

from the average pressure acting on the piston through the power stroke.

- Almost every medium and large bore highway diesel engine is manifold boosted; that is, it is turbocharged.
- An engine attempts to convert the potential heat energy of a fuel into useful kinetic energy: the degree to which it succeeds is rated by its thermal efficiency.
- That portion of the heat energy of a fuel not converted to kinetic energy is termed rejected heat that must be dissipated to the atmosphere by means of the engine cooling and exhaust systems.
- The diesel cycle is a four-stroke cycle consisting of four separate strokes of the piston occurring over two revolutions; a complete engine cycle is therefore extended over 720 degrees.
- The two-stroke diesel cycle enables every downstroke of a piston to be a power stroke, so in theory, it has the potential to produce more power than the four-stroke cycle. In practice, this is unobtainable largely because the engine cylinders must be scavenged of end gases and aspirated with a new charge of air in a fraction of the time.
- Ideally, engine fueling should be managed to produce peak cylinder pressures at somewhere around 10 to 20 degrees ATDC when the relative mechanical advantage provided by the crank throw position is low. This means that as cylinder pressure drops through the power stroke, throw mechanical advantage increases peaking at 90 degrees ATDC, providing a smooth unloading of force to the engine flywheel.
- For study purposes, the diesel engine has been divided into seven subcircuits: the powertrain, feedback assembly, engine housing components, lubrication circuit, cooling system, breathing system, and fuel management system.
- Diesel fuel is composed of roughly 85% carbon and between 12% and 15% hydrogen; these properties of carbon and hydrogen are similar to those in gasoline or engine oil. Diesel fuel is composed of less volatile fractions than gasoline, giving it a higher temperature flashpoint.

REVIEW QUESTIONS

1. A diesel engine has a bore of 4.85″ and a stroke of 5.15″. Which of the following correctly describes the engine?
   a. 10″ displacement
   b. 10 liter displacement
   c. Oversquare
   d. Undersquare

2. Which of the following accurately describes engine displacement?
   a. Total piston swept volume
   b. Mean effective pressure
   c. Peak horsepower
   d. Peak torque

3. Engine breathing in the two-stroke cycle is sometimes referred to as:
   a. Inertial
   b. Scavenging
   c. Exhaust blowdown

4. The tendency of an object in motion to stay in motion is known as:
   a. Kinetic energy
   b. Dynamic friction
   c. Inertia
   d. Mechanical force

5. When a modern diesel engine is running at optimum efficiency, the percentage of rejected heat would typically be:
   a. 20 %
   b. 40 %
   c. 60 %
   d. 80 %

6. What percentage of the potential heat energy of the fuel does the modern diesel engine convert to kinetic energy when it is operating close to optimum efficiency?
   a. 20 %
   b. 40 %
   c. 60 %
   d. 80 %

7. In which of the four strokes of the cycle is the cylinder pressure at its highest in a running diesel engine?
   a. Intake
   b. Compression
   c. Power
   d. Exhaust

8. Where does scavenging take place on a four-stroke cycle, diesel engine?
   a. BDC after the power stroke
   b. TDC after the compression stroke
   c. Valve overlap
   d. 10 to 20 degrees ATDC on the power stroke

9. Ideally, where should peak cylinder pressure occur during the power stroke?
   a. TDC
   b. 10 to 20 degrees ATDC
   c. 90 degrees ATDC
   d. At gas blowdown

10. Which of the seven subcircuits of the engine deals with the camshaft, valve trains, and mechanical actuation of injectors and fuel pumping apparatus?
    a. Powertrain
    b. Feedback assembly

c. Lubrication circuit
d. Engine housing circuit

11. A Cummins N series engine has a displacement of 855 cubic inches. Convert this to metric values.
    a. 11 liters
    b. 12 liters
    c. 14 liters
    d. 15 liters

12. Convert the displacement of a Caterpillar C10 engine to express it in cubic inches.
    a. 454 cubic inches
    b. 555 cubic inches
    c. 610 cubic inches
    d. 724 cubic inches

13. A Mack Truck's E7 engine has a total displacement of 728 cubic inches. Express this value in metric terms.
    a. 11.1 liters
    b. 11.9 liters
    c. 12.7 liters
    d. 14 liters

14. A six-cylinder engine has a bore of 4.875″ and a stroke of 6.5″. Calculate the total displacement and express in cubic inches.
    a. 530 cubic inches
    b. 610 cubic inches
    c. 728 cubic inches
    d. 855 cubic inches

15. A six-cylinder engine has a bore of 140 mm and a stroke of 152 mm. Calculate the total displacement and express in liters.
    a. 11.1 liters
    b. 12.7 liters

c. 14 liters
d. 14.8 liters

16. When running an engine at idle speed, which would typically be the optimum location to peak the cylinder pressure?
    a. 20 degrees BTDC
    b. 10 degrees BTDC
    c. TDC
    d. 15 degrees ATDC

17. When running an engine at rated speed and load, which would typically be the optimum location to peak the cylinder pressure?
    a. 30 degrees BTDC
    b. 15 degrees BTDC
    c. TDC
    d. 15 degrees ATDC

18. At which point in the engine cycle does the crank throw leverage peak?
    a. TDC
    b. 15 degrees ATDC
    c. 90 degrees ATDC
    d. BTDC

19. Which law states that all energy forms are convertible?
    a. The 1st law of thermodynamics
    b. The 2nd law of thermodynamics
    c. Boyle's law
    d. Charles's law

20. The energy of motion is known as:
    a. Inertia
    b. Kinetic
    c. Thermal
    d. Potential

Chapter 5

History of the Heat Engine

PREREQUISITE

Chapter 4

OBJECTIVES

After studying this chapter, you should be able to:

- Identify some of the key players in the history of technology.
- Outline the development of the diesel engine chronologically.
- Outline a pen picture history of Watt, Carnot, Diesel, and Cummins.
- Describe the first heat engine and the first jet engine.
- List the inventors of most motive power achievements.
- Identify the builder of the world's first automobile.
- Recognize the significant events that contributed to the evolution of the modern truck engine.
- Recount the role of Henry Ford in revolutionizing industrial process.
- Recount how a family chauffeur eventually became known as the "father of the highway diesel engine."
- Identify the inventors of the first electronic digital computer.
- Define the acronym ENIAC.

KEY TERMS

aeolipile

Carnot cycle

digital computer

heat engine

latent heat

power

reaction turbine

INTRODUCTION

The modern vehicle has reached a stage of technological development in which an operator rarely has to consider exactly what happens when the ignition key is turned and energizes the series of circuits that make it run. It is an action that many of us perform several times a day in a world in which a person may travel more in a single day than our forebearers of just 200 years ago did in a lifetime.

The modern vehicle is a result of humanity's pursuit of constant improvement. We may be able to identify the individual credited with the invention of the diesel engine, but it should always be remembered that he had the cumulative practical and technical achievements of all those who preceded him as a starting base for his research. The first diesel engines were crude and unreliable machines. In a hundred or so years, the observations, research, and sometimes lifelong efforts of thousands of engineers, technicians, and designers, most of whom are "nameless," have resulted in the technical excellence of the modern, highly reliable diesel engine.

Thousands of components are integral in the modern truck chassis, each with roles in various mechanical, hydraulic, pneumatic, electrical, and electronic circuits. Each had to be invented, manufactured, and systematically improved to its present stage of development. Each will surely be further improved as each year goes by. And as each year goes by, the rate of improvement accelerates. Consider the truck manufactured just 20 years ago and performance match it head to head with today's version. Aside from subjective considerations such as appearance, the contemporary vehicle will in most cases significantly outperform its predecessors just as surely as the 2025 model will outperform today's.

In inventing the diesel engine, Rudolph Diesel took a significant step forward in the progress of technology as applied to the modern truck. There were significant steps both before and after him. The argument has been made that the invention of the Otto cycle and diesel engines was inevitable after the theoretical foundation provided by the French scientist Beau de Rochas, but it would be difficult to make the argument that Diesel's engine of 1892 was not a pretty significant event. This chapter identifies some of the individuals in the history of technology whose accomplishments were important to the development of both the **heat engine** (any machine that converts the heat energy of a fuel to mechanical outcomes) and the modern diesel engine. In examining the history of technology until fairly recent times, it was usually possible to match an accomplishment with the name of the inventor. Today the thrust toward technological advance is something that large corporations must finance and "inventions" are usually the work of dozens of individuals and are *owned* by the funding company.

Within this list of innovators, there are people who may be described as mathematicians, scientists, engineers, chemists, technicians, physicists, and diesel mechanics with many different levels of education but with certainly one label in common—they were inventors. Most inventions result from an individual's refusal to accept the impossible.

In many ways, the history of technology is the history of modern man. Take any current truck, put it on an empty asphalt parking lot, and completely disassemble it. Disassemble every component and subcomponent until further disassembly is impossible. What you will now see are 100,000 or so manufactured components that would seem to have little obvious connection to a truck. Think about how each of those components was designed and manufactured, how each evolved. How many countries were sourced for the raw materials? How many individuals were involved in the manufacture? Now, think about the number of people through the ages that were involved in the evolution of that truck. You can start with the invention of the wheel. There is no chance the first wheel was as remotely round as a wheel on a vehicle today, the progressive efforts of thousands of minds helped make that happen.

Everything in technology is rational. In other sciences such as physics or astronomy, the more we learn, the more we discover we do not know. There are no mysteries in technology, just challenges that take a little longer to figure out. So the next time you are stymied with a troubleshooting problem on an engine, remember that the cause will be rational. Make it your challenge to discover and then explain the cause in terms that make sense to you. Think a little about what James Watt and Rudolf Diesel did and then refuse to accept that the concept of the unexplained can exist in technology.

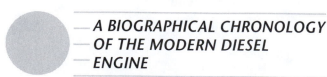

A BIOGRAPHICAL CHRONOLOGY OF THE MODERN DIESEL ENGINE

This collection of minibiographies and events is not intended to be a comprehensive history of technology but rather a listing of significant innovations. The innovators and some critical events are generally listed chronologically although some overlapping occurs in periods of hectic innovation such as at the end of the nineteenth century.

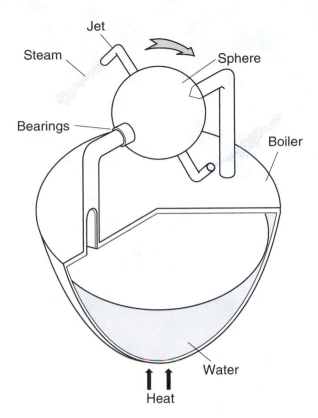

FIGURE 5–1 *Hero's aeolipile or reaction turbine.*
(Courtesy of Facts on File)

HERO OF ALEXANDRIA (AROUND A.D. 60) AKA HERON, GREECE

Noted for:

The aeolipile

Hero's formula (area of a triangle)

Hero was a Greek mathematician and inventor who authored a number of books, mostly on the subject of mathematics. His importance to technology is as the inventor of the first known heat engine somewhere around the year A.D. 60. Hero called his machine an **aeolipile** (Figure 5–1) but in our age we describe it as a **reaction turbine**. The aeolipile was a spherical vessel with an axial shaft through its center supported by bearings and mounted over a heat source. The vessel was fitted with two canted (angled) nozzle jets. When filled with water and heated, escaping steam from the jets produced thrust and resulted in rotary motion. This engine was rated at less than 1% thermal efficiency. It was the forerunner of all modern engines, but very especially the jet turbine, because it developed thrust. In one of his books, *Pneumatica*, Hero describes a number of devices including fire engines, water fountains, coin-operated machines, water clocks, and vari-

ous other steam-powered machines; it is doubtful whether any of these were actually built. In many ways, Hero was a visionary and thinker rather than a doer, much as Leonardo da Vinci a few centuries later.

GUNPOWDER AND THE ROCKET (THIRTEENTH CENTURY)

The invention of both gunpowder and the rocket is attributed to the Chinese. Fireworks were known to exist in China in 1232 and gunpowder was used as a propulsive agent in arrows and firearms shortly after. The first firearms using gunpowder appeared in Europe in the early part of the fourteenth century.

LEONARDO DA VINCI (1452–1519) REPUBLIC OF FLORENCE

Leonardo's genius manifested itself in a number of fields. He was a noted painter, draftsman, sculptor, architect, biologist, and engineer. As a young man he was apprenticed to a sculptor and established himself as a painter and cartoonist of some note. Today, he is known primarily for his artistic achievements and especially the *Mona Lisa*. Leonardo was a visionary and demonstrated a mechanical inventiveness centuries ahead of his time. His engineering accomplishments were theoretical and survive mainly in the form of drawings. Among these are rudimentary designs for various flying machines (including what appears to be a primitive sort of helicopter), ball bearings, and the smoke jack (chimney smoke driven, vaned rotor).

DENIS PAPIN (1647–1712) FRANCE/UNITED KINGDOM

Noted for:

Designing the first pressure cooker

Conceiving the first cylinder and piston steam engine

Both of Papin's inventions are, of course, closely related. The pressure cooker works by fitting a saucepan with a tightly fitting sealed lid, permitting the vessel's contents to rise to a temperature above normal boiling temperature, thus enabling shorter cooking times. A pressure relief valve prevents explosions (usually!). Anyone who has watched water boil in a saucepan with a lid will probably sooner or later figure that steam might be used to move a piston in a cylinder. Papin was the first to realize this and he sketched some rudimentary designs for steam engines, none of them as it turned out, very practical. He never actually built an engine, but it is known that Savery referenced his work.

THOMAS SAVERY (1650–1715)
UNITED KINGDOM

Noted for:

The first *useful* steam engine

The Marine Odometer

There is no doubt that a number of steam engines were built in the 1600 plus years between Hero's reaction turbine and 1698, but not one of them was capable of doing any real work. Thomas Savery was a military engineer, and in 1698 he patented an engine designed to pump water out of mines. This consisted of a closed vessel filled with water into which pressurized steam was piped: this forced the water to a higher level at which the steam was condensed by a sprinkler. The vacuum created by condensing the steam to a liquid again was used to draw still more water upward. The action was made more or less continuous by having two containing vessels in the same apparatus or engine. Savery built and sold his engines both for pumping out mines and supplying water to buildings. He later worked with Newcomen in developing his atmospheric engine.

THOMAS NEWCOMEN (1663–1729)
UNITED KINGDOM

Noted for:

Inventor and builder of the atmospheric steam (Newcomen) engine

Thomas Newcomen was an ironmonger who tackled the high cost of using horses to draw water out of mines. With his assistant John Cawley, a plumber, he developed the Savery engine by making it double-acting; that is, steam was used both to produce work first by pressure and subsequently as it condensed, by vacuum. Newcomen formed a partnership with Savery in his later years and together they built engines that pumped out mines all over Europe. Note the size of the engine in Figure 5–2.

JAMES WATT (1736–1819)
UNITED KINGDOM

Noted for:

The Watt steam engine

The centrifugal governor

The physics of power analysis

Units of power (BHP and Watt)

Sun and planet gearing

Crank and throw principle

The pressure gauge

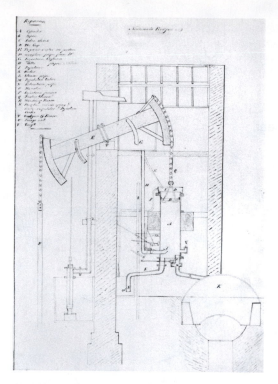

FIGURE 5–2   *Newcomen's 1712 engine. (Courtesy of Institute of Mechanical Engineers)*

James Watt was born just outside Glasgow, Scotland, the son of a successful ship and house builder (Figure 5–3). In childhood he was educated mostly by his mother and in his father's workshops, where at an early age he was entrusted with making models and repairing instruments. He was trained formally as an instrument technician for one year in London, following

FIGURE 5–3   *James Watt. (Courtesy of Institute of Mechanical Engineers)*

which he obtained employment at the University of Glasgow repairing mathematical instruments such as quadrants, compasses, and scales. This job was significant as it enabled him to continue his education and, in an informal way, meet many of the leading scientists and mathematicians of the day. One of these was Joseph Black who invented the concept of latent heat. Watt first focused on the steam engines of the day when asked to repair a Newcomen single chamber steam engine. He was puzzled by the inefficiency of the machine, realizing that the loss of **latent heat** (the heat energy required to change the state of a substance from a liquid to a gas) prevented it from achieving better than 1% thermal efficiency. Watt analyzed **power** production from a mathematical perspective. He then designed and patented a steam engine in 1765 with a separate condenser and expansion chamber (Figure 5–4). Until James Watt's 1765 engine, no heat engine had achieved better than 1% thermal efficiency; within 10 years, engines were built with thermal efficiencies of 4%. To put this achievement into perspective, it took another 100 years before a further quadrupling of thermal efficiency took place.

Watt worked for seven years as a land surveyor after patenting his steam engine and then moved to Birmingham, England, in 1744 where he spent the rest of his life. He formed a business partnership with Matthew Boulton, an iron works manufacturer, and together they would make steady improvements to the steam engine. Financially successful, Watt devoted most of the rest of his long life to inventing. Among his patents are the double-acting piston, the centrifugal governor, crank throw principle, sun and planet gearing, and the pressure gauge. Perhaps his most important contribution to technology from our twenty-first century perspective was that he wrote the language and mathematics to understand power production, which gave his followers the tools to develop the modern engine.

NICOLAS-JOSEPH CUGNOT (1725–1804) FRANCE

Noted for:

Designed and built the world's first automobile

Cugnot was a French military engineer who designed a two-piston, steam engine referencing the work of Papin but independently from Newcomen and Watt. His engine was the first to use high-pressure steam expansively without condensation. It was mounted in a tricycle-type carriage in which a single front wheel was responsible for driving and steering the vehicle. The vehicle was severely handicapped by the need for an on-board water supply and maintaining the high-pressure steam and therefore had little practical value. However, it was the first automobile and as such has an honored place in automotive history.

RICHARD TREVITHICK (1771–1833) UNITED KINGDOM

Noted for:

Building the first high-pressure steam engine
Designing and building the first steam locomotive

The son of a mine manager, Trevithick who performed poorly in school, was described by his own father as a loafer and remained semiliterate throughout his life. However, he displayed a remarkable ability to solve engineering problems and played a critical role in advancing the steam engine. He reportedly had little business sense and never profited from his genius. Trevithick's success with building steam engines and locomotives occurred largely because he focused on the materials used to construct pressure vessels, rails, and drivetrain components. His locomotives were designed primarily for on-site mine operations (Figure 5–5). He died in poverty and was buried in an unmarked grave.

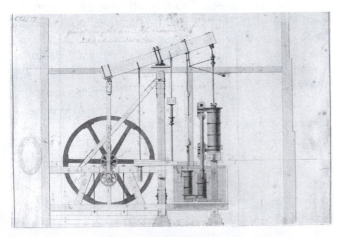

FIGURE 5–4 *Watt's condensing engine. (Courtesy of Institute of Mechanical Engineers)*

FIGURE 5–5 *Trevithick's locomotive of 1804. (Courtesy of Institute of Mechanical Engineers)*

GEORGE STEPHENSON (1781–1848) UNITED KINGDOM

Noted for:

Inventing and building the first updraft steam engine

Building the world's first commercial railway

Achieving the first land speed record—58 km/hr. (36 mph)

Stephenson was the son of a mechanic who operated a coal mine pump, atmospheric engine; he evidently had no formal education and started work at an early age. At 19 years of age he was operating a Newcomen engine, and he taught himself to read and write in his twenties. He was later taught mathematics by his only son who studied the subject in college. Stephenson first sought to improve the power of the steam engines of the day and he achieved this by inventing *steam blast* venting in which exhaust steam was redirected up the chimney, pulling air after it, and increasing the breathing draft. This innovation is said to have made the steam locomotive truly practical and began a countdown on the days the horse was to be used as a railway "engine." Next, Stephenson turned his attention to establishing a functional railway, connecting centers located some distance apart and capable of carrying both cargo and passengers. In 1825, the commercial railway age effectively began when 450 passengers were carried from Darlington to Stockton in England at a speed of (15 mph) 24 km/hr. Stephenson followed this up by connecting the cities of Liverpool and Manchester, (40 miles) 64 km apart by a railway line. A little later, he designed and built a locomotive named the *Rocket* to run on this track and take part in a challenge competing against three other entries; it won, achieving a then record-breaking speed of (36 mph) 58 km/hr. Perhaps Stephenson's greatest achievement was to demonstrate the viability of reliable engine-powered transportation; his locomotives *Locomotion*, *Planet*, and *Rocket* (Figure 5–6) heralded a boom in railway construction throughout the world.

Stephenson's "Rocket."

FIGURE 5–6 *Stephenson's locomotive. (Courtesy of Institute of Mechanical Engineers)*

SADI CARNOT (1796–1832) FRANCE

Noted for:

The Carnot cycle relating to heat engines

Carnot was the eldest son of the French revolutionary figure, Lazare Carnot, and educated by his father and at the Ecole Polytechnique in Paris. However, as a result of the Napoleonic wars, he spent most of his working life in the French military. He developed what we today describe as the Carnot cycle in the latter part of his short life, which he devoted to the study of physics.

The **Carnot cycle** relates to the ideal cyclical sequence of changes in fluids in a heat engine. During an operating cycle, an engine develops heat from a heat source, delivers the results of the heat expansion to a powertrain, rejects the heat not converted to mechanical force (to a heat sink/atmosphere), and then must produce work during compression at low temperature. Carnot asserted that the ratio of work output to the heat input (represented by the heat energy potential of the fuel) is equal to the difference between the temperatures of the heat source and rejected heat combined, divided by the temperature of the heat source. The Carnot cycle was key to the work of Beau de Rochas, Otto, Diesel, and others who developed the heat engine.

ALPHONSE BEAU DE ROCHAS (1815–1893) FRANCE

Noted for:

Originating the principle of the four-stroke cycle, internal combustion engine

In his theoretical works on heat engine theory, Beau de Rochas emphasized the importance of compressing the fuel/air mixture prior to ignition and detailed a four-stroke cycle, internal combustion engine, which he patented in 1862. Beau de Rochas was a theoretician only. He left subsequent development of the internal combustion engine to others such as Otto, Lenoir, and Diesel. However, without his scientific foundation, four-stroke cycle, internal combustion engines would not have replaced steam engines by the end of the nineteenth century.

ETIENNE LENOIR (1822–1900) BELGIUM

Noted for:

Building the first commercially successful internal combustion engine

The two-stroke cycle

The electric locomotive brake

Lenoir's engine was a converted double-acting steam engine with side valves to admit air/fuel mixture and discharge combustion end gases. The engine was fueled with coal gas (essentially methane) and could achieve about a 4% thermal efficiency. Lenoir also built the first automobile with an internal combustion engine, adapting his two-stroke cycle engine to run on liquid fuels extracted from coal tar.

ALLESSANDRO VOLTA (1745–1827) ITALY

Noted for:

Inventing the battery

Discovering methane gas

Identifying static electricity

The first known battery was that invented by the Italian physicist, Allessandro Volta in about 1800. Volta was an academic who taught at the University of Padua where he was known as a philosopher. He is said to have demonstrated his battery to Napoleon. The voltaic cell was later developed by the English chemist, John Daniell in 1836.

NIKOLAUS OTTO (1832–1891) GERMANY

Noted for:

Building the first four-stroke cycle, internal combustion engine

Otto built and attempted to patent the first four-stroke cycle engine in 1876, using the principles employed in the typical automobile engine of today. He was initially granted a patent for his engine but later

FIGURE 5–7 *1885 ad for an Otto engine. (Courtesy of Cummins)*

had it revoked when it was contested by Beau de Rochas. Notwithstanding this setback, Otto built thousands of engines, and we use his name today to describe the four-stroke cycle, spark-ignited engine principle: the Otto cycle (Figure 5–7).

RUDOLPH DIESEL (1858–1913) FRANCE/GERMANY

Noted for:

Inventing the four-stroke cycle, compression ignition engine (diesel cycle)

Rudolph Diesel was born in Paris and educated in France, England, and finally Germany, where at the Technische Hochschule in Munich he was known as a brilliant scholar (Figure 5–8). His first job was working on thermodynamics with the refrigeration engineer Carl von Linde and he joined the Paris-based firm in 1880.

FIGURE 5-8 *Rudolf Diesel.*

However, Diesel devoted much personal time to developing an internal combustion engine that would theoretically produce thermal efficiency values much greater than the Otto cycle engines. In 1890 he moved to Berlin and in 1892, he patented the compression ignition engine that today bears his name (Figure 5-9). From 1892 to 1897, Diesel worked with the MANN and Krupp companies producing a series of increasingly successful diesel engines, culminating in a single cylin-

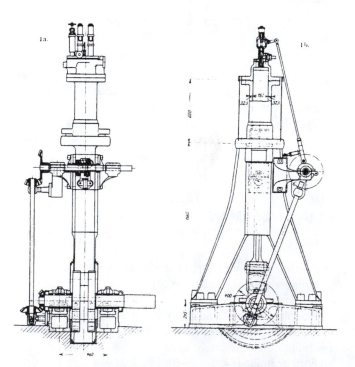

FIGURE 5-9 *Line drawing of Diesel's 1893 engine.* *(Courtesy of Cummins)*

der, coal dust fueled version that would produce 25 brake horsepower (BHP). From the moment that Diesel introduced his patent in 1892, his engine was an international success, and versions were built in France, the United Kingdom, and the United States, before the turn of the century.

Diesel was both a great linguist and salesman. In 1897, he decided to leave the development of the diesel engine to others and instead devote himself to popularizing it throughout the world. He lectured in many different countries and became extremely wealthy on the royalty proceeds of his 1892 patent. His economic success did not however, translate into any personal happiness. He suffered from bouts of depression and alcoholism and in 1913 mysteriously disappeared from the deck of a ship traveling from Dresden to London, in what may have been a suicide.

ROBERT BOSCH (1861–1942) GERMANY

Noted for:

The invention of the spark plug

The invention of the magneto

The invention of high-pressure oil injection for diesel engines

Robert Bosch was educated in the United States where he worked with Thomas Edison and his colleague, Siegmund Bergmann. Bosch returned to Stuttgart, Germany, where he founded Bosch GmbH in 1886, a firm that became Europe's largest auto parts manufacturer. While continuing to develop the company that bore his name, Bosch was active as an engineer and inventor, primarily in the auto electrical field. However, in 1927 he overcame a problem that had stunted the progress of the diesel engine for a number of years, namely that of introducing the fuel charge to a high-compression diesel engine. Bosch devised his port-helix metering, high-pressure liquid injection pump to overcome the problem of precise fueling of diesel engines running at higher speeds. His system and competitor's copies of it remain in use today with surprisingly few hydromechanical modifications and electronic management.

HENRY FORD (1863–1947) UNITED STATES

Noted for:

Inventing assembly line procedure

Establishing the world's second largest automobile manufacturer

The son of Irish immigrants, Ford dropped out of school before he turned 15 years old. He worked as a machinist's apprentice in Detroit and later set up his own machine shop on his father's farm, which evolved into an automobile building company. He also worked for a time as the chief engineer for the Edison Company in Detroit. Henry Ford is not known for technical innovation so much as the fact that he revolutionized factory component assembly procedure, introducing what is now known as the assembly line method. He established the Ford Motor Company in 1903 and in 1908 the Model T appeared. By 1913, mass production enabled him to sell it for $500. His business philosophy was to reduce unit cost and maximize sales.

HARRY RICARDO (1885–1974) UNITED KINGDOM

Noted for:

The Ricardo combustion chamber—side valve design

Fuel research and theories of ignition, combustion, and detonation

The indirect injected diesel engine

Stratified charge combustion

Smoke emission controls

Ricardo was born in London and studied as a mechanical engineer at Cambridge University. Although he is best known for the SI engine side valve combustion chamber that bears his name, Ricardo is important in the history of diesel technology for his research in the nature of combustion in the diesel engine; many of his conclusions are still in practice to this day. During the Second World War, Ricardo worked extensively on controlling smoke emission from engines powering British tanks, which was making them conspicuous to enemy gunners. He was the pioneer of emissions technology.

HANS LIST (1896–1996) AUSTRIA

Noted for:

Research on the scavenging of two-stroke cycle engines

Developing high-speed, direct-injected diesel engines

Author _The Internal Combustion Engine_, 1935

List was born in Graz in 1896, the son of a railway design engineer. He studied thermodynamics at the Graz Technical University and wrote his PhD on the control of diesel engines. When he was 50 years old, List started a consultancy engineering firm called AVL in his hometown of Graz, which rapidly became world renowned in addressing the combustion dynamics of diesel engines. Many current high-speed, direct-injected diesel engines had at least some of their engineering undertaken by List's AVL GmbH.

CLESSIE CUMMINS (1888–1968) UNITED STATES

Noted for:

Inventing the mechanically actuated, liquid fuel injector

Entering the first diesel-powered car in the Indianapolis 500

Building the Cummins engine company

Known as the "father" of the truck engine

Inventing the engine compression brake

Inventing the Cummins cycle engine—a four-stroke, three-barrel design

Clessie Cummins was the first of five children born to an Indiana farmer/cooper (barrelmaker). He attended thirteen different schools as his parents moved around. He left school after completing grade 8, and he remained contemptuous of formal education throughout his life. Cummins began his working life employed as a chauffeur for an Indiana banking family called the Irwins. He turned the Irwins's garage into a machine shop and with the financial help of his employers, introduced his first diesel engine to the marketplace in 1919, a 6 BHP, single-cylinder, kerosene-fueled engine (see Figure 5–10). Over the next few years, he and his

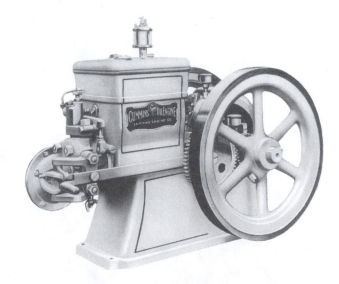

FIGURE 5–10 _Cummins 6 BHP, four-stroke cycle engine of 1919. (Courtesy of Cummins)_

company became associated with a number of publicity stunts, all of which had the objective of furthering Cummins's firm conviction that the diesel engine was the engine of the future. The first of these was staged in 1930, when a 1925 Packard, powered by a 50 BHP, four-cylinder Cummins U model, drove 792 miles from Indianapolis to the New York auto show on $1.38 of fuel. The following year he set the speed record for a diesel engine powered car in Daytona Beach (Figure 5–11) using a U model engine now rated at 86 BHP@1700 rpm of 101.4 mph; by 1934, Cummins had increased this record to 138 mph using a Model H engine powering a Duesenberg. Cummins also entered Cummins-powered Duesenbergs into the Indianapolis 500 with moderate success on the track but with great success in promoting diesel engines. The model H (Figure 5–12) had been designed for a truck application, however, and Cummins adopted the same publicity stunt methods of marketing the diesel as a truck engine, first hauling a 3½ ton load from coast to coast for $11.38 of fuel in the middle of the Depression and next setting the world endurance record for any vehicle driving 14,600 miles around the Indy 500 track. The math was simple: the truck averaged 43.4 mph at a cost of ½ cent of fuel per mile and only one refuel was required at the 10,000-mile point. Clessie Cummins spent his retirement years doing what he enjoyed most, inventing. One invention was the engine compression brake, variations of which are widely used in current trucks, while another was a four-stroke cycle, three- or five-barrel, opposed piston engine that never saw production but did serve as a testament to his creativity.

EVOLUTION OF THE COMPUTER (500 B.C.–TODAY)

The abacus is generally thought to be the earliest computing device. It is thought to be Babylonian (now modern Iraq) in origin (beginning around 500 B.C.) but its use spread throughout Europe and the Far East. It is still used in remote areas of the Far East today in place of a cash register. The abacus is a simple arithmetic calculator with racks of counting beads to represent numeric values. Expert practitioners can use them at speeds that compete with cash register input/output transactions.

The first real analogue computer, the slide rule, was invented by Edmund Gunter, an English mathematician, in 1620. The slide rule was used extensively in mathematics and engineering until replaced by electronic digital computer technology in the 1970s. Blaise Pascal, the French scientist and philosopher, is credited with conceiving the first mechanical digital calculating machine, built in 1642. Various forms of Pascal's digital calculator were used until the introduction of the modern electronic digital calculator.

Eckert and Mauchly (see next section) built the first automatic electronic digital computer and launched the computer revolution that has resulted in the affordable personal computer that in a mere 20 years has

Clessie Lyle Cummins
Daytona Beach, Florida, 1931

FIGURE 5–11   *Cummins at Daytona Beach in 1931. (Courtesy of Cummins)*

FIGURE 5–12 *Cummins H engine. (Courtesy of Cummins)*

changed the world more rapidly than any invention in history. A computer is required to manage every highway truck diesel engine sold in America today.

PRESPER ECKERT (BORN 1919) AND JOHN MAUCHLY (1907–1980) UNITED STATES

Noted for:

Inventing the electronic digital computer

During the Second World War, Eckert and Mauchly were contracted by the U.S. Army to accelerate the re-computation of artillery firing tables. The result was the world's first electronic **digital computer** that would handle data in coded formats. The research took place at the University of Pennsylvania. Eckert and Mauchly called their machine an Electronic Numeric Integrator and Calculator (ENIAC) and, far from being a mere prototype of the modern computer, it actually contained all of the high-speed circuitry used in today's computers.

SUMMARY

- Most technical achievements have been steps forward rather than leaps.
- Technical advances almost never occur in isolation of all other technical accomplishments.
- Creativity is a trigger for technical advances, but the concept must always be validated by practical results.
- Because the events occurred too long ago to have valid data, it is unknown what accomplishments Hero had at his disposal to build his ideas on.

- Between Hero and Watt, a period of 1,600 years, inventors built and ran thousands of different engines, but not one achieved better thermal efficiency than 1%.
- Within 10 years of Watt's mathematical analysis of power, engines that would produce thermal efficiencies of better than 4% were built.
- Diesel's engine relied heavily on the work of other inventors, especially Carnot and Beau de Rochas.
- Bosch solved the problem of high-pressure liquid fuel injection that made the modern diesel engine possible.
- Clessie Cummins is regarded as the "father of the highway diesel engine."

REVIEW QUESTIONS

1. Who invented the first heat engine?
   a. Otto
   b. da Vinci
   c. Hero
   d. Diesel
2. Which of the following properly describes the first heat engine?
   a. Internal combustion
   b. 2-stroke cycle
   c. 4-stroke cycle
   d. Reaction turbine
3. Who invented the term horsepower?
   a. Papin
   b. Watt
   c. Stephenson
   d. Diesel
4. Who built the world's first useful steam engine?
   a. Savery
   b. Newcomen
   c. Watt
   d. Carnot
5. Who designed the first centrifugal governor?
   a. da Vinci
   b. Watt
   c. Bosch
   d. Diesel
6. In what year did Diesel patent his compression ignition engine?
   a. 1864
   b. 1892
   c. 1896
   d. 1912
7. Who patented the first high pressure liquid fuel injection pump for a diesel engine?
   a. Otto
   b. Diesel
   c. Ricardo
   d. Bosch

8. Who is known as the father of the highway diesel engine?
   a. Diesel
   b. Bosch
   c. Cummins
   d. Roosa

9. What are the U.S. inventors Eckert and Mauchly known for?
   a. The first electronic digital computer
   b. The first turbocharger
   c. The first internal combustion engine
   d. The first dry cell battery

10. The world's first automobile was built by:
    a. Trivithick
    b. Watt
    c. Carnot
    d. Cugnot

Chapter

6

Power

PREREQUISITE

Chapter 4

OBJECTIVES

After studying this chapter, you should be able to:

- Understand the language of power as it applies to a truck diesel engine.
- Define the terms torque and power and describe what is required to produce each in an engine.
- Construct the formulae required to calculate power equations.
- Calculate brake power using actual engine data.
- Calculate indicated power using the PLANC formula.
- Define the term specific fuel consumption.
- Understand the term load in the context of a power analysis graph.

KEY TERMS

brake horsepower

brake specific fuel consumption (bsfc)

dynamometer

force

horsepower

indicated power (ip)

kilowatt (kw)

load

mechanical efficiency

power

rated power

rated speed

rejected heat

SAE power

specific fuel consumption (sfc)

torque

torque rise

torque rise profile

torsionals

work

INTRODUCTION

In the 1,600 years between Hero and Watt (see previous chapter), many heat engines had been designed and built, but not one had achieved better than 1% thermal efficiency and most a lot less. James Watt began a process of analyzing all the factors that contribute to an engine's ability to perform work and wrote the language and mathematics to describe it. It is important that the diesel technician acquire a basic understanding of power production. Although it is possible to operate diagnostic equipment such as **dynamometers** simply by matching specifications to readouts on test instruments, the technician will not be capable of properly analyzing test data without understanding some of the language and technology of power. The following attempts to simplify some complex concepts, and the interested learner may seek to further develop knowledge in this area by consulting reference texts and on-line research. To complicate the issue of understanding power, both standard and metric measurement systems are used by the industry; probably the best solution is for the technician to become familiar with both. At the time of writing, the engine OEM that designs an engine using the metric system exclusively will most often publish power analysis specifications using only the standard system, so it is more important to understand these first.

DEFINITIONS AND FORMULAE

To understand the concept of power, it is necessary to understand the language of power. The term power is very often misused, which adds to the difficulty in understanding it. This section contains a series of definitions and develops them into the formulae required to calculate power. It is important that the contents of this section be properly understood before tackling the later chapters on engine dynamometers and power analysis.

ENERGY

Energy is best defined as the capacity for producing work. Energy exists in a number of forms; kinetic, potential, electrical, thermal, chemical, and nuclear are its common forms. Kinetic energy is the energy of motion and any body in a state of motion is described as possessing kinetic energy. Air in its compressed state and

contained in a reservoir has the potential for creating motion and is therefore described as *potential energy*. The first law of thermodynamics states that energy can be neither created nor destroyed; however, the way in which energy manifests itself can be changed. In an internal combustion engine, the potential heat energy of a fuel is released when it is combusted, producing pressure that acts on a piston and drives it through its power stroke producing kinetic energy. Not all of the potential heat energy of the fuel can be successfully converted to kinetic energy. This energy is described as **rejected heat** and must be dissipated to atmosphere.

EFFECTS OF HEAT ENERGY

Heat is easily converted into mechanical energy. The sun's heat daily raises tons of water vapor high into the atmosphere, so all the mechanical energy of falling water whether as rain, in rivers, or glaciers stems directly from the sun's heat. In the heat engine, the heat released from the burning of fuel is converted into mechanical or kinetic energy.

FORCE

Force is generally defined in terms of the effects it produces, although it should always be remembered that force can be exerted with no result. If force is applied to a body at rest, it may be sufficient to cause the body to move; however, it might not. Using the standard system, force is measured in pounds. In the SI metric system, the Newton is the unit by which force is measured.

1N = force that produces acceleration of 1 meter/second on a mass of 1 kg

In a diesel engine, force is represented by cylinder pressure. This force acts on the sectional area of the piston crown and is transmitted to rotary motion by acting on the crank throw. The amount of force is controlled by the amount of fuel delivered to the engine cylinder because of the excess air factor in a diesel.

WORK

Work is accomplished when force produces a result. When the definition is applied to an engine, work is accomplished when force acting on the piston results in piston travel. In the following formula, standard values are always listed first:

$$\text{Work} \atop \text{(watt-second/joules)} = {\text{Force} \atop \text{(pounds/Newtons)}} \times {\text{Distance} \atop \text{(feet/meters)}}$$

To use a nonengine analogy to define work, if two persons of identical weight run exactly 100 meters, each has accomplished the same amount of work.

TORQUE

Torque is turning effort. In a typical engine, the force acting on the piston is transmitted to a crankshaft throw. The throw is a lever. Torque is the product of force on the torque arm or crank throw and its perpendicular distance from the shaft center. The greater the distance of the throw centerline from the crankshaft main centerline, the greater the leverage and the more potential torque. The ability to produce torque in an engine is directly related to its cylinder pressures. For instance, peak torque will always occur when cylinder pressures peak. A bicycle has a crankshaft driven by a pair of throws called pedals, offset 180 degrees. Peak torque is realized whenever muscle force acting on the pedals is at a maximum. High torque is required to propel a bicycle and the weight of its rider up a steep hill so gear selection must be made so that the required torque is within the capabilities of its engine, the cyclist.

$$\text{Torque} = \text{Force(pounds/Newtons)} \times \text{Leverage(feet/meters)}$$

POWER

Power is the rate of doing work. If two sprinters of equal weight run 100 meters, each will have accomplished the same amount of work. However, if one runs this distance in 20 seconds and the other in 10 seconds, the latter has twice the power.

$$\text{Power (watts)} = \frac{\text{Work (joules)}}{\text{Time (seconds)}}$$

James Watt coined the term **horsepower** before he developed his steam engine, and as steam engines of that time had at best 1% thermal efficiencies, the draft horse was more often the engine of choice. The draft horse was widely enough used that most people had at least an idea of what the animal was capable of. Watt observed the horse at work with a mathematician's eye and concluded: a medium-sized draft horse could raise a 330 lb. weight through a distance of 100 ft. in 60 seconds.

Subsequent observation has concluded that this is actually about 50% more than the average draft horse could sustain over a working day, but Watt obviously did not observe the performance of the horse for a prolonged period. So the foregoing data, expressed in a manner that can be used in a formula is:

$$330 \text{ lb} \times 100 \text{ ft.} = 33,000 \text{ lb.-ft./min.}$$

$$= 550 \text{ lb.-ft. per second} = 1\text{hp}$$

$$1\text{hp} = 33,000 \text{ lb.-ft. work in 1 minute}$$

$$= \frac{33,000}{60} \text{ lb.-ft.} = 550 \text{ lb.-ft. in 1 second}$$

Power is always related to time—it is simply the rate of doing work. A cyclist in competition will usually select a gear ratio that permits the pumping out of power strokes at the fastest rate manageable, knowing that each downstroke of a pedal from TDC to BDC translates into an actual distance on the road. If a race were to be held between a group of runners of identical weight, every runner who completed the race would have accomplished the same amount of work. The work accomplished by each runner would be the act of transporting their mass over a defined distance. However, one of the runners will have won the race by crossing the finish line ahead of the other runners. This runner has accomplished the race distance in the least amount of time and can be said to have demonstrated the most power.

POWER CALCULATIONS

In a diesel engine, as with the cyclist, the number of power strokes per second, in conjunction with the actual torque value, determines brake power. A dynamometer is a tool for testing engine power. It applies a resistance to turning effort and measures it; it therefore measures *torque*. When torque is factored with *time* (engine rpm), the brake power can be calculated. This is an automatic function of an engine dynamometer.

BRAKE POWER

This is the power measured at the flywheel of an engine. It is always lower than indicated power (calculated power) because of the amount of power consumed by the engine in overcoming internal friction and pumping losses—usually about 10%. The term is used to describe actual engine power over calculated power.

CALCULATING BRAKE POWER

Despite the fact that industry in both Canada and the United States has embraced the metric system, for instance, most engines are engineered using the metric system, it seems that it is part of trucking culture to use the standard system of weights and measurements. That means we express brake power in horsepower.

$$\textbf{Brake Horsepower} = \frac{\text{Force} \times \text{Distance} \times \text{Time}}{33,000 \text{ lb} \times 1 \text{ ft.} \times 1 \text{ min.}}$$

Distance must take into account that an internal combustion engine rotates and, therefore, the distance factor is not linear but circular; to accommodate this, 2π is used to construct the equation, which is simplified as follows:

$$2\pi = 2 \times 3.1416 = 6.2832$$

$$33,000 \div 6.2832 = 5252.10084$$

or, for calculation purposes, 5252

In the first brake power formula, the values time and distance are multiplied by each other and by the force value. In a diesel engine we express force as *torque*. However, we commonly express time and distance as one value, rpm. So for practical purposes, the brake power formula that we use becomes:

$$\text{Brake Horsepower Power} = \frac{\text{Torque (1 lb.-ft.)} \times \text{Time (rpm)}}{5252}$$

$$\text{Brake Power (kW)} = \frac{\text{Torque} \times \text{rpm}}{9429}$$

This formula can be used to calculate brake horsepower providing the torque value and rotational speed values are known (Figure 6–1). For instance, a Cummins 460E CELECT is specified to produce a peak torque value of 1,550 lb.-ft. at 1,200 rpm—to calculate the amount of power this engine produces at this speed and torque:

$$\text{BHP} = \frac{1550 \text{ lb.-ft.} \times 1200 \text{ rpm}}{5252}$$

$$= 354 \text{ BHP}$$

Using the metric system, brake power is measured in watts, so using the formulae explained earlier in this section:

1 Newton-meter = 1 joule

1 joule per second = 1 watt

CONVERSIONS

1 lb.-ft.	= 1.356 N·m
1 N·m	= 0.737 lb.-ft.
1 HP	= 0.746 kW = 746 watts
1 kW	= 1.341 HP
1 HP	= 2,545 Btu per hour
1 metric HP	= 4,500 kilogram/meters per minute
	= 0.9863 HP

Truck and bus diesel engines are usually classified by rating their actual or brake power either in hp or **kilowatt (kW)**, usually the former, measured at the flywheel (Figures 6–2 and 6–3). The brake power of the engine measured at the chassis wheels, such as would be displayed by a chassis dynamometer, would be somewhat less. There are other methods of reckoning power values some of which are listed later in this chapter.

INDICATED POWER (IP)

Indicated power is calculated power. It is useful for comparison purposes but obviously not as reliable of indicator of true power as brake power. The following formula is used to calculate indicated power:

P   MEP <u>(Pressure)</u>

L   <u>Length</u> of stroke

A   Piston cross sectional <u>area</u>

N   <u>Number</u> of power strokes per minute

C   Number of engine <u>cylinders</u>

$$\text{IP} = \frac{\text{PLANC}}{33,000}$$

Example:

A four-stroke cycle, six-cylinder diesel engine with a bore of 5.0″ and a stroke of 5.5″ is run at a speed of 2,000 rpm. The mean effective pressure obtained from an indicator diagram is specified as 200 psi. Calculate the indicated horsepower of the engine.

First, use the engine specifications to construct the data required for the PLANC formula:

P = (MEP) Calculated by determining the average cylinder pressure through the compression stroke, subtracted from the average cylinder pressure through the power stroke. This data is provided: 200 psi

$$\text{So, IP} = \frac{200 \times L \times A \times N \times C}{33,000}$$

$$L = \text{Stroke} = 5.5 \text{ Length in feet} = \frac{5.5}{12} = 0.458$$

$$\text{So, IP} = \frac{200 \times 0.458 \times A \times N \times C}{33,000}$$

A = Piston sectional area: Bore diameter = 5.0″, so radius = 2.5″ To calculate the area of a circle, the square of the radius is multiplied by 3.1416:

$$= (2.5 \times 2.5) \, 3.1416 = 19.635$$

$$\text{So, IP} = \frac{200 \times 0.458 \times 19.635 \times N \times C}{33,000}$$

N = number of power strokes per minute in one engine cylinder

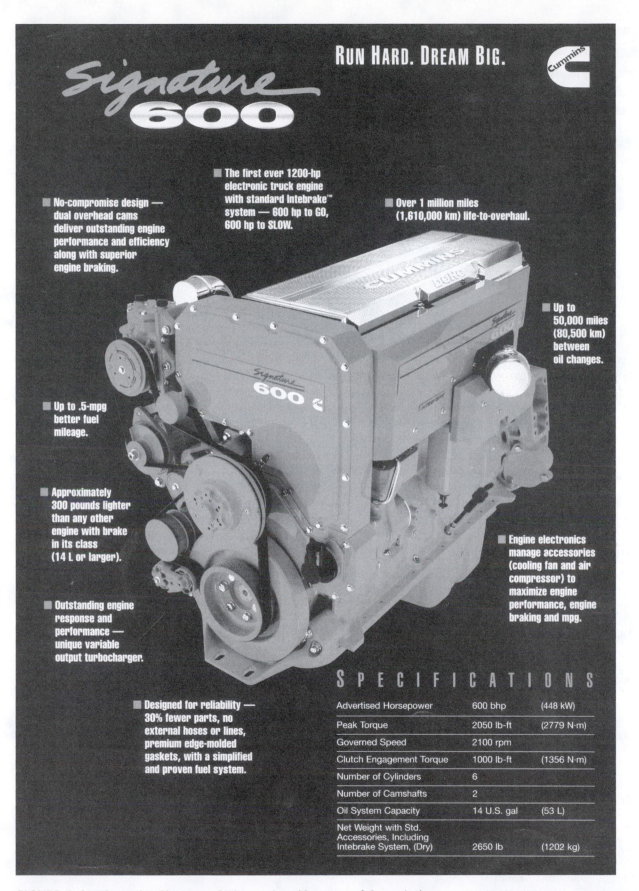

RUN HARD. DREAM BIG.

Signature **600**

■ No-compromise design — dual overhead cams deliver outstanding engine performance and efficiency along with superior engine braking.

■ The first ever 1200-hp electronic truck engine with standard Intebrake™ system — 600 hp to GO, 600 hp to SLOW.

■ Over 1 million miles (1,610,000 km) life-to-overhaul.

■ Up to 50,000 miles (80,500 km) between oil changes.

■ Up to .5-mpg better fuel mileage.

■ Approximately 300 pounds lighter than any other engine with brake in its class (14 L or larger).

■ Engine electronics manage accessories (cooling fan and air compressor) to maximize engine performance, engine braking and mpg.

■ Outstanding engine response and performance — unique variable output turbocharger.

■ Designed for reliability — 30% fewer parts, no external hoses or lines, premium edge-molded gaskets, with a simplified and proven fuel system.

SPECIFICATIONS

Advertised Horsepower	600 bhp	(448 kW)
Peak Torque	2050 lb-ft	(2779 N·m)
Governed Speed	2100 rpm	
Clutch Engagement Torque	1000 lb-ft	(1356 N·m)
Number of Cylinders	6	
Number of Camshafts	2	
Oil System Capacity	14 U.S. gal	(53 L)
Net Weight with Std. Accessories, Including Intebrake System, (Dry)	2650 lb	(1202 kg)

FIGURE 6–1   Cummins Signature Series engine. (Courtesy of Cummins)

DETROIT DIESEL

SERIES 50

8.5L
Automotive
250-315 BHP

General Specifications

Basic Engine	4 cycle
Model	6047GK60
Number of Cylinders	4 Inline
Air System	Turbocharged Air-to-Air Charge Cooling
Control	DDEC
Bore and Stroke	5.12 in x 6.30 in (130 mm x 160 mm)
Displacement	519 cu in (8.5 liters)
Compression Ratio	15.0 to 1
Dimensions: (approx.)	
Length	45 in (1143 mm)
Width	44.2 in (1123 mm)
Height	47.5 in (1207 mm)
Weight (dry)	2250 lbs (1061 kg)

(The Series 50 has the same Bore and Stroke and many other parts in common with the Series 60.)

Rated Power Output

Gross Power	250 BHP (187 kW) @ 2100 RPM
Peak Torque	780 lb ft (1058 N•m) @ 1200 RPM
Gross Power	275 BHP (205 kW) @ 2100 RPM
Peak Torque	890 lb ft (1207 N•m) @ 1200 RPM
Gross Power	300 BHP (224 kW) @ 1800 RPM
Peak Torque	1000 lb ft (1356 N•m) @ 1200 RPM
Gross Power	300 BHP (224 kW) @ 2100 RPM
Peak Torque	1000 lb ft (1356 N•m) @ 1200 RPM
Gross Power	315 BHP (235 kW) @ 1950 RPM
Peak Torque	1150 lb ft (1559 N•m) @ 1200 RPM
Gross Power	315 BHP (235 kW) @ 2100 RPM
Peak Torque	1150 lb ft (1559 N•m) @ 1200 RPM

Equipment Specifications

DDEC—Detroit Diesel Electronic Controls are standard on all Series 50 engines. This electronic unit fuel injector and engine management control system is the most advanced system available in the industry. DDEC includes state of the art diagnostics for critical engine functions.

Overhead Camshaft—This design optimizes intake and exhaust air passages in the cylinder head for easier breathing, and minimizes valve train losses by eliminating the need for push rods.

Short Ports—The cylinder head has very short intake and exhaust ports for efficient air flow, low pumping losses and reduced heat transfer.

Iron Crosshead Pistons—The top ring can be placed much closer to the top of the iron crosshead piston. This reduces the dead volume above the top ring and improves fuel economy.

Injector Rocker Arm with Ceramic Rollers—The cam follower roller in the Series 50 injector rocker arm is made of silicon nitride. The low wear properties of this ceramic makes it possible to operate at very high injection pressures while maintaining long life of the roller. High injection pressure is one way Detroit Diesel is able to meet the stringent particulate and smoke emission standards without aftertreatments.

Bearings—The Series 50 features large main and connecting rod bearings for long life

Eight Head Bolts per Cylinder—The head bolts provide a uniform load on the gasket and liner to reduce stress on the liner flange and block counterbore

High Efficiency Turbocharger—Combined with a pulse-recovery exhaust manifold, the high efficiency turbocharger provides an efficient transfer of energy for improved fuel economy. 315 HP has ceramic turbo resulting in more responsive and better performance.

Balance Shafts—The Series 50 engine has counter-rotating balance shafts which make it run smooth. The balance mechanism is gear driven, and is attached to the underside of the engine inside the oil pan. With the balance shafts, the Series 50 is as smooth as the six cylinder Series 60 engine.

Top Liner Cooling—The Series 50 features top liner cooling. This has been accomplished by machining a coolant channel high up on the block, so that the top of the liner is surrounded by coolant, resulting in longer ring life.

Photograph illustrates a typical automotive engine
Rating conditions of SAE: 77°F (25°C) and 29.31 in Hg (99 kPa) Barometer (Dry)

For a complete listing of standard and optional equipment, consult your distributor or authorized Detroit Diesel Corporation representative.

FIGURE 6–2   *DDC Series 50 engine. (Courtesy of DDC)*

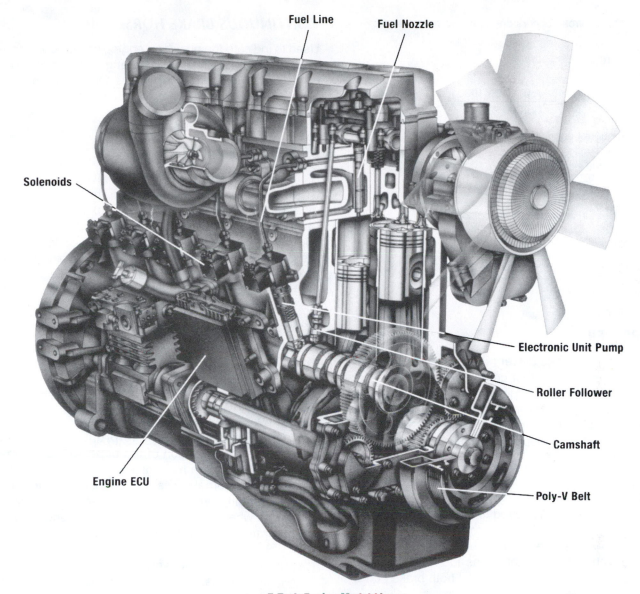

E-Tech Engine Model Lineup

Engine Family	Primary Applications	Engine Models	Operating Speed Range	Cruise RPM*	Torque Rise	Typical** Transmission
Econodyne®	Owner-operators; High GCW; Maximum performance	E7-460 E7-427 E7-400	1800-1200 rpm	1600 ± 50 rpm	25%	10, **13**, 15, 18 spds
	General purpose	E7-350			33%	9, **10** speeds
	Low cost/high MPG	E7-300			33%	9, **10** speeds
MaxiCruise™	Fleet highway; Heavy distribution; Bulk haul	E7-355/380 E7-330/350 E7-310/330	1800-1100 rpm	1500 ± 50 rpm	50+%	**T-2060A**; **9**, 10 spds
Maxidyne®	Vocational	EM7-300 EM7-275	1750-1020 rpm	1500 ± 50 rpm	58%	**T-2050/60/70/80**; T-14607B; 9, 10 speeds
	Automatic transmission	E7-300(A)	1950-1200 rpm	N/A	44%	Allison HD or CEEMAT

* Cruise RPM is the recommended engine cruising rpm in top gear at the Cruise Control Maximum mph setting.

** Transmissions listed in **bold** type are generally recommended for the type of engine or operation listed.

FIGURE 6–3   _Mack Trucks E-Tech power ratings. (Courtesy of Mack Trucks)_

So, a power stroke occurs once per two full rotations:

$$\frac{2000}{2} = 1000$$

So, $IP = \dfrac{200 \times 0.458 \times 19.635 \times 1000 \times C}{33,000}$

C = Number of engine cylinders = 6

So, $IP = \dfrac{200 \times 0.458 \times 19.635 \times 1000 \times 6}{33,000}$

Now the equation is complete, so solve it:

$$IP = \frac{200 \times 0.458 \times 19.635 \times 1000 \times 6}{33,000}$$
$$IP = 327 \text{ HP}$$

SAE POWER

SAE power is a brake power rating known to be especially reliable because test criteria and conditions are rigorously detailed. The engine must be tested in a controlled temperature environment with all parasitic loads attached including the primary transmission from which the torque load is applied and power measured.

FRICTION POWER

The amount of power required to overcome the friction of all the moving parts in an engine—friction power increases as engine size and rotational speeds increase and is affected by such things as lubricant type and temperature. It is easily calculated if both brake power and indicated power values are known:

$$FP = IP - BP$$

MECHANICAL EFFICIENCY

Mechanical efficiency is a measure of the pumping and friction efficiency of an engine and, like *friction power*, is a term more likely to be found in a textbook than in a listing of engine specifications. Usually expressed in percentage terms:

Mechanical Efficiency = BP ÷ IP

For example, an engine that has been tested to produce 400 BHP and has its indicated power calculated at 475 BHP would have a mechanical efficiency rating of 84%.

400 ÷ 475 = 84%

CONTINUOUS BRAKE HORSEPOWER

Used to indicate the power an engine can produce continuously without compromising engine longevity. The term is used by one diesel engine OEM but it means pretty much what its competitors refer to as brake power.

BRAKE POWER AT THE WHEELS

A chassis dynamometer is used to test power output in large dealerships. This means that the engine power must be transmitted through gear box and final drive transmissions, drive shafts, and wheel assemblies to a floor-mounted roller assembly. Actual brake power readings on a chassis dynamometer are typically 10% or more below readings taken on an engine dynamometer. Knowing the chassis being tested is important before accurately validating chassis dynamometer readings.

TORQUE RISE

Torque rise occurs between peak torque rpm and rated speed rpm. As a truck or bus engine is required to operate over a wide rpm range, high torque rise can be desirable. The term **torque rise profile** comes from the graphic representation of the upper portion of the fuel map often shown by engine OEMs on their sales literature. It is perhaps best understood as being the engine-operating range. An engine described as high torque rise when compared with the same engine with identical rated power but low torque rise:

- produces poorer fuel efficiency.
- requires a transmission with less gear ratios and/or less shifting.
- is described by truck drivers as being "more powerful."

Torque rise is often expressed as a percentage of rated speed. In other words, an engine with a nominal rated speed of 2,000 rpm and torque rise of 35% would produce peak torque at 1,300 rpm.

Generally, a truck engineered for a linehaul highway application requires fuel system programming for fuel economy and, therefore, does not require a high torque rise characteristic. Sure enough, the driver may be shifting gears a little more often while getting to a cruise speed, but once there it is desirable to operate the engine as efficiently as possible, in the sweet spot of the torque rise profile. This is usually somewhere close to the midpoint between peak torque rpm and rated rpm. However, a dump truck hauling aggregates in a constant stop-start application would require a much wider engine-operating rpm range, a high torque rise engine.

HIGH TORQUE RISE ENGINES

These are also known as constant horsepower engines. The two primary variations on an engine described as high torque rise over the conventional version of the same engine are found in the fuel system management and in the turbocharger geometry. In the high torque rise engine:

1. The fuel injected per cycle at peak torque is designed to exceed that injected per cycle at rated speed.
2. The turbocharger geometry is tailored to provide a constant rate of air flow per minute from the peak torque rpm (base of the torque rise profile) through to the **rated speed** (peak BHP rpm); this means that the mass of air flow delivered to the engine cylinder per cycle is greatest at peak torque.

SPECIFIC FUEL CONSUMPTION

Specific fuel consumption (sfc) is the fuel consumed per unit of work produced. This data described as **brake specific fuel consumption (bsfc)** is often diagrammed into power/fuel map/performance graphs to demonstrate fuel efficiency through the torque rise of the engine. If the data is constructed around brake power values (and it usually is when truck engine data is displayed) the specific fuel consumption is labelled as bsfc; when the data is constructed around indicated power specifications, the term indicated specific fuel consumption (isfc) is used,

LOAD

Load is usually defined as the ratio of power developed to the normal **rated power** at a specific rpm of an engine and is expressed as a percentage. In some instances, load is used to express the ratio of torque developed to that of peak torque and again, this is expressed as a percentage. When performing dynamometer testing, after the rated power/speed value is established, the load data can be useful in establishing whether abnormalities exist in the fuel map.

ENGINE CONFIGURATION, ENGINE SPEED, AND TORSIONAL FORCES

The number of cylinders in any engine will define the geometric interval (expressed in crank degrees) between firing pulses. This plus the factor of engine speed will define the actual time interval between the engine's firing pulses. Because the pressure developed in engine cylinders is seldom exactly synchronized with the throw vector angle of the crankshaft (see Chapter

4), there is usually a fractional acceleration of the crankshaft above mean crankshaft speed at each firing pulse followed by a fractional slow down just prior to the next firing pulse. These torsional forces (sometimes called **torsionals**) are amplified at slower engine rpms when the time interval between each power stroke is greatest and can affect the driven components in the drivetrain. As the latest generation of high-speed diesel engines tends to be programmed with *run-slow, gear-fast* governing, torsionals can be a problem. As a rule, prolonged running with high cylinder pressures at the *lower* rpm levels within the torque curve should be avoided. Most electronic engines cannot be lugged down (the programming will not permit it), but with high torque rise engines, there are consequences to high load operation in the lower portion of the torque rise curve. These consequences can involve premature drivetrain component (transmission, drive axle carrier) failures.

HOW MUCH POWER DO YOU NEED?

Specing out truck engines is a specialty and there are so many variables that they are difficult to outline in a way that is readily understood. But here is some of the hard math simplified a little—we will use the example of a truck hauling an 80,000 lb load and overlook some of the gearing complexities that could arise:

- It takes 255 BHP to move the truck down a level asphalt road at 60 mph with no wind.
- On a 1% uphill grade to maintain the same speed, another 65 BHP will have to be added, so now the engine requires 320 BHP.
- Back on the level highway, if you have a 5 mph headwind, another 16 BHP will have to be added to maintain that road speed of 60 mph. Should the headwind be 10 mph, another 36 BHP has to added.
- Parasitic losses such as an air-conditioning system (up to 15 BHP) and engine fan (viscous type: constant 4–6 BHP, on-off type up to 15 BHP but only on 2% of running time) all have to be factored into the BHP requirement.
- So running up a 3% grade (not that steep a hill) with a 20 mph headwind (not that strong a wind) is going to require a whopping 531 horsepower to maintain the same road speed. Throw in some of the common parasitic losses an engine has to sustain and you can see why some operators are targeting 600 BHP engines.

SUMMARY

- Energy exists in many forms; kinetic, potential, electrical, and chemical are some of its forms.
- Energy is the capacity for performing work.
- Force is usually defined in terms of the effects it could produce, despite the fact that the application of force does not necessarily result in work accomplished.
- Torque is defined as turning effort; that is, it is twisting force and as such may not necessarily result in work accomplished.
- Work is accomplished when force produces a result.
- Power is the rate of producing work; as such, it is always related to time.
- Brake power is power measured at the flywheel of an engine.
- Indicated power is calculated power using the engine specifications and does not factor in friction and pumping losses.
- When engine torque and rpm are known, brake power can be calculated.
- Indicated power can be calculated if mean effective pressure, stroke, bore, power strokes per minute, and the number of engine cylinders are known.
- Mechanical efficiency is the brake power specification divided by the indicated power specification expressed as a percentage.
- Brake power measured at the flywheel is typically around 10% more than brake power measured at the drive wheels of a truck.
- Torque rise is desirable in highway diesel engines; it occurs as an engine is lugged down from its rated power rpm to its peak torque rpm.
- Load is the ratio of power delivered to rated power at the same rpm, expressed as percentage.
- Running an engine in the lower rpm portion of the torque rise profile, especially high torque rise engines, can create torsional oscillations that can affect the entire drivetrain.

REVIEW QUESTIONS

1. Calculate the brake power of an engine producing 1,150 lb.-ft. of torque at the flywheel when it is turning at 2,000 rpm.
2. Use the PLANC formula to calculate the indicated power of a 6-cylinder, 4-stroke cycle engine run at 1,800 rpm using the following data:
   stroke: 6.0″
   bore: 5.5″
   MEP: 220 psi

3. Which of the following is required to produce peak torque in a diesel engine?
   a. Peak brake power
   b. Peak cylinder pressure
   c. Peak volumetric efficiency
   d. Peak mechanical efficiency
4. In most hydromechanical fuel systems, when maximum engine rpm is increased, the engine brake power potential will also increase.
   a. True
   b. False
5. Which of the following statements helps explain why an engine develops peak brake power at a higher rpm than peak torque?
   a. Cylinder pressures are highest
   b. More power strokes per second occur
   c. More time for combustion
   d. Better volumetric efficiency at high rpm
6. Which of the following is true of an engine described as having high torque rise when comparing it with its standard version with the same nominal rated power?
   a. It will produce higher brake power.
   b. It will have a higher maximum speed.
   c. It will provide a wider operating rpm range.
   d. It will be more fuel efficient.
7. A group of cyclists all weighing 80 kg (175 lb) and riding on bicycles of equal weight ride a 10 kilometer (6.4 miles) race and all finish it. Which of the cyclists has performed the most work?
   a. The winner
   b. All have performed equal work
   c. The loser
8. Usable power at the flywheel of an engine is known as:
   a. bsfc
   b. Brake power
   c. Indicated power
   d. Torque
9. When engine power is calculated rather than tested, it is known as:
   a. Mean effective pressure
   b. bsfc
   c. Indicated power
   d. Brake power
10. Which of the following expresses 1 HP correctly?
    a. 33,000 lb.-ft. of work per second
    b. 550 lb.-ft. of work per second
    c. 5,252 lb.-ft. of work per minute
11. Convert 25 lb.-ft. into Newton-meters.
    a. 19.8
    b. 21.3
    c. 33.9
    d. 38.1
12. A Cummings ISM engine, programmed to 330 HP, produces a peak torque value of 1350 ft.-lb at

1300 rpm. Convert the torque value to Newton-meters.
   a. 1007 N·m
   b. 1275 N·m
   c. 1830 N·m
   d. 2225 N·m
13. A Caterpillar C-16 is rated at 600 HP. Express this in kilowatts.
   a. 400 kW
   b. 448 kW
   c. 464 kW
   d. 488 kW
14. A DDC Series 50 engine is programmed to produce 235 kW at 2,100 rpm. Convert the kW to a hp value.
   a. 244 HP
   b. 285 HP
   c. 315 HP
   d. 325 HP
15. When comparing truck engine brake power measured at the flywheel to that measured at the drive wheels, which of the following should be true?
   a. No difference if transmission and drive carrier are functioning properly
   b. Brake power at the wheels greater than that at the flywheel by 10%
   c. Brake power can only be measured at the drive wheels
   d. Brake power at the flywheel exceeds that at the drive wheels by 10%

Chapter 7

Engine Powertrain Components

PREREQUISITE

Chapter 4

OBJECTIVES

After studying this chapter, you should be able to:

- Identify the engine powertrain components.
- Define the functions of the piston assembly.
- Identify trunk and articulating pistons and list their advantages and disadvantages.
- Outline the advantages of the Mexican hat, open combustion chamber design in modern direct-injected, low-emission diesel engines.
- Diagnose some typical piston failures and determine cause.
- Explain how piston rings act to lubricate the cylinder walls and seal the cylinder.
- Identify some commonly used diesel engine piston rings.
- Outline the conditions required to enable rings to seal most efficiently.
- Classify piston wrist pins by type.
- Describe the role of connecting rods and outline the stresses they are subject to.
- Identify common crankshaft throw arrangements and match to the appropriate cylinder block configurations.
- Outline the forces a crankshaft is subjected to under normal operation.
- Describe the materials, manufacturing, and surface hardening processes of typical heavy-duty crankshafts.
- Outline the causes of abnormal bending and torsional stresses a crankshaft may be subject to.
- Identify some typical crankshaft failures and their causes.
- Evaluate crankshaft condition visually, by precision measuring and electromagnetic flux inspection.
- Describe some common crankshaft reconditioning practices.
- Outline the procedure for an in-chassis, rod and main bearing rollover.
- Measure friction bearing clearance using Plastigage™.
- Define the term hydrodynamic suspension.
- Outline the roles played by the harmonic balancer and flywheel assemblies.
- Describe the principle of operation of a viscous-type harmonic balancer.
- Perform a ring gear removal and replacement on a flywheel.
- Outline the steps required to recondition a flat or pot type, heavy-duty flywheel.

KEY TERMS

anodizing	headland piston	Plastigage™
antithrust side	headland volume	quiescent
articulating piston	hone	ring belt
bearing shell	indirect injected (IDI)	ring groove
big end	keystone ring	rod eye
broach	keystone	small end
cam ground	lands	thrust bearing
cracked rods	lugging	thrust face
compression ring	major thrust side	torsion
compressional load	metallurgy	torsional stress
connecting rod	Mexican hat piston crown	trapezoidal ring
crosshead piston	minor thrust side	trapezoidal
crown	Ni-Resist™ insert	trunk type piston
direct injected (DI)	pin boss	valve pocket
friction bearing	piston pin	wrist pin
gas dynamics		

INTRODUCTION

This chapter addresses the group of engine components responsible for transmitting the gas pressures developed in engine cylinders to the engine power take-off mechanism, usually a flywheel. This group of components is defined as the engine powertrain (Figure 7–1), and includes:

pistons

piston rings

wrist pins

connecting rods

crankshafts

friction bearings

flywheels

vibration dampers/harmonic balancers

By definition, a piston is a usually circular plug that seals an engine cylinder bore and reciprocates within it; it is subject to the gas conditions within the cylinder on which it either imparts or receives force. The piston assembly consists of the piston; piston rings, which are used both to seal the cylinder and lubricate the cylinder walls; and a wrist pin, which connects the piston to the connecting rod (Figure 7–2). The upper face of the piston is called the crown, and it is exposed directly to the cylinder chamber and therefore the effects of combustion. A connecting rod links the piston assembly to a throw on the crankshaft. The crankshaft throw is offset from the centerline of the crankshaft. As the crankshaft is rotated, the piston reciprocates in the cylinder bore. In this way the linear reciprocating movement of the piston in the cylinder bore is translated into rotary movement at the crankshaft. In such an arrangement, piston *stroke* or travel distance is defined by the throw dimension: this is the distance from the centerline of the crankshaft main bearing to the centerline of the crank throw.

PISTON ASSEMBLIES

Diesel engine pistons absorb up to 20% of the heat of the combustion gases. A piston's ability to rapidly dissipate heat is essential, and high-output, turbocharged engines often have piston cooling jets that spray lubricating oil on the underside of the piston to help remove heat. Piston crown temperatures are always high, and the rings assist in cooling by conducting some of this heat to the cylinder walls. The crown geometry (the shape of the piston leading edge) has much to do with the gas dynamics (swirl and squish) produced during the compression stroke that will determine the fuel

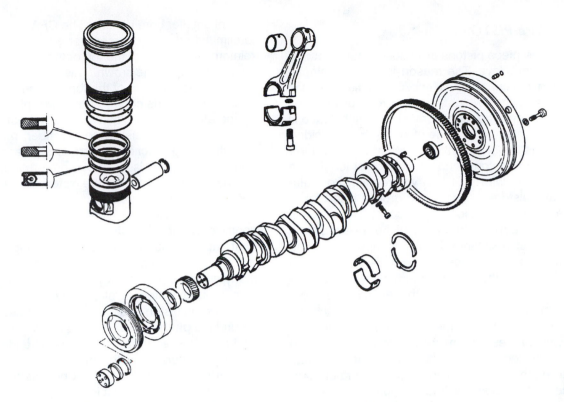

FIGURE 7–1 *Engine powertrain components. (Courtesy of Mack Trucks)*

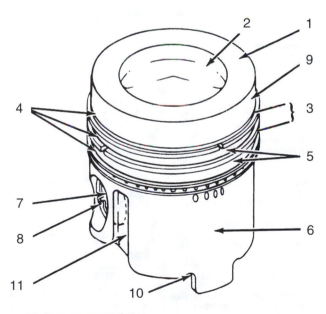

Piston nomenclature

1. Crown
2. Crater
3. Ring band
4. Ring grooves
5. Ring lands
6. Skirt
7. Pin bore
8. Snap ring groove
9. Top land
10. Cooling jet relief
11. Skirt relief

FIGURE 7–2 *Piston nomenclature. (Courtesy of Caterpillar)*

air mixing characteristics and ignition location within the cylinder. The Mexican hat piston crown design is common in low-emission truck DI engines, which tend to have low clearance volumes; at TDC, the piston may rise in the cylinder bore to a height that recesses are required to accommodate cylinder valve protrusion: the recesses are known as **valve pockets**.

It cannot be emphasized enough how critical it is for a piston to efficiently transfer the heat it is exposed to as combustion temperatures may exceed the melting point of the material from which it is made.

- Combustion temperatures rise to transient spikes of 2000°C (3630°F).
- Aluminum melts at 660°C (1220°F).
- Cast iron melts at 1540°C (2800°F).

Two basic piston designs are used in current truck and bus diesel engines. **Trunk type pistons**, which until the 1990s were the piston of choice in most diesels in the 200 BHP (150 kW) to 500 BHP to (375 kW) power range. These have given way more recently to articulating, two-piece pistons. Articulating pistons have been used by manufacturers for many years; many diesel technicians familiar with DDC engines will know them as crosshead pistons. There are advantages and disadvantages to both trunk and articulating designs.

TRUNK TYPE PISTONS

Single- or one-piece pistons are usually manufactured from aluminum alloys. The reason for using a single-piece piston in a current truck application diesel engine is to minimize piston weight, so cast-iron trunk pistons are seldom seen. The aluminum alloying substance is most often a small percentage of silicon that considerably toughens the aluminum. The low melting temperature of aluminum combined with a lack of toughness when compared with cast iron will require most aluminum trunk pistons to use a ring groove insert for at least the top compression ring groove. The insert is usually a **Ni-Resist™ insert**, a nickel bearing, iron alloy with great resistance to high temperatures and wear, but also with a coefficient of heat expansion nearly identical to that of aluminum. The Ni-Resist insert is molecularly bonded to the aluminum piston. Heat treatment also improves the metallurgical characteristics of aluminum. In some instances, hypereutectic (an alloying process) aluminum pistons are used to provide increased fatigue resistance. Another method of increasing the toughness and high-temperature performance of aluminum (without compromising its primary advantage of low mass light weight) is the ceramic fiber reinforced (CFA) process used on some current smaller bore diesels. The CFA process is used to eliminate the requirement for a Ni-Resist insert and reinforce the piston at the top of the ring belt extending up into the top of the crown. This permits both comparatively higher location of the top piston ring (an advantage for emissions control) and lower piston operating temperatures as the coefficient of heat transfer is higher than Ni-Resist. Another fiber reinforcement manufacturing practice known as squeeze cast, fiber reinforced (SCFR) is used by some manufacturers to toughen the crown area of the piston; this is an alumina fiber manufacturing process.

Aluminum pistons are light in weight and have a high coefficient of heat transfer, which means they dissipate heat rapidly. Aluminum also has a high coefficient of heat expansion and contraction, which dictates that many diesel engine trunk pistons are **cam ground**. A cam ground piston is slightly elliptical (oval) when cold—as the piston heats, the greater mass of material around the **pin boss** will expand more than the thinner skirt area between the pin bosses. The idea is that at running temperatures, the piston should expand to a circular shape. Aluminum trunk pistons are shaped to beef up the piston where it is most vulnerable, so apart from the reinforcement at the pin boss, they also have increased mass at the crown (Figure 7–3).

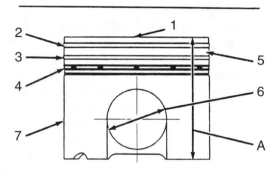

Piston Group.

(1) Crater. The piston is a symmetrical piston with an "on center" crater.
Thoroughly lubricate piston group 360° in zone (A) with clean engine oil prior to inserting into the block group.

(2) Top piston ring.
Install ring with side marked "UP-1" toward top of piston (yellow color stripe to right of ring end gap).
Clearance between ends of piston ring when installed in a cylinder liner with a bore size of 125.000 mm (4.9213 in) .... 0.625 ± 0.125 mm (.0250 ± .0050 in)
Increase in clearance between ends of piston ring for each 0.03 mm (.001 in) increase in cylinder liner bore size .... 0.09 mm (.004 in)

(3) Intermediate piston ring.
Install ring with side marked "UP-2" toward top of piston (green color stripe to right of ring end gap).
Width of groove in piston for intermediate ring (new) .... 3.061 ± 0.013 mm (.1032 ± .0005 in)
Depth of groove in piston for intermediate ring (new) .... 3.727 mm (.1467 in)

Thickness of intermediate ring (new) .... 2.990 ± 0.010 mm (.1177 ± .0004 in)
Clearance between groove and intermediate ring (new) .... 0.048 to 0.094 mm (.0002 to .0037 in)
Clearance between ends of piston ring when installed in a cylinder liner with a bore size of 125.000 mm (4.9213 in) .... 0.625 ± 0.125 mm (.0250 ± .0050 in)
Increase in clearance between ends of piston ring for each 0.03 mm (.001 in) increase in cylinder liner bore size .... 0.09 mm (.004 in)

(4) Oil regulating piston ring.
Oil ring spring ends to be assembled 180° from ring end gap (white colored portion of spring must be visible at ring end gap).
Width of groove in piston for oil ring (new) .... 4.033 ± 0.013 mm (.1588 ± .0005 in)
Depth of groove in piston for oil ring (new) ..... 3.727 mm (.1467 in)
Thickness of oil ring (new) .... 3.987 ± 0.013 mm (.1570 ± .0005 in)
Clearance between groove and oil ring (new) .... 0.020 to 0.072 mm (.0008 to .0028 in)
Clearance between ends of piston ring when installed in a cylinder liner with a bore size of 125.000 mm (4.9213 in) .... 0.55 ± 0.15 mm (.022 ± .006 in)
Increase in clearance between ends of piston ring for each 0.03 mm (.001 in) increase in cylinder liner bore size .... 0.09 mm (.004 in)

After piston rings have been installed, rotate rings so the end gaps are 120° apart.

(5) Crown assembly.

(6) Piston pin bore diameter in piston skirt .... 51.15 ± 0.15 mm (2.014 ± .006 in)

Thoroughly lubricate piston pin with clean engine oil prior to inserting into piston group and rod assembly.

(7) Piston skirt.

FIGURE 7–3 *Piston terminology and tolerance specifications. (Courtesy of Caterpillar)*

Aluminum trunk type pistons may also be surface treated to improve their wear characteristics. Some common methods are:

- Tin plating (tinning can also be used to repair scores on pistons)
- **Anodizing**
- Chrome plating (usually of crown)
- Squeeze cast fiber reinforced (more than just surface treatment, the structural integrity of the piston is increased)

Piston pins when not full floating are usually press fit to the piston boss and float on the rod eye. Heating the piston to 95°C (200°F) in boiling water facilitates pin assembly. Trunk pistons are prone to piston slap. This is the tilting action on the piston when the piston is thrust loaded by cylinder combustion pressure. It can be minimized by tapering the piston so that the outside diameter at the lower skirt slightly exceeds the outside diameter over the ring belt region. The ring belt region is exposed to more heat and expands more as the piston heats to operating temperatures.

Advantages of aluminum alloy, trunk type pistons:

- Lightweight. This reduces the piston mass and therefore the inertia forces that the connecting rod and crankshaft must sustain; tensile stressing of the connecting rod and crankshaft are significantly lower, permitting the use of lighter weight components throughout the engine powertrain.
- Cooler piston crown temperatures. The ability of aluminum alloy pistons to rapidly dissipate combustion heat permits lower crown temperatures.
- Quieter. Engines using aluminum alloy, trunk type pistons generally produce less noncombustion-related noise than comparatively configured and sized engines with articulating piston assemblies.

ARTICULATING PISTONS

Most engine OEMs are now manufacturing at least some engines equipped with articulating pistons, usually in their high power output models. The articulating piston (Figure 7–4) consists of a crown usually manufactured of cast-iron alloy and a separate skirt, usually (but not always) manufactured of aluminum alloy; a wrist pin links the two components to the connecting rod eye. This is not a new concept; DDC has used **crosshead pistons** for many years. The crosshead piston is a two-piece piston assembly consisting of a crown and a skirt united by a semifloating wrist pin; because both the crown and skirt assemblies are afforded some degree of independent movement (i.e., they pivot), the assembly articulates. DDC wrist pins are

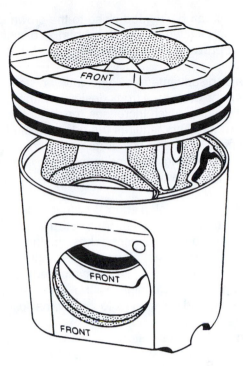

FIGURE 7–4 *Articulating piston. (Courtesy of Mack Trucks)*

bolted to the rod assembly, so they are described as semifloating. The true **articulating piston** assembly has a free-floating piston pin and therefore bearing surface with both the pin boss and the connecting rod eye. Otherwise, it is similar to the crosshead piston.

The **crown** is in most cases manufactured from a cast-iron alloy suitable for the high pressures and temperatures it is directly exposed to, while the skirt can be made out of a lighter material, usually an aluminum alloy. Two-piece, articulating pistons offer greater longevity and reduced tendency to piston slap, but the reason they have become the piston design of choice in recent engines has more to do with the higher combustion pressures, temperatures, and emissions standards required of today's high-efficiency, low-emission diesel engines. Engine designers are fueling diesel engines with ever higher injection pressures to ensure complete combustion of the fuel injected into the cylinder. This extends the duration (measured in small fractions of a second) of peak combustion temperature. Also, the volume in the cylinder above the top compression ring and below the crown leading edge tends to be unaffected by cylinder turbulence and the effects of cylinder scavenging so contains dead gas. This volume, **headland volume**, can be minimized by placing the top compression ring as close as possible to the crown leading edge. Because the tensile strength of cast iron is more than twice that of the aluminum alloy trunk piston and requires no groove insert, the upper ring groove can be located close to the crown leading edge. This is known as **headland piston** design and it has been proven to reduce smoking and improve fuel economy.

A major disadvantage of articulating piston assemblies over the aluminum trunk design is significantly increased weight, which increases tensional loading on the powertrain. This requires the use of beefed-up engine cylinder block and powertrain components, especially the connecting rods and crankshaft.

Advantages of articulating pistons:

- Reduced piston slap effect. When the piston crown is subjected to cylinder gas pressure on both the compression and power strokes, the thrust loading tends to cock (pivot off vertical centerline) the piston in its bore. This action is minimized with the articulating piston design as the skirt assembly is separate and not subject to the vertical load forces of the crown.
- Reduced thermal distortion. The cast iron used as the crown material is less subject to temperature distortion at combustion temperatures and the skirt is to some extent isolated from the crown, permitting cooler and more consistent running temperatures and therefore closer fit tolerances.
- Greater longevity. The superior toughness of the cast-iron crown can withstand more rigorous operating conditions and abuse.
- Improved emissions. Locating the top ring close to the piston leading edge is only possible with the superior toughness of cast iron that permits the headland piston crown design.
- Improved fuel economy. Despite the substantial increase in piston mass of articulating pistons over equivalent aluminum alloy trunk pistons, the greater toughness and temperature tolerance of cast-iron piston crowns permit higher cylinder pressures and temperatures. This increases the thermal efficiency of the engine and therefore the fuel economy.

Piston Thrust Faces

As cylinder gas pressure acts on a piston, there is a tendency for it to cock (pivot off a vertical centerline) in the cylinder bore because it pivots on the wrist pin. This action creates thrust surfaces on either side of the piston. The major thrust face is on the inboard side of the piston as its throw rotates through the power stroke. The minor thrust face is on the outboard side of the piston as its throw rotates through its power stroke (Figure 7–10, see pg. 91). The major thrust face is sometimes simply called the **thrust face** while the minor thrust face is called the antithrust face. It is important for the diesel technician to identify the thrust faces of a piston for purposes of failure analysis.

COMBUSTION CHAMBER DESIGNS

Most contemporary truck and bus diesel engines are direct injected. The fuel charge is therefore injected directly into the engine cylinder. In an **indirect-injected (IDI)** diesel engine, the fuel charge is injected into a cell connected to but not integral with the cylinder cavity, and in most cases ignition takes place within this cell. The cell is known by several names, which categorize type and location. Some examples are precombustion chamber, energy cell, and turbulence chambers. The truck diesel technician is only likely to have encountered IDI technology in the now-obsolete Caterpillar precombustion chambers used with their poppet nozzles. Today, while some small-bore diesel engines still use IDI engines, all of the medium- and heavy-duty truck engines manufactured in North America are direct injected.

Direct-injected (DI) engines use an open combustion chamber principle. In an open combustion chamber, the injector is usually located in the cylinder head and positioned over the piston crown. The shape of the piston crown therefore defines the type of combustion chamber. Since the mixing of the fuel charge with air takes place within the engine cylinder, the **gas dynamics** (swirl and turbulence) are critical in determining the mixing efficiency and the actual location where ignition occurs. Because the piston is the only moving component in the cylinder after the intake valves close, the geometry of the piston crown is critical in defining the gas dynamics. High-turbulence gas dynamics are those designed to produce more vigorous gas movement than **quiescent** gas dynamics, which use a lesser amount of swirl and turbulence. Most DI engines use one of three basic piston crown designs.

Mexican Hat

The **Mexican hat piston crown** design is by far the most common and the title perfectly describes its shape. The central area of piston crown is recessed below the piston leading edge, which forms a kind of wall, and in the center of the recess is a cone-shaped protrusion (Figure 7–10). This shape produces desirable turbulence through the compression stroke and the fuel injector is positioned so that it directs atomized fuel into the recessed cavity of the piston crown where the swirl effect is greatest. The bowl depth of Mexican hat piston crowns usually determines how much gas movement is generated. Deep bowl designs produce greater turbulence and are often used with fuel systems with lower peak injection pressures. Some engine OEMs describe their engines as having quiescent gas dynamics, meaning that relatively low turbulence is generated. A shallower Mexican hat bowl is used with quiescent systems which, in most cases, require the use of higher fuel injection pressures. A major advantage of quiescent cylinder dynamics is reduced after-burn (see Chapter 16) that lowers emissions, which is why late model engines use it. Engine OEMs frequently use the Mexican hat piston crown design because it provides desirable gas dynamics, low risk of fuel burnout on the piston below the injector, and long service life.

Mann Type (or "M" Type)

The Mann type piston crown is named after the German company that first designed it. It is usually used with trunk type pistons and consists of a spherical recess or bowl located directly under the injector, though not necessarily in the center of the piston crown. Depending on the depth of the recess, the Mann type combustion chamber generally produces high turbulence but is more vulnerable to localized burnout in the bowl.

Dished

The dished piston crown has a slightly concaved to almost flat design that produces low turbulence when compared with the previous types. It is more likely to be encountered in an IDI application in a small bore engine.

PISTON COOLING

The size of the piston, BMEP, and whether the engine is boosted all determine the need for piston cooling and the type used. Because combustion temperatures may peak at values that exceed the melting temperature of the cylinder materials, it is essential that the heat not converted to usable energy (rejected heat) is either exhausted directly or dissipated through the cylinder materials. Therefore, some of this heat must be transferred through the piston assembly. For obvious reasons, the cooling of aluminum, trunk type pistons is more critical than with articulating pistons. However, it should be remembered that a slightly misaimed piston cooling jet (Figure 7–5) can cause a rapid failure. Basically three methods of cooling pistons are used:

1. Shaker—Oil is delivered through the connecting rod to a cell in the underside of the crown; this oil is

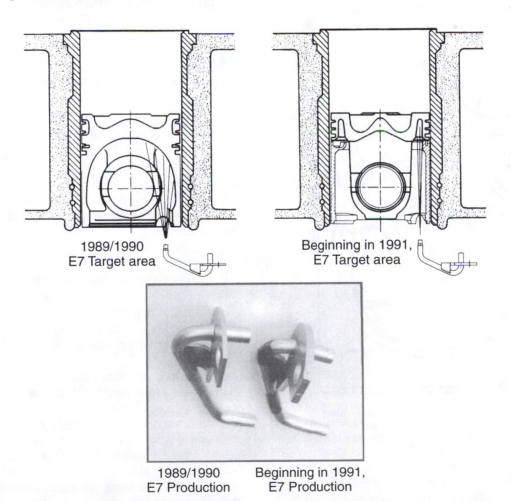

1989/1990
E7 Target area

Beginning in 1991,
E7 Target area

1989/1990
E7 Production

Beginning in 1991,
E7 Production

FIGURE 7–5 *Piston cooling jets. (Courtesy of Mack Trucks)*

distributed by the motion of the piston after which it drains to the crankcase.

2. Circulation—Oil is delivered through the connecting rod, through the wrist pin, and subsequently circulated through a series of grooves machined into the underside of the piston crown. It then drains back into the crankcase.

3. Spray—A stationary jet cylinder block mounted below the cylinder liner and fed by a lubricating oil gallery is directed at the underside of the piston. This oil cools the piston crown and may also lubricate the wrist pin. The jet must be precisely aimed on installation to be effective. This is usually accomplished using a clear perspex template that fits over the fire ring groove on the cylinder block deck and an aim rod inserted in the jet orifice; a direction or target window is scribed in the perspex template and the aim rod must be positioned within the window (Figure 7–6). The spray cooling method is used in most current turbocharged diesel engines.

Engines may use one or combinations of these piston cooling methods.

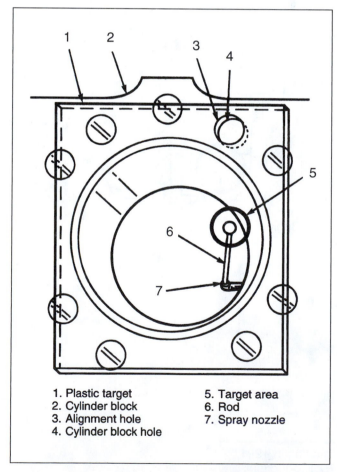

1. Plastic target
2. Cylinder block
3. Alignment hole
4. Cylinder block hole
5. Target area
6. Rod
7. Spray nozzle

FIGURE 7–6 *Spray nozzle targeting. (Courtesy of Mack Trucks)*

CAUTION: A slightly misaligned piston cooling jet can cost you an engine by torching the piston it is supposed to cool. Always check the cooling jet spray window and pay special attention to avoid clunking a cooling jet when installing piston/rod assemblies.

Some Facts

- The rate of heat flow is approximately three times greater in aluminum than cast iron. Therefore, aluminum will dissipate the heat it is exposed to much more quickly than cast irons and steels.
- Weight of aluminum is 0.097 lb. per cubic inch.
- Weight of cast iron is 0.284 lb. per cubic inch.

PISTON FIT PROBLEMS

- Excessive piston-to-bore clearance results in piston knocking against cylinder wall—especially noticeable when engine is cold.
- Too little piston clearance causes piston scoring and scuffing (localized welding)—the film of lube oil on the cylinder wall is scraped off.

PISTON RINGS

The function of piston rings is to seal the piston in the cylinder bore. Most pistons require rings to effectively seal, and those that do not are usually found in automobile racing applications using special piston materials and more importantly, run at high rpms that permit little *time* for cylinder leakage. Rings have three important functions:

1. Sealing: they are designed to seal compression and combustion gases within the engine cylinder.
2. Lubrication: they are designed to apply and regulate a film of lubricant to the cylinder walls.
3. Cooling: rings provide a path for heat to be transferred from the piston to the cylinder walls.

Piston rings are located in circumferential recesses in the piston known as **ring grooves**. Ring grooves are located between **lands**.

Piston rings may be broadly categorized as compression and oil control rings. **Compression rings** are responsible for sealing the engine cylinder and play a role in helping to dissipate piston heat to the cylinder walls. Oil control rings are responsible for lubricating the cylinder walls and also provide a path to dissipate piston heat to the cylinder walls. Piston rings are designed with an uninstalled diameter larger than the cylinder bore, so that when they are installed, radial pressure is applied to the cylinder wall. Until recently, most piston compression rings were manufactured from cast-iron alloys and tended to be more brittle than today's versions. Now, most piston rings are metallurgically more similar to stainless steel than cast iron and as a consequence, their susceptibility to fracture has diminished; in fact, many are highly flexible. The open graphite structure of the cast-iron rings gave them desirable self-lubricating properties but a not-so-desirable shorter service life. In most current diesels these have given way to much tougher and more flexible rings that often show little evidence of wear at engine overhaul. The wall section of the piston in which the set of rings is located is known as the **ring belt**.

RING ACTION

The major sealing force of piston rings is high-pressure gas. Piston rings have a small side clearance. The result of this minimal side clearance is that when cylinder pressure acts on the upper sectional area of the ring, it is first forced down into the land; this enables cylinder pressure to get behind the ring and drive it into the cylinder wall (see lower right side of Figure 7–7). Sealing efficiencies increase with cylinder pressure values.

The number of rings used is determined by the engine manufacturer and factors are bore size, engine speed, and engine configuration. Time is probably the major factor in determining the number of compression rings; the slower the maximum running speed of the engine, the greater the total number of rings because there is more time for gas blowby to occur. Most current truck and bus medium- and large-bore diesel engines have rated speeds that are in the 2,000 rpm range. Engine OEMs commonly use a three-ring configuration of two compression rings and a single oil control ring. However, four- and five-ring configurations can still be seen.

The top compression ring gets the greatest sealing assist from cylinder pressures. Gas blowby from the top compression ring is that used to seal the second compression ring, and so on. Gas that blows by all the rings enters the crankcase, so crankcase pressure values are often used as an indication of the overall health

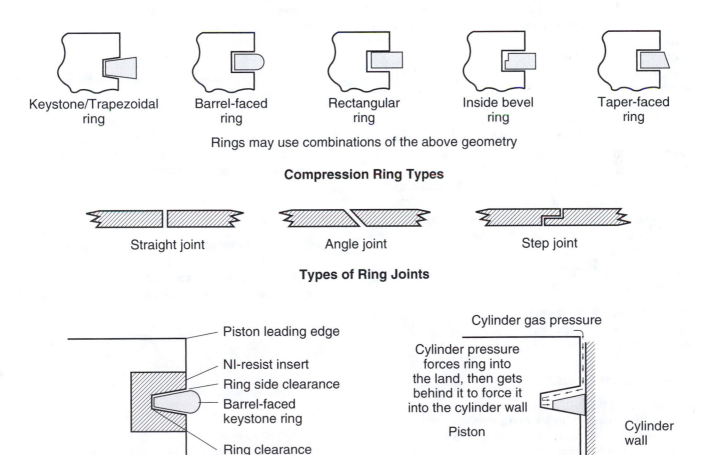

Keystone/Trapezoidal ring   Barrel-faced ring   Rectangular ring   Inside bevel ring   Taper-faced ring

Rings may use combinations of the above geometry

Compression Ring Types

Straight joint   Angle joint   Step joint

Types of Ring Joints

Piston leading edge
Nl-resist insert
Ring side clearance
Barrel-faced keystone ring
Ring clearance

Ring Geometry Terminology

Cylinder gas pressure

Cylinder pressure forces ring into the land, then gets behind it to force it into the cylinder wall

Piston

Cylinder wall

Compression Ring Action

FIGURE 7–7   Piston ring geometry and action.

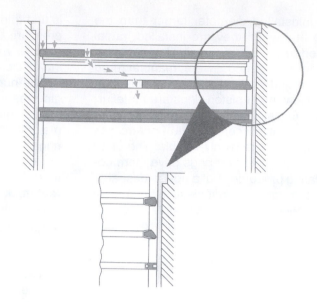

Top compression
• Full keystone
• Plasma

Intermediate compression
• Full keystone
• Taper face

Oil control
• One piece chrome
• Spring expander

FIGURE 7–8   *Ring stagger and pressure balance. (Courtesy of Navistar International Corp., Designer and manufacturer of International Brand diesel engines)*

of an engine. As a rule of thumb, gas pressure diminishes by around 50% beyond the top compression ring. Because the cylinder is sealed by rings, some cylinder leakage past the ring belt is inevitable, but the limiting factor is time. When an engine is operated at 2,000 rpm, one full stroke of a piston takes place in 15 milliseconds (0.015), so there is quite simply insufficient time for significant cylinder leakage to take place. Gas blowby is likely to be more pronounced when the engine is lugged, a running condition in which cylinder pressures are peaking and there is more time. It should be noted that all the piston rings play a role in controlling the oil film on the cylinder wall including the top ring, though it may be categorized as a compression ring, and that role becomes increasingly more important as noxious emissions standards become tougher and more rigorously enforced.

PISTON RING TYPES

There are many different types of piston ring categorized by function and geometry. When classified by function there are three types (Figure 7–8).

Compression Rings

Compression rings are designed with the primary objective of sealing cylinder compression and combustion pressures, though they also play a lesser role in controlling the oil film applied to the cylinder wall. There are many different designs, and a variety of materials are used. Malleable and cast irons have generally given way to highly alloyed steels containing molybdenum, silicon, chromium, vanadium, and other alloys.

Most lack the brittleness of the older cast-iron rings; on the contrary, they can be substantially deformed without fracturing. Compression rings are usually coated by tinning, plasma, and chrome cladding to reduce friction (Figure 7–9). Some ring coatings are break-in coatings designed to facilitate run-in; as these temporary coatings wear and end up in the crankcase lube, this should be taken into account when examining oil sample analyses. Sometimes the upper compression ring is known as the fire ring, but more often this term is used to describe the cylinder seal at the top of the liner that is often integral with the cylinder head gasket (Figure 7–10).

Combination Compression and Scraper Rings

Combination compression and scraper rings are designed both to assist in sealing combustion gases that have blown by the ring above it and to assist in controlling the oil film on the cylinder wall. Manufacturers

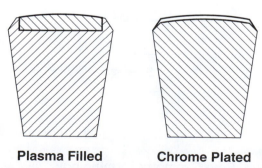

Plasma Filled   **Chrome Plated**

FIGURE 7–9   *Comparison of plasma-filled ring face and chrome-plated keystone rings. (Courtesy of Caterpillar)*

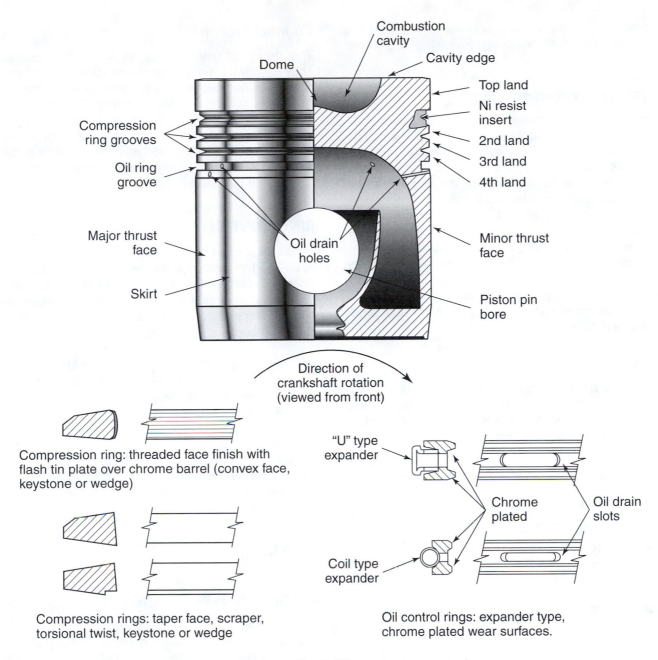

Piston Rings

FIGURE 7–10 *Piston and ring terminology. (Courtesy of Mack Trucks)*

who use this term are referring to a ring or rings located in the intermediate area of the ring belt, under the top compression ring and above the oil control ring(s).

Oil Control Rings

Oil control rings are designed to control the oil film on the cylinder wall, which has to be achieved with some precision; too much oil on the cylinder wall will end up in the combustion chamber, while too little will result in scoring and scuffing of the cylinder wall. The action of first applying and then wiping lubricant from the cylin-

der wall is also responsible for removing heat from the cylinder and transferring it to the engine lube. Most oil control rings use circumferential scraper rails forced into the cylinder wall by a circumferential expander, usually a coiled spring. They are sometimes known as *conformable* rings, because they will flex to conform to a moderately distorted liner bore.

PISTON RING GEOMETRY

Ring geometry describes the physical shape of the piston rings. Figure 7–7 shows some examples of ring geometry. Most piston rings use combinations of the characteristics outlined here. Following are some examples:

Keystone or Trapezoidal Rings

Keystone or **trapezoidal rings** are wedge-shaped and fitted to a wedge-shaped ring groove. The design is commonly used for the top compression ring because its shape allows gas pressure exerted on its upper sectional area to get behind the ring and act on its inner circumference to load it into the cylinder wall. Keystone rings also inhibit the buildup of carbon deposits due to the angled scraping action caused by the relative motion of the ring within its groove, so are less susceptible to sticking. They will also function effectively as they wear. These rings are used in most headland designed pistons. In fact, they have become the ring design of choice among medium- and large-bore diesel OEMs.

Rectangular Rings

The sectional shape of the ring is rectangular. Commonly used in the past when the ring material of choice was cast iron, this ring design is loaded evenly (that is, with relatively little twist) into the cylinder wall when subjected to gas pressure resulting in lower unit sealing pressures but greater longevity.

Barrel Faced

The outer face is "barreled" with a radius (rounded), usually with the objective of increasing its service life; there is no sharp edge to bite into the cylinder wall when the ring subjected to gas pressure twists within the groove. Keystone rings are often barrel faced. Figures 7–7 and 7–8 show examples of barrel-faced keystone rings.

Inside Bevel

A peripheral recess is machined into the inner circumference of the ring to facilitate cylinder gas getting behind the ring and causing it to twist within the groove. When a ring twists within its groove the result is usually high unit-sealing pressures; that is, the ring is allowed to bite into the cylinder wall to maximize the seal.

Taper Faced

The design is much the same as the rectangular ring with the exception that the outer face is angled, giving it a sharp lower edge. Once again, this enables the ring to achieve high unit-sealing pressures; that is, to bite into the cylinder wall when loaded with cylinder gas pressure.

Channel Section

This design is used exclusively for oil control rings. They usually consist of a channel sectioned or grooved ring with a number of slots to permit oil to be both applied and removed from the cylinder walls. Often an expander ring is used in conjunction with this design to improve its conformability: this is a coiled or trussed spring installed to a groove behind the face rails. See the channel section oil control rings shown in the lower right corner of Figure 7–10.

RING JOINT GEOMETRY

Piston rings must be designed so that when heated to operating temperatures, the ring does not expand sufficiently to permit the joint edges to contact each other. They must also seal with some efficiency when cold and cylinder pressures are low. Three types of joint design are used:

1. Straight. The split edges of the ring abut. This design has the disadvantage of affording the most potential for gas blowby at the ring joint. It is, however, the most commonly used.
2. Stepped. This design uses an L-shaped step at the joint and affords the least potential for cylinder leakage at the ring joint.
3. Angled. The ring is faced with complementary angles at the abutting joint and seals fairly efficiently at the ring joint.

Figure 7–7 shows some examples of ring joint geometry.

INSTALLING PISTON RINGS

Rings should be installed to the piston using the correct installation tool. Stretching rings over the piston by hand can crack the plating and cladding materials as can using some generic, multipurpose ring expanders. The engine OEM's special tool is usually the best bet, an example of which is shown in Figure 7–11. NEVER install a piston ring in which the coating appears cracked or chipped. Most rings must be correctly installed and that means they usually have an identified upside. OEM instructions should be consulted to avoid installing a ring upside down; most are marked, typically by using a dot to indicate the upside.

Ring End Gap

The ring end gap is checked by installing a new ring into the cylinder bore and measuring with thickness gauges. Spec is usually 0.003" to 0.004" per 1" of diameter.

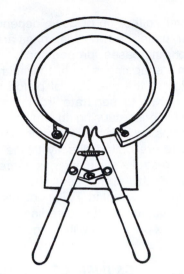

FIGURE 7–11 *Ring expander. (Courtesy of Mack Trucks)*

Ring Gap Spacing or Stagger

Observe OEM instructions. The gaps are usually offset by dividing the number of rings into 360 degrees, so if there were three rings, the stagger would be 120 degrees offset. However, there are other ways of doing this, and Figure 7–8 and Figure 7–12 demonstrate two different methods. The ring gaps are usually not placed directly over the thrust or antithrust faces of the piston.

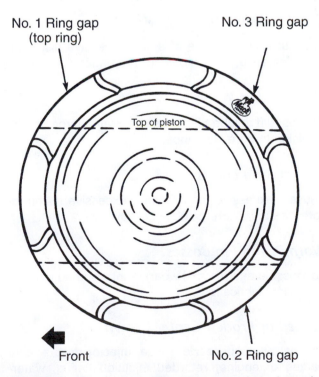

FIGURE 7–12 *Ring stagger. (Courtesy of Mack Trucks)*

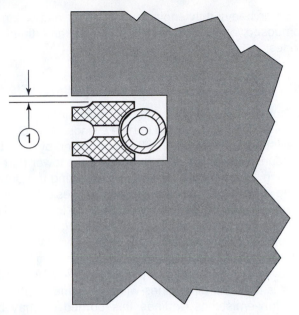

FIGURE 7–13 *Oil ring side clearance. (Courtesy of Mack Trucks)*

Ring Side Clearance

Ring side clearance is the installed clearance between the ring and the groove it is fitted to. The dimension is measured using thickness gauges. Figure 7–13 shows the ring side clearance dimension.

TECH TIP: Most piston rings have an upside that is often not easy to see at a glance. Check the OEM instructions for installing rings and identify the means each uses to identify the upside of their rings.

RING MATERIALS AND COATINGS

Base metals used are grey cast iron, die-formed stainless steels, and nodular cast irons. Nodular cast irons have tensile strengths exceeding 200,000 psi and are ductile to the extent they can be twisted without fracturing. Some cast-iron rings with a stainless steel (molybdenum) inlay have been used in diesel applications. Stainless steels are sometimes used in diesel oil control rings and these may be clad or chromium faced.

Rings coated with tin or electroplated with cadmium are less common in today's engines—plasma sprayed molybdenum is used to coat piston rings in engines designed for longevity. A Caterpillar example of plasma fill is shown in Figure 7–9. Molybdenum is harder than chrome, has a lower coefficient of friction, and due to its high melting point [2620°C (4750°F) versus 1766°C (3210°F) for chrome and 1538°C (2800°F) for cast iron], is highly scuff (localized welding) resistant.

Some heavy-duty engines use ceramic ring coatings composed of a mixture of aluminum and titanium oxides.

PISTON AND CYLINDER WALL LUBRICATION

Oil control rings are designed to maintain a precisely managed film of oil on the engine cylinder wall. On the downstroke of the piston, when not loaded by cylinder pressure, lubricating oil is forced into the lower part of the ring groove while the ring is contacting the upper ledge of the land. When the piston changes direction to travel upward, the ring is forced into the lower land of the ring groove, allowing the lubricating oil to pass around the ring to be applied to the liner wall. While the action of simultaneously applying and scraping oil from the cylinder walls ensures that the applied film thickness is minimal, all engines will burn some oil. In current low-emissions engines this burned oil may be minute, but it does occur.

WRIST PINS

Wrist or **piston pins** are used to connect the piston assembly with the connecting rod eye or small end. In the two-piece, articulating piston assembly, they also link the crown with the skirt; both crown and skirt are permitted independent movement on the wrist pin. The wrist pin serves to transmit loading to and from the piston assembly and the connecting rod while also acting as the axis for the angular movement of the connecting rod as the crankshaft rotates. The extent of the loads to be transmitted will determine whether the pin is solid or bored through; the weight of the wrist pin adds to the piston assembly mass, so it is engineered to be as light as possible while sustaining the loads it is subjected to. Maximum engine speed and peak cylinder pressures determine actual wrist pin design and material. In most cases, wrist pins are manufactured from mild steel and surface hardened, but middle alloy steels are used in some heavy-duty applications. Their bearing surfaces are lubricated by engine oil directed upward through a rifle bore in the connecting rod. Full floating piston pins are fitted to both the rod eye and the piston boss with minimal clearance, usually between 0.0025 mm and 0.0250 mm (0.0001″–0.001″).

Piston Pin Retention

All full-floating piston pins require a means of preventing the pin from exiting the pin boss and contacting the cylinder walls. Snap rings and plugs are used. When installing the internal snap rings used by most engine OEMs it is important to observe the installation instructions. These usually require that the split joint be installed perpendicular to the piston (on a vertical plane

to piston travel), either up or down depending on the manufacturer, usually down. Snap rings are subject to inertia, which increases proportionally with piston speed. When a snap ring is installed with the split joint at right angles to the direction of piston travel, they have been known to separate from their retention groove in the pin boss, causing an engine failure. Semi-floating wrist pins such as those used in crosshead piston assemblies are bolted directly to the connecting rod. In DDC two-stroke cycle engines, a press fit, sealing cap is used; this component should be vacuum leak-tested when assembled. Failure of this cap to seal will result in wrist pin lubricant bleeding to the cylinder walls, and air box pressures will charge the crankcase.

PISTON AND RINGS INSPECTION AND FAILURE ANALYSIS

When clamping a piston/connecting rod assembly in a vise, use brass jaws or a generous wrapping of rags around the connecting rod: the slightest nick or abrasion may cause a stress point from which a failure could develop. When attempting to diagnose the cause of a piston failure, the following may be used as a very general guideline:

Skirt Scoring

Causes: overheating, overfueling, improper piston clearance, insufficient lubrication, improper injector nozzle.

Cracked Skirt

Causes: excessive use of ether, excessive piston-to-bore clearance, foreign objects in cylinder.

Uneven Skirt Wear

Causes: dirt in lube oil, abrasives in air charge, too little piston-to-bore clearance.

Broken Ring Lands

Causes: excessive use of ether, excessive piston-to-bore clearance, foreign objects in cylinder.

Worn Piston Pin Bosses

Causes: old age (normal wear), contaminated lube oil, insufficient lubrication.

Burned or Eroded Center Crown

Causes: plugged nozzle orifice, injector dribble, cold loading of engine, retarded injection timing, water/coolant leakage into cylinder.

Fuel Injection/Ignition Timing Related Failures of Pistons

- Advanced engine timing causes excessive combustion pressures and temperatures: it may result in torching or blowing out the lower ring lands in severe cases. In less severe cases, erosion and burning of the top ring land or headland area result, evidenced by pitting. The latter condition is more common and is often the result of abuse by technicians occasioned by the belief it will increase engine power; it often does, but at a heavy cost, a significant reduction in engine longevity. Total failure will usually occur in one cylinder with the evidence in the remaining cylinders.
- Retarded engine timing will cause excessive cylinder temperatures and burning/erosion damage through the central crown area of the piston. An indicator of retarded injection timing is piston crown scorching under the injector nozzle orifii. The condition is seen much less frequently than advanced injection timing because it is seldom intentional and usually the result of technician error or component failure.

Torched Pistons

This term is used to describe a piston or set of pistons that have been overheated to such an extent that meltdown has occurred. Diagnosing a torched piston in isolation from the engine from which it was removed may be a game of guesswork rather than sound failure analysis practice. However, the cause of the condition must be unmistakably diagnosed before the engine is reassembled, or a recurrence of the failure will result. If only one piston is affected, the diagnosis may be simpler, but it is always important to remember that piston torching may be related to lubrication, cooling system, or fuel injection causes.

Cold Stuck/Gummed Rings

Causes: a condition in which the piston ring seizes in the ring groove when the engine is cold but frees and starts to seal the cylinder when the engine is at running temperature—caused by carbon/sludge buildup in the ring groove. The condition results in high crankcase pressure during engine warmup and fuel contaminated engine lubricating oil.

Hot Stuck Rings

Causes: crystallized carbon buildup resulting from the use of contaminated or improper fuel or the failure of oil control rings. The ring is usually captured stationary in the groove and the contact face blackened by cylinder blowby gas.

Scuffed Rings

Causes: usually related to an improper ring-to-cylinder bore fit such as would occur with an out-of-specification ring end gap, piston-to-bore clearance, or ring-to-groove side clearance. May also be caused by cylinder wall lubrication failure. Scuffing is a localized weld condition, the heat for which is created by friction resulting from metal-to-metal contact; it often results in the seizure of the piston in the liner bore.

Eroded/Glazed Rings

Causes: usually an indication that poorly filtered or unfiltered intake air has been ingested into the engine cylinders. Engines used in construction and aggregate hauling applications often ingest enough ultrasmall particles of abrasive silica sand through a properly functioning air cleaner system to glaze and erode rings and liners and shorten engine life. However, whenever this condition is identified, always test the engine intake system for leaks.

Broken Rings

Causes: improper installation or crystallized carbon buildup in a sector of the ring groove. A condition that is not often observed in the current generation of engines because brittle cast irons are seldom used as a ring material.

Plugged Oil Control Rings

Causes: low engine operating temperatures or broken down, sludged engine lubricant. The result of this condition is usually an engine that burns engine oil, because the oil applied to the liner walls by the piston and bearing throw-off is not controlled. The condition causes coincidental problems such as compression ring gumming or sticking.

REUSING PISTON ASSEMBLIES

This is not a common practice with aluminum alloy trunk type pistons, but it is becoming more common as articulating pistons with tougher, wear-resistant cast-iron crowns become commonplace. Attempt to observe OEM recommended practice, but routine replacement of pistons at an engine overhaul may not always be justified.

- Clean crystallized carbon out of the ring groove using a correctly sized ring groove cleaner, or if a used top compression ring can be broken (many cannot!), file square and use this. Then visually assess the condition of the groove.

- The ring groove is correctly measured with a new ring installed and thickness gauges. A typical maximum clearance spec would be 0.150 mm (0.006") but once again, always observe the OEM specification.
- Ring end gap is measured by inserting the ring by itself into the cylinder bore and measuring gap with thickness gauges. Typically, ring end gaps are 0.075 to 0.100 mm (0.003–0.004") per 2.5 cm (1") of cylinder diameter, so for a 5" cylinder bore end gaps would typically measure between 0.015" to 0.020". Remember, check OEM specifications and *always* measure new rings before installation.

Piston Thrust and Antithrust Side Identification

The piston thrust side is that half of the piston divided at the wrist pin pivot on the inboard side of the crank throw during the downstroke. For a typical engine that rotates clockwise viewed from the front, the major thrust side is the right side of the piston observed from the rear of the engine. The opposite side is known as the **antithrust side**. The terms **major thrust side** and **minor thrust side** are also used.

CONNECTING RODS

Connecting rods (also: conn rods) transmit the force developed in the cylinder and acting on the piston to the throw on the crankshaft (Figure 7–14). The end of the connecting rod that links to the piston wrist pin is known as either the **rod eye** or the **small end** while the end that links it to the crankshaft throw is known as the **big end**. Both the rod eye and big end have bearing surfaces. In this way the linear force that acts on the piston and drives it through its stroke can be converted to rotary force or torque by the crank throw, which rotates around the centerline of the crankshaft. Connecting rods used with crosshead type pistons have no rod eye bearing and instead have a saddle that bolts directly to the wrist pin; upper bearing action is therefore provided by the piston pin boss bearing. Most truck and bus diesel engines use two-piece connecting rods: the rod is usually forged in one piece, the big end cap being subsequently cut off, faced, and fastened (bolted) for machining.

For a number of years, **cracked rod** technology has been used in auto racing applications. This technology has recently been introduced in truck diesel engines. Cracked rods have a big end that is machined in one piece: after machining, the rod is frozen to subzero

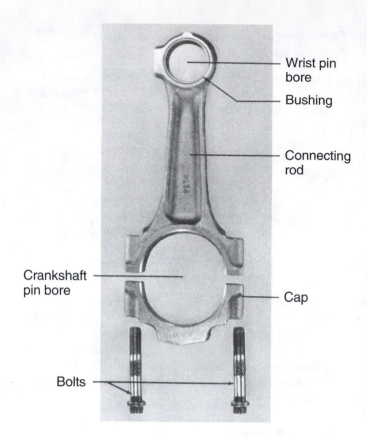

FIGURE 7–14   *Connecting rod terminology. (Courtesy of DDC)*

temperatures and then diametrically cracked through the big end. This produces an abutment of the rod and rod cap with rough-appearing mating faces but final-fit and align perfectly. Mercedes-Benz, Caterpillar, and Detroit Diesel are using cracked rods in some engines but their use could be more widespread.

Most rods use an I-beam section design but round section has also been used. The majority are rifle drilled from big to small ends to carry lubricant from the crank throw up to the wrist pin for purposes of both lubrication and cooling. **Trapezoidal** or **keystone** eye rods have become commonplace because they reduce bending stresses on the wrist pin by increasing the loaded sectional area: these have a wedge-shaped rod eye. Connecting rods are subjected to two types of loading: compressional and tensional.

COMPRESSIONAL LOADING

During the compression and power strokes of the cycle, the connecting rod is subjected to **compressional loading**. Another way of saying compression is squeezing. The extent of compressional loading on a rod can be calculated knowing the cylinder pressure and the piston sectional area values. Truck and bus

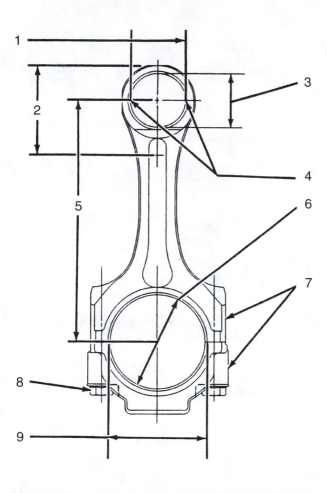

(1) Bore in connecting rod for piston pin bearing

NOTE: The connecting rod must be heated for installation of piston pin bearing. Do not use a torch.

(2) Distance rod may be heated to 175 to 260° C (347 to 500° F) to install the piston pin bearing

(3) Bore in bearing for piston pin

(4) Bearing joint must be at either location

(5) Distance between center of bearings

(6) Bore in bearing for crankshaft journal

(7) Rod cap and rod fastener

(8) Rod cap bolt head

(9) Big end bore dimension (minus friction bearing shell)

FIGURE 7–15 *Critical specification dimensions on a Caterpillar C-10 connecting rod. (Courtesy of Caterpillar)*

diesel engine connecting rods seldom fail due to compressional overloading of conn rods and when they do, it is usually coincidental with another failure such as hydraulic lock. Hydraulic lock is usually the result of a cylinder head gasket failure that has permitted coolant to leak into the cylinder.

TENSIONAL LOADING

Tensional loading is stretching force. At the completion of each stroke, the piston actually stops in the cylinder at either TDC or BDC. This reversal of motion occurs nearly 70 times per second when an engine is run at 2,000 rpm. The greater the mass of the piston assembly, the greater the inertial forces and therefore the tensile stress the rod and crank throw are subject to. This stress can be extreme in modern engines using heavy articulating piston assemblies. Tensile loading on connecting rods increases with engine rpm and the resulting piston speed increase. When an engine is overspeeded, the increased tensile loading on the conn rod can lead to failure (Figure 7–15).

CONNECTING ROD MATERIALS

Connecting rods are metallurgically complex due to the punishment they must sustain under normal operation. Desirable characteristics are a degree of elasticity, lightness, and the ability to absorb the compressional loads of the piston.

Medium carbon steel forgings such as SAE 1041 and SAE 1015 are used as are some alloy steel forgings such as chromium-molybdenum steel SAE 4140. Rods may be heat treated and some are shot peened.

CONNECTING ROD INSPECTION

1. Remove bearing shells (Figure 7–16), install rod cap, and torque to specification.

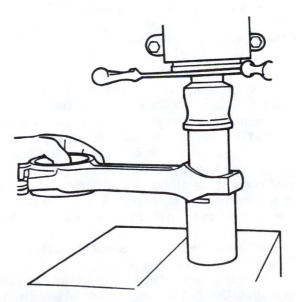

FIGURE 7–16 *Wrist pin bushing removal. (Courtesy of Mack Trucks)*

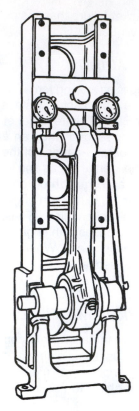

FIGURE 7–17   Connecting rod fixture. (Courtesy of Mack Trucks)

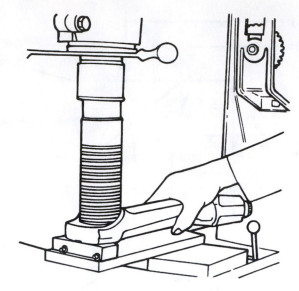

FIGURE 7–18   Burnishing a wrist pin bushing. (Courtesy of Mack Trucks)

2. Measure big and small end bores with snap or tele-scoping gauges used in conjunction with an out-side micrometer. Big end concentricity is critical and stretch is a result of tensile loading of the rod.

3. Rods should be checked for straightness and twist-ing in a mechanical rod fixture (Figure 7–17) or electronic rod gauge fixture, preferably the latter.

4. The oil passage in the bore should be blown out with compressed shop air and if necessary, cleaned with a nylon bristle, rifle brush.

5. Finally, the connecting rod should be electromag-netic flux tested for cracks. This process involves magnetizing the rod using an electromagnet de-signed for the purpose and then coating it with a fine magnetic flux, either dry or in a solution. A crack will interrupt the magnetic field, and the flux will concentrate at the flaw. When the magnetic flux is suspended in solution, it is coated in white pigment so that small cracks may be observed using black light (ultraviolet). Any components that have been magnetic flux tested should be de-magnetized before being reused.

6. When connecting rod eye bushings are replaced, the newly installed bushings should be sized with a burnishing **broach** as demonstrated in Figure 7–18.

Connecting rods should be handled with a great amount of care. When assembling the rod to the piston, a brass jaw vise and light clamping pressure should be used.

Slight nicks and scratches on connecting rods cre-ate stress focal points that can lead to a separation fail-ure. Most diesel engine OEMs suggest that connecting rods be electromagnetic flux tested *every* time they are removed from the engine; the cost of this procedure is small compared to the damage potential of a failure. When a connecting rod fails in a running engine, the re-sult is often a rod driven through the cylinder block casting. OEMs recommend that connecting rods that fail the inspection procedure be replaced. Any recondi-tioning procedure that involves removing material from the rod changes its weight and therefore alters the dy-namic balance of the engine.

Replacing Rods

As reconditioning of rods is not widely practiced when rebuilding today's diesel engines, connecting rods should be inspected according to the procedures out-lined by the manufacturer and if rejected, replaced. When replacing rods, they should be weight matched according to OEM technical service instructions. Be-cause most truck and bus diesels are relatively slow running compared with gasoline engines, some OEMs permit flexibility here, but the consequence of replacing a connecting rod with one of either greater or lesser weight is an unbalanced engine. Manufacturers usually code connecting rods to a weight class and typically each weight class will have a window of about ¼ ounce. When replacing defective connecting rods, al-ways match the weight codes.

Rod Cap Fasteners

Most OEMs require that these be replaced at each re-assembly. Moreover, they should be replaced with the correct OEM fastener and not cross-matched to an SAE-graded bolt. Thread deformation and stretching make these bolts a poor reuse risk, considering the consequences of fastener failure at the rod cap. The actual chances of failure are low, but remember, the cost of connecting rod failure is steep. If the work is being performed for a customer, recommend that they be replaced and leave the decision to reuse the fasteners to the customer.

Connecting Rod Bearings

Most engine OEMs use a single-piece bushing, press fit to the rod eye, and two-piece friction **bearing shells** at the big end. Rod eye bearings should be removed using an appropriately sized mandrel or driver (see Figure 7–16) and an arbor or hydraulic press. Using a hammer and any type of driver that is not sized to the rod eye bore is not recommended because the chances of damage are high. New one-piece bushings should be installed using a press and mandrel, ensuring that the oil hole is properly aligned, and then sized using a broach or **hone** as shown in Figure 7–18. Split big end bearings should also be installed respecting the oil hole location and the throw journal-to-bearing clearance should be measured. Bearing shells and rod eye bushings should both be installed to a clean dry bore: remove any packing protective coating from the bearings by washing them in solvent, followed by compressed air drying.

TECH TIP: Rod sideplay must be checked after a rod cap has been torqued to the rod. Cocking of the rod cap to rod fit can cause an engine to bind and damage the crankshaft. In applications in which cracked rods are used, this check is unnecessary as a perfect rod cap to rod fit can be assumed. Snapping the assembly fore and aft on the journal should produce a clacking noise indicating sideplay.

CRANKSHAFTS AND BEARINGS

Figure 7–19 shows some typical crank throw arrangements. A crankshaft is a shaft with offset throws or journals to which piston assemblies are connected by means of connecting rods. When rotated, the offset crank throws convert the linear, reciprocating movement of the pistons in the cylinder bores into rotary motion at the crankshaft. Crankshafts are supported by friction bearings at main journals and require pressurized lubrication at all times the engine is run to enable their hydrodynamic suspension within the bearing bores. Hydrodynamic suspension is the supporting of a rotating shaft on a fluid wedge of constantly changing, pressurized engine oil. In this way, direct shaft main journal-to-bearing bore contact is avoided; an oil film protects the friction bearing shell when the engine is stationary or cold cranked. Friction bearings are also used at the crank throw/connecting rod big end. The lubrication required for these bearings is supplied from oil passages routed through the crankshaft and oil holes located at each journal. Offsetting the oil hole angle from the crankshaft axis generally provides a thicker wedge of oil for the crankshaft to ride and therefore, most OEMs do this. Crankshafts are designed for dynamic balance and use counterweights to oppose the unbalancing forces generated by the pistons. These forces tend to diminish as the number of cylinders in an engine increases and companion throws (geometrically paired) contribute a counterbalancing effect (Figure 7–19).

Crankshafts are subjected to two types of force: bending forces and torsional forces.

BENDING FORCES

Bending stress occurs between the supporting main journals between each power stroke. Crankshafts are engineered to withstand the considerable bending forces that result from subjecting a crank throw to the

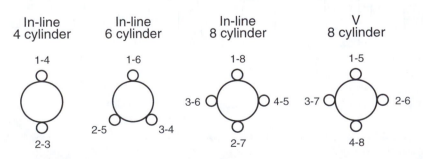

FIGURE 7–19 *Crank throw configurations.*

compression and combustion pressures developed in the cylinder. This normal bending stress takes place between the main journals at any time the crank throw between them is loaded by cylinder pressure, and therefore it peaks when engine cylinder pressures peak.

TORSIONAL FORCES

Torsion is twisting. **Torsional stresses** are the twisting vibrations that a crank is subject to that occur at high speed. Crankshaft torsional vibration occurs because a given crank throw while under compression (that is, driving the piston assembly attached to it upward on the compression stroke) will slow to a speed marginally less than average crank speed. Subsequently, this throw on receiving a power stroke will accelerate to a speed marginally greater than average crank speed. These twisting vibrations or oscillations take place at high frequencies and crankshaft design, materials, and hardening methods must take them into account. Torsional stresses tend to peak at crank journal oil holes at the flywheel end of the shaft. Torsional oscillations are amplified when an engine is run at slower speeds with high cylinder pressures because the real time duration between cylinder firing pulses is extended. This type of running (lower speed/high load) is known as **lugging**, but with many new generation, high torque rise engines, engine torsional oscillations can be projected through the entire drivetrain of the truck.

CRANKSHAFT CONSTRUCTION

Figure 7–20 is a technician's guide to crankshaft terminology. Understanding these terms is critical when measuring and machining crankshafts. Most crankshafts are made of steel forgings, but exceptionally special cast-iron alloys have been used. All crankshafts are tempered (heat treated) to provide a tough core with the required flexing characteristics to withstand the bending and torsional punishment they will be subjected to. An understanding of journal hardening procedures is important as the reconditionability of the crankshaft often depends on this.

Journal Surface Hardening Methods

• Flame hardening. Used on plain carbon steels and therefore seldom on current truck and bus engine applications. Consists of the application of heat followed by quenching with oil or water. The result is relatively shallow surface hardening and the actual hardness dependent on the carbon and other alloy content of the steel.

• Nitriding. Used on alloy steels. Involves higher temperatures than flame hardening and surface hardens to a greater depth—around 0.65 mm (0.025″)

• Induction hardening. Area to be hardened is enclosed by an applicator coil through which alternating current (AC) is pulsed heating the surface—tempering is achieved by blast air or liquid quenching. This process results in hardening to depths of up to 1.75 mm (0.085″), providing a much wider wear and machinability margin than the preceding methods. Most current diesel crankshafts are surface hardened by this method.

Removing Crankshaft from Cylinder Block

The cylinder block should be in an engine stand and in the inverted position. Once the main bearing caps and any other obstructions have been removed, a rigid

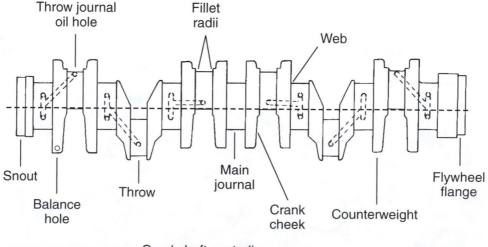

FIGURE 7–20   Crankshaft terminology. (Courtesy of Caterpillar)

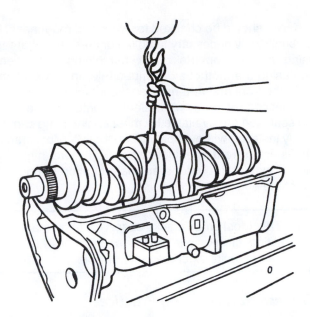

FIGURE 7–21 *Crankshaft removal from cylinder block. (Courtesy of Mack Trucks)*

crankshaft yoke should be fitted to the hoist: the yoke prongs should be fitted with rubber hose and linked to a pair of throws in the center of the crankshaft (for weight balance). Figure 7–21 demonstrates the process.

CRANKSHAFT FAILURES

A certain small percentage of crankshaft failures will result from manufacturing and design problems. Design problems usually only occur shortly after an engine series is introduced. No OEM wishes to have an engine labelled with a basic engineering fault, so design problems are usually rectified quickly.

Because research, development, and testing of new engine series is so thorough in today's engines, basic design flaws seldom occur even in newly introduced engine series. Metallurgical (**metallurgy:** the study of the properties and production of metals) and casting defects are also rare due to the advances in production technology. When crankshaft failures occur, the cause will most likely be attributable to one of the categories listed below.

Bending Failures

Abnormal bending stresses occur when:

- Main bearing bores are misaligned. Any kind of cylinder block irregularity causes abnormal stress over the affected crankshaft area.
- Main bearings fail or become irregularly worn.
- Main caps are broken or loose.
- Standard specification main bearing shells are installed where an oversize is required.

- The flywheel housing is eccentrically (relative to the crankshaft) positioned on the cylinder block. A possible outcome of failure to indicate a flywheel housing on reinstallation.
- Crankshaft is not properly supported either out of the engine before installation or while replacing main bearings in-chassis. The latter practice is more likely to damage the block line bore but may deform the crankshaft as well.

Bending failures tend to initiate at the main journal fillet and extend through to the throw journal fillet at 90 degrees to the crankshaft axis.

Torsional Failures

Excessive torsional stress can result in fractures occurring from a point in a journal oil hole extending through the fillet at a 45° angle or circumferential severing through a fillet. In an in-line six-cylinder engine, the #5 and #6 journal oil holes tend to be vulnerable to torsional failure when the crankshaft is subjected to high torsional loads.

Causes of crankshaft torsional failures:

- Loose, damaged, or defective vibration damper or flywheel assembly.
- Unbalanced engine-driven components: fan pulleys and couplings, fan assembly, idler components, compressors, and PTOs.
- Engine overspeed. Even fractional engine overspeeding will subject the crankshaft to torsional stresses that it was not engineered for that may initiate a failure. It should be remembered when performing failure analysis that the event that caused the failure and the actual failure may be separated by a considerable time span.
- Unbalanced cylinder loading. A dead cylinder or fuel injection malfunction of over- or underfueling a cylinder(s) can result in a torsional failure.
- Defective engine mounts. This can produce a "shock load" effect on the powertrain.

Spun Bearing(s)/Bearing Seizure

A lubrication-related failure caused by insufficient or complete absence of oil in one or all the crank journals. The bearing is subjected to high friction loads and surface welds itself to the affected crank journal. This may result in a spun bearing in which the bearing friction welds itself to the journal and rotates with the journal in the bore or alternatively continues to scuff the journal to destruction. When a crankshaft fractures as a result of bearing seizure, the surface of the journal is destroyed by excessive heat, and it fails because it is unable to sustain the torsional loading.

Causes of spun bearings and bearing seizure:

- Misaligned bearing shell oil hole.
- Improper bearing-to-journal clearance. Excessive clearance will result in excessive bearing oil throw-off, which starves journals farthest from the supply of oil. Insufficient bearing clearance can be caused by overtorquing, use of oversize bearing shells where a standard specification is required, and cylinder block line bore irregularities.
- Sludged lubricating oil causing restrictions in oil passages.
- Contaminated engine oil. Fuel or coolant in lube oil will destroy its lubricity.

Etched Main Bearings

Etched main bearings are caused by the chemical action of contaminated engine lubricant. Chemical contamination of engine oil by fuel, coolant, or sulfur compounds can result in high acidity levels that can corrode all metals. However, the condition is usually first noticed in engine main bearings. It may result from extending oil change intervals beyond that recommended. Etching appears initially as uneven, erosion pock marks, or channels.

CRANKSHAFT INSPECTION

Most truck and bus engine OEMs recommended that a crankshaft be magnetic flux tested at every out-of-chassis overhaul. Essentially this process requires that the component to be tested be magnetized and then coated with minute iron filings: the magnetic field force lines will bend into a crack or nick and the iron filings will tend to collect there. A crankshaft is magnetized using a circumferential electromagnet. The magnetic particles are coated in white pigment and suspended in fluid, allowing them to be sprayed over the crankshaft where they will collect over any flaw that breaks up the magnetic field force lines. Minute flaws may be identified by inspection with an ultraviolet or black light. Any component that has been magnetized for a magnetic flux examination must be demagnetized before it is reused. It should be noted that most small cracks observed by magnetic flux inspections are harmless. The following cracks require crankshaft replacement:

- Circumferential fillet cracks
- 45° cracks extending into critical fillet areas or journal oil holes

Evidence of failures in these locations *require* the replacement of the crankshaft; the consequence of not doing so could be total destruction of the engine. Following magnetic flux inspection, the crankshaft thrust surfaces should be checked for wear and roughness. It is usually only necessary to polish or dress these areas using the appropriate crank machining equipment and using the methods described later in this section. Also, the front and rear main seal contact areas should be checked for wear; wear sleeves, interference fit to the seal race are available for most engines and can be easily installed without specialized tooling. Most crankshafts in current medium- and large-bore highway diesel engines require no special attention during the life of the engine.

TECH TIP: Most small cracks observed when magnetic flux testing crankshafts are harmless: beware fillet cracks and cracks extending into oil holes.

Crankshaft Measuring Practices

Specifications listed here are *typical* maximums; remember, the OEM workshop manual should always be consulted when determining the serviceability of a specific crankshaft. Journals should be measured across axes 90 degrees apart in three places, usually at either end and in the center of the journal and the results matched to OEM specification. Using a measuring chart is the preferred method of organizing the data and retaining it.

- *Journal out of roundness:* 0.025–0.050 mm (0.001″–0.002″)
- *Journal taper:* 0.0375 mm (0.0015″)
- *Bent crankshaft:* Check in V-blocks with a dial indicator while smoothly rotating by hand—out of service (OOS) specifications differ widely due to differing crankshaft materials and lengths.

Polishing Crankshafts

The objective of polishing a crankshaft is to remove minor main and throw journal scratches, scores, and nicks from a crankshaft that has been measured and gauged to be within the OEM specifications. The use of oversized bearings would not normally result from a polishing procedure because no significant amount of journal surface material should be removed. The crankshaft should be mounted to the appropriate lathe mandrels with the journal oil holes plugged with flush port caps and rotated at low speed in a crank grinding lathe in the normal engine rotation direction. A long, narrow strip of low abrasive emery cloth (the type sold in rolls) wetted with diesel fuel or kerosene should be applied to the journals requiring the polishing for short periods and frequent inspection. When polishing the throw journals, the rotation will be eccentric, so a longer strip of emery cloth should be used. The crankshaft thrust faces may also have to be similarly dressed. This procedure is a little bit more tricky, and the emery cloth should be wrapped around an oversized, flat wood tongue depressor. Care should be exercised whenever hands have to be placed near rotating machinery. Thrust face scoring may require dressing by grinding, but before performing this procedure check on the availability of oversized thrust washers/bearings.

Reconditioning Crankshafts

While it should be made very clear that most truck and bus engine OEMs do not approve of the reconditioning of crankshafts, the practice is widespread. In most cases, the reconditioning procedures attempt not to compromise the original surface hardening but where the damage is severe, such as in the example of a spun bearing, reconditioning is mandated.

While journal surfaces can be rehardened, this practice is not so common. The reconditioning methods are simply outlined here because the general industry consensus is that they are bad practice and poor economics in the long term. There are four basic methods of crankshaft reconditioning:

1. Grinding to an undersize dimension. This may require that oversize bearings, not always supplied by the engine OEM, are available.
2. Metallizing surface followed by grinding to the original size.
3. Chroming surface to original size.
4. Submerged arc welding buildup followed by grinding to original size.

ROD AND MAIN BEARINGS

There are probably engineers who devote their entire careers to the study and development of **friction bearings** used in diesel engines. It makes sense for the truck technician to have a rudimentary appreciation of bearing construction and to be able to diagnose characteristic failures (Figure 7–22). Most of the truck and bus engine manufacturers make available excellent bearing failure analysis charts and booklets. These use high-definition color photography and examples of actual failures that make diagnosing failures a cinch. Color Figure C1 shows some examples of classic bearing failures.

Construction and Design

Two basic designs are used in current applications:

1. Concentric wall—uniform wall thickness
2. Eccentric wall—wall thickness is greater at crown than parting faces; also known as deltawall bearings

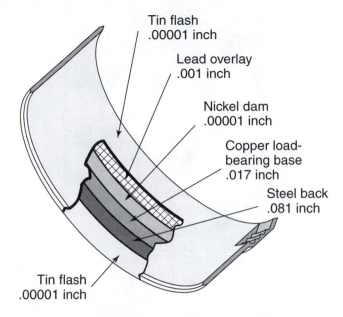

Tri-metal Bearing Construction

FIGURE 7–22   _Construction of a typical bearing. (Courtesy of Navistar International Corp., Designer and manufacturer of International Brand diesel engines)_

Materials

Rod and main friction bearings consist of a steel base or backing plate, onto which is layered copper, lead, tin, aluminum, and sintered combinations of metals often with a zinc or tin outer protective coating (Figure 7–23). Friction bearings are designed to have a degree of embedability, that is, permit small abrasives to penetrate the outer shell known as the overlay, to a depth in which they will cause a minimum amount of scoring to the crank journals.

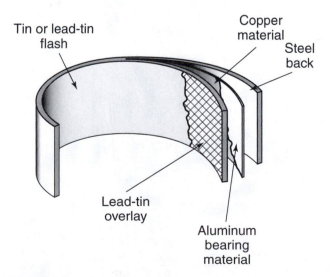

FIGURE 7–23   _Copper-bonded bearing construction. (Courtesy of Caterpillar)_

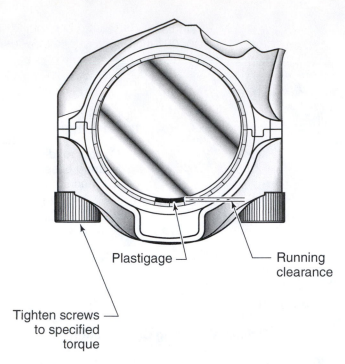

FIGURE 7–24   *Checking bearing clearance using Plastigage. (Courtesy of Mack Trucks)*

Bearing Clearance

The engine OEM's specification must be precisely observed. Bearing clearance is measured with Plastigage (Figure 7–24 and Figure 7–25); most manufacturers

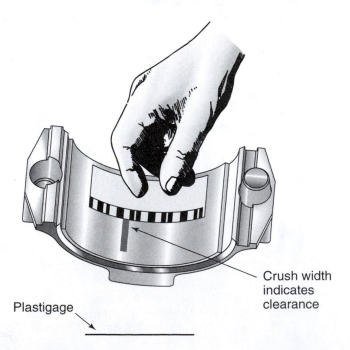

FIGURE 7–25   *Correct method of locating Plastigage. (Courtesy of Mack Trucks)*

make several oversizes (of bearings) to accommodate a small amount of crankshaft journal wear or machining. It is important not to assume that a new engine will always have standard size bearings. Typical bearing clearances in truck and bus engine applications run from 0.050–0.100 mm (0.002–0.004"). The ability to maintain hydrodynamic suspension of the crankshaft diminishes with increased bearing clearances and the resultant drop in oil pressure may require an in-chassis bearing rollover. Increased bearing clearance also increases oil throw-off, and this can result in excessive lube being thrown up onto cylinder walls.

The technician should not attempt to measure bearing clearance when the engine is in-chassis because the results will have little validity. The engine should be upside down and level so that there is no weight load acting on the retaining cap of the bearing being measured. The bearing clearance specifications should be consulted first, and Plastigage capable of measuring between the specified parameters selected. **Plastigage** is soft plastic thread that easily deforms to conform to whatever clearance space is available when compressed between a bearing and journal; the crushed width can then be measured against a scale on the Plastigage packaging. The less clearance available will result in the Plastigage being flattened to a wider dimension. A short strip of Plastigage should be cut and placed across the center of the bearing in line with the crankshaft. The bearing cap with the bearing shell in place should then be installed and torqued to specification in the incremental steps outlined in the workshop manual. Do not rotate the engine with the Plastigage in place. Next the bearing cap and shell should be removed and the width of the flattened Plastigage checked against the dimensional gauge on the packaging. If clearance is within specifications, carefully remove the Plastigage from the journal before reinstalling the cap and shell assembly.

TECH TIP: Always ensure that all Plastigage residue is removed from a bearing after measuring. Use a mild solvent and never an abrasive object to perform this.

Crankshaft Endplay

One of the main bearings is usually flanged to define crank endplay; this is known as the **thrust bearing** and it is available in several sizes to accommodate some thrust surface wear as well as some thrust face dressing in the crankshaft. Alternatively, split rings known as thrust washers may be used to control crank endplay. Endplay specifications would typically be in the 0.2 to 0.3 mm (0.008–0.012") range. Use a dial indicator to measure: force shaft fore and aft for measurement.

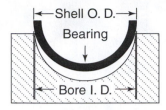 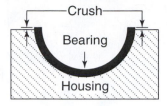

Bearing Free Spread   **Bearing Crush**

FIGURE 7–26 *Bearing spread and crush. (Courtesy of DDC)*

Bearing Retention

Bearings are retained primarily by *crush*: the outside diameter of a pair of uninstalled bearing shells slightly exceeds the bore to which it is installed (Figure 7–26). This creates radial pressure that acts against the bearing halves and provides good heat transfer. The bearing halves are also slightly elliptical to allow the bearing to be held in place during installation; this is known as bearing spread (Figure 7–26). Tangs in bearings (Figure 7–27) are inserted into notches in the bearing bore to minimize longitudinal movement, prevent bearing rotation, and align oil holes.

Bearing Removal and Installation

Workshop manuals should always be consulted and their procedures observed. This operation is very straightforward when performed out of chassis because the crankshaft is removed when installing the cylinder block side bearing shell and the technician is working above the engine with perfect visibility and accessibility for the remainder of the procedure.

For many novice truck technicians, their first experience of engine work will be performing what is known as an in-chassis bearing roll-over, a simple procedure

but one that can be messy, especially on a hot engine that may drip oil for hours. The procedure is practiced more often than necessary on today's engines because a slight drop in oil pressure is often mistakenly attributed to bearing wear. Many service facilities offer bearing rollover "specials" usually early in the spring season. It is not unusual when performing a bearing rollover to remove a set of bearings in near perfect condition: this is a testament to the quality of today's bearings (and other factors) compared with those of a generation ago.

In-chassis, main bearings are rolled out using a capped dowel (often " home" made by machining a bolt) inserted into the journal oil hole; ensure that the dowel cap does not exceed the thickness of the bearing, then rotate the crank to roll the bearing out, tang side first. Using a screwdriver or similar method will almost certainly result in scratch and score damage to the journal. Bearing halves are installed clean, dry, and with as little handling as possible. If a new bearing is coated with a protective film of grease or light wax, this should be removed using a bristle brush and solvent (do not use a solvent that has any residual oil component such as Varsol™), followed by air drying. The backing is always installed clean, dry, and preferably without finger contact—trace quantities of lubricant, moisture, or dirt will reduce its ability to transfer heat. The facing may be lightly coated with clean engine lubricant applied by finger. Most engine OEMs recommend avoiding the use of any type of grease, such as white lube (lithium-based grease) on the bearing face, because these lower the ability of the engine lubricant to hydrodynamically support the crankshaft on startup and may, in the crankcase, interact with the additives in the engine oil. After performing any procedure that involves the draining of lubricant from the oil galleries and passages that supply the bearings, it is good practice to prime the engine lubrication circuit using a remote pump before cranking.

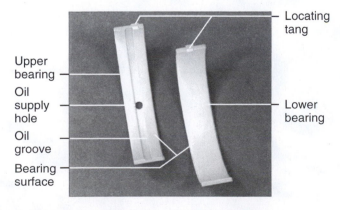

Upper bearing — Oil supply hole — Oil groove — Bearing surface — Locating tang — Lower bearing

FIGURE 7–27 *Upper and lower bearing shells. (Courtesy of DDC)*

VIBRATION DAMPERS/HARMONIC BALANCERS

Both these terms are used to describe the same component, but vibration damper is generally used in this text. A vibration damper is mounted on the free end of the crankshaft, usually at the front of the engine. Its function is to reduce the amplitude of vibration and add to the flywheel's mass in establishing rotary inertia. In other words, its primary function is to reduce crankshaft torsional vibration (there is a full explanation of crankshaft torsional vibration earlier in this chapter under crankshafts).

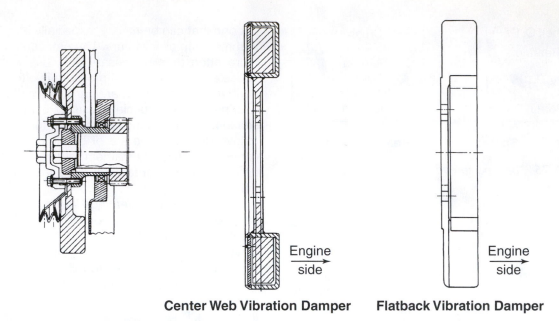

Center Web Vibration Damper   **Flatback Vibration Damper**

FIGURE 7–28   *Sectional view of a vibration damper. (Courtesy of Mack Trucks)*

A typical vibration damper consists of a damper drive or housing and inertia ring. (Figure 7–28 and Figure 7–29.) The housing is coupled to the crankshaft and using springs, rubber, or viscous medium, drives the inertia ring: the objective is to drive the inertia ring at *average* crankshaft speed. Viscous type harmonic balancers are most common in truck and bus diesels; the annular housing is hollow and bolted to the crankshaft. Within the hollow housing, the inertia ring is suspended in and driven by silicone gel. The shearing of the viscous fluid film between the drive ring and the inertia ring effect the damping action.

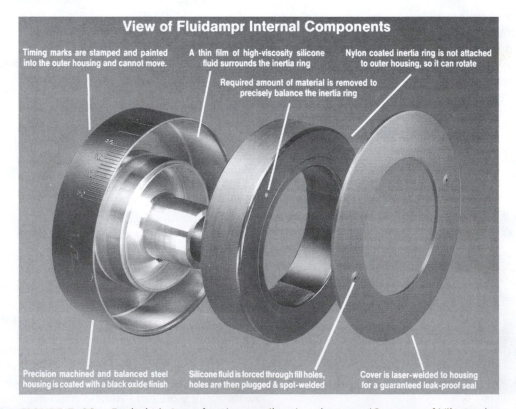

FIGURE 7–29   *Exploded view of a viscous vibration damper. (Courtesy of Vibratech IDEX)*

Most OEMs recommend the replacement of the harmonic balancer at each major overhaul but this is seldom observed due to the expense. Practice has shown these components to frequently exceed OEM projected expectations. The consequences of not replacing the damper when scheduled are economic as they may result in a failed crankshaft. The shearing action of the silicone gel produces friction, which is released as heat. This leads to eventual breakdown of the silicone gel, a result of prolonged service life or old age. Drive housing damage is another common reason for viscous damper failure, and this is probably caused by careless service facility practice in most cases.

Rubber type vibration dampers are not often observed on today's heavy truck diesel engines, suggesting that they are probably less effective. This type consists of a drive hub bolted to the crankshaft: a rubber ring is bonded both to the drive hub and the inertia ring. The rubber ring therefore acts both as the drive and the damping medium. The inherent elasticity of rubber enables it to function as a damping medium, but the internal friction generates heat, which eventually hardens the rubber and renders it less effective and vulnerable to shear failures.

INSPECTING A VISCOUS TYPE DAMPER

Because the critical functional element in a viscous type harmonic balancer is the silicone gel damping/drive medium sealed inside the drive ring, the typical service facility lacks the equipment to check the operating or dynamic effectiveness of these devices. However, if an engine balance irregularity is suspected and the OEM projected service life is known to have been exceeded, the damper should be replaced. If the harmonic balancer service life has not been exceeded, there are some external checks that the technician may perform to help diagnose the condition.

1. Visually inspect the damper housing, noting any dents or signs of warpage: evidence of either is reason to reject the component.
2. Using a dial indicator, rotate the engine manually, and check for damper housing radial and axial runout against the OEM specification. This is a low tolerance specification, usually 0.005″ (0.127 mm) or less.
3. Check for indications of fluid leakage, initially with the damper in place. Trace evidence of leakage justifies replacement of the damper.
4. Should the damper not be condemned using the above tests, remove it from the engine. By hand, shake the damper; any clunking or rattle is reason to replace it.
5. Next, using a gear hotplate or component heating oven, heat the damper to its operating tempera-

ture, usually close to the engine operating temperature, around 90°C (180°F)—this may produce evidence of a leak.
6. A final strategy is to mount the damper in a lathe using a suitable mandrel. It should be run up through the engine operating rpm range and monitored using a balance sensor and strobe light.

CAUTION: Recommend that a viscous type vibration damper be replaced at engine overhaul regardless of its external appearance. Explain that this is an OEM recommendation. If the customer declines, he has made the decision, not you. A failed vibration damper can cause crankshaft failure.

FLYWHEELS

The engine flywheel in the typical truck/bus diesel engine is normally mounted at the rear of the engine. It has three basic functions:

1. To Store kinetic energy (the energy of motion) in the form of inertia and both help smooth out the power pulses in the engine and establish an even crankshaft rotational speed.
2. To provide a mounting for engine output: it is the power takeoff device to which a clutch or torque converter is bolted.
3. To provide a means of rotating the engine by cranking motor during startup.

As an energy storage device, the flywheel plays a major role in dampening the torsional vibrations that act on the crankshaft during engine operation, and its mass helps rotate the engine between firing pulses. Flywheel mass (weight) depends on a number of factors, such as whether the engine is two- or four-stroke cycle, the number of engine cylinders, and the engine operating rpm range. Because the number of crank angle degrees between power strokes is half that on a two-stroke cycle compared to a four-stroke cycle of the equivalent number of cylinders, generally less flywheel mass is required.

six-cylinder, two-stroke cycle—frequency of power strokes: 60 crank angle degrees

six-cylinder, four-stroke cycle—frequency of power strokes: 120 crank angle degrees

The flywheel therefore stores kinetic energy in the form of inertia. Engines designed to be run at consistently high rotational speeds require less flywheel mass, and it should be noted that a heavy flywheel will inhibit rapid response to acceleration demand. While a four-stroke cycle, single cylinder diesel engine is unlikely to power any modern truck, it is worth taking a look at such an engine and noting its flywheel mass, a dominant characteristic of its appearance.

TYPES OF FLYWHEELS

The flywheels used on all North American-built trucks are categorized by the SAE by size, shape, and bolt configuration. Two basic geometric shapes are used: the *flat face* design and the *pot* design. Figure 7–30 shows a flat face truck flywheel. Most trucks equipped

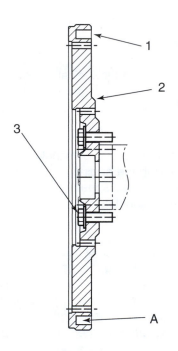

Typical Example

Make reference to Flywheel Runout for correct method of flywheel inspection

(1) Flywheel starter gear. Gear must be assembled against shoulder (A) of the flywheel. Maximum temperature of gear when shrinking in place
315°C (599°F)

(2) Flywheel.

(3) Apply 5P3413 Thread Sealant to bolts and torque to a torque of
300 ± 30 N•m (220 ± 22 lb.-ft.)

FIGURE 7–30 *Sectional view of a flywheel. (Courtesy of Caterpillar)*

with current medium- and large-bore highway diesel engines use one of two SAE flywheel sizes:

SAE # 4—accommodates a 15½″ clutch assembly

SAE # 5—accommodates a 14″ clutch assembly

FLYWHEEL CONSTRUCTION

Flywheels used on most truck and bus diesel engines are machined from cast iron or cast steel. Rim velocities are factored by the rotational speed (engine rpm) and the clutch diameter. However, rim stresses seldom generate metallurgical failures in truck and bus diesel engine flywheels due to their relatively low rotational speeds. Stresses tend to peak at the juncture of the hub with the rim, an area that is subject to torsional as well as centrifugal loads, but failures are rare.

RING GEAR REPLACEMENT

Shrunk fit to the outer periphery of the flywheel is the ring gear, which is the means of transmitting cranking torque to the engine by the starter motor during start-up. A worn or defective ring gear is removed from the flywheel by first removing the flywheel from the engine and then using an oxyacetylene torch to partially cut through the ring gear working from the outside. This is in most cases sufficient to expand the ring gear so that the removal can be completed using a hammer and chisel. Care should be taken to avoid heating the flywheel any more than absolutely necessary or damaging the flywheel itself by careless use of the oxyacetylene flame.

To install a new ring gear, place the flywheel on a flat, level surface, and check that the ring gear seating surface is free from dirt, nicks, and burrs. Ensure that the new ring gear is the correct one and if its teeth are chamfered on one side, that they will face the cranking motor pinion after installation. Next, the ring gear must be expanded using heat so that it can be shrunk to the flywheel. Most OEMs specify a specific heat value because ring gears are heat treated, and overheating will damage the tempering and substantially reduce the hardness. A typical temperature specification would be around 200°C (400°F). Because of its size, the only practical method of heating a ring gear for installation is using a rosebud type (high gas flow), oxyacetylene heating tip. To ensure that the ring gear is heated evenly to the specified temperature and especially to ensure that it is not overheated, the use of a temperature indicating crayon such as Tempilstick is recommended. When the ring gear has been heated evenly to the correct temperature, it will usually drop into position and almost instantly contract to the flywheel.

RECONDITIONING AND INSPECTING FLYWHEELS

Flywheels are commonly removed from engines for reasons such as clutch damage, leaking rear main seals, or leaking cam plugs, and care should be taken when both inspecting and reinstalling the flywheel and the flywheel housing. Flywheels should be inspected for face warpage, heat checks, scoring, intermediate drive lug alignment and integrity, and axial and radial runout using dial indicators, straightedges, and thickness gauges. Damaged flywheel faces may be machined using a flywheel resurfacing lathe to OEM tolerances: typical maximum machining tolerances range from 0.060" to 0.090" (1.50 mm–2.30 mm). It is important to note that when resurfacing pot type flywheel faces, the pot face must have the same amount of material ground away as the flywheel face; the consequence of machining only the clutch face is to have an inoperable clutch.

SUMMARY

- For purposes of study, the engine powertrain comprises those engine components responsible for delivering the power developed in the engine cylinders to the power takeoff mechanism, usually a flywheel.
- Trunk type pistons manufactured of aluminum alloys were widely used by truck and bus engine OEMs until the late 1980s because of their light weight and ability to rapidly transfer heat.
- Two-piece, articulating piston assemblies are more common in today's low-emission, extended operational life medium-, and large-bore highway diesel engines.
- Most aluminum alloy trunk type pistons used in truck diesel engines support the top compression ring with a Ni-Resist insert and are both cam ground and tapered.
- Most truck and bus engine OEMs use articulating pistons in their high power engines.
- The Mexican hat piston crown, open combustion chamber is most common in current direct-injected diesel engines.
- Engine oil is used to help cool pistons in three ways: the shaker, circulation, and spray jet methods.
- Piston rings seal when cylinder pressure acts on the exposed sectional area of the ring, which first forces it down into the land and then gets behind it to load the ring face into the cylinder wall.
- The efficiency with which piston rings seal a cylinder increases with increased cylinder pressure.

- Gases that manage to pass by the piston rings enter the crankcase and are known as blowby gases.
- The keystone ring design is the most commonly used top compression ring in today's highway diesel engines.
- Oil control rings are designed to apply a film of oil to the cylinder wall on the upstroke of the piston and "scrape" it on the downstroke.
- All the piston rings play a role in controlling the oil film on the cylinder.
- Full floating wrist pins have a bearing surface with both the piston boss and the connecting rod eye.
- Crosshead pistons articulate but have a semifloating wrist pin that bolts directly to the rod small end.
- Full floating wrist pins are retained in the piston boss by snap rings.
- DDC two-stroke cycle engines use press fit caps to seal the pin boss to seal the crankcase from the air box.
- Connecting rods are subjected to compressional and tensional loads in normal service operation.
- Most connecting rods will survive the life of the engine, but they should be dimensionally and crack inspected at each overhaul.
- Crankshafts must be designed to withstand considerable bending and torsional stress.
- Most medium- and large-bore highway diesel engines use induction-hardened crankshafts.
- Engine OEMs often do not approve of reconditioning failed crankshafts; however, the practice is widespread despite the risk of a subsequent failure.
- The friction bearings used in crankshaft throw and main journals are retained by crush.
- Harmonic balancers or vibration dampers consist of a drive plate, drive medium, and inertia ring.
- The viscous type damper is the most commonly used on today's truck and bus diesels: the hollow drive ring is bolted directly to the crankshaft and suspended in gelled silicone is the inertia ring. The shearing action of the silicone drive medium between the drive ring and the inertia ring effects the damping.
- The flywheel stores kinetic energy in the form of inertia to help smooth out the power pulses delivered to the engine powertrain.
- Flywheels are categorized by size and shape by the SAE.

REVIEW QUESTIONS

1. Which of the following engine conditions would be the likely cause of a tensile failure of a connecting rod?

a. Engine overspeed
b. High cylinder pressures
c. Prolonged lug down
d. Prolonged engine idling

2. Which of the following is a disadvantage of an articulating piston assembly when comparing it with an aluminum alloy trunk piston?
a. Piston slap control
b. Highest possible location of upper ring groove
c. Service life
d. Increased tensile loading on connecting rods

3. Where would a Ni-Resist insert be located?
a. Upper ring belt, trunk piston
b. Upper ring belt, crosshead piston
c. Trunk piston, pin boss
d. Exhaust valve seat

4. Under which of the following conditions would a piston ring seal more effectively?
a. High engine temps
b. Low engine temps
c. High cylinder pressures
d. Low cylinder pressures

5. A crankshaft has a fracture failure that appears to have initiated at a throw journal oil hole. Which of the following would be the more likely cause?
a. Broken main cap
b. Excessive cylinder pressure
c. Failed harmonic balancer
d. Spun rod bearing

6. Which of the following journal surface hardening methods would provide the highest machinability margin?
a. Flame hardening
b. Nitriding
c. Shot peening
d. Induction hardening

7. Rod and main journal bearing clearances are properly measured with:
a. Tram gauges
b. Dial indicators
c. Plastigage
d. Snap gauges

8. Which of the following components helps prevent air box pressure from charging the crankcase in DDC two-stroke cycle engines?
a. Crosshead pivot
b. Pin boss cap
c. Viscous seal
d. Crankcase pressure regulating valve

9. When observing a disassembled engine, the outer crown edge of all the pistons show signs of melting and erosion. This could be caused by:
a. Advanced timing
b. A dribbling injector
c. Retarded timing
d. High sulphur content fuel

10. When examining a set of pistons from a disassembled engine, scuffing is observed on both the thrust sides of all of the pistons. Which of the following would be the more likely cause?
a. High cylinder pressures
b. Overheating
c. Contaminated engine lube
d. High sulphur content fuel

11. Etched main bearings could be caused by all of the following conditions except:
a. Fuel in lube
b. Coolant in lube
c. Sulphur in lube
d. Particulate in lube oil

12. Which of the following conditions would be indicative of retarded injection timing in a diesel engine?
a. Erosion near center of piston crown
b. Upper land failure
c. Overall piston scuffing
d. Fractured rings

13. Which of the following conditions could cause carbon buildup on pistons?
a. Overfueling
b. Injector nozzle failure
c. Retarded injection timing
d. A, B, and C

14. Which of the following operating modes would be more likely to result in a compressional failure of a connecting rod?
a. Prolonged low idle running
b. Lugdown
c. Prolonged high idle running
d. Overspeeding engine

15. Which of the following would be a likely outcome of running an engine with a failed harmonic balancer?
a. Increased fuel consumption
b. Increased cylinder blowby
c. Crankshaft failure
d. Camshaft failure

16. What prevents a crosshead piston's wrist pin from contacting the cylinder walls?
a. Pin retainer
b. Piston boss cap
c. Connecting rod-to-wrist pin capscrews
d. Wrist pin-to-crown capscrews

17. The upper face of a piston assembly is called the:
a. Skirt
b. Boss
c. Trunk
d. Crown

18. *Technician A* states that a cam ground piston is circular when cold and expands to an elliptical shape at operating temperature. *Technician B* states that the reason the cam ground piston is elliptical when cold is to accommodate the thermal expansion of the greater mass of material at the

piston boss as it is brought up to operating temperature. Who is right?
a. A Only
b. B Only
c. Both A and B
d. Neither A nor B

19. The most commonly used piston crown design by OEMs in today's highway diesel engines is:
a. Mexican hat
b. Mann type
c. Dished
d. Barrel face

20. A piston in which the upper compression ring is located close to the leading edge of the piston is known as a:
a. Keystone ring
b. High top piston
c. Ni-Resist type
d. Headland piston

21. Which of the following is true of a CFA piston?
a. The ring belt reinforcement insert extends up into the leading edge of the piston.
b. It is a type of Ni-Resist insert.
c. It is only used on articulating pistons.
d. The ring belt operates at higher temperatures.

22. When crankshaft thrust surfaces are reground, which of the following would be true?
a. Undersize thrust bearings are required.
b. Oversize thrust bearings are required.

c. The thrust washers must be shimmed.
d. The thrust washers must be machined to a smaller dimension.

23. *Technician A* states that main bearing shells are usually interchangeable and may be installed in either the upper or lower position. *Technician B* states that the main bearing shell with an oil hole must always be installed in the main bearing cap. Who is right?
a. A only
b. B only
c. Both A and B
d. Neither A nor B

24. Excessive clearance at the crankshaft main bearings could result in which of the following conditions?
a. Aerated oil
b. Fluctuating oil pressures
c. High oil pressure
d. Low oil pressure

25. *Technician A* states that the most commonly used type of vibration damper used on truck diesel engines is the viscous type. *Technician B* states that the inertia ring drive medium on a viscous type vibration is a rubber compound. Who is right?
a. A only
b. B only
c. Both A and B
d. Neither A nor B

Chapter

8

Engine Feedback Assembly

PREREQUISITES

Chapters 4 and 7

OBJECTIVES

After studying this chapter, you should be able to:

- Identify the engine feedback assembly components.
- Describe the role of the engine timing gear train in managing engine functions.
- Describe the procedure required to time an engine gear train.
- Define the role of the camshaft in a typical diesel engine.
- Interpret the terminology used to describe camshaft geometry.
- Inspect a camshaft for wear and damage, selecting the appropriate tools.
- Outline the procedure required to remove and replace a set of block camshaft bearings.
- Identify the role valve train components play in running an engine.
- List the types of tappet/cam follower used in diesel engines.
- Inspect a set of push tubes or rods and evaluate their serviceability.
- Describe the role of the rocker assembly in the engine feedback assembly.
- Define the role of cylinder head valves and interpret the terminology used to describe them.
- Outline the procedure required to recondition a set of cylinder head valves, identifying valve margin and other critical wear/machining tolerances.
- Describe how valve rotators operate.
- Define the role of valve seat inserts and outline the servicing procedure.
- Perform a valve lash adjustment on a diesel engine using OEM specifications.
- Perform basic failure analysis on diesel engine cylinder valves.
- Outline the consequences of either too much or too little valve lash.

KEY TERMS

base circle	hunting gears	tappets
cam geometry	inner base circle	Tempilstick™
cam profile	lifters	train
camshaft	outer base circle	valve
companion cylinders	periphery	valve float
concept gear	ramps	valve margin
followers	rockers	valve train
helical gear	spur gear	

INTRODUCTION

The feedback assembly of an engine is the engine's mechanical management apparatus and includes timing and accessory drive gearing, the camshaft, tappets, valve and unit injector trains, and fuel pumping mechanisms. Its components are driven by and often timed to, the engine crankshaft. The drive mechanism for medium- and heavy-duty highway diesel engine feedback assemblies is almost always a gearset. In lighter duty engines pulleys and belts, chains and sprockets may be used as drives. A typical timing gear set and cover are illustrated in Figure 8–1 and Figure 8–2.

TIMING GEARS

Diesel engine timing gears are normally located at the front of the longitudinally mounted engines that power most commercial trucks and the transverse-mounted engines that power most buses. Timing gears are responsible for driving the camshaft and most of the engine accessories on a diesel-powered truck engine. The gear train is enclosed in a housing that permits engine oil lubrication of the rotating components. Some larger engines may locate the timing gear train in the rear or both front and rear. The gear ratios dictate the rotational speed and maintain a fixed location between the driven components in any given moment

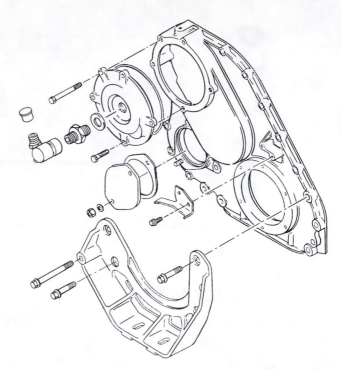

FIGURE 8–2   Timing gear cover. (Courtesy of Mack Trucks)

of engine operation. Timing gears are lubricated in two ways:

1. Splash: gear teeth rotated through the lubricant in the oil sump pick up oil and transfer it to other gears in the timing train before draining back to the sump.
2. Bearing spill: oil used to lubricate the shaft support bearings spills to the timing gear housing where it is circulated before returning to the oil sump.

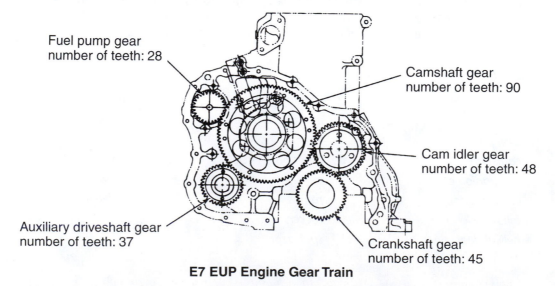

E7 EUP Engine Gear Train

FIGURE 8–1   Mack Trucks E-Tech engine timing gear train. (Courtesy of Mack Trucks)

TIMING GEAR CONSTRUCTION

Diesel engine gears are cast or forged alloys that are heat tempered and then surface hardened by a flame, nitriding, carburizing, or induction hardening process. The gear teeth are milled in manufacture to spur and helical designs: combinations of both are used in the gear trains of some current engines. The noise produced by the gear train is a factor and for this reason, helical cut gears tend to be more common. **Helical gears** have the advantage of increased tooth contact area lowering unit forces. The **spur gear** design offers the advantage of much lower thrust loads. The gears are commonly press fit to the shafts that they drive and are positioned on the shaft by means of keys and keyways.

TIMING GEAR INSPECTION, REMOVAL, AND INSTALLATION

Visual inspection should be sufficient to determine the condition of timing gears. Indications of cracks, pitting, heat discoloration, or lipping of the gear teeth usually require its replacement. Press fit gears require the use of a mechanical, pneumatic, or hydraulic puller to remove the gear from the shaft. When the shaft and gear can be removed from the engine, a shop air-over-hydraulic press should be used, ensuring that the usual safety precautions are observed and that care is taken to support the components on separation. When a gear has to be separated from a shaft while on the engine, a portable hydraulic press is usually required; while using the press, ensure that it is mounted in such a way that it will not damage either the cylinder block or other gears.

Install the new gear to the shaft by heating to the OEM's specified temperature. To heat the gear to the prescribed temperature use a thermostatically regulated oven or hot plate. If a bearing hot plate is used, ensure that it is large enough to fit the entire surface area of the gear. Typical specified temperatures are around 300°F (150°C). The consequence of overheating a gear will be to destroy the heat treatment/tempering and possibly the surface hardening, resulting in premature failure. Heat indicating crayon such as **Tempilstick**™ may be used to determine the exact temperature. At the required temperature, the gear can be dropped over the shaft and allowed to air cool. Use of a power press to install interference fit gears to shafts should be generally avoided as damage to the shaft may result: a hand-actuated arbor press is less likely to inflict damage. The engine gear train must be timed according to the OEM procedure to ensure the relative position of the camshaft and critical accessory drives are correct and balanced. Figure 8–3 shows timing seat indices aligned on a Caterpillar C-12 engine.

A number of gears are used in the engine gear train. If the camshaft is to be rotated in the same direction of rotation as the crankshaft, an idler gear must act as an intermediary between the crankshaft and camshaft gears. There is no requirement that the camshaft be turned in the same direction as the crankshaft, but this is the case in many engines.

Timing the Engine Gear Train

The timing procedure is simple but critical and therefore carefully outlined in the OEM technical support literature. As the procedure involves locating the engine powertrain position relative to that of the feedback assembly, the crankshaft should first be located in a specific position usually with the #1 cylinder at TDC. A procedure for locating true TDC is outlined in Chapter 14. Gears that are required to be timed to each other are indexed by stamped markings on the teeth that must be intermeshed. Idler gears may have a hunting tooth relationship with the crankshaft and camshaft gears, meaning that after timing the gear train, the engine may have to be rotated through a large number of revolutions before the timing indexes realign. **Hunting gears** are not uncommon in engine timing gear trains. Engine timing gear trains may be designed with one or more idler gears. Some engines use offset gear-to-shaft locating keys to trim the timing requirements; the procedure for determining the specific offset key required should be precisely observed.

TECH TIP: A quick check for proper timing gear alignment can be made by positioning the #1 crank throw in the TDC position on the compression stroke. This should result in a camshaft position in which the #1 valve cam profiles are both on their inner base circle with the overlap ramps exactly opposite to the followers.

Gear backlash should always be checked when the engine gear train has been assembled using a dial indicator or thickness gauge. A typical gear train backlash specification would be in the region of 200 mm (0.008″), but obviously the OEM data should be consulted. A backlash specification higher than the OEM specification is an indication of gear contact face wear and usually requires that a gear or gears be replaced. A lower than specified backlash factor often indicates an assembly or timing problem.

TIMING OVERHEAD CAMSHAFTS

The procedure for timing overhead camshafts varies and depends on the engine. We have chosen one of the trickier engine timing gearsets, that on a Cummins ISX series engine. This engine uses a double overhead

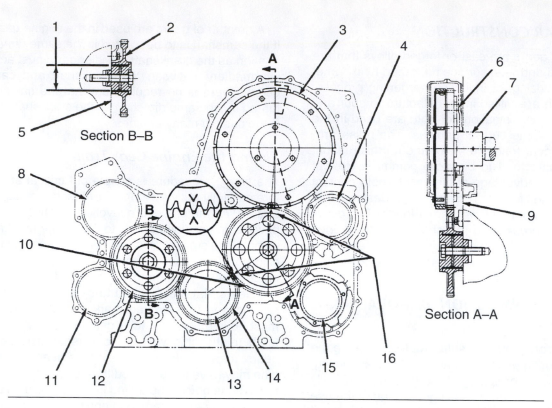

(1) Diameter of gear bore for bearing .... 60.163 ± 0.015 mm (2.3686 ± .0006 in)

(2) Gear assembly (Idler gear).

(3) Ring assembly. TDC occurs when the slot edge is in the center of the electronic control (speed sensor).

(4) Air compressor drive gear.

(5) Front face of the block.

(6) Front timing gear housing.

(7) Camshaft.

(8) Water pump drive gear

(9) Front face of block.

(10) Idler gear. Diameter of gear bore for bearing .... 74.452 ± 0.015 mm (2.9312 ± .0006 in)
Diameter of shaft for idler gear .... 69.321 ± 0.020 mm (2.7292 ± .0008 in)

(11) Oil pump drive gear.

(12) Idler gear.
Diameter of shaft for idler gear .... 55.047 ± 0.020 mm (2.1672 ± .0008 in)

(13) Crankshaft.

(14) Crankshaft gear. Heat gear to a maximum temperature of 316° C (601° F) for installation.

(15) Hydraulic pump drive gear.

(16) Align timing marks on idler gear with marks on crankshaft gear and camshaft gear as shown.

FIGURE 8–3   *Caterpillar C-12 timing gear components. (Courtesy of Caterpillar)*

camshaft arrangement as shown in Figure 8–4. Each camshaft has different functions. The left side camshaft is responsible for actuating the mechanical injectors used on these engines, while the right side camshaft takes care of valve actuation and engine braking. Two **concept gears** are used.

A concept gear uses coaxial springs between the drive hub and the gear to dampen gear train oscillations. In effect, a concept gear allows a zero lash condition to exist between intermeshing gears regardless of engine temperature and without the risk of gears radially loading each other, which would cause noise and damage. In the ISX engine, the lower concept gear is an idler gear with seven coaxial springs and the upper concept gear (four coaxial springs) is that used to drive

the injector camshaft and impart drive to the valve/brake camshaft. The procedure used to time the ISX gear train is briefly outlined in the following list, and it will help to consult Figure 8–4 while going through the procedure:

1. Pin the crankshaft and wedge both of the camshafts: this effectively locks both camshafts and the crankshaft into position. The wedges are Cummins special tools and must be used.

2. Now install the lower concept gear, shaft, and retainer plate, making sure that the thrust bearing is behind the gear. Install the gear shaft bolts followed by the gear fasteners and torque to specifications.

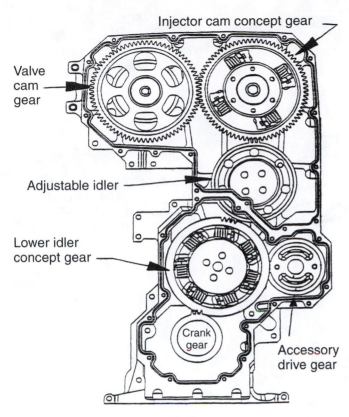

Injector cam concept gear

Valve cam gear

Adjustable idler

Lower idler concept gear

Crank gear

Accessory drive gear

FIGURE 8–4 *Cummins ISX timing gear train. (Courtesy of Cummins)*

3. The lower gear cover and seal can now be installed and torqued to specifications. Next, the vibration damper should be installed.

4. Both cam noses should have been cleaned with solvent leaving no oil residue. The two cam gears should be installed loosely (so they can be moved) on each cam nose. The injector cam concept gear should be installed with two gear screws backed out with the gear teeth aligned. Both cam gears should be snug on the camshaft nose tapers but not so snug that they cannot be rotated.

5. Place a 0.010″ feeler gauge between the teeth of the adjustable idler gear and its mesh point with the injector cam concept gear. Now using hand pressure only, move the adjustable idler gear into mesh toward the center of the engine so there is no air gap on either side of the feeler gauge: both cam gears may rotate slightly while you perform this step. This sets the lash between the adjustable idler gear and the injector camshaft concept gear. Lash at the lower idler concept gear is set by the spread of the lower idler concept gear teeth, which are constrained by the crank gear.

6. While holding the adjustable idler gear, torque the idler shaft fasteners to specification, and remove the feeler gauge.

7. Next, torque the injector cam concept gear fasteners to specification, followed by the valve cam gear fasteners. Remove the camshaft wedges. Lubricate the engine gear train with whatever lubricant is normally used in the engine.

CAMSHAFTS

The **camshaft** in most truck and bus diesel engines is gear driven by the crankshaft through one revolution per complete cycle of the engine. In a four-stroke cycle engine, to complete a full cycle the engine must be turned through two revolutions or 720 degrees, during which the camshaft would turn one revolution. Camshaft speed is therefore one-half engine speed. In a two-stroke cycle engine, the full cycle is completed in a single rotation or 360 degrees; camshaft speed and crankshaft (engine) speed is therefore geared so they are equal. The camshaft in a diesel engine actuates the **valve trains** (the term **train** can describe any components that ride the cam profile and are actuated by it) and ever more commonly, the injection pumping apparatus such as unit pumps, mechanically or electronically actuated unit injectors. The camshaft is supported at its journals by bushings or bearings that are in most cases pressure lubricated. Truck and bus medium- and large-bore diesel engine camshafts are gear driven by the engine crankshaft. **Cam geometry** or the profile outside of base circle will actuate the trains riding the profile and convert the rotary movement of the camshaft into reciprocating motion. Overhead camshafts are becoming more common in current truck diesel engines—Cummins ISX, Caterpillar C-15, and DDC Series 60 are examples. Valve train timing and unit injection pump or unit injector stroke are dictated by cam geometry and the camshaft gear timing to the engine crankshaft. In some applications, such as Cummins PT engines, this may be adjustable within a small window by the use of offset camshaft-to-camshaft gear locating keys or the shimming of cam follower housings. Figure 8–5 shows the sequence of camshaft-actuated events.

CONSTRUCTION AND DESIGN

Middle alloy steels are used with hardened journals and cams. Hardening is usually by nitriding or other hard facing processes, followed by finish grinding. Diesel engine camshafts are not usually reconditioned; resurfacing of the journals is possible but most camshaft failures are cam lobe related. Camshafts are supported by bearing journals within a longitudinal bore in

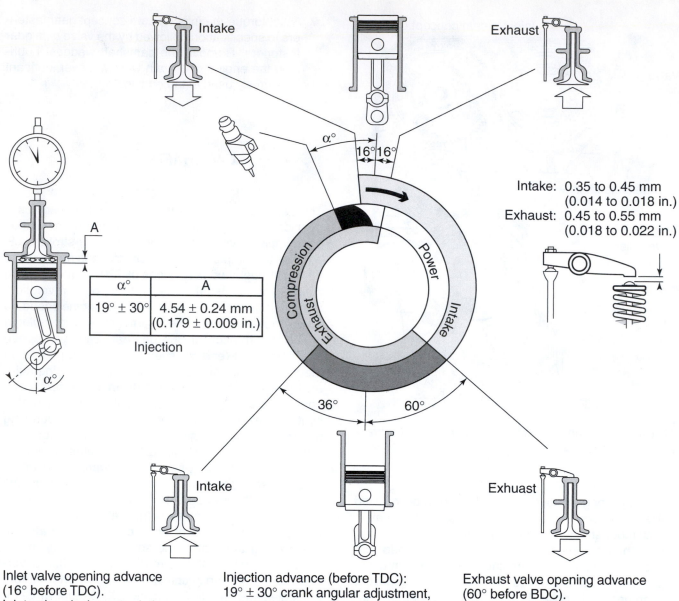

Intake

Exhaust

Intake: 0.35 to 0.45 mm
(0.014 to 0.018 in.)
Exhaust: 0.45 to 0.55 mm
(0.018 to 0.022 in.)

α°	A
19° ± 30°	4.54 ± 0.24 mm (0.179 ± 0.009 in.)

Injection

Intake

Exhuast

Inlet valve opening advance
(16° before TDC).
Inlet valve closing retardation
(36° after BDC).

Injection advance (before TDC):
19° ± 30° crank angular adjustment,
or 4.54 ± 0.24 mm (0.179 ± 0.009 in.)
linear adjustment.

Exhaust valve opening advance
(60° before BDC).
Exhaust valve closing retardation
(16° after TDC).

FIGURE 8–5   *Camshaft timing events. (Courtesy of Mack Trucks)*

the cylinder block or on a pedestal arrangement in the cylinder head in the case of an overhead cam. Cams are eccentrics machined to convert rotary motion to linear movement. The smallest radius of a cam concentric with the camshaft centerline is known as the **base circle** or **inner base circle** (IBC). The largest radial dimension from the camshaft centerline is known as **outer base circle** (OBC). The shaping of the profile that connects cam base circle with its outer base circle is described as ramping. A cam may be designed so that the larger percentage of its **periphery** (circumference) is base circle and in this case, the train that it is responsible for actuating will be unloaded for the larger percentage of the cycle; cams used to actuate engine

cylinder valves are of this design. Figure 8–6 shows a typical camshaft assembly. Alternatively, cams may be designed so that most of their periphery is outer base circle in which case, the train it actuates will be loaded for most of the cycle. Cummins PT injectors are actuated by cam profiles of this design. Figure 8–7 explains some cam terminology.

REMOVING AND INSTALLING THE CAMSHAFT FROM THE ENGINE

The procedure for removing the camshaft from a cylinder head-mounted overhead camshaft is relatively straightforward, usually only requiring the cam caps to

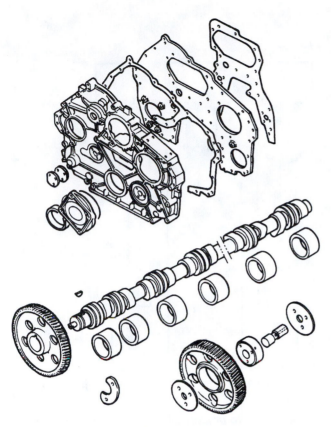

FIGURE 8–6 _Camshaft, bushings, and timing gear housing. (Courtesy of Mack Trucks)_

be released from the cam pedestals. When a camshaft has to be removed from a cylinder block, the tappet/follower assemblies will not permit the camshaft to be simply withdrawn while the engine is in an upright position. If the engine cannot be inverted, such as when removing the camshaft from an engine in-chassis, the

tappet/follower mechanisms must be raised sufficiently so that they do not obstruct the camshaft as it is withdrawn. Most engine OEMs provide special tools to perform this procedure—usually magnets on a shaft that lift the follower after which the shaft is locked to the train bore in the cylinder head. All the followers must be raised in this fashion so that the cam lobes and journals do not interfere with them as the camshaft is withdrawn from the block. When the correct tools are not available, the OEM special tools may by improvised by using coat hanger wire (mechanic's wire is not usually substantial enough) to first hook the tappet and then hold it up by bending the wire at the top. When withdrawing or installing the camshaft from a cylinder block bore, it is important that it never be forced because the result will almost certainly be damage to the camshaft or its support bearings. In certain engines, powertrain components, such as the crank web and crank throw, protrusions may interfere with the camshaft eccentrics requiring that the engine be rotated at intervals while the camshaft is being removed and installed.

Camshaft endplay is defined either by free or captured thrust washers/plates. Thrust loads are not normally excessive unless the camshaft is driven by a helical toothed gear in which instance there is more likely to be wear at the thrust faces. Endplay is best measured with a dial indicator: the camshaft should be gently levered longitudinally rearward then forward and the travel measured and checked to specifications. Figure 8–8 shows a camshaft used to actuate both valve and injector trains.

CAMSHAFT INSPECTION

1. Visual: pitting, scoring, peeling of lobes and scoring, wear, blueing of journals. Check drive gear keyway for distortion and key retention ability. Any visible deterioration of the hard surfacing on the journal or **cam profile** indicates a need to replace the camshaft. The inspection can be by touch; a fingernail stroked over a suspected hard surfacing failure can identify failing hard surfacing in its early stages better than the eye.
2. Mike the cam profile heel-to-toe dimension and check to OEM specifications. Mike the base circle dimension and check to specifications. Subtract the base circle dimension from the heel-to-toe dimension to calculate the cam lift dimension. Check to specifications. Cam lift can also be measured using a dial indicator with the camshaft mounted in V-blocks: zero the dial indicator on the cam base circle and rotate the camshaft to record the lift. Ensure that a dial indicator with sufficient total travel to measure the expected lift is used. When checking the cam lift dimension with the camshaft in-engine using a dial indicator, the

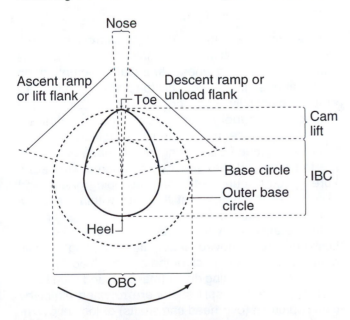

FIGURE 8–7 _Camshaft terminology._

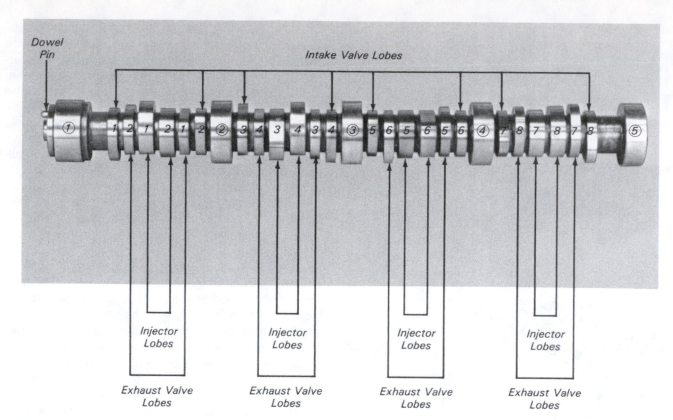

FIGURE 8–8 *Camshaft configuration appearance when a center lobe is used to actuate MUIs or EUIs. (Courtesy of DDC)*

measurement may not exactly coincide with the parameters required by the OEM, even when the cam profile is in sound condition, due to the lie of the camshaft in the support journals.

3. Check for cam lobe surface wear using a straight edge and thickness gauges sized to the maximum wear specification.
4. Install the camshaft in V-blocks (Figure 8–9) and check for shaft bending using a dial indicator.

Failure to meet the OEM specifications in any of the foregoing categories indicates that the camshaft should

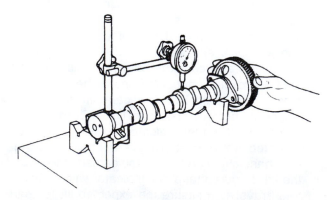

FIGURE 8–9 *Checking a camshaft on V-blocks. (Courtesy of Mack Trucks)*

be replaced. The technician should be aware that the smallest indication of a hard surfacing failure on the cam profile requires little time to advance to a total failure. Probably the most important inspection is the visual inspection that initiates the listed sequence.

CAMSHAFT BUSHINGS/BEARINGS

The camshaft is supported by pressure lubricated, friction bearings at main journals. The camshaft is subjected to loading whenever a cam is actuating the train that rides its profile. This loading can be considerable, especially when engine compression brake hydraulics and fuel injection pumping apparatus are actuated by the camshaft. Cam bushings are normally routinely replaced during engine overhaul, but if they are to be reused they must be measured with a dial bore gauge (Figure 8–10) or telescoping gauge and micrometer to ensure that they are within the OEM reuse specifications.

Interference fit, cylinder block located, camshaft bushings are removed sequentially starting usually from the front to the back of the cylinder block using a correctly sized bushing driver (mandrel) and slide hammer. Cam bearing split shells are retained either by cam cap crush (overhead camshafts) or lock rings. Interference fit cam bushings are installed using the

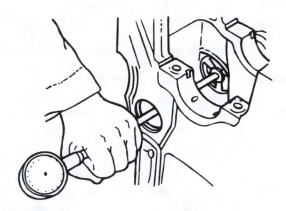

FIGURE 8–10 *Measuring cam bushing bore. (Courtesy of Mack Trucks)*

same tools used to remove them. Care should be taken to properly align the oil holes: when installing bushings to a cylinder block in-chassis where access and visibility are restricted, painting the oil hole location on the bushing rim with type correction fluid white-out may help align the bushing before driving it. Ensure that the correct bushing is driven into each bore. Bushings and their support bores vary dimensionally, and new bushings rarely survive being driven into a bore and removal. Bearing clearance is the camshaft bushing dimension measured with a dial bore gauge or telescoping gauge and micrometer minus the camshaft journal dimension measured with a micrometer. Prolonged use of certain types of engine compression brake can be especially hard on camshaft bearings.

VALVE/INJECTOR TRAINS AND ROCKER ASSEMBLIES

The **valve** and injector trains are responsible for transmitting the effects of cam geometry to action at the valve and injector assemblies. Followers convert the rotary movement of the camshaft into linear motion. This linear motion is converted to rotary movement once again at the rocker, which in its rocking action provides the force that opens valves and provides the pumping force for mechanical unit injectors (MUIs) and electronic unit injectors (EUIs). Rockers may be used in both cylinder block mounted camshaft engines (Figure 8–11) and those with overhead camshafts.

TAPPETS/LIFTERS/CAM FOLLOWERS

Tappets, lifters, and followers describe components that usually are positioned to directly ride, or at least be actuated by, a cam profile. The term **tappet** has a broader definition and is sometimes used to describe

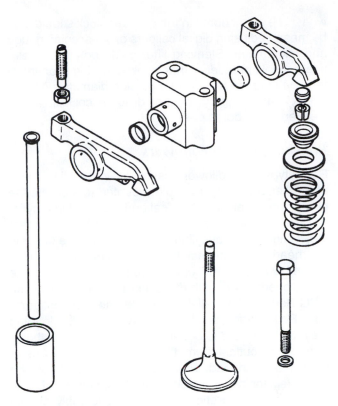

FIGURE 8–11 *Typical valve train assembly. (Courtesy of Mack Trucks)*

what is more often referred to as a **rocker** lever. Because North American engine and fuel system OEMs use all three terms, the technician should be familiar with them. In this text, the term tappet is not generally used to describe a rocker and the term **follower** is generally used to describe a component that rides the cam profile, except when describing specific OEM systems in which an alternative term is preferred. The function of cam followers is to reduce friction and evenly distribute the force imparted from the cam's profile to the train it is responsible for actuating. Diesel engines using cylinder block mounted camshafts use two categories of follower while those using overhead camshafts use either direct-actuated rockers or roller type cam followers:

Solid Lifters

Lifters are manufactured from cast iron and middle alloy steels and are usually located in guide bores in the cylinder block that allow them to ride the cam profile over which they are positioned; push rods are fitted to lifter sockets. The critical surface of a solid lifter is the face that directly contacts the cam profile. This face must be durable and may either be chemically hardened, cladded with a toughened alloy or have a disc of special alloy steel molecularly bonded to the face. Solid lifters should be carefully examined primarily for thrust face wear at engine overhaul but also stem and socket wear. The guide bores in the cylinder block should also be measured using digital calipers or telescoping gauge and micrometer. Sleeving lifter guide bores is a relatively simple procedure that involves boring to an interference fit to the new sleeve outside diameter. Always check that the lifters do not drag or cock in newly sleeved guide bores.

Roller Type Cam Followers

Roller type cam follower assemblies generally consist of a roller supported by a pin mounted to a clevis: the clevis can either be cylindrical and mounted in a cylinder block guide bore or be a pivot arm fitted to either the cylinder block or cylinder head. In the case of some overhead camshaft designs, a roller type, cam follower assembly and the rocker arm are integral. An example is the overhead camshaft design used to actuate the DDC Series 60, parallel port valve and EUI configuration. Roller type cam followers distribute the loading they are subjected to better than solid types and therefore tend to outlast them; the rollers are usually chemically hard surfaced. Roller contact faces should be inspected for pitting, scoring, and indications of hard surfacing disintegration. The roller assembly should also be checked for axial and radial runout (Figure 8–12).

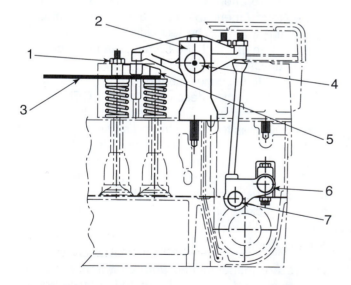

1. Adjusting screw and locknut
2. Rocker pedestal
3. Valve lash — valve stem to rocker pallet clearance
4. Rocker shaft
5. Valve bridge/yoke assembly
6. Cam follower pivot
7. Roller type cam follower/lifter

FIGURE 8–12 *Valve train components. (Courtesy of Caterpillar)*

PUSHRODS AND TUBES

Engines using cylinder block mounted camshafts require a means of transmitting the effects of cam action to the rocker assemblies in the cylinder head. Push tubes and pushrods act as intermediaries in the train and are located between the cam followers and rocker assemblies. Because they are subjected to shock loads that increase in proportion to engine rpm, they are manufactured from alloyed steels both to sustain these loads and minimize the mass of the train. The hollow push tube is more commonly used than solid pushrods, especially in cases where the camshaft is mounted low in the cylinder block, and they are required to be fairly long. A typical push tube is a cylindrical, hollow shaft fitted with a solid ball or socket at either end. Balls and sockets form bearing surfaces at the follower and rocker. This is especially important at the rocker end as the rocker moves through an arc while the push tube moves linearly. Balls and sockets are usually chemically hard surfaced. Push tubes are used over pushrods because the tubular shape provides nearly as high section modulus (shape characteristic related to rigidity) as a cylindrical rod and provide as much strength with less weight.

Pushrods are cylindrical and solid. They are used mostly in applications that permit them to be short and relatively low in mass.

Both pushrods and tubes seldom fail under normal operation but they are vulnerable to inaccurate valve lash adjustments and engine overspeeding. Ball and socket wear can be checked visually, but note that hard surface flaking/disintegration indicates that the tube or rod should be replaced. Next check:

1. Ball/socket to tube integrity: test by dropping onto a concrete floor from a height of 2″. A separating ball will always ring flat. NEVER reuse a push tube with a separating ball or socket.
2. Straightness: roll the push tube on a true flat surface such as a (new) toolbox deck. The slightest bend is reason to reject the rod or push tube.

CAUTION: It does not make economic sense to ever straighten a push tube or rod. The failure will recur—it is only a question of when. Even as a "temporary" repair, it makes little sense as the consequences of push tube failure can be much more expensive than the cost of replacing it.

ROCKER ARM ASSEMBLIES

Rocker assemblies transmit camshaft motion to the valves. They are used in both in-block camshaft and overhead camshaft configurations. The rocker pivots on a rocker shaft. When a cam **ramps** off its base circle, it acts on the train that rides the cam profile and "rocks" the rocker arm, thereby actuating the components on the opposite side of the rocker arm, either valves or unit injectors. Lubricating oil is normally ducted through the rocker shaft to supply each rocker arm. Rocker arm ratio may be used to amplify the cam lift dimension to increase valve or unit injector plunger travel. This requires that the distance from the centerline of the rocker shaft pivot bore to the pushrod side be less than its distance to the valve or unit injector side. Rocker ratio expresses the mechanical advantage ob-

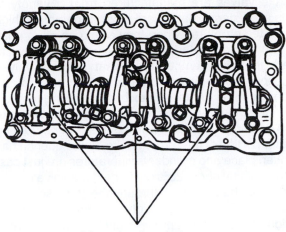

Mounting brackets

FIGURE 8–14 *Rocker assembly mounting brackets. (Courtesy of Mack Trucks)*

tained over the actual cam lift dimension. A rocker ratio of 2:1 would convert a cam lift dimension of 1″ into 2″ of valve opening travel. The rocker arms should be inspected for wear at the push tube end cup or ball socket, the pivot bore, and the pallet end. If the hard surfacing at either end shows signs of deteriorating, the rocker arm should be replaced. The pivot bore bearing can be replaced if it shows signs of wear as can the adjusting screw ball. The rocker shafts should be checked for straightness and wear at the rocker bearing race (Figure 8–13 and Figure 8–14).

CYLINDER HEAD VALVES

Cylinder head valves provide a means of timing and admitting air into the engine cylinders, and timing and exhausting end gases. In all current medium- and large-bore North American engines, they are mechanically actuated although this is expected to change in the near future with the introduction of camshaftless engines. The timing and lift of engine cylinder valves is defined by the actuating cam geometry. That is, the shape of the cam profile dictates how the valves perform.

VALVE DESIGN AND MATERIALS

Cylinder head valves are mushroom-shaped, poppet type valves. They are fitted to guides in the cylinder head and loaded by a spring or springs to seal to a seat. A disc-shaped, spring retainer locks the spring in position; split locks fitted to a peripheral groove at the top of the valve stem hold the spring retainer in position.

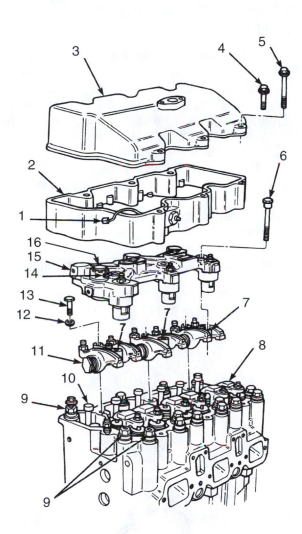

1. Wire connector	9. Head bolt with threaded
2. Riser	hole
3. Valve cover	10. Pushrod
4. Short capscrew	11. Rocker arm assembly
5. Long capscrew	12. Washer
6. Actuator capscrew	13. Capscrew
7. Rocker arm shaft	14. Connector
bracket	15. Actuator assembly
8. Valve yoke	16. Actuator solenoid

FIGURE 8–13 *Complete rocker assembly on a Jacobs brake equipped engine. (Courtesy of Mack Trucks)*

Intake valves are manufactured from a variety of middle alloy steels. Valves are most vulnerable to high cylinder temperatures when they are in an open position, that is, not seated. Intake valves do not have to sustain the high temperatures that exhaust valves must. However, they are actuated at high speeds and must have some degree of flexibility. Running temperatures of exhaust valves are much higher than intake valves. When the exhaust valve is first opened during the latter portion of the power stroke, flame quench has just taken place and cylinder temperatures are just past their peak. Only about 20% of the heat the valve is subjected to can be dissipated through the stem, so the exhaust valve must sustain high temperatures for the duration of the exhaust process until it seats. Exhaust valves are manufactured out of special ferrous alloys. They may be cladded at the head and often use chromium, nickel, manganese, tungsten, cobalt, and molybdenum to improve flexibility, toughness, and heat resistance. Exhaust valves are cooled primarily by dissipating heat to the valve seats and secondarily, by intake air at valve overlap and through the stem to guide contact area.

Valves machined with 45 degree seats have lower breathing efficiencies than those with 30 degree seats given identical lift. Valve seats with 45 degree seats are widely used in highway diesel applications simply because they have greater longevity. They can sustain greater dynamic loads, high seating forces and high temperatures. Valves machined with 30 degree seats tend to run hotter than those with 45 degree seats because they have less mass a their periphery. They also have less distortion resistance and lower unit seating force.

VALVE OPERATION

Current diesel engine valves rely on maximizing the seat contact area for cooling purposes and as a consequence, interference angles are seldom machined. An interference angle requires that the valve be machined at 0.5 degree to 1.0 degree more acute angle than the seat. This results in the valve seating with high unit pressures and inhibits the formation of carbon on the seat, but reduces valve to seat contact area. A majority of current diesel engines use valve rotators to minimize carbon buildup on the seat and promote even wear: an interference angle is never machined to valves using valve rotators.

Valve rotators use a ratchet principle or a ball and coaxial spring to fractionally rotate the valve each time it is actuated. Valve rotation should be checked after assembly by marking an edge of the stem and then tapping the valve stem with a light nylon hammer a number of times; the valve should visibly rotate each time the stem is struck.

Cylinder head poppet valves are seated by a spring or pair of springs. When a pair of springs is used, they are often oppositely wound to help cancel the coincidence of harmonies that may contribute to valve flutter. **Valve float** can occur at higher engine speeds when valve spring force is insufficient and may cause asynchronous (out of time) valve closures. Springs are a critical component of the valve train assembly and their importance is increased as engine rpm increases and reduces the real time periods between opening and closing.

INSPECTING VALVE SPRINGS AND RETAINERS

1. Check the keepers for wear. Inspect retainers.
2. Measure the spring vertical height and compare to specifications. Minor differences do not matter.
3. Test valve spring tension using a tension gauge (Figure 8–15). The valve spring tension must usually be within 10% of the original tension specification. If not, replace the valve spring.
4. Check the spring(s) for wear at each end.
5. Check that the spring is not cocked using a straight edge.
6. Check valve spring operation with the valve installed in the cylinder head. Remember that as the cylinder head valve seat material is ground away, the spring operating height is lengthened, reducing spring tension. Valve protrusion and recession must be checked to specification as shown in Figure 8–16 and Figure 8–17.

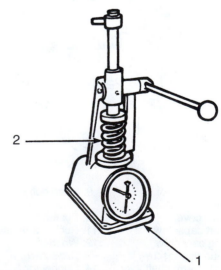

1. Valve spring tester
2. Valve spring

FIGURE 8–15 *Valve spring tension tester. (Courtesy of Mack Trucks)*

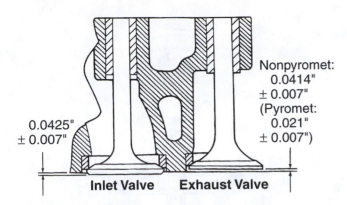

Nonpyromet:
0.0414"
± 0.007"
(Pyromet:
0.021"
± 0.007")

0.0425"
± 0.007"

Inlet Valve   Exhaust Valve

FIGURE 8–16 _Valve protrusion and recession dimensions on a Mack E-Tech engine. (Courtesy of Mack Trucks)_

FIGURE 8–17 _Measuring valve protrusion with a depth micrometer. (Courtesy of Mack Trucks)_

VALVE SERVICING

Use a valve spring compressor (Figure 8–18) to remove valves from the cylinder head. The valves should be tagged by location. Valves must first be cleaned on a buffer wheel, taking great care not to remove any metal. A glass bead blaster may be used to remove carbon deposits from the seat face, fillet, and head, but avoid blasting the stem. Valve inspection should begin with visually checking for dishing, burning, cracks, and pits; evidence of any of these conditions requires that the valve be replaced. Check the valve fillet for cracks and nicks—using a portable magnetic flux crack detector may help here—and again, reject the valve if cracks are evident in this critical area. Next, the valve should be measured. Using a micrometer, mark the valve stem at three points through the valve guide bushing sweep and check to specifications. Valve stem straightness must be checked with a vertical runout indicator. If the valve seat face is to be resurfaced, the valve margin must first be checked. The valve margin is a critical

specification and it should meet the minimum specification after grinding, or the valve will overheat and fail in operation. The split retainer grooves (keeper grooves: Figure 8–18) on the valve stem should also be checked for wear, nicks, and cracks.

Before servicing a set of valves, the previously outlined checks and measurements must be completed. It is pointless machining valves that will fail in any of the previous categories. First, dress the grinding stone using a diamond dressing tool. Adjust the valve grinder chuck to the specified angle and the carriage stop to limit travel so the stone cannot contact the stem. Run the coolant (soluble machine oil and water solution), and make a single shallow pass. Measure the **valve margin**. Valve margin is the dimension between the valve seat and the flat face of the valve mushroom. This specification is critical when machining valves, and it must exceed the minimum specified value when the grinding process has been completed. A valve margin that is lower than the specification will result in valve failures caused by overheating. When machining valves, make the minimum number of passes to obtain a valve seat face surface free of ridging and pitting; the final step is always to check the valve margin. In cases in which the valve has been loosely adjusted, the stem end may be slightly mushroomed: grind a new chamfer,

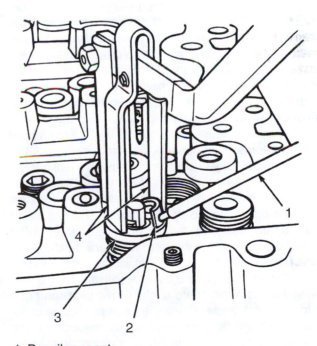

1. Pencil magnet
2. Valve retainer or keeper
3. Valve spring retainer
4. Valve spring compressor

FIGURE 8–18 _Removing a valve from a cylinder head. (Courtesy of Mack Trucks)_

taking care to remove as little material as possible. Too much chamfer will reduce the rocker-to-stem contact area and may damage the rocker pallet. When valves must be replaced, the new valves must be inspected, measured, and sometimes ground using the same procedure as with used valves.

VALVE SEAT INSERTS

Most current truck and bus diesel engines use valve seat inserts rather than integral valve seats, machined into the cylinder head. The primary advantages of valve seat inserts is that they can be manufactured from tough, temperature resistant material and then easily replaced when the cylinder head is serviced. Some cast-iron alloys toughened with nickel, chromium, and molybdenum are used, but the use of alloyed steels, some with stellite cladding, is more common. Valve seat inserts are press fit to a machined recess in the cylinder head—sometimes they are staked to position after installation. Since most of the heat of the valve must be transferred from the valve to the seat, it is essential that the contact area of the seat and the cylinder head be maximized.

Valve Seat Removal and Installation

Valve guides must be removed using a removal tool consisting of a sectored collet designed to expand into the valve seat after which it can be either levered out or driven out with a slide hammer. When installing new valve seats, the seat counterbore must be first cleaned out using a low abrasive emery cloth. The new insert should be inserted into the OEM specified driver, which has a pilot shaft that fits tight to the valve guide bore. A hammer should be used to drive the insert into its bore until it bottoms. Next check the concentricity of the valve seat with the valve guide bore using a dial gauge; this is a critical specification (Figure 8–19). In most cases the seat will have to be ground, even though new seats are finish ground. It is important that the valve grinders are serviced before fitting and machining the valve seats so that the new seats are machined to be concentric with the guides that will be put in service.

Valve seats are ground using a specified grit stone, dressed to the appropriate angle, with a pilot shaft that fits to the valve guide. The valve head protrusion or recess dimension specifications must be respected when grinding valve seats, and in cases in which the cylinder head deck surface has been machined, undersize valve seat inserts are available from most OEMs. Whenever a cylinder head deck has been machined, the valve stem height, which is dictated by the specific valve seat insert location, must be within specifications to ensure the correct valve spring dynamics.

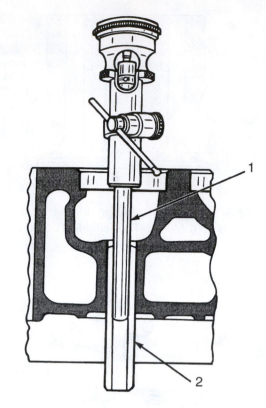

1. Arbor   2. Valve guide

FIGURE 8–19   *Measuring valve seat concentricity with a concentricity gauge. (Courtesy of Mack Trucks)*

VALVE LASH ADJUSTMENT

When valves are properly adjusted, there should be clearance between the pallet end of the rocker arm and the top of the valve stem. Valve lash is required because as the moving parts heat up, they expand, and if clearance were not factored somewhere in the valve train, the valves would remain constantly open by the time an engine reached operating temperature. Actual valve lash values depend on factors such as the length of the push tubes and the materials used in valve manufacture. Exhaust valves are subject to more heat and as a consequence, OEMs require a valve lash setting for exhaust values that is usually greater than the intake valve lash setting.

MALADJUSTED VALVES

Loose valve adjustment will retard valve opening and advance valve closing, decreasing the cylinder breathing time. Actuating cam geometry is designed to provide some "forgiveness" to the train at valve opening and valve closure to reduce the shock loading the train is subject to. When valves are set loose, the valve train is loaded at a point on the cam ramp beyond the intended point. The same occurs at valve closure when the valve is seated. High valve opening and closing velocities subject the valve and its seat to hammering that can result in cracking, failure at the head to stem fillet, and scuffing to the cam and its follower.

VALVE ADJUSTMENT PROCEDURE

The following steps outline the valve adjustment procedure on a typical four-stroke cycle, in-line six-cylinder diesel engine. Valves should always be adjusted using the OEM's specifications and procedure.

CAUTION: Shortcutting the engine OEM recommended valve (and injector) setting can result in engine damage unless the technician knows the engine well. Cam profiles are not always symmetrical and some engines may have camshaft profiles designed with ramps between base circle and outer base circle for purposes such as actuating engine compression brakes. Similarly, a valve rocker that shows what appears to be excessive lash when not in its setting position is not necessarily defective.

Six-cylinder engine firing order: 1–5–3–6–2–4
Companion cylinders or cylinder throw pairings:

1–6 at TDC
5–2 at 120 degrees BTDC
3–4 at 120 degrees ATDC

Cylinder throw pairings are called **companion cylinders**. In other words, when #1 piston is at TDC completing its compression stroke, #6 piston (its companion) is also at TDC having just completed its exhaust stroke. If the engine is viewed from overhead with the rocker covers removed, engine position can be identified by observing the valves over a pair of companion cylinders. For instance, when the engine calibration index indicates that the pistons in cylinders #1 and #6 are approaching TDC and the valves over #6 are both closed (lash is evident), then the point at which the valves over #1 cylinder rock (exhaust closing, intake opening) at valve overlap will indicate that #1 is at TDC having completed its exhaust stroke and #6 at TDC having completed its compression stroke. This method of orienting engine location is commonly used for valve adjustment.

Adjustment

1. Locate the valve lash dimensions. The lash specification for the exhaust valve(s) is usually (but not always) greater than that for the inlet valve.

2. The valves on current diesel engines should usually be set under static conditions and with the engine coolant 100°F (37°C) or less. Locate the engine timing indicator and the cylinder calibration indexes, 120 degrees apart; depending on the engine, this may be located on a vibration damper, any pulley driven at engine speed, or the flywheel.

3. Ensure that the engine is prevented from starting by mechanically or electrically no-fueling the engine. The engine will have to be barred in its normal direction of rotation through two revolutions during the valve-setting procedure requiring the engine to be no-fueled to avoid an unwanted startup.

4. If the engine is equipped with valve bridges or yokes (Figure 8–20 and Figure 8–21) that require adjustment, this procedure should be performed before the valve adjustment. To adjust a valve yoke, back off the rocker arm then loosen the yoke adjusting screw locknut and back off the yoke adjusting screw. Using finger pressure on the rocker arm (or yoke), load the pallet end (opposite to the adjusting screw) of the yoke to contact the valve; next, screw the yoke adjusting screw clockwise until it bottoms on its valve stem. Turn an additional one flat of a nut (⅙ of a turn), then lock to position with the locknut.

CAUTION: When loosening and tightening the valve yoke adjusting screw locknut, the guide on the cylinder head is vulnerable to bending. Most OEMs recommend that the yoke be removed from the guide and placed in a vise to back off and final torque the adjusting screw locknut.

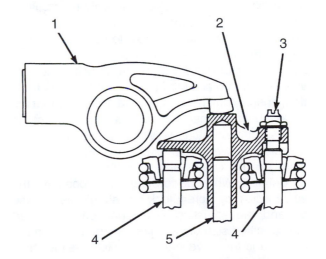

1. Rocker arm   4. Valve stem
2. Valve yoke   5. Yoke guide pin
3. Adjusting screw

FIGURE 8–20 *Valve bridge/yoke assembly. (Courtesy of Mack Trucks)*

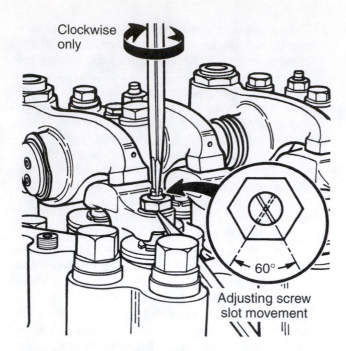

FIGURE 8–21   *Adjusting the valve bridge/yoke assembly. (Courtesy of Mack Trucks)*

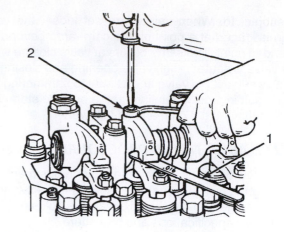

1. Thickness gauge   2. Adjusting screw and nut

FIGURE 8–22   *Adjusting intake valve lash on a Mack E-7 engine. (Courtesy of Mack Trucks)*

5. To verify that the yoke is properly adjusted, insert two similarly sized thickness gauges of 0.010" or less between each valve stem and the yoke. Load the yoke with finger pressure on the rocker arm and simultaneously withdraw both thickness gauges; they should produce equal drag as they are withdrawn. If the yokes are to be adjusted, they can be adjusted in sequence as each valve is adjusted. In some engines, valve yokes can only be adjusted with the rocker assemblies removed because it is impossible to access the yoke to verify the adjustment otherwise.

6. If the instructions in the OEM literature indicate that valves must be adjusted in a specific engine location, ensure that this is observed; the cams that actuate the valves may only have a small percentage of base circle. Setting valves requires the lash dimension between the rocker arm and the valve stem on the rocker arm and the valve yoke to be defined. When performing this procedure, the valve adjusting screw lock nut should be backed off and the adjusting screw backed out. Insert the specified size of thickness gauge between the rocker and the valve stem/yoke. Release the thickness gauge. Then turn the adjusting screw clockwise until it bottoms; turn an additional ½ flat of a nut ($\frac{1}{12}$ turn). Hold the adjusting screw with a screwdriver and with a wrench lock into position. Now for the first time since inserting the thickness gauge, handle it once again; withdraw the thick-

ness gauge. A light drag indicates that the valve is properly set. If the valve lash setting is either too loose or too tight, repeat the setting procedure. Do NOT set valves too tight. Set all the valves in cylinder firing order sequence rotating the engine 120 degrees between settings. It is preferable to begin at #1 cylinder and proceed through the engine in

1. Thickness gauge   2. Adjusting wrench

FIGURE 8–23   *Adjusting exhaust valve lash on a Mack E-7 engine equipped with a Dynatard engine brake. (Courtesy of Mack Trucks)*

firing order sequence (Figure 8–21, Figure 8–22, and Figure 8–23).

***

TECH TIP: Never attempt to adjust valve lash while simultaneously checking it with a feeler gauge. Allow the gauge blade to be clamped by the rocker, then release it. Make the adjustment at the rocker screw, then check the gauge blade drag. This should reduce the time you spend performing this simple engine maintenance adjustment.

Valves: Conclusion

Valves are normally but not always set when the piston is at TDC on the cylinder being set. When the piston is at TDC on the compression stroke, it can usually be assumed that the valves are fully closed. The procedure for setting valves in an engine is sometimes referred to as overhead setting or tune up. In most cases, the valve setting procedure is accompanied by procedures such as injector timing and/or lash setting/ train loading. When barring an engine over during a tune-up, ob-

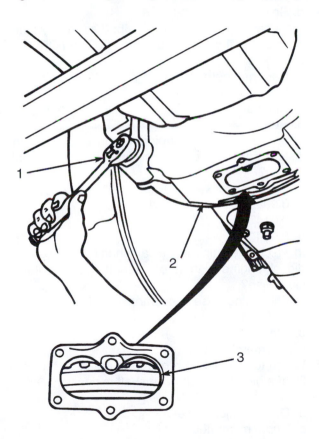

1. Ratchet and engine barring tool
2. Flywheel housing
3. Flywheel and engine position scale

FIGURE 8–24 *Barring an engine using a toothed barring tool while observing the engine position scale. (Courtesy of Mack Trucks)*

serve the correct direction of rotation and use an approved barring tool (Figure 8–24). Regardless of what is meant by overhead setting, it is critically important that each step in the procedure is performed in exactly the OEM-prescribed sequence.

OTHER FEEDBACK ASSEMBLY FUNCTIONS

The engine feedback assembly in many modern engines is responsible for actuating the pumping of fuel to injection pressure values. Most current engines are fueled by PT (Pressure-Time injectors), MUI, EUI, and EUP (Electronic Unit Pumps) high-pressure fueling apparatus, all of which are actuated by cam profile. As the means of actuating each type of pumping device is distinct to the fuel system, it is appropriate that this be covered under each fuel system. The injector pumping (MUI, PT, and EUI) train components are somewhat similar to the valve trains and depend largely on the location of the camshaft and the force required to actuate the pumps, which can be considerable. For instance, in engines with a block-mounted camshaft, the pushrods used in the injector train must be built to sustain much higher compressional loads than those used in the valve trains.

SUMMARY

- The engine feedback components incorporate the engine timing gear train, the camshaft, valve and unit injector trains, and in some cases the injection pumping apparatus. For purposes of study, the actuation mechanisms of the fuel injection components are examined in the sections dealing with specific fuel systems.
- Camshaft drive gears must be precisely timed with the crankshaft driven, engine gear train so that the events activated by the engine feedback assembly are synchronized with those in the engine.
- The camshaft drive gear is most often interference fit to the camshaft, positioned by a keyway.
- Camshaft gears are heat treated and when they are fitted to camshafts, it is essential that they are heated evenly to a precise temperature; overheating will cause premature failure.
- Camshafts may be rotated either with or oppositely to the direction of engine rotation.
- Camshaft gears may use spur or helical cut gear teeth; thrust loads are much higher when helical gears are used.
- Gear backlash must be measured using thickness gauges or dial indicators.

- Cam lift on block located camshafts may be inspected using a dial indicator mounted above the push tube or rod.
- Cam base circle or IBC is that portion of the cam periphery with the smallest radial dimension. Cam OBC is that portion of the cam periphery with the largest radial dimension.
- The critical cam dimensions may be checked on an overhead camshaft or an out-of-engine camshaft with a micrometer.
- A visual inspection of the camshaft should identify most cam failures but profile wear may be checked to specification using a straightedge and thickness gauges. The camshaft should be mounted in V-blocks to test for straightness.
- Out-of-engine camshafts should be supported on pedestals, on V-blocks, or hung vertically to prevent damage.
- Most diesel engine cam followers are of the solid or roller types, that is, they are not hydraulic.
- The cam's train consists of the series of components it is responsible for actuating.
- Most truck diesel engines with block-mounted camshafts use trains consisting of a follower assembly, push tubes, and a rocker.
- Most valve trains are adjusted with a lash factor to allow for expansion of the materials as the engine heats to operating temperature.
- Some injection pumping actuation trains are set at zero lash or even a slight load when the actuating cam profile is on its IBC.
- Rocker assemblies provide a means of reversing the direction of linear movement of the push tube or follower, and in some cases providing a mechanical advantage.
- Cylinder head valves are used to aspirate or breathe the engine cylinders: they are actuated by the cam geometry and time the air into, and end gases out of, the engine cylinders.
- Exhaust valves are often manufactured out of more highly alloyed steels than intake valves because they must sustain much higher temperatures.
- When reconditioning valves by regrinding, a critical specification is the valve margin.
- Most diesel engines do not use an interference angle to seat valves because the seating contact surface area is compromised and because of the fact that valve rotators are widely used.
- A 45-degree cut valve has higher seating force but lower gas flow than a 30-degree cut valve.
- Valve seats are usually interference fit to the cylinder head and finish ground concentric to the valve guide bore.
- When setting valve lash, the OEM specifications as to engine position for the valve being adjusted should be observed because the cam geometry on some engines is not clearly divisible into IBC and OBC sections.
- Valve lash should be set using thickness gauges; valves in current engines are set statically and cold.
- Loose valves cause lower cylinder breathing efficiencies and what is known as top end clatter, and may damage cam profiles.
- Valve yokes or bridges do not usually have to be adjusted as part of a routine valve adjustment: it is important that the bridge is properly supported to prevent the pedestal shaft from being damaged.

REVIEW QUESTIONS

1. The portion of the cam profile that is exactly opposite the toe is referred to as the:
   a. Nose
   b. Ramp
   c. Heel
   d. Sole
2. On a cam profile that is described as mostly inner base circle, the profiles between IBC and OBC are known as:
   a. Cam geometry
   b. Ramps
   c. Ridges
   d. Heels
3. Which of the following dimensions would be consistent with a typical engine timing gear backlash setting?
   a. 0.002″
   b. 0.008″
   c. 0.014″
   d. 0.024″
4. In which direction must a camshaft be rotated on a diesel engine with a crankshaft that is rotated clockwise?
   a. Either clockwise or counterclockwise depending on the engine
   b. Clockwise
   c. Counterclockwise
5. Which tool would be required to measure the cam lift of a block-mounted camshaft in position?
   a. Dial indicator
   b. Outside micrometer
   c. Depth micrometer
   d. Thickness gauges
6. A camshaft gear is precisely positioned on the camshaft using a(n):
   a. Interference fit
   b. Key and keyway
   c. Captured thrust washer
   d. Dial indicator

7. When grinding a valve face to remove pitting, the critical specification to monitor would be the:
   a. Shank diameter
   b. Stem
   c. Poppet diameter
   d. Margin

8. Which of the following is more likely to cause a valve float condition?
   a. Engine lugdown
   b. Operating in the torque rise profile
   c. Engine overspeed
   d. Operating in the droop curve

9. When grinding a new set of cylinder valve seats, which of the following should be performed before the machining?
   a. Installing the valves.
   b. Installing the rockers.
   c. Adjusting the valve yokes.
   d. Installing the valve guides.

10. The reason for using a pair of oppositely wound valve springs is to:
    a. Minimize valve dynamic flutter
    b. Double the closing force of two similarly wound valves
    c. Increase the valve closing velocity
    d. Diminish valve noise

11. *Technician A* states that the backlash in timing gearsets can be measured with a dial indicator. *Technician B* states that if the gear backlash is insufficient, the result can be gear whine. Who is right?
    a. A only
    b. B only
    c. Both A and B
    d. Neither A nor B

12. *Technician A* states that helical cut gears are often used in engine timing gear trains because they operate more quietly. *Technician B* states that shaft thrust loads are much higher when helical cut gears are used. Who is right?
    a. A only
    b. B only
    c. Both A and B
    d. Neither A nor B

13. *Technician A* states that when installing an interference fit gear to a shaft, the gear should not be heated above 400°F. *Technician B* states that if an oxyacetylene rosebud is not available, an interference fit gear should be pressed to the shaft using a hydraulic press. Who is right?
    a. A only
    b. B only
    c. Both A and B
    d. Neither A nor B

14. *Technician A* states that it is good practice to lubricate the inside bores of camshaft bushings to facilitate camshaft installation. *Technician B* states that if the cam bushing oil holes are not properly aligned an immediate camshaft failure could be the result. Who is right?
    a. A only
    b. B only
    c. Both A and B
    d. Neither A nor B

15. Which of the following methods is used to drive the camshaft on most heavy-duty truck diesel engines?
    a. Belt and pulley
    b. Timing chain and sprocket
    c. Gears
    d. Fluid coupling

16. A camshaft on a four-stroke cycle, truck diesel engine is driven at what speed in relation to the engine crankshaft speed?
    a. One-half engine speed
    b. Engine speed
    c. Two times engine speed
    d. Four times engine speed

17. A cam-actuated train that is loaded for most of the cycle would have geometry consistent with which of the following?
    a. Mostly OBC profile
    b. Mostly IBC profile
    c. Symmetrical OBC and IBC profiles

18. *Technician A* states that the camshaft on a two-stroke cycle engine must turn through one complete rotation per complete effective cycle of the engine. *Technician B* states that the camshaft on a two-stroke cycle engine rotates at one-half engine speed. Who is right?
    a. A only
    b. B only
    c. Both A and B
    d. Neither A nor B

19. *Technician A* states that to check cam lift, the camshaft must be removed from the engine to obtain a valid reading. *Technician B* states that to check for a bent camshaft, the camshaft must be removed from the engine and runout tested in V-blocks. Who is right?
    a. A only
    b. B only
    c. Both A and B
    d. Neither A nor B

20. *Technician A* states that valves machined with 45-degree seats breathe with higher efficiencies than those machined with 30-degree seats. *Technician B* states that valves machined with 30-degree seats run cooler than those machined with 45-degree seats. Who is right?
    a. A only   c. Both A and B
    b. B only   d. Neither A nor B

Chapter 9

Engine Housing Components

PREREQUISITES

Chapters 4, 7, and 8

OBJECTIVES

After studying this chapter, you should be able to:

- *Identify the components classified as engine housing components.*
- *Identify the types of cylinder block used in current truck diesel engines.*
- *Outline the procedure required to inspect a cylinder block.*
- *Measure an engine block to specifications using a workshop manual.*
- *Identify the types of cylinder liners used in truck diesel engines.*
- *Explain the procedure required to remove wet and dry liners.*
- *Describe the process required to remove seized dry liners from a block bore.*
- *Perform selective fitting of a set of dry liners to a cylinder block.*
- *Explain how cavitation erosion occurs on wet liners.*
- *Identify the types of cylinder heads used in truck diesel engines.*
- *Describe the component parts of a cylinder head.*
- *Define component creep and gasket yield.*
- *Explain the procedure required to test a cylinder head.*
- *Describe the role of the intake and exhaust manifolds.*
- *Describe the function of the oil pan in the engine.*
- *Determine the types of oil pan failures that may be repairable by welding.*

KEY TERMS

bubble collapse	exhaust manifold	sleeves
cavitation	fire rings	sump
crankcase	gasket	template torque
creep	intake manifold	wet liners
cylinder block	interference fit	yield point
cylinder head	liners	
dry liners	oil pan	

INTRODUCTION

The engine housing components are those components that enclose the internal engine components. The cylinder block acts as the central frame of the engine: all the other engine components are in some way attached to it including the remaining engine housing components. The components examined in this chapter are as follows:

1. Engine cylinder block and cylinder liners
2. Cylinder head assemblies
3. Intake and exhaust manifolds
4. Oil pans

ENGINE CYLINDER BLOCK

The engine **cylinder block** is the frame of the engine around which all the other components are assembled in much the same way that subcomponents are assembled around a truck frame (Figure 9–1). The cylinder block houses the engine cylinders and the engine crankcase. North American highway medium- and large-bore diesel engines all use in-line or V-cylinder configurations requiring a single crankshaft and therefore use cylinder blocks cast as a single unit. In all these engines, the cylinder block is manufactured essentially from a cast-iron alloy. Although aluminum and plastics may be used as a material for oil pans, and Caterpillar used an aluminum spacer deck in the original 3176 engine, cast-iron alloy remains the material of choice. The extent to which the cylinder block is alloyed has increased in recent years in the pursuit of lighter, stronger engines that last longer.

In most current highway diesel engines, the cylinder block is bored and fitted with cylinder **liners** or **sleeves**, which can be replaced when the engine is overhauled. The crankshaft is underslung from the cylinder block and supported in a cradle of main bearings in what is usually referred to as the **crankcase**.

The engine configuration of choice in medium- and large-bore track diesel engines has been the in-line, six-cylinder engine (Figure 9–2). Most of the truck engine OEMs have flirted with V configurations, usually V8s with V angles ranging from 90 degrees to a relatively tight 50 degrees. Only one larger bore, V8 engine has prevailed over the years in truck applications, Mack Trucks 90-degree V E9 engine. The story is somewhat different in bus and coach applications in which the relatively lower height of the V configuration makes it a good fit in the engine compartment located under the rear passenger housing. Here, DDC's range of 60-degree V engines were dominant until emissions legislation took them out of the market in the mid-1990s, especially in low-profile city buses.

Cylinder blocks may be bored to support the camshaft or camshafts and are cast with coolant passages and a water jacket. Even in a liquid-cooled engine, the cylinder block frame plays a major role in dissipating rejected heat to the atmosphere.

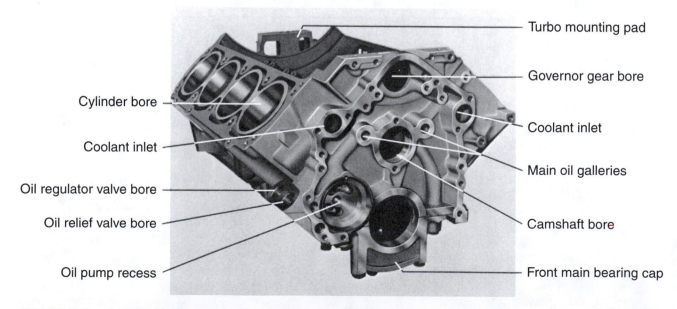

Cylinder bore

Coolant inlet

Oil regulator valve bore

Oil relief valve bore

Oil pump recess

Turbo mounting pad

Governor gear bore

Coolant inlet

Main oil galleries

Camshaft bore

Front main bearing cap

FIGURE 9–1 *Cylinder block terminology. (Courtesy of DDC)*

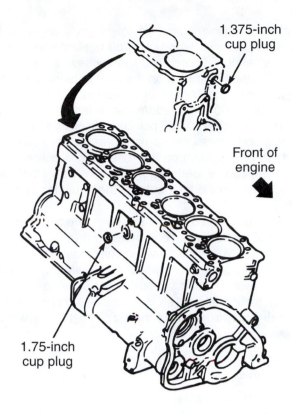

1.375-inch cup plug

Front of engine

1.75-inch cup plug

FIGURE 9–2 *In-line six-cylinder block showing the location of the cup plugs and installation method. (Courtesy of Mack Trucks)*

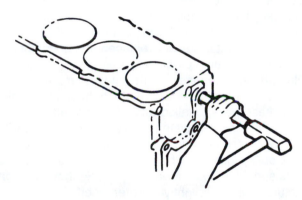

CYLINDER BLOCK MATERIALS

Most truck and bus cylinder blocks are manufactured from grey cast irons usually alloyed with at least some silicon to reduce brittleness of the block material. Most current medium- and large-bore highway diesel engines use cylinder liners, meaning that the block base materials are not directly subject to the engine cylinder pressures and temperatures. This permits the use of relatively uncomplicated cast-iron alloys as cylinder block materials. A recent trend has been to develop lower mass, higher strength cylinder block castings with the objective of reducing overall weight. For in-

stance, the high output, 15-liter Cummins ISX introduced in 1999 actually weighs less than most of the 12-liter diesel engines marketed through the 1990s.

A cylinder block:

• Incorporates bores for the piston assemblies
• Incorporates main bearing bores to mount the crankshaft
• May contain bores to mount the camshaft
• Incorporates coolant passages/water jacket
• Incorporates lubricant passages/drillings
• Incorporates mounting locations for other engine components

Cylinder blocks used in truck and bus CI engines can be categorized by type:

• Integral cylinder bore
• Wet sleeve/liner
• Dry sleeve/liner
• Combination wet/dry liners

Other factors such as two-stroke cycle and air cooled are also reflected in the block design.

Integral Cylinder Bore

Most automobile and small-bore engines use integral or parent bore cylinder blocks. They are less often seen in truck diesel engines, especially those of large-bore dimensions. When the integral cylinder bore is used in a medium-bore application, they are sometimes referred to as "throwaway blocks." This derogatory term came to be used because on some parent-bore designs, it was difficult to bore for sleeves due to the close proximity of the bore's cylinder to one another. The Caterpillar 3208 is an example of a parent cylinder block assembly. The initial cost savings in producing a block of this design are compromised at engine overhaul when the block requires either replacement or a boring/sleeving machining operation. The reason this design is seldom used as a diesel engine cylinder block is essentially that a diesel engine is designed with the premise that it will probably undergo reconditioning at least once in its operational life, whereas with the gasoline, spark-ignited engine this is not a foremost design consideration.

Recently, Freightliner/Mercedes-Benz have introduced a parent bore engine in which the cylinder bores have been induction hardened. The MB-900 engine, upper ring belt sweep in the cylinder bores has a unique, helical-striped, induction-hardening feature that is claimed to give the bores considerably extended service life compared with other parent bore engines. In the event the cylinder bores are worn to oversize, they can be bored and sleeved. Mercedes-Benz recommended that sleeving is performed only by them on these engines.

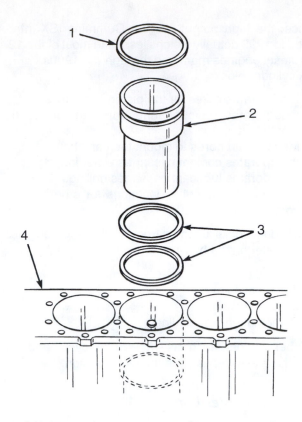

1. Crevice seal   3. Shims
2. Cylinder sleeve   4. Cylinder block

FIGURE 9–3   *Wet/dry liner components. (Courtesy of Mack Trucks)*

Wet Liner/Sleeve

The cylinder block is designed so that the water jacket is in direct contact with the liner in the block bore; the liner must therefore have a wall thickness sufficient to sustain the engine's peak combustion pressures (Figure 9–3). **Wet liners** transfer heat efficiently into the coolant and are easily replaced at overhaul. Their main disadvantage is that a seal must be maintained for the life of the liner and that an O-ring failure results in coolant contaminated engine lube.

Cavitation caused by vapor bubble implosion can shorten the life of wet liners but this is seldom a problem if the coolant chemistry is properly monitored. Despite the fact that a wet liner is constructed from a sizeable mass of cast iron or steel, it is relatively easily flexed. Try installing a telescoping gauge into the bore of a wet liner on the bench and locking it into position. Gentle hand pressure is sufficient to flex the liner for the gauge to unseat and drop to the bottom of the liner. Wet liners, when subject to cylinder pressures, which may be as high as 2,000 psi (18.78 MPa) expand outward into the wall of coolant that surrounds them then

contract, creating a sort of lower pressure void or bubble whose matter is boiled off coolant vapor. This "bubble" almost immediately collapses, causing the wall of coolant to impact on the liner exterior wall. This condition repeats itself at high frequency (17 times per second at 2,000 rpm) and has been tested to produce pressures of up to 60,000 psi. **Bubble collapse** results in cavitation unless the coolant provides protection in the form of a coating on the exterior of the liner wall. Cavitation can be identified by pitting/erosion that usually appears on the liner outside the thrust faces of the piston.

The O-rings used to seal wet liners are made from a variety of rubber type compounds, and it is important that the OEM installation recommendations be observed; they may be installed dry, coated in coolant, soap, engine oil, and various other substances. Wet liners are often alloyed so they possess characteristics metallurgically superior to the block casting, which increase service life.

Dry Sleeves

A thinner walled sleeve than the wet liner, the dry sleeve is installed into the block bore usually with a marginally loose fit and retained by the cylinder head. The dry sleeve does not transfer heat as efficiently as the wet liner, but it is easily replaced and does not present coolant sealing problems.

In older applications, **dry liners** were made of cast-iron material almost identical to that of the engine block; these were installed with a fractional interference to maximize heat transfer. Current dry liners tend to be alloyed for superior toughness, and because a liner's ability to transfer heat is dependent on maximizing its surface contact area in the block, they are manufactured with a slightly greater coefficient of heat expansion than cast iron and designed to be installed loose so they expand into the block bore when heated. If these are installed with the interference fit mandated in older liners, they will buckle in use, greatly reducing their service life. Precise measuring of the block id, the liner od, and selective fitting will ease installation and increase engine longevity. Loose fits of around 0.035 mm (0.0015″) are common.

Combination Wet/Dry Sleeves

The wet dry sleeve is designed so that the hottest part of the liner at the top is in direct contact with the coolant in the water jacket and the lower portion fits directly to the cylinder block with a fractional loose or interference fit. Consequently, the upper portion in direct contact with the water jacket must have considerably more mass because it has to contain the cylinder combustion pressures: the wet/dry liner also must seal the water

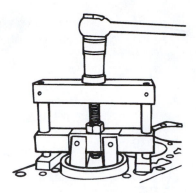

FIGURE 9–4   *Pulling a liner. (Courtesy of Mack Trucks)*

jacket and as with the wet liners, O-rings are usually the means. These are also known as midstop liners.

Cylinder Sleeve Removal

Sleeves should be removed with a puller and adaptor plate or shoe. The procedure is obviously more simple on wet liners. Dry sleeves often require the use of mechanical, hydraulic, or air-over-hydraulic pullers (Figure 9–4). When these fail to extract a seized liner, vertical arc weld runs may break a liner free, but great care has to be exercised because the block may become distorted. To use this method of removing a seized liner, first ensure that every critical engine component is moved well out of the way, then use two vertical-down runs with an E6010/11 electrode at 90 degrees from the piston thrust faces. Shock cool the welds with cold water and attempt to use the puller again. Avoid fracturing seized liners out of their bores as this practice almost always results in bore damage. As a last resort, liners can be machined out using a boring jig.

TECH TIP: Removing pistons from cylinder liners can be made much easier by removing the carbon around the cylinder wear ridge that forms in use (Figure 9–5). The carbon in the wear ridge can be removed using a flexible knife blade followed by emery cloth.

Block Serviceability Checks

1. Strip block completely including cup expansion and gallery plugs.
2. Soak in a hot or cold tank with the correct cleaning solution.
3. Check for scaling in the water jacket not removed by soaking. One OEM reports that 0.060″ scale buildup has the insulating effect of 4″ cast iron.
4. Check for excessive erosion around the deck coolant ports and fire ring seats.
5. Electromagnetic flux test the block for cracks at each out-of-chassis overhaul.

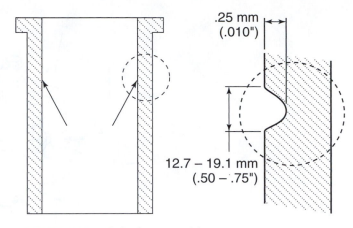

FIGURE 9–5   *Cylinder wear ridge.*

Final Inspection and Assembly

1. Check for deck warpage using a straightedge and thickness gauge. A typical maximum specification is approximately 0.004″ but always reference OEM tolerances.
2. Check the main bearing bore and alignment. Check the engine service history to ensure that the engine has not been previously line bored. A master bar is used to check alignment for a specific engine series. Check that the correct bar has been selected for the engine being tested and then clamp to position by torquing down the main caps minus the main bearings. The master bar should rotate in the cylinder block main bearing line bore without binding. Should it bind, the cylinder block should be line bored.
3. Check the cylinder sleeve counterbore for the correct depth and circumference. Counterbore depth should typically not vary by more than 0.001″. Counterbore depth can be subtracted from the sleeve flange dimension to calculate sleeve protrusion. Recut and shim the counterbore using the OEM recommended tools and specifications (Figure 9–6).
4. Check the cam bore dimensions and install the cam bushings with the correct cam bushing installation equipment: use drivers with great care as the bushings may easily be damaged and ensure that the oil holes are lined up before driving each bushing home.
5. Install gallery and expansion plugs. These are often interference fitted and sealed with silicone, thread sealants, and hydraulic dope.

LINER AND SLEEVE RECONDITIONING

Liner and sleeve reconditioning is not a common current practice due to the time limit constraints on the

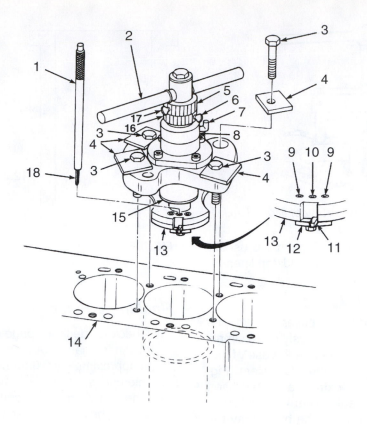

1. Cutter plate holder	7. Oil fill tube	13. Cutter plate
2. T-handle	8. Main housing	14. Cylinder block
3. Capscrews, M16 x 2 x 90	9. Hold-down capscrews	15. Main shaft
4. Special washers	10. Cutter bit adjuster	16. Lower depth-set collar
5. Upper depth-set collar	11. Cutter bit	17. Upper thumbscrew
6. Lower thumbscrew	12. Hold-down cap	18. Hex key wrench

FIGURE 9–6 *Counterbore cutting tool for use on an engine cylinder block with wet/dry liner. (Courtesy of Mack Trucks)*

technician and projected service life. It is probably good practice to advise customers against reconditioning liners citing the fact that it will probably cost them more in the long term. OEMs make oversize liners when the block bores are damaged and have to be rebored. Two basic methods of liner reconditioning are used:

Glaze Busting

When checked to be within serviceability specs, the liner should be deglazed. Deglazing involves the least amount of material removal. A power driven (heavy-duty electric drill with accurate rpm control) flex hone or rigid hone with 200 to 250 grit stones can be used for deglazing. The best type of glaze buster is the flex hone, typically a conical (Christmas tree) or cylindrically shaped shaft with flexible branches of carbon/abrasive balls. The objective of glaze busting is to machine away the cylinder ridge above the ring belt travel and re-establish the crosshatch. The drill should be set at 120 to 180 rpm and used in rhythmic reciprocating

thrusts: short sequences stopping frequently to inspect the finish produces the best results. A 60-degree crossover angle, 15 to 20 microinch crosshatch should be observed (Figure 9–7). When using a spring-loaded hone to deglaze liners, faster results are produced but there is also more chance of damaging the cylinder liner.

Honing

Honing is performed with a rigid hone, powered once again at low speeds by either a drill or overhead boring jig. The typical cylinder hone consists of three legs, which are static set to produce a radial load into the liner wall. The abrasive grit rating of the stones determines the aggressiveness of the tool—200 to 250 grit stones are typical. Overhead boring tools can be programmed to produce the required stroke rate for the specified crosshatch, but if using a hand-held tool, remember that a few short strokes with a moderate radial load tend to produce a better crosshatch pattern than

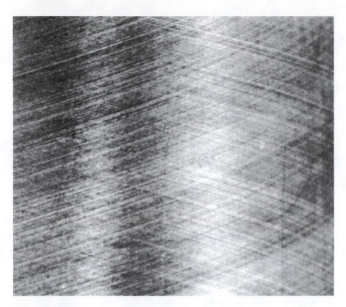

FIGURE 9–7 *Cylinder crosshatch pattern produced by a 150 to 250 grit stone to a depth of 15 to 20 microinches. (Courtesy of Mack Trucks)*

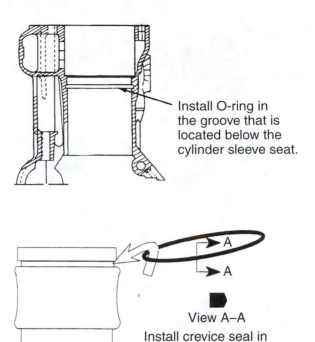

Install O-ring in the groove that is located below the cylinder sleeve seat.

View A–A
Install crevice seal in cylinder sleeve groove with pointed face toward the cylinder block and lubricate with ethylene glycol.

FIGURE 9–8 *Sealing a wet/dry liner. (Courtesy of Mack Trucks)*

many strokes with a light radial load. Once again, a 60-degree crossover angle, 15–20 microinch crosshatch should be observed. It should be clearly visible by eye as shown in Figure 9–7. When installing wet liners, observe the OEM installation procedure. Figure 9–8 shows the Mack Trucks method of sealing a wet/dry liner in the cylinder bore.

Selective Fitting of Liners

Selective fitting of liners to block bores is good practice whether or not the block bore has been machined. To selective fit a set of dry liners to a block, measure the inside diameter of each block bore across the North-South and East-West faces and grade in order of size. Next, get the set of new liners and measure the outside diameter of each, once again grading in order of size. Ensure that *every* measurement falls within the OEM specifications. Then fit the liner with the largest outside diameter to the block bore with the largest inside diameter and so on, down in sequence.

CYLINDER HEADS

Cylinder heads seal the engine cylinders and manage cylinder breathing. In diesel applications they are usually cast-iron assemblies machined with breathing tracts and ports, cooling and lubrication circuit manifolds, fuel manifolds, and injector bores. They may also support rocker assemblies and camshafts when over-

head camshaft design is used. There are several configurations used in diesel engines:

1. Multicylinder, single-slab casting
2. Multicylinder, multiple cast units
3. Single cylinder
4. Integral with block (requires use of valve cage assembly)

Only the first three are found in current diesel engines. Cylinder heads usually contain the valve assemblies: breathing ports, injector bores, and coolant and lubricant passages. On IDI applications, precombustion chambers are either integral or installed as units in the head.

Detail of cylinder heads is shown in Figure 9–9, Figure 9–10, Figure 9–11, and Figure 9–12.

CYLINDER HEAD DISASSEMBLY, INSPECTION, AND RECONDITIONING

1. Remove valves with a C-type spring compressor and tap valve with a nylon hammer to loosen keepers retainers.
2. Clean cylinder heads in a soak tank (preferably hot).

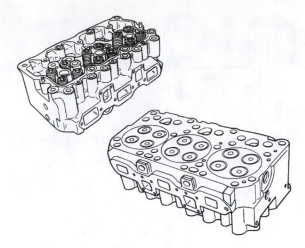

FIGURE 9–9   *Top and bottom views of a typical cylinder head. (Courtesy of Mack Trucks)*

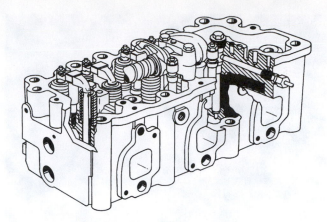

FIGURE 9–11   *Cutaway view of a cylinder head. (Courtesy of Mack Trucks)*

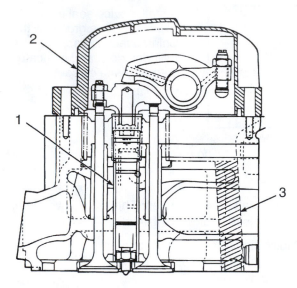

1. Large diameter nozzle sleeve to accommodate 22 mm nozzle
2. E7 EUP cylinder head cover
3. Pushrod hole angled at four degrees

FIGURE 9–10   *Sectional view of a Mack E-Tech cylinder head. (Courtesy of Mack Trucks)*

3. Dress head gasket surface with emery cloth or other fine grit, noncorrosive abrasive.
4. Electromagnetic flux test head for cracks. Optionally dye penetrant testing can be used, but this is messy and inaccurate.
5. Hydrostatic pressure test: cap and plug coolant ports then place head in a test jig and heat by running hot water through it—when hot, hydrostatically test using shop air at around 100 psi. Areas to watch are valve seats and injector sleeves.

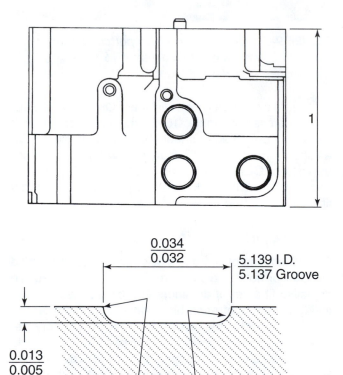

FIGURE 9–12   *Cylinder head height dimension. (Courtesy of Mack Trucks)*

6. If injector sleeves have to be replaced, use appropriate removal and installation tools and repeat pressure test after the operation.
7. Check for warpage using a straightedge and an appropriately sized thickness gauge—this specification will vary according to the size of the head.

8. Check valve guide bore to specs with a ball gauge. If in need of replacement, the guide must be pressed or driven out with the correct driver or mandrel. Damage to the guide bore may require reaming to fit an oversize guide. Integral guides can be repaired by machining for guide sleeves or knurling. Installation of new guides can be made easier by freezing (dry ice) or use of press fit lubricant.

9. New guides may require reaming after installation, but some OEMs use cladded guides that should *never* be reamed.

10. Check valve seats for looseness using a light, ball peen hammer and listening—a loose seat resonates at a much higher pitch. To recondition, select the appropriate mandrel pilot and insert into the guide, then match valve seat to the mandrel grinding stone. Stellite-faced seats require special grinding stones. Dress stone to achieve the required seat angle. Interference angles are seldom used in current applications because the valve's ability to transfer heat to the head is compromised. An interference angle is never cut when valve rotators are used.

New valve seats are installed with an **interference fit** and require the use of the correct driver; after installation the seat is usually knurled or staked in position. Most current diesels use alloy steel seats—cast iron does not require staking of seats.

After valve and seat reconditioning, check valve head height (valve protrusion) with a dial indicator.

11. Valve reconditioning. Clean valves with wire wheel or glass bead blaster, then check for stretching, cupping, burning, or pitting. When refacing (dressing) valves, ensure that the valve margin remains within OEM specifications. Check valve seating using Prussian blue (aka machinist's blue). Lapping is not required if the grinding has been done properly.

12. Check the valve springs for straightness, height, and tension using a right-angle square, tram gauge, and tension gauge.

13. Valve rotators. Positive rotators (roto coil) can be checked after hand assembly by tapping valve open with a nylon hammer after assembly to simulate valve train action—they should rotate.

CYLINDER HEAD GASKETS AND CYLINDER HEAD TORQUE DOWN PROCEDURE

Most current cylinder head gaskets are integral, meaning the fire rings and grommets are manufactured integrally with the gasket plate. If not, great care should be taken to ensure that the **fire rings** (sealing rings at the liner flange) and grommets are properly positioned while the cylinder head is being installed to the block. The head **gasket** must be properly torqued to ensure that its **yield point** is attained for proper sealing. Yield point means that a malleable gasket is crushed to conform to the profile designed to produce optimum sealing. Properly torquing a cylinder head means observing the OEM torque increments and sequencing. An engine cylinder head that is torqued to the correct specification without observing the incremental steps can damage the gasket by deforming it. The objective of torquing cylinder head gaskets is to ensure that the required amount of clamping force is obtained and the gasket conforms to its engineered yield shape.

Most head gaskets require no applied sealant and in fact may fail if sealants are used. Head gaskets are designed to seal engine components in a region where both the temperature and pressure are at their highest; they must do this and at that same time accommodate a large amount of component **creep**. Creep is the relative movement of clamped engine components due to different coefficients of heat expansion or different mass. For instance, multiple cast-iron cylinder heads clamped to a cast-iron cylinder block will expand and arrive at their operating temperature well before the cylinder block. The opposite occurs on cooling. The head gasket is engineered to seal under all the operating conditions of the engine. Cylinder head fasteners should be lightly lubed with engine oil before installation. Oil should never be poured into the block threads because a hydraulic lock may result. The torque sequence for a Mercedes-Benz MB-906 cylinder head is shown in Figure 9–13.

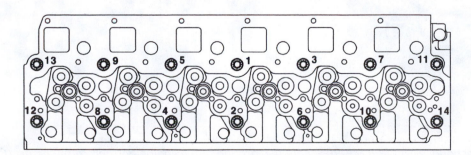

FIGURE 9–13 *Cylinder head torque sequence. (Courtesy of Mercedes-Benz)*

Cylinder Head Installation

On in-line multicylinder heads, it is usually required that separate heads be aligned with a straightedge across the intake manifold faces before torquing. Failure to observe torquing increments and sequencing can result in cracked cylinder heads, failed head gaskets, and fire rings that will not seal. Because of the large number of fasteners involved, a click type torque wrench should be used. Some OEMs require that a **template torque** method be used: this requires setting a torque value first and then turning a set number of degrees beyond that value using a template or protractor. Cylinder head bolts should be installed lightly oiled. Excessive quantities of oil should be avoided because the excess can drain into the bolt hole and cause a hydraulic lock.

TECH TIP: Installing a cylinder head onto an engine that uses nonintegral fire rings in the cylinder head gasket can be made easier by using four guide studs inserted into cylinder bolt holes. This reduces the chances of a fire ring misalignment occurring during head installation.

INTAKE AND EXHAUST MANIFOLDS

In a CI engine, the **intake manifold** is required to deliver air only to the engine cylinders and is bolted to the cylinder heads enclosing the intake tracts. Because of the almost universal use of turbochargers on diesels, intake manifold design is less complex than that for naturally aspirated SI engines. This usually means that the runners that extend from the intake plenum can be of unequal lengths and this does not compromise engine breathing due to turbocharging. A tuned intake manifold is one in which the shape and length of each runner is similar and designed to establish optimum gas dynamics for engine breathing. A single box manifold fed off the turbocharger compressor pipe can often meet a boosted engine's breathing requirements. Intake manifolds can either be wet (coolant ports) or dry; the latter is generally used in truck and bus applications. Materials used are aluminum alloy or cast iron, but some OEMs are experimenting with plastics and carbon-based fibers. The gaskets used are usually fiber based and must be able to accommodate component creep and the peak boost pressures. When a single section cast aluminum manifold is bolted to a multicylinder head configured engine, it is critical that the cylinder heads be aligned with a straightedge before they are torqued down. Figure 9–14 shows a sectional view of an engine: note the location of the engine housing components.

EXHAUST MANIFOLD

The function of the **exhaust manifold** is to collect cylinder end gases and deliver them to the turbocharger or in the case of a naturally aspirated engine, directly to

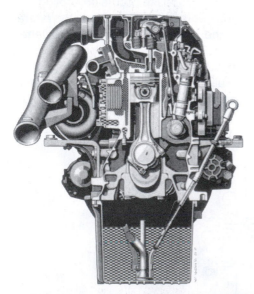

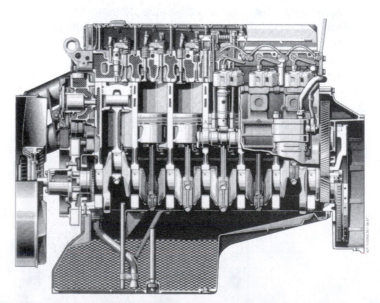

FIGURE 9–14 *Sectional view of an MB-906: note the location of the engine housing components. (Courtesy of Mercedes-Benz)*

the exhaust piping. They are usually manufactured in single or multiple sections of cast iron. The exhaust manifold assembly is bolted to the cylinder head and in engines using multiple heads, it should be aligned before torquing.

Most current diesel engine exhaust manifolds are described as tuned, meaning that exhaust gas is routed to the turbocharger with minimal flow resistance The gas dynamic of exhaust manifolds is discussed in more detail in Chapter 12. The term tuned is applied to any exhaust system designed with at least some accommodation for the exhaust gas flow: if the exhaust manifold and piping is properly designed, as each slug of cylinder exhaust is discharged, it will not "collide" with that from another cylinder but instead be timed to unload into its tailstream.

Exhaust manifold gaskets in truck/bus engine applications are usually of the embossed steel type, although occasionally fiber gaskets are used. They are almost always installed dry. Embossed steel gaskets must be evenly torqued and are designed to be used once only. They can accommodate a large amount of component creep and high temperatures while maintaining a seal. In many cases the fastening hardware such as studs, bolts, and nuts is manufactured from highly alloyed steels to accommodate high temperatures and the high thermal expansion/contraction rates.

OIL PANS OR SUMPS

The **sump** is a reservoir usually located at the base of the engine cylinder block enclosing the crankcase. **Oil pans** are manufactured from cast aluminum, stamped mild steel, and various plastics and synthetic fibers. The oil pan acts as the reservoir used to collect the lubrication oil that gravity causes to drain to the crankcase and from which the oil pump pickup can recycle it through the lubrication circuit. Oil pans can act as a sort of boom box and amplify engine noise, so they are usually designed to minimize noise. The oil pan also plays a role in dissipating lube oil heat to atmosphere but the effectiveness of this role obviously has a lot to do with the material from which it is manufactured. Aluminum will achieve this objective more effectively than a fiber-reinforced plastic. Also this heat exchanger role of the oil pan can produce problems when running in midwinter or subarctic conditions and cause gelling and sludging. Some off-highway trucks use a deep oil pan design to accommodate a scavenging pump. Scavenging pumps ensure that the engine lubrication circuit is provided with adequate oil when the vehicle is operating on steep grades.

Oil pans are usually located in the air flow under the frame rails and are vulnerable to damage from objects on the road, from rocks on rough terrain to small animals on the highway. Most highway diesel engines have oil pans that can be removed from the engine while it is in chassis. It is advisable to drain the oil sump before removing it. Oil pans that seal to the engine block using gaskets can be difficult to remove, especially where adhesives have been used to ensure a seal. A 4 lb rubber mallet may make removal easier. Avoid driving screwdrivers between the oil pan and its block mating flange because the result will be damage to the oil pan flange and the cylinder block mating flange.

TECH TIP: A pneumatic gasket scraper is a great way to remove gasket residue from the engine block oil pan flange face. Care must be taken to avoid gouging the mating surfaces because some pneumatic scrapers can be aggressive.

Oil pans that use rubber isolator seals tend not to present removal problems. After removal, the oil pan should be cleaned of gasket residues and washed with a pressure washer. Inspect the oil pan mating flanges, check for cracks, and test the drain plug threads. Cast aluminum oil pans that fasten both to the cylinder block and to the flywheel housing are prone to stress cracking in the rear; with these oil pans always meticulously observe torque sequences and values. Cast aluminum oil pans can be successfully repair welded without distortion using the tungsten inert gas (TIG) process providing the cracks are not too large and the oil saturated area of the crack is ground clean. It is more difficult to execute a lasting repair weld on a stamped steel oil pan due to the distortion that results. Steel oil pans may rust through; in which case they should be replaced rather than welded. Whenever an oil pan has indications of porosity resulting from corrosion, it should always be replaced. The technician should always be aware of the fact that a failure of the oil sump or its drain plug on the road can result in the loss of the engine.

SUMMARY

- The engine cylinder block can be considered to be the main frame of an engine, the component to which all others are attached.
- Most truck diesel engines use either wet or dry liners to make engine overhaul easier and faster, and extend engine life.
- Dry liners are fitted to the cylinder block fractionally loose or with a fractional interference fit.

- Dry liners do not transfer heat from the engine cylinder to the coolant in the water jacket as efficiently as wet liners.
- Selective fitting of dry liners to cylinder block bores ensures the best liner-to-bore fit.
- Wet liners have much greater mass than dry liners because they must fully support combustion pressures.
- Wet liners transfer heat to the water jacket efficiently because they are surrounded by the coolant.
- Wet liners must seal the water jacket using O-rings.
- Engine cylinder blocks should be boiled in a tank, have every critical dimension measured, especially deck straightness and line bore, and be magnetic flux tested at every major engine overhaul.
- The cylinder head houses the valve train assemblies, the injectors, and the engine breathing passages.
- The engine cylinder block should be pressure tested hydrostatically by first heating through with hot water.
- Common locations for cracks in cylinder heads are the injector bore tubes and the valve seats.
- Cylinder heads must be torqued in sequence to ensure even clamping pressure.
- Torquing cylinder heads in increments is designed to achieve the cylinder head gasket yield point by evenly achieving the required clamping force.
- Cylinder head bolts should be lightly lubed with engine oil before installation.
- The intake manifold is responsible for directing intake air into the cylinder head intake tracts.
- Due to the fact that most truck diesel engines are turbocharged, intake manifolds tend to be simple in design.
- Most truck diesel engines use the term tuned to describe their exhaust manifolds because they feed the turbine housing of the turbocharger and the gas dynamic is critical for scavenging.

REVIEW QUESTIONS

1. Which of the following cylinder block designs would be the most common in medium- and heavy-duty truck diesel engines?
   a. Eight-cylinder, 90-degree V-configuration
   b. In-line six cylinder
   c. Six-cylinder, 60-degree V-configuration
   d. In-line eight cylinder
2. The main reason for bench pressure testing a diesel engine cylinder head is to:
   a. Test cylinder gas leakage
   b. Check for air leaks
   c. Check for exhaust leaks
   d. Check for coolant leaks
3. When selective fitting a set of dry liners to a cylinder block, which of the following statements should be true?
   a. The liner with the largest od is fitted to the bore with the largest id.
   b. The liners are installed with an interference fit.
   c. The liner with the smallest od is fitted to the bore with the largest id.
   d. The liners are installed with a fractionally loose fit.
4. Which of the following is the recommended method of pulling dry liners from a cylinder block?
   a. The arc welding method
   b. A puller and shoe assembly
   c. Heat and shock cool method
   d. Fracture the liner in position
5. Which of the following tools is recommended for checking cylinder block line bore?
   a. Straightedge and thickness gauges
   b. Dial bore gauge
   c. Dial indicator
   d. Master line bore bar
6. When boiling out a cylinder block, why is it critical that all the scale in the water jacket be removed?
   a. Scale is an effective insulator.
   b. Scale may contaminate the engine lubricant.
   c. Scale can accelerate coolant silicate drop-out.
   d. Scale causes cavitation of wet liners.
7. Which of the following tools should be used to check a cylinder block deck for warpage?
   a. Master bar
   b. Dial indicator
   c. Laser
   d. Straightedge and thickness gauges
8. Which of the following tools should be used to check a cylinder head for warpage?
   a. Master bar
   b. Dial indicator
   c. Laser
   d. Straightedge and thickness gauges
9. *Technician A* states that all the machined surfaces of a cylinder block should be checked to specification with a straightedge and thickness gauges. *Technician B* states that cylinder blocks should be boiled in a soak tank at every major overhaul. Who is right?
   a. A only
   b. B only
   c. Both A and B
   d. Neither A nor B
10. When cavitation damage is evident on a set of wet liners, which of the following is more likely to be responsible?

a. Lubrication breakdown
b. High cylinder pressures
c. Coolant breakdown
d. Cold engine operation

11. When cylinder block counterbores are machined, which of the following would necessarily be required?
a. Oversize fire ring
b. Shims
c. Undersize liner flange
d. Cylinder head resurfacing

12. *Technician A* states that it is good practice to magnetic flux test a cylinder block for cracks at each major overhaul. *Technician B* states that before performing a magnetic flux test, the cylinder block should be boiled in a soak tank. Who is right?
a. A only
b. B only
c. Both A and B
d. Neither A nor B

13. Most diesel engine cylinder heads are manufactured from:
a. Cast aluminum alloy
b. Cast-iron alloys
c. Composite fibers

14. Once a cylinder head gasket fire ring has been torqued to its yield point, it should:
a. Not be reused after removal
b. Be immediately heated to operating temperature
c. Deform and no longer seal effectively

15. *Technician A* states that all cylinder heads on current engines should be hot torqued after a rebuild procedure. *Technician B* states that cylinder head bolts should be soaked in engine lubricating oil before installation. Who is right?
a. A only
b. B only
c. Both A and B
d. Neither A nor B

Chapter 10

Engine Lubrication Systems

PREREQUISITES

Chapters 4, 7, 8, and 9

OBJECTIVES

After studying this chapter, you should be able to:

- Describe the objectives of an engine lubrication circuit.
- Outline the function of the main components in a typical diesel engine lubrication circuit.
- List the properties of a heavy-duty engine oil.
- Define the term hydrodynamic suspension and describe how this principle is used in a typical diesel engine.
- Interpret the terminology used to classify lubrication oil.
- Interpret API classifications and SAE viscosity grades.
- List some of the properties and ingredients of synthetic engine oils.
- Identify the components used in a diesel engine lubricating system.
- Replace and properly calibrate a lube oil dipstick.
- Describe the two types of oil pumps commonly used on diesel engines and outline the operating principles of each.
- Perform the measuring procedures required to determine the serviceability of an oil pump.
- Describe the operation of an oil pressure regulating valve.
- Define the term positive filtration.
- Outline the differences between full flow and bypass filters.
- Service a set of oil filters.
- Outline the role of an oil cooler in the lubrication circuit.
- Test an oil cooler core using vacuum or pressure testing.
- Identify the methods used to measure oil pressure in current diesel engines.
- Outline the procedure for taking an engine oil sample for analysis.
- Interpret the results of a laboratory oil analysis.

KEY TERMS

American Petroleum Institute (API)

blotter test

boundary lubrication

bundle

bypass filter

bypass valve

centrifugal filter

dry sump

fire point

flash point

fluid friction

full flow filter

gerotor

hydrodynamic suspension

inhibitors

lamina

lubricity

oil cooler

positive filtration

pour point

relief valve

shear

spectrographic testing

synthetic oil

thick film lubrication

tribology

viscosity

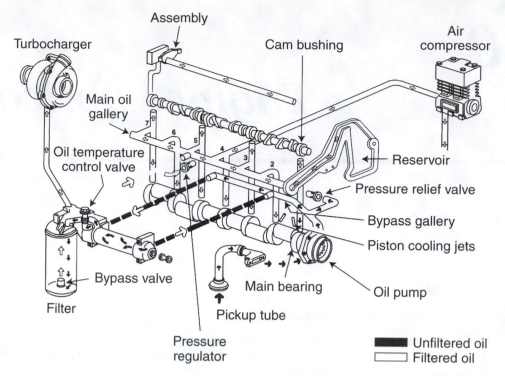

FIGURE 10–1 *Navistar lubrication circuit. (Courtesy of Navistar International Corp., Designer and manufacturer of International Brand diesel engines)*

INTRODUCTION

All medium- and large-bore diesel engines currently used on North American highways use a pressure lubrication system to supply the bearings and moving components with engine oil. Engine oil is the medium used in lubrication circuits, and it is especially formulated to fulfill the lubrication and service life requirements of diesel engines. Engine lubricating oil temperatures run somewhat higher than engine coolants, and the actual oil temperature is probably a more reliable indicator of true engine running temperature. The basic components required of a pressure lubrication system are as follows (Figure 10–1 and Figure 10–2):

- **Lubricant:** a petroleum-based, liquid medium used to reduce friction, hydrodynamically support shafts, help seal pistons, and act as a cooling medium.
- **Sump:** storage space located usually to enclose the crankcase but exceptionally, a remotely located tank in what is called a **dry sump** system.
- **Pump:** responsible for moving the oil through the lubrication circuit to supply the system oil passages and bearings.
- **Filter:** truck and bus diesel engine lubrication systems require multistage filtration systems to remove particulates from the engine oil.

- **Oil cooler:** a heat exchanger that uses engine coolant as its medium. Heat picked up by the engine oil is transferred to the coolant from which it can be dissipated to atmosphere.

FRICTION

Friction is a common element in our lives that we simply take for granted. For instance, we can walk up a steep hill without slipping because of high friction between the soles of our feet and the ground surface. We also accept that when that same hill is covered with packed snow, we can ski down it. In the first instance, the coefficient of friction is high, and in the second it is low. Another way of thinking about the coefficient of friction is as a means of rating the aggressiveness of friction surfaces. Rubber soles have a higher coefficient of friction than skis. Lubricants are designed to reduce friction between surfaces that are, or could be, in contact with one another.

ENGINE LUBRICATING OIL

The main functions of a diesel engine lubricating system are:

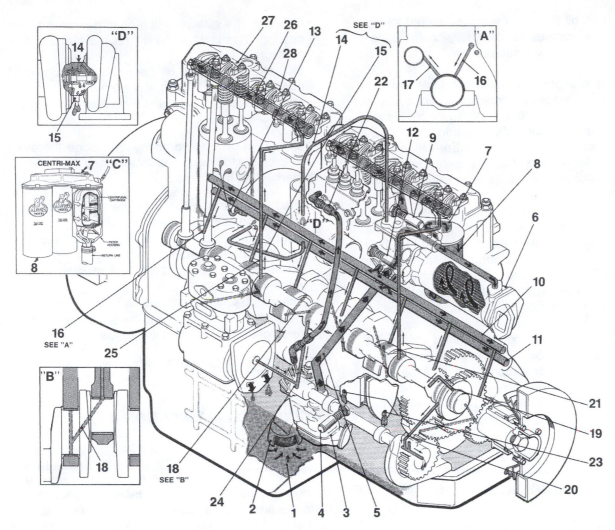

1. Oil pump
2. Oil pump inlet tube
3. Oil pump
4. Oil pressure relief valve (located in pump)
5. Oil pump discharge line to oil cooler
6. Lube oil cooler
7. Oil filter mounting adapter
8. Oil filter(s) (see inset "C")
9. Oil filter bypass valve
10. Main oil gallery
11. Piston oil cooling gallery
12. Oil pressure relief valve (piston cooling gallery)
13. Piston cooling oil spray nozzle
14. Lube feed to turbocharger (external)
15. Turbocharger lube oil return (external)
16. Oil supply from main oil gallery to crankshaft main bearings (see inset "A")
17. Oil flow from main bearings to camshaft bearings (see inset "A")
18. Main bearing and connecting rod bearing lube oil passages (see inset "B")
19. Line to injection pump drive assembly from camshaft bushing number 1
20. Line to auxiliary shaft front bushing from camshaft bushing number 1
21. Oil supply from camshaft bushing number 2 to rocker arm shaft
22. Line to injection pump (external) and governor from camshaft bushing number 4
23. Fuel injection pump lube oil drains
24. Line to auxiliary shaft rear bushing and compressor from camshaft bushing number 4
25. Oil supply from camshaft bushing number 5 to rocker arm shaft
26. Rocker arm shaft
27. Valve rocker arm
28. Oil supply from rocker arm shaft to rocker arm tip

FIGURE 10–2   _Mack Truck lubrication circuit. (Courtesy of Mack Trucks)_

1. Lubrication. The primary task of an engine oil is to minimize friction and act as a medium to support the **hydrodynamic suspension** of the crankshaft and camshaft.
2. Act as a sealant to enable the piston and ring assembly to seal compression and combustion gases from the crankcase.
3. Coolant. Heat generated by combustion and friction must be dissipated to atmosphere via heat exchangers.
4. Cleaning agent. Condensed by-products of combustion gases end up in the engine crankcase and can combine to form harmful liquids (acids) and particulates (sludge).

5. Maintain an oil film even when subjected to high thrust loads, such as the compressional loading a piston/rod assembly is subjected to.

Lubricating oils are petroleum products that are complex mixtures made up of many different fractions. Like gasoline and diesel fuel, engine lubricant is elementally composed of about 85% carbon and 15% hydrogen. The fractioned compounds are refined from petroleum and asphalt bases and subsequently mixed. The refining processes of petroleum are presented in a little more detail in Chapter 17.

The theoretical action of an engine oil is to form a film between moving surfaces so that any friction that results occurs in the oil itself, that is, it is **fluid friction**. Fluid friction generates considerably less heat than dry friction. The lubricating requirements of an engine oil can be classified as **thick film lubrication** and **boundary lubrication** (thin film). Thick film lubrication occurs where mating tolerances between components are wide. Boundary lubrication is required where mating tolerances are narrow, such as when pressure is applied to one of the components: a breakdown of boundary lubrication will result in metal-to-metal contact. A quality engine oil should be capable of performing both thick film and boundary lubrication. Because engine lube oil is usually petroleum based it is flammable.

THE PRINCIPLE OF HYDRODYNAMIC SUSPENSION

When a shaft is rotated within friction bearings such as those used to support an engine crankshaft, and that shaft is stationary, a crescent-shaped gap is formed on either side of the line of direct contact due to the clearance between the journal and the inside diameter of the friction bearing. A static film of oil prevents shaft-to-bearing contact when stationary. When the shaft is rotated and the bearing is charged with engine oil under pressure, a crescent-shaped wedge of lubricant is formed between the journal and its bearing; oil is introduced to the bearing where the shaft clearance is greatest, usually at the top. This wedge of oil is driven ahead of the direction of rotation in a manner that permits the shaft to be "floated" on a bed of constantly changing, pressurized oil. This principle is used in most crankshaft and camshaft dynamics and is known as hydrodynamic suspension.

In hydrodynamic lubrication, keeping a liquid film between moving surfaces is done mechanically, by pumping the lubricant. In the case of a rotating shaft and a stationary friction bearing, the shaft acts as a pump to maintain the lubricant film. The result is that the shaft journal floats on a film of oil the thickness of which depends on:

• Oil input: the rate at which oil is delivered to the bearing. This is why you will have a problem if oil pressure is low
• Oil leakage rate: the oil that spills from a bearing during operation. This is why we make the association between worn engine main bearings and low oil pressure

The thickness of the hydrodynamic wedge therefore depends on the following four factors, any one of which will change that thickness:

1. Load increase: causes oil to be squeezed out of the bearing at a faster rate
2. Temperature increase: causes oil leakage rate from bearing to increase
3. Lower viscosity oil: flows with less resistance causing more leakage
4. Changing shaft speed: reduce speed (remember, the shaft is the "pump") and the film becomes thinner; increase speed and it thickens

Figure 10–3 shows how hydrodynamic lubrication functions.

ENGINE OIL CLASSIFICATION AND TERMINOLOGY

To fully understand the terms and codes used on the label of any engine oil probably requires the learning of a tribologist, that is, an expert in **tribology**, the study of friction, wear, and lubrication. However, every technician should have a rudimentary understanding of the codes and terms used to describe engine oils. The following section introduces some of the basic language of lubricants.

Viscosity

The **viscosity** rating of an oil usually describes its resistance to flow. High viscosity oils have molecules with

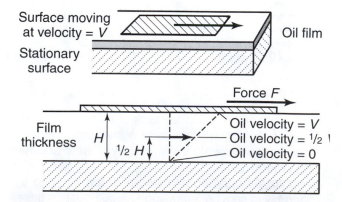

FIGURE 10–3   *Concept of hydrodynamic viscosity.*

greater cohesion ability. However, properly defined, viscosity denotes resistance to **shear**. When two moving components are separated by engine oil, the **lamina** (portion of the oil film closest to each metal surface) on each moving component should have the least fluid velocity while the fluid in the center has the greatest fluid velocity. Shear occurs when the lamina fluid velocity is such that it is no longer capable of adhering to the surface of the moving components.

Viscosity Index

The viscosity index (VI) is a measure of an oil's tendency to thin out as temperature increases. Temperature affects viscosity; viscosity is considerably reduced with an increase in temperature and to a smaller extent, as fluid pressure is increased. The greater the VI, the less of an effect temperature will have on the actual viscosity. In other words, oils that show relatively small viscosity changes with changes in temperature can be said to have high VI.

SAE Numbers and Viscosity

The viscosity of automotive engine oils is graded by the Society of Automotive Engineers (SAE). These gradings and the recommendations that accompany them are listed later in Chapter 10. These gradings specify the temperature window within which the engine oil can function to provide adequate shear resistance under boundary lubrication conditions.

Multiviscosity Oils

Multiviscosity engine oils now tend to be the lubricating oil of choice among all the truck and bus engine OEMs. They have been around for many years but initially, some manufacturers took a cautious approach before finally recommending them. They have the advantage over straight grade oils of being able to provide proper lubrication to the engine over a much wider temperature range. In other words, they have a relatively flat viscosity to temperature curve. Multiviscosity oils are produced by special refining processes and the addition of VI improvers. They possess good cold cranking characteristics and show comparatively small variations of viscosity over their nominal operating range. Synthetic oils, when marketed with an SAE grading, usually greatly exceed the nominal grading window.

Lubricity

Two oils with identical viscosity gradings can possess different lubricity. The **lubricity** of an oil properly describes its flow characteristics. Lubricity is also affected by temperature: hotter oils flow more readily, colder oils less readily. In comparing two engine oils, the one that has the lowest frictional resistance to flow, can be said to possess the greater lubricity. In thick film fluid lubrication, flow friction is determined by the fluid's viscosity, that is, its resistance to shear. In thin film or boundary lubrication, flow friction is determined by the lubricity of the fluid. Sulfur compounds add to the lubricity of diesel engine oils. Synthetic oils tend to have high lubricity with low sulfur content.

Flash Point

Flash point is the temperature at which a flammable liquid gives off enough vapor to ignite momentarily. The **fire point** of the same flammable liquid is usually about 10°C higher and is the temperature at which a flammable liquid gives off sufficient vapor for continuous combustion. The flash point value has some significance when assessing a diesel engine lubricating oil because a large portion of the cylinder wall is swept by flame every other revolution in a four-stroke cycle engine and every revolution in a two-stroke cycle engine. However, actual cylinder wall temperatures are significantly lower than the temperatures of the combustion gases and the oil is only exposed to them for very short periods of time. Most diesel engine lubricating oils have flash points of 400°F (205°C) or higher.

Pour Point

The temperature at which a lubricant begins to gel or simply ceases to flow is known as the **pour point**. Engine lubricating oils formulated for extreme cold weather operation have pour point additives or depressants that act as "antifreeze" for lubrication oils. Pour point is an important engine oil specification especially due to the tendency of many operators to use a multigrade engine oil viscosity nominally not suited for midwinter conditions in the northern part of the continent (Figure 10–4).

Inhibitors

The term **inhibitors**, when applied to an engine lubricant, refers to the additives that protect the oil itself against corrosion, oxidation, and acidity. Their objective is to make the oil less likely to partake in reactions with the contaminants that find their way into the crankcase, such as combustion by-products, moisture, and raw fuel.

Ash

Ash in lubricant is mineral residue that results from oxide and sulfate incineration. High ash levels in oil analyses are often the result of high temperature operation.

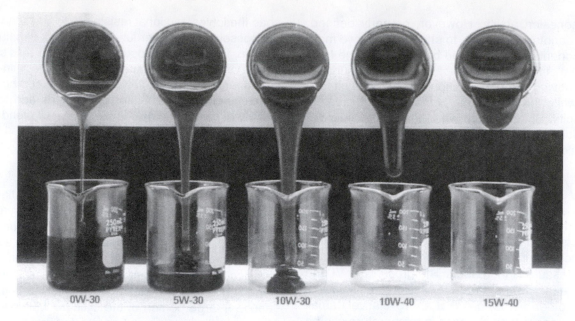

FIGURE 10–4 *Demonstration of cold weather flow ability of different oil viscosity grades. (Courtesy of Esso/Imperial Oil)*

Film Strength

Most mineral lubricants possess adequate film strength to prevent seizure and galling of contacting metal surfaces, but high-speed, high-output truck diesel engines usually contain additives to improve film strength and cohesiveness of the oil.

Detergents

Detergents are added to engine oils to prevent the formation of deposits on internal engine components. These function to keep soluble oxidation products from becoming insoluble. Polymeric detergents and amine compounds are used.

Dispersants

Dispersants are added to engine oils to help keep insoluble oxidation products in suspension and prevent them from coagulating into sludge and deposits. When sludge and deposits do form in the crankcase, the capability of the dispersant content in the engine oil has been exceeded, often an indication that the oil change interval should be reduced.

OIL CONTAMINATION AND DEGRADATION

When oil becomes contaminated, it can lead to complete engine failure. Contaminated engine oil has some characteristic tattletales. These can be quickly identified by visual and odor testing, and confirmed, if required, by oil analysis. The following are some common contaminants:

Fuel

When fuel contaminates engine oil, the oil loses its lubricity and it appears thinner and blacker in color. The condition is usually easy to detect because small amounts of fuel in oil can be recognized by odor. When fuel is found in significant quantities in engine oil, the cylinder head(s) are the likely source.

Coolant

Coolant in the engine lubricant gives it a milky, cloudy appearance when churned into the oil. After settling, the coolant usually collects at the bottom of the oil sump. When the drain plug is removed, the heavier coolant exits first as long as sufficient time has passed since running the engine. When coolant is found in the engine oil, the cylinder head(s) and cylinder head gasket are the most likely source.

Oil Aeration

Aerated engine oil, also known as foaming, can severely affect its ability to properly lubricate the engine. Aerated oil can be caused either by the chemical characteristics of the oil itself (such as detergent additives) or by contaminants such as water; surface tension is a factor. The conditions that aerate oil in the crankcase are the rotary action of the crankshaft, sucking of air into the oil pump inlet, and the free fall of oil into the crankcase from the oil pump **relief valve** and cylinder walls. Antifoaming additives inhibit oil aeration. Silicone polymers are used to prevent foaming in diesel engine lubricants.

Cold Sludge

Cold sludge is caused by oil degeneration that occurs during prolonged engine low load operation at low temperatures. It settles in the engine crankcase and can accelerate engine wear rates. When a diesel engine is operated with its coolant thermostat(s) removed, the result can be the formation of cold sludge.

API CLASSIFICATIONS

The **American Petroleum Institute**, usually known as **API**, classifies all engine oil sold in North America. There are two main classifications designated by the prefix letters S and C. The S classes of engine oil designate those oils suitable for passenger cars and light trucks. The S represents SI or spark ignited. The C classes of engine oils designate oils suitable for heavy-duty trucks, buses, and industrial and agricultural equipment. The C represents CI or compression-ignited engines. C category and the most recent S category oil classifications are listed and described here because many fleets use engine oils that claim to be suitable for both C and S classifications.

In most cases, OEMs have specific requirements for engine lubricants that should be observed because failure to do so could result in higher hydro-carbon (HC) emissions.

CA—for Light-Duty Diesel Engine Service. Service typical of diesel engines operated in mild or moderate duty with high quality fuels. Occasionally has included gasoline engines in mild service. Oils designed for this service were widely used in the late 1940s and 1950s.

CB—for Moderate-Duty Diesel Engine Service. Service typical of diesel engines operated in mild to moderate duty, but with lower quality fuels that require more protection from wear deposits. Occasionally has included gasoline engines in mild service. Oils designed for this service were introduced in 1949.

CC—for Moderate-Duty Diesel and Gasoline Engine Service. Typical of lightly supercharged diesel engines operated in moderate to severe duty and has included certain heavy-duty gasoline engines. Oils designed for this service were introduced in 1961 and used in many trucks, in industrial and construction equipment, and farm tractors. These oils provide protection from high-temperature deposits in lightly supercharged diesels and also from rust, corrosion, and low-temperature deposits in gasoline engines.

CD—for Severe-Duty Diesel Engine Service. Service typical of supercharged diesel engines in high-speed, high-output duty requiring highly effective control of wear and deposits. Oils designed for this service were introduced in 1955 and provide protection from bearing corrosion and from high-temperature deposits in supercharged diesel engines when using fuels of a wide quality range.

CD-II—Service typical of two-stroke cycle diesel engines requiring highly effective control over wear and deposits. Oils designated for this service also meet all performance requirements of API Service Category CD.

CE—Service typical of turbocharged or supercharged heavy-duty diesel engines manufactured since 1983 and operated under both low-speed, high-load and high-speed, high-load conditions. Oils designed for this service may be used when previous API Engine Service Categories for diesel engines are recommended.

CF—May be used in diesel engines exposed to fuels that may have a sulfur content greater than 0.05%. Effective in controlling piston deposits, wear, and bearing corrosion. May be used where API Service Category CD is recommended.

CF-2—Service typical of two-stroke cycle engines that require effective control over cylinder and ring-face scuffing and deposits. May be used where API Service Category CD-II is recommended.

CF-4—Service typical of on-highway, heavy-duty truck applications. Designated for multigraded oils and introduced in 1991. Oils meeting this category will also meet API Service Category CE.

CG-4—For use in high-speed, four-stroke cycle diesel engines used on both heavy-duty on-highway and off-highway applications (with less than 0.05% wt. sulfur fuel). CG-4 oils provide effective control over high-temperature piston deposits, wear, and soot accumulation. Effective in meeting 1994 exhaust emission standards and may also be used where API Service Categories CD, CE, and CF-4 are specified.

CH-4—For use in high-speed, four-stroke cycle diesel engines used in on- and off-highway applications that are fueled with fuels containing less than 0.05% sulfur. Supercedes CD, CE, CF, and CG category oils.

CI-4—Introduced in September 2002 for use in high-speed, highway diesel engines meeting 2004 exhaust emission standards, implemented in October 2002 by EPA agreement with engine OEMs. CI-4 oil is especially formulated for low-emission engines, using fuels containing less than 0.05% sulfur and are designed to sustain engine durability when exhaust gas recirculation (EGR) devices are used. CI-4 engine oil supercedes CD, CE, CF, CG, and CH category oils. It meets the requirements of heavy duty engine oil (HDEO) standard PC-9.

PC-10—This oil is scheduled for introduction in 2007, and it will probably be known as CJ-4.

Oils formulated for use in gasoline-fueled engines are categorized in much the same way. The most recent engine oils formulated for service in gasoline engines areas follows:

SH   1993

GF-2   1996

GF-3   2001

GF-4   2004

There are still some engine oils formulated that claim to meet both the needs for SI gasoline fueled and diesel fuel engines, but this is not as common as it was a decade ago.

SAE VISCOSITY GRADES

The following list the SAE engine oil grades and the recommended temperature operating ranges. The W denotes a winter grade lubricant (Figure 10–5).

Multigrade engine oils:

0W-30 Recommended for use in arctic and sub-arctic winter conditions.

Outside temperature °C

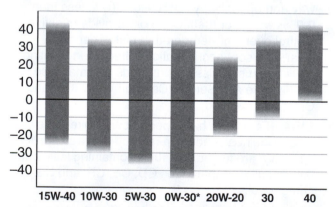

Choose the Right Grade of Oil for Your Climate

* Part-synthetic Essolube XD-3 Extra DW-30.
 Note 1: Detroit Diesel allows 15W-40 multigrade use in two-cycle engines only below 0°C.
 Note 2: Multigrade oils are recommended for 4-stroke modern diesel engines.

FIGURE 10–5   Oil grade and climate. (Courtesy of Esso/Imperial Oil)

5W-30 Recommended for winter use where temperatures frequently fall below 0°F (–18°C) and seldom exceed 60°F (15°C).

5W-40 Recommended for severe-duty winter use where temperatures frequently fall below 0°F (–18°C).

5W-50 Recommended for severe-duty winter use in Arctic and sub-Arctic conditions where temperatures frequently fall below 0°F (–18°C). A synthetic oil viscosity grade.

10W-30 Recommended for winter use where temperatures never fall below 0°F (–18°C).

10W-40 Recommended for severe-duty winter use where temperatures never fall below 0°F (–18°C).

15W-40 Recommended for use in climates where temperatures never fall below 15°F (–9°C). Despite this and diesel engine OEM recommendations that support the use of lighter multigrades in winter conditions, this is by far the most commonly used viscosity grade for truck and bus engines year round, often in climates that have severe winters. A true 15W-40 engine oil will freeze to a greaselike consistency in subzero conditions, making the engine almost impossible to crank; additionally, engine wear is accelerated during the warmup phase of operation. However, many oils with the nominal 15W-40 grading are actually formulated to perform effectively at temperatures of 0°F (–18°C).

20W-40 Recommended for use in high performance engines in climates where temperatures never fall below 20°F (–6°C).

20W-50 Recommended for use in high performance engines in climates where temperatures never fall below 20°F (–6°C).

Straight grades:

10W Recommended for winter use in climates where temperatures never fall below 0°F (–18°C) and never exceed 60°F (15°C).

20W-20 Recommended for use in climates where temperatures never fall below 20°F (–6°C).

30 Recommended for use in climates where temperatures never fall below 32°F (0°C). This grade of oil is the most widely used truck and bus diesel engine in operations that persist in using straight grade oils.

40 Recommended for severe-duty use in climates where temperatures never fall below 40°F (4°C).

50 Recommended for severe-duty use in climates where temperatures never fall below 60°F (15°C). Often used to disguise engine problems, especially oil burners and leakage.

SYNTHETIC OILS

Most diesel engine OEMs approve of the use of synthetic lubricants and often recommend them for severe-duty applications such as extreme cold weather operation, providing the oil used meets their specifications. Synthetic lubricants are largely petrochemically derived, but some plant and coal-sourced additives are also used. Synthesis is the process of creating compounds in laboratory, so while conventional lubricants are obtained from petroleum crude oils by distillation or other refining methods, synthetics are manufactured or "synthesized" in chemical plants by reacting components. Some examples are poly-alpha-olefins, diesters, polyesters, and silicone fluids. Only the very high price of synthetic lubricants is limiting their use in the trucking industry. Most of the indicators suggest that they substantially outperform conventional lubricants. It should also be noted that, while most of the engine OEMs approve of the use of **synthetic oils**, they have not, up until now endorsed increased service intervals when they are used.

WHAT OIL SHOULD BE USED?

Each engine OEM has specific requirements for the engine oils it wants you to put in its engines. These requirements are outlined in tests that engine oils are subjected to prior to an approval. These tests correlate to a standard that oil manufacturers then attempt to meet in the oil they formulate: you see these standards written on the containers in which engine oil is sold. Here are some of them:

- Caterpillar Series 3, TO-2, and I-R
- Cummins CES 20071 and 2076
- Mack EO-M and EO-M Plus
- Volvo VDS standard

Oil refiners attempt to meet as many standards as possible when marketing engine oils for the obvious reason that they can service more potential buyers with a single product. The problem is, oil is like a brew and you do not necessarily improve it by throwing more additives in. For instance, if your preference is to drink bourbon and water, you probably will not think it a better drink if some gin, vodka, and rum were added to it. For this reason, a general purpose heavy-duty engine oil is not necessarily better than the OEM-labeled oil, although of course it will be both cheaper and more advertised.

The best choice for the engine is to use the manufacturer-recommended engine oil. After all, the research and performance profiles have all been performed on the engine using this oil. This may not be the best choice for the pocketbook, and there are general

purpose diesel engine oils that have been proved to perform well in engines over time.

You should also know that engine OEMs usually do not "engineer" their own oils. This development is done by specialist oil additive companies contracted by the manufacturer. Developing the right additive package for an engine oil is a science in itself and is usually the result of extensive testing by lubrication engineers as tribologists.

LUBRICATION SYSTEM COMPONENTS

Engine lubricant must be retained in a reservoir, pumped through the lubrication circuit, filtered, cooled, and have its pressure and temperature monitored. The group of components that performs these tasks is known as the lubricating circuit components. These components vary little from one engine to another but OEM service literature should be consulted before servicing and reconditioning any components.

SUMP—OIL PAN

The sump is a reservoir usually located at the base of the engine cylinder block enclosing the crankcase (Figure 10–6). Oil pans are manufactured from cast

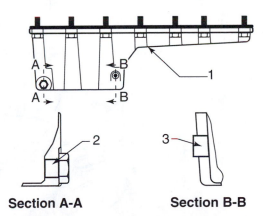

Section A-A   **Section B-B**

(1) Oil pan.
 Oil pan is shown in rear sump mounting position. An alternative front sump mounting position is available.
(2) Assemble gasket with seam against oil pan.
(3) Apply 9S3263 thread lock to plug threads, and tighten to a torque of
$$80 \pm 11 \text{ N•m} (60 \pm 8 \text{ lb. ft.})$$

FIGURE 10–6 *A Caterpillar C-10 oil pan. (Courtesy of Caterpillar)*

aluminum, stamped mild steel, and various plastics and fibers. The oil pan acts as the reservoir used to collect the lubrication oil, which gravity causes to drain to the crankcase, and from which the oil pump pickup can recycle it through the lubrication circuit. A more complete description of an oil pan is provided in Chapter 9.

TECH TIP: Observe the torque sequence when fitting an oil pan, especially those designed to bolt both to the engine cylinder block and the flywheel housing. The consequences of not doing so can be leaks at the pan gasket or stress cracks to the oil pan.

DIPSTICKS

The dipstick is a rigid band of hardened steel that is inserted into a round tube to extend into the oil sump. Checking the engine oil level is performed daily by the vehicle operator, so its location is always accessible. In cab-over-engine (COE) chassis, the dipstick must be accessible without raising the cab, so it may be of considerable length. It is crucial that the correct dipstick is used for an engine. When replacing a missing or defective dipstick, replace the engine oil and filters, installing the exact OEM-specified quantity. Run the engine for a couple of minutes then shut down and leave for 10 minutes. Dip the oil sump with the new dip-

stick and scribe the high level graduation with an electric pencil. Measure the distance from the high level to low level graduations on the old dipstick and then duplicate on the replacement. Remember, the consequences of low or high engine oil levels can be equally serious, so ensure that this operation is performed with precision. Some electronically managed engines have an oil level sensor that signals a low oil level condition to the ECM, which can then initiate whatever failure strategy it is programmed for.

TECH TIP: To obtain an accurate oil level reading, ensure that an engine has been shut down for at least 5 minutes before reading the dipstick indicated level.

OIL PUMP

Engine oil pumps are usually of the positive displacement type and have pumping capacities that greatly exceed the requirements of the engine. They are gear driven and usually located in the crankcase close to the oil they pump, though in some Cummins and Caterpillar applications they are external. Most oil pumps located in the crankcase are driven by a vertical shaft and pinion engaged with a drive gear on the camshaft. A pickup is located close to but not contacting the base of the oil pan (Figure 10–7). Two basic types of oil pumps are used.

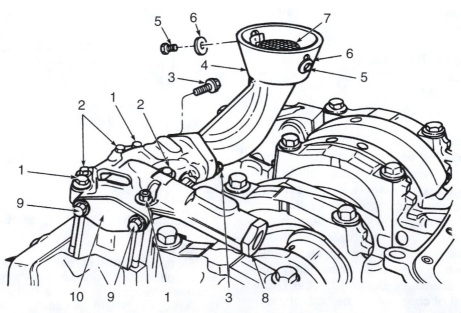

1. Capscrews	4. Sump	7. Screen	10. Plate
2. Capscrews	5. Capscrews	8. Relief valve cap	
3. Capscrews, 12-point	6. Washer	9. Capscrews	

FIGURE 10–7 *View of an oil pump and pick-up on an inverted engine. (Courtesy of Mack Trucks)*

External Gear

External gear pumps consist of two meshed gears, one driving the other within a housing machined with an inlet (suction) port and outlet (charge) port. As the gears rotate, the teeth entrap inlet oil and force it outside between the gear teeth and the gear housing to the outlet port. Where the teeth mesh in the center, a seal is formed that prevents any backflow of oil to the inlet. This is by far the most common oil pump design on current engines.

Gear type engine oil pumps seldom malfunction. When they show evidence of wear, the underlying reason is usually found in contaminated engine lube. To inspect a gear type pump, remove the cover plate and with the gears in the housing, use a thickness gauge to check the gear to housing radial clearance. If this exceeds the OEM specification, the gears should be replaced. Next, the gear-to-cover plate clearance (gear member axial clearance) should be measured using a straightedge and thickness gauges: replace the gears if the measurement exceeds specifications. Before reassembling a gear type pump, clean thoroughly using solvent and compressed air drying, ensuring that the pickup pipe and inlet screen are not overlooked. Reassemble using engine oil as assembly lubricant and check the drive gear teeth (Figure 10–8, Figure 10–9, and Figure 10–10).

Gerotor

Gerotor type oil pumps use an internal crescent gear pumping principle. An internal impeller with external crescent vanes is rotated within a internal crescent gear also known as a rotor ring. The inner rotor or impeller has one less lobe than the rotor ring. The result is that as the inner rotor is driven within the outer rotor, only one lobe is engaged at any given moment of operation. In this way, oil from the inlet port is picked up in the crescent formed between two lobes on the impeller and as the impeller rotates, forced out through the outlet port as the lead lobe once again engages. The assembly is rotated within the gerotor pump body.

Gerotor pumps tend to wear most between the lobes on the impeller and on the apex of the lobes on the rotor ring. These dimensions should be checked to OEM specifications using a micrometer. The rotor ring-to-body clearance should be checked with a thickness gauge sized to the OEM maximum clearance specification and the axial clearance of the rotor ring and impeller measured with a straightedge and thickness gauges.

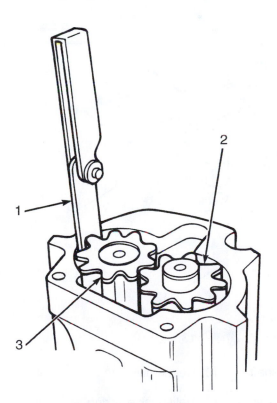

1. Thickness gauge   3. Pump driven gear
2. Pump driven gear

FIGURE 10–8 *Checking gear side clearance. (Courtesy of Mack Trucks)*

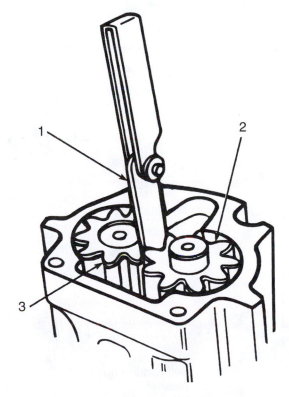

1. Thickness gauge   3. Pump driven gear
2. Pump driven gear

FIGURE 10–9 *Checking gear backlash. (Courtesy of Mack Trucks)*

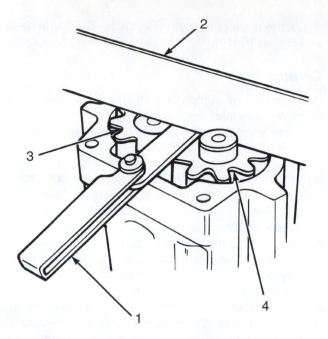

1. Thickness gauge   3. Pump driven gear
2. Straightedge   4. Pump driven gear

FIGURE 10–10 *Checking gear end clearance.*
(Courtesy of Mack Trucks)

Scavenge Pumps/Scavenge Pickups

Scavenge pumps are used in the crankcase of some off-highway trucks required to work on inclinations that could cause the oil pump to suck air. They are designed with a pickup located at either end of the oil pan.

PRESSURE-REGULATING VALVES

Pressure-regulating valves are responsible for defining maximum system oil pressure. Most are adjustable. Typically, an oil pressure regulating valve consists of valve body with an inlet sealed with a spring-loaded, ball check valve. Other types of poppet valve are also used but the principle is the same. The regulating valve body is plumbed in parallel to the main oil pump discharge line (Figure 10–11). When oil pressure is sufficient to unseat the spring-loaded check ball, it unseals permitting oil to pass through the valve and spill to the oil sump; this action causes the pressure in the oil pump discharge line to drop. The regulating pressure value is adjusted by setting the spring tension of the spring usually by shims or adjusting screw. Some

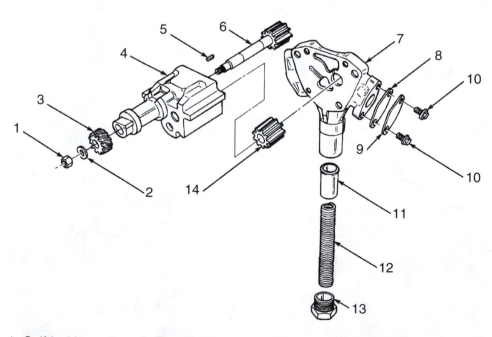

1. Self-locking nut
2. Washer
3. Drive gear
4. Housing
5. Key
6. Pumping gear and shaft assembly
7. Oil pump housing cover
8. Gasket
9. Inlet flange plate
10. Capscrews
11. Gasket
12. Relief valve spring
13. Relief valve cap
14. Idler gear

FIGURE 10–11 *Exploded view of an oil pump and integral pressure relief valve.*
(Courtesy of Mack Trucks)

OEMs use color-coded springs to define the oil pressure values.

TECH TIP: When testing pressure-regulating valve operation, always use a master gauge to accurately verify pressures.

FILTERS

The role of oil filters in the diesel engine lubrication system is to remove and hold contaminants while providing the least amount of flow restriction in the lubrication circuit. Filters use several different principles to accomplish this objective. The term **positive filtration** is used to describe a filter that operates by forcing all the fluid to be filtered through the filtering medium. Most engine oil filters use a positive filtration principle. It should be noted that filters function at higher efficiencies when the engine oil is at operating temperature. Filters work to clean the engine oil by using the following methods:

Mechanical Straining

Mechanical straining is accomplished by forcing the lubricant through a filtering medium, which if greatly enlarged, would have the appearance of a grid. The sizing of the grid openings defines the size of the particle that may be entrapped by the filtering medium. Most current diesel engine oil filters make use of mechanical straining in conjunction with other principles. Straining media would include rosin-impregnated paper often pleated to increase the effective area, and cotton fibers.

Absorbent Filtration

These filters work by absorbing or sucking up engine contaminants as a sponge would. Effective absorbent filtering media include cotton pulp, mineral wools, wool yarn, and felt. These filters not only absorb coarse particles but may also remove insoluble oxidized particulate, moisture, and acids.

Adsorbent Filtration

Filters adsorb by holding (by adhesion) the molecules of dissolved substances or liquids to the surface of the filtering medium. Adsorbent filtering media include charcoal, Fuller's earth, and chemically treated papers. Because adsorbent filters may act to remove oil additives, they are often used only where low-additive engine oils are specified.

Filter Types and Efficiencies

Most current filters are spin-on disposable cartridges. Older engines may have permanent canisters enclosing a replaceable element; the canister was mounted to the filter pad with a long threaded shaft that extended through the length of the canister. They are seldom seen today as most OEMs made conversion adapters so that disposable spin-on cartridges could replace them. Filters are categorized by the manner in which they are plumbed into the lubrication circuit:

Full Flow

The filter mounting pad is usually plumbed into the lubrication circuit close to the oil pump outlet, and all of the oil exiting the pump is forced through the filter. The filtering media is usually rosin or otherwise treated paper or cotton fiber. In most cases, these filters employ a mechanical straining principle, so the particles entrapped by the filtration media are those too large to pass through it. All **full flow filters** used on current truck and bus diesel engines use positive filtration and are rated to entrap particulate sized between 25 and 60 microns. **Bypass valves** located on full flow filter mounting pad(s) protect the engine should a filter become plugged. In this event, the oil exiting the oil pump would be routed around the plugged filter directly to the lubrication circuit.

Bypass

Bypass filters are used to complement the full flow filters on current highway diesel engines. These are plumbed in parallel in the lubrication circuit, usually by porting them into the main engine oil gallery. They filter more slowly, but are rated to entrap particles down to 10 microns in size. Two types are used:

Luberfiner Filter. These are large canister-type filters designed to entrap much smaller particles than full flow filters. Luberfiner filters are supplied from any point in the engine lubrication circuit and return the filtered lubricant directly to the oil sump. They may serve an additional role as an oil cooler as they are often mounted in the air flow. Luberfiner filters are large volume filters that substantially increase the amount of engine oil required: this is important to remember when servicing the engine because more oil will be required. A replaceable filter element is installed to a permanent canister; after replacing a filter element, the canister should be purged of air after engine startup.

TECH TIP: Always crack the nut on the exit fitting of a luberfiner filter after each oil change to purge the air. Failure to do this can result in an air-locked filter.

Centrifugal Filter. **Centrifugal filtration** or centrifuging (Figure 10–12) is used to entrap smaller particles than most full flow filters so they are usually, but not always, of the bypass type. The filter consists of a canis-

FIGURE 10–12 *A filter configuration showing two full flow filters and a centrifugal bypass filter. (Courtesy of Mack Trucks)*

ter within which a cylindrical rotor is supported on bearings. The filter is plumbed into the lubrication circuit so that the rotor is charged with engine oil at lube system pressure. It exits the rotor via two thrust jets, angled to rotate the assembly at high velocity. The centrifuge forces the engine oil through a stationary cylindrical filtering medium wrapped outside the rotor. The filtering medium is usually a rosin-coated paper element. The filtered oil drains back to the oil sump.

TECH TIP: Bypass filters boast high filtering efficiencies, and failure to observe scheduled maintenance can result in plugged filters.

CAUTION: The OEM-specified oil capacity of an engine may not include the volume of oil retained in the bypass filters. Purge bypass filters where required, and check oil level after running the engine following the oil change.

Filter Bypass Valves

Filter mounting pad bypass valves operate in much the same way as the oil pressure regulating valves, except that their objective is to route the lubricant around a restricted full flow filter to prevent engine damage by oil starvation. When a filter bypass valve is actuated and the check valve unseated, instead of spilling the oil to the crankcase it reroutes the oil directly to the lubrication circuit, effectively shorting out the filter assem-

bly. So, when a bypass value trips, unfiltered oil is circulated through the lubrication circuit. Note the location of the filter bypass in Figure 10–13.

Replacing Filters

Filters are removed using a band, strap, or socket wrench. Ensure that the filter gasket and seal (if used) are removed with the filter. Precautions should be taken to capture oil that spills when the filter is removed. Filters mounted vertically usually spill their contents onto the mounting pad assembly. Disposable filters and elements are loaded with toxins and must be disposed of in accordance with federal and local jurisdiction legislation pertaining to used engine oils and filters.

Most OEMs require that new oil filters be primed. If this is not done, in some cases, the lag required to charge the oil filters on startup is sufficient to generate a fault code. Priming an oil filter requires that it be filled with new engine oil on the inlet side of the filter until it is just short of the top of the filter; this will take a little while as the oil must pass through the filtering media to fill the outlet area inside the element. The sealing gasket should be lightly coated with engine oil. OEMs are

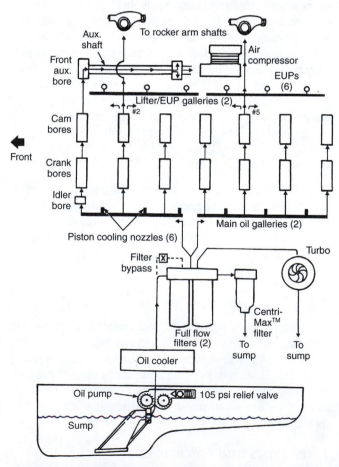

FIGURE 10–13 *Schematic showing oil flow through a lubrication circuit.*

usually fairly specific about how much the filter should be tightened and caution against overtightening. In most cases, the filter should be tightened by rotating it one-half to a full turn after the gasket and filter pad mounting face contact.

OIL COOLERS

Oil coolers are heat exchangers consisting of a housing and a bundle (element/core) through which coolant is pumped and around which oil is circulated. Oil temperatures in diesel engines run higher than coolant temperatures typically around 110°C (230°F). However, the engine coolant reaches its operating temperature more rapidly than the oil and plays a role in heating the oil to operating temperature in cold weather startup/warmup conditions. Two types of oil cooler are in current use:

Bundle

The **bundle**-type oil cooler is the most common design. It consists of a usually cylindrical "bundle" of tubes with headers at either end, enclosed in a housing. Engine coolant is flowed through the tubes and the oil is spiral circulated around the tubes by helical baffles. The assembly is designed so that the oil inlet is at the opposite end to the coolant inlet. This arrangement means that the engine oil at its hottest is first exposed to the coolant at its coolest, slightly increasing cooling efficiency (Figure 10–14, and Figure 10–15).

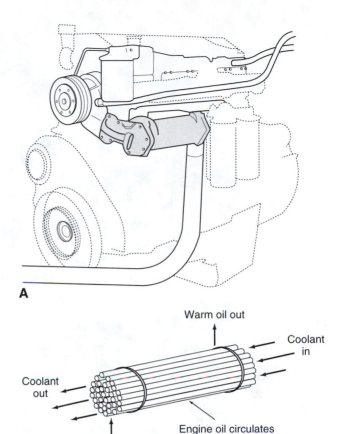

FIGURE 10–14   A. _Oil cooler location. (Courtesy of Mack Trucks)_ B. _Oil flow through a bundle type oil cooler._

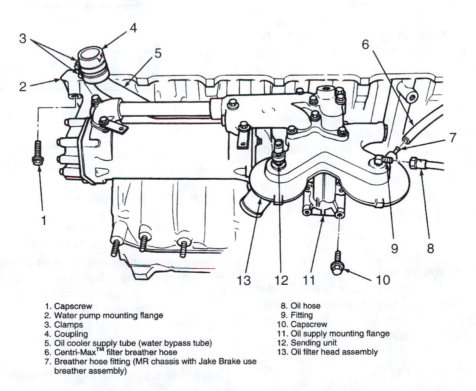

1. Capscrew
2. Water pump mounting flange
3. Clamps
4. Coupling
5. Oil cooler supply tube (water bypass tube)
6. Centri-Max™ filter breather hose
7. Breather hose fitting (MR chassis with Jake Brake use breather assembly)
8. Oil hose
9. Fitting
10. Capscrew
11. Oil supply mounting flange
12. Sending unit
13. Oil filter head assembly

FIGURE 10–15   _External view of oil cooler and filter mounting pad showing the coolant and lubricant plumbing. (Courtesy of Mack Trucks)_

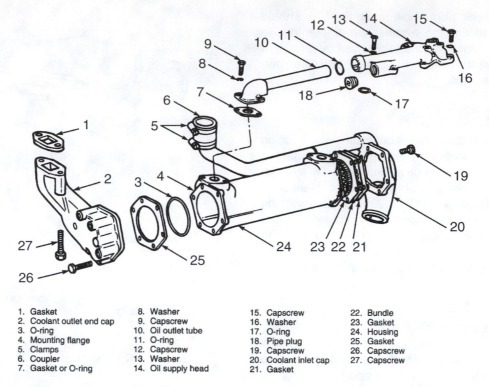

1. Gasket
2. Coolant outlet end cap
3. O-ring
4. Mounting flange
5. Clamps
6. Coupler
7. Gasket or O-ring
8. Washer
9. Capscrew
10. Oil outlet tube
11. O-ring
12. Capscrew
13. Washer
14. Oil supply head
15. Capscrew
16. Washer
17. O-ring
18. Pipe plug
19. Capscrew
20. Coolant inlet cap
21. Gasket
22. Bundle
23. Gasket
24. Housing
25. Gasket
26. Capscrew
27. Capscrew

FIGURE 10–16   *Exploded view of a bundle-type oil cooler. (Courtesy of Mack Trucks)*

The consequence of a failed cooler bundle or the header O-rings is oil charged to the cooling circuit. Most OEMs prefer that bundles be leak tested by vacuum because this tests the assembly in the direction of fluid flow in the event of a leak. One header should be capped with a dummy plate and the other fitted with the evacuation adaptor: load to the OEM vacuum value and leave for the required amount of time to observe any drop off.

Alternatively, the bundle can be pressure tested using regulated shop air pressure and a bucket of water. This is known as reverse flow testing. Whenever oil has leaked into the coolant circuit, the engine cooling system must be flushed with an approved detergent and water with the engine run at its operating temperature for at least 15 minutes. Figure 10–16 is an exploded view of a bundle-type oil cooler.

Plate Type

In the plate-type oil cooler, the oil circulates within a series of flat plates and the coolant flows around them within a housing assembly. They have lower cooling efficiencies than bundle element coolers, but they are usually easier to clean and repair (Figure 10–17).

OIL PRESSURE MEASUREMENT

Of all the engine monitoring devices used on an engine, the oil pressure is one of the most critical. Loss of

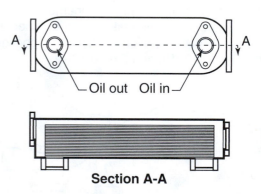

Section A-A

With SAE 30-W oil, the oil flow through the cooler assembly is 110 liter/min (29 U.S. gpm). The maximum water flow through the cooler assembly is 190 liter/min (50 U.S. gpm).

FIGURE 10–17   *A plate-type oil cooler. (Courtesy of Caterpillar)*

engine oil pressure will, in most cases, cause a nearly immediate engine failure. Back in the days when few of an engine's operating conditions were monitored and displayed to the operator, there was always a means of signalling a loss of oil pressure. Several types of sensors are used in today's engines.

Variable Capacitance (Pressure) Sensor

Most current truck and bus diesel engines managed electronically use variable capacitance type sensors.

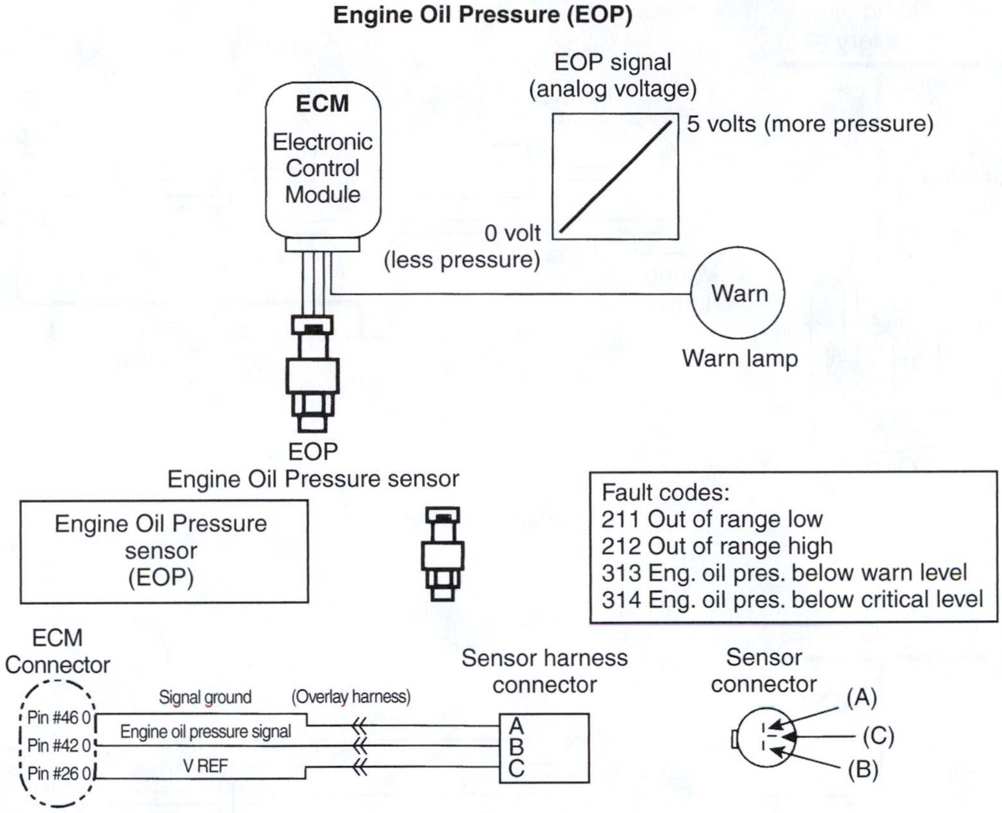

Engine Oil Pressure (EOP)

FIGURE 10–18　*A Navistar oil pressure sensing circuit. (Courtesy of Navistar International Corp., Designer and manufacturer of international Brand diesel engines)*

These are supplied with reference voltage. Oil pressure acts on a ceramic disc and moves it either closer or farther away from a stationary steel disc varying the capacitance of the device and thus the voltage signal value returned to the ECM. The ECM is responsible for outputting the signal that activates the dash display or gauge. It should be noted that engine oil pressure usually has to fall to dangerously low levels before programmed failure strategies are triggered (Figure 10–18).

Piezoelectric

Oil pressure can be measured using a piezoelectric sensor. Certain crystals become electrically charged when exposed to pressure and produce a small voltage. The voltage increases with pressure increase, and the signal is sent to the ECM or a voltmeter-type display gauge.

Bourdon Gauge

A flexible, coiled, bourdon tube is filled with oil under pressure. The bourdon tube will attempt to uncoil and straighten incrementally as it is subjected to pressure. This action of somewhat bending the bourdon tube rotates a gear by means of a sector and pinion. A pointer is attached at the gear across a calibrated scale and provides a means of reading it. A bourdon gauge is also known as a mechanical gauge.

Electrical

Engine oil pressure acts on a sending unit diaphragm, which in turn, moves a sliding wiper arm across a variable resistor, which incrementally grounds out a feed from the electric gauge. The gauge is a simple armature and coil assembly that receives its feed from the vehicle ignition switch.

TECH TIP: Whichever means is used to display oil pressure, none should be used as the only means of diagnosing a low oil pressure complaint. Use a good quality master gauge, usually a bourdon gauge with a fluid-filled display dial.

Oil Temperature Management

As EPA noxious emission standards become more stringent, there is an ever greater requirement to manage the combustion temperatures, and engine oil temperatures are one of the most accurate indicators of engine temperature. Whenever engine oil is used as a

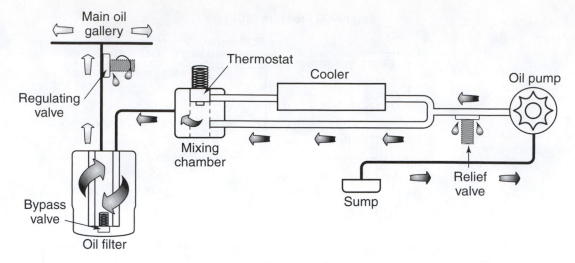

Warmup Mode

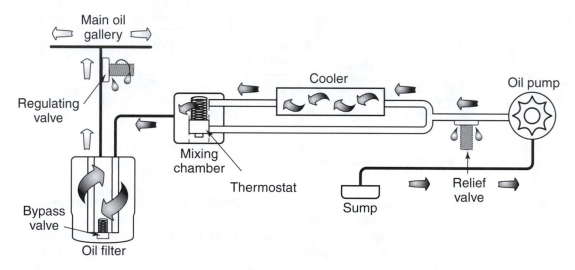

Operating Temperature

FIGURE 10–19   *Navistar temperature control circuit. (Courtesy of Navistar International Corp.,
Designer and manufacturer of international Brand diesel engines)*

hydraulic medium, its performance is to some extent dependent on its operating temperature. Figure 10–19 shows a Navistar temperature control circuit that enables the engine oil to bypass the oil cooler while the engine is warming up.

INTERPRETING OIL ANALYSES

Oil analysis is performed to determine:

- the viscosity
- the presence of coolant

- dirt contamination
- abnormal wear contamination

The engine oil temperature should be close to operating temperature. The best time to take a routine preventative maintenance (PM) sample is during an oil change. The technician taking the sample should avoid taking the sample from either the beginning or end of the drainoff: midway through the runoff is preferred. Precautions should be taken to avoid burns from the oil while obtaining the sample. Also avoid contaminating the sample after it has been taken. Label the container and complete the engine identification and data form that will accompany the sample to the laboratory. If the objective of the sample is diagnosis of a condition, the oil for the sample may be syringed from the oil pan—again the oil should be at running temperature.

When used engine oil is analyzed, the objective is to produce a report card on the rate of engine wear, the suitability of the oil change interval to the application, and perhaps head off imminent engine problems before they become major problems. Oil-testing laboratories use primarily blotter and spectrographic testing to produce reports that reference the engine OEM limit specifications. A blotter test kit is available for non-laboratory testing of engine oil. A **blotter test** is a relatively crude means of testing some of an oil's characteristics. A change in the oil's viscosity can be determined using a flow comparator. A significant increase in the engine oil's viscosity can be caused by prolonged high temperature operation, aerated lubricant, or coolant contamination. A decrease in viscosity is normally attributable to fuel contamination. It should be noted that all truck diesel engine oil is affected to some extent by fuel contamination; however, this should not be sufficient to cause any significant alteration in viscosity. Next, a measured drop of used oil is dropped on a test paper or blotter. The darkness of the sample relative to that on a code chart is used to indicate the amount of soot, dirt, and other suspended material in the sample. On some blotter test kits, the acidity can also be read on the test paper, by a change of color. Laboratories perform more comprehensive testing usually involving the use of spectrochemical analyses. **Spectrographic testing** produces more specific results and is used to identify metallic and organic contaminants in oil. A quantity of the sample is vaporized, subjected to ultraviolet radiation, and the extent to which its constituent molecules react to the radiation is used to identify and quantify them.

Oil sample analyses must be interpreted comparatively not only with the OEM maximum specifications but also with the vehicle's service application. They are most meaningful when interpreted by a fleet equipped with identical engines in units engaged in similar operating modes. The following list includes some of the elements found on a typical oil analysis report.

Al—aluminum. A constituent of many friction bearings, pistons, turbo impeller housings, Roots blowers.

B—boron. An engine coolant SCA (supplemental cooling additive) and an engine oil additive.

Cr—chromium. A common piston ring coating.

Cu—copper. A constituent of many oil cooler bundles, friction bearings, engine bushings, and thrust washers. Copper levels tend to be higher when an engine is new (100 ppm) and should drop down to lower levels (20 ppm) after 100,000 miles (160,000 km).

Fe—iron. Most of the engine components are iron-based alloys. High Fe readings could be an indication of rapid wear on almost any engine component exposed to the lube oil; engine cylinder and valve train components should be suspected first.

Pb—lead. A constituent of most friction bearings.

Si—silicon. Silicon dioxide from airborne fine sands and dusts ingested by the engine intake system often indicate malfunctioning air filter or intake ducting leaks. Silicon is a constituent of new engine oil in which it serves as an antifoam agent, so the silicon level reported by the laboratory is corrected to read only externally acquired silicon.

Na—sodium. Found in some lubricants as an additive, so the laboratory reading must be corrected to read acquired sodium. When the source is other than the engine oil, the culprit is usually coolant SCAs.

Sn—tin. A constituent of most friction bearings and occasionally used to coat pistons.

Soot—also known as total solids, soot is a form of fine carbon particle, a combustion byproduct that is not readily identifiable in routine lab testing; in fact, some laboratories report total suspended particles as soot level. Fuel soot is present in all engine oil. It is nonabrasive but may increase the oil's viscosity.

TECH TIP: Whenever a high reading in an oil analysis chart cannot be explained, contact the National Service Operations Department of the engine OEM before taking any other action. Their experience of failure feedback from across the continent may produce a simple explanation of the problem that could avoid an unnecessary engine teardown.

Guidelines for Servicing Lubricating Circuit Problems

- When investigating a low oil pressure complaint, first check the oil sump level. Next, check the appearance and odor of the oil.
- Low oil pressure complaints must be verified with a master gauge.
- High sump levels caused simply by excessive oil can aerate the lubricant, causing low pressure and fluctuations or surging.
- After an engine rebuild, it is good practice to pressure prime the engine lubrication system with an external pump.
- Most OEMs recommend priming full flow filters on all turbocharged engines. Oil filters should be primed by filling with new engine oil poured into the inlet side of the oil filter. One OEM (Caterpillar) prefers that oil filters not be primed before installation, due

TABLE 10–1   *API Classification Correlation with OEM Tests*

	CG-4	CH-4	CI-4
Fuel sulfur	Low (0.05%)	Both	Low
Soot thickening	Mack T-8	Mack T-8E	Mack T-8E
Piston deposits		Caterpillar 1K	Caterpillar 1K
Low sulfur—Al	Caterpillar 1N		Caterpillar 1N
Low sulfur—Fe		Caterpillar 1P	Caterpillar 1R
Follower wear	GM 6.5L	GM 6.5L	GM 6.5L
Slider wear		Cummins M11	Cummins M11 EGR
Bearing corrosion	L-38	Mack T-9	Mack T-10
Ring zone wear		Mack T-9	Mack T-10
Oil aeration	HEUI EOAT	HEUI EOAT	HEUI EOAT
Oxidation	Seq IIIE	Seq IIIE	Seq IIIE

Key:

HEUI = hydraulically-actuated, electronic unit injector

EOAT = engine oil aeration test

to the risk of dirt contamination caused by careless handling of the filter during priming and improper priming technique. It is not necessary to prime by-pass filters, although it may be necessary to purge air from them as the engine is started up after an oil change.

- When taking an oil sample, draw off midway through the sump runoff, never at the beginning or end. If an oil sample is required outside of the normal oil change interval, use a syringe. Oil samples should be taken shortly after an engine has been run to obtain the most reliable results.
- Oil change intervals are usually determined primarily by mileage in linehaul applications in conjunction with OEM recommendations. In fact, mileage rather than engine operating hours is used as the indicator of service intervals in most highway vehicles.

with EGR systems. Any EGR operates by dumping end gas back into the engine cylinders to "dilute" the oxygen in the air charge to the cylinder, which increases the acidity of the engine oil. CI-4 oils are designed to help keep the acidity level in the engine oil in check. Table 10–1 shows some of the differences between the current API classifications and the OEM tests they meet.

Want to know more? Try these Web sites:

http:///www.api.org
   American Petroleum Institute

http://www.dieselnet.com
   EPA diesel emissions news

http://www.lubrizol.com
   Lubrizol home page

API OIL CATEGORY COMPARISONS

So what is the difference in the three most recent API categories? Can you damage an engine by using a CG-4 oil instead of a CI-4 oil? The answers are not that easy, because the oils are formulated to meet individual OEM test criteria. However, the argument for using a CI-4 oil for any engine that specifies it is a strong one because CI-4 oil has been formulated for use in diesel engines

EXTENDED OIL CHANGES

Servicing trucks costs money. When oil change intervals can be extended, it saves money. Engine OEMs in conjunction with those who research their oil requirements are continually working on developing engine oils that will sustain longer service life before degradation, because their customers are asking for this feature. So you can expect that over the next few years, engine oils will be introduced that claim longer service life.

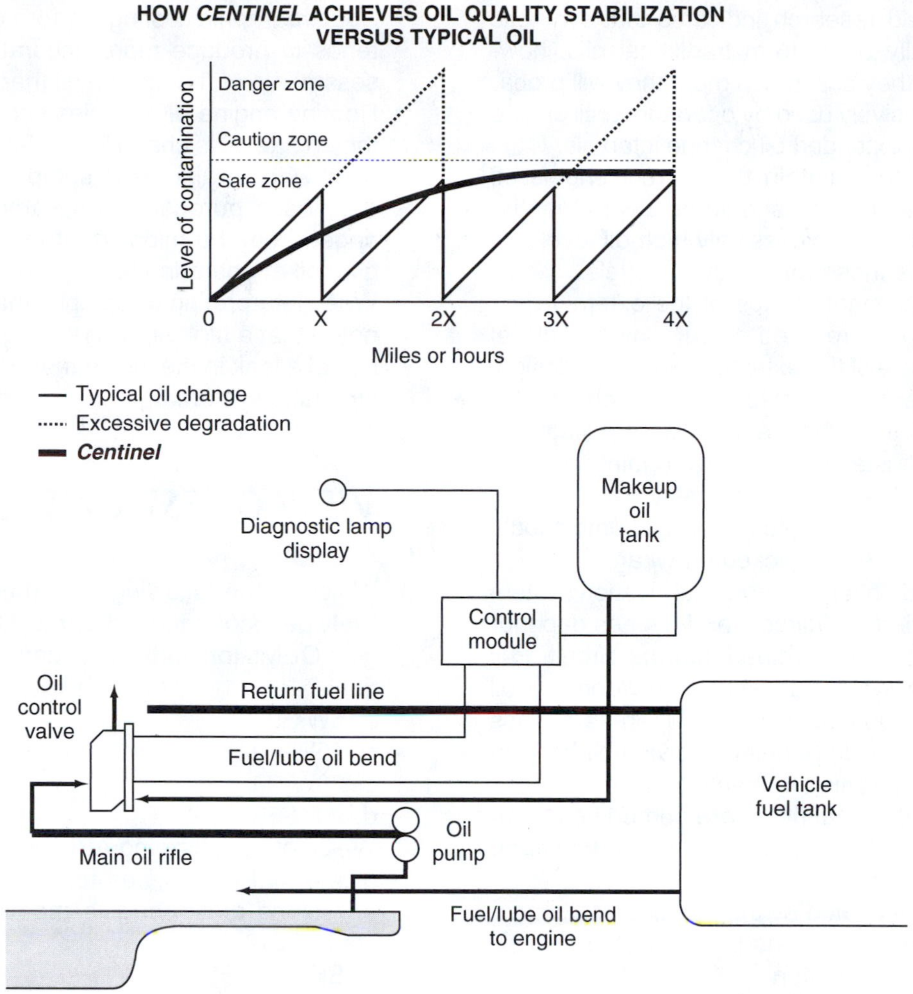

HOW *CENTINEL* ACHIEVES OIL QUALITY STABILIZATION VERSUS TYPICAL OIL

FIGURE 10–20   *The Centinel oil management system. (Courtesy of Cummins)*

Service intervals can be extended by a replenishment system such as the Cummins Centinel Oil Management System. Cummins claims that oil change intervals can be extended up to 300,000 miles using Centinel. In Centinel, the lubrication circuit is monitored by a control module. The control module can use oil from a makeup tank to replenish spent oil in the sump. The spent oil is removed from the circuit, mixed with fuel at a low concentricity, and combusted in the engine cylinders. Figure 10–20 shows the components and operation of the Centinel System.

SUMMARY

- A diesel engine lubricant performs a number of roles including those of minimizing friction, supporting hydrodynamic suspension, cooling, and cleaning.
- Viscosity describes a liquid fluid's resistance to shear.

- Lubricity describes the flow characteristics of a liquid fluid. Lubricity in engine oils is affected by temperature, with hot oils tending to flow more readily than cold oils.
- Most diesel engine OEMs recommend the use of multigrade oils over straight grades and approve the use of synthetic engine oils, especially for operation in conditions of severe cold.
- The pour point of an engine oil is the temperature at which the oil starts to gel: oils formulated for winter use have pour point depressant additives.
- Sludged oil is usually a result of oil degradation caused by prolonged low load, cold weather operation.
- Ash in used engine oils is mineral residue caused by oxide and sulfate incineration. High ash levels are indicative of high temperature operation.
- Oil aeration can be caused by high oil sump levels.
- When interpreting API oil classifications, the S prefix denotes an oil formulated for spark-ignited engines and the C denotes an oil formulated for compression-ignition engines.

- Most of the field research indicates that synthetic oils substantially outperform traditional oils; however, because they cost much more they will probably not be extensively used by operators until engine OEMs endorse extended oil change intervals.
- It is important to maintain the correct engine oil level, and the technician should be aware that the consequences of an excessively high oil level can be as severe as those for low sump levels.
- Positive displacement pumps of the external gear and gerotor types are used as oil pumps in diesel engines; most are of the external gear pump design.
- Oil pumps are designed to pump much greater volumes of oil than that required to lubricate the engine. An adjustable, pressure-regulating valve defines the peak system pressure.
- When an engine is overhauled, the oil pump should be disassembled and checked for wear.
- The filters used on a heavy-duty lubrication system may be classified as full flow and bypass depending on how they are plumbed into the circuit. Full flow filters are located in series between the oil pump and the lubrication circuit whereas bypass filters are arranged in parallel receiving oil from an oil gallery and returning it directly to the oil sump.
- Most OEMs prefer that filters are primed before installation; filters should always be primed by filling from the inlet side only.
- Filters are usually rated by their mechanical straining specification in microns but they also filter by absorption and adsorption.
- Oil coolers are heat exchangers used on most heavy-duty, highway diesel engines; the cooling medium used is engine coolant.
- The common type of oil cooler is the bundle, which consists of a housing within which coolant is pumped through a cylindrical "bundle" of copper tubes in one direction while engine oil is pumped around the tubes in the opposite direction.
- Most OEMs prefer that bundle elements be vacuum tested for leaks before assembly; pressure testing may also be used.
- Engine oil pressure measurement is, in most modern engines, performed by a variable capacitance type sensor that signals the management computer. The dash gauge or display is therefore a computer output.
- Engine oil pressures usually have to fall to very low levels before programmed electronic failure strategies are triggered.
- Laboratory testing of used engine oil samples is a relatively inexpensive method of tracking engine wear rates and providing early warning alerts to engine malfunctions. It should be adopted as part of a regular PM schedule.
- Spectrographic testing performed by laboratories tends to produce more accurate and precise assessments of the engine oil than blotter tests.
- Routine engine oil samples should be taken when the engine oil is changed, preferably at the midflow point when draining the sump. When extracted for diagnostic purposes, the sample should be syringed from the midpoint of the sump with the engine oil at operating temperature.
- When interpreting oil sample analyses, it should be noted that a high silicon reading is often an indication of a leak in the air cleaner or suction side of the air intake system.

REVIEW QUESTIONS

1. Which of the following SAE multigrade oils would likely be recommended by most highway diesel engine OEMs for North American summertime conditions?
   a. 5W-30
   b. 15W-40
   c. 20W-20
   d. 20W-50
2. Which of the following API classifications would indicate that the oil was formulated for a minimum emissions diesel engine meeting 2004 emission standards?
   a. SF
   b. CC
   c. CI
   d. CG
3. Which of the following conditions could result from a high crankcase oil level?
   a. Lube oil aeration
   b. Oil pressure gauge fluctuations
   c. Friction bearing damage
   d. All of the above
4. When investigating a low oil pressure complaint, which of the following should be performed first?
   a. A rod and main bearing rollover
   b. Installing a master gauge to verify the complaints
   c. Testing the oil for fuel contamination
   d. Checking the crankcase oil level
5. A full flow oil filter has become completely plugged. What is the likely outcome?
   a. A bypass valve diverts the oil around the filter.
   b. Engine lubrication ceases.
   c. Oil pump hydraulically locks.
   d. Engine seizure
6. Which of the following statements would usually correctly describe the relative temperatures of the

coolant and lubricant when the engine is at operating temperature?

a. Coolant temperatures are higher than engine oil temperatures.

b. Engine oil temperatures are higher than coolant temperatures.

c. Coolant and engine oil temperatures should be equal.

7. Which type of oil pump is most commonly used in current highway diesel engines?

a. External gear

b. Plunger

c. Centrifugal

d. Vane

8. Which of the following is usually the preferred OEM method of testing an oil cooler bundle?

a. Pressure test using shop air and a bucket of water

b. Vacuum test

c. Die penetrant test

9. When interpreting a used engine oil analysis profile, which of the following conditions would be most likely to cause high silicon levels?

a. Oil cooler bundle disintegration

b. Plugged air cleaner

c. Turbocharger bearing failure

d. Air cleaner perforation

10. Which of the following correctly describes crude petroleum?

a. Vegetable-based oil

b. Fossil fuel

c. Asphalt

11. Engine oil service classifications are standardized by which of the following organizations?

a. SAE

b. API

c. ASTM

d. EPA

12. _Technician A_ states that the operating temperature of an engine oil should be within 5°F of that of the engine coolant operating temperature. _Technician B_ states that engine oil temperatures are probably a more accurate means of assessing actual engine operating temperature than the coolant temperature reading. Who is right?

a. A only

b. B only

c. Both A and B

d. Neither A nor B

13. An auxiliary oil pump designed to feed oil to the lubrication circuit on a vehicle that operates at extreme working angles is known as a(n):

a. Gerotor pump

b. Scavenging pump

c. Vane pump

d. Emergency pump

14. Oil that has a milky, clouded appearance is probably contaminated with:

a. Fuel

b. Engine coolant

c. Dust

d. Air

15. Which of the following statements correctly describes viscosity?

a. Resistance to flow

b. Resistance to shear

c. Lubricity

d. Breakdown resistance

16. _Technician A_ states that an oil cooler operates by having coolant pumped through the bundle tubes and oil circulated around the tubes. _Technician B_ states that a leaking oil cooler core must always be replaced. Who is right?

a. A only

b. B only

c. Both A and B

d. Neither A nor B

17. _Technician A_ states that 15W-40 is the most commonly used truck engine oil. _Technician B_ states that 15W-40 oil is not an appropriate all-season oil, because winter temperatures are often much lower than the SAE recommended ambient temperature window for this oil. Who is right?

a. A only

b. B only

c. Both A and B

d. Neither A nor B

18. _Technician A_ states that the device used to signal oil pressure values to the ECM on most electronically managed engines uses a variable capacitance principle. _Technician B_ states that oil sump level can be signalled to the ECM using a thermistor. Who is right?

a. A only

b. B only

c. Both A and B

d. Neither A nor B

Chapter

11

Engine Cooling Systems

PREREQUISITES

Chapters 4, 7, 8, and 9

OBJECTIVES

After studying this chapter, you should be able to:

- Describe the cooling system components and their principle of operation.
- Define the terms conduction, convection, and radiation.
- Identify the three types of coolant used in current highway diesel engines and the relative merits and disadvantages of each.
- Outline the properties of a heavy-duty antifreeze and supplemental cooling additive package.
- Calculate the boil and freeze points of a coolant mixture.
- Mix coolant using the correct proportions of water, antifreeze, and SCAs.
- Outline the causes of wet liner cavitation and the steps required to minimize it.
- Recognize the degree to which coolant system scaling can insulate and outline the steps required to eliminate it.
- List the performance and economic advantages claimed for extended life coolants.
- Describe the requirements of a heavy-duty radiator.
- Identify the types of heavy-duty radiators including downflow, crossflow, and counterflow.
- Test a radiator for external leakage using a standard cooling system pressure tester.
- Describe the process required to repair radiators.
- Test the operation of a radiator cap.
- List the different types of thermostats in use and describe their principle of operation.
- Describe the role of the coolant pump.
- Describe the process required to recondition a coolant pump.
- Define the role of the coolant filters and their servicing requirements.
- List the types of temperature gauges used in highway diesel engines.
- Describe how a coolant level warning indicator operates.
- Define the roles played by the shutters and engine fan in managing engine temperatures.
- Diagnose basic cooling system malfunctions.

KEY TERMS

antifreeze	fanstat	refractometer
cavitation	headers	rejected heat
conduction	heat exchanger	shutterstat
convection	hydrometer	single pass
counterflow radiator	kinetic energy	supplemental cooling system additives (SCA)
crossflow radiator	pH	
diesel coolant additive (DCA)	propylene glycol (PG)	thermatic fan
double pass	radiation	thermostat
downflow radiator	radiator	total dissolved solids (TDS)
ethylene glycol (EG)	ram air	
extended life coolant (ELC)	refraction	

INTRODUCTION

Cooling systems dissipate a percentage of engine **rejected heat**. Rejected heat is that percentage of the potential heat energy of a fuel that the engine is unable to convert into useful **kinetic energy** (the energy of motion) and therefore must dissipate to atmosphere, either in the exhaust gas or indirectly using the engine cooling system. If an engine is operating at 40% thermal efficiency, 60% of the potential heat energy is rejected. Approximately half of the rejected heat is discharged in the exhaust gas, which leaves the engine cooling system responsible for dissipating the other half to atmosphere. This task is complicated by the extremes of the North American climate because it is necessary to manage a consistent engine-operating temperature at all engine speeds and loads to ensure optimum performance and minimum emissions. Liquid cooling systems are universal in North American truck and bus applications, and only they will be addressed in this section. The Deutz engine company of Germany manufactures air-cooled engines in the 200 to 500 BHP power range, but in North America, their engines are generally only found in agricultural and off-highway applications.

The functions of a diesel engine liquid cooling system are to:

- Absorb heat from engine components.
- Transfer the absorbed heat by circulating the coolant.
- Dissipate the heat to atmosphere by means of heat exchangers.
- Manage engine operating temperatures.

Combustion heat can be dissipated by the cooling system to atmosphere in three ways:

1. **Conduction:** the transfer of heat through solid matter, such as the transfer of heat through the cast-iron material of a cylinder block.
2. **Convection:** the transfer of heat by currents of gas or liquids, such as in the movement of ambient air through an engine compartment.
3. **Radiation:** transfer of heat by means of heat rays not requiring matter, such as a fluid or solid. The turbine housing of a turbocharger radiates a considerable amount of heat.

Cooling systems are both sealed and maintained under pressure. By confining a liquid under pressure, its boil point is increased. Most cooling systems are designed to manage coolant temperatures at just below their boil points. The chemistry of the **antifreeze** and its concentration in the coolant will define the actual boil point of a coolant. Most antifreezes double as antiboil agents. When an engine is approaching an overheat condition, the coolant will first boil at the location within the system where the pressure is lowest. In most cases, boiling will occur first at the inlet (suction side) of the system water pump. Figure 11–1, Figure 11–2, and Figure 11–3 illustrate some typical diesel engine cooling circuits.

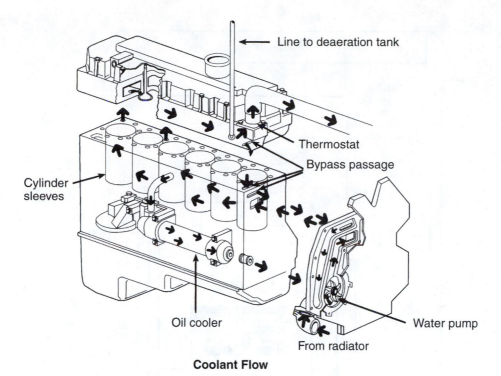

Coolant Flow

FIGURE 11–1   Coolant flow through a typical in-line, six-cylinder engine. (Courtesy of Navistar International Corp., Designer and manufacturer of International Brand diesel engines)

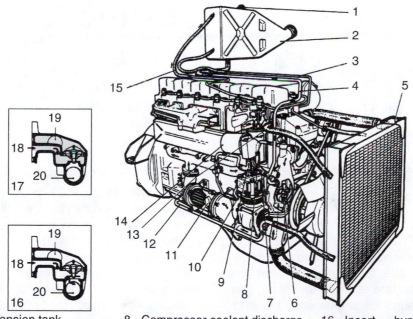

1. Coolant expansion tank pressure cap
2. Coolant expansion tank
3. Radiator gas relief hose
4. Feed line from coolant expansion tank water pump (suction side)
5. Radiator
6. Water pump impeller
7. Water conditioner discharge line

8. Compressor coolant discharge line
9. Coolant conditioner supply
10. Coolant feed to air compressor cylinder head
11. Oil cooler
12. Heater core supply line
13. Heater core discharge line
14. Coolant conditioner
15. Cylinder head gas relief line

16. Insert — bypass circuit (thermostats closed)
17. Insert — radiator circuit (thermostats open)
18. Thermostat housing inlet
19. Thermostat housing outlet to radiator inlet
20. Thermostat housing outlet to water pump inlet

FIGURE 11–2   Cooling system components. (Courtesy of Mack Trucks)

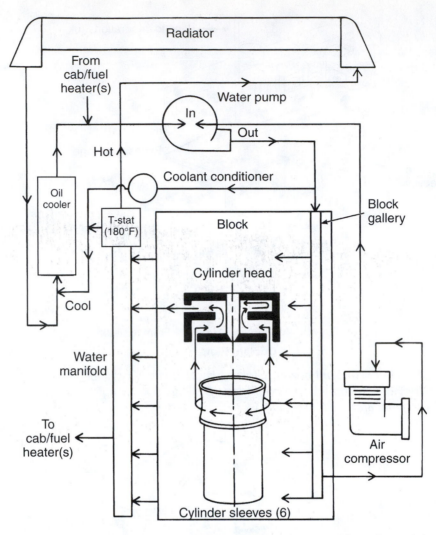

FIGURE 11–3 *Cooling system flow schematic. (Courtesy of Mack Trucks)*

ENGINE COOLANT

Water-based coolant is the medium used to absorb engine rejected heat, transfer that heat to a **heat exchanger**, and then dissipate it to atmosphere. The coolant is circulated through the engine water jacket to absorb the heat of combustion in the manner shown in Figure 11–3. Engine coolant is a mixture of water, antifreeze, and supplemental cooling additives (SCA). If the only objective of diesel engine coolant was to act as a medium to transfer heat, pure water would accomplish this more efficiently than any currently used antifreeze mixture. However, water possesses inconvenient boil and freeze points, poor lubricating properties, and promotes oxidation and scaling activity.

Almost all current truck and bus engines use a mixture of **ethylene glycol (EG), propylene glycol (PG)**, or carboxylate type extended life antifreeze, plus water as coolant; alcohol-based solutions are no longer used as they evaporate at low temperatures. A properly formulated diesel engine coolant is always made up of the correct proportions of water, antifreeze, and supplemental coolant additives or SCAs. When the antifreeze properties of the coolant are EG- or PG-based, the SCAs require monitoring and routine replenishing. Extended life coolant (ELC) is low maintenance in that the coolant life is 6 years, with only one SCA charge required in that period.

Water expands about 9% in volume as it freezes and it can distort or fracture containers it is housed in even when this container is a cast-iron engine block. Water occupies the least volume when it is in the liquid state and close to its freezing point of 0°C or 32°F. As water is heated from a near freezing point to a near boiling point it expands approximately 3%; a 50/50 mixture of water and ethylene glycol will expand even more, approximately 4% through the same range. Consequently, cooling systems must be designed to accommodate the expansion and contraction of the cooling medium while it is in the liquid state. Most importantly, antifreeze is also antiboil.

The mixture of water, antifreeze, and SCAs that is referred to as engine coolant should perform the following:

1. Corrosion protection. Corrosion inhibitors in both the antifreeze and the SCA package protect the metals, plastics, and rubber compounds in the engine cooling system.
2. Freeze protection. The degree of freeze protection of the coolant is directly related to the proportion of antifreeze in the mixture.
3. Antiboil protection. The degree of antiboil protection of the coolant is again directly related to the proportion of antifreeze in the mixture.
4. Antiscale protection. The antifreeze should contain antiscale additives that prevent hard water mineral deposits from adhering to the cooling system heat transfer surfaces.
5. Acidity protection. A pH buffer is used to inhibit the formation of acids in the coolant, which would result in corrosion.
6. Antifoam protection. Prevents aeration of the coolant that could be caused by the action of pumping and flowing it through the cooling circuit.
7. Antidispersant protection. This prevents insoluble matter from coagulating and plugging cooling system passages.

Except for the antifreeze and antiboil characteristics of glycol-based coolant mixture, the remaining protection additives deplete with engine operation, and they must be evaluated and restored at appropriate maintenance intervals. Both propylene and ethylene glycol are petrochemical products. EG has been used as the standard antifreeze for some time, but the Federal Clean Air Act and OSHA have both come to regard EG as a toxic hazard. PG in its virgin state is said to be less toxic than EG, and for that reason it is gaining acceptance as a base antifreeze ingredient. However, leaks and spillage of both EG- and PG-based coolants should be regarded as dangerous to mammals (including humans) and plant life. The chemical characteristics of engine coolant are altered in use and the toxicity increases.

Ethylene glycol is derived from ethylene oxide, which in turn is produced from ethylene—a basic petroleum fraction. Propylene glycol is derived from propylene oxide, which is produced from propylene, another basic petroleum fraction. When mixed in a solution with water, both PG and EG are described as aqueous, meaning they mix pretty well with it. Extended life coolants are claimed to have lower toxicity but OSHA has insufficient data to rule on PG and ELC. The following chart compares the freeze points of PG and EG:

CONCENTRATION OF ANTIFREEZE BY PERCENT VOLUME	FREEZE POINT OF COOLANT			
	EG		PG	
0 (water only)	32°F	0°C	32°F	0°C
20	16°F	–0°C	19°F	–7°C
30	4°F	–16°C	10°F	–12°C
40	–12°F	–24°C	–6°F	–21°C
50	–34°F	–37°C	–27°F	–33°C
60	–62°F	–52°C	–56°F	–49°C
80	57°F	–49°C	–71°F	–57°C
100	–5°F	–22°C	–76°F	–60°C

Propylene and ethylene glycol based coolants should *never* be mixed. The mixture in itself will not cause any engine or cooling system problems, but it will be impossible to determine the antifreeze mixture strength with either a refractometer or a **hydrometer**. If a mixture of EG and PG is known to have taken place and the coolant, for whatever reasons, cannot be immediately replaced, use a refractometer with an EG and a PG scale and average the two readings. However, the cooling system should be drained and refilled with either aqueous PG or EG when practical to avoid problems later on. ELC is only sold premixed and is dyed a red color. Only ELC premix should be added to the cooling system, or in conditions of extreme cold, ELC concentrate. ELC is incompatible with PG and EG.

MEASURING COOLANT MIXTURE STRENGTH

Standard antifreeze hydrometers are calibrated for measuring ethylene glycol mixtures and even when measuring these, tend to be inaccurate and require calculation to temperature correct the reading. The Truck Maintenance Council (TMC) and most OEMs recommend the use of a **refractometer** to test the antifreeze strength of a coolant. A refractometer should have an accuracy variable within 7°F of the actual freeze point of the coolant throughout the range of temperatures found in North America. A refractometer designed for measuring truck diesel engine coolant will have a PG and EG calibration scale. Ensure that the correct scale is used for the antifreeze being tested. A refractometer measures **refraction** in a liquid: the refractive index of antifreeze increases with an increase in concentration. Refractometer readings are only valid in the context of a specific type of antifreeze.

SUPPLEMENTAL COOLANT ADDITIVES

Supplemental coolant additives (SCA) are a critical ingredient of the coolant mixture. The actual SCA package recommended by an engine manufacturer will depend on whether wet or dry cylinder liners are used, the materials used in the cooling system components, and the fluid dynamics (high flow/low flow) within the cooling system. The operator may wish to adjust the SCA package to suit a specific operating environment or set of conditions. Abnormally hard water, for instance, requires a greater degree of antiscale protection. Depending on the manufacturer, SCA may be added to a cooling system in a number of ways. The practice of installing the SCA in the system coolant filter is less commonly used today because it resulted in generally higher levels of additives than required, and an excess of additives can create problems. Most OEMs suggest testing the SCA levels in the coolant, followed by adding SCA, to adjust to the required values. Never dump unmeasured quantities of SCAs into the cooling system at each PM. **Diesel coolant additive (DCA)** is one manufacturer's brand name for its additive solution.

Cavitation

SCAs are especially important to controlling cavitation on engines that use wet liners. **Cavitation** is caused by vapor bubble collapse, which results in pitting of the exterior of the liner wall across the thrust faces of the piston. When the liner is subjected to the gas pressure of combustion, it first expands then contracts, forming a low pressure vapor bubble. The vapor bubble almost immediately collapses (implodes), resulting in high velocity coolant impacting on the liner wall. This has been proved to produce surface pressures up to 60,000 psi, which can cause pitting and erosion. The characteristic will normally appear on the exterior of the liner wall across the thrust faces of the piston. This liner pitting can be controlled by adding molybdates and nitrite to coolant, which act to form an invisible but very tough, protective oxide film on the liner wall. The vapor bubbles will still form, but the liner is shielded by the oxide film. The bad news is that this film breaks down as the coolant ages.

Scaling

The Cummins engine company reported a number of years ago that the scale buildup of $\frac{1}{16}''$ (1.5 mm) had the equivalent insulating effect of 4″ (100 mm) of cast iron. Scaling is caused by hard water mineral deposits (especially magnesium and calcium) adhering to the surfaces of the cooling system where temperatures are highest. Scale formation on wet liners is an especially serious condition because it seldom forms in an even layer, and as it insulates, that is, inhibits the liner's ability to transfer heat, the resulting high temperatures may cause buckling and other distortions to the liner. Left unchecked, scale buildup insulates engine components designed to transfer heat, resulting in overheating and subsequent failure (Figure 11–4).

There are commercially available de-scalants that may work to remove minor scale buildup followed by a cooling system flush, but most often, by the time scale buildup has progressed to the point that it is causing an engine to overheat, the engine must be disassembled and the cylinder block and heads boiled in a soak tank.

TESTING SCA LEVELS

Generally OEMs recommend that the coolant SCA level be tested at each oil change interval. Additionally, whenever there is a substantial loss of coolant and the system has to be replenished, the SCA level should be tested. Each test provided by an OEM is designed to monitor the SCA package required for their product and cannot generally be used for other OEMs' product. Also, the test kits usually consist of test strips that must be stored in air-tight containers and that have expiration dates that should be observed. Coolant test kits permit the technician to test for the appropriate SCA concentration, the pH level, and the total dissolved solids (TDS). The **pH** level determines the relative acidity or alkalinity of the coolant. Acids may form in engine coolant exposed to combustion gases, or in some cases, when cooling system metals (ferrous and copper based) degrade. The pH test is a litmus test in which a test strip is first inserted into a sample of the coolant, then removed, and the color of the test strip is indexed to a color chart provided with the kit. The optimum pH window is defined by each OEM, but normally falls between 7.5 and 11.0 on the pH scale. Higher acidity readings (below 7.5 on the pH scale) in tested coolant are indications of corrosion of ferrous and copper metals, coolant exposure to combustion gases, and in some cases, coolant degradation. Higher-than-normal alkalinity readings indicate aluminum corrosion and possibly that a low silicate antifreeze is being used where a high silicate antifreeze is required.

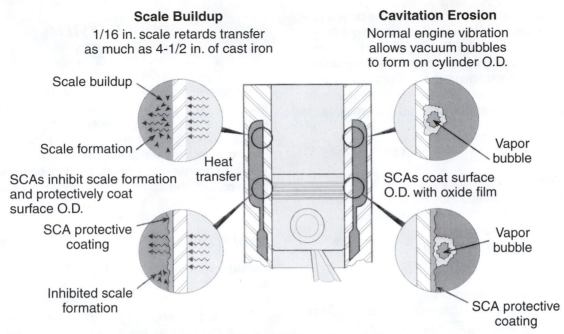

Scale Buildup
1/16 in. scale retards transfer
as much as 4-1/2 in. of cast iron

Scale buildup

Scale formation

SCAs inhibit scale formation
and protectively coat
surface O.D.

SCA protective
coating

Inhibited scale
formation

Heat
transfer

Cavitation Erosion
Normal engine vibration
allows vacuum bubbles
to form on cylinder O.D.

Vapor
bubble

SCAs coat surface
O.D. with oxide film

Vapor
bubble

SCA protective
coating

Supplemental Coolant Additives Protect Engine Surfaces

FIGURE 11–4   How SCAs help scaling and cavitation problems in wet liners. (Courtesy of Navistar International Corp., Designer and manufacturer of International Brand diesel engines)

Testing for **total dissolved solids (TDS)** requires using a TDS probe, which essentially measures the conductivity of the coolant by conducting a current between two electrodes. It should be noted that distilled water does *not* conduct electricity; the ability of water to conduct electricity increases with its TDS content. The TDS test is performed by inserting the probe into the top radiator tank. A reading higher than the OEM specified TDS measured in parts per million is an indication the condition of the coolant may be conducive to scale buildup.

BLENDING HEAVY-DUTY COOLANT

Ideally, coolant for use in heavy-duty diesel engines should not be blended in the engine cooling system, but in a container before pouring or pumping it into the engine. Good quality water (known to be not excessively hard and to have no iron content; it should not appear reddish) should always be poured into the container first, followed by the correct proportions of antifreeze and finally the SCA package. Mix the solution before adding to the engine cooling system. When adding coolant to compensate for a low coolant level, some OEMs manufacture premixed, heavy-duty coolant. This shifts the responsibility of mixing coolant from the technician because the water, antifreeze, and SCA are pre-

mixed in the correct proportions, it also eliminates the problems associated with poor quality water.

High Silicate Antifreeze

High silicate concentrations are required to protect aluminum components exposed to the engine coolant. However, many OEMs require that low silicate coolant be used in their engines. High silicate and low silicate antifreeze should not be mixed. Generally, high silicate antifreeze should not be used except when required by the engine OEM. ELCs do not use silicates, nitrates, borates, phosphates, and amines to inhibit scaling, but instead use a carboxylate base that, at least according to the manufacturer, significantly outperforms the complex chemical brew required of an EG or PG coolant.

EXTENDED LIFE COOLANTS

Extended life coolants (ELC) use an EG base and promise a service life of up to 600,000 miles (960,000 km) or six years with one additive recharge at 300,000 miles (480,000 km) or three years. This compares with a typical service life of two years during which up to 20 recharges of SCA would be required for conventional EGs and PGs. ELCs are so far only available as premixed solution to ensure that the water quality is at the required level. The pricing of ELCs by quantity is generally comparable with EG and PG, but because of reduced cooling system maintenance and extended service life, is fast becoming the coolant of choice of the engine OEMs. No test kits are required to monitor the ELC SCA level. Notable among the advantages claimed for ELCs are:

- Greatly extended service life—six years or 600,000 miles
- No inhibitor testing required
- Improved water pump longevity due to much lower TDS content (TDS are often abrasive)
- Reduced hard water scaling
- Improved cavitation protection
- Improved corrosion protection
- Improved heat transfer ability
- No gelling problems: no silicates are used in ELC, which eliminates the problem of silicate dropout responsible for sludging EG and PG
- Improved aluminum corrosion inhibitors
- Better high temperature performance than EGs and PGs

ELCs may be used in engine cooling systems that have previously used either EG or PG solutions. The EG and PG should be drained from the cooling system, which should then be flushed with clean water. The ELC can then be installed. If a coolant filter is used, replace the existing filter with an SCA-free filter. For mid-winter Northern United States and Canadian operation, the ELC premix should be strengthened with concentrate, check with the ELC manufacturer. The Caterpillar Engine Company was the first to endorse the use of ELCs in its products, and most of the other engine OEMs have now followed.

TECH TIP: Always use a refractometer to test the freeze protection of coolant solution and ensure that the correct scale is used for the type of antifreeze being tested.

WARNING: When running a truck in severe winter conditions, the thermostats will close cycle coolant through the engine and in the radiator, which is exposed to frigid ram air, icing can occur if there is not adequate freeze protection. Ensure that coolant freeze protection accounts for the lowest ambient temperature with a margin of at least −10°F.

CAUTION: After coming into contact with any antifreeze or coolant solution, wash the affected skin areas immediately and thoroughly.

COOLING SYSTEM COMPONENTS

The components used to store, pump, condition, and manage engine coolant flow and temperature are known as the cooling system components (Figure 11–5). These components vary little from one diesel engine manufacturer to the next. However, when servicing and repairing cooling system components, always consult the appropriate service literature.

RADIATORS

Radiators are heat exchangers. The engine power rating usually dictates the frontal area of a radiator, typically 3 to $4^{2}''$ (squared) per BHP unit. The cooling medium is **ram air**, that is, ambient air forced through the radiator core as the truck is driven down the highway. Vehicle speed and ambient temperatures obviously determine the radiator's efficiency as a heat exchanger. Fan shrouds improve air flow through the radiator core and the efficiency of the fan.

Radiator Materials and Construction

Most truck diesel engine radiators are currently fabricated mainly from copper and brass components, but the use of aluminum and plastics is increasing. Radiators typically consist of bundled rows of round or elliptical tubes through which the coolant is flowed and to which are connected fins that increase the sectional area to which the ram air is exposed. The tubes are usually brass and the fins are copper in truck applications but OEMs are experimenting with aluminum, widely used in automobile radiator construction, due to lower cost and lighter weight. Copper, brass (an alloy of copper and zinc), and aluminum all have high coefficients of heat transfer and are ideal as base material for radiators. Aluminum is more susceptible to corrosion than copper and brass, both from within (coolant breakdown, poor water quality) and outside (salt, both ambient and road salt). Plastics are being increasingly used in the construction of radiator tanks replacing metal tanks; plastic tanks are usually crimped onto the main radiator core, enclosing the **headers**.

All radiators are equipped with a drain valve located at the lowest point in the assembly, inlet and outlet piping, and a filler opening sealed with a radiator cap. Most use a **single pass**, downflow principle, which requires that the cooling tubes run vertically from the top tank to the bottom tank. Radiators are classified by their flow characteristics. The following types may be found in current truck applications:

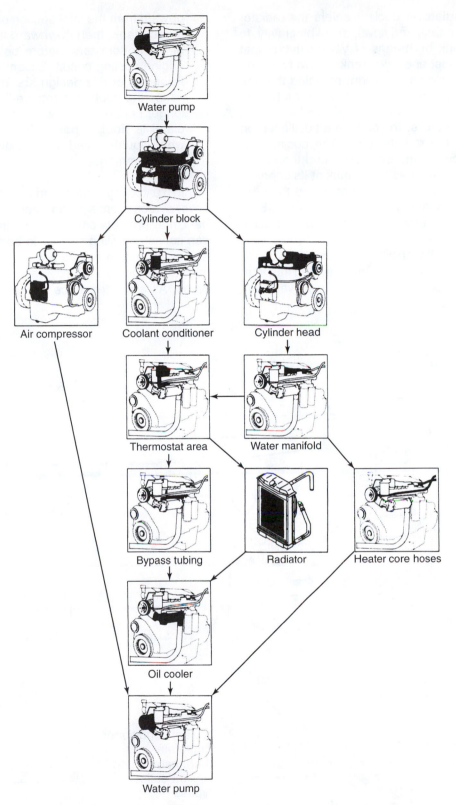

FIGURE 11–5 *Cooling system components. (Courtesy of Mack Trucks)*

- **Downflow radiators:** coolant enters the radiator through the top tank and flows, aided by gravity, to the bottom tank by means of vertical tubes that connect the upper and lower tanks. Downflow radiators are a single pass design, meaning that the coolant is routed from the top tank to the bottom and then exits (Figure 11–6).
- **Crossflow radiators:** the tanks are positioned on either side of the radiator core. The coolant enters through one of the side tanks and then flows through horizontal tubes to the tank at its opposite side. This design affords a lower profile than the downflow design, and it is used by chassis OEMs using low-nose, aerodynamic designs of trucks. Flow is single pass.
- **Counterflow radiators:** the coolant usually enters through a bottom tank that is divided into inlet and outlet sections. The coolant flows vertically upward from the inlet section of the bottom tank to the top tank, then downward to the outlet section of the bottom tank before being returned to the engine cooling circuit. Essentially, the cooling efficiencies of this design are improved mainly because the coolant is retained in the radiator for a longer period. The flow of coolant through the radiator is **double pass**. The design was popular when liquid-cooled, charge-air heat exchangers were common.

Air in a cooling system can cause many problems. It severely compromises the coolant's ability to transfer heat, may promote corrosion, and in extreme instances can cause a cooling system to become air bound. An air-bound system occurs when air is entrapped in the inlet to the coolant pump, and it effectively loses its prime. As a consequence, most cooling systems are

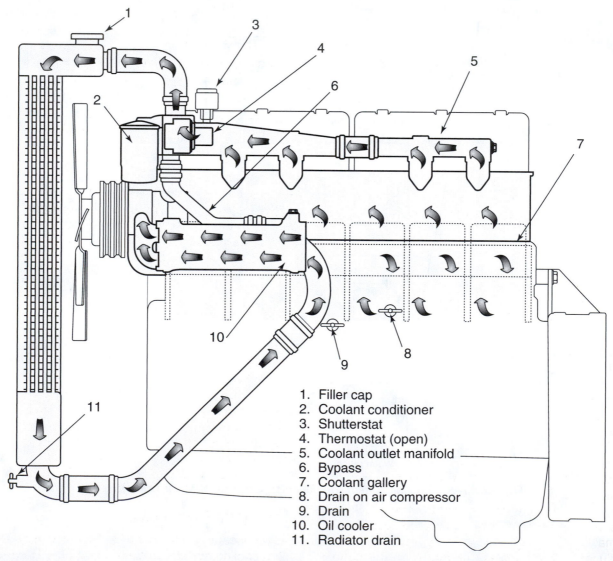

1. Filler cap
2. Coolant conditioner
3. Shutterstat
4. Thermostat (open)
5. Coolant outlet manifold
6. Bypass
7. Coolant gallery
8. Drain on air compressor
9. Drain
10. Oil cooler
11. Radiator drain

FIGURE 11–6 *Cooling system flow with a downflow radiator. (Courtesy of Mack Trucks)*

designed to limit aeration problems. Some radiators have a divided top tank in which coolant enters the lower section from the cylinder head and the upper section from a standpipe; the two sections are separated by a baffle. This design tends to reduce cooling system aeration problems. Also, vent lines help to de-aerate the cooling system. The radiator top tank stores reserve coolant volume and accommodates the thermal expansion of coolant.

Servicing and Testing Radiators

Radiators in highway trucks tend to be left in the chassis with no maintenance until they fail, either because they leak or fail to adequately cool. An often overlooked PM practice is the external cleanliness of the radiator; buildup of road dirt and summer bugs can severely compromise the radiator's ability to effectively cool. Radiators should be cleaned externally using either a low-pressure steamer or regular hose, detergent, and a soft nylon bristle brush. Never use a high-pressure washer because this will almost certainly result in damaging the cooling fins.

Leaks are more often the result of external damage than corrosion failure: they are indicated by white or reddish streaks at the location of the leak. Radiators are commonly pressure tested at around 10% above normal operating pressure, but ensure that the OEM test specifications are consulted, especially where plastic tanks are used. Pressure testing may help identify the locations of leaks. It is important that radiator leaks are promptly repaired. If the leak has been caused by external damage and the radiator is in otherwise good condition, the radiator may be repaired by shorting out the affected tubes. This usually involves removing the top and bottom tanks and plugging the damaged tube(s) at the headers. If a leaking tube is accessible, a soldering repair may be possible. Soldering radiators is at best a risky business and before beginning, assess how the heat will affect any nearby soldered joints. Low melt point solder has little structural strength and although it may be used to seal a hairline crack or small impact leak, it should not be used otherwise. Silver solder is a preferred solder repair medium, but more heat is required to apply it than 50/50 lead-tin solder.

Radiator Flushing and Major Repairs

Most commercially available in-chassis radiator descaling solutions are a poor risk as they are seldom 100% effective and may dislodge scale, which will subsequently plug up elsewhere in the cooling circuit. For the same reasons, reverse flow flushing of the cooling system makes little sense. When OEMs recommend radiator flushing, it is generally performed in the normal direction of flow often aided by a cleaning solution.

Major radiator repairs should be referred to a radiator specialty shop. A scaled radiator falls into the classification of major repair. An ultrasonic bath will remove most scale rapidly and effectively.

A properly equipped radiator repair shop will also be able to determine the extent of repairs required and whether recoring is necessary. One of the problems of performing radiator repairs without the proper test equipment is the inability to test the radiator until it is assembled and reinstalled in the chassis. Removing and installing radiators can be a labor intensive operation in some chassis.

RADIATOR CAP

Radiators are usually equipped with a pressure cap whose function it is to maintain a fixed operating pressure while the engine is running. This cap is additionally equipped with a vacuum valve to admit surge tank coolant (or air) into the cooling circuit (the upper radiator tank) when the engine is shut down to accommodate coolant thermal contraction. Radiator caps permit pressurization of a sealed cooling system. For each 1 psi (7 KPa) above atmospheric pressure, coolant boil point is raised by 3°F (1.67°C) at sea level; for every 1,000 feet of elevation, the boil point decreases by 1.25°F (0.5°C). System pressures are seldom designed to exceed 25 psi (172 KPa) and more typically they range between 7 psi (50 KPa) and 15 psi (100 KPa).

Radiator caps are identified by the pressure required to overcome the cap spring pressure and unseat the seal: when this occurs, the coolant is routed to a surge tank. The surge tank coolant level is always at its highest when the engine is running hottest. As the engine cools, the pressure within the cooling system drops, and when it falls to a "vacuum" value of ±¼ psi, the radiator cap vacuum valve is unseated, which allows coolant from the surge tank to be pulled back into the radiator.

CAUTION: Great care should be exercised when removing a radiator cap from the radiator: if the system is pressurized, hot coolant may escape from the filler neck with great force. Most filler necks are fitted with double cap lock stops to prevent the radiator cap from being removed in a single counterclockwise motion: if the radiator is still pressurized, the cap will jam on the intermediate stops. Never attempt to remove a radiator cap until the cooling system pressure is equalized.

Testing Radiator Caps

Radiator caps can be performance tested using a standard cooling system pressure testing kit. The radiator cap should first be installed to an appropriately sized adapter on the hand pump, then pump to the seal crack value (this should exceed the cap rated value by 1 psi [7 KPa]). Next, release the pressure and once again recharge to the exact rated pressure value of the cap and observe the pump gauge: pressure drop-off should not exceed 2 psi over 60 seconds.

WATER PUMPS/COOLANT PUMPS

Water pumps are usually nonpositive, centrifugal pumps driven directly by a gear or by belts. When the engine rotates the coolant pump, an impeller is driven within the housing, creating low pressure at its inlet, usually located at or close to the center of the impeller. The impeller vanes throw the coolant outward and centrifugal force accelerates it into the spiralled pump housing and out toward the pump outlet. Because the cooling system pressure at the inlet to the coolant pump is at its lowest, boiling always occurs at this location first. This will very rapidly accelerate the overheating condition as the pump impeller will be acting on vapor. Coolant pumps are the main reason that engine coolants should have some lubricating properties, because they are vulnerable to abrasion damage when the coolant TDS levels are high (Figure 11–7, Figure 11–8, and Figure 11–9).

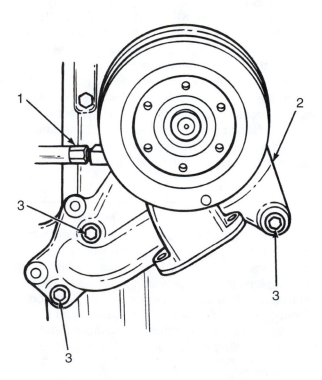

1. Air compressor line   3. Capscrew
2. Water pump

FIGURE 11–7   Exterior view of a water pump and drive pulley. (Courtesy of Mack Trucks)

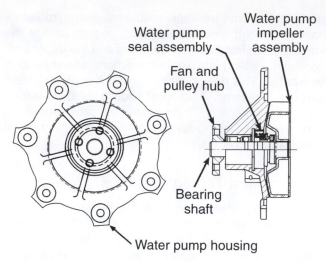

Water Pump Sub Assembly

FIGURE 11–8   Sectional view of a water pump and housing. (Courtesy of Navistar International Corp., Designer and manufacturerr of International Brand diesel engines)

Coolant pumps fail for the following reasons:

- Overloading of the bearings and seals caused by misalignment or tight drive belts
- High TDS levels in the coolant that erode the impeller
- Mineral scale buildup on the pump housing
- Overheating—boiling usually occurs first at the inlet to the water pump, so a system that is not properly sealed, or hot shut downs, can cause vapor lock.

Inspecting, Replacing, Rebuilding Coolant Pumps

A defective coolant pump should first be removed from the engine and analyzed for the cause of the failure to avoid a repetition. Although they were commonly rebuilt by the technician a generation ago, this is seldom the case now. When defective, they are replaced as a unit with a rebuilt/exchange unit. Rebuilding of water pumps is usually performed at a rebuild center equipped with the proper equipment by persons who specialize in the process. Although the technician who reconditions one or two coolant pumps a year may not be able to compete with the time of the specialist rebuilder, it is certainly possible to perform the work to the same standard. A slide hammer is usually required to remove the pulley from the impeller drive shaft, and an arbor press is generally preferred over power presses both for disassembly and reassembly. The pump is one of the more simple subcomponents of the engine and essentially consists of a housing, impeller, impeller shaft, bearings, and seals. When a pump is rebuilt, inspect

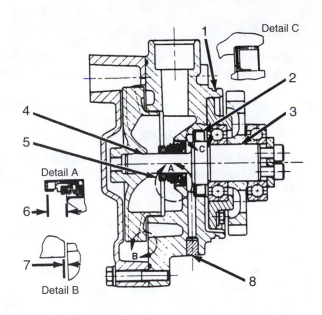

(1) Seal groove. Prior to installation lubricate bore in engine front cover, where seal groove and seal will slide, with engine oil.

(2) Lip type seal. Install as shown.

(3) Shaft diameter on drive gear end ..... 29.987 ± 0.008 mm (1.1806 ± .0003 in)

(4) Shaft diameter on impeller end .... 15.912 ± 0.006 mm (.6265 ± .0002 in)

(5) Seal assembly.

(6) Distance from surface in pump housing to top of seal assembly must be ...... 12.83 ± 0.13 mm (.505 ± .005 in)

(7) Clearance between impeller and pump housing .... 1.50 ± 0.50 mm (.059 ± .020 in)

(8) Filter. Install filter flush with pump housing.

FIGURE 11–9 _Sectional view of a water pump and subcomponents. (Courtesy of Caterpillar)_

the components thoroughly; in many cases, especially where plastic impellers are used, only the housing and shaft are reused. Examine the shaft seal contact surfaces for wear. When gear-driven coolant pumps are reconditioned, pay special attention to the drive gear teeth. The OEM instructions should be observed, and where ceramic seals are used, great care is required to avoid cracking them during installation. A critical specification is the impeller-to-housing clearance and failure to meet this will reduce pumping efficiency.

FILTERS

Coolant filters are usually of the spin-on cartridge type connected in parallel to coolant flow (Figure 11–10). Corrosion inhibitors may also be packaged within the OEM coolant filter, a good reason to avoid overservic-

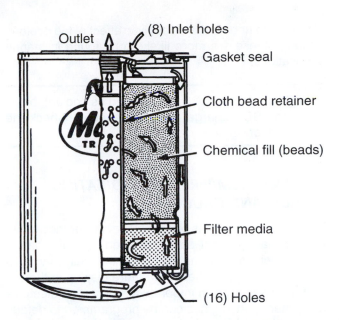

FIGURE 11–10 _Cutaway view of a coolant filter. (Courtesy of Mack Trucks)_

ing. When coolant filters are required to be changed, check the type of shutoff mechanism used: some are automatic, others have manual shutoff valves (Figure 11–11). New filters do not require priming. Flow through the filter is consistent with that of most other engine filters in that the coolant enters the canister through outer ports and exits through a central single port. Because some coolant filters are loaded with the SCA charge, always ensure that the correct filter is installed; it is important to observe both the OEM recommendation and the type of coolant used in the cooling system. Some coolant filters are equipped with a zinc

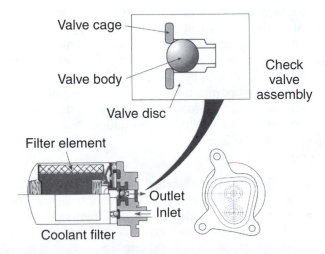

FIGURE 11–11 _Coolant filter with a mounting pad check valve. (Courtesy of Navistar International Corp., Designer and manufacturer of International Brand diesel engines)_

electrode to negate the electrolytic effect of the coolant, although these are more likely to be found in marine applications.

CAUTION: Where ELC is being used as coolant, blank filters (no SCA charge) may be required. Check the OEM or fleet service schedules.

COOLANT TEMPERATURE INDICATING CIRCUIT AND GAUGES

Coolant temperature is often a primary reference to the engine management software when factoring timing and air/fuel ratio parameters on electronically managed engine systems, so the temperature display to the operator is secondary in importance. All current electronically managed engines can be programmed to default to failure strategies, which may include engine shutdown based on the coolant temperature readings. Thermistors are almost universally used to sense the temperature of the cooling system as well as other engine fluid temperatures, including ambient air, boost air, and lubricating oil. The following methods are used to sense coolant temperature:

Thermistors

Thermistors are solid state, semiconducting devices whose internal resistance varies with temperature change. They are supplied with a specific reference voltage and output a signal based on temperature. Negative temperature coefficient (NTC) thermistors tend to be more commonly used: the internal resistance in an NTC thermistor decreases as temperature rises. Positive temperature coefficient (PTC) thermistors function oppositely: the internal resistance increases as temperature rises. Thermistors are commonly used to read coolant temperature on electronically controlled engines. The reference voltage is an ECM output: the thermistor returns an electrical signal representative of the coolant temperature to the ECM. This signal is used in processing to help factor fueling logic and as an output for the dash digital display or gauge.

Electric

Electric sensors use a bimetal arm in conjunction with a resistor fed with a modulated electrical signal from the temperature gauge: when the bimetal arm is heated, the greater linear expansion of one of the bimetal strips causes it to bend one way and as it cools and contracts, bend in the opposite direction. Connected to the bimetal strip is a wiper, which short-circuits the current flow through the resistor to ground, thereby altering the gauge value.

Expansion

An expansion-sensing gauge consists of a tube filled with a liquid that expands as it is heated and in expanding, activates the gauge indicator needle. This gauge tends not to be used much in today's engines.

Coolant Level Indicators

Most current electronically managed engines have radiators equipped with low coolant level warning systems. Most operate using the same principles. The ECM outputs a signal to a probe (or sensor), usually located in the top radiator tank, which grounds through the coolant. When the probe fails to ground through the coolant, a low coolant level warning is generated; the outcome depends on how the ECM has been programmed (this is a customer data program option). In most cases, a lag (somewhere around 5 to 12 seconds) is required before the ECM resorts to a programmed failure strategy. This may be simply to alert the operator, ramp down to a default rpm/engine load, or shut down the engine after a suitable warning period. Some radiators are equipped with two probes used to signal low and dangerously low coolant levels.

THERMOSTATS

Thermostats function as a type of automatic valve that will sense changes in engine temperature and regulate coolant flow to maintain an optimum engine-operating temperature. To function effectively, a thermostat must:

- Start to open at a specified temperature
- Be fully open at a set number of degrees above the start to open temperature
- Define a flow area through the thermostat in the fully open position
- Permit zero coolant flow or a defined small quantity of flow when in the fully closed position

The cooling system thermostat is normally located either in the coolant manifold or in a housing attached to the coolant manifold (Figure 11–12). Its primary function is to permit a rapid warmup of the engine: when the engine has attained its normal operating temperature, the thermostat opens and permits coolant circulation. As the thermostat defines the flow area for circulating the coolant, there may be more than one. A heat-sensing element actuates a piston attached to the seal cylinder. When the engine is cold, coolant is routed to the coolant pump to be recirculated through the engine. When the engine heats to operating temperature, the seal cylinder gates off the passage to the coolant pump and routes the coolant to the radiator. The heat-sensing element consists of a hydrocarbon or wax pellet into which the actuating shaft of the thermo-

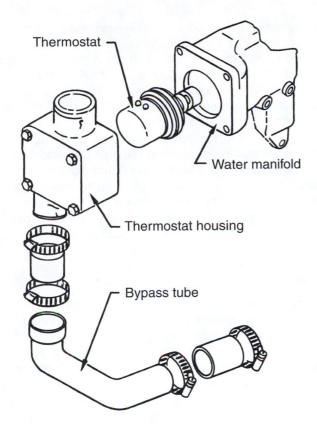

Thermostat

Water manifold

Thermostat housing

Bypass tube

FIGURE 11–12 *Exploded view of a thermostat housing. (Courtesy of Mack Trucks)*

stat is immersed. As the hydrocarbon or wax medium expands, the actuating shaft is forced outward in the pellet, opening the thermostat. Thermostats can be full blocking or partial blocking (Figure 11–13).

Top Bypass Thermostat

The top bypass thermostat simultaneously controls the flow of coolant to the radiator and the bypass circuit. During engine warmup, all of the engine coolant is directed to flow through the bypass circuit. As the temperature rises to operating temperature, the thermostat begins to open and coolant flow is routed to the radiator, increasing incrementally with temperature rise.

Poppet or Choke Thermostats

Poppet type thermostats control the flow of coolant to the radiator only, and the bypass circuit is open continuously. Flow to the radiator is discharged through the top of the thermostat valve.

Side Bypass or Partial Blocking Thermostat

The side bypass thermostat functions similarly to the poppet type. It has a circular sleeve below the valve that moves with the valve as it opens: this serves to partially block the bypass circuit and direct most of the flow to the radiator.

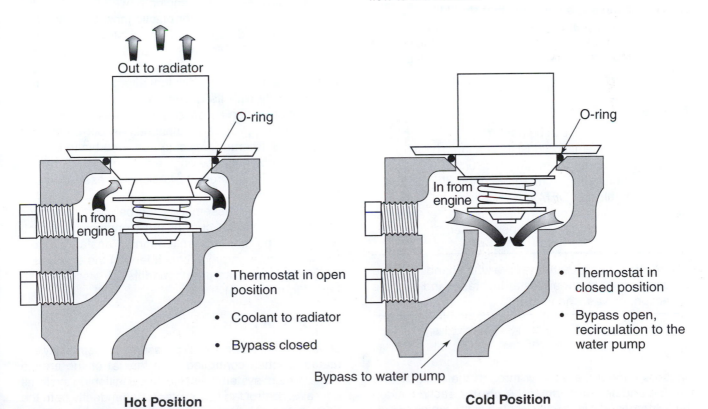

Out to radiator

O-ring

In from engine

- Thermostat in open position
- Coolant to radiator
- Bypass closed

Hot Position

O-ring

In from engine

- Thermostat in closed position
- Bypass open, recirculation to the water pump

Bypass to water pump

Cold Position

FIGURE 11–13 *Sectional view showing thermostat operation. (Courtesy of Navistar International Corp., Designer and manufacturer of International Brand diesel engines)*

Vented and Unvented Thermostats

Vented thermostats have a small orifice in the valve itself or a notch in the seat; usually this must be positioned in an upright position on installation. The function of the vent orifice is to help de-aerate the coolant by routing air bubbles out of the bypass circuit. Positive de-aeration type systems usually require nonvented thermostats.

Bypass Circuit

The term *bypass circuit* describes the routing of the coolant before the thermostat opens, that is, through the engine cylinder block and head. The flow of bypass coolant permits rapid engine warmup to the required operating temperature.

Running Without a Thermostat

This practice is not recommended and engine OEMs may void warranty. It also violates the EPA requirements regarding tampering with emission control components. Removing the thermostat invariably results in the engine running too cool. This can cause vaporized moisture in the crankcase to condense and cause corrosive acids (the chemical formula for sulfuric acid is H_2SO_4) and sludges in the crankcase. Additionally, low engine running temperatures will increase the emission of HC. Conversely, on engines that should use top bypass or partial bypass type thermostats may overheat when the thermostat is removed as most of the coolant will be routed through the bypass circuit with little being routed in the radiator.

Testing Thermostats

Testing a thermostat can be performed using a specialized tool that essentially consists of a tank, heating element, and accurate thermometer. Such a test device can be manufactured from an open top electric kettle. Always consult the OEM specifications for testing a thermostat, and remember, there is a difference between start to open and fully open temperature values.

CAUTION: Exercise extreme care when handling close to boiling water in a thermostat test tank, and use eye protection, gloves, and tongs.

SHUTTERS

Shutters control the air flow through the radiator and into the engine compartment. A **shutterstat** manages the system, and it is usually located in the coolant manifold. Shutters are sets of louverlike slats that pivot on shafts and are interconnected to rotate in unison from a fully open to a closed position in much the same manner as a set of Venetian blinds. The shutterstat is a temperature actuated control mechanism. It receives a feed of system air pressure, which it will allow to pass through until the coolant temperature reaches a predetermined value specified on the shutterstat. The shutter assembly is mounted on the radiator and is usually spring loaded to the opened position. When air from the shutterstat is delivered to the shutter cylinder, the plunger extends to actuate a lever and close the shutters. When no air is available at the shutter cylinder, the shutters are held open by spring force. During engine warmup, the shutters would normally be held closed once there was sufficient air pressure in the system.

COOLING FANS

OEMs use two basic operating principles of engine compartment cooling, requiring the use of either suction or pusher fans. Suction fans pull outside air into the engine compartment whereas pusher fans do the opposite and push heated air out. Highway vehicles that receive ram air assistance most commonly use suction fans. In fact, depending on the variables of engine size, radiator design, and vehicle application, ram air is often sufficient to perform cooling 95% of operating time.

Fan design is important, and fans should draw the least possible amount of engine power: 6 BHP is typical of modern fiberglass or plastic fans, but twice that may be required to rotate a large steel fan. Consequently, many OEMs have adopted these lighter fan blades over steel or aluminum. Many current designs use flexible pitch fan blades that alter a fan's efficiency proportionally with its driven speed; this permits efficient fan operation at low engine speeds. Fan assemblies must be precisely balanced. An out-of-balance fan (a small fragment missing from one blade is sufficient) will unbalance the engine driving it, which can be severe enough to promote torsional stress failures of the crankshaft.

Because a fan assembly when being driven draws engine power, most current truck engines use lightweight, temperature-controlled fans. Fanstat, air/electric/oil pressure engaged clutch fans of the on-off type are used as well as thermomodulated, viscous drive fans (Figure 11–14).

On-Off Fan Hubs

On-off fan hubs use air, oil pressure, or electrically actuated clutches controlled by a fanstat or the engine management system. Vehicle air-conditioning systems may also control fan hub cycles independently from the ECM using electric-over-pneumatic switching. However, increasingly this type of switching can be handled by the vehicle data bus: see Chapter 35. A

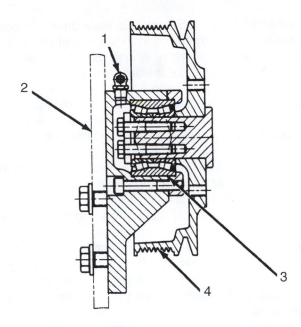

1. Fitting. Fill seal cavity with 2S3230 Bearing Lubricant.
2. Fan drive mounting plate
3. O-ring seal
4. Poly-V belt pulley

FIGURE 11–14 *Sectional view of a Poly-V fan drive. (Courtesy of Caterpillar)*

fanstat is usually located in the water manifold: it is a temperature-triggered switch that can either lock up or freewheel the fan hub using an electrical signal or chassis system air pressure to control the clutch. Most on-off fan hubs are spring loaded to the engaged mode, and the pneumatic or electrical signal disengages the fan clutch to permit it to freewheel. This provides a locked fan hub in the event of control circuit malfunction. When an air pressure-engaged, spring-released fan hub is used, the fan hub must be mechanically locked in the event of control circuit failure. On-off types of fan hub leach the least amount of engine power when the coolant temperatures are below the trigger value, which is usually designed to be 90% to 95% of operating time, but ambient temperatures and the requirements of the vehicle air-conditioning system may change this ratio.

When engine oil is used as the hub drive coupling medium (some bus applications), the fanstat controls the flow of oil to the hub assembly. At the fanstat nominal value, oil pressure directed to the fan hub acts as a fluid coupling so there is always a percentage of slippage. Some electronically managed engines directly control the fan cycle and use it as a retarding mechanism as well as a cooling aid. This is another operation that may be switched by the chassis data bus.

Thermatic Viscous Drive Fan Hubs/Thermomodulated Fans

Thermatic viscous drive fan hubs are integral units with no external controls: they use silicone fluid as a drive medium between the drive hub and the fan drive plate. The hub assembly consists of three main subcomponents: a drive hub (input section), a driven fan drive plate (output section), and the control mechanism. There is no mechanical connection between the drive hub and the driven member. During minimum slip operation, torque is transmitted through the internal friction of the silicone drive fluid in the working chamber that couples the input and output sections. A wiper attached to the driven member continually wipes the fluid, and centrifugal force returns it to a supply chamber where an open valve cycles it back to the working chamber. As ambient temperature drops, a bimetal temperature-sensing strip contracts to close the valve that supplies the silicone drive medium to the working chamber, and the fluid is trapped in the supply chamber. An advantage of the viscous drive fan hub is that it is capable of infinitely variable drive efficiencies depending on the temperature of the bimetal strip.

WARNING: Winter fronts, which are installed on the hood grill to limit the amount of ram air driven through the engine compartment, can load a fan off its axis creating an unbalancing effect whenever the fan is engaged. A winter front should not be necessary on most engines produced after 1990. When one is installed, ensure that it is approved by the truck chassis OEM. Never completely close a winter front.

Fan Shrouds

Fan shrouds are usually molded fiber devices bolted to the radiator assembly and may partially enclose the fan, providing a small measure of safety if the fan is engaged when the engine is running and the hood is open and exposed to the technician. Shrouds play an important role in shaping the air flow through the engine compartment and a missing or damaged shroud can result in temperature management problems. In hot weather conditions, fan efficiencies can be dramatically lowered by a defective or missing shroud, so they should be examined at each inspection.

Fan Belts and Pulleys

Fan pulleys use external V or poly-V grooves and internal bearings of the roller, taper roller, and bushing types. The belt tension should be adjusted using a belt tensioner. The consequences of maladjusted belts are:

- Too tight: excessively loads the bearings and shortens bearing and belt life

• Too loose: causes slippage and destroys belts even more rapidly than a too tight adjustment

Belts should be inspected periodically as part of a PM routine. Replace belts when glazed, cracked, or nicked. Replacing belts with early indications of failure costs much less in the long run than the breakdowns that may be caused by belts that fail in service.

Water Manifolds

The water manifold acts as a sort of main artery for the cooling system. It is usually a cast-iron or aluminum assembly fitted to the cylinder heads and may house the thermostat(s). Figure 11–15 shows a typical water manifold.

COOLING SYSTEM LEAKS

Cooling system leakage is common, and the system should be inspected by the operator daily. Cold leaks may be caused by contraction of mated components at joints, especially hose clamps; cold leaks often cease to leak at operating temperatures. Many fleet operators replace all the coolant hoses after a prescribed in-service period regardless of their external appearance

to avoid the costs incurred in a breakdown. Silicone hoses are more expensive than the rubber compound type but they usually have longer service life. Silicone hoses require the use of special clamps, and these are sensitive to overtightening so torque to the required specification. Pressure testing a cooling system will locate most external cooling system leaks. A typical cooling system pressure-testing kit consists of a hand-actuated pump and gauge assembly calibrated from zero to ±25 psi (170 kPa) plus various adapters for the different types of fill neck and radiator cap. Some are capable of vacuum testing.

Internal leaks can be more difficult to locate. When coolant is present in the engine oil, it appears as a milky sludge before settling to the bottom of the oil pan. It is the first fluid to exit the oil pan when the drain plug is removed. When the leak is sourced to a failed wet liner seal, to locate the engine cylinder, remove the oil pan, allow the oil to drip for a period, then pressurize the cooling system using standard hand pump apparatus and place a sheet of cardboard under the engine. Wet liner O-ring seals may leak either cold or hot and even this method is not surefire.

To determine if combustion gases are leaking into the cooling system, remove the piping to the upper radiator tank and thermostat, fill the engine with coolant

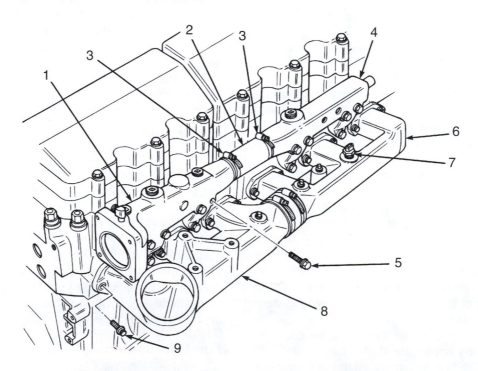

1. Coolant manifold, front section
2. Couple
3. Clamp
4. Coolant manifold, rear section
5. Capscrew

6. Air inlet manifold, rear section
7. Inlet air temperature sensor (sending unit)
8. Air inlet manifold, front section
9. Capscrew, 12-point

FIGURE 11–15 *Two-section coolant manifold on a six-cylinder engine with a two-cylinder head design: the intake manifold is located below. (Courtesy of Mack Trucks)*

(water will do), disconnect the water pump drive (if possible), and run the engine. The appearance of bubbles indicates that combustion gases are leaking into the coolant.

WARNING: The use of cooling system agents that claim to stop leaks should generally be avoided even in a situation that might be described as an emergency. They may work temporarily but in doing so, they have been known to plug thermostats, radiator/heater cores, and oil cooler bundles. Generally, they cause more trouble than they cure.

STRAY VOLTAGE DAMAGE

Stray voltage grounding through engine collant can result in electrolytic action that can cause considerable engine damage in unbelievably short periods of time. This electrolytic damage ranges from pinholes in heat exchangers to erosion pitting failures of cast-iron liners and cylinder blocks. The incidence of stray voltage damage has become more commonplace due to the increase in electrical and electronic components combined with the introduction of nonconducting components such as plastic radiator tanks. Chassis static voltge buildup can also discharge through engine coolant. One fleet hauling both tankers and flatbed trailers, using tractors with identical powertrains, reported that engine failures attributed to coolant electrolysis were only occurring in the tractors assigned to flatbeds. Trailer technicians will know that tankers have a bonded ground circuit whereas flatbeds do not. Use the following method to identify stray voltage and eliminate it.

Testing for Stray Voltage

Stray voltage can be AC (alternator diode bridge leakage) or DC, but is more likely to be the latter. Use a DMM on autorange to perform the following tests, checking for DC first.

1. Run the engine and turn on all the vehicle electrical loads.
2. Place the negative DMM probe directly on the battery negative terminal and the other into the coolant at the neck of the radiator without touching metal.
3. Record voltage reading. 0.1 V-DC is OK, the maximum acceptable to most OEMs is 0.3 V-DC. If higher, you must locate the leakage source. 0.5 V-DC leakage is capable of eating out a cast-iron engine block.
4. Sequentially shutdown each electrical component, checking the voltage reading. When the leaking component or circuit has been identified, repair its

ground. Attempting to ground the coolant at the radiator will not repair the problem.

COOLING SYSTEM MANAGEMENT

Most truck engine cooling systems are managed with the objective of opening the thermostat first, followed by the shutterstat, and finally the fan will engage. Because the fan consumes power (around 6 HP at rated speed), most OEMs go with the premise that it is desirable for it to be driven as little as possible. However, some will elect to open shutters before opening the thermostat to allow the thermostat full control of engine temperature and avoid the sudden changes of engine compartment temperature caused by cycling the shutters. **Thermatic fans** sense underhood or engine compartment temperature. An underhood temperature of 155°F can be generally reckoned to an engine coolant equivalent of 190°F.

SUMMARY

- Approximately 50% of the rejected heat of combustion is transferred to the engine cooling system, which is responsible for transferring it to atmosphere.
- The cooling system uses the principles of conduction, convection, and radiation to transfer heat from the coolant to atmosphere.
- A diesel engine cooling system has four main functions: to absorb combustion heat, to transfer the heat using coolant to heat exchangers, to dissipate the heat from the heat exchangers to atmosphere, and to manage engine-operating temperatures.
- The main components of a diesel engine cooling system are the water jacket, coolant, a coolant pump, a radiator, thermostat(s), filter(s), shutters, temperature sensing circuit, and a fan assembly.
- Water expands in volume both as it freezes and as it approaches its boil point. The engine cooling system must accommodate this change in volume.
- Engine coolant is a mixture of water, antifreeze, and supplemental coolant additives.
- Three types of diesel engine coolant are in current use: ethylene glycol (EG), propylene glycol (PG), and extended life coolant (ELC).
- A properly formulated diesel engine coolant should protect against freezing, boiling, corrosion, scaling, and foaming, and should inhibit acid buildup.

- The correct instrument for reading the degree of antifreeze protection of a coolant is a refractometer.
- Mixing coolant solutions should be performed in a container outside of the engine cooling system and then added.
- Supplemental coolant additives (SCAs) are a vital component of engine coolant. The SCA levels must be routinely tested in EG and PG coolants because they deplete in service.
- ELC is claimed to have a service life of 600,000 miles or six years, during which a single SCA recharge is required.
- Where possible, the radiator is located in the air flow at the front of the chassis to optimize ram air cooling effect.
- Most radiators are of the single pass, downflow type.
- Radiator caps are equipped with a pressure valve used to define the cooling system operating pressure and a vacuum valve to prevent hose collapse as the system cools and pressure drops.
- Coolant pumps are either belt or gear driven and are of the nonpositive, centrifugal type.
- Coolant pumps are lubricated by the coolant and tend to be vulnerable to high TDS levels, which are abrasive.
- Care should be taken when servicing coolant filters because some are charged with SCAs and an excess of SCAs can cause coolant problems just as depleted SCAs.
- The commonly used coolant temperature sensor on today's electronically managed engines is the thermistor, which is a temperature sensitive, variable resistor.
- Coolant level sensors ground a reference signal into the coolant in the top radiator tank and trigger an alert when the ground circuit is broken for a programmed period.
- Thermostats are used to manage the engine temperature to ensure optimum performance, fuel economy, and minimum noxious emissions.
- Thermostats route the coolant through the bypass circuit to permit rapid engine warmup.
- Shutters control the air flow through the engine compartment by means of louverlike slats that open and close like Venetian blinds.
- Shutters are controlled by a shutterstat, which is located in the water manifold and uses chassis system air pressure to open and close them.
- Shutters are usually designed to fail in the fully open position.
- Most engine compartment fans are temperature controlled, either directly on the basis of coolant temperature measured by a fanstat or indirectly based on engine compartment temperature.

- Flex blade, fiberglass fans are designed to alter their pitch based on their driven rotational speed, permitting higher fan efficiencies at low rpm.
- Viscous type, thermatic fans sense underhood temperatures and are driven by a fluid coupling designed to produce minimum slip at their nominal operating temperature.
- Fan assemblies are driven by V or poly-V belts whose tension should be adjusted using a belt tension gauge to avoid bearing and slippage problems.
- Testing a cooling system for external leaks is performed using a hand-actuated, pressure testing kit consisting of a pump, pressure gauge, and a variety of fill neck and radiator cap adaptors.

REVIEW QUESTIONS

1. Which type of diesel engine coolant is regarded as the most toxic?
   a. EG
   b. PG
   c. Pure water
   d. ELC
2. Which of the following cooling media would transfer heat most efficiently?
   a. EG
   b. PG
   c. Pure water
   d. ELC
3. Wet liner cavitation is caused by:
   a. Aerated coolant
   b. Combustion gas leakage
   c. Air in the radiator
   d. Vapor bubble collapse
4. The cooling system hoses on an engine collapse when the unit is left parked overnight. Which of the following is the likely cause?
   a. This is normal.
   b. Defective thermostat
   c. Improper coolant
   d. Defective radiator cap
5. If a radiator cap pressure valve fails to seal, which of the following would be the likely outcome?
   a. Coolant boil-off
   b. Cooler operating temperatures
   c. Higher HC emissions
   d. Cavitated cylinder liners
6. Which would be the warmest portion of a typical downflow type radiator when the engine is at operating temperature?
   a. The top tank
   b. The surge tank

c. The bottom tank
d. The center of the core

7. The temperature rating of a thermatic viscous fan hub is a nominal 155°F. At approximately what equivalent coolant temperature will this produce minimum slip drive as temperatures rise?
   a. 155°F
   b. 165°F
   c. 190°F
   d. 225°F

8. In a typical diesel-powered highway truck, which of the following is true?
   a. Coolant temperatures run cooler than lube oil temperatures.
   b. Coolant temperatures run warmer than lube oil temperatures.
   c. Coolant temperatures should be equal to lube oil temperatures.

9. Which of the following properly describes the operating principle of a typical coolant pump on a highway diesel engine?
   a. Positive displacement
   b. Centrifugal
   c. Constant volume
   d. Gear type

10. Which of the following controls the drive efficiency of a thermatic, viscous drive fan hub?
    a. Bimetal strip
    b. Viscosity of the silicone drive medium
    c. Fanstat
    d. Solenoid

11. When the thermostat routes the coolant through the bypass circuit, what is happening?
    a. The coolant is cycled through the radiator.
    b. The coolant is cycled through the engine.
    c. The coolant is cycled through the filter.

12. Fiberglass fans with flexible pitch blades are designed to flow air at greatest efficiency at:
    a. Low speeds
    b. All speeds
    c. High speeds

13. Coolant silicate drop-out usually causes:

a. Scaling
b. Gooey sludge
c. High TDS readings

14. In the event of an engine overheating, where is the coolant likely to boil first?
    a. Engine water jacket
    b. Top radiator tank
    c. Inlet to the coolant pump
    d. Thermostat housing

15. The correct instrument for checking the degree of antifreeze protection in a heavy-duty diesel engine coolant is a:
    a. Hydrometer
    b. Refractometer
    c. Spectrographic analyzer
    d. Color coded test coupon

16. *Technician A* states that some ELC has an approved service life of up to 6 years. *Technician B* states that water should never be added to ELC even when it is distilled. Who is right?
    a. A only
    b. B only
    c. Both A and B
    d. Neither A nor B

17. *Technician A* states that when a shutterstat management system fails, the shutters are usually left in the open position. *Technician B* states that if the chassis system air pressure failed, the shutters would close. Who is right?
    a. A only
    b. B only
    c. Both A and B
    d. Neither A nor B

18. *Technician A* states that a double pass radiator offers higher cooling efficiencies than a similarly sized single pass radiator. *Technician B* states that crossflow radiators are often used in aerodynamic low hood and grill designs. Who is right?
    a. A only
    b. B only
    c. Both A and B
    d. Neither A nor B

Chapter
12

Engine Breathing

PREREQUISITES

Chapters 4, 7, 8, and 9

OBJECTIVES

After studying this chapter, you should be able to:

- Identify the intake and exhaust system components.
- Describe how intake air is routed to the engine's cylinders and exhaust gases are routed out the tailpipe.
- Define the term positive filtration.
- Outline the operating principle of an air precleaner.
- Service a dry, positive air cleaner.
- Perform an inlet restriction test.
- Outline the operation of a Roots blower on a two-stroke cycle engine.
- Identify the subcomponents on a truck diesel engine turbocharger.
- Outline the operating principles of an exhaust gas-driven, centrifugal turbocharger.
- Troubleshoot common turbocharger problems and perform some basic failure analysis.
- Define the role of a charge air cooler and the relative efficiencies of each type.
- Test a charge air heat exchanger for leaks.
- Relate valve configurations and seat angles to breathing efficiency and cylinder gas dynamics.
- Describe how a pulse type exhaust manifold can boost turbocharger efficiency.
- Outline the function of a pyrometer on a truck diesel engine.
- Describe the role of the exhaust silencer and its operating principles.
- Comprehend the nature of sound dynamics and how combustion noise is minimized in an engine.
- Understand some basic exhaust gas emissions chemistry.
- Describe the operation of a diesel engine catalytic converter and EGR system.

KEY TERMS

air box
catalyst
catalytic converter
C-EGR
charge air cooling
compressor housing
crossflow valve configuration
diffuser
exhaust gas recirculation (EGR)
heat exchanger
I-EGR
impeller
induction circuit

inlet restriction gauge
intake circuit
manometer
naturally aspirated
opacity meter
oxidation catalyst
palladium
parallel port valve configuration
platinum
positive filtration
pulse exhaust
pyrometer
rejected heat

resonation
rhodium
Roots blower
scavenge
sound absorption
substrate
supercharger
tip turbine
tuned exhaust
turbine
turbocharger
volute

INTRODUCTION

Because of the nearly universal use of turbochargers and increasing use of exhaust gas recirculation (EGR) on current commercial vehicle diesel engines, information that could be handled separately under the headings *"Intake Systems"* and *"Exhaust Systems"* will be dealt with in this one chapter on *"Engine Breathing"* to facilitate the study of the component common to both intake and exhaust systems. A turbocharger charges the intake manifold at pressures above atmospheric; turbocharged engines are commonly described as providing manifold boost. The manifold-boosted CI engine cylinder is *charged* with air rather than induced. In a naturally aspirated engine, air or an air/fuel mixture is drawn into the engine cylinder by the below atmospheric pressure created in the engine cylinder on the downstroke of the piston on the first stroke of the four-stroke cycle. This air or air/fuel mixture is therefore induced into the cylinder. The correct term to collectively describe the components responsible for delivering breathing air in **naturally aspirated** engines is **induction circuit**. The preferred term for collectively describing the breathing air delivery components in an engine with manifold boost is air **intake circuit**.

The function of the air intake system in a diesel engine is to supply a charge of clean, cool air to the engine cylinders for combustion, cooling, and scavenging. Diesel engines are designed for lean burn operation. Essentially this means that the air charged to the engine cylinder will always substantially exceed that required to combust the fuel. Engine life expectations have lengthened dramatically during the past decade, in part due to the design of air intake system components.

The function of the exhaust system is to minimize both engine noise and noxious emissions while restricting the exhaust gas flow to a minimum degree. The turbocharger is a component common to both intake and exhaust systems; its function is to use the exhaust system to recapture some of the engine **rejected heat** to pressurize the air delivered to the engine cylinders. Manifold boosted diesel engine breathing requires that the gas dynamics of the intake and exhaust systems work to complement each other, especially during the very critical period of valve overlap (Figure 12–1). Diesel engine exhaust systems are increasingly using postcombustion, emission controls. **Oxidation catalysts** are used by some manufacturers, and many engines meeting EPA 2004 emissions are using cooled EGR.

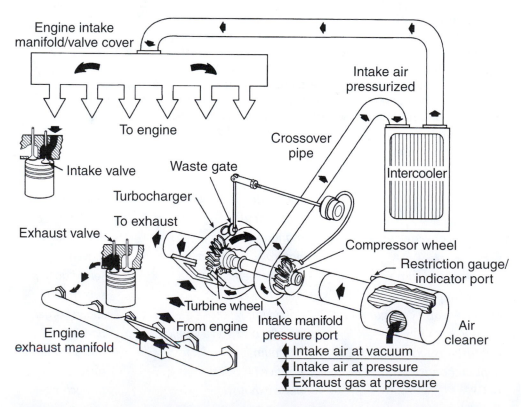

FIGURE 12–1   *Schematic showing engine breathing circuit. (Courtesy of Navistar International Corp., Designer and manufacturer of International Brand diesel engines)*

 **AIR INTAKE SYSTEM COMPONENTS**

The air intake system components are those responsible for delivering ambient air to the engine's cylinders. This air will always be filtered and in most truck diesel engines it will be pressurized well above atmospheric pressure values. Because pressurizing the air charge increases temperature, it reduces its density. To counter this, most turboboosted engines use some form of heat exchanger to cool the boost air before it is directed into the engine cylinders. Because the turbocharger is driven by exhaust gas heat and boosts the intake charge, it is considered as both an intake and exhaust system component (Figure 12–2).

ENGINE BREATHING COMPONENTS ON A TYPICAL FOUR-STROKE CYCLE TRUCK DIESEL ENGINE

Intake system components:

1. Air cleaner system consisting of a main filter and a precleaner

2. Intake ducting/piping
3. Turbocharger and turbocharger controls
4. Charge air cooling circuit
5. Intake manifold
6. EGR system
7. Valve porting and intake tract design

Exhaust system components:

1. Valve configuration
2. Exhaust manifold
3. Turbocharger and turbocharger controls
4. EGR system
5. Exhaust piping
6. Pyrometer
7. Engine silencer
8. Catalytic converter

ENGINE BREATHING COMPONENTS ON A TYPICAL TWO-STROKE CYCLE, TRUCK OR BUS DIESEL ENGINE

Intake system components:

1. Air cleaner system consisting of a main filter and a precleaner

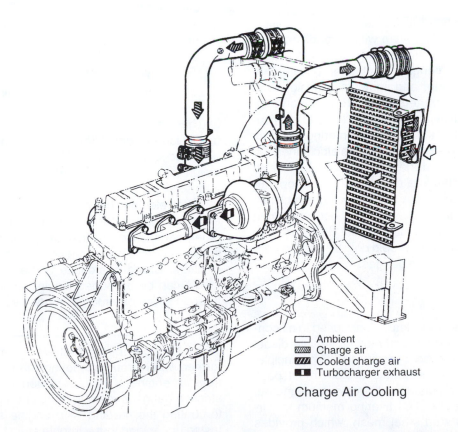

☐ Ambient
▨ Charge air
▧ Cooled charge air
■ Turbocharger exhaust

Charge Air Cooling

FIGURE 12–2   *Air flow through turbocharged, charge air cooled, in-line six-cylinder engine. (Courtesy of Mack Trucks)*

2. Intake ducting/piping
3. Turbocharger and turbocharger controls (on some applications)
4. Roots blower
5. Charge air cooling circuit
6. Intake manifold
7. Air box
8. Liner intake ports

Exhaust system components:

1. Exhaust valves
2. Turbocharger
3. Engine silencer
4. Catalytic converter (some engines)
5. Exhaust piping

AIR CLEANERS

The function of the air cleaner system on a highway diesel engine is to filter airborne particulates from the air that will be delivered to the engine cylinders. Airborne dirt can be highly abrasive and when it finds its way past the air cleaner system, it can destroy an engine in a very short period.

Precleaners

Precleaners are required in trucks operating in dusty conditions or in North American winter conditions, especially where highways are salted. They may triple the service life of the main engine air cleaner. Located before the main air cleaner and sometimes designed to receive ram air, precleaners generate a cyclonic air flow with the objective of separating heavier particulates by centrifugal force from the air entering the main cleaner. Depending on the design of the air cleaner, the particulate removed from the air stream can either be discharged or collected in a dust bowl. Trucks operating in conditions in which they are subjected to high levels of larger airborne matter such as ash and grain chaff may use precleaner screens, but these are vulnerable to plugging so they should only be used in conjunction with inlet restriction gauges.

Dry, Positive Filters

Dry, **positive filtration** air cleaners are used in all contemporary North American commercial vehicle diesel engines. Because they use a positive filtration principle, *all* of the air entering the intake system must pass through the filtering media, usually resin-impregnated, pleated paper elements. The filtering medium is surrounded by a perforated steel mesh, which provides the element with a limited amount of structural integrity. If a visible dent is evident in the surrounding mesh, the filter will probably not seal in the canister and should be replaced. Filtering efficiencies are high throughout the speed and load range of the engine, usually better than 99.5% and highest just before replacement is required. Some dry element filters are two stage and may eliminate the need for a precleaner; these induce a vortex flow and use centrifugal force to separate heavier particulate, which is then discharged by an ejector valve.

Dry paper element filters are designed to last for as long as 12 months in a linehaul application and should not be serviced unless the inlet restriction specification exceeds the OEM maximum or a used oil analysis report indicates that dirt may be bypassing the air cleaner. On-board inlet restriction gauges are not accurate instruments. If such a gauge indicates that the filter element is responsible for high inlet restriction before the regular service interval is achieved, check the reading using a water-filled manometer. Never remove an air filter from its canister unnecessarily. Every time a filter is serviced, some dust will be admitted beyond the filter assembly, no matter how much care is taken.

Checking filter restriction with a trouble light is a common practice that finds little favor with OEMs because whether the light is visible or not visible through the filtering medium depends more on the filtering medium materials than the degree it might be plugged up. The only valid reason for inspecting a dry filter with a trouble light is to locate a perforation. OEM specifications should always be consulted but typical *maximum* inlet restriction specs are:

15″ H_2O vacuum for NA engines
25″ H_2O vacuum for boosted engines

The practice of removing and attempting to clean a dirt-laden, dry filter on the shop floor should be avoided. A dirt-laden filter weighs many times the weight of a new filter and dropping it onto a concrete floor to shake dust free is usually sufficient to damage the filter element (by crumpling the perforated mesh) enough to prevent it from sealing in the housing. Another malpractice that invariably ends up costing more in the long term is reverse compressed air blowing out of filters. Sharp particulate embedded in the filtering media can be dislodged leaving an enlarged opening through which other larger particles may pass.

TECH TIP: One of the first tests that should be performed when troubleshooting black smoke emission from an engine is inlet restriction. If the on-board restriction gauge is reading high, fit an H_2O **manometer** to confirm the reading. The engine should ideally be tested under load, but a throttle snap to high idle should give a close indication.

Laundering Dry Element Filters

When a dry filter element has become plugged the obviously ideal solution is to replace it. However, dry filter elements are expensive and in some operations, such as on construction sites, mining, and aggregate hauling, these may become restricted in a couple of working days even when precleaners are used. Professional laundering is perhaps a second best but necessary option to replacement. The laundering process usually requires that the filter element be soaked in a detergent solution for a period of time, followed by reverse flushing with low pressure clean water. The element is next dried with warm air and inspected for perforations in the element and for seal/gasket integrity. OEM testing of professionally laundered filter elements indicates that filtering efficiencies are reduced with each successive laundering.

Inlet Restriction Gauges

Inlet restriction gauges are resettable gauges mounted either on the filter canister or remotely in the vehicle dash. They provide a readout in inches of water vacuum in the same way the manometer does. Although these instruments are not noted for their accuracy, they provide an indication of when filter service is required.

Oil Bath Filters—Viscous Impingement Oil Filters

Low operating efficiencies have made these types of nonpositive filters things of the past. They consist of a mesh-filled canister with a sump filled with engine oil. Air flow is cycloned through the canister and wets down a cylindrical mesh with engine oil to which dirt particles attach themselves. This principle works effectively to remove larger particles from the air stream and has higher efficiencies when induced air flow is highest, that is, at rated speed. They possess low filtering efficiencies at low-load, low-speed operation. Oil bath filters are serviced by replacing the sump oil and washing down the mesh with solvent.

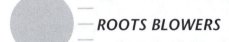

ROOTS BLOWERS

The **Roots blower** (also known as a **supercharger**) is a gear-driven, positive displacement air pump used to **scavenge** the engine cylinders in DDC two-stroke cycle engines. It produces relatively low-peak pressures compared to those produced by turbochargers. As a scavenging component, the primary function of the Roots blower is to displace end gases from the engine cylinders. It may also produce a small amount of manifold boost. All DDC two-stroke cycle engines used in automotive applications (that is, highway trucks and buses) use Roots blowers for cylinder scavenging; these engines may, in addition, use a turbocharger to produce manifold boost. Under conditions of high engine loading and therefore peak turbocharger efficiency, the turbocharger may be capable of performing cylinder scavenging. In some nonautomotive applications of DDC two-stroke cycle engines, the required scavenging may be performed exclusively by turbocharger(s) but only where engine loading is required to be consistently high, such as in a genset.

Roots blowers are used as scavenging pumps because of their positive displacement pumping principle. They pump air to an air box, which in turn supplies air to the cylinders by means of intake ports in the cylinder liner. A positive displacement pump will displace a constant slug volume per cycle. This means that the Roots blower will unload a constant slug volume of air per rotation so the volume of air charged to the **air box** is directly dependent on the driven speed. This gives the Roots blower somewhat higher efficiencies (than a turbocharger) as a pump when engine rpm and loads are low. Turbochargers are driven by engine-rejected heat energy and therefore tend to have low efficiency when engine loads are low, plus a lag between power demand and turbo response.

CONSTRUCTION

The DDC Roots blower housing is manufactured from an aluminum alloy casting. Within this housing are a pair of intermeshing, spiral-fluted rotors supported by bearings located at either end of the assembly. One of the rotors receives gear-driven input, and it drives the other. Each rotor has three spiralled lobes timed to intermesh, so one spirals clockwise and the other counterclockwise. The rotor drive may be on either side, depending on the model. The rpm ratio of the blower to that of the engine also varies and depends on factors such as whether a turbocharger and intercooler are used. The fuel pump and water pump are also driven by blower drive shaft, the water pump at one end and the fuel pump at the other end of the assembly.

Lubrication

The DDC Roots blower is supported by pressure-lubricated roller bearings that ensure that the rotors have close clearance in operation but no contact. As a consequence, the rotors themselves require no lubrication, because in normal operation, they have no metal-to-metal contact. Lip type seals prevent engine oil from bypassing the rotor shaft and being pumped into the engine. Evidence of engine oil on the rotor lobes is an indication that the blower shaft seals are failing. The oil to lubricate the blower gears, bearings, governor, and fuel pump drive is ported from the main oil gallery and will drain back to the crankcase via ducting. On in-line engines, oil drain-off from the rocker housing drains to the blower housing and plates and lubricates the moving components aided by a slinger.

Diagnosing Roots Blower Problems

In order to function effectively as a scavenging pump on a two-stroke cycle diesel engine, the Roots blower must be capable of displacing up to 50 times the total engine cylinder swept volume per cycle. When attempting to diagnose an engine breathing problem, all the factors that may affect breathing must be tested, including air inlet restriction, air box pressure, valve lash, turbocharger operation and controls, and exhaust restriction. When inspecting the blower itself, the following should be checked:

1. Rotor leading edges for evidence of contact indicating a possible bearing problem or wear indicating the presence of abrasives in the air stream.
2. Traces of engine oil streaked over the rotors are evidence of failing seals, a condition that should be repaired immediately because the consequence of a total seal failure is an engine runaway. Worn bearings, high manifold boost (turbocharged engines), or a restricted crankcase breather can all contribute to seal failures.
3. Blower drive shaft. The blower drive shaft is a flexible coupling that is designed as shear point should anything jam the blower rotors. The shaft should flex under torsion approximately ⅜″ when the rotors are turned. Excessive flexing or zero flex are reasons to replace the blower drive shaft.

TURBOCHARGERS

Turbochargers are almost universally used on North American medium- and large-bore, highway diesel engines. By definition, a **turbocharger** is an exhaust gas-driven, centrifugal pump that "recycles" some of the rejected heat from the engine's cylinders. Turbochargers may be driven to speeds exceeding 200,000 rpm in certain race car engine applications. In most highway applications of the technology, the turbocharger is used to deliver a pressurized charge of air to the engine's cylinders; in short, it increases the oxygen density in the air charge. In larger, off-highway diesel engines, the ex-

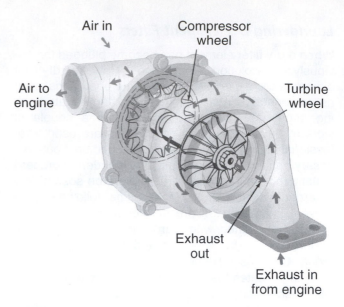

FIGURE 12–3 *Gas flow through a turbocharger. (Courtesy of Schwitzer)*

haust gas-driven **turbine** may be used with a fluid coupling to drive reduction gearing connected to the engine crankshaft; in this instance, the turbocharger assists in driving the crankshaft (Figure 12–3).

PRINCIPLES OF OPERATION

A turbocharger is an exhaust gas-driven air pump consisting of a turbine and an **impeller** mounted on a common shaft. The shaft is floated (hydrodynamically suspended) on friction bearings supplied with pressurized lube oil. The turbine wheel is subject to engine exhaust gas energy (heat) and is driven within a turbine housing through which the exhaust is routed. The impeller is enclosed in a separate **compressor housing** and acts on intake system air, pumping it through the charge side of the intake system. The exhaust gas that drives the turbine and the intake air the impeller acts on do not come into contact.

The greater the engine rejected heat value (this increases somewhat proportionally with engine output), the greater the exhaust gas heat energy. This means that the gases acting on the turbine vanes will expand more and the result will be turbine rotational speed. The exhaust gas is routed to flow into a **volute** (snail-shaped, diminishing sectional area) and exit it to expand on the turbine vanes. Alternatively, a nozzle may be used instead of a volute, but the principle is the same. The volute flow area or the nozzle ring size helps define maximum rotor speed. It is important to underline the fact that turbocharger rotational speed is factored mainly by exhaust gas heat and not exhaust gas pressure.

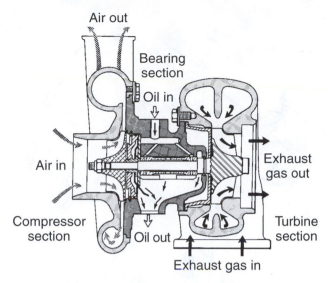

FIGURE 12–4 *Sectional view of a turbocharger, the lubrication circuit, and air flow. (Courtesy of Schwitzer)*

Turbochargers are designed for optimum performance when the engine is fully loaded and operating at an rpm in the torque rise profile (in high torque rise engines, turbo optimum efficiency occurs close to peak torque). If the load demanded of an engine is reduced but the rpm maintained, the weight of the exhaust gas remains fairly constant (as does manifold pressure),

but the reduced exhaust gas heat results in considerably reduced turbine speeds. The geometry of the turbocharger always determines the specific engine rotational speed at which the turbocharger operates at peak efficiency.

Filtered intake air is admitted to the compressor housing, pulled in by the impeller (compressor wheel), which is attached to the turbo shaft and driven by the turbine. This accelerates the air to high velocities. High velocity air flows radially outward to a diffuser. The **diffuser** is designed to convert the kinetic energy (energy of motion) of the intake air into pressure. The diffuser may be a volute or a blade type (Figure 12–4). Blade type diffusers have higher efficiencies.

Construction

Turbochargers used in truck diesel engine applications can sometimes wind out at peak rpm of up to 130,000 rpm, commonly perform at 70,000 rpm, and must sustain temperatures well over 1,200°F (650°C), so the materials used in turbo construction must be able to withstand high temperatures and centrifugal forces. Figure 12–5 shows the components used in a typical turbocharger. The compressor components are often manufactured from aluminum alloys. Turbine housings must be able to sustain rotor burst and are manufactured out of austenitic (a particular crystalline structure

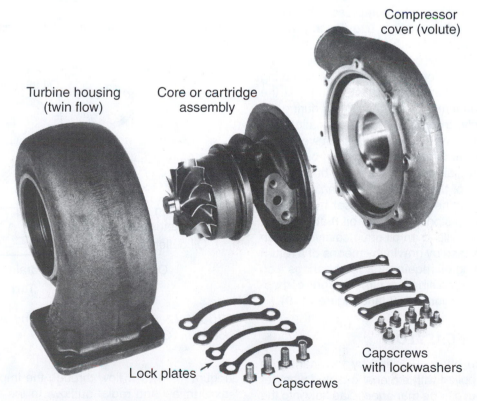

FIGURE 12–5 *Turbocharger components. (Courtesy of Schwitzer)*

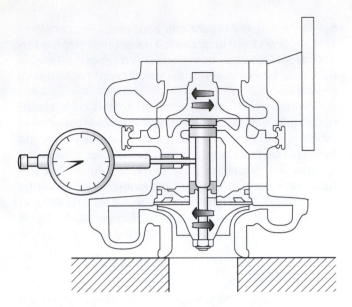

FIGURE 12–6   Checking turbine shaft radial play.

of iron and carbon) Ni-Resist™ cast iron, which has high hot strength and resistance to oxidation. The turbines are manufactured from nickel or cobalt steels and ceramics. The turbo shaft is usually an alloy steel to which the turbine is either welded or bolted and the impeller is bolted. Sometimes the turbo shaft, turbine, impeller, and bearing assembly are manufactured in a cartridge for ease of replacement. When a turbocharger is recored, this shaft assembly is replaced as a unit and the hub, turbine, and compressor housings reused. All current turbochargers use floating friction bearings. Turbo bearings are designed to rotate in operation and do so at about one-third shaft speed with a film of oil on either side of the bearing. The operating principle is that of friction bearings, meaning that the turbine shaft is hydrodynamically suspended during operation. Accordingly, the radial play specification is a critical determination of bearing and/or shaft wear. Figure 12–6 shows a method of measuring turbine shaft radial play using a dial indicator. A thrust bearing defines the axial play of the turbine shaft, and gas pressure on both the turbine and compressor housings is sealed by a pair of piston rings. The oil that pressure feeds the bearings spills to an oil drain cavity and then drains to the crankcase by gravity by means of a return hose. The lubricating oil required for the bearings also plays a major role in cooling the turbocharger components (Figure 12–7, Figure 12–8, and Figure 12–9).

TURBOCHARGER GAS FLOW

Turbochargers are often classified by the manner in which they are supplied with exhaust gas and whether turbocharger output can be managed. Gas flow into the turbine housing is radial and with an axial outflow (see

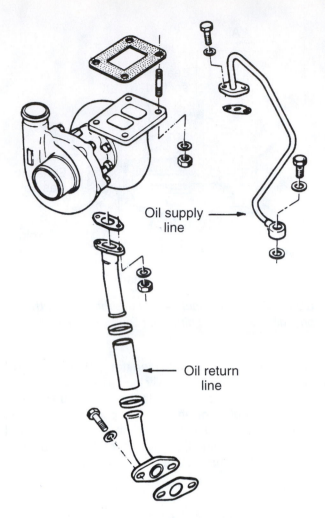

FIGURE 12–7   Turbocharger lubrication lines. (Courtesy of Mack Trucks)

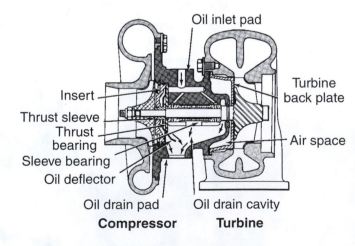

FIGURE 12–8   Turbocharger bearing and thrust sleeve location. (Courtesy of Schwitzer)

Figure 12–4). Air flow through the impeller housing is axial inflow and radial outflow. In the simplest type of turbocharger design, the turbine housing intake tract

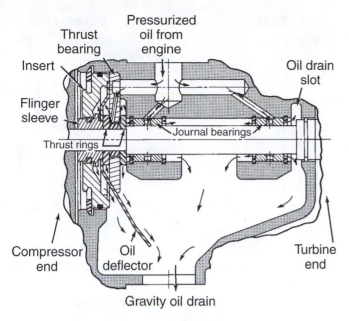

FIGURE 12–9 *Oil flow through a turbocharger. (Courtesy of Schwitzer)*

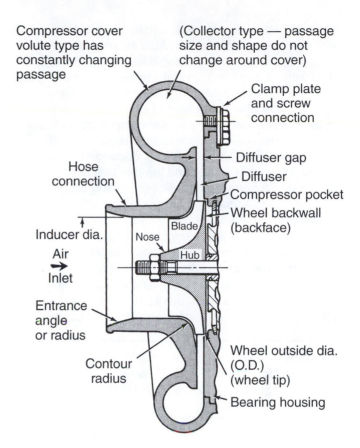

FIGURE 12–10 *Sectional view of a turbocharger compressor housing. (Courtesy of Schwitzer)*

or throat is undivided, meaning that all the cylinders feed gas into a single passage. In a double flow design, the turbine throat is divided and each half feeds one-half of the turbine wheel circumference. Another variation is the twin flow in which two passages are fed from a common throat for the full rotation of the turbine wheel. V-configured engines commonly use either the double or twin flow turbine housing design (Figure 12–10).

Common Manifold Turbocharging

A single pipe manifold is flange mated to the exhaust tract of each engine cylinder with the turbocharger turbine housing flange mounted to the center of the assembly in a manner that the exhaust gas is routed through the turbine housing. This manifold design creates certain problems with the gas dynamics and produces inconsistencies in the effective heat per exhaust slug discharged into the manifold; simply, each exhaust slug volume discharge from cylinders 1 and 6 on an in-line, six-configured engine must travel considerably farther before they enter the turbine housing tract than those discharged at cylinders 3 and 4 located at the center of the engine.

Pulse/Tuned Exhaust Manifold

OEMs of many current diesel engines often used **pulse/tuned exhaust** manifolds that use geometrically tuned pipes to direct the exhaust gas from each cylinder almost directly into the inlet tract of the turbine housing. This minimizes gas flow interference and ex-

haust slug pressure variables that reduce the efficiency of common manifold systems. Pulsed manifolds tend to diminish turbo lag (turbocharger response time) and increase low engine load performance.

TURBOCHARGER GEOMETRY AND PERFORMANCE

Unlike most other diesel engines, turbocharger efficiencies in highway truck diesel engines tend to be designed to be highest at peak torque, rather than at rated speed. A highway diesel engine described as *high torque rise* (increase in torque as engine rpm decreases with a relatively constant power curve) is mainly different from the same model not described as such by the fuel injection setup and the turbocharger geometry. Constant geometry turbochargers route all exhaust gas through the turbine housing all the time and cannot change the volute flow area. Constant geometry turbochargers used with high torque rise engines produce an approximately constant rate of flow per minute from the peak torque rpm through to the rated speed rpm. Therefore, the mass flow of air per cycle will be greatest at the peak torque speed. A turbocharger that is matched to the engine to produce optimum efficiency at peak torque will produce lower efficiencies at higher rpms. Also as engine speed increases, the real time available for charging the cylinders with air and injecting fuel diminishes, providing the turbo with a self-regulating overspeed capability. The torque curve of a turbocharged engine quickly falls off in the lower speed range as the compressor efficiency plunges; fuel economy is also adversely affected.

Turbocharger performance is dependent on actual exhaust gas heat values and the gas flow dynamics so they are to a certain extent self-regulating with respect to air requirements of the engine. Many truck diesel engines use constant geometry turbochargers that are precisely matched to the engine's requirements and do not require any type of external controls. Variable geometry turbochargers are managed externally by using controls in the form of wastegating or nozzle ring control. Figure 12–11 and Figure 12–12 show the gas flow routing on single, twin, and double flow turbochargers.

WARNING: Mismatching of turbochargers may result in high engine cylinder pressures that result in engine failure or oppositely, to low power, smoking, and high noxious emissions.

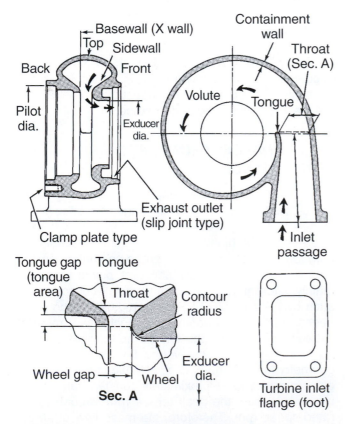

FIGURE 12–11 Sectional view of a turbocharger turbine housing. (Courtesy of Schwitzer)

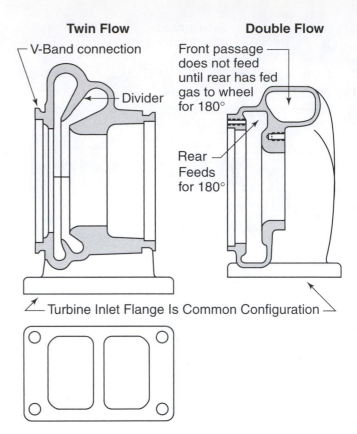

FIGURE 12–12 Sectional view of a divided turbine housing. All housing types can have a water passage around the gas passage for cooling. (Courtesy of Schwitzer)

WASTEGATES

The idea of a wastegate is to increase both the operating and load range of a turbocharger. For instance, if you wished to increase the manifold boost at low load and rpm operation, a higher capacity turbocharger could be used. The problem is, a high capacity constant geometry turbocharger would produce too much boost and create too high cylinder pressures at rated speed and loads. Wastegating allows a turbocharger to behave like a high capacity output turbo under low engine loads and like a low capacity turbo under conditions of high engine loading. The payoff is much faster response (yes, reduced turbo lag) and of course reduced engine emissions.

When a diesel engine uses a variable geometry turbocharger, the wastegate manages turbo output using manifold boost to actuate the gating or bypass mechanism. In this arrangement, a wastegate actuator is used to control the wastegate valve: the wastegate valve position determines how much exhaust gas is routed around the turbocharger in any given moment of engine operation. A wastegate actuator is a canister with a rod. One end is attached to the wastegate and the other to a bellows inside the canister. An internal spring loads the actuator to close the wastegate valve. In the closed position, all of the exhaust gas is routed through the turbine housing. On the other side of the bellows, manifold boost acts in opposition to the spring, that is, it attempts to open the wastegate valve and allow some of the exhaust gas to bypass the turbine housing.

This operation can be made more complex by allowing the ECM to take over and manage some of the wastegate functions. For instance, on a Cummins ISX engine, the turbocharger has a dual entry port. The ISX system is designed so that end gas from the front three cylinders is constantly routed through the turbine housing, and the wastegate can only bypass directly to exhaust end gas from the rear three cylinders. However, the ECM uses a combination of solenoids (inlet and vent) and manifold boost to precisely manage the pressure acting on wastegate actuator bellows to provide four levels of boost. Table 12–1 shows how this works.

As EPA requirements for reductions in noxious emissions and customer demands for increased power and fuel economy continue, you can expect to see the engine management computer play a larger role in managing turbocharger output. Constant geometry turbochargers are becoming increasingly rare today.

Two-Stage or Series Turbocharging

Series turbocharging requires use of two turbochargers in series. The terms primary and secondary turbochargers (or low/high pressure turbochargers) designate the roles played by the two units used. This technology was last used by Cummins in their NTC–475 in the early 1980s and will be used by Caterpillar in 2004 ACERT engines.

TABLE 12–1   *ISX Four-Step Wastegate*

STEP	INLET SOLENOID	VENT SOLENOID	RESULT
1	Energized	De-energized	Maximum air pressure to wastegate. Minimum boost.
2	De-energized	De-energized	One step of boost. The default position.
3	Energized	Energized	Two steps of boost.
4	De-energized	Energized	Minimum pressure to wastegate. Maximum boost.

Paralleling

Paralleling describes the use of multiple turbochargers usually to charge each bank of a V configuration engine. Parallel turbocharger configurations are not used on any current truck diesel engines, but they are used extensively in larger, off-highway applications.

Compounding

This word is often incorrectly used to describe two-stage turbocharging. The turbine is connected to a fluid coupling connected to reduction gearing with output shaft connected to the engine crankshaft. In other words, the turbocharger helps directly drive the crankshaft. It is not used in any current North American truck/bus engines.

Turbocharger Precautions

1. Avoidance of hot shutdown: allow at least 5 minutes of idling before shutting down the engine after prolonged loading on a dynamometer or road test.
2. Prelubing: directly pour oil onto the turbine shaft through the oil supply flange when installing a turbocharger before connecting the lube supply line.
3. Some OEMs require that the oil filters are primed when replacing them at service intervals.

Turbocharger Failures

1. Hot shutdown: extreme temperature failures producing warped shafts and bores.
2. Turbocharger overspeed: this may be caused in a number of ways, including fuel rate tampering, high-altitude operation with a defective altitude compensation fuel control, and mismatching of complete turbocharger units.
3. Air intake system leaks: allows dirt to enter the compressor housing, causing the vanes to be eroded wafer thin.
4. Lubrication-related failure: caused by abrasives in the engine oil, improper oil, broken down oil, or restricted oil supply.

CHARGE AIR HEAT EXCHANGERS

The act of compressing the intake air charge by the turbocharger typically produces air temperatures of 260°F (125°C) at a 70°F (20°C) ambient temperature and proportionally more at higher ambient temperatures. The objective of charge air coolers, whatever their cooling medium, is to cool the air pressurized by the turbocharger as much as possible while maintaining the pressure. As intake air temperature increases, air density diminishes and the oxygen charge in the cylinder is reduced. The result is lower power and higher cylinder temperatures as the proportion of rejected heat increases. To minimize this loss of power potential, the boosted air charge is cooled using one of several types of heat exchanger. As a rule, a 1°C increase in intake air temperature will produce a 2°C increase in exhaust gas temperature. It should also be noted that because the performance of charge air coolers greatly influences combustion temperatures and the density of the cylinder air charge, they are an integral component in the vehicle's noxious emission control system. Most North American boosted diesel engines use some type of heat exchanger to cool intake air.

AIR-TO-AIR HEAT EXCHANGERS

These have the appearance of a coolant radiator and are often chassis mounted in front of the radiator (Figure 12–13). As the vehicle moves down the highway, ambient air is forced through the fins and element tubing. Ram air is therefore the cooling medium. Cooling efficiencies are highest when the vehicle is travelling at higher speeds. They are least efficient when the air flow through the cooler is minimal, and while the engine fan may assist, air-to-air cooling may not be suitable for construction applications. Under optimum conditions, air-to-air heat exchangers have efficiency ratios that better any liquid-cooled heat exchangers. Optimum conditions means maximum air flow through the element so usually the vehicle must be travelling at full speed down the highway. Thermal efficiency is increased by the air-to-air method of **charge air cooling** because where liquid-cooled heat exchangers are

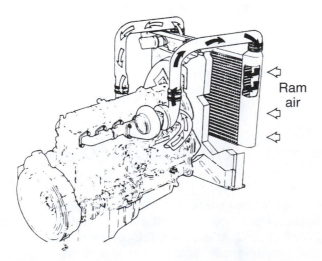

FIGURE 12–13   Location of an air-to-air, charge air cooler. (Courtesy of Mack Trucks)

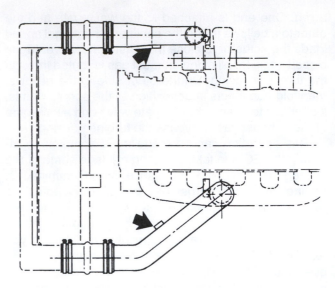

FIGURE 12–14   Overhead view of an air-to-air, charge air cooler. (Courtesy of Mack Trucks)

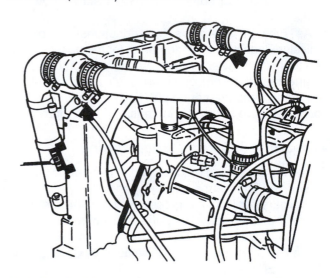

FIGURE 12–15   Charge air cooler plumbing. (Courtesy of Mack Trucks)

used, actual intake air temperatures are reduced to values only slightly higher than the coolant temperature values (Figure 12–14 and Figure 12–15).

TESTING AIR-TO-AIR COOLERS

When an engine is operating at close to rated speeds and loads, air enters a charge air cooler at around 300°F. If the heat exchanger is doing its job properly, the air should exit at around 110°F when ambient temperatures are 75°F. Charge air coolers are designed so that they can leak small amounts of air without any performance drop-off. When testing the air leak-off rate, plug the inlet and outlets of the charge air cooler and pressurize to the OEM recommended test value. Some examples of OEM recommended test values:

Engine OEM	psi Start Test Value	psi Drop-off over 15 Secs.
Caterpillar	30 psi	5 psi
Cummins	30 psi	7 psi
Detroit Diesel	25 psi	5 psi

If the pressure drop-off exceeds specifications, drop the pressure down to 5 psi and hold. Then attempt to locate the leak using a soap and water solution.

AFTERCOOLERS AND INTERCOOLERS

These normally refer to **heat exchangers** that use liquid engine coolant as the cooling medium. The two terms are synonymous and the one used merely indicates the OEM preference. Boosted air is forced through an element containing tubing through which coolant from the engine cooling system is pumped. Cooling efficiencies tend to be lower than with air-to-air heat exchangers due to the heat of the cooling medium; they continue to be used in applications in which low air flow through the engine housing rules out the use of an air-to-air exchanger.

TIP TURBINE HEAT EXCHANGERS

The **tip turbine** type of heat exchanger uses both ambient air and engine coolant as cooling media. Boosted air is delivered by a main duct to a heat exchanger through which it is forced before entering the engine cylinders. Engine coolant is circulated through tubing within the heat exchanger. However, a small amount of boosted air exiting the impeller housing of the turbocharger is ported off to drive a tip turbine. The tip turbine is an air-driven centrifugal pump whose function it is to force fresh, filtered ambient air through the heat exchanger element. This air does not come into direct contact with the charge air but acts to cool both it and the engine coolant. Tip turbine, charge air cooling is popular in construction and stop/start applications in which air flow through the engine housing would be insufficient to use air-to-air exchangers (Figure 12–16).

Testing for Boost Circuit Leakage

When leaks occur on the boost side of the turbocharger, they are accompanied by a loss of power and

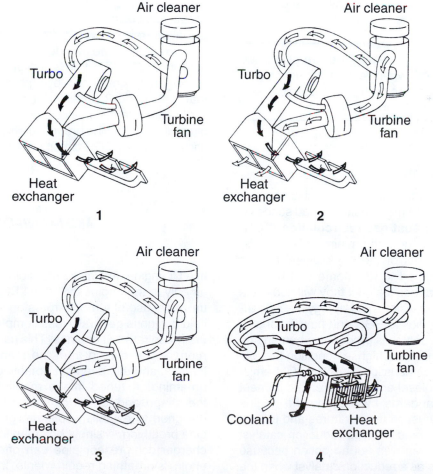

FIGURE 12–16 *Tip turbine operation. (Courtesy of Mack Trucks)*

sometimes a high pitch turbo whistle. The following procedure is recommended to locate leaks downstream from the turbocharger compressor and before the intake manifold:

1. Clean off the intake ducting and the exterior of the charge air cooler with a pressure washer from a distance to ensure that the heat exchanger fins are not damaged.
2. Inspect the exterior of the charge air cooler and the intake ducting for obvious indications of damage. Replace any suspect hoses.
3. Remove the rubber flex hoses at both the turbocharger compressor outlet and the intake manifold, making sure that no dirt is allowed to enter the intake system. Close off the hoses at either end using plug and hose clamp.
4. There is usually an aperture located in the charge air cooler designed specifically for pressure testing: it is sealed by a threaded pipe plug. If there is not such an aperture, one may have to be tapped into one of the plugs clamped at either end of the circuit. Fit an air pressure regulator to the aperture.
5. Set the air pressure regulator to 5 psi (35 kPa): this should be sufficient to locate most leaks. Use a soap and water solution in a spray bottle or apply with a paint brush to locate leak(s).

CAUTION: Never regulate the air pressure at a value above 30 psi (200 kPa) when testing for boost side leaks because this may result in personal injury and damage to the components under pressure.

EXHAUST GAS RECIRCULATION

Until the introduction of the EPA 2004 emission standards, truck diesel engine manufacturers had successfully avoided using **exhaust gas recirculation (EGR)** systems. No longer. To meet the minimum NO_x emissions requirements, most of the truck diesel engine manufacturers have opted to use some sort of EGR system; only Caterpillar has stated it they will not. For a full understanding of emissions technology, see Chapter 45; here we briefly look at EGR and how it works.

Oxides of nitrogen, or NO_x is produced when engine combustion temperatures are high, a condition that occurs during lean burn combustion. Until the 2004 emission standards, truck diesel engines were able to meet the NO_x emission standards by allowing the engine ECM to manage combustion temperatures and fuel injection timing: using a rhodium type reduction catalyst like those on automobiles was not an option because rhodium only functions as a reduction catalyst when the air-fuel ratio is stoichiometric or richer. Diesels run lean.

So, most diesel engine manufacturers had no option but to turn to EGR. An EGR system routes some of the exhaust gas back into the intake system. This "dead" gas is used simply to occupy some space in the cylinder, to reduce the extent of lean burn that is conducive to NO_x formation. A better way of saying this is that EGR dilutes the intake charge of oxygen. It is not popular for the following reasons:

- Putting end gas back into the engine cylinder in place of fresh filtered air adds wear-inducing contaminants and increases engine oil acidity, both of which reduce engine longevity.
- EGR reduces engine power.
- EGR badly affects fuel economy.

It is not difficult to see why EGR is not popular with the engine OEMs, but it finds even less favor among their customers.

In most cases, truck diesel engines using EGR cool the end gas using a heat exchanger before routing it back into the engine cylinders. The ECM controls the cycling of the EGR gas. Cooled EGR systems have produced the acronym **C-EGR**. Mack Trucks uses C-EGR in highway applications but has opted to use internal exhaust gas recirculation or **I-EGR** in engines powering vocational trucks. I-EGR functions by using the valve actuation train to permit some of the combustion end gas to remain in the cylinder. It accomplishes the same objective as C-EGR, that is, it dilutes the cylinder charge of oxygen by simply occupying space. It can be argued that both C-EGR and I-EGR essentially defeat what took years of engineering to accomplish, the diesel engine with high engine breathing efficiency. Figure 12–17 shows a cooled EGR circuit on a diesel engine.

INTAKE MANIFOLD DESIGN

In a CI engine, the intake manifold is required to deliver air only to the engine cylinders. Because of the almost universal use of turbochargers on diesels, intake manifold design is generally less complex than that for naturally aspirated SI engines. This usually means that the runners that extend from the plenum are of unequal lengths, and this does not notably compromise engine breathing. A tuned intake manifold is one in which the shape and length of each runner is similar and designed to establish optimum gas dynamics for engine breathing. A single box manifold fed off the turbocharger compressor pipe can often meet a boosted engine's breathing requirements. Intake manifolds can either be wet (coolant ports) or dry; the latter is gener-

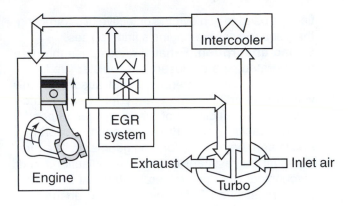

FIGURE 12–17 *C-EGR: Layout of a cooled exhaust gas recirculation system on a diesel engine*

ally used in truck and bus applications. Materials used are aluminum alloy or cast iron, but some OEMs are experimenting with plastics and carbon-based fibers.

Where a tuned intake manifold is used in a boosted CI engine, it is designed to work with a tuned exhaust manifold and establish ideal gas dynamics for effective scavenging during valve overlap. In full authority, electronically managed engines, a thermistor is located in the intake manifold; this thermistor is often a primary reference for determining air/fuel ratio (AFR).

VALVE DESIGN, CONFIGURATION, AND BREATHING

Most current diesel engines use multivalve configurations consisting of two inlet and two exhaust valves. Two basic breathing configurations are used, crossflow and parallel port. Until recently, most engines used **crossflow valve configuration**. Crossflow breathing locates both sets of valves transversly, meaning that in- or outflow of the inboard valve can interfere with that from the outboard valve. Breathing efficiencies can be enhanced by using parallel port configurations, which permit each of the valves to be responsible for an equal amount of gas flow without crossflow interference. **Parallel port valve configuration** enhances both cylinder charging and scavenging, but requires a more complex camshaft assembly (or dual camshafts) and to some extent, compromises gas velocity and swirl effect. Sometimes this is desirable; in older high-swirl engines, fuel sometimes was thrown into the cylinder wall causing it to condense and only partially combust. Many newer engines use quiescent cylinder gas dynamics (lower swirl) and much higher injection pressures to prevent fuel condensing in the cylinder. Valve seat angle also affects cylinder breathing. Valve seats are usually cut and machined at either 30 degrees or 45 degrees.

Gas flow is generally around 20% greater using a 30-degree valve seat angle and the same lift as a 45-degree valve seat angle. However, a 45-degree seat is used more due to higher seating force and distortion resistance.

EXHAUST SYSTEM COMPONENTS

The exhaust system is required to perform the following:

- Assist cylinder scavenging.
- Minimize engine noise.
- Minimize engine noxious gas emission.
- Exhaust heat, noise, and end gases safely to atmosphere.

EXHAUST MANIFOLD

The exhaust manifold collects cylinder end gases and delivers them to the turbocharger. The exhaust manifold is usually manufactured in single or multiple sections of cast iron. Most diesel engine exhaust manifolds are "tuned" to deliver to the turbocharger exhaust gas with minimal and balanced flow resistance from each cylinder. The term tuned is used to denote any exhaust system designed with at least a little consideration for the exhaust gas flow. If the exhaust manifold and piping is properly designed as each slug of cylinder exhaust is discharged, it will not "collide" with that from another cylinder but instead will be timed to unload into its tailstream. The pulsed manifolds discussed earlier in this section are an example of a tuned exhaust system. Exhaust backpressure factors in a diesel engine are the turbocharger turbine assembly, the catalytic converter (if fitted), and the engine silencer. Exhaust manifold gaskets in truck/bus engine applications are usually of the embossed steel type, although occasionally fiber gaskets are used; they are usually installed dry.

PYROMETER

Normally found at a specified distance down line from the turbocharger, a **pyrometer** can provide the driver with engine loading information useful for shifting. Thermocouple pyrometers are used. These consist of two dissimilar insulated wires (often pure iron and constantin [55% copper, 45% nickel]) connected at each end to form a continuous circuit. The two junctions are known as hot end, located where temperature reading is required, and reference end connected to a millivolt meter. Whenever the junctions are at different temperatures, there is current flow, the voltage increasing with difference in temperature. The practice of locating one pyrometer at the exhaust exit to each cylinder to indicate cylinder balance is not a current one in truck and bus engines.

EXHAUST PIPING

Where mild steel piping, flex piping, and clamps are used, the steel is corrosion protected usually by galvanizing (zinc coating). However, it is becoming more common to use stainless steels, which provide the exhaust system components with at least double the service life.

Stainless steel band clamps have been used for many years and are an effective means of sealing flex pipe to straight pipe as they yield to shape at installation; they are deformed at installation and should not be reused. The function of exhaust piping is to collect the engine exhaust gases unloaded by the turbine housing, deliver them first to the engine silencer/catalytic converter, and then to atmosphere clear of the tractor cab structure and the trailer structure. If rain caps are used on vertical exit pipes, they should be mounted transversely, opening away from the chassis. They should never be mounted longitudinally facing either forward or rearward.

ENGINE SILENCERS

Sound is produced by the firing pulses in engine cylinders as sonic nodes and antinodes and the function of the silencer or muffler is to modify engine noise to legally acceptable levels. Essentially, this requires the scrambling of the frequency of the sonic nodes and antinodes produced by engine firing pulsations. Before the rigorous sonic emissions standards of the 1990s, a turbocharger in conjunction with piping design was in some cases able to meet the minimum required standard, but this is no longer so. Pressure waves caused by the unloading of each slug of exhaust gas into the piping can have an amplifying effect on engine sound, and the turbocharger is capable of smoothing these pressure waves and enabling a simpler muffler design.

Muffler volume in today's truck engine systems is typically almost five times total exhaust slug volume per cycle. Exhaust slug volume is always somewhat greater than total swept volume in manifold-boosted engines. Engine silencers use two basic principles to achieve their objective of dampening sound:

1. **Resonation.** Resonation requires reflecting sound waves back toward the source, thereby multiplying the number of sound emission points. Separate chambers connected by offset pipes as well as baf-

fles are used to achieve the objective of scrambling the frequency of the engine's firing pulses.
2. **Sound absorption.** Exhaust gases pass through a perforated pipe enclosed in a canister filled with sound-absorbing material. Sound absorption involves converting sonic energy into heat by friction, and the efficiency of these devices depends on the packing density of the sound-absorbing media and the pipe perforation geometry. The media must be resistant to pressure pulsations and mechanical vibration; mineral and metal wools are used, most often basalt wool. Sound absorption mufflers generally have improved flow resistance characteristics over comparable resonator principle mufflers.

Truck and bus mufflers use one or a combination of the foregoing principles to meet sonic emission requirements.

CATALYTIC CONVERTERS

Until 2002 truck and bus engines met noxious gas emission requirements by carefully managing the combustion process, usually using comprehensive monitoring and an on-board computer. A few OEMs use **catalytic converters** on current engines, but this is necessarily a single stage device incorporating just an oxidizing catalyst. However, exhaust gas treatment processes are becoming more commonplace as EPA emissions legislation becomes more stringent, so the truck technician should possess a basic understanding of how these systems operate.

By definition, a **catalyst** is a substance that enables a chemical reaction without itself undergoing any change. An automobile catalytic converter is normally described as a two-stage, three-way convertor. The two stages refer to the objectives of the catalytic converter. When cylinder end gas contains HC and CO, the oxidizing stage of the catalytic converter attempts to oxidize these to H_2O and CO_2. Where oxidized nitrogen compounds (known collectively as NO_x) are present in the exhaust gas, a reduction stage attempts to reduce these to elemental nitrogen and oxygen. The three-way denotes the catalysts.

Platinum and **palladium** are both oxidation catalysts; **rhodium** is a reduction catalyst. All three are classified as noble metals and are sensitive to lead contamination. They are expensive. Oxidizing catalysts enable what can be termed catalytic afterburning, sometimes assisted by the introduction of fresh air into the reaction; peak temperatures run in the range of 800°C to 1,000°C.

The commonly used reduction catalyst is rhodium and when oxidized nitrogen compounds have been formed it attempts to reduce them to elemental oxygen and nitrogen. However, rhodium only functions as a reduction catalyst when the cylinder burn is managed close to stoichiometric ratios. This makes it of little use in lean burn diesel applications. Other reduction catalysts exist, but they are expensive. Because of the lean burn characteristic of diesel engines, NO_x emission continues to be a problem, so the introduction of multistage catalytic converters using a new technology reduction stage may not be too far away. Current research indicates that selective catalytic reduction (SCR) using water/urea injection can effectively reduce NO_x. The urea (crystallized nitrogen compounds) reduces to gaseous ammonia that reacts with NO_x compounds, producing harmless nitrogen and water. Bosch hopes to have a system ready for truck diesel engines by 2005. In most cases, a combination muffler/catalytic converter assembly will replace the engine silencer unit.

The noble metal catalysts are thinly coated on aluminum oxide or granulate monolith **substrate** (catalytically inert material onto which active catalysts are coated) providing optimum use of catalyst surface area combined with minimal flow resistance. Catalytic converters only operate efficiently when at operating temperatures and do little to limit HC emissions during startup/warmup; computer-controlled variable timing and minimal fueling attempt to achieve this. To date, CI truck and bus engines use only oxidation catalysts.

Prolonged cold temperature engine operation (that is, engine loading is light or idle), especially where ambient temperatures are low, can cause diesel catalytic converters to plug. When this occurs, a full load run on a dynamometer for 30 minutes may burn off the deposits that restrict the converter. In some cases, the catalytic converter will have to be replaced.

OTHER MEANS USED TO CONTROL NOXIOUS EMISSIONS

Particulate traps have been used by some OEMs, especially in special applications such as having to run a truck inside a building. These are essentially soot filters usually using a ceramic filtration medium. It should be noted that effective cooling of intake air lowers combustion temperatures, making it less likely that the nitrogen in the air is oxidized during the combustion process. A full account of emissions control in diesel engines is provided in Chapter 45.

Opacity Meters (aka Opacimeters)

An **opacity meter** is a light extinction, exhaust smoke testing tool that has become the aftermarket field monitoring tool of choice for jurisdictions wishing to enforce smoke emission standards. The tool is fitted to the exhaust stack and consists of a light emitter that is projected through the exhaust gas at a sensor; the percentage of emitted light that is picked up at the sensor is read as "percent opacity."

Smoke opacity readings of 5% or less are not easily observed. When using an opacity meter, it is important to meticulously observe the test procedures if any belief is to be given to the results.

A more complete account of using opacity meters is provided in Chapter 45.

SUMMARY

- Most current trucks use a dry, positive filter system.
- Trucks operated in environments with airborne particulates such as grain chaff dust, construction, and road dust should use some kind of precleaner to extend the air cleaner element service life.
- Dry, positive type filters function at optimum efficiency just before their service life is completed.
- It is important not to overservice air filters, because every time the canister is opened, some dust will find its way downstream from the filter assembly; many will last for up to a year in a linehaul application.
- Air inlet restriction should be tested with a water manometer.
- Charge air heat exchangers cool the turbo-boosted air charge and therefore increase its density: the result is more oxygen molecules in the intake charge delivered to the cylinder.
- Air-to-air charge air coolers boast higher cooling efficiencies than the liquid medium coolers but must have adequate air flow: this makes them ideal for use in highway applications but not in high-load, low road speed vocational applications.
- Because of the almost universal use of turbochargers in truck diesel engines, intake manifold designs remain relatively uncomplicated.
- Valve configuration affects both the cylinder breathing efficiency and the cylinder gas dynamics.
- Parallel port valve configurations generally produce better and more balanced cylinder breathing efficiency but also produce lower swirl cylinder dynamics.
- Valve seats cut at a 45-degree angle produce higher flow restriction and higher seating force than valves cut at a 30-degree angle, assuming identical lift.
- Turbochargers represent an exhaust backpressure factor, but their objective is to recapture some of the engine rejected heat by using it to pressurize the intake charge to the cylinders.
- Turbochargers are driven by the *heat* in the exhaust gas, so the more heat, the faster the turbine speeds.

- Turbochargers in truck diesel engine applications may wind out at up to 130,000 rpm with mean running speeds in the 70,000–80,000 rpm range.
- Truck diesel engine OEMs are using variable geometry turbochargers to increase the efficient operating range and reduce the turbo lag duration.
- Turbocharger radial and axial runouts should be routinely inspected at engine PM intervals.
- Hot shutdowns are a primary cause of turbocharger failures. After prolonged high load operation and especially after a dynamometer test, a cool-down period of at least five minutes is required.
- Sound is transmitted in sonic waves producing nodes and antinodes. An exhaust silencer operates by scrambling (altering the frequency) of these nodes and antinodes, thereby altering its nature.
- Engine silencers use resonation and sound absorption principles to alter the frequency of the sound emitted from the engine; most use a combination of both principles.
- The exhaust silencer volume on truck diesel engines is typically five times the total cylinder exhaust slug volume through a complete cycle.
- A catalyst is a substance that enables a chemical reaction without itself undergoing any change.
- A catalytic converter on a truck diesel engine is of the single stage, oxidizing type.
- Catalytic converters used on diesel engines are vulnerable to plugging up when an engine is operated for prolonged periods under light or idle loads.
- An opacity meter (or opacimeter) is a light extinction tester that operates by aiming a light beam at a sensor that records the amount of light sensed.

REVIEW QUESTIONS

1. The tool of choice to accurately check the inlet restriction of a dry air filter would be a:
   a. Pyrometer
   b. Trouble light
   c. Water manometer
   d. Restriction gauge
2. A typical maximum specified inlet restriction for an air filter on a turbocharged diesel would be:
   a. 25" water
   b. 25" mercury
   c. 25 psi
   d. 25 kPa
3. Which of the following type of filter has the highest filtering efficiencies?
   a. Centrifugal precleaners
   b. Oil bath
   c. Dry, positive

4. Which of the following should be performed first when checking an engine that produces black smoke under load?
   a. Injection timing
   b. Plugged particulate trap
   c. Fuel chemistry analysis
   d. Air filter restriction test
5. Which of the following would best describe the catalytic converter fitted to a diesel engine?
   a. Two stage, three-way
   b. Single stage, oxidizing
   c. Single stage, reduction
6. Which of the following components would play the largest role in minimizing NO_x emission in a highway diesel engine?
   a. Catalytic converter
   b. Air-to-air charge air cooler
   c. Particulate trap
   d. Turbocharger
7. In current electronically managed highway diesel engines, which of the following would be a primary reference to the ECM in determining the air/fuel ratio (AFR)?
   a. Pyrometer
   b. Intake manifold thermistor
   c. Inlet restriction sensor
   d. Vehicle speed sensor
8. Which of the following correctly describes the opacimeter used to measure exhaust smoke emission?
   a. Light extinction tester
   b. Filtration tester
   c. Constant volume sampling (CVS) tester
   d. Flame ionization tester
9. Removing an engine coolant thermostat would likely produce an increase in which of the following noxious emission levels?
   a. NO_x
   b. HC
   c. Sulfur dioxide
   d. Carbon dioxide
10. The constant geometry turbocharger used with a high torque rise highway diesel engine is usually designed to produce optimum efficiency at which rpm?
    a. Rated
    b. Top engine limit (TEL)
    c. Peak torque
    d. High idle
11. *Technician A* states that an air filter functions at optimum efficiency when it is new and that by the time it requires replacing, it has low filtering efficiencies. *Technician B* states that oil bath filters are seldom used with today's highway diesel engines because they have low filtering efficiencies. Who is right?

a. A only
b. B only
c. Both A and B
d. Neither A nor B

12. _Technician A_ states that a Roots blower is a positive displacement air pump, which volumetrically displaces air proportionally with rpm increase. _Technician B_ states that Roots blowers are used to aspirate two-stroke cycle engines and are not found on current highway diesel engines. Who is right?
    a. A only
    b. B only
    c. Both A and B
    d. Neither A nor B

13. _Technician A_ states that the term manifold boost is used to describe breathing on a turbocharged engine. _Technician B_ states that turbine speeds on truck diesel engines can exceed 100,000 rpm. Who is right?
    a. A only
    b. B only
    c. Both A and B
    d. Neither A nor B

14. _Technician A_ states that using a light bulb inside an air filter element is a reliable means of determining whether the filter is plugged. _Technician B_ states that air filters are best left untouched between service intervals unless the filter restriction gauge indicates an out-of-specification reading. Who is right?
    a. A only
    b. B only
    c. Both A and B
    d. Neither A nor B

15. _Technician A_ states that the first item to be checked when an engine produces black smoke should be the air filter inlet restriction gauge. _Technician B_ states that whenever an engine produces black smoke the air filter should automatically be changed. Who is right?

a. A only
b. B only
c. Both A and B
d. Neither A nor B

16. Which of the following methods would identify a hole or tear in an air filter element?
    a. Light bulb test
    b. Air inlet restriction gauge
    c. Manometer test
    d. Regulated air pressure test

17. When is a diesel engine turbocharger capable of highest rotational speeds?
    a. At peak engine rpm
    b. When rejected heat is greatest
    c. At peak torque
    d. When exhaust manifold pressure is greatest

18. Muffler volume in a truck diesel engine would typically be equivalent to:
    a. Engine displacement volume
    b. Five times the total exhaust slug volume
    c. The air cleaner volume
    d. Ten times the total exhaust slug volume

19. When an engine is described as high torque rise, at which rpm is the turbocharger likely to be operating at highest efficiency under load?
    a. Peak torque
    b. Rated speed
    c. High idle speed
    d. Top engine limit

20. _Technician A_ states that the function of a turbocharger wastegate is to divert exhaust gas around the turbine directly into the exhaust system. _Technician B_ states that a wastegated turbocharger can improve turbine efficiencies at lower loads and rpms. Who is right?
    a. A only
    b. B only
    c. Both A and B
    d. Neither A nor B

Chapter
13
Engine Retarders

PREREQUISITES

Chapters 4 through 12. Some knowledge of the content of Sections II and III will also help.

OBJECTIVES

After studying this chapter, you should be able to:

- Identify the types of engine brakes used in trucks.
- Describe the operating principles of each type of engine brake and the relative advantages and disadvantages of each.
- Outline the control mechanisms used to manage each type of retarder system.
- Interpret the electrical schematics used in the electronic and electric controls of engine brakes.
- Interpret a schematic representation of the hydraulic circuit of a typical internal engine compression brake.
- Describe how the hydraulic actuation of internal engine compression brakes is timed.
- Describe the operation of a Mercedes-Benz constant throttle valve brake.
- Describe the pneumatic controls used to manage external engine compression brakes and engine-mounted hydraulic retarders.
- Outline the differences in automatic and manual control of the Caterpillar BrakeSaver.

KEY TERMS

brake fade

BrakeSaver™

coefficient of friction

CTV brake

exhaust brake

external compression brake

Intebrake

internal compression brake

Jacobs brake

Jake brake

retarder

INTRODUCTION

Engine compression brakes are widely used on North American truck diesel engines. A large percentage of vehicle braking requires application pressures of 20 psi or less in a typical air brake system; in other words, less than a fifth of the brake system's potential is being used. The objective of engine compression brakes is to relieve the vehicle braking system of some of the light-duty braking, especially in instances that require prolonged light-duty braking such as on an extensive downgrade.

A typical truck braking system is air actuated and converts the kinetic energy (energy of motion) of the vehicle to heat energy: this takes place at the brake foundation assembly, and its mechanisms are drums (or rotors), brake shoes (or calipers), and actuating apparatus that converts the potential energy of compressed air into mechanical force. When the brakes are applied, the friction facings on the brake shoes or calipers are loaded into a rotating drum or rotor and resultant resistance or friction retards the movement. The friction produces heat so the brake system must be designed to sustain high temperatures and dissipate braking heat rapidly. The **coefficient of friction** describes the aggressiveness of friction materials. It is desirable to maintain a consistent coefficient of friction in foundation brake components for obvious reasons. However, coefficient of friction is altered by changes in temperature and the addition of a lubricant such as water. When foundation brake systems are subjected to the most amount of use, they are required to dissipate the most amount of heat to atmosphere. Failure to dissipate this heat can greatly increase foundation brake temperatures causing both a reduction in the coefficient of friction of the critical friction surfaces at the shoe and drum and an expansion of the drum dimension. When a tractor-trailer combination has to run downhill for prolonged periods, even though brake application pressures are relatively light, excessively high foundation brake temperatures result, causing a reduction in braking efficiency at a time when it is most needed, a condition known as **brake fade**.

A supplementary braking system such as an engine compression brake or driveline retarder can support the main vehicle braking system, enhance peak braking performance, and in cases in which engine braking is ECM managed, reduce some of the driver-actuated braking.

The engine compression brake on which most current versions are modelled was invented by Clessie Cummins after he had officially retired from his position as CEO of the company that bears his name. The Jacobs Manufacturing Company, perhaps better known for its drill chucks, has manufactured engine compression brakes since its introduction of Cummins's first engine brake in the late 1950s to the present day. Jacobs manufactures an engine brake for most of the major North American diesel engines and has established universal brand identification: "Jake brake" is used to refer to any engine compression brake by many drivers. Trucks may also use hydraulic retarders, which may be located anywhere in the driveline, but are usually integral with the engine or transmission. Hydraulic driveline retarders use a torque absorption principle that can be compared to that used in a torque converter. Electric driveline retarders are also currently available and one of them is marketed by Jacobs. The Caterpillar BrakeSaver, which is coupled to the engine is covered in this chapter because it is an engine component rather than a transmission or driveline component.

The term **retarder** can be applied to any component or system that effects braking action, that is, retards or slows motion. However, the primary vehicle braking system is seldom referred to as a vehicle retarder system, whereas engine compression brakes and driveline brakes are commonly described as retarders. Compression and driveline brakes do no more than supplement the primary braking system: the retarding powerflow is necessarily confined to those axles directly linked to the drivetrain, that is, through the wheels on the drive axles. The objective of the engine brake system is identical to that of the main vehicle braking system—namely, converting the energy of vehicle motion to heat energy. In the engine compression brake, kinetic energy (that of vehicle motion) is converted first to the potential energy of compressed air and then released into the exhaust system as heat.

BRAKE CYCLE CONTROLS

With the increased use of electronics in truck chassis systems, the way engine brake cycles are managed has changed considerably over the past few years. In older systems, control of the brake retarding cycle was left entirely to the driver, whereas in computer-controlled systems, the chassis management electronics are playing an ever-larger role. An obvious use of a computer-controlled engine brake is by the cruise control system to help maintain a consistent road speed. However, as multiplexing becomes more of a factor on our trucks and we see the emergence of collision warning, electronic lane tracking, and smart braking systems, this opens the door for increased use of engine braking. Engine braking makes sense both from safety and maintenance perspectives. It has become much more than a way of extending foundation brake, friction face service life.

PRINCIPLES OF OPERATION

The term *engine brake* can be used to describe retarder devices that use three distinct operating principles:

1. Internal engine compression brakes
2. External engine compression brakes
3. Hydraulic engine brakes

INTERNAL ENGINE COMPRESSION BRAKES

In the four-stroke cycle diesel engine, the piston is required both to perform work and receive work. On its upward travel during the compression stroke, it compresses the cylinder air charge (performs work), following which it is forced through its downstroke by expanding cylinder gases (receives work). The **internal compression brake** operates by making the piston perform the work of compressing the air charge on the compression stroke and then negating (cancelling) the power stroke by releasing the compressed cylinder gases to the exhaust system (gas blowdown) somewhere around TDC. This has the effect of reversing the engine function, converting it from an energy-producing pump to an energy-absorbing compressor. All engine compression brakes use this principle of operation; the mechanisms used to actuate, time, and control the brakes vary. The braking powerflow of the vehicle begins at the drive axle wheels, extends through the transmissions and driveshafts, through the engine powertrain, and the kinetic energy is converted to pressure; the potential energy of the compressed air is then dumped into the exhaust system. Braking efficiencies are highest when engine rpm is highest.

EXTERNAL ENGINE COMPRESSION BRAKES

The operating principle of the **external engine compression brake** is not that different from the internal compression brake. External engine compression brakes are also known as **exhaust brakes**. They consist of a housing located downstream from the turbine housing discharge in the exhaust system. Within the housing is a valve that when actuated chokes off the exhaust discharge. This restricts the exhaust gas flow and once again reverses the role of the engine, converting it into an energy absorbing pump; however, in the internal engine compression brake, the effective pumping stroke is the compression stroke, whereas in the exhaust brake, the effective retarding stroke becomes the exhaust stroke. Therefore, as each

succeeding exhaust slug is unloaded into the exhaust manifold, the pressure rises and braking efficiency increases. Retarding efficiency is to some extent diminished by pressure loss to the intake during valve overlap. Operation of the exhaust brake is managed so that the engine is not fueled during braking and as with the internal engine compression brake, braking efficiencies are highest when the engine rpm is in its higher range.

HYDRAULIC ENGINE BRAKES

Many types of electric and hydraulic driveline brakes exist, but these are more appropriately dealt with when studying transmission and driveline components. However, the **Caterpillar BrakeSaver**™ is a hydraulic retarder that is coupled to the rear of the engine and uses engine lubrication oil as its medium. A rotor is coupled directly to the engine crankshaft, so it rotates at any time the engine is running. The rotor is turned within the BrakeSaver housing, which is coupled to the flywheel housing and to which the transmission is mounted. The engine flywheel is bolted through to the crankshaft, but because of the BrakeSaver, the starter motor must crank the engine by means of a ring gear on a ring gear plate mounted to the crankshaft behind the BrakeSaver rotor. The rotor is therefore driven by the crankshaft between the BrakeSaver housing and a stator. When the BrakeSaver is actuated, the housing is charged with pressurized engine oil. Vaned pockets on the rotor mean that when the BrakeSaver is charged with oil, the rotor encounters fluid resistance defined by the stator geometry and the rotor rotational speed. Once again, retarding efficiencies are greatest when rotational speed is greatest. The oil that acts as the BrakeSaver hydraulic medium is engine oil, supplied from the engine sump by a section of the oil pump dedicated to charging the BrakeSaver.

ENGINE BRAKE OPERATION AND CONTROL CIRCUITS

Engine brakes typically use electric control switches, which means that they may be actuated either directly by the driver or by the engine management ECM in "smart" cruise applications. When engine brakes are used in hydromechanically managed engines, the switching of the engine brake is usually electrical but the retarding effect must be actuated either hydraulically or pneumatically. In such engines, a control circuit would require that a series of switches be closed before engine braking can be effected. The first of this series of switches would be the driver control switch, which can be proportional depending on the system. The next would be located at the clutch; this must be necessarily fully engaged to operate engine braking. The final essential switch in the series (though any number of others may be installed) would be to ensure that the accelerator was not depressed; in other words, engine fueling is at zero (current engines) or at least at a minimum in older engines to limit the quantity of raw fuel dumped into the exhaust system. Current regulations require that no uncombusted fuel is discharged into the exhaust system, so the rules of operation have changed in recent years. However, engine brakes used in many current engines are managed or monitored by the engine electronics. When an engine brake is electrically switched to the "on" position, the result depends on the type of engine brake used. A description of some of the common engine brakes follows:

WILLIAMS EXHAUST BRAKES

Williams exhaust brakes are classified as external engine compression brakes (Figure 13–1). Control of the brake is electric over pneumatic. The electrical circuit required to actuate the engine brake consists of three switches, all of which must be closed: a control switch (dash mounted), a clutch switch (clutch must be fully engaged), and an accelerator switch (accelerator must be at zero travel). The sliding gate exhaust brake uses a pneumatically activated gate, actuated by chassis system pressure. The air supply to close the gate is controlled by an electrically switched pilot valve. An aperture in the sliding gate permits a minimal flow through the brake gate during engine braking. The butterfly valve version operates similarly (Figure 13–2 and Figure 13–3).

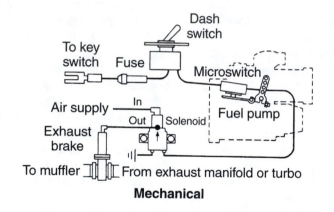

Mechanical

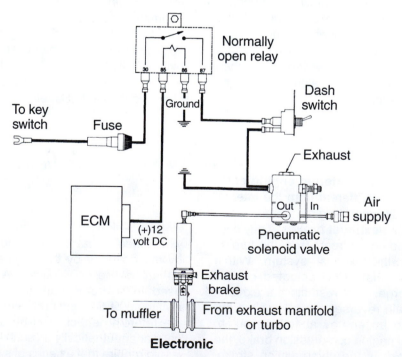

Electronic

FIGURE 13–1 *Control and actuation circuits on Williams mechanical and electronic exhaust brakes. (Courtesy of Williams)*

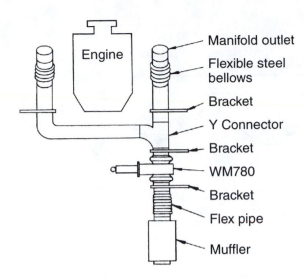

FIGURE 13–2   *Layout of an exhaust brake on a V-configured engine. (Courtesy of Williams)*

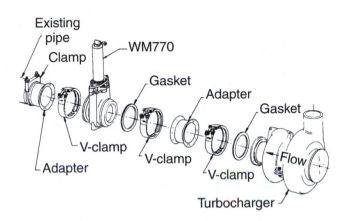

FIGURE 13–3   *Exhaust brake components. (Courtesy of Williams)*

JACOBS COMPRESSION BRAKES

The Jacobs family of engine compression brakes is classified as internal compression brakes. People in the trucking industry use the term **Jake brake** to refer to any internal engine compression brake and indeed, this manufacturer dominates the marketplace by designing its engine brakes for most of the major OEMs. However, there are other manufacturers of similar devices and more evidence of this can be seen in imported truck engines. Jacobs also manufactures driveline retarders.

As an internal engine compression brake, a **Jacobs brake** operates by having the piston perform the work of compressing the cylinder air charge on the compression stroke and then negating the power stroke by opening the exhaust valves somewhere around TDC and releasing the cylinder compressed air to the exhaust. Jacobs engine brakes use electric switching of a hydraulic actuating circuit, so they adapt readily to elec-

tronic management, becoming an integral component in "smart" cruise control systems.

All Jacobs brakes hydraulically actuate the opening of the exhaust valve(s) in the engine cylinder performing the braking. The manner in which this hydraulic circuit is timed and actuated varies with the engine; however, the mechanical force required to actuate the hydraulic circuit is always provided by the engine camshaft. This means that the timing of the exhaust valve opening is also governed by cam profile. The Jacobs brake was first designed for Cummins PT engines, so its operation on this engine is described first.

The Jacobs brake on a Cummins PT engine is managed electrically from a circuit that requires that switches at the clutch, throttle arm on the PT pump, and dash located control switch all be closed. This energizes a three-way solenoid valve located in the rocker housing that charges the Jacobs hydraulic circuit with pressurized engine lubricating oil. The solenoid acts as an electrically controlled pilot valve that merely controls the flow of oil into the hydraulic circuit of the engine brake controls. This oil flows through the hold-down control piston and when the PT injector rocker arm loads the plunger into the cup, oil from the hold-down piston control valve is trapped between it and the hold-down piston, preventing the PT injector plunger from lifting and therefore metering and pumping. This oil remains entrapped in this circuit until the three-way solenoid valve is de-energized. When the three-way solenoid is energized, oil also flows to the slave piston control valve and feeds a circuit, which has a master piston located directly above the push tube end of the injector rocker and a slave piston located at the exhaust valve bridge. The oil pressure first loads the master piston contacting the adjusting screw of the injector rocker. When the injector rocker travels upward (actuated by cam profile), the oil is trapped in the circuit by the slave control piston and the oil pressure is driven upward, acting on the slave piston located over the exhaust valves. The slave piston acts on the exhaust valve bridge, opening them and causing gas blowdown to the exhaust manifold.

A version of the Jacobs brake is available for most North American-built engines and the principles used to actuate each are similar. For instance, on engines using mechanical unit injectors (MUIs) and electronic unit injectors (EUIs) both of which have cam-actuated pumping strokes, the Jacobs brake hydraulic timing and actuation is almost identical to that described on the Cummins PT engine. On Caterpillar and Mack engines using hydraulic injector nozzles, the Jacobs brake hydraulic timing/actuation must be triggered by valve train movement over another cylinder under the same cylinder head.

An example of an engine brake actuated by movement from an adjacent exhaust valve is shown in

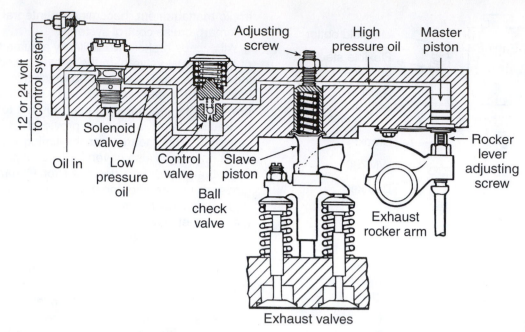

The blowdown of compressed air to atmospheric pressure prevents the return of energy to the engine piston on the expansion stroke, the effect being a net energy loss since the work done in compressing the cylinder charge is not returned during the expansion process.

Exhaust blowdown of the braking cylinder is accomplished by utilizing the pushrod motion of an exhaust valve of another cylinder during its normal exhaust cycle as follows:

1. Energizing the solenoid valve permits engine lube oil to flow under pressure through the control valve to both the master piston and the slave piston.

2. Oil pressure causes the master piston to move down, coming to rest on the corresponding exhaust rocker arm adjusting screw.

3. The exhaust rocker pushrod begins upward travel (as in normal exhaust cycle), forcing the master piston upward and directing high pressure oil to the slave piston of the braking cylinder. The ball check in the control valve traps high pressure oil in the master/slave piston circuit.

4. The slave piston (under the influence of the high pressure oil) moves down, momentarily opening the exhaust valves while the engine piston is near its top dead center position, releasing compressed cylinder air to the exhaust manifold.

5. Compressed air escapes to the atmosphere, completing a compression braking cycle.

The level of engine braking is controlled by using the solenoid to turn each housing ON or OFF. The above figure shows the relationships between master pistons, slave pistons, and control valves within the housing.

FIGURE 13–4   *Jacobs compression brake schematic (on Caterpillar or Mack) and operating description. (Courtesy of Jacobs Vehicle Systems)*

Figure 13–4. In this circuit, once oil is trapped in the engine brake hydraulic circuit, the master piston is forced inboard by exhaust rocker upward (opening) movement that creates pressure rise. Pressure in the engine brake actuation circuit acts on the sectional area of the slave piston, which in turn acts on the exhaust valve bridge on the cylinder being braked. The slave piston thereby forces the valve bridge downward, opening both exhaust valves, dumping cylinder pressure created by the piston compression stroke.

Figure 13–5 and Figure 13–6 show respectively sectional and exploded views of a typical internal engine compression brake. Remember that all internal engine compression brakes use the same general principles although you should expect to see minor differ-

ences in the hardware. Figure 13–7 shows the setup of a Jacobs brake used on a current Mack Trucks E-Tech engine.

PROGRESSIVE STEP ENGINE BRAKING

Engine braking can be used most effectively when the greatest number of levels (steps) of braking is available. Until recently, internal engine compression brakes have operated on steps that consist of braking those cylinders located under one cylinder head. For instance, an in-line six-cylinder engine with two cylinder heads would have two steps, one with three cylinder heads, three steps. In the Cummins ISX engine, engine braking can be achieved in six steps, meaning that under a braking

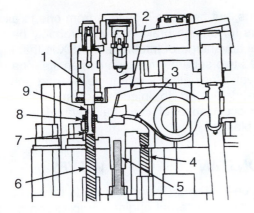

1. Jake slave piston
2. Rocker arm
3. Valve yoke
4. Inboard exhaust valve
5. Valve yoke pin (in cylinder head)
6. Outboard exhaust valve
7. Valve yoke hollow adjusting screw
8. Yoke adjusting screw jam nut
9. Exhaust valve actuating pin (activates outboard exhaust valve only, for jake operation)

FIGURE 13–5 *Jacobs brake on a Mack E-Tech engine. (Courtesy of Mack Trucks)*

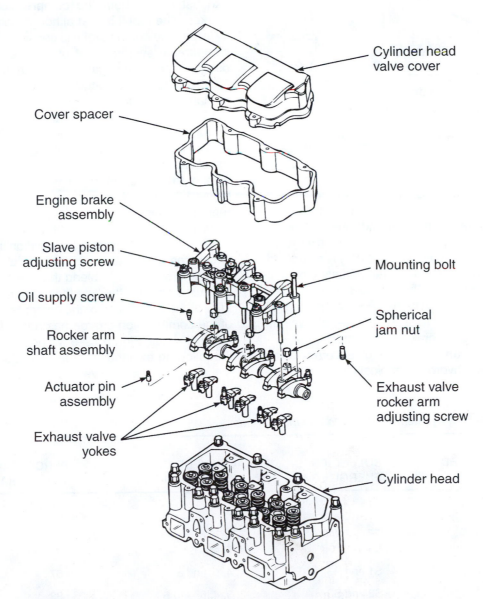

Cylinder head valve cover

Cover spacer

Engine brake assembly

Slave piston adjusting screw

Oil supply screw

Rocker arm shaft assembly

Actuator pin assembly

Exhaust valve yokes

Mounting bolt

Spherical jam nut

Exhaust valve rocker arm adjusting screw

Cylinder head

FIGURE 13–6 *Exploded view of Jacobs compression brake components. (Courtesy of Mack Trucks)*

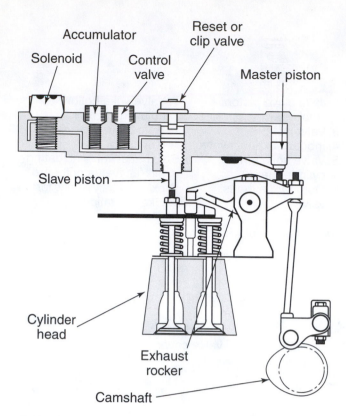

FIGURE 13–7 *Sectional view of Jacobs engine brake components.*

cycle, one cylinder progressing to all six can be used. Cummins call this Intebrake and it is achieved using engine oil as the hydraulic medium and three solenoids. When one cylinder is selected for braking, only the solenoid between cylinders one and two (S1) is energized. The solenoid between two and three actuates, braking for cylinders two and three. The solenoid between cylinders four and five (S2) actuates engine braking on cylinders four, five and six (S3). Table 13–1 shows how the Intebake progressive braking is effected.

InteBrake is still an internal engine compression brake and functions hydromechanically like other ver-

sions, but because it options from one to all six cylinders for engine braking, it will probably be copied. A real advantage of Intebrake on ISX is that this engine uses two separate function camshafts, one dedicated to valve actuation and engine braking and the other one dedicated to injector actuation. This reduces the occurrence of cam train-related engine-braking problems.

CONSTANT THROTTLE VALVES

Mercedes-Benz (Freightliner) has introduced a variation on the internal engine compression brake called the constantly open throttle valve or **CTV brake**. Small valves (the CTVs) are fitted into the engine cylinder head. These values allow some cylinder leakage to the exhaust during both the compression and exhaust strokes. The result is that although some braking ability is lost, the engine brake is considerably quieter than comparable systems that use the exhaust valves to dump compression gas into the exhaust system. A CTV is fitted to each engine cylinder and the system is managed electrohydraulically using engine oil as the medium. When activated, the CTVs remain open throughout the braking cycle. Actuation of the CTV brake requires that the accelerator pedal is at zero travel and the clutch is fully engaged. The system is designed to cease retarding when engine rpm drops below 1,100 rpm (factory preset), but this value can be reprogrammed up or down depending on application. Figure 13–8 shows an MB-906 CTV.

When the CTV circuit is actuated on the compression stroke, the constantly open throttle valve permits cylinder compressed air to bleed through the valve into the exhaust circuit. By the time the piston moves away from TDC at the completion of the compression stroke, most of the compression charge has been dumped into the exhaust circuit. Figure 13–9 will help guide you through the following explanation. After the electrical actuation of the solenoid slide valve (4) by the control module, the

TABLE 13–1   *ISX Intebrake Stepped Braking Options.*

NO. OF BRAKED CYLINDERS	SOLENOIDS ENERGIZED	ACTUAL BRAKED CYLINDER NUMBER	PERCENTAGE OF TOTAL BRAKING
1	S1	1	17
2	S2	2 and 3	33
3	S3	4, 5, and 6	50
4	S1 and S3	1, 4, 5, and 6	67
5	S2 and S3	2, 3, 4, 5, and 6	83
6	S1, S2 and S3	1, 2, 3, 4, 5, and 6	100

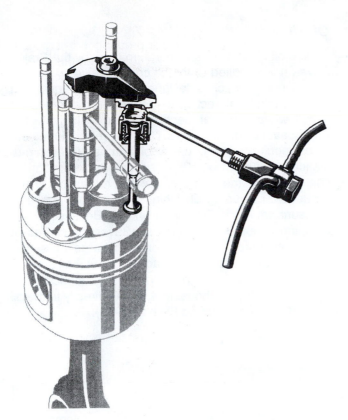

FIGURE 13-8 *Location of a CTV on the engine. (Courtesy of Mercedes-Benz)*

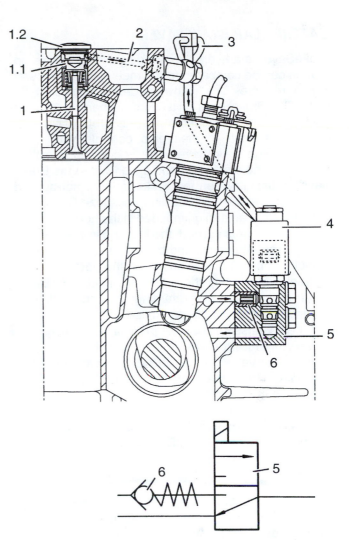

1.	Constantly-open throttle valve	2.	Oil port
1.1.	Constantly-open throttle piston	3.	Oil line
		4.	Solenoid slide valve
1.2.	Constantly-open throttle valve cover	5.	3/2-way valve
		6.	Pressure holding valve (0.2 bar)

FIGURE 13-9 *Components in a CTV engine brake system. (Courtesy of Mercedes-Benz)*

3/2-way valve spool (5) opens, allowing engine oil to be directed into the cylinder head to each CTV holding each open. Unintentional closure (fluttering) of the CTV by cylinder backpressure is prevented by a pressure-holding valve (6). To take the system out of braking cycle, the solenoid slide valve is de-energized, closing the oil supply at the 3/2 valve spool routing the oil back to the sump. This causes oil pressure in the pressure chambers above the CTV pistons (1.1) to collapse, and the CTVs close to seal the cylinder by spring force. The normal engine cycle resumes at this point.

The constant throttle valve retarder is internal engine compression braking with a difference. By not using the engine camshaft to manage the engine braking, camshaft problems associated with engine braking are eliminated. The system is also notably quiet in operation.

OTHER INTERNAL RETARDERS

Although they may be referred to as Jake brakes, there are other internal engine compression brakes available on diesel engines marketed in North America, many of them manufactured offshore. Volvo makes an engine brake for its VED-12 engine that uses two stages so it can be classified as an internal/ external engine retarder. Pacific Diesel Brake, known as PACBrake, also

offers an aftermarket engine brake. Any engine brake described as an internal engine compression brake uses almost identical principles, so the only factor that changes is the means of managing the braking cycle. For instance, the engine electronics are increasingly used to control engine retarding.

TECH TIP: Engine brake operation can be tough on camshafts on some engines, especially older ones. Where low power complaints and repeated top end adjustments are required on an engine, suspect camshaft failure and check cam lift to specifications.

CATERPILLAR BRAKESAVER

BrakeSaver is a hydraulic retarding device whose operation can be compared to a torque converter operating in reverse. The following description is of the Caterpillar BrakeSaver on a hydromechanical 3406 engine. This engine-mounted hydraulic retarder system also lends itself to electronic controls with minor changes. The BrakeSaver rotor is coupled directly to the rear of the crankshaft and driven within the BrakeSaver housing. When the BrakeSaver is not applied, the housing is not charged with pressurized engine oil, and the rotor can turn unimpeded in the housing. When the BrakeSaver is applied, the housing is charged with pressurized engine oil, and the rotor's ability to turn is modulated by fluid resistance and the pocket or vane geometry, producing the retarding effect. The Brake-Saver offers modulated retarding; the control circuit is electrically or manually controlled and pneumatically actuated. Chassis system air pressure is reduced by a pressure-reducing valve, which drops the pressure to a maximum value of 50 psi (345 kPa): this feeds the manual control valve and the solenoid control valve. Both the automatic solenoid valve and the manual control valve use air to control the flywheel-mounted oil control valve. The function of the oil control valve is to meter the flow of oil to the BrakeSaver housing.

Control Valves

The automatic control circuit requires an electrical circuit consisting of the ignition switch, mode selection switch, clutch switch, and accelerator switch. When the automatic option is toggled on the mode selection switch (manual/automatic), the BrakeSaver is actuated whenever the accelerator is released and the clutch and accelerator switches are closed. The automatic control circuit is "on/off". The steering column-mounted (usually) manual control valve meters the air delivered to the oil control valve. This controls the flow of engine oil to the BrakeSaver housing and thus the braking effort. A pneumatic two-way check valve prioritizes delivery of the actuation signal at the highest pressure to the oil control valve.

Oil Flow

Engine oil is the hydraulic medium used by the Brake-Saver. It is supplied to the oil control valve by the engine oil pump, which is designed with a front and a rear section. The front section feeds the engine lubrication circuit while the rear section feeds exclusively the BrakeSaver oil control valve. BrakeSaver-equipped engines require larger capacity oil sumps. When pneumatically actuated by either the automatic or manual control valve, the oil control valve routes oil to the BrakeSaver housing at a high rate of flow and at a pressure value of ±70 psi (480 kPa). Because of the pumping action within the BrakeSaver, pressures at the outlet increase 50% over the charge pressure. When the BrakeSaver is not actuated, a small amount of oil is pumped through the housing to lubricate the seals. Oil exiting the housing is routed directly to the oil cooler (Figure 13–10 and Figure 13–11).

SUMMARY

- Engine brakes are designed to complement the main vehicle brake system, not replace it.
- An engine compression brake's ability to absorb power is low compared to the brake capacity of the vehicle; however, a large percentage of vehicle braking uses less than one-fifth of the total vehicle capacity, so engine brakes can greatly extend the life of the vehicle foundation brakes.
- Most engine brakes are controlled by an electric circuit that requires that a series of switches is closed: at minimum the control switch, a clutch switch, and an accelerator or governor switch must be closed.
- Internal compression brakes use the electric control circuit to manage the brake's hydraulic circuit.
- The operating principle of an internal engine compression brake is to make the piston perform the work of compressing the cylinder air charge and then negate or cancel the power stroke by releas-

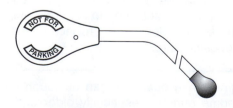

Manual Control Lever

BrakeSaver Control Selector

BrakeSaver Control Air Pressure

BrakeSaver Oil Pump

FIGURE 13–10   Caterpillar BrakeSaver controls. (Courtesy of Caterpillar)

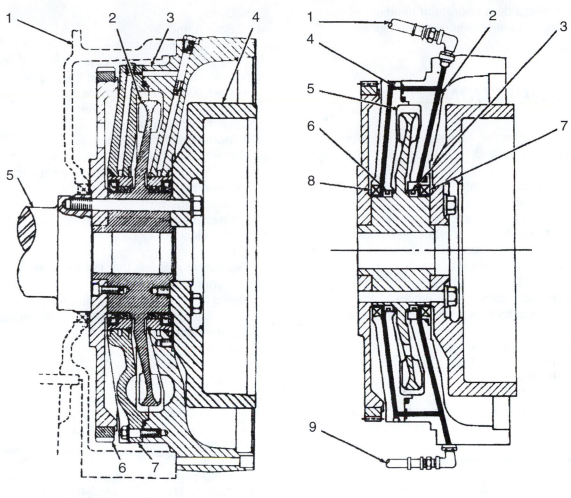

BrakeSaver Components

1. Flywheel housing 2. Rotor 3. BrakeSaver housing
4. Flywheel 5. Crankshaft flange 6. Ring gear plate
7. Stator

1. Oil Line 2. Orifice 3. Piston-type ring seal
4. Orifice 5. Chamber 6. Piston ring seal
7. Lip-type seal 8. Lip-type seal 9. Oil line

FIGURE 13–11   *Sectional view of the Caterpillar BrakeSaver. (Courtesy of Caterpillar)*

ing the cylinder charge by opening the exhaust valves at TDC on the compression stroke. This changes the engine's role from that of a power-producing pump to that of a power-absorbing pump.

- The mechanical force used to time and actuate the hydraulic circuit of a Jacobs type internal engine compression brake is rocker movement. This may be the movement of the PT, MUI, or EUI rocker, or in cases in which hydraulic injectors are used, movement of an exhaust valve rocker over a different cylinder.

- The engine external or exhaust brake operates by choking down the exhaust flow: the retarding stroke of the piston is therefore the exhaust stroke.

- Exhaust brakes are managed electrically and actuated pneumatically.
- Some engines use both an internal engine compression brake in conjunction with an exhaust brake enabling both upward strokes of the piston to be retarding strokes. This increases engine braking capacity significantly.
- The BrakeSaver uses the principle of a torque converter in reverse to absorb energy and retard vehicle speed.
- The BrakeSaver uses engine oil as its retarding medium. The engine oil is provided to the hydraulic brake from a larger capacity oil sump by a dedicated section of the oil pump.

- All engine retarders operate at optimum efficiency when engine rotational speeds are highest.

REVIEW QUESTIONS

1. Which type of engine brake uses the compression stroke of the piston as its retarding stroke?
   a. Internal compression brake
   b. External compression brake
   c. Hydraulic retarder
2. Which type of engine brake uses the exhaust stroke of the piston as its retarding stroke?
   a. Internal compression brake
   b. External compression brake
   c. Hydraulic retarder
3. Which type of internal engine compression brake does not use the camshaft to open the exhaust valves?
   a. Caterpillar BrakeSaver
   b. Jacobs or Mack E-Tech
   c. ISX Intebrake
   d. Mercedes-Benz CVT
4. The effective braking cycle on most internal engine compression brakes fitted to hydromechanical engines is actuated by which means?
   a. Electric over hydraulic
   b. Electric over pneumatic
   c. Pneumatic over hydraulic
   d. Hydraulic over electric
5. When will any kind of engine brake usually operate at peak efficiency?
   a. At idle rpm
   b. At peak torque rpm
   c. At rated speed rpm
   d. At the highest rpm
6. *Technician A* states that on most internal engine compression brakes, the effective braking stroke of the piston is the compression stroke. *Technician B* states that on some engine brake systems, both the compression and exhaust strokes can deliver retarding force. Who is right?
   a. A only

   b. B only
   c. Both A and B
   d. Neither A nor B
7. The operating principle of an internal engine compression brake is to convert the engine into an:
   a. energy-absorbing compressor
   b. energy-releasing machine
   c. inertia-absorbing device
   d. inertia-producing device
8. The energy of vehicle motion is ultimately converted to what energy form when an engine compression brake is used?
   a. Kinetic energy
   b. Heat energy
   c. Chemical energy
   d. Potential energy
9. *Technician A* states that the force used to actuate the hydraulic circuit in a typical Jacobs brake is delivered by an electrically actuated solenoid. *Technician B* states that Jacobs brake solenoids act as a pilot switch. Who is right?
   a. A only
   b. B only
   c. Both A and B
   d. Neither A nor B
10. When an internal engine compression brake is actuated, where does gas blowdown occur?
    a. At valve overlap
    b. Close to BDC following the power stroke
    c. At the beginning of the exhaust stroke
    d. Close to TDC at the completion of the compression stroke
11. Which of the following engine brakes offers progressive step braking using from one to all six cylinders?
    a. Cummins Intebrake
    b. Mercedes-Benz CTV
    c. Mack Trucks Jacobs on E-Tech
    d. Caterpillar BrakeSaver
12. When a Caterpillar BrakeSaver is used, which of the following is also true?
    a. External exhaust brake principles apply.
    b. A larger capacity engine oil sump is used.
    c. It must be controlled by the vehicle ECM.
    d. Retarding efficiency can never be modulated.

Chapter

14

Engine Removal, Disassembly, Cleaning, Inspection, and Reassembly Guidelines

PREREQUISITES

Chapters 2 through 13

OBJECTIVES

After studying this chapter, you should be able to:

- Describe the procedure required to remove an engine from a typical truck chassis.
- Disassemble an engine for reconditioning.
- Outline the process of cleaning and inspecting engine components.
- Understand the importance of systematically tagging components and connectors.
- Describe some of the key reconditioning procedures.
- Develop good inspection and failure analysis habits.
- Describe the procedure required to reassemble a diesel engine.
- Outline some of the reassembly steps that require special attention or precautions.

KEY TERMS

air conditioning (A/C)

anaerobic sealant

buttress

cab over engine (COE)

diamond dowels

flywheel housing concentricity

high-pressure washer

magnetic flux test

master bar

soak tank

spreader bar

Tempilstick

INTRODUCTION

Most of the procedures and terms used in this chapter have been used elsewhere in this textbook. The objective of this chapter is to put some of the content together in sequences typical of disassembly, cleaning, inspection, and reassembly of a truck diesel engine. It should never be used as a replacement for OEM literature, and remember that every engine and for that matter, every truck chassis, presents its own small and larger distinct problems that can only be addressed using product-specific guidelines.

REMOVAL OF AN ENGINE FROM A VEHICLE

Details of this procedure obviously vary from one vehicle to another and depend to large extent on the specific engine and whether it is hydromechanically or electronically managed. The following is intended as a rough set of guidelines and the sequencing of the steps required to remove an engine from a highway tractor chassis.

GETTING READY

1. It is good practice before beginning to pressure wash or steam clean the engine and engine compartment to remove road grime, grease, and oil. This will provide a clean work area that will enable more precise inspection and promote a more professional work environment.

2. Park the vehicle on a clean, level surface and ensure that the engine compartment can be accessed by the hoist required to lift the engine. Engage the vehicle parking brakes, block wheels, and ensure that there is sufficient bench space or a mobile steel cart on which removed components can be placed. Ensure that some adhesive labels and masking tape are at hand so that electrical wiring and fluid hoses can be identified. Failure to do this can considerably lengthen the reinstallation time. If the truck is of the **cab-over-engine (COE)** design, ensure that the cab lift hydraulics operate properly and that the positional locks are in place. With the COE chassis, the cab must usually be elevated to its extreme travel position, so check that the mechanical stops are in place and that the extended lift rams are not supporting the weight of the cab for the duration of the job.

3. If the engine is electronically managed, download the engine/chassis identification data and the customer data programmable options using a PC or ProLink and retain on a diskette or printout. Disconnect the main battery leads at the battery terminals.

WARNING: On electronic engines with a nonvolatile RAM component, this data is lost the instant the batteries are disconnected. Check the OEM instructions before disconnecting the batteries.

---

4. Remove the radiator cap. To avoid injury when removing a radiator cap, rotate it counterclockwise to the first stop, but do not depress. This will allow any pressure to bleed from the cooling system. After the pressure has completely dissipated, press the cap downward and continue turning to remove. Open the cooling system drain cocks usually located on the lower radiator tank and somewhere on the engine block to drain the coolant. If the coolant is to be reused, store in a sealed container: do not leave in drainage tubs exposed to the shop atmosphere. If the coolant is to be replaced (and it really should be at an engine overhaul), ensure that the used coolant is disposed of in an environmentally safe manner consistent with federal and local jurisdiction regulations.

5. Shut off the air supply to any engine pneumatic controls such as shutterstats, fan controls, puff limiters, air-controlled exhaust brakes, and any other pneumatic apparatus in the engine compartment that may have to be removed or disconnected.

6. Remove the oil pan plug and drain the engine oil. Remove the oil, fuel, and coolant filters and dispose of them and the engine oil in an environmentally safe manner consistent with federal and local jurisdiction regulations.

REMOVING THE ENGINE FROM THE CHASSIS

1. Disconnect the piping/ducting to charge air heat exchangers such as aftercoolers, air-to-air coolers, and tip turbine assemblies, labelling any plumbing that could be a problem during reassembly. Remove the intake ducting, capping the turbocharger intake and exhaust porting. Ensure that all intake ducts, air hoses, and the air filter assembly are capped to prevent contaminants from entering the intake system.

2. On vehicles with **air conditioning (A/C)**, it is often possible to remove the engine without losing the refrigerant charge. Remove the condenser from the heat exchanger cluster at the front of the engine housing and fully support it so that it is not

hanging by its hoses. Remove the A/C compressor, again without disconnecting its plumbing and fully support it somewhere where it is not going to hinder further work in the engine compartment. If it the refrigerant must be discharged, connect the appropriate recovery station and evacuate the system, plugging any open hoses.

3. Usually the radiator should be removed. In most cases, the radiator tie rods obstruct access to the engine, so these at least should be removed. To remove the radiator, when possible, leave the air-to-air type, charge air coolers attached to the rad assembly and attach chains to the radiator tie-rod upper brackets. Support the assembly using a shop crane (chain hoist or cherry picker). Next, disconnect the upper and lower radiator hoses. Inspect the hoses for cracks and general overall condition to note whether replacement is required: this should be done on removal of the engine from the chassis so that quotes are reliable and parts can be ordered when necessary. Remove the radiator lower mounting support bolts and rubber insulators. Lift radiator/air-to-air charge air cooler assembly from the chassis using the shop crane. Lift slowly and carefully to avoid damaging any components.

4. When the assembly is out of chassis, separate the two heat exchangers. Carefully, place the radiator upright on a flat surface. Use extreme care when handling heat exchanger cores—they are easily damaged. Remove the fan shroud, brackets, and other hardware if the radiator is to be recored.

5. Remove fan and fan hub assembly.

6. Clearly mark all the electrical leads before disconnection. Do the same with fuel lines, all linkages, oil lines, and water hoses. Cap all disconnected fuel lines to keep out dust.

7. Locate the rear engine mounts. These are either on the engine flywheel housing or on the transmission bell housing. When the rear engine mounts are located on the flywheel housing, the transmission must be fully supported. Position a mechanical transmission jack or blocks under the transmission. In some vehicles it may be necessary to completely separate the transmission from the engine *before* attempting to remove the engine.

8. Loosen and remove the capscrews that hold the transmission bell housing to the engine flywheel housing.

9. Obtain or fabricate a suitable **spreader bar** and lifting chain so that the engine can be lifted without lift chain contact to critical engine components. Fit engine lifting eyes to the OEM recommended location on the engine. Move into position the hoist to be used to lift the engine from the chassis. Attach lifting chain hooks to the engine lifting eyes.

CAUTION: Never attempt to lift an engine using only rope slings or cable slings. Never support the engine on its oil pan and never place any kind of sling around the oil pan.

---

10. After the hoist is chained to the engine lift eyes, apply a slight load to the lifting chain.

11. Remove the engine mounting bolts. Perform a thorough visual inspection to ensure that everything that has to be uncoupled is and that nothing is obstructing the planned removal path of the engine.

12. Separate the engine from the transmission, taking care not to force anything or overload the transmission input shaft. Use the hoist to remove the engine from the chassis. This may be a two-person operation, depending on the chassis. If unfamiliar with the chassis, ensure that someone is around to assist, even if it means just having an extra pair of eyes.

13. Remove external accessories as required: alternator, starter, power steering pump, oil, coolant and fuel filter mounting pads, and anything else on the block that might obstruct the engine overhaul mount stand. Bolt the engine into the overhaul stand and do not release the weight from the hoist until the engine stand mounting plate bolts are tight.

14. Remove the hoist chain hooks and then the lifting eyes from the engine unless they were originally on the engine.

15. Remove the turbocharger oil supply and return lines. Remove the turbocharger and assess its condition.

16. Remove the crankcase breather assembly if fitted.

TECH TIP: Always work on an engine with the assumption that it may have to be reinstalled by some one else. Tag every electrical wire and hose coupling. Work cleanly and methodically. Use several containers to put fasteners in and label each container by component and location.

 ENGINE DISASSEMBLY

In most cases, the engine will be removed from the location of the chassis before it is disassembled. If an engine is to be rebuilt in a general area of the shop floor, it is essential to ensure that no work, such as welding, be undertaken in close proximity to the engine work.

1. On engines that use integral or unit high-pressure injection pumps, remove the injection lines and any electrical wiring by separating the connectors. On engines using integral hydraulic injectors accessible without removing the valve covers, remove the injectors, capping the leakoff and feed ports. On some engines the high-pressure pipe is connected to a cylindrical injector through a port in the rocker housing. The two nuts that locate and seal this connector pipe must be backed off and the pipe removed before any attempt is made to remove the injector.
2. Remove the exhaust manifold assembly.
3. Disconnect the manifold boost sensor plumbing, the electrical wiring, and devices such as puff limiters, where applicable, from the intake manifold assembly, then remove the intake manifold.
4. Remove the water manifold assembly, complete with the thermostat housing when possible.
5. Remove the vibration damper and thoroughly inspect to assess its serviceability. Next, remove the crank pulley and crankshaft hub assembly. Usually this requires the removal of the hub retaining capscrew(s) and then using a universal "T" puller (Figure 14–1).

FIGURE 14–1 *Removing a pulley using a T-puller. (Courtesy of Caterpillar)*

WARNING: Exercise extreme caution when pulling the hub from the crankshaft—damaging a crankshaft is likely to more than erase any profit that might be earned on the engine overhaul. Always consult the engine OEM service literature.

6. Remove the engine timing gear cover. If the timing gear cover has thrust buttons fitted, back these off before removing. On engines where the timing indicator is fastened by the timing gear cover screws, remove them first and ensure that they are not damaged.
7. Remove the oil cooler assembly.
8. Disconnect any plumbing to the water pump not already removed, then remove the water pump itself.
9. Remove the clutch assembly from the flywheel and assess its serviceability.
10. Remove the flywheel assembly using a suitable hoist.
11. Remove the flywheel housing and any attached components, checking for sensor wires. Check for eccentricity at the locating dowel holes.
12. Remove all accessory drive components not already removed, such as power steering pumps, air compressors, and so forth. Remove accessory drive pulleys using a suitable puller (Figure 14–1).
13. Remove the oil filler and dipstick tubes.
14. Remove the mechanical tachometer drive, if fitted.
15. If the engine is fitted with electronic unit pumps (EUPs), remove them and store them as per OEM recommendations. If the engine has a high-pressure injection pump, remove any electrical wiring from the rack actuator housing (electronically managed, port-helix metering), transducer module, or governor housing. Disconnect the throttle arm and fuel stop arm from hydromechanical pumps. Check that all the plumbing is disconnected from both sides of the pump and remove any support brackets: loosen the pump mounting capscrews, and, using a smooth motion, pull the injection pump away from its mounting flange. Pressure-time, common rail (PT) pumps and CAPS pumps can be removed in a similar fashion. The various types of transfer pumps used with the electronic unit injector systems (EUIs) are relatively straightforward to remove, but the OEM literature should be consulted whenever hydromechanical or electronic fuel pumping apparatus is removed from the engine. When performing a complete engine overhaul, it may be part of the procedure to recondition or replace with rebuilt/exchange fuel injection components, so check with the OEM and shop work order instructions.
16. Remove the rocker housing covers.

17. Remove the rocker shaft assemblies including valve bridges where used. On overhead camshaft engines, remove the camshaft assembly.

TECH TIP: It is recommended by most OEMs that rocker arm shaft assemblies and valve bridges/yokes be tagged for position to help maintain the same wear surfaces if they are to be reused.

---

18. Remove the injectors, using the appropriate puller. When cylindrical hydraulic injectors are used, make sure that all the plumbing is clear of the injector seat before attempting to remove them. Where EUIs, electrohydraulic injectors, and HEUIs are used, ensure that all electrical and hydraulic connections are removed. Check the injector nozzle gasket: if it is not on a removed injector, remove from the injector bore with a magnet or small tapered rod. Do not leave in the cylinder head because this will cause problems on reassembly.
19. Remove the push tubes or rods where fitted. It is good practice to inspect each for serviceability on disassembly. Some OEMs recommend that the push tubes be tagged for position in the engine.
20. Remove fuel manifolds (internal charge and return lines) where fitted, jumper pipes, and crossover pipes.
21. Remove all the cylinder head capscrews and washers. Remove the cylinder head(s).
22. Remove the cylinder head gasket and the sealing grommets and fire rings if not integral with the cylinder head gasket. Inspect the cylinder head gasket for signs of failure, paying special attention to the fire rings.
23. Remove the oil pan capscrews and the oil pan.
24. Remove the oil pump, which may be located in the crankcase or outside fastened to the engine cylinder block.
25. Before removing each piston and connecting rod assemblies, scrape carbon deposits (ring ridge) from the upper inside wall of each cylinder liner using a flexible knife.
26. Plan to remove the pistons on an in-line six-cylinder engine in companion pairs; that is, 1 and 6, 2 and 5, 3 and 4.
27. First, remove the connecting rod bearing capscrews and separate the cap from the rod by tapping with a light-duty nylon hammer.
28. Remove each connecting rod and piston assembly carefully, guiding the rod bottom end clear of the piston cooling oil nozzle and avoiding contact with the liner wall. Arrange the piston assemblies sequentially on a bench, making note of any unusual characteristics.

29. Remove the cylinder block located piston cooling nozzles. Oil will be trapped in the gallery that supplies the piston cooling jets, so be prepared to capture it. If the engine is rotated to an angle on the stand, remove either 1 or 6 first and allow the oil in the gallery to drain.
30. Disassemble the engine timing gear train. Remove accessory drive gears, idler gears, and an in-block camshaft gear and camshaft as an assembly. Rotating the engine upside down will ease the removal of an in-block camshaft because this will drop the cam follower/lifter assemblies out of the way so they do not interfere with the cam lobes and journals as the shaft is withdrawn.
31. Remove the cam follower/tappets/valve lifter assemblies. If the camshaft is to be reused, inspect and tag individual lifters for reassembly to position. Where cam follower housings are used, tag for position. In engines where the cam follower housing shimming defines engine timing, it is advisable to measure and record the shim (gasket thickness) with a micrometer even when the engine is to be retimed on reassembly.
32. Disassemble any remaining gears in the engine gear train and remove any auxiliary or balance shafts remaining in the engine cylinder block.
33. Remove the crankshaft rear seal housing assembly.
34. Remove all the camshaft bushings using the correct sized driver and slide hammer.
35. With the engine still inverted, remove the main bearing cap capscrews, the cap brackets, and the main bearing caps by gently tapping with a nylon hammer. Tag the bearings by location for inspection and failure analysis purposes.

CAUTION: Some lightweight cylinder blocks use transverse buttress screws in the main caps.

---

36. Fit a crank-lifting yoke to the crank throws on a pair of companion cylinders (preferably 2 and 5 on an in-line six): the yoke should have rubber conduit installed over the throw hooks to prevent damage to the throw journals. Lift the crankshaft free from the cylinder block using a hoist.

TECH TIP: Due to the weight of the crankshaft, extreme care must be observed during removal. Lift the crankshaft straight up to avoid damage. No scratches, nicks, burrs, or any other kinds of distress are permitted on the main or throw journals and their fillets.

---

37. Remove the cylinder liners using a suitable puller. Wet liners are retained by O-rings and do not

usually require a great amount of force to separate them from the block. Dry liners and combination wet/dry liners may require considerable force and hydraulic and air over hydraulic pullers with the appropriated adapter plates are usually required. In cases where even hydraulic pullers fail to move the liner, a single vertical down welding bead using a 3.2 mm (⅛") deep penetration electrode (such as E6010 or E6011) followed by the immediate application of cold water will often work. Whenever such a practice has to be resorted to, it should be recognized that a mistake may destroy the cylinder block. Always ensure that the welding is performed well separated from any engine components. And always use a vertical down technique to avoid overheating the liner.

38. All expansion (frost) press fit plugs and threaded oil passage plugs must be removed from the cylinder block before cleaning.

TECH TIP: Although cleaning and inspection is covered in the following section, it is important that an engine is not simply ripped apart. Technicians should get in the habit of inspecting every component as it is removed and tagging it when it is important that it is reinstalled in the same location.

CLEANING AND INSPECTING COMPONENTS

The cleaning and inspection of an engine is a critical stage in the engine reconditioning procedure. If the reason for the engine rebuild is a failure, ensure that the cause is identified before any attempt at reassembly is made.

1. Use a gasket scraper to remove all the gasket material and heavy dirt from the cylinder block. Install the cylinder block into a preferably heated, **soak tank** using a heavy-duty alkaline soak cleaner for a period of one to two hours. One OEM reports that 1/16" of coolant scale has the insulating properties of 4" of cast iron. It is therefore important not to skip this procedure.

CAUTION: Use extreme care and wear protective clothing when working with alkaline solutions.

---

2. Remove the cylinder block from the cleaning or soak tank. Thoroughly flush the cylinder block using a shop **high-pressure washer**, ensuring that all particles and sludge produced by the block boiling are removed.

3. Visually inspect the cylinder block, checking all the coolant passages and ensuring they are clean and unobstructed. Ream or drill out if necessary to dislodge any deposits.

4. Check to see that there are no casting fins or residues that might obstruct coolant flow by removing any casting irregularities with a pry bar.

5. Run a cylindrical wire brush in all the oil passages to ensure they are unobstructed.

6. Flush the oil passages with air and solvent.

MAGNETIC FLUX TEST

It makes sense to **magnetic flux test** engine cylinder blocks, crankshafts, and all connecting rods at every major engine overhaul. These processes are neither expensive nor time consuming and as the consequences of a single warranted engine failure out of 20 overhauls will demolish the profits of the other 19, it is shortsighted to overlook it. Remember, this expense is passed on to the customer, and if it is to be passed over, it is the customer who should make the decision: get a refusal in writing. The usual drawback is not having the equipment on site, but in most cases, the equipment is accessible and machine shops usually pick up and deliver.

ENGINE REASSEMBLY GUIDELINES

Although engines may be disassembled with little or no reference to a workshop manual (not recommended practice), they should be assembled precisely according to the sequencing in the OEM workshop manual. Maybe the only excuse for not observing this practice could be made by the technician who continually overhauls the same engine series and who keeps abreast of every OEM technical service bulletin (TSB). Even for the experienced engine overhaul technician, using the manual as a guide is simply good practice. For the novice engine overhauler, the workshop manual should guide every move: it is important that short cuts are never experimented with. As every experienced diesel technician will acknowledge, there are a few short cuts to some procedures, but these must involve a zero risk factor, and that requires an intimate knowledge of the engine that only comes with experience. To provide a general guide to an engine assembly procedure would simply reverse the general guidelines for disassembly (Photo Sequence #1 shows the disassembly/reassembly of an engine), and the point is engines cannot be successfully rebuilt according to general guidelines but

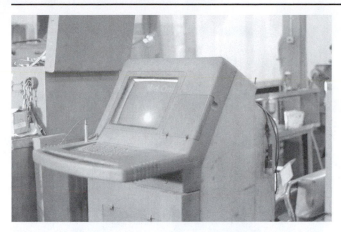

PS1–01   Prior to working on the engine, download the engine and chassis service and repair history from the engine ECM and the proprietary data hub. It is good practice to download a printout of all customer data programming at the beginning of the procedure.

PS1–02   When anticipating the possibility of major engine work, take the time to pressure wash the engine exterior and the engine compartment.

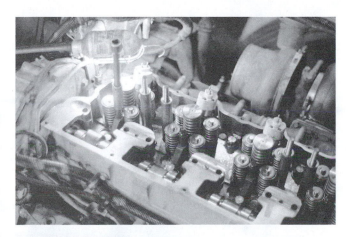

PS1–03   Preliminary diagnosis of an engine problem with the rocker covers removed.

PS1–04   Once it has been determined that the engine must be removed from the chassis, the hood and radiator assembly should be removed to enable proper access to the engine compartment.

PS1–05   Often an engine ends up partially disassembled during the troubleshooting process while it is still in the chassis: ensure that the disassembly of components is performed methodically and the work area is well organized.

PS1–06 *This engine had the oil pan and the cylinder head removed before it was determined that the engine would have to be removed from the chassis: always ensure that lifting eyes are securely fitted to the engine cylinder block before attempting to remove the engine. The hoist must be rated to easily handle the weight of the engine as the removal process usually requires some levering and twisting.*

PS1–07 *The process of resurfacing the journals on an induction hardened crankshaft using a crank grinding lathe.*

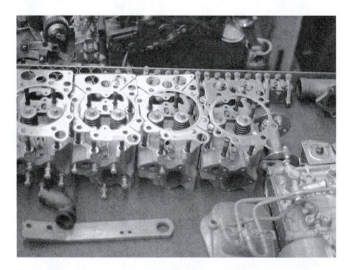

PS1–08 *The reassembly area should be well organized so that the assembly can proceed methodically as outlined in the OEM service literature.*

PS1–09 *The engine cylinder block should be mounted in an engine stand with a rotating turntable for reassembly.*

must be assembled in the precise sequence outlined by the OEM. However, the following lists some general precautions that the engine rebuilder may encounter when assembling engines.

O-RINGS ON WET LINERS

Use the lubricant recommended by the manufacturer. This may be antifreeze, engine oil, silastic, or nothing at all. Using an inappropriate lubricant may chemically interact with the "O" material and cause it to swell, contract, or disintegrate, generating a premature failure.

Use of Anaerobic Sealants (Silicone, Silastic/RTV, Etc.)

The use of **anaerobic sealants** has become popular in recent years. Only use where specified, and do not overuse as it may plug up apertures.

NOTE: Many sealants do not cure when in contact with any type of oil or grease.

Test the Cylinder Block Line Bore

This check is performed most easily using a **master bar**, which is clamped into the block line bore without the bearings in place. The test is used to determine whether the cylinder block line bore must be machined. After torquing the main bearing caps, the master bar should rotate by hand: binding indicates that line boring is required.

Cylinder Block Fretting and Warpage

This should be checked to specification using a straightedge and thickness gauges. Today's high brake power, lightweight engines make this a critical check because they are more subject to torque twist.

Sleeve Height Protrusion Specification

This is a critical specification. The protrusion variance of the liners under one cylinder head determines how evenly each liner is clamped to the cylinder block.

A liner protrusion error can damage a cylinder head and/or cause an immediate head gasket failure.

Piston Cooling Jets

Always check the directional aim using the appropriate tooling, usually a perspex template and rod. Some cooling jets may be bent to correct aim, some may not: check OEM literature.

Heat Shrinking Gears to Shafts

A bearing hotplate may be used to heat interference fit gears to the OEM specified temperature. A kitchen toaster oven (dedicated to the purpose of heating engine components) has an accurate thermostat and works better: the gear is more evenly heated. **Tempilstick** crayon may be used to check the temperature of a heated component. Do not use an oxyacetylene torch to heat components because the steel may carburize. Gears should not be heated for periods exceeding 45 minutes.

Buttress Screws on Main Bearing Caps

Some engines use **buttress** screws on some of the main bearing caps to add rigidity to the block assembly. These are usually installed after the main bolts have been torqued: do not omit them. Cylinder blocks for the high-horsepower engines of today are made from castings of higher strength and lower weights, enabling them to flex more than those of a generation ago, which is the reason the buttress screws are now engineered into some engines.

Diamond Dowels

Diamond dowels work to retain the alignment of mated components much better than cylindrical dowels. However, all the critical alignment checks should still be performed. For instance, if a flywheel housing has been removed, the housing should be radially indicated (dial indicator) for its concentricity to the crankshaft even when diamond dowels are used. If they are used, there is much less chance of an eccentric measurement (Figure 14–2).

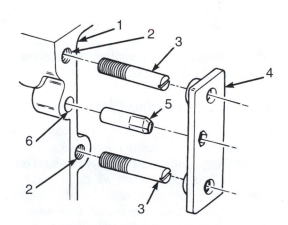

1. Engine cylinder block
2. Flywheel housing bolt holes
3. Guide stud
4. Diamond dowel alignment / installation fixture
5. Diamond dowel
6. Diamond dowel interference fit bore

FIGURE 14–2 *Diamond dowel installation. (Courtesy of Mack Trucks)*

Rear Cam Bushings

When performing an in-chassis engine overhaul, when the cam bushings are removed, remember that the rear cam bushing in some engines may not be removed without unseating the rear cam plug. Because the replacement of the rear cam plug involves the removal of the transmission/clutch assembly and flywheel housing—procedures that are unlikely to have been factored into the estimate—it is advisable to leave the rear cam bushing in place. Record the fact that it was not replaced on the hardcopy of the work order.

Connecting Rod Weight Classifications

Connecting rod weights are coded by manufacturers. When replacing connecting rods, it may be possible to either increase or decrease the weight class by one increment, but not more. The consequence of an overweight or underweight rod is to unbalance the engine and generate a premature failure. Always check OEM technical literature.

Ring End Gap

A procedure often overlooked by experienced diesel technicians probably due to the reliability of machining and packaging today (that is, the nominal dimension is usually correct), checking the ring end gap specification takes little time and should be done. Insert the ring into the cylinder bore and measure the gap with thickness gauges.

Piston Wrist Pin Retainer Snap Rings

These should be installed flat side out and with the gap either up or down (varies with OEM, but usually down) and never crossways to the direction of piston travel. The inertia forces on acceleration or deceleration acting on the snap ring have been known to unseat them.

Checking Main Bearing Clearance

This procedure is designed to be performed with the engine upside down using Plastigage™; in other words, when the engine is being overhauled outside of the chassis. The specification will differ and have less relevance when performed with the engine in-chassis. Identify the OEM specification first, then select the appropriate plastigage color code. When making the measurement, ensure that the main bearing cap is torqued to specification.

Rod Side Play

This is a critical specification: slightly cocked rod caps can damage crankshafts and promote rapid bearing failure. The use of a pair of equal-sized thickness gauges close to the specification value inserted on one side of the cap may assist with alignment problems while torquing the rod cap. Cracked rods used on some engines make this check unnecessary.

Pressure Testing Cylinder Heads

It is preferable to run hot water through cylinder heads for about 10 minutes to heat them before hydrostatically testing them. Hot water and shop air regulated at 100 psi should locate any leakage problems. Consult with an experienced technician to check whether that engine series has a track record of cylinder head problems and where leakage problems occur.

Cylinder Head Warpage

Check the cylinder for head warpage with a straightedge and thickness gauges. This test is critical in today's low-weight, high-brake power engines especially those using single slab castings on 6-cylinder engines.

Measure and Recut the Cylinder Head Fire Ring Groove

Measure to specification and recut when necessary. This is not a complex procedure but it does require the correct tooling.

Valve Margin

When dressing valves, remember that the valve margin specification is critical, and if it cannot be met, the valve should be replaced.

Valve Interference Angle

Valves are almost never designed with an interference angle in today's diesel engines because this would compromise the valve's ability to transfer heat to its seat due to the reduced direct contact surface area. When valve rotators are used, an interference angle is never machined into the valve.

Cylinder Head Alignment

When installing multiple heads on an engine block deck, align the heads before torquing using a straightedge. Perform this even when the cylinder block has cylinder head alignment dowels. This enables improved sealing and lower stress loads on the manifolds.

Most current cylinder head gaskets are of the integral design; that is, the fire rings and sealing grommets are incorporated in the head gasket template. However, in nonintegral gasket designs, especially those installed on a cylinder block deck that is angled (such as on a V engine), always use alignment dowels and eyeball the gasket components before finally decking the cylinder head.

Setting Valves and Injectors

When setting cylinder valves and cam-actuated injectors, always use the correct engine locations to perform the adjustment. Novice technicians may observe experienced technicians taking shortcuts that save time and do not endanger the engine. However, until a technician is completely familiar with a specific engine series, always go by the book. The general procedure for setting valves is outlined in Chapter 8.

Injector Installation

When installing hydraulic, electrohydraulic, MUIs, EUIs, and HEUIs, always use the OEM recommended procedure. Lubricating the O-rings with the correct medium is critical. Generally, the use of antiseize compound on cylindrical injectors with O-ring seals is never approved and where the technician may believe he is making the next removal of that set of injectors easier, their removal may come a lot sooner than anticipated. Use the correct lubricant medium for the O-rings and accept that occasionally injectors do seize in their bores.

Torque High Pressure Injection Pipe Nuts

Overtorquing high pressure injection pipe nuts at the injector and the injection pump can ridge the nipple seat and more significantly, collapse the aperture, reducing flow area; this reduction in flow area will create lack of power problems. Torquing injector lines using a line wrench socket and torque wrench is good practice and will avoid damaging the high-pressure pipe nipple and seat.

VERIFYING TRUE TDC

When a piston reaches the top of its travel on its upstroke, it remains stationary while the crank throw turns through its apex before beginning the downstroke. True top dead center is the exact midpoint between the moment the piston stops moving upward and the moment it begins its downstroke. A timing indicator must be set at true TDC and not simply at any point where the piston is at the top of its travel to ensure that components such as injection pumps are accurately timed to the engine. The following list gives the procedure for determining true TDC for a typical four-stroke cycle, in-line, six-cylinder engine.

1. Locate the fixed engine timing marker and rotating calibration scale: this may be on the harmonic balancer pulley, any other pulley driven at engine speed, or on the flywheel. The procedure may be performed on either 1 or 6 engine cylinder. Ensure that the fuel system is no-fueled either mechanically or electrically, whichever is appropriate. Manually bar the engine in its normal direction or rotation to locate cylinders 1 and 6 at indicated TDC using the fixed timing indicator or by making one out of mechanic's wire and clamping it close to where the fixed timing indicator is to be positioned. The flywheel (using a gear and ratchet barring tool) or crank hub (using a barring fixture) should be used to rotate the engine.

2. If the cylinder heads are installed, remove the 1 or 6 injector and install in its place a dial indicator fitted with an extension probe long enough to contact the piston crown. In cases where the cylinder heads are removed from the engine, the dial indicator may be positioned on the cylinder block with the probe contacting the piston crown. Next, zero the indicator at the highest point of piston travel by barring the engine slightly both sides of TDC.

3. Cut a 4″ strip of masking tape and place it on the flywheel or pulley calibration scale.

4. If the dial indicator was properly zeroed, barring the engine in either direction (BTDC and ATDC) will produce a negative scale reading on the indicator. Pick an arbitrary value such as that represented by a complete rotation of the dial indicator (0.050″ or 0.100″: for the purposes of this explanation of the procedure, 0.050″ is used), then bar the engine BTDC until the chosen value is exceeded by 0.030″, that is, until the indicator reads 0.080″. Next, bar the engine back in the normal direction of rotation until 0.050″ is read at the dial indicator. With a pencil, draw a line under the fixed or temporary timing indicator. Then reverse bar the engine until 0.050″ is read at the dial indicator. Draw a second line under the fixed or temporary timing marker. The reason for turning the engine 0.030″ beyond the selected value is to eliminate backlash factors while performing this procedure.

5. Next, place a third mark exactly between the first two. One way of accurately performing this is with a knife. Neatly cut the masking tape at both of the two lines drawn and fold back one end of the masking tape so that the two cut ends align. Rotate the engine until the fold crease is positioned under the fixed or temporary timing marker. When the engine is rotated back to the exact midpoint between the two lines, it will be positioned at true TDC.

6. With the engine located at true TDC, the tape can be removed and the fixed timing marker adjusted to the TDC point on the engine calibration scale.

INDICATING A FLYWHEEL HOUSING

Any time a flywheel is removed from the cylinder block, the housing inner flange face concentricity with the crankshaft should be checked using a dial indicator. The maximum tolerance for the crankshaft to flywheel housing eccentricity is low, typically around 0.012″ (0.3 mm) total indicated runout (TIR), and the consequences of installing a flywheel housing that exceeds the allowable specification are severe. When this occurs the drive axis is broken, a condition that can result in clutch, engine mount, transmission, and engine failures. The flywheel housing to crankshaft concentricity is preserved by interference fit cylindrical or diamond dowels. These may be relied on to properly realign the flywheel housing on the cylinder block at each reinstallation, but the specification is so critical, it should be checked. The following procedure outlines a method of checking **flywheel housing concentricity**, and a couple of typical strategies for correcting an out-of-specification condition.

1. Ensure that the engine is properly supported. Indicating a flywheel housing may be performed with the engine in-chassis or out; obviously a procedure such as this is made rather easier if the engine is out of chassis. Locate the flywheel inner flange face to crankshaft concentricity TIR tolerance in the OEM technical literature. Either mechanically or electrically ensure the engine fuel system is no-fueled.
2. Mount the flywheel housing to the engine cylinder block using the dowels to align the assembly and snug the flywheel fasteners at about half of the OEM specified torque value.
3. Using fabricators chalk, mark the flywheel flange face with indices in the following rotational positions: NE, NW, SW, SE, or N W S E.
4. Next, thoroughly clean the inside face of the flywheel housing with emery cloth and fix a magnetic base dial indicator to any position on the crankshaft, setting the probe to contact the flywheel housing inside face. Using an engine barring tool, rotate the engine in its normal direction of rotation until the indicator probe is positioned at any one of the chalk indexes. Now set the indicator at zero.
5. Bar the engine through a full rotation, recording the indicator reading at each of the indexes. The indicator should once again read zero when the revolution is complete. If the readings were as follows: NE:0, NW: minus 0.003″, SW: minus 0.005″, SE: plus 0.004″, the TIR would be the highest negative value (0.005″) added to the highest positive value (0.004″) giving a reading of 0.009″. If the OEM TIR maximum specification were 0.012″, this reading would be within it (Figure 14–3).

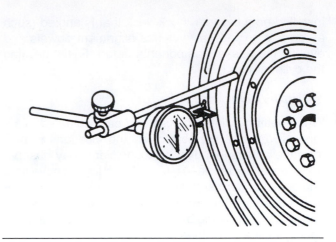

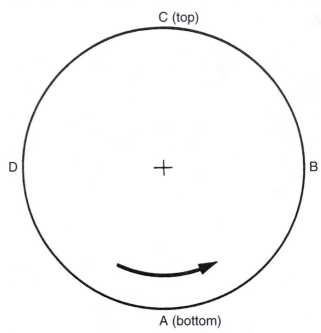

FIGURE 14–3 *Caterpillar recommended method for checking flywheel housing radial concentricity. (Courtesy of Caterpillar)*

If the flywheel housing to crankshaft concentricity is outside of the specification, it must be reset. The following outlines this procedure for an engine that aligns the flywheel housing using cylindrical locating dowels. Begin by removing the dowels. Check the availability of oversize locating dowels with the parts department. Loosen the flywheel mounting capscrews until they are barely snug, permitting the flywheel housing to be moved fractionally when struck with a rubber mallet. Perform the flywheel housing concentricity sequence as in the foregoing list. Adjust the position of the flywheel housing by tapping with the rubber mallet on the basis of the readings of the dial indicator. Repeat the procedure until the readings meet the required specification. When they do, fully torque the flywheel

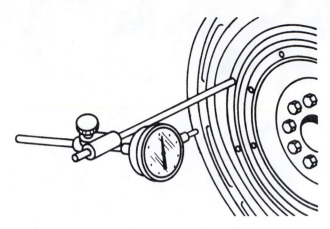

FIGURE 14–4 *Checking flywheel housing axial concentricity. (Courtesy of Caterpillar)*

housing capscrews to specification. Next, the locating dowel holes in the cylinder block and flywheel housing will have to be reamed to an oversize specification factoring in the interference fit requirement. Drive the oversize dowels into the reamed holes. Next, check the flywheel housing axial concentricity as shown in Figure 14–4. If this is out of specification, machining is required.

Reinstall Engine to Chassis

If an adequate amount of care was taken in removing the engine from the chassis, the reinstallation should be straightforward. Usually this can be expedited in less time than the removal as the components should be clean and the hoses and connectors tagged. Never hesitate to obtain assistance, especially in snub nose conventional chassis with an engine compartment that extends into the cab.

SUMMARY

- Most of the procedures outlined in this chapter are covered in greater detail elsewhere in the book. The content in this chapter is intended to be used as a very general guideline for the novice technician.
- Ensure the work area around the chassis and the engine during rebuild is clean, organized, and uncluttered.

- The importance of tagging components, lines, linkages, and connectors cannot be overemphasized.
- Always work on an engine so that if another technician were to take over the rebuild at any point in the sequence, he or she would have no problem in determining where they are.
- Remember to note the condition of every engine component as it is removed for failure analysis, especially until a definitive diagnosis of the cause of the failure has been made.
- When the reason for the engine rebuild is a failure, never begin the engine reassembly before defining the failure.
- Always use the OEM technical service literature when undertaking engine overhaul.

REVIEW EXERCISES

The contents of this chapter are entirely practical. Simply, it is the application of some of the theory dealt with in previous chapters. At this point it is probably essential to take down and reassemble a truck diesel engine and put to practice some of the theory covered thus far. The following are some tasks to prepare you for the experience.

1. Identify the engine to be worked on and obtain the requisite OEM literature.
2. Review the disassembly procedure of any diesel engine according to the OEM workshop manual procedure. Ensure that writing paper and a pencil are available and record any observations, especially those procedures that are not common to all engines.
3. Review the OEM recommended cleaning and testing procedure for all the engine components including the cylinder heads, cylinder block, connecting rods, and crankshaft. List all those procedures requiring the use of specialized equipment.
4. Review the OEM recommended reassembly procedure for the same engine, once again noting those procedures that are distinct to that particular engine.
5. Review the OEM requirements for initial startup of the engine following an engine overhaul. Make a list of each step in the procedure. Take special notice of the recommended break-in procedure.

Chapter

15

Engine Run-In and Performance Testing

PREREQUISITES

Chapters 4, 6, 7, 8, 9, 10, 11, 12, 13, and 14. Some reference to Sections 2 and 3 of this textbook is desirable.

OBJECTIVES

After studying this chapter, you should be able to:

- *Understand the basics of power analysis and dynamometer testing.*
- *Outline the OEM requirements of engine run-in after rebuild.*
- *Describe the process required to check out both the engine and vehicle before a chassis dynamometer test.*
- *Outline personal and equipment safety while operating an engine or chassis dynamometer.*
- *Describe the objectives of run-in and performance testing on a dynamometer.*
- *Interpret the data on a dynamometer test profile.*
- *Describe how a heavy-duty truck is installed in a chassis dynamometer test bed.*
- *Outline the objectives of a chassis dynamometer performance and engine run-tests.*

KEY TERMS

brake horsepower (BHP)

chassis dynamometer

droop curve

dynamometer

dyno

engine dynamometer

engine longevity

linehaul

peak torque

prelubricator

preventative maintenance (PM)

rated speed

run-in

torque

torque rise profile

INTRODUCTION

This chapter discusses the running-in and testing of truck and bus diesel engines. The procedures outlined here are designed to complement those produced by the manufacturers of engine test apparatus and the engine OEMs and not to replace them. In the larger truck facility, the chassis dynamometer is commonly used as a tool for loading an engine with the objective of diagnosing malfunctions, and it may not always be necessary to connect all of the instrumentation the **dyno** is equipped with. Also, most electronically managed engines are capable of replicating the dynamometer data. However, it is strongly recommended that all the available dynamometer instrumentation is used for every dyno test regardless of what data is generated by the vehicle management system. This will verify the ECM data and help develop a dynamometer performance data base, which can be referenced in future testing of similar engine/chassis configurations.

When using the road test for engine break-in and testing, the technician should be always aware that the primary responsibility is to safely operate the vehicle on the road. The road test is definitely the least preferred method of engine testing, but nevertheless one that many repair and service operations must use if they cannot access a dynamometer. It makes sense to develop a set of guidelines to be observed by technicians when road testing at a service operation to ensure some consistency of test conditions; this can be useful as a comparative performance reference. If the data generated by dynamometer and road testing is recorded to a database, it will not be long before sufficient data is retained to provide an invaluable reference for diagnosing engine conditions.

The terms used in this chapter concerning power terminology are covered in an earlier chapter of this textbook. The terms and principles concerning engine governing, electronic management systems, and diagnostic instruments are covered in the later sections. In other words, to fully understand all of the concepts introduced in this chapter, you will have to reference other sections of the textbook.

REBUILT ENGINE RUN-IN PROCEDURE

When OEMs talk about **engine longevity**, they are referring to the projected operational service life of the engine from manufacture to first overhaul and thereafter, between engine overhauls. Actual engine longevity is determined by a large number of factors including the type of chassis it powers, the physical conditions under which it is operated, preventive maintenance practices of the operator, and the skills of the driver. There is no doubt that in recent years, expected engine life of electronically managed, highway diesels has probably more than doubled. Engine longevity can be measured either in engine hours or by highway miles completed. In most truck operations, highway miles are used to evaluate engine life. This can be misleading as it would not take into account factors such as prolonged idling, power takeoff (PTO) operation, and the conditions of the operating environment. For instance, a truck engine used for short pick-up and delivery runs in a large city will have a much tougher life than one used in a long distance, linehaul application. The same would apply in the comparison between the city bus and the intercity, highway coach. The type of service a vehicle is subjected to is always a factor. However, it is not unreasonable to expect a service life of 3,000,000 *linehaul* miles or around 5,000,000 km from a current truck engine. Within this life span, one major out-of-chassis and two in-chassis overhauls would be expected at appropriate intervals. **Linehaul** refers to terminal-to-terminal, long distance highway travel, usually regarded as the least punishing operating condition for every chassis system. The same engine in a truck hauling aggregates out of a quarry and running them short distances might be lucky to achieve a sixth of that mileage and that would require conscientious **preventative maintenance (PM)** and a knowledgeable operator.

A critical factor in the expected operational life of a rebuilt engine is the initial **run-in** following the overhaul. After a complete overhaul or major repair job involving the installation of piston rings, pistons, cylinder liners/ sleeves, or bearings, the engine must be "run-in" before release for service. The procedure for run-in varies depending on the method used and that largely depends on the equipment available. Engine dynamometer, chassis dynamometer, or highway run-in methods may be used. The preferred engine run-in method would definitely be the engine dynamometer, but time constraints make this method impractical in many operations. The chassis dynamometer is probably the most commonly used method of engine break-in but dynamometer equipment is very expensive and only larger truck service facilities could justify the expense. The least preferred method is the highway run-in. Many rebuilders have no choice but to use this method of engine break-in, but even when the operator is knowledgeable it is difficult to fulfill the proper requirements of the break-in procedure on a highway road test with little control over load and limited monitoring instrumentation.

PREPARING AN ENGINE FOR THE FIRST STARTUP

Regardless of the method to be used for engine run-in, the engine must be properly prepared before starting for the first time. The following list may be used as a general guide:

1. Lubrication System: The lubricating oil film on the rotating parts and bearings of an overhauled engine is usually insufficient for proper lubrication when the engine is started up for the first time after an overhaul. Install new oil filters. Fill a pressure **prelubricator** (usually an electrically or pneumatically actuated external oil pump) with the OEM-recommended oil and connect the supply line to the main oil gallery. Prime the engine lubrication system with sufficient oil. There are various points on the engine into which the pressure line may be tapped, but if no other is apparent, the oil gauge line may be disconnected and the pressure tank applied at that point. Remove the oil level dipstick, and check crankcase level. Add oil, if necessary, to bring it to the full mark on the dipstick. DO NOT OVERFILL! When using a prelubricator, it is unnecessary to prime the oil filters. Some OEMs prefer that the oil filters are never primed due to the risk of contaminating the oil during the procedure. When priming filters, pour the oil into the inlet side (outer annulus) of the filter and never into the outlet side (center) (Figure 15–1).

2. Turbocharger: Disconnect the turbocharger oil inlet line and pour approximately half a liter (one pint) of clean engine oil into the turbo, ensuring that the bearings are lubricated for the startup. Reconnect the oil line.

3. Air Intake System: Check the integrity of the air intake system, checking all the hose clamps, support clamps, piping, charge air cleaner, and the air cleaner element(s). Always replace the air cleaner element after an engine overhaul.

4. Cooling System: Fit a new coolant filter and if required, separate conditioning additives. Fill the cooling system with the recommended coolant mixture. Ensure that all or at least most air is purged from the cooling system; remove a plug from the water manifold during filling to allow air to escape.

5. Fuel System: Install new fuel filters, priming them as required, with the correct grade of filtered fuel. Next, prime the fuel system by actuating a hand pump or external priming pump. Avoid priming a fuel system by charging the fuel tanks with compressed air; the practice can be particularly dangerous when ambient temperatures are high.

6. Electrical System: Ensure that the batteries hold a proper level of charge. This is especially critical with some electronically managed engines in which the ECM requires a specific minimum operating voltage.

7. Initial Startup: Crank the engine with no-fuel for 15 seconds; check for leaks. Next, start the engine, run for 2 minutes, and then shut down. Check for leaks. Do not run for a period longer than 2 minutes.

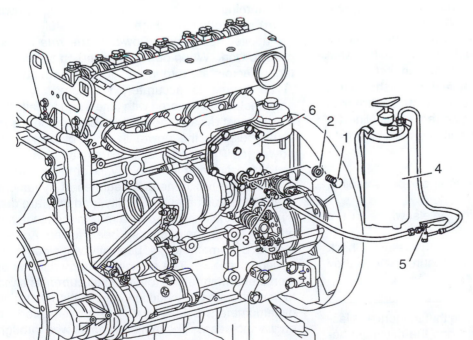

1. Bleed screw
2. Seal
3. Prelubricator nipple
4. Prelubricator
5. Prelubricator valve
6. Oil cooler

FIGURE 15–1 *Prelubricator set up to prime an MB-900. (Courtesy of Mercedes-Benz)*

TECH TIP: Do not allow the engine to run for more than two minutes during the initial startup test. The run-in objective after an engine rebuild is to apply a moderate load to the engine until it achieves its normal operating temperature and then to run the engine somewhere close to full load for at least fifteen minutes to seat the rings. Idling an engine for any length of time after an overhaul can glaze the liner walls and make it impossible for the piston rings to seat.

CAUTION: Avoid priming a dry fuel system by pressurizing fuel tanks with shop air whether regulated or unregulated. In hot weather conditions, potentially explosive mixtures of air and diesel fuel are formed, especially when the tank fuel levels are low.

ENGINE RUN-IN PROCEDURE

Regardless of the run-in method, the technician performing the procedure must be observant throughout to detect and act on any problems that may develop. Instrumentation displaying engine and chassis functions must be monitored constantly and all readings recorded. If a computerized dynamometer with a printout capability is used, also record the cab instrument display and data from the chassis bus. Compare the downloaded data where possible. During dynamometer or road testing on electronically managed systems, use a notebook PC or ProLink to snapshot data as required.

If an engine develops any abnormal running characteristics during run-in, it should be immediately shut down. Investigate and correct the problem before continuing the run-in procedure. The technician should use all his or her senses during the entire startup/break-in process. The following should act as a guideline for conditions that may develop and require immediate investigation:

- Any unusual noises such as knocking, scraping, squealing, or combustion knock.
- Any vibration in the engine or elsewhere in the drivetrain.
- Any significant drop in engine oil pressure. Check the OEM parameters for oil pressure values—they vary widely.

NOTE: The low oil pressure value that triggers an electronic alert is often much lower than the minimum acceptable normal oil pressure value.

- A rise in coolant temperature that exceeds the maximum specified by the OEM, typically around 200°F (95°C) (maximum values vary—check OEM specifications).
- A rise in engine oil temperature that exceeds the maximum specified by the OEM, typically 240°F (115°C) (maximum values vary—check OEM specifications).
- An exhaust temperature that exceeds maximum limit for the specific engine being tested, as measured by a chassis pyrometer or a dynamometer pyrometer. If the latter is used, ensure that it is positioned in the correct location for the specification reading (12″ downstream from the turbine exhaust outlet is typical), and the probe is not contacting any metal components.
- Any oil, coolant, fuel, manifold boost, or air intake system leaks.
- Any nonengine chassis system malfunction. When using a chassis dynamometer or road test to make engine break-in easier, never forget that while the engine may be the focus of the testing, every chassis system is getting a thorough workout.

DYNAMOMETER TESTING OBJECTIVES

A **dynamometer** is primarily designed to measure power, a term fully defined in an earlier chapter. Essentially, the dynamometer is an instrument that applies turning resistance to the torque output (twisting effort) of another machine and accurately measures the applied resistance. Power is the *rate* of accomplishing work. When power is tested on a dynamometer, its factors are torque and time. The **torque** output of an engine is accurately measured by the dynamometer and factored with time (rpm) to calculate its power: most dynamometers take care of the math and usually display the power in units of **brake horsepower (BHP)** or kW. For a full explanation of power characteristics and terminology, refer to Chapter 6.

There are two general categories of dynamometer used in the truck and bus facility, and they are defined by the method used to apply a resistance to the turning effort of the engine or chassis being tested. An electromotive dynamometer is basically an electric motor turned in reverse. The engine is coupled to the dynamometer armature and rotates it: as current is switched to flow through the induction coils of the electromotive dynamometer, resistance to the turning effort of the engine increases. The more current flowed through the dynamometer coils, the stronger the electromagnetic field produced and the greater the amount of torque required by the engine. The second category of dynamometer is hydraulic. The hydraulic medium used is usually water but other types of hydraulic media are also used. The critical component in the hydraulic dynamometer is a load cell or multiple load cells. These use a principle similar to that used by a hydraulic driveline retarder such as the Caterpillar brake saver: water is flowed through the cell and acts on an impeller. Inlet and outlet valves are used to control the flow of the hydraulic medium into and out of the load cell and define the torque required to rotate the impeller. All dynamometers measure torque. Most also measure rotational speed. When both torque and rpm are known, brake horsepower can be calculated using the equations introduced in Chapter 6:

$$BHP = \frac{T \times rpm}{5252}$$ T = Torque expressed in lb.-ft.

or

$$kW = \frac{T \times rpm}{9429}$$ T = Torque expressed in N·m

$$Torque = \frac{BHP \times 5252}{rpm} \quad or \quad T = \frac{kW \times 9429}{rpm}$$

When performing chassis dynamometer testing of diesel engines whether for purposes of diagnosis or to perform an engine run-in routine, all of the gauges and instrumentation should be connected to the engine being tested. Loading an engine down on a dynamometer is the ultimate performance test for an engine, and it should be as thoroughly monitored as possible. It takes a little extra time to connect all the required gauges and instruments, but they will assist to ensure accurate diagnoses and complete reports. Electronic printouts do not look professional when half the data categories are left blank.

CHASSIS DYNAMOMETER TESTING

Because of the time required to mount and dismount an engine to an engine dynamometer test bed, the chassis dynamometer is used for most truck engine diagnosis and testing. **Chassis dynamometers** use a roller or rollers to receive input. The truck chassis therefore must be located so that its drive axle(s) are properly aligned and contacting the rollers, and then be chained into position. Determine that the dynamometer is in calibration before starting the test. When double roller chassis dynamometers are used, periodic testing by running a sin-

gle drive axle chassis on each roller separately, will provide an indication of the condition of each absorption unit and the display instrumentation. The following outlines the procedure to be followed when running a truck chassis on a dynamometer.

1. Check all the chassis fluid levels including engine oil, fuel, coolant, transmission, and drive axle carrier. Check the tire pressure and the wheel lug integrity. Inspect suspension components, driveline and brake adjustment, ensuring the chassis rolls freely with brakes released.
2. If the dynamometer is of the double roller type, measure the truck chassis drive axle spread and set the rollers on the dynamometer to the required spread (Figure 15–2 and Figure 15–3). Then engage the roller brake to lock the roller(s) to enable the truck to be driven onto the dynamometer test bed. Next, drive the truck onto the dynamometer roller(s), release the brake, and align by running

FIGURE 15–2 *Double roller, chassis dynamometer. (Courtesy of Superflow Corp.)*

FIGURE 15–3 *Front axle tie-downs to dynamometer test bed. (Courtesy of Superflow Corp.)*

up the "road speed" to around 10 mph (16 km/hr) in one of the low range gears for 15 seconds or so, then shut down the engine. Chock the front wheels using four wheel wedges and install the safety chains; when performing the latter, be sure to securely load down (but not overload) the safety chains. Generally, single roller dynamometers require the truck to be fastened down with less force than double roller versions. Check on the procedure outlined by the dynamometer manufacturer.

3. All chassis operating on a dynamometer must be provided with adequate cooling air. This is almost impossible to achieve in the modern, sound insulated dynamometer room. A large fan can assist, but it does not come close to inducing the air flow transfer rate of a tractor driving down the highway at 60 mph (100 km/hr). To make matters worse, air-to-air charge air coolers have generally replaced coolant medium heat exchangers, and while these boast higher efficiencies in highway operation, they have lower operating efficiencies when ambient air transfer rates are low, such as operation in a mining pit or an enclosed chassis dynamometer. Fit the exhaust gas extractor pipes to the truck exhaust pipes.

4. Connect all the dynamometer gauges (these are usually of superior quality and accuracy) to the vehicle dash instruments, and when possible, additionally monitor the engine running data using a ProLink or a PC loaded with the appropriate software. Do not omit connecting monitoring gauges with the belief that it will save time. In the event of a malfunction, critical diagnostic data may be unavailable. Typical dynamometer gauges include: crankcase pressure, air inlet pressure, manifold boost pressure, fuel subsystem charging pressure, coolant temperature, coolant pressure, oil temperature, oil pressure, ambient air temperature (engine compartment), exhaust temperature (pyrometer), and exhaust back pressure (Figure 15–4, Figure 15–5, and Figure 15–6).

5. If the engine is hydromechanically managed, ensure that the accelerator linkage is properly set. This should be tested by having someone fully depress the accelerator pedal with the engine stopped while observing the throttle arm/fuel control lever. Normally, a visible travel breakover is required, typically ±0.125″ (3 mm). With electronically managed engines, ensure that the TPS calibration programming is correct.

6. On certain vehicles with front axle/rear axle drives, the transfer case to front axle drive will have to be eliminated. In some cases, this might involve the removal of the propeller shaft. Check with the OEM technical service literature on the appropriate test method.

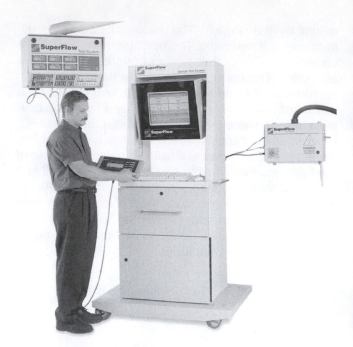

FIGURE 15–4   *Data display console used on a truck chassis dynamometer system. (Courtesy of Superflow Corp.)*

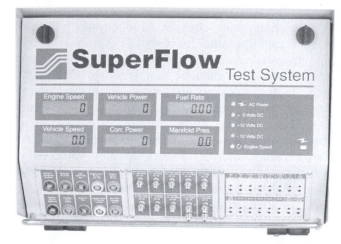

FIGURE 15–5   *Data display console. (Courtesy of Superflow Corp.)*

7. Remove any stones lodged in the tire grooves. Switch on the exhaust extraction system and the ambient air transfer system. The fumes produced by burning tire rubber while dyno testing are harmful and should not be inhaled. Wear hearing protection.

8. Obtain the OEM test data for dynamometer testing and take particular note of the engine **torque rise profile** (often identified by the recommended gear shift points), rated speed rpm low, high idle speeds and **droop curve**. Either obtain or make up a

Superflow's model SF-602 includes:

1. Proven Superflow rollset and dynamometer
2. Sensor input box with pivoting boom
3. Hand-held controller
4. Computer console with high-performance
   Windows™ computer, 17" monitor, color printer
   and WinDyn™ software
5. Interconnect box
6. Gravimetric fuel-flow measurement system.

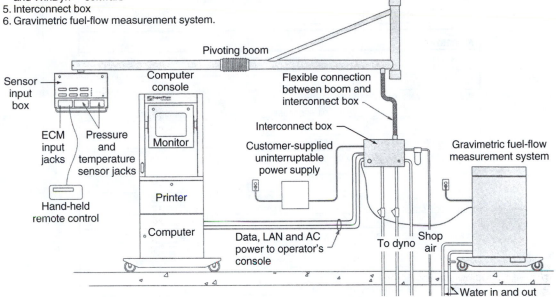

FIGURE 15–6   *Chassis dynamometer circuit schematics. (Courtesy of Superflow Corp.)*

dynamometer test profile and record the running data while testing. It is good practice to do this even when a dynamometer printout is available. Familiarization with all critical maximum and minimum running parameters for the engine being tested will help in conducting a thorough engine test (Figure 15–7).

9. Start the engine and while stationary, test idle and high idle engine speed. On a hydromechanically governed engine, if there is any irregularity with either parameter, cease the test and adjust the governor/set engine overhead as necessary before resuming. On a tandem drive axle unit, be sure to engage the interaxle differential lockout, especially when testing on a double roller dyno test bed. Smoothly shift through the gear ratios (skip shifting is acceptable) with no applied dynamometer load until the chassis is being run in the OEM recommended ratio for testing, usually 1:1. Generally, overdrive ratios are not used. Record the road speed at both the base **(peak torque)** and peak **(rated speed)** rpms of the torque rise profile.

10. Run at the rated speed and load down the engine to approximately 50% of the engine rated power value until the engine coolant and/or oil are at operating temperature. Monitor all the engine data parameters, but pay particular attention to the coolant and exhaust pyrometer temperatures.

WARNING: In testing using the double roller type of chassis dynamometer, the two bogie axles must never exceed "5 miles per hour" difference in road speed. With the engine run at a given test speed, the two dynamometer "Load" application (one for each set of axle rollers) buttons must be applied evenly, while carefully monitoring the "MPH" display for each axle, ensuring that they are within 5 mph and are preferably identical. Application of the "Load" and "Unload" control buttons should always be balanced. Remember, whenever vehicles are equipped with a power divider lockout control, the lockout should be engaged during testing.

11. Never exceed a maximum limit. If excessively high temperatures are observed and the dynamometer ambient temperature is not suspected to be the cause, check the boost air temperature. Install a calibrated temperature gauge downstream from the charge air cooling core, preferably at the intake manifold. Temperatures should generally not exceed 150°F (65°C), but check OEM specifications.

12. When the engine is at operating temperature and the displayed data indicates that its systems are properly functional, the dynamometer test sequence may proceed. Full load rated power or "governed speed" is normally tested first. Running the engine rpm held at the rated speed (not higher than rated

ENGINE TEST REPORT

Date: _____   Unit Number: _____
Repair Order Number: _____   Model Number: _____
PROM I.D.: _____   Max. N/L RPM: _____
Rated F/L RPM: _____
Idle RPM: _____

A. PRESTART

1. PRIME LUBE OIL SYSTEM	2. PRIME FUEL OIL SYSTEM	3. FILL COOLING SYSTEM

B. START-UP AND IDLE FOR 30 SECONDS

START_____ STOP _____ OIL PRESSURE _____ WATER TEMPERATURE _____

C. WARM-UP — 5 MINUTES   START _____   STOP _____

RPM MAX. SPEED	LOAD 50%	OIL PRESSURE	WATER TEMPERATURE
1. LUBE OIL LEAKS	2. FUEL OIL LEAKS	3. COOLANT LEAKS	4. LOOSE BOLTS

D. RUN-IN — 5 MINUTES   START _____   STOP _____

RPM MAX. SPEED	LOAD 75%	OIL PRESSURE	WATER TEMPERATURE

E. FINAL RUN-IN — 20 MINUTES   START _____   STOP _____

RPM MAX. SPEED	LOAD 100%	CRANKCASE PRESSURE AT F/L	EXHAUST BACK PRESSURE AT F/L
LUBE OIL PRESS. AT F/L	LUBE OIL TEMP. AT F/L	FUEL OIL TEMP. AT F/L	FUEL OIL PRESSURE AT F/L
WATER TEMP. AT F/L	TURBO BOOST PRESS. AT F/L	LUBE OIL PRESSURE AT IDLE	IDLE RPM

REMARKS:
_____
_____

OK _____   Reject _____   Dynamometer Operator _____   Date _____

FIGURE 15–7   *Typical dynamometer test profile recommended by DDC for engine testing. (Courtesy of DDC)*

speed), load down the roller(s) until the accelerator pedal is at full travel, and at the point at which the engine is unable to sustain the rated speed rpm, record the brake power value. Ensure that the air compressor is unloaded, the A/C system is switched off, and that the engine fan is disengaged so that the results of the test are as accurate as possible. Check twice. Allow power readings to stabilize before recording. Next, perform the same test at 100 rpm below rated speed. Dropping the engine rpm in 100 rpm increments, test the engine power descending through the torque rise profile to its base rpm. This may be (but does not have to be) the "peak torque" rpm.

13. If the objective of the dynamometer test is to run-in a rebuilt engine rather than to diagnose engine performance, the engine should be run at full power at 100 rpm increments through the torque rise profile for a period not less than 20 minutes. This should be sufficient to properly seat the rings.

14. The technician undertaking the dynamometer test should constantly monitor the engine data displays for abnormal readings, especially those indicating temperature values. It is not unusual for

engines using air-to-air charge air coolers to approach overheat values when performing full load chassis dynamometer testing due to low air flow factors.

WARNING: After running a full load dynamometer test on an engine, allow the engine to run at no-load idle speed for a period of at least 5 minutes following the test. This provides the turbocharger with adequate oil for lubrication and cooling during the shutdown period and permits engine cooling system flow to assist with cooling the engine down when it is at its hottest.

TECH TIP: Chassis dynamometer test results vary from dynamometer to dynamometer. It is up to each service facility to establish acceptable at-the-wheel horsepower ranges on the dynamometer they use for testing. Generally, BHP output at the wheels should approximate 80% to 85% of the OEM engine brake power specification for tandem drive axle chassis and 85% to 90% of the OEM brake power specification for single drive axle chassis. The variables in the truck drivetrain are always a factor. Variations from one dynamometer to another can be due to variables such as the altitude of the location, wheel roller diameter, ambient temperature, and humidity. The type and condition of the tires on a vehicle will cause traction variations and small roller dynamometers generally produce more tire slippage than dynamometers equipped with larger diameter rollers. Single, large diameter roller dynamometers used on tandem drive trucks produce less slippage and generally more accurate results than multiple roller types. The specific values obtained from chassis dyno tests are most useful for comparisons to data obtained from testing similarly configured chassis on the same dynamometer.

WARNING: Never perform a dynamometer test with mismatched duals, recapped, or snow tread tires mounted on the vehicle drive axles. Tread separation may occur when using recapped tires while snow tread tires can produce erratic dynamometer test readings. Using a set of "slave" tires (used solely for dyno testing) when testing a chassis equipped with inappropriate tires is an option.

CAUTION: When a truck is run on a chassis dynamometer, ensure that no person is in the test room. Stones and rocks lodged in the tire treads can free up and be propelled behind the truck at high velocity. It is also important to ensure that no equipment is located behind the truck chassis.

ENGINE DYNAMOMETER TESTING

The **engine dynamometer** method of checking engine power is more accurate than the chassis dyno method because it removes all the variables involved when transmitting power through a transmission, drive shafts, drive axle carriers, and wheel assemblies. This method is ideal for engine run-in because it affords the test technician perfect observation of the engine throughout the test sequence, something not possible when the engine is in-chassis. Two types of engine dynamometer are used: the test bed type, which is a permanent fixture usually isolated in an insulated, soundproof room often with a separate observation room, or the portable unit, which is coupled directly to the engine to be tested. The main disadvantage of engine dynamometer testing is that it adds considerable time to an engine overhaul to fit and subsequently remove the engine from the dynamometer test bed, and test conditions are far removed from the more realistic chassis dyno test. In this respect, the portable unit usually requires less time to fit to the engine, but these can seldom be safely used in a large truck garage due to the noise levels emitted and the potential for danger when operating an engine under high loads in a room not dedicated to the purpose.

Before using an engine dynamometer for run-in or testing, the technician should become familiar with the procedure outlined previously for checking power using a chassis dynamometer. The first step is to obtain the OEM dynamometer test data, connect and properly phase and align the drive shaft, connect the appropriate gauges, plumb in a cooling tower, and check all the engine fluid levels.

1. With the engine no-fueled, crank until an oil pressure value is observed on the oil gauge. Then start the engine.
2. Check the engine idle and high idle rpms and on hydromechanical engines ensure that the dyno fuel control linkages produce full travel plus the required breakover factor. On electronic engines ensure that the throttle position sensor is properly calibrated and programmed to the engine ECM.
3. Start the engine, and run at ±1,000 rpm with no load for 5 minutes to check for oil pressure, noises, vibration, leaks, and drive shaft runout. Shut down engine, and check and correct oil and coolant levels.
4. Run the engine at a midpoint rpm within the torque rise profile, and load the engine at around 50% of its rated power value. Operate the engine under this half-load condition until the coolant and oil

PHOTO SEQUENCE 2
Chapter 15 Chassis Dynamometer Installation and Test Procedure

PS2–01 The chassis dynamometer to be used for this test sequence is a double roller, water brake model. Note that it is housed in a bay isolated from the main truck shop. The bay is also soundproofed.

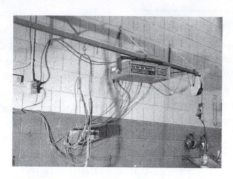

PS2–02 The dynamometer display console is designed to slide on a rail, so it can be positioned to give the operator optimum view of the readouts from the cab during the test.

PS2–03 The truck should be backed onto the dynamometer test bed on to the double rollers: the bogie axle should be measured and the roller spread adjusted prior to backing the truck over the rollers.

PS2–04 When backing the truck up over the dynamometer rollers, the roller brake must be actuated.

PS2–05 Next, the truck chassis must be chained to the deck from both front and rear. Secure the rear chain with a moderate tensile load.

PS2–06 The front axle is secured to the dynamometer test bed by means of a chain assembly that holds the front tires on either side.

PS2–07 The dynamometer test sequence should be performed as per the OEM test procedure and the safety guidelines outlined in this chapter.

PS2–08 Following the test sequence, the engine should be permitted a cool-down period of at least 5 minutes prior to removal from the dynamometer test bed.

PS2–09 Releasing the rear hold-down chain prior to removing the truck from the dynamometer test bay.

PHOTO SEQUENCE 3
Chapter 15 Engine Dynamometer Test Procedure

PS3–01   *The engine to be tested should be mounted in an appropriate cradle on the test bed deck. Plumb in the cooling tower (located on the left).*

PS3–02   *It is good practice to connect all the dynamometer test and monitoring instrumentation to the engine for a test sequence.*

PS3–03   *The engine mounted from the dynamometer test bed viewed from the rear.*

PS3–04   *The engine monitoring and test instrumentation may have to be read from within the test cell.*

PS3–05   *or from an adjacent observation room.*

PS3–06   *The view of the dynamometer observation room panel display and the observation window. The engine should be run through the OEM test sequence and the results downloaded on a hard printout or to a diskette. Use the chassis serial number as the file ID.*

PS3–07   *Following the test sequence, the engine should be allowed to idle for at least 5 minutes before shutting down. Ensure that the engine has cooled completely before removing the cooling tower plumbing and the test instrumentation.*

temperatures are each in the operating temperature window.

5. Set the accelerator/throttle to produce the rated speed rpm and smoothly load the engine until it is unable to sustain the rated speed: determine the peak power setting and then test at descending increments of 100 rpm engine power through the torque rise profile.

6. Monitor and record all the engine running data during the test sequence including: crankcase pressure, air inlet pressure, manifold boost pressure, fuel subsystem charging pressure, coolant temperature, coolant pressure, oil temperature, oil pressure, exhaust temperature (pyrometer), and exhaust back pressure.

7. Run under conditions of peak loading for a minimum period of 20 minutes when the dyno test is being performed to run-in the engine.

WARNING: After running a full load dynamometer test on an engine, allow the engine to run at no-load idle speed for a period of at least 5 minutes after the test to provide the turbocharger with adequate oil for lubrication and cooling during the shutdown period and permit the engine cooling system to assist with cooling the engine down when it is at its hottest.

ROAD TESTING

Before performing a road test, the technician should ensure that the qualifications for operating a heavy-duty truck or bus on the highway required by the state or provincial jurisdiction are met. It should be emphasized that although the engine may be the chassis system being tested, all the vehicle systems are subjected to a performance workout and it is any vehicle operator's responsibility to ensure that they are both safe and functional. When diagnosing an engine malfunction, the road test usually takes second place to a chassis dynamometer test. And when the objective is the run-in of a rebuilt engine, the road test is the least preferred method after engine and chassis dyno loading, simply because there are too many variables and unknown factors that may arise during a road test.

When running-in an engine, it should be loaded close to its maximum for at least some of the procedure and therefore the truck should be coupled to a preferably loaded trailer. Using the vehicle braking system to achieve this can seriously overheat the foundation brakes.

When road testing an electronically managed engine, the operator should be thoroughly familiar with the characteristics of the management system and its programming. Before the road test, scan the ECM for active and historic codes, download any tattletales, and the customer and proprietary data programming that might impact on the diagnosis. During the road test, connect a ProLink or notebook PC with the appropriate software to the ATA J1708/J1939 connector to scan and snapshot, when required, vehicle running data. If a digital dash display is installed in the truck, determine what data may be displayed during the test that would be most useful in assessing the engine condition. The technician should never forget while road testing that the primary responsibility is to safely operate the vehicle and the distractions of the dash and electronic displays used to diagnose must not interfere with that responsibility. The following is intended to act as a guide to road test performance whether the objective is engine run-in or malfunction diagnosis:

1. Perform a pretrip inspection or circle test. Most state and provincial jurisdictions require this test to be performed by the driver of any heavy-duty highway vehicle before operating it on a road. This test at minimum would require that the following be checked and corrected before running the vehicle on a highway: vehicle lighting, tire inflation, tire condition, brake adjustment, air system pressure buildup time and values, audible air leaks, mudflap condition, suspension condition, fifth wheel or pintle hook integrity, and load securement.

2. If the objective of the road test is to run-in a newly rebuilt engine, perform the sequence outlined earlier in this chapter entitled "Preparing the engine for the first startup." If the objective is to diagnose a malfunction, the engine fluid levels should be checked.

3. Next, the technician should become familiar with the setup of the primary drivetrain components, making a mental note of the engine rated speed, torque rise profile, and any customer-programmed data such as progressive shifting, road speed limit, and governor type that could be a factor while performing the road test. Note the recommended gear shift points (often identified on the visor or elsewhere in the cab) when mechanical engine/transmissions are to be tested.

4. Start the engine and check the idle and no-load high-idle speeds, correcting the values as required. Warm the engine up with the vehicle stationary but avoid idling for longer than 5 minutes, especially where the engine has been newly rebuilt. The test procedure outlined for chassis dyno testing should be reviewed because a road test attempts to repli-

cate some of this. Road traffic largely dictates how thoroughly a road test can be performed, so try to select a highway where traffic is known to be light.

5. Once on the highway, bring the vehicle up to speed and test the performance through the torque rise profile. It really helps if the operator is not forced to maintain a consistent road speed, but this factor will be determined by the amount of road traffic. A road test of 30 minutes after the specified engine operating temperature has been reached is usually sufficient to break in a rebuilt engine (seat the rings). When diagnosing engine malfunctions, the road test duration will depend on the data revealed during the procedure.

WARNING: After running a road test on an engine, allow the engine to run at no-load idle speed for a period of at least 5 minutes after the test to provide the turbocharger with adequate oil for lubrication and cooling during the shutdown period and permit the engine cooling system to assist with cooling the engine down when it is at its hottest.

SUMMARY

- The entire content of this chapter involves practical test procedure and to properly comprehend it, some exposure to the technology is required, even if it is just observation.
- Before operating a truck chassis on a dynamometer, it is advisable to acquire some driving skills. Consult a fleet driver-trainer or experienced technician for some advice.
- Engine dynamometers present a certain potential for danger that can be minimized by properly aligning the engine on the test bed and always using safety guards and shrouds around rotating components such as fans and drive shafts.
- Chassis dynamometers must be properly harnessed to the test bed. Consult the dynamometer operating manual and an experienced operator before running the engine.
- Before road testing vehicles, ensure that you hold a valid license to operate the vehicle on a highway and that you fully understand how to operate the vehicle.
- Whether dynamometer or road testing, ensure that all the appropriate observation instrumentation is connected; in the case of most current electronic engines this means that data is duplicated but this duplication also corroborates the data. Apart from reducing the scope of data on which an analysis

must be made, it looks unprofessional to display dynamometer printouts with half of the critical data displayed as NA (not available).

REVIEW EXERCISES

1. A Caterpillar C-15 engine is engine dynamometer tested to produce 1,800 lb.-ft. of torque at 1,200 rpm. Calculate the BHP.
2. A Mack Trucks E9 engine produces 1,400 lb.-ft. of torque at 1,900 rpm. Calculate the BHP value.
3. Obtain an OEM engine power and torque performance chart. Using calculation, confirm that the nominal BHP value is correct at three points in the torque rise profile.
4. Use a college dynamometer or visit a local service facility with the objective of observing the procedure for mounting an engine to a dyno test bed.
5. Visit a facility with a chassis dynamometer and observe the procedure for harnessing a truck to the test bed.
6. Using an OEM dynamometer performance test profile for either a chassis or engine dynamometer, record the preparation and test sequence.

REVIEW QUESTIONS

1. *Technician A* states that a truck rebuilt diesel engine run-in procedure usually requires that the engine be operated at medium to low loads for the first 1,000 miles (1,600 km) of operation. *Technician B* argues that it is usually essential to run a rebuilt diesel engine at maximum load for a short period after a rebuild to seat the piston rings. Who is right?
   a. A only
   b. B only
   c. Both A and B
   d. Neither A nor B
2. *Technician A* states that an engine dynamometer is likely to produce a brake power rating closer to the OEM specified power than a chassis dynamometer test. *Technician B* states that the chassis dynamometer produces brake power readings at the wheels that are always lower than OEM rated power. Who is right?
   a. A only
   b. B only
   c. Both A and B
   d. Neither A nor B
3. If all three options were available and time is not a consideration, which of the following options would

be the preferred method of running-in a rebuilt engine?
a. Engine dynamometer
b. Chassis dynamometer
c. Road test

4. *Technician A* states that when performing dynamometer testing of an engine, the altitude of the test location must be considered when analyzing the power. *Technician B* states that ambient temperature variations have a greater effect on brake power produced than altitude. Who is right?
a. A only
b. B only
c. Both A and B
d. Neither A nor B

Section

2

Hydromechanical Diesel Fuel Injection Systems

Section 2 begins with some basic fuels chemistry, truck fuel subsystems, and a look at the general objectives of a fuel management system. Each hydromechanical fuel system found on diesel engines is then explored, focusing on principles of operation. While many systems in this section are being gradually phased out, they should not be overlooked by the student truck technician as most of the fundamental principles of electronically managed fuel systems are sourced from them. Certain hydromechanical components such as hydraulic injectors and fuel subsystem components that are found in electronic management systems are dealt with in this section. The section concludes with some basic approaches to troubleshooting and failure analysis of diesel engines and their fuel systems.

Chapter
16

Chemistry and Combustion

OBJECTIVES

After studying this chapter, you should be able to:

- Understand basic chemistry and its application to fuel systems.
- Define elements, mixtures, and compounds.
- Describe a simple chemical reaction and chemical bonding.
- Outline the structure of an atom.
- Define the states of matter and the conditions that predetermine them.
- Describe the properties of common elements, mixtures, and compounds.
- Outline the dynamics of a combustion reaction in an engine cylinder.
- Define the conditions required for a stoichiometric reaction.
- Calculate air/fuel ratio.
- Describe the actual processes of combustion in a diesel engine cylinder.
- Describe the dynamics of detonation.

KEY TERMS

afterburn	diesel knock	matter
air/fuel ratio (AFR)	direct injection (DI)	mixture
atom	electron	monatomic
balanced atom	element	neutron
chemical bonding	fluidity	particulate matter (PM)
compound	hypothesis	propagate
condensation	ion	proton
covalent bonding	kinetic	stoichiometric
cylinder gas dynamic	kinetic molecular theory	sublimation
detonation	lambda	triatomic
diatomic	mass	vaporization

INTRODUCTION

It is important for any mechanical technician to have a fundamental understanding of chemistry for purposes of explaining fuel composition, combustion dynamics in engines, and for developing an ability to work with electricity and electronics. A knowledge of basic chemistry can be an especially useful diagnostic tool for the technician specializing in engines and fuel systems. This chapter briefly addresses basic chemistry, atomic structure, combustion dynamics, and the language of fuel technology, all topics that have individually been the subject of many textbooks. The technician is encouraged to explore this subject matter to a greater depth using other specialized texts.

Chemistry is the science that seeks to understand the composition of **matter** (physical substance in general), the **elements** (any substance that cannot be resolved into simpler substances), the **compounds** (substances containing two or more elements combined in definite proportions and held together by chemical force) they form, and the reactions they undergo. It is one of the oldest sciences, and it evolved from the observations of thinkers in times well before the modern age. The chemist would observe natural phenomena, hypothesize (attempt to reason) its causes, then confirm the **hypothesis** (supposition) by experiment. When it seemed that a hypothesis could be conclusively proven by documented observation of cause and effect, a law was established.

Of course, the history of science is generally littered with hypotheses that became irrelevant as knowledge progressed, so a theory of physics or chemistry today may not necessarily be so tomorrow.

solved into simpler substances. They consist of minute particles known as *atoms* which, for purposes of study, are considered to be indivisible. An atom is the smallest particle of an element that can take part in a chemical reaction. The atoms of any one element are exactly alike and possess identical **mass** (quantity of matter a substance contains: weight). The name of an element is always a single word such as oxygen or hydrogen (Figure 16–1 and Figure 16–2). All elements have a short form symbol usually derived from the Latin name for the substance. Sometimes the symbol is obvious because the Latin word for an element is similar to the English word (much of the English language is derived from Latin), such as that for the element oxygen, which is O. However, the symbol for the element iron is Fe derived from the Latin word for iron, which is ferrum. Atomic number identifies the number of **protons** in an atom of the element.

SOME COMMON METALLIC ELEMENTS	ATOMIC NUMBER	SOME COMMON NONMETALLIC ELEMENTS	ATOMIC NUMBER
Iron – Fe	26	Hydrogen – H	1
Sodium – NA	11	Carbon – C	6
Magnesium – Mg	13	Helium – HE	2
Aluminum – Al	13	Sulfur – S	16
Nickel – N	28	Silicon – SI	14
Rhodium – Rh	35	Selenium – Se	34
Silver – Ag	47	Oxygen – O	8
Zinc – Zn	30	Nitrogen – N	7
Gold – Au	79	Argon – Ar	18
Platinum – Pt	78	Radon – Rn	86

BASIC CHEMISTRY

The building blocks of all matter are **atoms**. All atoms are electrical. Electrical charge is a component of all atomic matter. This section examines some key definitions required to develop an understanding of atomic theory. Some of this information is revisited in Chapter 29 on electrical fundamentals.

ELEMENTS

An element is any one of more than 100 substances, most naturally occurring, that cannot be chemically re-

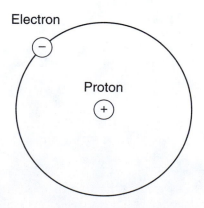

FIGURE 16–1   *Hydrogen atom.*

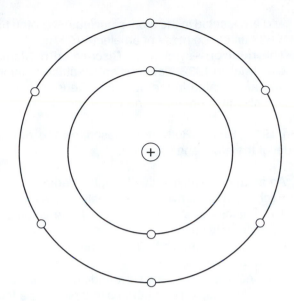

FIGURE 16–2 *Oxygen atom.*

MIXTURES

A **mixture** is composed of two or more elements and/or compounds, all of which retain their own characteristics and identity. Air is an example of a mixture; it is composed of 23% oxygen and 76% nitrogen by mass (plus about 1% of inert gases). Both the main constituent elements retain their own identity and may partake in reactions independently of each other. The oxygen in air for instance, is commonly involved in reactions that do not involve nitrogen. The properties of a mixture depend on the substances in it; the compound water is a pure substance with identifiable characteristics, dissimilar from any other pure substance. However, salt water is a mixture of salt and water; it boils at a higher temperature and freezes at a lower temperature than pure H_2O.

CHEMICAL BONDING

Chemical bondings are interactions that account for the association of atoms into molecules, ions, and crystals. When atoms approach one another, their nuclei and **electrons** interact and distribute themselves in such a way that their combined energy is lower than it would be in an alternative arrangement. Whenever the total energy of a group of atoms is lower than the sum of energies of its component atoms, they are chemically bonded; bonding energy accounts for the overall lowering of energy.

The number of bonds an atom can form is called its valency (or valence number). The valency of an atom is simply the number of unpaired electrons in its valence shell. The valence shell is the outermost shell of elec-

trons. A water molecule consists of an oxygen atom with a valency of two combined with two hydrogen atoms, which have a valency of one. An electrovalent or ionic (**ion:** an atom with either an excess or deficiency on electrons) bond occurs when an electron is transferred from one neutral atom to another—the resultant charge results in the atoms being held together by electrostatic attraction.

COMPOUNDS

A compound is a substance composed of two or more elements combined in definite proportions and held together with chemical force. A compound is composed of identical molecules, made up of atoms of two or more elements. Most commonly occurring materials are mixtures of different chemical compounds. Pure compounds can usually be obtained by physical separation processes such as filtration and distillation. Compounds themselves can be broken down into their constituent elements by chemical reactions.

Carbon atoms are unique among the chemical elements in their ability to form covalent bonds with each other and with other elements. **Covalent bonding** occurs when two electrons are shared by two atoms. Because of its atomic structure, carbon is more likely to share electrons than to gain or lose them. There are ten times as many carbon compounds as compounds of all other elements combined; common carbon compounds are formed with hydrogen, oxygen, and nitrogen.

MOLECULE

A molecule is the smallest particle of a compound that can exist in a free state and take part in a chemical reaction. When an H_2O molecule (Figure 16–3) is chemically reduced, it ceases to have the properties of water and it possesses those of elemental oxygen and hydrogen.

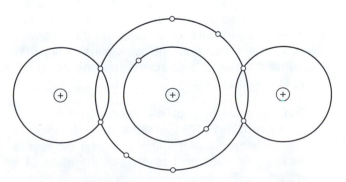

FIGURE 16–3 *Water molecule: observe the shared electrons.*

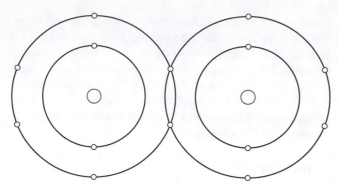

FIGURE 16–4   *Oxygen molecule: oxygen in the lower atmosphere is mostly diatomic.*

The oxygen found at sea level in the earth's atmosphere combines to form **diatomic** molecules, that is, two oxygen atoms to a molecule (Figure 16–4).

ATOMIC STRUCTURE

Atomic theory seeks to explain both the composition of matter and the laws that apply to chemical combination. It was conceived in 1801 by a chemist, John Dalton, who essentially confirmed theorizing by some great thinkers in the era around 400 years B.C., suggesting that matter could not be subdivided indefinitely without altering the properties of that substance. Dalton's atomic theory suggested:

1. Each element is made up of minute, indivisible particles called atoms.
2. Atoms cannot be created or destroyed.
3. Every atom that a specific element is composed of is identical.
4. Atoms of different elements are different.
5. Atoms of one element may combine with other elements to form compounds.

Dalton's atomic theory developed into our current notions of the structure of an atom by the scientists Thomson, Rutherford, and Chadwick in the early part of the twentieth century; these scientists identified the following subatomic particles:

1. The electron—symbol e–. Discovered by J. J. Thompson in 1897. Electrons carry a negative charge and orbit in the shells around the atom's nucleus. An electron has $1/1837$ of the mass of a proton.
2. The proton—symbol p+. Discovered by Ernest Rutherford in 1911. Protons carry a positive charge

and are located in the atom's nucleus. A proton has 1,837 times the mass of an electron.
3. The **neutron**—symbol n°. Discovered by James Chadwick in 1932. Neutrons are electrically neutral and are located in the atom's nucleus. A neutron has slightly more mass than a proton.

In 1911, Ernest Rutherford described the nuclear model of the atom and asserted:

1. An atom consists mostly of empty space.
2. Each atom has a minute, extremely dense nucleus.
3. The nucleus is surrounded by a large volume of nearly empty space.
4. Most of the mass and *all* of the positive charge is located in the nucleus.
5. The nearly empty space surrounding the nucleus is sparsely occupied by electrons that possess a fraction of the mass but a negative charge that balances the positive charge of the protons.

A hydrogen atom (Figure 16–1) consists of a single proton in the nucleus and one electron in its orbital shell; the electron orbits the nucleus in much the same manner Earth orbits our sun. The electrical force attracting the electron toward the positive charge of the nucleus is balanced by the mechanical force acting outward on the rotating electron; this ensures that the electron remains in its orbital shell and is not drawn into the nucleus. All atoms are electrical. An electrically **balanced atom** is one in which there is an equal number of electrons and protons (Figure 16–5 and Figure 16–6). While electrons can move from atom to atom in many substances, protons do not, and they are held in the nucleus of the atom. An atom with either an excess or deficit of electrons is known as an ion. An ion is an atom in an electrically unbalanced state.

Current atomic theory suggests that, in much the same way protons and neutrons make up atomic nuclei, these particles are themselves made up of

Copper atom

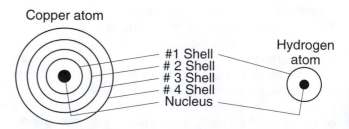

Copper has three full shells: #1, #2, and #3. Its # 4 shell has a single electron.

FIGURE 16–5   *Copper atom: a copper atom has four shells, three of which are full. (Courtesy of Freightliner Corp.)*

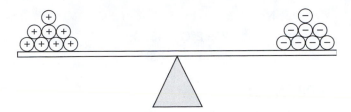

FIGURE 16–6 *Atomic charge balance: if an atom has eight protons it will want to have eight electrons.*

quarks. Quarks are therefore subatomic particles that are theorized to be the fundamental constituents of all matter. However, the study of quarks probably goes beyond what today's technician needs to know of atomic structure.

STATES OF MATTER

In physics, matter is defined as anything that has mass and occupies space. Matter can be generally classified into one of three states or phases within the earth's atmosphere: solid, liquid, or gas. Water is the only substance that is familiar in all three states, which we describe as ice, water, and steam (Figure 16–7). A solid does not readily change in size or shape when subjected to physical forces. Like a solid, a liquid has a definable volume but it has **fluidity** and will change its shape if poured from one container to another. A gas generally expands to fill the volume of the vessel in which it is contained.

On Earth's surface, under a given set of temperature and pressure conditions, a substance will be in one of

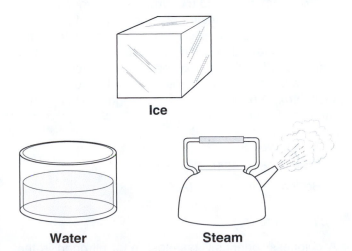

Ice

Water   Steam

FIGURE 16–7 *States of matter: H_2O is the only substance readily observed in all three states.*

these states; and most substances can exist in any of the three states. The differences between solids, liquids, and gases can be explained in terms of **kinetic molecular theory; kinetic** refers to motion. Kinetic molecular theory states that all matter consists of molecules that are constantly in motion and their extent of motion in a substance will be greater at high temperatures and less at cooler temperatures. A certain amount of attraction exists between all molecules in a substance, and whenever repulsive forces are weaker than the intermolecular attractive forces, the molecules stick together. However, as temperature increases, so does the molecular motion.

In a solid, intermolecular attractive forces overcome the disruptive thermal (heat) energy of the molecules and they are bound together in an orderly arrangement called a crystal. The molecules in a crystal vibrate due to their thermal energy and again, the amount they vibrate is determined by temperature or the amount of retained heat energy.

As a solid is subjected to heat, its molecules vibrate with increasing energy until they are capable of overcoming the intermolecular attractive forces. At this point the substance melts or changes to the liquid state and acquires fluidity. A change of state works in reverse; the freeze point of a substance is the same as its melt point.

Kinetic molecular theory also explains the change of state from that of a liquid to a gas known as **vaporization**. As heat is applied to a liquid, some molecules acquire sufficient thermal energy to overcome the intermolecular attraction and break free from the surface of the liquid; these molecules are converted to a gaseous state. If heat continues to be applied to the liquid, more molecules will move away from the liquid until the substances' boil point is achieved and all the molecules escape the liquid state. When vaporization is reversed, it is known as **condensation**. The distance between molecules of a gas are large in comparison with that of a liquid because the intermolecular forces are weak. A gas molecule will move until it either strikes another gas molecule or the walls of the vessel that contains it. The effect of many gas molecules striking a container wall is known as pressure.

It is possible for a substance to pass directly from the solid to the gaseous state in a process known as **sublimation**. Dry ice (CO_2) sublimates at atmospheric pressure and temperature.

PLASMA AND OTHER STATES OF MATTER

At very high temperatures, atoms may collide with each other with a force that results in some electrons being jolted free of the nuclei. A mixture containing these positive and negative ions (ion is an atom that has lost or gained one or more electrons) is not defined as a gas but as plasma. Scientists commonly refer to a plasma state as being the fourth state of matter and in fact, most of the known universe is in this state. In the universe, matter may be found in forms that cannot be grouped into these previously described states of matter. Dying stars are theorized to collapse into forms in which matter is so dense that their gravitational force pulls in all matter and radiation. Because no light can escape from a black hole little is known of them.

STATES OF MATTER—CONCLUSION

As far as the technician is concerned, a basic understanding of what we know as the three states of matter should suffice. A practical example of some of the preceding information can be observed in what happens in a typical injection pulse. Fuel directly injected to a diesel engine cylinder is atomized. Atomized fuel is in the liquid state. This fuel, exposed to the heat of compression in the engine cylinder, must be vaporized, or in other words, changed in state before it can be combusted. If liquid fuel contacts the piston crown, it can result in a failure, because more heat energy is required to vaporize it, retarding its ignition. Gases condensing in the exhaust gas stream will be observed as white smoke. Particulate or minute solids in the exhaust gas stream will be observed as black smoke. In diagnosing these, and many other conditions, the technician will benefit from a basic understanding of chemistry.

PROPERTIES OF SOME COMMON ELEMENTS

Each element has a special identity and set of characteristics that make it unique, in terms of both its behavior and appearance. Some of the characteristics of common elements are outlined below.

HYDROGEN (H)

Hydrogen is a colorless, odorless, tasteless flammable gas, the simplest member of the family of chemical elements with a nucleus consisting of one proton, orbited by one electron. Hydrogen atoms are reactive and combine in pairs forming diatomic (combinations of two atoms to a molecule) molecules: H_2. Hydrogen is the most abundant element in the known universe accounting for about 75% of matter, although only around 1% of all matter on Earth. It is present in all animal and vegetable substances in the form of compounds, most often carbon. It also constitutes 11% of the mass of water. Combustion of hydrogen with oxygen produces temperatures in the region of 2,600°C (4,700°F).

Atomic No.: 1
Melting point: −259.2°C (−435°F)
Boiling point: −252.8°C (−423°F)

CARBON (C)

Carbon is a nonmetallic element that exists in a number of forms and combines to form compounds more readily then any other element. About 0.2% of the Earth's crust is comprised of carbon. Elemental carbon exists in three forms: diamonds, graphite, and carbon black, which includes charcoal and coal. Pure diamond is the hardest substance known, is a poor conductor of both electricity and heat, and is used in industrial applications as a cutting and drilling medium. Graphite has many opposite characteristics notably, it is a conductor of heat and electricity and commonly used as a lubricant. When iron is alloyed with small quantities of carbon, *steel* is formed.

Atomic No.: 6
Melting point: 3,550°C (6,420°F)
Boiling point: 4,827°C (8,721°F)

OXYGEN (O)

A colorless, odorless, tasteless gas, oxygen is the most plentiful element in the Earth's crust. Oxygen is present in air (23% by mass), water (86%), and in the Earth's crust (47%). Almost all atmospheric oxygen is the result of photosynthesis. During respiration, animals and some plants process atmospheric oxygen to form carbon dioxide; by photosynthesis, green plants assimilate carbon dioxide in the presence of sunlight to produce oxygen. Most of the gaseous oxygen found in the atmosphere consists of molecules of two atoms known as **diatomic** oxygen (O_2).

Monatomic (O) and **triatomic** (O_3) oxygen (ozone) are more predominant in the upper strata of the atmosphere where ozone filters the sun's ultraviolet radiation. Commercial oxygen is produced by fractional distillation of liquid air. Oxygen has a valence of 2 and forms a large range of covalently bonded compounds. Molten oxygen forms a blue liquid and solid oxygen is attracted by a magnet.

Atomic No.: 8
Melting point: −218°C (−361°F)
Boiling point: −183°C (−297°F)

NITROGEN (N)

A colorless, odorless, tasteless gas, nitrogen is the most plentiful element in the Earth's atmosphere; the mixture *air* contains approximately 76% nitrogen by mass. Nitrogen is a constituent of all living matter. Chemically, nitrogen is fairly inert at ambient temperatures but during combustion it will combine to form various oxides, especially in high-temperature, lean burn conditions. The automotive industry identifies those various oxides collectively as NO_x and they are subject to EPA standards and regulation as noxious emission.

> Atomic No.: 7
> Melting point: $-210°C$ ($-346°F$)
> Boiling point: $-196°C$ ($-320°F$)

SULFUR (S)

A nonmetallic element, sulfur is also one of the more reactive elements. Pure sulfur is a tasteless, odorless, brittle solid that is a pale yellow in color and a poor conductor of electricity. Sulfur is perhaps important to the diesel technician mostly because it is a constituent of petroleum. Sulfur appears prominently in the residual oil fractions (those left over after distillation and other refining practices) of a diesel fuel that provide the fuel with its lubricating properties. Current United States and Canadian statutory requirements of on-highway diesel fuels permit no more than 0.05% sulfur content. This has lowered the lubricity of the fuel to a degree that has presented problems with some fuel injection systems, especially those engineered before the current standards. Ultralow sulfur (ULS) fuel, which has a maximum sulfur content of 0.005%, is being advocated for use as highway diesel fuel. Sulfur is oxidized to sulfur dioxide (SO_2) in the combustion process. Sulfur dioxide is a colorless and poisonous gas regarded by the EPA as a noxious emission and one that readily forms acidic compounds such as H_2SO_4 (sulfuric acid). Sulfur is also a constituent of lubricating oils.

> Atomic No.: 16
> Melting point: $113°C$ ($235°F$)
> Boiling point: $445°C$ ($832°F$)

IRON (FE)

A metallic element that makes up about 5% of the Earth's crust, iron is the second most common metal after aluminum. It combines with many other elements to form hundreds of minerals. Iron almost always contains small amounts of carbon, which modifies its properties. When alloyed with small amounts of carbon, steel is produced—the most widely used of all metals. Elements other than carbon can also be added to iron in the alloying process to modify the characteristics of a steel. Common alloying metals are nickel, chromium, molybdenum, tungsten, vanadium, titanium, and manganese.

> Atomic No.: 26
> Melting point: $1,535°C$ ($2,795°F$)
> Boiling point: $3,000°C$ ($5,432°F$)

ALUMINUM (AL)

A lightweight, silver-white colored metal that makes up about 8% of the Earth's crust, aluminum is therefore the most common metal. Aluminum never occurs in the metallic form naturally, but its compounds are present in most rocks, vegetation, and animals. Modern commercially produced aluminum is obtained from bauxite and isolated by electrolysis. Pure aluminum tends to be soft and weak, but when alloyed with small amounts of silicon and iron, it becomes harder and tougher. It is an excellent conductor of heat and electricity.

> Atomic No.: 13
> Melting point: $660°C$ ($1,220°F$)
> Boiling point: $2,467°C$ ($4,473°F$)

PROPERTIES OF SOME MIXTURES AND COMPOUNDS

The way in which mixtures and compounds behave when they undergo chemical reactions can be explained by the constituent elements of which they are composed. The following section describes some of the mixtures and compounds that occur in or result from combustion reactions in an internal combustion engine. Many of the by-products of combustion described here are referred to in more detail in Chapter 45.

AIR

Air is a mixture of oxygen and nitrogen. This mixture contains a group of gases with nearly constant concentrations and another group that are variable. Atmospheric gases of more or less consistent concentration by volume are:

Nitrogen (N_2)	78.084%
Oxygen (O_2)	20.946%
Argon (Ar)	0.934%
Neon (Ne)	0.0018%
Helium (He)	0.000524%
Methane (CH_4)	0.0002%
Krypton (Kr)	0.000114%
Hydrogen (H_2)	0.00005%
Nitrous oxide (N_2O)	0.00005%
Zenon (Xe)	0.0000087%

Of gases present in variable concentrations, water vapor, ozone, and carbon dioxide are of principal importance and are critical to the maintenance of life on the planet. Water vapor is the source for all forms of precipitation and is also both an absorber and emitter of infrared radiation. Carbon dioxide is also an absorber and emitter of infrared radiation and is also critical to the process of photosynthesis. Ozone (triatomic oxygen, O_3) is found principally in the upper strata of the atmosphere, 10 to 50 kilometers (6–30 miles) above the Earth's surface. It shields Earth from most cosmic radiation waveforms. Photochemical reactions between nitrogen oxides (NO_x) and hydrocarbons (HC) can produce ozone in quantities large enough to cause respiratory problems in animals including humans.

Concentrations of these gases in air are:

Water vapor (H_2O) 0–7%
Carbon dioxide (CO_2) 0.01–0.1%
Ozone (O_3) 0.01%

CARBON DIOXIDE (CO_2)

Carbon dioxide is a compound. It is a colorless gas with a sharp odor and a sour taste. It is an oxidized form of carbon, produced in a number of ways but most notably in respiration by animals, as a by-product of the combustion of carbon-containing substances (such as a HC fuel); in the fermentation of vegetable matter; and by plants in the photosynthesis process. The presence of carbon dioxide in the atmosphere permits the retention of radiant energy (primarily from our sun) received by the Earth. Carbon dioxide is commonly used as refrigerant and in fire extinguishers. It is also, by mass, the primary product of the combustion of hydrocarbon fuels. However, carbon dioxide is not classified or regulated as a noxious emission, although the excess of it in our atmosphere contributes to global warming. In its solid form it is known as dry ice.

WATER

Water is a compound. It can be readily observed in its three physical states: ice, water, and steam. It is vital to sustenance of all life forms. In its liquid state it is colorless, tasteless, and odorless. An oxygen atom has six electrons in its valence shell; when bonded to two hydrogen atoms, it shares an electron from each hydrogen atom and fills its valence shell. Like carbon dioxide, it is a by-product of the combustion of a HC fuel (the result of oxidizing hydrogen), so it is emitted from any internal combustion engine in the form of steam.

Melting point: 0°C (32°F)
Boiling point: 100°C (212°F)

CARBON MONOXIDE (CO)

Carbon monoxide is a compound of carbon and oxygen. It is a highly toxic, colorless, odorless, and flammable gas. Concentrations of 0.3% can be lethal within 30 minutes.

Carbon monoxide is produced in the stoichiometric or slightly fuel rich combustion zone. It is associated with incomplete combustion in SI engines but tends not to be an important factor as in CI engines due to the considerable excess air factors in diesel engines. Carbon monoxide's toxicity results from its absorption by red blood cells in preference to oxygen. CO poisoning initially produces headaches, dizziness, nausea, and culminates in respiratory failure.

Melting point: −192°C (−314°F)
Boiling point: −199°C (−326°F)

SULFUR OXIDES (SO_X)

Sulfur dioxide (SO_2) emission is the primary product of oxidizing sulfur impurities in hydrocarbon fuels. Sulfur tends to be present in the residual oil fractions constituent in diesel fuel and legislation limiting the sulfur content in on-highway diesel fuel to 0.05% (from 0.335%) has diminished the diesel engines' responsibility for producing this emission. Sulfur dioxide is a *compound* of sulfur, is odorless and nonflammable, and is a source of sulfuric acid (H_2SO_4). Sulfur dioxide smog tends to be more of a problem in geographic areas where burning of solid fuels and heavy oils is not regulated. This type of smog is especially worsened by dampness and high concentrations of suspended particulate.

NO_X

By mass, nitrogen represents the largest ingredient in the engine cylinder during the combustion process in any engine aspirated with air. At ambient temperatures and pressures, nitrogen is relatively inert, that is, unlikely to become involved in chemical reactions. However, subjected to engine cylinder heat and pressure, nitrogen may become involved in the oxidation process, producing compounds of nitrogen known collectively as NO_x:

1. Nitric oxide (NO)—a colorless, odorless gas that in the presence of oxygen will convert to NO_2
2. Nitrogen dioxide (NO_2)—a reddish orange gas that has corrosive and toxic properties
3. Nitrous oxide (N_2O)—a colorless, odorless gas better known as laughing gas

NO_x emission is generally the result of a number of fuel/engine parameters, the most important of which are actual reaction temperatures and the excess air factor. Generally but not consistently, high combustion temperatures tend to produce higher NO_x emission. NO_x is a major contributor to photochemical smog because it reacts with HC to form O_3 molecules or ozone.

UNBURNED HYDROCARBONS

Unburned hydrocarbons (UHC) consist of any emitted unburned fuel fractions and represent the range of fractions in the fuel. They include paraffins, olefins, and aromatics (see Chapter 17). The least volatile elements of a fuel are more likely to result in UHC emissions.

PARTIALLY BURNED HYDROCARBONS

A result of low-temperature combustion partially burned hydrocarbons (PHCs) are substances (aldehydes, ketones, and carboxylic acids) resulting from quenching (extinguishing the flame front) before a molecule is completely combusted.

PARTICULATE MATTER

Any liquid or solid matter that can be detected in light extinction test apparatus, such as a smoke opacimeter, is classified as **particulate matter** (PM). The term is most often applied to emitted ash and carbon soots that are in the solid state.

Diesel engines, even when operating under conditions of optimum efficiency, can emit carbon soot particulate sized 1μ or less in small quantities.

COMBUSTION

Despite the fact that human society has used fire in one form or another since prehistoric times, no person was able to explain it in appropriate chemical terms until Antoine Lavoisier in 1783. Lavoisier proved that combustion, the rusting of metals, and the breathing of animals all involved the combining of oxygen with other chemicals.

When a fuel is heated to its ignition temperature in the presence of oxygen, a chemical reaction takes place in which the heat energy contained in the fuel is liberated, resulting in a large volume of hot gases. In the oxidation reaction that results from igniting a fuel charge in an engine cylinder, oxygen molecules attack hydrocarbon molecules and if they succeed in penetrating their central nuclei (that is, completely combust/oxidize the fuel), produce the compounds H_2O (in the form of steam) and CO_2 (carbon dioxide). Despite

much research concerning the chemical processes that take place during combustion in an engine cylinder, knowledge of the thermodynamic process is largely limited to initial and end states and the intermediate pressure rise profile because it occurs within a minute time frame and at high temperatures. Combustion calculations are simplified when the reaction ingredients are limited to a hydrocarbon fuel and pure oxygen. But combustion in an engine cylinder uses the oxygen available in the ambient air mixture, so proportionally, the largest ingredient of the reaction is always the element nitrogen. Ideally, the elemental nitrogen aspirated by the engine should remain inert and be exhausted, unaffected by the oxidation of the fuel. However, under some circumstances the nitrogen is oxidized, producing a number of oxides (known collectively as NO_x) that are identified as a noxious emission. The tendency to involve nitrogen in the oxidation reaction taking place in the cylinder generally increases as combustion temperatures increase and the fuel concentration in the mixture decreases, that is, under lean burn conditions.

Compression pressures in diesel engines range generally from 27 atms/400 psi (2,750 kPa) to 48 atms/700 psi (4,862 kPa), but in most current truck and bus **direct injection (DI)** engines, actual pressures typically would be close to the middle of that range. This produces cylinder temperatures at the end of the compression stroke approximating 500°C, around twice the minimum required to ignite the fuel. Peak combustion pressures of between two and four times the compression pressure values result, depending on the actual engine and how it is being fueled. Peak cylinder pressures would be managed (by the fuel system) to occur somewhere between 10 to 20 degrees ATDC to optimize the relationship between cylinder pressure and the mechanical advantage of the crank throw vector angle (see Chapter 6) (Figure 16–8).

CYLINDER GAS DYNAMICS

Piston crown design, valve configuration, manifold boost value, and the engine breathing manifolds design all play a role in determining the **cylinder gas dynamics**, that is, how injected fuel is dispersed, mixed, and subsequently combusted in the cylinder. The intent is to create cyclonic turbulence (swirl) in the cylinder as the piston is driven upward through its stroke (squish), and current piston design theory suggests that the Mexican hat crown best enables this. The manner in which cylinder gas behaves under compression and through the expansion or power stroke will govern both the engine's performance efficiency and noxious

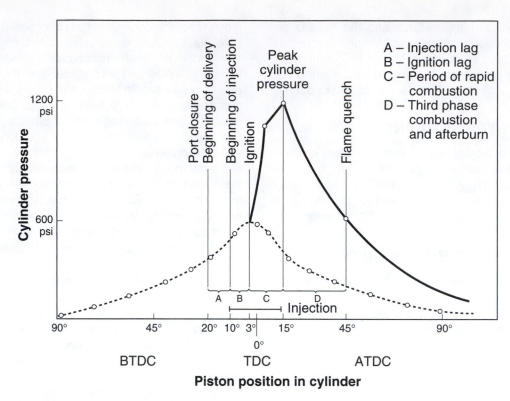

FIGURE 16–8   *Pressure volume curve in a diesel engine.*

emissions. The manner in which cylinder gas behaves during breathing (intake and exhaust strokes) determines how effectively combustion end gases are expelled and a new air charge is induced. Because truck and bus diesel engines are almost always turbo boosted, volumetric efficiencies exceed 100% by a wide margin in most performance operating phases. In fact, it would be true to say that all diesel engines are designed to operate with an excess air factor. The actual excess air percentage is usually highest when the injected fuel quantity is lowest (engine loads are lightest). The excess air factor may exceed 600% at low idle and drop to values that only slightly exceed the stoichiometric requirement when operating in the upper portion of the torque rise profile at full power loads.

STOICHIOMETRY

The term **stoichiometric** is derived from the Greek school of philosophy, which, among other things, preached the avoidance of excess. In chemistry, this has come to mean the relationship between the relative quantities of substances involved in a reaction. In motive power technology, managing a stoichiometric burn ratio means controlling fueling so that the air in the engine cylinder is precisely that required to completely oxidize the fuel. No more, no less. A more comprehensive definition of stoichiometric is: the actual ratio of the reactants in any reaction (not necessarily just a combustion reaction) to the exact ratios required to

complete the reaction. The stoichiometric ratio or **lambda** (λ) factor is therefore dependent on the actual chemical composition of the fuel to be burned: when this is known, the stoichiometric air-to-fuel ratio may be calculated. Many SI automobile engines are electronically managed to run at stoichiometric air/fuel ratios. This is not a requirement of the diesel engine, although it should be noted that the ignition location in the cylinder will occur where the mixture proportions are close to stoichiometric.

> = greater than
< = smaller than

$$\lambda = \frac{\text{actual air supplied}}{\text{stoichiometric requirement}}$$

$\lambda > 1$ lean burn

$\lambda < 1$ rich burn

$\lambda = 1$ stoichiometric AFR

CALCULATING AIR/FUEL RATIO

Gasolines and diesel fuels are refined from crude oil petroleum. Petroleum on a mass basis consists of the following elements:

Carbon 84–87%

Hydrogen 11–15%

Sulfur 0–2%

So on this mass basis, petroleum products whether they are gasoline, diesel fuel, or engine lubricating oil are not going to possess widely differing stoichiometric requirements in combustion reactions. That is, the theoretical **air/fuel ratio (AFR)** will not change that much when different petroleum products are combusted. As soon as the exact composition of the fuel is known, the stoichiometric air requirement to combust it can be calculated using the following data:

To oxidize:

1 kg carbon (C) requires 2.66 kg of oxygen (O)

1 kg hydrogen (H) requires 8.0 kg of oxygen (O)

1 kg sulfur (S) requires 1.0 kg of oxygen (O)

Air is a mixture composed of 23% oxygen and 76% nitrogen by mass. So 1 kg of air would contain 0.23 kg of oxygen. So:

$$\frac{1}{0.23} = 4.35$$

Therefore 1 kg of O would be contained in 4.35 kg of air.

Example:

1. A hypothetical diesel fuel contains by mass 86% carbon, 13% hydrogen, and 1% sulfur. What _mass_ of air would be required to completely oxidize 1 kg of this fuel?

O required = $(2.66 \times .86) + (8.00 \times .13) + (1.00 \times .01)$
           carbon   hydrogen   sulfur

Air required = $4.35 \{(2.66 \times .86) + (8.00 \times .13) + (1.00 \times .01)\}$

Air required = $4.35 (2.29 + 1.04 + .01)$

Air required = 4.35×3.34

Air required = 14.5 kg

∴ AFR = 14.5:1

2. A hypothetical gasoline contains by mass 86% carbon and 14% hydrogen. Calculate the stoichiometric ratio for this fuel.

Air required = $4.35 \{(2.66 \times .86) + (8.0 \times .14)\}$

Air required = $4.35 (2.29 + 1.12)$

Air required = 4.35×3.41

Air required = 14.8 kg

∴ AFR = 14.8:1

SOME FACTS

In combustion:

1 lb carbon combines with oxygen and releases 14,540 Btu.

1 lb hydrogen combines with oxygen and releases 62,000 Btu.

Air/fuel ratio is normally expressed as a ratio of masses. AFR by mass of diesel fuels and gasolines falls within a fairly narrow window ranging from 14.5:1 up to about 15.2:1 depending on the exact chemical constituents of the fuel. AFR by volume ranges from around 10,000:1 up to around 14,000:1.

THE ACTUAL COMBUSTION CYCLE IN A DIESEL ENGINE

The following briefly outlines Harry Ricardo's (engine designer/combustion specialist) description of the diesel engine combustion cycle:

IGNITION DELAY OR IGNITION LAG

Ignition delay or ignition lag occurs between the _events_ of the start of injection (injector nozzle opening) and the moment ignition occurs. Although some chemical reaction begins the moment the first liquid fuel droplet enters the cylinder, the moment of ignition is normally considered to occur either when a visible flame or measurable pressure rise occurs.

In a diesel engine, ignition is not fixed by any factor as easily controlled as the ignition spark in an SI engine. The duration of the ignition delay is defined by the evaporation rate of the fuel, which is factored by the ignition qualities of the fuel and the actual temperature of the air in the cylinder.

In a DI diesel engine, the term air/fuel ratio cannot be used to compare the conditions in the cylinder with those evident in the premixed charge present in an SI engine because as long as the fuel is not completely evaporated, the complete range of fuel/air ratios must be present from zero (no fuel) to infinity (no air within the fuel droplets). Ignition will occur where local air/fuel ratio is most favorable. That is going to be where the mixture conditions are somewhere close to the stoichiometric requirement.

PERIOD OF RAPID COMBUSTION

In this phase, the fuel that evaporated and mixed during the ignition delay period is burned, so the rate and duration of *rapid combustion* are closely associated with the length of the delay period. Generally, as ignition delay is prolonged (for whatever reasons), the rate and resultant pressure rise increase in the second phase. In modern diesel engines with electronically managed variable timing, ignition timing can be controlled so that the period of rapid burning produces peak cylinder pressure at an ideal crank angle under all operating conditions from idle speed/load up to rated speed/load. However, in many older engines with static injection timing (fuel delivery timing could neither be advanced or retarded), this period of high cylinder pressure could occur out of phase with the mechanical dynamics of the crank throw angle.

THIRD PHASE OF COMBUSTION

The third phase of the combustion cycle begins at the moment of peak cylinder pressure (wherever that happens to occur) and ends when combustion is measurably complete; that is, the available fuel has been oxidized. Under conditions of an extended fuel delivery pulse (high engine loads), some portion of the fuel will be injected into the cylinder during this third phase, so the burn rate will be influenced by the rate of injection as well as the mixing rate. Generally, engines are designed with cylinder gas dynamics to enable rapid mixing of fuel and air during the third phase so that the combustion process is completed as early as possible.

AFTERBURN PHASE

Afterburn in the diesel combustion cycle is a period in which any unburned fuel in the cylinder may find oxygen and burn. Most modern engines incorporate this phase in the third phase of combustion because with improvements in cylinder gas dynamics and computer-controlled injection timing, it should not be a significant factor in the present day diesel engine.

DETONATION

Detonation describes the phenomenon that the diesel technician describes as "**diesel knock**" and a driver of any car knows as "ping." In the diesel engine, it is identified by sound caused by intense pressure rise that vibrates the cylinder walls.

Following the ignition of the fuel charge as the flame **propagates** (spreads) through the combustion chamber, the portion of the charge farthest from the primary flame front is subject to both radiated heat and compression caused by the gas expansion of the primary flame front. The heat and compression may cause unburned portions of the charge to ignite before the arrival of the primary flame front. This multiple flame front condition causes an abnormally high rate of combustion and the resultant pressure rise in the cylinder known as diesel knock. When the terms combustion and explosion are used to describe oxidation reactions in an engine cylinder, the physical results of the reaction are the same; however, the term explosion is used to describe an oxidation reaction that takes place at higher speeds. Detonation is therefore correctly described as combustion occurring at explosive rates. Its causes relate generally to the fuel chemistry, prolonged ignition lag, and advanced injection timing, but generally, detonation is seldom a significant problem in contemporary electronically managed diesel engines using North American fuels. It was, and still is, a problem in diesel engines fueled by hydromechanical injection, and the condition can be observed when starting a cold engine during winter conditions.

SUMMARY

- An element is any one of more than a hundred substances that cannot be chemically resolved into simpler substances.
- Elements consist of minute particles known as atoms.
- A mixture is composed of two or more elements and/or compounds, all of which retain their own characteristics and identity.
- A compound is composed of two or more elements combined in definite proportions and held together by chemical force.
- A molecule is the smallest particle of a compound that can exist in a free state and take part in a chemical reaction.
- Electrons carry a negative charge and orbit in shells around the atom's nucleus.
- Protons carry a positive charge and are located in the atom's nucleus.
- Neutrons are electrically neutral and are located in the atom's nucleus.
- Matter can be classified into three states: solid, liquid, or gas.
- Water is the only substance that is familiar in all three states: ice, water, and steam.
- Hydrogen is the simplest of the chemical elements and one of the most reactive.
- Carbon exists in a number of forms and combines to form compounds more readily than any other element.

- Most fuels are elementally composed of carbon and hydrogen.
- The products of combustion of an HC fuel are water and carbon dioxide.
- When nitrogen is oxidized in the combustion process it forms several compounds known collectively as NO_x.
- A stoichiometric combustion reaction occurs when the exact proportions of the reactants (fuel and air) are present.
- Stoichiometric ratios for typical diesel fuels and gasolines range between 14.5:1 and 15:1.
- The moment of ignition in an engine cylinder occurs when there is visible flame or measurable pressure rise.
- Diesel knock is a detonation condition.

REVIEW QUESTIONS

1. Which of the following statements correctly describes air?
   a. A mixture of nitrogen, oxygen, and inert gases
   b. A compound of nitrogen, oxygen, and inert gases
   c. A molecule of nitrogen, oxygen, and inert gases
   d. An element of nitrogen, oxygen, and inert gases
2. A cubic centimeter of a typical diesel fuel would require approximately what volume of air to completely oxidize it?
   a. 12 cc
   b. 15 cc
   c. 1,000 cc
   d. 12,000 cc
3. The smallest particle of a compound that can exist in the free state and retain the chemical identity of the compound is known as:
   a. A quark
   b. A nucleus
   c. An atom
   d. A molecule
4. The theory that explains the atomic differences in the three states of matter is known as:
   a. Covalent bonding
   b. Kinetic molecular theory
   c. Atomic theory
   d. Plasma theory
5. The most abundant element found in the known universe is:
   a. Oxygen
   b. Carbon
   c. Hydrogen
   d. Nitrogen
6. The most abundant element found in the Earth's crust is:

   a. Oxygen
   b. Carbon
   c. Hydrogen
   d. Nitrogen
7. Oxygen has a valence of:
   a. 0
   b. 2
   c. 4
   d. 8
8. Proportionately, the largest ingredient in an in-engine cylinder, combustion reaction is:
   a. HC fuel
   b. Oxygen
   c. Water
   d. Nitrogen
9. The term used to describe a combustion reaction in which the exact proportions of fuel and air are present is:
   a. Stoichiometric
   b. Perfect burn
   c. Afterburn
   d. Gas blow-down
10. Combustion in an engine cylinder that takes place with more than one flame front is correctly known as:
    a. Stoichiometric
    b. Normal
    c. Detonation
    d. Afterburn
11. Calculate the stoichiometric requirement of a fuel composed of 84% carbon, 15% hydrogen, and 1% sulfur when combusted in pure oxygen.
12. Calculate the stoichiometric requirement of a fuel composed of 84% carbon, 15% hydrogen, and 1% sulfur when combusted in ambient air.
13. How many protons would a balanced atom of each of the following elements possess?
    a. Oxygen _____
    b. Hydrogen _____
    c. Nitrogen _____
    d. Sulfur _____
14. How many electrons would a balanced atom of each of the following elements possess?
    a. Oxygen _____
    b. Hydrogen _____
    c. Nitrogen _____
    d. Sulfur _____
15. The atomic number of an element indicates the number of _____ in an atom of the element.
    a. Electrons
    b. Neutrons
    c. Protons
    d. Quarks
16. *Technician A* states that the stoichiometric requirement of a gasoline combusted in air at sea level must be precisely 14.7:1. *Technician B* states that

the stoichiometric requirements of both diesel fuel and gasoline are similar and range between 14.5:1 and 15.1:1. Who is right?

a. A only
b. B only
c. Both A and B
d. Neither A nor B

17. *Technician A* states that an atom with an equal number of electrons and protons has no charge. *Technician B* states that an atom with a deficit of electrons is known as an ion. Who is right?

a. A only
b. B only
c. Both A and B
d. Neither A nor B

18. *Technician A* states that most electronically managed diesel engines are run at precise stoichiometric ratios of air and fuel after the engine is at running temperature. *Technician B* states that most diesel engines are managed to run at rich fuel-to-air ratios. Who is right?

a. A only
b. B only
c. Both A and B
d. Neither A nor B

19. *Technician A* states that hydrogen is an inert element. *Technician B* states that carbon can be classified as a reactive element. Who is right?

a. A only
b. B only
c. Both A and B
d. Neither A nor B

20. *Technician A* states that the end products of both an explosion and a combustion reaction are similar. *Technician B* states that an explosion is an oxidation reaction that occurs in much less time than what would be described as combustion in an engine cylinder. Who is right?

a. A only
b. B only
c. Both A and B
d. Neither A nor B

Chapter 17

Diesel Fuel

PREREQUISITE

Chapter 16

OBJECTIVES

After studying this chapter, you should be able to:

- Interpret the language used to describe hydrocarbon fuels.
- Define those terms that specifically apply to diesel fuel.
- Describe how the cetane number of a diesel fuel is determined.
- Outline the minimum requirements of a North American diesel fuel.
- Calculate how much ignition accelerator is required to restore original CN value.
- Determine the calorific or heating value of a fuel.
- Understand the problems associated with storing fuel.
- Identify degraded diesel fuel.
- Explain the effects of contaminated or degraded fuel on a typical fuel subsystem.
- Explain how an in-cylinder combustion catalyst works.
- Define the terms cloud point and pour point and their importance in cold weather operation.
- Outline the constituents of a typical aftermarket diesel fuel conditioner.

KEY TERMS

AFR	distillate	microorganism growth
ash	fire point	natural gas
ASTM	flame front	octane rating
boil point	flame propagation	oil window
bomb calorimeter	flash point	oxidation stability
calorific value	fossil fuel	photochemical smog
catalyst	fractions	pour point
cetane number (CN)	gasoline	specific gravity
cloud point	heat energy	stoichiometric ratio
compressed natural gas (CNG)	ignition accelerators	sulphur content
crude oil	kerosene	viscosity
diesel fuel	liquefied petroleum gas (LPG)	volatility

FUEL TERMINOLOGY

Before studying fuel chemistry and combustion some fundamental terminology is required. This terminology is used by the fuel refiner and supplier to describe the fuel supplied. This chapter builds on the basics of chemistry introduced in the previous chapter and many of the terms introduced there are applied here. It is probably important, if not essential, that the diesel technician have a fundamental understanding of fuel chemistry and combustion. This knowledge can prove to be a useful tool when troubleshooting engine malfunctions. You do not have to memorize most of the terms listed here; just use them as a reference to interpret the contents of this chapter.

AFR—air-to-fuel ratio. The *actual* ratio of air to fuel in a combustion reaction. Distinct from stoichiometric AFR, which describes the proportions of the reactants in a combustion reaction required to complete the reaction. Normally expressed by mass (weight).

Ash—Diesel fuels normally contain a certain quantity of suspended solids or soluble metallic compounds such as sodium and vanadium. The fuel's ash content can affect injector, fuel pump, and any engine components subjected to high temperatures such as piston rings, exhaust valves, and turbochargers.

ASTM—American Society for Testing and Materials. Organization that classifies diesel fuel (and other fuels) to a standard.

ASTM #1D Fuel—Fuel recommended for use in high speed, on-highway diesel engines required to operate under variable load and variable speeds. Minimum CN must be above 40. In theory, the ideal fuel for highway truck and bus diesel engines but in practice, it not as often used as #2D fuel because it has less heat energy by weight making it less economical.

ASTM #2D Fuel—Fuel recommended for use in high speed, on-highway diesel engines required to operate under constant loads and speeds. Like #1D fuel, the minimum CN is required to be above 40. Widely used in highway truck operations because it produces better fuel economy than #1D fuel due to its higher calorific value albeit at the expense of slightly inferior performance.

Boil point—the temperature at which a liquid vaporizes. When applied to liquid hydrocarbon fuels, it becomes a measure of volatility.

Calorific value—heat energy. The potential heat energy of a fuel is measured in Btu (British thermal units), joules, or calories.

Catalyst—a substance that enables a chemical reaction without itself undergoing any change.

Cloud point—the temperature at which the normal paraffins in a fuel become less soluble and begin to precipitate as wax crystals. When these become large enough to make the fuel appear cloudy, this is termed the "cloud point." Cloud point exceeds the pour point by 3°C (5°F) to 15°C (25°F) in HC fuels.

CN (cetane number)—a measure of the ignition quality of a diesel fuel defined in some detail later in this section.

CNG—compressed natural gas. See Natural gas.

Crude oil—raw petroleum. It consists of a mixture of many kinds of hydrocarbon compounds of differing molecular weights and small quantities of organic compounds such as sulfur. Crude oil is distilled and cracked in the refining process to produce residual oils, distillates, and fractions, blended to manufacture fuels, oils, and tars.

Diesel fuel—term used to describe distillate petroleum compounds and fractions formulated for use in on-highway CI engines. They are generally composed of fractions from the paraffin (the most volatile range used in a diesel fuel), naphthene, and aromatic (the least volatile range found in a diesel fuel) series of crude oil fractions and are graded by the ASTM. Highway-use diesel fuel has a distillation range between 150°C (300°F) to 290°C (550°F) and specific gravity values ranging from 0.78 to 0.86. (See further references to diesel fuel in this text.)

Distillate—term sometimes used to describe diesel fuel formulated for on-highway use. The term diesel fuel is preferred.

Fire point—the temperature at which a liquid hydrocarbon fuel evaporates sufficient flammable vapor to burn continuously in air. Fire point generally exceeds flashpoint by about 10°C in hydrocarbon fuels.

Flame front—the forward boundary of the reacting zone in-cylinder combustion. Usually luminous depending on the fuel used.

Flame propagation—the way in which a fuel combusts inside the engine cylinder as determined by the manner the flame front spreads; dependent on cylinder gas dynamics, the actual AFR, temperature, and fuel chemistry.

Flash point—the temperature at which a liquid hydrocarbon fuel evaporates sufficient flammable vapor to momentarily ignite when a flame is brought near its surface.

Fractions—a portion of a mixture separated by distillation or a cracking procedure such as hydrocracking or catalytic cracking. Most fuels are carefully balanced brews of combustible petroleum fractions with a range of volatility. Each fraction possesses distinct characteristics.

Gasoline—the group of liquid petroleum fuels blended for use in SI engines. The actual composition of gasolines varies according to the crude oil source, refining processes, and blend requirements. Typically they have a volatility range that extends from 35°C (90°F) to 210°C (400°F) and a specific gravity that ranges from 0.70 to 0.78.

Kerosene—made up of heavier fractions than gasoline, kerosene is widely used in heating oil and jet fuel. Kerosene typically has a distillation range between 150°C (300°F) and 270° (510°F) and specific gravity values ranging from 0.78 to 0.85.

LPG—liquified natural gas. See *Natural gas*.

Microorganism growth—airborne bacteria and fungi commonly enter vehicle and storage tanks through their venting systems. When water is present in the bottom of the tank, bacteria may reside in it and feed off the fuel hydrocarbons. The metabolic waste from such microorganisms is acidic and may corrode fuel injection components. Fungal growth can plug fuel filters when pumped through the fuel subsystem. It is good practice to fill truck on-board tanks before parking the vehicle to minimize water condensation problems; draining tanks daily (at the sump tap) also helps. However, when the problem is sourced at the storage tank, chemical treatment of the fuel is required.

Natural gas—the gaseous product of petroleum either suspended above liquid crude oil or dissolved in it, in which case it becomes the first product to be separated in the distillation process. Natural gas is composed primarily of methane CH_4 with a lesser amount of ethane C_2H_6, propane C_3H_8, and butane C_4H_{10}. Propane and butane are extracted from natural gas and stored as liquids under pressure. Both are used as an automotive fuel and are usually known as liquefied petroleum gas or LPG.

Octane rating—octane rating is a measure of the antiknock quality of a fuel, usually a gasoline. *Knock* in an actual engine depends on complex combustion reaction phenomena and on the engine design. However, as a method of classifying gasoline, the ASTM has standardized two methods, the Motor method (MON) and the Research method (RON). Both relate the antiknock performance of a test gasoline to an actual fuel. Two primary reference fuels are used: iso-octane with "ideal" antiknock characteristics and assigned an octane number of 100 and heptane with poor antiknock characteristics and assigned an octane number of 0.

A mixture of these two primary reference, pure HC fuels are then used as the means for grading the actual performance of a gasoline. Therefore, a mixture of 90% isooctane and 10% heptane would theoretically have identical antiknock characteristics as a gasoline sold at the pumps as "90 octane."

The actual test conditions used by Motor and Research methods differ. The Motor method testing operates at higher speed and inlet mixture temperatures than the Research method. The Research method is generally the better indicator of fuel antiknock quality for engines operating at full throttle and low speeds, whereas the Motor method is the better indicator at full throttle and high speeds. Federal regulations in the United States and Canada require the posting of the *average* of Research Octane Number (RON) and Motor Octane Number (MON) at dispensing pumps. This is expressed (R + M)/2.

A gasoline's tendency to knock can be decreased by the addition of antiknock additives, metallo-organic compounds that precipitate in the burn and slow the combustion rate. Tetraethyl lead (TEL) and tetramethyl lead (TML) were used until legislated out of use. Currently potassium and oxygenated compounds are used as octane boosters. Manganese compounds are still used in some Canadian gasolines to increase octane number, but these are not permitted in the United States.

Finally, when the octane rating of a gasoline is greater than 100, it is based on the milliliters of tetraethyl lead required to be added to the reference fuel isooctane to produce the improved rating. In North America, where leaded fuel is still available, it has a pump octane rating of 88. Unleaded fuels are available at the pumps with (R+M)/2 of between 87 and 93 generally. Higher ratings are used for performance and specialty applications.

Oxidation stability—the products of oxidizing stored diesel fuel can result in deposits, filter plugging, and lacquering of fuel injection equipment. Antioxidants in the fuel inhibit the condition.

Photochemical smog—photochemical smog results from the photochemical reaction of hydrocarbons and NO_x with sunlight in the lower atmosphere. It manifests itself as a brownish haze and the ozone that results may cause reduced visibility, plant damage, eye irritation, and respiratory distress. A photochemical reaction occurs when a physical substance absorbs visible, infrared, or ultraviolet radiation.

In the case of photochemical smog the radiation source is the sun; its ingredients are HC and NO_x from combustion processes and subjecting it to still air and sunlight will cause the required molecular decomposition, producing ozone. Photochemical smog is primarily a problem in urban metropolis with ample sunshine and low air movement. Of these, the metropolis of greater Los Angeles, California, is most notable—having an urban population of 16.2 million inhabitants (Demographica.com) effectively walled-in on its east side by the San Gabriel mountains, limiting air movement that is generally inclined to move from west to east. Chapter 45 gives a full account of photochemical smog.

Pour point—as fuel temperature drops below the cloud point, paraffin wax crystals increase in size and the pour point generally denotes the lowest temperature at which the fuel can be pumped. Pour point is generally 3°C (5°F) to 15°C (25°F) below cloud point. Pour point depressant additives are required for extreme cold weather operation in No 1D and No 2D on-highway fuels. Pour point depressants have no effect on cloud point.

Specific gravity—the specific gravity of a liquid is the weight of a volume of the liquid compared to the weight of the same volume of water. The specific gravity of a petroleum base fuel is a direct measure of its heating value (calorific value).

At 15°C (60°F) a diesel fuel measured to a specific gravity of 0.85 would release 139.50 Btu per gallon when combusted.

Stoichiometric ratio—the stoichiometric ratio is an expression of the exact ratio of the reactants required for a chemical reaction to take place. In the internal combustion engine, the specific stoichiometric ratio of a fuel at sea level depends on the chemistry of the fuel and not on the conditions of combustion. It is a ratio of masses.

Sulfur content—on recommendation of the API, the EPA effected a maximum sulfur content of 0.0015% by weight effective in October 1993 in the United States and October 1994 in Canada. This fuel is known as low sulfur (LS) fuel. Some jurisdictions have adopted the use of ultra low sulfur (ULS) fuel that contains a maximum of 0.005% sulfur. Many want ULS adopted as the standard highway diesel fuel.

Viscosity—a measure of a liquid's resistance to sheer, a value that generally decreases as temperature increases. It also influences its resistance to flow and therefore its fluidity. The fluidity of a liquid is graded by seconds Saybolt universal (SSU)—a measure of its ability to flow through a defined flow area against time at 100°F (39°C). The SSU rating is used for comparing the viscosity of diesel fuels; N° ID is typically 34.4 SSU, N° 2D typically 40 SSU, and N° 4 125 SSU (preheating required). Viscosity and volatility in diesel fuels are closely associated. Fuel viscosity directly affects fuel injection pump service life.

Volatility—the tendency of a liquid to vaporize. The volatility rating of a fuel is more important in SI engines as it determines the vapor-to-air ratio at the time of ignition. The volatility rating of diesel fuel is critical in summer operation when its higher fractions tend to boil off reducing CN.

PETROLEUM

The word petroleum is derived from the Latin words petra meaning rock, and oleum meaning oil. Petroleum is broadly used to describe hydrocarbon fossil fuels found in the upper strata of the Earth's crust ranging from solid tars through crude oil to natural gas. Somewhere around 70% of the energy consumed in North America and 40% worldwide is derived from petroleum products.

Most of this fuel is extracted from crude oil and the natural gases that are often contained in the proximity of petroleum deposits or boils off from them at surface pressures and temperatures. Crude oils are usually black, but many reflect a yellow or greenish tint. They range in density from light, very fluid liquids of high volatility to viscous tars of lower volatility.

FORMATION OF PETROLEUM

The origin of the carbon and hydrogen that are the elemental components of petroleum is in the organic materials that made up the primordial (existing at the beginning) Earth. These elemental components passed through an organic phase, mostly single-celled plants and algae and mostly located in aquatic environments. Many such simple organisms were known to have been abundant 570 million years ago; rapid burial of these organisms preserved them in sedimentary rock and enabled a series of biological, physical, and chemical changes that permitted them to evolve to what we know as petroleum. Put simply, this evolution requires heat, pressure, and eons of time. Petroleum is therefore a **fossil fuel**, an unrenewable energy source with limited reserves. It has never been duplicated in a commercial laboratory.

Crude oils are loosely classified by their content, which indicates exactly what can be extracted from them in refining processes. There are three types:

1. Asphalt-based crudes: These are usually black in color, and generally the higher the fluidity of a crude oil, the better its quality, meaning that a larger percentage of lighter (by mass) fractions can be extracted from it. In fact, the market value of a crude oil is often determined by the number of gasoline-suitable fractions that can be refined from it.
2. Paraffin-based crudes: These often reflect a greenish tint. When refined, they produce paraffin waxes and lubricating oils. They tend to be marketed at lower values than asphalt-based crudes.
3. Mixed base crudes: These combine the characteristics of asphalt-based and paraffin-based crudes. Accordingly, they are usually priced in between.

The area of the Earth's crust in which a crude oil is formed plays a role in defining its properties. A minimum temperature of 50°C (120°F) is required to enable the natural formation of crude petroleum.

Oils formed at lower temperatures tend to be heavier and located in the upper strata of the **oil window**. The oil window is the area of the Earth's crust in which petroleum oils can be formed and extends from a depth of 1,500 meters (5,000 ft.) up to 6,500 meters (20,000 ft.). Oils formed in the deeper portion of the oil window at temperatures up to 175°C (350°F) tend to be lighter and commercially more valuable. Petroleum located at depths below the oil window is usually in the form of natural gas.

Oil may also be obtained from oil shales and oil sands. In these, organic matter has been reduced to kerogen, an intermediate stage between the organic matter and oil. Oil shale and oil sands originate from the fine-grained sand deposits below ancient lakes and oceans. The kerogens in oil shales and sands can be processed to produce around 100 liters (27 gallons) per ton when heated to 350°C (660°F). Both Canada and the United States have large reserves of oil shale and sands. However, because of the energy required to process them to petroleum oils, at this moment in time it has proved more economical to import petroleum shortfalls.

REFINING PETROLEUM

The refining of crude petroleum begins with two processes: vaporization and condensation. A crude oil is made up of many different hydrocarbon compounds each with different characteristics. Each has distinct boil and condensation temperatures permitting relatively easy separation. These are known as fractions.

Boiled-off vapors are passed through a distilling column. The distilling column is a vertical cylinder containing a series of stacked vertical trays. Rising vapors bubble upward through holes in the trays and through any liquid that has already condensed. As these vapors rise they cool, condensing on the trays. Condensed liquid is permitted to spill from the trays to external containers. Each batch of liquid drawn off from the distillation column is called a "cut" or "fraction."

The fractions that are drawn off at high temperatures are called heavy fractions and those at low temperatures, light fractions. Straight run distillation therefore grades the fractions extracted from a crude petroleum by cut points or boil temperature. The gas taken off the top of the distilling column (or tower) is natural gas and in descending order of boil temperature, gasoline, naphtha, kerosene, light gas oil, and heavy gas oil. The liquid remaining at the bottom is known as residuum. After removal from the column, the cuts are purified (Figure 17–1).

THERMAL AND CATALYTIC CRACKING

Most highway fuels are a "brew" of fractions carefully blended by the refiners. Racing gasolines are highly complex petroleum brews made from formulae that are sometimes carefully kept secret from competitors. The gasolines and diesel fuel purchased at the pumps are a blend of fractions with a range of volatility designed to produce balanced combustion. The higher or lighter fractions predominate in gasolines, and while a diesel fuel possesses some lighter fractions it also contains heavier fractions with higher lubricity and calorific value than those found in a gasoline. The lighter fractions are more highly valued than the heavier ones.

A problem inherent in the distilling of crude petroleum is that it produces too few of the lighter, more in-demand fractions. Cracking describes processes by which heavier fuel oils may be chemically modified by dividing its heavy molecules into smaller light molecules. Two types of cracking are used. Thermal cracking involves subjecting crude petroleum to high temperatures and pressures in a cylindrical tower. Catalytic cracking is a more efficient method of accomplishing the same objective. A cracking tower is a reaction chamber that is pumped full of cracking stock (low grade crude) and catalysts; the contents of the cracking tower are then subjected to pressure and heat. The cracking products are then drawn off from the reaction chamber and separated in a fractioner by density. The fractioner residuum (what is left in the cat cracker after the cracking process) is called cycle oil. This can sometimes run through a catalytic cracker again.

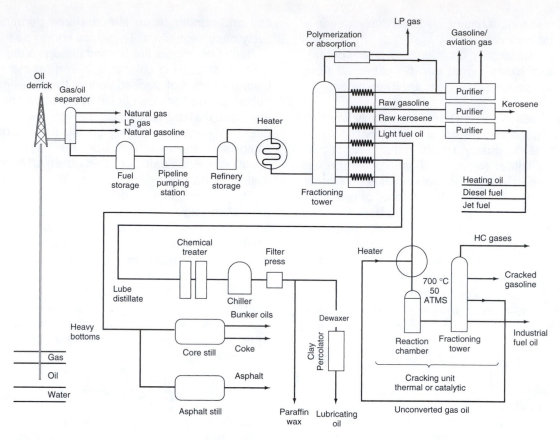

FIGURE 17-1 *Petroleum refining processes.*

Hydrocracking is a catalytic cracking process undertaken in the presence of hydrogen. It is used to produce the higher volatility, lighter fractions required in gasolines and light distillates such as diesel fuel. Before fuels are blended, their composite fractions must be purified. This generally refers to the removal of salt, sulfur, and water impurities (Figure 17–2 and Figure 17–3).

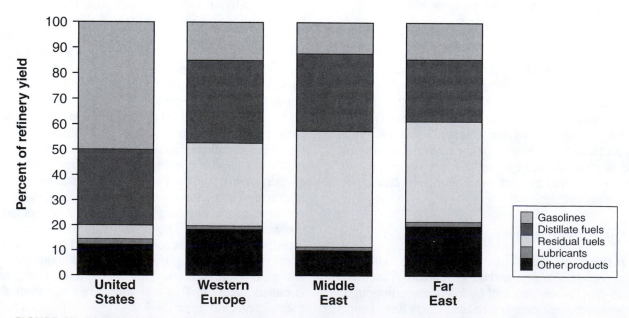

FIGURE 17-2 *Petroleum products as a percent of refinery yields. (Courtesy of Chevron Research)*

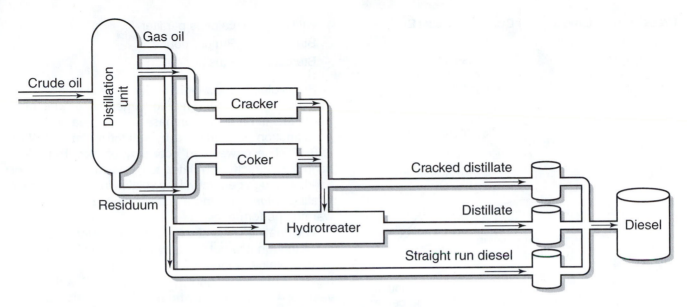

FIGURE 17–3   _Diesel fuel blending. (Courtesy of Chevron Research)_

PETROLEUM PRICING

Crude oil is marketed on pricing based on several benchmark crudes on a per barrel price. Posted price is rounded off to the nearest ten cents and depending on worldwide market conditions, is close to the Organization of Petroleum Exporting Countries (OPEC) price. Spot price is expressed to the nearest cent and continuously changes reflecting market supply and demand.

Spot price fluctuations generally are high averaged to produce a posted price. Posted price governs actual pump pricing (taxation excluded!).

DIESEL FUEL CHARACTERISTICS

Because we have become accustomed to readily available, high quality, uncontaminated fuels in North America, we are often slow to attribute a fuel system or engine problem to the actual fuel being used. Filling a set of vehicle tanks with contaminated or poor quality (low grade) fuel can create performance conditions that last in the vehicle fuel system well beyond the fuel that caused them. When this happens, the cause is usually the purchase of fuel from a source in which the fuel is retained for prolonged periods in storage tanks (most fuel degrades in contact with air). As North American trucking activity extends into Central and South America and fuel is purchased in those countries, the chances of fuel contamination-related problems are going to increase.

CETANE NUMBER

Cetane number (CN) is a measure of the ignition quality of a diesel fuel. Cetane number by ASTM definition is the percentage by volume of a test fuel consisting of _cetane_ (ideal ignition quality, CN 100) mixed with heptamethylnonane (poor ignition quality, CN 0) required to match the fuel to be classified. A mixture of 45% cetane with 55% heptamethylnonane would have a CN of 45. ASTM N° 1D and N° 2D diesel fuels for on-highway use must have a minimum CN of 40 in North America. The CN directly determines the ignition delay phase of the diesel combustion cycle and it is not uncommon for diesel fuel refiners to adjust CN seasonally. However, increasing CN generally reduces fuel density and therefore, the fuel mileage. As the CN of a diesel fuel increases, its ignition temperature decreases (Figure 17–4).

IGNITION ACCELERATORS

Ignition accelerators (aka: cetane improvers) are fuel additives that increase the CN value of a fuel. They are used as a scientifically preferable option to adding alcohol-based fuel conditioner to tanks. An ignition accelerator should only be added to fuel after testing and on the recommendation of the fuel supplier.

Cyclohexanol nitrate when added to fuel at a 0.2% concentration will raise CN by 7 points. It should always be added to a known quantity of fuel and that usually means full tanks on either on-board or ground storage tanks. Other alkyl nitrates are also used as ignition accelerators.

Engine performance factors, which are influenced by the ignition quality of the fuel are cold-starting,

Pressure (and temperature) Cycle in a Diesel Engine

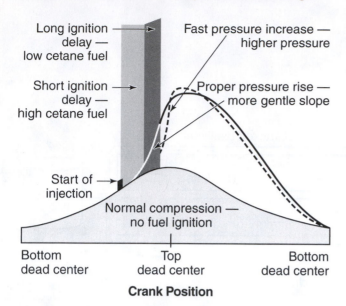

FIGURE 17–4 *Cetane number versus ignition delay.* (*Courtesy of Chevron Research*)

engine warmup, combustion roughness, acceleration, and exhaust smoke density. The ignition quality requirement of an engine is governed largely by engine-operating speeds, atmospheric temperature, and altitude. Increasing the ignition quality over the required level will have the effect of advancing the ignition timing and this will not generally improve engine performance and can compromise engine life.

TECH TIP: Ignition accelerators do not improve fuel, they merely lower the ignition temperature and thereby reduce ignition lag. They should only be used after analysis of fuel by the fuel supplier and then only exactly as prescribed. Excessive quantities of ignition accelerator in fuel has the effect of advancing ignition timing, which can cause engine damage.

HEATING VALUE (CALORIFIC VALUE)

An important property of a diesel fuel is the **heat energy** it releases during combustion. This value is used to reckon the thermal efficiency of an engine's ability to produce power. The gross heat of combustion at a constant volume is determined by a **bomb calorimeter** test, which measures the amount of heat released by burning a known quantity of fuel outlined by ASTM standard D240. Heating values may be expressed as:

J/kg	Joules per kilogram
J/L	Joules per liter
cal/g	calories per gram
cal/L	calories per liter
Btu/lb	Btu per pound
Btu/gal.	Btus per gallon

Both volume and weight bases have significance to the engine designer and equipment user because diesel fuel is normally purchased by volume while fuel consumption is expressed by weight as in: kg/kW or lb/BHP. The power and economy of diesel engines cannot be directly correlated to the volatility of the fuel used, although generally, less volatile fuels have higher heating values and more volatile fuels, better startup/ warmup performance.

The heating value of a fuel directly relates to its density although it is not the only determining factor of its heating value (Figure 17–5). The specific gravity of highway grade 1D and 2D fuels at 15°C with a CN value of 40 or better vary from 0.870 to 0.780 with high CN rating corresponding with lower specific gravity. Perhaps more important, especially to fleet operators, is the fact that 1D fuel generally possesses a lower calorific value than 2D fuel. In practice, this results in better fuel-to-mileage figures for 2D fuels over 1D and this fact has made it the fuel of choice among truck fleet operations.

LOW SULFUR AND ULTRALOW SULFUR FUELS

Sulfur is present in most crude petroleums and is more prominent in the heavier residual fractions from the refining process. When combusted, sulfur in diesel fuel is oxidized to form sulfur dioxide (SO_2), which combines with water to form sulfuric acid (H_2SO_4). Sulfur dioxide produced from the combustion of diesel fuel is recognized as environmentally hazardous and the sulfur content of diesel fuels has progressively been legislated to lower levels in recent years. Today, low sulfur fuel is classified as diesel fuel containing 0.05% sulfur or less. All 1D and 2D diesel fuel currently sold in North America at the pumps as *clear* fuel is low sulfur. Low sulfur fuel has been in use since 1993 and because of a documented problem with its lubricating properties or lubricity, it has caused failures in certain older hydromechanical fuel injection pumps and even some reported premature failures in newer electronically managed fuel systems. In some cases, engine and fuel system manufacturers have recommended the use of additives that specifically enhance the lubricity of the fuel. When using a lubricity-enhancing additive, ensure that only the OEM-recommended product is added to the tank in the correct proportions. This procedure should be undertaken only when the vehicle tanks are full; that is, when the quantity of fuel to be treated is known exactly.

Ultralow sulfur (ULS) fuels are currently being used in some jurisdictions in North America; these are clas-

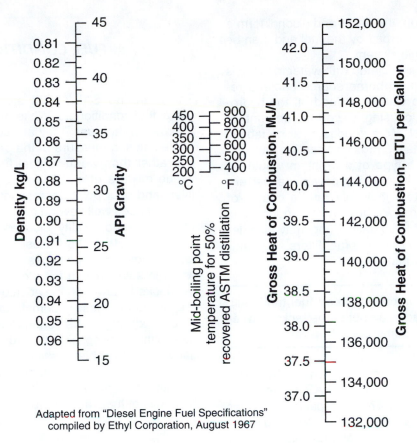

Adapted from "Diesel Engine Fuel Specifications"
compiled by Ethyl Corporation, August 1967

FIGURE 17–5 *Heating value relationships. (Courtesy of Chevron Research)*

sified as fuels that contain 0.005% sulfur or less. ULS fuel is available in some Western European countries where it is used voluntarily by operators. Environmentalists are lobbying to have ULS adopted as the standard diesel fuel on American highways.

FUEL STORAGE, FUEL DETERIORATION, AND PERFORMANCE

The fuel chemistry of both gasolines and diesel fuels is adjusted seasonally by the refiner/supplier largely because what is required of a fuel varies with temperature. To further complicate matters, a truck hauling a load of fruit from Miami, Florida to Edmonton, Alberta can be subject to temperature extremes in excess of 30°C (85°F) to −17°C (0°F) within the span of a single trip. Generally, fuel purchased directly from the refiner/supplier presents fewer problems than fuel that is bulk stored in reservoirs by operators.

Whenever fuel is drawn out of a tank it is replaced by air. This air always contains some percentage of mois-

ture and in areas of high humidity, this can be extreme. Cooling temperatures cause this vaporized moisture to condense as water, which because of its greater weight than the fuel, settles at the bottom of the tank. Most vehicle and storage tanks contain some water, and draining the tanks periodically (just the water trap sump of the tank: ensure that fuel is settled before attempting to do this) helps to prevent excessive water buildup. Both storage reservoirs and vehicle fuel tanks locate the pickup tube in a position so that trace quantities of water present in the bottom of the tank are not drawn out.

Water presents two main problems while it is actually in a fuel tank. However, it presents a host of others, should it actually be drawn out and pumped through a fuel system. Some trucks carry fuel in a pair of cylindrical tanks located at either side of the chassis that have a single pickup tube located in one of the tanks; the tanks are connected by a crossover pipe at the lowest point. If an appreciable quality of water were present in such an arrangement, that water would tend to collect in the crossover pipe, which becomes vulnerable to freeze-up given its location in the air flow. The common preventative and antidote for this condition is to add methyl hydrate (alcohol) to the tank. While this will prevent water freeze-up, the water and alcohol form a solution that can be absorbed by the fuel and then be pumped through the fuel system.

The other problem associated with water in tanks is microorganism growth. Airborne source bacteria reside in the tank sump water and feed off the hydrocarbons in the fuel causing a certain amount of degradation of the fuel. More serious, their metabolic waste is acidic and has a corrosive effect on fuel system components. This type of a problem would be more manifest in the emergency power, genset engine and its fuel system that only runs a couple of times a year than in truck or bus engine applications that are likely to run at least daily. Fungi may also develop in fuel tanks and can plug fuel filters and damage in-tank, transfer pumps.

TECH TIP: It is good practice to keep fuel tanks full. This keeps moisture-laden air out of the tanks and reduces water-in-fuel problems.

Hot weather presents other problems. The year-round average CN value of 1D and 2D ASTM fuels sold in North America is around 47. The legal minimum for both fuels is 40. It makes sense that fuel supplied in the northern states and Canada in winter has higher mean CN values than that sold in the same geographic areas in the summer. However, when fuel is stored and exposed to high temperatures, its more volatile fractions are boiled off, effectively lowering the CN value at a time of year when the supplier has already lowered the CN because the seasonal requirement has dropped.

Cloud point and pour point are important diesel fuel characteristics in frigid temperature conditions. The cloud point value indicates the point at which filters can become plugged, especially the secondary filters in some current applications designed to entrap particulate sized between 2μ and 4μ. At the fuel's pour point temperature, it is on the verge of gelling. Pour point is the lowest temperature a fuel can be pumped through the fuel system. The viscosity of the fuel is temperature related. High viscosity fuel can result in excessive pump resistance and filter damage. Low fuel viscosity may result in excessive wear in fuel injection apparatus and account for leaks. As temperature-conditioned ASTM 2D fuel is, in practice, the fuel used by a majority of operators year-round, fuel heaters are sometimes used to help compensate for deficiencies in cloud point, pour point, and viscosity ratings.

Generally, on-board fuel tanks and stationary reservoirs should be maintained as full as possible to minimize the quantity of moisture-laden air the fuel is exposed to. It is good practice for fleets to require refilling of vehicle tanks at the point of returning from a journey and before overnight parking.

FUEL CONDITIONERS

There are no SAE/ASTM standards that apply to diesel fuel conditioners. While most engine OEMs disapprove of their use, they are commonly sold in their dealerships, a measure of the profitability of the product rather than worth. Diesel fuel conditioners are a vague mixture of cetane improvers, cleansing additives, and pour point depressants in an alcohol base, usually priced well beyond the cost of their ingredients. In a freeze-up, reality dictates that the addition of methyl hydrate is the fastest way out of the problem and in those circumstances, the negatives of putting water in solution with the alcohol are understandably overlooked. However, the addition of methyl hydrate to the tank under these circumstances is arguably preferable to pouring in the contents of a "fuel conditioner" can, in that they specifically address the problem. Similarly, when the objective is to improve the CN value of the fuel, then adding a cetane improver in a measured quantity is a more scientific and cheaper strategy. Most diesel fuels are carefully balanced by the refiner and the use of additives should be avoided whenever possible. Under circumstances such as the effect of reduced lubricity of 0.05 low sulfur fuel on some fuel systems, the OEM will recommend an additive to specifically manage the lubricity problem. Aftermarket fuel conditioners generally create more problems than they solve but the reality is that there are occasions when they have to be used. The technician should recognize the problems that can be caused by fuel conditioners and use them only when there is no other option. Diesel fuel conditioner should always be used according to the manufacturer's recommendations (Figure 17–6).

CERIUM DIOXIDE (CEO$_2$)

Cerium oxide is an in-combustion catalyst that tends to increase the reactiveness of an HC fuel in the process of being oxidized, enabling more complete combustion to occur at lower temperatures. This fact can help to reduce both HC and NO$_x$, plus increase fuel mileage. A number of aftermarket suppliers use in-combustion catalysts with the claim that it increases both engine power and fuel economy. The technology has been around for over 40 years and research suggests that it works at least to some extent. Because some fleets are using these devices, they are addressed here in introductory detail.

CLASSIFICATION	TYPE	FUNCTION
Contaminant Control		
Biocides	Boron compounds, ethers of ethylene glycol, quaternary amine compounds	Inhibit growth of bacteria and fungi—prevent filter clogging
Demulsifiers and Dehazers	Surface-active material that increase water/oil separation	Improve separation of water and prevent haze
Rust and Corrosion Inhibitors	Organic acids, amines and amine phosphates	Prevent rust and corrosion in fuel systems, pipelines, and storage facilities
Fuel Stability		
Metal Detectors	Chelating agents	Inhibit gum formation
Oxidation Inhibitors	Alkyl amines	Minimize oxidation, gum, and precipitate formation
Dispersants	Polymeric amine surfactants	Prevent agglomeration and disperses residue
Engine Performance		
Detergents	Polyglycols, polyether amines	Prevent injector deposits Increase injector life
Dispersants	Basic nitrogen polymeric amine surfactants	Peptize injector deposits Increase filter life
Cetane Improvers	Alkyl nitrates	Increase cetane number
Smoke Suppressants	Overbased barium compounds	Minimize exhaust smoke
Fuel Handling		
Pour Point Depressants	Polymeric compounds	Reduce pour point and improve low-temperature fluidity properties
Could Point Depressants	Polymeric compounds	Reduce cloud point and improve low temperature filterability
De-icers	Low molecular weight alcohols	Reduce freezing point of small amounts of water to prevent fuel line plugging
Antifoam	Silicone and nonsilicone surfactant	Minimizes the formation of fuel foam

FIGURE 17.6   _Diesel additive classification. (Courtesy of Chevron Research)_

TABLE 17–1   *Minimum Fuel Specification for ASTM 2D Low Sulfur Fuel*

PROPERTY	SPECIFICATIONS	ASTM METHOD
Aromatics, % volume	35% maximum	D 1319
Distillation, 90% point, °C (°F)	378°C (640°F) maximum	D 86
Flashpoint, °C (°F)	52°C (125°F) minimum or legal	D 93
Cetane number	40 minimum	D 613
Viscosity, cSt* @ 40°C (104°F)	1.9–4.1	D 445
Sediment and water, Volume %	0.05 maximum	D 1796
Ash, weight %	0.01 maximum	D 482
Total sulfur, weight %	0.05 maximum	D 2622
Copper strip corrosion @ 100°C (212°F)	No. 3 maximum	D 130
Ramsbottom carbon residue on 10% residuum, %	0.35 maximum	D 524
Cloud point, °F	*See* Note 1	D 2500

Source: Courtesy of TMC RP304B

*Kinematic viscosity, centistokes.

Note 1: Cloud point no higher than −12°C (10°F) above the tenth percentile minimum ambient temperature for the area use. Tenth percentile minimum ambient temperature for the U.S. are shown in appendices of ASTM Standard D975-81, Diesel Fuel Oils.

The typical setup involves passing intake air (or a portion of it) through a canister containing cerium dioxide. Minute particles of cerium dioxide are thereby metered to the engine cylinders through the intake system. In combustion, the cerium dioxide gives up one of its oxygen molecules, which participates in the oxidation reaction while also acting as a combustion catalyst.

$$CeO_2 => CeO+[O]$$

Cerium dioxide induction apparatus is generally approved by North American jurisdictions, and testing seems to indicate no negative and possible positive, emissions effects. Among the claims:

- Increased fuel-to-mileage economy
- Reduced HC emission
- Greatly reduced NO_x emission
- Prolonged engine lube life
- Increased power output

EPA REQUIREMENTS OF DIESEL FUELS

The EPA standards pertaining to diesel fuels become ever more stringent as every year passes. Diesel engine particulate emissions have been reduced over

95% in a period of 25 years due both to the mandated improvements in engine technology and in the fuels used to power them. Table 17–1 outlines the requirements of 2D diesel fuel for the year 1997.

SUMMARY

- Crude petroleum is the base used for diesel fuel and many other HC fuels.
- Petroleum is a nonrenewable, fossil fuel with limited reserves.
- Crude petroleum must be refined to separate the fractions used to formulate fuels identified as diesel fuel, gasoline, kerosene, and heavy furnace oil.
- Several different cracking processes may be used to obtain the lighter fractions constituent in diesel fuels and gasolines.
- Diesel fuels and gasolines are a brew of many different petroleum fractions.
- Gasolines are usually more chemically complex brews than are diesel fuels.
- The ignition quality of a diesel fuel is rated by its CN.
- Minimum CN values for on-highway fuels are legislated in North America.
- The minimum CN for a fuel sold as 1D or 2D is 40.

- There is a correlation between fuel density and its heating value.
- Temperature-conditioned (by the refiner) 2D fuel is the common, on-highway fuel.
- Fuel deteriorates chemically and can be subject to microorganic contamination when stored for prolonged periods.
- Addition of methyl hydrate and other alcohols to fuel tanks should be corrective and not precautionary because it may cause other problems.
- Low sulfur diesel fuel with 0.05% sulfur or less is mandated for on-highway use.
- ULS diesel fuel with 0.005% sulfur or less is currently mandated in some North American jurisdictions and environmentalists are advocating its adoption as the standard highway diesel fuel.
- Low sulfur fuels generally have lower lubricity and have produced failures in older hydromechanical fuel systems and even some newer EUI systems.

REVIEW QUESTIONS

1. As the octane rating value of a gasoline increases, what can be said of its burn rate?
   a. It is not affected.
   b. It decreases.
   c. It increases.
2. As the CN value of a diesel fuel increases, what happens to its ignition temperature?
   a. It is not affected.
   b. It decreases.
   c. It increases.
3. The temperature at which a fuel evaporates sufficient flammable vapor to momentarily ignite when a flame is brought close to its surface is known as:
   a. Flash point
   b. Fire point
   c. Boil point
4. What is the minimum CN requirement of 1D or 2D on-highway diesel fuel?
   a. 30
   b. 40
   c. 45
   d. 50
5. Which of the following specific gravity values would correspond to that of a typical 2D diesel fuel?
   a. 0.710
   b. 0.840
   c. 0.970
   d. 1.170
6. Which of the following diesel fuel additives is a commonly used cetane improver?
   a. Iso-propyl alcohol
   b. Kerogen
   c. Kerosene
   d. Cyclohexanol nitrate
7. The temperature at which a diesel fuel begins to form paraffin wax crystals is known as:
   a. Cloud point
   b. Pour point
   c. Flash point
8. The lowest temperature at which a diesel fuel can be pumped through a fuel system is known as:
   a. Cloud point
   b. Pour point
   c. Gel point
9. If an additive were required to prevent freeze-up of a fuel crossover pipe, which of the following would be the preferred option?
   a. Add 8 oz. of fuel conditioner to the tanks.
   b. Add 8 oz. of methyl hydrate to the tanks.
   c. Add 8 oz. of gasoline to the tanks.
   d. Add 8 oz. of cetane improver to the tanks.
10. Current diesel fuel formulated for on-highway use is required to have what maximum sulfur content?
    a. 0.005%
    b. 0.05%
    c. 0.335%
    d. 0.5%
11. *Technician A* states that 2D diesel fuel will, in most cases, produce slightly better fuel mileage from a highway truck engine than a 1D fuel. *Technician B* states that 2D usually has more heat energy than a 1D fuel. Who is right?
    a. A only
    b. B only
    c. Both A and B
    d. Neither A nor B
12. *Technician A* states that all 1D and 2D fuels must be sold with a CN number of exactly 40. *Technician B* states that refiners and fuel suppliers seasonally adjust ignition quality of fuels. Who is right?
    a. A only
    b. B only
    c. Both A and B
    d. Neither A nor B
13. Which of the following best describes the term octane rating when applied to a fuel?
    a. Volatility
    b. Burn rate
    c. Antiknock resistance
    d. Ignition temperature
14. Which of the following describes cerium dioxide when used on a diesel engine application?
    a. Combustion catalyst
    b. Exhaust catalyst
    c. Emission control device
    d. Antioxidant
15. *Technician A* states that a catalyst is a substance that enhances or enables a chemical reaction

without itself undergoing any change. *Technician B* states that the laws of thermodynamics dictate that a catalytic reaction must involve a change in the state of the catalyst. Who is right?

a. A only
b. B only
c. Both A and B
d. Neither A nor B

16. *Technician A* states that denser fractions of crude petroleums tend to command higher prices than lighter fractions. *Technician B* states that diesel fuel tends to be more volatile than gasoline. Who is right?

a. A only
b. B only
c. Both A and B
d. Neither A nor B

Chapter
18

Fuel Subsystems

PREREQUISITE

Chapter 17

OBJECTIVES

After studying this chapter, you should be able to:

- Identify fuel subsystem components on a truck or bus chassis.
- Define the functions of internal and external fuel tank components.
- Troubleshoot a fuel sending unit.
- Define the role of primary and secondary fuel filters.
- Service primary and secondary fuel filters.
- Describe the three ways water can be suspended in fuel.
- Explain how a water separator works.
- Service a water separator.
- Define the principles of operation of a transfer or charge pump.
- Prime a fuel subsystem.
- Test the suction side of the fuel subsystem for inlet restriction.
- Test the charge side of the fuel subsystem for charging pressure.

KEY TERMS

canister	fuel filter	positive displacement
cartridge	fuel heater	primary filter
centrifuge	fuel subsystem	prime mover
charging circuit	fuel tank	secondary filter
charging pressure	gear pump	section modulus
charging pump	Hg manometer	sending unit
clockwise (CW)	inlet restriction	suction circuit
coalesce	micron (μ)	transfer pump
crossover	pickup tube	venting
emulsify	plunger pump	water separator

INTRODUCTION

The diesel **fuel subsystem** on a commercial vehicle is best defined as the group of components responsible for fuel storage and its transfer to the injection pumping apparatus. While injection pumping mechanisms differ greatly from manufacturer to manufacturer, the fuel subsystems that supply them tend to have much in common. Many of the problems that challenge the diesel technician at the novice level focus on the fuel subsystem. The principal components described in this chapter are often dismissed as being straightforward and exempted from study.

However, a thorough knowledge of how these components interact and how they affect the performance of the fuel injection apparatus, is essential. **Fuel tanks, fuel filters, water separators, transfer pumps, fuel heaters**, and their interconnecting plumbing are examined in this chapter (Figure 18–1 and Figure 18–2).

FUEL SUBSYSTEM OBJECTIVES

In the fuel system shown in Figure 18–1 and Figure 18–2, there exists a clear divide between the suction side and the charge side represented by the fuel transfer pump. Most fuel subsystems are of this type and the terms **suction circuit** and **charging circuit** are used to describe each. A **primary filter** is most often located on the suction side of the transfer pump while the **secondary filter** is located on its charge side. However, there are some fuel systems, notably those manufactured by Cummins, in which all movement of fuel through the fuel subsystem is under suction. When such a fuel system uses multiple filters, the terms primary and secondary tend not to be used. This section attempts to generally describe fuel subsystems and troubleshooting methods. The characteristics of particular fuel subsystems are described in sections dealing with proprietary systems.

FUEL TANKS

Fuel is stored on commercial vehicles in fuel tanks. The location of the fuel tank in a typical diesel fuel system is shown in Figure 18–3. Many diesel fuel management systems are designed to pump much greater quantities of fuel through the system than that required for actually fueling the engine.

This excess fuel factor varies from a minimal amount to values exceeding 60% of pumped fuel; it is used to lubricate and to cool high-pressure injection compo-

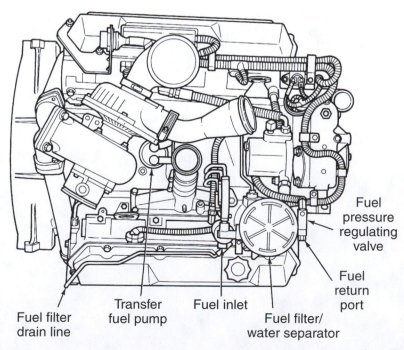

Fuel filter drain line   Transfer fuel pump   Fuel inlet   Fuel filter/ water separator   Fuel pressure regulating valve   Fuel return port

FIGURE 18–1   Location of fuel subsystem components. (Courtesy of Navistar International Corp., Designer and manufacturer of International Brand diesel engines)

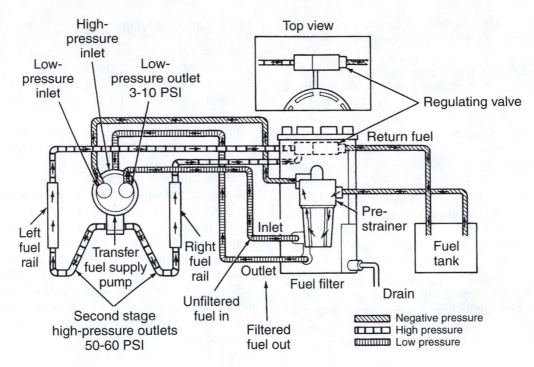

FIGURE 18–2 *Fuel subsystem schematic. (Courtesy of Navistar International Corp., Designer and manufacturer of International Brand diesel engines)*

nents, especially those exposed directly to the extreme temperatures of engine cylinders. As a cooling medium, the fuel transfers heat from the injection pumping unit(s) to the fuel tank. This provides the fuel tank(s) with a role as heat exchanger.

Engine and fuel injection system manufacturers tend not to manufacture most of the fuel subsystem compo-

nents but often make specific recommendations to the truck/bus OEM, which may or may not be observed. The fuel tank is always a chassis OEM-supplied component. Figure 18–4 shows a typical truck fuel tank.

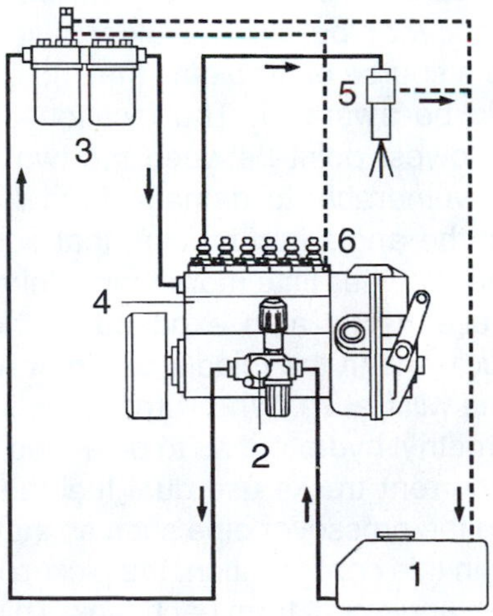

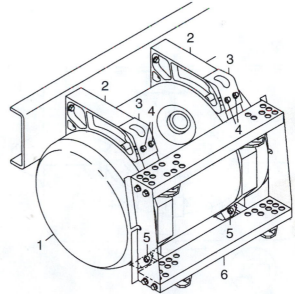

1. Fuel tank
2. Fuel supply pump
3. Fuel filter
4. Fuel injection pump
5. Injection nozzle
6. Overflow valve, overflow restriction

FIGURE 18–3 *Bosch fuel subsystem layout. (Courtesy of Robert Bosch Corporation)*

1. Fuel tank
2. Fuel tank bracket
3. T-bolt guard plate
4. $\frac{3}{8}$-16 Bolts and nuts with washers
5. $\frac{5}{16}$-20 Bolts and nuts with washers
6. Fuel tank step assembly

FIGURE 18–4 *Typical fuel tank with cab access steps. (Courtesy of Freightliner)*

Essentially, a vehicle fuel tank will function most effectively as a heat exchanger if the following is true:

1. Located in the air flow. Truck fuel tanks tend to be mounted in cradle brackets bolted to the outboard side of laddered frame rails (Figure 18–4). This ensures fairly good air flow around the tank into which heat removed from the cylinder head by the fuel can be dissipated.
2. Cylindrically shaped. A cylindrically shaped vessel helps maximize the surface area of the tank exposed to the air flow. This shape additionally has higher **section modulus** (relates the shape of a vessel or beam to rigidity) than a rectangular-shaped vessel permitting thinner wall thickness: that means, an overall lighter fuel tank.
3. Aluminum construction. The coefficient of heat transfer of aluminum is high, enabling heat to be transferred efficiently from the fuel to atmosphere. Aluminum is less susceptible than steel to water corrosion at the base of the tank and is much lighter.
4. Maintained 25% full or better. In a fuel system that circulates fuel through the system at a high rate such as the Cummins HPI-TP system, the fuel can heat up to temperatures where its lubricity is compromised when the tanks are near empty. It is good practice in such systems to maintain the tank level at better than 25% full.

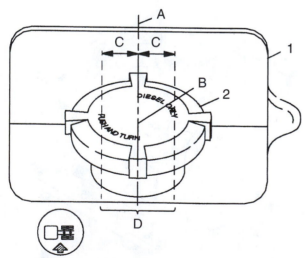

A. Centerline of fill door opening
B. Centerline of fuel cap
C. 1 inch (25 mm)
D. Fuel cap centerline to be within this area.

1. Fill Door Opening
2. Fuel Cap

FIGURE 18–5 Fuel tank cap. (Courtesy of Freightliner)

FUEL TANK SENDING UNITS

Most commercial truck fuel subsystems use remote (from the tank) fuel transfer pumps and not assemblies that incorporate the **sending unit** and a transfer pump. Therefore, the fuel sending unit is an integral assembly flange fitted to the tank. It consists of a float and arm connected to a variable resistor whose function it is to control current flow to a cab gauge proportionally with fuel tank level.

Some dual fuel tank truck chassis where the tanks are connected underneath by a **crossover** pipe (a crossover pipe connects a pair of cylindrical fuel tanks, saddle mounted on the outside of opposing frame rails) may have a single sending unit located in one of the tanks. It is generally preferable to locate a sending unit in each tank and provide a dash gauge for each, providing the operator with some advance warning of a crossover pipe restriction. Fuel sending unit problems can be diagnosed with a digital multimeter (DMM) in resistance mode, by moving the float arm through its sweep and observing readings.

DUAL TANKS

Most heavy highway trucks use multiple fuel tanks (usually two) to increase on-board fuel quantity and therefore range and also to evenly distribute fuel weight. It should be noted that 100 gallons of a typical diesel fuel weighs between 700 and 730 pounds. To maintain even weight distribution as fuel is consumed, a y-type **pickup tube** pulls fuel from both tanks simultaneously; assuming the tanks are of equal volume, fuel level is automatically equalized. Each tank requires a fuel cap (Figure 18–5) and must be filled separately when no crossover pipe is used. Older style crossover pipes can be a source of problems that the diesel technician should be aware of. The location of the pipe slung at the lowest point between the two fuel tanks makes them vulnerable to damage from road debris and animals; the angle iron bracket that supports the crossover line provides little more than minimal protection. The crossover is also exposed to the air flow under the truck and in the middle of winter, water present in the line will freeze. When crossover lines freeze up, alcohol (methyl hydrate) has to be added to the fuel tanks. Most current trucks use dual fuel tank designs that eliminate the crossover pipe such as that shown in Figure 18–6; in this configuration, the pickup lines draw in a parallel arrangement from each tank. This arrangement is also known as a Y-type pickup.

PICKUP TUBES

Fuel pickup tubes are positioned so that they draw on fuel slightly above the base of the tank and thereby avoid picking up water and sediment. Pickup tubes are

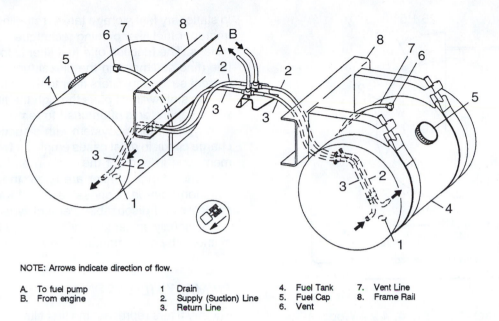

NOTE: Arrows indicate direction of flow.

A.	To fuel pump	1. Drain	4. Fuel Tank	7. Vent Line
B.	From engine	2. Supply (Suction) Line	5. Fuel Cap	8. Frame Rail
		3. Return Line	6. Vent	

FIGURE 18–6 *Dual fuel tank arrangement that eliminates the fuel crossover pipe. (Courtesy of Freightliner)*

quite often welded into the tank; in this case, if they fail the tank may have to be replaced. Fuel pickup tubes seldom fail but when they do it is usually by metal fatigue crack at the neck; this results in no fuel being drawn out of the tank by the transfer pump whenever the fuel level is below the crack.

VENTING

Currently, most jurisdictions in North America permit **venting** of diesel fuel tanks to atmosphere. Therefore, as fuel is pumped out of on-board tanks it is replaced by ambient air drawn in through vent valves. Similarly, on refueling, fuel tank vapors are expelled to atmosphere. In hot weather conditions, some of the lighter fuel fractions may be boiled off perhaps causing a slight reduction in the CN value. However, this boil-off seldom occurs at a rate sufficient to significantly compromise the fuel except in instances in which a tank of fuel is retained for a prolonged period in extreme heat. Boiled-off fuel fractions from diesel fuel have thus far not been considered a noxious HC emission problem of significance by current emission legislation. Fuel tank vents should be routinely inspected for restrictions and should be protected from ice buildup. A plugged fuel tank vent will rapidly shut down an engine, creating a suction side **inlet restriction** value the transfer pump will not be capable of overcoming.

TECH TIP: To check for the presence of water in fuel tanks, first allow the fuel tanks to settle then insert a probe (a clean aluminum welding rod) lightly coated

with water detection paste through the fill neck until it bottoms in the base of the tank; withdraw the rod and examine the water detection paste for a change in color. This test will give some idea of the quantity of water in the tank by indicating the height on the probe where the color has changed. Trace quantities (just the tip of the probe changes color) in fuel tanks are not unusual and will not necessarily present any problems.

FUEL TANK REPAIR

Generally, OEMs recommend that a defective fuel tank be replaced rather than repaired. However, corroded and punctured fuel tanks are going to be repaired regardless of OEM recommendation so the following suggestions are offered, not to endorse fuel tank repair practice but to promote some safety awareness. It is essential to recognize the explosion hazard represented by diesel fuel vapors.

If a metal fuel tank must be welded, all fittings and plugs should be removed and the tank high pressure steamed until the temperature exceeds 200°F (95°C) throughout or for a minimum of one hour; it should also be remembered that in some jurisdictions, the practice of steaming fuel storage vessels to atmosphere is prohibited. Fuel tanks sectioned with internal baffles must be directly subjected to the steam within each section. After steaming, the tank should be filled with nitrogen

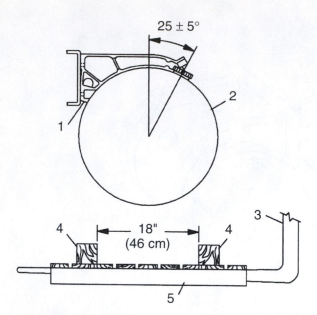

25 ± 5°

1. Fuel tank bracket
2. Fuel tank
3. Lift-truck forks
4. 4 × 4 Wooden block
5. Wooden pallet

FIGURE 18–7   Fuel tank mounting geometry required to prevent overfill. (Courtesy of Freightliner)

gas before repair welding and/or weld patching. When a gasoline tank is to be repair welded, it should be steamed then evacuated with a vacuum pump to a moderate vacuum before filling with nitrogen gas.

TECH TIP: When evacuating fuel tanks, check with the tank OEM for the safe vacuum test value. A high vacuum value may collapse a fuel tank; as a rule, cylindrical tanks will withstand higher vacuum test values than square section tanks.

The practice of welding full fuel tanks in position, while arguably safer than welding on an empty tank filled with fuel vapor, should never be undertaken. A single mistake or metallurgical defect in the tank wall will result in the death of the welder and any other person in the vicinity (Figure 18–7).

FUEL FILTERS

Diesel fuel injection equipment is manufactured with minute clearances and impurities in fuel, which if not removed by the fuel subsystem, can cause premature failures. Most dirt found in fuel is a result of conditions in stationary fuel storage tanks, refueling practices, and improper fuel filter priming techniques by service technicians. The function of a fuel filter is to entrap particulate (fine sediment) in the diesel fuel, and while some current secondary filters filter to the extent that water in its free state will not pass through the filtering media, a water separator is often used to remove H_2O.

A typical fuel subsystem with a suction circuit and a charge circuit in most cases employ a two-filter arrangement, one in each of the suction and charge circuits. Two basic types of filter are used: the currently more common spin-on, disposable **cartridge** type and the **canister** and disposable element type. Spin-on filters are obviously easier to service and are the filter design of choice by most manufacturers (Figure 18–8).

PRIMARY FILTERS

Primary filters represent the first filtration stage in a typical two-stage filtering fuel subsystem. Primary filters are therefore usually under suction, plumbed in series between the fuel tank and the fuel transfer pump. They are designed to entrap particulate sized larger than 20–30 μ (1μ = one millionth of a meter) depending on the fuel system, and they achieve this using media ranging from cotton threaded fibers, synthetic fiber threads, and resin-impregnated paper.

SECONDARY FILTERS

Secondary filters represent the second filtration stage in two-stage filtering. In a typical fuel subsystem, the secondary filter is charged by the transfer pump, and this enables use of more restrictive filtering media. The secondary filter would therefore normally be located in series between the transfer or **charging pump** (the pump responsible for pulling fuel from the fuel tank and charging the fuel injection components) and the fuel injection apparatus. In some diesel fuel subsystems using two-stage filtering, a primary and secondary filter may be both located on the same circuit, usually the charge circuit. In such cases, both filters are mounted on the same base pad with the primary filter feeding the secondary filter; such an arrangement is more likely to be found in off-highway applications of diesel engines. Current secondary filters may entrap particulate sized as small as 1 μ (**micron:** one millionth of a meter) but filtering efficiencies of 2–4 μ are more common.

Water in its free or emulsified state will not be pumped through many of the current secondary fuel filters. This results in the filter plugging on water and shutting down the engine by starving it for fuel. Secondary filters use a variety of media including chemically treated pleated papers and cotton fibers.

Fuel Filter
Two-stage box-type filter

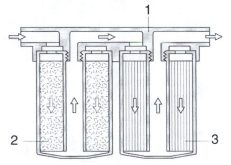

Multistage Filter
With spiral V-form filter element

1. Filter cover with mounting
2. Coarse filter
3. Fine filter

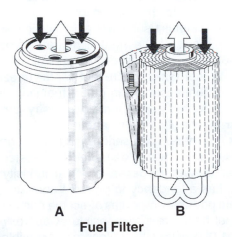

A   B
Fuel Filter

A. Easy-charge filter
B. Spiral V-form filter element

FIGURE 18–8 Types of fuel filters. (Courtesy of Robert Bosch Corporation)

In a fuel subsystem that is entirely under suction, such as some Cummins systems, the terms primary and secondary are not used to describe multiple filters when fitted to the circuit. Because every filtering device used in the fuel subsystem is under suction, the inlet restriction specification is critical and, if exceeded, would result in a loss of power caused by fuel starvation.

SERVICING FILTERS

Most fuel filters are routinely changed on preventative maintenance schedules that are governed by highway miles, engine hours, or calendar months. They are seldom *tested* to determine serviceability. When filters are tested, it is usually to determine if they are restricted (plugged) to the extent they are reducing engine power by causing fuel starvation.

Primary filters or a filter under suction should be tested for inlet restriction with a mercury (**Hg**) filled **manometer**. A manometer is a clear tubular column formed in a U shape around a calibration scale marked off in inches: the column is then filled with either mercury or water (usually colored with dye for ease of reading) to a zero point on the calibration scale. When the manometer is connected to a fluid circuit, it will produce a reading according to the pull (vacuum circuit) or pressure acting on the fluid in the column. Actual inlet restriction values vary considerably with the fuel system and specifications should always be referenced. The Hg manometer should be plumbed between the filter mounting pad and the transfer pump. Transfer pumps are usually **positive displacement** (unload a constant slug volume of fluid per cycle), so they pump more fuel proportionally with rpm increase; this means valid test results can be obtained without loading the engine. When testing circuit restriction on a fuel subsystem that is entirely under suction, the specifications are likely to be fairly exacting. Circuit restriction specification readings on some Cummins systems should be 4–6″ Hg with a new filter but 7″ Hg is the maximum specification; if this value is exceeded, it will result in fuel starvation. The technician should note that most filters function with optimum efficiency just before they completely plug; in other words, at the end of their service lives.

Secondary filters are usually charged by the transfer pump. Testing **charging pressure** (the pressure downstream from the charging/transfer pump) is normally performed with an accurate, fluid-filled pressure gauge plumbed in series between the transfer pump and injection pump apparatus; it is not generally used as a method of determining the serviceability of a secondary filter. Secondary filters tend to be changed by preventative maintenance schedule rather than by testing or when they plug on water or midwinter fuel waxing and shut down an engine.

PROCEDURE FOR SERVICING SPIN-ON FILTER CARTRIDGES

Much dirt is placed in diesel fuel systems by technicians using improper service techniques. Most diesel service technicians realize that sets of replacement filters should be primed, that is, filled with fuel before installation, but few concern themselves about the source of the fuel. Filters should be primed with filtered fuel. Shops performing regular engine service should have a reservoir of clean fuel; any process that requires the technician to remove fuel from vehicle tanks will probably result in it becoming contaminated at least to some extent, however much care is exercised. The container used to transport the fuel from the tank to the filter should be cleaned immediately before it is filled with fuel. Paint filters (the paper cone-shaped type) can be used to filter fuel. The inlet and outlet sections of the filter cartridge should be identified. The filter being primed should be filled only through the inlet ports usually located in the outer annulus (ring) of the cartridge and never directly into the outlet port, usually located at the center. Some manufacturers prefer that only the primary filters be primed before installation during servicing. After the primary filter(s) has been primed and installed, the secondary filter should be installed dry and primed with a hand primer pump.

REPLACEMENT PROCEDURE

1. Remove the old filter cartridge from the filter base pad using an appropriately sized filter wrench. Drain the fuel to an oil disposal container.
2. Ensure that the old filter cartridge gasket(s) have been removed. Wipe the filter pad gasket face clean with a lint-free wiper.
3. Remove the new filter cartridge from the shipping wrapping. Fill the filter cartridge with clean, filtered fuel poured carefully into the inlet section. The inlet ports are usually located in the outer annulus of the cartridge. Fuel poured into the filter inlet ports passes through the filtering media and fills the center or outlet section of the filter; this method will take a little longer because it requires some time for the fuel to seep through the filtration medium.
4. The fuel oil itself should provide the gasket and/or O-ring and mounting threads with adequate lubricant; it is not necessary or good practice to use grease or white lube on filter gaskets.
5. Screw the filter cartridge **CW (clockwise:** right-hand threads are used) on to the mounting pad; after the gasket contacts the pad face, a further rotation of cartridge is usually required. In most cases, hand tightening is sufficient but each filter manufacturer has its own specific recommendations on the tightening procedure and these should be referenced.

TECH TIP: When a hand primer pump is fitted to a fuel subsystem, externally prime only the primary filter, ensuring that all the fuel is poured through the inlet side only. Install the secondary filter dry and prime using the hand primer pump.

WARNING: When removing filter cartridges, ensure that the gasket is removed with the old filter. A common source of air in the fuel subsystem is double gasketing of the primary filter. Double gasketing will usually produce a leak at the secondary filter.

WATER SEPARATORS

Most current diesel engine-powered highway vehicles have fuel subsystems with fairly sophisticated water removal devices. Water appears in diesel fuel in three forms: free state, emulsified, and semiabsorbed. Water in its free state appears in large globules and, because of its greater weight than diesel fuel, readily collects in puddles at the bottom of fuel tanks or storage containers.

Water emulsified in fuel appears in small droplets; because these droplets are minutely sized, they may be suspended for some time in the fuel before gravity takes them to the bottom of the fuel tank. Semiabsorbed water is usually water in solution with alcohol, a direct result of the methyl hydrate (type of alcohol added to fuel tanks as deicer or in fuel conditioner) added to fuel tanks to prevent winter freeze-up. Water that is semiabsorbed in diesel fuel is in its most dangerous form because it may **emulsify** (the fine dispersion of one liquid into another) in the fuel injection system where it can seriously damage components.

Generally, water damages fuel systems for three reasons. Water possesses lower lubricity than diesel fuel, has a tendency to promote corrosion, and its different physical properties affect the pumping dynamics. Diesel fuels are compressible at approximately 0.5% per 1,000 psi; water is less compressible at approximately 0.35 per 1,000 psi. Fuel injection pumping apparatus is engineered to pump diesel fuel, and if water with its lower lubricity and compressibility is pumped through the system, the resultant pressure rise can cause structural failures, especially at the sac/nozzle area of the fuel injectors.

Water separators have been used in diesel fuel systems for many years. These were most often fairly crude devices that used gravity to separate the heavier water from the fuel. However, over the past two decades as both injection pumping pressures have steadily risen and the consumer's expectation of engine longevity greatly increased, water separators have developed accordingly. Often a water separator will combine a primary filter and water separating mechanism into a single canister. Many of these combination primary filter/water separators are manufactured by aftermarket suppliers such as Racor, CR, Davco, Dahl, and others. These use a variety of means to separate and remove water in free and emulsified states; they will not remove water from fuel in its semiabsorbed state.

Water separators use combinations of several principles to separate and remove water from fuel. The first is gravity. Water in its free state or emulsified water that has been **coalesced** (where small droplets come together to combine into larger droplets) into large droplets will, because of its heavier weight, be pulled by gravity to the bottom of a reservoir or sump. Some water separators use a **centrifuge** to help separate both larger globules of water and emulsified water from fuel; the centrifuge subjects fuel passing through it to centrifugal force, throwing the heavier water to the sump walls where gravity can pull it into the sump drain. A centrifuge acts to separate particulate from the fuel in the same manner. Fuel directed through a fine resin-coated, pleated paper medium passes through the medium with greater ease than water. Water entrapped by the filtering medium can collect and coalesce in large enough droplets to permit gravity to pull it down into the sump drain. In many cases, aftermarket water separator/fuel filters are designed to replace the fuel system OEM's primary filter; in others this unit may work in conjunction with the primary filter. When installing an aftermarket water separator/filter unit on the suction side of a fuel subsystem, it is good practice to locate the manufacturer's maximum restriction specification and test that it is not being exceeded.

A mercury (Hg) manometer is the appropriate test tool. In cases in which the entire fuel subsystem is under suction, the consequences of exceeding the restriction specification are generally more severe, the result being fuel starvation to the engine. Servicing water separator units is a simple process but one that should be undertaken with a certain amount of care because it is easy to contaminate the fuel in the separator canister either by priming it with unfiltered fuel or by permitting dirt to enter when the canister lid is removed. Most aftermarket water separators have a clear sump through which it is easy to observe the presence of water. All water separators are equipped with a drain valve; the purpose of this valve is to siphon water from the sump. Water should be routinely removed from the sump using the drain valve. The filter elements used in combination water separator/primary filter units should be replaced in most instances with the other engine and fuel filters at each full service. However, some manufacturers claim their filter elements have an in-service life that may exceed the oil change interval by two or more times. Whenever a water separator is fully drained, it should be primed before attempting to start the engine.

TECH TIP: To troubleshoot the source of air admission to the fuel subsystem, a diagnostic sight glass can be used; it consists of a clear section of tubing with hydraulic hose couplers at either end, and it is fitted in series with the fuel flow. However, the process of uncoupling the fuel hoses will always admit some air into the fuel subsystem, so the engine should be run for a while before reading the sight glass.

Some current systems use a water-in-fuel (WIF) sensor to alert the operator of fuel contamination.

FUEL HEATERS

In recent years it is more common to find trucks equipped with fuel heaters. In fuel systems in which fuel is flowed through the injection system circuitry at a rate much higher than that required for fueling the engine, constant filtering of fuel removes some of the wax and therefore some of its lubricity even when the appropriate seasonal pour point depressants are present.

Pour point depressants tend not to have too much effect on the cloud point of a fuel, which is its first stage of waxing. This is a condition to which ASTM N° 2D fuels are more prone than N° 1D. However, there is some debate about the use of fuel heaters, and the fuel system/engine manufacturer should always be consulted when fitting such a device. One engine manufacturer warns that its warranty is voided if electric element type fuel heaters are used in its system.

There are two types of fuel preheaters in current use:

1. Electric element type. An electric heating element uses battery current to heat fuel in the subsystem. This type offers a number of advantages, most notable of which is that the heater can be energized before startup so that cranking fuel is warmed up. Electric element fuel heaters may be thermostatically managed so that fuel is only heated as much as required and not to a point that compromises some of its lubricating properties.

2. Engine coolant heat exchanger type. This type of fuel heater consists of a housing within which coolant is circulated in a bundle (heat exchanger core) and over which the fuel is passed. A disadvantage of this type is that the engine cooling system must be at operating temperature before the fuel can be heated.

Fuel heaters exist that use both electric heating elements and coolant medium heat exchangers and furthermore, manage the fuel temperature. Optimumly, fuel temperature should be managed not to exceed 90°F (32°C). Once fuel exceeds this temperature its lubricating properties start to diminish and the result is reduced service life of fuel injection components.

FUEL CHARGING/TRANSFER PUMPS

Fuel charging or transfer pumps are positive displacement pumps driven directly or indirectly by the engine. A positive displacement pump displaces the same volume of fluid per cycle and therefore, fuel quantity pumped increases proportionately with rotational speed; similarly, if a positive displacement pump unloads to a defined flow area, pressure rise can be said to be proportional with rpm increase. On most truck and bus fuel systems, charging/transfer pumps are of the plunger or gear types. These pumps are responsible for all movement of fuel through the fuel subsystem (Figure 18–9 and Figure 18–10).

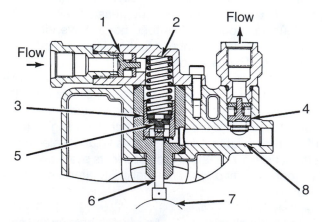

Fuel transfer pump
1. Inlet check valve
2. Spring
3. Piston assembly
4. Outlet
5. Piston check valve
6. Tappet assembly
7. Cam
8. Passage

FIGURE 18–9   Caterpillar transfer pump. (Courtesy of Caterpillar)

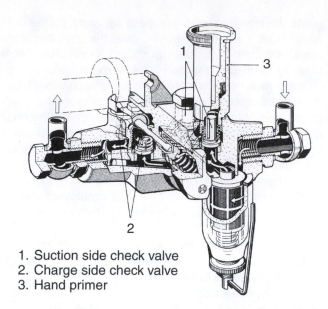

1. Suction side check valve
2. Charge side check valve
3. Hand primer

FIGURE 18–10   Bosch charging pump with integral hand primer and primary filter. (Courtesy of Robert Bosch Corporation)

PUMPING PRINCIPLE

In describing pump operation in the fuel subsystem, in common with most truck OEMs we use the terms suction circuit and charge circuit. Fuel movement through a fuel subsystem is created by a positive displacement pump, also known as a **prime mover**. The way this works is that as the pump forces fuel into the charge circuit, lower-than-atmospheric pressure is created upstream from the pump inlet. Atmospheric pressure acting on the fuel in the tank then forces fuel from the tank toward the pump inlet. In this way, fuel is moved through the fuel subsystem.

PLUNGER-TYPE PUMPS

Plunger-type pumps are often used with port helix metering injection pumps. They are usually flange mounted to the injection pump cambox and driven by a dedicated cam on the pump camshaft. Single-acting and double-acting plungers may be used, with the latter type specified in higher output engines requiring more fuel. A single-acting **plunger pump** (a pump with a single reciprocating element such as a bicycle pump) has a single pump chamber and an inlet and outlet valve. Fuel is drawn into the pump chamber on the inboard stroke and pressurized on the outboard or cam stroke.

A double-acting pump has twin chambers each equipped with its own inlet and outlet valve. On the cam stroke, a two-way plunger charges the pump chamber while admitting fuel to the other. On the return stroke, the pump retraction spring reverses the process (Figure 18–10).

GEAR-TYPE PUMPS

Gear pumps are also commonly used as transfer pumps, especially on most electronically managed engines. These are normally driven from an engine accessory drive and are located wherever convenient. Gear pumps usually have an integral relief valve that defines the system charging pressure. Fuel injection systems designed to be charged at pressure values higher than typical, tend to use gear-type transfer

pumps instead of cam-actuated plunger pumps. In instances in which a gear pump feeds an injection system with no main filter in series, a filter mesh is sometimes incorporated to protect injection pumping apparatus; when gear pumps are used, there is a small chance that gear teeth cuttings can be discharged into the system. A majority of the full authority, electronic management fuel systems use gear-type transfer pumps.

HAND PRIMER PUMPS

A hand primer pump may be a permanent fixture to a fuel subsystem located on the fuel transfer pump body or a filter mounting pad (Figure 18–10 and Figure 18–11). A hand primer pump can be a useful addition to the technician's tool kit, and it can be fitted to a fuel subsystem when priming is required. The function of a hand primer pump is to prime the fuel system whenever prime is lost. Typically, they consist of a hand-actuated plunger and use a single-acting pumping principle. On the outward stroke, the plunger exerts suction on the inlet side, drawing in a charge of fuel to the pump chamber; on the downward stroke, the inlet valve closes and fuel is discharged to the outlet. When using a hand primer pump, it is important to purge air downstream from the pump on its charge side. Some fuel subsystems mount a hand primer to the transfer pump housing. Some newer fuel systems have self-contained, electric priming pumps whose function is to prime the system after servicing.

Operating principle (single-acting)

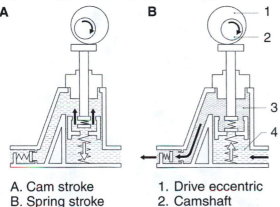

A. Cam stroke
B. Spring stroke

1. Drive eccentric
2. Camshaft
3. Pressure chamber
4. Suction chamber

Operating principle (double-acting)

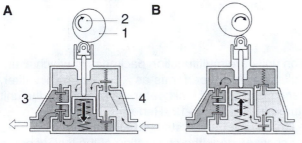

A. Cam stroke
B. Spring stroke

1. Drive eccentric
2. Camshaft
3. Pressure chamber
4. Suction chamber

FIGURE 18–11   Action of single- and double-acting plunger pumps. (Courtesy of Robert Bosch Corporation)

PRIMING A FUEL SYSTEM

Although priming a fuel system is a relatively simple procedure, it is usually advisable to consult the appropriate service manual.

Most OEMs prefer that the technician avoid pressurizing air tanks with regulated air pressure to prime a diesel fuel system. It should be remembered that diesel fuel contains volatile fractions and the act of pressurizing a fuel tank with air pressure will vaporize some fuel. Air exiting an air nozzle creates friction and the potential for ignition. Avoid this practice in extreme hot weather conditions.

RECOMMENDED PROCEDURE

1. When a vehicle runs out of fuel and it is determined that the fuel subsystem requires priming, remove the filters and fill with filtered fuel.

2. Locate a bleed point in the system—often on an in-line injection pump system, this will be at the exit of the charging gallery—and crack open the coupling. A fuel manifold outlet coupling in the cylinder head should be opened in most other systems.

3. Next, if the system is equipped with a hand primer pump, actuate it until air bubbles cease to exit from the cracked open coupling. If the system is not equipped with a hand primer pump, fit one upstream from the transfer pump and actuate until air bubbles cease to exit from the cracked open coupling.

4. Retorque the coupling. Crank the engine for 30-second segments with at least 2-minute intervals between crankings until it starts; this will allow for starter motor cool-down. In most diesel engine systems, the high pressure circuit will self-prime once the subsystem is primed.

Refueling

It is good practice to refuel tanks on returning from a trip; filling the tanks with fuel expels the air present in the tank that replaced the fuel as it was consumed. When tanks are left in a near empty condition for any length of time (overnight is long enough), the moisture in the air condenses and contaminates the fuel.

CAUTION: When refueling tanks, many drivers and technicians overlook the fact that diesel fuel vaporizes and combines with air to form combustible mixtures that only require an ignition source to cause an explosion. Diesel fuel is less volatile than gasoline but it should always be handled with care, especially in the heat of summer.

SERVICE INTERVALS

During the past fifteen years, most truck innovations focused on a dramatic shifting from hydromechanical to electronic systems control. Probably the most significant advances next will be innovations designed to reduce the frequency of service intervals. Some engine OEMs are making lubricating oil monitoring systems available that will significantly extend the engine oil change intervals, and because that more often than not defines service intervals, the remainder of the filter package will require upgrading. The engine OEMs are making superior filter packages available (an example is DDC Davco) and aftermarket companies such as Racor that specialize in the manufacture of filters with the claim that they perform better and much longer than the original filters. Many of the aftermarket fuel fil-

FIGURE 18–12   *Racor filter/separators. (Courtesy of Parker Hannefin)*

tration units are multifunction packages in which a single filter assembly performs as a high efficiency filter, water separator, and fuel heater. An example is the unit featured in Figure 18–12. Reducing the number of filters on the fuel subsystem requiring service lowers labor costs as does the eliminating of the loss of prime that occurs after filter service. Figure 18–13 features DDC's Pro-Chek device. Detroit Diesel Corporation's Fuel Pro filter shown in Figure 18–14 and Figure 18–15 is a single filter system that replaces the primary and secondary filters and also performs as a fuel heater and water separator.

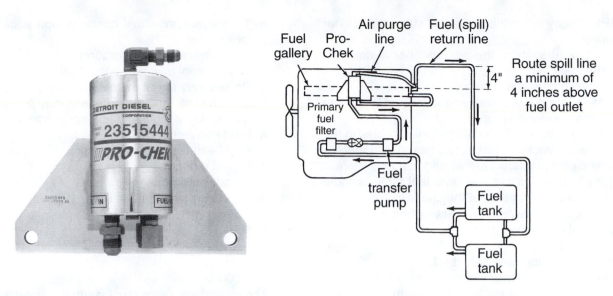

FIGURE 18–13   *DDC's Pro-Chek for Series 60 engines. (Courtesy of Detroit Diesel Corp.)*

Features

1 Self-Priming Port
If engine loses prime—just *spin off cap, pour in fuel, and restart engine with clean "filtered" fuel.*

2 Clear Cover
See when to change filters
See air leaks in fuel
5 year guarantee

3 Secondary Filter (primary location) 5-micron media with waterproof coating.

4 Fuel Heater Thermostatically controlled. Turn key to "accessory", or "on" position and preheat fuel at filter for faster winter starts.

5 Check Valve Large inlet (1/2 NPTF) may replace more restrictive fittings.

6 Drain Valve
Brass "self-venting" ball valve.

7 Aluminum Cylinder
Acts as a fuel cooler.

FIGURE 18–14   *DDC's Fuel Pro features. (Courtesy of Detroit Diesel Corp.)*

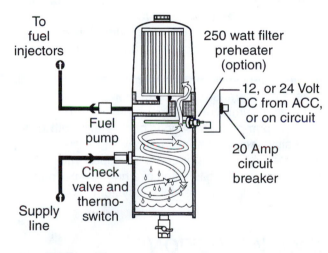

FIGURE 18–15   *Fuel Pro operating principles. (Courtesy of Detroit Diesel Corp.*

SUMMARY

- The fuel subsystem is defined as the group of components responsible for fuel storage and its transfer to the injection pumping apparatus.
- The typical fuel transfer system can be divided into a suction circuit and a charge circuit, separated by a transfer or charge pump.
- Some fuel systems may locate a transfer pump in the tank itself whereas others such as Cummins

HPI-TP may retain the entire fuel subsystem under suction.

- In a typical system, the primary filter is normally under suction.
- The secondary filter entraps smaller-sized particulates than a primary filter and is subject to system charging pressure.
- Aluminum alloy, cylindrical fuel tanks located in the air flow on truck chassis act as heat exchangers.
- Most diesel fuel tanks are vented to atmosphere.
- Many highway trucks use dual fuel tanks, saddle mounted on either side of the chassis.
- Many current secondary filters will plug on water and shut down the engine.
- Water may be found in fuel in three forms: free state, emulsified, and absorbed.
- Fuel system inlet restriction values are tested on the suction side of the fuel subsystem using a Hg manometer.
- A common source of air in the fuel subsystem is double-gasketing of a filter under suction.
- Many fuel subsystems are equipped with a water separator.
- A water separator is designed to remove free state and emulsified water from fuel.
- Two types of fuel heater are in current use: the electric element and coolant medium, heat exchanger types.
- Diesel fuel systems commonly use one of two different fuel transfer or charge pumps: reciprocating plunger pumps and gear pumps.
- Gear pumps and plunger pumps are both of the positive displacement type.
- Some fuel subsystems are equipped with a hand primer pump; the function of a hand primer pump is to purge air from the fuel subsystem.

REVIEW QUESTIONS

1. On the typical truck diesel fuel subsystem, which of the following is subject to the lowest pressure in the circuit?
   a. Fuel heater
   b. Primary filter
   c. Secondary filter
   d. Charging gallery
2. A secondary filter is located:
   a. Upstream from the transfer pump
   b. On the charge side of the transfer pump
   c. In the fuel rail
   d. In the return gallery
3. The main reason for filling vehicle fuel tanks before overnight parking is:
   a. To prevent moisture condensation within the tank
   b. To minimize fuel evaporation
   c. To help cool down on-board fuel
   d. Drivers may forget the next morning
4. Besides fuel storage, the fuel tank may play an important role in a high-flow fuel system as a(n):
   a. Heat exchanger
   b. Fuel heating device
   c. Ballast equalizer
   d. Aerodynamic aid
5. Approximately how much does 100 gallons of 2D diesel fuel weigh?
   a. 500 lb
   b. 730 lb
   c. 1,000 lb
   d. 2,000 lb
6. The correct instrument for testing the low-pressure side of most fuel subsystems for restriction is:
   a. Diagnostic sight glass
   b. H_2O manometer
   c. Hg manometer
   d. Accurate pressure gauge
7. The correct instrument for testing fuel subsystem charging pressure is:
   a. Diagnostic sight glass
   b. H_2O manometer
   c. Hg manometer
   d. Accurate pressure gauge
8. The correct instrument to use for checking for the admission of air to a fuel subsystem is:
   a. Diagnostic sight glass
   b. H_2O manometer
   c. Hg manometer
   d. Accurate pressure gauge
9. Which type of fuel charging/transfer pump is commonly used by electronically managed engine/fuel systems?
   a. Gear pumps
   b. Centrifugal pumps
   c. Diaphragm pumps
   d. Plunger pumps
10. What type of pumping principle is used by the typical hand primer pump?
    a. Single-acting plunger
    b. Double-acting plunger
    c. Rotary gear
    d. Cam-actuated diaphragm
11. *Technician A* states that when servicing fuel filters on a fuel subsystem with a primary and secondary filter, a transfer pump located, hand primer pump can be used to prime both filters. *Technician B* states that if a transfer pump located, hand primer pump is used to prime the secondary filter after installation, it reduces the chances of dirt contamina-

tion beyond downstream from the filter. Who is right?

a. A only
b. B only
c. Both A and B
d. Neither A nor B

12. Charging pressure rise is usually directly related to which of the following?

a. Increased rpm
b. Engine load
c. Peak power
d. Throttle position

13. *Technician A* states that when hydrostatically testing a truck aluminum fuel tank, the regulator should be set at 250% of the working pressure. *Technician B* states that the fuel tank vent valve should not be removed before performing a hydrostatic test. Who is right?

a. A only
b. B only
c. Both A and B
d. Neither A nor B

14. *Technician A* states that the presence of water in fuel tanks can be detected using a probe and water detection paste. *Technician B* states that the presence of water in the fuel subsystem can usually be detected by a WIF sensor. Who is right?

a. A only
b. B only
c. Both A and B
d. Neither A nor B

15. *Technician A* states that some water separator units use a centrifuge to separate water from fuel. *Technician B* states that because fuel is heavier than water, fuel can be easily separated from the water because it always settles under it. Who is right?

a. A only
b. B only
c. Both A and B
d. Neither A nor B

16. *Technician A* states that gear pumps are of the positive displacement type. *Technician B* states that most diesel engine fuel subsystems use gear pumps as transfer pumps. Who is right?

17. If the fuel tank vent valve on a diesel-powered vehicle is plugged, which of the following would be the likely outcome when running the engine?

a. The tank would explode.
b. The engine would shut down due to fuel starvation.
c. The tank would implode.
d. The engine would run away due to excess fuel.

18. *Technician A* states that diesel engine fuel tank caps are fitted with a two-way valve that permits both fuel vapor seepage and air admission to the tank. *Technician B* states that most truck diesel engines use a vane-type transfer pump located in the fuel tank to pump fuel to the transfer pump. Who is right?

a. A only
b. B only
c. Both A and B
d. Neither A nor B

19. *Technician A* states that diesel fuel can combine with air to form potentially explosive mixtures. *Technician B* states that diesel fuel is likely to degrade more quickly during the winter than in the heat of summer. Who is right?

a. A only
b. B only
c. Both A and B
d. Neither A nor B

20. At refueling on the completion of a journey, fuel is steaming in the tanks. *Technician A* states that this condition is a result of normal operation in some fuel systems. *Technician B* states that such a condition can be a result of running the tanks low on fuel in a high-flow fuel system. Who is right?

a. A only
b. B only
c. Both A and B
d. Neither A nor B

Chapter 19

Hydromechanical Injection Principles

PREREQUISITE

A thorough understanding of the four-stroke cycle diesel engine

OBJECTIVES

After studying this chapter, you should be able to:

- *Interpret the contents of later chapters dealing with hydromechanical and electronic engine management.*
- *Understand the objectives of a fuel management system.*
- *Define timing and explain the need to vary it for optimum performance.*
- *Define metering and its application in a fuel system.*
- *Explain atomization and the droplet sizings required for a direct-injected diesel engine.*
- *Describe the factors that determine emitted droplet sizing.*
- *Explain the overall objectives of an engine fuel system.*
- *Describe the relationship between cylinder pressure and crank throw to crank axis angle.*
- *Relate how the fuel system manages engine cylinder pressures.*

KEY TERMS

after top dead center (ATDC)

atomization

before top dead center (BTDC)

compression ignition (CI)

crank angle

crank axis

crank throw

direct injection (DI)

electronic management

hydromechanical

indirect injection (IDI)

injection rate

lever

mechanical advantage

metering

orifice

orifii

timing

top dead center (TDC)

units of linear measurement: micron, millimeter, centimeter, and meter

units of pressure: atms, psi, kPa, and MPa

OVERVIEW OF DIESEL FUEL INJECTION PRINCIPLES

Before beginning the study of **hydromechanical** (term used to describe a diesel engine that is managed without a computer) and **electronic management** (refers to any system that is managed by computer) of the diesel engine (Figure 19–1), a thorough understanding of engine components and principles of operation is required. In any **compression ignition (CI)** engine, a power stroke will only take place if the fuel system is accurately phased to the engine and performs the following five objectives:

1. TIMING

Fuel delivery **timing** is critical during all engine operating phases. Typically, fuel is injected into the engine cylinder slightly before the piston completes its compression stroke. While older engines tended to have hard (nonvariable) timing values, newer engines have variable timing features that can be managed mechanically, hydraulically, or electronically. Variable timing is required in today's engines to produce optimum performance and minimal noxious emissions. **Indirect injection (IDI)** engines, which are not often seen today in truck applications, require timing delivery much advanced of that in **direct injection (DI)** engines due to the much larger droplet sizes emitted from their injectors. The larger droplets emitted from IDI engines rely on turbulence created in prechambers to help rip them

down to a smaller size suitable for combustion in the short amount of time available. DI engines with high-pressure injection (and therefore smaller emitted droplet sizes) have injection timing typically around 20 degrees **before top dead center (BTDC)**.

2. PRESSURIZING

The fuel system must be capable of pressurizing the fuel sufficiently to open the injector nozzles and deliver fuel to the engine cylinders correctly prepared for combustion. The means used to pressurize the fuel to the required injection pressures varies with the type of fuel system. In most cases, the fuel injection pumping apparatus is actuated mechanically, typically by a cam profile located either on the main engine camshaft or on a dedicated pump camshaft. The aggressiveness of the cam flank in conjunction with engine rpm, will play a big role in defining what peak system pressure values are. In some more recent engines, the pumping to injection pressures is by precisely controlled hydraulic pressure. Injector nozzle opening pressure in current engines range from 200 to 350 atms (3,000 to 5,500 psi). Peak injection pressures in current CI engines can exceed 2,000 atms (30,000 psi).

3. METERING

Metering is the precise control of fuel quantity. The only factor that controls the output of a diesel engine is the amount of fuel put into it. Gasoline-fueled, spark-ignited engines define peak output by the amount of air that can be induced into the engine cylinders so the

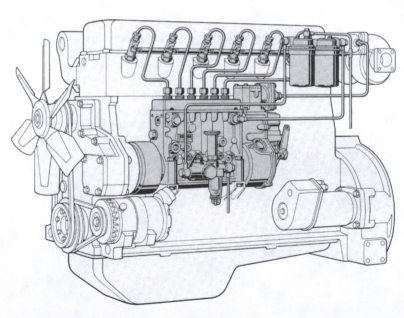

FIGURE 19–1   *A typical hydromechanical fuel system layout.*
(Courtesy of Robert Bosch Corporation)

size of the throttle bore is the defining factor. Diesel engines operate with excess air, that is, under any load or operating condition, there is always more air present than the minimum required to completely combust the fuel. Because of this excess air factor, most truck- and bus-sized diesel engines given unlimited fuel will accelerate at a rate of up to 1,000 rpm per second. Diesel engines must have a means of precisely metering fuel into the engine's cylinders. Metering in the engines of today is achieved by several distinct means; while diesel engine manufacturer's engines have many components and operating principles that are common, each fuel system that manages these engines tends to be distinct. Most current engines have fuel systems that are managed electronically (by a computer).

Injection rate is the fuel quantity injected per **crank angle** degree.

4. ATOMIZATION

Atomization of the fuel delivered to the cylinder means preparing the fuel charge for combustion given the conditions within the cylinder and the time dimension it is to take place in. In today's DI diesel engines, the fuel is required to be atomized within a droplet sizing range of between 10 and 100μ (microns); the reasons for those precise sizings are dealt with in Chapter 16 on combustion. The degree to which the fuel exiting the nozzle **orifii** (the plural of **orifice**) is atomized depends on:

• the pressure of the fuel supplied to the nozzle orifii.
• the flow area or sizing of the nozzle orifii.

It should be emphasized that atomized fuel is in the liquid state and, therefore, it must be vaporized and ignited before it can be combusted (oxidized) in the engine cylinder. The smaller the atomized droplets, the more rapidly they will respond to the heat within the cylinder to undergo the change of state required.

Indirect-injected engines use injector nozzles that provide a lesser degree of atomization (the emitted droplets are larger) and depend on prechamber turbulence to help reduce the emitted droplet sizing; that is, turbulence in the prechamber helps rip the fuel droplets emitted from the injector into smaller droplets. Because this takes time, fuel injection timing in IDI engines is significantly advanced when compared with that in DI engines.

5. DISTRIBUTION

In multicylinder engines:

a. The fuel system must be phased (sequenced) to deliver the fuel to each engine cylinder at the cor-

rect time and in the correct firing order. Correct fuel system phasing is required to balance the engine output. If in a given engine position, ignition is set to occur at 3 degrees BTDC, it must occur in that position in each cylinder in the engine.

b. Dispense the fuel to the correct area of the combustion chamber so that the fuel droplets combust at the appropriate time (spray dispersion). The position of the fuel injector is critical in ensuring that the injected fuel droplets are directed to the correct area of the combustion chamber in DI engines. For instance, if an injector was seated on double sealing washers (a not uncommon occurrence), the nozzle would be effectively too high in the cylinder; the result would be to inject fuel above the combustion bowl on a Mexican hat piston crown, often causing fuel to condense on the piston headland and cylinder walls. Fuel exiting the nozzle orifii should be vaporized before it comes into contact with the cylinder hardware, which because of its comparatively lower temperature, may cause vaporized fuel to condense. Fuel that condenses on the piston can cause erosion (pitting).

The five objectives listed above can be met using a variety of very different fuel delivery mechanisms and while the hydromechanical principles used to achieve them evolved in the late 1920s (Robert Bosch, Harry Ricardo), they have been fine tuned to deliver the high thermal efficiencies and low noxious emissions required in diesel engines of the twenty-first century (Figure 19–2).

UNITS OF PRESSURE AND LINEAR MEASUREMENT

Although the tendency in North America is to speak and think using English (standard) unit values of measurement, most truck and bus engines and fuel systems engineered during the past two decades have used the si (Système internacional d'unités) or metric system. Most North American-built diesel fuel injection equipment has always used the metric system. The contemporary diesel technician should be prepared to readily convert values from standard to metric systems and vice versa. It is ironic that the manufacturer who engineers machinery using the metric system entirely then converts the specifications to standard values for the technical service manual specifications. The trend today is to present specifications using both metric and standard values. Manufacturers of fuel injection apparatus often express pressure values in units of atmospheric pressure known as *atms*. This has the

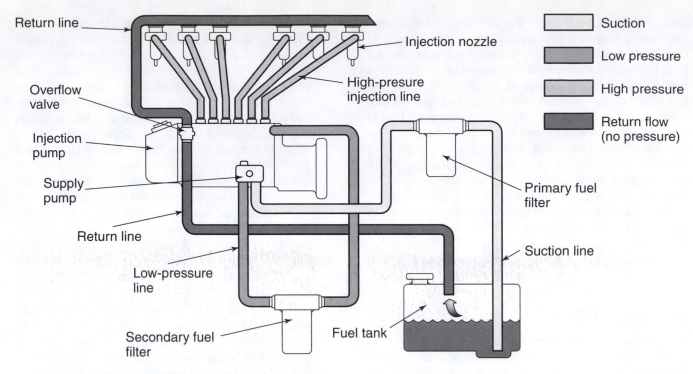

FIGURE 19–2   *A Mack Trucks fuel system schematic.*

advantage of being easily converted either to the metric or standard systems. In Europe, units of *bar* (10⁵ Newtons per square meter) are used to express units of atmosphere: one unit of bar is not precisely equivalent to one unit of atm. Most fuel injection test equipment is calibrated atms and bar. Consult Chapter 2 for a more comprehensive look at measurement systems. Here are some units of measurement used in this chapter:

1 **atm** = 14.7 **psi** = 101.3 **kPa** (kilopascals)

200 **atms** = 2,940 psi = 20.26 **MPa** (megapascals)

1 **micron** = millionth of a meter

1 **millimeter** = 0.0394 inches

1 **centimeter** = 0.394 inches

1 **meter** = 39.4 inches

ENGINE MANAGEMENT OBJECTIVES

The fuel system manages *engine* output. The overall objective of any engine whether it powers a lawnmower, a ship, or a highway truck is to transmit the force developed in its cylinders as smoothly and evenly as possible to the power takeoff device, usually a flywheel. To achieve this objective, the fuel system must manage fueling to complement the geometry of the engine powertrain components or most specifically, that of the crankshaft.

First, it is important to understand the mechanical relationship of the **crank throw** (the crankshaft connecting rod journal) with the crankshaft axis; it is a **lever**. When a piston is at **TDC (top dead center)**, the crank throw-to-**crank axis** (center point of crank main journal) angle is zero so the leverage factor is therefore also zero. In this position, the force of any cylinder pressure cannot be transmitted to the engine powertrain. As the piston is driven down through its stroke, the crank throw axis angle (and leverage) increases.

Remembering that the objective is to *smoothly* and *evenly* transmit force developed in the engine cylinders to the engine flywheel, ideally the fuel system should manage cylinder pressures to peak when the crank throw angle offers low **mechanical advantage** (but not zero, which would be true when the throw and piston are at TDC) and diminish as the crank throw is driven to a 90 degree angle with the conn rod in which position mechanical advantage (leverage) peaks. With this in mind, most spark-ignited and diesel cycle engines are engineered to attempt to produce a peak cylinder pressure in the 10- to 20-degree **after top dead center (TDC)** range, at any phase of load or speed operation. To enable this, engines will advance either the ignition timing (SI engines) or the injection timing (CI engines) as engine speed increases and the real time available for combustion diminishes (Figure 19–3).

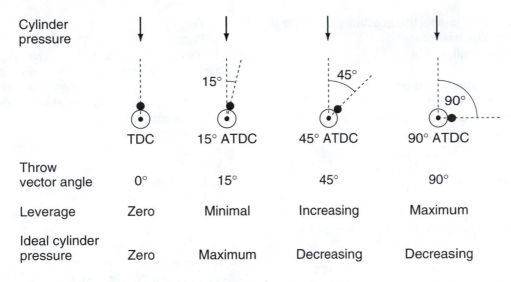

FIGURE 19–3   *Relationship between cylinder pressure and crank pressure.*

Hydromechanically managed engines attempt to manage this optimum relationship between cylinder pressure and mechanical advantage but most will only achieve it in the torque rise profile. Electronically managed engines generally come closer to achieving this objective, but even in these, optimum performance can often be compromised by noxious emission control criteria.

Finally, the fuel system must precisely manage engine fueling within minute time dimensions. A complete engine effective cycle is 720 degrees and the fueling pulse even when it is at maximum usually occupies less than 40 crank angle degrees. A four-stroke cycle diesel engine rotating at 2,000 rpm produces approximately 17 power strokes in each of its cylinders every second so the pumping, metering, and injection activity is generally measured in milli and microseconds.

SUMMARY

- In any CI engine, a power stroke only takes place if the fuel system is accurately phased to the engine.
- The fuel system must be timed to the engine.
- The fuel system must be capable of pressurizing the fuel to values exceeding 30,000 psi.
- Fuel delivered to the engine cylinders must be precisely metered.
- Injection rate refers to fuel injected per crank angle degree and is usually governed by the pump actuating mechanism geometry, usually a cam.

- The fuel system must be capable of atomizing fuel to the precise dimensions required of the specific fuel system.
- Atomized droplet sizing is determined by the injector nozzle orifii sizing and the pressure it is subjected to.
- Atomized fuel is in the liquid state.
- The fuel system must deliver the fuel to the correct cylinder at the correct time to ensure balanced power output.
- The relationship between the crank throw and the crank axis is that of a lever.
- The fuel system is responsible for managing engine cylinder pressures.
- To optimize the relationship between cylinder pressure and mechanical advantage, cylinder pressure should ideally peak between 10 and 20 degrees ATDC regardless of speed or load conditions.
- A four-stroke cycle engine run at 2,000 rpm will produce approximately 17 power strokes per second in each of its cylinders.

REVIEW QUESTIONS

1. The precise control of injected fuel quantity is known as:
   a. Timing
   b. Metering
   c. Atomization
   d. Spray dispersion

2. Which two factors define the emitted droplet sizings from a nozzle assembly?
   a. Pressure and orifice size
   b. Load and rpm
   c. Accelerator position and governor
   d. Cylinder pressure and crank angle
3. Given unlimited fuel a truck engine may accelerate at a rate up to:
   a. 60 rpm per second
   b. 120 rpm per second
   c. 120 rpm per minute
   d. 1,000 rpm per second
4. If a four-stroke cycle engine is run at 1,000 rpm, approximately how many power strokes per second will occur in any one of its cylinders?
   a. 1,000
   b. 8
   c. 17
   d. 32
5. Crank throw leverage will peak in which location?
   a. 10 degrees BTDC
   b. TDC
   c. 10–20 degrees ATDC
   d. 90-degree rod to throw angle
6. An ideal fuel system manages fueling to produce peak cylinder pressure at which location to produce optimum performance on any given engine cycle?
   a. TDC
   b. 15 degrees ATDC
   c. 90 degrees ATDC
   d. BDC
7. The specification for an injector nozzle opening pressure reads 265 atms. Convert this to standard US value.
   a. 26,500 psi
   b. 389.5 psi
   c. 3,895 psi
   d. 2,650 psi
8. If a four-stroke cycle engine is idled at 600 rpm, how many power strokes will take place per second in any one of its cylinders?
   a. 1.3
   b. 3
   c. 5
   d. 30
9. In which engine location does the crank throw have the least amount of leverage?
   a. TDC
   b. 15 degrees ATDC
   c. 90 degrees ATDC
   d. 120 degrees BTDC
10. Which of the following diesel fuel injection terms is used to describe the amount of fuel injected per crank angle degree?
    a. Metering
    b. Injection rate
    c. Timing
    d. Phasing
11. *Technician A* states that an advantage of variable fuel delivery timing is better control over the noxious exhaust gas emissions. *Technician B* states that another reason for variable fuel delivery timing mechanisms on engines is to produce optimum performance. Who is right?
    a. A only
    b. B only
    c. Both A and B
    d. Neither A nor B
12. *Technician A* states that the term metering refers to the control of ignition timing in a diesel engine fuel system. *Technician B* states that the term injection *rate* refers to the speed of the injection pulse in real time values. Who is right?
    a. A only
    b. B only
    c. Both A and B
    d. Neither A nor B
13. *Technician A* states that the degree of atomization is factored by pressure the injection nozzle is subjected to and the sizing of the nozzle orifii. *Technician B* states that atomized droplets are in a gaseous state. Who is right?
    a. A only
    b. B only
    c. Both A and B
    d. Neither A nor B
14. Which of the following correctly describes the unit of measurement known as *micron*?
    a. A thousandth of an inch
    b. A millionth of a yard
    c. A thousandth of a meter
    d. A millionth of a meter
15. A direct injected diesel engine governed at a rated speed of 2,000 rpm would require injected atomized fuel droplets closest to which of the following sizes?
    a. 0.001″
    b. 0.0001″
    c. 50 microns
    d. 500 microns
16. Caterpillar C-16 injection pressures peak at 30,000 psi. Convert this to metric units and round to the nearest of the following values.
    a. 123 kPa
    b. 123 MPa
    c. 210 kPa
    d. 210 MPa
17. *Technician A* states that indirect-injected diesel engines generally require smaller atomized droplets of fuel delivered to the engine cylinders than equivalent output direct-injected engines. *Technician B* states that indirect-injected diesel engines have the

diesel fuel injected to the intake tract in the cylinder head. Who is right?

a. A only
b. B only
c. Both A and B
d. Neither A nor B

18. *Technician A* states that in most IDI diesel engines, the injector nozzle injects fuel to a prechamber. *Technician B* states that injection timing in DI engines is considerably advanced when compared to injection timing in IDI engines. Who is right?

a. A only
b. B only
c. Both A and B
d. Neither A nor B

19. In a four-stroke cycle, six-cylinder diesel engine, the injector nozzles are designed to open at 10 degrees BTDC on each engine cylinder. In what en-gine position will #1 piston be when the injector nozzle opens on #5 cylinder?

a. 100 degrees BTDC
b. 80 degrees ATDC
c. 130 degrees BTDC
d. 110 degrees ATDC

20. *Technician A* states that a hydraulic injector seated on double seal washers causes fuel to be injected above the combustion bowl in an engine that uses Mexican hat piston crowns. *Technician B* states that an injector nozzle located either too high or too low in the cylinder of a direct-injected engine can cause the fuel to condense on the cylinder hardware before ignition occurs. Who is right?

a. A only
b. B only
c. Both A and B
d. Neither A nor B

Hydraulic Injector Nozzles

PREREQUISITE

Chapter 19

OBJECTIVES

After studying this chapter, you should be able to:

- Identify four types of injector nozzles.
- Identify the subcomponents of a nozzle assembly.
- Describe the injector nozzle's role in system pressure management.
- Describe the hydraulic principles of operation of poppet, pintle, multi-orifii and electrohydraulic nozzles.
- Define nozzle differential ratio.
- Describe a valve closes orifice (VCO) nozzle.
- Explain the difference between a low and high spring injector.
- Bench (pop) test a hydraulic injector nozzle.
- Disassemble, ultrasonically bath, and reassemble an injector.
- Test a nozzle for forward leakage.
- Test nozzle back leakage.
- Set injector nozzle opening pressure (NOP).
- Evaluate the serviceability of a hydraulic injector nozzle.

KEY TERMS

atomization

back leakage

cavitation

chatter

direct injection (DI)

electrohydraulic

forward leakage

high spring injector

hydraulic injectors

indirect injection (IDI)

leakoff lines/pipes

low spring injector

mechanical injectors

micron or μ

multi-orifii nozzle

nozzle differential ratio

nozzle opening pressure (NOP)

nozzle seat

orifice nozzle

peak pressure

pencil injector nozzle

pintle nozzle

pop tester

poppet nozzle

popping pressure

residual line pressure

sac

spindle

valve closes orifice (VCO) nozzle

INTRODUCTION

When the truck technician refers to an injector, the nozzle holder assembly is generally being described. However, the term is also used perhaps somewhat incorrectly to describe components such as mechanical unit injectors (MUIs), electronic unit injectors (EUIs), and hydraulically actuated, electronic unit injectors (HEUIs), that, while they may incorporate a hydraulic nozzle, also perform the metering, timing, and pumping functions of the diesel fuel system. For the sake of clarity, this text refers to integral injector assemblies whose function it is to define **nozzle opening pressure (NOP)** and atomize fuel as **hydraulic injectors**. It should be noted that one OEM refers to these devices as **mechanical injectors** (Cummins).

Hydraulic injectors are normally classified by nozzle design. They are simple hydraulic switch mechanisms whose functions are to atomize and inject fuel into the engine cylinders. There are four basic types of injector nozzles, two of which are effectively obsolete in medium- and heavy-duty truck and bus engine applications. However, both are briefly described in this chapter. Poppet and pintle nozzles are best suited to indirect-injected engine applications and therefore are not used in any current North American medium- and heavy-duty engines. **Multi-orifii nozzles** are used in most **direct-injected (DI)** diesel engines using in-line, port helix metering injection pumps, and as an integral subcomponent of mechanically and electronically controlled unit injection systems. All current, high-speed diesel engines found in highway applications use either hydraulic or **electrohydraulic** nozzles with the exception of the open nozzle designs used in Cummins hydromechanical and electronic common rail systems. Open nozzle injectors are covered in later sections of this textbook. There are almost no similarities in operating principle between the closed hydraulic nozzles addressed in this chapter and open nozzle injectors.

NOZZLE OPENING PRESSURE (NOP)

An injector nozzle is a hydraulic switch. One of its primary functions is to define the pressure required to trigger its opening. The term **popping pressure** is also used to describe the opening pressure of a nozzle. The actual NOP value is defined by the mechanical spring tension of the injector spring. This spring tension loads the nozzle valve onto its seat and therefore determines the hydraulic pressure required to unseat the valve. Most injectors incorporate a means of adjusting the injector spring tension so that the NOP value can be set to specification. The spring tension adjustment

mechanism is either shims or an adjusting screw and locknut. The NOP value is always one of the first performance specifications to be evaluated when testing injector nozzles on a bench test fixture **(pop tester)**.

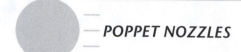

POPPET NOZZLES

Poppet nozzles were used by Caterpillar in the past on **indirect-injection (IDI)** engines using a precombustion chamber. They are perhaps the most simple of the hydraulic injector nozzles and as such were the cheapest to manufacture. Poppet nozzles are not easily reconditioned; the technician is normally required to bench test the nozzle for the correct NOP value and observation of spray pattern and either reject or accept it for continued service.

Poppet nozzles use an outward or forward opening valve principle. This means that all the fuel pumped to the injector ultimately ends up in the engine cylinder eliminating the necessity for the leakoff lines (return circuit) used with most other nozzle types. Nozzle opening pressure parameters generally range between 500 and 1,800 psi (35–125 atms); for instance, the Caterpillar poppet nozzles had NOPs of around 800 psi (55 atms).

Poppet nozzles operate as hydraulic switches to define the NOP and atomize the fuel. The nozzle spring loads the poppet valve onto its seat. Hydraulic line pressure (from the injection pump) acts on the sectional area of the seated upper portion of the poppet valve, and when the injection pump delivers pressure rise sufficient to unseat the poppet valve, fuel passes around the poppet to exit from the nozzle's single orifice.

When a precombustion chamber design is used, the engine cylinder clearance volume is generally too large to generate compression temperatures in cold weather sufficiently high to ignite the fuel charge (Figure 20–1). Glow plugs were often designed into IDI engine systems as a cold weather starting aid. It should be remembered by technicians who come across older precombustion chamber engines that the use of ether should be avoided whenever a glow plug circuit is energized. The result of ether usage in IDI engines is explosion damage to the prechamber.

PINTLE NOZZLES

Like poppet nozzles, **pintle nozzles** are used mainly in IDI engine applications and seldom in truck/bus-sized CI engines of the recent past. The pintle nozzle body is

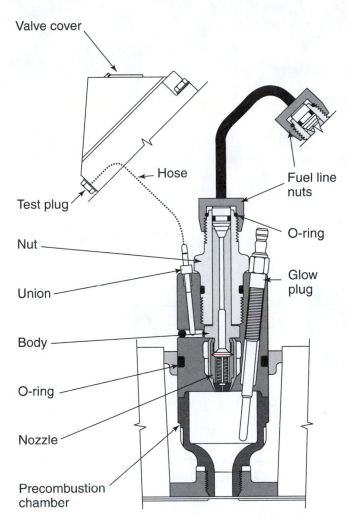

Valve cover

Hose

Test plug

Fuel line nuts

Nut

O-ring

Union

Glow plug

Body

O-ring

Nozzle

Precombustion chamber

FIGURE 20–1 *Poppet injector and prechamber assembly.*

designed with a single exit orifice; the valve pintle protrudes through this single orifice even when in the fully open position. This means that the fuel exiting the nozzle is forced around the pintle producing a somewhat conical (cone shaped) spray pattern. The valve body and pintle valve are lapped together in manufacture often to a tolerance of around 2μ (μ: micron or one millionth of a meter), which does not permit the components to be interchanged.

Pintle nozzles generally achieve better **atomization** (reduction of a liquid to minute droplets) than poppet nozzles and generally much greater service life as they are easily reconditioned. Nozzle opening pressure parameters range between 100 and 150 atms (1,470–2,200 psi) with peak system pressures seldom exceeding 500 atms (7,350 psi).

ACTION

Fuel is delivered to the pintle injector from the fuel injection pump by means of a high-pressure pipe. The

nozzle valve is held in the closed position mechanically by spring pressure either acting directly on the pintle valve or transmitted by means of a **spindle** (shaft that transmits spring force to the nozzle valve). Fuel is ducted through the injector assembly to the pressure chamber within the nozzle; this permits the line pressure to act on the sectional area of the pintle valve exposed to the pressure chamber. Whenever the hydraulic pressure generated by the injection pump acting on the sectional area of the pintle valve exposed to the pressure chamber is sufficient to overcome the spring pressure loading the valve on its seat, it retracts, permitting fuel to flow past the seat and exit through the injector orifice. Fuel will continue to flow through the nozzle orifice for as long as hydraulic pressure exceeds the mechanical force represented by the injector spring. The length of a fueling pulse is determined by the fuel injection pump and when line pressure collapses, the spring reseats the nozzle valve, ending injection.

The pintle nozzle uses an inward opening valve principle requiring the use of **leakoff lines/pipes** to return fuel to the tank. The leakage takes place at the pintle valve-to-nozzle body clearance and tends to increase as these matched components age.

The shaping of the spray pattern emitted from a pintle nozzle depends on the pintle design. Throttling nozzles are designed to emit less fuel at the beginning of the injection pulse and may be identified by their conically shaped pintle valves (Figure 20–2 and Figure 20–3).

ORIFICE NOZZLES

As stated in the introduction to this chapter, most current truck and bus diesel engines use closed hydraulic injector nozzles; these nozzles are, in almost all cases, of the multi-orifii type. Multi-orifii nozzles appear in truck engines in traditional, pump-line-nozzle configurations and they are an integral subcomponent in EUIs and HEUIs; they are also used with electronic unit pump (EUP) fueled engines. **Orifice nozzles** may have a single orifice or several orifii, almost always the latter. Multi-orifii and the closely related electrohydraulic nozzles are required in most high-speed DI diesel engines because only these produce the necessary degree of atomization, droplets sized between 10μ to 100μ. Atomized droplet sizings of 10μ or less can result in erratic ignition or no ignition at all. Atomized droplet sizings 100μ or more will produce extended ignition lag and are too large to be completely combusted (oxidized) in the available time window of high-speed CI

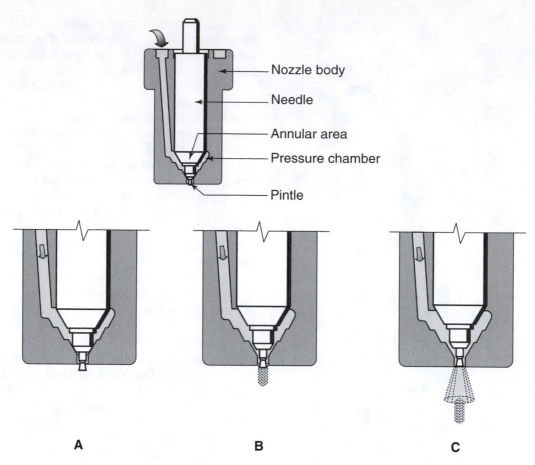

FIGURE 20–2 *Pintle nozzle operation: terminology.* A. *nozzle seated;* B. *opening;* C. *open.*
(Courtesy of Robert Bosch Corporation)

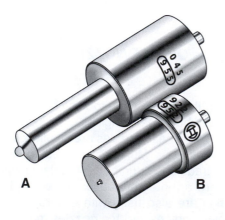

FIGURE 20–3 A: *Multi-orifii nozzle;* B; *pintle nozzle.*

engines with rated speeds of 1,800–2,600 rpm. The time dimension window within which combustion of the fuel must take place is defined by engine rpm. If an engine is run at 2,000 rpm, it takes exactly 0.030 seconds or 30 milliseconds for a single revolution of the engine; this means that a piston travels from TDC to 90 de-

grees ATDC, more or less the period available for combustion, in 7.5 milliseconds. Top end engine speeds dictate the combustion or burn duration window and, therefore, the maximum atomized droplet sizing.

Orifice nozzles use an inward opening valve principle and that requires leak-off lines. The leakage is that occurring at the nozzle valve-to-nozzle body clearance and measured on the injector test bench as **back leakage**. NOP parameters in current engines range from 120 to 400 atms (1,800–5,880 psi) with peak system injection pressures ranging from 2 to 10 times the NOP value. Multi-orifii nozzles used in hydromechanical engines typically have NOPs in the region of 200 atms or 3,000 psi. The multi-orifii nozzles used in EUIs and HEUIs generally have NOP values closer to the higher range of the typical parameters, around 350 atms (5,000 psi).

ACTION

The multi-orifii nozzle is usually dowel positioned in the injector assembly to ensure that the spray pattern is directed to a specific location of the engine cylinder. Fuel from the injection pump is delivered to the nozzle holder and then ducted to an annular recess in the upper nozzle valve body. Either single or multiple fuel ducts extend from the annular recess to the pressure chamber, meaning that the pressure here will be the same as that in the high-pressure pipe. The nozzle valve is loaded to a closed position on its seat, either directly by the injector spring or indirectly by spring pressure relayed to the nozzle valve by means of a spindle. This spring tension is adjustable by either a screw or by shims and will set the NOP value. When line pressure is driven upward by the injection pump, pressure increases in the nozzle pressure chamber, and when this pressure is sufficient to overcome the mechanical force of the injector spring, the nozzle valve retracts permitting the fuel to flow past the seat into the **sac** and exit the nozzle orifii. At the center of the nozzle valve seat, a single duct connects to the nozzle sac, a spherical chamber into which the nozzle orifii are machined. The sac hydraulically balances the fuel exiting the orifii, so that droplets start to exit from each orifice at approximately the same moment at the beginning of the fuel pulse.

The injection pulse continues for as long as the nozzle valve remains open; depending on the specific means used to generate injection pressures, maximum fueling pulses can extend from 25 to 50 crank angle degrees. The fueling pulse ends when the pressure is collapsed by the metering/pumping element and the pressure in the nozzle assembly drops to a value below NOP.

The emitted droplet sizing from the nozzle depends on pressure and flow area. The flow area is represented by the sizing of the nozzle orifii and therefore remains constant. However, the pressure values vary considerably, extending from a value lower than NOP up to the **peak pressure** (the highest pressure attainable in a fuel injection system) value. As pressure increases, the droplet sizing decreases. The more prolonged the injection pulse (that is, the longer the duration of the effective pump stroke), the higher the circuit pressure at injection pump port opening (end of delivery) and the smaller the atomized droplet sizing. The reduction in droplet sizing that occurs as the fueling pulse is extended is generally favorable for the complete combustion of the fuel, except for the short period of fueling that takes place *after* pump effective stroke completion and before nozzle valve closure. As the pressure collapses in the pump element and injection circuit, emitted droplet sizing increases until the moment the nozzle valve actually closes, so it is desirable for this pressure collapse to occur as rapidly as possible (Figure 20–4).

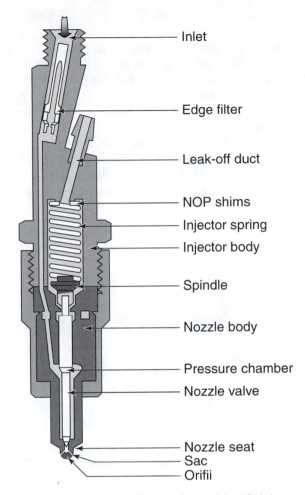

Inlet

Edge filter

Leak-off duct

NOP shims

Injector spring

Injector body

Spindle

Nozzle body

Pressure chamber

Nozzle valve

Nozzle seat
Sac
Orifii

FIGURE 20–4 *Sectional view of a multi-orifii injector nozzle.*

ELECTROHYDRAULIC INJECTORS

Electrohydraulic injectors have been introduced recently and are used in conjunction with common rail, diesel fuel injection systems such as those used in Mack Trucks light-duty engines, Cummins ISB, and the GM/Isuzu engines. Electrohydraulic injectors are similar to multi-orifii hydraulic injectors in most ways but are controlled electrically rather than hydraulically. When the unit is electrically energized, it uses the hydraulic pressure in the rail to effect the opening of the nozzle. The electrohydraulic injector (Figure 20–5) can be subdivided as follows:

- Nozzle assembly
- Hydraulic servo system
- Solenoid valve

OPERATION

Referencing Figure 20–5, fuel at rail pressure is supplied to the high-pressure connection (4), to the nozzle through the fuel duct (10), and to the control chamber (8) through the feed orifice (7). The control chamber is connected to the fuel return (1) via a bleed orifice (6) that is opened by the solenoid valve. With the bleed orifice closed, hydraulic force acts on the valve control plunger (9) exceeding that at the nozzle-needle pressure chamber (located between the shank and the needle of the nozzle valve). Although the hydraulic

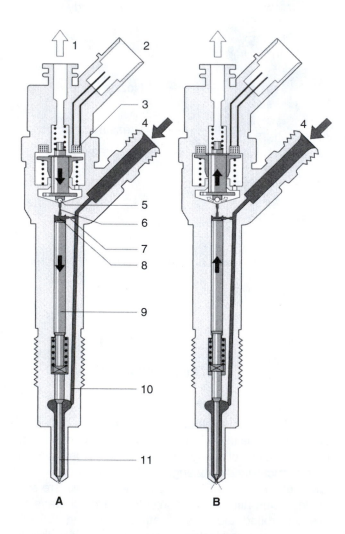

FIGURE 20–5 *Electrohydraulic injector. (Courtesy of Robert Bosch Corporation)*

Injector (schematic)

A Injector closed (at-rest status),
B Injector opened (injection).
1 Fuel return,
2 Electrical connection,
3 Triggering element (solenoid valve),
4 Fuel inlet (high pressure) from the rail,
5 Valve ball,
6 Bleed orifice,
7 Feed orifice,
8 Valve control chamber,
9 Valve control plunger,
10 Feed passage to the nozzle,
11 Nozzle needle.

pressure values acting on the top of the nozzle valve and that in the pressure chamber are identical, the sectional area at the top of the nozzle valve is greater. As a result, the nozzle needle is loaded into its seated position, meaning that the injector is closed.

When an ECM signal triggers the injector control solenoid valve, the bleed orifice opens. This immediately drops the control-chamber pressure and, as a result, the hydraulic pressure acting on the top of the nozzle valve (11) also drops. When hydraulic force acting on top of the nozzle valve drops below the force on the nozzle-needle pressure shoulder, the nozzle valve retracts and allows fuel to pass around the seat to be injected through orifii into the combustion chamber. The hydraulic assist and amplification factor are required in this system because the forces necessary for rapid nozzle valve opening cannot be directly generated by an electrically actuated solenoid valve. Fuel used as hydraulic media to open the nozzle valve is in addition to the injected fuel quantity, so this excess fuel is routed back to the tank. In addition to this fuel, some leak-by fuel losses occur at the nozzle-valve to body clearance and the valve-plunger guides clearance.

Injector-Operating Phases

When the engine is stopped, all injector nozzles are closed, meaning that their nozzle valves are loaded onto their seats by spring pressure. In a running engine, injector operation takes place in three phases.

Injector closed. In the at-rest state, the solenoid valve is not energized and therefore the nozzle valve is loaded onto its seat by the injector spring combined with hydraulic pressure (from the rail) acting on the sectional area of the valve control plunger. With the bleed orifice closed, the solenoid valve spring forces the armature ball check onto the bleed-orifice seat. Rail pressure builds in the injector control chamber, but identical pressure will be present in the nozzle pressure chamber. Given equal pressure acting on the larger sectional area of the nozzle control plunger (which is mechanically connected to the nozzle valve) and in the nozzle pressure chamber, this pressure and the force of the nozzle spring combine to load the nozzle valve on its seat holding the injector closed.

Nozzle opening. When the injector solenoid valve is energized by the ECM injector driver, it is actuated at high pressure, typically around a 90 V spike at a current draw of around 20 A. Force exerted by the triggered solenoid now exceeds that of the valve spring, and the armature opens the bleed orifice. Almost instantly, the high-level pickup current to the solenoid is reduced to a lower hold-in current flow required by the solenoid. Hold-in current is less than pickup current for the simple reason that the magnetic field air gap is reduced when the solenoid valve is fully actuated. As the bleed orifice opens, fuel flows from the valve-control chamber into the cavity above it and out to the return circuit. This collapses the hydraulic pressure acting on the valve control plunger that was helping to hold the nozzle valve closed. Now, pressure in the valve-control chamber is much lower than that in the nozzle pressure chamber that is maintained at the rail pressure. The result is the force that was holding the nozzle valve closed collapses and the nozzle valve opens, beginning the injection pulse.

The nozzle needle opening velocity is determined by the difference in the flow rate through the bleed and feed orifices. When the control plunger reaches its upper stop, it is cushioned by fuel generated by flow between the bleed and feed orifices. When the injector nozzle valve has fully opened, fuel is injected into the combustion chamber at a pressure very close to that in the fuel rail.

Nozzle closing. When the solenoid valve is de-energized by the ECM, its spring forces the armature downward and the check ball closes the bleed orifice. The closing of the bleed orifice creates pressure buildup in the control chamber via the input from the feed orifice. This pressure should be the same as that in the rail and now it exerts an increased force on the nozzle valve control plunger through its end face. This force, combined with that of the nozzle spring, exceeds the hydraulic force acting on the nozzle valve sectional area and the nozzle valve closes, ending injection. Nozzle valve closing velocity is determined by the flow through the feed orifice. The injection pulse ceases the instant the nozzle valve seats.

Nozzle Hole Geometry

Common Rail injectors use both sac-chamber nozzles and what are generally referred to as valve closes orifice (VCO) nozzles. Both sac and VCO nozzles are covered in a little more detail later in the chapter but VCO nozzles tend to be required in engines meeting North American emissions standards. Most Bosch electrohydraulic injectors use a 4 mm nozzle valve diameter. Both sac-hole and seat-hole nozzles, have input edges of each orifice rounded by hydroerosive (HE) machining. Hydroerosive machining helps prevent edge wear caused by the abrasive particles in the fuel and carbon coking that can reduce flow and disrupt the

spray geometry. For reasons of strength, the nozzle tip is conically shaped.

NOZZLE DIFFERENTIAL RATIO

Nozzle differential ratio describes the geometric relationship between the sectional area of the **nozzle seat** and that of the pressure chamber or valve shank. Nozzle valves are opened by hydraulic pressure acting on the sectional area of the valve subject to the pressure chamber; when this overcomes the spring pressure that loads them on their seat, they retract. However, the instant the nozzle valve unseats, hydraulic pressure is permitted to act over the whole sectional area of the nozzle valve. The whole sectional area of the nozzle valve is the sectional area of both the seat and that of the pressure chamber combined (Figure 20–6).

When pressure rise begins in the injection line circuit before the opening of the nozzle valve, the hydraulic circuit being acted on is closed (it is sealed at the nozzle seat). The instant the NOP value is achieved, the nozzle valve unseats and opens the circuit. This must result in a drop in line pressure. However, because at the moment the nozzle valve opens, the sectional area over which the hydraulic pressure acts is increased by that of the seat sectional area, less pressure is now required to hold the nozzle open. Nozzle differential ratio must be sufficient to prevent nozzle closure when this pressure drop occurs. After NOP, fuel passing around the seat fills the sac and pressure rise resumes because of the restriction represented by the minute sizing of the nozzle orifii. Nozzle differential ratio means that nozzle closure at the end of injection occurs at a value somewhat below the NOP value. It also helps define the specific **residual line pressure** (pressure that dead volume fuel is retained at in the high-pressure pipe) value.

The bottom line is that more pressure is required to unseat a nozzle valve than that required to hold it off its seat due to nozzle differential ratio.

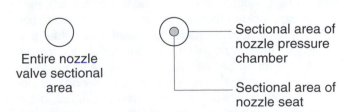

Entire nozzle valve sectional area

Sectional area of nozzle pressure chamber

Sectional area of nozzle seat

FIGURE 20–6 *Nozzle differential ratio: sectional areas of a nozzle valve.*

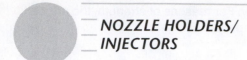

NOZZLE HOLDERS/ INJECTORS

Injectors are simply mounting devices for hydraulic nozzles. They come in many different shapes and sizes for a variety of reasons; for instance, long stem nozzles are easier to cool. Older hydraulic injectors tended to be of the high spring design. **High spring injectors** located the spring in the upper portion of the holder and relayed the spring tension to the nozzle value by means of a spindle. Spring tension was altered directly by an adjusting screw and setting the NOP value on the pop tester was a quick and easy procedure. A disadvantage of the high spring design was at NOP, the nozzle valve slammed open until it was mechanically prevented from further inboard travel. Because of its high opening velocity, the spindle was driven into the spring, which rebounded sufficiently to hammer the nozzle valve from its open position back into the seat, interrupting the injection pulse. More recent hydraulic injector nozzles have tended to use a low spring design that eliminates the spindle and thus reduces the mass of moving parts; this has minimized rebound interference of the injection pulse. However, **low spring injectors** generally use shims acting on the spring to define the NOP value, which extends the time required to set the NOP value on the test bench.

PENCIL-TYPE INJECTOR NOZZLES

Pencil injector nozzles are a type of multi-orifii injector nozzle and, in fact, share common operating principles with a couple of small exceptions. They are seldom found in any current truck engine applications. Caterpillar has used them in some of its hydromechanical engines. The pencil nozzle assembly is cylindrical and has the approximate appearance of a pencil. Within the nozzle body, the nozzle valve extends through nearly the full height of the injector assembly, and the injector spring acts directly on top of the nozzle valve shaft. Adjustment is by means of an adjusting screw and locknut located at the top of the assembly. Some pencil nozzles have no leakoff lines; fuel that bleeds by the nozzle valve during injection is accumulated in a chamber above the valve, and pressure equalization with line pressure occurs after nozzle closure. Pencil nozzles are most often damaged in the process of removal and service. When making any adjustments, use the recommended mounting fixture and observe the specified torque procedure.

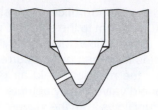

FIGURE 20–7   *Valve closes orifice (VCO) nozzle.* *(Courtesy of Robert Bosch Corporation)*

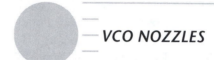

VCO NOZZLES

Valve closes orifice (VCO) nozzles have eliminated the sac. The function of the sac in the nozzle is to provide balanced fuel dispersal, which is especially important in keeping the ignition lag time consistent. However, at the completion of the injection pulse, the volume of fuel in the sac was essentially wasted fuel that added to HC emission. At the instant of nozzle closure, the sac and the nozzle orifii would contain fuel that would be vaporized due to the heat of combustion but at best, only partially combusted. Current injector nozzle designs have either substantially reduced the sac volume or eliminated it entirely. A true VCO nozzle has orifii that extend directly from the seat. VCO nozzles are most often used on electronically managed injection systems that use high NOP values, which can to some extent compensate for the compromising of the balanced fuel dispersal offered by nozzle sacs. Figure 20–7 shows a VCO nozzle design.

NOZZLE TESTING AND RECONDITIONING

Injector nozzles are probably subjected to more abuse than any other engine component. They are exposed to temperature peaks of 2,550°F (1,400°C) outside and to pressures exceeding 30,000 psi (2,000 atms) internally. Manufacturers' service intervals are seldom respected and field diagnostics tend to be poor. Today's diesel technician is seldom required to fully recondition nozzle assemblies due to the high cost of labor and service tooling versus the relatively low cost of nozzle replacement. However, most manufacturers include the injectors in preventive maintenance schedules and all that is required in terms of equipment is a simple bench test fixture and ultrasonic bath. Servicing a set of injectors is a safe and easy procedure, but some precautions are required.

CAUTION:

- Eye protection should be worn when working with or near high-pressure fluids.
- High-pressure atomized fuel is extremely dangerous and no part of the body should be in danger of contacting it.
- Never touch the nozzle assembly on the pop tester when it is at any pressure above atmospheric—and remember that the entire injector assembly is under pressure.
- Diesel fuel oil is a known carcinogen, that is, a cancer-causing agent. Hands should always be washed after contact with it.

REMOVAL OF INJECTORS FROM THE CYLINDER HEAD

The OEM recommended method of removing an injector from a cylinder head must be adhered to and failure to do so can result in failed injectors. Before attempting to remove injectors, thoroughly clean the surrounding area, remembering that any dirt around the injector bore can end up in the engine cylinder below it. Thin-bodied, pencil-type nozzles are vulnerable to any side load force and the correct pullers should always be used. Additionally, whenever pencil-type nozzles are removed from a cylinder head, they must be tested in a test fixture before reinstallation because they are so easily damaged during the removal process. Many injector assemblies are flanged, in which case the hold-down fasteners and clamps should first be removed and the injector levered out using an injector heel bar. The injector heel bar is 8" to 12" (20–30 cm) in length to prevent the application of excessive force to the injector flange. Where cylindrical injectors are used, a slide hammer and puller nut that fits to the high-pressure inlet of the injector should be used to pull the injector. With certain cylindrical hydraulic injectors, the high-pressure delivery pipe fits to a recess in the injector through the cylinder head; the high-pressure pipe must be backed away from the cylinder head before attempting to remove the injector, or both the injector and the seating nipple on the high-pressure pipe will be damaged. When removing injectors from cylinder heads, make a practice of removing the injector nozzle washer at the same time; these have an id (inside diameter) less than the cylinder injector aperture id, so providing the piston is not at TDC, they can usually be removed either by inserting an O-ring pick or, in cases

of difficult-to-remove washers, the tapered end of the injector heel bar and jamming it into the washer. It is not good practice to reuse injector washers. Both steel and copper washers harden in service. The copper washers can be annealed by heating to the point they shimmer green, then quenched with water. Steel washers are a poor reuse risk and no attempt should be made to anneal them.

When the injector has been removed, use plastic caps to seal the cylinder head injector bores, the injector inlet, and high-pressure pipe nipples. Ensure that a set of injectors is marked by cylinder number and properly protected in an injector tray; wrapping the injectors in shop rags will do if no injector tray is available.

Occasionally, an injector may seize in a cylinder head and defy any normal means used to remove it. Once the injector becomes damaged (for instance, the line threads that fit the injector to the puller are destroyed), the technician has no option but to remove the cylinder head; never risk damaging a cylinder head attempting to remove an injector. With the cylinder head removed, a seized injector can usually be removed with a punch and hammer.

TECH TIP: The reality of the trucking industry today is that it is economically unwise to recondition injector nozzles. It is a labor-intensive procedure that is not justified in modern diesel engines that have exacting requirements for atomized droplet sizing. If a customer seeks your advice, recommend that a failed nozzle be replaced rather than reconditioned.

TESTING

OEM specifications and procedures should be consulted before placing the nozzle to be tested in the nozzle test fixture. A typical procedure for testing nozzle assemblies would be as follows:

Assessment Procedure

Perform the following tests in sequence. At the first failure, cease following the sequence and proceed to the next section.

1. Clean the injector externally with a brass wire brush (Figure 20–8).
2. Locate the manufacturer's test specifications.
3. Mount the injector in the bench test fixture. Build pressure slowly using the pump arm, watching for external leakage.
4. Bench test the NOP value and record. Use three discharge pulses and record the average value. Sticking or a variation in NOP value that exceeds 10 atms (150 psi) fails the nozzle.

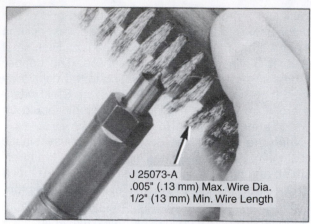

J 25073-A
.005" (.13 mm) Max. Wire Dia.
1/2" (13 mm) Min. Wire Length

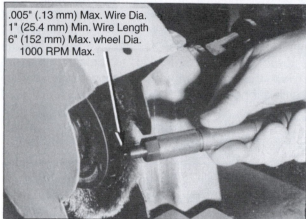

.005" (.13 mm) Max. Wire Dia.
1" (25.4 mm) Min. Wire Length
6" (152 mm) Max. wheel Dia.
1000 RPM Max.

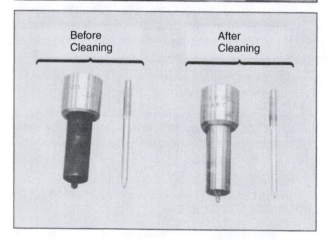

Before Cleaning   After Cleaning

FIGURE 20–8 *Cleaning the exterior of the nozzle assembly.* Top: *Brass wire brush;* middle: *brass wire wheel;* bottom: *before and after. (Courtesy of Mack Trucks)*

5. Test **forward leakage** by charging to 10 atms (150 psi) below the NOP value and holding the gauge pressure at that value while observing the nozzle. Any leakage evident at the tip orifii fails the nozzle.
6. Check back leakage by observing pressure drop from a value 10 atms (150 psi) below NOP. Pressure drop values should typically be in the range of 50–70 atms (700–1,000 psi) over a 10-second test

period. Pressure drop that is less than OEM specification indicates too little valve-to-body clearance (possibly caused by valve-to-body mismatch). A rapid pressure drop exceeding the OEM specification indicates excessive nozzle valve-to-body clearance, a condition usually caused by wear.

7. Once again, actuate the bench fixture pump arm and observe the nozzle spray pattern, checking for orifii irregularity. Ignore nozzle **chatter** (rapid pulsing of the nozzle valve); this can be regarded as a test bench phenomenon due to *slow* rate of pressure rise and inability to drive pressure much above NOP. In some modern injectors, the intensity of chatter noise is reduced due to the use of double seats (Figure 20–9).

Reconditioning

The function of the assessment procedure is to determine whether the nozzle should be reconditioned or placed back in service. Reconditioning in the service shop of today seldom means regrinding of the valve

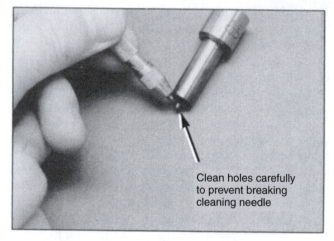

Clean holes carefully to prevent breaking cleaning needle

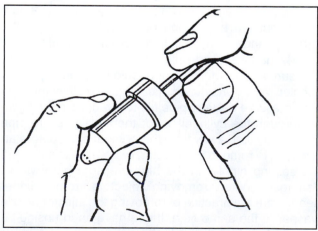

FIGURE 20–9 Top: *Cleaning nozzle orifii;* bottom: *performing valve-to-nozzle slide test. (Courtesy of Mack Trucks)*

nozzle seat or reaming nozzle orifii due to high labor and tooling costs in comparison with the cost of nozzle replacement. The following simple procedure probably summarizes what most operations interpret as reconditioning:

1. A nozzle that fails the inspection procedure should first be removed from the injector. Disassemble the injector and remove the nozzle, separating the nozzle valve from the nozzle body. Ensure that nozzle valves remain matched with the nozzle bodies they are removed from because they are not interchangeable.

2. Prepare the nozzle valve and body for ultrasonic cleaning (Figure 20–10). An ultrasonic bath filled with clean soapy water should be used. In other words, change the cleaning solution regularly. Alternatively diesel test oil can be used, but it is not as effective. A five-minute, ultrasonic bath is generally sufficient. This should be followed by dehumidification either in a dehumidifier or an oven. If a dehumidifier is not available, fuel test oil rather than soapy water should be used as the ultrasonic soaking medium. Remember, lapped components are not interchangeable so if cleaning multiple nozzle assemblies, do not mismatch components.

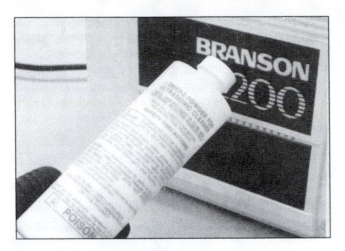

CAUTION: Ultrasonic baths emit sound waves that are outside the frequencies detected by the human ear and have been known to trigger seizures in those susceptible to epilepsy. It is good practice to vacate a room in which ultrasonic cleaning is being performed even though it represents to most people nothing more than an irritating buzzing.

3. Test or replace the injector spring. Reassemble the injector assembly using diesel fuel as an assembly lubricant. Never use engine oil or any grease-based lubricant.

4. As a general rule, set the NOP value at 10 atms (150 psi) above the recommended specification if no parts have been replaced. Add another 5 atms (75 psi) if using a new spring, and a further 5 atms (75 psi) if using a new nozzle assembly. After 100 engine hours, the nozzle will function at the specified NOP value. In other words, when reassembling an injector with a new nozzle assembly and a new spring, the NOP would be set 20 atms (300 psi) above the recommended specification. Always consult the appropriate service literature for the NOP setting procedure.

- Screw type: back off the locknut, then turn the adjusting screw CW to increase or CCW to decrease NOP.
- Shim type: add or subtract shims acting on the injector spring.

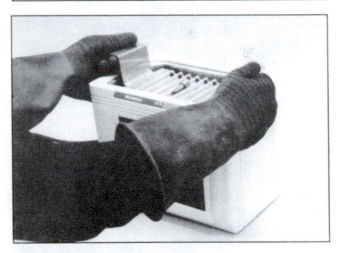

FIGURE 20–10 *Ultrasonic bath procedure: Top: follow instructions; middle: fill tank with solution; bottom: lower tray with the nozzle assemblies into the tank. (Courtesy of Mack Trucks)*

5. Check forward leakage.
6. Check back leakage.
7. Check back leakage pipe restriction.
8. Observe spray pattern and check for nozzle orifii irregularity, preferably using a template.

A critical function of the nozzle washer (spacer) is to ensure optimum droplet penetration into the engine cylinder. It is not a mere joint gasket. Spacer thickness is critical and determines the nozzle protrusion dimension. Use of an incorrectly sized washer, two washers, or no washer can result in rapid engine failures.

REINSTALLATION OF INJECTORS

The injector bore should be cleaned before installation using a bore reamer if necessary to remove carbon deposits. Blow out the injector bore using an air nozzle. Where injector sleeves are used, service/replace as per OEM instructions. Most diesel engine OEMs prefer that injectors are installed dry; lubricants such as Never-Seize™ can hinder the ability of the injector to transfer heat to the cylinder head. It is good practice to turn the engine over and pump fuel through the high-pressure pipe before connecting it to the injector; this will purge the pipe of air and possible debris that may have intruded into the line while it was disassembled. Always torque the nut on the high-pressure pipe with a line socket; failure to do this can result in damage to both the line nut and the nipple and seat.

WARNING: Whenever fuel has spilled over an engine, ensure that the engine is pressure washed afterwards.

HIGH-PRESSURE PIPES

In most hydromechanical fuel injection systems and some electronically managed systems, fuel is delivered from the pumping elements in the fuel injection pump to the nozzle holders by high-pressure pipes. These are subject to very high pressure values and are manufactured from alloyed steels with substantial wall thickness. Pressure wave reflection caused by pressure fluctuations and high-speed switching must also be accommodated. Failures are caused by pipe erosion, **cavitation** (vapor bubble collapse), and technician abuse-related problems such as overtorquing and bending. High-pressure pipes should never be repair welded or braised when worn through due to the extreme pressures they are required to contain. Weld or braising repairs can penetrate into the worn-away area and reduce the flow area within the high-pressure pipe. They may also completely plug the pipe and when a pump is capable of generating pressures exceeding 30,000 psi, something has to let go. Perhaps the most important reason that high-pressure pipes should never be repair welded is that if the weld is unseated by the pipe pressure in operation, it can be propelled with sub-

stantial force and have the potential to injure anyone in the vicinity.

In multicylinder engines, it is critical that the high-pressure pipes are of equal length and identical bores. Small variations in length or bore size can substantially alter injection timing and unbalance engine performance. High-pressure pipe nuts at both the injector and injection pump ends should always be torqued to specification. When attempting to locate a cylinder misfire, technicians sequentially crack the high-pressure pipe nuts in a procedure known as "shorting out injectors." In practice, where no torque wrench is used, the result is usually overtorquing—this has the effect of crushing the sealing nipple of the high-pressure pipe and reducing flow through the line. The nipple sealing seats on high-pressure pipes should be inspected for ridging and aperture (bore) size at each removal. It should also be noted that this condition can result in low power and/or unbalanced fuel delivery.

TECH TIP: Always use the engine OEM's specification, a line socket, and torque wrench to torque line nuts. A common cause of low-power complaints is collapsed nipple seats due to overtorquing, which reduces the line flow area.

SUMMARY

- Injector nozzles are simple hydraulic switching mechanisms.
- Direct injected engines usually require the use of multi-orifii or electrohydraulic nozzles.
- Direct-injected engines require atomized droplets within the 10–100 micron range.
- Currently, multi-orifii or electrohydraulic nozzles are used in most truck and bus engine applications.
- Injector spring tension defines the NOP value.
- Injector spring tension is set either by adjusting screws or by shims.
- Locating dowels are used to position a nozzle in the injector body to ensure correct spray dispersal in the engine cylinder.
- Nozzle sac volumes have either decreased or been eliminated (VCO nozzle) to address legislated emissions standards.
- Nozzle chatter can usually be ignored when pop testing injector nozzles; it is a bench test condition.
- Injector nozzles are seldom fully reconditioned in current trade practice.
- Torquing of high-pressure pipe union nuts at both the injection pump and injector ends is critical to avoid damaging the nipple seat and aperture.
- High-pressure pipe bore should be checked to specification at the seats.

REVIEW QUESTIONS

1. Which type of injection nozzle would be found on most direct-injected truck diesel engines?
   a. Poppet
   b. Pintle
   c. Throttling
   d. Multi-orifii
2. When injector back leakage is bench checked, which of the following is being tested?
   a. NOP
   b. Nozzle seat seal
   c. Valve-to-body clearance
   d. Injector spring
3. When nozzle forward leakage is bench checked, which of the following is being tested?
   a. NOP
   b. Nozzle seat seal
   c. Valve-to-body clearance
   d. Injector spring
4. Setting the injector NOP value is accomplished by:
   a. Adding or subtracting shims
   b. An adjusting screw
   c. Both a and b
   d. Neither a nor b
5. An injector spring that is fatigued would likely produce which of the following conditions?
   a. Higher NOP
   b. Higher NOP and retarded injection timing
   c. Nozzle chatter
   d. Lower NOP and advanced injection timing
6. Replacing a single high-pressure injection pipe on a multicylinder engine by one of greater length would likely have what effect on injection timing in the affected cylinder?
   a. Advance
   b. Retard
   c. None
   d. Decrease the fuel pulse width
7. _Technician A_ states that nozzle chatter while bench testing an injector is an indication of a fatigued injector spring. _Technician B_ states that a high-injector back leakage rate is an indicator of excessive nozzle valve-to-nozzle body clearance. Who is right?
   a. A only
   b. B only
   c. Both A and B
   d. Neither A nor B
8. What is the distinguishing feature of a VCO nozzle?
   a. The sac is eliminated.
   b. High spring location
   c. Low spring location
   d. Used only in IDI engines
9. Which of the following injector types would be more likely to require a spindle?
   a. High spring
   b. Low spring
   c. Poppet
   d. VCO
10. Which type of injector nozzle would be incorporated in an EUI?
   a. Pintle
   b. Poppet
   c. Multi-orifii
   d. Pencil
11. _Technician A_ states that injection nozzles are designed so that the pressure required to open an injector nozzle valve is greater than the pressure required to hold it in the open position. _Technician B_ states that after NOP, the nozzle line pressure steadily decreases until the nozzle closes at the end of the injection pulse. Who is right?
   a. A only
   b. B only
   c. Both A and B
   d. Neither A nor B
12. Nozzle differential ratio is a ratio of:
   a. Nozzle seat sectional area and total nozzle sectional area
   b. The high-pressure pipe sectional area and the orifii flow area
   c. The nozzle pressure chamber sectional area and the high-pressure pipe sectional area
   d. The ratio of mechanical spring force and hydraulic pressure
13. Which of the following factors does most to prevent secondary injections?
   a. Pressure wave reflection
   b. Nozzle differential ratio
   c. The nozzle sac
   d. The pressure chamber sectional area
14. Which of the following is another way of stating the NOP specification?
   a. Not operating properly
   b. Residual line pressure
   c. Peak pressure
   d. Popping pressure
15. Most diesel engine OEMs recommend that injectors are installed into the cylinder head injector bore:
   a. Dry
   b. Coated with engine oil
   c. Coated with Never-Seize
   d. Coated with lubriplate
16. _Technician A_ states that when a hydraulic multi-orifii injector is used in an electronically managed fuel system, nozzle opening and closing is switched hydraulically. _Technician B_ says that the opening and closing of an electrohydraulic injector is controlled by a solenoid. Who is right?

a. A only
b. B only
c. Both A and B
d. Neither A nor B

17. Excessive fuel returned into the leak-off lines from a set of hydraulic injector nozzles is an indication of:
    a. Nozzle seat leakage
    b. Wear in the nozzle body-to-nozzle valve fit
    c. Wear in nozzle seat
    d. A weak injector spring

18. *Technician A* states that the reason for using VCO nozzles in many current diesel engines is to reduce HC emissions. *Technician B* states that an advantage of reducing the nozzle sac volume is to minimize the amount of fuel in the cylinder that cannot be properly combusted. Who is right?
    a. A only
    b. B only
    c. Both A and B
    d. Neither A nor B

19. *Technician A* states that the injector nozzles used on current EUIs and HEUIs use the same operating principles as those on any hydraulic pump-line-nozzle system. *Technician B* states that EUP fuel systems use standard multi-orifii, hydraulic nozzles. Who is right?
    a. A only
    b. B only
    c. Both A and B
    d. Neither A nor B

20. Which of the following values would be a typical NOP parameter for a multi-orifii nozzle used on an electronically managed engine?
    a. 1,500 psi (102 atms)
    b. 3,000 psi (205 atms)
    c. 5,000 psi (340 atms)
    d. 12,000 psi (816 atms)

Chapter 21

Port-Helix Metering Injection Pumps

PREREQUISITES

Chapters 18, 19, and 20

OBJECTIVES

After studying this chapter, you should be able to:

- Identify the major components in a typical port-helix metering injection pump.
- Explain the principles of operation of an in-line, port-helix metering injection pump.
- Define the terms effective stroke, port closure, port opening, NOP, residual line pressure and peak pressure.
- Understand how the rotary motion of the camshaft is converted to the reciprocating motion required of the injection pump elements.
- Explain how the pump element components create injection pressures.
- Define metering and the factors that control it.
- Explain the operation of aneroid devices, altitude compensators, and variable timing/timing advance mechanisms.
- Time an injection pump to an engine using the appropriate OEM method.
- Relate event sequencing in the injection pump to that in the engine combustion chamber.

KEY TERMS

altitude compensator	double helix	port closure (PC)
aneroid	dual helices	port opening
atm	effective stroke	pump drive gear
barometric capsule	flutes	pump-line-nozzle (PLN)
barrel	fuel rate	register
boosted engine	helix/helices	retraction collar/piston
calibration	high-pressure pipes	retraction spring
cambox	ignition lag	spill timing
camshaft	injection lag	static timing
charging pressure	lower helix	tappets
charging pump	metering recesses	transfer pump
control rack	phasing	upper helix
control sleeve	plunger	vial
dead volume fuel	plunger geometry	
delivery valve	plunger leading edge	

INTRODUCTION

The first high-pressure liquid fuel injection to a high-speed diesel combustion chamber was developed in 1927 by Robert Bosch. This evolved into the fueling apparatus used by Caterpillar, Mack, Navistar, and other truck engine manufacturers into the 1990s. The principles of pumping and metering a fuel charge have changed little since Bosch's first 1927 high-pressure injection pump. However, the port-helix (helix: scroll shaped) metering injection pump, emissions certified for the 1990s, will, in most cases, be managed electronically rather than hydromechanically, and the hard (hard: not changeable) delivery timing window defined solely by the geometry of actuating cam profile and helix geometry has given way to variable timing devices. This has permitted the technology of the in-line, port-helix metering pump to live on into the electronic age at least for a short while. In most cases, when the truck technician uses the term **pump-line-nozzle (PLN)**, an in-line port-helix metering injection pump is usually being referred to. Despite the fact that a few of these injection pumps are still in use, their days are numbered. This chapter describes the port-helix metering pumps engineered and manufactured by Bosch (Figure 21–1 and Figure 21–2), Delphi-Lucas, and Caterpillar and used on Caterpillar, Mack, Cummins, Navistar, and other current highway and off-highway diesel engine applications.

TECHNICAL DESCRIPTION

The typical port-helix metering pump used to fuel a truck (or bus) CI engine is in-line configured and flange mounted to an engine accessory drive to be driven through one complete rotation (360 degrees) per complete engine cycle (720 degrees). The internal pump components are housed in a frame constructed of cast aluminum, cast iron, or forged steel. The engine crankshaft drives the injection pump by means of timed reduction gearing. The gear-driven pump drive plate is connected to the injection pump **camshaft** (the shaft fitted with eccentrics designed to actuate the pump elements), so rotating the pump drive plate rotates the pump camshaft. The camshaft is supported by main bearings and rotates within the injection pump **cambox**. The cambox is the lower portion of the injection

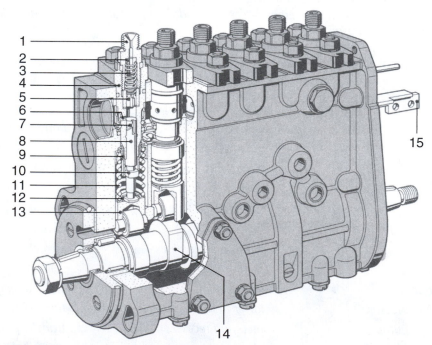

1. Delivery valve	6. Inlet and spill port	11. Plunger return spring
2. Filler piece	7. Helix	12. Spring seat
3. Delivery valve spring	8. Pump plunger	13. Roller tappet
4. Pump barrel	9. Control sleeve	14. Actuating cam
5. Delivery valve	10. Plunger control arm	15. Control rod

FIGURE 21–1   *In-line, port-helix metering injection pump components.*
(Courtesy of Robert Bosch Corporation)

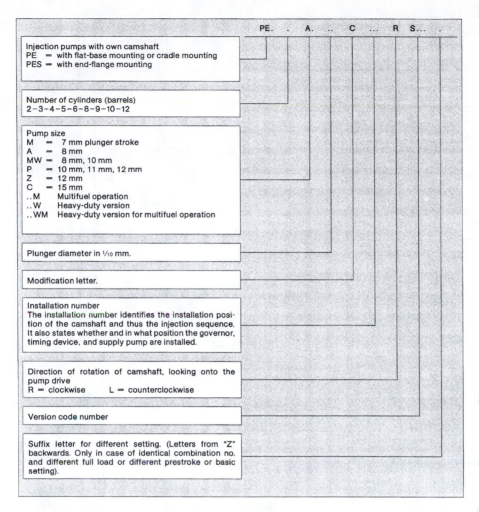

FIGURE 21–2 _Interpreting a Bosch injection pump serial number. (Courtesy of Robert Bosch Corporation)_

pump that houses the camshaft, **tappets**, and an integral oil sump (Figure 21–3 and Figure 21–4).

The camshaft is designed to have one cam profile dedicated to each engine cylinder. Riding each cam profile is a tappet assembly driving a pump element consisting of a **plunger** and a **barrel**. The barrel is stationary and drilled with two ports in its upper portion, which are exposed to the fuel charging gallery.

The fuel gallery is charged with low pressure fuel, typically between 1 to 5 atms (**atm:** unit of atmospheric pressure) (15–75 psi). This permits fuel to flow into and through the barrel ports when they are not obstructed by the plunger. The plunger reciprocates within the barrel; it is loaded by spring pressure to ride its actuating cam profile. Therefore, actual plunger stroke is constant. Plunger-to-barrel tolerance is close, the components being lapped in manufacture to tolerance of 2μ to 4μ. Fuel quantity to be delivered in each stroke is controlled by managing plunger **effective stroke** (Figures 21–5 through 21–8). The plunger is milled with a vertical slot or cross and center drillings and helical

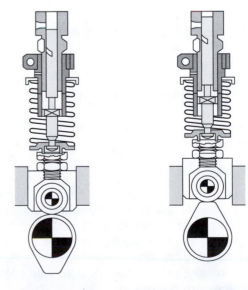

CAM Base Circle   **CAM Outer Base Circle**

FIGURE 21–3 _Actuating a port-helix type pump element. (Courtesy of Robert Bosch Corporation)_

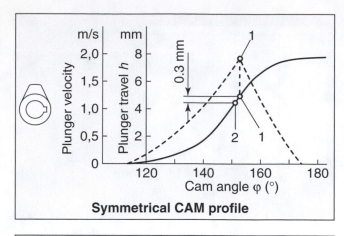

Symmetrical CAM profile

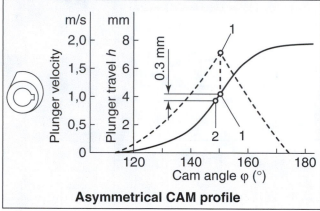

Asymmetrical CAM profile

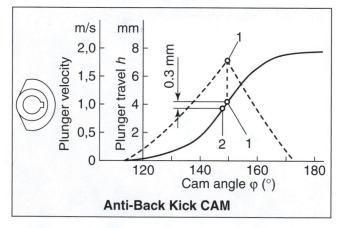

Anti-Back Kick CAM

FIGURE 21–4   Actuating cam geometry. (Courtesy of Robert Bosch Corporation)

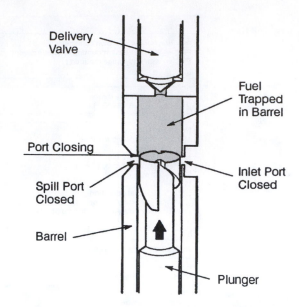

FIGURE 21–5   Port closure. (Courtesy of Mack Trucks)

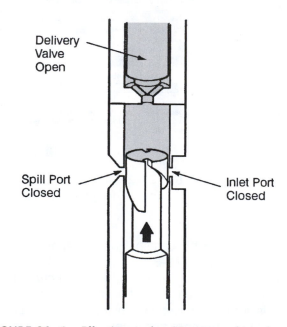

FIGURE 21–6   Effective stroke. (Courtesy of Mack Trucks)

recesses. The function of the vertical slot or cross and center drillings is to maintain a constant conduit between the pumping chamber above the plunger and the helical recesses. In other words, whatever pressures exist in the pumping chamber must also exist in the helical recess. Initially, only plunger designs with a **lower helix** or **helices** are discussed.

Effective stroke describes the delivery stroke. The delivery stroke begins when the plunger is forced upward by cam profile and the **plunger leading edge**

(uppermost part of the plunger) traps off the spill port(s).

As the plunger rises through its stroke in the barrel after trapping off the spill port, rapid pressure rise occurs, creating the required injection pressures. The precise moment that begins effective stroke is known as **port closure** (Figure 21–5). It is of critical importance to the diesel technician because its precise setting is used to control ignition timing. As pressure rises in the pump chamber, it acts first on a **delivery valve**, next on the fuel confined in the high-pressure pipe transmitting fuel to the injector nozzle, and finally delivering a fuel

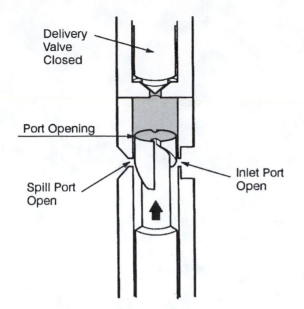

FIGURE 21–7 *Port opening. (Courtesy of Mack Trucks)*

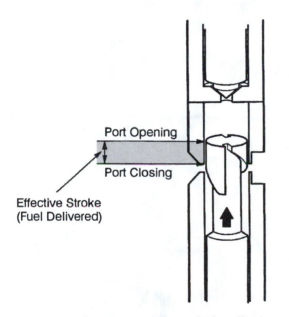

FIGURE 21–8 *Plunger travel through the effective stroke. (Courtesy of Mack Trucks)*

pulse to the engine cylinder. Effective stroke ends at **port opening** (Figure 21–7). This is the precise moment that the upward travel of the plunger exposes the helical recess(es) to the spill port. High-pressure fuel is spilled back to the charging gallery causing a rapid collapse of pressure in the pump chamber, line, and nozzle. The injection pulse ceases when there is no longer sufficient pressure to hold the delivery and nozzle valves open. Port opening always occurs while the plunger is moving in an upward direction, that is, not at plunger TDC or beyond. This is required because the pressure in a port-helix pump element is designed to

rise through the delivery stroke, thereby producing smaller atomized droplets from the injector toward the end of effective stroke; however, at the point of port opening, regardless of the length of the effective stroke, pump pressure should collapse as rapidly as possible and minimize the larger droplets emitted from the injector as pump pressure falls to a value below NOP.

The length of plunger effective stroke depends on where the plunger helix **registers** (vertically aligns) with the spill port. **Control sleeves** lugged to the plunger permit the plunger to be rotated while reciprocating. Rotating the plunger in the bore of the barrel changes the location of register of the spill port with the helix. Therefore, plunger effective stroke depends entirely on the rotational position of the plunger. In multiple cylinder engines, the plungers must be synchronized to move in unison to ensure balanced fueling at any given engine load.

The control sleeves are tooth meshed to a governor **control rack**, which when moved linearly, rotates the plungers in unison (Figure 21–9). This is important. It means that in any linear position of the rack, all of the plungers will have identical points of register with their spill points, resulting in identical pump effective strokes. The consequence of not doing this would be to unbalance the fueling of the engine, that is, deliver different quantities of fuel to each cylinder (Figure 21–10).

Engine shutdown is achieved by moving the control rack to the no-fuel position. The rotational position of the plungers is now such that the vertical slot will be in register with the spill port for the entirety of plunger travel; the plunger will merely displace fuel as it travels

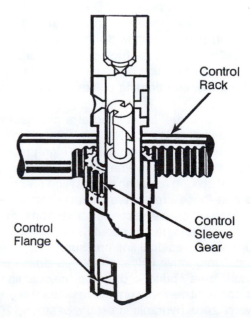

FIGURE 21–9 *Control rack and sleeve gear. (Courtesy of Mack Trucks)*

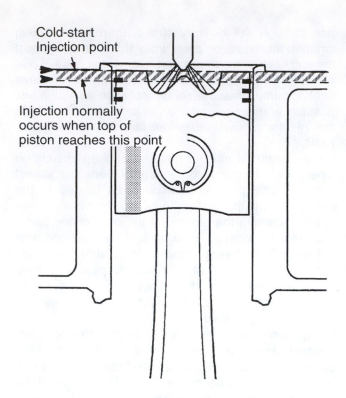

FIGURE 21–10   Cold-start, retarded injection timing.
(Courtesy of Mack Trucks)

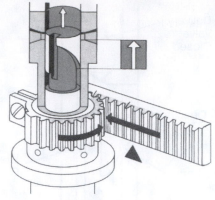

Full Fuel
Maximum Effective Stroke

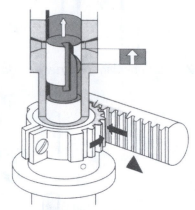

Medium Fuel
Partial Effective Stroke

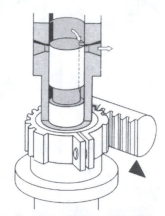

No Fuel Rotational Position
No Effective Stroke

FIGURE 21–11   Rack position and relationship to fuel
delivery quantity. (Courtesy of Robert Bosch Corporation)

upward, with no pumping action possible. In other
words, as the plunger is driven into the pump chamber,
the fuel in the chamber will be squeezed back down the
vertical slot to exit through the spill port and return to
the charging gallery (Figure 21–11).

Most port-helix metering injection pumps use delivery valves to reduce the amount of work required of
each pump element per cycle. Delivery valves function
to isolate the high-pressure circuit that extends from
the injection pump chamber to the seat of the nozzle
valve. Fuel retained in the **high-pressure pipes** (pipes
that connect the injection pump elements with hydraulic
injectors) between pumping pulses is known as **dead
volume fuel**.

Dead volume fuel is retained at a pressure necessarily somewhat below the NOP value; the high-speed
hydraulic switching that occurs in the high-pressure circuit creates pressure wave reflections, which may have
the effect of spiking (causing surges) line pressures. To
ensure that these pressure spikes do not exceed the
NOP value and cause secondary injections, the dead
volume fuel is retained at about ⅔ of the NOP value;
this is known as residual line pressure.

The delivery valve is loaded into its closed position
on its seat by a spring and by the residual line pressure. If, for whatever reason, the residual line pressure
value were zero, hydraulic pressure of around 20 atms
(300 psi) would have to be developed in the pump element to overcome the mechanical force of the spring.

This mechanical force is therefore compounded when
the residual line pressure acts on the sectional area
represented by the delivery valve and establishes the
pressure value that must be developed in the pump
chamber before it is unseated. The delivery valve
flutes provide a means of guiding the valve in its
bore while permitting hydraulic access between the

retraction collar or retraction piston (both terms are used) and the pump chamber (plunger and barrel assembly). The retraction collar seals the pump chamber from the dead volume fuel, which is retained at a higher pressure value. Consequently, when the delivery valve is first unseated, it is driven upward in its bore by rising pressure in the pump chamber and acts as a plunger driving inward onto the fuel retained in the high-pressure pipe (Figure 21–12 and Figure 21–13).

The moment the retraction collar clears the delivery valve seat, fuel in the injection pump chamber and that in the high-pressure pipe unite and the injection pump plunger is driven into a volume of fuel that extends from

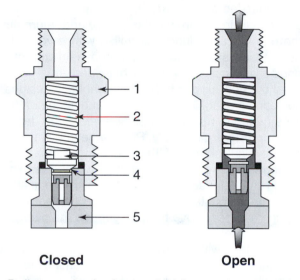

Closed	Open

1. Delivery valve body   4. Valve seat
2. Delivery valve seat   5. Valve holder
3. Delivery valve core

FIGURE 21–12 *Typical delivery valve assemblies in closed and open positions. (Courtesy of Robert Bosch Corporation)*

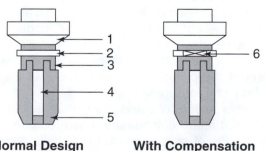

Normal Design	With Compensation Siphon

1. Delivery valve seat   4. Delivery valve guide
2. Retraction collar/piston   flutes
3. Annular piston   5. Guide flute
       6. Retraction collar siphon

FIGURE 21–13 *Delivery valve core terminology. (Courtesy of Robert Bosch Corporation)*

the plunger to the nozzle valve seat. Rising pressure subsequently unseats the injector nozzle valve (NOP) and forces atomized fuel into the engine cylinder.

At port opening, the effective pump stroke ends beginning a rapid pressure collapse as fuel spills from the barrel spill port. When there is insufficient pressure in the pump chamber to hold the nozzle valve in its open position, spring pressure overcomes hydraulic pressure and it seats, sealing the nozzle end of the high-pressure pipe. Almost simultaneously, the delivery valve begins to retract. The instant the retraction collar passes the delivery valve seat, it hydraulically seals the pump end of the high-pressure circuit. However, after sealing the pump end of the high-pressure circuit, the delivery valve must travel farther before it seats, increasing the volume available for dead volume fuel storage. This causes a drop in line pressure and defines the residual line pressure value. In short, the volume available for fuel storage in the high-pressure pipe is increased by the swept volume of the retraction collar. This volume of fuel is known as dead volume fuel and is retained at residual line pressure. Retraction collar swept volume is matched to the length of the high-pressure pipe to achieve a precise residual line pressure value so that pipe length should not be altered. Most delivery valves are of the constant volume type described here. However, some fuel systems have a constant pressure delivery valve that consists of a forward delivery valve in the forward flow direction and a pressure holding valve (spring and ball) in the reverse flow direction. This rather more complex valve design helps minimize the effects of pressure wave reflection, which cause wear and cavitation.

INJECTION PUMP COMPONENTS

The following is a list of subcomponents that assembled, will form a typical port-helix metering, injection pump.

PUMP HOUSING

The pump housing is the frame that encases all the injection pump components and is a cast aluminum, cast-iron, or forged steel enclosure. The pump housing is usually flange mounted by bolts to the engine cylinder block to be driven by an accessory drive on the engine gear train. In some offshore applications of in-line, port-helix metering injection pumps, the pump assembly is cradle mounted on its base, in which case, it is driven by an external shaft from the timing gear train.

CAM BOX

The cam box is the lower portion of the pump housing incorporating the lubricating oil sump and main mounting bores for the pump camshaft. Camshaft main bearings are usually pressure lubed by engine oil supplied from the engine crankcase, and the cambox sump level is determined by the positioning of a return port. In older injection pumps, the pump oil was isolated from the main engine lubricant, and the oil was subject to periodic checks and servicing.

CAMSHAFT

The camshaft is designed with a cam profile for each engine cylinder and supported by main bearings at the base of the pump housing. It is driven at one-half engine rotational speed in a four-stroke cycle engine by the pump drive plate, which is itself, either coupled directly to the **pump drive gear** or to a variable timing device. Camshaft actuating profiles are usually symmetrical (that is, geometrically similar on both sides of the toe) and mostly inner base circle (IBC: the smallest radial dimension of an eccentric) in profile, though asymmetrical (the geometry of each cam ramp or flank differs) and mostly outer base circle (OBC: the largest radial dimension of an eccentric) designs are used. For a full explanation of cam geometry, reference Chapter 8.

TAPPETS

The tappets are arranged to ride the cam profile and convert the rotary motion of the camshaft to the reciprocating action required of the plunger. A **retraction spring** is integral with the tappet assembly. This is required to load the tappet and plunger assembly to ride the cam profile and it is necessarily large enough to overcome the low pressure (vacuum) established in the pump chamber on the plunger return stroke. This low pressure can be considerable when plunger effective strokes are long but it enables a rapid recharge of the pump chamber with fuel from the charging gallery. The time dimension within which the pump element must be recharged decreases proportionately with pump rpm increase.

BARREL

The barrel is the stationary member of the pump element, located in the pump housing so its upper portion is exposed to the charging gallery. This upper portion of the barrel is drilled with diametrically opposed ports known as inlet and spill ports that permit through flow of fuel to the barrel chamber to be charged. Because it contains the spill ports, both its height and rotational position in relation to the plunger are critical. Barrels are often manufactured with upper flanges so that their relative heights can be adjusted by means of shims, and fastener slots permit axial movement for purposes of **calibration** and **phasing** (the procedures of calibration and phasing are covered in more detail later in this chapter).

PLUNGER

Plungers are the reciprocating (something that reciprocates, moves backward and forward such as in the action of a piston in an engine cylinder) members of the pump elements and they are spring loaded to ride their actuating cam's profile. Plungers are lapped to the barrel in manufacture, to a clearance close to 2 μ, ensuring controlled back leakage directed toward a viscous seal consisting of an annular groove and return duct in the barrel. Each plunger is milled with a vertical slot, helical recess(es), and annular groove. In current truck engine applications, a lower helix design is generally used but both **upper helix** and dual helix designs are sometimes used. The positioning and shape of the helices (plural of helix) on a plunger are often described as the **plunger geometry**. Plunger geometry describes the physical shape of the **metering recesses** machined into the plunger, and this defines the injection timing characteristics. The function of the vertical slot is to ensure a constant hydraulic connection between the pump chamber above the plunger and the plunger helical recess(es).

A plunger with a lower helix will have a constant beginning, variable ending of delivery timing characteristic, whereas an upper helix design will be of the variable beginning, constant ending type. **Double helix** designs have both an upper and a lower helix and a variable beginning and variable ending of delivery; this geometric design tends not to be often used in highway diesel engines. Another perhaps more common plunger design is the **dual helices** design—identically shaped helices are machined into the plunger, diametrically opposite each other. Plungers with diametrically opposed helices are used in many modern high-pressure injection pumps to provide hydraulic balance to the pump element; specifically, it prevents the side loading of the plunger into the barrel wall that may occur at high-pressure spill-off.

A further feature of some plungers is a start retard notch. Start retard notches are milled recesses in the leading edge of plungers with lower helix geometry. The start retard notch is usually on the opposite side of the vertical slot from the helix and in a position that would correlate close to a full-fuel effective stroke; the governor of the injection pump is designed to permit the start retard notch to register with the spill port only at cranking speeds (under 300 rpm) and with the accelerator fully depressed. The objective of the start retard notch on a lower helix design plunger is to retard the injection pulse until there is a maximum amount of heat in the engine cylinder, usually when the piston is close to TDC. The instant the engine exceeds 300 rpm, it becomes no longer possible for the start retard notch to register with the spill port (Figure 21–10).

RACK AND CONTROL SLEEVES

The rack and control sleeves permit the plungers in a multicylinder engine to be rotated in unison to ensure balanced fuel delivery to each cylinder. Plungers must therefore be timed either directly or indirectly to the control rack. The rack is a toothed rod that extends into the governor or rack actuator housing. The rack teeth mesh with teeth on plunger control sleeves, which are either lugged or clamped to the plunger. It must be possible to rotate the plungers while they reciprocate to permit changes in fuel requirements while the engine is running. Linear movement of the rack will rotate the plungers in unison, alter the point of register of the helices with their respective spill ports, and thereby control engine fueling. Timing of the plungers to the control rack is the means used to adjust the effective stroke in individual pump elements in a procedure known as calibration. This timing procedure is effected either directly, by adjusting the relationship of the plunger with the rack or indirectly, by adjusting the rotational position of the barrel and therefore its relationship with the plunger geometry. The means of calibrating an injection pump depends on the make and model.

DELIVERY VALVES

Delivery valves isolate the high-pressure circuit that extends from the injection pump chamber to the seat of the injector nozzle valve. They act as somewhat like check valves. Because they seal before they seat, they permit line pressure to drop to a residual value well below the NOP value, and this helps prevent secondary injections. The delivery valve is machined with a seat, retraction collar, and flutes to guide it in the bore of the delivery valve body (Figure 21–13).

CHARGING PUMPS

The terms **charging pump** and **transfer pump** are used interchangeably, depending on the OEM. The charging pump is responsible for all fuel movement in the fuel subsystem. In truck applications using port-helix metering injection, the charging pump is normally a plunger pump, flange mounted to the fuel injection pump and actuated by a dedicated eccentric on the in-

jection pump camshaft. The charging or transfer pump is responsible for producing **charging pressure**. Charging pressures range from 1 to 5 atms (15–75 psi) depending on the system. The role of charging/transfer pumps is dealt with in greater detail in the context of the fuel subsystem, the subject of Chapter 18.

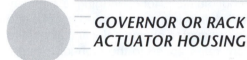

GOVERNOR OR RACK ACTUATOR HOUSING

Either a governor or rack actuator housing must be incorporated to a port-helix metering injection pump. This acts as the control mechanism for managing fueling. As both the subject of governors and the rack actuators used in electronically managed versions of port-helix metering injection pumps are dealt with in detail elsewhere in this textbook, only a brief description is provided here. The only factor that determines the output of a diesel engine is the amount of fuel metered into its cylinders. Given unlimited fuel, a diesel engine is capable of accelerating at a rate of 1,000 rpm per second until self-destruction occurs. To prevent such an "engine runaway" and to provide the engine with a measure of protection from abuse to ensure that it meets the manufacturer's expectations of longevity, fuel quantity delivered to the engine cylinders must be precisely managed by a governor under all operating conditions. The fuel control mechanism on a port-helix metering injection pump is the rack. Therefore, the governor or rack actuator controls engine fueling by precise positioning of the rack.

Hydromechanical governors are not so often used to manage medium- and heavy-duty, on-highway diesel engines of today but when they are, they consist of a self-contained housing mounted to the rear of the injection pump. In a simple mechanical governor, the vehicle accelerator linkage connects to a throttle arm or governor input lever located on the side of the governor housing. Enclosed within the governor housing, a weight carrier is mounted to the rear of the camshaft. Flyweights are mounted in the weight carrier, loaded inboard by spring force. As the camshaft rotates, centrifugal force acting on the flyweights thrusts them outward. The faster the rotational speed, the more centrifugal force is generated. Countering the centrifugal force generated by the flyweights is an accumulation of spring forces. The amount of spring force is usually variable and is increased as the accelerator arm is moved. A thrust collar acts as an intermediary (that is, something in between two other components) between the mechanical force of the spring(s) and the centrifugal force generated by the flyweights. The fuel control mechanism or rack is connected to the thrust collar.

Spring force acting on the thrust collar attempts to increase engine fueling. Centrifugal force acts in opposition to the spring force and attempts to diminish engine fueling.

Maximum engine speed is defined by ensuring that at that set speed, the centrifugal force generated by the flyweights will overcome any amount of spring force the governor can counter it with. Even the simplest mechanical governor suitable for managing a highway truck engine will define idle speed, rated speed, and the droop curve (graduated engine defueling as engine speed rises from rated speed to the high-idle speed) leaving intermediate speed selection to the operator. A simple governor may also provide torque rise fuel enhancement and an **aneroid** (light pressure sensing device) to sense manifold boost and limit fueling until manifold pressure achieves a predetermined value.

Two main categories of hydromechanical governor exist. The first is the LS or limiting speed governor, which defines idle and high-idle engine speeds, leaving the intermediate speed ranges to the operator. The second category is the variable speed governor in which throttle arm position defines an engine rpm value; the governor then attempts to maintain this speed while engine loading varies. Hydromechanical governors are not capable of managing today's high-power engines that must achieve statutory and customer requirements of low emission operation and high fuel economy. They have been seldom used since 1991.

When the port-helix metering injection pump is managed by a computer, engine governing depends on how the ECM is programmed. The governor housing attached to the rear of the injection pump is replaced by a rack actuator housing consisting of switched ECM output devices and sensors. The housing contains a rack actuator, either a linear magnet (proportional solenoid) controlled directly by the ECM or an electric-over-hydraulic device (engine oil acts as the hydraulic medium) also ECM controlled. The rack actuator sets rack position on ECM command and nothing more. A rack position sensor reports the exact rack position to the ECM. Additionally, the rack actuator housing may house other sensors to report rotational speed (pulse wheel fitted to the rear of the camshaft), engine position, and timing data to the ECM. Rack actuator housings are not governors. Where a rack actuator housing is fitted to an in-line, port-helix metering injection pump, the engine governing functions are undertaken by the software programmed into the fuel management ECM.

Electronic governing offers infinitely more control over engine fueling and it allows OEMs to meet statutory noxious emissions requirements and achieve better fuel economy, but it additionally can be easily programmed/reprogrammed with customer data to tailor the engine for varying engine and chassis applications. This subject matter is dealt with in detail elsewhere in this text in chapters relating to partial authority, electronic engine management.

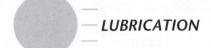

LUBRICATION

The lower portion of the port-helix metering injection pump is lubricated by engine oil. In older versions, the injection pump lubricating circuit was often isolated from that of the engine and the lube level would be checked and replenished through a dipstick located in the cambox. Contemporary injection pumps tend to be plumbed into the main engine lubricating circuit. The camshaft main bearings are often pressure lubed directly from an engine oil gallery, and the remainder of the cambox components are splash lubricated from the oil held in the sump.

The upper portion of the injection pump is lubricated by the diesel fuel being pumped through the charging gallery, so the lubricity of the diesel fuel is critical. Compromising the lubricating qualities of the fuel may cause premature failure and/or leakage and some manufacturers have reported such problems following the introduction of low-sulfur, low-lubricity fuels mandated by EPA standards. It is critical for the operation of both the fuel injection pump and the engine it manages that the fuel used to lubricate the upper portion of the pump never comes into contact with the engine oil in the cambox. Plunger-to-barrel clearance is lapped in manufacture to a minute tolerance and fuel that leaks by the recessed metering areas bleeds to an annular belt in the barrel. A duct connects the annular belt in the barrel to the charging gallery permitting bleed-by fuel to be routed there. This is known as a viscous seal. Trace leakage of a viscous seal can rapidly cause engine oil contamination and lead to lubricant breakdown. Viscous seals fail due to plunger side loading (single helix design), prolonged usage wear, and fuel contaminants—especially water. Fuel that bleeds by the metering recesses serves to guide the plunger true in the barrel bore minimizing metal-to-metal contact. When studying the results of engine oil analysis, it should be remembered that failure of the injection pump's viscous sealing ability is probably the least likely source of a fuel-in-engine oil condition.

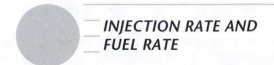

INJECTION RATE AND FUEL RATE

Injection rate is defined as fuel quantity injected per crank angle degree; it is therefore determined by the geometry of the pump element actuating cam profile. Most port-helix metering injection pumps use a symmetrical cam profile whose periphery (outer surface) is mostly inner base circle; this design holds the plunger at the bottom of its travel for most of the cycle. Occasionally asymmetrical cams and cams whose periphery is mostly outer base circle are used.

The shaping of the cam contour affects not only injection rate but also pump chamber breathing and cooling. A cam profile that is mostly outer base circle is sometimes known as a back-kick cam, which will prevent the engine from being started in reverse rotation.

Fuel rate is a term usually used to define the actual rate of delivering fuel to an engine cylinder, so it is factored by both cam profile and engine rpm.

TIMING ADVANCE AND VARIABLE TIMING MECHANISMS

Older port-helix metering injection pumps were usually directly driven by reduction gearing from the engine camshaft gear. Such a system would dictate that the **static timing** value (port closure) occur at the same number of crank angle degrees BTDC regardless of engine load or speed when lower helix geometry is used. Fuel economy and noxious emission considerations led to the development of first, mechanical advance mechanisms and more recently, electronically managed, variable timing.

Mechanical timing advance mechanisms are actuated using a set of flyweights and eccentrics to advance the drive angle of the pump camshaft in relation to the pump drive gear, using a spiral gear on a shaft. In other words, the position of the fuel injection pump relative to that of the engine is advanced in direct proportion to the centrifugal force generated by the weight carrier. Port closure was therefore advanced as engine rpm increased. The extent of the advance offered by mechanical advance mechanisms can be as little as 3 degrees crank angle and seldom more than 10 degrees crank angle (Figure 21–14).

Electronically managed variable timing devices such as those used by Mack and Caterpillar employ a variable timing coupling mechanism between the pump drive gear and the pump camshaft. This intermediary was designed to establish a variable timing window of up to 20 degrees crank angle, managed by the engine electronics. The static timing value (that is, port closure) specified is usually the most retarded parameter in the variable timing window. For example, an injection pump timed at 7 degrees BTDC with a 20-degree variable timing mechanism would permit the ECM to select any port closure value between 7 degrees BTDC and

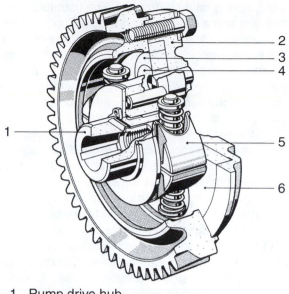

1. Pump drive hub
2. Advance assembly housing
3. Advance eccentric
4. Compensating eccentric
5. Flywheel assembly
6. Adjusting disc

FIGURE 21–14 Mechanical timing advance mechanism. (Courtesy of Robert Bosch Corporation)

27 degrees BTDC. Injection pump static timing (port closure) specs of as little as 4 degrees BTDC are used to limit combustion heat and therefore NO_x emission; substantially retarded injection timing does not generally enhance either performance or fuel economy. In fact, the tradeoff of retarded injection timing is an increase of HC emission but all emissions legislation is about keeping the exhaust noxious emissions within a window of acceptability. Both Mack and Caterpillar manage their variable timing devices in different ways, which are discussed in later chapters dedicated to these systems.

ANEROIDS

By definition, an aneroid is a low-pressure sensing device. In application, it is used on a turbocharged diesel engine to measure manifold boost and limit fueling until the boost pressure achieves a predetermined value.

Such devices are known variously as puff limiters, turbo-boost sensors, AFC valves and smoke limiters, and all seek to accomplish the same objective. They typically consist of a manifold within which is a diaphragm; boost air is piped from the intake manifold to act on the diaphragm. Such devices are used on most current **boosted engines** (turbocharged engines).

When an aneroid is used on an in-line, port-helix metering pump, it is usually a mechanism consisting of a manifold, spring, and control rod. The manifold is fitted with a port and a steel line connects it directly to the engine intake manifold. In this way, boost pressure is delivered to the aneroid manifold where it will act directly on the diaphragm within. Attached to the diaphragm is a linkage connected either directly or indirectly to the fuel control mechanism (rack); a spring loads the diaphragm to a closed position, which limits fueling by preventing the rack from moving into the full-fuel position. When manifold boost acting on the diaphragm is sufficient to overcome the spring pressure, it acts on the linkage permitting the rack full travel and thus maximum fueling. Such systems are easily and commonly shorted out by operators in the mistaken belief that aneroid systems reduce engine power. The emission of a "puff" of smoke from the exhaust stack at each gear shift point is an indication that the aneroid/boost sensing mechanism has been tampered with. Modern aneroids are designed to be more difficult to tamper with but nevertheless, aneroids are still commonly tampered with on hydromechanical engines (Figure 21–15 and Figure 21–16).

ALTITUDE COMPENSATOR

An **altitude compensator** device contains a **barometric capsule** that measures barometric pressure and on this basis downrates engine power at higher altitudes to prevent overfueling. They are required when running at higher altitudes because the oxygen density in the air charge decreases with an increase in altitude and, unless the fuel system is aware of this, it effectively overfuels the engine. The critical altitude at which some measure of injected fuel quantity deration is required is 1,000 ft., but in older ingines this may not actually occur until 3,000 ft. in altitude.

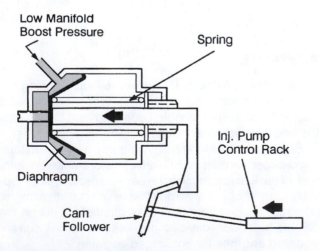

FIGURE 21–15 *Aneroid operation at low boost. (Courtesy of Mack Trucks)*

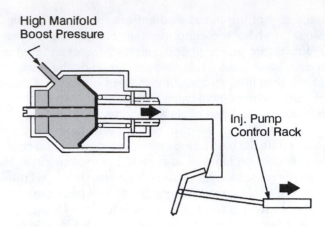

FIGURE 21–16 *Aneroid operation at high boost. (Courtesy of Mack Trucks)*

TIMING INJECTION PUMPS TO AN ENGINE

Port-helix metering injection pumps are timed to the engine they manage by phasing port closure on the #1 cylinder (usually in North American-engineered engines; check the specification—for a variety of reasons, the pump may be required to be timed to the #6 cylinder) to a specific engine position. All injection pumps must be accurately timed to the engine they will fuel. This usually means the phasing of pump port closure to a specific number of degrees BTDC on the cylinder to be timed to a specification that seldom can be outside of 1 degree crank angle and may have to be within ¼ degree crank angle. Methods used to time the injection pump to the engine vary by OEM. Actual spill timing of the pump to the engine is a procedure that has become effectively obsolete; however, it is important that the diesel technician understand this procedure, which can be used in the absence of OEM tooling.

TECH TIP: Most injection pumps on North American engines are timed to the #1 engine cylinder but not all, so watch for those that are not. The #6 cylinder (on an in-line 6) is the next most common but assume NOTHING. Always check the specifications in the work shop manual.

SPILL TIMING PROCEDURE

Before beginning the **spill timing** procedure, check the engine (and OEM chassis) manual for positioning of the fuel control lever, stop fuel lever, brake valve, and gear shift lever position. See Photo Sequence 4.

PHOTO SEQUENCE 4
Chapter 21 Spill Timing for an In-Line, Port-Helix Metering Injection Pump

PS4–01   Check the pump port closure specification on the injection pump ID plate, noting which engine cylinder the pump is timed to. Locate the engine in the correct position for the spill timing procedure by observing the valve rocking action on the companion cylinder to the one being timed. Position the engine to ± 20 before the port closure specification value.

PS4–02   The injection pump in the photograph is timed on its #1 cylinder. Remove the hydraulic nut on the #1 high-pressure pipe and move the line out of the way; it will probably be necessary to remove some line clamps to enable this.

PS4–03   Remove the delivery valve body and the delivery valve core and spring: ensure that no dirt gets into the exposed injection pump chamber.

PS4–04   Insert a spill tube (an old injection high-pressure pipe, cut and shaped to a gooseneck) into the delivery valve body in place of the delivery valve. Actuate the hand primer pump to ensure that fuel exits from the spill tube. Capture the fuel exiting the spill tube in a container.

PS4–05   Next, manually bar the engine slowly in the correct direction of rotation . . .

PS4–06   . . . while actuating the hand primer pump. The exact point of injection pump, spill cutoff is achieved when the stream of fuel exiting the spill tube breaks up into droplets. This occurs when the leading edge of the plunger rises in the barrel bore to begin to cut off the spill port.

PHOTO SEQUENCE 4
Chapter 21 Spill Timing for an In-Line, Port-Helix Metering Injection Pump (continued)

PS4–07 Spill cutoff. Be sure that when the hand primer pump is actuated, droplets of fuel still exit the spill tube. If no fuel exits the spill tube, the engine has been barred past the point of injection pump port closure: if this is so, back engine up at least 20 degrees before the port closure specification and repeat the previous steps.

PS4–08 After locating the port closure on the pump, check the engine calibration plate. On this engine, the calibration plate and pointer are located on the vibration damper: in some engines the calibration plate is on the flywheel. If the engine location reading is within 1 crank degree of the port closure specification, assume the engine to be properly timed. If not . . .

PS4–09 . . . remove the pump drive gear cover plate located directly in front of the injection pump.

PS4–10 Loosen the pump drive gear to pump drive plate fasteners: this should be loose enough to permit the engine to be rotated independently of the gear pump.

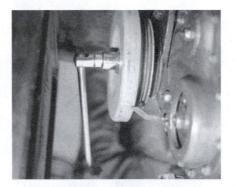

PS4–11 Carefully bar the engine to the correct port closure location on the engine calibration plate: the objective is to rotate the engine while the injection pump remains stationary in its location port closure at #1 pump element.

PS4–12 With engine in the correct location, torque the pump drive gear to pump drive plate fasteners. Then repeat the sequence outlined in steps 4 through 8 to confirm that the injection pump is properly timed to the engine.

PS4–13 When the port closure specification has been confirmed, remove the spill tube and carefully reassemble the injection pump delivery valve, reconnecting the high-pressure pipe.

1. Check the injection pump data plate for port closure value. Manually bar engine in the direction of rotation to position the piston within the #1 cylinder on its compression stroke. Locate the engine calibration scale, usually to be found on the front pulley, vibration damper, or on the flywheel. Position the engine roughly 20 degrees before the port closure specification.

2. Remove the high-pressure pipe from the delivery valve on the injection pump #1 cylinder. Unscrew the delivery valve body and remove the delivery valve core and spring. Replace the delivery valve body and install a spill tube. A discarded high-pressure pipe neatly cut and shaped to a goose neck will suffice. When the hand primer pump is actuated, the charging gallery will be pressurized. The amount of pressure created by the hand primer pump will be insufficient to open the delivery valves, so it will exit through the spill tube fitted to the #1 cylinder. The fuel should exit in a steady stream, and it should be captured in a container held under the spill tube. Next, slowly and smoothly bar the engine in its direction of rotation observing the stream of fuel exiting the spill tube.

    When the plunger leading edge rises to trap off the spill port, the steady stream of fuel exiting the spill tube will break up first into droplets and then cease as the plunger passes the spill port.

    The objective of this step is to locate the pump precisely at port closure—this means that the flow area at the spill port should exist but be minimal; two to six drops per 10 seconds should be within the specification window. Ensure that the pump has not been barred past port closure cutting the fuel off altogether as it is impossible to determine how much beyond port closure the plunger has travelled.

3. Next, check the engine calibration scale. The specification is typically required to be within 1 degree crank angle of the port closure (PC) specification. If this is so, the pump can be assumed to be correctly timed to the engine. If it is not, proceed as follows.

4. Remove the accessory drive cover plate. Loosen the fasteners that couple the pump drive gear to the pump drive plate. Bar the engine to the correct PC specification position. By uncoupling the pump drive plate from the pump drive gear, it is hoped that the pump will remain stationary at PC on the #1 cylinder while the engine is barred independently. When the engine is in the correct position, torque the fasteners that couple the pump drive gear to the pump drive plate. Back the engine up roughly 20 degrees before the PC specification and then repeat steps 1 through 3.

Every diesel technician should be acquainted with the foregoing procedure even though he/she may seldom practice it. Spill timing can also be performed using compressed air as the test medium; the setup procedure is the same as just listed, but regulated compressed air is ported into the charging gallery. Instead of a spill tube, a flexible hose is connected to the pump element used for the spill timing procedure and immersed in a water-filled glass jar; when air is supplied to the charging gallery, it will pass through the barrel spill ports and exit through the flexible hose producing bubbles in the glass jar. As the pump is rotated toward the port closure location and the plunger leading edge starts to trap off the spill port, the stream of bubbles will turn into smaller bubbles produced less frequently. Using air to spill time injection pumps is probably the least recommended method. The air supply should always be equipped with an air dryer/filter assembly; and even when it is so fitted, the danger of moisture and other air suspended contamination does not warrant the risk of using this method. OEMs prefer alternative methods of static timing of port-helix metering injection to be used that do not require the removal and disassembly of the delivery valve. Following are some of the alternatives.

1. Timing pin. This is probably the simplest method and least likely to present problems. The engine is located to a specific position by inserting a timing bolt usually in the cam gear but sometimes in the flywheel. Similarly, the injection pump is pinned by a timing tool to a specific location. The injection pump is always removed and installed with the timing tools in position. It goes without saying that the timing tools must be removed before attempting to start the engine.

2. High-pressure pump. This involves connecting a high-pressure timing pump into the circuit. This portable electric pump charges the charging gallery at a pressure in excess of the 20 atms required to crack the delivery valves, which causes them all to open; consequently the injector leakoff system and pump gallery return must be plugged off. Next, the high pressure pipe on the #1 cylinder is removed at the pump, and a spill pipe discharging into the portable pump sump is fitted. The procedure then replicates that used to spill time the injection pump—the engine must be correctly located well before the PC specification (to eliminate engine gear backlash variables) and then barred to spill cutoff.

    If the settings are out of specification, they are rectified by altering the coupling location of the pump drive gear to the pump drive plate.

3. Electronic. This method tends to be used with later generation hydromechanical and electronically

managed, port-helix metering injection pumps. The static timing value is checked with an electronic timing tool that senses the positioning of a raised notch located on the pump camshaft. The procedure is outlined in some detail using a specific injection pump and engine in Chapter 37.

Injection pump-to-engine timing is critical. It should be checked each time a pump is reinstalled to the engine and as a step in performance complaint troubleshooting. Pump timing may be checked using a diesel engine timing light. This consists of a transducer that clamps to the high-pressure pipe and signals a strobe light when it senses the line pressure surge that occurs when the delivery valve opens. The test is not an accurate one and the values read on the timing light should not be confused with the pump manufacturer's port closure specification because the pressure rise pulse used to trigger the light occurs after port closure. The test has some validity as a comparative test (using data obtained from other engines in the same series, timed at the same port closure value) and is a good means of verifying the operation of timing advance mechanisms.

TECH TIP: Although approved by manufacturers of older automotive diesel engines, using a timing light is not considered to be a sufficiently accurate check of larger diesel engine timing. Limit the use of a timing light to verify the operation of timing advance mechanisms.

Removing a Port-Helix Metering Injection Pump from an Engine

Probably close to half the port-helix metering injection pumps that are removed from engines have nothing wrong with them. This is due to poor field diagnostics caused by a low level of understanding of the operating principles of injection pumps combined with inaccurate interpretation of service literature. Another factor is a tendency of field technicians to black-box fuel pump technology as the responsibility of the pump room technician and as a result, package up problems for someone else to repair. The correct location to diagnose most fuel injection pump problems is with the pump on the truck engine. If the OEM recommended on-board tests are performed before removing the fuel injection pump, then if the pump has to be removed from the engine, the job of the pump room technician has been made easier. When it has been determined that the injection pump is responsible for a fuel problem and it must be removed, the procedure should be as follows:

1. Power wash the injection pump and surrounding area of the engine.

2. If the fuel pump and engine are equipped with locking/timing pins, position the engine to install them.
3. Remove the fuel supply and return lines. Remove the lubricating oil supply and return lines.
4. Remove the accelerator and fuel shutoff linkages (if equipped) or the electronic connector terminals.
5. Disconnect the high-pressure pipes from the delivery valves. Ensure that the high-pressure pipes can be moved away from the delivery valves without forcing or bending them; this usually means releasing insulating and support clamps. Cap both the high-pressure pipe nipples and the delivery valves.
6. Unbolt the fasteners at the pump mounting flange. Depending on the pump, the pump may have to be removed separately from the variable timing/advance timing mechanism. Some pumps may require that the pump drive gear be separated from the pump drive plate or the variable timing device from the front of the pump drive gear, requiring the removal of the timing gear cover plate. Most pumps should slide back easily after this but care should be taken not to support the weight of the injection pump on its drive gear.

Reinstallation

Essentially, the foregoing procedure is reversed but the port closure timing should be confirmed using the OEM-recommended method (outlined earlier in this section). When installing the pump, ensure that it is installed with the pump drive gear/plate correctly registered with the engine timing gear. Resistance can be an indication that the pump is being installed a tooth out of register and contacting a dowel or key; never force an injection pump into position. When resistance is encountered, remove the pump and check both the pump position and its drive mechanism for problems.

CAUTION: If resistance is encountered while installing a fuel injection pump, remove and check for the cause. Forcing a pump home on its mounting flange using the fasteners can damage the pump drive and will almost certainly result in an out-of-time pump.

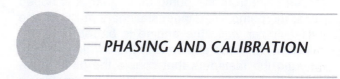

PHASING AND CALIBRATION

Calibration and phasing of port-helix metering injection pumps is the responsibility of the manufacturer trained pump technician; these operations require the use of specialized tooling, a comparator bench, and the manufacturer's technical data. However, the diesel technician should be capable of defining the procedures.

PHASING

Phasing the injection pump sets the phase angle between the individual pump elements, essentially to ensure that port closure on each occurs at a precise spacing. On a six-cylinder engine this would be exactly 60 degrees. Phasing the injection pump ensures balanced, multicylinder engine timing. It is checked on the comparator bench using a degree wheel (protractor located on the comparator drive turret) and adjusted by varying either the barrel height or shimming the plunger tappets. The critical specification is the *lift to port closure*, a measure of the positioning of the plunger in relation to cam lift. Depending on the type of pump used, either the tappet height or the barrel height is adjusted to ensure that port closure in each pump element is phased to occur at a precise interval. A typical method of phasing a port-helix pump is outlined in the following procedure. A dial indicator assembly fitted with a spill tube is mounted in place of the delivery valve on the cylinder to be phased; when a hand primer pump is actuated and the pump rotated, fuel will be cut off by the leading edge of the plunger at port closure. The dial indicator probe rides on top of the pump plunger and thus its position relative to the moment of port closure can be determined. An adjustment would require the raising or lowering of either the barrel height (by means of split shims) or the raising or lowering of the tappet height (by means of phase angle shims or an adjusting screw). For a pump required to fuel a six-cylinder engine, port closure in each of the six cylinders should be phased to be exactly 60 degrees apart. Using the degree wheel as a reference, port closure in each cylinder would be set in engine firing sequence at the required interval. Phasing can be also checked/ adjusted using the spill method in conjunction with the degree wheel on the comparator bench turret. It should be noted that primary phasing is the responsibility of the camshaft geometry and a failing cam profile will result in a phasing problem.

CALIBRATION

Calibrating the injection pump dynamically balances fuel delivery quantity output from the individual pumping elements. The injection pump is mounted to a comparator bench with sufficient power to handle the torque required to drive the pump when it is delivering full fuel loads. The output from each pump element is measured in calibrated **vials** (graduates/burets) and often displayed on a video monitor or cathode ray tube

(CRT). In a typical test sequence, fuel output is recorded through a sequence of specified rotational speeds and simulated fuel demands. Fuel quantity output in a pump element is adjusted by one of two methods. In certain fuel pumps, the barrels are manufactured with upper flanges that are located by means of eccentric stud holes; these permit some rotation of the barrels in their mounting bores. This rotation turns the spill ports relative to the plunger geometry and therefore alters plunger effective stroke. Another method used to balance fueling in port-helix pump elements is the use of control sleeve lock collars to clamp the plungers. When the lock collar is loosened, the plunger can be rotated within and retorqued to adjust the delivered fuel quantity.

BENCH TESTING

Comparator bench testing is a specialty. It requires manufacturer training and a properly equipped fuel injection pump repair facility to develop the techniques required to test fuel injection pumps. It also requires the manufacturer's service literature and specifications. The term *comparator* is used to describe the test bench because the performance output values of the pump being tested are *compared* directly with the manufacturer's specifications. A typical comparator bench procedure requires a test sequence that includes:

1. Phasing
2. Full-fuel quantity calibration
3. Peak torque rpm fuel quantity calibration
4. Droop calibration
5. High-idle rpm calibration
6. Idle speed calibration
7. Cranking fuel calibration
8. Start retard fueling
9. Aneroid operation and adjustment
10. Supply pump charging pressure

TECH TIP: Never be tempted to perform any adjustment on an in-line port-helix metering pump that should be performed on a comparator bench. Both specialized training and equipment are required to perform internal pump adjustments, and the cost of attempting to adjust fuel pumps outside of the manufacturer specifications can be the price of a replacement engine.

CRITICAL SYSTEM PRESSURE VALUES

The following is a review of some of the system values in a typical port-helix metering injection pump system. Much of the language of fuel injection systems stems from this original high-pressure injection pump system and is used in many of the systems that have evolved from it.

Charging Pressure

Charging pressure is generated by the transfer pump. It is the pressure in the injection pump charging gallery usually between 1 to 5 atms (15–75 psi), and it varies according to the system.

Delivery Valve Crack Pressure

When there is no residual line pressure in the high-pressure pipe, pressure equivalent to about 300 psi (20 atm) is required to crack the delivery valve that is held closed only by the delivery valve spring. This plus the residual line pressure is required to crack the valve when the engine is running.

Residual Line Pressure

Residual line pressure is the pressure the dead volume fuel is retained at in the high-pressure pipe. It is usually about two-thirds of the NOP value.

Nozzle Opening Pressure

Also known as popping pressure, NOP is the pressure at which the nozzle valve unseats in a hydraulic injector. NOP parameters range between 2,200 and 5,100 psi (150 to 375) atms in current bus and truck multi-ori-fii nozzles. It denotes the beginning of injection.

Peak Pressure

Peak pressure is the highest pressure a system can generate. In most systems, the pressure increases as the injection pulse is prolonged, so peak pressure will be attained only at full fuel. Peak pressure parameters range between 2 and 10 times NOP values.

DELIVERY, INJECTION, AND COMBUSTION

The function of any port-helix metering injection pump is to manage the fueling of an engine. This section attempts to match the critical events in the injection pump with the critical events in the engine; in other words, to describe how the fueling pulse effects combustion in the engine cylinders (Figure 21–17).

INJECTION LAG

Injection lag is the time measured in crank angle degrees between port closure in the injection pump and

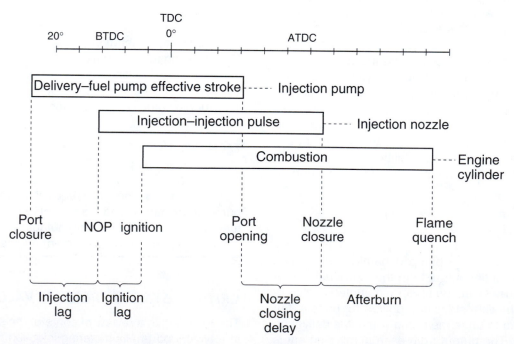

FIGURE 21–17 *Graphic showing the phasing of the events of delivery, injection, and combustion.*

NOP, accounted for primarily by the time required to raise pump chamber pressure first to the residual line pressure value, then to the NOP value. Secondary causes of injection lag are:

1. The elasticity of the high-pressure pipe.
2. The compressibility of the fuel. This is approximately 0.5% per 1,000 psi (67 atms) but it must be accounted for when the system is engineered.

Injection lag can be regarded by the diesel technician as a constant.

IGNITION LAG

Ignition lag is the time period between the events of NOP and ignition of the fuel charge. It is usually measured in crank angle degrees at any given engine rpm and load. Liquid droplets exiting the nozzle orifii must be vaporized and ignited. This time period is variable and depends on the ignition quality of the fuel (CN value) and the actual compression temperature. Because it is variable, so will be the amount of fuel in the cylinder at ignition. In cold weather startup conditions, this can be excessive and can result in a detonation condition known as diesel knock.

COMBUSTION

The duration of combustion depends on the length of the injection pulse (that is, the total quantity of fuel injected to the cylinder and the point in crank angle degrees when injection ceases) and the engine rpm. The rate of injection is optimumly designed so that cylinder pressure and crank angle leverage are synchronized to deliver power to the flywheel as smoothly as possible.

NOZZLE CLOSURE LAG

Nozzle closure lag is the time period between the end of injection pump delivery and actual nozzle closure. This varies, increasing as the injection pressure pulse increases.

AFTERBURN

Afterburn is the normal combustion of fuel in the engine cylinder after injection nozzle closure. Its duration depends on the length of the injection pulse, the actual quantity of fuel in the cylinder, and many other factors including droplet sizing just before nozzle closure. Afterburn duration is managed to be as short as possible in later versions of hydromechanical, port-helix injection pumps by delivering higher peak pressures and using smaller injector nozzle orifii sizing, the result of which are smaller droplets of fuel that oxidize more quickly.

SUMMARY

- High-pressure, liquid fuel injection to a diesel engine has been used since Robert Bosch designed the first system in 1927.
- The hydromechanical pumping apparatus using inline port-helix metering injection pumps has changed little since 1927.
- Management of the port-helix metering injection pump has evolved from hydromechanical governing to electronic governing.
- Most in-line, port-helix metering injection pumps are flange mounted to the engine block and gear driven at camshaft speed.
- The port-helix metering pump is driven through one full rotation (360 degrees) per full effective cycle of the engine (720 degrees in a four-stroke cycle).
- The pump camshaft is supported by main bearings and driven in the cambox, which also acts as a lubrication sump.
- Pump element actuating tappets are spring loaded to ride the cam profiles.
- Cam geometry dictates the pump element activity.
- There is a pump element dedicated to each engine cylinder.
- A pump element consists of a stationary barrel and a reciprocating plunger.
- The plunger is milled in manufacture with a metering recess known as a helix or scroll.
- Plunger rotational position determines the point of register of the barrel spill port and the helix.
- The plungers are rotated in unison by a toothed rack meshed to slotted control sleeves themselves lugged to the plungers.
- Plunger effective stroke begins at port closure and ends at port opening.
- Delivery valves separate the pump elements from each high-pressure pipe and act to retain dead volume fuel at pressure values approximating two-thirds NOP.
- Delivery valves are designed to seal before they seat.
- Delivery valves increase the volume available for dead volume fuel storage in the high-pressure pipe by the swept volume of the retraction collar.
- Injection rate is a term defined as fuel injected per crank angle degree.
- Most current port-helix metering injection pumps have a variable timing mechanism that acts as an intermediary between the pump drive gear (on the engine) and the pump camshaft coupling.
- Hydromechanically managed injection pumps often incorporate an aneroid device and an altitude compensator to prevent more fuel being injected to a engine cylinder than there is oxygen to burn it.

- In-line port-helix metering injection pumps must be accurately timed to the engine.
- Timing an injection pump to an engine synchronizes the activity in the pump to that in the engine.
- The manufacturer's procedure and specifications must be strictly adhered to when timing an injection pump to an engine.
- The diesel technician should understand how to spill time an injection pump to an engine even though the practice is considered generally obsolete.
- It is critical that the diesel technician understand basic combustion technology and the relationship between the activity in the injection pump, injection nozzle, and the engine cylinder.
- Peak pressure is the highest pressure value that an injection pump is designed to produce.

REVIEW QUESTIONS

1. The beginning of the injection pump effective stroke is known as:
   a. Port closure
   b. Port opening
   c. Injection lag
   d. Afterburn
2. The ending of injection pump effective stroke is known as:
   a. Port closure
   b. Port opening
   c. NOP
   d. Flame quench
3. The device used to rotate the injection pump plungers in unison is known as a:
   a. Barrel
   b. Tappet
   c. Rack
   d. Control sleeve
4. Charging pressures in a port-helix metering injection pump are created by a(n):
   a. Supply pump.
   b. Accumulator
   c. Vane pump
   d. Centrifugal pump
5. What component replaces the governor housing at the rear of a port-helix metering injection pump if it is managed by a computer?
   a. Electronic governor
   b. Rack actuator housing
   c. ECM
   d. Module
6. Which of the following will define injection rate?
   a. Engine speed
   b. Fuel demand

   c. Cam profile geometry
   d. NOP
7. Static timing can also be referred to as:
   a. Port closure timing
   b. NOP timing
   c. Point of ignition
   d. Completion of injection
8. Which of the following components would limit engine fueling at high altitudes?
   a. Barometric capsule
   b. Aneroid
   c. Governor
   d. Variable timing
9. A port-helix metering injection pump managing a direct-injected diesel engine would typically be port closure timed to what position on the engine?
   a. 50 degrees BTDC
   b. 50 degrees ATDC
   c. 20 degrees BTDC
   d. 20 degrees ATDC
10. A common cause of *diesel knock* is:
    a. Prolonged injection lag
    b. Afterburn
    c. Prolonged ignition lag
    d. Preignition
11. *Technician A* states that when static timing a port-helix metering pump to an engine, the pump #1 element must always be timed to the #1 engine cylinder. *Technician B* states that the spill timing method is seldom recommended by OEMs to time current port-helix metering pumps to engines. Who is right?
    a. A only
    b. B only
    c. Both A and B
    d. Neither A nor B
12. When a port-helix metering injection pump that requires to be locked into position by a timing pin is to be installed to an engine, what else must also be done?
    a. The pump should be high-pressure timed.
    b. The engine should be locked into position by a timing bolt.
    c. Timing should be checked with a pulse actuated timing light.
    d. Spill timing.
13. At what speed is a port-helix metering injection pump driven on a four-stroke cycle engine in relation to engine speed?
    a. ¼ engine speed
    b. ½ engine speed
    c. Engine speed
    d. 2 times engine speed
14. 360 degrees of pump rotation is equal to how many degrees of engine rotation?
    a. 120 degrees

b. 180 degrees

c. 360 degrees

d. 720 degrees

15. Which type of cam geometry is used to actuate the high-pressure pump elements in most in-line, port-helix metering injection pumps?

a. Symmetrical of mostly base circle design

b. Asymmetrical of mostly outer base circle design

c. Antikick back

d. Double lobe

16. *Technician A* states that the phasing of an in-line, port-helix metering pump involves the correct spacing of port closure in each pump element through its effective cycle and is usually performed with a degree wheel (protractor) on the comparator bench drive turret. *Technician B* states that calibration of an in-line, port-helix metering injection pump can be performed on the engine with the appropriate governor adjusting tools. Who is right?

a. A only

b. B only

c. Both A and B

d. Neither A nor B

17. What is the function of the flutes machined into a typical delivery valve core?

a. Guide the valve in the body

b. Permit the valve to seal before it seats

c. Define the residual line pressure

d. Define valve core travel in the valve body

18. What is the function of the retraction collar/piston on the delivery valve core?

a. Guide the valve in the body

b. Permit the valve to seal before it seats

c. Limit pressure wave reflection

d. Limit back leakage to the pump chamber

19. Which of the following does most to define the actual residual line pressure value in a running diesel engine using a port-helix metering PLN fuel system?

a. The swept volume of the retraction collar/piston

b. The mechanical force of the delivery valve spring

c. The dead volume fuel

d. The length of the high-pressure pipe

20. An electronically managed port-helix metering injection pump has ECM-managed variable timing with a 20-degree window, statically timed at 8 degrees BTDC. What is the timing window within which the ECM can select port closure?

a. 28 degrees BTDC to 8 degrees BTDC

b. 8 degrees BTDC to 12 degrees ATDC

c. 8 degrees BTDC to 28 degrees ATDC

d. 12 degrees ATDC to 8 degrees BTDC

Chapter
22

Detroit Diesel
Mechanical Unit
Injection

OBJECTIVES

After studying this chapter, you should be able to:

- Identify a DDC mechanical engine by interpreting the specification plate.
- Describe the layout and components in a Detroit Diesel Corp. (DDC) fuel subsystem used on 53, 71, and 92 series engines.
- Outline the principles of operation of a DDC mechanical unit injector.
- Identify the governor types used on highway applications of DDC engines.
- Describe how MUI effective stroke is varied to control injected fuel quantity.
- Describe the components that link the MUIs with the governor assembly.
- Outline the procedure for performing DDC tune up.
- Test system charging pressure to specification.
- Perform MUI fuel timing and cylinder head valve adjustment.
- Outline the procedure for balancing the MUIs.
- Perform basic fuel system troubleshooting.
- Perform a DDC two-stroke cycle engine tune up.

KEY TERMS

bell crank	governor differential lever	Roots blower
buffer screw	governor gap	spill deflector
bushing	governor weight forks	starting aid screw
clevis	jumper pipes	tailored torque (TT)
control rack	mechanical unit injector (MUI)	throttle delay
control rod	metering recesses	timing dimension
crown valve nozzle	needle valve nozzle	unit injector
effective stroke	plunger	valve bridges
follower	rack clevis	
gear pump	rack lever	

INTRODUCTION

Detroit Diesel Corp. hydromechanical, two-stroke cycle engines have not been EPA certifiable for use on North American highways since 1991. However, they have been the bus and coach power plant of choice for nearly 50 years and due to the tendency of transit corporations to recondition components many times over, these engines will be with us for a few more years. Detroit Diesel two-stroke cycle engines enjoyed some popularity as a truck engine; this popularity peaked in the 1970s (I6–71, 8V–71, and 12V–71) and declined through the 1980s (8V–92). Detroit Diesel was owned by General Motors until 1987 when the division was purchased by Penske Corporation; in 2001, Daimler-Chrysler assumed ownership.

In a **mechanical unit injector-**fueled engine, each engine cylinder has its own **unit injector**, essentially a rocker-actuated pumping and metering element and hydraulic injector nozzle combined in one unit. Each mechanical unit injector or **MUI** has fuel delivered to it at charging pressure, variable between 30 and 70 psi (200 kPa–470 kPa) depending on engine speed. Charging pressure is generated by a positive displacement **gear pump**, driven by the **Roots blower** drive shaft, and responsible for all movement of fuel through the fuel subsystem. Engine output is controlled by a mechanical governor (in most truck and bus applications) that regulates fuel quantity by controlling the unit injector fuel racks.

DDC mechanical unit injection system engines were engineered using the English measurement system and most of the technical specifications are listed by DDC in standard-only values. As a consequence, standard system values are used where specifications are referred to in this chapter.

ENGINE IDENTIFICATION

See Figure 22–1. The following lists governor acronyms used in truck and bus applications:

    LS—Limiting speed
    VS—Variable speed
    DW—Double weight
    SW—Single weight
    TT—Torque tailored

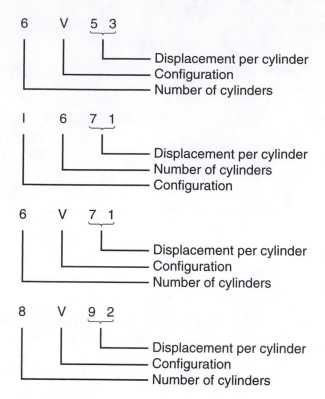

FIGURE 22–1   *DDC engine identification.*

FUEL SYSTEM COMPONENTS

The layout and most of the fuel system components are almost identical for 53, 71, and 92 series engines. The components listed in the fuel subsystem are for typical applications; some differences may occur in specific chassis requirements of the fuel subsystem such as in bus, coach, vocational, and truck applications due to such factors as distance of engine from fuel tanks and controls.

FUEL FILTERS

Most DDC two-stroke cycle engines with MUIs use a typical diesel fuel filter configuration consisting of two filters, one of which is under suction and the other under charging pressure produced by the gear pump.

Primary Filters

The primary filter is located in series between the fuel tank and the engine-mounted gear pump; this filter is under suction. Older applications used nondisposable canisters within which was a disposable element. This element was surrounded by a cloth enclosure and the flow of fuel was from the outside of the element to the inside, with filtered fuel exiting the center of the assembly. Most newer engines use disposable spin-on cartridges. Older engines may be retrofit with filter mounting pads that use the spin-on cartridges for ease of servicing and higher filtering efficiencies. The filter mounting pads are fitted with an inlet check valve to prevent fuel siphoning (drain back) to the tank.

Secondary Filters

The secondary filter is located in series between the gear pump and the inlet port in the cylinder head fuel manifold. Secondary filters on most current engines are disposable, spin-on filter cartridges, although some of the older canister and element filters can still be seen. Current spin-on filter cartridges entrap particulate down to 1 micron and inhibit water from passing through the filtering media to the extent that they will plug and starve the engine for fuel.

GEAR PUMP

The transfer pump is an external gear pump. It is an engine-driven, positive displacement pump responsible for charging the unit injectors with pressures between 30 and 70 psi (200–470 kPa). It is driven by the Roots blower drive shaft.

FUEL LINES AND FUEL MANIFOLD

Hydraulic hose is used to connect the fuel tank with the filters, pump, and deliver fuel to the fuel manifolds. In 53, 71, and 92 Series, the fuel manifolds are drillings within the cylinder head(s) that deliver fuel to the unit injectors at charging pressure and return fuel back to the tank(s). A restriction fitting is located at the exit port of each return fuel manifold. This defines the flow area that establishes the charging pressure window. A relief valve in the transfer pump prevents the charging pressure from exceeding 70 psi.

JUMPER PIPES

Jumper pipes connect the fuel manifolds in the head with the unit injectors; one charges the unit injector, the other returns fuel to the return manifold. Fuel is cycled through the system whenever the engine is running; in other words, all of the fuel delivered to the MUIs is not used for fueling the engine. Fuel cycled through the MUI circuitry is used for cooling and lubricating the internal components of the assembly.

TECH TIP: The number one cause of oil dilution by fuel in DDC 2-stroke engines is improperly installed jumper pipes. Consult DDS service literature and inspect for crossed trheads and leaks behind the pipe nuts.

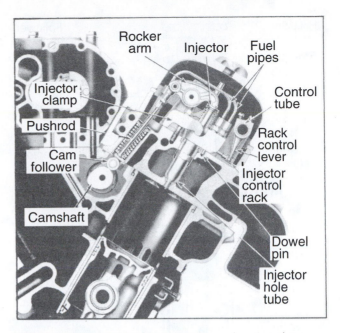

FIGURE 22–2　*DDC engine cutaway showing the location of the MUI. (Courtesy of DDC)*

DDC MECHANICAL UNIT INJECTORS

A mechanical unit injector combines a complete pumping, metering element and a hydraulic injector nozzle in a single, cam-actuated unit. There are two types differentiated by nozzle valve design, **crown valve nozzle** and **needle valve nozzle**. However, the crown valve type with NOP values of 800 psi can be considered obsolete and is not discussed here. The needle valve type was introduced in the early 1970s and is almost universal in DDC mechanical engines in current use (Figure 22–2 and Figure 22–3).

DESIGN

The **plunger** and **bushing** within the unit injector can be compared to the pumping element in an in-line, port-helix metering pump. The bushing is stationary and machined with upper and lower ports, 180 degrees offset. The plunger reciprocates within the bushing. It is actuated by the engine camshaft by a cam profile that is mostly inner base circle and an injector train consisting of a **follower**, rocker, and pushrod. The plunger is milled (a machining process) with helical **metering recesses** and is center and cross drilled.

The unit injector follower is lug connected to the plunger and the injector follower spring serves to load the plunger to ride its actuating cam profile. A gear positioned over the upper portion of the plunger is tooth-

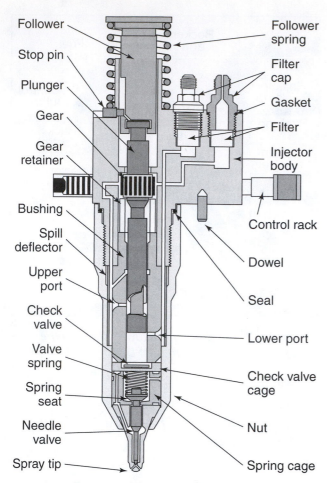

FIGURE 22-3 *Sectional view of a DDC mechanical unit injector.*

meshed with the **control rack**, permitting the plunger to be rotated within the bushing when the rack is moved linearly. Surrounding the bushing is a **spill deflector**; this is a stellite alloy sleeve whose function is to prevent high-velocity spilled fuel from eroding the injector body. Below the pumping element formed by the plunger and bushing is a needle valve assembly; ducting connects the pump chamber with the nozzle assembly. The needle valve is a simple multi-orifii, hydraulic injector nozzle. Spring pressure will determine NOP values that will be within the range of 2,200–3,400 psi.

OPERATION

For most of the cycle when the injector train is riding on the inner base circle of its actuating cam, the MUI plunger will be retracted by the injector spring, which also serves to load the injector actuating train, extending from the rocker to the cam follower. Fuel flows into the MUI from the supply jumper pipe, passes through the lower bushing port charging the pump chamber, passes up through the plunger center and

cross drillings, charging the recessed metering helices, and exits at the upper bushing port. From the upper bushing port, the fuel flows through ducts to exit the MUI by means of return jumper pipes.

Actual plunger stroke is defined by cam profile geometry and therefore will not vary. At each rotation of the camshaft, each MUI plunger is driven through a single stroke that begins when the cam ramps off its base circle toward peak lift. Cam peak lift represents the maximum point of downward travel of the plunger into the bushing pump chamber. After peak lift, the cam ramps back toward cam base circle and the injector spring lifts the plunger back into its retracted position.

The term **effective stroke** is used to describe that portion of the plunger stroke where fuel is actually being pumped. The actual fuel quantity injected to the engine's cylinders fueling is therefore the result of the MUI effective stroke. The length of the effective stroke depends on the rotational position of the plunger, which is controlled by the MUI rack. Specifically, it depends on where the recessed metering helices register with the upper and lower ports machined into the bushing. For an effective stroke to occur, both bushing ports must be closed for some part of the downward stroke of the plunger. When the injector cam rotates off base circle, the cam ramp loads the injector train and the rocker arm drives the plunger through its downward stroke within the bushing. As the plunger begins to move, its leading edge first closes the lower bushing port. As the plunger descends through its stroke, fuel in the pump chamber is displaced, passing through the plunger center and cross drillings and exiting the upper bushing port. Fuel continues to be displaced by the plunger until the helical edge of the plunger metering recess closes off the upper bushing port; because the lower bushing port has already been closed, the fuel is trapped in the pump chamber, and the effective stroke begins. Effective stroke or pumping stroke will continue until the lower edge of the metering recess is exposed to the lower bushing port; at this point, the high-pressure fuel in the pump chamber spills, exiting the pump chamber through the now exposed lower bushing port (Figure 22–4).

The length of the effective stroke depends on the rotational position of the plunger and the points at which the plunger metering recesses register with the upper and lower bushing ports. Because an effective stroke can only take place when both bushing ports are closed off, to shut down the engine, the plunger rotational position must be such that one of the bushing ports is always exposed. To obtain no-fuel, the plunger must be rotated to a position of register where the lower bushing port is exposed before the upper bushing is closed off. At no-fuel, the fuel in the MUI pump chamber is displaced through the entire plunger stroke; as the plunger descends into the pump chamber, fuel is forced up

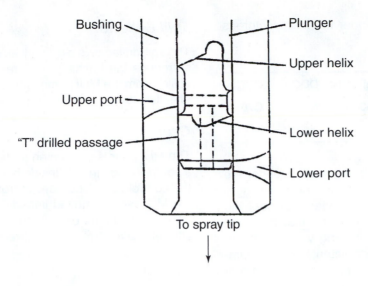

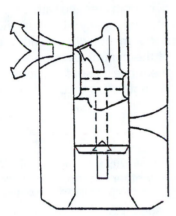 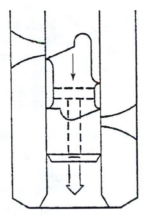 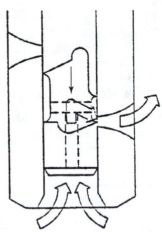

Injector Operation: The plunger descends, first closing off the lower port and then the upper. Before the upper port is shut off, fuel being displaced by the descending plunger may flow up through the "T" drilled hole in the plunger and escape through the upper port.

After the upper port has been shut off, fuel can no longer escape and is forced down by the plunger and sprays out the tip.

As the plunger continues to descend, it uncovers the lower port, so that fuel escapes and injection stops. Then the plunger returns to its original position and awaits the next injection cycle.

FIGURE 22–4 *Effective stroke in an MUI with variable beginning and variable ending plunger geometry (Courtesy of DDC)*

through the plunger center and cross drillings to exit at the bushing ports.

The geometry (shape) of the metering recesses/helices determines the injection timing characteristics in the unit injector. The term injection timing is used to denote the points at which the injection pulse begins and ends; these points are usually specified in engine crank angle degrees. Depending on application, these may be:

• Variable beginning, variable ending of injection pulse

• Constant beginning, variable ending of injection pulse
• Variable beginning, constant ending of injection pulse

UNIT INJECTOR IDENTIFICATION

A circular identification tag pressed into the unit injector body identifies the class number. Unit injectors with no line under the manufacturer name are generally those with the obsolete crown valve nozzles. A needle valve nozzle would have this line under the manufacturer's name:

Manufacturer ID	GM	Reliabilt	DDC	DDC
Code	N60	B60	N70	C70
Nozzle type	Needle	Crown	Needle	Needle

In the first example in the table, the line under the manufacturer name identifies the injector as having a needle valve nozzle. The letter codes indicates the plunger specific plunger geometry; for instance, a C would identify the plunger as having variable beginning, variable ending timing characteristics. The number following the letter designation is the comparator bench specification for fuel output per 1,000 strokes measured in cubic millimeters. The digits engraved on the nozzle valve identify the number of orifii, orifii sizing, and orifii spray angle.

MUI SUBCOMPONENTS

The technician should be familiar with the components that collectively make up the assembly. The following lists the components and briefly describes their function in a needle valve nozzle MUI.

Follower Assembly

The follower is the actuating mechanism for the MUI plunger. The follower is directly linked to the plunger by a slotted lug. A stop pin limits the upward travel of the plunger. The MUI spring surrounds the follower. It holds the follower (and therefore the plunger) in the raised position and loads the injector train.

Plunger

The plunger is the moving component of the pumping element. It reciprocates within the bushing, which is held stationary in the MUI body. The plunger is milled with helically shaped metering recesses for purposes of varying effective stroke, and center and cross drilled, to provide a hydraulic connection between the pumping chamber below the plunger and the metering recesses. Plungers are lapped to bushings in manufacture. Lapped components are not interchangeable.

Bushing

The MUI bushing is a cylindrical housing within which the plunger reciprocates and with which it forms the pump element. The bushings are drilled with diametrically opposed upper and lower ports and are positioned in the MUI by dowels.

Spill Deflector

The spill deflector is a cylindrical, stellite alloy sleeve that surrounds the bushing and prevents high velocity fuel spilled from the MUI pump chamber at the completion of effective stroke from eroding the unit injector body.

Gear

The MUI gear is connected to the plunger by means of a flat machined into its inside bore and tooth meshed to the control rack. The gear therefore rotates whenever the MUI rack is moved linearly. In this way, the gear provides a means of rotating the plunger (required to alter effective stroke) while it reciprocates.

Control Rack

Each MUI control rack is tooth meshed with the plunger gear, so when it is moved linearly, the plunger is rotated. The control rack is manufactured with a **clevis** into which the foot of the control lever on the control tube is placed. The linear position of the control rack is controlled by these control levers that extend from the control tube.

Check Valve

In the event of the needle valve being unable to seat due to carboning or any other reason, the check valve prevents cylinder gases from entering the MUI circuitry beyond the nozzle assembly.

Needle Valve Nozzle Assembly

The DDC needle valve assembly uses all the same operating principles as any other multi-orifii nozzle assembly (this subject matter is handled in detail in Chapter 20). As with any hydraulic injector nozzle, an MUI needle valve nozzle is held closed and seated by the nozzle valve spring. Fuel pressures developed in the MUI pump chamber below the plunger are ducted to act on the pressure chamber sectional area of the nozzle valve. When that pressure is sufficient to overcome the nozzle valve spring tension, the needle valve retracts, allowing high-pressure fuel from the MUI pump chamber to pass around its seat to a sac and exit through nozzle orifii, beginning the injection pulse. When the pump element effective stroke ends, fuel starts to spill from the pump chamber, exiting the lower bushing port. The moment that fuel pressure acting on the sectional area of the needle valve is insufficient to hold the valve spring retracted, it closes, ending the injection pulse.

INJECTOR TO GOVERNOR LINKAGES

The output of the mechanical unit injectors is governor controlled by means of a set of linkages that mechanically connect the governor with the MUI. Governor **control rods** extend from the **governor differential lever** (twin arm lever that pivots on a fulcrum) in the governor housing and connect to a control tube lever by means of a clevis and pin; linear movement of the governor control rods will by this means be converted into rotary movement of the control tube, which runs lengthwise through the upper cylinder head. Extending from the control tube are the **rack levers**. When the control tube is rotated by the control rods, the rack levers rotate with it. The feet of the rack levers connect to the MUI fuel racks by means of a clevis, so once again, rotary movement is converted to linear movement to position the racks. The mechanical relationship between the MUI rack and the rack lever is adjusted by an adjusting screw (or pair of screws in older engines) on the rack lever to control tube clamp.

Therefore, the rack levers link the individual MUIs with the control tube. When the control tube is rotated by the governor control rods, the rack levers extending from the control tube move the unit injector racks linearly. A critical procedure in DDC engine tune up is ensuring that the unit injectors are balanced; that is, that the point of register of the helices with the bushing ports in each unit injector in the engine is identical. This is achieved by correctly setting the control tube adjusting screw (newer engines) or adjusting screws (older engines). The result of balancing is equal fueling of each cylinder.

GOVERNOR

A simple mechanical governor is used in most truck and bus applications. Governor components consist of a set of engine-driven flyweights, a thrust collar, **bell crank**, spring set, and differential lever. As in most mechanical governors, centrifugal force exacted by the flyweights attempts to diminish engine fueling while spring force moderated by speed control lever position attempts to increase engine fueling. The thrust collar acts as an intermediary between the spring force defined by precise adjustment and moderated by the speed control lever position and the centrifugal force that directly correlates to engine speed. Extending from the governor housing are the control rods that link the MUI fuel control linkages with the governor assembly. Both limiting speed and variable speed governors are used. The governor type should always be identified prior to engine tune up as the procedure varies with governor type.

DETROIT DIESEL TUNE UP

Although it is not an objective of this textbook to address specific workshop procedure, the DDC tune up is one instance in which the workshop text might benefit from some enhancement. The following is intended to be used as a guide to DDC tune up of two-stroke-cycle engines and it should always be used in conjunction with DDC workshop texts, specifications, and service bulletins. Initially, the engine and its governor type should be identified.

DWLS—double weight, limiting speed (mobile equipment)

SWLS—single weight, limiting speed (mobile equipment)

SWVS—single weight, variable speed (industrial, marine)

DWDRG—double weight, dual range governor (mobile/highway)

The tune up procedure does not require a readjustment of every setting—it is a procedural method of checking for possible change in the settings.

The following description is loosely based on V–92 or 71 Series tune up procedure, and it should be remembered that the procedures will differ somewhat from engine to engine.

GENERAL TUNE UP GUIDELINES

DDC prefers that all tune up adjustments be made with the engine at 160–185°F coolant temperature. As many of these adjustments have to be made with the engine stopped, it may have to be run between adjustments to maintain the required temperature. Ensure before every startup, that when the stop lever is engaged, the injector racks move to the no-fuel position. The consequence of not doing this could result in engine overspeed or even a runaway.

If the engine classification has a **TT** suffix, it is a **tailored torque**, highway application fuel squeezer engine with a Belleville spring retainer governor. The spring retainer nut must be backed out until there is approximately 0.060″ clearance between the washers and the retainer nut (Figure 22–5).

ADJUST EXHAUST VALVE BRIDGES

Adjusting exhaust **valve bridges** procedure is required when a cylinder head is removed/overhauled/replaced or if there is a reason to suspect that it should be

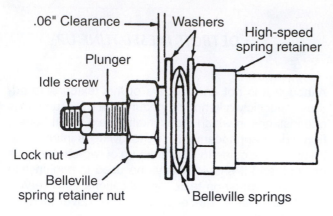

FIGURE 22–5 *Belleville washer location. (Courtesy of DDC)*

checked. Remove the bridge assembly from the cylinder head (essential: attempting to do this on the engine can bend the bridge guide stem) and mount it in a soft-jawed vise; loosen the locknut on the adjusting screw. Remove the bridge from the vise and install on the bridge guide; turn the adjusting screw CW until it contacts the valve stem then rotate an extra ⅛ of a turn. Hold with locknut finger tight, remove the bridge, and install once again in a soft-jawed vise; hold the adjusting screw stationary and torque the locknut to specification (25 lb.-ft.). Reinstall the bridge saddle on the bridge guide. To check, cut two pieces of 0.0015" shimstock (or thickness gauges) so they fit under the bridge saddle. Load the bridge onto the guide with finger pressure and check for even drag as the two pieces of shimstock are simultaneously withdrawn from either side of the bridge. Ensure that all the valve bridges are properly positioned when the rocker assembly is reinstalled.

VALVE LASH ADJUSTMENT

Unless there is substantial reason to suspect the valve bridge adjustment such as excessive valve clatter, this adjustment can be omitted from the routine tune up. Valve lash in a DDC two-stroke cycle engine can be set/checked in one engine rotation. Check the engine cylinder firing sequence. Engine rotation is always determined by observing the engine from the front. In a V configuration engine, the cylinder banks are oriented by observing the engine from the rear. Following is the cylinder firing sequence for an 8-V configured engine.

CW = RH–IL 3R 3L 4R 4L 2R 2L 1R

CCW = LH–IL IR 2L 2R 4L 4R 3L 3R

It is not necessary to align timing marks to set the valves on these engines; observing the action of the injector actuating train will ensure that the valves are closed. Precise positioning of the engine on the timing indices is not required. A remote starter switch can be used to bunt the engine to each valve set position.

WARNING: Ensure that the governor speed control lever is in the idle position and the stop lever is secured in the no-fuel position before rotating an engine during tune up regardless of whether the rotation is performed by hand or using the vehicle starter motor.

Bunt the engine using an auxiliary switch and the cranking motor to the required position for adjusting each set of valves. Adjustment should be performed when the injector rocker is fully actuated. Check data in the workshop manual for the lash specifications. DDC prefer the use of go-no-go thickness gauges. Remember also that cold lash settings differ from hot lash settings (Figure 22–6, and Figure 22–7).

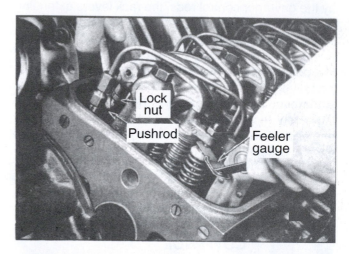

FIGURE 22–6 *Adjusting valve bridges. (Courtesy of DDC)*

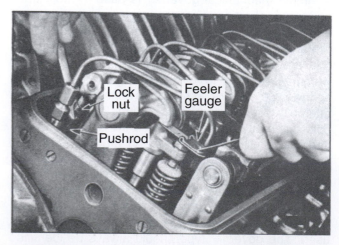

FIGURE 22–7 *Adjusting valves on a four-valve cylinder head. (Courtesy of DDC)*

FUEL INJECTOR TIMING

The injector **timing dimension** specification can be found stamped on the valve/rocker cover plate. A timing pin with the correct dimension stamped on its flange should be selected. The engine may be rotated in the same manner as when performing the valve adjustment, and there is no reason why both adjustments cannot be set simultaneously, observing correct engine position and sequencing. Set the timing dimension when the valves in the cylinder are fully depressed. Insert the dowel in the timing gauge into the injector timing hole located in the upper flange of the MUI. Smear a drop of engine oil on the injector follower flange and then adjust the injector pushrod so that when the timing tool is rotated, it should gently wipe the oil film from the follower. The injector-actuating train consists of a solid pushrod that connects to the rocker by a pivot pin and clevis; the pushrod connects to the clevis by screw threads and a locknut. An adjustment is therefore performed by backing off the locknut and either reducing or lengthening the pushrod total length to set the specified plunger travel. All the injectors are set in the same way; that is, setting actual plunger travel by adjusting the MUI actuating train dimension (Figure 22–8).

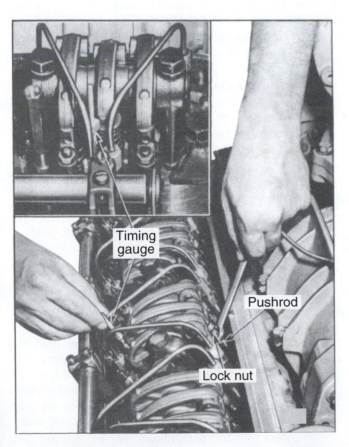

FIGURE 22–8 Fuel injector timing. (Courtesy of DDC)

GOVERNOR GAP ADJUSTMENT

Governor gap adjustment is a critical step to engine performance and if not properly set, the result will be a lack of power and fueling flat spots. The procedure varies somewhat according to engine series and governor type so check DDC service literature.

Governor Gap Adjustment Procedure— Engine Stationary

1. Remove the governor cover and back out the **buffer screw** and the **starting aid screw**, if equipped.
2. Next, bar the engine to position the governor weights in a horizontal position, then insert a **governor weight fork** (the common practice of using a screwdriver to spread the weights is not recommended) down between the inner lobe of either one of the low-speed governor weights (the larger weights) and the weight riser shaft. Use the governor weight fork to force the weights outward to their fully extended position against the stop on the weight carrier shaft.
3. The preceding step will have forced the low-speed spring cap against the high-speed plunger. Select the correct thickness gauge and insert it at the governor gap; there should be a slight drag. Adjust if required.
4. Reinstall the governor cover. If the governor gap setting was checked cold, the adjustment should once again be checked with the engine at operating temperature.

Governor Gap Adjustment Procedure— Engine Running

DDC has technical service bulletins on the subject of governor gap adjustments with the engine running (Figure 22–9) but typically the procedure would be as follows:

1. Nonfuel squeezer engines: the engine should be run between 800 and 1,000 rpm and the governor gap should be 0.0015". When using shimstock cut with mechanics shears or scissors, ensure that the edge has not been distorted to a much greater-than-specified dimension.
2. Fuel squeezer (TT) engines: the engine should be run between 1,100 and 1,300 rpm and the governor gap should be 0.002" to 0.004".

WARNING: When the engine is run with the governor cover removed, great care should be taken not to over-rev the engine.

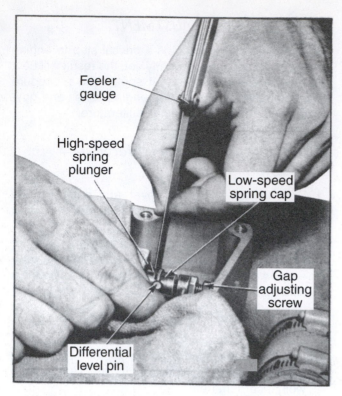

Feeler gauge

High-speed spring plunger

Low-speed spring cap

Gap adjusting screw

Differential level pin

FIGURE 22–9 *Governor gap adjustment. (Courtesy of DDC)*

SET INJECTOR RACKS

Setting injector racks is a critical adjustment that must be precisely executed to balance fueling to each engine cylinder. Erratic setting of the MUI racks during tune up is a main cause of hunting, a rhythmic fluctuation in engine rpm. The exact linear position of each MUI rack is determined by its control lever adjustment. The governor's location on the engine determines which injector rack should be set first; this is always that closest to the governor.

1. Disconnect all linkages connected to the governor speed control lever.
2. Back out the idle speed adjusting screw until there is no tension on the low-speed spring (LS governors). When backed out approximately ½", the low-speed spring will have little or no tension, permitting the closing of the low-speed gap without risk of bending the fuel rods or damaging the yield link spring mechanism (used with **throttle delay** mechanisms). Failure to back out the idle speed adjusting screw will result in inaccurate fuel rack settings.
3. The throttle delay mechanism should have been removed when setting the governor gap; ensure that it remains off. Also, back out the buffer screw approximately ⅝" as it was before setting the governor gap.

4. On TT engines, the Belleville spring retainer nut should have 0.060" clearance and on engines with a starting aid screw, this should be left backed out.
5. Loosen the adjusting screw (inner and outer adjusting screws on older, two-screw rack lever adjusters) of each MUI rack control lever on the control tube. Ensure that the foot of the rack lever is not binding into the MUI **rack clevis**. More recent model engines use a single adjusting screw and locknut with a torsion spring instead of the two-screw rack adjusters, to prevent jamming a whole bank into the full-fuel position in the event of a single MUI rack seizure.
6. On V-configuration engines, remove the clevis pin from the right bank fuel rod connection to the control tube, leaving the left bank control tube connected to its fuel rod.
7. On engines with VS mechanical governors, the speed control lever should be pinned into the full-fuel position while setting the injector racks and the stop lever should be held in the run position. On LS mechanical governors, load the speed control lever into its full-fuel position with light finger pressure.
8. Next, with both LS- and VS-governed engines, observing the previous step, turn CW the inner adjusting screw (two-screw type) or adjusting screw (one-screw type) on the injector closest to the governor until the rack control clevis can be observed to kick up; at this point, a slight increase in effort required to turn the screwdriver should be detected. On the single-adjustment screw-type, turn screw CW on further ⅛ turn, then lock in position with the locknut. On the older two-screw adjustment rack levers, turn the inner screw a further ⅛ turn CW, then screw in the outer screw until it bottoms. Next, alternately tighten each of the screws by turning ⅛ turn CW, finally torquing to specification (24–36 lb.-in.).
9. Next, on LS governor engines, move the speed control lever through its travel arc, or likewise, the stop lever on VS engines through its arc; if any drag is detected, the rack should be readjusted.
10. On all engines, the first MUI rack to be adjusted is the one closest to the governor; this becomes the master rack once adjusted according to the procedure outlined in the previous step. The remaining rack control levers are adjusted as follows:
    • In-line engines: remove the fuel rod clevis pin from the control tube actuating arm, then hold the injector control racks in the full-fuel position using the control tube. Adjust each MUI rack lever in sequence moving away from the master rack down the cylinder bank. Check to ensure that each rack clevis has equal spring back when tested with the screwdriver. Never readjust the master rack once this is properly set.

• V-configuration engines: remove the clevis pin from the fuel rod at the left bank injector control lever. Reinstall the clevis pin in the fuel rod in the right bank, reconnecting it with the governor. Adjust the #1 right bank cylinder injector rack exactly as outlined for the left bank #1 cylinder. To verify that both #1 left and #1 right racks are balanced, connect the left bank fuel rod again, then move the speed control lever (LS governor) to maximum speed or the stop lever (VS governor) to the run position. Check for drag or binding at the clevis pin connecting each bank. Next, check the spring back at the rack clevis on both the #1 left and #1 right cylinders; this should be identical. If not, readjust the #1 right cylinder. This must be correct before proceeding.

Now remove the fuel rod clevis pins and readjust each bank of injector racks sequentially moving away from the "master" on each bank (Figure 22–10).

WARNING: If the inner adjusting screw (two-screw type) or the adjusting screw (one-screw type) are turned in too far in the previous step, the stop control lever (VS governor) or speed control lever (LS governor), both of which are being held in by finger pressure, will be felt to move. This happens when the MUI rack is forced beyond full-fuel position; that is, it is loaded beyond its normal travel. If this happens, back off the screw and start again.

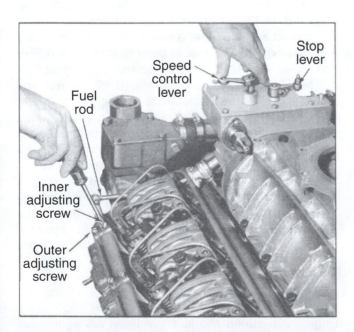

FIGURE 22–10   *Setting the injector racks. (Courtesy of DDC)*

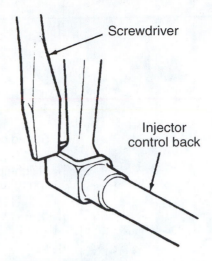

FIGURE 22–11   *Checking rack clevis spring back. (Courtesy of DDC)*

TECH TIP: Using a long thin screwdriver held between the thumb and index finger, apply light pressure to the injector rack clevis, tilting it. When the screwdriver is removed, the clevis should *spring back* to its original position. If this does not happen, repeat this step. The idea is to have each rack lever lightly load its injector rack into its full-fuel position with the end objective being that every MUI installed in the engine produced the identical metered fuel quantity in any given governor position.

---

11. When all the injector racks have been set, install the fuel rod clevis pins. Check for drag on each clevis pin and once more recheck spring back at rack clevises (Figure 22–11).
12. Finally, secure clevis pins with cotter pins and screw the idle adjust screw until it projects ³⁄₁₆″ from the locknut; this will allow the engine to start.

ADJUST MAXIMUM NO-LOAD (HIGH-IDLE) ENGINE SPEED (LS GOVERNORS)

The specification should be stamped on the engine option plate on the rocker cover. Proceed as follows:

1. Ensure the buffer screw is still backed out. On LS governors loosen the spring retainer locknut, then back off (CCW) the high-speed spring retainer nut approximately 5 full rotations. The engine should be running and at operating temperature with no load on it.
2. Next, using an accurate tachometer, hold the speed control lever in the maximum speed position then rotate the high-speed spring retainer nut CW until the specified maximum no-load rpm is achieved (Figure 22–12).

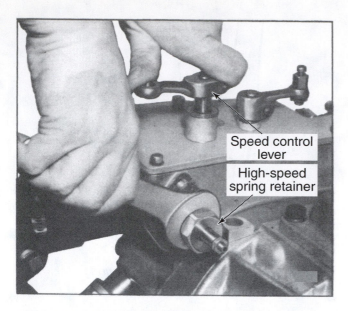

FIGURE 22–12 *Adjusting the maximum no-load speed. (Courtesy of DDC)*

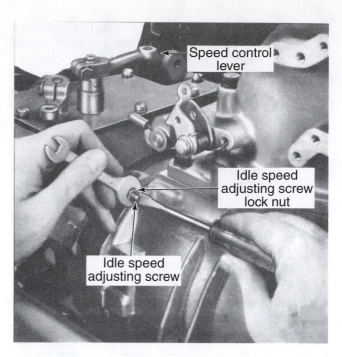

FIGURE 22–13 *Adjusting idle speed. (Courtesy of DDC)*

- TT engines: the Belleville springs must be readjusted following a no-load maximum speed adjustment. With these engines, set the maximum no-load speed 150 rpm above specification before adjusting the Belleville springs.
- VS governor: shims are used on the bellcrank end of the governor spring to set the maximum no-load speed (spring tension is set by shims rather than a nut). Consult the workshop manual for specifications. This method is also used on some 53 Series with LS governors.

ADJUST IDLE SPEED

The engine should still be at operating temperature and the buffer screw backed out. Check the idle speed specification. Loosen the locknut on the idle screw and rotate either CW or CCW to set idle speed 15 rpm below the recommended value. It may be necessary to use the buffer screw to reduce engine rpm roll/hunting while performing this operation. However, once set, lock the idle screw with the locknut and back out the buffer screw (Figure 22–13).

BELLEVILLE SPRING ADJUSTMENT (TT ENGINES ONLY)

This is a very critical adjustment and fractional inaccuracies can result in greatly reduced power output. Reference Figure 22–5.

1. Idle drop method—The engine tune up should have been performed up to this stage. Use an accurate tachometer for rpm readings. Disconnect the accelerator linkages. The engine should be at running temperature. Refer to DDC service literature to obtain the idle-drop specifications, peak brake horsepower, and rated engine speed values. Set the initial idle speed by using the idle adjust screw. Then with the speed control lever in the idle position, turn the Belleville spring retainer nut CW until the specified idle drop value is obtained exactly. This must be performed with precision, and it should be noted that a difference of 1 rpm off the specified idle drop value can result in a 2 to 3 BHP power loss. Then, lock the Belleville spring retainer nut. Next, lower the idle speed to the required specification using the idle adjust screw.

2. Chassis dynamometer method—This adjustment can be made on a chassis dynamometer, but it is not recommended due to the variables of dynamometer calibration, chassis driveline efficiency, air density, tire slippage, and the fuel CN. It does set the peak BHP value, however, and may be preferred by some operators. Consult the engine specifications, and remember that engine power at the flywheel and engine power at the wheels are entirely different.

ADJUST BUFFER SCREW

Engine idle speed should have been set at a value below specification in a previous step (15 rpm) and the engine should be at operating temperature. Turn

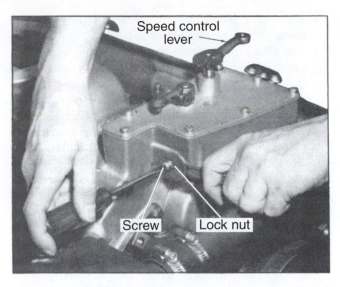

FIGURE 22-14   *Adjusting the buffer screw. (Courtesy of DDC)*

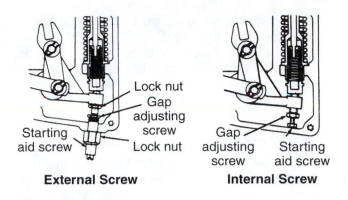

External Screw   **Internal Screw**

Injector	Gage Setting*	Tool Number
S80	.345"	J 24889
N60	.345"	J 24889
N65	.385"	J 24882
N70	.385"	J 24882
N75	.385"	J 24882
N80	.385"	J 24882
N90	.454"	J 23190
C65	.385"	J 24882
C80	.385"	J 24882

FIGURE 22-15   *Adjusting the starting aid screw. (Courtesy of DDC)*

the buffer screw CW so that it just begins to contact the governor differential lever; this should have the effect of fractionally increasing engine speed and leveling off engine rpm roll. Use an accurate tachometer and ensure that the idle speed does not exceed the required specification. Next, check the high-idle speed and if this has increased by more than 25 rpm, back off the buffer screw. Lock up the buffer screw when rpm values are set to specification (Figure 22–14).

ADJUST STARTING AID SCREW (DWLS-TURBO-BOOSTED ENGINES)

The starting aid screw is designed to minimize fuel injected during cranking to reduce exhaust smoke. It is necessary because the turbocharger impeller restricts air flow to the cylinders until the engine fires, and exhaust heat and flow act to rotate the turbine. Adjust the external (internal on some governors) adjusting screw on the governor as follows:

1. With the engine not running, the stop lever in the run position, and the speed control lever in the idle position, select the correct gauge as per DDC technical literature. The setting can be measured at any convenient cylinder location.
2. When the starting aid screw is properly adjusted, the gauge should have an end clearance of 1/64″. Lock up the starting aid screw (Figure 22–15).

ADJUST THROTTLE DELAY CYLINDER

If the tune up is being performed on an engine equipped with a throttle delay device, it must be first removed to avoid interference in the governor or injector control linkage. Used on engines with and without turbos, is purpose is to retard movement of fuel control tube and injector racks on acceleration. It is a crude smoke-limiting device that functions by preventing the control tube from snapping the MUI racks immediately to full-fuel when requested by the operator. It also prevents drivers from overrevving the engine while upshifting. Any time the injector racks are moved toward the no-fuel position, free movement of the throttle delay piston is ensured by air drawn into the throttle delay cylinder. However, movement of the piston toward the rack to no-fuel position allows oil from the reservoir to enter and displace the air. Sudden fuel demand such as that required for acceleration causes the racks to move toward full-fuel position, but the throttle delay piston must first displace the oil in the throttle delay cylinder through an orifice. This momentarily retards the inboard travel of the MUI racks toward the full-fuel position.

1. Insert the correct gauge between the injector body and the shoulder on the injector rack clevis on the injector closest to the throttle delay cylinder. Move the speed control lever to full-fuel and hold in that position, causing the injector rack to move against the gauge.
2. Insert DDC pin gauge (J255580.072") in the cylinder oil fill hole. Rotate the throttle delay lever until

further movement is limited by the piston coming up against the pin gauge.

3. Carefully hold the throttle delay piston in this position and tighten the U-bolt nuts while exerting light pressure on the throttle delay lever in its direction of actuation.

CONCLUSION

Adjusting the throttle delay cylinder (when equipped) completes the DDC two-stroke cycle tune up. It is important to check that when accelerator linkage is reconnected to the speed control lever, it moves to the maximum speed position when the accelerator pedal is fully depressed.

A DDC tune up sets all the critical fuel system adjustments. The first time the technician undertakes DDC tune-up procedures, it is essential that the appropriate DDC service literature and specifications be consulted and that the novice technician is guided by one with some DDC tune-up experience. Although the chances of a runaway are rare, the technician should be aware of how to trip the blower air gates (these cut intake air to the engine) and take the precaution of having a CO_2 fire extinguisher close by.

SUMMARY

- DDC mechanical unit injector fuel systems on 53, 71, and 92 series engines have been the bus engine of choice for half a century.
- Ownership of the DDC Corporation has transferred from General Motors to Penske Corp. and more recently to DaimlerChrysler.
- The series number in all the older hydromechanical DDC engines denotes the swept volume per cylinder, so to calculate total engine displacement, the series number should be multiplied by the number of cylinders.
- The hydraulic principles used on DDC mechanical unit injector (MUI) fuel systems relate closely to those on a port-helix metering pump.
- The DDC MUI fuel subsystem consists of a fuel tank and a primary filter under suction, and a gear-type transfer pump feeding a secondary filter and the cylinder head charging gallery.
- Jumper pipes connect the DDC MUIs with the cylinder head charge and return galleries.
- The MUI is an integral pumping, metering, and injection unit.
- MUIs are actuated by a dedicated cam profile on the engine camshaft(s).
- The MUI actuating train consists of an injector cam, lifter, pushrod, and rocker.

- The MUI bushing is machined with an upper and lower port, diametrically offset.
- The MUI plunger is milled with helices which, dependent on helix geometry, may offer constant beginning, variable ending; variable beginning, constant ending; or variable beginning, variable ending, delivery timing characteristics.
- Metering of fuel is accomplished by rotating the plunger within the bushing bore by means of a gear, which is itself tooth meshed to the MUI control rack.
- A needle valve hydraulic nozzle assembly is connected to the MUI pump chamber by a duct; this atomizes fuel directly to the engine cylinder similar to any multi-orifii, hydraulic injector nozzle.
- A throttle delay mechanism limits snap throttle surge fueling to diminish smoking at shifting.
- Various governors are used on DDC mechanical engines, depending on application, but common for on-highway applications is the mechanical LS type.
- A complete DDC overhead adjustment is known as a tune up.
- Fuel rack adjustment balances MUI fueling.
- Fuel timing adjustment on MUIs phases the plunger stroke with the bushing ports.

REVIEW QUESTIONS

1. Which of the following correctly describes the type of fuel transfer pump used on DDC highway, two-stroke engines?
   a. Plunger
   b. Diaphragm
   c. Vane
   d. Gear
2. Charging pressure values to DDC mechanical unit injectors would approximate which of the following?
   a. 17 psi
   b. 70 psi
   c. 800 psi
   d. 2,300 psi
3. Which of the following correctly describes the component(s) that feed and return fuel to DDC mechanical unit injectors?
   a. High-pressure pipes
   b. Annular charge galleries
   c. Jumper pipes
   d. Rail pipes
4. Which of the following is most responsible for developing injection pressures in a DDC mechanical unit injector?
   a. Charge pump
   b. Rack actuator

c. Control tube

d. Injector cam

5. Which of the following will determine the amount of fuel injected per crank angle degree in a DDC mechanical unit injection system?
   a. Rack position only
   b. Engine rpm
   c. Cam profile only
   d. Rack position and cam profile

6. Which of the following is *not* a function of a DDC mechanical unit injector?
   a. Converts charging to injection pressures
   b. Meters fuel quantity
   c. Atomizes fuel
   d. Determines injection rate

7. Effective stroke in a DDC mechanical unit injector occurs when:
   a. Both upper and lower bushing ports are exposed.
   b. Both upper and lower bushing ports are covered.
   c. The plunger is on its downstroke.
   d. The plunger is on its upstroke.

8. When considering the injection timing characteristics of a DDC mechanical unit injector, which of the following is true?
   a. Variable beginning, constant ending
   b. Constant beginning, variable ending
   c. Variable beginning, variable ending
   d. All of the above

9. What would be the total swept volume of a DDC 8V–92?
   a. 8 liters
   b. 736 cc
   c. 736 cu. in.
   d. 460 BHP

10. Which of the following conditions would cause a hunting engine roll on a two-stroke cycle DDC engine?
    a. Weak high idle spring
    b. Belleville spring adjustment
    c. Injectors out of balance
    d. Seized control tube

11. NOP parameters in DDC mechanical unit injectors with needle valve nozzles would be:
    a. 600–850 psi
    b. 2,200–3,400 psi
    c. 10,000–12,000 psi
    d. Up to 25,000 psi

12. *Technician A* states that hunting is a condition that may result from the erratic setting of MUI racks during tune up. *Technician B* states that engine hunting can always be removed by adjusting the buffer screw CW until the rpm smooths out. Who is right?
    a. A only
    b. B only
    c. Both A and B
    d. Neither A nor B

13. *Technician A* states that the exhaust valves on DDC two-stoke cycle engines are set with identical lash to the intake valves. *Technician B* states that the adjusting screw on DDC valve bridges is used to set valve lash. Who is right?
    a. A only
    b. B only
    c. Both A and B
    d. Neither A nor B

14. If a single MUI is seized in the full-fuel position in a DDC two-stroke cycle engine, which of the following devices would do most to prevent an engine runaway?
    a. 2-screw type, rack lever adjuster
    b. Single-screw type, rack lever adjuster
    c. Throttle delay cylinder
    d. A limiting speed governor

15. *Technician A* states that the length of a DDC MUI effective plunger stroke is determined by the rotational position of the plunger in the bushing. *Technician B* states that plunger rotational position is determined by the linear position of the rack. Who is right?
    a. A only
    b. B only
    c. Both A and B
    d. Neither A nor B

Chapter 23

Caterpillar—Mechanical Unit Injection

OBJECTIVES

After studying this chapter, you should be able to:

- *Describe the layout and components in a Caterpillar fuel subsystem supplying 3114, 3116, and 3126 mechanical unit injector (MUI) equipped engines.*
- *Outline the principles of operation of a Caterpillar mechanical unit injector.*
- *Describe how MUI effective stroke is varied to control injected fuel quantity.*
- *Describe the components that link the MUIs with the governor assembly.*
- *Outline the procedure for performing Caterpillar top end adjustments.*
- *Test MUIs for fueling balance using an infrared thermometer.*
- *Test system charging pressure to specification.*
- *Locate TDC on #1 piston.*
- *Perform MUI fuel timing and cylinder head valve adjustment.*
- *Outline the procedure for synchronizing the MUIs.*
- *Perform basic fuel system troubleshooting.*

KEY TERMS

energized-to-run (ETR) latching solenoid

fuel ratio control (FRC)

infrared thermometer

link and lever assembly

mechanical unit injector (MUI)

synchronizing position

timing bolt

INTRODUCTION

Truck and bus technicians familiar with Detroit Diesel Corporation (DDC) **mechanical unit injection (MUI)** systems will require little introduction to the Caterpillar **MUI** system. However, there are some critical differences and Caterpillar technical literature must be referenced when performing its version of the tune up sequence. Caterpillar MUIs can be found on 3114, 3116, and 3126 engines. Most technicians will see the 3126 version of this engine with Caterpillar's HEUI fuel system, a computer-controlled, hydraulically actuated unit injector system; this is covered later on in this text. Currently the 3116 is available exclusively to General Motors Corporation (Figure 23–1). To interpret the numeric engine series code, the first digit (3xxx) means Caterpillar, the next two digits denote the displacement volume per cylinder (x12x = 1.2 liters) and the fourth digit (xxx6) denotes the number of cylinders.

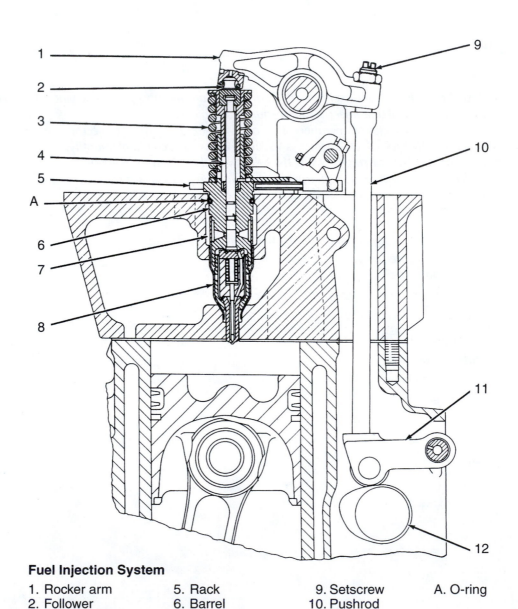

Fuel Injection System

1. Rocker arm	5. Rack	9. Setscrew	A. O-ring
2. Follower	6. Barrel	10. Pushrod	
3. Tappet spring	7. Fuel gallery	11. Lifter	
4. Plunger	8. Sleeve	12. Cam	

FIGURE 23–1 *Cutaway view of a Caterpillar MUI engine. (Courtesy of Caterpillar)*

FUEL SUBSYSTEM COMPONENTS

Fuel from the fuel tank is pulled from the fuel tank by a plunger-type transfer pump; a primary fuel filter may be fitted between the fuel tank and the transfer pump. The transfer pump is integral with the governor housing.

The transfer pump charges a secondary or main fuel filter and the unit injector assemblies. The charge side of the transfer pump is pressure protected by an outlet check valve and a pressure relief valve. When equipped, the hand primer pump is located on the secondary filter mounting pad; its function is to purge air from the fuel subsystem responsible for charging the unit injectors. Fuel is delivered from the fuel subsystem to drilled passages in the cylinder heads that intersect annular galleries around the cylindrical bore of the unit injectors.

The fuel transfer pump is a cam-actuated plunger pump located in the front housing of the governor. Its actuating cam is located on the governor drive shaft and it provides the pump plunger with the required reciprocating action working in conjunction with a retraction spring. Inlet and outlet check valves aspirate the pump chamber; these close when the engine is shut down. Outlet pressure should test at a minimum of 200 kPa (29 psi) at high idle (Figure 23–2 and Figure 23–3).

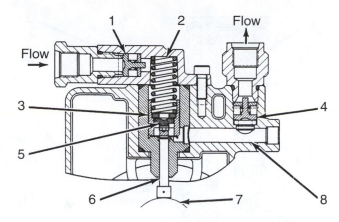

1. Inlet check valve
2. Spring
3. Piston assembly
4. Outlet check valve
5. Piston check valve
6. Tappet assembly
7. Cam
8. Passage

FIGURE 23–3 *Caterpillar plunger-type fuel transfer pump. (Courtesy of Caterpillar)*

PRINCIPLES OF OPERATION OF CATERPILLAR MECHANICAL UNIT INJECTORS

The mechanical unit injectors are responsible for metering, pumping to injection pressure values, and atomizing fuel directly to the engine cylinders. The hydraulic principles of the Caterpillar mechanical unit injector relate closely to those of the in-line, port-helix metering pump and a hydraulic, multi-orifii injector nozzle. In a mechanical unit injector (Figure 23–4), the pump is actuated by a dedicated cam profile on the engine camshaft, the high-pressure pipe is eliminated, and the hydraulic injector nozzle is integral with the assembly. The mechanical unit injector is installed within a sleeve press-fit to the cylinder head; this is directly exposed to engine coolant passages. The sleeve should only be serviced with Caterpillar special tools.

The injector cam profile actuates a train consisting of a roller-type lifter, pushrod, and rocker arm. The rocker arm acts on the tappet assembly of the mechanical unit injector using a "floating button" as an intermediary; this prevents side loading of the tappet by the rocker arm. The plunger is linked to the tappet and reciprocates within a stationary barrel. It is milled with a helix that can be classified as a lower helix design, resulting in a constant beginning, variable ending of fuel delivery timing. With this design of helix, the beginning of the effective pumping stroke is determined by the angular location of the cam. The barrel is machined with an upper port and a lower port diametrically offset. Metering of fuel is accomplished by rotating the plunger in the

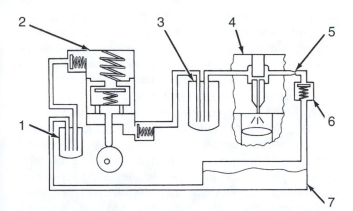

1. Screen
2. Fuel transfer pump
3. Main filter
4. Cylinder head
5. Pressure regulating orifice
6. Check valve
7. Fuel tank

FIGURE 23–2 *Schematic of Caterpillar fuel subsystem layout used with MUI fueled engines. (Courtesy of Caterpillar)*

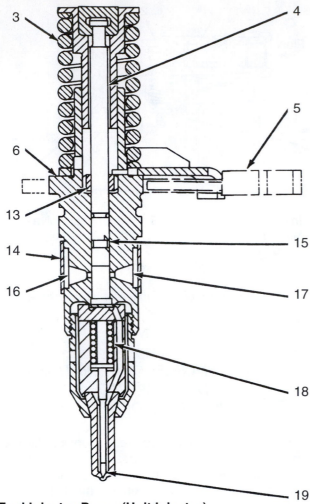

Fuel Injector Pump (Unit Injector)

3. Tappet spring
4. Plunger
5. Rack
6. Barrel
13. Gear
14. Sleeve filter

15. Helix
16. Lower port
17. Upper port
18. Spring
19. Check (needle valve)

FIGURE 23–4   Caterpillar MUI. (Courtesy of Caterpillar)

barrel bore and altering the point of register of the spill ports with the helix. The plunger is rotated by means of a gear that is tooth meshed to the rack. The gear collars the plunger in a manner that permits plunger rotation while it reciprocates.

EFFECTIVE STROKE

Plunger actual stroke is determined entirely by cam profile and, therefore, remains constant. Control of fuel quantity depends on the rotational position of the plunger, specifically the point of register of the helix, with the upper barrel port. When the actuating cam ramps off its base circle, the plunger descends within

the barrel and first, its leading edge will close off the upper barrel port. As the barrel continues to descend, it displaces fuel below it in the pump chamber, forcing it out of the lower barrel port. Effective stroke begins when the lower bushing port is closed by the plunger leading edge. Pressure rise occurs in the pump chamber below the plunger. Connected to the pump chamber by a duct is the pressure chamber sectional area of the needle valve in the hydraulic injector nozzle at the base of the mechanical unit injector.

At nozzle opening pressure (NOP), sufficient pressure rise has occurred to overcome the spring pressure that loads the needle valve on its seat causing it to retract. Fuel passes around the needle valve seat and is forced out through the nozzle orifii. Effective pumping stroke will continue as long as both the upper and lower barrel ports are closed off by the plunger. It ends at the moment the upper barrel port is exposed by the recessed helix milled into the plunger, permitting fuel to be spilled to the return circuit. The resultant pressure collapse enables the injector nozzle spring to close and seat the needle valve, ending the injection pulse. Plunger descent continues until the injector train is maximumly extended by the cam nose. As the injector cam ramps toward its base circle, the plunger retracts creating low pressure, enabling rapid recharging of the pump chamber.

FUEL RACK CONTROL

The governor is mechanically connected to the rack shaft assembly by means of a **link and lever assembly** (Figure 23–5). When the governor demands fuel, the injector rack must be moved linearly away from the injector body rotating the plunger CCW. The governor output shaft, therefore, moves to rotate the shaft, causing the rack levers to move the injector racks simultaneously outboard. To no-fuel the engine, the injector racks must be forced inboard, causing the plungers to move CW. A torsion spring at each rack lever assembly allows the rack control linkage to return to a no-fuel position should one injector seize in a fuel-on position.

Power setting of the #1 cylinder injector is made with the fuel setting screw on the rack lever assembly. As the fuel setting screw is turned, shaft position relative to the governor link is altered. Adjusting screws on each clamp and lever assembly permit synchronization of the remainder of the unit injectors to that on the #1 cylinder. Synchronization of the unit injector racks is essential to balance engine fueling (Figure 23–6 and Figure 23–7).

FUEL RATIO CONTROL

Turbocharged engines use a **fuel ratio control (FRC)** to limit engine smoke during conditions of low boost. Essentially, the FRC limits fueling to the cylinders until

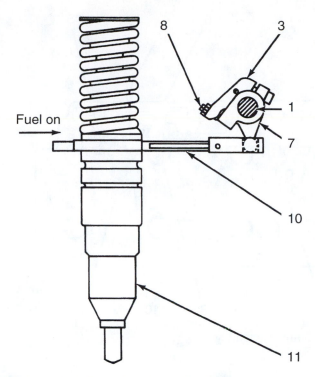

View from A-A from Previous Illustration

1. Shaft	8. Synchronization screw
3. Clamp	10. Rack
7. Lever assembly	11. Injector

FIGURE 23–5   MUI fuel-on position. (Courtesy of Caterpillar)

there is sufficient air in them to combust it. Boost air is ported from the engine intake manifold to the FRC inlet port. At low boost, when the governor output shaft moves to the fuel-on position attempting to draw the unit injector racks outboard, the limit lever contacts the set screw on the FRC lever. As manifold boost increases, it acts against the FRC diaphragm and overcomes the FRC spring pressure; this permits the retainer shaft to move outward enabling the FRC lever and the limit lever to rotate CW and the rack shaft to rotate toward full-fuel.

FUEL SHUTOFF SOLENOID

The fuel shutoff solenoid used in off-highway applications is an **energized-to-run (ETR) latching solenoid** (latching solenoid: a pilot valve that stays in the energized position even when not electrically energized and remains so until the circuit it controls is shut down). When the engine is used in trucks, the shutoff solenoid does not latch. A spring-loaded plunger inside the solenoid acts on a lever assembly located in the front housing of the governor. This lever assembly forces the governor output shaft to the fuel-off position when the solenoid armature is released at engine shutdown.

At start-up, the solenoid armature is energized to latch in the run position permitting the governor output shaft to move to the fuel-on position. Engines with latching solenoids can be manually shut off by

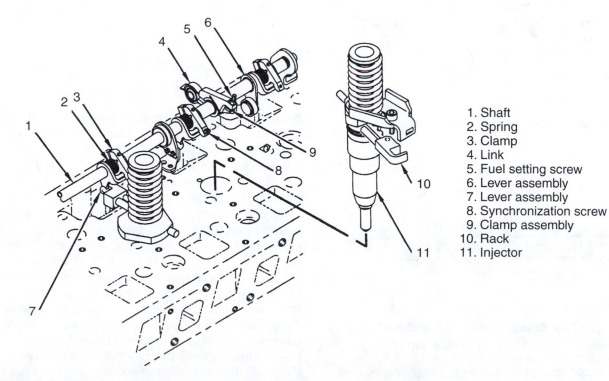

1. Shaft
2. Spring
3. Clamp
4. Link
5. Fuel setting screw
6. Lever assembly
7. Lever assembly
8. Synchronization screw
9. Clamp assembly
10. Rack
11. Injector

FIGURE 23–6   MUI location and rack control assembly. (Courtesy of Caterpillar)

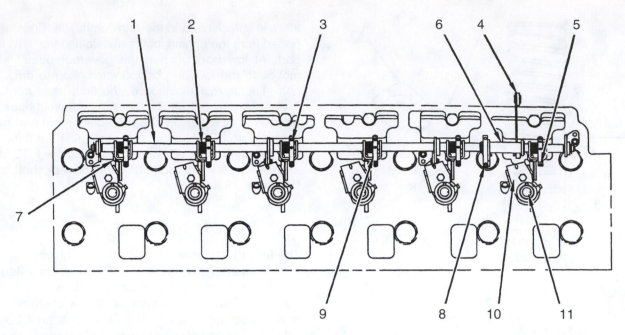

1. Shaft
2. Spring
3. Clamp
4. Link
5. Fuel setting screw
6. Lever assembly
7. Lever assembly
8. Synchronization screw
9. Clamp assembly
10. Rack
11. Injector

FIGURE 23-7   Overhead view of the MUIs and the rack control linkages. (Courtesy of Caterpillar)

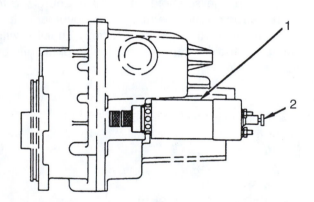

Fuel Shutoff Solenoid (Latching Type)

1. Solenoid   2. Button

FIGURE 23-8   Location of fuel shutoff solenoid. (Courtesy of Caterpillar)

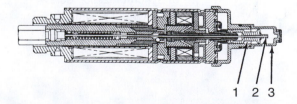

Governor Fuel Shutoff Solenoid (Latching Type)

1. Knob   2. Rod   3. Plastic cap

FIGURE 23-9   The ETR, latching solenoid. (Courtesy of Caterpillar)

depressing the override button. This solenoid cannot be manually latched in the run position for starting (Figure 23-8 and Figure 23-9).

GOVERNORS

Six different governor types are used on 3116, 3114, and 3126 engines. Each is identified sequentially by Roman numerals. Governors for Caterpillar mechanical unit injector fueled engines are covered only briefly in this section. Proprietary literature (SEN6454) and test equipment (IU7326 Governor Test Stand) are required to service and adjust these governors. The governor is gear driven by the engine camshaft, rotating a weight carrier and flyweight assembly. When the flyweights are rotated, centrifugal force exacted will be proportional to rotational speed and thus to engine rpm. Opposing the outward thrust of the flyweights is spring force, which uses a riser (thrust collar) as an intermediary. Fuel request from the operator is delivered to the governor by means of a control lever. This input is relayed to the governor output shaft, but the governor will moderate engine fueling when rated load or a lug condition is attained. The only on-engine adjustments to

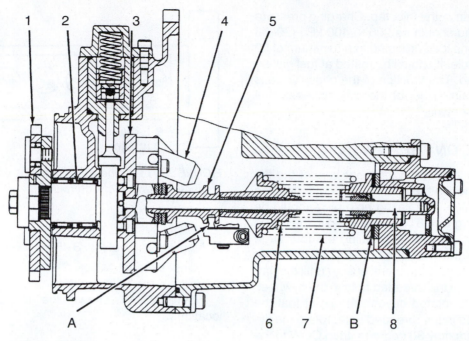

1. Governor drive gear
2. Shaft
3. Flyweight carrier
4. Flyweights
5. Riser
6. Low idle spring
7. High idle spring
8. Shaft
A. Pin
B. Shims

FIGURE 23–10 *Sectional view of one of the six governors used on Caterpillar MUI-fueled engines. (Courtesy of Caterpillar)*

the governor are the final low and high idle settings (Figure 23–10).

3114, 3116, AND 3126 ENGINE TESTING AND ADJUSTMENTS

Caterpillar service literature should be consulted when performing all engine and testing procedures. Technical service bulletins should also be referenced.

TESTING INJECTORS

Caterpillar recommends a preliminary test for mechanical unit injectors using an **infrared thermometer**. When performing this test, the engine should be operated at *low idle* and the temperature at the exhaust manifold ports measured. Low temperature at one of the ports can indicate no-fuel from an injector, while excessively high temperature is evidence of overfueling. The difference between cylinders should not be greater than 70°C (158°F).

With the valve cover removed and the engine idling, each unit injector may be tested individually by moving its rack to the fuel-on position; this should immediately result in combustion knock in the cylinder being tested. If this combustion knock does not result, there may be a problem with the injector assembly, the fuel supply to it, or the seal between the injector and sleeve.

Out-of-Engine Injector Inspection

Out-of-engine injection inspection requires the use of a Caterpillar test fixture. It should be performed by a Caterpillar dealership only.

GOVERNOR ADJUSTMENT

The only on-engine adjustments of the governor are the final low and high idle settings. Both settings should be trimmed to specification on a running engine by loosening the locknut and adjusting the screw against throttle arm stops.

FUEL PRESSURE TEST

A fuel pressure test measures the charging pressure to the engine cylinder head. To check the filter, the pressure gauge should be installed at the taps in the filter mounting pad. A restricted filter causes lower pressure at the outlet tap than at the inlet tap. Charging pressure to the cylinder head should be 200 to 400 kPa (29–58 psi) at rated rpm and load dropping to a minimum of 50 kPa (7 psi) at low idle. It should be tested at fuel gallery in the cylinder head. The function of the check valve is to prevent fuel draining out of the cylinder head fuel gallery during shutdown.

LOCATING TDC ON THE #1 PISTON

The timing hole is located on the front face of the flywheel housing either on the right or left side depending on engine application. The timing hole plug should be removed and **timing bolt** 8T0292 inserted in the timing hole. Next, bar the engine CCW using the four large bolts at the front of the crankshaft (do not use the eight small bolts on the front of the crank pulley) until the timing bolt engages with the threaded hole in the flywheel.

If the flywheel is rotated beyond the point that the timing bolt engages in the threaded hole, rotate the flywheel CW approximately 30 degrees, then CCW to the timing point; this is important to ensure that gear backlash is not a factor when determining the exact TDC position. Insert the timing bolt into the threaded hole. The valve cover should be removed and the rocker arms on the #1 cylinder should both show lash indicating that the cylinder valves are closed. If so, then the #1 piston is at TDC on the compression stroke. If the rocker arms on #1 evidence no lash, the timing bolt must be removed, the engine rotated 360 degrees and the timing bolt reinstalled.

FUEL TIMING, VALVE SETTING, AND INJECTOR SYNCHRONIZATION

3116/3126 crankshaft positions for fuel timing and valve setting:

3116/3126	VALVE LASH
Intake valves	0.38 + 0.08 mm (0.015" + 0.003")
Exhaust valves	0.64 + 0.08 mm (0.025" + 0.003")

The firing order is 1 5 3 6 2 4.
With #1 piston at TDC on compression stroke:

Adjust injectors: 3–5–6

Adjust intake valves: 1–2–4

Adjust exhaust valves: 1–3–5

Rotate engine 360 degrees.
With #1 piston at TDC on exhaust stroke:

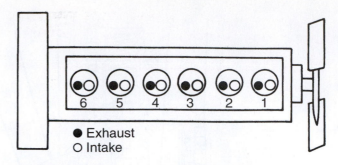

● Exhaust
○ Intake

FIGURE 23–11   *The valve locations on 3116 and 3126 engines. (Courtesy of Caterpillar)*

Adjust injectors: 1–2–4

Adjust intake valves: 3–5–6

Adjust exhaust valves: 2–4–6

Figure 23–11 shows the intake/exhaust value locations.

INJECTOR SYNCHRONIZATION

Injector synchronization is the setting of all the control racks on a set of unit injectors on the engine so that each delivers the same amount of fuel to each cylinder. This is performed by setting each injector rack to the same position using a 3.5 mm calibration block while the control linkage is in a fixed position called **synchronizing position**. Synchronizing position is when the injector at #1 cylinder is at fuel shutoff. Because #1 injector is the reference point for the other injectors, no synchronizing adjustment is made to the #1 injector rack. An injector should always be synchronized whenever it has been removed and reinstalled or replaced. When the #1 injector has been removed and reinstalled or replaced, all the injectors must be synchronized. Figure 23–12 shows the synchronizing tool.

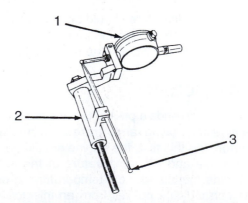

1. 6V6106 Dial indicator   3.   Lever (part of 1U6679
2. 1U6679 Indicator group   indicator group)

FIGURE 23–12   *Fuel synchronization tool used to set Caterpillar MUIs. (Courtesy of Caterpillar)*

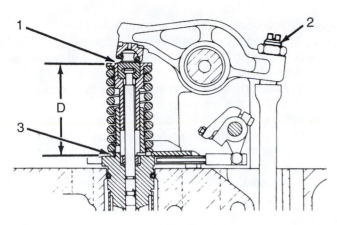

1. Tappet
2. Adjustment screw
3. Shoulder
D. Fuel timing dimension

FIGURE 23–13 *The fuel timing dimension setting.* *(Courtesy of Caterpillar)*

FUEL TIMING

The fuel timing dimension is a critical tune-up adjustment that sets the unit injector tappet dimension using the injector actuating train. Essentially, this process positions the plunger at a specific location in the barrel while the injector cam is on base circle; this "times" the activity in the unit injector with that in the engine and ensures that the beginning of injection occurs at a consistent crank angle in each cylinder. Figure 23–13 shows the fuel timing dimension (D), which is set using a timing pin.

SUMMARY

- The hydraulic principles used on Caterpillar MUI fuel systems relate closely to those on a port-helix metering pump.
- System hardware and layout of Caterpillar MUI fuel systems have many similarities with DDC mechanical unit injection systems.
- The Caterpillar MUI fuel subsystem layout consists of a fuel tank and optional primary filter under suction and a cam-actuated plunger-type transfer pump feeding a main filter and the cylinder head charging gallery.
- The MUIs are cylindrical and located in bores in the cylinder head.
- The MUI is an integral pumping, metering, and injection unit.
- MUIs are actuated by a dedicated cam profile on the engine camshaft.
- Charge fuel from the transfer pump enters the MUIs directly from the cylinder head fuel gallery by means of inlet ports located in exterior annuli.

- The MUI actuating train consists of an injector cam, roller-type lifter, push rod, and rocker.
- The MUI barrel is machined with an upper and lower port, diametrically offset.
- The MUI plunger is milled with a helix classified as lower helix type with a constant beginning, variable ending of delivery timing.
- Metering of fuel is accomplished by rotating the plunger within the barrel bore by means of a gear that is itself tooth-meshed to the MUI control rack.
- A needle valve hydraulic nozzle assembly is connected to the MUI pump chamber by means of a duct; this functions to atomize fuel directly to the engine cylinder similar to any multi-orifii, hydraulic injector nozzle.
- A fuel ratio control (FRC) device limits fueling under engine operating conditions of low boost.
- The fuel shut off device used by Caterpillar on its MUI systems is an energized-to-run (ETR) latching solenoid.
- Six different governors are used on this series of engine, depending on application.
- The only on-engine governor adjustments are the final low and high idle adjustments.
- The procedure required to balance MUI fueling is termed injector synchronization by Caterpillar.
- The fuel timing adjustment performed on Caterpillar MUIs phases the plunger stroke with the barrel ports and, therefore, defines the beginning of injection.

REVIEW QUESTIONS

1. Caterpillar MUIs are actuated:
   a. Hydraulically
   b. Electrically
   c. Electronically
   d. Mechanically
2. What component prevents side loading of Caterpillar MUI tappets by the rocker arm?
   a. Injector link
   b. Floating button
   c. Tappet spring
   d. Toothed collar
3. The Caterpillar MUI actuating train uses what type of lifter?
   a. Roller
   b. Solid
   c. Hydraulic
   d. Pneumatic
4. What component in a Caterpillar MUI-fueled engine dictates the length of the plunger stroke?

a. injector cam
b. Rack
c. Governor
d. Crankshaft rpm

5. Altering a Caterpillar MUI effective stroke is accomplished by:
   a. Rotating the plunger in the barrel
   b. Rotating the barrel within the MUI body
   c. Varying plunger stroke
   d. Varying cam lift

6. In the event of an MUI seizure in a fuel-on position, what component permits the remaining MUIs to return to a no-fuel position?
   a. Governor spring
   b. Torsion spring
   c. MUI tappet spring
   d. Hydraulic charging pressure

7. What event occurs at the precise moment that begins effective stroke?
   a. The lower barrel port is covered.
   b. The upper barrel port is covered.
   c. The plunger begins its downstroke.
   d. The plunger completes its downstroke.

8. Synchronizing MUIs is accomplished by:
   a. Setting the MUI timing dimension
   b. Adjusting screws on the clamp and lever assemblies
   c. Adjusting screws on the injector train rockers
   d. Setting the governor link

9. What component is used on Caterpillar MUI systems to limit fueling under conditions of low manifold boost?
   a. ETR solenoid
   b. Aneroid
   c. FRC device
   d. AFC valve

10. When testing for a malfunctioning MUI using an infrared thermometer, Caterpillar suggests that the maximum temperature difference between cylinders be no greater than:
    a. 25°C (77°F)
    b. 70°C (158°F)
    c. 100°C (212°F)
    d. 150°C (302°F)

Chapter 24

Cummins PT

PREREQUISITES

Chapters 18 and 19

OBJECTIVES

After studying this chapter, you should be able to:

- *Define Cummins pressure-time (PT) system theory.*
- *Describe the principles of operation of PT injectors.*
- *Describe the function of each PT injector subcomponent.*
- *Define the role of the two critical flow areas established at the PT injector.*
- *Outline the principles of operation of a Cummins PT pump.*
- *Describe the function of each PT pump subcomponent.*
- *Interpret PT fueling circuit schematics under several typical operating modes.*
- *Identify the different types of PT injector and the methods used to adjust them.*
- *Outline the Cummins PT engine timing procedure used on N, L, and K series engines.*
- *Outline the principles of operation of a step timing control (STC) injector.*
- *Install and prepare a PT pump for startup.*
- *Test and adjust the PT pump and injectors on-engine.*
- *Outline the role of the PT PACE/PACER, electronic, road speed governing.*

KEY TERMS

air/fuel control (AFC) circuit

balance orifice

calibrating orifice

cam follower housing timing

common rail system

flow area

governor barrel

governor button

governor plunger

metering orifice

offset camshaft key timing

PACE and PACER

partial authority

plunger link

pressure-time (PT) fuel system

PT control module (PTCM)

pulsation damper

shutdown solenoid

step timing control (STC)

top stop injector

INTRODUCTION

Until the aftermath of World War II, diesel engine manufacturers had been generally content to use port-helix metering fuel injection systems; there were a few exceptions but these tended not to survive. However, with port-helix metering injection, a disproportionate percentage of the total engine cost must be allocated to the fuel management apparatus and engine designers began to explore alternative systems. The Cummins engine company introduced its **pressure-time (PT) fuel system** in the early 1950s and by 1970, PT-managed engines were the North American market sales leaders in the 200–500 BHP range. By 1980, Cummins PT engines outsold all others worldwide in the 200 BHP and larger market, manufactured out of facilities in the United States, Britain, and Germany, supported by casting and subassembly plants in half a dozen other countries. This success continued until the 1994 North American EPA noxious emission legislation rendered the PT-managed engine ineligible for on-highway certification.

Unlike some of the other hydromechanical fuel injection systems, the PT system did not readily lend itself to electronic management and Cummins developed its own full authority electronic management systems to replace the PT **common rail system** (a fuel system in which a fuel pump supplies injectors from a common pipe or accumulator). However, it continues to be important for the truck/bus diesel technician to understand the hydromechanical PT systems first, due to the number of these engines sold before 1994, and more importantly, as the system has evolved into electronically managed derivatives. PT systems continue to be used on off-highway engines and could be until 2007 when on- and off-highway emission standards will be identical (Figure 24–1).

PT SYSTEM THEORY

Power output of any diesel engine is essentially determined by the amount of fuel metered into its cylinders. The Cummins PT system manages fueling using a set of principles quite distinct from those used in port helix metering systems. The sequence of the two letters in the acronym PT is important because Cummins also

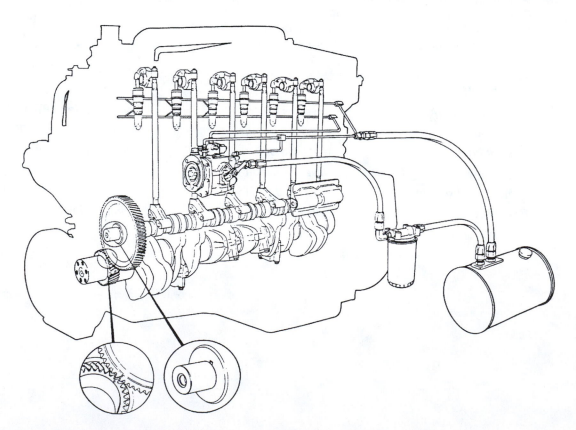

FIGURE 24–1   *Layout of a Cummins PT fuel system. (Courtesy of Cummins)*

uses the acronym TP representing time/ pressure; in each case, the first letter of the acronym denotes the control variable. In the case of the PT system, that would be the pressure factor.

A simple hydraulic equation constructed to determine a volume of flow requires the following data:

1. The fluid pressure
2. The flow time
3. The flow area

Changing any of these values in this hydraulic equation will change the outcome, a specific volume of flow. In the Cummins PT system, the critical **flow areas** (flow area: the most restricted sectional area of a hydraulic circuit under flow) are designed to remain constant, so the variables in the hydraulic equations are:

1. *Flow time*—variable, but dictated by engine speed, which has little relationship with fuel quantity required. The time period diminishes inversely with rpm increase.
2. *Fluid pressure*—variable and precisely managed by the system to control engine output.

The critical component in terms of managing the fueling is the PT pump; this is *not* an injection pump. It is best regarded as a flow control device. The PT pump feeds all the injectors by means of a common rail; that is, a single outlet pipe from the PT pump that supplies a rail gallery in the cylinder heads. The injectors are cylindrical and receive and return fuel through exterior annuli separated by O-rings. As the PT injectors are mounted in parallel and the only way in which the rail fuel can pass into the cylinder head return gallery is by passing through the PT injectors, their inlet orifii will collectively identify the flow area the PT pump unloads to. This flow area remains constant (they are accurately sized to the application on the test bench) so it can be said that the PT pump directly manages the rail pressure. Rail pressure values are used to manage engine output through the entire rpm and load ranges of the engine; they must, therefore, be widely variable. Typically they range from 8 psi to 210 psi.

The PT system is dealt with component by component. However, it is important to bear in mind the system objectives while doing this. Actual quantity of fuel to metered per cycle depends on the following five factors:

Rail Pressure

The PT pump is a flow control device responsible for managing rail pressure. Actual rail pressure will be factored by the quantity of fuel unloaded into the single rail pipe that supplies all of the PT injectors and the flow area as defined collectively by the PT injectors.

Balance Orifice Sizing

All of the fuel unloaded by the PT pump to the rail is routed through the PT injectors mounted in parallel. The **balance orifice** is the inlet orifice of each PT injector; each balance orifice is accurately sized because collectively, they define the flow area for factoring the actual rail pressure parameters. Although the term balance orifice is used in this text, Cummins also uses the term **calibrating orifice** to describe this precisely sized inlet to the PT injector.

Engine Speed

Engine speed defines the real time available for metering and pumping in the PT injector. It also determines the PT gear pump (a subcomponent of the PT pump) pressure; this increases proportionally with a rise in rpm given a constant flow area and accordingly influences rail pressure values.

Metering Orifice Sizing

The PT injector is essentially a pumping device responsible for converting rail pressure values to injection pressure values. The larger percentage of fuel entering the PT injector circulates (for purposes of lubing and cooling) and exits to the return gallery in the cylinder head, from which it is routed back to the fuel tank. The percentage of fuel to be used for fueling the engine is that forced through the **metering orifice** within the PT injector in the time period factored by the cam geometry (the PT injector pumping stroke is actuated by cam profile) and engine rpm.

Injector Timing

The pumping action of the PT injector is actuated by a dedicated cam profile (located between the valve cams in the cylinder) and injector train, consisting of a cam follower, push tube, and rocker assembly. Timing adjustments are critical and marginal inaccuracies can cause both fueling problems and physical damage to the injector and injector train.

The PT fuel system uses a common rail fueling principle. The pump is a flow control device that manages rail pressure. It feeds PT injectors positioned in parallel extending from the common rail. Fuel exiting the rail must pass through the PT injectors where a portion is metered for injection and the remainder routed to the return manifold. Fuel is flowed through a circuit in the PT injectors at any time the engine is running. The fuel used for metering is diverted from this circuit within the PT injector.

THE PT INJECTOR

The PT injector is a rocker-actuated pumping element responsible for converting rail pressures to injection pressures. It also defines a couple of critical flow areas essential for metering. In mobile diesel engine applications, it is cylindrically shaped and has two exterior annuli separated by O-rings. When installed in the injector bore in the cylinder head, the lower annular area in which the balance orifice is located is subjected to rail pressure. The balance orifice is the injector inlet orifice. In this way fuel enters the PT injector, circulates within it, and exits through an orifice in the upper exterior annulus, to be returned to the fuel tank.

The PT injector plunger extends through the length of the body and is seated in the cup at the base of the assembly. The cup orifii are sealed only when the plunger is seated. Due to the geometric design of the injector cam lobe (more than half its periphery is outer base circle), the plunger is loaded into the cup for the larger portion of the cycle. When seated, fuel will circulate through the PT injector, the flow volume dictated by whatever value the rail pressure happens to be at a given moment. When the injector cam unloads the injector train (that is, it ramps off OBC and moves toward IBC), the injector spring will lift the plunger off its seat in the cup. As the plunger is lifted, it exposes the metering orifice, permitting fuel flow through it into the cup.

Metering continues until the injector cam profile passes over the IBC of the cam profile onto the ramp toward OBC. This ramp drives the injector train (follower, push tube, and rocker) and therefore, drives the plunger downward into the cup, first purging air from the cup and next, acting on whatever amount of fuel was metered to pump it through the cup orifii. This action creates the required injection pressures as the fuel metered into the cup must be forced out through the cup orifii, directly into the engine cylinder. The sequence ends with the plunger being loaded into the cup with full mechanical crush (considerable force: equivalent to 380 lb of linear force), this being required first, to seal the PT injector from the engine cylinder, and second, to enable effective heat transfer from the cup to the plunger. The cup orifii are sized to produce an average emitted droplet sizing of 50 m when the engine is run at rated speed and loads, which Cummins considers ideal for PT engines.

PT injectors use an open nozzle design. This means that during metering, air from the engine cylinder (under compression) can siphon into the cup. As engine compression pressures exceed rail pressure by a considerable margin, this could cause fuel flow reversal. To prevent this from occurring, a check ball is located at the metering orifice, which is designed to engage any time the pressure in the cup exceeds the metering pressure. Although this interferes with the metering process, it tends only to present a problem at low engine speeds. The minute sizing of the cup orifii means that at higher speeds, there is insufficient time during metering for cylinder pressures to siphon to the injector cup. The check ball also engages whenever a full charge of fuel is metered into the cup. In this instance, the plunger acts directly on the fuel metered into the cup before it has closed off the metering orifice. Once again, the check ball (the term lock-off ball is also used) engages whenever cup pressure exceeds metering pressure, regardless of the reason.

Typically, only about 35% of the fuel exiting the PT pump rail pipe actually fuels the engine. The remainder is used for lubrication and very importantly, taking heat away from the PT injectors. This heat is substantial and the fuel subsystem should be designed to dissipate it; cylindrical aluminum tanks placed in the air flow and maintained at better than 25% full help this process. The specific percentage of the rail volume used to fuel the engine is known as the fuel rate.

Cam profile will determine discharge rate. Cam profile, engine speed, and the sizing of the cup orifii will determine the actual droplet sizing. The ideal 50 m would be obtained at rated speed and load. In injector comparator testing, fuel rate is monitored rather than total flow through the injector. It would be incorrect to describe PT injectors as unit injectors because they do not manage metering. Their function is to create the injection pressure and to atomize the fuel. The two flow areas they define are critical for determining engine fueling. For instance, if a set of low flow balance orifii were installed in a set of injectors, rail pressure parameters would go up; that is, the flow area the PT pump unloads to would be reduced so the pressure would increase. It is important to remember that all of the rail fuel must pass through the PT injectors; there is no other connection between the cylinder head rail manifold and the return manifold (Figure 24–2 and Figure 24–3).

PT INJECTOR COMPONENTS

The main components of PT injectors are identified here. Step timing control (STC) injectors have some additional components and these are described later in this chapter.

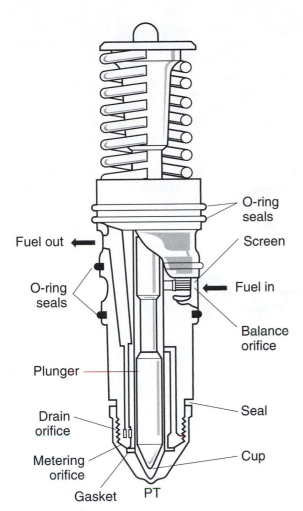

Fuel out ←

O-ring seals

Screen

Fuel in

O-ring seals

Balance orifice

Plunger

Seal

Drain orifice

Metering orifice

Cup

Gasket   PT

FIGURE 24–2   _An older PTC-type injector. (Courtesy of Cummins)_

Body

a. Provides mounting for other components
b. Provides fuel routing within the body

Plunger and Cup

The plunger and cup:

a. Form the high-pressure pump element
b. The cup provides space for fuel storage during metering.
c. The cup orifii sizing will determine droplet sizing and spray dispersal characteristics. The plunger and cup are lapped in manufacture, therefore not interchangeable.

Plunger Spring

a. Lifts plunger for metering
b. Loads the injector train

Check Ball

The check ball prevents fuel flow reversal that could be caused by preinjection pressures or seepage of compression pressures into the cup. The check ball, also known as a lock-off ball, is designed to isolate the cup pump chamber at any time cup pressure exceeds metering pressure.

Balance (Calibrating) Orifice Plug

a. Provides a means of customizing fuel flow through the injector, which permits standardization of the injector body to a number of engines
b. Balances injector sets to an engine by defining the flow area to which the rail unloads

The balance orifice is sized by burnishing, a procedure that must be performed using Cummins service literature, tooling, and a PT injector comparator.

Metering Orifice

The metering orifice is the passage within the PT injector through which fuel must pass to enter the cup. It is exposed to the cup whenever the plunger is lifted. When the plunger is lifted, some of the fuel that is flowing through the PT injector diverts through the now-exposed metering orifice to flow into the cup. The metering orifice defines the critical flow area that determines the actual quantity of fuel to be injected.

Plunger Link

The **plunger link** is a short rod that is inserted into an aperture on the PT injector follower that helps prevent side-loading of the plunger (which moves linearly) by the actuating rocker arm (which moves radially). They are surface-hardened devices and should be replaced when evidence of surface hardening failure is observed.

THE PTG-AFC PUMP

Only the PTG-AFC pump is discussed in this text. PTR pumps are not used in highway applications of PT technology and PTG pumps were replaced by PTG-AFC pumps many years ago. The PTG-AFC pump has the following functions:

1. Responsible for all movement of fuel in the system
2. Manages rail pressure
3. Provides limiting speed (LS) governing

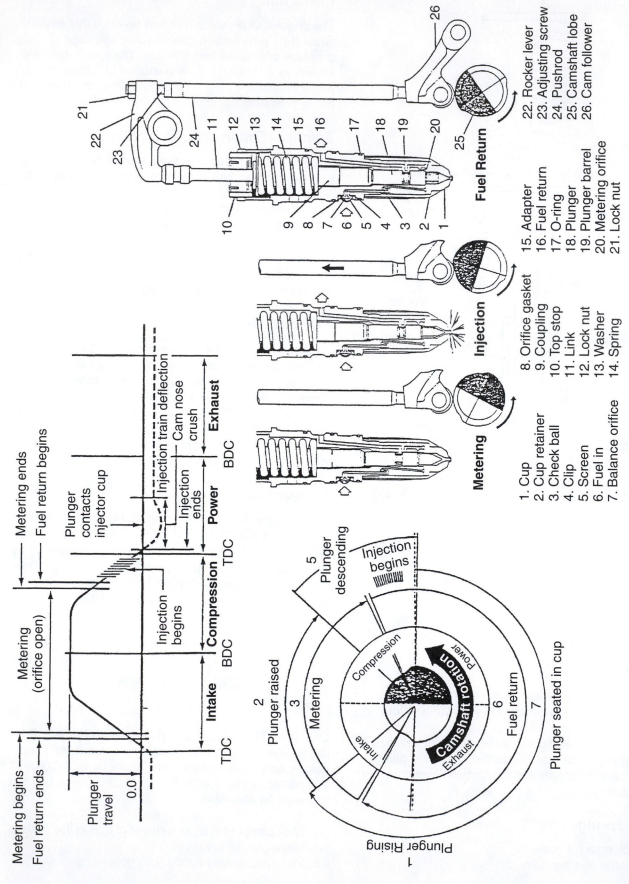

22. Rocker lever
23. Adjusting screw
24. Pushrod
25. Camshaft lobe
26. Cam follower

15. Adapter
16. Fuel return
17. O-ring
18. Plunger
19. Plunger barrel
20. Metering orifice
21. Lock nut

8. Orifice gasket
9. Coupling
10. Top stop
11. Link
12. Lock nut
13. Washer
14. Spring

1. Cup
2. Cup retainer
3. Check ball
4. Clip
5. Screen
6. Fuel in
7. Balance orifice

Fuel Return

Injection

Metering

FIGURE 24–3   *The metering and injection cycle of a PT injector. (Courtesy of Cummins)*

374

4. Limits fueling when manifold boost is low (internal aneroid)
5. Provides for ignition key engine shutdown

The PT pump is a precise flow control device that manages rail pressures, the control factor in the Cummins PT fueling equation. It is best studied by first identifying the subcomponents and subsequently analyzing performance schematics (Figure 24–4).

PT PUMP COMPONENTS

It is critical for the technician to understand the subcomponents of the PT pump and how they interact with each other in order to effectively troubleshoot PT system malfunctions.

Gear Pump

The gear pump is located at the rear of the PT pump and is encased within a cast-iron housing. It is responsible for all movement of fuel in the system. Therefore,

the entire fuel subsystem is under suction. In early PT pump versions, the fuel filter mounting pad was integral with the pump; in contemporary versions, the filter pad is always remote mounted. Remote positioning of the filter mounting pad facilitates troubleshooting the admission of air into the fuel subsystem (a not uncommon problem) by permitting the insertion of a diagnostic sight glass in series between the gear pump and the filter pad. The gear pump is positive displacement, that is, through each cycle, it unloads a constant slug volume into the outlet. Because it unloads to a defined flow area, pressure rise will be proportional to its rotational speed. It is driven at engine rpm. The pressure values produced by the pump always exceed rail pressure.

Pulsation Damper

The gear pump produces pressure in pulses, that is, there is a surge each time a slug of fuel is unloaded by the gear teeth into the ducting supplying the governor assembly. To smooth these pressure pulses, a **pulsation damper** consisting of a steel disc supported on either side by a pair of O-rings within an aluminum housing is mounted directly onto the gear pump housing. A fractured pulsation damper disc would produce erratic/fluctuating rail pressures, the result of which is engine rpm surging.

Internal Filter

The internal filter is a cylindrical core consisting of a mesh and a magnet. It is located downstream from the gear pump and its purpose is to minimize the possibility of a gear tooth cutting passing into the rest of the PT pump circuitry. It can be replaced without disassembling the pump but is not considered part of routine service maintenance and, in fact, should be left untouched.

Governor Assembly

The governor assembly (Figure 24–5) is driven at approximately two times the PT pump-driven speed, therefore smaller weights can be used. The governor employs all the principles of a typical mechanical governor. In any mechanical governor, centrifugal force generated by rotating flyweights attempts to diminish engine fueling at speeds beyond governor break (rated speed), while spring force attempts to increase engine fueling. At engine high-idle speed, the centrifugal force generated by the flyweights _must_ be capable of overcoming any of the spring forces that are applied against it to prevent engine overspeed. At any engine speed, the position of the **governor plunger** within the **governor barrel** is determined by the balance between flyweight force acting on one end of the plunger and spring force acting on its opposite end. The governor

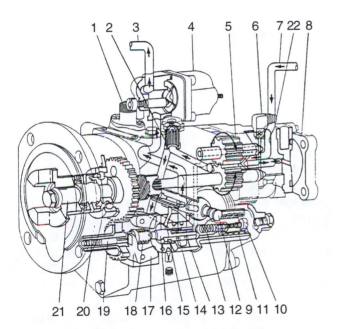

1. Tachometer drive shaft
2. Filter screen
3. Supply to injector rail
4. Shutdown solenoid
5. Gear pump
6. Check valve elbow
7. Fuel from tank
8. Pulsation damper
9. Throttle shaft
10. Idle adjust screw
11. High-speed spring
12. Idle spring
13. Gear pump pressure
14. Fuel manifold pressure
15. Idle pressure
16. Govenor plunger
17. Govenor weights
18. Torque spring
19. Weight assist plunger
20. Weight assist spring
21. Main shaft
22. Bleed line

FIGURE 24–4   _PT pump components. (Courtesy of Cummins)_

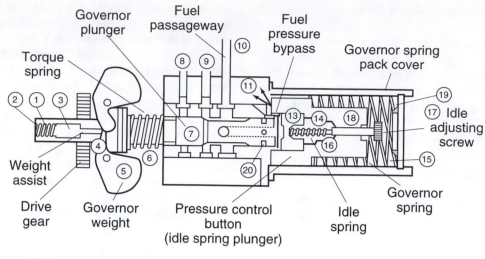

Governor plunger
Torque spring
Fuel passageway
Fuel pressure bypass
Governor spring pack cover
Weight assist
Drive gear
Governor weight
Pressure control button (idle spring plunger)
Idle spring
Idle adjusting screw
Governor spring

1. Weight assist spring
2. Weight assist spring shims
3. Weight assist plunger
4. Governor weight carrier
5. Governor flyweights
6. Torque control spring
7. Governor plunger
8. Idle passage
9. Main passage
10. Supply passage
11. Bypass passage
12. Idle plunger guide
13. Governor button
14. Idle spring
15. Governor spring
16. Idle spring seat washer
17. Idle spring adjusting screw
18. Idle screw retention spring
19. Governor (high idle) shims
20. Overspeed dump hole

FIGURE 24–5 *PTG governor. (Courtesy of Cummins)*

plunger is machined with a recessed annulus, which is cross and center drilled; it is driven by a radial lug that rests between the flyweight feet and therefore rotates within the stationary governor barrel. As speed increases, the plunger is forced down the governor barrel. Fuel from the gear pump enters the governor barrel through the supply passage at any time the engine is running.

The supply passage is always exposed to the recessed annular area of the governor plunger. The cross and center drillings hydraulically connect the plunger annulus with the **governor button**, which is spring loaded to contact the inboard end of the plunger. The sizing of the recess in the governor button will determine how much fuel pressure is required to overcome the spring pressure applied to the opposite side of the governor button and separate the button from the plunger to spill fuel back into the bypass passage for recirculation through the PT pump circuitry. The smaller the surface area of the governor button recess, the more pressure is required in the governor barrel before button separation and spill occurs. A governor button with a small recess has the effect of increasing supply pressure downstream of the governor assembly.

Fuel is spilled to the bypass passage continually; the amount spilled will determine the actual pressure in the plunger annular recess. Fuel may also exit the plunger recess through the idle passage and the main passage. At idle speeds, both the idle and main passages are exposed to the plunger annulus; as the speed increases and centrifugal force drives the governor plunger inboard, the idle passage ceases to register with the plunger recessed annulus. Through torque rise, the torque control spring becomes a factor; this spring helps define torque rise rpm. Beyond rated speed, the governor plunger has been driven inboard sufficiently that the flow to the main passage begins to diminish, reducing the flow potential to the rail.

As engine speed progresses through the droop curve (graduated fuel deration that occurs as the engine rpm increases from the rated to the high idle speeds), the flow area to the main passage continues to diminish, meaning that most of the supply fuel passes through the plunger cross and center drillings, acts on the governor button, and spills to the bypass passage. At the end of the droop curve or high idle, transverse drillings in the governor plunger called overspeed dump holes, register directly with the bypass passage and most of the fuel entering the governor barrel is recirculated to the gear pump intake. The tension of the governor spring assemblies determines how much centrifugal force (this correlates to an rpm value) is required to drive the plunger to a position where the main passage is taken out of register with the plunger annular recess. Governor spring tension is set with shims and defines the high-idle speed. The sizing of the governor button recess moderates the actual supply pressure value and affects engine fueling throughout the speed range of the engine. The governor button is loaded against the inboard end of the plunger by the idle spring. Final adjusting of this spring tension is performed dynamically (engine running) and sets the engine idle speed.

THROTTLE SHAFT ASSEMBLY

The throttle shaft assembly receives fuel from two passages in the governor assembly and supplies fuel to the AFC circuit and the no-air set screw. It is actuated by a throttle lever (accelerator lever) with a spring breakover; the throttle lever is clamped to the throttle shaft. The throttle shaft is set to move through an arc of 28 degrees ± 1 degree. Moving the throttle shaft will set a flow area from the main passage to the AFC circuit by setting the extent of register of the throttle shaft fuel orifice with the main passage. When the throttle lever travel is measured with a protractor and exceeds the required 28 degrees ± 1 degree, the complaint is usually that of low power, and it is an indication that the throttle arm stops have been tampered with.

When the throttle shaft is in the idle position, a small quantity of fuel flows through the throttle shaft fuel orifice; this fuel is defined as throttle leakage. The idle passage feeds an eccentric in the throttle assembly and bypasses the throttle fuel orifice. As the throttle shaft is rotated, the throttle shaft fuel passage is brought into full register with the main passage. Flow area through the throttle fuel orifice is set by a fuel adjusting screw within the throttle shaft bore; it is used to adjust rail pressure on the pump calibration stand.

No-Air Set Screw

The no-air set screw defines the maximum flow area from the throttle assembly to the rail *before* a predetermined manifold boost value has been achieved. It should only be adjusted on the PT pump calibration fixture.

AFC Circuit

The **air/fuel control (AFC) circuit** is an acceleration smoke control mechanism that replaced the external aneroid device used in earlier PT pump models. It is a

requirement of all turbo-boosted engines that fuel delivery is modulated to the actual amount of air charged into the engine cylinders. In a Cummins PT engine, intake manifold pressure is piped to the AFC manifold on the PT pump, where it acts on an AFC diaphragm to which the AFC plunger is attached. A spring acting on the AFC plunger loads it into its closed position. When manifold boost reaches a predetermined value, usually 15 psi, it acts on the diaphragm to overcome spring tension to drive the AFC plunger inboard, exposing the AFC ducting and increasing the fuel flow potential to the rail. Unlike most aneroid devices that are on-off, the PTG-AFC increases flow to the rail in proportion to manifold boost increase. The AFC circuit is usually fully open by the time 25 psi of manifold boost is attained, regardless of peak boost on the engine, which is usually around 35 psi.

Shutdown Solenoid

The **shutdown solenoid** is an energized-to-run device that permits the engine to be shut down using the ignition key; it acts by gating the rail. When the solenoid is not energized, a spring loads a disc into a closed position that gates the supply passage from the AFC ducting to the rail pipe. When the solenoid is energized, the disc is pulled into the solenoid permitting flow. A manual override is incorporated in the device in the event of a chassis electrical failure. When turned CW, the shutdown solenoid override displaces the shutdown disc by forcing it back against the spring, thus opening the rail circuit.

PT PUMP CIRCUIT SCHEMATICS

Probably the best way to understand the operation of a PT pump is to interpret the following sequence of schematics that show how every internal pump component responds to any given set of running conditions. The technician who is serious about accurately troubleshooting Cummins PT systems should etch these schematics into memory where they will be invaluable.

IDLE SPEED

(Figure 24–6.)

1. Fuel is pulled through the fuel subsystem by the gear pump.
2. Pressure waves caused by the gear pump are smoothed by the pulsation damper.

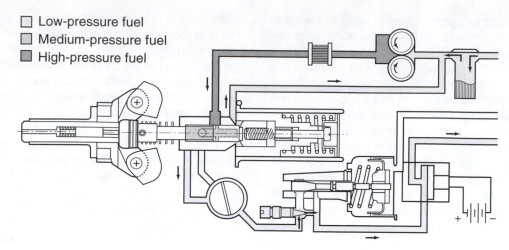

Low-pressure fuel
Medium-pressure fuel
High-pressure fuel

FIGURE 24–6   *PTG-AFC circuit schematic: idle speed. (Courtesy of Cummins)*

3. Fuel is forced through an internal filter consisting of a mesh and a magnet.
4. Fuel enters the governor assembly through the supply port in the governor barrel. The governor weights generate little centrifugal force at idle speed and, therefore, the governor plunger (which rotates within the governor barrel) is in a position in which both the idle and main ports are in register with the recessed annular area of the plunger—fuel at supply pressure will exit through the idle and main ports. Because the governor plunger is center and cross drilled, supply pressure will also act on the governor button, separating it, and recirculate fuel to the bypass circuit.
5. Fuel from the governor assembly arrives at the throttle assembly from the idle and main passages. However, the throttle is in the idle fuel position, which locates the throttle fuel orifice out of register with the main port; a small quantity of fuel, known as throttle leakage, is permitted to pass through the throttle fuel orifice. Fuel from the idle passage is permitted to bypass the throttle shaft.

6. Fuel from the throttle assembly is ducted to the AFC circuitry. However, at idle speed there is little or no manifold boost to act on the AFC diaphragm; accordingly the AFC plunger is in the closed position. The flow area defined by the no-air set screw setting is the total flow area feeding the rail pipe.
7. Fuel passes around the no-air set screw and exits the PT pump through the solenoid to which the rail pipe is connected.

NORMAL ENGINE OPERATION—OPERATION IN TORQUE RISE RPM

(Figure 24–7.)

1–3. As in the previous schematic, but pressure within the defined flow area represented by the passage feeding the governor barrel will be higher as the

Low-pressure fuel
Medium-pressure fuel
High-pressure fuel

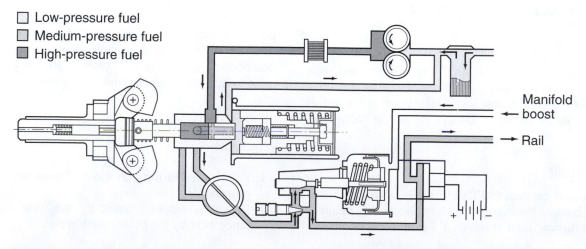

Manifold boost

Rail

FIGURE 24–7   *PTG-AFC circuit schematic: normal operation. (Courtesy of Cummins)*

positive displacement gear pump is being rotated at a higher speed.

4. Fuel enters the governor assembly through the governor barrel supply port and circulates in the annular recess of the governor plunger. However, due to the higher rpm and greater centrifugal force generated by the flyweights, the governor plunger has been driven inboard into the barrel sufficiently to have taken the idle passage out of register with the plunger recessed annulus. Supply fuel exits the governor barrel through the main passage and also acts on the governor button, spilling fuel to the bypass for recirculation. This action moderates supply pressure. The torque spring is now a factor; this helps resist further inboard plunger travel toward a diminished fuel position.

5. Fuel from the governor is ducted to the throttle assembly by the main passage. As the engine is running somewhere within the torque rise rpm, the throttle shaft is positioned so there is some degree of register of the fuel orifice with the main passage, defining a flow area.

6. Fuel flows from the throttle assembly to the AFC circuitry. As before, it flows up to and around the no-air set screw, but now there is sufficient manifold boost acting on the AFC diaphragm to overcome the AFC spring and drive the AFC plunger inboard sufficiently to permit flow through the AFC circuit, increasing flow area to the rail.

7. Fuel from the AFC circuitry passes through the shutdown solenoid and out to the rail pipe.

HIGH-SPEED GOVERNING

(Figure 24–8.)

High-speed governing begins at governor break, that is, as the engine rpm enters the droop curve. High-speed governing is a common running condition for Cummins PT engines in which the accelerator is held in the full-travel position by the operator.

1–3. Supply pressure is higher than in previous schematics due to higher rpm but otherwise the operating conditions are the same.

4. Fuel enters the governor assembly through the governor barrel supply port—but now the centrifugal force exacted by the flyweights has driven the governor plunger inboard to the extent that the main passage flow area is diminished. More fuel is being spilled by passing through the center and cross drillings acting on the governor button to enter the bypass circuit.

5. For the engine to be run in this condition, the accelerator would necessarily be fully depressed, allowing the throttle shaft fuel orifice full register with the main passage.

6. Fuel from the throttle assembly flows to the AFC circuitry. However, until the rpm penetrates well into the droop curve, the engine will be sufficiently fueled to generate enough rejected heat to ensure that the turbocharger can maintain manifold boost in excess of 15 psi. This permits at least some flow through the AFC circuit.

COMPLETE HIGH-SPEED GOVERNING

(Figure 24–9.)

Engine rpm has risen above high idle; that is, the engine is being run at an rpm beyond the droop curve. In most cases, this means that the engine is being driven by the drivetrain of the vehicle—a condition that would occur when running on a prolonged downhill. The schematic shows what is occurring in the PT pump circuitry when the engine speed exceeds the high-idle speed.

1–3. Supply pressure being dependent on rpm is at its highest.

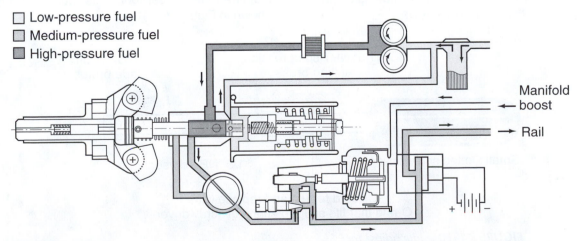

□ Low-pressure fuel
▨ Medium-pressure fuel
■ High-pressure fuel

Manifold ← boost

→ Rail

FIGURE 24–8 *PTG-AFC circuit schematic: high-speed governing. (Courtesy of Cummins)*

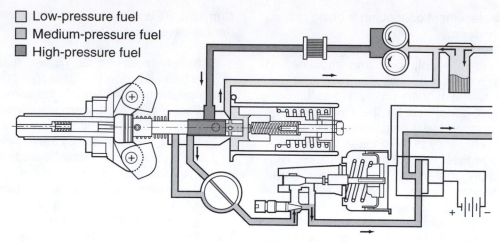

☐ Low-pressure fuel
☐ Medium-pressure fuel
■ High-pressure fuel

FIGURE 24–9 *PTG-AFC circuit schematic: complete high-speed governing. (Courtesy of Cummins)*

4. Fuel enters the governor barrel through the supply port in the governor barrel. In this running condition, the governor flyweights are maximumly extended and have driven the governor plunger inboard against the governor spring (high-speed spring) to the extent that:
   • The main passage has been entirely or almost entirely taken out of register with the recessed plunger annulus.
   • The plunger has been driven inboard to the extent that a transverse drilling through the plunger has extended beyond the governor barrel allowing most of the supply fuel to be spilled directly to the bypass.

   The rpm at which complete high-speed governing occurs is determined by the spring tension of the governor (high-idle) spring, which is set by shims.

5. At complete high-speed governing, so little fuel is being injected into the engine's cylinders, that manifold boost will drop below the 15 psi required to hold the AFC ducting open. All of the fuel that exits the PT pump to the rail must do so by passing around the no-air set screw.

PT PUMP TESTING AND ADJUSTMENTS

Cummins PT pumps are calibrated on a comparator bench—a procedure that should be performed using the Cummins service literature and specifications. This procedure is the specialty of the diesel pump room specialist and is not within the scope of this textbook. The following outlined procedures are those that the truck diesel technician would use to set up and diagnose problems occurring in a PT fuel system.

DYNAMIC IDLE ADJUST TOOL

After the PT pump has been calibrated on the comparator bench or if an exchange pump has been installed on an engine, the final adjustment to idle speed should be set to specification with the engine running. Install an idle adjust tool. This tool can be made as shown in Figure 24– 10.

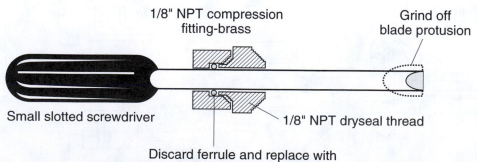

1/8" NPT compression
fitting-brass

Grind off
blade protusion

Small slotted screwdriver

1/8" NPT dryseal thread

Discard ferrule and replace with
O-ring that fits snug to shank of screwdriver

FIGURE 24–10 *Idle adjust tool.*

Dynamic-Idle Adjustment

This should be set whenever a PTG-AFC is replaced. The idle adjust tool should be installed to the access plug on the governor spring assembly cover. The engine should then be started and run until there is no performance evidence of air in the PT pump circuit, usually about 5 minutes. The idle adjust screw is then turned CW to increase the idle speed or CCW to decrease it; this procedure sets the spring tension that acts on the governor button and the higher this is, the less easily it spills from the governor barrel assembly. Always consult Cummins specifications before setting the idle speed; this setting will impact on engine fueling overall when maladjusted. The idle speed specification is stamped on the control parts list (CPL) plate (Figure 24–11).

High-Idle Adjustment

On PTG-AFC pumps, the high-idle spring tension is set by shims. To adjust the high-idle speed, the governor spring cover should be removed. Next remove the snap ring, spring assembly retainer, and shim pack. When adding or subtracting shims, 0.001″ shim thickness is approximately equivalent to 2 rpm.

FUEL SUBSYSTEM INLET RESTRICTION

The entire fuel subsystem is under suction in the Cummins PT system so ensuring the inlet restriction values are within specification is critical. These values should be tested with a mercury (Hg) manometer or gauge reading Hg (inches of mercury).

Specifications:

With a new filter: 4–6 ″Hg

Typical in-service specification: 7 ″Hg

Replace filter when specification exceeds: 8 ″Hg

When inlet restriction readings exceed specification after the filter has been replaced, any component between the PT pump and the pickup tube inlet in the tank may be responsible. Start by checking filter gaskets and remember that internal failure of the fuel hoses can cause the problem.

TECH TIP: Installation of an aftermarket filter/water separator assembly not specified for a Cummins PT engine can cause high inlet restriction readings resulting in low engine power. Check that any aftermarket devices fitted to the fuel system are specified for a Cummins PT system.

ADMISSION OF AIR TO THE FUEL SUBSYSTEM

This condition is characterized by rough engine running overall and an erratic rpm roll at all no-load speeds. Check using a diagnostic sight glass installed in series between the gear pump and the filter mounting pad. In some older applications, the filter mounting pad assembly was integral with the pump, so the sight glass would have to be located upstream from the filter. After installation, the engine should be run for about 5 minutes to purge the air that was admitted when installing the sight glass. To analyze what is observed through the sight glass, the engine should be run at a high idle speed; with good light focused on the sight glass, the fuel should appear transparent. A cloudy appearance is an indicator that air is being pulled into the fuel system; only in cases in which a large quantity of air is being admitted to the suction side of the fuel system will actual bubbles be observed.

CHECKING THROTTLE ARM LINKAGES

The PT throttle is set to travel 28 degrees ± 1 degree on the comparator bench. The travel stops should not be adjusted on-engine. The technician investigating a low power complaint should first check the accelerator linkage by having someone fully depress the accelerator in the cab; a visible breakover should be observed at the throttle arm (⅛ inch). If not, adjust the accelerator linkage to obtain the required throttle breakover

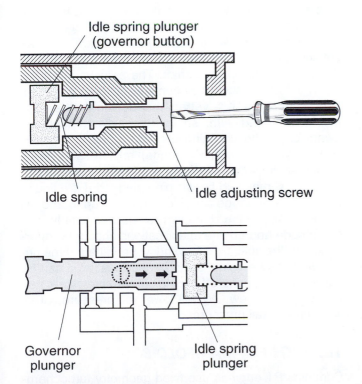

Idle spring plunger (governor button)

Idle spring

Idle adjusting screw

Governor plunger

Idle spring plunger

FIGURE 24–11   _The idle adjustment procedure._ _(Courtesy of Cummins)_

ensuring that it returns to the idle position after each adjustment.

Should this first test perform to specification, remove the accelerator linkage from the throttle arm and use a protractor to check that the throttle arm moves through the required 28-degree arc. Throttle lever travel should be set on the PT pump comparator bench. Tampering with the throttle stop settings almost always results in reducing flow at the throttle shaft bore at full travel. Should an out-of-specification throttle arm travel reading be observed, the pump should be recalibrated.

TESTING PT RAIL PRESSURE

The following tests are used by the technician to diagnose PT system malfunctions arising from incorrect rail pressure values. When performing these tests, Cummins service literature should be consulted, especially when using the chassis dynamometer as a diagnostic tool.

Snap Rail Test

The snap rail test is a fast means of checking that the rail pressure is at least approximately achieving specification. It is outlined in some of the older Cummins service literature with the caution that it is inclined to produce erratic results. The test works by "surprising" the governor into producing a surge of rail pressure. It has validity because it is a means of roughly checking rail pressure without the use of a dynamometer. Locate the specification for peak rail pressure on the engine to be tested. Insert a fluid-filled pressure gauge calibrated to read up to 250 psi in the rail pipe plug in the shutdown solenoid assembly. Start the engine and idle. Check the base rail pressure value. Next, "snap" the throttle arm to the full-speed position observing the pressure gauge. It should produce a spike reading that correlates to a percentage of the specified peak rail value before the governor responds. Check Cummins service literature for specifications.

Should a PT pump fail this test, it should not in itself condemn the pump for removal; proceed to the following test.

Rated Load and Speed Rail Test

This test should be performed on a dynamometer, preferably in conjunction with full monitoring instrumentation hookup. A fluid-filled pressure gauge capable of reading to at least 250 psi should be fitted to the rail pipe plug in the shutdown solenoid and positioned so it can be read while running the dynamometer. Manifold boost should also be monitored while performing this test. Incrementally load the engine at 100 rpm steps through the torque rise profile (see Chapter 15),

recording the rail pressure values. Peak rail pressure should read to within 2% (about 4 psi) of specification.

FUEL RATE ADJUSTMENT

This tests the fuel quantity pumped through the system that is actually used for fueling the engine. It is measured to specification by the technician at rated speed and load on an engine or chassis dynamometer in pounds per hour. The fuel rate adjustment screw is located within the throttle shaft assembly behind a ball seal. The ball seal should be removed by drilling a small hole through the seal, then gently levering it out of the throttle shaft. The fuel rate adjusting screw is either a slotted or Allen head and should be turned CW to decrease fuel rate, CCW to increase it. This procedure should only be performed using a dynamometer and the appropriate Cummins test instrumentation. Actually, it should perhaps only be performed on the calibration bench; it is outlined here because it is one of many tampering abuses this pump is subjected to and which the technician must be aware of when troubleshooting the system. It should be especially noted that backing the fuel rate screw too far out of the throttle shaft (at least 0.250" depth into the shaft bore is required) can result in the screw being forced out of the shaft.

ADJUSTING THE NO-AIR SET SCREW

This setting defines the maximum flow area from the pump circuitry to the rail when there is insufficient manifold boost to have activated the AFC circuit. The no-air set screw is located behind the throttle cover plate, directly above the throttle shaft. The adjustment of the no-air set screw should be confined to the comparator bench but again, it is yet another setting that is commonly tampered with, so the technician should be aware of it. The checking procedure seeks to determine a specific (within a range) rail pressure for a loaded engine at a set engine speed with the AFC signal (that is, manifold boost) pipe plugged. Fit a liquid-filled pressure gauge capable of reading up to 250 psi to the rail plug in the shutdown solenoid assembly. The pump code and engine specifications should be referenced for the specification window permitted. If the no-air rail pressure value is out of specification it can be adjusted by backing off the no-air set screw lock nut. Rotating the screw CW will lower rail pressure, CCW will increase rail pressure.

TESTING PEAK MANIFOLD BOOST

Cummins PT engines use fixed geometry turbochargers so testing manifold boost is a relatively easy procedure. A dynamometer MUST be used. Attempting to test manifold boost on a road test will produce generally meaningless data. Fit a good quality, liquid-filled pressure gauge (calibrated to 50 psi) to an intake manifold port and position so that it can be read while operating the dynamometer. Reference Cummins service literature for specifications. Peak manifold boost will be observed at rated speed and load.

PT PUMP INSTALLATION

The PTG pump is flange mounted to an accessory drive rotated at engine rpm; it is not timed to the engine. When removing the pump from the engine, all the plumbing ports should be sealed with dust caps. When reinstalling the pump, the flange gasket should be replaced and the drive coupling spider inspected for serviceability. The pump is often installed to the rear of the compressor housing in truck applications.

1. Install the coupling spider to the drive lugs, place a new flange gasket in position, and torque the four mounting bolts.
2. Connect the fuel and AFC plumbing. Connect the shutdown solenoid wiring. Connect the tachometer cable when fitted.
3. Remove the priming plug (a standard pipe plug located at the upper deck of the PT pump) and fill the pump housing with filtered fuel. Replace and prime the fuel filter.
4. Attach the accelerator linkage. Adjust to ensure the throttle arm moves through the required 28-degree arc and produces a breakover of 1/8 inch. Set the accelerator pedal stop.

5. Start the engine. There will usually be evidence of air in the system for about 5 minutes but generally it will self-purge.

PT INJECTOR INSTALLATION AND ADJUSTMENT

The engine firing order is 1-5-3-6-2-4.

Short cuts should never be used when performing Cummins PT overhead adjustment. It should also be noted that when the A, B, and C (1-6, 2-5, and 3-4 on older engines) calibration indices on the accessory drive pulley are aligned with the calibration notch on the engine block, no pair of pistons is at TDC. See Chart A.

Always consult Cummins technical data when setting the overhead adjustments. The following is a brief description of the methods used.

PRETORQUE

Injector cam lobe must be on OBC (outer base circle: the largest radial dimension of a cam profile), that is, _not_ as shown in Chart B. This method sets the actual loading of the plunger into the cup, a linear force of around 380 lb. This is done by torquing the injector train adjusting screw on its actuating rocker to a specific value. The torque applied to the adjusting screw varies between 45 lb.-ft. to 90 lb.-ft. depending on the rocker housing casting material (aluminum/cast iron) and whether the adjustment is performed hot or cold.

CHART A

ENGINE POSITION		PISTON LOCATION				
A or 1–6	1–6	90 degrees ATDC	2–5	30 degrees BTDC	3–4	150 degrees BTDC
B or 2–5	2–5	90 degrees ATDC	3–4	30 degrees BTDC	1–6	150 degrees BTDC
C or 3–4	3–4	90 degrees ATDC	1–6	30 degrees BTDC	2–5	150 degrees BTDC

CHART B _Top end set positions—rotate the engine CW._

ENGINE POSITION	SET INJECTOR	SET VALVES
A or 1–6 (#5 on compression)	3	5
B or 2–5 (#3 on compression)	6	3
C or 3–4 (#6 on compression)	2	6
A or 1–6 (#2 on compression)	4	2
B or 2–5 (#4 on compression)	1	4
C or 3–4 (#1 on compression)	5	1

DIAL INDICATOR

The injectors should be set with the engine location as shown in Chart B. That is, each injector is set when its actuating cam is at a specific location of its base circle. This adjustment requires that plunger free travel is set to precise dimension. A dial indicator is located in position to ride on the plunger flange and the rocker is actuated using a rocker actuating claw. The travel of the plunger is then set by adjusting the rocker adjusting screw to an exact value; 0.170″ ± 0.001″ is typical for N series, but again, always consult the OEM specifications. Setting PT injectors by this method should ensure that the required loading of the plunger into the cup on OBC is achieved and the method is preferred by Cummins.

TOP STOP

Top stop PT **injectors** are adjusted as shown in Chart B; in other words, as with the dial indicator method, they are set on a specific location of IBC (inner base circle or the smallest radial dimension of the cam profile). However, these are actuated by a train that Cummins wants set with zero lash. Zero lash is identified by Cummins as having a slight load on the injector actuating train. The setting is achieved by torquing the train adjusting screw to 5–6 lb.-in. It is good practice to initially overtorque the specification by about 50%, back off, then retorque to specification. Correctly set, the setting can be checked by attempting to rotate the injector link—it should rotate but with some difficulty.

STC INJECTOR SETTING

Step timing control (STC) injectors can be set using an OBC method. It is not straightforward, however, and Cummins Service Bulletin #3666006–00 of May 1991 should be consulted.

LOCATING CYLINDER MISFIRE

This process is the equivalent of shorting out hydraulic injectors by cracking the union nuts on high-pressure pipes. A tool called a rocker claw or rocker lever is used to perform this test on a PT-fueled engine. This tool has an arced yoke that fits under the rocker shaft and with the help of a ¾″ or 19 mm wrench, levers the injector rocker into its fully actuated position. This action holds the PT injector plunger seated in the cup, preventing it from metering and pumping. An idling engine may be tested for cylinder misfire using this test; each cylinder should be cut out (the term *shorted* is often used) in sequence using the rocker claw. If rpm drops each time a PT injector is cut out, that injector can be

presumed to be functioning properly. If the engine rpm does not drop when an injector is shorted out, that injector should be removed and tested.

CAUTION: Care should always be taken when working above an exposed rocker assembly on a running engine.

CAM FOLLOWER HOUSING TIMING/OFFSET CAMSHAFT KEY TIMING

Actuation of the fuel system pumping pulse by the engine timing gear train must be precisely phased in any diesel engine. This process essentially times the events that manage the fuel system pumping pulse to a specific engine position. Cummins achieves this using two methods depending on the engine series. The **offset camshaft key timing** (the camshaft gear is set to a specific axial position on the camshaft by the offset of the locating key) is used on L-, M- and K-series (Figure 24–12), whereas **cam follower housing timing** is used on N-series. To perform cam follower housing setting on N-series, PT engines, a Cummins engine timing fixture is used. The timing fixture consists of a bracket that mounts to the cylinder head (Figure 24–13). The timing fixture is fitted with two dial indicators and probe assemblies. One probe is positioned so that it can be inserted through the injector bore in the cylinder head to contact the piston crown and the second probe is positioned to ride on the injector push tube. In this way, one probe is set to read engine position while the other reads camshaft position.

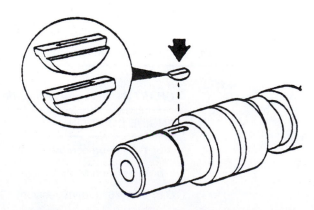

FIGURE 24–12 *The cam timing procedure used on some PT engines showing the offset camshaft key method. (Courtesy of Cummins)*

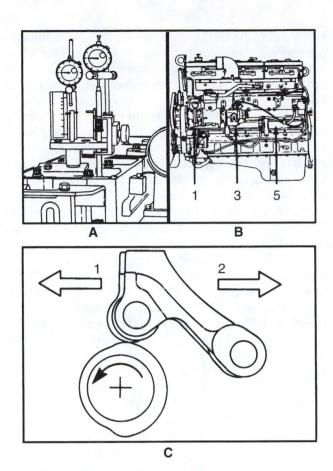

FIGURE 24–13 A: *Injection timing tool in position; the tool determines injector pushrod travel in relation to piston travel.* B: *Location of the cam follower housing and reference cylinders.* C: *1. Advance 2. retard.* *(Courtesy of Cummins)*

An adjustment is made by either adding or subtracting gaskets that seal the cam follower housing and cylinder block. Seen from the front of the engine, the camshaft rotates CCW (viewed from the front); the camshaft and cam follower assemblies are positioned on the left side of the engine (viewed from the rear). The result is that if shims are added to the cam follower housing, the cam follower rollers (which actuate the injector train) will register with any portion of the cam geometry slightly earlier. Similarly, if shims are removed from the cam follower housing, the cam follower rollers will move farther inboard and away from the camshaft, registering with any portion of the cam geometry fractionally later. Therefore, in N-series engines, shimming up the cam follower housing advances the injector train actuation timing, whereas subtracting shims, retards timing.

The procedure is outlined in Cummins technical service literature and is straightforward, but the method

must be meticulously adhered to. The procedure should be undertaken at each major overhaul; measuring and replicating the shim (gasket) thickness on disassembly is not considered sufficiently accurate; however, it is a good starting point when beginning the timing procedure. The offset camshaft gear achieves the same objective; it phases fuel injection actuation with a precise engine position. Several offset keys are available and their function is to alter the axial relationship between the camshaft gear and the camshaft.

STEP TIMING CONTROL

A characteristic of PT, common rail injection is to have a variable beginning and constant ending of the injector fueling pulse. This translates into the end of injection always occurring in exactly the same engine location regardless of engine load or speed. The Cummins PT fueling window was originally engineered for optimum performance around the rated speed and load operating zone. This meant that when the engine was operated at low speeds and lighter loads, the beginning of the injection pulse was effectively retarded. In fact, because the beginning of injection depends directly on how much fuel has been metered into the cup, the lighter the engine load, the more retarded the beginning of injection becomes. In an effort to rectify this problem, Cummins introduced the **step timing control** or **STC** system.

STC is both hydraulically actuated and managed. It attempts to "advance" the beginning of injection whenever engine loading is light. Although the term "advance" is used in describing the STC system, perhaps the objective of STC should more properly be described as getting injection pulse within a *normal* range. Figure 24–14 shows the components used in an STC injector.

The STC injector components are shown in Figure 24–14.

STC hardware consists of PTD top-stop style injectors fitted with tappets and an STC control valve, plus an interconnecting hydraulic circuit (Figure 24–15). The STC tappet is located at the top of the PT injector; when this tappet is charged with engine oil, it effectively lengthens the PT plunger. The STC actuating cam profile has greater lift than non-STC injector cams.

As a result, when the injector train is actuated with the STC tappet charged, the plunger both begins its downstroke earlier, then bottoms into the cup before peak cam lift is attained. When the plunger bottoms

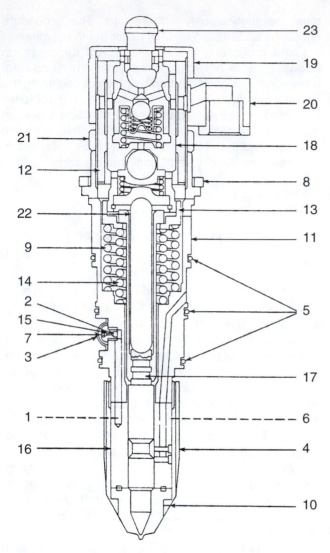

1. Check ball
2. Injector gasket
3. Screen retainer
4. Cup retainer
5. O-ring seal
6. Roll pin
7. Filter screen
8. Washer
9. Plunger spring
10. Injector cup
11. Injector adapter
12. Stop screw
13. Spring retainer
14. Plunger spring
15. Metering orifice plug
16. Barrel
17. Plunger
18. STC tappet
19. Tappet stop cap
20. Tappet top stop
21. Lock nut
22. Lock nut
23. Injector link

FIGURE 24–14   *STC injector components. (Courtesy of Cummins)*

into the cup, the STC actuating oil is trapped in the tappet so the oil pressure increases until a load check ball is unseated. This occurs at between 1,100 and 1,500 psi, spilling the oil from the tappets into the rocker housing, from which it is returned to the sump.

Whenever the STC tappets are charged with engine oil by the STC control valve, the timing operates in advanced mode. This occurs at light engine loads. When the STC control valve sets normal timing mode, the STC tappets no longer charge with oil, so the first portion of the injector cam flank travel into the PT injector is responsible for collapsing the STC tappet and the PT plunger is then actuated mechanically rather than hydraulically.

The STC control valve monitors actual rail pressure and uses predetermined values to differentiate between normal and advanced modes. Most commonly, the critical value is 53 psi. In other words, when rail pressure is 53 psi or less, the STC control valve will operate in advance mode and charge the STC tappets. As rail pressure rises beyond the critical 53 psi parameter, the STC control valve switches to normal operation. Normal operation continues until rail pressure decreases to 25 psi; at this point, the system switches to advanced mode once again.

Direct fuel feed (DFF) injectors are a later version of the STC injector. DFF injectors are designed to prevent metering when the engine is in a motoring condition (throttle arm at zero travel, chassis momentum driving engine). This reduces noxious emission, cup orifii carboning, and of course, wasted fuel.

PACE AND PACER

PACE and **PACER** are Cummins PT system, partial authority computer control systems. A **partial authority** engine management system is one in which a hydromechanical fuel system is adapted to computer controls; Chapter 32 on computerized engine management should be consulted to understand the operating principles of ECM-managed engines. A brief account is provided here.

The PACE/PACER management system adapts the PT system for some electronic controls but leaves most of the key components of the system operating as if there were no computer controls. The primary feature of PACE/PACER is road speed governing. In fact, PACE and PACER are not true acronyms but instead refer to the road speed governing feature. A major disadvantage of the hydromechanical PT system in truck

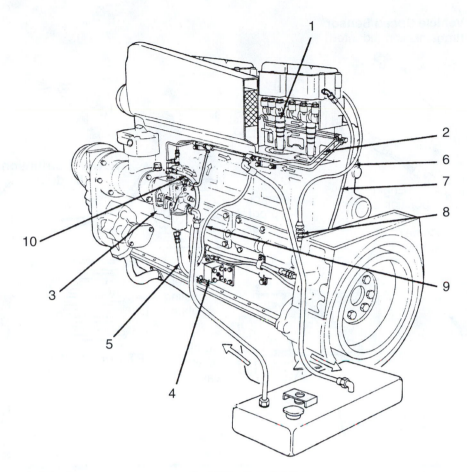

1. Injector
2. Injector drain line
3. Fuel pump
4. STC control valve
5. STC oil supply line
6. STC oil outlet line
7. C-brake sensing line/oil drain line
8. Check valve
9. STC fuel drain line
10. STC fuel pressure sensing line

FIGURE 24–15   *STC circuit layout. (Courtesy of Cummins)*

applications was the tendency of drivers to operate the engine in the governor droop curve rpm. That is, drive the vehicle with foot to the floor, resulting in an engine speed somewhere above rated and below high idle. When an engine is run in the droop curve it will usually run at lower-than-optimum efficiency and produce higher levels of noxious emissions. The PACE/PACER essentially governs the operator out of the droop curve.

The PT fuel system remains largely intact. PACE/PACER is mastered by a **PT control module (PTCM)** supplied with inputs from a transmission tailshaft-mounted vehicle speed sensor (VSS), a brake switch sensor, clutch switch sensor, and engine position sensor. The PTCM processes input data and controls fueling by means of an electronic fuel control valve assembly mounted on top of the PT pump. This unit is equipped with a throttle bypass valve positioned in parallel with the PT throttle shaft, which is capable of routing fuel alternately (bypassing the rail) when energized. Its function is solely to limit fueling; the PTCM can thereby define a maximum road speed or a set PTO (power takeoff) mode speed.

The PACE/PACER system is a very early example of electronic management of a highway truck diesel engine and there are not too many examples of the technology on the roads today. The foregoing description is included here to emphasize that the standard PT system is left largely intact. In short, PACE/PACER is a hydromechanically controlled PT fuel system with electronic governing of road speed, cruise control, and PTO operation. Should the PACE/PACER system fail, the PT system continues to operate but at slightly reduced power (Figure 24–16).

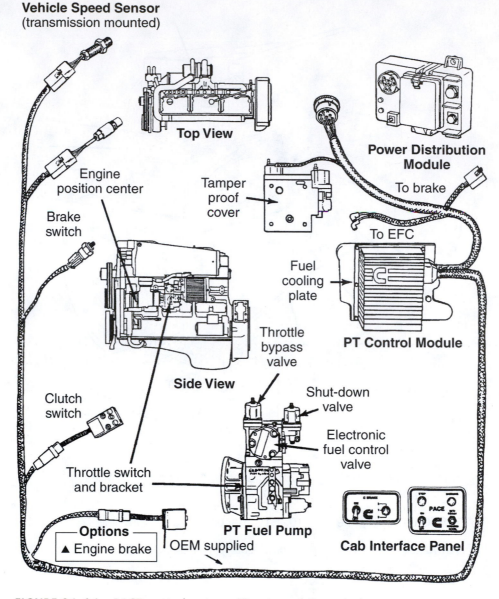

FIGURE 24–16 *PACE control system. (Courtesy of Cummins)*

SUMMARY

- A hydraulic equation constructed to calculate a volume of flow is factored by the time of the flow, the flow area, and fluid pressure.
- The Cummins PT system uses the pressure variable in a simple hydraulic equation to manage fueling.
- Two critical flow areas are defined by the PT injector: the balance orifice and the metering orifice.
- The collective restriction of the balance orifii in the rail (six in a six-cylinder engine) defines the flow area required to factor rail pressure.

- The PT injector metering orifice defines the flow area for determining the actual volume of fuel metered to the engine cylinder.
- PTD top-stop injectors have plunger travel set internally and are adjusted to be actuated by a zero lash injector train.
- PTD-STC-DFF injectors are used on later version PT engines.
- Cummins identify both a zero lash and OBC method of setting PTD-STC-DFF injectors.
- The PTG-AFC pump is a flow control and governing device that manages rail pressure values to control engine fueling.

- The PTG-AFC pump uses mechanical governing and most versions are of the limiting speed type with enhanced torque rise fueling and reduced fueling in conditions of low manifold boost.
- A PTG-AFC pump should be diagnosed on the engine before removal.
- The technician should learn how to identify common tampering abuses of PTG-AFC pumps.
- Cummins PT injectors have a variable beginning (dependent on metered fuel quantity) and constant ending (when plunger bottoms in cup) fuel delivery characteristic resulting in retarded injection in low load/speed operation.
- STC injectors use hydraulic tappets to advance injection timing when rail pressure values are lower than a specified value.
- Control of the STC tappets is the function of the STC control valve.
- PACE/PACER is a partial authority computerized management system that adapts PT hydromechanical fueling for electronic governing of maximum road speed, cruise control, and PTO operation.

REVIEW QUESTIONS

1. The PT injector is actuated by:
   a. Spring pressure
   b. Hydraulic pressure
   c. Mechanical pressure
   d. Electrical pressure
2. Which of the following represents the critical flow area when factoring PT rail pressures?
   a. Collective restriction of the balance orifii in the rail
   b. Collective restriction of the metering orifii in the rail
   c. Outlet orifice in the PT pump solenoid
   d. PT pump housing pressurizing valve
3. When considering the actual quantity of fuel delivered to the cylinders of a PT-fueled engine, which of the following represents the critical flow area?
   a. Balance orifice
   b. Metering orifice
   c. No-air screw
   d. AFC valve
4. Which of the following represents the correct maximum fuel inlet restriction specification for a PTG pump?
   a. 8 inches H_2O
   b. 8 inches Hg
   c. 8 psi
   d. 15 psi
5. Removing shims/gaskets from the cam follower housing assembly would have what effect on PT injector timing?
   a. Advance
   b. Retard
   c. No difference
6. When setting an injector using the pretorque method, which of the following should be true?
   a. Injector cam profile is on outer base circle.
   b. The piston is at TDC completing its compression stroke.
   c. The injector cam profile is on inner base circle.
7. Which would be the correct amount of travel of the throttle arm on a PTG pump?
   a. 15 degrees
   b. 28 degrees
   c. 90 degrees
   d. 170 degrees
8. When using the dial indicator method of setting PT injectors, which of the following should be true?
   a. Injector cam profile is on outer base circle.
   b. Injector cam profile is on inner base circle.
   c. The piston is at TDC completing its compression stroke.
9. When setting 0 lash (top-stop) D injectors, which of the following would result in a correct adjustment?
   a. Click-type torque wrench set at 0–1 lb.-in.
   b. Dial-type torque wrench indicating 0–1 lb.-in.
   c. Dial-type torque wrench indicating 6 lb.-in.
   d. Click-type torque wrench set at 72 lb.-in.
10. After diagnosing fluctuating rail pressure values on a PTG-AFC-fueled engine, which of the following components would most likely be at fault?
    a. Solenoid
    b. Throttle arm
    c. AFC diaphragm
    d. Pulsation dampener
11. When checking for air in a PTG pump, which would be the location of choice for an in-line diagnostic sight glass?
    a. In the rail
    b. Between gear pump and filter pad
    c. Between filter pad and separator
    d. Between separator and tank
12. When considering how a PT pump is coupled to an engine, which of the following would be true?
    a. Pump timed to engine, driven at crank speed
    b. Pump not timed to engine, driven at crank speed
    c. Pump timed to engine, driven at cam speed
    d. Pump not timed to engine, driven at two times crank speed
13. The PTG governor weight carrier is driven at:
    a. Camshaft speed
    b. Crankshaft speed
    c. More than two times crank speed
    d. Pump driven speed

14. Which of the following would increase the high-idle value in a PTG pump?
    a. Subtract shims acting on the maximum speed spring.
    b. Add shims acting on the maximum speed spring.
    c. Turn CW the maximum speed spring adjusting screw.
    d. Turn CCW the maximum speed spring adjusting screw.

15. Which of the following would increase the idle speed value on a running Cummins PT engine?
    a. Add shims acting on the idle spring pack.
    b. Subtract springs acting on the idle spring pack.
    c. Turn CW the idle spring adjust screw.
    d. Turn CCW the idle spring adjust screw.

16. Approximately what average percentage of fuel exiting the PTG-AFC pump actually fuels the engine?
    a. 10%
    b. 35%
    c. 65%
    d. 100%

17. If a set of lower flow injectors were installed in a Cummins engine, rail pressures would:
    a. Increase
    b. Decrease
    c. Not be affected

18. Under which running condition would engine compression pressures be most likely to interfere with the metering process in a PT-fueled engine?
    a. Operation in the droop curve
    b. Rated speed
    c. Peak torque
    d. Idle speed

19. *Technician A* states that the AFC circuit helps minimize smoking while the engine is being accelerated. *Technician B* states that the AFC circuit can help reduce overfueling when running the engine at high altitudes. Who is right?
    a. A only
    b. B only
    c. Both A and B
    d. Neither A nor B

20. *Technician A* states that when the engine calibration index C aligns with the timing notch, the companion pistons #3 and #4 are at TDC. *Technician B* states that when the engine calibration index B aligns with the timing notch with #3 piston on its compression stroke, the PT injector on cylinder #3 can be adjusted. Who is right?
    a. A only
    b. B only
    c. Both A and B
    d. Neither A nor B

Chapter 25

Rotary Distributor Pumps

PREREQUISITES

Familiarity with Section 1 of textbook and Chapters 18, 19, and 20

OBJECTIVES

After studying this chapter, you should be able to:

- *Identify the two categories of rotary distributor pump likely to be found on small-bore diesel engines.*
- *Identify the main components of an inlet-metering, distributor injection pump.*
- *Describe the operating principles of an opposed-plunger, inlet-metering injection pump.*
- *Explain how fuel is routed, pumped and metered from the fuel pump to the injector during an inlet-metering, opposed-plunger injection pump cycle.*
- *Outline the circuits in an inlet-metering, opposed-plunger rotary distributor pump.*
- *Identify the main components of a sleeve-metering, distributor injection pump.*
- *Describe the operating principles of a sleeve-metering, distributor injection pump.*
- *Explain how fuel is routed, pumped, and metered from the fuel pump to the injector during a sleeve-metering, distributor injection pump.*
- *Outline the circuits in a sleeve-metering, distributor injection pump.*
- *Recognize and describe the operating principles of Stanadyne Roosa, Delphi Lucas CAV, and Bosch rotary injection pumps.*

KEY TERMS

cam plate
distributor head
distributor plunger
distributor rotor
hydraulic head
inlet-metering
internal cam ring
opposed plungers
sleeve metering
thrust collar

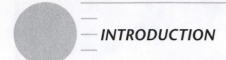

INTRODUCTION

A significant disadvantage of a port-helix metering injection pump is up-front cost due to its requirement for close tolerance manufacture and a pump/metering element for each cylinder on the engine. This cost factor becomes increasingly noticeable in small-bore engines in which the fuel system alone (pump/lines/nozzles) can represent up to 35% of the total market cost of the engine. This fact gave manufacturers plenty of incentive to look for options to Robert Bosch's 1927 fuel pump. Over the years, many alternate designs of diesel fuel pumping apparatus have been used to manage diesel engines, but few have lasted. Many of these alternate designs used rotary distributor injection principles, the objectives of which were to reduce complexity, cost, and the number of pumping elements.

Rotary distributor injection pumps have a significant initial cost savings advantage over equivalent port-helix metering pumps, but they also have the disadvantage of reduced longevity. Over the years, rotary distributor pumps have used a number of different operating principles, but only a couple of designs have survived. Today, rotary distributor pumps are mostly associated with small-bore diesel engines, and it has been many years since they have been used on American built truck engines in the medium- to large-bore range. Larger bore engines can be fueled by rotary distributor pumps, but this use tends not to be the case in North America where there is a requirement for good longevity and minimal downtime, both disadvantages of rotary distributor pumps. Because rotary distributor pumps are simple in construction, they have been popular in geographic regions where fuel quality is less reliable and necessitates frequent pump disassembly and cleanup. Many distributor pumps can be disassembled for cleaning without requiring complex comparitor bench adjustments following reassembly.

The first practical design for a rotary distributor pump was engineered by Vernon Roosa, a diesel mechanic employed by the City of New York, where he maintained and repaired diesel-driven generators. Roosa patented his design in 1941, but it took him some time to generate any interest in his invention. He tried to interest both American Bosch (now Ambac) and CAV (now Delphi/Lucas CAV) in his patent. Following the conclusion of World War II, a machine shop best known for the manufacture of aircraft engine components for Pratt and Whitney, the Hartford Machine and Screw Company, decided to test out the Roosa design and brought its inventor to Hartford to assist in the research and development. At the same time, CAV in the United Kingdom started to work with the Vernon Roosa blueprints, resulting in the simultaneous introduction of the Roosa pump (manufactured by Hartford Machine and Screw Company) and a CAV version of the pump in the year 1952. The CAV DPA version pump became the biggest selling fuel injection pump ever manufactured worldwide, and it is still used today in areas of the world in which noxious emissions and fuel economy are subject to less regulation than in North America.

Although the Roosa design for a rotary distributor injection pump was not the first, it was easily the most successful and accordingly, it generated a succession of copycat designs. Only one of these, the Bosch VE, lasted into the 1990s and is used primarily on small automotive diesel engines. The attraction of the rotary distributor injection pump was low cost, simplicity, and easy serviceability. The Hartford Machine and Screw Company is now part of Stanadyne Corporation and therefore, the Roosa injection pump is today known as a Stanadyne pump. Currently, the two versions of the Roosa pump (Stanadyne and Delphi/ Lucas/ CAV) and the Bosch design are still seen with partial authority management electronics.

In this chapter, two distinct types of rotary distributor pumps are addressed. It is important that you do not confuse them because they use very different operating principles. We begin by looking at the Roosa pump, which is generically described as an **inlet-metering**, opposed-plunger, distributor injection pump, then examine Bosch hydromechanical distributor pumps. Truck technicians should be aware that, while they are unlikely to come across either of these injection pumps in current highway trucks, in rural areas where the truck shop does all the diesel repair work, having some knowledge of these pumps can be useful.

OPPOSED-PLUNGER, INLET-METERING INJECTION PUMPS

In terms of operating principles, the Roosa injection pump and its close relative, the CAV DPA are nearly identical. Although the DPA significantly outsells Stanadyne/Roosa pumps worldwide, this is not true in North America, where both GM and Ford have used the Stanadyne pump. For purposes of describing the operation of the pump, the DB 2 version of the Roosa pump is primarily referenced, but remember that there are a number of different versions within the Stanadyne Roosa and Delphi/Lucas/ CAV families of opposed-plunger pump.

ROOSA DB2

When the Roosa DB2 pump was first introduced, it was marketed as a Roosa-Master pump. Older technicians might often use this term today. The DB2 injection pump is an opposed-plunger, inlet-metering, distributor type, diesel injection pump. It was designed for low cost production and simplicity. A typical DB2 pump has a total of around 100 component parts and only four main rotating members. There are no spring-loaded components, none are lap-fitted in manufacture, and there are no ball bearings or gears. The pump has a single pumping chamber in which two **opposed plungers** are actuated by an **internal cam ring**. Because these injection pumps are used primarily with small-bore, low-cost diesel engines, low initial cost is a feature of the pump.

The geometry of the **hydraulic head** determines the distribution of fuel between cylinders, and because fuel flow can be preset, lengthy periods on the fuel pump test bench are avoided. DB2 pumps are self-lubricated and contain roughly the same number of component parts regardless of the number of cylinders served. They may be mounted in any position on the engine.

MAIN PUMP COMPONENTS

The operation of an inlet-metering, opposed-plunger distributor pump is not complicated. First, it is necessary to be able to identify the major components and circuits of the pump. A cutaway version of the DB2 version of the Roosa pump which we use as our primary reference for study is shown in Figure 25–1.

Using Figure 25–1, note that the drive shaft directly engages with the **distributor rotor** in the hydraulic head. In other words, the distributor rotor rotates within the stationary hydraulic head. It is difficult to see in Figure 25–1, but the drive end of the DB2 rotor incorporates two pumping plungers that oppose each other in a single chamber. These plungers are actuated by the

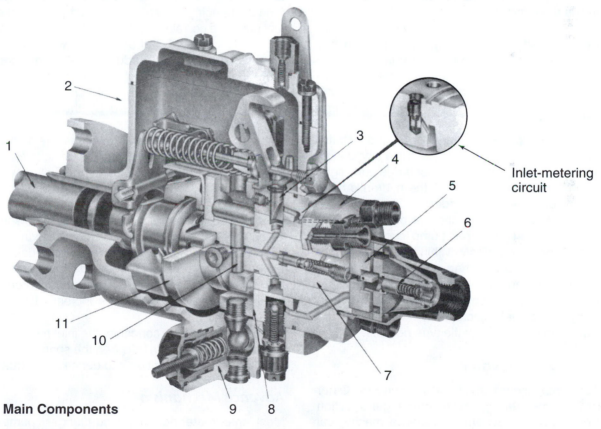

Main Components

1. Drive shaft
2. Housing
3. Metering valve
4. Hydraulic head assembly
5. Transfer pump blades
6. Pressure regulator assembly
7. Distributor rotor
8. Internal cam ring
9. Automatic advance (optional)
10. Pumping plungers
11. Governor

FIGURE 25–1 _Cutaway of a Stanadyne DB2 injection pump. (Courtesy of Stanadyne Automotive Corporation)_

internal cam ring. (See Figure 25–3 for a closer look at the pump chamber components.)

OVERVIEW OF OPERATING PRINCIPLES

The DB2 injection pump is driven through one complete rotation per complete engine cycle. This means that it is rotated at camshaft speed, that is, to turn the pump through one full 360 degree rotation, the engine has to be turned through two complete rotations or 720 degrees. The pump chamber is formed between the pump plungers, and they are actuated toward each other simultaneously by an internal cam ring. The plungers ride on rollers and shoes that are carried in slots at the drive end of the rotor. The number of cam lobes normally equals the number of engine cylinders and in turn determines how many effective strokes will occur during each pump cycle.

Pump Components

A fuel transfer pump is located at the rear of the rotor. The pump is a positive displacement vane type so it will displace fuel in proportion to rotational speed. The assembly is enclosed in the end cap, which also houses the fuel inlet strainer and the transfer pump pressure regulator. The face of the regulator assembly is compressed against the liner and distributor rotor and forms an end seal for the transfer pump. The injection pump is designed so that end thrust acts against the face of the transfer pump pressure regulator. The distributor rotor incorporates two charging ports and a single axial bore with one discharge port. This discharge port aligns with stationary distribution ports in the hydraulic head dedicated to each engine cylinder. As the distributor rotor rotates, it is brought in and out of register with each discharge port once per pump cycle. The hydraulic head contains the bore within which the distributor rotor rotates, the metering valve bore, charging ports, and the hydraulic head discharge fittings. High-pressure injection pipes connect these discharge fittings to hydraulic fuel injectors. In most cases, the hydraulic fuel injectors use pintle type nozzles.

Governing and Metering

The DB2 pump uses a mechanical governor. Other versions of inlet-metering, opposed-plunger injection pumps use hydraulic governors. When a mechanical governor is used, speed regulation is achieved by using centrifugal weights to measure the driven speed of the pump (to sense engine speed). This is opposed by variable spring force, which is regulated by the driver's accelerator position (this inputs a fuel demand request). In the governor weight carrier, centrifugal force is transmitted through a sleeve to the governor arm and from there uses a linkage to the metering valve.

All fuel that is allowed to enter the pump chamber formed by the two opposed plungers must pass through the metering valve. The metering valve rotates in bore and defines a flow area to the duct that feeds the pump chamber. Depending on its rotational register with this duct, fuel flow to the pump chamber can be managed from no fuel graduating up to full fuel. The no-fuel rotational position of the metering valve is achieved by moving a dedicated shutoff solid linkage by an independently operated shutoff lever (outside of the governor housing) or by an electrical solenoid.

Hydraulic Governing

Many versions of the Delphi Lucas CAV pumps use hydraulic governing. Hydraulic governing simplifies the injection pump and reduces its size. In hydraulically governed versions of this pump, the metering valve moves longitudinally in its bore instead of rotationally. Because the fuel transfer pump is a positive displacement vane type pump, it displaces fuel proportionately to engine rpm. If this fuel is unloaded into a defined (and unchanging) flow area, it makes sense that engine fuel pressure correlates with rpm. The metering valve is cross and center drilled and milled with a graduated metering recess. This graduated metering recess can be brought into and out of register with the metering duct that directs fuel to the pump chamber: its longitudinal position is determined by the forces that act above and below it.

Below the metering valve, fuel pressure acts against the sectional area of the valve, attempting to force it upward, tending to reduce fuel delivery. Fully upward, no fuel is metered to the pump chamber and shutdown occurs.

Opposing the hydraulic force acting below the metering valve is spring force. This force is variable and determined primarily by the accelerator pedal position. As the operator pushes harder on the accelerator pedal increasing the pedal angle, the spring force acting against the hydraulic pressure is increased, resulting in more fuel delivery.

So, in any running condition, the inlet metering valve establishes equilibrium between the spring and the hydraulic forces that position it to define a fuel quantity.

Advance Mechanism

Most inlet-metering, opposed-plunger pumps are equipped with an automatic advance device that can advance fuel injection timing. This can be either a hydraulic or electrically actuated mechanism that advances or retards the pumping cycle. The **internal cam ring** is responsible for actuating the pumping plungers. When the rollers contact the internal cam ramps, the plungers are forced inboard against each other, pressurizing the fuel in the pump chambers in between them. If the cam ring is rotated toward the direction of rotor rotation, an advance is achieved. An external lug protrudes from the cam ring into the advance chamber below it, permitting a window of advance.

In the case of hydraulic advance, spring force acts on one side of the cam ring lug, positioning it in the most retarded location. An advance piston is located on the other side of the lug, and fuel pump pressure from the vane pump is allowed to act on this piston. Because a vane pump is positive displacement, when it unloads to a defined flow area, pressure rise will be proportional to pump rotational speed. When this fuel pressure acts on the advance piston, it overcomes the spring pressure opposing it to move the cam ring to advance fuel delivery timing. The result is that any advance achieved will be entirely speed sensitive. The same thing can be accomplished electrically in later versions of the pump by using a linear proportioning solenoid controlled by the engine ECM.

FUEL FLOW

Now that you have some understanding of the pump operation, we can take a look at exactly how the fuel is routed through the pump during its cycle. It makes sense to use both the text and the figures to understand how this is achieved. Reference Figure 25–2, a fuel flow schematic of the DB2 pump.

Movement of fuel through the fuel subsystem is the responsibility of the transfer pump, which is the vane pump integral with the injection pump. Fuel is pulled from the main fuel tank through a filter or filters into the pump inlet. At the pump inlet, fuel is pulled through the inlet filter screen by the transfer pump. Some fuel is bypassed through the pressure regulator and routed back to the suction side.

Fuel under transfer pump pressure flows through the center of the transfer pump rotor shaft, past the rotor retainers into an annular groove on the rotor. It next passes through a connecting passage in the head to the advance cylinder, up through a radial passage, and then through a duct to the metering valve. The rotational position of the metering valve, directly controlled by the governor, regulates fuel flow to the radial charging passage, which incorporates the hydraulic head charging ports.

As the rotor revolves, the two rotor inlet passages register with charging ports in the hydraulic head,

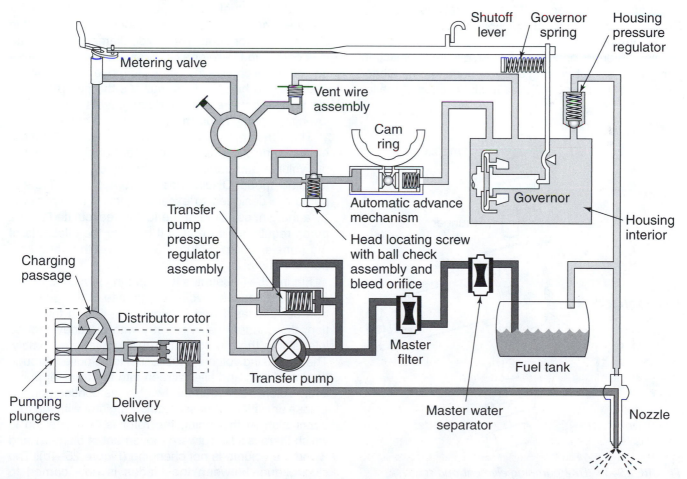

FIGURE 25–2 *Fuel flow schematic of a Stanadyne DB2 pump. (Courtesy of Stanadyne Automotive Corporation)*

allowing fuel to flow into the pumping chamber. Fuel flowing into the pump chamber spreads the pumping plungers outward. The extent to which they spread outward depends entirely on the fuel quantity metered into the pump chamber: this fact tends to mean injection timing is advanced proportionate with load. With further rotation, the inlet passages move out of register and complete the metering phase. Next, the discharge port of the rotor is brought into register with one of the hydraulic head distribution outlets that connect the pump with an individual fuel injector. During the register of the discharge port with the rotor, the pumping phase take place. The plunger-actuating rollers contact the lobes of the internal cam ring, forcing the opposed plungers inboard into the pump chamber. The pumping element is shown in Figure 25–3. The higher the fuel load in the pump chamber, the more advanced the timing of this moment that begins delivery. Fuel trapped in the pump chamber between the plungers is pressurized and unloaded through the discharge port in the hydraulic head during the delivery phase. The discharge port is connected by a high-pressure pipe to a hydraulic injector nozzle located in the engine cylinder head. When fuel pump pressure exceeds the nozzle-opening pressure (NOP), the injector nozzle valve unseats and fuel is injected. Most applications using the Roosa pump use indirect injection and pintle type nozzle assemblies.

Inlet-metering, opposed-plunger, distributor injection pumps are self-lubricating and self-priming. The lubricity of diesel fuel accomplishes the lubrication of the in-

ternal components of the pump. As fuel at transfer pump pressure reaches the charging ports, slots on the rotor shank allow fuel and any entrapped air to flow into the pump housing cavity. An air vent passage in the hydraulic head connects the outlet side of the transfer pump with the pump housing. This permits air and some fuel to bleed back to the fuel tank by means of a return line. Bypass fuel fills the pump housing, lubricates the internal components, acts as coolant, and purges the housing of small air bubbles. The pump is therefore designed to operate with the housing charged with fuel. During operation there should be no air within the pump.

INLET-METERING DISTRIBUTOR PUMP CIRCUITS

There are nine subcircuits in an inlet-metering, opposed-plunger distributor injection pump. Technicians should understand the role of each circuit, as it will help isolate problems during troubleshooting.

Transfer Pump Circuit

The vane type fuel transfer pump consists of a stationary liner and spring-loaded vanes or blades carried in slots in the rotor shaft. The inside diameter of the liner is eccentric to the rotor axis. Rotation creates centrifugal force that causes the blades to move outward in the rotor slots to hug the liner wall. As the pump rotates, the volume between the blade segments is varied, allowing the pump to discharge fuel to the outlet.

The transfer pump uses a positive displacement operating principle, which means that transfer pump output volume and pressure increase in direct proportion to pump speed. Displacement volumes of the transfer pump are designed to exceed injection requirements by a margin, so some of the fuel is recirculated by the pump regulator that routes it back to the inlet side of the transfer pump. Figure 25–4 shows the vane type transfer pump.

Figure 25–4 illustrates the pumping principle. Radial movement causes a volume increase in the quadrant between blades 1 and 2 (Figure 25–4a). In this position, the quadrant is in registry with a kidney-shaped inlet slot in the top portion of the regulator assembly. The increasing volume causes fuel to be pulled through the inlet fitting and filter screen into the transfer pump liner. Volume between the two blades continues to increase until blade 2 passes out of register with the regulator slot. At this point, the rotor is in a position in which there is little outward movement of blades 1 and 2 and the volume is not changing (Figure 25–4b). The slug of fuel between the blades is now carried to the bottom of the transfer pump liner.

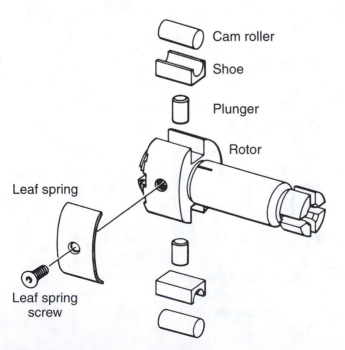

FIGURE 25–3 *DB2 pumping element and rotor. (Courtesy of Stanadyne Automotive Corporation)*

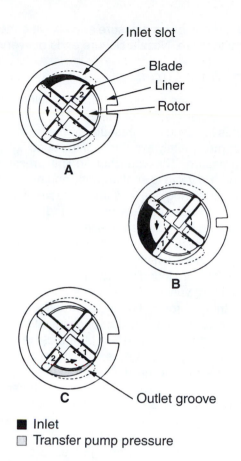

FIGURE 25–4 *Roosa vane type transfer pump. (Courtesy of Stanadyne Automotive Corporation)*

As blade 1 passes the edge of the kidney-shaped groove in the lower region of the regulator assembly (Figure 25–4c), the eccentric liner wall compresses blades 1 and 2 in an inward direction (Figure 25–4a). This action reduces the volume available for fuel storage but not the quantity, so the fuel is pressurized. Pressurized fuel is unloaded through a groove in the regulator assembly and directed into a channel on the rotor leading to the hydraulic head. As the rotor continues with its rotation, volume between blades continues to decrease, pressurizing the fuel in the quadrant, until blade 2 passes the groove in the regulator assembly.

Regulator Assembly Circuit

The regulator prevents excessive vane pump pressures in the event of an engine overspeed condition. Fuel from the discharge side of the transfer pump forces the regulator piston against the regulator spring. As pump speed and therefore flow increases, the regulator spring is further compressed until the leading edge of the regulator piston starts to expose the pressure-regulating slot. Because fuel pressure acting on the regulator piston is opposed by the regulator spring, delivery pressure of the transfer pump is controlled by:

- regulator spring tension
- flow area defined by the regulating slot

Metering Circuit

Inlet metering refers to the fact that only fuel permitted to pass through the metering valve enters the pump chamber. The fuel delivered to the pump chamber during each pump cycle is the metered fuel quantity and defines the injection pulse duration. The metering valve rotates in the bore. It receives fuel from the charging passage. Because it is milled with a scrolled metering recess, it directs this fuel to the opposed plunger injection pump through a charging passage. In other words, the rotational position of the metering valve controlled by the governor defines the flow area to the duct that feeds the pump chamber. This flow area must accommodate all the fuel requirements of the engine from no fuel to full fuel. The no-fuel rotational position of the metering valve is achieved by moving a dedicated shut-off linkage or by an electrical solenoid.

Charging Circuit

The charging circuit determines the exact quantity of fuel that enters the pumping chamber. As the rotor revolves (Figure 25–5), the two inlet passages in the rotor register with charging passage ports. Fuel from the transfer pump, controlled by the opening of the metering valve, flows into the pumping chamber thereby forcing the plungers apart.

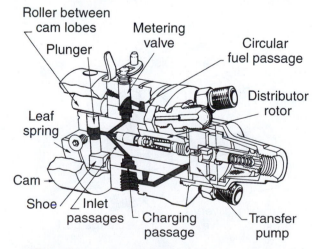

FIGURE 25–5 *Charging cycle. (Courtesy of Stanadyne Automotive Corporation)*

The plungers move outward only to the extent required to accommodate the fuel required for injection on the following stroke. If only a small quantity of fuel is charged to the pumping chamber, for instance that required to idle the engine, the plungers spread only a short distance. When engine fuel load is high, the plungers are forced outward further to accommodate the higher volume of fuel. Maximum plunger travel is limited by a leaf spring that contacts the edge of the roller shoes. Only when the engine is operating at full load will the plungers move to the most outward position.

Discharge Circuit

As the cycle continues, the inlet passages are moved out of register with the charging ports. Next, the rotor discharge port is brought into register with one of the outlets in the hydraulic head that will direct fuel to an injector. The plunger-actuating rollers then contact the cam profiles in the internal cam ring, driving the shoes against the plungers and forcing them inboard into the pump chamber. This action begins the high-pressure pumping phase called the discharge cycle shown in Figure 25–6.

The beginning of injection varies according to the required fuel load. Assuming that the internal cam ring remains stationary, the higher the fuel load admitted to the pump chamber, the more advanced the resultant injection timing. During the discharge stroke, fuel trapped in the pump chamber between the plungers is forced through the axial passage of the rotor to the discharge port and out through the injection line to the injector nozzle. Fuel delivery continues until the plunger-actuating rollers are ramped beyond the highest point of the cam lobe and are allowed to move outward. Pressure in the axial passage is allowed to

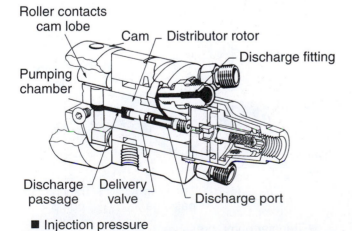

Roller contacts
cam lobe
Cam — Distributor rotor
— Discharge fitting
Pumping
chamber
Discharge / Delivery   — Discharge port
passage   valve

■ Injection pressure

FIGURE 25–6 *Discharge cycle. (Courtesy of Stanadyne Automotive Corporation)*

collapse, dropping line pressure and allowing the injector nozzle to close. Nozzle closure ends delivery.

Delivery Valve Circuit

Delivery valves are an option on most inlet-metering, opposed-plunger distributor pumps. The delivery valve accelerates injection line pressure drop after injection to a value approximately two-thirds of the specified nozzle closing pressure. This reduction in line pressure permits the nozzle valve to seat abruptly and minimizes the fuel injected during the collapse phase of injection. When larger droplets of fuel are injected to the engine cylinder during the collapse phase at the end of the injection pulse, there is insufficient time available to properly combust them, which causes higher HC emissions. Delivery valves are found on most highway applications.

The delivery valve is located in a bore in the center of the distributor rotor. This type of delivery valve requires no seat and uses just a mechanical stop to limit travel. Sealing is accomplished by the close tolerance fit between the valve and bore. When injection begins, fuel pressure moves the delivery valve slightly out of its bore, adding its swept volume displacement to the spring chamber. Because at this moment the discharge port is already exposed to a hydraulic head outlet, the retraction volume and plunger displacement volume are discharged under high pressure to the nozzle. Delivery ends when pressure on the plunger side of the delivery valve drops as the cam rollers ramp over the high point of the cam profile. Next, the rotor discharge port is closed off completely and a residual line pressure is sealed in the high-pressure injection pipe. The delivery valve seals only while the discharge port is open: once the port is closed, residual line pressure is maintained by the close tolerance fit of the hydraulic head and rotor.

Fuel Return Circuit

Fuel under transfer pump pressure is discharged into a vent passage in the hydraulic head. Flow through the passage is restricted by a vent wire assembly to prevent excessive return fuel that could cause undue pressure loss. The actual amount of return fuel is controlled by the size of wire used in the vent bore assembly. The smaller the wire the greater the return flow and vice versa. Vent wires are available in several size options to meet differing specifications. The vent wire assembly can be accessed by removing the governor cover. The vent passage is located behind the metering valve bore and connects with a short vertical passage containing the vent wire assembly, leading to the governor housing.

Any air entering the transfer pump is routed to the vent passage as shown. Both air and fuel then flow from the housing to return to the fuel tank via the return line. Housing pressure is maintained by a spring-loaded ballcheck return fitting in the governor cover of the pump.

Mechanical Governor

The governor used can be classified as a variable speed governor, also known as an all-speed governor. This type of governor functions to maintain desired engine speed (requested rpm) and will do so within a certain window, as load changes.

Speed sensing is performed by the flyweights as shown in Figure 25–1 and indicated by arrow 11. Movement of the flyweights acting against the governor thrust sleeve rotates the metering valve in its bore by means of the governor arm and linkage hook. Rotation of the metering valve varies the register of the metering valve scroll to the passage from the transfer pump, thereby controlling how much fuel is metered to the pump element. Centrifugal force, which directly correlates with rotational speed, forces the flyweights outward, moving the governor thrust sleeve against the governor arm, actuating the linkages that rotate the metering valve. Force of the flyweights acting on the governor arm is balanced by the governor spring force. This spring force is variable and controlled by the manually positioned throttle lever. The throttle lever connects to the vehicle accelerator linkage.

Load decrease. A variable speed governor defines a specific rpm value. In the event of a load reduction on the engine, engine speed would tend to increase. This would in turn increase the centrifugal force produced by the flyweights, which would act to rotate the metering valve clockwise to reduce engine fueling.

Load increase. When the load on the engine is increased, engine speed initially tends to drop. As engine speed reduces, centrifugal force generated by the weights drops, permitting the spring forces that oppose it to rotate the metering valve in the counterclockwise direction, increasing fuel metered to the pumping element. Engine speed at any point in the operating range of the engine is dependent on the combination of forces that act on the governor thrust lever. As with any mechanical governor, centrifugal force acting on the thrust lever will result in speed reduction (less fuel) and spring forces acting on the thrust lever will attempt to increase speed (more fuel).

Governor operation summary. A light idle spring is provided for more sensitive regulation when flyweight centrifugal force is low, such as when the engine is operating close to idle speeds. The limits of throttle travel are set by adjusting screws to define low-idle and high-idle speeds. A light tension spring on the linkage as-sembly takes up slack in the linkage joints and permits the shutoff mechanism to close the metering valve without having to overcome the governor spring force, which means that only a little force is required to rotate the metering valve to the closed position to shut the engine down.

The function of any variable speed governor is to attempt to hold engine rpm at a consistent value (based on a given throttle position) and let the governor adjust fueling to hold that rpm constant as engine load varies. Positioning the throttle lever simply defines a request for an rpm value, not a fuel quantity, which would be the case in a limiting speed governor.

Advance Circuit

Opposed-plunger, inlet-metering distributor pumps permit the use of a simple, direct-acting hydraulic mechanism powered by fuel pressure from the transfer pump to rotate the internal cam within the pump housing. This changes the phasing of injection pump stroke with that of the engine that drives the injection pump and varies delivery timing. The advance mechanism defaults to the most retarded position when there is no or low pressure developed by the vane-type transfer pump. As vane pump pressure increases, fuel pressure acting on the advance piston overcomes the advance spring pressure and rotates the internal cam ring to advance the start of fuel delivery. Total movement of the cam is limited by the piston length. A trimmer screw is provided to adjust the advance spring tension, which essentially determines how much fuel pressure is required (that is, the specific rpm) to begin the advance. It can be incorporated at either side of the advance mechanism and may be adjusted on the test bench while running. Because the extent of advance depends on vane pump pressure, it is speed sensitive. Figure 25–7 shows the advance circuit components and operating principle.

Advancing fuel injection timing compensates for inherent injection lag and greatly improves high-speed engine performance. Beginning the injection pulse earlier when the engine is operating at higher speeds ensures that peak cylinder combustion pressures are developed when the piston is ideally positioned in its downstroke to optimize torque transfer to the crankshaft. Advancing the beginning of the injection pulse results in the plungers completing their pumping stroke earlier.

OPPOSED-PLUNGER, INLET-METERING PUMP SUMMARY

The worldwide success of Roosa's injection pump design is a testament to its low initial cost, ease of maintenance, and ability to manage a range of

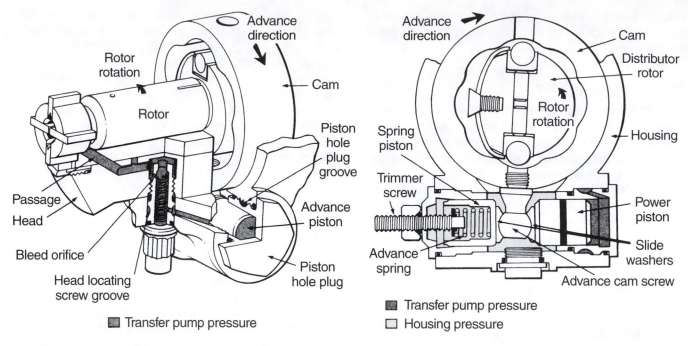

FIGURE 25–7 *Advance circuit components and operating principle. (Courtesy of Stanadyne Automotive Corporation)*

engines. Though not seen in anything but light-duty trucks, opposed-plunger, inlet-metering, rotary distributor pumps are found in many small diesel engine applications. They are especially prominent in agricultural equipment. The design lends itself to some limited enhancement by using partial authority computer controls, usually governing (ECM control of metering valve) and advance functions: in these, the critical components of the fuel pump are retained. However, with today's expectations for minimum emissions, better fuel economy, and higher longevity, this category of fuel injection pump has some limitations and can be regarded as a disappearing technology, at least in North America.

SLEEVE-METERING, SINGLE PLUNGER DISTRIBUTOR PUMPS

Bosch hydromechanical, **sleeve-metering** distributor injection pumps use a single plunger pumping element. In describing this technology, the popular VE pump is used as our primary reference. The VE pump has been used in a number of off-shore small-bore engines including Volkswagen and was an option on the Cummins B series engine. These pumps manage fueling for engines with up to six cylinders. A VE pump has four primary circuits:

1. Fuel-supply pump
2. High-pressure pump

3. Governor
4. Variable timing

The VE distributor pump has been used in passenger cars, commercial vehicles, agricultural tractors, and stationary engines.

SUBASSEMBLIES

The VE distributor pump uses only one pump cylinder and a single plunger. It is designed to fuel multicylinder engines. The general layout of a typical VE fuel system is shown in Figure 25–8.

All movement through the fuel subsystem is the responsibility of a vane-type transfer pump integral with the VE pump assembly. Once fuel from the vehicle tank enters the VE pump, fuel is pressurized to injection pressures and routed to the engine cylinders by means of high-pressure pipes. Injection pressures are created by a single plunger type pump. Fuel delivered by the pump plunger is routed by a distributor groove to the outlet ports, which connect to hydraulic injectors located at each engine cylinder. The VE distributor pump housing contains the following subcircuits:

- High-pressure (injection) pump with distributor
- Mechanical (flyweight) governor
- Hydraulic timing device
- Vane-type, fuel supply pump
- Shutoff device
- Engine-specific add-on modules

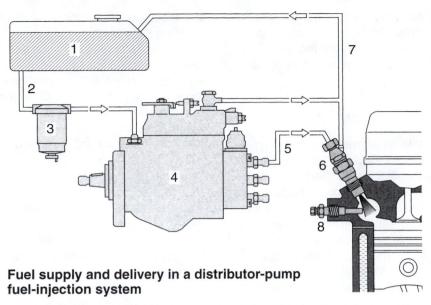

Fuel supply and delivery in a distributor-pump fuel-injection system

1. Fuel tank
2. Fuel line (suction pressure)
3. Fuel filter
4. Distributor injection pump

5. High-pressure fuel injection line
6. Injection nozzle
7. Fuel return line (pressureless)
8. Sheathed-element glow plug

FIGURE 25–8 *VE distributor pump fuel system layout. (Courtesy of Robert Bosch Corporation)*

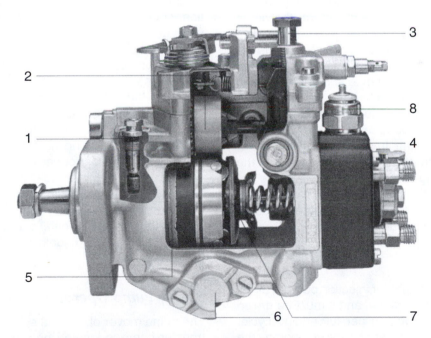

1. Pressure-control valve
2. Governor assembly
3. Overflow restriction
4. Distributor head with high-pressure pump

5. Vane-type fuel supply pump
6. Timing device
7. Cam plate
8. Electromagnetic shutoff valve

FIGURE 25–9 *Location of pump subassemblies. (Courtesy of Robert Bosch Corporation)*

Figure 25–9 shows a cutaway view of the subcircuits used on a typical VE pump. Add-on modules, some of which are described later in the chapter, allow the pump to be adapted to the requirements of specific diesel engines.

DESIGN AND CONSTRUCTION

The pump drive shaft is supported by bearings in the pump housing and drives the vane-type fuel supply pump. A roller ring is located inside the pump at the end of the drive shaft although it is not connected to it. Figure 25–10 shows the arrangement of the pump drive shaft and components in the front section of the pump. A rotating-reciprocating movement is imparted to the distributor plunger by means of a **cam plate** driven by the input shaft that rides on the rollers of the roller ring. The plunger moves inside the distributor head, which is itself bolted to the pump housing. Located in the distributor head are the electric fuel shutdown, screw plug with vent screw, and the delivery valves. If the distributor pump is also equipped with a mechanical fuel shutdown, it is mounted to the governor cover.

The governor assembly, which includes flyweights and the control sleeve, is driven by the drive shaft. The governor linkage, made up of control, starting, and tensioning levers, pivots in the housing. The governor shifts the position of the control sleeve on the pump plunger and, in this way, defines plunger effective stroke. Located above the governor mechanism is the governor spring that connects with the external control lever by means of the control-lever shaft, itself held in bearings in the governor cover.

The control sleeve is used to control pump output. The governor assembly is located at the top of the pump, and it contains the full-load adjusting screw, the overflow restriction or the overflow valve, and the engine speed adjusting screw. The variable timing device is located under the pump assembly and it functions to advance pump timing based on fuel pressure developed by the internal vane pump.

Pump Drive

The sleeve-metering, distributor injection pump is direct driven by the engine it manages and it must be driven through one complete rotation per full engine cycle. This means that on a four-stroke cycle engine, the pump is driven through one complete revolution per two crankshaft revolutions, in other words at camshaft speed. In common with most other injection pumps, sleeve-metering injection pumps must be precisely timed to the engine it manages.

The injection pump can be driven by toothed timing belts, a pinion, gear wheel, or chain. The direction of rotation can be either clockwise or counterclockwise, depending on the engine manufacturer requirements. The fuel delivery outlets in the **distributor head** are supplied with fuel in rotational geometric sequence and each is identified with a letter, beginning with A and following through with B, C, D, and so on, up to the total number of engine cylinders. This is done to avoid confusion with engine cylinder numbering. VE distributor pumps will fuel an engine with up to six cylinders.

FUEL SUBSYSTEM

The fuel subsystem of a sleeve-metering distributor injection pump is also known as the low-pressure circuit. It consists of a fuel tank, fuel lines, fuel filter, vane-type fuel supply pump, pressure control valve, and overflow restriction.

The vane-type supply pump is responsible for all movement of fuel in the fuel subsystem. It pulls fuel from the fuel tank and routes it through a filter before it enters the injection pump. The vane pump is positive displacement in operating principle, so the volume of fuel it pumps is directly related to rotational speed. A pressure control valve ensures that injection pump internal pressure is managed as a function of vane pump speed. This valve sets a defined internal pressure for any given speed, meaning that pump internal pressure rises directly in proportion to engine speed. In operation, some fuel flows through the pressure-regulating valve and is routed back to the suction side of the vane pump: some fuel also flows through the overflow restriction located at the top of the governor housing, and this fuel is routed back to the fuel tank. By flowing excess fuel through the injection pump, fuel is used for cooling and venting the injection pump housing. In some applications, an overflow valve is fitted instead of the overflow restriction. The interaction between the supply pump, pressure control valve and overflow restriction valve is shown in Figure 25–10.

In some applications, the vane pump exerts insufficient suction to pull fuel through the fuel subsystem, and in these applications, a presupply pump is required. This pump is usually located in or close to the fuel tank.

Supply Pump Operation

The prime mover of the fuel subsystem, the vane-type transfer pump is located on the injection pump drive shaft. The pump impeller assembly is concentric with the shaft and lugged to it by means of a Woodruff key. The impeller assembly rotates within an eccentric liner. When the drive shaft rotates, centrifugal force throws the four vanes in the impeller outward against the wall of the eccentric liner. Fuel enters the impeller assembly through an inlet passage and a kidney-shaped recess

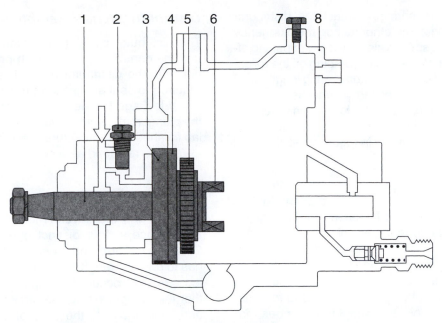

1. Drive shaft
2. Pressure control valve
3. Eccentric ring
4. Support ring
5. Governor drive
6. Drive shaft dogs
7. Overflow restriction
8. Pump housing

FIGURE 25–10 *Interaction of fuel supply pump, pressure control valve, and overflow restriction valve. (Courtesy of Robert Bosch Corporation)*

in the pump housing and charges the cavity formed between the vanes and the liner wall. As the pump rotates, fuel between adjacent vanes is forced into the upper or outlet, kidney-shaped recess and from there, directed to the injection pump circuitry. Some discharge fuel is also directed to the pressure control valve, a spring-loaded spool valve, which options fuel to the return circuit when pressure exceeds a specified value.

DEVELOPING INJECTION PRESSURES

Fuel injection pressures are produced by the high-pressure stage of injection pump assembly. High-pressure fuel is then routed to delivery valves and from there to injection nozzles located at each engine cylinder by means of high-pressure pipes.

Distributor Plunger Drive

Rotary movement of the drive shaft is transferred to the **distributor plunger** by a coupling unit as shown in Figure 25–11. Here, drive lugs or dogs on the cam plate engage with the recesses in a yoke, located between the end of the drive shaft and the cam plate. The cam plate is loaded onto the roller ring by a spring, so when it rotates, the cam lobes riding on the ring rollers con-

vert the rotational movement of the drive shaft into a rotating-reciprocating movement of the cam plate.

The distributor plunger is locked into position relative to the cam plate by a pin. The distributor plunger is actuated upward through its stroke by cams on the cam plate while a pair of symmetric return springs force it back downward. Because the plunger is actuated by cam profile and loaded to ride that profile by springs, its actual stroke does not vary.

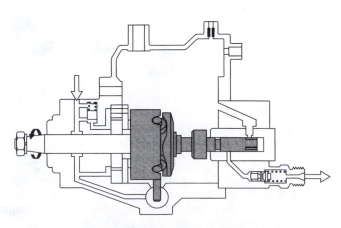

FIGURE 25–11 *High-pressure pump components. (Courtesy of Robert Bosch Corporation)*

The plunger return springs contact the distributor head at one end, and at the other, act on the plunger by a link element. These springs also have a dampening effect and can prevent the cam plate jumping off the rollers during sudden speed change. Return spring length must be carefully matched so that the plunger is not side-loaded in the pump bore. (See the lower portion of Figure 25–12.) An exploded view of the high-pressure pumping components is shown in Figure 25–12.

Cam Plates

The cam plate and its cam contour help define fuel injection pressure values and injection duration, along with pump-driven speed that determines plunger actuation velocity. Because of the different requirements of each type of engine, fuel injection factors produced by an injection pump are distinct to each engine. That means a specific cam plate profile is required for each engine type and technicians should remember that cam plates are generally not interchangeable because of the engine-specific machining of each.

Distributor Head Assembly

The distributor plunger, the distributor head bushing, and the control collar are each precisely fitted by lapping into the distributor head. These components are required to seal at injection pressure values. Some internal leakage losses do occur and serve to lubricate the plunger. In common with other components that are lap finished in manufacture, the distributor head should always be replaced as a complete assembly.

Metering

Metering and the development of injection pressures takes place in four distinct phases shown in Figure 25–13. In a four-cylinder engine, the distributor plunger has to rotate through 90 degrees for a complete pumping stroke to occur. A complete pump stroke means that the plunger has to be stroked from BDC to TDC and back again. In the case of a six-cylinder engine, the plunger has to complete a pumping stroke in 60 degrees of pump rotation.

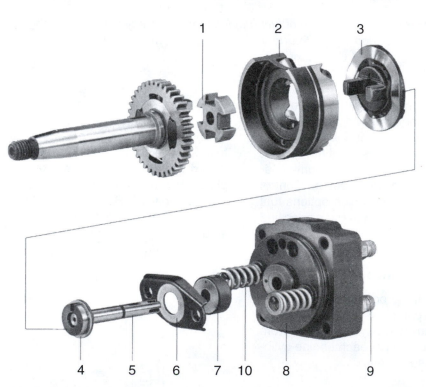

Generates the high pressure and distributes the fuel to the respective fuel injector.

1. Yoke
2. Roller ring
3. Cam plate
4. Distributor plunger foot
5. Distributor plunger

6. Link element
7. Control collar
8. Distributor head flange
9. Delivery valve holder
10. Plunger return spring

4–8. Distributor head

FIGURE 25–12   *Exploded view of the high pressure pump components. (Courtesy of Robert Bosch Corporation)*

Use Figure 25–13 and correlate the text with the four phases of a complete pumping stroke. As the distributor plunger is forced down from TDC to BDC, fuel flows through the open inlet passage and passes into the pumping chamber located above the plunger. At BDC, plunger rotational movement takes the plunger out of register with the inlet passage and exposes the distributor slot for one of the outlet ports as shown in Figure 25–13A. The plunger now reverses direction and is driven upward to begin the working stroke. Pressure rise is created in the pump chamber above the plunger and when sufficient, opens the delivery valve and forces fuel through the high-pressure pipe to the injector nozzle. The working (delivery) stroke is shown in Figure 25–13B. The delivery stroke is complete when the plunger transverse cutoff bore (cross drilling) protrudes beyond the metering sleeve, collapsing the pressure. When collapse is initiated, pressure drops in the high-pressure line and pump chamber until there is no longer sufficient pressure to hold the nozzle valve

A Inlet passage closes.
At BDC, the metering slot (1) closes the inlet passage, and the distributor slot (2) opens the outlet port.

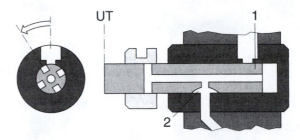

B Fuel delivery.
During the plunger stroke toward TDC (working stroke), the plunger pressurizes the fuel in the high pressure chamber (3). The fuel travels through the outlet port passage (4) to the injection nozzle.

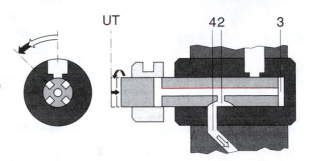

C End of delivery.
Fuel delivery ceases as soon as the collar (5) opens the transverse cutoff bore (6).

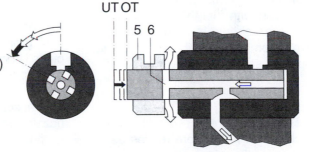

D Entry of fuel.
Shortly before TDC, the inlet passage is opened. During the plunger's return stroke the BDC, the high-pressure chamber is filled with fuel and the transverse cutoff bore is closed again. The outlet port passage is also closed at this point.

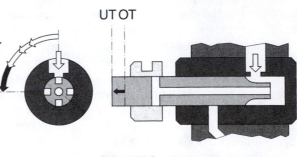

OT = TDC
UT = BDC

FIGURE 25–13 *Distributor plunger stroke and delivery phases. (Courtesy of Robert Bosch Corporation)*

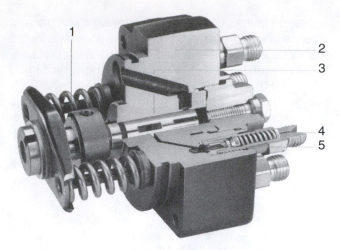

1. Control collar
2. Distributor head
3. Distributor plunger
4. Delivery valve holder
5. Delivery valve

FIGURE 25–14   Distributor head with pump chamber. (Courtesy of Robert Bosch Corporation)

open. Injection ceases the instant the nozzle valve closes.

As the plunger continues to move upward to TDC, fuel spills backward through the cutoff bore to the pump housing. During this collapse phase, the inlet passage is again exposed, ready for the next pump working cycle as shown in Figure 25–13C. As the plunger is forced back down by the return springs from TDC to BDC, the transverse cutoff bore is taken out of register by the plunger rotational movement, and the pump chamber is again charged with fuel through the now-exposed inlet passage as shown in Figure 25–13D. Figure 25–14 shows a complete distributor head and high-pressure pumping assembly in cutaway view.

Delivery Valve

The delivery valve seals the high-pressure pipe from the injection pump chamber. It therefore retains dead volume fuel (static fuel in the pipe) at a value well above that in the pump chamber but comfortably below that required to open the injector nozzle. This static pressure is known as residual line pressure and it ensures precise closure of the injector nozzle at the end of injection and ensures that a stable pressure is maintained in the high-pressure pipe between injection pulses, regardless of injected fuel quantity.

One delivery valve is used per engine cylinder. The delivery valve is a spring-loaded plunger. It is opened by delivery pressure developed in the injection pump chamber and closed by a return spring. Between injection pulses, the delivery valve remains closed. Its function can be most simply described by stating that it separates the high-pressure pipe from the distributor

head outlet port for the larger portion of the cycle when no fuel is being pumped to a given engine cylinder.

Delivery valve operation.   The construction of each delivery valve is identical to that used on port-helix metering pumps: you may wish to consult Chapter 21 in which the operation of delivery valves is covered in greater detail. The valve core consists of a stem, seat, retraction piston, and flutes. Subject to pressure rise created during the injection pump delivery stroke, the piston is hydraulically opened when pump chamber pressure exceeds the combined forces acting to load it on its seat: spring force and line pressure. Initially, the delivery valve acts as a hydraulically actuated plunger, compressing the fuel retained in the high-pressure pipe. Once the retraction piston of the delivery valve exits above the seat in the delivery valve body, fuel in the pump chamber is united with that in the high-pressure pipe and the valve can be considered open.

Following the injection pump delivery stroke, line pressure is collapsed by the spill taking place at the control sleeve: the instant there is insufficient hydraulic pressure to hold the delivery valve open, it begins to retract toward its seat. However, the moment the retraction piston passes the delivery valve seat, dead volume fuel quantity is defined, the injector nozzle having sealed. Because the delivery valve has to retract further before it physically seats, the defined dead volume fuel is given a fraction more space, so the pressure is further reduced. This reduction in pressure establishes the residual line pressure value, usually about two-thirds of that required to open the injector nozzle.

Delivery valve with return-flow restriction.   Pressure drop-off to a precise value in the high-pressure pipes is desirable at the end of injection. However, the high-speed, high-pressure switching that the high-pressure pipe is subject to creates pressure waves that are reflected between the delivery valve and injector nozzle seat. This pressure wave reflection causes local spiking of line pressure and may cause undesirable nozzle opening known as secondary injection or vacuum phases in the high-pressure pipe, causing cavitation. Using a delivery valve with a restriction bore that is only a factor in the direction of return (spill) fuel flow can minimize pressure wave reflection. The return-flow restriction consists of a valve plate and a pressure spring arranged so that the restriction is only effective in the return direction when it dampens pressure spikes and vacuum phases.

Constant-Pressure Valve

Another means of dealing with the problems associated with pressure wave reflection (pressure spikes and vacuum phases) is to use a constant-pressure type

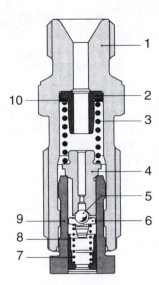

1. Delivery valve holder
2. Filler piece with spring locator
3. Delivery valve spring
4. Delivery valve plunger
5. Constant pressure valve
6. Spring seat
7. Valve spring (constant pressure valve)
8. Setting sleeve
9. Valve holder
10. Shims

FIGURE 25–15 *Constant pressure delivery valve. (Courtesy of Robert Bosch Corporation)*

delivery valve. These are usually found on high-speed engines using direct-injection (DI) engines. Constant-pressure valves relieve the high-pressure pipe pressure by means of a single-acting, nonreturn valve set to a specific pressure. The specific pressure would define the residual line pressure. A diagram of a constant-pressure delivery valve is shown in Figure 25–15.

GOVERNING

Bosch sleeve-metering, rotary distributor pumps are available with both variable speed and limiting speed governing options. These are usually described by Bosch using the British terms:

Variable speed governor = all speed governor

Limiting speed governor = min-max governor

A full description of governor principles of operation is provided in Chapter 26 so it is not repeated here. Instead, we look at how the governor functions to control the positioning of the sleeve-metering collar under various operating conditions. In addressing the governors used on Bosch sleeve-metering distributor pumps, we reference a variable speed governor. Remember, regardless of governor type, the position of the sleeve-metering collar defines the injection pump effective stroke, so engine output is entirely dependent on this.

Governor Design

The governor assembly is attached to the governor drive shaft located in the governor housing. It consists of a flyweight carrier, **thrust collar**, and tensioning lever. When the flyweights rotate, they are forced outward due to centrifugal force. This radial outward movement is converted to axial movement of the thrust collar or what Bosch describes as a sliding sleeve. Thrust collar travel is allowed to act on the governor lever assembly, which is made up of a start lever, tensioning lever, and adjusting lever. As in any mechanical governor, spring forces defined at the governor lever assembly act in opposition to the centrifugal force produced by the flyweights. Centrifugal force tends to reduce fueling/rpm, while spring force tends increase fueling/ rpm. Therefore the interaction of spring forces and centrifugal force acting on the thrust collar (sliding sleeve) defines the positioning of the governor lever assembly. The governor lever assembly controls the position of the control sleeve or collar. As we learned earlier, the position of the control sleeve determines the plunger effective stroke that defines the quantity of fuel to be delivered. The governor assembly on a VE distributor pump is shown in Figure 25–16.

Startup Fueling

When the engine is stationary, the flyweights and thrust collar are in their initial position. This results in the start lever being forced into the start position by the starting spring, moving the control sleeve on the distributor plunger to its start fuel position. When the engine is cranked, the distributor plunger travels through a complete working stroke before the cutoff cross-drilling is exposed to end delivery. The result is a full-fuel delivery pulse.

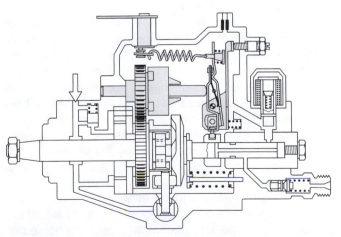

FIGURE 25–16 *VE governor assembly. (Courtesy of Robert Bosch Corporation)*

Low-Idle Operation

Once the engine is running with the accelerator pedal released, the engine speed control lever shifts to the idle position. This positions the control sleeve to provide a short plunger effective pumping stroke and represents the lowest fuel delivery condition of a running engine. Idle speed can be adjusted independent of the accelerator pedal setting, and can be increased or decreased if temperature or load conditions require it.

Operation Under Load

During actual operation, the driver requests the required engine speed by accelerator pedal angle. If higher engine speeds are required, the driver pushes harder on the pedal, increasing the angle and therefore, the governor spring force. If a lower engine speed is required, he reduces the pedal angle. At any engine speed above idle, the start and idle springs are compressed completely, so do not influence governing. Governed speed becomes the responsibility of the governor spring.

Under load, the driver sets the accelerator pedal at a specific position, increasing the pedal angle. If a higher engine speed is required, the pedal angle has to be further increased. As a result, the governor spring is tensioned increasing the spring force available to counter the centrifugal force produced by the flyweights. This acts through the thrust collar governor levers to shift the control sleeve toward the full-fuel direction, increasing the injection pump effective stroke. As a result, injected fuel quantity increases and engine speed rises.

The control collar remains in the full-fuel position until equilibrium is established once again between the centrifugal force generated by the flyweights (now greater) and the governor spring forces that oppose it. Should engine speed continue to increase, the flyweights extend further, resulting in governor thrust collar movement that forces the control sleeve toward the no-fuel direction, trimming back fueling. The governor can reduce delivery fueling to no-fuel, ensuring that engine speed limitation takes place. During operation, assuming the engine is not overloaded, every position of the engine speed control lever relates to a specific engine speed: the governor manages that speed by having the ability control the control sleeve in any position between full fuel and no fuel. In this way, the governor maintains desired speed. The speed at which the governor responds to a change in engine load in order to maintain desired engine speed is known as *droop*.

If during engine operation, engine load increases to the extent that when the control sleeve is in full-fuel position, engine speed continues to drop, the engine can be assumed to be overloaded and the driver has no option but to downshift.

Governor Break

During the downhill operation of a vehicle, the engine is driven by vehicle momentum, and engine speed tends to increase. This results in the governor flyweights moving outward, causing the governor thrust collar to press against the tensioning and start levers. Both levers react by changing position and pushing the control sleeve toward the no-fuel position until a reduced fuel equilibrium is established in the governor that corresponds to the new load/speed condition. At engine overspeed, the governor can always no-fuel the engine. With a variable-speed governor, any control sleeve position can be set by the governor in order to maintain desired speed.

VARIABLE TIMING DEVICE

A hydraulically actuated timing device is located under the main pump housing at right angles to the pump longitudinal axis as shown in Figure 25–17. The variable timing device is a speed-sensitive advance mechanism that defaults to the most retarded delivery position in the absence of hydraulic (fuel) pressure.

The timing device housing is closed with a cover on either side. A passage is located in one end of the timing device that allows fuel from the vane pump to enter. This fuel is allowed to act on the sectional area of the advance piston. On the opposite side of the piston, spring force opposes the hydraulic pressure of the fuel. The piston is connected to the roller ring by

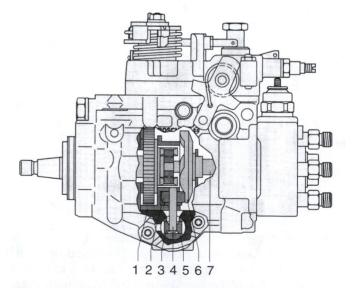

1. Roller ring
2. Roller ring rollers
3. Sliding block
4. Pin
5. Timing-device piston
6. Cam plate
7. Distributor plunger

FIGURE 25–17   *VE variable timing device. (Courtesy of Robert Bosch Corporation)*

means of a sliding block and pin, enabling piston linear movement to be converted to rotational movement of the roller ring.

Timing Device Operation

The timing device is held in its initial or default position by the timing device spring as shown in Figure 25–18a. When the engine is started, the pressure control valve regulates fuel pressure so that it is exactly proportional to engine speed. As a result, this engine-speed-dependent fuel pressure is applied to the end of the timing device piston in opposition to the spring force acting on the other side of it.

At a predetermined vane pump fuel pressure, the advance piston overcomes the spring preload and shifts the sliding block and the pin that engages with the roller ring. The roller ring is rotated, moving its relative position with the cam plate, which results in the rollers lifting the rotating cam plate earlier. This action means that the actuation of the injection plunger is advanced. The maximum advance angle achievable in Bosch sleeve-metering distributor injection pumps is limited by timing piston linear movement and is usually 24 crank angle degrees. Figure 25–18b shows

the maximum timing angle location of the advance piston.

ADD-ON MODULES AND SHUTDOWN DEVICES

Bosch sleeve-metering, distributor injection pumps are available with a variety of add-on modules and shutdown devices. Because of the modular construction of the pump, these supplementary devices can be added to optimize engine torque profile, power output, fuel economy, and exhaust gas composition. A brief description of add-on modules and how they impact on engine operation follows. Figure 25–19 is a schematic that shows how these add-on modules interact with the basic distributor pump.

Torque Control

Torque control relates to how fuel delivery is managed in respect to engine speed and the engine load requirement characteristic. Generally, engine fuel requirement increases when a request for higher engine speed is made, that is, the driver pushes his foot harder on the

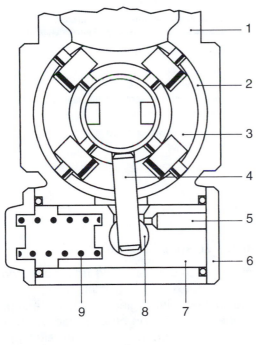
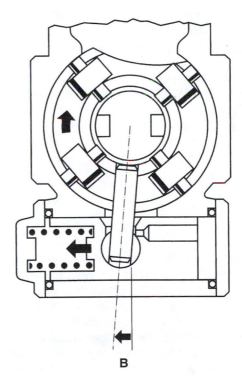

A

B

1. Pump housing
2. Roller ring
3. Roller ring rollers
4. Pin
5. Passage in timing device piston

6. Cover
7. Timing device piston
8. Sliding block
9. Timing device spring

A. Initial position
B. Operating position

FIGURE 25–18 *Timing device operating principle. (Courtesy of Robert Bosch Corporation)*

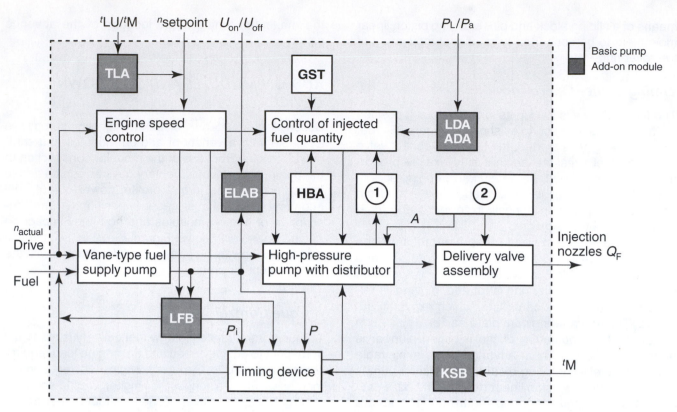

FIGURE 25–19　*VE add-on modules. (Courtesy of Robert Bosch Corporation)*

LDA Manifold pressure compensator.
Controls the delivery quantity as a function of the change-air pressure.

HBA Hydraulically controlled torque control.
Controls the delivery quantity as a function of the engine speed (not for pressure-charged engines with LDA).

LFB Load-dependent start of delivery.
Adaptation of pump delivery to load. For reduction of noise and exhaust gas emissions.

ADA Altitude pressure compensator.
Controls the delivery quantity as a function of atmospheric pressure.

KSB Cold start accelerator.
Improves cold start behavior by changing the start of delivery.

GST Graded (or variable) start quantity.
Prevents excessive start quantity during warm start.

TLA Temperature-controlled idle speed increase.
Improves engine warmup and smooth running when the engine is cold.

ELAB Electrical shutoff device.

A Cutoff port, n_{actual} Actual engine speed (controlled variable), $n_{setpoint}$ Desired engine speed (reference variable). Q_F Delivery quarterly, t_M Engine temperature, t_{LU} Ambient air temperature, P_L Change-air pressure, p_i Pump interior pressure.

①Full-Load Torque control with governor lever assembly.
②Hydraulic full-load torque control.

accelerator pedal. Fueling should then level off as actual engine speed approximates desired engine speed. Note that the same position of the fuel control sleeve will result in slightly more fuel delivery at higher engine speeds than at lower rpms due to the throttling effect that occurs at the distributor plunger cutoff port. This means that if the fuel trim settings were set to produce maximum torque at low engine speeds, the engine would be overfueling at high engine speeds, resulting in smoking and possible engine overheat. Conversely, if the fuel trim settings were set to produce optimum performance at rated speed, the engine would not be able to develop sufficient power at lower-than-rated speeds. Getting fueling quantities optimized throughout the engine operating range is known as torque control.

Positive Torque Control

Positive torque control would be required on a pump that delivers excess fuel at higher speeds. Positive torque control limits engine fueling in the upper portion of torque rise. This can be achieved by using a lower tension spring in the delivery valve, altering the cutoff port geometry producing a throttling effect, or using additional torque control springs in the governor.

Negative Torque Control

Negative torque control would be required in an engine that overfueled in the lower speed ranges but produced satisfactory performance at speeds closer to the rated speed of the engine. Negative torque control limits fueling in the lower portion of torque rise speed. It achieves this by governor spring pack trim or hydraulically, using pump housing fuel pressure.

ANEROID

The manifold pressure compensator (known by Bosch as an LDA, a German acronym) is a simple aneroid device. As such, it reacts to the manifold boost pressure generated by the turbocharger and essentially limits fueling until there is sufficient air in the engine cylinder to properly combust it.

Charge air is ported directly to the LDA assembly. The LDA is divided into two separate airtight chambers divided by a diaphragm. Manifold boost is applied to one side of the diaphragm while spring force is applied to the other. The diaphragm is connected to the LDA sliding pin, which has a taper in the form of a control cone. This is contacted by a guide pin through which levers act on the full fuel stop setting for the control sleeve.

At lower engine loads, there is insufficient manifold boost developed to effect movement on the diaphragm so the spring remains in control, with the result that the LDA limits maximum fueling. When engine load is increased and a predetermined manifold boost pressure is achieved, the LDA spring pressure is overcome, allowing the control sleeve to permit a longer pump effective stroke.

ALTITUDE-PRESSURE COMPENSATOR

The objective of an altitude-pressure compensator (known by Bosch as ADA, a German acronym) is to limit engine fueling when the vehicle is operated at high altitudes and the air density is reduced. A reduction in air density means that a lower number of oxygen molecules are charged to the engine cylinder per cycle resulting in overfueling that can cause smoking. All altitude compensators are deration devices, that is, they reduce engine fueling and power output at altitude to eliminate smoking. The construction of an ADA is somewhat similar to that of the LDA. The major difference is that the ADA is equipped with a barometric capsule that connects to a vacuum system somewhere on the vehicle, for instance, a power-assisted brake system circuit will do.

ADA Operation

Atmospheric pressure is applied to the upper side of the ADA diaphragm. A reference pressure from the barometric capsule is applied to the lower chamber in the ADA. The two chambers are separated by a diaphragm. Should a drop in atmospheric pressure occur such as would be experienced by driving a vehicle up a mountain, the barometric capsule (constant pressure) would be at higher pressure than that on the opposing side of the diaphragm. This pressure would cause the diaphragm to move the sliding bolt vertically away from the lower fuel stop and reduce engine fueling.

ENGINE SHUTOFF

Diesel engines are shut down by no-fueling them. Most Bosch sleeve-metering distributor pumps are shut down using a solenoid-operated shutoff (ELAB). A few of these pumps are equipped with a mechanical shutoff device.

Electrical Shutoff Device

Electrical shutoff is desirable because the solenoid can be energized by the vehicle ignition circuit. On the Bosch distributor pump, the fuel shutoff solenoid (acronym ELAB is German) is installed in the distributor head. When the engine is running, the solenoid is energized. When energized, the solenoid shutoff valve is positioned open, allowing fuel to pass into the injection pump chamber. When the ignition circuit is opened (turned off), current flow to the shutdown solenoid winding is cut, its magnetic field collapses, and spring force closes the shutdown valve. This seals off the inlet passage to the injection pump chamber, resulting in no-fuel delivery. The shutoff solenoid is energized to run and does not latch. In the event of unwanted interruption of the electrical circuit, the engine will shut down.

Mechanical Shutoff Device

The mechanical shutoff device is located in the governor cover and has an outer and inner stop lever. The outer lever can be actuated by the driver from inside the vehicle. When the shutdown cable is actuated, both inner and outer levers swivel around a common pivot, causing the inner stop lever to push against the start lever of the governor-lever mechanism. This in turn, moves the control sleeve to the no-fuel position, therefore the distributor plungers cutoff port remains open throughout plunger stroke and no fuel can be pressurized.

SUMMARY

- There are two categories of hydromechanical rotary distributor pump likely to be found on small-bore diesel engines today: the inlet-metering, opposed-plunger injection Roosa type pump manufactured by Stanadyne or by Delphi Lucas CAV, and the sleeve-metering, rotary injection pump manufactured by Bosch.

- The main components of an inlet-metering, distributor injection pump are the vane pump, rotor and hydraulic head assembly, internal cam ring, opposed plungers, inlet metering valve, governor assembly, and timing advance.
- An opposed-plunger, inlet-metering injection pump is driven through one full revolution per full cycle on the engine or two revolutions.
- An opposed-plunger, inlet-metering injection pump has a single pump chamber actuated by a pair of opposed plungers: fuel quantity metered into the pump chamber determines the outward travel of the plungers, advancing fuel injection timing in respect of load.
- Fuel is moved through the fuel subsystem by a vane-type transfer pump, integral with the pump housing: this fuel is then routed through the fuel pump circuitry, directed to the metering valve and from there, to the pumping chamber. Pressure rise to injection pressures is developed in the pump chamber and unloaded to the rotor, from which it is distributed to the injectors.
- The main components of a Bosch sleeve-metering, distributor injection pump are a vane-type transfer pump, cam plate, single plunger actuated pump chamber, distributor head, delivery valves, control sleeve, governor assembly, and advance mechanism.
- A sleeve metering, distributor injection pump uses axial movement of a sleeve on the pumping plunger to alter the effective pumping stroke: the plunger both rotates and reciprocates to pressurize and distribute fuel to outlets in the distributor head.
- Movement of fuel through the fuel subsystem in a sleeve-metering, distributor injection pump is the responsibility of a vane-type transfer pump. Actual pump plunger stroke is defined by the actuating cam geometry, and the position of the metering or control sleeve determines the actual fuel pumped.
- The operating principles of the two types of distributor pump presented in this chapter are distinct in the way in which they pump, meter, and distribute fuel, and it is important not to confuse the two systems during study.

REVIEW QUESTIONS

In answering these questions, remember two different pump systems are studied in this chapter: the first 10 questions (1–10) address inlet-metering, opposed-plunger distributor pumps, the next 10 (11–20) address the Bosch sleeve-metering distributor pump.

1. If a four-stroke cycle diesel engine fueled by an inlet-metering, opposed-plunger fuel pump is rotated through a complete cycle, how many times is the pump rotated?
   a. One-half revolution
   b. One revolution
   c. Two revolutions
   d. Four revolutions

2. What actuates the plungers of an inlet-metering, opposed-plunger fuel injection pump to create fuel injection pressures?
   a. Metered fuel quantity
   b. Charging fuel pressure
   c. External cam profile
   d. Internal cam ring

3. In an inlet-metering, opposed-plunger fuel pump, what causes the outward movement of the opposed plungers in the pump chamber during metering?
   a. Centrifugal force
   b. Metered fuel quantity
   c. Spring force
   d. Pump housing backpressure

4. How is an inlet-metering, opposed-plunger fuel pump lubricated?
   a. Dedicated supply of engine lubricating oil
   b. Pressurized engine lubrication oil
   c. Fuel from the vane pump
   d. Prelubed on assembly

5. *Technician A* says that when a mechanical governor is used on an inlet-metering, opposed-plunger fuel pump, metering valve rotational position determines how much fuel is metered to the pump element. *Technician B* says that the metering valve determines the pressure value developed by the vane-type transfer pump. Who is correct?
   a. A only
   b. B only
   c. Both A and B
   d. Neither A nor B

6. When an inlet-metering, opposed-plunger fuel pump fuels an automotive eight-cylinder diesel engine, how many delivery valves would be used?
   a. One
   b. Two
   c. Four
   d. Eight

7. *Technician A* says that on an inlet-metering, opposed-plunger fuel pump, plunger actual stroke is constant regardless of fuel requirement. *Technician B* says that in the same pump, plunger effective stroke is determined by vane pump pressure. Who is correct?
   a. A only
   b. B only
   c. Both A and B
   d. Neither A nor B

8. How is the timing advance mechanism on a hydro-mechanical, inlet-metering, opposed-plunger fuel pump actuated?
   a. Hydraulically by the vane pump fuel pressure
   b. Hydraulically by engine oil pressure
   c. Pneumatically by manifold boost
   d. Manifold vacuum

9. What limits the maximum fuel metered to the pump chamber in an inlet-metering, opposed-plunger fuel pump?
   a. Vane pump pressure
   b. Metering valve flow area
   c. Leaf spring stops
   d. Roller outboard travel

10. *Technician A* says that pintle type injector nozzles are usually used with an inlet-metering, opposed-plunger fuel pump system. *Technician B* says that this type of fuel system is more likely used on indirect injected diesel engines rather than direct-injected engines. Who is correct?
    a. A only
    b. B only
    c. Both A and B
    d. Neither A nor B

11. When a Bosch sleeve-metering, rotary distributor pump is used to fuel a four-stroke cycle engine, at what speed is the fuel pump driven?
    a. Engine camshaft speed
    b. Engine crankshaft speed
    c. Two times engine speed
    d. Four times engine speed

12. *Technician A* says that Bosch sleeve-metering, rotary injection pumps can be used to fuel high-speed diesel engines of up to eight engine cylinders. *Technician B* says that when this type of pump is used on a V-configured engine, two distributor heads are used. Who is correct?
    a. A only
    b. B only
    c. Both A and B
    d. Neither A nor B

13. When a Bosch sleeve-metering, rotary distributor pump is used on a six-cylinder, four-stroke cycle diesel engine how many delivery valves would be required?
    a. One
    b. Two
    c. Four
    d. Six

14. *Technician A* says that the pumping plunger used on a Bosch sleeve-metering, rotary distributor pump reciprocates. *Technician B* says that the pumping plunger used in this fuel system rotates. Who is correct?
    a. A only

b. B only
c. Both A and B
d. Neither A nor B

15. What determines plunger stroke in a Bosch sleeve-metering, rotary distributor pump?
    a. Position of the sleeve-metering collar
    b. Cam geometry
    c. Rotational speed of the pump
    d. Metered fuel quantity

16. *Technician A* says that the cam geometry used on a Bosch-sleeve metering, rotary distributor pump is specific to the engine that it fuels. *Technician B* says referring to the same pump, that the cam ring rotates at engine camshaft speed. Who is correct?
    a. A only
    b. B only
    c. Both A and B
    d. Neither A nor B

17. How is timing advance achieved in a Bosch hydro-mechanical, sleeve-metering, rotary distributor pump?
    a. Centrifugal force generated by flyweights
    b. Hydraulic pressure generated by the vane pump
    c. Spring force relayed by the governor
    d. Mechanical force from the accelerator linkage

18. *Technician A* says that a variable speed governor on a Bosch sleeve-metering, rotary distributor pump functions by defining a specific fuel quantity that corresponds to throttle lever angle. *Technician B* says that a variable speed governor will attempt to maintain a specific rpm value as engine load changes. Who is correct?
    a. A only
    b. B only
    c. Both A and B
    d. Neither A nor B

19. *Technician A* says that most Bosch sleeve-metering, rotary distributor pumps use a shutoff solenoid known as an ELAB. *Technician B* says that an ELAB does not latch so in the event of a loss of electrical power, the fuel system will shut the engine down. Who is correct?
    a. A only
    b. B only
    c. Both A and B
    d. Neither A nor B

20. Which of the following could address the problem of a Bosch sleeve-metering, rotary distributor pump fueled engine that functioned well at peak torque but was overfueling and smoking when run at rated speed.
    a. Positive torque control
    b. Negative torque control
    c. Trim back maximum speed stop screw
    d. Replace the aneroid

Bearing Failure Analysis

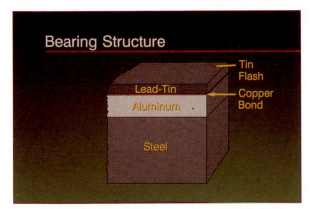

Bearing Structure   **C1-1**

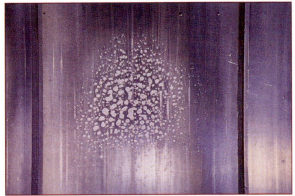

High mileage wear: lead-tin bond weakens in high load area   **C1-2**

Sliding contact stress fatigue caused by normal, high mileage wear   **C1-3**

High mileage, normal wear: wear in high load areas with some cavitation   **C1-4**

Misaligned loading: wear occurs on opposite sides   **C1-5**

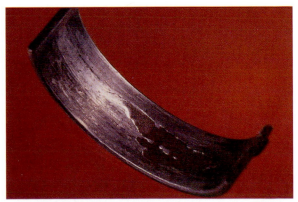

Thermal growth (overheating) of aluminum layer causes bond separation   **C1-6**

Courtesy of Caterpillar Inc.

Valve Failure Analysis

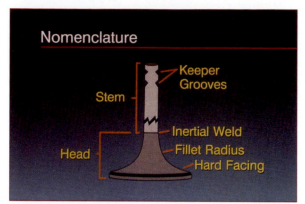

Valve Terminology   **C2-1**

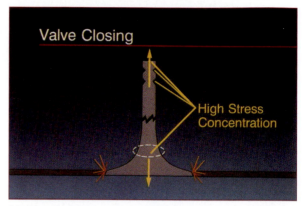

Valve closing stress points   **C2-2**

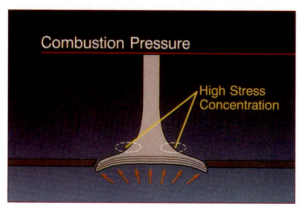

Stress during combustion   **C2-3**

Thermal cracks caused by overheating   **C2-4**

Cupping or dishing: also caused by overheating   **C2-5**

Valve leakage effects caused by ash and carbon build-up valve face   **C2-6**

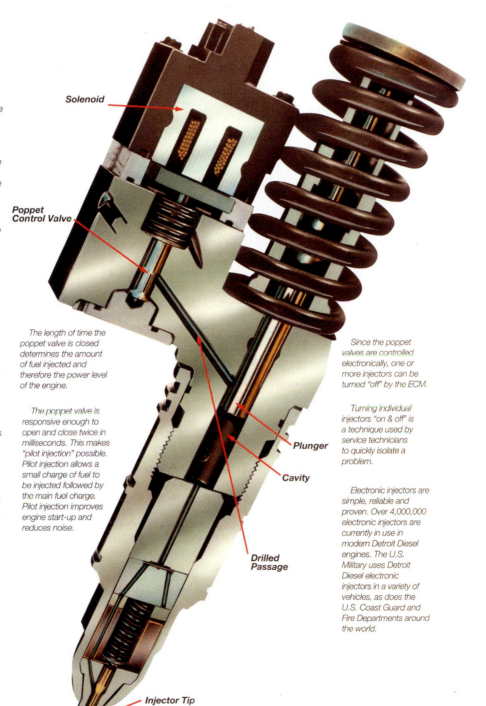

The rocker arm, operated by the camshaft, depresses the plunger. The cavity below the plunger is continuously filled with diesel fuel by the fuel pump. A drilled passage connects the cavity below the plunger to the poppet control valve.

When the poppet control valve is open, no pressure is created below the plunger and no fuel is injected into the cylinder.

When the poppet control valve is closed, diesel fuel is trapped in the cavity. The downward motion of the plunger pressurizes the fuel and injection takes place through the injector tip.

The poppet valve is controlled by a simple solenoid, which in turn is controlled by the Electronic Control Module (ECM).

The poppet valve can be opened or closed at any time. Therefore, engine timing can be varied over a wide range, improving start-up, smoke control and overall performance.

The length of time the poppet valve is closed determines the amount of fuel injected and therefore the power level of the engine.

The poppet valve is responsive enough to open and close twice in milliseconds. This makes "pilot injection" possible. Pilot injection allows a small charge of fuel to be injected followed by the main fuel charge. Pilot injection improves engine start-up and reduces noise.

Since the poppet valves are controlled electronically, one or more injectors can be turned "off" by the ECM.

Turning individual injectors "on & off" is a technique used by service technicians to quickly isolate a problem.

Electronic injectors are simple, reliable and proven. Over 4,000,000 electronic injectors are currently in use in modern Detroit Diesel engines. The U.S. Military uses Detroit Diesel electronic injectors in a variety of vehicles, as does the U.S. Coast Guard and Fire Departments around the world.

Solenoid

Poppet Control Valve

Plunger

Cavity

Drilled Passage

Injector Tip

Caterpillar EUI Injection Cycle

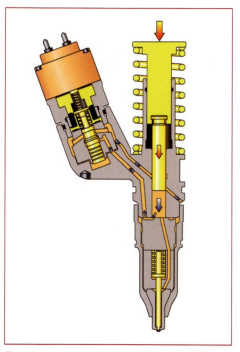

Pre-Injection

C4-1

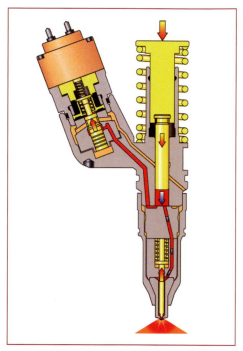

Injection

C4-2

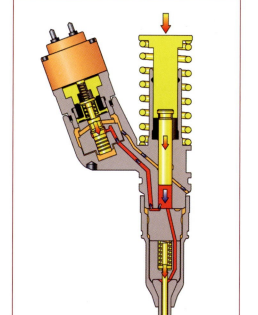

Spill

C4-3

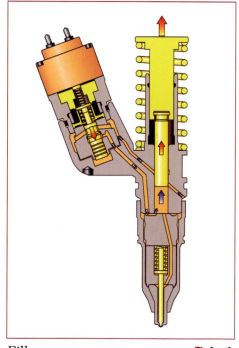

Fill

C4-4

Comparison between Caterpillar EUI and HEUI Systems

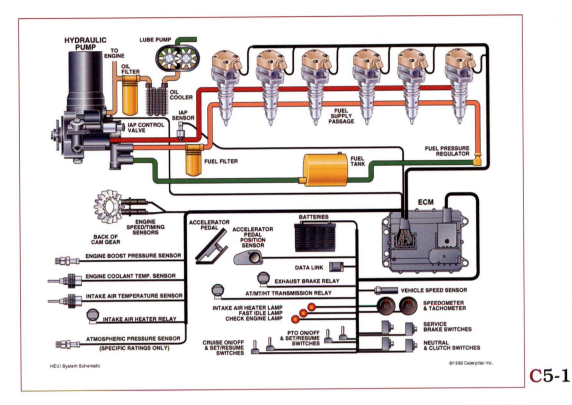

HEUI System Schematic

©1999 Caterpillar Inc.

C5-1

EUI System Schematic

©1998 Caterpillar Inc.

C5-2

The HEUI Injection Cycle

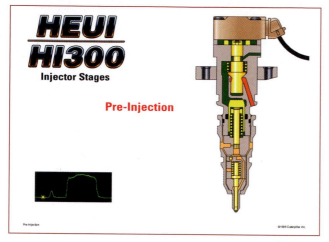

C6-1

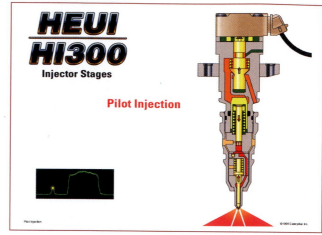

C6-2

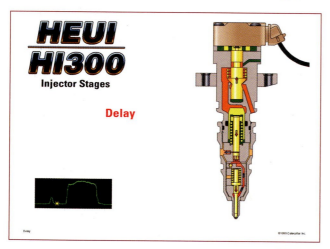

C6-3

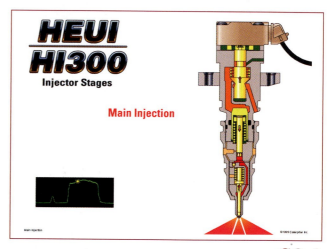

C6-4

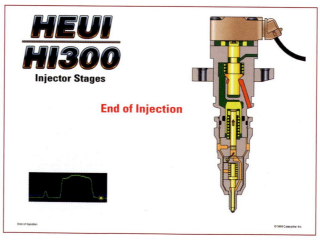

C6-5

Courtesy of Caterpillar Inc.

HEUI Injection Control and Pulse

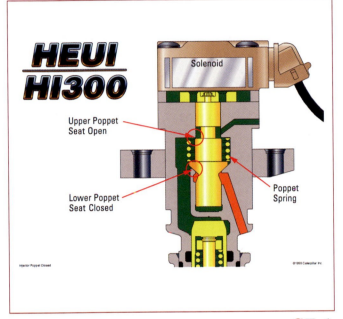

HEUI HI300

Upper Poppet Seat Open

Lower Poppet Seat Closed

Poppet Spring

Injector Poppet Closed

©1999 Caterpillar Inc.

C7-1

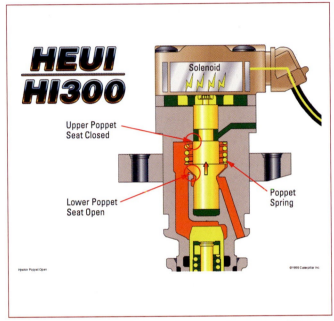

HEUI HI300

Solenoid

Upper Poppet Seat Closed

Lower Poppet Seat Open

Poppet Spring

Injector Poppet Open

©1999 Caterpillar Inc.

C7-2

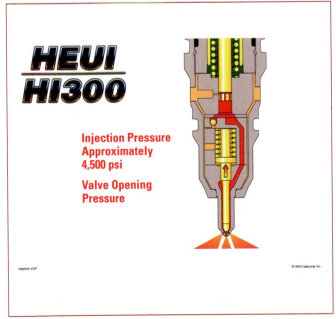

HEUI HI300

Injection Pressure Approximately 4,500 psi

Valve Opening Pressure

Injection VOP

©1999 Caterpillar Inc.

C7-3

HEUI HI300

Injection Pressure Below 4,000 psi

Valve Closing Pressure

Injection VCP

©1999 Caterpillar Inc.

C7-4

Courtesy of Caterpillar Inc.

Cummins HPI-TP
Fuel System Schematic

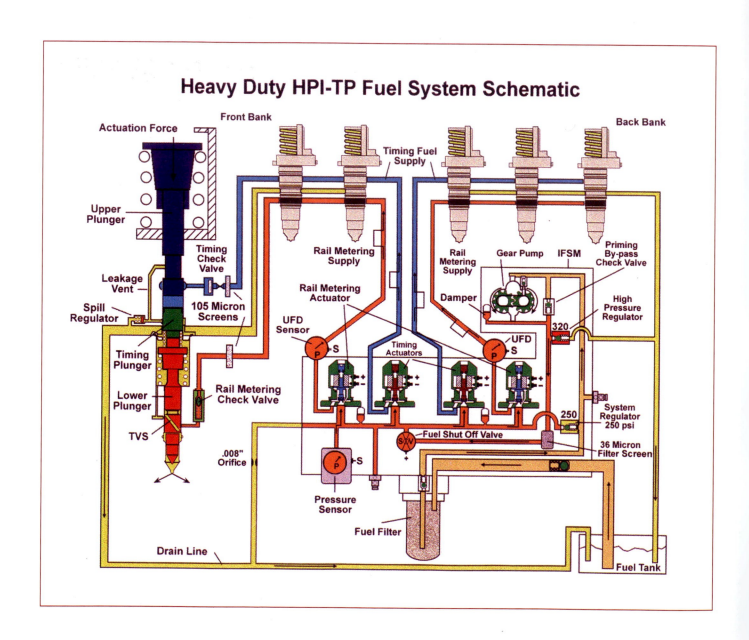

Heavy Duty HPI-TP Fuel System Schematic

Front Bank

Back Bank

Actuation Force

Timing Fuel Supply

Upper Plunger

Timing Check Valve

Rail Metering Supply

Rail Metering Supply

Gear Pump

IFSM

Priming By-pass Check Valve

Leakage Vent

Rail Metering Actuator

Damper

High Pressure Regulator

105 Micron Screens

UFD Sensor

UFD S

Spill Regulator

320

Timing Plunger

Timing Actuators

Rail Metering Check Valve

Lower Plunger

System Regulator 250 psi

TVS

250

Fuel Shut Off Valve

36 Micron Filter Screen

.008" Orifice

Pressure Sensor

Fuel Filter

Drain Line

Fuel Tank

Chapter
26
Governors

PREREQUISITES

A good understanding of diesel engine fundamentals and fuel management

OBJECTIVES

After studying this chapter, you should be able to:

- *Outline the reasons why a governor is required on a diesel engine.*
- *Define the functions of hydromechanical governors on a diesel engine.*
- *Describe the principles of operation of a simple mechanical governor.*
- *Outline the characteristics of five different types of governors.*
- *Classify governors by management mode.*
- *Interpret governor performance terminology.*

KEY TERMS

automotive governor	hydromechanical governing	power takeoff (PTO)
droop	idle speed	rated speed
droop curve	isochronous governor	throttle
electronic governor	limiting speed (LS) governor	thrust collar
governor spring	mechanical governor	top engine limit (TEL)
high-idle speed	motoring	torque rise
hunting	overspeed	torque rise profile
hydraulic nonservo governor	peak torque	variable speed (VS) governor
hydraulic servo governor		

INTRODUCTION

The output of a diesel engine aspirated with excess air (usually above the stoichiometric requirement) is only limited by the fuel put into it. Given unlimited fuel, diesel engines will accelerate at rates of up to 1,000 rpm per second, often to self-destruction. The primary function of a diesel engine governor is to sense engine speed and limit fueling when the engine is run at its specified maximum speed. However, the **electronic governors** used on today's truck and bus engines manage engine output based on command input from the vehicle operator and engine/chassis input data to an electronic control module (ECM), which then plots a fueling profile (fuel quantity, injection timing, and pilot pulse) from programmed software instructions. The ECM computations are subsequently converted to outputs by switching apparatus such as injector drivers, which control the duty cycle of fuel injectors.

Today's ECMs may be programmed with comprehensive fuel maps, engine malfunction strategy, and a wide variety of customer data options, but all the language used to describe engine governing conditions evolved from James Watt's 1786 centrifugal governor and most of it is still in use today.

James Watt invented the **mechanical governor** because he wanted to regulate the speed of his steam engines. Speed sensing was accomplished by driving a set of flyweights in a carrier at a speed proportional to engine speed; the flyweights pivoted in the carrier and were loaded into their most retracted position by a spring. As the carrier was rotated, the centrifugal force exacted from the flyweights would act against the applied spring force; the spring tension could then be set so that at a specified maximum speed, centrifugal force would overcome the spring force to act directly on a fuel control mechanism to limit fueling. Most truck and bus governors classified as "hydromechanical" use variants of Watt's governor.

This chapter introduces the basics of governor terminology and some of the principles of operation of simple mechanical governors. The diesel technician should be familiar with this terminology because it is used in performance testing and assessment. At the end of the chapter, there is a glossary of governor terminology that gives more detailed explanations of terms used within the text than those found in the main glossary: it is recommended that these terms be referenced when studying this chapter.

GOVERNOR TYPES

Governors are classified by type; that is, by operating principles. The oldest type, the mechanical governor, was most common on hydromechanical engines until 1990. Every governor managing a truck or bus engine has the benefit of an operator or driver who regulates engine output according to the requirements of the trip being undertaken. The accelerator therefore represents an essential input to the governor. The governor must also know exactly how fast the engine is rotating in any given mode of operation; so another essential input to the governor is a means of precisely determining engine rpm. In vehicle engine governors, these two inputs would be regarded as essential. A governor controls how the engine is fueled. It limits fueling at the highest intended engine rpm to prevent overspeed, but it is also required to define the lowest no-load rpm of the engine to enable it to idle without any input from the driver's accelerator pedal. The governor must also provide a means of no-fueling the engine to shut it down. Also most modern governors have many other features such as the ability to provide extra fuel over the normal engine operating rpm (**torque rise:** identified to the operator by the upper and lower shift rpms), graduated fuel deration at the top engine speeds, and excess fuel/timing adjustments for startup. Each governor type listed in this section is simply a means of grouping governors into categories. Not every mechanical governor is the same. In fact, some are very basic whereas others may have an extensive array of optional features.

MECHANICAL GOVERNORS

Speed sensing is by means of rotating flyweights driven at a speed proportional to engine rpm; in a mechanical governor, centrifugal force generated by the rotating flyweights acts directly on the fuel control mechanism. (See Figure 26–1, Figure 26–2, and Figure 26–3.) The **governor spring** tension is set to oppose the centrifugal force and will define the **top engine limit (TEL)** or **high-idle speed** (the highest speed at which an engine is designed to run). The **thrust collar** (aka thrust washer or thrust sleeve) acts as an intermediary between the spring force and centrifugal force and is connected to the fuel control mechanism. Governor spring tension is usually designed to be variable and increases with accelerator pedal travel.

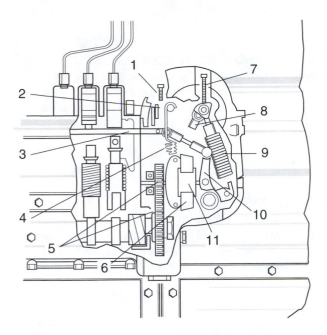

1. Low-idle adjusting screw
2. Spring-loaded stop assembly
3. Fuel rack
4. Shutoff spring
5. Gears
6. Governor weight
7. High-idle adjusting screw
8. Low-idle stop lever
9. Governor spring
10. Terminal lever
11. Thrust bearing and sliding sleeve

FIGURE 26–1   *Cutaway view of a simple mechanical governor. (Courtesy of Caterpillar)*

- Centrifugal force (produced by rotating the flyweights) acts on the thrust sleeve to attempt to decrease fueling.
- Spring force (often variable and increased with accelerator pedal travel) acts on the thrust sleeve to attempt to increase fueling.

HYDRAULIC GOVERNOR—SERVO TYPE

In a **hydraulic, servo-type governor**, speed sensing is accomplished once again by rotating flyweights driven at a speed proportional to engine speed, and opposing this force is the governor spring. However, the intermediary between the centrifugal and spring forces is a control valve that feeds oil (usually pressurized oil from the engine lubricating circuit) to a servo, which moves the fuel control mechanism, so the actual force used to move the fuel control mechanism is hydraulic. This design permits the use of smaller flyweights and more sensitive response to speed and load fluctuations.

HYDRAULIC GOVERNOR—NONSERVO

In a **hydraulic nonservo governor**, speed sensing is accomplished by using a positive displacement engine-driven transfer pump that unloads to a defined (and constant) flow area, so fuel pressure rises proportionally with engine speed. This type of governor is used in inlet metering fuel systems, and speed regulation is achieved by gating or diverting fuel from the inlet metering apparatus at a specified fuel pressure. It is not used on any current medium- and heavy-duty truck diesel engines.

PNEUMATIC

Pneumatic governors are used on naturally aspirated diesel engines and therefore seldom in truck and bus applications. Fueling is factored by manifold vacuum, which is regulated by a **throttle** valve controlled by an operator. Pneumatic governors are common in older marine and agricultural applications.

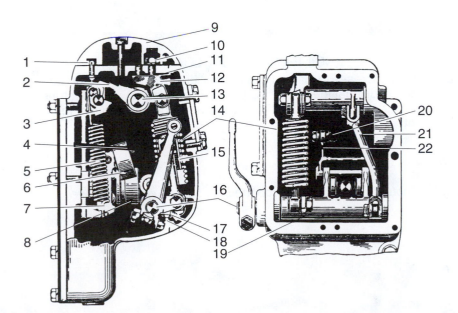

1. Low-idle adjusting screw
2. Low-idle stop lever
3. Stop
4. Small governor weight
5. Shutoff spring
6. Large governor weight
7. Spindle assembly
8. Thrust bearing and sliding sleeve
9. Top cover
10. Capscrew
11. Shims
12. Anchor
13. Stop lever shaft
14. Governor spring
15. Small spring
16. Terminal lever shaft
17. Throttle shaft control lever
18. Setscrew (earlier engines)
19. Terminal lever
20. Link
21. Rod
22. Lever

FIGURE 26–2   *Sectional and overhead views of a simple mechanical governor and components. (Courtesy of Caterpillar)*

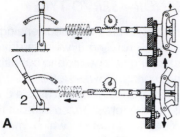

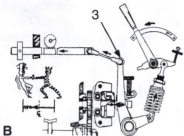

GOVERNOR SPRING AND WEIGHT FORCES

A. The force generated by the rotating governor weights is illustrated by the figures 1 and 2. Figure 1 shows half speed and Figure 2 shows high speed. As the engine speed increases the weight force increases. The spring force varies, depending on how much it is stretched.

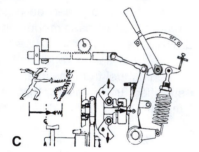

FORCES BEFORE STARTING THE ENGINE

B. When the governor control is moved to the "fuel on" position, the governor spring is stretched, and the lever (3) pushes the fuel rack toward the increased fuel position, since the weights are not rotating. As the engine is started, the speed will increase and the governor weights move outward to supply the opposing force.

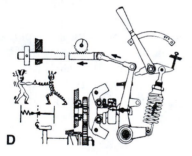

FORCES BALANCED AT HIGH IDLE

C. As the governor spring force moves the fuel rack to increase the fuel, the engine speed increases and the weight force increases. The forces oppose each other and rapidly assume a balance, whereby the correct amount of fuel is injected to balance the forces at high speed.

LOAD BEING APPLIED

D. When the engine is running at high idle speed and a load is applied, the engine speed decreases momentarily, with a corresponding decrease of governor weight force. The governor spring takes immediate advantage of this to move the rack to increase fuel.

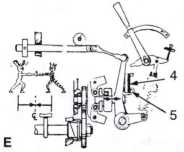

ENGINE UNDER FULL LOAD

E. A further increase in load results in additional movement of the rack to provide additional fuel. The stop pin (5) is in contact with the torque spring (4) or stop. As long as the load remains constant the fuel remains constant.

FIGURE 26–3 *Governor operation graphics: note the position of the rack and the figures representing weight and spring force. (Courtesy of Caterpillar)*

ELECTRONIC GOVERNING

Electronic governing is commonly used to govern today's truck and bus diesel engines. Engine speed sensing is by means of a reluctor type, inductive sensor, which signals an electronic control module (ECM) that plots a fueling profile and then effects it by switching actuators. Today's trucks and buses are drive-by-wire, meaning that there is no direct mechanical connection between the accelerator pedal and the fuel control mechanism. In other words, the accelerator position is one of a number of sensor inputs to the ECM. Electronic governing is used by all the diesel engine electronic management systems including Detroit Diesel DDEC, Cummins CELECT and IS (Interact System), Caterpillar and ADEM, Mack Trucks V-MAC, Navistar HEUI, and Volvo VECTRO. Each of these systems is handled in some detail elsewhere in this textbook.

GOVERNOR CLASSIFICATION

Any governor, hydromechanical or electronic, is classified by how it manages engine performance. The governor classifications found on contemporary truck and bus diesel engines fuel management systems follow:

LIMITING SPEED (LS)

A **limiting speed (LS) governor** (aka min-max) sets the engine **idle speed**, defines the high-idle speed, and permits fueling between those values to be controlled by an operator (driver). Limiting speed governors are the most common in commercial vehicle applications. One of their advantages is to make the diesel engine respond to accelerator input in much the same manner as the SI engine responds to throttle control. A governor classified as limiting speed will in most cases provide excess startup fuel, define a **torque rise profile**, define **droop curve**, and be capable of no-fueling the engine for shutdown. A mechanical limiting speed governor is sometimes known as an **automotive governor** (Figure 26–4).

VARIABLE SPEED (VS)

A **variable speed (VS) governor** (aka all-speed) sets engine idle speed, defines high idle and *any* speed in the intermediate range depending on accelerator pedal position. A given amount of accelerator pedal travel corresponds to an engine rotational speed; as engine loading either increases or decreases, the governor manages fueling to attempt to maintain that engine speed. Hydromechanical variable speed governors were common in many Mack Trucks and Caterpillar applications and others where **PTO** (**power takeoff:** where the engine is used to drive auxiliary equipment) management was a consideration. From the driver

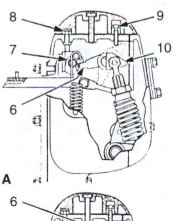
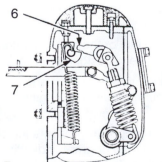

HIGH-AND LOW-IDLE ADJUSTMENTS

A. When the low-idle stop lever (6) is in contact with the spring-loaded stop (7) as illustrated, the engine operates at low-idle. The screw assembly (8) controls the low-idle setting. When the lever (10) contacts the high-idle screw (9) which is also adjustable, the engine operates at the high-idle speed.

GOVERNOR SHUTOFF POSITION

B. When the governor control is in the shutoff position, it causes the stop lever (6) to pivot the spring-loaded stop (7) and it will hold the fuel rack in the "fuel off" position.

FIGURE 26–4 *Governor adjustments and shutoff position. (Courtesy of Caterpillar)*

perspective, the VS governor takes a little getting used to. Most of today's electronic management systems can be toggled (as customer data programming) to either LS or VS mode. A governor classified as variable speed usually provides excess startup fuel, defines a torque rise profile, defines droop curve, and is capable of no-fueling the engine for shutdown.

ISOCHRONOUS

Isochronous governing is only required when driving a generator. In this application the engine must respond instantly to load changes with zero **droop** (*no* rpm fluctuation when engine load changes) or the (electrical) frequency will alter. However, the term is being used to describe an option in diesel engine electronic management systems. In this instance, isochronous governing mode would be used to manage PTO fueling while stationary and one OEM uses the term to describe engine fueling at an electronically managed, default (when critical input signals are lost) rpm.

GOVERNOR TERMINOLOGY

All-speed governor: British term used to describe a variable speed governor, but it is also used by Mack Trucks and Bosch.

Deadband: term used to describe the sensitivity of a governor. It is the speed window around set speed in which fueling correction is made by the governor.

Droop: transient (of short duration) speed variation from set speed when engine load changes.

Droop curve: expressed as a percentage of high-idle speed, the droop curve is the graduated fuel deration between rated load and speed (peak power) rpm and the high-idle speed rpm. Droop curve is usually between 5 and 20% in hydromechanical highway diesel engines.

Governor cutoff: speed at which governor cuts off fueling.

Governor flight path: see droop curve.

Governor spring: force used to counter centrifugal force developed by the flyweights in a mechanical governor.

High idle: (WOT or top engine limit). The maximum no-load speed of an engine.

Hunting: rhythmic change in engine speed often caused by unbalanced fuel delivery in multicylinder engines.

Hydromechanical governing: refers to engines that are governed without the use of computers (ECMs).

Idle: any no-load running speed of an engine but usually refers to low idle, the lowest speed the engine is designed to run at, usually with no input from the speed control mechanism.

Min-Max: British term for a limiting speed governor; also used by Bosch.

Motoring: running an engine at 0 throttle, with chassis momentum driving engine.

Overrun: the inability of a governor to keep the engine speed below the high-idle speed when it is rapidly accelerated.

Overspeed: any speed above high idle.

Peak torque: rpm at which the engine develops peak torque, often located at the base of the torque rise profile.

Rated speed: the rpm at which peak power is achieved from a diesel engine.

Road speed governing: any governor system in which engine fueling is moderated by a predetermined road speed value.

Sensitivity: ability to respond to maintain a set rpm without rpm fluctuation as load changes.

Speed drift: in which engine speed rises above or below set speed often in surges. Differentiated from hunting by the fact that it is not rhythmic.

Stability: ability to maintain set rpm.

TEL: see top engine limit.

Throttle: air flow to the intake manifold control mechanism used in SI gasoline and diesel engines with pneumatic governors. The term is commonly used to describe the speed control/accelerator/fuel control mechanism in a diesel engine.

Thrust collar: in a mechanical governor, the intermediary between the centrifugal force exacted by the flyweights and the spring forces that oppose it; the thrust collar in a governor is usually connected to the fuel control mechanism. aka thrust washer and thrust bearing.

Top engine limit: high idle or the fastest no-load speed at which an engine is designed to run.

Torque rise: rpm range through which engine can maintain near-maximum torque. The engine-operating range.

Torque rise profile: a graphic representation of engine fueling on a fuel map or graph indicating the torque rise window. Often, but not necessarily, begins at peak torque and ends at rated load speed.

Underrun: a governor's inability to maintain the engine low idle speed when rpm is quickly dropped.

Work capacity: a measure of a mechanical governor's ability to produce the centrifugal force required to move the fuel control mechanism. The work capacity of a mechanical governor driven at engine rpm exceeds that of one driven at camshaft speed assuming the same flyweight mass.

WOT: wide open throttle. A term used mainly on SI automotive engines and less often on diesels; means *full-fuel request.*

SUMMARY

- Speed sensing in many hydromechanical governors is performed by rotating flyweights at a speed proportional to engine rpm and measuring the centrifugal force exacted.
- Most current truck and bus diesel engines manufactured since 1991 are equipped with electronic governors.
- Electronic governor terminology has evolved from hydromechanical governor technology.
- James Watt invented the first centrifugal governor.
- A mechanical limiting speed governor is sometimes referred to as an automotive governor.
- On a variable speed governor, accelerator position commands a specific engine speed. As engine load increases and decreases, the governor adjusts engine fueling to attempt to maintain that engine speed.
- Mack Trucks and Caterpillar have traditionally used VS governors on their hydromechanical engines, which accounts for the unique driveability characteristics of their product. VS governors are usually an option from other engine OEMs.
- Road speed governing is commonly specified in current truck engine applications.
- In a basic mechanical governor system, centrifugal force exacted by flyweights will attempt to diminish engine fueling while spring force, often moderated by operator demand, attempts to increase fueling; however at a predetermined maximum engine speed value, the centrifugal force must always overcome the spring force.

REVIEW QUESTIONS

1. Rhythmic fluctuation in engine rpm is known as:
   a. Governor overrun
   b. Governor underrun
   c. Dieseling
   d. Hunting

2. An engine run at a speed exceeding rated but less than high-idle speed would be operating:
   a. In the torque rise profile
   b. At peak torque
   c. In the droop curve
   d. In a overrun condition

3. Which type of hydromechanical governor would most commonly be used to manage a highway truck or bus diesel engine?
   a. Mechanical, limiting speed
   b. Hydraulic, limiting speed
   c. Mechanical, variable speed
   d. Hydraulic, variable speed

4. Speed sensing in mechanical governors is performed by:
   a. A pulse wheel
   b. Matching centrifugal force to spring force
   c. Precise determination of fuel pressure

5. A mechanical, limiting speed governor has a fatigued governor spring. What would be the likely consequence?
   a. Erratic low-idle speed
   b. Hunting
   c. Lower high-idle speed
   d. Engine runaway

6. Under which of the following conditions would a properly functioning mechanical governor on a truck or bus diesel engine NOT be capable of controlling engine rpm?
   a. Loaded, running downhill
   b. Unloaded, running level
   c. Loaded, running level
   d. Unloaded, running uphill

7. Which of the following governors would sometimes be known as an automotive governor?
   a. Hydraulic, servo type
   b. Hydraulic, nonservo
   c. Mechanical LS
   d. Mechanical VS

8. When an engine is rapidly accelerated and momentarily exceeds the nominal high-idle speed, the condition is known as:
   a. Runout
   b. Governor cutoff
   c. Speed drift
   d. Overrun

9. *Technician A* states that governor rated speed and governor high-idle speed should occur at the same rpm in a truck engine, hydromechanical governor. *Technician B* states that governor droop curve is typically between 5% to 20% in truck hydromechanical engines. Who is right?

a. A only
b. B only
c. Both A and B
d. Neither A nor B

10. *Technician A* states that a limiting speed governor is designed to produce accelerator response close to that of an automobile throttle. *Technician B* states that a limiting speed governor is sometimes known as an automotive governor. Who is right?
    a. A only
    b. B only
    c. Both A and B
    d. Neither A nor B

11. *Technician A* states that a variable speed governor provides zero droop performance. *Technician B* states that a variable speed governor usually provides faster response to a load change than an isochronous governor. Who is right?
    a. A only
    b. B only
    c. Both A and B
    d. Neither A nor B

12. *Technician A* states that an isochronous governor could be used on a vehicle engine providing it had a fluid clutch. *Technician B* states that an isochronous governor is normally used to manage a genset. Who is right?
    a. A only
    b. B only
    c. Both A and B
    d. Neither A nor B

13. *Technician A* states that a high-speed diesel engine given unlimited fuel will accelerate to self-destruction. *Technician B* states that an engine fueled on engine oil can also run away to self-destruction. Who is right?
    a. A only
    b. B only
    c. Both A and B
    d. Neither A nor B

14. *Technician A* states that most computer-managed engines still use hydromechanical governors. *Technician B* states that electronic governors can be programmed to function in either LS or VS modes. Who is right?
    a. A only
    b. B only
    c. Both A and B
    d. Neither A nor B

15. *Technician A* states that when a governor high-idle speed is increased, an increase in brake horsepower is a more likely result than a decrease. *Technician B* states that drivers are more likely to complain about a lack of power on an engine with a short torque rise profile than one with high torque rise. Who is right?
    a. A only
    b. B only
    c. Both A and B
    d. Neither A nor B

Chapter

27

Alternate Fuels

PREREQUISITES

A good understanding of Section 1 and chapters 16, 17, and 18.

OBJECTIVES

After studying this chapter, you should be able to:

- Identify some alternatives to diesel fuel that may be viable in commercial vehicle engines.
- List some possible power plant alternatives to the diesel engine.
- Describe the characteristics of biodiesel fuels.
- Outline the properties of B100 and B20 fuels.
- List some of the means of using electricity to power heavy-duty highway, commercial vehicles.
- Describe how diesel electric hybrid engines are used as a power plants in city transit bus applications.
- Identify some types of fuel cells currently used and being considered for use in highway vehicles.
- Outline the operating principles of a proton exchange membrane (PEM) fuel cell.
- Identify some of the advantages and disadvantages of gaseous fuels in medium- and heavy-duty commercial vehicle applications.
- List the reasons why hydrogen may become the fuel of the future.
- Identify the advantages and disadvantages of alcohol-based fuels used in medium- and heavy-duty commercial vehicle applications.

KEY TERMS

alcohol

auxiliary power unit (APU)

biodiesel

B20

B100

compressed natural gas (CNG)

hydrogen

liquefied petroleum gas (LPG)

parallel hybrid

propane

proton exchange membrane fuel cells (PEM)

regenerative braking

INTRODUCTION

In the first edition of this book in a section on alternate fuels, it was suggested that diesel fuel might not be around in the foreseeable future as a highway truck fuel. We could say that again, but it is just as unlikely now as it was then that diesel fuel could be replaced as the primary fuel used by commercial vehicles in anything less than 10 years after the innovation of a viable technology. Currently, there are no alternate fuel technologies on the horizon that could compete with diesel fuel. This fact means that changes will focus on the fuel itself and the way diesel engines combust it. However, it is valuable to look at potential challengers to diesel fuel, some of which are actually used and others that are just talked about. Most of the alternate fuel technologies are marginal at best and do not pose any immediate threat to the diesel engine.

The most unlikely scenario is that some revolutionary power plant technology will replace the diesel engine in the immediate future. The diesel engine has been around for a hundred or so years and is supported by a massive engineering, production, fueling, and repair infrastructure that would take years to replace. Although the use of diesel fuel in some jurisdictions is statutorily limited, usually within the limits of some especially polluted city, the consequence has been to put the alternate fuel (CNG/ propane/alcohol-based) under the microscope, and the emission and performance results have not been as good as expected. And despite all of the recent talk about fuel cells, these have generated a lot more talk than electricity, and it is unlikely that the fuel cell will ever be a viable technology to replace the diesel engines in our commercial trucks. It is more likely to be developed as a transit bus power plant. In the next decade or so, we are likely to see a continued effort led by the EPA and CARB to improve the chemical characteristics of diesel fuel in an effort to reduce or eliminate its more toxic combustion byproducts.

For two decades, transit corporations have used gaseous and alcohol-based fuels in jurisdictions that limit the use of diesel fuel. The downside has been decreased engine life and increased fuel mileage and maintenance costs. These disadvantages arise partly because few engines used in this way have been designed from scratch as an "alternate" fuel engine. Instead, an existing commercial diesel engine has been modified to combust the alternate fuel. In such engines fueled by **alcohol**-based fuels, **liquefied petroleum gas (LPG)**, or **compressed natural gas (CNG)**, the fuel charge will not autoignite at diesel compression temperatures, so a means of igniting the fuel must be provided. The result is that either a spark ignition system has to be incorporated or a small shot of diesel fuel has to be injected as an ignition pilot. When an alcohol-based fuel is used in pumping hardware designed for diesel fuel, the low lubricity of the alcohol necessitates frequent overhauls of expensive fuel pumping and injection apparatus.

In this chapter, we look at some friendlier diesel fuels, at least from the environmental perspective, and at some of the technologies that are, or may be, able to replace the diesel engine. We also examine fuel cells, because, after all, everybody is talking about them and they are finding their way into some intracity transit buses, and they could even be used in a limited way on trucks, say, as an **auxiliary power unit (APU)**. An APU is used on a truck to provide heating and electrical power when the engine is not running.

BIODIESEL

We begin with a fuel, **biodiesel**, that actually has a small track record of success as a diesel fuel, largely because it combusts in the diesel engine without any significant changes having to be made to the engine. Biodiesel has some real limitations because it is best suited to operation in warm climates, but it is cleaner-burning than petroleum-based fuels. Mixed with standard diesel fuel at a 20% concentration improves its temperature operating range while lowering harmful emissions.

In the past, biodiesel conjured images of hippies concocting fuels from McDonald's waste cooking fat, an operation that could take place in a barn or perhaps a kitchen. The definition of biodiesel is a little hazy because the term can be applied to a number of home-made fuels as well as those refined under more stringent guidelines. In describing biodiesel in this chapter, we use ASTM standard D6751 and the U.S. Department of Energy (DOE) research rather than referencing some of the cruder do-it-yourself technologies. This is not to denigrate those who manufacture their own fuel (official recognition of biodiesel as a viable technology has probably been hastened by them), but rather to state that fuels from this source will not be used to power highway trucks. By our definition, biodiesel is fuel produced from farm products. Its base is vegetable oil and alcohol. And, yes, the commercially sold vegetable oil can be sourced from recycled restaurant greases, so it has great appeal to those who advocate renewable energy practices.

The chemical composition of biodiesel is given in Table 27–1. In the table, biodiesel manufactured to

TABLE 27–1 *Biodiesel Compared with Highway Diesel Fuel*

FUEL PROPERTY	DIESEL	BIODIESEL
Specific gravity	0.85	0.88
Caloric value (heat energy)	131,295 Btu/gal	117,093 Btu/gal
Density (lb per gal at 15°C)	7.079	7.328
Water percentage of volume	0.05	0.05
Carbon (by weight)	87%	77%
Hydrogen (by weight)	13%	12%
Oxygen (by weight)	0%	13%
Sulfur (by weight)	0.05%	0 to 0.0025%
Flash point	60–80°C	100–170°C
Cetane number	40 to 55	48 to 60
Pour point	−35 to −15°C	−15 to 16°C

Source: Data from U.S. DOE, Feb. 2002.

TABLE 27–2 *Biodiesel Emissions Compared with EPA Certified Diesel Fuel*

EMISSION	B100	B20
Carbon monoxide	−43%	−12%
Hydrocarbons	−56%	−11%
Particulates	−55%	−18%
Nitrogen oxides	+ 6%	+ 1.2%
Air toxics	−80 to − 90%	−20%

Source: Data from U.S. DOE, Feb. 2002.

ASTM D6751 is compared to a highway diesel fuel ASTM 975 manufactured to 2002 standards.

COMBUSTING BIODIESEL

Biodiesel burns more cleanly than petroleum-based diesel fuel. Most of the harmful emissions produced by combusting petroleum diesel fuel are significantly reduced with the exception of oxides of nitrogen (NO_x). NO_x emission is slightly increased over diesel fuel because biodiesel is an oxygenate fuel; that is, like alcohol, there is an oxygen component in the fuel. Perhaps the biggest disadvantage of biodiesel is its inability to flow at colder temperatures. This means that its primary markets are going to be in warmer climates. Biodiesel becomes a more viable fuel when it is mixed at a 20% ratio with petroleum-based diesel fuel. This mixture is known as **B20**. Biodiesel B20 is able to retain some of the advantages of biodiesel **B100** (100% vegetable base) and is probably more practical as a future commercial vehicle fuel. B20 and B100 are compared in Table 27–2.

COSTS AND AVAILABILITY

Biodiesel is new, expensive, and limited to seasonal use in much of the United States and Canada. These factors have meant that the supply infrastructure has been slow to grow. Biodiesel sells at between $1.10 and $1.85 per gallon (June 2003). Add a further $.50

tax and it is expensive compared with current #2 diesel fuel. Therefore consumers tend to opt for cheaper B20 over B100. However, the DOE is actively working with the biodiesel industry to reduce the cost of the product. Combine this effort with EPA efforts to make highway fuel cleaner (therefore more expensive!) reducing the cost differential, and the future for biodiesel may be bright. For the time being, the DOE suggests that it is best to avoid any biodiesel that does not meet the ASTM D6751 standard.

POWER AND FUEL ECONOMY

Using B100 in place of diesel fuel will reduce power by about 10%. And, quite simply, there is less heat energy in biodiesel, therefore you will use more of it. For every gallon of diesel fuel consumed, approximately 1.1 gallons of B100 will have to be consumed. The difference is much less noticeable when B20 is used. Most data indicate that the reduction in power and fuel economy is around 2%.

WARNING: B100 gels in anything but midsummer conditions. It should not be considered a year-round fuel for engines.

CAUTION: Some OEMs do not approve of biodiesel use in their fuel systems. Using biodiesel can void warranty. If considering the use of either B100 or B20, consult the OEM and ensure that it is approved.

ELECTRICITY

Most of the automobile and some of the bus OEMs are producing a limited range of electrically powered and hybrid or dual-powered vehicles. The limitations of electrically powered vehicles means that they are an unlikely truck power plant in the near future. Electrically powered vehicles can be categorized as follows.

Battery powered. The mass and the space occupied by the battery packs are just the beginning of the problems of battery driven, electrically powered vehicles. The challenge to industry is to be able to build batteries of much larger storage capacity combined with much lower weights. Nevertheless, cars, vans, and light trucks powered by nickel metal hybrid battery packs are being manufactured in limited numbers, mostly for intracity use.

Electric motor, gas turbine driven. The thrust toward higher efficiency power plants has been driven by the Europeans who have the best reasons for developing fuel efficiency, namely, astronomically high fuel prices. They are engineering and manufacturing long-range, light vehicles that use a superefficient gas turbine fueled with kerosene to drive an electric motor.

Electric motor, hard-wire powered. Used for decades to power trolley buses and streetcars, these power plants have obvious limitations and will probably never be a viable truck fueling technology. Electrically powering large commercial vehicles is not generally as environmentally friendly as some would like to believe, depending on the means used to generate the electricity. Fossil fuels and nuclear reactors are used to produce electricity in many parts of North America and, even if the source is hydroelectric, some would argue that there is an ecological cost.

Fuel cells. These generate electricity and we focus on how they operate in the next section. They have been appearing on our roads as a city bus power plant, and Freightliner has offered a fuel cell auxiliary power unit (APU) on certain trucks.

Parallel hybrid. Dual power plant or **parallel hybrid** vehicles overcome some of the disadvantages of the electric motor as a vehicle engine by using a traditional internal combustion engine in conjunction with the electric motor. Several designs are available in the marketplace, and the New York City Metropolitan Transportation Authority (NYC-MTA) has committed to parallel hybrid technology by purchasing 125 Orion hybrid buses. The Orion buses use a Lockheed Martin engineered powertrain, consisting of a Cummins ISB (6 cylinder, 5.9 liter) 275 BHP diesel engine to generate electrical power to AC electric motors located on the drive trucks. The Cummins ISB 275 engine (shown in Figure 27–1) on its own would have insufficient power and torque to drive a full-size bus, but in a par-

FIGURE 27–1   *Cummins ISB 5.9 liter engine. (Courtesy of Cummins)*

allel hybrid arrangement becomes an ideal power source. The engines used are equipped with a particulate trap and catalytic converter and in testing with the MTA have produced 50% less NO_x, 98% less CO, and 200% better fuel economy than diesel-powered buses.

A bonus of a diesel electric hybrid drive is achieved with **regenerative braking**. When the drive electric motor magnetic field is reversed, it applies retarding torque and generates electricity used to charge the batteries. Diesel-hybrid drive is not a new concept; for many decades most railway locomotives have been powered by a diesel/electric powertrain. It is simply new on our highways.

FUEL CELLS

A fuel cell is an electrochemical device in which the energy of a chemical reaction is converted into electricity. It does nothing more than produce electricity. If a vehicle is to be powered by a fuel cell, the fuel cell simply produces electricity, and electric motors are used to drive the vehicle. Fuel cells are not new, having been invented by Sir William Grove in 1839. After their invention, not a whole lot was done with fuel cells until NASA used them to provide some of the electrical requirements of their spacecraft in the 1960s. Figure 27–2 shows the power equation of a fuel cell.

Fuel cells are not "free" energy as some have described them; they need **hydrogen** to function, which creates problems. Commercial hydrogen is usually produced electrolytically from water, a process that requires large amounts of electricity. Further, safe storage of highly explosive hydrogen on a vehicle is not easy. The option is to use fuel reforming to extract the hydrogen from hydrocarbon (HC) fuels such as gasoline, natural gas, or alcohols. The fuel cells themselves do not discharge like a battery; they will run as long as supplied with hydrogen fuel. The claim that fuel cells are zero-emission energy producers is only true when the fuel provided is pure hydrogen. Most current fuel cells used in small vehicles require an HC fuel such as gasoline as the hydrogen source, using a reforming process. Reforming is the separating of H_2 from the HC

fraction molecule in the gasoline or other HC fuel. When a fuel cell is fueled with reformed hydrogen, the results are not as environmentally friendly as some might claim because the carbon byproduct of the process has to be considered.

So, although hydrogen is the most abundant element in the known universe, to produce it on Earth either takes lots of electricity (the production of which usually causes noxious emissions) or it has to be reformed from a fossil fuel, the greater percentage of which is carbon. These factors are likely to make the fuel cell a very limited mobile technology in the immediate future.

HOW A FUEL CELL WORKS

A fuel cell functions on the principle of the thermodynamic reversibility of the electrolysis of water. For many years we have produced hydrogen and oxygen by the electrolysis of water, and, as mentioned before, it does take lots of electricity. When supplying hydrogen and oxygen to the two electrodes of an electrolysis cell, a potential difference is created, and electric current begins to flow. Join several cells together and you create a multicell fuel cell. The electrochemical reaction is:

$2H_2 + O_2$ produces $2H_2O$ + electricity

Figure 27–3 demonstrates the operating principle of a fuel cell: hydrogen and oxygen are electrochemically reacted to produce water and electricity.

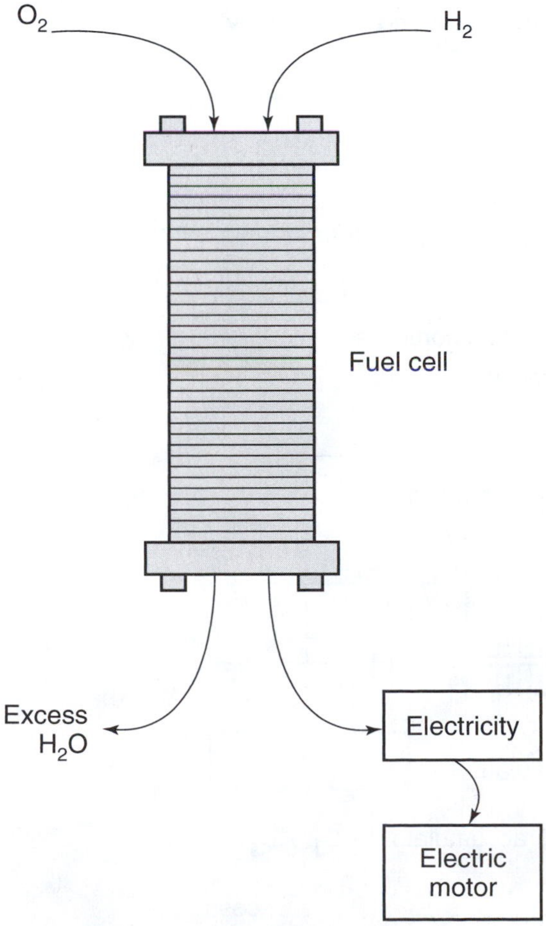

FIGURE 27–2   *Fuel cell power equation.*

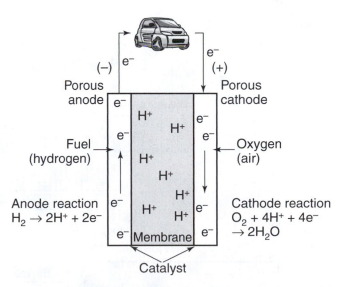

Overall electrochemical reaction
$2H_2 + O_2 \rightarrow 2H_2O$ + electricity

FIGURE 27–3   *Fuel cell chemical reaction.*

TYPES OF FUEL CELLS

There are a number of different types of fuel cells. We look briefly at some of them and in more detail at one currently being tested in city buses that may perhaps have a future as a short-range, small truck power plant. Each type of fuel cell is distinguished by the electrolyte material used.

Alkaline fuel cells.   Alkaline fuel cells have been used by NASA in the Apollo and space shuttle programs. To function they must be provided with pure hydrogen and pure oxygen, substances that are stored and readily available on a spacecraft, but difficult to store on a truck or automobile.

Phosphoric acid fuel cells.   Phosphoric acid fuel cells are suitable only in stationary power generation applications due to the corrosive nature of the liquid electrolyte and high operating temperatures, usually in the range of 200°C. They have been used as electrical generators in large buildings with a need for a power source independent from the grid such as hospitals and hotels.

Solid oxide fuel cells.   Solid oxide fuel cells operate at the highest temperatures, between 700° and 1,000°C, but because of this, will tolerate relatively impure fuels such as the hydrocarbon gas obtained from the gasification of coal. These fuel cells have a relatively simple design and are thought to be most suitable for large stationary power generators. It is unlikely that there would be any mobile vehicle application in the near future.

Proton exchange membrane fuel cells.   **Proton exchange membrane (PEM) fuel cells** have created some excitement in the automotive industry because they are the most viable mobile fuel cell technology. The PEM fuel cell has been used as a primary power source in some vehicles, and Freightliner chose this technology for its optional APU unit. A PEM fuel cell uses a solid polymer membrane (a thin plastic film) as the electrolyte. If the fuel cell can be supplied with pure hydrogen (as opposed to reforming it from a fossil fuel), then the only emissions that have to be considered are those used to produce the pure hydrogen. The Ballard fuel cell being used to power city buses uses a PEM principle. Figure 27–4 shows how a PEM functions.

Ballard Power Systems has led the way in developing fuel cells for automotive use and has partnership liaisons with several OEMs. The Ballard Mark 900 fuel cell is being used in trial applications in city buses in Chicago, Vancouver, and Palm Springs. More recently, Ballard has agreed to supply fuel cells to Daimler-Chrysler for European trials in a number of cities, in-

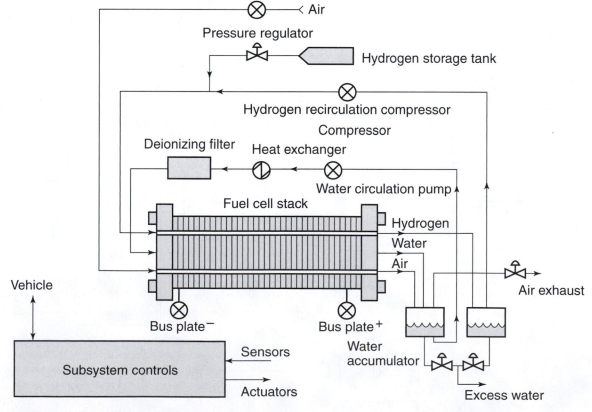

FIGURE 27–4   PEM fuel cell principle.

cluding Paris, Barcelona, Reykjavik, and Hamburg, beginning in 2005. These types of research and development trials are laying the groundwork for the fuel cell to become a viable technology within 10 years or so. However, its use may be confined to city buses.

FUEL ALTERNATIVES FOR ENGINES

Generally, where engines and fueling apparatus have been successfully adapted for fueling with any of the alternatives to diesel fuel, their use is confined to short haul and intracity transit applications. The thrust toward maximizing engine efficiency and finding alternatives to diesel fuel and gasoline has taken a different path in North America than in Japan and Western Europe. The objective of lower reliance on imported fossil fuels has driven development in Japan and Western Europe due to the much higher cost of fossil fuels in those geographic areas. It should be stated that a large portion of this high cost is a result of government taxation of fuels. Development of low emissions engines has been the primary objective of North American OEMs, driven by clean air legislation. We now look at the characteristics of some of the possible alternatives to the gasoline and diesel fuel we use today.

HYDROGEN

Hydrogen could be the fuel of the future. It is the most abundant element in the universe and it oxidizes (combusts) producing no noxious emissions. There are no current commercial vehicle engine designs in production, but most of the major manufacturers have experimented with using hydrogen as a fuel. If you read the earlier section in this chapter on fuel cells, you know that the uses of hydrogen are not limited to burning it, it can be used to produce electricity using fuel cells.

A rocket engine runs on hydrogen fuel reacted with oxygen. When you watch a rocket launch, the clouds of white smoke you see emitted are composed of vaporized water or steam. Combusted hydrogen combines with oxygen to form water (H_2O), so if the oxygen is pure, the result is a zero toxic emissions engine. However, if hydrogen is reacted with the oxygen in air, then oxides of nitrogen can result from the combustion reaction. Because of its explosiveness and difficulty in the storage/refueling process, hydrogen is unlikely to be combusted in mobile engines in the near future. For instance, a hydrogen tank similar in size to one 100-gallon-truck diesel fuel tank would provide the truck with only enough fuel for a single 100-mile trip. If hydrogen is ever to be adopted as commercial vehicle

fuel, it will more likely be to fuel advanced versions of the fuel cells described in the previous section.

Hydrogen Combustion Products

Hydrogen has the simplest atomic structure of any element. It consists of one proton and a single electron. When hydrogen is oxidized or combusted, two hydrogen atoms bond with an oxygen atom to form H_2O or water. The water formed from reacting hydrogen with oxygen is so pure that astronauts on space shuttles drink the "waste" from the liquid hydrogen fuel that powers their rocket engines.

Sources of Hydrogen

Despite its abundance, hydrogen atoms only exist naturally in combination with other elements on Earth. That means that it has to be stripped away from whatever other atoms it is combined with. And that takes energy. In most cases the energy required is electricity (if water is the source) and lots of it is needed. Producing hydrogen by electrolysis occurs when a lead-acid battery is overcharged.

Advantages
High heating value, three times that of diesel fuel
Clean burning—oxidizes to H_2O (water)
Lightweight
Abundant

Disadvantages
Expensive to produce
Low boiling point −423°F (−252°C)
Difficult to store safely on a vehicle
Highly explosive

WARNING: Hydrogen is highly explosive. Special training is required when working with hydrogen-fueled vehicles. Leaks may result in life-threatening situations.

COMPRESSED NATURAL GAS

There are numerous currently produced, commercial vehicle engine designs using CNG, (chemically composed mostly of methane), but note the following list of disadvantages. The CNG-fueled engine has been used in intracity transit vehicle applications with limited success. As a fuel, many transit corporations experimented with CNG in the early 1990s, but most did not like the results. It is unlikely that CNG will replace diesel fuel as the fuel of the twenty-first century. Often low pressure, plumbed city gas is compressed into vehicle tanks, but this procedure is usually slow and can take a number of hours. Part of the problem with using CNG as a fuel for commercial engines is that no engines have been designed specifically for CNG. In most cases, diesel engines have been adapted to burn it. CNG does not autoignite as does diesel fuel, so a means of igniting the fuel charge has to be provided. This characteristic means that either a spark ignition system or diesel pilot must be used.

Advantages

Nontoxic

Abundant with good distribution infrastructure

Technology already exists for use as a vehicle fuel

High octane (would permit high compression ratio)

Relatively clean burning—less HC, CO, and NO$_x$

Disadvantages

Low calorific value compared with diesel fuel

Bulky storage tanks of large volume

Reduced peak power performance

Refueling can be a problem—very slow

Range limitations

LIQUEFIED PETROLEUM GAS

LPG (chemically composed mostly of methane) or **propane** is a byproduct of the process of extracting crude petroleum from the ground. As with CNG, there are numerous existing proven designs, but none of them have prevailed beyond some limited, usually intracity, light duty, commercial vehicle designs. Propane does not autoignite under compression pressures so, as with CNG, when used to fuel an engine designed as a diesel, a means of igniting the fuel charge must be provided. Ignition can be by spark or pilot pulse.

Advantages

Equals fuel weight-to-mileage range of gasoline

Technology already exists for use as a vehicle fuel

High octane

Clean burning—emits less HC, CO, and NO$_x$

Increased engine longevity

Disadvantages

Heavier than diesel fuel by 20%

Must be stored at −260°F (−162°C) or less to remain liquid

Lower heating value than diesel fuel

Reduced peak power performance

ALCOHOL-BASED FUELS (METHANOL/ETHANOL)

There are some current alcohol-fueled engines used in commercial vehicles, mostly intracity buses. Engine longevity can be drastically reduced when alcohol fuels are used in existing diesel engines adapted for alcohol fueling because of the low lubricity of **alcohol**. Its use in commercial vehicles has been confined to short-haul operations within cities where local jurisdictions have banned or discouraged the use of diesel fuel.

Advantages

Produced from natural gas, coal, or vegetation (garbage)

Adapts readily to current liquid fuel delivery technologies

High octane rating—up to 110

Used in current bus diesel applications

Reduced HC emissions

Technology exists for use in diesel engines

Disadvantages

Half the calorific value of diesel fuel

Low lubricity (a disadvantage for use in current diesel engines)

Low CN—requires spark or piloted ignition

Poor cold weather performance (does not readily vaporize)

Higher acidity promotes corrosion and lube oil breakdown

SUMMARY

- Biodiesel, hydrogen, propane, compressed natural gas, and alcohol-based fuels are some possible alternatives to diesel fuel that may be viable in commercial vehicle engines. Although some of these are currently used, none are likely to entirely replace diesel fuel in the near future.
- Electrical power both in the form of fuel cells and hybrid drive power plants are some possible alternatives to the diesel engine.
- Biodiesel fuels are refined from vegetable oils and waste cooking grease. They are currently used in warmer geographic locations of the United States where they have been proved to perform well.
- Biodiesel is marketed as B100, a pure vegetable oil-based stock, and B20, a fuel composed of 80% petroleum-based diesel and 20% biodiesel.

- Direct electricity is used to power heavy-duty highway commercial vehicles in limited applications. It is used through an overhead or rail-based grid to power transit light rail and trolley buses. This type of application is obviously limited by the range of the electrical grid.
- Diesel electric hybrid engines are used as a power plant in many city transit bus applications, notably by the MTA in New York City. This technology uses a small-bore diesel engine to drive a generator, and it uses batteries to drive electric motors located on the drive axles.
- The fuel cell currently used and being considered for much more extensive use in highway vehicles is the proton exchange membrane (PEM) fuel cell.
- The operating principles of a PEM fuel cell are simple: hydrogen and oxygen are reacted to produce electricity and water. The procedure is the reverse of electrolysis.
- Gaseous fuels have the advantage of producing lower emissions when combusted in an engine, but generally have lower power and present storage difficulties on board a vehicle. This limits range as there is not a refueling infrastructure in place.
- When used in diesel engines, alcohol-based fuels produce lower emissions, but because of the low lubricity of alcohol, they tend to wear out the fuel delivery systems when existing diesel engine fuel injection systems are adapted for alcohol delivery.

REVIEW QUESTIONS

1. Which of the following is a correct description of a B20 fuel?
   a. 80% biodiesel cut with 20% petroleum base diesel
   b. 80% petroleum-based diesel cut with 20% biodiesel
   c. Biodiesel rated with a CN of 20
   d. Petroleum-based diesel with an ignition temperature of 200°C.
2. What would a B100 fuel be composed of?
   a. 100% petroleum-based diesel fuel
   b. 100% alcohol-based stock
   c. Even mixture of petroleum and vegetable-based stock
   d. 100% vegetable-based stock
3. Which of the following makes biodiesel less suitable for use in cooler geographic locations?
   a. Pour point
   b. Cetane number
   c. Sulfur content
   d. Particulate emissions
4. Which of the following noxious emissions tends to increase when the combustion of biodiesel is compared with the combustion of petroleum-based diesel fuel?
   a. Sulfur oxides
   b. Particulate matter
   c. Hydrocarbons
   d. Nitrogen oxides
5. What is the fuel required by a fuel cell?
   a. Nitrogen only
   b. Hydrogen and oxygen
   c. Hydrogen only
   d. Water
6. What results from the electrochemical reaction in a fuel cell?
   a. Electricity only
   b. Water only
   c. Hydrogen and oxygen
   d. Electricity and water
7. Which type of fuel cell is thought to have the most potential as a power plant in highway commercial vehicle applications?
   a. Proton membrane exchange
   b. Alkaline
   c. Solid oxide
   d. Phosphoric acid
8. Which of the following has the most amount of heat energy by mass (weight)?
   a. Propane
   b. Compressed natural gas
   c. Hydrogen
   d. Diesel fuel
9. When propane or compressed natural gas is used in an internal combustion engine designed for diesel fuel, what must also be provided?
   a. Particulate trap
   b. Means of igniting the fuel
   c. Cooled EGR
   d. Distributor
10. *Technician A* states that a disadvantage of alcohol as an internal combustion engine fuel is its low octane rating. *Technician B* states that an advantage of LPG over gasoline as an internal combustion engine fuel is its high calorific value. Who is correct?
    a. A only
    b. B only
    c. Both A and B
    d. Neither A nor B
11. When a regenerative braking cycle is engaged in a diesel-electric hybrid drive system, which of the following should be true?
    a. The diesel engine turns in reverse.
    b. The diesel engine compression retarder is actuated.

c. The AC drive motor reverses.
d. Electric motor magnetic field is reversed.

12. Which of the following fuels has the most heat energy (calorific value/Btus by mass)?
    a. Alcohol
    b. Compressed natural gas
    c. Propane
    d. Diesel fuel

13. When hydrogen is burned using pure oxygen, what end gas results from the combustion process?
    a. Hydrogen
    b. Oxygen
    c. Water
    d. Carbon dioxide

14. Which of the following fuels is often a byproduct of crude petroleum?
    a. Biodiesel
    b. Propane
    c. Alcohol
    d. Hydrogen

15. Which of the following elements is the most abundant in the known universe?
    a. Iron
    b. Oxygen
    c. Nitrogen
    d. Hydrogen

Chapter
28

Failure Analysis, Troubleshooting, and Diagnoses Strategies

OBJECTIVES

After studying this chapter, you should be able to:

- *Define the terms failure analysis and troubleshooting.*
- *Analyze exhaust gas smoke emission by color.*
- *Relate some typical engine performance malfunctions to smoke color.*
- *Identify step-by-step sequential troubleshooting practices.*
- *Troubleshoot some typical engine and fuel system failures.*
- *Identify common operator and technician abuses of engine and fuel systems.*
- *Develop a checklist for tackling lack of power complaints specific to an engine or chassis system.*
- *Perform rudimentary failure analysis on failed engine components including pistons, rings, rod and main friction bearings, cylinder sleeves/liners, and crankshafts.*
- *Relate a failed component to the conditions that caused the failure.*

KEY TERMS

arcing	failure analysis	scuffing
black smoke	glazing	sequential troubleshooting chart
blue smoke	gumming	spalling
cavitation	lacquering	torched piston
cylinder leakage test	master gauge	troubleshooting
dynamometer	master pyrometer	white smoke
etching	pitting	

INTRODUCTION

The term **troubleshooting** is generally used to describe noninvasive (meaning that as little of the engine is disassembled as possible) methods of determining the cause of an engine problem. These methods can vary from educated guesswork to highly structured procedures, such as would be mandated when diagnosing most electronic engine management systems. Most manufacturers make troubleshooting guides available for purposes of diagnosing common engine problems. These guides provide a means of helping the technician to work methodically and not overlook an obvious possible cause of a problem. Troubleshooting electronically managed engines is necessarily a much more structured procedure requiring the use of **sequential troubleshooting charts**. These require the technician to test circuits and components in a step-by-step procedure, in which the results of a test in one step determine the routing to the next step in the sequence. Skipping or inaccurately performing just one of the steps in a sequential troubleshooting chart can render the whole procedure meaningless. When troubleshooting hydromechanical engines and fuel systems, there is no substitute for a thorough knowledge of the specific engine system. The following information is provided as a general guideline and avoids addressing manufacturers' specific product problems; most OEMs do this in their own technical literature.

The term **failure analysis** normally refers to component analysis methods used to determine the cause of a failure. In most cases, the failure analysis takes place after the engine or other component has been disassembled with the objective of avoiding recurrence. Failure analysis often relies on learned knowledge of the specific engine. What follows is a general look at typical engine component problems and their causes.

SMOKE ANALYSIS

An engine can emit from its exhaust noxious gases, water, and carbon dioxide. Providing these emissions remain in the gaseous, state they will not be observed exiting the exhaust piping. For exhaust gas to be identified as smoke, the state of the emission must be a liquid or a solid. When exhaust smoke can be described as white, the emission is in the form of condensing liquid droplets from which light reflects or refracts, making it appear white to the observer. The condensing liquid could be engine coolant, fuel, or, if cold enough, the water that is a normal product of combustion.

When smoke appears to be black, the state of the emission is solid; specifically, particulate solids through which light will not pass at all, making it appear black to the observer. **Blue smoke** is normally associated with engine oil emission. This is usually classified as a condensing liquid emission from which light is mainly reflected rather than refracted, due to the usually dark color of engine oil.

As outlined in a previous chapter, there are three states of matter—solid, liquid, and vapor. The technician analyzing the smoke emitted from a diesel engine is making an observation on the state of the emission. It is important to note that the absence of observable smoke emission does not mean that the engine is not producing noxious emissions. It means only that engine emissions are in a gaseous state. Gaseous emissions can only be analyzed using costly exhaust gas analysis equipment. Nevertheless, observable smoke emitted from an engine tells a story and the technician should have some ability to interpret the causes of a smoking engine.

BLACK SMOKE

The term **black smoke** is used to describe anything from a greyish tinge, in exhaust smoke, to heavily sooted emission. It is the result of the incomplete combustion of fuel and therefore has many causes.

Insufficient Combustion Air

The causes for insufficient combustion air are air starvation caused by any defect in the air intake system from the air filter, turbocharger, boost air/heat exchanger through to intake valve problems; clogged cylinder sleeve ports (two-stroke cycle engines); restricting emergency stop gate, and so on. Check for a problem that can be read electronically first in electronically managed engines. Test the intake air inlet restriction using a water manometer. The specifications should always be checked to the OEM values, but typical maximum values will be close to:

15″ H_2O vacuum—naturally aspirated engines
25″ H_2O vacuum—boosted engines

Exhaust System Restriction

Exhaust system restriction causes are turbocharger failure, collapsed exhaust system piping, internal failure of engine silencer or catalytic converter, or a clogged particulate trap.

Excess Fuel/Irregular Fuel Distribution

Excess fuel/irregular fuel distribution is caused by injection timing, defective pump-to-injector piping, injector nozzle failure, variable timing control device failure, unbalanced fuel rack settings, lugging engine (operation at high loads at speeds below peak torque rpm), incorrect governor settings, inoperative or tampered with manifold boost management system, or defective barometric capsule (altitude compensator).

Improper Fuel Grade

When fuel is stored for prolonged periods, the more volatile fractions evaporate, altering the fuel's chemical characteristics. Additionally, fuel suppliers seasonally adjust fuel to accommodate temperature extremes. The use of used engine lube/fuel mixers often produce smoking. Heavier residual oils may not vaporize, and if they do, there is insufficient time to properly combust them. This practice may cause high acidity in the exhaust gas (high sulfur), which may rust out exhaust systems unusually quickly. Ensure that the engine lube/fuel mixer permits the engine to meet the emission standards for the year the engine was manufactured.

BLUE SMOKE

Blue smoke is usually caused by lube oil getting involved in the combustion process.

Some possible causes are turbocharger seal failure, roots blower seal failure, pullover of lube oil from oil bath air cleaner sump, worn valve guides, ring failure, glazed cylinder liners, high oil sump level, excessive big end bearing oil throw-off, low-grade fuel, fuel contaminated with automatic transmission fluid (ATF), or engine lube placed in fuel tanks as an additive.

WHITE SMOKE

White smoke is caused by condensing liquid in the exhaust gas stream. Temperature usually plays a role when white smoke is observed, both ambient temperature and the engine-operating temperature. Remember that water is a natural product of the combustion of any hydrocarbon fuel and in mid winter conditions it is normal for some of this to condense in the exhaust gas. However, when white smoke is a problem, the following are some of the possible causes.

Cylinder Misfire

Check for a problem that can be read electronically first on electronically managed engines; perform an electronic cylinder cutout test using a PC or ProLink if that option is in the software diagnostics. With nonelectronic engines, short out injectors in sequence using a rocker claw/lever for MUI and PT injectors, and by cracking (loosening the line nut) high-pressure pipes in PLN systems.

CAUTION: When cracking high-pressure pipes, ensure that suitable eye protection is worn and that spilled fuel is not in danger of igniting.

Low Cylinder Compression Pressure

The procedure for checking cylinder compression pressures and cylinder leakage is outlined later in this section.

Low CN Fuel

Low CN fuel may produce white smoke. Test the fuel specific gravity using a fuel hydrometer and compare to the fuel supplier's specification. Consult the fuel supplier as to the correct additive and proportions to correct a CN deficiency in a storage tank.

Air Pumped through the High Pressure Injection Pump Circuit

Check the fuel subsystem because this is usually the source of the air, depending on the type of system. This condition is normally accompanied by rough engine operation.

Coolant Leakage to Cylinders

Confirm that the emission truly is coolant. Engine coolant has an acrid, bittersweet odor that is very noticeable. Locate the source. Some possibilities are injector cup failure, head gasket (fire ring) failure, cracked cylinder head, cracked wet liner flange, or **cavitation** perforation of wet liner. Coolant leakage may be difficult to locate; if the failure is not immediately evident, drop the oil pan, pressurize the cooling system, and observe. If necessary, place clean cardboard under the engine with the cooling system pressurized and leave for a while to attempt to identify the source; this may save an unnecessary engine disassembly. It is often easier to identify an internal coolant leak with the engine intact rather than disassembled, so explore all the options with the engine assembled first.

Low Combustion Temperatures

Low combustion temperatures may be the result of extreme low temperature conditions or a fault in the cooling system management system, such as a defective thermostat, fan drive mechanism, or shutters.

SMOKE EMISSION MEASUREMENT

The Environmental Protection Agency (EPA) sets maximum smoke density standards measured with a light extinction test instrument called an opacitymeter or opacimeter. Smoke opacity is defined as the fraction of light transmitted from a source that is prevented from reaching a sensor; it is expressed as a percentage value. Some jurisdictions in the United States and Canada choose to enforce these standards with heavy fines. Smoke emission testing is covered in Chapter 45.

 TROUBLESHOOTING

As a general rule when troubleshooting a system complaint, investigate the possible causes that can be eliminated easily (and inexpensively) before proceeding to those that require more time. Component disassembly should generally be avoided until all other investigative options have been pursued. This practice will help avoid presenting a customer with an inflated bill for diagnosing a simple complaint. It makes sense to develop some guidelines for troubleshooting engine problems in a service facility. A set of guidelines will help the technician to strategize the troubleshooting procedure. Engine OEMs produce some excellent troubleshooting charts for their own products; Figure 28–1 reproduces a Navistar chart.

The troubleshooting strategy used by the technician will depend on whether the engine is hydromechanically or electronically managed, and on the engine OEM. It should be noted that the troubleshooting procedure required of most electronically managed engines is structured and sequential; it must be adhered to. This subject is dealt with in more detail in the next section of this textbook. The following lists some typical engine problems with their possible causes and some solutions; the focus is on hydromechanical problems.

LOW OIL PRESSURE

Verify the problem:

1. Check the oil sump level.
2. Install a **master gauge** (an accurate, fluid-filled gauge).
3. Investigate oil consumption history.
4. Determine the cause.

Some possible causes and suggested solutions:

- Restricted oil filter or oil cooler bundle: change the oil and filter(s). If the problem persists, clean or re-

place the oil cooler bundle (core), and check or clean the filter and oil cooler bypass valves.
- Contaminated lube (fuel): detect by oil analysis or by odor. Engine lube contaminated with fuel can have a darker appearance and feel thin to the touch. If the cause is fuel, locate the source. This may be difficult and the procedure varies with the type of fuel system and the routing of the fuel to the injector.
Pressure testing the fuel delivery components may be required. Porosity in the cylinder head casting, failed injector O-ring seals, leaking fuel jumper pipes, and cracked cylinder head galleries are some possible causes. Perform repairs as required, then service the oil and filters.
- Excessive crankshaft bearing clearance: inspect the bearings to determine the cause. Visually check the crank journals to determine whether removal is required. Replace the bearings, ensuring that the clearance of the new bearings is checked.
- Excessive camshaft or rocker shaft bearing clearance: replace the bearings.
- Pump relief valve spring fatigued/stuck open: clean the valve and housing, replacing parts as necessary. Check the bypass/diverter valves in the oil cooler and filter mounting pad.
- Oil pump defect: recondition or replace the oil pump.
- Oil suction pipe defect: replace the oil suction pipe.
- Defect in oil pressure gauge or sending unit: replace the oil pressure gauge or sending unit.
- Broken-down (chemically degraded) lube oil: change the oil and filters.

HIGH OIL CONSUMPTION

Verify the condition by monitoring oil consumption and analyzing exhaust smoke. Some possible causes and suggested solutions:

- Excess cylinder wall lubrication: high oil sump level, excessive connecting rod and big end bearing oil throwoff, plugged oil control/wiper rings, oil pressure too high, or oil diluted with fuel. This type of problem usually requires an engine disassembly to diagnose and repair.
- External oil leaks: steam or pressure wash the engine, then load on a chassis dynamometer or road test to determine the source.
- High oil temperatures: malfunctioning boost air heat exchanger, lug down engine loading, overfueling, incorrect fuel injection timing, or problem with the oil cooler or engine cooling system.
- Piston ring fit abnormality: usually caused by wear. Replace the rings.
- Piston ring failure: determine the cause. Check for other damage, then replace the rings.

INTERNATIONAL	DT–466E AND THE INTERNATIONAL 530E HARD START / NO START & PERFORMANCE ENGINE DIAGNOSTICS	Date		Miles		Hours	Technician
		Eng. S/N		VIN			Unit #
		Eng. HP		Ambient Temp.		Coolant Temp.	

◄ HARD START / NO START DIAGNOSTICS ►

1. SUFFICIENT CLEAN FUEL
- Free of Water–Icing and clouding
- Correct grade of fuel

Method	Check
Visual	

2. VISUAL INSPECTION
- Inspect for leaks
- Inspect for loose connections, etc.

Fuel	Oil	Coolant	Electrical	Air
Method		Check		
Visual				

3. CHECK ENGINE OIL LEVEL
- Check engine crankcase oil level
- Check for contaminants (fuel, coolant)
- Correct Grade/Viscosity
- Miles/Hours on oil, correct level
- Check oil pressure on dash gauge

Method	Check
Visual	

4. INTAKE/EXHAUST RESTRICTION
REFER TO FIGURE A ON REVERSE SIDE
- Inspect hoses and piping
- Check filter minder
- Inspect exhaust system

Method	Check
Visual	

Perform Test 7 if EST is not available or inoperative

5. EST TOOL – FAULT CODES
REFER TO FIGURE B ON REVERSE SIDE
- Install Electronic Service Tool

Active	
Inactive	

☐ See Electronic Diagnostic Form for codes

6a. EST – ENGINE OFF TESTS
- Select "Engine Off" test from diagnostic test menu

Faults Found	

☐ Repair fault codes before continuing

6b. EST–INJECTOR "BUZZ TEST"

NOTE: "Engine Off Test" must be performed first, in order to gain access to the Injector "BUZZ TEST"

- Select "Injector Test" from "The Engine Off Tests" menu

Faults Found	

☐ See Electronic Diagnostic Form for codes

Perform Test 7 if EST is not available or inoperative

7. STI BUTTON – FLASH CODES
REFER TO FIGURE F ON REVERSE SIDE
- Depress and hold "Engine Diagnostics" switch, then turn the ignition switch to the "ON" position.

Faults Found	

☐ Refer to Electronic Diagnostic form, if fault code(s) set

8. EST TOOL – DATA LIST
- Select and enter the following data as the first 3 lines in a custom data list
- Monitor the data while cranking the engine for 20 seconds minimum

Data	Spec	Actual
Bat. Voltage	7 Volts min.	
Eng. RPM	150 RPM min.	
ICP Pressure	800 PSI min.	

☐ If voltage is low, refer to ECM diagnostics
☐ If no RPM is noted, recheck fault codes
☐ If ICP pressure is low, refer to Test 10

Perform Test 9 if EST is not available or inoperative

9a. ECM VOLTAGE
REFER TO FIGURE C ON REVERSE SIDE
- Check while cranking the engine
- Measure with DVOM
- Breakout box pins 57+ & 40–

Instrument	Spec	Actual
DVOM 57+ & 40–	7 volts minimum	

☐ If voltage is low, refer to ECM diagnostics

9b. ENGINE CRANKING RPM
REFER TO FIGURE C ON REVERSE SIDE
- Minimum 150 RPM engine cranking speed for 20 seconds
- Breakout box pins 34+ & 46– with Fluke 88

Instrument	Spec	Actual
Fluke 88 34+ & 46–	150 RPM minimum	

☐ If no RPM is noted, recheck fault codes

9c. INJECTION CONTROL PRESSURE
REFER TO FIGURE D OR E ON REVERSE SIDE
- Minimum 150 RPM engine cranking speed for 20 seconds
- Measure with breakout box: pins 27+ & 46– or breakout "Tee" signal (green) & ground (black)

Instrument	Spec	Actual
DVOM 27+ & 46–	1 Volt Minimum	

☐ If ICP pressure is low, refer to Test 10

10. LOW ICP PRESSURE TEST

NOTE: Perform this test if ICP Pressure was low in Test 8 or 9C.

REFER TO FIGURE G ON REVERSE SIDE
- Remove EOT sensor and check for oil in reservoir and reinstall EOT
- Remove high pressure hose from oil manifold
- Attach adapter and ICP sensor to hose
- Monitor pressure while cranking the engine

Instrument	Spec	Actual
EST	800 PSI min.	
DVOM	1 Volt min.	

☐ If pressure is within specifications, check for high pressure oil leakage. Refer to EGES–145 Sec. 2.2
☐ If pressure is still low, verify that pump is rotating
☐ If pressure is still low, replace IPR and retest

11. FUEL PUMP PRESSURE
REFER TO FIGURE K ON REVERSE SIDE
- Measure at bleeder valve on filter header
- Minimum 150 RPM cranking speed for 30 seconds

Instrument	Spec	Actual
0–160 PSI Gauge	20 PSI minimum	

☐ If pressure is low, replace fuel filter, clean fuel strainer and retest.

☐ If pressure is still low, perform Transfer Pump Restriction Test 2B (of Performance Diagnostics)

PERFORM TESTS IN SHADED AREAS IF EST TOOL IS NOT AVAILABLE OR ATA CODES ARE NOT TRANSMITTED

EGED–150 MARCH 1995 © NAVISTAR INTERNATIONAL TRANSPORTATION CORPORATION

FIGURE 28–1 *International Truck troubleshooting strategy chart. (Courtesy of Navistar International Corp., Designer and manufacturer of International Brand diesel engines)*

Injector P/N		Turbocharger P/N	
Engine Family Rating Code			
Complaint			

◄——————— PERFORMANCE DIAGNOSTICS ———————►

ALL TESTS SHOULD BE PERFORMED WITH ENGINE AT OPERATING TEMPERATURE

1. CHECK ENGINE OIL LEVEL
- Check engine crankcase oil level
- Check for contaminants (fuel, coolant)
- Correct Grade/Viscosity

Method	Check
Visual	

2. SUFFICIENT FUEL/PRESSURE
REFER TO FIGURE K ON REVERSE SIDE
- Drain sample from tank(s)
- Inspect fuel for contamination
- Measure fuel pressure at fuel filter bleeder
- Measure pressure at high idle

Instrument	Spec	Actual
0–160 PSI Gauge	20 PSI minimum @ High idle	

☐ If pressure is low, replace fuel filter, clean fuel strainer and retest.
☐ If pressure still low, proceed with step 2B.

2b. TRANSFER PUMP RESTRICTION
REFER TO FIGURE L ON REVERSE SIDE

NOTE: Perform this test only if fuel pressure is low.

- Measure at fuel filter inlet @ High idle.

Instrument	Spec	Actual
0–30" Vacuum Gauge	Less than 8 " Hg.	

☐ If restriction is high, check for blockage between pump and fuel tank
☐ If restriction < 8 " Hg., refer to EGES–145 Sec. 2.3 for additional diagnostics.

3. EST TOOL – FAULT CODES
REFER TO FIGURE B ON REVERSE SIDE
- Install Electronic Service Tool

Active	
Inactive	

☐ See Electronic Diagnostic Form for codes

4a. EST – ENGINE OFF TESTS
- Select "Engine Off Test" from diagnostic test menu

Faults Found	

☐ Repair fault codes, before continuing

4b. EST–INJECTOR "BUZZ TEST"

NOTE: "Engine Off Test" must be performed first, in order to gain access to the Injector "BUZZ TEST"

- Select "Injector Test" from "The Engine Off Tests" menu

Faults Found	

☐ See Electronic Diagnostic Form for codes

Perform Test 5 if EST is not available or inoperative

5. STI BUTTON – FLASH CODES
REFER TO FIGURE F ON REVERSE SIDE
- Depress and hold "Engine Diagnostics" switch, then turn the ignition switch to the "ON" position.

Faults Found	

☐ Refer to Electronic Diagnostic form if fault code(s) set

6. INTAKE RESTRICTION
REFER TO FIGURE H ON REVERSE SIDE
- Measure at high idle and no load
- Use manometer or magnehelic gauge

Instrument	Spec	Actual
Manometer or Magnehelic Gauge	12.5" H_2O	

7a. EST–ENGINE RUNNING TEST
- Select "Engine Running" test from the diagnostic test menu

Faults Found	

☐ Refer to Electronic Diagnostic form if fault code(s) set

7b. EST TOOL–INJECTOR TEST
(CYLINDER CONTRIBUTION)

NOTE: "Engine RUNNING Test" must be performed first, in order to gain access to the "INJECTOR TEST"

- Select "Injector Test" from "Engine Running" test menu

Faults Found	

☐ Refer to Electronic Diagnostic form if fault code(s) set

Tests 8, 9 & 10 to be performed at Full load

8. FUEL PRESSURE (FULL LOAD)
REFER TO FIGURE K ON REVERSE SIDE
- Measure fuel pressure at fuel filter bleeder.
- Measure pressure at full load rated speed.

Instrument	Spec	Actual
0–160 PSI Gauge	20 PSI minimum	

☐ If pressure is low, replace fuel filter, clean fuel strainer & retest
☐ If pressure is still low, perform Test 2B.

---

9. ICP PRESSURE
REFER TO FIGURE D OR E ON REVERSE SIDE
- Monitor ICP pressure and engine RPM with the EST tool in data list mode

Or use breakout "TEE" and DVOM
- Refer to EGES–145 for specifications

Data	Spec	Actual
Low Idle	PSI/Volts	
High Idle	PSI/Volts	
Full Load	PSI/Volts	

☐ If pressure is low or unstable, disconnect ICP sensor and retest
☐ If problem is resolved, refer to ICP diagnostics
☐ If pressure is still low or unstable, replace IPR and retest

10. BOOST PRESSURE
REFER TO FIGURE I ON REVERSE SIDE
- Monitor boost pressure and engine RPM with the EST tool in data list mode

Or use dash tach and 0–30 PSI gauge and "T", if EST tool is not available
- Measure pressure at full load rated speed
- Refer to EGES–145 for specifications

Spec	Actual
PSI @ RPM	

11. CRANKCASE PRESSURE
REFER TO FIGURE J ON REVERSE SIDE
- Measure at road draft tube with orifice tool (ZTSE–4039)
- Measure at High Idle no load RPM

Instrument	Spec	Actual
0 to 60" H20 Magnehelic Gauge	< 6" H_2O	

STOP

IF GUIDELINE DATA WAS OBTAINED DURING THE FIRST 11 TESTS, ENGINE OPERATION IS SATISFACTORY. NO FURTHER TESTING IS REQUIRED

12. WASTEGATE ACTUATOR TEST
- Apply regulated air to actuator
- Inspect for leakage
- Inspect actuator for movement

Instrument	Spec	Actual
0 to 60 PSI Gauge	28–32 PSI	

13. EXHAUST RESTRICTION
- Visually inspect exhaust system for damage
- Measure at a point 3 to 6 inches after turbo outlet
- Measure at full load and rated speed

Instrument	Spec	Actual
Manometer or Magnehelic Gauge	0–35" H_2O	

14. VALVE CLEARANCE
- Engine off: Hot or Cold

Instrument	Spec	Actual
Feeler Gauge		

FIGURE 28–1 (Continued)

- Turbocharger seal failure: recondition the turbocharger (recore or replace).
- High oil sump level: this causes aeration. Determine the cause of high oil level. Check for presence of fuel and engine coolant in oil.
- Glazed cylinder liners or sleeves: caused by improper break-in procedure or prolonged engine idling. Replace cylinder liners/sleeves or use a glaze-buster to machine crosshatch.
- Worn cylinder head valve guides or seals: measure to specification and recondition the cylinder head if required.

HIGH OIL TEMPERATURE

Some possible causes and suggested solutions:

- Insufficient oil in circulation: check sump level, pump pressure, and lubrication circuit restrictions. Repair as required.
- High water jacket temperatures: test cooling system performance.
- Plugged/failed oil cooler: disassemble and inspect galleries, bundle, and diverter valve for restrictions and scaling. Service the cooling system afterward to prevent a recurrence.
- Oil badly contaminated: submit sample for analysis and repair cause.
- Engine lugdown: provide some driver training and identify the transmission shift points. Drivers familiar with older engines may not have been adequately retrained to operate today's fuel efficient engines, which may be programmed with short torque rise windows.

HIGH COOLANT TEMPERATURE

Accurately verify the condition:

1. Check coolant level in the radiator.
2. Install a master gauge to verify the problem.
3. Observe exhaust pyrometer readings and compare to specification.

Some possible causes and solutions:

- Incorrect mixture: the correct mixture for EGs is usually an equal proportion of water and antifreeze with somewhere between 3% and 6% coolant conditioner. Increasing the antifreeze proportion with both EG and PG reduces cooling efficiency.
- Aerated coolant: usually caused by combustion gases entering the cooling circuit through a defective cylinder head or failed fire ring/cylinder head gasket. Heat is not transferred efficiently through a gas. Verify the condition: check for bubbles at the water manifold.
- Fan clutch: test fanstat/thermal fan operation using a master gauge and loading the engine until it reaches the temperature required to cycle the fan.
- Radiator: check for internal and external flow restrictions. Clean externally. Internal restrictions usually require the removal of the radiator and either recoring or the use of specialized reconditioning equipment.
- System does not seal: this permits the coolant to boil at a lower temperature and cause boilover. Test with a cooling system pressure tester. Check and repair/replace the defective radiator cap and/or pressure relief valves.
- Improper air flow: incorrectly sized fan or missing/damaged radiator shroud may significantly reduce flow through the radiator and engine compartment.
- Loose pump and fan drive belts: this will cause a reduction in coolant and/or air flow. Check visually and with a belt tension gauge.
- Coolant hoses: coolant hoses should be changed every couple of years. They often fail internally while appearing sound externally. A collapsed hose may cause a significant flow restriction.
- Restricted air intake: may cause high engine temperatures. Test inlet restriction with a water manometer. Typical maximum specifications are 15" H_2O for naturally aspirated engines and 25" H_2O for boosted engines.
- Exhaust restriction: causes high engine temperatures. Uncouple the exhaust piping from turbo or remove the exhaust manifold to see if the condition is corrected. Check the exhaust silencer and if equipped, the catalytic converter for internal collapse.
- Shunt line failure: a restriction in the shunt line from the radiator top tank to the water pump inlet may cause a boil condition at the inlet, reducing coolant flow and causing overheating.
- Thermostat: test out of engine for opening value and full open position using a boiler and a thermometer.
- Water pump: remove and check for a loose or damaged impeller and check the impeller to housing clearance to specification. Rebuild or replace as required.
- Boost air heat exchanger: an air flow or internal restriction will cause a rise in engine temperatures. Test air flow through the engine compartment. Winter grill covers impede air flow through the heat exchangers and may unevenly load the fan.
- High-altitude operation: cooling system efficiency diminishes as altitude increases. Larger cooling system capacity is required for high-altitude operation.

- Lugging: operating an engine at high loads and lower speeds means high temperatures and reduced coolant flow. Driver training is required.
- Overfueling: this raises the amount of rejected heat and may exceed the cooling system's ability to handle it. Recalibrate engine fueling to specification.
- Fuel injection timing: depending on the type of fuel system, both retarding and advancing fuel injection timing may result in engine overheating. Fuel injection timing is commonly tampered with; fuel injection timing should be set to OEM specification. Greatly diminished engine life is the consequence of minor injection timing adjustments, whereas major component failure is the consequence of more dramatic adjustments.

LOW COOLANT TEMPERATURE

Accurately verify the condition:

1. Check cooling system temperature management components such as shutter and fan operation.
2. Install a master gauge to verify the problem.
3. Observe exhaust pyrometer readings and compare to specification.

Some possible causes and solutions:

- Thermostat: a thermostat that is stuck in the open position will cause the engine to run cool. Remove the thermostat and check its start to open and fully open temperature specifications.
- Air vent valve: if stuck in the open position, this may cause low coolant temperatures when the engine is under light loads.
- Prolonged idle or light load operation: when little fuel is being used by the engine, there is less rejected heat for the cooling system to handle, and operating temperatures may be lower.

CYLINDER COMPRESSION PROBLEMS

Overall low or unbalanced cylinder compression can create problems of excessive blowby, low power, and vibration. Cylinder pressure may be checked using a compression tester. In diesel engines, the compression test should be performed when the engine is close to operating temperature by first removing all of the injectors, fitting the compression test gauge to the injector bore in one of the cylinders, and then cranking through five rotations. The batteries should be fully charged during this test. Test each cylinder sequentially and match the test results to specifications. The results should be within the engine OEM cylinder compression

parameters and normally within 10% of each other. Compression testing will locate which cylinder(s) is at fault, but not the cause of the problem. A cylinder leakage or air test kit sometimes stands a better chance of identifying the cause of a problem. The cylinder leak test consists of a pressure regulator and couplings that fit to the injector bores. Once again the engine should be close to its operating temperature at the beginning of the test. The test is performed with the piston in the cylinder to be tested at TDC and regulated air pressure is delivered to the cylinder. The test may indicate valve seal leaks, head gasket leaks, leaks to the water jacket, leaking piston rings, and defective pistons. The **cylinder leakage test** regulator delivers air at a controlled volume and pressure, then measures the percentage of leakage.

HIGH EXHAUST PYROMETER READINGS

Verify the condition:

1. Install a **master pyrometer** (do not rely solely on the vehicle pyrometer).
2. Chassis **dynamometer** test using a full instrumentation readout. See Chapter 15 on engine testing.

TECH TIP: A pyrometer is an operator's instrument as much as it is the technician's. A complaint of high pyrometer readings without other symptoms may be an indication that the operator requires some driver training. The truck driver should regard a pyrometer reading approaching its maximum as a signal to downshift gear ratios.

Some possible causes and solutions are:

- Air inlet restriction: test inlet restriction value to specification using a manometer.
- Flow restricted, boost air heat exchanger: check air flow through the hood intake grill. A winter grill cover can limit air flow through the heat exchangers located at the front of the chassis. Check the operation of a tip turbine if equipped.
- High engine load, low air flow operation: engines using air-to-air boost air cooling *require* air flow through the heat exchanger. A dump truck with air-to-air boost cooling operating in a pit in extreme heat may produce high pyrometer readings even when operated skillfully.
- Fuel injection timing: check to specification.
- Overfueling: look for other indicators of an overfueling condition and check fuel settings to specification.

SUDDEN ENGINE STOPPAGE

Some possible causes and solutions:

- Electrical failure: use the appropriate electrical circuit troubleshooting to locate the problem. Check the most obvious causes first.
- Electronic failure: access the on-board diagnostics for clues to the shutdown.
- No fuel: refuel tanks and use an appropriate priming method to get the engine running again.
- Air in fuel system: air may be pulled into the fuel subsystem from any of a number of locations; common causes are the fuel filter sealing gaskets, double gaskets (caused by failure to remove a used gasket during servicing), and a failed filter mounting pad assembly.
- Water in fuel system: most current secondary filters will plug on water. Replace and prime filters. Locate the source of the water.
- Plugged fuel line: sequentially work through the fuel subsystem circuit to locate the failed component. An internally failed fuel hose can be difficult to locate because an internal flap may require pressure to block flow.
- Lubrication failure: caused by a failed lubrication circuit component. A lubrication circuit failure will rapidly develop into a seized engine. Attempt to manually bar the engine over. A lubrication failure often results in a complete engine disassembly.

ENGINE RUNS ROUGH

This describes a condition in which the engine produces an inconsistent rpm roll, cylinder misfire, or surging at a specific rpm or throughout the engine-operating range.

Some possible causes and solutions:

- Air in fuel system: often accompanied by white smoke emission. Locate the source of the air, checking the suction side of the fuel subsystem first.
- Leak or restriction on the charge side of the fuel subsystem: repair a leak if evident. Use a pressure gauge to check charging pressure to specification.
- Injection nozzle, PT, or MUI failure(s): this results in a cylinder misfire condition. Locate the affected cylinder by shorting out the injectors sequentially. Crack the high-pressure pipe nuts at the pump on hydraulic injector systems and load the injector rocker in its downstroke position on PT and MUI systems. When checking electronically managed engines, use the electronically driven diagnostics and never mechanically short out an injector.

- CN value of fuel is too low: this occurs when fuel is stored for prolonged periods either in the vehicle or base storage tanks and the fuel deteriorates chemically or biochemically.
- Fuel cloud point is high: when fuel begins to wax, it can restrict filters and pumping apparatus. Analyze the fuel and add an approved cloud point depressant as required.
- Advanced injection timing: a common tampering problem. Check injection timing and reset to specification if required.
- Unbalanced MUI fueling: produces a hunting condition. Balance the MUI racks according to the OEM procedure. If the condition persists after the tune up, check that the MUI fuel codes are matched.
- Bent or broken MUI push rod: usually a cylinder misfire accompanied by mechanical clatter. Replace defective injector train components.
- Maladjusted valves: perform an overhead adjustment.
- Cylinder leakage: caused by a failed fire ring(s), a crack in the cylinder head/block, or valve leakage. Perform a cylinder leakage test appropriate for the engine.

ENGINE BLOWBY

Because rings create the seal of a piston in a cylinder bore, some blowby is inevitable, due to end-gap requirement. Normally, blowby is a specification determined by the OEM. Remember, piston rings seal most efficiently when cylinder pressures are highest, and engine rpm is high when there is less time for gas blowby.

Some causes and solutions of excessive engine blowby are:

- Cracked head or piston: diagnose cause and recondition engine.
- Worn stuck or broken piston rings: diagnose cause and recondition engine.
- Failure of a wrist pin cap's ability to seal in DDC mechanical two-stroke cycle engines: reseal and vacuum test the wrist pin caps.
- Glazed liner/sleeve inside wall (due to improper break-in practice or prolonged idling): diagnose cause and recondition engine.
- Poor quality or degraded engine lube: change the oil and filters.

High crankcase pressures can also be caused by:

- Air compressor (plugged discharge line, air pumped through oil return line): test air compressor operation.

- Turbocharger (seal failure allows turbine housing to leak pressurized air through to the turbine shaft and back through the oil return piping): test turbocharger operation and recondition or replace as required.

ENGINE WILL NOT CRANK

Determine whether the problem is related to the cranking circuit (electrical or pneumatic problem) or a mechanical engine condition. Focus on the cranking circuit first and eliminate any obvious causes; in most cases, the cause of a failure to crank condition will be found here. Attempt to bar the engine over by hand. If it cannot be turned over, check the engine externally and both the transmission and PTO. If engine seizure is suspected, remove all the injectors and again attempt to bar the engine over by hand before beginning to disassemble the engine. If disassembly is required, carefully remove and label components so that the cause of the failure can be determined.

TECH TIP: Use OEM software-guided troubleshooting whenever you can, even when you know the problem is not electronic. These programs are designed to prevent technicians from overlooking a critical step in a troubleshooting procedure.

LACK OF POWER

"Lack of power" probably appears more frequently on shop work orders than any others. In many cases, the lack of power is more closely associated with driver expectations than a genuine engine problem. A chassis dynamometer with a data printout is usually required to convince a skeptical truck driver that his equipment is functioning properly. Although the technician can eliminate many causes of low power, realistically, when performing this work for a customer, it is essential to test that the engine performs to specification on a chassis dynamometer when the work is completed.

It makes a lot of economic sense for shops working frequently on low-power complaints to prepare a strategy check sheet that addresses the specific equipment worked on; this ensures that the work is performed sequentially (eliminating obvious/quick to perform tasks first) and enables each technician to pursue the steps in more or less the same manner. A successful method of developing a "lack of power checklist" is to devote a technicians' shop meeting to brainstorming troubleshooting strategies and sequences for a specific engine/fuel system. Apart from the fact that the resulting checklist will incorporate the input of the entire workshop team, this exercise will have the added benefit of "educating" the less experienced technicians. In short,

the time devoted to the meeting will be paid for many times over by the time saved avoiding inconsistent and inaccurate troubleshooting practices.

There are too many variables based on engine manufacturer and fuel system type to provide definitive lack of power troubleshooting strategies. The following is a general list of some causes of low power and their solutions:

- Restricted fuel filters: replace the filters. Check the vehicle fuel tanks for contamination.
- Restricted air inlet system: test the inlet restriction value using a manometer. Remember, air filter elements are best tested without removal from the canister—do NOT replace unless they fail an inlet restriction test or have exceeded the in-service time limit.
- Leaks in the boost air circuit: test for leakage using an OEM-approved method such as direct application of a soap solution with the engine under load, ether spray at idle, or others. Small leaks in charge air cooler cores can be difficult to locate and may require removal and pressure testing. When dynamometer testing, always fit instrumentation to read manifold boost.
- Low manifold boost—turbocharger problem: visually inspect the turbocharger, checking radial play and endplay—check for rotation drag caused by carbon deposits (coking).
- Restricted exhaust system: this condition is usually accompanied by poor engine response. Check piping, engine silencer(s), and catalytic converters when equipped. Engine silencers can fail internally when baffles and resonator walls collapse; test backpressure value to specification.
- Low fuel subsystem charging pressure: fit a pressure gauge and test fuel pressure at idle and high-idle speeds referencing the specifications.
- Valve lash maladjustment: often accompanied by smoking and top end clatter/valve clatter. Adjust valves to specification.
- Defective boost fueling device: these are known variously as "aneroids," turbo-boost sensors, puff limiters, LDAs (Bosch), AFC (air/fuel control) valves, and FARC (fuel/air ratio control) valves. They usually have a trigger specification (predetermined manifold boost value), and their function is simply to limit fueling until there is enough air in the cylinders to properly combust it. Some permit a graduated increase of fueling, proportional to increase in manifold boost, whereas others simply switch at a predetermined boost value, limiting fueling until it is achieved. Replacing a properly functioning boost fueling control device with one of lower trigger value will not cure a lack of power complaint or increase engine power, although it will

cause puff smoking at shifting, waste fuel, and contaminate the atmosphere. Use the OEM method of testing the device, and ensure that the correct one is fitted. Some use an intake manifold sensor and system air pressure to actuate a governor-located pilot plunger—these must be precisely set to specification.

- Governor maladjusted: usually a result of repeated tampering with components that should only be adjusted on a pump comparator bench. Performance test the engine on a dynamometer and reset the governor, removing if required, the fuel injection pump assembly if the governor is integral.

- Contaminated fuel: fuel is abused unknowingly by both drivers and technicians whenever they add any substance to vehicle fuel tanks other than diesel fuel. Some of these additives may create conditions that result in lack of power complaints. There are certainly seasonal conditions that mandate the addition of alcohols to fuel tanks (crossover pipe freeze-up), but drivers and technicians should be educated to understand that dumping additives into fuel tanks, especially aftermarket additives of dubious chemistry, should be avoided. Purchasing #1D fuel is much cheaper than purchasing low grade #2D fuel and paying for the problems it creates when operating equipment in extreme winter conditions.

- Defective fuel: not a common problem in North America and difficult to diagnose without the use of specialized equipment. Check with the fuel supplier. Testing specific gravity may identify a fuel in which the lighter fractions have boiled off lowering the CN but the exact original values must be known. Defective fuel problems are usually the consequence of storing a fuel for prolonged periods or use of fuels outside the season in which they were purchased.

ENGINE VIBRATION

While driveline vibrations are not uncommon, true engine vibrations are not often a problem. When investigating a vibration complaint, the technician should eliminate all possible causes in the driveline behind the engine first. In doing this, it should be remembered that the clutch may disengage the driveline from the transmission and beyond from the engine, but that most of the mass of the clutch is rotated with the engine, engaged and disengaged.

- Cylinder misfire: see section "Engine Runs Rough."
- Loose vibration damper: check bolts for shear damage and damper fastener holes for eccentricity. Visually inspect damper for other damage. Retorque and test operation.

- Defective vibration damper: viscous-type dampers should be replaced at each engine overhaul but seldom are—visually inspect. The slightest external defect is reason to replace the unit. To dynamically test a vibration damper, a lathe and dynamic balance apparatus is required—a procedure beyond the scope of most shops. Visually inspect rubber drive ring dampers.

- Defective external driven component: every driven engine accessory is capable of unbalancing the engine. Most air compressors marketed today are balanced units and do not require timing; but some single cylinder compressors must be timed to the engine.

Some common accessories that may cause an unbalanced engine condition when they or their bearings fail are fan assembly (broken/damaged blades, failed bearings), water pump, accessory drive bearings, idler pulley bearings, alternator, and others.

SOOT IN INLET MANIFOLD

Some soot in the intake manifold is normal in most diesel engines; it usually indicates an engine operated at low loads and speeds for prolonged periods. In such operating conditions, the valve overlap duration is at a maximum in real time values and manifold boost is minimal or nonexistent. Excessive soot in the intake manifold may be an indication of imminent turbocharger failure or an injection timing problem.

MECHANICAL ENGINE KNOCK AS DIFFERENTIATED FROM COMBUSTION KNOCK

- Bottom end knock: produces an easily recognizable low frequency thump. The cause is big end or connecting rod journal bearing failure, a condition that rapidly develops into a crankshaft failure if not attended to. Replace the bearings and thoroughly inspect and measure the throw journals.

- Failed crankshaft: diagnose the problem in-chassis by visual inspection of journals and their bearings and the critical crankshaft stress points.

- Damaged gears: often identified by a high-frequency whine. Pull the timing gear cover and replace failed components, remembering that debris from the failure may have been pumped through the entire lubrication circuit.

- Failure of feedback circuit component: check rocker trains and camshaft. Repair components as required, once again remembering that debris from the failure may have been pumped through the entire lubrication circuit.

COMBUSTION KNOCK

Sometimes known as *diesel knock*. If the noise is rhythmic, it may be that it is being produced from one engine cylinder, which may be verified by comparing cylinder temperatures with an infrared thermometer. Some causes and solutions:

- Fuel injection timing: usually produces erratic knock, amplified at higher loads and rpms. Check fuel injection timing to specification.
- Air in fuel: usually accompanied by white smoke emission. Check for air admission on the suction side of the fuel subsystem.
- Low grade fuel: fuel with a low CN (caused by additive contamination, prolonged storage, nonhighway ASTM grade, etc.) will extend the ignition lag phase. This results in excess fuel in the cylinder at the time of ignition, rapid pressure rise, and detonation. Analyze the fuel quality using the fuel supplier/refiner's recommended procedure.
- MUIs out of balance: a single MUI delivering excess fuel due to a maladjusted rack can produce combustion knock. Check overhead adjustment. Check individual MUI racks for binding.

 — FAILURE ANALYSIS

Failure analysis usually refers to the diagnosing of a failure after component disassembly. Failure analysis is always made much easier when the disassembly is organized and methodical. It really helps to label and tag components on removal. Never begin to reassemble a failed engine until the exact cause of the failure has been determined. Most of the engine OEMs make failure analysis guides available with precision photography that can make failure analysis a little more scientific than guesswork.

This section deals with some common failure modes of engine components, but it should be remembered that it is by no means comprehensive.

PISTONS

When disassembling an engine, always tag pistons for location on removal, even when the primary cause of the failure has been determined.

- Excessive combustion pressure: caused by overfueling and ether abuse (Figure 28–2 and Figure 28–3).
- Advanced timing (in CI engines): outer edge of piston melts—erosion of top ring land above the Ni-

FIGURE 28–2 *Piston failure due to high cylinder pressure. (Courtesy of Mack Trucks)*

FIGURE 28–3 *Blown-out ring lands due to high cylinder pressure. (Courtesy of Mack Trucks)*

resist insert. The erosion is caused by fuel dispersal/condensation outside of the combustion bowl (Figure 28–4 and Figure 28–5).

- Retarded ignition/detonation (in SI and CI engines): failure occurs near the center of piston crown and, in cases where the extent of retardation is minor, pitting is evident. In more severe cases, the pitting can rapidly eat through a piston. The pitting is caused by injected fuel condensing on the piston crown before being combusted.
- Piston **scuffing**—metal-to-metal contact.

FIGURE 28–4 *Major piston failure due to advanced timing. (Courtesy of Mack Trucks)*

FIGURE 28–5 *Early indication of an advanced timing failure of a piston. (Courtesy of Mack Trucks)*

FIGURE 28–6 *Torched piston due to advanced timing or excessive overheating. (Courtesy of Mack Trucks)*

FIGURE 28–7 *Complete crown melt-down caused by excessive cylinder temperatures. (Courtesy of Mack Trucks)*

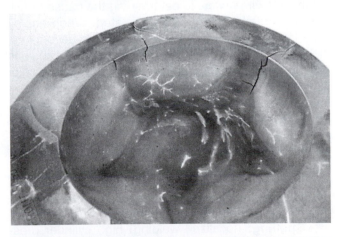

FIGURE 28–8 *Piston failure due to overheating causing cavity edge cracks. (Courtesy of Mack Trucks)*

On thrust sides only—caused by high cylinder pressures: high boost pressures, overfueling, lug down.

Scuffing overall—caused by inadequate cylinder wall lubrication, lube oil contamination or overheating, or insufficient ring end gap.

- Carbon buildup: overfueling, retarded injection timing, inappropriate fuel, injector nozzle failure.
- Crown **pitting** (pocked erosion): retarded injection timing, injector nozzle failure/dribbling, moisture from intake air condensing on piston crown.
- **Lacquering:** high sulphur content in fuel.
- **Torched piston:** a characteristic of an advanced timing condition when the result is a torched set of pistons. Can also be caused by extreme overheating. When a single piston is torched, it is usually caused by a failure to dissipate heat. An example of this would be a misaimed cooling jet. Figure 28–6, Figure 28–7, and Figure 28–8 show some examples of piston failure due to extreme overheating.

TECH TIP: Number pistons by cylinder when disassembling an engine, even when it is known they are to be replaced—it may help diagnose a failure.

PISTON RINGS

Once again, tag all piston rings for location on the piston and on the engine in sequence as they are removed on disassembly of the engine. Some common causes of piston ring failures are:

- Land failures: excessive manifold boost, ether abuse, overfueling.
- Sticking/stuck rings: lacquering and **gumming** caused by high sulphur fuel or contaminated fuel, overfueling, or prolonged lugdown operation. A failure characteristic noted in engines fueled with mixed used engine oil and fuel.
- Broken rings: overfueling, ring erosion followed by fracture caused by dirt.
- Cylinder scuffing: insufficient ring end gap.
- Seized rings: high operating temperatures, overfueling, contaminated fuel.

TECH TIP: Always check that piston rings were originally installed correctly when removing them from piston assemblies. Place the rings with the pistons they were removed from.

FRICTION BEARINGS

Despite some OEM's opinions to the contrary, it does not make economic sense to reuse connecting rod and main bearings. When bearings are removed from an engine, whether at overhaul, bearing rollover, or other engine problem, they should be numbered and examined with a view to determining the cause of failure. Many failures are lube oil related. When engine lube becomes contaminated, the bearings become especially vulnerable. Improper installation practices can also shorten a bearing's life. An example of a color bearing failure analysis guide is shown in Figure C1.

- **Etching:** caused by chemical contamination of lube—usually fuel, coolant, or sulphur.
- **Glazing:** caused by broken down lube or improper lube.
- **Lacquering:** orange/brown film on bearing caused by poor quality lube or high sulphur content in lube.
- **Pitting:** particulate in lube oil.
- **Spalling:** (localized welding) high cylinder pressures caused by high manifold boost, overfueling,

lugging or detonation—less commonly a result of excessively high lube oil temperatures.

Figure 28–9 shows a broad range of typical engine bearing failures.

TECH TIP: Obtain a color bearing failure analysis guide from the OEM whose product you are working on. This is an excellent way for rookie technicians to begin to perform accurate failure analyses and avoid recurrent failures (such as that shown in Figure C1).

- **Arcing:** incorrect chassis welding practices.
- Low oil pressure is often related to bearing wear and can be the rationale for a bearing rollover.
- Bottom end failures: spalling increases the bearing id and the resultant hammering will increase the tensional loading on the throw/conn rod assembly.
- Lack of crush on a bearing will result in poor heat dissipation and a failure characterized by heat discoloration.

CRANKSHAFTS

Modern crankshafts seldom fail. On a newly introduced engine series, they may fail according to a pattern; always report crankshaft failures to the manufacturer whether or not the equipment is under warranty. Metallurgical failures are infrequent. Subjected to abuse, crankshafts will fail, but more often, the abuse will manifest itself with the failure of a subsidiary component. Crankshaft failures can be categorized by the type of loading that initiates the failure.

- Torsional failures: caused by a loose or defective crank vibration damper or flywheel, out-of-balance engine-driven components, engine lugdown, or engine overspeed. Characterized by fractures that initiate at the journal oil holes or run circumferentially through the journal fillets.
- Bending failures: caused by misaligned main bearing bores, main bearing failure or broken/loose main bearing caps. Characterized by cracks that initiate at the main journal fillet and extend at a 45-degree angle through the throw journal fillet.

CYLINDER SLEEVES AND LINERS

If the liners are to be reused, ensure that they are identified so that they are reinstalled to the same bore they were removed from. Identify manufacturer size codings when used. When disassembling an engine for the first

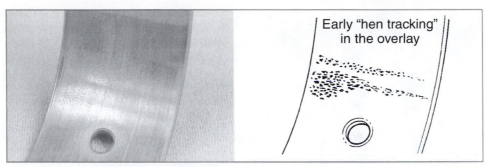

A. Overlay fatigue, normal wear: early stages

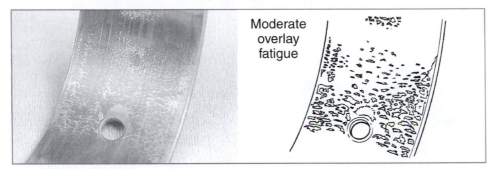

B. Overlay fatigue, normal wear: intermediate stage

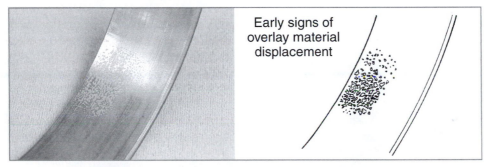

C. Orange peel appearance, normal wear: early stages

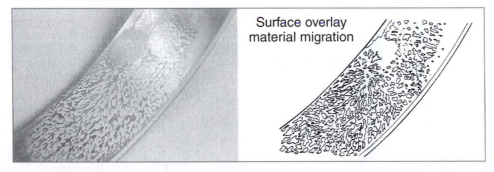

D. Orange peel appearance, normal wear, high mileage condition, intermediate stage

FIGURE 28–9   *Engine bearing distress analysis. (Courtesy of Federal Mogul Corporation)*

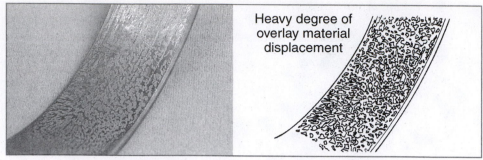

E. Orange peel appearance: advanced suction erosion condition
caused by high mileage

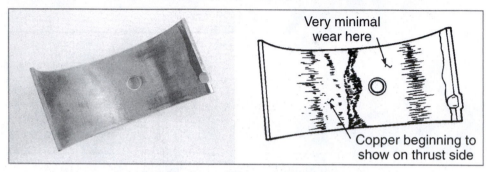

F. High mileage, normal wear

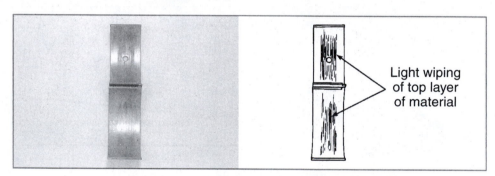

G. Lack of lube: early stages

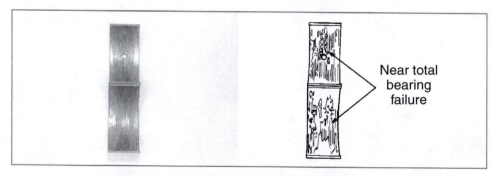

H. Lack of lube: advanced stage

FIGURE 28–9   *(Continued)*

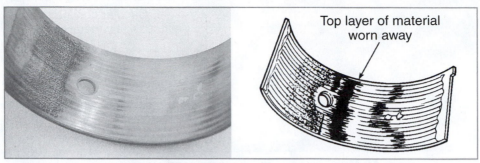

I. Coolant contaminated lube

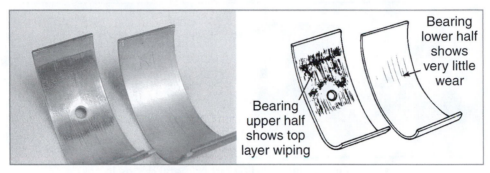

J. Fuel contaminated lube: early stages

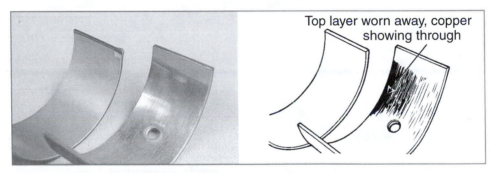

K. Fuel contaminated lube: advanced stage

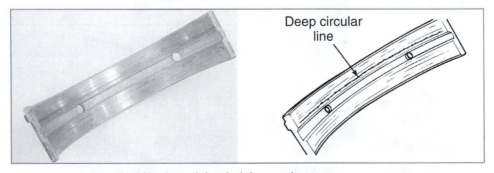

L. Metal particles in lube: early stages

FIGURE 28–9   *(Continued)*

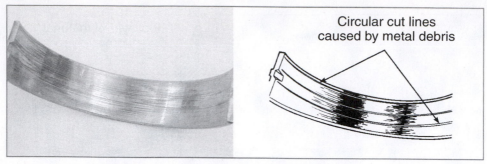

M. Metal particles in lube: advanced stage

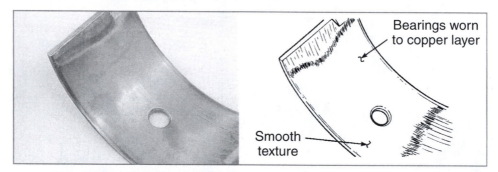

N. Crankshaft journals damaged by dirt or metal debris in lube

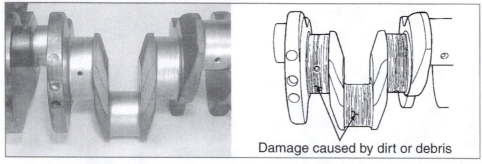

O. Airborne dirt in oil: abrasive action wear to a smooth texture

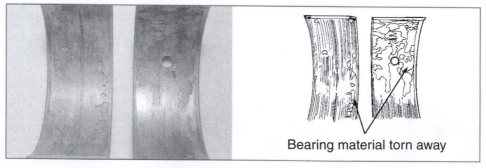

P. Hot short rod bearing failure caused by high shear loads
and extreme temperatures

FIGURE 28–9   (Continued)

time, do not make the assumption that oversize sleeves and liners have not been used.

DRY SLEEVES

Reference Chapter 9 for the procedure used to remove dry sleeves. When the correct pullers are used, sleeves can usually be removed without damaging them. Label them with a permanent marker pen by cylinder sequence.

Scoring

Tight fit of sleeve od to bore, causing sleeve distortion. Tight fit of piston assembly to sleeve id.

Scuffing

Sleeve-to-bore fit is too loose, causing poor heat transfer—usually accompanied by discoloration.

CI engine cylinder dry sleeves and wet liners are routinely replaced at a scheduled engine overhaul without much inspection. Correct installation practices should be carefully observed to ensure maximum engine life. It is common practice to install current sleeves 0.0005" to 0.0015" loose; this means that in some cases, the sleeve can be pressed into its cylinder bore entirely by hand. Interference-fit sleeves are seldom used in current diesel engines and machining an interference will cause early failure because the sleeve will buckle. Selective fitting of sleeves to bores should be practiced; this requires measuring the od of each new sleeve and each bore id, then getting the largest sleeve to the largest bore, down to the smallest sleeve to the smallest bore.

Counterbore depth should be inside the specification window and usually within 0.001" of each other under one cylinder head. Recutting counterbores and shimming are common practices, but these procedures usually require special tools to properly perform them. When honing cylinder dry sleeve bores in the engine block, ensure that the machining residues of the procedure are completely removed. This is especially important when undertaking the procedure in chassis with the crankshaft in place.

Cavitation

Evidence of cavitation indicates problems with the engine coolant chemistry. If reusing a slightly cavitated wet liner, rotate it 90 degrees from its original location. Cavitation is caused by liner pulsing during expansion strokes. This acts on the coolant it is in contact with in the water jacket, forming a vapor bubble, which implodes with considerable force. Subjecting a wet liner to repeated bubble collapse erodes it, causing pitting. This pitting is usually observed on the liner exterior at the piston thrust faces. It may eventually perforate the liner. Continued cavitation problems on wet liners can usually be cured by using ELC.

Liner O-Ring or O-Ring Seat Failure

Check seating areas thoroughly and also check that correct sealing medium has been used. The sealing medium used on the O-ring during installation depends on the "O" material. This can be dry (nothing), antifreeze (EG, PG, etc.), soap solution, OEM sealants, and so forth. It is important to remember that what may be referred to as a rubber can actually be one of a number of different rubber, silicone, and petrochemical compounds, so always observe the OEM instructions for the specific engine.

Coolant in Combustion Chamber

This condition is evidenced by white smoke emission and the characteristic bittersweet odor of vaporized coolant: a cracked wet liner flange can leak coolant into the combustion chamber.

Wet liners should be installed by hand and subsequently checked with a plug gauge. This is machined to within 0.0015≤ of the liner id and ensures that it is not distorted. Liner flange protrusion of a set of liners clamped under one head should normally be within 0.001" of each other; check OEM specification.

TECH TIP: Always label cylinder liners numerically on removing them from the cylinder block even when they are going to be replaced. You never know when failure analysis evidence may be required.

WARNING: Removing a seized cylinder sleeve from an engine block should be performed in a manner that does not distort or damage the cylinder block bore—fracturing the sleeve in sections almost always results in damage to the block bore. If a sleeve cannot be removed using a hydraulic puller, try gently applying oxyacetylene heat on the lower portion of the ring belt sweep on the sleeve, cool with a small quantity of water, then attempt to pull out hydraulically again.

 COMMON ABUSES OF ENGINES AND FUEL SYSTEMS

- Cold-startup procedure: excessive use of ether, excessive engine speeds when oil temperatures are low, excessive engine loading when engine temperature is low.

- Excessive engine idling: this can glaze cylinder walls or break down engine lube (rings do not seal well when cylinder pressures are lower); prolonged idling at low temperatures may cause fuel contamination of engine lube.
- Overfueling: causes high cylinder pressures (which can reduce engine life or cause outright mechanical failure), increases noxious emissions, reduces engine longevity, and costs money through wasted fuel.
- Engine overspeed: can cause valve float, which may result in dropping a valve, and increases tensional loading on connecting rods and crank throws, which may result in complete engine failure.
- Advancing fuel injection timing: a very common abuse, the objective being to improve engine performance; the consequences of this practice vary from greatly reducing engine life at best, to rapid mechanical engine failure.
- Fuel additive abuses: these include adding to fuel moth balls, gasoline, kerosene, alcohols, aftermarket fuel conditioners, engine oil, and ATF.
- Fuel heaters are not generally recommended by OEMs as the fuel's lubricity is compromised by heating it. The preferred method of running an engine through arctic, subarctic, Canadian, and Northern United States midwinter conditions is to use #1D fuel, temperature conditioned by the refiner.
- Hot shutdown: can damage turbocharged engines especially. The turbine bearings are vulnerable because the turbine shaft is still rotating after the oil flow has ceased. Additionally, the heat from the hot turbine exhaust cooks the oil at the bearings, rendering them baked dry for the next startup.

TECH TIP: When diagnosing engine performance complaints and component failures, try to keep an open mind and avoid making assumptions before developing a foundation for them. *Never* indulge in coffee table diagnosis, that is, diagnosing a condition based on a nontechnical person's report. In most cases, a truck driver can be considered a nontechnical person. *Do* consult experienced technicians when in doubt, especially where the results of a troubleshooting sequence indicate that the next step is to begin the disassembly of a component.

SUMMARY

- The term *troubleshooting* describes a systematic procedure used to diagnose an engine or fuel system complaint.

- Troubleshooting practice attempts to identify the cause of a malfunction to determine the extent of repair required and often to provide a customer with a cost estimate for the repairs.
- Failure analysis, by definition, is the examination of failed components to attempt to determine the cause of a breakdown.
- The diesel technician should aim to develop sound failure analysis skills to prevent repeated failures.
- Failure analysis is usually performed on fully disassembled components or subcomponents.
- The diesel technician should be able to analyze exhaust smoke emission and use this knowledge to help diagnose engine and fuel system malfunctions.
- White exhaust smoke indicates the presence of condensing liquid in the exhaust gas.
- Black exhaust smoke indicates the presence of particulate in the exhaust gas.
- Blue smoke emission is generally associated with an engine oil-burning condition.
- An opacimeter is a light extinction, test instrument used by transportation regulation enforcement agencies to test exhaust smoke density to EPA standards.
- Work methodically through sequential troubleshooting procedures. NEVER skip a step on the basis of an assumption.
- Keep an open mind. When seeking advice, seek the advice of an expert.

REVIEW QUESTIONS

1. The tool of choice to accurately check dry air filter restriction would be a:
   a. Pyrometer
   b. Trouble light
   c. H_2O manometer
   d. Dash restriction gauge
2. A typical maximum inlet restriction specification on a turbo-boosted diesel engine would be:
   a. 25″ H_2O
   b. 25″ Hg
   c. 25 psi
   d. 250 KPa
3. Which of the following conditions would be LEAST likely cause black smoke emission?
   a. Restricted air filter
   b. Retarded injection timing
   c. Plugged particulate trap
   d. Excessive use of cetane improver

4. Which of the following devices helps prevent engine overfueling at high altitudes?
   a. PT-AFR assembly
   b. Aneroid
   c. Barometric capsule
   d. Puff limiter

5. When observing a disassembled engine, the outer edge of all the pistons have evidence of melting and erosion. This could be likely caused by:
   a. Advanced timing
   b. A dribbling injector
   c. Retarded timing
   d. High sulphur content fuel

6. When examining a set of pistons from a disassembled engine, scuffing is observed on the thrust sides of all of the pistons. Which of the following would be the more likely cause?
   a. High cylinder pressures
   b. Overheating
   c. Contaminated engine lube
   d. High sulphur content fuel

7. Which of the following components when worn may cause lower oil pressures?
   a. Main bearings
   b. Valve guides
   c. Oil control rings
   d. a, b, and c

8. Which of the following might result in glazed cylinder sleeves?
   a. High cylinder pressures
   b. Prolonged engine idling
   c. Ether usage
   d. Engine lugging

9. Removing an engine coolant thermostat would most likely result in an increase of which noxious emission?
   a. HC
   b. NO_x
   c. SO_2
   d. CO_2

10. Etched main bearings could be caused by all of the following conditions *except:*
    a. Fuel in lube
    b. Coolant in lube
    c. Sulfur in lube
    d. Particulate in lube oil

11. Evidence of cavitation erosion on a set of wet liners would most likely be caused by:
    a. High cylinder pressures
    b. Incorrect coolant chemistry
    c. Piston spalling
    d. Liner O-ring failure

12. Which of the following conditions would be indicative of retarded injection timing in a diesel engine?
    a. Erosion near center of piston crown
    b. Upper land failure

c. Overall piston scuffing
d. Fractured rings

13. Which of the following conditions could cause carbon buildup on pistons?
    a. Overfueling
    b. Injector nozzle failure
    c. Retarded injection timing
    d. a, b, and c

14. Which of the following operating modes would be most likely to result in a tensile conn rod failure?
    a. Overfueling
    b. Lugdown
    c. Ether abuse
    d. Overspeeding engine

15. Which of the following operating modes would be more likely to produce valve float?
    a. Overfueling
    b. Lugdown
    c. Ether abuse
    d. Overspeeding engine

16. *Technician A* states that an excessive amount of oil in the crankcase can produce a low oil pressure problem in a diesel engine. *Technician B* states that the first thing to check when oil pressures fluctuate is the crankcase oil level. Who is right?
    a. A only
    b. B only
    c. Both A and B
    d. Neither A nor B

17. *Technician A* states that a small dent in a viscous-type vibration damper is not a concern unless there is visible leakage. *Technician B* states that some compressors have to be timed to the engine or a vibration will result. Who is right?
    a. A only
    b. B only
    c. Both A and B
    d. Neither A nor B

18. There is evidence of soot in the intake manifold of a boosted truck diesel engine. Which of the following conditions would be most likely to contribute to this condition?
    a. Engine lugdown
    b. High-idle operation
    c. Rated speed operation
    d. Running at low speed with low loads

19. *Technician A* states that white smoke emission is often the result of air starvation to the engine. *Technician B* states that blue smoke is associated with an engine that is burning oil in its cylinders. Who is correct?
    a. A only
    b. B only
    c. A and B
    d. Neither A nor B

20. An aluminum trunk-type piston has failed with a hole burned through the center of the crown. *Technician A* states that this might be due to a retarded timing condition. *Technician B* states that such a failure could occur caused by a dribbling injector nozzle. Who is right?
    a. A only
    b. B only
    c. Both A and B
    d. Neither A nor B

Section 3

Computerized Truck Management Systems

Section 3 is dedicated to electronic management systems. It begins with a review of the fundamentals of electricity and electronics and proceeds to the study of the basics of computer and communications technology. Vehicle electronic systems and their input circuits are introduced in Chapter 32 followed by chapters devoted to the electronic service tools (ESTs) and wiring repair techniques. Next, each specific type of current truck engine management system is examined either by type or by manufacturer. The section concludes with a look at the causes of noxious emissions and the methods used to test and control them in highway diesel engines.

Chapter
29

Electricity and Electronics

PREREQUISITES

Chapter 16

OBJECTIVES

After studying this chapter, you should be able to:

- Define the terms electricity and electronics.
- Describe atomic structure.
- Outline the properties of conductors, insulators, and semiconductors.
- Describe the characteristics of static electricity.
- Define what is meant by the conventional and electron theories of current flow.
- Describe the relationship between electricity and magnetism.
- Define what is meant by an electrical circuit and the terms voltage, resistance, and current flow.
- Identify different types of electrical circuit used in chassis electrical systems.
- Perform simple electrical circuit calculations using Ohm's law.
- Identify the characteristics of DC and AC.
- Describe some methods of generating a current flow in an electrical circuit.
- Apply Ohm's law to series, parallel, and series-parallel circuits.
- Describe Kirchhoff's first and second laws and calculate voltage drop in circuit components.
- Define the term capacitance and identify some types of capacitor used in electrical circuits.
- Define the operating principles of coils and transformers.
- Describe some types of electrical waveforms and their application in electronic circuits.
- Define the term pulse width modulation.
- Describe some types of semiconductors used in truck electronics and define the properties of N and P type semiconductors.
- Outline the operating principles and applications of diodes and transistors.
- Describe the optical spectrum and identify some commonly used optical components.
- Explain what is meant by an integrated circuit.
- Define the role of AND, OR, NOR, and NOT gates in electronic circuits.
- Interpret a truth table.
- Explain why the binary numeric system is used in computer electronics.
- Define the terms bits and bytes.
- Describe how data can be transmitted using electronic circuits.

KEY TERMS

alternating current (AC)	AND gate	bit
ampere	anode	byte
ampere-turns (At)	atom	capacitance
amplification	binary system	capacitor

cathode	electronics	potential difference
charge differential	farad	proton
closed circuit	field effect transistors (FET)	pulse width modulation (PWM)
coils	gates	reluctance
condenser	Integrated circuit (IC)	resistance
conductance	insulators	silicon-controlled rectifier (SCR)
conductors	ion	semiconductor
conventional theory	Kirchhoff's laws	series circuits
coulomb	light emitting diode (LED)	shell
current	modulation	static electricity
cycles	neutron	thyristor
Darlington pair	NOT gate	transistors
data processing	nucleus	truth table
dielectric	ohm	valence
direct current (DC)	Ohm's law	voltage
electricity	open circuit	voltage drop
electromagnetism	OR gate	watt
electron	parallel circuits	wavelength
electron theory	photonic semiconductors	zener diode

INTRODUCTION

This chapter reviews some of the basic electrical principles required to understand how an electronically managed engine functions. As it encompasses a range of principles that have been the subject of many textbooks, the approach here is that of reviewing the material, so you may find that you need to use other texts. There is no doubt that the modern truck technician should have a thorough grasp of the basics of electricity and electronics, which has become the single most important subject area. The days when many truck technicians could avoid working on an electrical circuit through an entire career are long past.

LEARNING ELECTRICITY

Mechanical technicians learn best by a hands-on approach. This fact tends to make electricity more difficult to learn because you cannot see or touch it. Some learning programs insist on beginning with the study of the science of electricity before progressing to its more practical aspects. Other programs teach practical electricity, the need-to-know material first, and introduce its theoretical aspects later. Both methods work. The approach used in this chapter is to begin by introducing some of the science of electricity and then progress toward applied electricity. You can skip content within the chapter or use it in any sequence you wish.

Most of the contents of this chapter are used throughout the remainder of the book. You can perform many repairs on truck electrical and computer control systems without understanding the building blocks of electrical theory, but when you encounter problems that fall outside the scope of troubleshooting software or flow charts, if you have good electrical instincts (you can think electrically), you are more likely to get to the root of that problem. So whichever way you study electricity, make sure you do not underestimate its importance. This chapter is key to understanding later chapters on computer technology. Try to learn to love electricity, even if you initially find it a real turnoff.

In this chapter, the theoretical aspects of electricity beginning with atomic structure are dealt with first, followed by a look at electrical circuits and components. Toward the end, some of the concepts behind information processing are introduced. Theoretical electricity begins with the study of atomic structure. If you find this confusing, skip forward and get back to it later when you feel more comfortable with some the practical aspects of electricity.

BRIEF HISTORY

All matter has some electrical properties, but it is only comparatively recently in history that humans have been able to make it work for them. The first type of electricity to be identified was **static electricity** by the Greeks more than 2,000 years ago when they observed that amber rubbed with fur would attract lightweight objects such as feathers. The Greek word for amber (a translucent, yellowish resin, derived from fossilized trees) is **electron** from which the word electricity is derived. Probably the next significant step forward occurred toward the end of the sixteenth century when the English physicist William Gilbert (1544–1603) made the connection between electricity and magnetism. Shortly after, Benjamin Franklin (1706–1790) proved the electrical nature of thunderstorms in his famous kite experiment, established the terms positive and negative, and formulated the conventional theory of current flow in a circuit. From this point forward, progress accelerated. In 1767, Joseph Priestly established that electrical charges attract with a force inversely proportional to distance and in 1800, Alessandro Volta invented the first battery. Michael Faraday (1791–1867) opened the doors of the science we now know as **electromagnetism** when he published his law of induction that simply states that a magnetic field induces an electromotive force in a moving conductor. Thomas Edison (1847–1931) invented the incandescent lamp in 1879 but perhaps even more importantly, built the first central power station and distribution system in New York City in 1881. This station provided a means of introducing electrical power into industry and the home. The discovery of the electron by J. J. Thomson (1856–1940) in 1897 introduced the science of electronics and quickly resulted in the invention of the diode (1904) and triode (1907).

DEFINITIONS

Electricity is a form of energy. This energy form can be observed in charged particles such as electrons or protons either statically, as accumulated charge, or dynamically, as current flow in a circuit. **Electronics** is a branch of electricity that addresses the behavior of flows of electrons through solids, liquids, gases, and across vacuums. However in automotive technology, the terms electronics and electronic engine management are generally used to describe systems that are managed by computers.

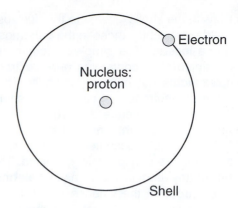

FIGURE 29–1   *Hydrogen atom.*

ATOMIC STRUCTURE AND ELECTRON MOVEMENT

An **atom** is the smallest particle of a chemical element that can take part in a chemical reaction. An atom is usually made up of **protons**, **neutrons**, and electrons. Protons and neutrons form the center or **nucleus** of each atom while the electrons orbit the nucleus, in a manner similar to that of planets orbiting the sun in our solar system. The simplest atom is hydrogen. It has a single electron orbiting a nucleus consisting of a single proton: it has no neutron. A hydrogen atom is shown in Figure 29–1.

Electrons orbit the nucleus of an atom in a concentric ring known as a shell. Electrons have a negative electrical charge. Protons and neutrons form the nucleus of all atoms around which the electrons orbit. Protons have a positive electrical charge whereas neutrons have no electrical charge. The nucleus of an atom comprises 99.9% of its mass. The number of protons in the nucleus is the atomic number of any given element: the sum of the neutrons and protons is the atomic mass number. In a balanced or electrically neutral atom, the nucleus is surrounded by as many electrons as there are protons.

All electrons are alike. All protons are alike. The number of protons in the nucleus of an atom identify it as a specific element. Electrons have 0.0005 of the mass of a proton. Under normal conditions electrons are bound to the positively charged nuclei of atoms by the attraction between opposite electrical charges.

Any atom may possess more or fewer electrons than protons. Such an atom would be described as negatively (an excess of electrons) or positively (a net deficit of electrons) charged and known as an **ion**. The concentric orbital **shells** (a shell is an orbital path) of an atom proceed outward from the nucleus of an atom.

The electrons in the shells closest to the nucleus of an atom are held most tightly: those in the outermost shell are held more loosely. The simplest element, hydrogen, has a single shell containing one electron. The most complex atoms may have seven shells. The maximum number of electrons that can occupy shells one through seven are in sequence: 2, 8, 18, 32, 50, 72, 98. The heaviest elements in their normal states have only the first four shells fully occupied with electrons: the outer three shells are only partially occupied. The outermost shell in any atom is known as its **valence**. The number of electrons in the valence will dictate some basic characteristics of an element. The chemical properties of atoms are defined by how the shells are occupied with electrons.

An atom of the element helium whose atomic number is 2 has a full inner shell. An atom of the element neon with an atomic number of 10 has both a full first and second shell (2 and 8): its second shell is its valence (Figure 29–2). Other more complex atoms that have eight electrons in their outermost shell, even though this shell might not be full, will resemble neon in terms of their chemical inertness. Remember that an ion is any atom with either a surplus or deficit of electrons. Free electrons can rest on a surface or travel through matter (or a vacuum) at close to the speed of light. Electrons resting on a surface cause it to be negatively charged. Because the electrons are not moving, that surface is described as having a negative static electrical charge. The extent of the charge is measured in **voltage** or **charge differential**. A stream of moving electrons is known as an electrical current. For instance, if a group of positive ions passes in close proximity to electrons resting on a surface, they will attract the electrons by causing them to fill the "holes" left by the missing electrons in the positive ions. Current flow is measured in **amperes**: one ampere equals 6.28 $\times 10^{18}$ electrons (a **coulomb**) passing a given point per second.

A number of factors such as friction, heat, light and chemical reactions can "steal" electrons from a surface, and when this occurs, the surface becomes positively charged. If the positive ions remain at rest, the surface will have a positive static electrical charge differential. Every time a person walks across a carpet, electrons are "stolen" from the carpet surface and this has an electrifying effect (electrification) on both the substance from which electrons are stolen (the carpet) and the moving body that performs the theft. When the moving body has accumulated a sufficient charge differential (measured in voltage), the excess electrons will be discharged through an arc and balance the charge.

Electrification results in both attractive and repulsive forces. In electricity, like charges repel and unlike charges attract. When a plastic comb is run through hair, electrons are stolen by the comb, giving it a negative charge. The comb may subsequently attract small pieces of paper as shown in Figure 29–3. The experiment always works better on a dry day because electrons can travel more easily through humid air and the accumulated charge will dissipate rapidly. Two balloons rubbed on a woolen fiber will both acquire a negative charge and therefore tend to repel each other. An atom is held together because of the electrical tendency of unlike charges attracting and like charges repelling each other. Positively charged protons hold the negatively charged electrons in their orbital shells and because like electrical charges repel each other, the electrons do not collide. All matter is composed of atoms. Electrical charge is a component of all atoms. When an atom is balanced, that is, the number of protons matches the number of electrons (Figure 29–4), the atom can be described as being in an electrically neutral state. So all matter is electrical in essence. The phenomena we describe as electricity concerns the behavior of atoms that, for whatever reason, have become unbalanced or electrified.

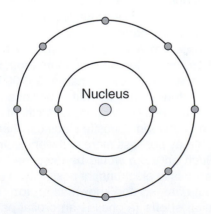

FIGURE 29–2 Atomic structure of a neon atom: outer shell is full.

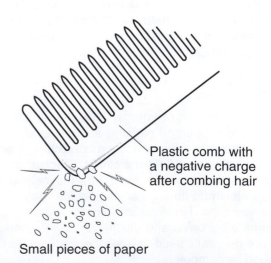

Small pieces of paper

FIGURE 29–3 Unlike charges attract.

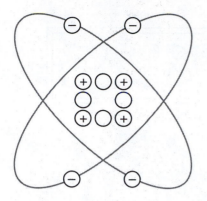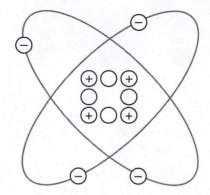

FIGURE 29–4   *Balanced atom: the number of protons and electrons are equal.*

CONDUCTORS AND INSULATORS

The ease with which an electron moves from one atom to another determines the conductivity of the material. **Conductance** is the ability of a material to carry electric current. To produce current flow, electrons must move from atom to atom as shown in Figure 29–5. Materials that readily permit this flow of electrons from atom to atom are classified as **conductors**. A conductor is generally a metallic element that contains fewer than four electrons in its outer shell. Examples of conductors include copper, aluminum, gold, silver, iron, and platinum.

Materials that inhibit or perhaps prevent a flow of electrons are classified as **insulators**. An insulator is a nonmetallic substance that contains more than four electrons in the outer shell. Examples of insulators include glass, mica, rubber, and plastic.

Semiconductors are a group of materials that cannot be classified either as conductors or insulators: they have exactly four electrons in their outer shell. Silicon is an example of a semiconductor.

Conductive metals even when in an electrically neutral state contain vast numbers of moving electrons that move from atom to atom at random. When a battery is placed at either end of a conductor, such as copper

wire, and a complete circuit is formed, electrons are pumped from the more negative terminal to the more positive until either the charge differential ceases to exist or the circuit is opened. The number of electrons does not change.

CURRENT FLOW

Current flow only occurs when there is a path and a difference in electrical potential: this difference is known as charge differential and is measured in voltage. Charge differential exists when the electrical source has a deficit of electrons and therefore is positively charged. Because electrons are negatively charged and unlike charges attract, electrons flow toward the positive source.

Initially it was thought that current flow in an electrical circuit had one direction of flow, that is, from positive to negative. This idea is known as the **conventional theory** of current flow. When the electron was discovered, scientists revised the theory of current flow and called it **electron theory**. In studying electricity, the technician should be acquainted with both conventional and electron theories of current flow. A conductor such as a piece of copper wire contains billions of neutral atoms

Conductor

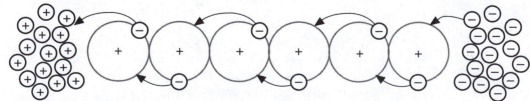

FIGURE 29–5   *Electron flow through a conductor.*

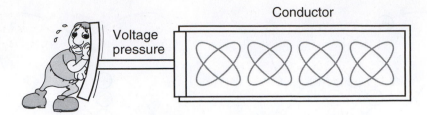

FIGURE 29-6   *Voltage or charge differential is the pressure that causes electrons to move.*

whose electrons move randomly from atom to atom, vibrating at high frequencies. When an external power source such as a battery is connected to the conductor, a deficit of electrons occurs at one end of the conductor and an excess of electrons occurs at the other end: the negative terminal repels free electrons from the conductor's atoms while the positive terminal attracts free electrons. This results in a flow of electrons through the conductor from the negative charge to the positive charge. The rate of flow depends on the charge differential (or **potential difference**/voltage). The charge differential or voltage is a measure of electrical pressure. Figure 29-6 shows the effect of voltage on electron flow. The role of a battery, for instance, is to act as a sort of electron pump. In a closed electrical circuit, electrons move through a conductor, producing a displacement effect close to the speed of light.

The physical dimensions of a conductor are also a factor. The larger the sectional area (measured by wire gauge size), the more atoms there are over a given sectional area, therefore the more free electrons: therefore as wire size increases, so does the ability to flow more electrical current through the wire. The rate of electron flow is called **current** and it is measured in amperes. A current flow of 6.25 billion billion electrons per second is equal to one ampere as shown in Figure 29-7.

HOLE THEORY

Rather than saying that the conventional theory of current flow is wrong and electron flow is right, you can think about current flow by comparing it with what happens in a ticket line at a movie theatre. As each person buys a ticket, the people remaining in the line move a step forward. This concept can be compared to what happens to each electron in an energized electrical circuit—it moves forward. However, as the line moves forward, the gap moves backward. You can compare this gap to the hole left by an electron in the electrical circuit as it moves forward. There is nothing different in what is happening in the movie theatre line, just whether you think of its movement from the point of view of a person in it or the gap.

Terminals

The terminal from which electrons exit in an electrical device is known as the **anode** or positive terminal. The terminal through which electrons enter a electrical component is known as the **cathode** or negative terminal.

MAGNETISM

The phenomenon of magnetism was first observed in lodestone (magnetite) and the way ferrous (iron based) metals reacted to it. When a bar of lodestone was suspended by string, the same end would always rotate to point toward the Earth's North Pole. The molecular the-

Conductor

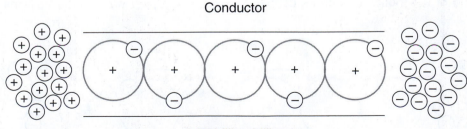

6.25 billion billion
electrons per second = one ampere

FIGURE 29-7   *The rate of electron flow is measured in amperes.*

ory of magnetism tends to be the most widely accepted: in most materials, the magnetic poles of the composite molecules are arranged randomly so there is no magnetic force. In certain metals such as iron, nickel and cobalt, the molecules can be aligned so that their north or N poles all point in one direction and their south or S poles point in the opposite direction. In lodestone, the molecules align themselves naturally. Some materials have good magnetic retention, which means that when they are magnetized they retain their molecular alignment. Other materials are only capable of maintaining their molecular alignment when positioned within a magnetic field; when the field is removed, the molecules disarrange themselves randomly and the substance's magnetic properties are lost. All magnetism is essentially electromagnetism in that it results from the kinetic energy of electrons. Whenever an electric current is flowed through a conductor, a magnetic field is created. When a bar-shaped permanent magnet is cut in two, each piece assumes the magnetic properties of the parent magnet with individual north and south poles. Figure 29–8 shows some magnetic fields and how separate magnetic fields react when in proximity with one another.

The term **reluctance** describes resistance to the movement of magnetic lines of force. Reluctance can be reduced by using permeable (susceptible to penetration) materials within magnetic fields. The permeability of matter is rated by giving a rating of 1 for air, generally considered to be a poor conductor of magnetic lines of force. In contrast, iron is ascribed a permeability factor of 2000, and certain ferrous alloys may have values exceeding 50,000.

The force field existing in the space around a magnet can be demonstrated when a piece of cardboard is placed over a magnet and iron filings are sprinkled on top of the cardboard. The pattern produced as the filings arrange themselves on the cardboard is referred to as flux lines. Flux lines are directional and exit from the magnet's north pole and enter through the south pole. Flux lines do not cross each other in a permanent magnet. Flux lines facing the same direction attract while flux lines facing opposite directions tend to repel.

The flux density (concentration) determines the magnetic force. A powerful magnetic field exhibits a dense flux field whereas a weak magnetic field exhibits a low density flux field. Flux density is always greatest at the poles of a magnet.

ATOMIC STRUCTURE AND MAGNETISM

In an atom, all of the electrons in their orbital shells also spin on their own axes, in much the same way the planets orbit the sun. They rotate axially each producing magnetic fields. Because of their axial rotation, each electron can be regarded as a minute permanent magnet. In most atoms, pairs of electrons spinning in opposite directions produce magnetic fields that cancel each other out. An atom of iron has 26 electrons, 22 of which are paired. In the second from the outermost

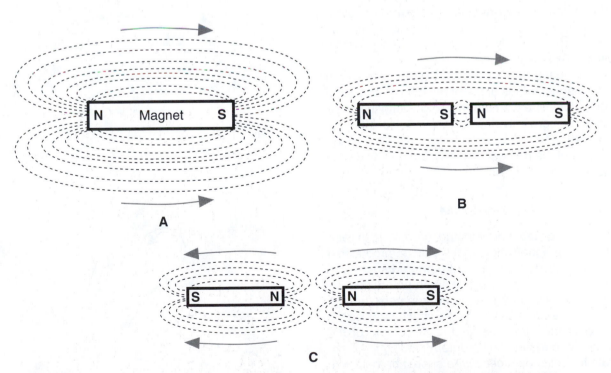

FIGURE 29–8 *Magnetic principles: A. All magnets have poles. B. Unlike poles attract each other. C. Like poles repel.*

shell, four of the eight electrons are not paired, meaning that they rotate in the same direction and do not cancel each other out. This fact accounts for the magnetic character of the metal iron. A large number of vehicle electrical components use electromagnetic principles in some way or other, and magnetic field strength is the reason that **coils** and other electromagnetic devices contain iron cores.

ELECTROMAGNETISM

Electrical current flow through a conductor such as copper wire, creates a magnetic field surrounding the wire. This effect can be observed by passing a copper wire through which current is flowing, lengthwise over a compass needle: the needle will deflect from its North-South orientation when this occurs. Any magnetic field created by electrical current flow is known as electromagnetism. Study of the behavior of electromagnetic fields has proved the following:

- Magnetic lines of force do not move when the current flowed through a conductor remains constant. When current flowed through the conductor increases, the magnetic lines of force will extend farther away from the conductor.
- The intensity and strength of magnetic lines of force increase proportionally with an increase in current flow through a conductor: similarly they diminish proportionally with a decrease in current flow through the conductor.
- The right-hand rule is used to denote the direction of the magnetic lines of force: the right hand should enclose the wire with the thumb pointing in the direction of conventional current flow (positive to negative), and the finger tips will then point in the direction of the magnetic lines of force as shown in Figure 29–9 and Figure 29–10.

USING ELECTROMAGNETISM

When wire is coiled and electric current is flowed through it, the magnetic field that would be produced if the wire were straight combines to form a larger magnetic field identified by north and south poles. This effect can be amplified by placing an iron core through the center of the coil (Figure 29–11), which reduces the reluctance of the magnetic field. The polarity of the electromagnet created can be determined by the right-hand rule for coils: the coiled wire should be held with the fingers pointed in the direction of conventional current flow (positive to negative) and the thumb will point

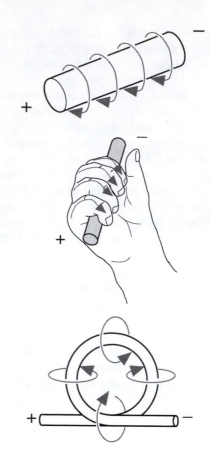

FIGURE 29–9   *Electromagnetic field characteristics.*

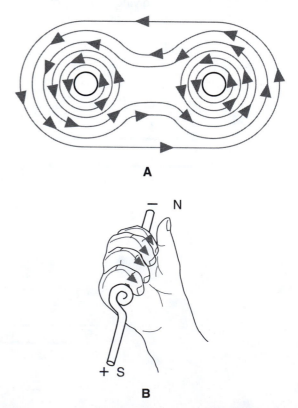

FIGURE 29–10   A. *Magnetic lines of force join together and attract each other;* B. *the right-hand rule.*

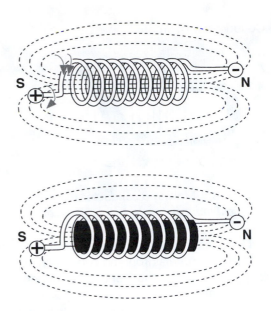

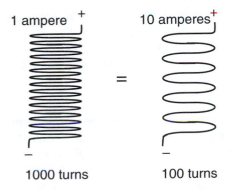

FIGURE 29–11 Magnetic field characteristics of a coiled conductor.

1 ampere ⁺   10 amperes ⁺

=

1000 turns   100 turns

FIGURE 29–12 Magnetic field strength is determined by the amount of amperage and the number of coils.

to the north pole of the coil (check out Figure 29–10 again). Electromagnetic field force is often described as magnetomotive force (mmf). The mmf is determined by two factors:

1. The amount of current flowed through the conductor.
2. The number of turns of wire in a coil

Magnetomotive force is measured in **ampere-turns (At)**. Ampere-turn factors are the number of windings (complete turns of a wire conductor) and the quantity of current flowed (measured in amperes). Referencing Figure 29–12, if a coil with 100 windings has 1 ampere of current flowed through it, the result will be a magnetic field strength rated at 100 At. An identical magnetic field strength rating could be produced by a coil with 10 windings with a current flow of 10 amperes. The actual field strength would have to factor in reluc-

tance. In other words, the actual field strength of both the foregoing coils would be increased if the coil windings were to be wrapped around an iron core.

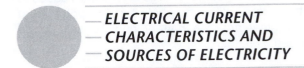

ELECTRICAL CURRENT CHARACTERISTICS AND SOURCES OF ELECTRICITY

Electrical current is classified as either direct current (DC) or alternating current (AC). The technician should understand the basic characteristics of both because both are used in vehicles. Although most of the chassis electrical circuit uses DC, an alternator and an inductive pulse generator both produce AC.

DIRECT CURRENT

Electrical current can be made to flow in two directions through a conductor. If the current flows in one direction it is known as **direct current**, usually abbreviated as **DC**. The current flow may be steady or have a pulse characteristic. DC can be produced in a number of manners outlined later in this section. DC has many applications and it is extensively, if not quite exclusively, used in highway vehicles through most of the chassis electrical circuits.

ALTERNATING CURRENT

Alternating current or **AC** describes a flow of electrical charge that cyclically reverses due to a reversal in polarity at the voltage source. It is usually produced by rotating a coil in a magnetic field. AC is used in vehicle AC generators (perhaps more often called alternators) and in certain sensors on modern vehicles: it is also better suited than DC for transmission through power lines. The frequency at which the current alternates is measured in **cycles**. A cycle is one complete reversal of current from zero though positive to negative and back to the zero point. Frequency is usually measured in cycles per second.

SOURCES OF ELECTRICITY

There are a number of ways of producing electricity and modern trucks use most of them in some way or another.

Chemical

Batteries are a means of producing DC from a chemical reaction. In the lead acid battery, a potential difference is created by the chemical interaction of lead and lead peroxide submerged in sulfuric acid electrolyte. When a circuit is connected to a charged battery (one in which there is a charge differential), the battery will pump electrons through the circuit from the negative terminal through the circuit load to the positive terminal. This process will continue until the charge differential ceases to exist. When the charge differential ceases to exist, there is no difference between the potential at either terminal and the battery can be said to be discharged.

Static Electricity

The term static electricity is somewhat misleading because it implies that it is unmoving. Perhaps it is more accurately expressed as frictional electricity because it results from the contact of two surfaces. Chemical bonds are formed when any surfaces contact, and if the atoms on one surface tend to hold electrons more tightly, the result is "theft" of electrons. Such contact produces a charge imbalance by pulling electrons of one surface from that of the other. As electrons are pulled away from a surface, the result is an excess of electrons in one surface (result = negative charge) and a deficit in the other (result = positive charge). The extent of the charge differential is measured in voltage. While the surfaces with opposite charges remain separate, the charge differential will exist. When the two polarities of charge are united, the charge imbalance will be canceled. Static electricity is an everyday phenomenon as described in examples earlier in this chapter. It usually involves voltages of more than 1,000 V and perhaps rising to as much as 50,000 V. A fuel tanker trailer towed by a highway tractor steals electrons from the air as it is hauled down the highway (as does any moving vehicle) and can accumulate a significant and potentially dangerous charge differential. This charge differential, which can be as high as 40,000 V, must be neutralized by grounding before any attempt is made to load or unload fuel that could otherwise be ignited by an electrostatic arc.

Electromagnetic Induction

Current flow can be created in any conductor that is moved through a magnetic field or alternatively by a mobile magnetic field and a stationary conductor. The voltage induced increases both with speed of movement and the number of conductors, so densely wound conductors tend to produce higher voltage values. Generators, alternators, and cranking motors all use the principle of electromagnetic induction. Figure 29–13 shows what happens when a conductor is moved through a magnetic field.

Thermoelectric

Electron flow can be created by applying heat to the connection point of two dissimilar metals. The pyrome-

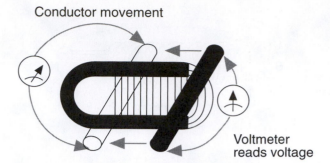

FIGURE 29–13   Moving a conductor through a magnetic field induces an electrical charge differential.

ter used to measure exhaust gas temperature consists of two dissimilar metals (iron and constantin, a copper-tin alloy) joined at the "hot" end and connected to a sensitive voltmeter at the gauge end. As temperature increases at the hot end, the reading will increase at the display gauge, which functions like millivoltmeter.

Photoelectric

When light contacts certain materials such as selenium and cesium, electron flow is stimulated. Photoelectric cells are used as sensors that can control headlight beams and automatic daylight/night mirrors.

Piezoelectric

Some crystals, notably quartz, become electrified when subjected to direct pressure, the potential difference increasing with pressure increase. Piezoelectric sensors are used as detonation sensors used on SI electronically controlled engines.

ELECTRICAL CIRCUITS AND OHM'S LAW

The German physicist Georg **Ohm** (1787–1854) proved the relationship between electrical potential (pressure or voltage), electrical current flow (measured in amperes), and the **resistance** to the current flow. In order to understand the behavior of electricity in an electrical circuit, it is necessary to understand the terminology used to describe its characteristics. When introducing electricity, comparisons are often made between electrical circuits and hydraulic circuits: these analogies are used in the following explanations.

VOLTAGE

You can describe voltage in a number of different ways. Charge differential, electrical pressure, and potential difference are all correct ways of describing voltage. Using a hydraulic analogy, voltage in an electrical circuit can be compared to fluid water pressure. Similar to pressure in the hydraulic circuit, voltage may be present in an electrical circuit without any current flow. You measure voltage using a digital multimeter (DMM) on the V-DC or V-AC settings. You can measure voltage in both a closed (energized) and an open electrical circuit.

RESISTANCE

Resistance is the opposition current flow. The resistance to the flow of electrons through a circuit is measured in ohms, the symbol for which is Ω. The resistance to the flow of free electrons through a conductor results from the innumerable collisions that occur and generally, the greater the sectional area of the conductor (wire gauge size), the less resistance to current flow, simply because there are more available free electrons.

Once again, using the hydraulic analogy, the resistance to fluid flow through a circuit would be defined by the pipe internal diameter or flow area. In an electrical circuit, resistance generally increases with temperature because of collisions between free electrons and vibrating atoms. As the temperature of a conductor increases, the tendency of the atoms to vibrate also increases and so does the incidence of colliding free electrons. The categories of resistance we see in vehicle electrical circuits are:

- Load—could mean lights, motors, anything that is powered electrically.
- Corrosion—eats away from outside in reducing the effective surface contact area
- Wiring—size and material should provide as low resistance to the electron flow

Resistance in an electrical circuit is measured with a DMM set in ohms.

∞ = infinite resistance, out of limit, open circuit

OL = infinite resistance, out of limit, open circuit

0 Ω = zero or minimum circuit resistance, complete or **closed circuit**

There can be both voltage and resistance in an open electrical circuit. For current to flow in a circuit, both voltage (charge differential) and a physical circuit (resistance) must first exist: when such a circuit is closed, a path for current flow is provided.

CURRENT

Current flows when the conditions of voltage and a closed circuit are met. Current is specifically the flow of electrons measured in amperes. One ampere is equal to 6.28×10^{18} electrons passing a given point in an electrical circuit in one second—or one coulomb. If the comparison with a hydraulic circuit is used once again, amps can be compared to gallons per minute (gpm). If current flow is to be measured in an electric circuit, the circuit must be electrically active or closed, that is, actively flowing current.

CLOSED CIRCUIT

A complete electrical circuit is an arrangement that permits electrical current to flow. At its simplest this arrangement would require a power source, a load, and a means of connecting supply and return paths to the power source. A circuit is described as closed when current is flowing and open when it is not. Figure 29–14 shows the simplest type of electrical circuit.

SERIES CIRCUITS

A **series circuit** may have several components such as switches, resistors, and lamps, but they are connected so that there is only one path for current flow thorough the circuit, such as that in Figure 29–14. In this figure, if the element in the light bulb were to fail, the circuit would open and no current would flow.

PARALLEL CIRCUITS

A **parallel circuit** is one with multiple paths for current flow meaning that the components in the circuit are connected so that current flow can flow through a component without having first flowed through other components in the circuit. Figure 29–15 shows an example of a parallel circuit using three paths for current flow each through a separate load.

SERIES-PARALLEL CIRCUIT

Many circuits are constructed using the principles of both series and parallel circuits and are known as series-parallel circuits.

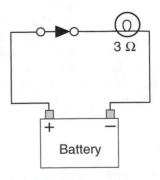

FIGURE 29–14 _Simplified series light circuit with 3 ohms resistance._

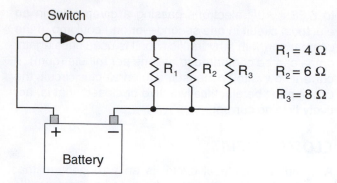

FIGURE 29–15   *A parallel circuit with different resistances in each branch.*

$R_1 = 4\ \Omega$
$R_2 = 6\ \Omega$
$R_3 = 8\ \Omega$

ELECTRICAL CIRCUIT TERMINOLOGY

The following terminology is used to describe both normal and abnormal behavior in electrical circuits.

Short Circuit

Short circuit is used to describe what occurs in an electrical circuit when a conductor is placed across the connections of a component and some or all of the circuit current flow takes a shortcut. Short circuits are generally undesirable and can quickly overheat electrical circuits. Electricity will generally choose to flow through the shortest possible path in order to complete a circuit.

Open Circuit

The term **open circuit** describes any electrical circuit through which there is no current flow. A switch is used in electrical circuits to intentionally open them. An electrical circuit may also be opened unintentionally, and this might occur when a fuse fails, a wire breaks, or connections corrode.

Ground

The term ground represents the point of a circuit with the lowest voltage potential. In vehicles, ground or chassis ground is integral in the electrical circuit as it provides the return electrical path for all components in the circuit. Technicians used to working on vehicles always use the term ground to mean chassis ground. To power a light bulb in a vehicle, a single wire can supply one terminal on the bulb with a feed from the positive terminal of the battery and the return path is completed by connecting the other bulb terminal directly to the chassis ground.

OHM'S LAW

Ohm's law tells us that an electrical pressure of 1 V is required to move 1 A of current through a resistance of 1 Ω.

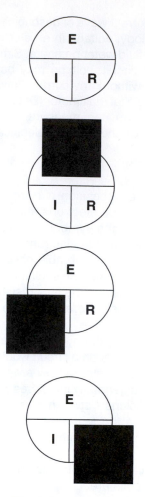

FIGURE 29–16   *Ohm's law formula graphic.*

I = intensity = current in amperes

E = EMF (electromotive force) = pressure in voltage

R = resistance = resistance to current flow in Ω

$$E = I \times R \qquad R = \frac{E}{I} \qquad I = \frac{E}{R}$$

Ohm's law can also be expressed in the units of measurement used in the following formulas:

$$V = A \times \Omega \qquad \Omega = \frac{V}{A} \qquad A = \frac{V}{\Omega}$$

See Figure 29–16. Better still, memorize it.

Ohm's Law Applied to Series Circuits

In a series circuit, all of the current flows through all of the resistances in that circuit. So the sum of the resistances in the circuit would define the total circuit resistance. If a series circuit were to be constructed with 1 ohm and 2 ohm resistances (Figure 29–17), total circuit resistance (Rt) would be calculated as follows:

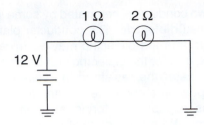

FIGURE 29–17   *Series circuit calculation.*

$Rt = R_1 + R_2$

$Rt = 1\,\Omega + 2\,\Omega$

$Rt = 3\,\Omega$

If the power source in the foregoing series circuit is a 12 V battery, then when the circuit is closed, current flow can be calculated using Ohm's law as follows:

$$I = \frac{E}{Rt}$$

$$I = \frac{12\,V}{3\,\Omega}$$

$$I = 4 \text{ amps}$$

Ohm's Law Applied to Parallel Circuits

According to Kirchhoff's law of current (see the next section), current that is flowed through a parallel circuit divides into each path in the circuit. When the current flow in each path is added, the total current will equal the current flow leaving the power source. When calculating the current flow in parallel circuits, each current flow path must be treated as a series circuit or the total resistance of the circuit must be calculated before calculating total current. When performing calculation on a parallel circuit, it should be remembered that more current will always flow through the path with the least resistance. If a parallel circuit is constructed with 2 Ω and 6 Ω resistors in parallel fed by a 12 V power source (Figure 29–18), total current can be calculated by treating each current flow path separately as follows:

$$I_1 = \frac{12\,V}{2\,\Omega} = 6 \text{ amps}$$

$$I_2 = \frac{12\,V}{6\,\Omega} = 2 \text{ amps}$$

$$I_t = 8 \text{ amps}$$

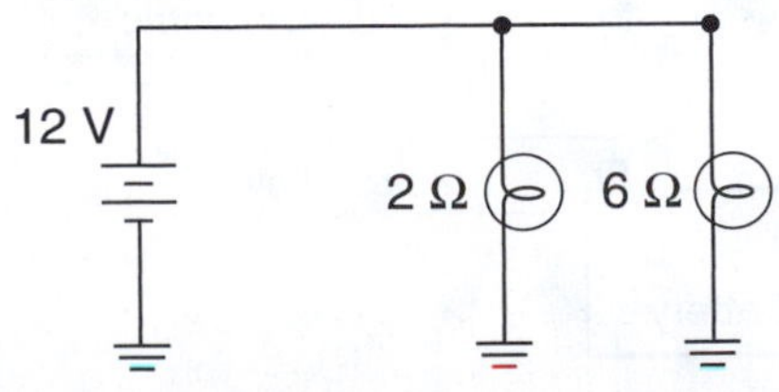

FIGURE 29–18   *Parallel circuit calculation.*

An alternative method would be to calculate Rt using the following formula:

$$Rt = \frac{R_1 \times R_2}{R_1 + R_2}$$

$$Rt = \frac{6\,\Omega \times 2\,\Omega}{6\,\Omega + 2\,\Omega} = \frac{12\,\Omega}{8} = 1.5\,\Omega$$

then

$$I = \frac{E}{R}$$

$$I = \frac{12\,V}{1.5\,\Omega} = 8 \text{ amps}$$

Remember, the information presented in this section is presented as a review guideline and the technician should be fully conversant with electrical circuit calculations. Many vehicle electrical circuits are of the series-parallel type, that is, they combine the characteristics of both the series circuit and the parallel circuit. When performing circuit analysis and calculation, it helps to visualize the circuit in terms of paths for current flow.

Kirchhoff's Law of Current (Kirchhoff's First Law)

Kirchhoff's law of current states that the current flowing into a junction or point in an electrical circuit must equal the current flowing out. High resistance anywhere in a closed circuit will choke down total current flow.

Kirchhoff's Law of Voltage Drops (Kirchhoff's Second Law)

Kirchhoff's law of **voltage drops** states that voltage will drop in exact proportion to the resistance and that the sum of the voltage drops must equal the voltage applied to the circuit. Measuring or calculating voltage drop is frequently performed by technicians troubleshooting electrical circuits. If a series circuit is constructed with a 12 V power source and 2 Ω and 6 Ω resistors, the voltage drop across each resistor when the circuit is closed and subject to a current flow of 1.5 A can be calculated:

Voltage drop for R_1:

$E_1 = I \times R_1$

$E_1 = 1.5\,A \times 2\,\Omega$

$E_1 = 3\,V$

Voltage drop for R_2:

$E_2 = I \Omega R$

$E_2 = 1.5 \, A \times 6 \, \Omega$

$E_2 = 9 \, V$

Total voltage drop through the circuit should equal that of the sum of the two calculations:

$E_1 + E_2$ = source voltage

$3 \, V + 9 \, V = 12 \, V$

TECH TIP: The importance of voltage drop testing cannot be overemphasized in troubleshooting vehicle electrical systems. Voltage drop verifies active or energized electrical circuits. More detail on voltage drop testing is provided in Chapter 34.

POWER

Just as in the internal combustion engine, the unit for measuring electrical power is the **watt** (named for James Watt, 1736–1819) usually represented by the letter P. In engine technology, power is defined as the rate of accomplishing work and therefore it is always factored by time. Remember, the definition of an ampere is 6.28×10^{18} electrons (1 coulomb) passing a point in a circuit per *second*. So the formula for calculating electrical power is:

$P = I \times E$ (spells "pie")

Using the data from the previous formula in which the circuit voltage was 12 V and the current flow was 1.5 A

$P = 1.5 \times 12 = 18 \, W$ = power consumed

One HP = 746 watts, so calculated values can be compared to standard values used to rate power-producing or power-absorbing components.

CAPACITANCE

The term **capacitance** is used to describe the electron storage capability of a commonly used electrical component known as a **capacitor**. Capacitors, which are also called **condensers**, all do the same thing—they store electrons. The simplest type of capacitor would

consist of two conductors separated by some insulating material called **dielectric**. The conductor plates could be aluminum and the dielectric may be mica (silicate mineral): the greater the dielectric properties of the material, the greater the resistance to voltage leakage. When a capacitor is connected to an electrical power source, it is capable of storing electrons from that power source. When the capacitor's charge storage capability is reached, it ceases to accept electrons from the power source. The charge is retained in the capacitor until the plates are connected to a lower voltage electrical circuit. At this point, the stored electrons are discharged from the capacitor into the lower potential (voltage) electrical circuit. Figure 29–19 shows the operating principle of a capacitor.

As a capacitor is electrified, for every electron removed from one plate, one is loaded onto the other plate. The number of electrons in a capacitor is identical when it is in both the electrified and neutral states.

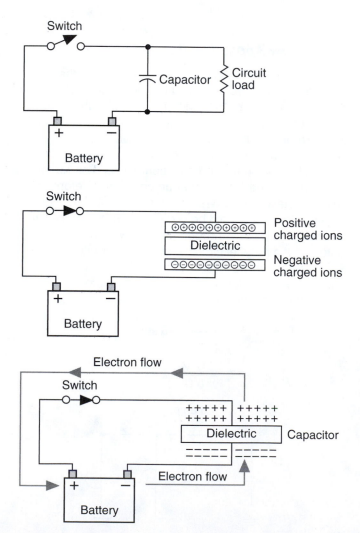

FIGURE 29–19 *Operating principle of a capacitor.*

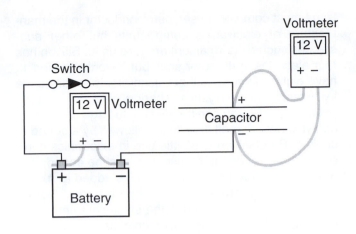

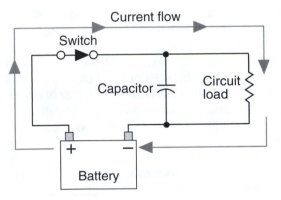

FIGURE 29–20 *Current flow with a fully charged capacitor.*

What changes is the location of the electrons. The electrons in a fully charged capacitor (Figure 29–20) will in time "leak" through the dielectric until both conductor plates have an equal charge: at this point, the capacitor would be described as being in a discharged condition. The ability to store electrons is known as capacitance, and this is measured in **farads** (named after Michael Faraday (1791–1867), the discoverer of the principle). One farad is the ability to store 6.28 × 10^{18} electrons at a 1 volt charge differential. Most capacitors have much less capacitance so they are rated in picofarads (trillionths of a farads) and microfarads (millionths of a farad).

1 farad = 1F

1 microfarad = $1\mu F$ = 0.000001F

1 picofarad = $1\rho F$ – 0.000000000001F

TYPES OF CAPACITORS

Many capacitors used in electric and electronic circuits are fixed value capacitors: these are coded by capacitance and voltage rating. Some capacitors have a variable capacity. These may have a combination of fixed and moving conductor plates, and the capacitance is

varied by a shaft that rotates the moving plates. In some variable capacitance type capacitors, the dielectric may be air. One truck engine manufacturer used to use a variable capacitance type sensor to signal throttle position. Electrolytic type capacitors tend to have much higher capacitance ratings than nonelectrolytic types, and they are polarized so they must be connected accordingly in a circuit. The dielectric in this type of capacitor is the oxide formed on the aluminum conductor plate.

When working on electrical circuits, it should be noted that capacitors can retain a charge for a considerable time after the circuit current flow has ceased. Accidental discharge can damage circuit components.

Capacitors are used extensively in electronic circuits performing the following roles:

- AC-DC filter: steadies a DC voltage wave offset sometimes caused by exposure to sunlight
- Power supply filter: smoothes a pulsating voltage supply into a steady DC voltage form
- Spike suppressant: when digital circuiting is switched at high speed, transient (very brief) voltage reductions can occur. Capacitors can eliminate these spikes or glitches by compensating for them.
- Resistor-capacitor circuits (R-C circuits): circuits that incorporate a resistor and a capacitor and one used to reshape a voltage wave or pulse pattern from square wave to sawtooth shaping or modify a wave to an alternating pattern.

CAUTION: Because capacitors store a charge differential, they can discharge when given the opportunity, even when a circuit is supposedly open because a battery is disconnected. This can be dangerous in circuits such as air bag deployment circuits that are designed to be actuated by capacitors in event of the vehicle batteries being destroyed in a collision.

 ## COILS AND TRANSFORMERS

When electrons are moved through a conductor, an electromagnetic field is created surrounding the conductor. When the conductor is a wire wound into a coil, the electromagnetic field created is stronger. Coils are the basis of electric solenoids and motors. They are used in electronic circuits to shape voltage waves because they tend to resist rapid fluctuations in current flow.

Like capacitors, coils can be used in electrical circuits to reshape voltage waves. Also, the energy in the

Input   Output

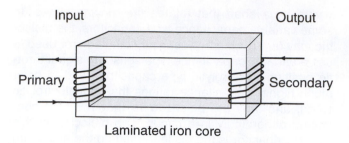

Primary   Secondary

Laminated iron core

FIGURE 29–21   Transformer operating principle.

electromagnetic field surrounding a coil can be induced to any nearby conductors. If the nearby conductor happens to be a second coil, then a current flow can be induced in it. The principle of a transformer is essentially that of flowing current through a primary coil and inducing current flow in a secondary or output coil as shown in Figure 29–21. Variations on this principle would be coils constructed with a movable core which permits their inductance to be varied, thereby altering frequency.

Transformers of various different types are used in vehicle electrical circuits, but they all fall into three general categories.

1. *Isolation transformers.* In an isolation transformer, the primary and secondary coils have the same number of windings, producing a 1:1 input to output ratio. Their objective is to "isolate" one portion of an electrical circuit from another: secondary voltage and current equal the primary.
2. *Step-up transformer.* The objective of a step-up transformer is to multiply the primary coil voltage by the winding ratio of the primary coil versus that of the secondary coil. For instance, if the primary to secondary coil winding ratio is 1:10, 12 V through the primary coil will induce 120 V through the secondary with a similarly proportional drop in current flow. Examples of step-up transformers are automotive ignition coils and the injector driver units used on some diesel EUI systems.
3. *Step-down transformer.* Step-down transformers function oppositely from step-up transformers. The primary coil winding ratio exceeds that of the secondary coil, resulting in a diminished output voltage and increased output current.

SEMICONDUCTORS

Semiconductors are a group of materials with exactly four electrons in their outer shell. As such, they cannot be classified either as insulators or conductors. Silicon is the most commonly used semiconductor in the manufacture of electronics components but other substances such as germanium are also used. Silicon has four electrons in its outer shell but it would "prefer" to have eight: this means that silicon atoms readily unite in clusters called crystals, sharing electrons in their outer shells. Silicon can be grown into large crystals by applying heat to melt the silicon followed by a period of cooling. Pure silicon is of little use in electronics components. For silicon to be useful it must be doped, that is, have small quantities of impurities added to it. The doping agents are usually phosphorus and boron. The doping intensity defines the electrical behavior of the crystal. After doping, silicon crystals may be sliced into thin sections known as wafers.

The type of doping agent used to produce silicon crystals defines the electrical properties of the crystals produced. A boron atom has three electrons in its outer shell. The outer shell is known as the valence and an atom with three electrons in its outer shell is known as trivalent. A boron atom in a crystallized cluster of silicon atoms will produce an outer shell with seven electrons instead of eight. This "vacant" electron opening is known as a hole. The hole makes it possible for an electron from a nearby atom to fall into the hole. In other words, the holes can move, permitting a flow of electrons. Silicon crystals doped with boron (or other elemental atoms with three electrons in the outer shell—trivalent) forms P-type silicon.

A phosphorus atom has five electrons in its outer shell. It is pentavalent. In the bonding between the semiconductors and the doping material, there is room for only eight electrons in the center shell. Even when the material is in an electrically neutral state, the extra electron can move through the crystal. When a silicon crystal is manufactured using a doping material with five electrons in the outer shell (pentavalent) it forms N-type silicon or semiconductor material. Like silicon, germanium also has four electrons in its outer shell. Figure 29–22 shows a germanium atom in a crystalized cluster with shared electrons.

A pure crystal of silicon or germanium is, electrically, of no use. To provide a semiconductor crystal with useful characteristics, it must be doped with an impurity. The doping of a semiconductor crystal will always define its electrical characteristics. In Figure 29–23, germanium is used once again to show how the element forms a P-type semiconductor when doped with arsenic and an N-type semiconductor when doped with boron. Semiconductor materials are classified as P-type and N-type. In different ways, P-type and N-type silicon crystals may permit an electrical current flow. In the P-type semiconductor, current flow is occasioned by a deficit of electrons while in the N-type semiconductor, current flow is occasioned by an excess of electrons. Whenever a voltage is applied to a semiconductor, electrons flow

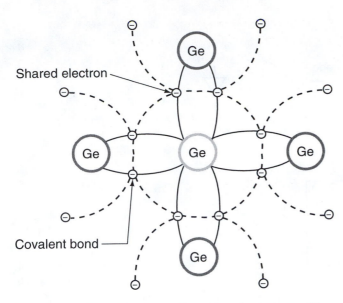

FIGURE 29–22 *Structure of a germanium atom with four electrons in the outer shell. Shown in a crystallized cluster with shared electrons.*

toward the positive terminal and the "holes" move toward the negative terminal.

DIODES

The suffix -ode literally means terminal. For instance, it is used as the suffix for cathode and anode. The word diode means literally, having two terminals. In the previous section it was explained how both P-type and N-type semiconductor crystals can conduct electricity: the actual resistance of each type is determined by either the proportion of holes or surplus of electrons. When a chip is manufactured using both P- and N-type semiconductors, electrons will flow in only one direction. The diode is used in electronic circuitry as a sort of one-way check valve that will conduct electricity in one direction (forward) and block it in the others (reverse). Figure 29–24 shows how a diode operates in forward and reverse bias modes.

The positive terminal (+) is called the anode and the negative terminal (–) the cathode. As an electrical one-way check valve, diodes will permit current flow only when correctly polarized. Diodes are used in AC generators (alternators) to produce a DC characteristic from AC. They are also used extensively in electronic circuits.

Diodes may be destroyed when subjected to voltage or current values that exceed their rated capacity. Excessive reverse current may cause a diode to conduct in the wrong direction and excessive heat can melt the semiconductor material. Diode voltage and current specifications are written as follows:

V_F = forward voltage I_F = forward current
V_R = reverse voltage I_R = reverse current

Numerous types of diodes play a variety of roles in electronic circuits. The following list gives examples of some of the more common types:

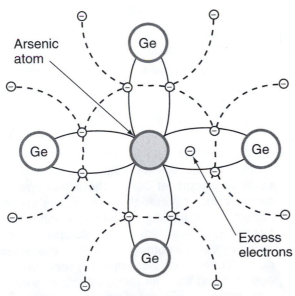

Germanium doped with arsenic to form an N-type semiconductor

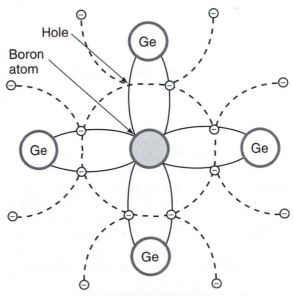

Germanium doped with boron to form a P-type semiconductor

FIGURE 29–23 *Doped germanium semiconductor crystals of the N- and P-types.*

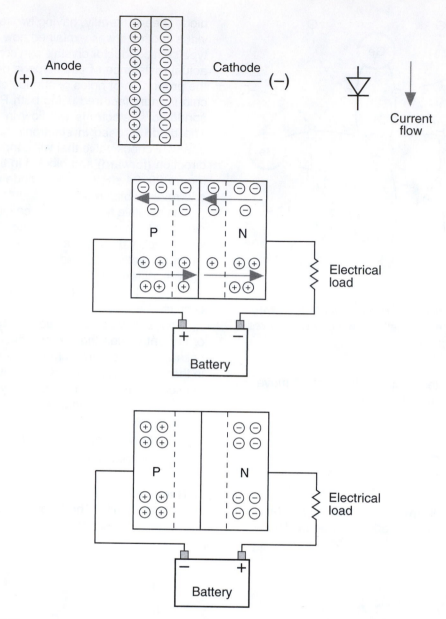

FIGURE 29–24   *Diode operation and diodes in forward and reverse bias.*

- Small signal. Small signal diodes are used to transform low current AC to DC (rectification), perform logic data flows and absorb voltage spikes.
- Power rectifier. Power rectifier diodes function in the same manner as small signal diodes but are designed to permit much greater current flow. They are used in multiples and often mounted on a heat sink to dissipate excess heat caused by high current flow.
- Zener. A **zener diode** functions as a voltage sensitive switch. Named after its inventor (Clarence Zener, who invented it in 1934), the zener diode is designed to block reverse bias current but only

up to a specific voltage value. When this reverse breakdown voltage is attained, it conducts the reverse bias current flow without damage to the semiconductor material. Zener diodes are manufactured from heavily doped semiconductor crystals. They are used in electronic voltage regulators and in other automotive electronic circuitry. Zener diodes are rated by breakdown voltage (V_Z) and this can range from 2 V to 200 V. A schematic of a Zener diode is shown in Figure 29–25.
- Light-emitting diode. All diodes emit electromagnetic radiation when forward biased, but diodes

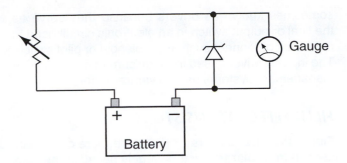

FIGURE 29–25   *Zener diode.*

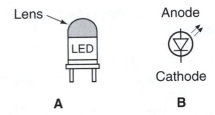

FIGURE 29–26   *Light-emitting diode (LED).*

TRANSISTORS

Transistors are three-terminal semiconductor chips that are used extensively in electronic circuits. Transistors can be generally grouped into bipolar and field effect categories. Many of their functions are either directly or indirectly associated with circuit switching and in this capacity they can be likened to relays in electrical circuits. Figure 29–27 shows a transistor-operating principle.

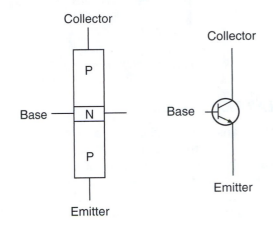

manufactured from some semiconductors (notably gallium arsenide phosphide) emit it at much higher levels. **Light-emitting diodes (LEDs)** may be constructed to produce a variety of colors and are commonly used for digital data displays. For instance a digital display with seven linear LED bars arranged in a bridged rectangle could display any single numeric digit by energizing or not energizing each of the seven LEDs. LED arrangements constructed to display alpha characters are only slightly more complex. LEDs convert electrical current directly into light (photons) and therefore are highly efficient as there are no heat losses. Figure 29–26 (*a*) shows an LED and Figure 29–26 (*b*) shows how one is represented on a schematic.

- Photo diode. All diodes produce some electrical response when subjected to light. A photo diode is designed to detect light and therefore has a clear window through which light can enter. Silicon is the usual semiconductor medium used in photo diodes.

Diodes operate in either forward or reverse bias.

- Forward bias: a positive voltage is applied to the P-type material and a negative voltage to the N-type material permitting current flow.
- Reverse bias: a positive voltage is applied to the N-type material and a negative voltage to the P-type material, meaning that current flow is blocked.

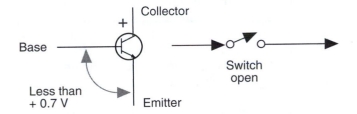

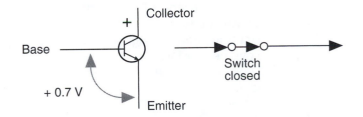

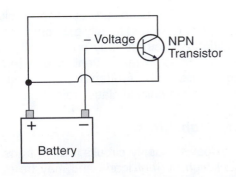

FIGURE 29–27   *Transistor-operating principle.*

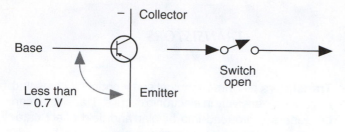

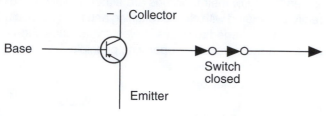

FIGURE 29-28   *Action of a PNP transistor.*

BIPOLAR TRANSISTORS

A bipolar transistor functions as a sort of switched diode with three terminals known as collector, emitter, and base each connected to semiconductor media in what is known as a silicon sandwich. The sandwich may either be NPN or PNP. The middle of the sandwich always acts as a *gate* capable of controlling the current flow through all three layers: the base is fairly thin and it has comparatively fewer doping atoms than that of the semiconductor material on either side of it. Bipolar transistors are designed so that a small emitter-base current will "ungate" the transistor and permit a larger emitter-collector current flow. Figure 29-28 shows the action of a PNP transistor.

Facts about Bipolar Transistor Operation

- The base emitter junctions will not conduct until the forward bias voltage exceeds ± 0.6 V.
- Excessive current flow through a transistor will cause it to overheat or fail.
- Excessive voltage can destroy the semiconductor crystal media.

Small Signal Switching Transistors

Small signal transistors are used to amplify signals. They may be designed to fully gate current flow in their off position and others may both amplify and switch. These are used in many vehicle electronically controlled circuits and are often the ones you get to test and build into circuits in class projects.

Power Transistors

Used in power supply circuits, power transistors may conduct high current loads and may be mounted on heat sinks to enable them to dissipate heat. They are sometimes known as drivers because they serve as the final or output switch in an electronic circuit used to control a component such as a solenoid or pilot switch. The injector drivers used in many current diesel engine management systems use power transistors.

FIELD EFFECT TRANSISTORS

Field effect transistors or **FETs** are more commonly used than bipolar transistors largely because they are cheaper to manufacture. They may be divided into junction type and metal-oxide semiconductors. The FET has three terminals: source, drain, and gate. The source supplies the electrons like the emitter in a bipolar transistor. The drain collects the current so it can be compared to the collector in the bipolar transistor. The gate creates the electrostatic field that "switches" the FET and permits electron flow from the source to the drain so it can be compared to the gate in a bipolar transistor.

Both junction type and metal-oxide FETs are controlled by a very small input or gate voltage. They are used in many smart switch multiplexing circuits used today in trucks. An FET functions like a relay with no moving parts. It is connected into a circuit that uses the FET like a switch. Figure 29-29 shows the operating principle of an FET.

Facts about FETs

- Gate-channel resistance is very high so the device has almost no effect on external components connected to the gate.
- Almost no current flows in the gate circuit because the gate-channel resistance is so high. The gate and channel form a "diode," and so long as the input signal reverse biases this diode, the gate will show high resistance.

THYRISTOR

Thyristors are three-terminal, solid-state switches. A small current flow through one of the terminals will switch the thyristors on and permit a larger current flow between the other two terminals. Thyristors are switches, so they are either in an on or off condition. They are classified by whether they switch AC or DC current. Some thyristors are designed with two terminals only: they will conduct current when a specific trigger or breakdown voltage is achieved, functioning somewhat like a diode.

Silicon-Controlled Rectifiers

Silicon-controlled rectifiers (SCRs) are similar to a bipolar transistor with a fourth semiconductor layer added as shown in the upper section of Figure 29-30.

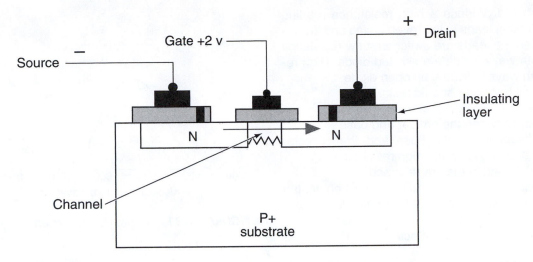

FIGURE 29–29   *FET transistor: an FET uses a positive charge to the gate, which then creates a capacitive field to permit electron flow. It acts like a relay with no moving parts.*

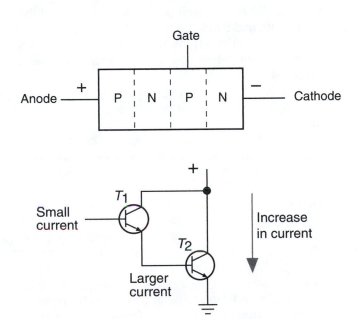

FIGURE 29–30   *A forward direction SCR and Darlington pair relationship.*

They are used to switch DC. When the anode of an SCR is made more positive than the cathode, the outer two PN junctions become forward biased: the middle PN junction is reverse biased and will block current flow. However, a small gate current will forward bias the middle PN junction, enabling a large current to flow through the thyristor. SCRs will remain *on* even when the gate current is removed: the *on* condition will remain until the anode-cathode circuit is opened or reversed biased. SCRs are used for switching circuits in vehicle electronic and ignition systems.

Darlington Pairs

A **Darlington pair** (named after the inventor) consists of a pair of transistors wired so that the emitter of one supplies the base signal to a second, through which a large current flows. The objective once again is to use a very small current to switch a much larger current. This type of application is known as **amplification**. Darlington pairs are used extensively in vehicle computer control systems. Figure 29–30 shows a Darlington pair relationship.

SUMMARY OF TRANSISTOR OPERATION

A transistor has three operating states:

- Cutoff: reverse bias voltage is applied to the base terminal. In cutoff state, no current will flow through the semiconductor.
- Conduction: bias voltage between the base and emitter has increased and switched the transistor to its on state. In its on state, the transistor is conducting, permitting electron flow. Output current is proportional to base current.
- Saturation: when collector-to-base voltage drops off to a near zero value by voltage drop across the collector, the transistor enters saturation state.

TESTING SEMICONDUCTORS

Diodes and transistors are normally tested using a DMM. The semiconductor being tested should be isolated from the circuit. The digital multimeter should be set to the diode test mode.

Diodes should produce a high resistance reading with the DMM test leads in the reverse bias and low resistance when the leads are switched. Low resistance readings both ways indicate a shorted diode. High resistance both ways indicates an open diode.

A functional transistor should test:

• Continuity between the emitter and base
• Continuity between the base and the collector when tested one way and high resistance
• When DMM test leads are reversed
• High resistance in either direction when tested across the emitter and collector terminals

Testing semiconductors is discussed in Chapter 33 in more detail with diagrams.

PHOTONIC DEVICES

Photonic semiconductors emit and detect light or photons. Photons are produced when certain electrons excited to a higher-than-normal energy level return to a more normal level. Photons behave like waves and the distance between the wave nodes and antinodes (wave crests and valleys) is known as **wavelength**. Electrons excited to higher energy levels emit photons with shorter wavelengths than electrons excited to lower levels. Photons are not necessarily visible and it is perhaps important to note that they may only truly be described as light when they are visible. Phototransistors are used in conjunction with an LED and slotted beam interrupter to accurately signal shaft/rotor speeds in vehicle electronics.

THE OPTICAL SPECTRUM

All visible light is classified as electromagnetic radiation. The specific wavelength of a light ray defines its characteristics. Light wavelengths are specified in nanometers, that is, billionths of a meter. The optical light spectrum includes ultraviolet, visible, and infrared radiation. Photonic semiconductors either emit or can detect near-infrared radiation, so near-infrared is usually referred to as light. As the computer age advances, technology is taking advantage of the ultrahigh frequencies of light waves and using them increasingly. Figure 29–31 shows the optical spectrum.

OPTICAL COMPONENTS

Optical components may conduct, refract, or modify light. For instance, an audio or data CD retains data

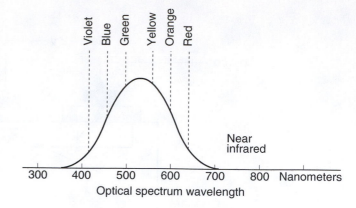

FIGURE 29–31   *The optical spectrum.*

that has to be read optically, usually by a laser. Some of the more common optical components are:

• Filters. Filters transmit only a narrow band of the spectrum and block the remainder.
• Reflectors. In much the same way a mirror functions, reflectors bend back a light beam, or at least most of it.
• Beam Splitters. Beam splitters transmit some of the optical wavelength and reflect back the rest of it.
• Lenses. Lenses bend light waves. They are often used in conjunction with semiconductor light sources and detectors. They are often used to collect and focus light onto a detector.
• Optical Fibers. Increasingly used to transmit digital data by pulsing light, optical fibers are thin, flexible strands of glass or plastic that conduct light: the light travels through a core surrounded by conduit or cladding. The use of fiber optics is growing rapidly, and it is only a matter of time before they become extensively used in vehicle technology.
• Electromagnetic Spectrum. The electromagnetic spectrum is shown in Figure 29–32 so that you have an idea of the relative frequencies of the wavelengths. The higher the frequency, the higher the intensity of data that can be relayed (reason for shift to fiber optics) and we use frequency shift analysis in technologies such as the Doppler radar used in truck collision warning systems.
• Solar Cells. A solar cell consists of a PN or NP silicon semiconductor junction built onto contact plates. A single silicon solar cell may generate up to 0.5 V in ideal light condition (bright sunlight), but output values are usually lower. Like battery cells, solar cells are normally arranged in series groups, in which case the output voltage would be the sum of cell voltages, or in parallel, where the output current would be the sum of the cell currents. They are used as battery chargers on vehicles.

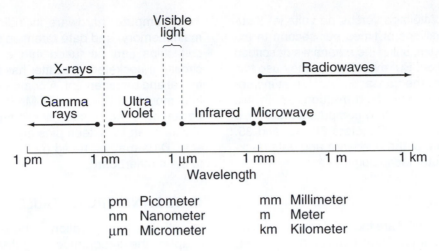

FIGURE 29–32 _The electromagnetic spectrum._

pm	Picometer
nm	Nanometer
μm	Micrometer

mm	Millimeter
m	Meter
km	Kilometer

USING ELECTRONIC SIGNALS

Simple electronic circuits can be designed to transmit relatively complex data by switching subcircuit components on or off, making use of the binary voltage or no-voltage features that are the essence of computer transactions. The idea is that a series of input signals work to produce a defined output signal. Signal processing includes filtering, amplification, and rectification performed by the devices presented earlier in this chapter. Digital circuits handle pulsed waveforms that can do things like add, subtract, and multiply and compare and perform logic processing. An example of a pulsed waveform is **pulse width modulation (PWM)**, used frequently in truck digital electronics both for input signaling (Caterpillar TPS) and to drive electronic unit injector (EUI) control cartridges. Gates are used to route signals. If a circuit consisting of a power source, light bulb, and switch is constructed, the switch can be used to "pulse" the ON/OFF time of the bulb. If you know Morse code, think of what can be expressed using the two Morse signals, a dot and a dash. This pulsing can be coded into many types of data such as alpha or numeric values. Pulses are controlled immediate variations in current flow in a circuit: ideally, the increase or decrease in current would be instantaneous. If this were to be so, the pulse could be represented graphically as in Figure 29–33.

However, true pulse shaping results in graduated rise when the circuit is switched to the ON state and graduated fall when the circuit is switched to the OFF state. Waves are rhythmic fluctuations in circuit current or voltage: they are often represented graphically and are described by their graphic shapes. Some waveforms are shown in Figure 29–34. You can see these

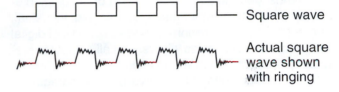

FIGURE 29–33 _Square sine waves._

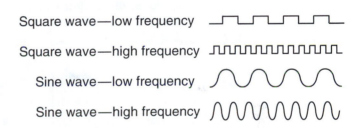

FIGURE 29–34 _Waveforms._

waveform types in vehicle electronic circuits when you use scopes to display them.

The term signals is used to describe pulses and waveforms that are shaped to transmit data. The mechanisms and processes used to shape data signals is **modulation**. In vehicle electronics, the term modulation is more commonly used in reference to digital signaling. Electronic noise is unwanted pulse or waveform interference that can scramble signals. All electrical and electronic components produce electromagnetic fields that may generate noise. All electronic circuits are vulnerable to magnetic and electromagnetic field effect.

COMPUTER BASICS

Computers process information (data). They do this using simple but long-winded methods of saying yes and no (presence of voltage versus no voltage) thousands sometimes millions of times per second to express simple messages, using the means we described in the previous section. But even though they use this long-winded process, the transactions occur at incredible speeds, that is, at very high frequencies. In this chapter, we begin to look at how computers function, a subject that is continued in Chapters 30, 32, and 35. We also look at some of the hardware and data packaging used in computer technology.

INTEGRATED CIRCUITS

Integrated circuits or **I/Cs** are the hardware, the building blocks of computers. They consist of resistors, diodes, and transistors arranged in a circuit on a *chip* of silicon. The number of electronic components that comprise the I/C varies from a handful to hundreds of thousands, depending on the function of the chip. Integrated circuits have innumerable household, industrial, and automotive applications and are the basis of digital watches, electronic pulse wipers, and all computer systems. Figure 29–35 shows a typical I/C.

Integrated circuits fall into two general categories. Analog integrated circuits operate on variable voltage values: electronic voltage regulators are a good vehicle example of an analog I/C. Digital integrated circuits operate on two voltage values only, usually presence of voltage and no voltage. Digital I/Cs are the basis of

most computer hardware including processing units, main memory, and data retention chips. Integrated circuit chips can be fused into a motherboard (main circuit) or socketed: the latter has the advantage of removal and replacement. A common chip package used in computer and vehicle ECMs is the DIP (dual in-line package), a rectangular plastic enclosed I/C with usually fourteen to sixteen pins arranged evenly on either side. DIPs may be fused (not removable) or socketed to the motherboard.

GATES AND TRUTH TABLES

In outlining the operation of transistors earlier in this chapter, the importance of **gates** was emphasized. Gates are the electronically controlled switching mechanisms that manage the operating mode of a transistor. Digital integrated circuits are constructed by using thousands of gates. In most areas of electronics, gates can be either open or closed: in other words in-between states do not exist. The terms used to describe the state of a gate is on or off. In a circuit, these states are identified as presence of voltage or no voltage. By saying yes or no through a number of channels, we can actually say quite a bit, as shown by what we can get a single byte of information to represent (shown a little later in this chapter). The best way to learn to understand the operation of gates in a digital circuit is to observe the operation of some electromechanical

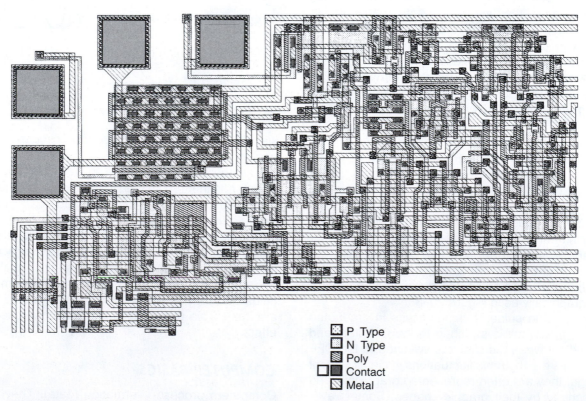

P Type
N Type
Poly
Contact
Metal

FIGURE 29–35 *Integrated circuit.*

switches in some simple electrical circuits such as the ones described next, but because these switches change state and therefore outcome on the simple basis of voltage ON, voltage OFF, they adapt ideally to electronic digital circuits.

AND Gates

In Figure 29–36, a power source is used to supply a lightbulb in a series circuit: in the circuit, there are two pushbutton switches that are in the normally open state. In such a circuit, the light bulb will only illuminate when both switches are closed. This kind of switch is known as an **AND gate**.

The operation of the foregoing circuit can be summarized by looking at the circuit and coming to some logical conclusions: a table that assesses a gated circuit's operation is often called a **truth table**. A truth table applied to the foregoing circuit, an AND gate would read as follows:

SWITCH A	SWITCH B	OUTCOME
Off	Off	Off
Off	On	Off
On	Off	Off
On	On	On

A truth table is usually constructed using the digits _zero_ (0) and one (1) because the **binary system** (outlined in detail following this section on gates) is usually used to code data in digital electronics. A truth table that charts the outcomes of the same AND switch would appear as follows:

SWITCH A	SWITCH B	OUTCOME
0	0	0
0	1	0
1	0	0
1	1	1

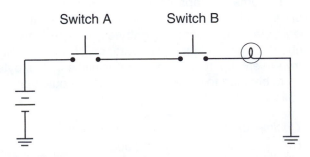

FIGURE 29–36 _AND gate: both switches are of the normally open type._

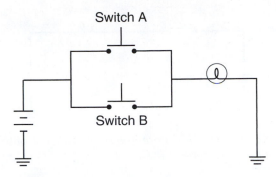

FIGURE 29–37 _OR gate: both switches are of the normally open type._

OR Gates

If another circuit is constructed using two normally open electromechanical switches but this time they are arranged in parallel, the result would be an **OR gate**. A schematic of an OR gate circuit is shown in Figure 29–37. A truth table constructed to represent the possible outcomes of this circuit would appear as follows:

SWITCH A	SWITCH B	OUTCOME
0	0	0
1	0	1
0	1	1
1	1	1

NOT or Inverter Gates

A **NOT gated** circuit switch can be constructed by using a push button switch that is in the normally closed state. This means that circuit current flow is interrupted when the button is pushed. In the series circuit shown in Figure 29–38 current will flow until the switch opens the circuit. A truth table constructed to graphically represent the outcomes of the NOT gated circuit shown in Figure 29–38 would look like the following:

IN	OUT
0	1
1	0

While the examples presented here used electromechanical switches for ease of understanding, in digital electronics, circuit switching is performed electronically using diode and transistor gates. AND, OR, and NOT gates are three commonly used means of producing an outcome that depends on the switching status of components in the gate circuit. If you under-

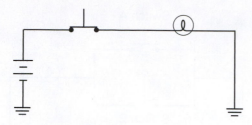

FIGURE 29–38   *NOT gate: the single switch is of the normally closed type.*

stood how transistors function earlier in this chapter, you should now begin to understand how they can function as the switches in gate circuits.

GATES, TRUTH TABLES AND BASIC DATA PROCESSING

Figure 29–39 shows some simple two input logic gates and the outcomes that can be produced from them. Logic gates are the basis of **data processing**.

Gates are normally used in complex networks in which they are connected by buses to form a logic circuit. Bits of data can be moved through the logic circuit or data highway to produce outcomes that evolve from inputs and retained memory. By massing hundreds of thousands of logic circuits, the processing of information becomes possible. A computer works simply by processing input data and generating logical outcomes based on the input data.

AND gate

A	B	Out
0	0	0
0	1	0
1	0	0
1	1	1

NAND gate

A	B	Out
0	0	1
0	1	1
1	0	1
1	1	0

OR gate

A	B	Out
0	0	0
0	1	1
1	0	1
1	1	1

NOR gate

A	B	Out
0	0	1
0	1	0
1	0	0
1	1	0

FIGURE 29–39   *Logic gates and truth tables showing the switching outcomes.*

BINARY SYSTEM BASICS

The binary system is an arithmetic numeric system using two digits: it therefore has a base number of 2. The base number of a numeric system is the number of digits used in it. The decimal system is therefore a ten base number system. The binary numeric system is often used in computer electronics as it directly correlates to the on or off states of switches and circuits. In computer electronics, the binary system is the primary means of coding data using the digits 0 and 1 to represent alpha, numeric, and any other data. Numeric data is normally represented using the decimal system: again, this is simple coding. When we see the digit 3 it is a representation of a quantitative value that most of us have been accustomed to decoding since early childhood. If decimal system values were coded into binary values, they could look like this:

Decimal digit	Binary digit
0	0
1	1
2	10
3	11
4	100
5	101
6	110
7	111
8	1000
9	1001
10	1010

In digital electronics a **bit** is the smallest piece of data that a computer can manipulate: it is simply the ability to represent one of two values either *on* (1) or *off* (0). When groups of bits are arranged in patterns they are collectively described as follows:

4 bits = nibble

8 bits = **byte**

Most data is referred to quantitatively as bytes. Computer systems are capable of processing and retaining vast quantities of data, so in most cases millions (megabytes) and billions (gigabytes) of bytes are described. A byte is eight bits of data, so it has the ability to represent up to 256 data possibilities. If a byte were to be used to code numeric data, it could appear as follows:

Decimal digit	Binary coded digit
0	0000 0000
1	0000 0001
2	0000 0010

3	0000 0011
4	0000 0100
5	0000 0101
6	0000 0110
7	0000 0111
8	0000 1000

and so on.

A number of methods are used to code data. If you are familiar with some computer basics, you may be acquainted with some ASCII (American Standard Code for Information Interchange) codes. This coding system has its own distinct method of coding values and would not be compatible with other coding systems without some kind of translation. Because on-off states can so easily be used to represent data, most digital computers and communications use this technology: it also is used in optical data processing, retention, and communications. Digital signals may be transmitted in series, one bit at a time, which tends to be slower, or in parallel, which is much faster. If the numbers zero through three had to be transmitted through a serial link, they would be signaled sequentially as shown in Figure 29–40.

If the same numbers were to be transmitted through a parallel link, they would be outputted simultaneously as shown in Figure 29–41.

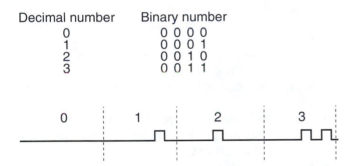

Decimal number	Binary number
0	0 0 0 0
1	0 0 0 1
2	0 0 1 0
3	0 0 1 1

FIGURE 29–40   _Sequential switching of binary coded numbers: serial link._

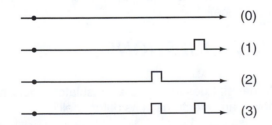

FIGURE 29–41   _Parallel link: transmission time is greatly reduced._

CONCLUSION

If you have read through this chapter, you will see that electricity has some complex aspects. However, it may comfort you to know that there are only three types of faults that can occur in an electrical circuit:

- opens
- shorts
- high resistance

When you understand the operating principles of the electrical circuits you are working on, it makes finding, then classifying electrical circuit faults that much easier. Most of the remainder of this textbook requires some understanding of the electrical and electronic basics introduced in this chapter. Chapter 34 uses some of the theory you have learned in this chapter and gets you to use it working on actual truck electrical circuits. Remember, you are not going to become an electrician overnight; it is a process that takes some time. But if you really want to understand what makes the wheels roll on any recent model truck, you have no choice but to become comfortable working with electricity and electronics.

TECH TIP: When confronted with the most frustrating type of electrical problems, remember that the cause of the problem will be classified into one of three categories: open, short, or high resistance.

SUMMARY

- All matter is composed of atoms.
- Electrical charge is a component of all atoms. When an atom is balanced, it can be described as being in an electrically neutral state.
- All matter is electrical in essence. What we call _electricity_ concerns the behavior of atoms that have become, for whatever reason, unbalanced or electrified.
- Electricity may be defined as the movement of free electrons from one atom to another.
- Current flow is measured by the number of free electrons passing a given point in an electrical circuit per second.
- Electrical pressure or charge differential is measured in _volts_, resistance in _ohms_, and current in _amperes_.
- If a hydraulic circuit analogy were used to describe an electrical circuit, _voltage_ is equivalent to fluid pressure, _current_ to the flow in gpm and _resistance_ to flow restriction

- The magnetic properties of some metals such as iron are due to electron motion within the atomic structure.
- There is a direct relationship between electricity and magnetism.
- Magnetomotive force (mmf) is a measure of electromagnetic field strength: its unit is ampere-turns.
- Ohm's law is used to perform circuit calculations on series, parallel, and series-parallel circuits.
- In a series circuit, there is a single path for current flow and all of the current flows through each resistor in the circuit.
- A parallel circuit has multiple paths for current flow: the higher the resistance in each path the less the current flow through it.
- Kirchhoff's law of voltage drops states that the sum of voltage drops through resistors in a circuit must equal the source voltage.
- When current is flowed through a conductor, a magnetic field is created.
- Reluctance is resistance to the movement of magnetic lines of force: iron cores have permeability and are used to reduce reluctance in electromagnetic fields. Air has high reluctance.
- Capacitors are used to store electrons: they consist of conductor plates separated by a dielectric.
- Capacitance is measured in farads: capacitors are rated by voltage and by capacitance.
- When current is flowed through a wire conductor, an electromagnetic field is created: when the wire is wound into a coil, the electromagnetic field strength is intensified.
- The principle of a transformer can be summarized by describing it as flowing current through a primary coil and inducing current flow in a secondary or output coil.
- Transformers can be grouped into three categories: isolation, step-up, and step down.
- Data can be transmitted electronically by means of electrical waveforms.
- Semiconductor elements have four electrons in their outer shells.
- Silicon is the most commonly used semiconductor material. Germanium is another.
- Semiconductors must be doped to provide them with the electrical properties that can make them useful as electronic components.
- After doping, semiconductor crystals may be classified as having N or P electrical properties.
- Diodes are two terminal semiconductors that often function as a sort of electrical one-way check valve.
- Zener diodes are commonly used in vehicle electronic systems: they act as voltage sensitive switches in a circuit.

- Transistors are three-terminal semiconductor chips.
- Transistors can be generally grouped into bipolar and field effect types.
- Essentially, a transistor is a semiconductor sandwich with the middle layer acting as a control gate: a small current flow through the base-emitter will ungate the transistor and permit a much larger emitter-collector current flow.
- Many different types of transistors are used in vehicle electronic circuits, but their roles are primarily concerned with switching and amplification.
- The optical spectrum includes ultraviolet, visible, and infrared radiation.
- Optical components conduct, reflect, refract, or modify light. Fiber optics are being used increasingly in vehicle electronics as are optical components because of the higher frequency of light.
- Integrated circuits consist of resistors, diodes, and transistors arranged in a circuit on a chip of silicon.
- A common integrated circuit chip package used in computer and vehicle electronic systems is a DIP with either fourteen or sixteen terminals.
- Many different chips with different functions including a CPU are often arranged on a primary circuit board, also known as a motherboard.
- Gates are the foundation of digital integrated circuits.
- Gates are switched controls that channel flows of data through electronic circuitry.
- AND, OR, and NOT gates are examples of gates used in electronics: they produce outcomes based on the switching status of the input circuit.
- The binary numeric system is a two-digit arithmetic system often used in computer electronics as it directly correlates to the on or off states of switches and circuits.
- A bit is the smallest piece of data that a computer can manipulate: it has the ability to show one of two states, either ON or OFF.
- A byte consists of eight bits.
- A byte of data can represent up to 256 pieces of coded data simply by changing the two states of each of the eight bits of which it is comprised.
- Never forget that there are only three types of electrical circuit faults: opens, shorts, and high resistance.

REVIEW QUESTIONS

1. A material described as an insulator would have how many electrons in its outer shell?
   a. Less than 4
   b. 4
   c. More than 4

2. Which of the following is a measure of electrical pressure?
   a. Amperes
   b. Ohms
   c. Voltage
   d. Watts
3. Which of the following units of measurement quantitatively expresses electron flow in a circuit factored with time?
   a. Coulombs
   b. Watts
   c. Farads
4. How many electrons does the element silicon have in its outer shell?
   a. 2
   b. 4
   c. 6
   d. 8
5. Who originated the branch of electricity generally described as electromagnetism?
   a. Franklin
   b. Gilbert
   c. Thomson
   d. Faraday
6. Which of the following elements could be described as being electrically inert?
   a. Oxygen
   b. Neon
   c. Carbon
   d. Iron
7. Which of the following is a measure of charge differential?
   a. Voltage
   b. Wattage
   c. Amperage
   d. Ohms
8. An element classified as a semiconductor would have how many electrons in its outer shell?
   a. Less than 4
   b. 4
   c. More than 4
   d. 8
9 Which of the following describes resistance to movement of magnetic lines of force?
   a. Reluctance
   b. Inductance
   c. Counter electromotive force
   d. Capacitance
10. Use Ohm's law to calculate the current flow in a series circuit with a 12 V power source and a total circuit resistance of 6 ohms.
11. Calculate the power consumed in a circuit through which 3 amperes are flowed at a potential difference of 24 V.
12. A farad is a measure of:

a. Inductance
b. Reluctance
c. Charge differential
d. Capacitance
13. The term pulse width modulation refers to:
   a. Waveforms shaped to transmit data
   b. Unwanted voltage spikes
   c. Electronic noise
14. To form a P-type semiconductor crystal, the doping agent would be required to have how many electrons in its outer shell?
   a. 3
   b. 4
   c. 5
   d. 8
15. To form an N-type semiconductor crystal, the doping agent would be required to have how many electrons in its outer shell?
   a. 3
   b. 4
   c. 5
   d. 8
16. The positive terminal of a diode is correctly called a(an):
   a. Electrode
   b. Cathode
   c. Anode
   d. Emitter
17. Which of the following terms best describes the role of a typical transistor in an electronic circuit?
   a. Check valve
   b. Relay
   c. Rectifier
   d. Filter
18. When testing the operation of a typical transistor, which of the following should be true?
   a. High resistance across the emitter and base terminals
   b. Continuity across the emitter and collector terminals
   c. Continuity across the base and emitter terminals
19. What would be the outcome in a OR gate if one of two switches in the circuit was closed?
   a. Off
   b. On
20. How many different data codes could be represented by a byte?
   a. 2
   b. 8
   c. 64
   d. 256
21. A potentiometer is a:
   a. Two-terminal, mechanically variable resistor
   b. Three-terminal, mechanically variable resistor
   c. Two-terminal, thermally variable resistor
   d. Three-terminal, thermally variable resistor

22. Which of the following devices can be used to block AC and pass DC?
    a. Rheostat
    b. Inductor
    c. Capacitor
    d. Thermistor
23. Which of the following components is typically used in a half-wave rectifier?
    a. Rheostat
    b. Transformer
    c. Diode
    d. Zener diode
24. An N-type semiconductor crystal is doped with:
    a. Trivalent atoms
    b. Pentavalent atoms
    c. Carbon atoms
    d. Germanium atoms
25. A PWM (pulse width modulated) signal is:
    a. Analog
    b. Digital
    c. Either analog or digital
    d. Neither analog or digital
26. Which one of the following can store energy in the form of an electric charge?
    a. Thermocouple
    b. Induction coil
    c. Potentiometer
    d. Capacitor
27. Which of the following can store energy in the form of an electromagnetic charge?
    a. Thermocouple
    b. Induction coil
    c. Potentiometer
    d. Capacitor
28. In which of the following devices is polarity always important?
    a. Coil
    b. Capacitor
    c. Resistor
    d. Diode
29. Which of the following is a three-terminal device?
    a. Zener diode
    b. Transistor
    c. Capacitor
    d. Diode
30. Which of the following is a two-terminal device?
    a. Potentiometer
    b. Thermocouple pyrometer
    c. NPN transistor
    d. PNP transistor

Chapter 30

Digital Computers

PREREQUISITE

Chapter 29

OBJECTIVES

After studying this chapter, you should be able to:

- Outline a brief history of computers from the abacus to the present day.
- Describe the hardware components in a basic computer system.
- Understand commonly used computer terminology.
- Comprehend the operating principles of the key components of a typical PC system.
- Describe the different types of data retention media and their appropriate application.
- Differentiate between magnetically retained, electronically retained, and optically encoded data.
- Outline the four stages of a computer processing cycle.
- Describe the role played by the CPU in the processing cycle.
- Define the role of the system clock in synchronizing processing activity.
- Identify common PC peripherals and their role in the system.
- Describe the main types of computer stations.
- Outline the role played by computers and their impact on the trucking industry.

KEY TERMS

analog	format	operating system
baud	gigabyte	optical disks
BIOS	hypermedia	output
bit	I/C	peripherals
boot-up	interface	personal computer (PC)
bus	kilobyte	PROM
byte	laptop computer	RAM
cache	laser	ROM
CD-ROM	mainframe	serial port
CPU	megabyte	soft copy
CRT	microprocessor	software
cursor	modem	storage media
data	monitor	system clock
data processing	motherboard	terminal
decoding	mouse	UPC
diskettes	multitasking	volatile memory
DOS	nodes	

INTRODUCTION

Today's diesel or automobile technician is required to have a working knowledge of **personal computers** or **PCs** as well as the computers that manage vehicle systems. Truck diesel engine and chassis OEMs use the PC as their primary diagnostic electronic service tool (EST). Since the early 1980s, PCs have filed and organized the business activity of the trucking industry, both in fleet and service/repair operations. From the early 1990s, the PC has played an ever-increasing role on the shop floor where it has been used to read, diagnose, guide sequential troubleshooting, and reprogram vehicle on-board computers. Often the PC is networked to a mainframe for purposes of logging repairs or reprogramming proprietary data to on-board vehicle computer systems. By definition, a computer is a device used for information processing, arithmetic calculation, and information storage.

There is no doubt that apprentice truck technicians with basic computer skills learn the technology of vehicle electronic management systems much more quickly than those without those skills. Technicians with 20 years experience before the electronics revolution tended to adapt to the changes and challenges less readily than those less set in their ways. This fact has opened up many opportunities for younger technicians to advance at an accelerated rate in the truck and bus repair and maintenance industries.

This chapter dealing with computer technology in a general sense provides a broad introduction to computer technology in preparation for study of specific computer systems later. It begins with a glossary of computer terminology; those with little exposure to the world of computers should reference this to interpret the text that follows.

COMPUTER TERMINOLOGY

In most cases, only the more common mode of expression is defined. For instance, the acronym CRT is more commonly used than its source cathode ray tube, so only CRT is defined. There is a good chance that you are already familiar with most of these terms so just quickly scan them.

Acronym—a word formed by the initial letters of other words.

Analog—the use of physical variables such as voltage or length to represent values.

Application—software programs that direct computer processing operations.

ASCII—American Standard Code for Information Interchange. Widely used data coding system used on PCs.

ATM—asynchronous transfer mode. A method of transmission and switching that can handle vast amounts of data at high speed.

AUTOEXEC.BAT—a batch file loaded into the DOS kernel that governs boot-up protocol.

Bandwidth—the rate at which data can be transmitted by any given method.

Baud—times per second that a data communications signal changes and permits one bit of data to be transmitted.

Baud rate—the speed of a data transmission.

Bay—a location in the computer housing/system unit designed to accommodate system upgrades.

BIOS—Basic Input/Output System. When a computer is booted, the CPU looks to the BIOS chip for instructions on how to interface between the disk-operating system and the system hardware.

Bit—a binary digit that can represent one of two values, on or off: presence of voltage or no voltage. The smallest piece of data a computer can manipulate. There are 8 bits to a byte.

Boot—the process of loading an operating system into RAM or main memory.

Boot-up—the process of loading an operating system to RAM, electronically reloading a system program, or resetting a computer.

BPS—bits per second. A measure of the speed at which data can be transferred.

Bridge—the software and hardware used to connect nodes (PC stations) in a network.

Buffers—memory locations used to store processed data before it is sent to an output device.

Bus—an electronic connection. Transmission lines that connect the CPU, memory, and the input/output devices. The term is increasingly used to mean "connected."

Byte—unit of measure of computer data, comprised of 8 bits. Used to quantify computer memory.

Cache—high-speed RAM located between the CPU and main memory used to increase processing efficiency.

CAD—computer-assisted design. The commonly used industrial component design tool.

Cage—a computer housing location accommodating two or more bays.

CAM—computer-assisted machining. Programmable computer managed machining.

Cartridge tape—sequential data storage medium, currently often used for PC data backup.

CD-ROM—an optical data disk that is read by a laser in the same way an audio CD is read.

Chip—a complete electronic circuit that has been photoinfused to semiconductor material such as silicon; also known as integrated circuit (I/C), microchip.

Clipboard—temporary storage location for data during cut and paste and program transfer operations.

Clock speed—the measure of computer processing speed measured in megahertz (MHz) or millions of cycles per second.

Cluster—the smallest data storage unit on a diskette.

Coaxial cable—type of wiring used to transmit signals with almost unlimited bandwidth but unable to carry two-way signals.

Control unit—the part of the CPU responsible for fetching, decoding, executing, and storing.

Conventional memory—the first data logged into RAM used primarily to retain the operating system.

Coprocessor—a chip or CPU enhancement designed for specific tasks such as mathematical calculation.

CPU—central processing unit. The computer subcomponent that executes program instructions and performs arithmetic and logic computations (Figure 30–1).

CRT—cathode ray tube. The commonly used acronym for "monitor."

Cursor—the underline character or arrow that indicates the working location on the screen display.

Cybernetics—the science of computer-controlled systems.

Cyberspace—a term commonly used to describe virtual reality technology.

DAT—digital audio tape. High-density data storage tape written to by a helical scan head.

Data—raw information.

Database—a data storage location or program.

Data compression—a means of reducing the physical storage space for data by coding it.

Datahub—the hub of a network system. Used by most truck OEMs to log data such as warranty status, repair history, and proprietary programming of on-board ECMs.

Data link—the connection point or path for data transmission in network devices.

Data processing—the production and manipulation of information by a computer.

Decoding—a CPU control unit operation that translates program instructions.

Device drivers—software used to control input and output devices.

Digital signals—data interchange/retention signals limited to two distinct states: combinations of ones and zeros into which data, video, or human voice must be coded for transmission/storage and subsequently reconstructed.

Digitizing—the process used to convert data to digital format.

Diskettes—magnetically written to portable data storage media for PCs.

DOS—disk-operating system. Set of software commands that govern computer operations and enable functional software programs to be run.

Download—data transfer from one computer system to another; often used to describe proprietary data transfer when reprogramming vehicle ECMs.

Downlink—the transmission signal from a communications satellite to an Earth receiver or the receiver itself.

Dumb node—a network node with no independent processing or data retention capability.

DVD-ROM—digital video disk–read-only memory. Provides up to 25 times the data storage capacity of a CD-ROM; offers high quality digital images for multimedia.

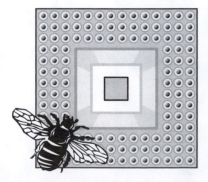

FIGURE 30–1 CPU size matched with a bumblebee: the CPU is the brain of the computer processing cycle.

E-mail—networking software that allows communications between PCs using modems network cards, and the phone system.

Execute—effect an operation or procedure.

Expansion board—a circuit board added to a computer system to increase its capability.

Fetching—CPU function that involves obtaining data from memory.

Fiber optics—the transmission of laser light waves through thin strands of fiber used to digitally pulse data more cheaply and at higher speeds than copper wire.

Fields—specific items of information.

File—a collection of related data.

Fixed disk—a data storage device used in PCs and mainframes consisting of a spindle and multiple stacked data retention platters.

Flash programming—term that is most often used to describe programming to nonvolatile RAM but is also used to describe magnetically retained programming (such as EEPROM) in some vehicle ECMs.

Flash RAM—nonvolatile RAM.

Floptical—a diskette that combines magnetic and optical technology to achieve high-density storage capability.

Font—typeface size and appearance.

Format—1. to alter the appearance or character of a program or document; 2. to prepare data retention media to receive data by defining tracks, cylinders, and sectors, a process that removes any data.

Function keys—numerical keys prefixed by an F that act as program commands and shortcuts.

Gigabyte—a measurement of memory capacity; a billion bytes.

Graphical user interface (GUI)—user friendly software that cues an operator through a procedure by using icons. MS Windows is an example (Figure 30–2).

Groupware—software that allows multiple users to work together by sharing information.

GUI—see *Graphical user interface.* Pronounced "gooey."

Handshake—establishing a communications connection.

Hard copy—computer-generated data that is printed rather than retained on disk.

Hard disk—see *Fixed disk.*

Hard drive—see *Fixed disk.*

Hardware—computer equipment excluding software.

Helical scan—technology used to write data at high density on tape instead of longitudinally.

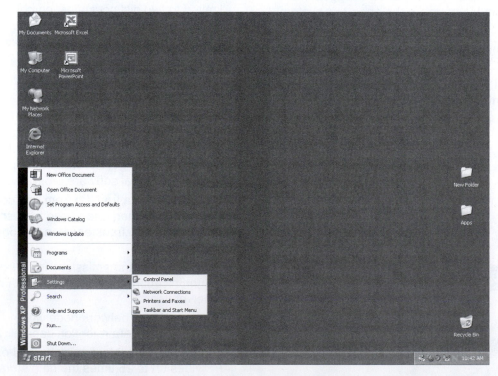

FIGURE 30–2 A graphical user interface such as MS Windows makes the computer easier to use: the icons represent different processing options.

Host computer—a main computer that is networked to other computers or nodes.

Hypermedia—a multimedia presentation tool that permits rapid movement between screens to display graphics, video, and sound.

I/C—integrated circuit. An electronic circuit that has been photoinfused on a semiconductor chip such as silicon.

Icons—pictorial/graphical representations of program menu options displayed on-screen such as those used with Windows software.

Ink jet printer—nonimpact printer that uses a nozzle to shoot droplets onto a page.

Input—the process of entering data into a computer system. Input devices would describe hardware such as a keyboard or sensors, on a vehicle system.

Integrated circuit—see *I/C*.

Interface—the point or device where an electronic interaction occurs. Separate vehicle system ECMs often require interface hardware and software to permit electronic interaction and avoid duplication of common input hardware.

Internet—the global, computer multimedia network communications system accessed through the phone system by means of a modem.

Intranet—an internal computer network designed for organizational communications using Internet protocols.

Kernel—the resident portion of a disk or program-operating system.

Keyboard—the data entry device used on PC systems enabling alpha, numeric, and command switching (Figure 30–3).

Kilobyte—a quantitive unit of data consisting of 1,024 bytes.

LAN—local area network. A private network used for communications and data tracking within a company or institution aka: Intranet.

Laptop computer—a portable PC.

Laser—any of many devices that generate an intense light beam by emitting photons from a stimulated source. Used in computer technology to read and write optically retained data and produce printed matter. The word is actually an acronym: **l**ight **a**mplification by **s**timulated **e**mission of **r**adiation.

Laser printer—a common PC printer device that aims a laser beam at a photosensitive drum to produce text or images on paper.

LCD—liquid crystal display. Flat panel display screen.

Local bus—an expansion bus that connects directly to the CPU.

Logical processing—data comparison operations by the CPU.

Log-on—an access code used in network systems (such as truck OEM data hubs) used for security and identification.

Machine cycle—the four steps that make up the CPU processing cycle: fetch, decode, execute, and store.

Mainframe—large computers that can process and file vast amounts of data. In the trucking industry, the data hubs to which dealerships are networked are mainframes.

Main memory or RAM—electronically retained data pipelined to the CPU. Data must be loaded into RAM for processing.

Master program—the resident portion of an operating system. In a vehicle ECM, the master program for system management would be retained in ROM.

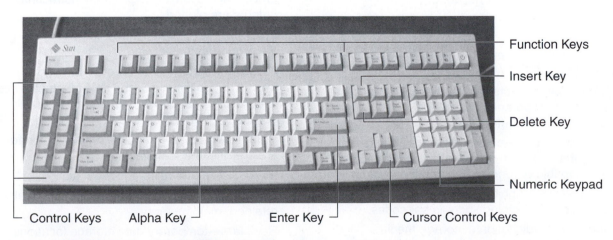

Function Keys

Insert Key

Delete Key

Numeric Keypad

Control Keys   Alpha Key   Enter Key   Cursor Control Keys

FIGURE 30–3   *A typical PC keyboard.*

Megabytes—one million bytes. Often abbreviated to "meg."

Megahertz—a measure of frequency: one million cycles per second. The system clock is speed rated in megahertz.

Memory address—the location of a byte in memory.

Menu—a screen display of program or processing options.

Microprocessor—a small processor. Sometimes used to describe a complete computer unit.

Microwaves—radio waves used to transmit voice, data, and video. Limited to line of sight transmission to distances less than 30 km.

MIPS—millions of instructions per second. Rating of processing speed.

Modem—a communications device that converts digital output from a computer to the analog signal required by the phone system.

Monitor—the common output screen display used by a computer system: a CRT.

Motherboard—the primary circuit board in the computer housing to which the other components are connected.

Mouse—input device that controls the cursor location on the screen and switches program options.

Multimedia—the combining of sound, graphics, and video in computer programs.

Multitasking—operating systems that permit two or more programs to be run at the same time. Most vehicle ECMs possess multitask capabilities.

Nanosecond—one billionth of a second.

Network—a series of connected computers designed to share data, programs, and resources.

Nodes—dumb terminals (no processing capability) and PCs connected to a network.

Notebook computer—briefcase-sized PC designed for portability.

Numeric keypad—numeric-only input keys such as on a ProLink EST.

OCR—optical character recognition. Scanners that read type by shape and convert it to a corresponding computer code.

Operating system—core software programs that manage the operation of computer hardware and make it capable of running functional programs.

Optical codes—graphic codes that represent data for purposes of scanning such as bar codes.

Optical disks—digital data storage media consisting of rigid plastic disks on which lasers

have burned microscopic holes; the disk can then be optically scanned (read) by a low-power laser. Audio CD and CD-ROM are examples.

Optical memory cards—digital data storage media the size of a credit card capable of retaining large amounts of data.

OS—see *Operating system*.

Output—the result of any computer processing operation.

Output devices—components controlled by a computer that effect the results of processing. The CRT and printer on a PC system and the injector drivers on a diesel engine are all classified as output devices.

Parallel ports—peripheral connection ports for devices that require large volume data transmission such as printers and disk drives.

Password—an alpha, numeric, or alphanumeric value that either identifies a user to a system or enables access to data fields for purposes of download or reprogramming.

PC—personal computer.

Peripherals—input and output devices that support the basic computer system such as the CRT and the printer.

Pipelining—rapid sequencing of functions by the CPU to enable high-speed processing.

Pixels—picture elements. A measure of screen display resolution; each dot that can be illuminated is called a *pixel*.

Port—connection socket used to link a computer with input and output devices.

POST—power on self-test. A BIOS test run at boot-up to ensure that all components are operational.

Processing—the procedure required to compute information in a computer system; input data is processed according to program instructions and outputs are plotted.

Program—set of detailed instructions that organize the programming activity.

PROM—programmable read-only memory. A means of qualifying general ROM data with system specific information; can be a removable chip socketed to the motherboard.

Proprietary OS—OS that are privately owned and are specific to a manufacturer or operator.

Protocols—set of rules and regulations.

RAM—random access memory. Electronically retained, main memory.

Registers—temporary data storage locations in the CPU.

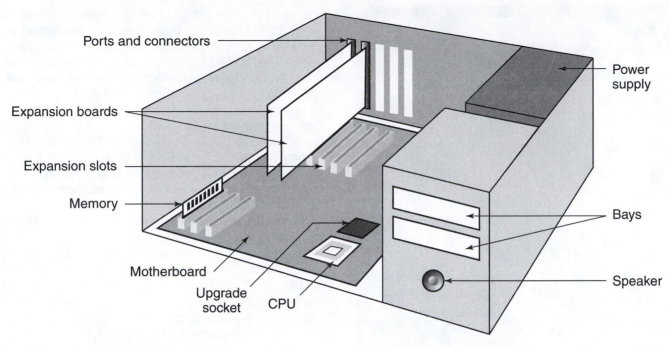

Ports and connectors
Expansion boards
Expansion slots
Memory
Motherboard
Upgrade socket
CPU
Power supply
Bays
Speaker

FIGURE 30–4   *PC system unit components.*

ROM—read-only memory. Data retained either magnetically or by optical coding and designed to be both permanent and read-only.

Router—connection between two networked computers or two vehicle system ECMs.

Scrolling—the moving of lines of data up or down on a display screen.

Sector—a pie-shaped section of a disk or a section of a track.

Sequential storage—storage of data on media, such as magnetic tape, where data is read and written sequentially.

Serial port—port connection that transfers data one bit at a time and therefore more slowly than a parallel port. A mouse is connected to a serial port.

SPRAM—synchronous dynamic RAM. High-speed RAM used with high-speed processors.

Soft copy—data retained electronically or on disk as opposed to being on paper.

Software—programs and the instructions within a program that organize the activity of a computer to produce/process outcomes.

Solid state storage—volatile storage of data in RAM chips.

Sound card—multimedia card capable of capturing and reproducing sound.

Spreadsheet—software that enables numeric data organization and calculation.

Star network—network set up to operate from a central hub computer.

Static RAM—a RAM chip with medium to large volume memory retention and high access speeds.

Storage media—any nonvolatile data retention device; floppy disks, data chips, CD-ROM, and PCMCIA cards are examples.

System board—see *Motherboard*.

System clock—generates pulses at a fixed rate to time/synchronize computer operations.

System unit—the main computer housing and its internal components (Figure 30–4 and Figure 30–5).

Teleconferencing—audio/video communication using computers, a camera, and modem linkages.

Terabyte—a trillion bytes.

Terminal—a computer station or network node.

Tower computer—a PC housed in an upright case.

Trackball—the cursor control device often used in notebook computers consisting of a rotatable ball integral with the keypad.

Typeface—the design appearance of alpha characters.

UPC—universal product code. The commonly used commercial bar code designed for optical scanners.

USB—universal serial bus. Permits the connection of external devices such as speakers, scanners, modems, and additional memory storage devices.

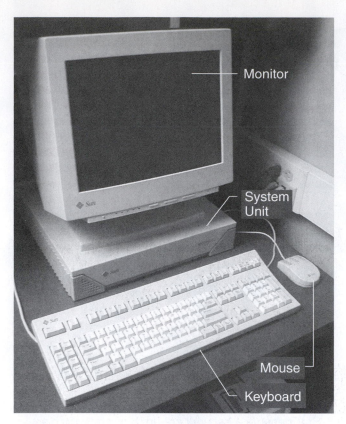

FIGURE 30–5   *Devices that make up a simple PC system.*

User ID—a password used for security and identification on multiuser systems.

Volatile memory—RAM data that is only retained while a circuit remains switched on.

Windows—The most commonly used GUI software for structuring the operating system architecture; undergoes version/generation upgrades every couple of years. Manufactured by Microsoft Corporation.

Wireless—the use of microwave or radio frequencies to transmit analog or digital signals.

Word processing—using a computer to produce text/mainly text documents and files.

WORM—write once, read many. Optical disk that can be written to once (permanently) and read many times.

A BRIEF HISTORY OF COMPUTERS

The first computer is generally believed to be a simple counting and calculation tool called an abacus. The abacus is thought to have had its origins in Babylonia some 2,500 years ago. It survives as a calculating instrument in remote parts of East Asia and the Middle East and as a child's toy in the West. An expert practitioner can compete with electronic calculation in both speed and accuracy. The abacus would be classified as an analog computer. An analog device operates on data represented by numbers or voltage values, which it processes to output a physical analogy of a mathematical problem to be solved. A slide rule, much used until the widespread acceptance of the electronic multitasking calculator in the 1960s, is another example of an analog computer.

After the abacus, the next significant events in the development of the computer took place in Europe during the seventeenth century. In 1620, Edmund Gunter, an English mathematician, invented the slide rule, and in 1642, the French scientist, Blaise Pascal, introduced the first mechanical digital calculating machine. These significant innovations, plus the ideas of many scientists and mathematicians who followed, paved the road to the development of the modern computer.

Most modern computers are digital. The digital computer works with data in discreet form, that is, expressed as digits of binary codes. Binary coding is explained in Chapter 29. The digital computer processes data by counting, listing, and rearranging *bits* of data in accordance with detailed program instructions. The results of processing are then translated into outputs that may be displayed on a screen, printed onto paper, or used to actuate other peripherals. Peripherals are input and output devices that are part of the computer system but not integral with main housing.

The invention of the first electronic digital computer is attributed to Presper Eckert and John Mauchly of the University of Pennsylvania in 1946. By the 1960s the computer had developed into a business tool used for data storage and arithmetic processing. The computers of the 1960s were physically vast, power-hungry machines full of vacuum tubes that required considerable maintenance.

The modern computer evolved from these beginnings. The typical home computer system of the twenty-first century has more computing power than that on board *Apollo 11* (the first manned moon landing craft). Today, computers have a role in driving almost every aspect of our lives.

COMPUTER HARDWARE

A computer system is made up of a number of pieces of hardware. The following components would be typical:

COMPUTER/SYSTEM HOUSING

The computer/system housing is a desktop case or tower constructed of plastic or steel. The system housing contains:

Motherboard

The motherboard is the main circuit board or deck to which the internal chips, cards, and other components are attached (Figure 30–6).

Central Processing Unit

The central processing unit (CPU), also known as the microprocessor unit (MPU), is the "brain" of the computer system and the primary indicator of its processing speed. It is attached to the motherboard by a multipin connector. The functions of the CPU are those of decision making, addressing and data transfer, timing and logic control, arithmetic and logic operations, fetching data and program instructions from memory, decoding instructions, and responding to control signals from input and output devices (I/O). Microprocessors contain a complex of many electronic circuits photoinfused on a silicon chip. The CPU individual subcircuits are referred to as blocks and collectively as the architecture of the microprocessor.

System Internal Buses

Buses are simply connection and data transfer devices. System internal buses link the main modules in a system as follows:

> Control bus—bidirectional. It transmits and receives operational commands to all parts of the computer system.
>
> Address bus—unidirectional. It connects the CPU to the memory banks and I/O ports. The CPU selects the address or location in memory to which data is targeted.
>
> Data bus—bidirectional. It acts as a data conduit to and from the memory banks, the CPU, and the I/O ports.

Word Size

CPUs are often described by word size such as 64 bit processor. Word size describes the number of bits that can be processed simultaneously. The larger the word size, the faster the processor.

DATA RETENTION HARDWARE—MEMORY CATEGORIES

The term *memory* is used to describe digital data generally in computer terminology. Memory or data can be retained using a number of different means, determined by factors such as whether the information must be accessed at high speeds by the processor and whether it needs to be retained permanently (Figure 30–7). This section categorizes memory by general type:

Random-Access Memory (RAM)

Random-access memory (RAM) is often referred to as "main memory" or "primary storage." It is retained electronically in semiconductor chips. RAM data includes the computer operating system software (a set of protocols that coordinate computer operations), program application software (direct procedure for the specific task(s) performed), and the data currently being processed. RAM data is volatile. Because it is electronically retained, it is lost when the electrical circuit supplying it is opened. Some vehicles have on-board computers with RAM capability described as both volatile RAM and nonvolatile RAM. Vehicle volatile RAM can be likened to main memory in a PC; it is switched by the vehicle ignition circuit. Some vehicles, especially those with first generation computerized system management electronics, had a nonvolatile RAM

FIGURE 30–6   *A look at the inside of a PC.*

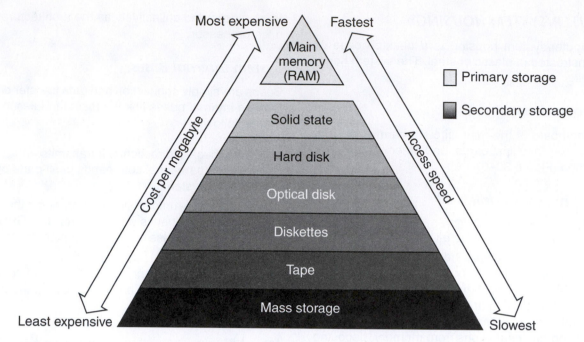

FIGURE 30–7   *Access speeds of data retention media.*

memory component; this data was electronically retained but supplied by a circuit directly from the vehicle battery after the ignition circuit was opened. Fault codes and failure mode strategy were written to nonvolatile RAM and "kept alive" until either the battery was disconnected or the ECM reset.

In the PC processing cycle, cache is used to retain frequently consulted program instructions or data. When the CPU requires data, it checks cache first, enabling faster execution than retrieval from (slower) main memory.

RAM data on PCs can be categorized as follows:

• Conventional memory. Conventional memory is the first logged in RAM at boot-up, and it consists of operating system data and programs.
• Upper memory. Upper memory is used to control the computer's integral hardware, input and output devices, and some peripherals.
• Extended memory. Extended memory is used to retain program and data; multiple programs may be logged in extended memory, permitting multitasking or driving more than one quite separate operations simultaneously.
• Expanded memory. Expanded memory was built into a memory expansion board and used on older PCs. It was driven by software called expanded memory manager (EMM) and resulted in slower processing speeds. Current PC systems tend to use increased extended memory capability and eliminate expanded memory.

Read-Only Memory

Read-only memory (ROM) chips retain data, protocols (rules and regulation), and instructions that once written, cannot be changed. In other words, the information can be read and used but never altered (although it can be destroyed or corrupted). The boot-up procedure in a PC system is used repeatedly and does not change so it makes sense that it is written to ROM. Truck and other vehicle computers (ECMs) have major portions of the total requisite management data written to ROM. A diesel engine series may power a truck, bus, generator, boat, or a grader; the master program for engine management would be written to ROM. This data in itself may not be sufficient to run the engine without further data qualification. However, it would contain all the common running characteristics required to run the engine in each of the very different applications. ROM describes nonvolatile data retention media; it is usually retained magnetically in current systems.

Programmable Read-Only Memory

Programmable read-only memory (PROM) is written to a chip or set of chips (card) and socketed to the PC or ECM motherboard. In a vehicle engine management system, PROM is often used to define application. For instance, the diesel engine designed for use in many different applications and manufactured with an ECM ROM common to all those applications would require function-specific running data before it would actually manage an engine. The data contained in the PROM chip or card would define the actual run parameters of the engine.

Continuing to use the example of an engine ECM, PROM data would qualify that contained in ROM to a specific chassis application. One truck OEM describes their PROM card as a "personalty module"; an excellent description of the actual role played by this type of data. PROM chips may often be removed and replaced when the engine application requires redefining.

Electronically Erasable Programmable Read-Only Memory

Electronically erasable programmable read-only memory (EEPROM) is used in most current truck engine and chassis management ECMs. It can be written to by the on-board software, receive customer and proprietary data programming, log driving and performance analyses, and retain codes manifestly and covertly. The type of data programmed to EEPROM would include customer data, such as a road speed limit, cruise parameters, transmission ratios, final drive ratios, tire rolling radii, and failure mode strategy—data that may be required to be changed a number of times during the vehicle/engine life. It would also log data such as the fuel map profiles and covert documentation of failure/abuse conditions—data that could only be read or rewritten directly by the OEM mainframe via networking.

Secondary Data Storage Media

Secondary data storage or auxiliary storage devices are means, such as diskettes, used for writing, storing, and reading data. The earliest home PCs used cassette tapes on which data was magnetically encoded; these presented some limitations imposed by the speed at which the tape could be driven through the reading heads and the sequential storage characteristic of cassette tape. Sequential data storage works fine to retain an album of music in which the notes and songs are replayed in sequence, but severely limits access speed when specific units of data logged on different locations on the tape are required. These were superceded by 5¼" floppy diskettes and then by 3½" disks. The 3½" floppy diskette is encased in rigid plastic and only the internal circular disk is "floppy"; this is nylon plastic coated with metal oxide. It is encoded with data magnetically in the same way an audiotape is encoded. It is read by a rotating drive mechanism and read heads.

Hard disks also retain data magnetically. They consist of multiple rigid platters coated with metal oxide that permit data to be magnetically written to the surface. On a hard disk assembly, the read-write heads, the drive mechanism, and data retention platters are incased in an airtight, sealed case. Hard disks are usually a permanent fixture within a PC housing. They rotate at high speeds (7,200 rpm) to receive and unload data (Figure 30–8).

The cylinder method reduces the movement of the read/write head (thereby saving time) by writing information down the disk on the same track of successive surfaces.

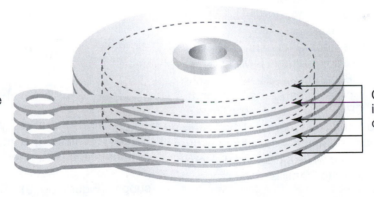

Cylinder 5 is comprised of all track 5s.

The clearance between a disk head and the disk surface is about 10 millionths of an inch. With this small difference, contamination such as a smoke particle, or human hair could render the drive unusable. Hard disk drives are sealed to prevent contamination.

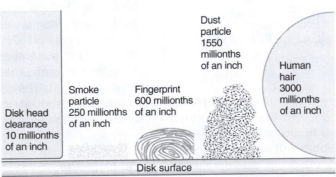

FIGURE 30–8 Hard disk drive. (Courtesy of Shelly, Cashman and Waggoner)

Data may also be stored magnetically on a new generation of cartridge tapes. In home PC applications these are often used for data backup. Tape is sequential storage media; data must be read in the sequence it was recorded in, so even when the data density of the tape is high (and it is, in current DAT-digital audiotape, up to 60,000 bpi), it has limited applications in current computers.

Optical disks are being used increasingly to store data in media better known as CD-ROM and DVD-ROM. A laser is used to burn microscopic holes on the surface of a rigid plastic disk. The disk can then be scanned by a low power laser that reflects light off the encoded disk surface to read digital data. This read-only format is being widely used in the automotive industry. Because of the amount of data that a CD can retain, many computer programs are now sold in this format. Optical disks have the additional advantage that the data written to them cannot be damaged by magnetic force.

Optical memory cards are about the size of a credit card and may contain data equivalent to 1,600 pages of text. Their use is increasing. They retain data for read out and can be written to with the correct equipment.

Memory Speed

Students and technicians alike are often curious as to access and processing speeds in computers. While this is not need-to-know information, to some it falls into the category of nice to know. You should also remember that with each generation of computer technology, these speeds increase dramatically, so what is said today does not apply tomorrow.

The speed of memory is usually measured in nanoseconds (billionths of a second). Current RAM chips may have access speeds of 8 nanoseconds while cache may have access times of less than 2 nanoseconds. Registers designed into CPU chips have access times of as little as a single nanosecond and represent the fastest type of memory. A vehicle ROM has access times in the 100 to 250 nanosecond range while a PC hard disk produces access times in the 10 to 20 millisecond range.

ELECTRONICALLY REPRESENTED DATA

In a digital computer, binary digits are used to represent data. The digits 0 and 1 are the only numbers used to represent billions of data possibilities. These digits represent the on (1) or off (0) status we talked about in Chapter 29. Each on or off digital value is called a bit. The word bit is derived from the words binary digit.

A byte comprises 8 bits. A byte is capable of representing 256 data possibilities. ASCII is a commonly used secondary data storage code. ASCII is an 8-bit storage code used on PC systems. Observe how the following alpha, numeric, and symbolic values are coded in ASCII.

SYMBOL	ASCII
0	01100000
1	01100001
2	01100010
3	01100011
4	01100100
5	01100101
6	01100110
7	01100111
8	01101000
9	01101001
A	01000001
B	01000010
C	01000011
D	01000100
E	01000101
F	01000110
G	01000111
H	01001000
I	01001001
J	01001010
K	01001011
L	01001100
M	01001101

SUMMARY OF COMPUTER OPERATION AND THE PROCESSING CYCLE

The system unit in a typical computer consists of a housing containing a motherboard, CPU, electronic and magnetic data retention media, data readers, co-processors, connection slots and ports, and a power supply (Figure 30–9). The motherboard is a circuit board; most of the system unit components are either permanently connected or socketed to the motherboard. In a typical PC, the CPU is contained in a single integrated circuit called a *microprocessor*. The CPU contains the control/command unit and the arithmetic/logic unit (ALU). The control/command unit in the CPU operates by sequencing a four-part processing cycle below and illustrated in Figure 30–10.

1. Fetching
2. Decoding
3. Executing
4. Storing

1. Notebook Laptop

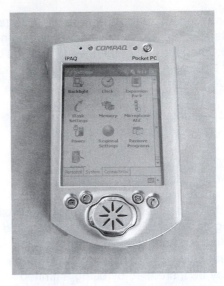

2. Personal Digital Assistant (PDA)

3. Tablet PC

4. Desktop PC

FIGURE 30–9   *Types of computers.*

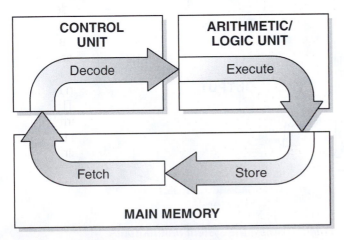

FIGURE 30–10   *The processing cycle.*

The control unit uses a system clock to synchronize all computer operations; the system clock generates electronic pulses at a fixed rate specified in megahertz (MHz). The CPU ALU incorporates the electronic circuitry required to perform arithmetic and logic data manipulation. Arithmetic operations include addition, subtraction, multiplication, and division. Logical operations consist primarily of data comparison and sequential strategizing. The CPU contains registers, memory storage locations that may temporarily store certain types of data. A critical factor of computer processing speed is word size: this indicates the number of bits that a CPU can process simultaneously.

Memory generally refers to data quantitively that can be either retained electronically or stored magnetically.

RAM refers to main memory or primary storage and retains data electronically. Electronically retained data is volatile; when the computer is switched off it is dumped. After booting-up a computer, RAM is loaded with the following data:

1. The operating system
2. The application program instructions
3. The data currently being processed

Some computers have a nonvolatile RAM component known as nonvolatile random-access memory (NV-RAM), keep-alive memory (KAM), or flash RAM, depending on the application. RAM is installed using a single in-line memory module (SIMM) or double in-line memory module (DIMM) on boot-up. Processing efficiency may be increased by using a limited amount of high-speed RAM called cache. The type of data stored in cache is frequently consulted program instructions and data; cache is located between the CPU and main memory.

Read-only memory describes chips that retain magnetically encoded data designed to be permanent. ROM chips are often fused to a motherboard and are not removable. PROM chips also retain data magnetically; however, these are often designed for removal and replacement. Electronically erasable PROM (EEP-ROM) is data retention that can be both read and written to, either by the computer system it is integral with, or it may receive programming from a networked source or data retention media.

The speed of memory is measured in nanoseconds. Registers in the CPU chip have access times of less than a single nanosecond; ROM memory on chips has access times of between 50 and 250 nanoseconds,

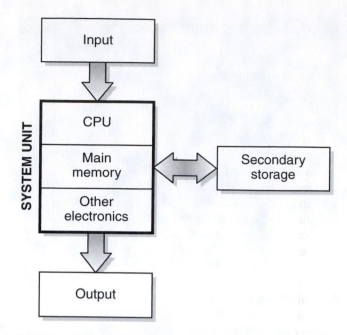

FIGURE 30–11   *Processing from input to outcome.*

while accessing data on a typical PC hard disk drive takes between 10 and 20 milliseconds.

Computer processing efficiency may be increased by using a coprocessor, a chip or card that functions to perform specific tasks. The computer power supply depends on the computer's functions (Figure 30–11 and Figure 30–12). A PC typically receives an electrical supply at 110 to 120 volts AC and transforms this to an operating value ranging from 3 to 12 volts DC. A vehicle ECM is usually supplied with 12 volts DC and transforms this to a lower voltage value, usually 5 volts

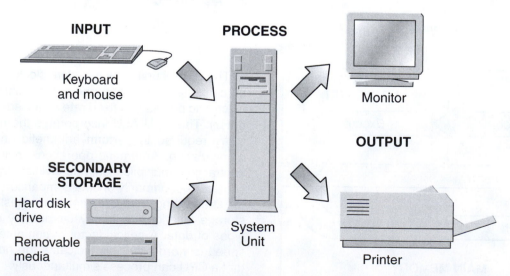

FIGURE 30–12   *A computer system consists of input devices, the system unit, output devices, and secondary storage devices.*

within the system. Operating voltage values tend to be lower in more recent computers, the objective being to reduce heat. PCs normally incorporate a cooling fan to remove heat from system components. Diesel-powered vehicle ECMs are sometimes mounted on a heat sink or heat exchanger through which fuel is routed on its way to the injection pumping apparatus.

Expansion slots are used in PCs to connect in devices that increase computing power or capability. Ports are used to connect peripherals such as printers, modems, and keyboards. Parallel ports are capable of transmitting data 8 bits (1 byte) at a time using cable that has 8 data lines. Printers and disk drives use parallel port connections. Serial ports transmit data 1 bit at a time and therefore function more slowly than parallel ports. Keyboards, a mouse, and communication devices (modem) use serial ports. A bay is simply a location within the system unit used to store hardware such as disk and CD drives. Two or more bays stacked vertically can be called a *cage*.

A machine language instruction is the driving command in binary data from which the CPU interprets and generates outputs. The instruction set consists of commands from which the computer generates outputs. Computers may use serial processing in which the CPU processes a single instruction at a given moment until the instruction is completed. It then begins the execution of the next instruction and sequences processing one instruction at a time until the program is completed. Parallel processing uses multiple CPUs, enabling the execution of several sets of instructions simultaneously. Massively parallel processors (MPP) are used in supercomputers where thousands of CPUs function simultaneously.

SUMMARY

- The first computer is generally accepted to be the abacus.
- The birth of the modern digital computer took place in 1946 at the University of Pennsylvania by Eckert and Mauchly.
- A computer system requires hardware and software to function.
- In a PC system, the system housing encases most of the components in either a desktop horizontal case or a vertical tower.
- Within the system housing, most of the internal components are connected to a main circuit board called a motherboard.
- The CPU is often referred to as the *brain* of a computer system.
- RAM is electronically retained and can be called main memory.

- Cache is a type of high-speed RAM used to park frequently consulted operating data requiring fast access.
- Commonly used data storage media include cassette tapes, floppy disks, hard drive disk platters, optically encoded disks (CDs), and optically encoded cards.
- Memory access speed is measured in nanoseconds.
- Cache access speed can be less than a single nanosecond while ROM chip access speed in current truck ECMs falls into the 50 to 250 nanosecond range.
- Data is represented electronically in computer systems in the form of bits and bytes.
- A bit is the smallest unit of data that can be handled by a computer: it is simply an *on* or *off* condition used to represent digital values. There are 8 bits to a byte.
- A byte is capable of representing 256 data possibilities.
- The four-part processing cycle consists of fetching, decoding, executing, and storing.
- Serial ports transmit data 1 bit at a time; transmission is therefore relatively slow.
- Parallel ports transmit data a byte at a time and consequently function at higher speed.

REVIEW QUESTIONS

1. In which of the following is data logged electronically?
   a. RAM
   b. ROM
   c. PROM
   d. EEPROM
2. Which of the following components is sometimes referred to as the brain of a computer system?
   a. CRT
   b. VDT
   c. CPU
   d. CD-ROM
3. Which scientist is credited with having invented the first mechanical digital computer?
   a. Hero
   b. Gunter
   c. Pascal
   d. Watt
4. What year was the world's first electronic digital computer introduced?
   a. 1960
   b. 1946
   c. 1642
   d. 2500 B.C.

5. Which of the following data retention categories is often referred to as main memory?
   a. RAM
   b. ROM
   c. PROM
   d. EEPROM

6. If data is read optically, how is it encoded on the storage media?
   a. Magnetically
   b. Electronically
   c. With a laser

7. Memory access speeds are usually specified using which units?
   a. Kilobytes
   b. Microseconds
   c. Gigabytes
   d. Nanoseconds

8. How many bits are there in a byte?
   a. 8
   b. 64
   c. 256
   d. 1024

9. How many different data possibilities could be contained in a single byte?
   a. 8
   b. 64
   c. 256
   d. 1024

10. The main circuit board in a computer is often referred to as a:
    a. Motherboard
    b. ROM card
    c. CPU chip
    d. BIOS

11. The word used in computer terminology to describe communication rules and regulations is:

   a. Acronyms
   b. Buffers
   c. Hypermedia
   d. Protocols

12. *Technician A* states that the keyboard on a PC station is an *input* to the computer. *Technician B* states that the printer can also be described as an *input*. Who is right?
    a. A only
    b. B only
    c. Both A and B
    d. Neither A nor B

13. Which of the following PC components would not be classified as a peripheral?
    a. Modem
    b. Printer
    c. Scanner
    d. CPU

14. *Technician A* states that RAM data is retained electronically. *Technician B* states that RAM is the main factor in defining a computer's processing speed. Who is right?
    a. A only
    b. B only
    c. Both A and B
    d. Neither A nor B

15. *Technician A* states that digital CDs are magnetically encoded with data and close contact with a magnetic field can corrupt the data. *Technician B* states that a 3½″ floppy disk retains data magnetically and close contact with a magnetic field can corrupt the data. Who is right?
    a. A only
    b. B only
    c. Both A and B
    d. Neither A nor B

Chapter 31

Networking and Communications in the Trucking Industry

PREREQUISITE

Chapters 29 and 30.

OBJECTIVES

After studying this chapter, you should be able to:

- Define the terms networking *and* telecommunications.
- Understand modern communications systems.
- Identify the hardware required to enable handshake connections between a PC and a computer network system.
- Demonstrate a basic understanding of the Internet and its capabilities.
- Describe the technology and hardware that enables a telephone to operate.
- Comprehend how North American telephone voice and data transactions take place.
- Outline the transmission media used by the telecommunications industry.
- Describe how a telecommunications satellite functions in a geosynchronous orbit.
- List some of the ways in which the trucking industry uses communications technology.
- Describe how GPS adapts trilateration geometry to geographically locate vehicles.

KEY TERMS

baud rate

cybernetics

data hub

downlink

dumb node

fiber optics

geosynchronous orbit

global positioning satellite (GPS)

handshake

Internet

Intranet

local area network (LAN)

modem

network

nodes

search engine

star network

transponder

trilateration

uplink

INTRODUCTION

This chapter introduces some of the communications technology that the diesel technician should be aware of to work effectively. Along with **cybernetics** (the science of computer control systems), today's trucking industry has widely embraced the full spectrum of data communications. It is used to download ECM system management programming from a **data hub** (focal point of a communications system sometimes known as a network star) using a **modem** (telecommunications device that translates digital signals to analog for transmission) and PC system through to the geographic tracking by satellite of fleet vehicles. A majority of today's computer systems may be networked via the **Internet** (global multimedia communications network) and various in-house, **Intranet** (an internal network using Internet protocols) systems that make communications within a company easier. This technology is introduced in this section with a reminder that its use is projected to expand rapidly in the next few years and with it, the diesel technician's requirement to understand it.

NETWORKING

A **network** in its simplest form would be a **"handshake"** connection between two computers (Figure 31–1). Handshake is the term used to describe the establishing of a communications link. More often, a network refers to a series of interconnected computers, usually PCs, for purposes of sharing data, software, and hard resources. A large office with several computer stations may network them to permit data exchange and perhaps have a single printer to serve all the stations. The means of connecting the computers in a network varies, but if it extends beyond a single building, it usually involves a modem connection into the telephone system. Cable and wireless connections are also used.

A modem is a device used for converting the digital output of the computer to the analog signals required for transmission by the phone system. The speed at which a modem can transmit data is specified by its **baud rate**. Baud rate indicates the number of times per second one bit can be transmitted. The baud rate specification is critical when considering the purchase of any equipment that must be networked. Truck engine OEMs always specify a minimum baud rate for networking computers with their mainframe systems and today usually recommend high-speed (large bandwidth) connections.

A **star network** is one that is constructed around a central computer with multiple terminals or **nodes**. Node is a term used to describe each user station. The central computer may be a mainframe. Often a star network will use **dumb nodes**, that is, stations that consist of a screen and keyboard but no processing or data storage capability independent of the central computer.

THE INTERNET

The Internet is a network of networks. It now connects most of the countries in the world and is said to be used by over one and a half billion people worldwide, a number that grows daily. It is very appropriately referred to as the world wide web or www. The following list demonstrates some of the uses of the Internet system:

1. Sending e-mail (electronic mail). This is a highly efficient means of messaging users served by the Internet. It has the advantage of nearly immediate transfer and is made easier without the protocols (and delay) of hard mail. Users must have an e-mail address and a minimal amount of computer skills, which should include some keyboarding ability.
2. File transfer. Files may be transferred from the Internet to a user PC system using a graphical user

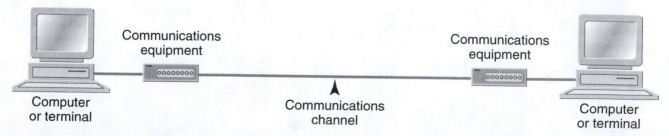

FIGURE 31–1 Basic model of a communication system.

interface (GUI) and dragging the file title from the Internet menu to the user menu.

3. News groups. Internet users may belong to a newsgroup, essentially a discussion group made up of Internet users with common interests. Messages on a newsgroup are titled and may consist of a single phrase or be pages long. Message titles may be browsed and only those of interest downloaded for reading.

4. Search databases. Most government agencies and educational institutions make extensive quantities of information available on the internet. This makes accessing information, especially detailed, subject-specific data easy, certainly much easier than attempting to locate it by phoning bureaucrats or searching a public library with its limited "hard" resources. A variety of **search engines** are used on the internet. Search engines will hunt down data on the web using a key word or phrase.

5. Play games. Many different games may be played either on-line with other Internet users or downloaded to a PC and retained.

6. Research a company. Most U.S. and Canadian companies have a web page if for no other reason than to maintain a profile on the Internet. The extent of information accessible varies with the company.

7. Go shopping. Thousands of products and services are advertised on the Internet and may be purchased using a credit card number. Although most are probably reputable, "buyer beware" should be exercised at all times.

8. Find a job. More and more companies list job vacancies on the Internet. There are search engines dedicated exclusively to job search on the Internet.

9. Receive on-line news. Magazines, newspapers, and other news services all make information available on the web and depending on the service, it may be up-to-date. Sports scores throughout the world are an example of data that is almost constantly updated. The scores are usually accompanied by stories that can be read or downloaded for printing.

10. PR (public relations). Almost every company, government agency, and organization in North America and Western Europe realizes the importance of the Internet as a communications tool. Connect to the White House, for instance, and the user may select anything from a pictorial tour to messages from the president and vice-president. A presence on the Internet is probably essential for any major company transacting business in North America.

LANS AND INTRANET SYSTEMS

Local area network (LAN) and Intranet are terms that usually describe private communication networks run by a corporation, government agency, or institution to handle the specific communications needs of their operation (Figure 31–2). Access is usually restricted with passwords required for system entry and fields within the system. The protocols used are Internet derived and while general access from the Internet may be to some extent blocked, access *to* it is usually enabled. A LAN could be housed in a single room and consist of two or three nodes or it may be spread over the country and incorporate thousands of nodes.

THE TELECOMMUNICATIONS SYSTEM

The telephone systems are responsible for both the hard (wiring/switching/decoding) and soft (microwave/

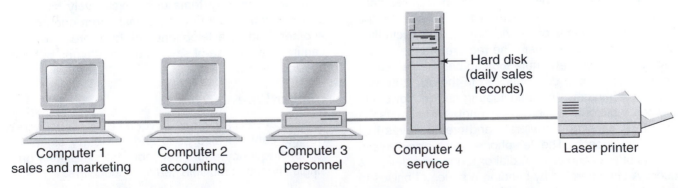

Computer 1
sales and marketing

Computer 2
accounting

Computer 3
personnel

Computer 4
service

Hard disk
(daily sales
records)

Laser printer

Local Area Network

FIGURE 31–2 *A LAN consists of multiple PCs connected to one another for purposes of sharing hardware and information.*

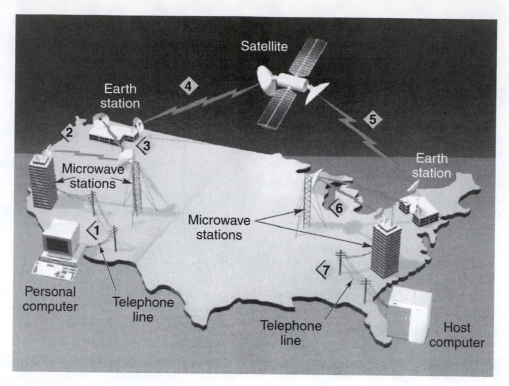

Satellite

Earth station

Microwave stations

Earth station

Microwave stations

Personal computer

Telephone line

Telephone line

Host computer

FIGURE 31–3   The use of telephone lines, microwave transmission, and a communications satellite allows a PC to communicate with a large host computer, star network, or data hub. (Courtesy of Shelly, Cashman and Waggoner)

radio waves, and so forth) linkages that enable networking (Figure 31–3). Of the telecommunications systems in use in the world today, the telephone system is probably the most used. The telephone system is continually being expanded, largely to handle the immense growth in the data communications traffic over the past decade.

A telephone consists of a transmitter, a receiver, and a dial or pushbutton mechanism. The telephone transmits voice or sound waves by having them act to vibrate a thin metal diaphragm backed by carbon granules through which current flows; when a sound wave acts to pulse the diaphragm inward, the carbon granules pack closely, permitting higher current flow. Therefore, the current flow depends on sound waves. This fluctuating current flows to a receiver. The receiver earpiece consists of an annular armature located between a permanent magnet and a coil; attached to the armature is a plastic diaphragm. When fluctuating current flows into the coil, it becomes an electromagnet. This attracts the armature, causing it to vibrate and create waves that duplicate speech. The telephone must also have a means of targeting calls. A dial or touch-tone device is used. A dial operates by sending a series of pulses to the switching network; when the number five is dialed, current from the source set is interrupted five times. The dialing creates a pattern of interruptions that are used to target the telephone signal to the correct location. With

a touch-tone device, each numeric button produces a different *tone* or frequency, which is electronically decoded to target the telephone call.

Most current telephone switching is performed digitally. Incoming signals from a telephone set are first converted into digital signals, patterns of on-off pulses. These pulses are then processed by computer to search the shortest (and least overloaded) path to connect the target telephone.

TRANSMISSION MEDIA

Early telephone systems used exclusively hard wire transmission media to carry signals from one point to another. Today, a telephone call to a foreign country can involve the use of several different types of transmission media.

Copper Wire

Most local calls are carried over pairs of copper wires. Coaxial cable is also used and can carry signals of higher frequency and allow more transmissions on a single line using multiple coaxial conductors.

Microwave

Microwave is the modulation of high-frequency radio signals. These radio signals are transmitted between

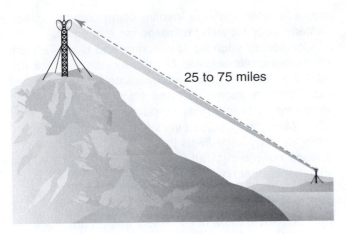

25 to 75 miles

FIGURE 31–4 *Microwave transmission.*

antennae located on towers that are within sight of one another. Microwaves travel only in straight lines and do not refract from the ionosphere layer of the atmosphere. Because they only travel short distances, multiple relays must be used. Microwaves have frequencies of many gigahertz, permitting large numbers of simultaneous transmissions on each frequency band. Micro-

wave accounts for the major portion of long distance transmissions transacted overland in North America (Figure 31–4).

Satellites

Telecommunications satellites parked in **geosynchronous orbit** (see later on in this section for a full explanation of geosynchronous orbit and telecommunications satellites) handle a large portion of transcontinental transmissions. The telecommunications satellite acts as a relay station for microwave signals receiving a signal from a ground base and retransmitting it to another Earth-based station.

Cell Phones

Each geographic region is divided into "cells," each with its own transmitter-receiver. The cells are arranged so that the frequencies used in adjacent cell locations are different although the same frequencies are reused many times over throughout the whole system. If a call is made from a cell phone in a car travelling over the boundary of one cell to another, the phone automatically switches frequencies permitting a continuous phone conversation (Figure 31–5).

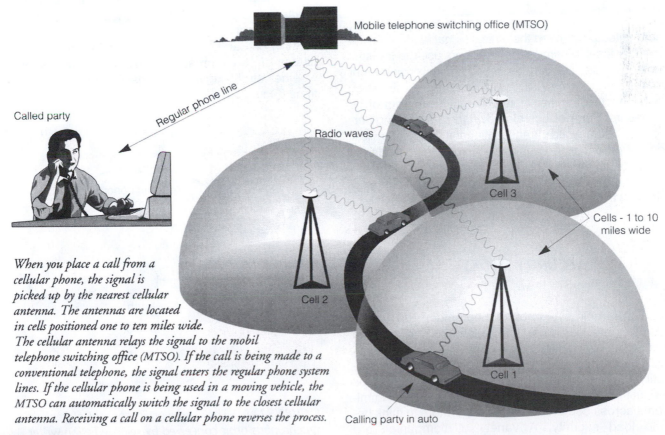

Mobile telephone switching office (MTSO)

Called party

Regular phone line

Radio waves

Cell 3

Cells - 1 to 10 miles wide

Cell 2

Cell 1

When you place a call from a cellular phone, the signal is picked up by the nearest cellular antenna. The antennas are located in cells positioned one to ten miles wide. The cellular antenna relays the signal to the mobil telephone switching office (MTSO). If the call is being made to a conventional telephone, the signal enters the regular phone system lines. If the cellular phone is being used in a moving vehicle, the MTSO can automatically switch the signal to the closest cellular antenna. Receiving a call on a cellular phone reverses the process.

Calling party in auto

FIGURE 31–5 *Cellular phone operation. (Courtesy of Shelly, Cashman and Waggoner)*

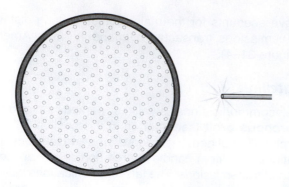

FIGURE 31–6   *The fiber optic cable (right) can transmit as much information as the 1,500-pair copper wire cable (left).*

FIBER OPTICS

Fiber optic technology is beginning to dominate the telecommunications industry because of the vast volume of transmissions it can handle. Optical fibers use light as the transmitting medium rather than radio waves. Because light has a frequency of around 100 TH$_z$ (100 trillion hertz), it has the potential to handle (theoretically) almost infinite numbers of transmissions. Telephone signals are first digitized then fed to semiconductor lasers, which produce pulses of light to a conduit of extremely thin glass fiber. At the receiving end of the optical fiber, photo detectors convert the optical signal to an electrical signal for local transmission. The fibers used to transmit the optical signals can be made almost transparent and the transmission volume can be increased simply by upgrading the transmission and receiving electronics and lasers. **Fiber optics** already handle a growing percentage of the world's transcontinental telecommunications—up-front expense is really the only thing hindering its rate of growth. In time, continental long distance services and then local systems will adopt the technology (Figure 31–6).

SECURITY ISSUES

Since the tragic events of 11 September 2001, truck operators are looking at vehicle location tracking and security in an entirely new light. Until that date, truck security usually meant theft avoidance. The realization that a truck could be used as a weapon of mass destruction or as a means of transporting illegal immigrants across borders has forced operators to think beyond load security. A variety of companies are competing for the business of addressing truck secu-

rity, and most of these involve using communication systems coupled with fast response strategies.

Companies such as Qualcomm and Corp Ten are marketing wireless panic buttons for use by truck drivers. An alarm is transmitted by a satellite-based, asset management, monitoring and tracking company and made available to clients on password protected Web sites. More sophisticated vehicle tracking technology is being used by large fleets that can afford telecommunications networking, but this use is not that new. The major operators have asset protection systems in place because they cannot afford not to have it. Cheaper satellite communications technology-based options are now beginning to be used. These use an automatically actuated, panic button signal to a telecommunications satellite when a vehicle deviates from a preprogrammed, dispatch route.

SATELLITE COMMUNICATIONS TECHNOLOGY

A satellite is by definition, a small body that orbits a larger astronomical object; by this definition, the Earth's moon is a natural satellite. The first artificial satellite was launched into orbit in 1957 by the Soviet Union. It was called *Sputnik I*. Today, thousands of satellites orbit the Earth; all but one are artificial, that is, manufactured on Earth. These satellites orbit the planet at altitudes of 160 kilometers or higher and are used for weather study, spying, navigation, and communications (Figure 31–7).

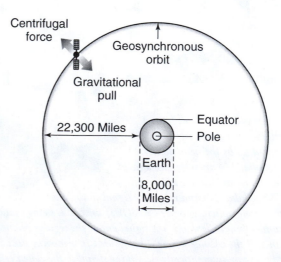

FIGURE 31–7   *The Clarke orbit: concept of geosynchronous orbit conceived by science fiction writer Arthur C. Clarke (2001: A Space Odyssey) in 1945.*

A communications satellite is often placed in geo-synchronous orbit. The time that it takes for a satellite to orbit the Earth once correlates exactly to its altitude. A satellite parked exactly 22,300 miles (35,900 km) above the Earth's surface on the equator will be positioned at a fixed point above the Earth's surface throughout the rotation. Such a satellite is said to be in geostationary orbit. If its altitude were to increase slightly, its orbit would cause it to fall behind the fixed point—if it descended below the geosynchronous altitude, the reverse would occur. The first telecommunications satellite was launched by the United States in 1961; it was named *Telstar*, and it became the subject of a popular hit musical recording of the day. To maintain a geosynchronous orbit, satellites must have on-board rockets and a fuel supply, usually hydrazine, to adjust the orbit periodically. The power supply required to run the on-board electronics of a communications satellite is obtained from solar cells or nuclear-fueled thermoelectric generators. On-board computers control the satellite's operation and orbit and manage the filtering, storing, and relaying of signals from and to ground stations (Figure 31–8, Figure 31–9, and Figure 31–10).

To avoid signal interference, telecommunications satellites in geosynchronous orbit are parked 2 degrees apart from each other on the Earth's equator.

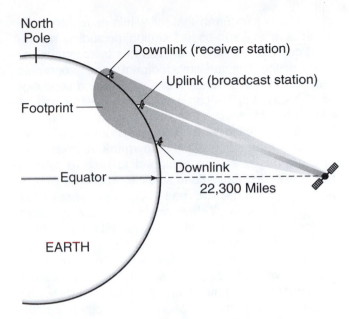

FIGURE 31–9   *Operating principle of a communications satellite in geosynchronous orbit.*

Therefore, there is limited parking space. The life of such a satellite is often governed by the fuel supply required to run the rockets that maintain the park position. The satellite receives ground signals from **up-links** (ground-to-satellite communications hardware

FIGURE 31–8   *Communications satellite. (Courtesy of Shelly, Cashman and Waggoner)*

FIGURE 31–10   *Earth stations use large dish-shaped antennae to communicate with satellites.*

and software) on Earth that fall within its footprint. An uplink is any Earth-based station (including mobile units) capable of relaying signals to a telecommunications satellite. The satellite's footprint is the geographic area on Earth over which it can receive and send signals. There are 32 telecommunications satellites parked with footprints directed on the American continents. Typically, a satellite receives a signal from an uplink, filters it, and rebroadcasts it to a **downlink**. A downlink is an Earth-based station equipped to receive signals from a telecommunications satellite. In such a way, a radio signal may be relayed through several satellites and ground stations. Television networks are big users of telecommunications satellites and will be until there is a fiber optics network capable of replacing it. For instance, a baseball game televised in Seattle may be watched live in Montreal; the signal is relayed through a number of telecommunications satellites, uplinks, and downlinks. Telecommunications technology has been used by the trucking industry since the late 1980s. Satellite communications technology has been used mainly by major fleets.

The television networks and cable network providers are the major consumers of telecommunications satellite air time. Consequently, air time is most expensive during prime time television hours and least expensive in the early hours of the morning. When a comprehensive fiber optics network is in place on the North American continent this will change; television networks will abandon satellite communications as will the telephone companies. Then the cost of air time for satellite communications will drop and satellite communications technology will become affordable for more than just the largest operators.

Satellite communications technology tracks a vehicle by placing an uplink, sometimes called a **transponder**, on the mirror brackets or cab roof of a truck. The uplink beams a signal to a telecommunications satellite, which may then relay it across the continent if necessary, targeting a downlink at a data hub. Any data that can be transmitted digitally can be sent in this manner.

VEHICLE NAVIGATION

Vehicle navigation systems vary in complexity (and cost) but a comprehensive system consists of:

• Geographic positioning
• Target destination
• Route mapping
• Route guidance
• Map display

The detail that the navigation system offers in each of the foregoing categories depends on the stored data (usually on CD) the system provider offers. Real detail (such as street addresses) tends to be pretty expensive because of the cost of collating data, but maybe it is not so expensive when compared with the cost of a lost truck driver navigating Manhattan in the rush hour. **Global positioning satellites (GPS)** are the key to vehicle navigation technology.

TRILATERATION

To understand how GPS functions, you have to understand a simple geometric principle called **trilateration**. Suppose you are completely lost somewhere in America, you pull into a truck stop to ask someone where you are, and the answer is that they are not sure, but they do know for sure they are exactly 490 miles from Cleveland. At this point you can get a map of the United States, find Cleveland, and draw a circle with a radius on a scale that equals 490 miles. You ask the next trucker, and you get the answer that he does not know exactly, but he has come directly from Houston and has traveled 710 miles. Now you can go back to the map and draw a circle with a radius representing 710 miles around Houston. At this point you still cannot say exactly where you are until you approach a third trucker and ask him the same question. He informs you that he is exactly 520 miles out of Jacksonville. Again, you draw a circle with a radius representing 520 miles around Jacksonville. Now if you look at your map (Figure 31–11), you will see that the three circles you have drawn all intersect at one point and you can determine your exact position.

FIGURE 31–11   *Position location by trilateration.*

GPS LOCATION

If the world were flat, trilateration could be used to determine position. For the principle to work in three-dimensional space, you have to use spheres instead of circles, so in order to pinpoint exact position, four GPS satellites are required. If the GPS receiver can only locate three satellites, then its software creates a *virtual* sphere, a dummy sphere that allows it to use the coordinates established by the three satellites it has located. This means that your position can be calculated but not your altitude.

Measuring Distance

GPS satellites broadcast radio signals that a GPS receiver can detect. The receiver then calculates how long the signal has taken to travel from the satellite to the receiver. For a GPS system to function, 24 satellites are required so that there is a minimum of four on the horizon at any time or location, during the day. The signals received tend to be weak, so this technology works better driving through the countryside than it does in the middle of a large city.

Mapping

The raw positional data produced by the GPS receiver unit on your car or truck can next be connected to the geographic information retained in the memory of the GPS ECM. In a top-of-the-line system, the navigational display may produce color map readouts and provide detail on street addresses. The retained data may be on CD-ROM, but the system works best with a telecommunications link found on some trucks that can provide live data on such things as traffic congestion and road construction.

Navigational Aids to GPS

Depending on the sophistication of the system, the effectiveness of a GPS unit can be improved by inputting data from the chassis electronics (road speed) and dedicated sensors such as a gyrometer to signal change of direction data. A top-of-the-line system uses Doppler effect (frequency shift analysis) to constantly monitor directional and vehicle speed data.

SUMMARY

- Computers drive all communication systems today.
- The connection of a computer either to another single computer or to a network of computers can be referred to as a handshake.
- A network is a series of interconnected computers for purposes of data sharing, communication, and hardware sharing.
- The Internet is a network of networks with millions of users spread all over the world. The Internet is used by private individuals, government agencies, and corporations.
- The telephone system is responsible for both the hard and soft links that enable networking.
- Most local telephone system transactions use pairs of copper wires. Most long distance telecommunications traffic currently uses microwave transmission with multiple relays.
- Fiber optics technology is positioned to replace both the local and long distance transmission media. It already handles a growing percentage of the world's transcontinental telecommunications transmissions.
- Most telecommunications satellites are parked in a geosynchronous orbit.
- Telecommunications satellites generally use microwave transmissions.
- An uplink is an Earth base that broadcasts a signal to a satellite. A downlink is an Earth base that receives a satellite signal.
- The uplink/downlink device on a truck is usually called a transponder.
- GPS technology is becoming widely used in the trucking industry.
- GPS works by adapting trilateration geometry to four spheres.

REVIEW QUESTIONS

1. Which of the following could be described as being a network of networks?
   a. The Internet
   b. Intranet
   c. E-mail
   d. LAN
2. Which of the following components is used to condition the output of a computer for transmission through the telephone system?
   a. CPU
   b. Transponder
   c. CRT
   d. Modem
3. What is the transmission medium used by the telephone companies for local telephone service?
   a. Microwave
   b. Copper wire
   c. Radio waves
   d. Telecommunications satellites
4. Which of the following properly describes the character of fiber optics transmissions?
   a. Analog
   b. Pulses of light
   c. Digital
   d. Nodes and antinodes

5. Which altitude from the Earth's surface would describe a geosynchronous orbit?
   a. 160 miles
   b. 22,300 miles
   c. 35,900 miles
   d. 56,700 miles

6. The first telecommunications satellite was called:
   a. *Sputnik I*
   b. *Apollo XI*
   c. *Telstar*
   d. *Discovery*

7. The Earth base station responsible for broadcasting microwave signals to a telecommunications satellite is called a(n):
   a. Modem
   b. Uplink
   c. Downlink
   d. Footprint

8. The uplink/downlink component used on trucks equipped with GPS technology is often called a(n):
   a. Aerial
   b. Transponder
   c. Modem
   d. Footprint

9. *Technician A* states that a telecommunications satellite orbits the Earth at a distance of 35,900 miles. *Technician B* states that all telecommunications satellites use fuel to maintain a fixed orbit on the Earth's equator. Who is right?
   a. A only
   b. B only
   c. Both A and B
   d. Neither A nor B

10. *Technician A* states that telecommunications satellites in geosynchronous orbit are required to park at about 2 degrees apart from one another. *Technician B* states that the geographic zone over which a satellite's signals are effective is known as its footprint. Who is right?
    a. A only
    b. B only
    c. Both A and B
    d. Neither A nor B

11. *Technician A* states that the telephone system can only handle analog transactions. *Technician B* states that most telecommunications systems in North America rely on hard wire mediums for at least some of the transmission. Who is right?
    a. A only
    b. B only
    c. Both A and B
    d. Neither A nor B

12. *Technician A* states that a LAN is sometimes known as a network of networks. *Technician B* states that the Internet is owned by the Microsoft Corporation. Who is right?

   a. A only
   b. B only
   c. Both A and B
   d. Neither A nor B

13. What is a transponder?
    a. Uplink/downlink
    b. Satellite dish
    c. Footprint
    d. CB antenna

14. Which of the following means the same thing as network star?
    a. The Internet
    b. The Intranet
    c. Remote node
    d. Data hub

15. Which of the following broadcast mediums relies on modulating of high-frequency radio waves?
    a. Hard wire telephone system
    b. Microwave
    c. Fiber optics

16. The software tool capable of finding information on the Internet and on databases is known as:
    a. A search engine
    b. An operating system
    c. A mouse
    d. A scanner

17. *Technician A* states that microwaves travel in geometrically straight lines. *Technician B* states that the point-to-point communications distance of microwaves is indefinite and a signal should be able to travel halfway around the world. Who is right?
    a. A only
    b. B only
    c. Both A and B
    d. Neither A nor B

18. Which of the following mediums should be able to transmit the most data given a similar radial diameter of each?
    a. Aluminum wire
    b. Copper wire
    c. Coaxial cable
    d. Fiber optic cable

19. What is the minimum number of GPS satellites a vehicle GPS system must locate to identify both vehicle position and altitude?
    a. One
    b. Two
    c. Three
    d. Four

20. To calculate a specific position on a map using trilateration geometry, what is the minimum number of coordinates required?
    a. One
    b. Two
    c. Three
    d. Four

Chapter 32

Vehicle Computer
Systems

PREREQUISITES

Chapters 29 and 30

OBJECTIVES

After studying this chapter, you should be able to:

- Understand the language of computerized truck engine management systems.
- Describe the circuit layout of an electronically managed truck engine.
- Identify the differences between partial authority and full authority electronic engine management.
- Outline the stages of a computer processing cycle.
- Describe the data retention media used in vehicle ECMs.
- Describe the role played by the various memory components in a truck ECM.
- Identify the command and monitoring input circuits on a vehicle electronic system.
- Define the principles of operation of thermistors, variable capacitance sensors, Hall effect sensors, potentiometers, induction pulse generators, and piezotresistive sensors.
- Describe how an ECM processes inputs and uses programmed data to generate outputs.
- Identify current computer-controlled engines by OEM and engine series.
- Define the role played by the injector driver unit in a typical full authority engine management system.
- Differentiate between customer and proprietary data reprogramming.
- Describe the processes used to reprogram a truck engine ECM with proprietary data.

KEY TERMS

actuator	foreground computations	SAE J standards
algorithm	fuzzy logic	SAE J1587
background computations	hard parameter	SAE J1708
central processing unit (CPU)	injector driver	SAE J1939
chopper wheel	microprocessor	sampling
cybernetics	multiplexing	soft parameter
download	parameter	strategy
electronically erasable, programmable read-only memory (EEPROM)	potentiometer	threshold value
	PROM	tone wheel
engine/electronic control module (ECM)	pulse wheel	upload
	RAM	
engine/electronic control unit (ECU)	reprogram	
	ROM	

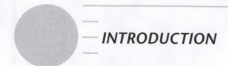

INTRODUCTION

Most highway truck, and increasingly, bus and off-highway equipment, use computers to manage engine and other on-board system functions. They are cybernetically controlled. **Cybernetics** is the science of computer control systems. The technician is required to troubleshoot system problems and reprogram **parameters** (values) using either generic reader-programmers or personal computers (PCs). On-board vehicle computers are referred to as **engine/electronic control modules (ECM)** or **electronic/engine control units (ECU)**. ECM is the acronym used in this text except when discussing a product whose OEM chooses to use the acronym ECU.

ECMs normally contain a **microprocessor**, data retention media, and often, the output or switching apparatus. The ECM can be mounted on the component to be managed (an engine management ECM is often located on the engine) or alternatively, inside the vehicle cab. Increasingly as this technology develops, several system ECMs can be mastered by a single vehicle management ECM. Some OEMs use the term **multiplexing** to refer to a vehicle management system using multiple, interconnected ECMs on a data bus. A more detailed introduction to multiplexing appears in Chapter 35. A truck and bus technician is required to have a basic understanding of both vehicle and personal computers to interact effectively with today's technology. This chapter introduces the essentials of electronic management of vehicle systems (Figure 32–1). The terminology introduced in Chapter 30 is used extensively in this chapter. Computerized system management can be summarized as a set of electronically connected components that enable an information-processing cycle comprising three distinct stages:

1. Data input
2. Data processing
3. Outputs

DATA INPUT

Data is simply raw information. Most of the data to be inputted to a truck diesel engine ECM comes from monitoring sensors, such as rpm signals, and command sensors, such as the throttle position sensor. Most of this data is in analog format, such as voltage values, and it must be converted to a digital format or digitized to enable processing by the ECM. Figure 32–2 demonstrates some typical input circuit devices.

DATA PROCESSING

A **central processing unit (CPU)** contains a control unit that executes program instructions and an arithmetic logic unit (ALU) to perform numeric calculations and logic processing such as comparing data. Random-access memory **(RAM)** is data that is electronically retained in the ECM—this data can be manipulated by the CPU. Input data and magnetically retained data in **ROM** (read-only memory), **PROM** (programmable read-only memory), and EEPROM are transferred to RAM for processing. RAM can be called primary storage or main memory. Because RAM data is electronically retained, it is lost when its circuit is opened. Figure 32–3 is a simplified schematic of an ECM showing some of its basic functions.

OUTPUTS

The results of processing operations must be converted to action by switching units and **actuators**. In most (but not all) truck/bus management ECMs, the switching units are integral with the ECM. In an ECM managing a full authority diesel engine, the **injector driver** unit would be one of the primary outputs to be switched. ECM commands would be converted to an electrical signal that would determine the pulse width or duty cycle of an electronic fuel injector (EUI) and define its effective pumping stroke. Note the output drivers in Figure 32–2 and Figure 32–3.

SAE HARDWARE AND SOFTWARE PROTOCOLS

Among the OEMs of truck and bus engines in the United States, there has been a generally higher degree of cooperation in establishing shared electronics hardware and software protocols than in the automobile manufacturing segment of the industry. To some extent, this cooperation has been orchestrated by the SAE and ATA, but it is a cooperation that has been necessary due to the fact that a Caterpillar engine could be specified to any truck chassis and have to electronically interact with a Fuller transmission and Wabco ABS/ATC (antilock brake system/automatic traction control). Until recently, separate **SAE J standards** (surface vehicle recommended practice) controlled the hardware and software protocols that enabled this type of data exchange using a chassis data bus. Truck multiplexing protocols are covered by the following J standards:

- **SAE J1587.** First generation, multiplexing data exchange protocols used in data exchange between heavy-duty, electronically managed systems.
- **SAE J1708.** First generation, multiplexing hardware compatibility protocols between microcomputer systems in heavy vehicles.

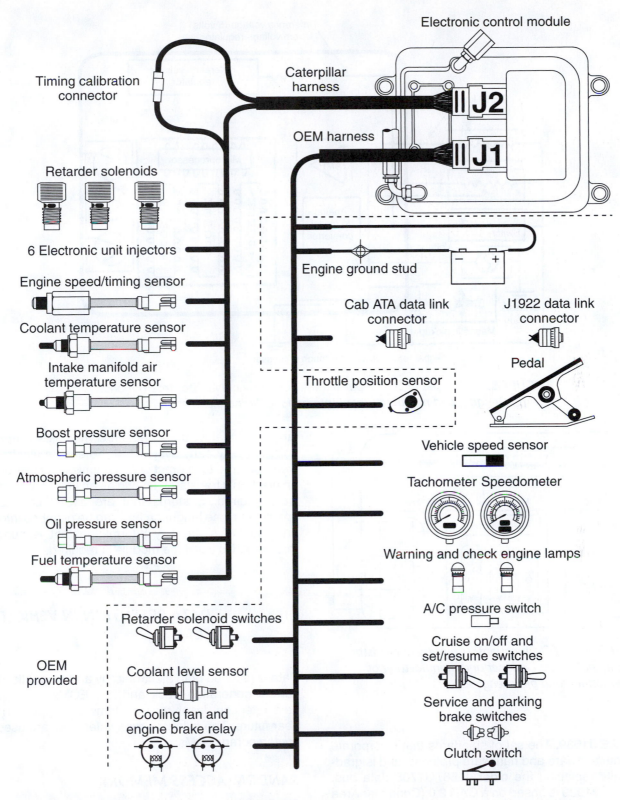

FIGURE 32–1 *Schematic of a typical full authority, diesel engine management system. (Courtesy of Caterpillar)*

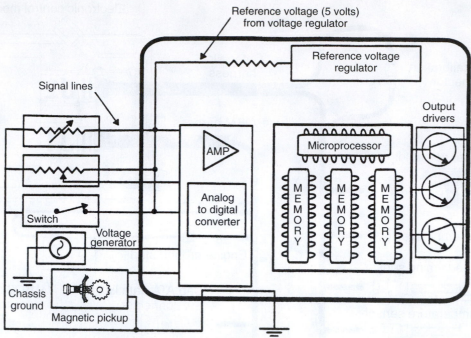

Reference voltage (5 volts)
from voltage regulator

Reference voltage
regulator

Signal lines

Output
drivers

AMP

Microprocessor

Analog
to digital
converter

MEMORY   MEMORY   MEMORY

Switch

Voltage
generator

Chassis
ground

Magnetic pickup

Signal return provided through processor

FIGURE 32–2   *Types of ECM input signals. (Courtesy of Navistar International Corp., Designer and manufacturer of International Brand diesel engines)*

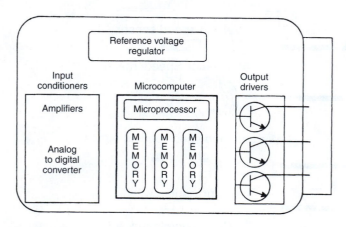

Reference voltage
regulator

Input
conditioners   Microcomputer   Output
drivers

Amplifiers   Microprocessor

Analog
to digital
converter   MEMORY   MEMORY   MEMORY

FIGURE 32–3   *ECM functions. (Courtesy of Navistar International Corp., Designer and manufacturer of International Brand diesel engines)*

• **SAE J1939.** The set of standards that incorporate both software and hardware protocols and is gradually replacing the older J1587/J1708 data bus. SAE J1939 is based on a CAN 2.0 (Controller Area Network generation 2.0) platform that is the subject matter of Chapter 35.

Data Exchange

For data exchange to take place between two separate electronic control systems, they must "speak" the same

language, that is, use common communication protocols. In many early systems, if the engine electronics were required to interact with transmission electronics, an electronic translator, known as an interface module, was required. Systems that are J1939 compatible speak the same language and can be plumbed into the data bus to share any information that is pumped through it. No interface module is required.

DATA RETENTION IN VEHICLE ECMs

Data is retained both electronically and magnetically in current generation truck and bus ECMs, but optical data retention, laser read systems will appear in the near future. The following data categories are used by contemporary ECMs:

RANDOM-ACCESS MEMORY

The amount of data that may be retained in RAM is a primary factor in quantifying the computing power of a system. It is also known as main memory because the CPU can only manipulate data when it is retained electronically. At startup, RAM is electronically loaded with the vehicle management operating system instructions and all necessary running data retained in other data categories (ROM/PROM/EEPROM) that the CPU must access at high speed.

RAM data is electronically retained, which means that it is always to some extent volatile; that is, data storage could be described as temporary—if the circuit is opened, RAM data is lost. Most truck ECMs use only fully volatile RAM; in other words, when the ignition circuit is opened, all RAM data is dumped.

Data such as coolant temperature, oil pressure, and boost pressure are also logged into RAM and, providing the values are within a threshold window defined in ROM/PROM/EEPROM, they only have significance at that specific moment; they can therefore be safely discarded when the ignition key is opened. This category of input data is continually monitored by a process known as **sampling**. Most RAM in truck engine management systems is volatile RAM, meaning that nothing is retained when the ignition switch is opened. However, a second type of RAM is used in some truck and many automobile systems, which have no EEPROM capability. This is nonvolatile RAM (NV-RAM) (or KAM—keep-alive memory) in which data is retained until either the battery is disconnected or the ECM is reset, usually by depressing a computer reset button that temporarily opens the circuit. Codes and failure **strategy** (action sequence) would be written to NV-RAM and retained until reset.

READ-ONLY MEMORY

ROM data is magnetically retained and is designed not to be overwritten; it is "permanent" though it can be corrupted (rare), and it is susceptible to damage when exposed to powerful magnetic fields. Low-level radiation, such as that encountered routinely driving on a highway from police radar and high-tension electrical wiring, will not affect any current ECMs. Most of the data retained in the ECM is logged in ROM. The master program for the system management is loaded into ROM. Production standardization is permitted by constructing ROM architecture so that it is written with common requisite data for a number of different systems. For example, identical ROM chips can be manufactured to run a group of different engines in a series—but to actually make an engine run in a specific chassis application, the ROM data would require further qualification from data loaded into PROM and EEPROM.

ROM contains all the protocols (rules and regulations) to master engine (or the system) management including all the hard **threshold values** (limits). The term **hard** is often used to describe **parameters**/values that are interpreted by the ECM in rigid terms. For instance, the temperature at which the ECM is programmed to identify an engine overheat condition is always a fixed value: at this specific value, a code and

failure strategy are effected. Engine overspeed would be another example of a hard parameter. A **soft parameter**/value is one that would be interpreted by the ECM with a cushion around the value, according to the input of sensors monitoring the engine and the conditions it is running under. When an engine or system is run under open loop conditions, and most current truck diesel engine are, the processing path to produce outcomes is sometimes known as **fuzzy logic**.

PROGRAMMABLE ROM

PROM is magnetically retained data—usually a chip, set of chips, or card socketed into the ECM motherboard. PROM can sometimes be removed and replaced. PROM function is to qualify ROM to a specific chassis application. In the earliest truck engine management systems, programming options, such as idle shutdown time, could only be altered by replacing the PROM chip; customer programmable options are written to EEPROM in current systems where they can be easily altered. Some OEMs describe the PROM chip as a "personality module," an appropriate description of its actual function of trimming or fine tuning the ROM data to a specific application.

ELECTRONICALLY ERASABLE PROM

The **electronically erasable programmable read-only memory (EEPROM)** data category contains customer data programmable options and proprietary data that can be altered and modified using a variety of tools ranging from a generic reader programmer to a mainframe computer. It is usually magnetically retained in current systems. A generic reader/programmer, such as Pro-Link 9000 equipped with the correct cartridge, can often be used to rewrite customer options such as tire rolling radius, governor type (LS or VS), cruise control limits, road speed limit, and others. Only the owner password is required to access the ECM and make any required changes.

Proprietary data is more complex both in character and methods of altering—it would contain data, such as the fuel map, in an engine system. The procedure here normally requires accessing a centrally located mainframe computer (usually at the OEM headquarters location) via a modem and the phone system, **downloading** the appropriate files to a PC and subsequently to a diskette and finally **reprogramming** (altering or rewriting the original data) the ECM from the diskette. Some proprietary reader/programmers can act as the interface link between the vehicle ECM and the mainframe, but the procedure is essentially the same. Caterpillar incorporates PROM and EEPROM on a single card and describes it as a personality module. This was replaceable in early ECMs but not in current versions.

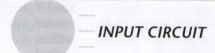

INPUT CIRCUIT

ECM inputs can be divided into sensor inputs and switched inputs. They are covered in this chapter in some detail because there is a high degree of input component commonality among the OEMs. These inputs can also be further described by grouping them into system monitoring and system command inputs.

SENSOR CIRCUITS

Anything that signals input data to a computer system can be described as a sensor. Sensors may be simple switches that an operator toggles open or closed to ground a reference voltage, modulate a reference voltage, or be powered up either by V-Ref or require power-up outside of the V-Ref circuit.

Sensors Using a Reference Voltage to Produce an Analog or Digital Signal

These sensors receive a constant reference voltage (usually 5 V) and the ECM compares the return signals with values logged in its data retention banks. There are four types:

1. Thermistors. **Thermistors** precisely measure temperature. Resistance through a thermistor either decreases as temperature increases (NTC—negative temperature coefficient) or vice versa (PTC—positive temperature coefficient). The ECM receives temperature data from thermistors in the form of analog voltage values. The coolant temperature sensor, ambient temperature sensor, and oil temperature sensor are usually thermistors. These temperature sensors are all fairly critical inputs to the ECM when factoring injection timing and injection fuel quantity data. Figure 32–4 is a schematic of a thermistor.

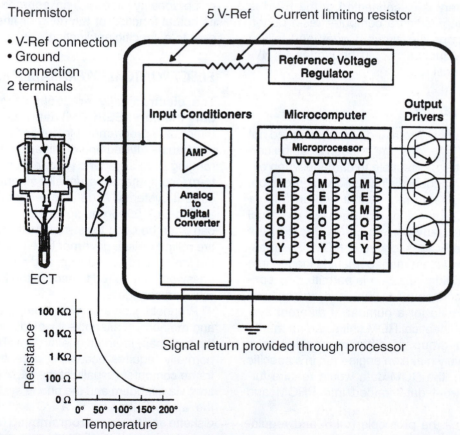

The chart indicates resistance of a thermistor decreases as temperature increases. Output of thermistor is not linear.

Thermistor Engine Coolant Temperature (ECT)

FIGURE 32–4   An engine coolant temperature sensor of the NTC thermistor type. (Courtesy of Navistar International Corp., Designer and manufacturer of International Brand diesel engines)

2. Variable capacitance (pressure) sensor. These are supplied with reference voltage and usually designed to measure pressure or linear position values. The medium whose pressure is to be measured acts on a ceramic disc and moves it either closer or farther away from a steel disc; this varies the capacitance of the device and thus the voltage value returned to the ECM.

Variable capacitance-type sensors are used for oil pressure sensing, MAP sensing (turbo-boost pressure), BARO (barometric pressure sensing), and fuel pressure sensing. Figure 32–5 shows an engine oil pressure sensor that uses a variable capacitance electrical operating principle. A variable capacitance principle has been used in the past by one OEM in a TPS.

3. Potentiometers. The **potentiometer** is a three-wire (V-Ref, ground, and signal) variable resistor. Once again, these receive a reference voltage and output a signal proportional to the motion of a mechanical device. They can be referred to as a voltage divider. As the mechanical device moves, the resistance is altered within the potentiometer. Throttle position sensors are commonly potentiometers. Figure 32–6 shows a TPS that uses a potentiometer principle of operation.

4. Piezoresistive pressure sensor. This type of pressure sensor is often used to measure manifold pressure, especially where greater accuracy is required. In trucks it is used as a manifold boost sensor and it is sometimes referred to as a Wheatstone bridge sensor. A doped silicon chip is formed in diaphragm shape so that it measures 250 microns around the outside and reduces to about 25 microns at the center: this means that it forms a sort of drumskin that is 10 times thinner in its center. This permits the diaphragm to flex at the center when subjected to pressure. A set of sensing resistors is formed around the edge of a vacuum chamber over which the diaphragm is stretched. When subjected to pressure, the diaphragm is deflected, causing the resistance of the sensing resistors to change in proportion to the increase in pressure.

An electrical signal proportional to pressure is produced by connecting the sensing resistors into a Wheatstone bridge circuit in which V-Ref is used to supply a constant DC voltage value across the bridge. When no pressure acts on the silicon diaphragm, all the sensing resistance will be equal and the bridge can be said to be balanced. When pressure causes the silicon diaphragm to deflect, resistance across the sensing resistors increases,

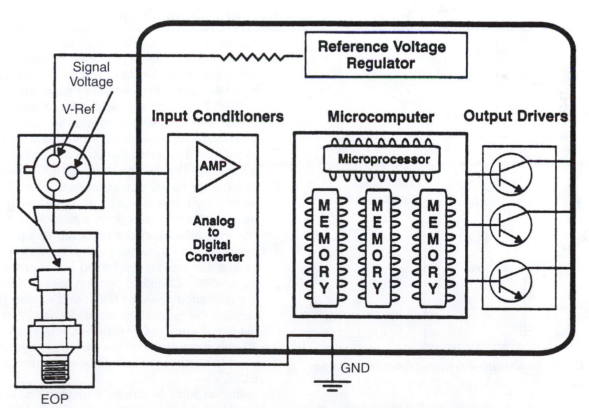

Variable Capacitance Sensor (Engine Oil Pressure Sensor)

FIGURE 32–5 *An engine oil pressure sensor of the variable capacitance type. (Courtesy of Navistar International Corp., Designer and manufacturer of International Brand diesel engines)*

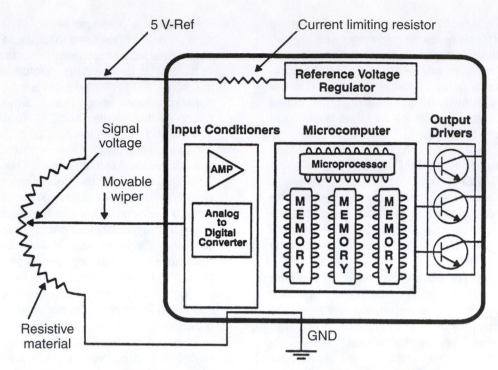

5 V-Ref

Current limiting resistor

Reference Voltage Regulator

Signal voltage

Input Conditioners   **Microcomputer**

Output Drivers

AMP

Microprocessor

Movable wiper

Analog to Digital Converter

MEMORY   **MEMORY**   **MEMORY**

Resistive material

GND

Potentiometer (Variable Resistance Voltage Divider)

FIGURE 32–6   An accelerator position sensor of the potentiometer type. (Courtesy of Navistar International Corp., Designer and manufacturer of International Brand diesel engines)

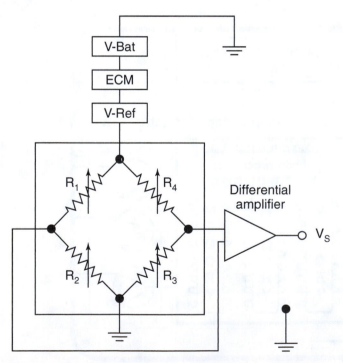

V-Bat

ECM

V-Ref

R_1 R_4

Differential amplifier

V_S

R_2 R_3

FIGURE 32–7   Piezoresistive sensor: Wheatstone bridge circuit.

unbalancing the bridge, creating a net voltage differential that can be relayed to the ECM as a signal. Figure 32–7 shows the operating principle of a typical piezoresistive sensor.

Signal-Generating Sensors

1. Hall effect sensors. Hall effect sensors generate a digital signal as timing windows or vanes on a rotating disc pass through a magnetic field. The disc is known as a **pulse wheel** or **tone wheel**, terms that are both used to describe the rotating member of the induction pulse generator, so care should be taken to avoid confusion. The frequency and width of the signal provides the ECM with speed and position data. The disc incorporates a narrow window or vane for relaying position data. The Hall effect sensor outputs a digital signal by blocking a magnetic field with a vane from a semiconductor sensor. Hall effect sensors are used to input engine position data for purposes of event timing computations such as the beginning and duration of the pulse width. Camshaft position sensor (CPS), timing reference sensor (TRS), and engine position sensors (EPS) are examples. Figure 32–8 shows the signal output of a Hall effect sensor and Figure 32–9 is a schematic of the Hall effect sensor when used to signal camshaft position information to an ECM.

2. Induction pulse generator. A toothed disc known as a reluctor, pulse wheel, or tone wheel (slang but commonly used: **chopper wheel**) with evenly spaced teeth or serrations is rotated through the magnetic field of a permanent stationary magnet.

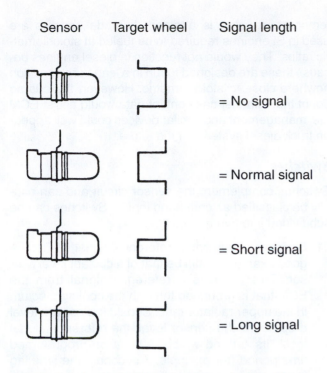

Sensor   Target wheel   Signal length

= No signal

= Normal signal

= Short signal

= Long signal

FIGURE 32–8 *Hall effect operating principle. (Courtesy of Navistar International Corp., Designer and manufacturer of International Brand diesel engines)*

As the field builds and collapses, AC voltage pulses are generated and relayed to the ECM. Reluctor-type sensors are used variously on modern truck chassis in applications such as ABS wheel speed sensors, vehicle speed sensors (VSS) located in the transmission tailshaft, or front wheel and engine speed sensors. Figure 32–10 shows the operating principle of an induction pulse generator.

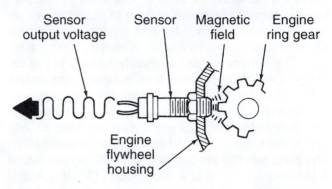

Sensor output voltage   Sensor   Magnetic field   Engine ring gear

Engine flywheel housing

FIGURE 32–10 *Engine rpm sensor of the induction pulse generator type. (Courtesy of Navistar International Corp., Designer and manufacturer of International Brand diesel engines)*

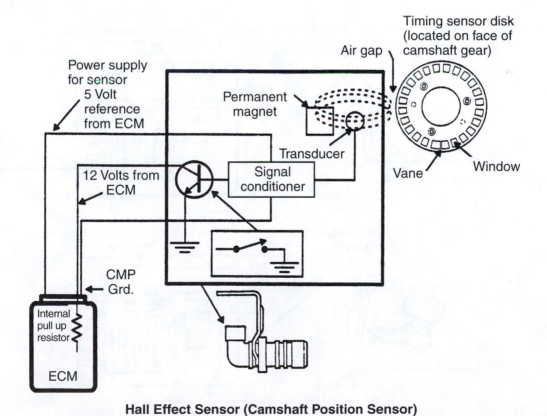

Timing sensor disk (located on face of camshaft gear)

Air gap

Power supply for sensor 5 Volt reference from ECM

Permanent magnet

Transducer

12 Volts from ECM

Signal conditioner

Vane

Window

CMP Grd.

Internal pull up resistor

ECM

Hall Effect Sensor (Camshaft Position Sensor)

FIGURE 32–9 *Camshaft position sensor of the Hall effect type. (Courtesy of Navistar International Corp., Designer and manufacturer of International Brand diesel engines)*

3. Galvanic sensors. Although not used on current diesel engines and gasoline-fueled engines are seldom found on today's trucks, the truck technician should have rudimentary understanding of a lambda (l) sensor (also known as O_2 sensor or exhaust gas sensor).

The lambda sensor is the closed loop driver on many SI engines; that is, it is the primary engine reference monitor used to condition AFR. It is used in most engines required to be fueled at close to stoichiometric ratios. The most common lambda sensor is a galvanic device that produces an electric current by chemical action. The lambda sensor is usually located in the exhaust piping exposed to the engine exhaust gas on one side and ambient air on the other. The device is constructed of a zirconium dioxide ceramic material with gas permeable platinum electrodes exposed on one side to the exhaust gas and on the other, to ambient air. The zirconium dioxide begins to conduct oxygen ions at a temperature value of around 300°C (550°F). At this temperature, as the oxygen proportion in the exhaust gas and that in the ambient air differ, a small voltage is generated due to the electrolytic properties of zirconium dioxide. The greater the difference of the oxygen proportions, the higher the voltage produced. These voltages range between 0.1 V and 1.0 V after the trigger

temperature value is attained. Lambda sensors are used in SI engines required to be fueled at stoichiometric ratios. They would not function in diesel engines because these are designed to run in a lean burn condition nowhere close to stoichiometric. However, there is no doubt that exhaust gas content data would assist ECM fuel management and similar devices could well appear on truck diesel systems of the future.

Switches

Switches complement the sensor circuit and can usually be classified as command inputs. Switches can be subdivided into two groups:

1. Switches grounding a reference signal (V-Ref). A good example would be that of a coolant level sensor. This receives a reference signal from the ECM that is grounded through the cooling medium in the upper radiator tank. Should the coolant level drop below the sensor level, the reference signal loses its ground which, after a preprogrammed time period (for example, 8 seconds, the time lag is necessary because otherwise a more temporary loss of ground caused by braking will open the circuit), will trigger the ECM to react with whatever action it is programmed with—electronic malfunction alert or engine shutdown. Figure 32–11 and

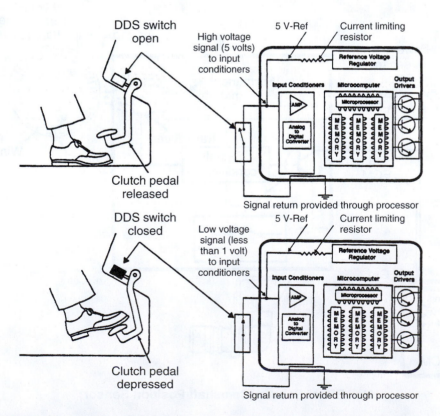

FIGURE 32–11 *Switch-type input sensors. (Courtesy of Navistar International Corp., Designer and manufacturer of International Brand diesel engines)*

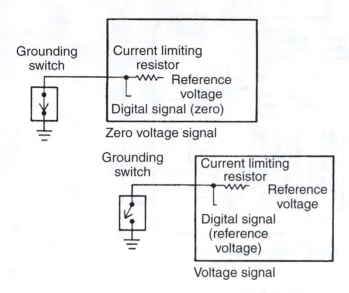

FIGURE 32–12 Grounding-type switch operation. (Courtesy of Navistar International Corp., Designer and manufacturer of International Brand diesel engines)

Figure 32–12 show examples of switches used in a typical electronically managed engine circuit.

2. Manual switches control electrical circuit activity. Many versions of this type of switch are used by the operator to control vehicle functions. Some examples would be the ignition key, engine retarder mode switches, and the cruise control switches. Switches that are controlled by the driver are sometimes called command switches.

ECMS AND THE PROCESSING CYCLE

Diesel engine management ECMs have four basic functions:

1. Regulating reference voltage (V-Ref)
2. Input conditioning, amplification, and ADC
3. Processing
4. Managing output drivers

ECM is the generic term for the unit housing the complete computer assembly and often some, or all, of the switching apparatus. The engine management (or vehicle management) ECM generally receives command and monitoring data from an electronic subcircuit that operates at an electrical pressure of 5 V or less. The ECM also manages reference voltage (V-Ref) at a specific value both to power up and to benchmark input signals. The electronic subcircuit consists primarily of sensors located variously on the engine and chassis. Because the data they input is mostly in the form of ana-

log voltage values, these must be converted to digital codes for processing by an ADC (analog to digital converter) integral with the ECM. Weak signals may be conditioned or strengthened before processing. Preparing input signals for processing are ECM functions categorized as input conditioning. Figure 32–13 outlines the role of V-Ref and signal conditioning in an ECM.

Processing in the ECM involves scanning the programmed fuel map data (in ROM, complemented or finally defined in PROM and EEPROM), engine and chassis monitoring sensors (such as engine coolant temperature [ECT], vehicle speed sensors [VSS], and so on) and command inputs (throttle position sensor [TPS], cruise control, engine brake, and so on), and subsequently plotting an actual fuel quantity to be delivered.

The results of CPU logic and arithmetic processing must next be converted to action by means of *outputs*. ECM processing may occur at different frequencies classified as **foreground** and **background computations**. Foreground operations would include response to a critical command input such as the TPS whose signal must be processed immediately to generate the appropriate outcome. The monitoring of engine oil temperature is often classified as a background operation. Although obviously valid, this signal does not require as immediate an adjustment to operating strategy as that from the accelerator. ECM outputs are switching functions. These switching functions are usually driven at system chassis voltage or in some instances, they are spiked to higher values. So the ECM manages the performance of hydromechanical fuel pumping apparatus. When the engine management ECM is bussed into other system ECMs (transmission, ABS, ATC) or there is a single vehicle management ECM managing multiple systems, input circuits decrease, and switching mechanisms tend to increase in complexity. Figure 32–14 is a simplified example of the processing cycle in a vehicle electronic system.

ECMs can be located anywhere on the vehicle chassis. Engine management ECMs are often mounted on the engine close to the devices to be switched. While engine-mounted ECMs require better shock and vibration insulation, they have certain advantages in being close to both the sensors and actuators and, where heat is a problem, can be mounted on a heat exchanger with diesel fuel acting as the cooling medium. Another option is to locate the ECM under the dash inside the vehicle cab, reducing the requirement for shock and vibration insulation.

The following are examples of some OEM acronyms used to refer to engine and fuel system controllers:

FIC—Fuel injection control module
EEC—Electronic engine control
ECI—Electronically controlled injection
ECU—Electronic control unit

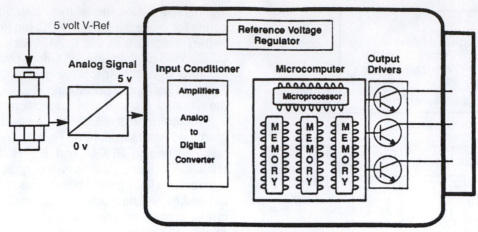

Electronic Control Module Signal Conditioning

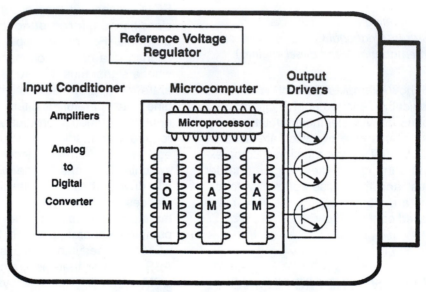

Electronic Control Module Microprocessor Memory

FIGURE 32–13   *ECM input signal conditioning and data retention. (Courtesy of Navistar International Corp., Designer and manufacturer of International Brand diesel engines)*

The term **algorithm** is used to describe the sequence of processing events in an ECM, from the arrival of a signal to the stitching of an outcome.

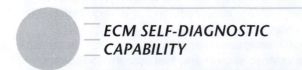

ECM SELF-DIAGNOSTIC CAPABILITY

A major benefit of the manufacturers of truck and bus electronics systems' compliance with SAE J1939 software and hardware data bus protocols is the fact that current systems can be read by most electronic service tools in use, using the SAE message identifiers (MIDs), system identifiers (SIDs), parameter identifiers (PIDs), and failure mode indicators (FMIs). ESTs that only read systems are known in the automobile service industry as *scan tools*, primarily because they have read-only capability. Currently, many of the automobile electronic management systems can also be programmed with the type of data referred to as customer data programming in the truck industry, but their ESTs still tend to be referred to as scan tools. The same tool might be referred to as a reader/programmer in the truck service facility, because truck ESTs had some programming ability from the onset. Self-diagnostic capability in truck and bus electronics systems can be grouped as on-

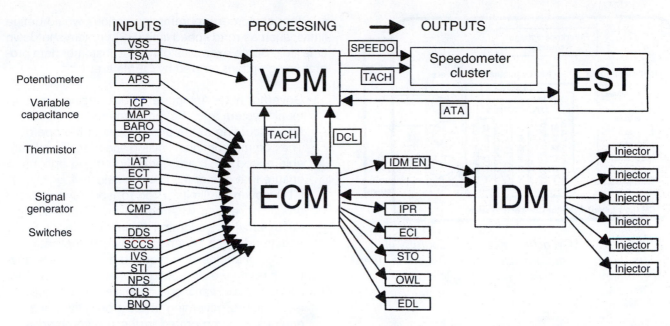

FIGURE 32–14 *The processing cycle of a International Trucks HEUI, two-module controller system. (Courtesy of Navistar International Corp., Designer and manufacturer of International Brand diesel engines)*

board in the form of blink or flash codes or digital data displays, and EST-read such as reader/programmers and PC-based systems. The use of these instruments is covered in Chapter 33 and in chapters dealing with each specific system.

TIME AND ELECTRONICALLY MANAGED ENGINES

Until the widespread use of electronics, engine technicians generally referred to timing and events within the engine cycle in terms of crank angle degrees. This tendency has shifted as the data readouts produced by the on-board electronics in many systems tend to use time as well as crank angle to report engine events such as EUI response time and duty cycle. It may help the technician to have a basic understanding of some of the time dimensions within which engine events take place especially when it comes to making sense of some the data displayed by electronic service tools. If an engine rotates at a typical idle speed of 600 rpm, it turns through 10 complete rotations in one second. When an engine rotates at 2,000 rpm, one rotation of the engine takes place in 30 milliseconds (0.030 seconds), the engine rotates through 33 revolutions in a second and the typical fueling pulse of an EUI lasts about 3 milliseconds (0.003 seconds). It should also be noted that a fuel pulse width (PW) of 3 milliseconds at 2,000 rpm results in fuel being injected for around 35 crank angle degrees, whereas a PW of 3 milliseconds at 600 rpm translates into fuel being injected for only around 10 crank angle degrees. The reason that a

small variance in a factor such as injector response time is significant is that it can dramatically impact on engine operation.

OUTPUT CIRCUIT

The switching apparatus used with each electronic management system is what truly differentiates each system from any other. Although the switching and pumping apparatus in full authority engine management systems have certain similarities regardless of OEM system, it is more appropriate to address this technology by manufacturer rather than generally in this section. Full authority engine management systems use EUIs, HEUIs, EUPs, and common rail actuators so the driver unit is obviously a primary output. However, there are significant differences from system to system requiring that each be examined separately. The output circuit always begins in the ECM. Figure 32–15 shows the relationship of the ECM switching unit or output drivers within the processing cycle.

Those systems classified as partial authority tend to have more complex switching apparatus, especially those that were designed to meet 1994 EPA emissions standards. These used an in-line, port-helix metering injection pump, and outputs included mechanisms for rack control and variable timing. Again, it is more appropriate to examine each OEM system individually, so this is done in later chapters.

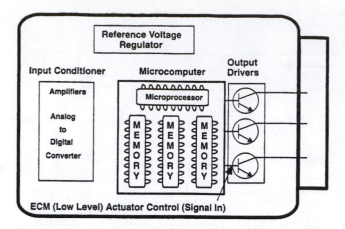

FIGURE 32–15   ECM actuator control. (Courtesy of Navistar International Corp., Designer and manufacturer of International Brand diesel engines)

ECM PROGRAMMING

Vehicle ECM programming can be divided into two general categories, customer data programming and proprietary data programming. Customer data programming generally refers to any information that may have to be changed on a routine basis. It is usually concerned with day-to-day running parameters that the vehicle owner has ownership of. Proprietary data programming is that data that the engine manufacturer has ownership of. Typical proprietary data concerns the core system management algorithms that may require alteration in the lifetime of the system or perhaps may be vulnerable to corruption.

CUSTOMER DATA PROGRAMMING

Most OEMs have designed their systems so that a wide range of running parameters can be altered to accommodate a change in chassis application and driver or owner preference. Usually all that is required to alter "customer data" is the ECM customer password. Fleets may sometimes not provide their drivers with the vehicle ECM password. Additionally, some fleets use an alpha password to either prevent or make difficult customer data programming changes with a generic reader programmer (ProLink 9000). Many customer data programming options are electronically toggled; for instance, the owner may choose variable speed (VS) governing over limiting speed (LS) governing so the VS governing option would be selected on the menu. "Toggling" means the either/or selection of a pair of options; it would not be possible, for instance, to elect to have no governing.

Other customer programming options would define limits, such as road speed limit (RSL) or idle shutdown time limit. Some examples of typical customer data programming options offered by OEMs are:

- Governor type. Any current truck engine management system can be programmed to governor type according to the preferences of the operator. The common options, limiting speed (LS) and variable speed (VS), are normally toggled options to ensure that one or the other is selected.
- PTO governing/limits. Precise isochronous (zero droop) governing was, before the electronic age, an expensive governing option required in any engine driving a generator, in which application, a change in engine load required an immediate governor response in fueling to maintain a constant speed to avoid variations in the electrical frequency. Isochronous governing in electronically managed engines is incorporated in the programmed software and the term is used by some OEMs to describe the operation of the engine when driving PTO equipment such as pumps and compressors.
- Cruise control parameters. The upper and lower limits are programmable. Actual values are programmed by the driver operating the vehicle using dash switches. Smart cruise (flexible parameter) programming available in many current systems makes cruise control capable of producing superior fuel-to-mileage figures than hard (rigid parameter) cruise control. Smart cruise programming can also "reward" a driver who improves on projected vehicle fuel economy by increasing governed road speed.
- Road speed limit. Any observer of trucks driven on our highways will note that this is a popular option with large fleets that capitalize on the fuel-to-mileage and safety benefits of having a vehicle run at a lower maximum road speed. It is a programming option that most owner-operators choose not to take advantage of.
- Critical shutdown sensors. A coolant level sensor can be programmed to a number of failure strategies if its ground (through the liquid coolant) is lost for a predetermined period. The failure strategies could vary from a simple warning alert to the driver, to ramp-down to a default rpm/maximum load, or to shutdown the engine. The ECM could be programmed not to shut the engine down in a vehicle routed through mountainous terrain.
- Tire size. Tire size must be programmed using a tire specification manual that defines a tire's rolling radius. This is usually programmed to the ECM in revolutions per mile or revolutions per kilometer. It is important to note that two nominal 22.5″ tires of different tread code ratings (even if they are manu-

factured by the same manufacturer) may have quite different dynamic rolling radii. Incorrectly programmed tire data will result in false road speed data signalled to the ECM and coincidentally, inaccurate speedometer and mileage information.

Additional examples of typical customer data programming options offered by OEMs include:

- Transmission ratio
- Carrier ratio
- Progressive shifting limits
- Idle speed
- Torque rise profile
- Peak brake power
- Idle shutdown duration

Customer data programming must fall within parameters defined in the ECM's data retention media. For instance, if the maximum high-idle speed in a given engine is defined by its management ECM as 2,000 rpm, this value cannot be reprogrammed to a higher value using the preceding methods.

TECH TIP: Customer data programming fields may be password protected to prevent unwanted rewriting of critical parameters. Unless properly authorized, avoid obliging fleet driver requests for reprogramming data in such fields as progressive shifting and road speed limit or you may lose a customer.

PROPRIETARY DATA PROGRAMMING

The technician enables proprietary data programming but does not actually perform the reprogramming. Proprietary data includes any ECM-programmed data that the OEM does not want the customer to alter. A good example would be the fuel map. There are circumstances in which this might have to be rewritten or altered, but this would be accomplished by the OEM. Also, critical ECM files can become damaged or corrupted and may require proprietary data reprogramming. Proprietary data programming normally takes place in three distinct stages: downloading, programming the ECM, and uploading verification.

Downloading

Downloading requires accessing the OEM's centrally located mainframe via a PC, modem, and appropriately driven software, or the OEM's reader/programmer instrument. The vehicle is identified electronically, usually by its VIN, and the system driver software routes itself to the correct mainframe location of the system files. The mainframe computer can be viewed as a sort of filing cabinet. The function of the driver software is to locate the system files that pertain to the specific vehicle identified by its VIN. Having located these files, they are next downloaded to the PC's electronic memory from which they can be copied to a diskette or directly to the ECM. This operation does not require that the vehicle be electronically connected to the PC; in fact, it could be out on the highway at the time of the transaction.

Programming the ECM

The system files that were copied to the diskette can now be transferred to the system ECM. A portable PC (a laptop computer would be the tool of choice) can be connected to the vehicle ECM by means of the ATA data connector, six or nine pin Deutsch, and adhereing to some menu-driven sequencing, the downloaded files can be transferred to the ECM. This procedure, if successful, will generate a verification file that will be created on the diskette from which the reprogramming files were copied.

Before reprogramming, certain requirements usually must be met. For instance, the vehicle ignition circuit must usually be closed with the engine stationary. All fault codes would normally have to be cleared before reprogramming. Additionally, it would make sense to obtain a printout of all customer programmable options *before* reprogramming because these may be erased by the process depending on the system. Customer data would have to be reprogrammed after the proprietary data reprogramming.

Uploading

The term **upload** usually describes the transfer of data from one computer system to another, and in this sequence of events, it is the final step in reprogramming. The verification file created by the vehicle ECM following reprogramming must, within a certain period of time, be uploaded to the OEM's centrally located mainframe computer. This would involve accessing the mainframe via modem in the same manner described in the download procedure and transferring the verification file. Uploading verification files is a record-keeping procedure; it basically informs the mainframe record file that the reprogramming was successfully undertaken.

Proprietary data reprogramming is a simple set of procedures. These require some familiarity with PCs. However most of the sequencing is menu driven and OEMs have designed the procedures to be as user friendly as possible. As each new software version is released, these procedures become progressively more simple.

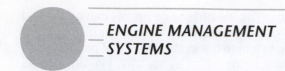

ENGINE MANAGEMENT SYSTEMS

Management systems for current truck and bus fuel systems tend to be classified by the degree of control they have over the fueling pulse.

FULL AUTHORITY SYSTEMS

Full authority systems use electronic unit injectors (EUIs) or electronic unit pumps (EUPs), both of which are cam actuated, or hydraulically actuated electronic unit injectors (HEUI). The term full authority indicates that the system ECM has full control of the fueling pulse, but in the case of EUI and EUP systems, this necessarily occurs within a timing window established by the physical dimensions of the actuating cam profile, usually a maximum of about 40 crank angle degrees. The HEUI system is "full authority" in the truest sense in that the plunger is actuated hydraulically, high-pressure engine lube is the medium and is therefore not limited to any hard window. Examples of full authority management systems are:

Bosch CR (used on GM-Duramax, Mack CR, and Cummins ISB)

Bosch EUP (used on V-MAC III, M-B 900, and M-B 4000)

Caterpillar ACERT-EUI used on C-11, C-13, and C-15

Caterpillar ACERT-HEUI used on C-7 and C-9

Caterpillar ADEM with Delphi EUI (used on C-10 and C-12)

Caterpillar ADEM with Cat EUI (used on C-15 and C-16)

Cat ADEM-HEUI (used on 3126)

Cummins CAPS (used on ISC and ISL)

Cummins CELECT Plus (used on ISM and N-14)

Cummins HPI-TP (used in ISX)

DDC-DDEC I (used on DDC Series 92)

DDC-DDEC II–V (used on DDC 50 and 60 Series)

International—Caterpillar HEUI (used on 466E, 530E, 444E, and 6.0)

Volvo VECTRO with Delphi EUI (used on VED-12)

PARTIAL AUTHORITY SYSTEMS

Partial authority systems are ECM-managed systems in which existing hydromechanical fueling apparatus has been adapted for computerized control. These are rapidly becoming obsolete and are not used on current systems. The extent to which the ECM can control fueling varies considerably, ranging from a limited function as a vehicle road speed governor, to complex management of vehicle functions and engine fueling. Some partial authority management systems offer comprehensive on-board diagnostics, extensive customer data programming options, and proprietary data reprogramming capability. Examples of partial authority management systems are:

Cat	PEEC (3406)
Cummins	PT PACE/PACER (N14)
Mack	V-MAC I and V-MAC II (E-7)

SUMMARY

- Almost all current on-highway trucks use computers to manage the engine and usually other chassis systems.
- A truck with multiple ECM-managed systems may electronically connect them to a vehicle management ECM capable of mastering all the systems using a chassis data bus.
- A vehicle ECM information processing cycle comprises three stages: data input, data processing, and outputs.
- RAM or main memory is electronically retained and therefore volatile.
- The master program for engine management in an engine controller ECM is usually written to ROM.
- PROM data is used to qualify the ROM data to a specific chassis application.
- Some OEMs describe their PROM and EEPROM component as a "personality module."
- EEPROM gives the ECM a read/write/erase memory component used for logging fault codes, covert codes, failure strategy management, and customer and proprietary data programming.
- Multiplexing is the term used to describe a system in which two or more ECMs are connected on a data bus to reduce input hardware and optimize vehicle operation.
- Input data may be categorized as command data and system monitoring data.
- Thermistors precisely measure temperature and may operate on either an NTC or PTC principle.
- Variable capacitance-type sensors are often used to measure pressure values.
- Piezoresistive sensors using a Wheatstone bridge are also used for pressure measurement signalling.
- Throttle position sensors are usually of the potentiometer type.
- Hall effect sensors generate a digital signal and are used to signal shaft position data.

- Induction pulse generators are used to input shaft rotational speed data.
- Truck engine management ECMs are responsible for regulating reference voltage, conditioning input data, processing, and managing outputs.
- The ECM processes data at different frequencies that are known as foreground and background computations.
- Engine management ECMs can be mounted on the engine itself or in a remote location such as under the dash.
- Customer data programming includes those parameters that may require changing on a day-to-day basis as well as those that may have to be reprogrammed if the vehicle application changes or critical components are changed.
- Proprietary data reprogramming is usually performed in these three stages: downloading system files, reprogramming the ECM, and uploading verification.
- The term partial authority management describes a system in which existing hydromechanical fueling apparatus has been adapted for computerized management of an engine.
- Full authority engine management is used to describe the systems designed for management by the computer.
- Full authority systems usually provide the ECM with a broader fueling window and therefore more control.

REVIEW QUESTIONS

1. The acronym that describes the internal computer component that executes program instructions is the:
   a. CRT
   b. RAM
   c. CPU
   d. ROM
2. Which of the following data retention media is electronically retained?
   a. RAM
   b. ROM
   c. PROM
   d. EEPROM
3. To which of the following memory categories would the master program for engine management be written in a typical truck engine ECM?
   a. RAM
   b. ROM
   c. PROM
   d. EEPROM
4. To which of the following memory categories would customer data programming be written from an electronic service tool?
   a. RAM
   b. ROM
   c. PROM
   d. EEPROM
5. Which of the following components conditions V-Ref?
   a. ECM
   b. Voltage regulator
   c. EUI
   d. Personality module
6. A thermistor is responsible for signalling data concerning:
   a. Pressure
   b. Temperature
   c. Rotational speed
   d. Rotational position
7. A Hall effect sensor is usually responsible for signalling data concerning:
   a. Pressure
   b. Temperature
   c. Fluid flow
   d. Rotational position
8. Induction pulse generators are used to input data concerning:
   a. Pressure
   b. Temperature
   c. Rotational speed
   d. Altitude
9. Which of the following components is used to signal accelerator pedal travel in most truck and bus TPS units?
   a. Pulse generator
   b. Potentiometer
   c. Thermistor
   d. Galvanic sensor
10. Which of the following input components would more likely use a variable capacitance-type sensor?
    a. CPS
    b. TRS
    c. VSS
    d. MAP
11. In a full authority diesel engine management system, EUI switching is performed by:
    a. A personality module
    b. Injector drivers
    c. V-Ref
    d. FIC module
12. Which of the following systems would be classified as a partial authority diesel fuel management system?
    a. CELECT Plus
    b. Caterpillar/International Trucks HEUI

c. DDEC V
d. V-MAC II

13. In which of the following systems is the injection pumping stroke actuated hydraulically?
a. CELECT Plus
b. Caterpillar/International HEUI
c. DDEC III
d. V-MAC II

14. *Technician A* states that V-Ref and the injector driver voltage are usually managed by the ECM to the same voltage values. *Technician B* states that V-Ref is controlled by the vehicle battery and is not conditioned by the ECM. Who is right?
a. A only
b. B only
c. Both A and B
d. Neither A nor B

15. *Technician A* states that in some of the earliest engine management ECMs, the only method of changing customer options, such as idle shutdown duration and transmission ratios, was to replace the ECM PROM chip. *Technician B* states that in most current vehicle management systems, customer programmable options are written to ROM. Who is right?
a. A only
b. B only
c. Both A and B
d. Neither A nor B

16. *Technician A* states that in most computer systems, RAM data is lost whenever the computer is shut down. *Technician B* states that some systems write data to a special type of RAM that is non-volatile. Who is right?
a. A only
b. B only
c. Both A and B
d. Neither A nor B

17. *Technician A* states that engine overspeed rpm is usually programmed to the ECM as a hard value. *Technician B* states that the engine oil/coolant overheat temperature value was usually programmed to the ECM as a soft value because it would depend on the ambient temperatures. Who is right?
a. A only
b. B only
c. Both A and B
d. Neither A nor B

18. *Technician A* states that it is not unusual for the vehicle maximum speed to be programmed at a lower speed than maximum cruise speed. *Technician B* states that unless the ECM is programmed for soft cruise, setting maximum cruise speed at a value above maximum vehicle speed can reduce fuel economy. Who is right?
a. A only
b. B only
c. Both A and B
d. Neither A nor B

Chapter
33
Electronic Service Tools

PREREQUISITES

Chapters 29 and 32

OBJECTIVES

After studying this chapter, you should be able to:

- *Define the acronym EST.*
- *Identify the different types of EST in current usage.*
- *Identify the levels of access and programming capabilities of each EST.*
- *Understand the dangers of electrostatic discharge and using inappropriate circuit analysis tools.*
- *Describe the type of data that can be accessed by each EST.*
- *Identify the categories of data that may be read using on-board flash codes.*
- *Perform some basic electrical circuit diagnosis using a DMM.*
- *Identify the function codes on a typical DMM.*
- *Perform tests on common input circuit components such as thermistors and potentiometers.*
- *Describe the full range of uses of a generic reader-programmer.*
- *Update reader-programmer software cartridges by replacing PROM chip(s) and data cards.*
- *Define the objectives of snapshot test.*
- *Outline the importance of completing each step when performing sequential troubleshooting testing of electronic circuits.*
- *Interpret the SAE J1587/1939 codes for MIDs, PIDs, SIDs, and FMIs using the included interpretation charts.*

KEY TERMS

active codes	ground strap	parameter identifier (PID)
ATA connector	Hall effect probe	parity
baud rate	handshake	ProLink 9000
blink codes	historic codes	reader/programmers
breakout box	inactive codes	root mean square (rms)
breakout T	J1587	SAE J1587
continuity	J1708	SAE J1708
current transformers	J1939	SAE J1939
digital multimeter (DMM)	message identifier (MID)	snapshot test
electronic service tool (EST)	meter resolution	stop bits
failure mode identifier (FMI)	open circuit	subsystem identifier (SID)
flash codes		thermistors

INTRODUCTION

The acronym **EST (electronic service tool)** is generally used in the trucking industry to cover a range of electronic service instruments ranging from on-board diagnostic/malfunction lights to sophisticated computer-based communications equipment. The use of generic ESTs and procedures is reviewed in this chapter; proprietary ESTs are designed to work with an OEM's specific electronics and are not introduced in this text. The proprietary EST is rapidly becoming a thing of the past, with OEMs generally opting for low-cost, powerful PCs whose hardware and software can be cheaply upgraded.

6-Pin J1587/1708 connector

A = Data bus, dominant high (+)
B = Data bus, dominant low (−)
C = Battery positive
D = Dummy
E = Ground
F = Battery negative

FIGURE 33–1   J1587/1708 Connector cavity pin assignments.

ESTs capable of reading ECM data are connected to the on-board electronics by means of Society of Automotive Engineers (SAE)/American Trucking Association (ATA) J1587/J1708 six-pin or J1939, nine-pin Deutsch connectors in all current systems. Data connectors are used by all the truck engine electronics OEMs and often are referred to as **ATA connectors**. Common connectors and the adherence by the engine electronics OEMs to SAE data bus software protocols enable proprietary software of one manufacturer to at least read the parameters and conditions of its competitors. This means that if a Cummins-powered truck has an electronic failure in a location where the only service dealer is DDC, some basic problem diagnosis can be undertaken using the DDC electronic diagnostic equipment. ATA connectors are wired as indicated in Figure 33–1 and Figure 33–2.

There are six categories of electronic service tools as follows:

ON-BOARD DIAGNOSTIC LIGHTS

Blink or **flash codes** (OEMs use both terms) are an on-board means of troubleshooting using a dash or ECM-mounted electronic malfunction light or check engine light (CEL). Usually only **active codes** (ones that indicate a malfunction at the time of reading) can be read in truck (engine) ECMs but some also read historic codes.

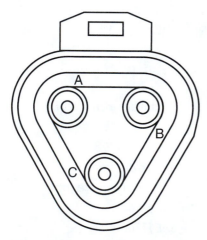

3-Pin J1939 connector

A = CAN busline, dominant high (+)
B = CAN busline, dominant low (−)
C = CAN ground

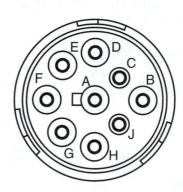

9-Pin J1939 connector

A = Battery negative
B = Battery positive
C = J1939 CAN busline, dominant high (+)
D = J1939 CAN busline, dominant low (−)
E = CAN ground
F = J1587 busline, dominant high (+)
G = J1587 busline, dominant low (−)

FIGURE 33–2   J1939 Connector cavity pin assignments.

DIGITAL MULTIMETERS

The **digital multimeter (DMM)** is an essential diagnostic instrument that should be part of every truck technician's tool chest. It is used to troubleshoot electronic circuits and test components.

SCANNERS

Scanners are read-only tools capable of reading active and **historic** or **inactive** (logged but not currently indicating a malfunction) **codes** and sometimes system parameters but little else. They are obsolete as a truck/bus diesel engine diagnostic tool, but the term is sometimes used to describe a reader/programmer.

GENERIC READER/PROGRAMMERS

Generic **reader/programmers** are microcomputer-based ESTs designed to read and reprogram all proprietary systems in conjunction with the appropriate software cartridge. These are usually tough and portable. **ProLink 9000** has become the industry standard although its use is in decline as electronic systems become more sophisticated.

PROPRIETARY READER/PROGRAMMERS

These are usually PC-based test instruments packaged by the OEM for use exclusively on its system. They have the advantage of offering an optimum degree of user friendliness and the disadvantage of being system specific and high in cost. Some examples are:

Cummins-Compulink, Echeck

Cat-ECAP

DDC-DDR Programming Station

Proprietary reader-programmers are rapidly becoming obsolete, being replaced by PC-based diagnostic instruments.

PERSONAL COMPUTER (PC)

Most truck engine OEMs have made the generic PC and its operating systems (DOS/Windows platform) their diagnostic and programming tool of choice; the remainder are expected to follow, which makes sense.

PCs are cheap, easily upgradable, and have vast computing power when compared with proprietary reader/programmers. OEMs can upgrade/update software more easily either by CD or on-line with access to their data hubs. PCs are connected to a vehicle ECM through the ATA data connector; a serial link or inter-

face module may also be required, depending on the system and the production year.

Truck engine OEMs have chosen to use MS Windows-driven programs and offer comprehensive courses on their own management systems, which usually include a thorough orientation of PCs. PCs are addressed in some detail in Chapter 30.

ACCESSING THE DATA BUS

Most truck electronic systems are connected to a data bus, a communications backbone that will be studied in more detail later in this chapter and again in Chapter 35. The term multiplexing is used to describe system-to-system data transactions, but electronic service tools (ESTs) must access the chassis data backbone in order to read, diagnose, and reprogram system control modules. Older truck systems used a **J1708** data bus that used **J1587** communications protocols. The J1708 data bus is accessed using the six-pin Deutsch connector shown in Figure 33–1. Current trucks use a more sophisticated data backbone known as **J1939**. The J1939 standard covers both hardware and software protocols and is accessed by an EST using a nine-pin Deutsch connecter shown in Figure 33–2.

Because there are usually a number of electronic systems connected to a truck data backbone, after connecting an EST to the data bus, the specific system you want to work with has to be selected. Each major electronic system on the data bus is assigned a message identifier or MID and you must select this before you can access the information in that system. It actually sounds more complicated than it is. After connecting an EST to the data bus, all of the MIDs on that chassis will be displayed. If you want to work on the engine MID, you use the EST to select this from the menu.

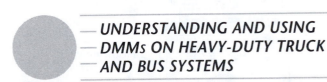

UNDERSTANDING AND USING DMMs ON HEAVY-DUTY TRUCK AND BUS SYSTEMS

A DMM is simply a tool for making electrical measurements. DMMs may have any number of special features but essentially, they measure electrical pressure or volts, electrical current quantitive flow or amps, and electrical resistance or ohms. A good quality DMM with minimal features may be purchased for as little as $100; as the features, resolution, and display quality increase in sophistication, the price increases proportionally.

Because most electronic circuit testing _requires_ the use of a DMM, this instrument should replace the analog multimeter and circuit test light in the truck/bus technician's toolbox. Reliability, accuracy, and ease of use are all factors that should be considered when selecting a DMM for purchase. Some options the technician may wish to consider are a protective rubber holster (will greatly extend the life of the instrument!), analog bar graphs, and enhanced resolution. This section deals with the practice of using DMMs and a knowledge of basic electricity is assumed.

DMM SUBCOMPONENTS

Figure 33–3 shows the location of the various features and subcomponents on a typical DMM. When learning how to get the best use of a DMM, you should first understand the function of each feature.

Input Terminals

The input terminals are located at the base of most DMMs as shown in Figure 33–4. The DMM shown has four input terminals into which lead jacks are inserted. The black lead is used for ALL DMM measurements, and is inserted into the common jack. To make voltage and resistance measurements, the common and volts/rpm/ohms/diode jacks would be selected, so the red lead would be inserted into the volts/rpm/ohms/diode jack. The two jacks to the left of these are used only for series current measurements and these are self-explanatory, one for milliamps, the other for amps. Remember that mixing leads up can blow the internal DMM fuses. These fuses are expensive so blowing a couple after purchasing a DMM can be regarded as a learning experience, and after that, as stupidity.

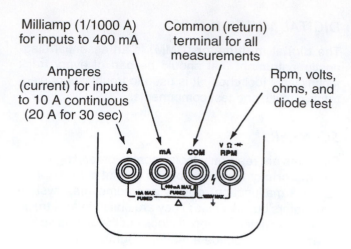

Milliamp (1/1000 A) for inputs to 400 mA

Amperes (current) for inputs to 10 A continuous (20 A for 30 sec)

Common (return) terminal for all measurements

Rpm, volts, ohms, and diode test

FIGURE 33–4   *DMM input terminals. (Courtesy of International Navistar Corp., Designer and manufacturer of International Brand diesel engines)*

Rotary Switch

The DMM rotary switch, shown in Figure 33–5, is used to set the DMM function. The DMM can make AC and DC voltage measurements, measure circuit resistance, test diodes, and make series amperage measurements. Most current DMMs will autorange, that is, determine what the measurement value is and select the appropriate display scale. Technicians working daily with DMMs should only consider units with autoranging capability.

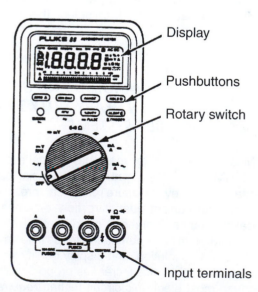

Display

Pushbuttons

Rotary switch

Input terminals

FIGURE 33–3   *Fluke 88 DMM. (Courtesy of International Navistar Corp., Designer and manufacturer of International Brand diesel engines)*

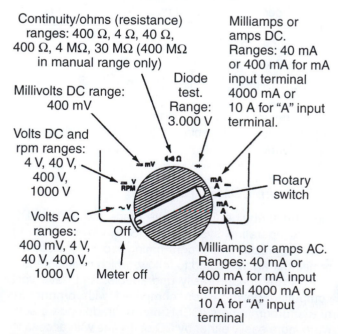

Continuity/ohms (resistance) ranges: 400 Ω, 4 Ω, 40 Ω, 400 Ω, 4 MΩ, 30 MΩ (400 MΩ in manual range only)

Millivolts DC range: 400 mV

Volts DC and rpm ranges: 4 V, 40 V, 400 V, 1000 V

Diode test. Range: 3.000 V

Milliamps or amps DC. Ranges: 40 mA or 400 mA for mA input terminal 4000 mA or 10 A for "A" input terminal.

Volts AC ranges: 400 mV, 4 V, 40 V, 400 V, 1000 V

Off

Meter off

Rotary switch

Milliamps or amps AC. Ranges: 40 mA or 400 mA for mA input terminal 4000 mA or 10 A for "A" input terminal

FIGURE 33–5   *DMM rotary switch. (Courtesy of International Navistar Corp., Designer and manufacturer of International Brand diesel engines)*

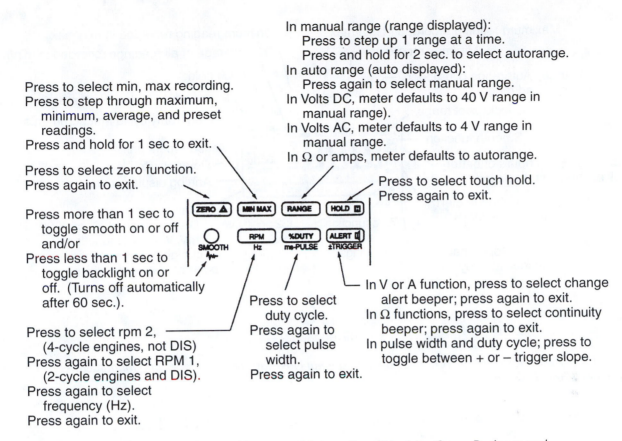

Press to select min, max recording.
Press to step through maximum,
    minimum, average, and preset
    readings.
Press and hold for 1 sec to exit.

Press to select zero function.
Press again to exit.

Press more than 1 sec to
    toggle smooth on or off
    and/or
Press less than 1 sec to
    toggle backlight on or
    off. (Turns off automatically
    after 60 sec.).

Press to select rpm 2,
    (4-cycle engines, not DIS)
Press again to select RPM 1,
    (2-cycle engines and DIS).
Press again to select
    frequency (Hz).
Press again to exit.

In manual range (range displayed):
    Press to step up 1 range at a time.
    Press and hold for 2 sec. to select autorange.
In auto range (auto displayed):
    Press again to select manual range.
In Volts DC, meter defaults to 40 V range in
    manual range).
In Volts AC, meter defaults to 4 V range in
    manual range.
In Ω or amps, meter defaults to autorange.

Press to select touch hold.
Press again to exit.

Press to select
    duty cycle.
Press again to
    select pulse
    width.
Press again to exit.

In V or A function, press to select change
    alert beeper; press again to exit.
In Ω functions, press to select continuity
    beeper; press again to exit.
In pulse width and duty cycle; press to
    toggle between + or – trigger slope.

FIGURE 33–6   *DMM pushbuttons. (Courtesy of International Navistar Corp., Designer and manufacturer of International Brand diesel engines)*

Pushbuttons

The pushbuttons on a DMM are used to select or fine-tune meter operations. On the DMM shown, when a pushbutton is depressed, a symbol is displayed on the data display and sometimes a beeper will sound. Figure 33–6 shows the pushbuttons used on a Fluke 99 DMM. In most cases, when the rotary switch setting is changed, the pushbutton settings will revert to default status.

Data Display

The data display is the means used to read the measurements made by the DMM. Figure 33–7 shows the data display used on a high-end DMM. Many current DMMs have the ability to display both digital and analog readings. The analog display is simply an analog representation of the digital reading made by the meter. The digital display should be used for stable input values while the analog display should be used for frequently changing inputs. The acronym OL used in the display means overloaded or out of range.

RESOLUTION

Resolution denotes how fine a measurement can be made with the instrument. Digits and counts are used to describe the resolution capability of a DMM. A 3½ digit meter can display three full digits ranging from 0 to 9 and one-half digit that displays either 1 or is left blank. A 3½ digit meter therefore displays 1,999 counts of resolution. A 4½ digit DMM can display up to 19,999 counts of resolution.

However, many DMMs have enhanced resolution so the meter's reading power is usually expressed in counts rather than digits. For instance, a 3½ digit meter may have enhanced resolution of 4,000 counts. Basically, **meter resolution** is expressed in counts rather than digits. For example, a 3½ digit or 1,999 count meter will not measure down to 0.1 V when measuring 200 V or higher. However, a 3,200 count meter will display 0.1 V up to 320 V, giving it the same resolution as a 4½ digit, 19,999 count meter until the voltage exceeds 320 V.

ACCURACY

Accuracy denotes how close the displayed reading on the DMM is to the actual value of the measured signal. This is usually expressed as a percentage of the reading. An accuracy rating of ±1% in a DMM reading a voltage value of 10 V would mean the actual value could range between 9.9 V and 10.1 V. DMM accuracy can be extended by indicating how many counts the

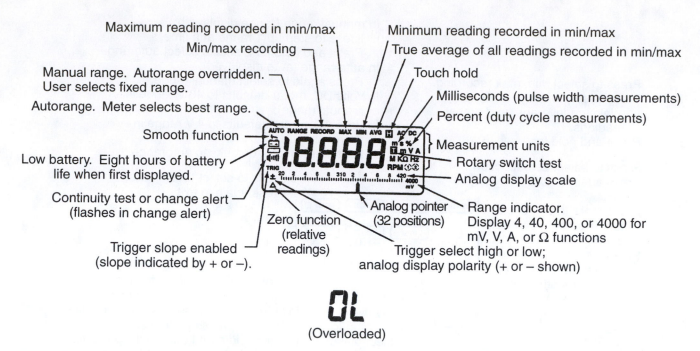

Maximum reading recorded in min/max
Minimum reading recorded in min/max
Min/max recording
True average of all readings recorded in min/max
Manual range. Autorange overridden. User selects fixed range.
Touch hold
Autorange. Meter selects best range.
Milliseconds (pulse width measurements)
Percent (duty cycle measurements)
Smooth function
Measurement units
Low battery. Eight hours of battery life when first displayed.
Rotary switch test
Analog display scale
Continuity test or change alert (flashes in change alert)
Zero function (relative readings)
Analog pointer (32 positions)
Range indicator. Display 4, 40, 400, or 4000 for mV, V, A, or Ω functions
Trigger slope enabled (slope indicated by + or –).
Trigger select high or low; analog display polarity (+ or – shown)

(Overloaded)

FIGURE 33–7 *DMM data display. (Courtesy of International Navistar Corp., Designer and manufacturer of International Brand diesel engines)*

right display digit may vary. So an accuracy rating of ± (1% + 2) would mean that a displayed voltage of 10 V could have an actual value range between 9.88 V and 10.12 V.

Analog multimeters have accuracy ratings that vary between 2% and 3% of full scale. DMM accuracy ratings range between ± (.07% + 1) to ± (0.1% + 1) of the *reading*.

OPERATING PRINCIPLES

In any electrical circuit, voltage, current flow, and resistance can be calculated using Ohm's law. A DMM makes use of Ohm's law to measure and display values in an electrical circuit. A typical DMM has the following selection options, chosen by rotating the selector to one of the following:

off: Shuts down the DMM
V~:   Enables AC voltage readings
V–:   Enables DC voltage readings
mV-:   Enables low-pressure DC voltage readings
Ω: Enables component or circuit resistance readings
α: Enables **continuity** (a circuit capable of being closed) testing. Identifies an open/closed circuit.
A~:   Checks current flow (amperage) in an AC circuit
A–:   Checks current flow (amperage) in a DC circuit

MEASURING VOLTAGE

Checking circuit supply voltage is usually one of the first steps in troubleshooting. This is performed in a vehicle DC circuit by selecting the V– setting and checking for voltage present or high/low voltage values. Most electronic equipment is powered by DC voltage. For example, home electronic apparatus such as computers, televisions, and stereos use rectifiers to convert mains AC voltage to DC voltage.

The waveforms produced by AC voltages can either be sinusoidal (sine waves) or nonsinusoidal (sawtooth, square, ripple). A DMM displays the **root mean square (rms)** value of these voltage waveforms. The rms value is the effective or equivalent DC value of the AC voltage. Meters described as "average responding" give accurate rms readings only if the AC voltage signal is a pure sine wave; they will not accurately measure nonsinusoidal signals. DMMs described as "true-rms" measure the correct rms value regardless of waveform and should be used for nonsinusoidal signals.

A DMM's ability to measure voltage can be limited by the signal frequency. The DMM specifications for AC voltage and current will identify the frequency range the instrument can accurately measure. Voltage measurements determine:

• Source voltage
• Voltage drop
• Voltage imbalance
• Ripple voltage
• Sensor voltages

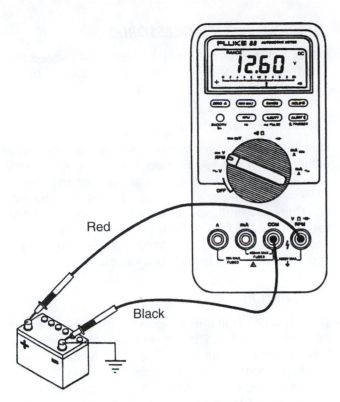

FIGURE 33–8 *DMM setup for making DC voltage measurements. (Courtesy of International Navistar Corp., Designer and manufacturer of International Brand diesel engines)*

Figure 33–8 shows a DMM set up to read DC voltage.

MEASURING RESISTANCE

Most DMMs measure resistance values as low as 0.1 Ω and some measure high resistance values up to 300 MΩ (megaohms). Infinite resistance or resistance greater than the instrument can measure is indicated as "OL" or flashing digits on the display. For instance, an **open circuit** (one in which there is no path for current flow) would read "OL" on the display.

Resistance and continuity measurements should be made on open circuits only (Figure 33–9). Using the resistance or continuity settings to check a circuit or component that is energized will result in damage to the test instrument. Some DMMs are protected against such "accidental" abuse and the extent of damage will depend on the model. For accurate low-resistance measurement, test lead resistance, typically between 0.2 Ω and 0.5 Ω depending on quality and length, must be subtracted from the display reading. Test lead resistance should never exceed 1 Ω.

If a DMM supplies less than 0.3 V DC test voltage for measuring resistance, it is capable of testing resistors isolated in a circuit by diodes or semiconductor junctions, meaning that they do not have to be remov-

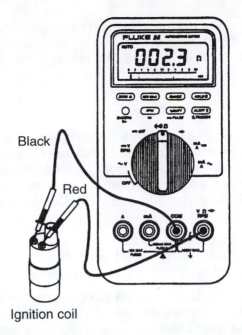

Black

Red

Ignition coil

FIGURE 33–9 *DMM setup for measuring resistance. (Courtesy of International Navistar Corp., Designer and manufacturer of International Brand diesel engines)*

ed from the circuit board. Resistance measurements determine:

- Resistance of a load
- Resistance of conductors
- Value of a resistor
- Operation of variable resistors

Continuity is a quick resistance check that distinguishes between an open and a closed circuit. Most DMMs have audible continuity beepers that beep when they detect a closed circuit, permitting the test to be performed without looking at the meter display. The actual level of resistance required to trigger the beeper varies from model to model. Continuity tests determine:

- Fuse integrity
- Open or shorted conductors
- Switch operation
- Circuit paths

Diode Testing

A diode is an electronic switch that can conduct electricity in one direction while blocking current flow in the opposite direction. Diodes are commonly enclosed in glass cylinders; a dark band identifies the cathode or blocking terminal. Current flows when the anode is more positive than the cathode. Additionally, a diode will not conduct until the forward voltage pressure reaches a certain value, 0.3 V in a silicon diode. Some meters have a diode test mode. When testing a diode with the DMM in this mode, 0.6 V is delivered through the device to indicate continuity; reversing the test leads should indicate an open circuit in a properly functioning diode. If both readings indicate an open circuit condition, the diode is "open." If both readings indicate continuity, the diode is shorted.

MEASURING CURRENT

Current measurements are made in series, unlike voltage and resistance readings, which are made in parallel. The test leads are plugged into a separate set of input jacks and the current to be measured flows through the meter (Figure 33–10). Current measurements determine:

- Circuit overloads
- Circuit operating current
- Current in different branches of a circuit

When the test leads are plugged into the current input jacks and they are used to measure voltage, this causes a direct short across the source voltage through a low value resistor inside the DMM called a current shunt. A high current flows through the meter and if not adequately protected, both the meter and the circuit can be damaged. A DMM should have current input fuse protection of high enough capacity for the circuit being tested. This protection is mainly of importance when working with high pressure (220 V +) circuits.

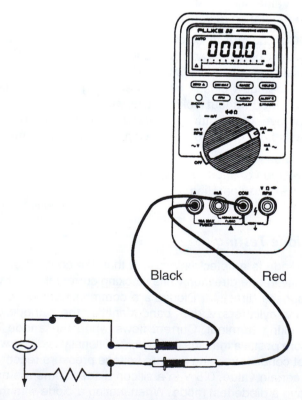

FIGURE 33–10 *DMM setup for measuring current flow. (Courtesy of International Navistar Corp., Designer and manufacturer of International Brand diesel engines)*

CURRENT PROBE ACCESSORIES

When making current measurements that exceed the DMM's rated capacity, a current probe can be used. There are two types:

Current Transformers

Current transformers measure AC current only. Their output is 1 mA per 1 A. Therefore, a current flow of 100 A is reduced to 100 mA, which can be handled by most DMMs. The test leads would be connected to the mA and Common input jacks and the meter function switch set to "mA AC." This is not a very accurate circuit test and is used for ballpark reckoning only. There are no applications for this tool in truck electrical systems.

Hall Effect Probe

The output of a **Hall effect probe** is 1 mV per ampere. It measures AC or DC. The test leads are connected to the V and Common input jacks. The DMM function switch should be set to the V or mV scale, selecting V AC for AC current and V DC for DC current measurements; this is not especially accurate. The Hall effect probe may be used to measure cranking motor current draw.

DMM Features

When considering a DMM for purchase, the following features should be considered:

- Fused current inputs
- Use of high energy fuses (600 V +)
- High-voltage protection in resistance mode
- Protection against high-voltage transients
- Insulated test lead handles
- CSA and UL approval
- Ammeter required
- Autorange capability
- Cost of replacement fuses

SOME TYPICAL DMM TESTS

Always perform tests in accordance with truck and bus OEM specifications; never jump sequence or skip steps in sequential troubleshooting charts. Most DMM tests on truck and bus electronic systems will be used in conjunction with a generic reader/programmer or PC. The following tests assume the use of a Fluke 88 DMM.

Engine Position (Fuel Injection Pump Camshaft), Cam, and Crank Position Sensors

Hall effect sensors:

a. Cycle the ignition key, then off.
b. Switch meter to measure V DC/rpm.

c. Identify the ground and signal terminals at the Hall sensor. Connect the positive (+) test lead to the signal terminal and the negative (–) test lead to the ground terminal. Crank the engine.

At cranking speeds, the analog bar graph should pulse; at idle speeds or above, the pulses are too fast for bar graph readout.

d. Press the duty cycle button once. Duty cycle can indicate square wave quality with poor quality signals having a low duty cycle. Functioning Hall sensors should have a duty cycle of around 50% depending on the sensor. Check to specifications.

Potentiometer-Type TPS

Resistance test:

a. Key off.
b. Disconnect the TPS.
c. Select Ω on the DMM. Connect the test probes to the signal and ground terminals. Next, move the accelerator through its stroke while observing the DMM display.
d. The analog bar should move smoothly without jumps or steps. If it steps, there may be a bad spot in the sensor.

Voltage test:

a. Key on, engine off.
b. Set the meter to read V DC. Connect the negative lead to ground.
c. With the positive lead, check the reference voltage value and compare to specifications. Next, check the signal voltage (to the ECM) value through the accelerator pedal stroke. Check values to specification. Also observe the analog pointer; as with the resistance test, this should move smoothly through the accelerator stroke.

NOTE: This test will not work on those systems that digitize the signal to produce a PWM input to the ECM.

Magnetic Sensors

Magnetic or variable reluctance sensors function similarly to a magneto. The output is an AC voltage pulse whose value rises proportionally with rotational speed increase. Voltage values range from 0.1 V up to 5.0 V depending on the rotational speed and the type of sensor. Vehicle speed sensors (VSS), engine speed sensors (ESS), and ABS wheel speed sensors all use this method of determining rotational speed. Test using V AC switch setting and locating test leads across the appropriate terminals.

Min/Max Average Test for Lambda (λ) (O_2) Sensors

a. Key on, engine running, DMM set at V DC. Select the correct voltage range.
b. Connect the negative test lead to a chassis ground and the positive test lead to the signal wire from the lambda sensor. Press the DMM min/max button.
c. Ensure the engine is warm enough to be in closed loop mode (100 mV–900 mV O_2 sensor output). Run for several minutes to give the meter time to sample a scatter of readings.
d. Press the min/max button slowly three times while watching the DMM display. A maximum of 800 mV and a minimum of less than 200 mV should be observed. The average should be around 450 mV.
e. Next, disconnect a large vacuum hose to create a lean burn condition. Repeat steps c and d to read the average voltage. Average voltage should be lower, indicating a lean condition.
f. The same test can be performed using propane enrichment to produce a rich AFR condition and therefore higher voltage values.
g. Lambda sensor tests can be performed while road testing the vehicle; 450 mV normally indicates stoichiometric fueling ($\lambda = 1$) but check to specifications.

Thermistors

Most **thermistors** used in computerized engine systems are supplied with V-Ref (5 V) and have a negative temperature coefficient (NTC), meaning that as the sensor temperature increases, its resistance decreases. They should be checked to specifications using the DMM ohmmeter function and an accurate temperature measurement instrument.

OEMs seldom suggest random testing of suspect components. The preceding tests are typical procedures. Circuit testing in today's computerized engine management systems is highly structured and part of sequential troubleshooting procedure. It is important to perform each step in the sequence precisely; skipping a step can invalidate every step that follows.

BREAKOUT BOXES AND BREAKOUT Ts

The DMM is often used in conjunction with a **breakout box** or **breakout T**. Breakout devices are designed to be teed into an electrical circuit to enable circuit measurements to be made on both closed (active) and de-energized circuits. The objective is to access a circuit with a test instrument without interrupting the circuit. A breakout T normally describes a diagnostic device that is inserted into a simple two- or three-wire circuit such as that used to connect an individual sensor, while a breakout box accesses multiple wire circuits for diagnostic analyses of circuit conditions. Most of the electronic engine management system OEMs use a breakout box that is often inserted into the interface connection between the engine electronics and chassis electronics harnesses. The face of the breakout box displays a number of coded sockets into which the probes of a DMM can be safely inserted to read circuit conditions. Electronic troubleshooting sequencing is often structured based on the data read by a DMM accessing a circuit. A primary advantage of breakout diagnostic devices is the fact that they permit the reading of an active electronic circuit, for instance, while an engine is running.

WARNING: When a troubleshooting sequence calls for the use of breakout devices, always use the recommended tool. Never puncture wiring or electrical harnesses to enable readings in active or open electronic circuits. The corrosion damage that results from damaging wiring conduit will create problems later on, and the electrical damage potential can be extensive.

Diagnostic Connector Dummies

Diagnostic connector dummies are used to read a set of circuit conditions in a circuit that has been opened by separating a pair of connectors. The dummies are manufactured by the electrical/electronic connector manufacturer as a means of accessing the circuitry with a DMM without damaging the connector sockets and pins. Ensure that correct dummies are used.

WARNING: The terminals in many connectors are especially vulnerable to the kind of damage that can be caused by attempting to insert DMM probes, paper clips, and other inappropriate devices. Even more important, remember always that it is possible to cause costly electrical damage by shorting and grounding circuits in a separated electrical connector.

TECH TIP: When performing a multiple-step, electronic troubleshooting sequence on a large multiterminal connector, photocopy the coded face of the connector(s) from the service manual and use it as a template. The alphanumeric codlings used on many connectors can be difficult to read and using a template is a good method of orienting the test procedure.

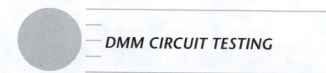

DMM CIRCUIT TESTING

Many truck technicians learn to use DMMs on-the-fly, that is, they acquire their understanding on an as-

needed basis, with the consequence that most do not properly understand the potential of a high-end DMM. It does not make sense to spend $500 on a complex DMM that will never be used to perform tasks that a DMM one-sixth of the cost will handle. This next section introduces some simple tasks that can be performed by low-end DMMs using components readily available in introductory electricity/electronics labs.

TESTING RESISTORS

Most carbon resistors used in electronic circuits are color coded. Although individual resistors would seldom be tested in the real world of the truck shop, testing them in a teaching/ learning setting can be a great way of getting to know the ohmmeter function of your DMM. You can grab a pile of resistors and check the color-coded resistance value against the actual measurement. First you have to understand the color codes:

Color	Value
Black	0
Brown	1
Red	2
Orange	3
Yellow	4
Green	5
Blue	6
Violet	7
Gray	8
White	9

In the above list, note that the darkest colors like black and brown are used for the lowest numeric values (0 and 9) moving up to white, which is given the highest numeric value. These codes are standardized by the Electronics Industries Association (EIA) and are also used for capacitors. In addition, the colors gold and silver are used to rate tolerance or the percentage amount that similarly rated resistors can differ from each other and still function within specification. Gold indicates 5% and silver 10% tolerance. Figure

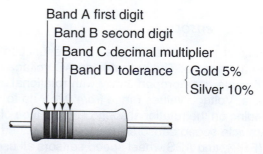

Band A first digit
Band B second digit
Band C decimal multiplier
Band D tolerance ⎰ Gold 5%
⎱ Silver 10%

FIGURE 33–11 *Interpreting bands on a carbon resistor.*

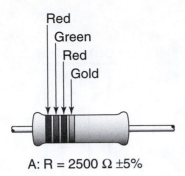

A: R = 2500 Ω ±5%

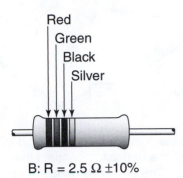

B: R = 2.5 Ω ±10%

FIGURE 33–12   Interpreting band resistance codes.

33–11 shows the significance of each band used on a resistor.

We can look at two actual examples. Using Figure 33–12, in Figure 33–12A the arrangement is:

1st stripe =   red = 2

2nd stripe =   green = 5

3rd stripe =   red = multiplier to power of 2 or 10^2, so you *add* 2 zeros = 2500 Ω

4th stripe =   gold = tolerance rated at within 5%

At first glance, Figure 33–12B does not appear to be unlike the previous resistor:

1st stripe =   red = 2

2nd stripe =   green = 5

3rd stripe =   black = multiplier is zero so the value is 25 Ω

4th stripe =   silver = tolerance rated at within 10%

TESTING SEMICONDUCTORS

Figure 33–13 and Figure 33–14 outline some methods of testing semiconductors that can be helpful in learning how to use a DMM to test circuit components and to understand how diodes and transistors operate.

Testing a Diode

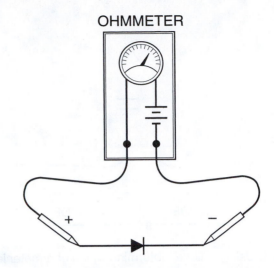

Step 1. Connect the ohmmeter leads to the diode. Notice if the meter indicates continuity through the diode or not. Now use the diode test mode on the DMM.

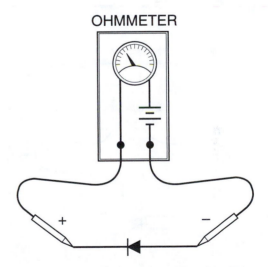

Step 2. Reverse the diode connection to the ohmmeter. Notice if the meter indicates continuity through the diode; the ohmmeter should indicate continuity through the diode in only one direction. (Note: If continuity is not indicated in either direction, the diode is open. If continuity is indicated in both directions, the diode is shorted.)

FIGURE 33–13   Testing diodes. (Courtesy of Utah Technical College)

Testing A Transistor

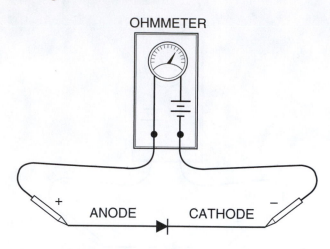

Step 1. Using a diode, determine which ohmmeter lead is positive and which is negative. The ohmmeter will indicate continuity through the diode only when the positive lead is connected to the anode of the diode and the negative lead is connected to the cathode.

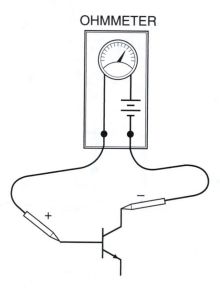

Step 2. If the transistor is an NPN, connect the positive ohmmeter lead to the base and the negative lead to the collector. The ohmmeter should indicate continuity. The reading should be about the same as the reading obtained when the diode was tested.

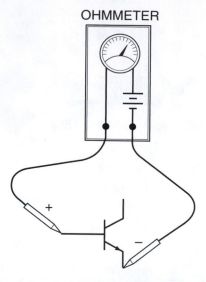

Step 3. With the positive ohmmeter lead still connected to the base of the transistor, connect the negative lead to the emitter. The ohmmeter should again indicate a forward diode junction. (Note: If the ohmmeter does not indicate continuity between the base-collector or the base-emitter, the transistor is open.)

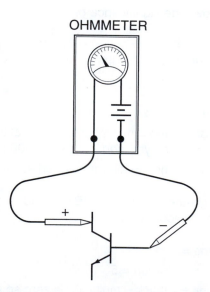

Step 4. Connect the negative ohmmeter lead to the base and the positive lead to the collector. The ohmmeter should indicate infinity or no continuity.

FIGURE 33–14 Testing transistors. (Courtesy of Utah Technical College)

OHMMETER

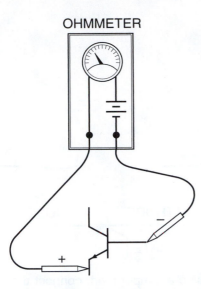

Step 5. With the negative ohmmeter lead connected to the base, reconnect the positive lead to the emitter. There should again be no indication of continuity. (Note: If a very high resistance is indicated by the ohmmeter, the transistor is "leaky" but may still operate in the circuit. If a very low resistance is seen, the transistor is shorted.)

OHMMETER

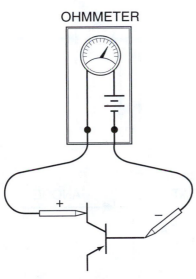

Step 6. To test the PNP transistor, reverse the polarity of the ohmmeter leads and repeat the test. When the negative ohmmeter lead is connected to the base, a forward diode junction should be indicated when the positive lead is connected to the collector or emitter.

OHMMETER

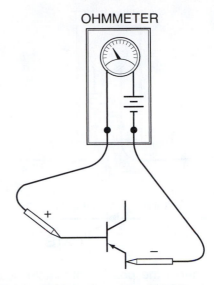

Step 7. If the positive ohmmeter lead is connected to the base of a PNP transistor, no continuity should be indicated when the negative lead is connected to the collector or the emitter.

Testing an SCR

OHMMETER

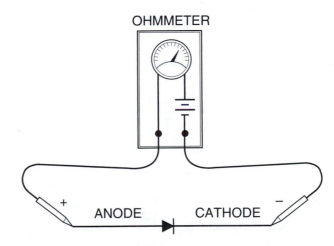

ANODE   CATHODE

Step 1. Using a junction diode, determine which ohmmeter lead is positive and which is negative. The ohmmeter will indicate continuity only when the positive lead is connected to the anode of the diode and the negative lead is connected to the cathode.

FIGURE 33–14 _(Continued)_

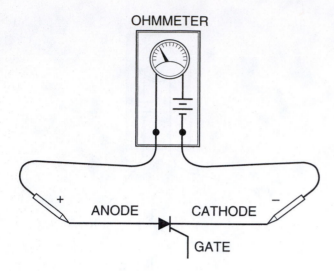

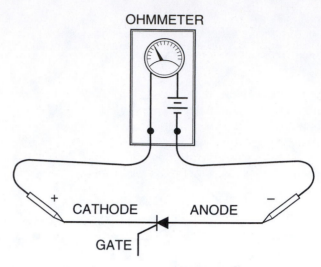

Step 2. Connect the positive ohmmeter lead to the anode of the SCR and the negative lead to the cathode. The ohmmeter should indicate no continuity.

Step 3. Using a jumper lead, connect the gate of the SCR to the anode. The ohmmeter should indicate a forward diode junction when the connection is made. (Note: If the jumper is removed, the SCR may continue to connect, or it may turn off. This will be determined by whether the ohmmeter can supply enough current to keep the SCR above its holding current.)

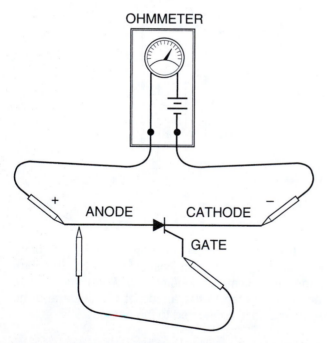

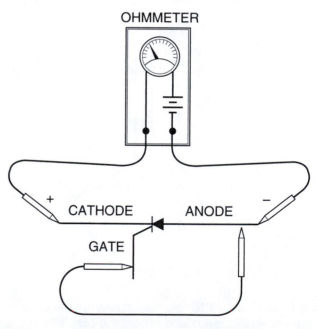

Step 4. Reconnect the SCR so that the cathode is connected to the positive ohmmeter lead and the anode is connected to the negative lead. The ohmmeter should indicate no continuity.

Step 5. If a jumper is used to connect the gate to the anode, the ohmmeter should indicate no continuity. (Note: SCRs designed to switch large currents [50 amperes or more] may indicate some leakage current with this test. This is normal for some devices.)

FIGURE 33–14   *(Continued)*

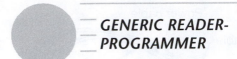

GENERIC READER-PROGRAMMER

The ProLink 9000 is an example of a generic reader/programmer. This tool was, for a number of years, the industry standard, electronic service tool capable of reading most current systems with the correct software cartridge or data card. These microprocessor-based test instruments will do the following (depending on the system):

- Access active and historic codes; erase historic (inactive) codes
- View all system identification data
- View data on engine operation with the engine running
- Perform diagnostic tests on system subcomponents such as EUI testing
- Reprogram customer data parameters on engine and chassis systems
- Act as a serial link to connect the vehicle ECM via modem to a centrally located mainframe for proprietary data programming (some systems only)
- Snapshot system data parameters to assist intermittent fault finding solutions

The advantages of reader/programmers are numerous, the most notable of which are:

- User friendly. Technicians require very little time to develop a working knowledge of these tools. The best method of becoming familiar with them is simply to use them. Menus can be selected, data can be scrolled through. Stuck? Simply unplug and reconnect. This can be done at any time without sustaining physical or electronic damage.
- Universality. These tools were for a while the industry standard and can be used extensively even on systems that require the use of PCs for more detailed diagnostic and programming functions. They will read most of the truck/bus engine and other on-board systems (and also automobile systems) simply by inserting the correct software cartridge or data card.
- Easily updated. Cartridges are updated simply by opening them up, removing and discarding out-of-date PROM chips or cards, and inserting new one(s). At PROM chip replacement, the RAM chip should also be momentarily disconnected from the motherboard to dump its nonvolatile content. The newer multiprotocol cartridges are updated by data card.
- Tough. The tough plastic housing really does withstand the rough treatment some of these instruments are likely to receive in the typical service facility.

The disadvantages of generic reader-programmers are:

- Display window is limited to four lines.
- Impossible to input alpha passwords (some systems) or long-winded procedure of inputting an alpha password on the numeric keypad.
- As system management systems become more complex, the limitations of ProLink 9000 become more evident. PC-based systems are more often required to properly troubleshoot electronic problems.

Generic reader/programmers are connected to the on-board ECM by means of the SAE/ATA J1708/1939 Data Link—that is, the ATA data connector. Both six- and nine-pine Deutsch connectors are used as shown in Figure 33–1 and Figure 33–2.

RS-232 SERIAL PORT

This port is located on the right side of the reader-programmer head. There are three choices when this menu item is selected:

1. Printer output
2. Terminal output
3. Port setup

Four parameters need to be set up for the device to communicate with a printer or PC terminal:

1. **Handshake**—this refers to how data will be transmitted between two electronic components. There are two commonly used methods: BUSY and XON/XOFF. Select the method specified on the printer specifications.
2. **Baud rate**—the speed at which data is transmitted. Both the output device and the receiving device must agree on transmission speed.
3. **Parity**—this is the even or odd quality of the number of 0s and 1s. The menu options are none, odd, or even.
4. **Stop bits**—the last element in the transmission of a character. The output and receiving devices must agree on how many stop bits to expect to determine the end of a transmission character.

TECH TIP: ProLink will transmit to a printer or PC terminal any data that it can read itself from a system ECM. Printouts of data are especially helpful when analyzing the causes of a condition (Figure 33–15 and Figure 33–16).

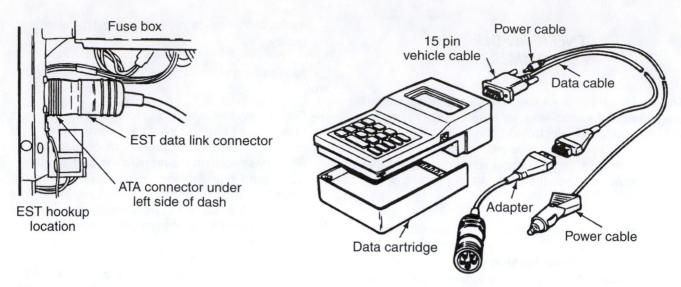

FIGURE 33–15 ProLink data connection hardware. (Courtesy of International Navistar Corp., Designer and manufacturer of International Brand diesel engines)

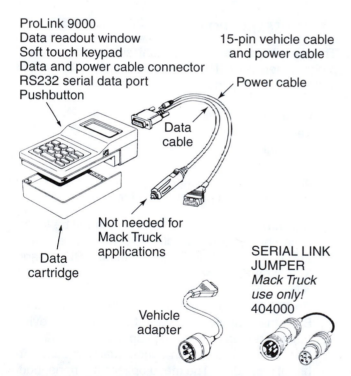

FIGURE 33–16 ProLink head, software cartridge, and cables. (Courtesy of MPSI)

UPDATING A PROLINK CARTRIDGE

Depending on when the software cartridge was made, the update method varies. Older cartridges are updated by replacing the PROM chip or chips using the following procedure:

1. Ensure that during this procedure, there is no danger of a static charge being unloaded into the cartridge; wear a **ground strap**.

2. Remove the cartridge from the ProLink 9000 reader/programmer. Remove the four screws fastening the cartridge cover plate, then separate the motherboard from the cartridge casing. Place on a clean wooden bench.

3. Using a chip lever, carefully remove the PROM chip(s) (one or two depending on the system) from the motherboard. Using the schematic in the update chips packaging, install the updated chips. Special care is required to avoid bending/damaging the connecting prongs of the new chips when installing them. Next, it may be required to temporarily remove the RAM chip in the same manner as the PROM chips. This should be momentarily separated from the motherboard (there is a non-volatile memory segment in some systems) and then carefully reinstalled. On other systems, removing the battery temporarily from the motherboard will accomplish the same objective. Ensure that all the chips are fully installed in their sockets.

4. Reassemble the cartridge. Note the date of the PROM transfer and the software version somewhere on the cartridge exterior. Then install in the ProLink 9000 tool.

5. Power up the ProLink by connecting to an appropriate vehicle ECM. Variously scroll through the menus to ensure everything is working.

NOTE: When system cartridges or data cards are updated, they will in most cases read and function with all system software/hardware versions that preceded it. Figure 33–17 shows the fuse location in a ProLink software cartridge; the fuse may be removed for testing.

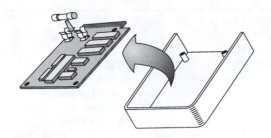

FIGURE 33–17 *Location of fuse on the ProLink circuit board. (Courtesy of MPSI)*

More recently manufactured ProLink cartridges are updated by replacing the data card, a credit card-sized data retention card that is inserted into a slot in the multi-protocol (MCP) cartridge.

ELECTRONIC TROUBLESHOOTING

Electronically managed engines are capable of malfunctioning hydromechanically, so the technician must not assume that every problem is electronically based. The major OEMs produce excellent sequential problem-solving technical manuals; to use such manuals effectively, the technician should be capable of using a DMM, have a basic understanding of electrical and electronic circuitry, and be capable of interpreting schematics. Sequential troubleshooting takes place in stages. It is critical that the instructions in each stage are precisely undertaken before proceeding to the next; subsequent stages will generally be rendered meaningless if a stage is skipped. Some OEMs have PC-based sequential troubleshooting programs for their systems, which simplifies the sequencing of the troubleshooting stages and saves a lot of page turning.

It cannot be emphasized how important it is to use the correct tools. Generally, test light circuit testers should never be used, nor should analog test meters.

When the use of a breakout box is mandated, make sure that it is used. Also, it should be remembered that electrostatic discharge can damage microprocessor components, so it is good practice to wear a ground strap when opening up any housing containing a microprocessor. When testing separated Deutsch and Weatherpac-type connectors with a DMM, ensure that the sockets have been identified before probing with the test leads. Use socket adapters where necessary to avoid damaging terminal cavities. Never spike wires when testing circuits.

SNAPSHOT TESTS

Most systems will accommodate a **snapshot test** readout from the ECM to facilitate troubleshooting intermittent problems that either do not generate codes or log a code with no clear reason. Snapshot mode troubleshooting can be triggered by a variety of methods (codes, conditions, manually) and record data frames before and after the trigger; these can be recorded to a reader/programmer instrument (ProLink) or a PC (laptop) while the vehicle is running. Because frames can be snapshot both before and after the trigger, it is possible to analyze a wide spectrum of data that may have contributed to a problem; each data frame can be examined individually after the event. The portability of the ProLink unit lends itself to this test.

SAE/ATA J1587, J1708, J1939 CODES AND PROTOCOLS

The following is a partial listing of SAE/ATA codes that have been adopted by all the North American truck engine/electronics OEMs. Generally, **SAE J1587** covered common software protocols in electronic systems, **SAE J1708** covered common hardware protocols, and the more recent **SAE J1939** covers both hardware and software protocols. The acceptance and widespread usage of these protocols enables the interfacing of electronic systems manufactured by different OEMs on truck and bus chassis, plus provides any manufacturer's software to at least read other OEMs electronic systems. The truck technician who works with multiple OEM systems may find it easier to work using these codes rather than use the proprietary codes.

The acronym **MID (message identifier)** is used to describe a major vehicle electronic system, usually with independent processing capability. Here are some of the common MIDs.

MESSAGE IDENTIFIER (MID)	DESCRIPTION
128	Engine Controller
130	Transmission
136	Brakes—Antilock/Traction Control
137–139	Brakes—Antilock, Trailer #1, #2, #3
140	Instrument Cluster
141	Trip Recorder
142	Vehicle Management System
143	Fuel System
162	Vehicle Navigation
163	Vehicle Security
165	Communication Unit—Ground
171	Driver Information System
178	Vehicle Sensors to Data Converter
181	Communication Unit—Satellite

The acronym **PID (parameter identifier)** is used to code components within an electronic subsystem.

PARAMETER IDENTIFIER (PID)	DESCRIPTION
65	Service Brake Switch
68	Torque Limiting Factor
70	Parking Brake Switch
71	Idle Shutdown Timer Status
74	Maximum Road Speed Limit
81	Particulate Trap Inlet Pressure
83	Road Speed Limit Status
84	Road Speed
85	Cruise Control Status
89	PTO Status
91	Percent Accelerator Pedal Position
92	Percent Engine Load
93	Output Torque
94	Fuel Delivery Pressure
98	Engine Oil Level
100	Engine Oil Pressure
101	Crankcase Pressure
102	Manifold Boost Pressure
105	Intake Manifold Temperature
108	Barometric Pressure
110	Engine Coolant Temperature
111	Coolant Level
113	Governor Droop
121	Engine Retarder Status

PARAMETER IDENTIFIER (PID)	DESCRIPTION
156	Injector Timing Rail Pressure
157	Injector Metering Rail Pressure
164	Injection Control Pressure
167	Charging Voltage
168	Battery Voltage
171	Ambient Air Temperature
173	Exhaust Gas Temperature
174	Fuel Temperature
175	Engine Oil Temperature
182	Trip Fuel
183	Fuel Rate
184	Instantaneous MPG
185	Average MPG
190	Engine Speed

The acronym **SID (subsystem identifier)** is used to identify the major subsystems of an electronic circuit. The SIDs listed here relate to cruise control.

SUBSYSTEM IDENTIFIERS (SID) COMMON TO ALL MIDS	DESCRIPTION
242	Cruise Control Resume Switch
243	Cruise Control Set Switch

SUBSYSTEM IDENTIFIERS (SID) COMMON TO ALL MIDS	DESCRIPTION
244	Cruise Control Enable Switch
245	Clutch Pedal Switch
248	Proprietary Data Link
250	SAE J1708 (J1587) 1939 Data Link

The following SIDs are used by engine and fuel system controllers:

SUBSYSTEM IDENTIFIERS (SID) FOR MID 128, 143	DESCRIPTION
01–16	Injector Cylinder #1 through #16
17	Fuel Shutoff Valve
18	Fuel Control Valve
19	Throttle Bypass Valve
20	Timing Actuator
21	Engine Position Sensor
22	Timing Sensor
23	Rack Actuator
24	Rack Position Sensor
29	External Fuel Command Input

FAILURE MODE INDICATORS

Because the diagnostic software used on trucks today is designed to be user friendly, you certainly do not have to remember specific MIDs, PIDs, SIDs and FMIs (**failure mode identifiers**) because in most cases, the software is designed to route you through the troubleshooting architecture without you having to memorize a bunch of meaningless numeric values. If you work on engines often enough, you will come to know that a 128 MID identifies the engine electronics, but it is more important that you are aware that MIDs are used to identify separate and integral, on-board electronic systems. Every MID has some processing capability. When you want to communicate with any MID on board the truck, first you have to be routed to the MID controller ECM. At that point, the subcircuit electronic troubleshooting kicks in and you are led through the PIDs and SIDs until you have identified a specific malfunctioning component.

If every engine OEM were allowed to come up with coding ways in which one of their components could fail, there would be a scary number of options, most overlapping or slightly differentiating similar failure modes. The best thing about SAE failure mode indicators or FMIs is that any electronic component can only fail in one of twelve ways. That means that when the diagnostic path has been completed, the failed component must be assigned one of the twelve FMIs. That makes life simple when troubleshooting one OEM's equipment with another's software and you are confined to those MIDs, PIDs, SIDs and FMIs. The following is a *complete* listing of FMIs, all twelve, plus those numbered 13 and 14, which are not actually failure modes.

FAILURE MODE IDENTIFIERS (FMI)	DESCRIPTION
0	Data valid but above normal operating range
1	Data valid but below normal operating range
2	Data erratic, intermittent, or incorrect
3	Voltage above normal or shorted high
4	Voltage below normal or shorted low
5	Current below normal or open circuit
6	Current above normal or grounded circuit
7	Mechanical system not responding properly
8	Abnormal frequency, pulse width, or period
9	Abnormal update rate
10	Abnormal rate of change
11	Failure mode not identifiable
12	Bad intelligent device or component
13	Out of calibration
14	Special instructions

SUMMARY

- The ESTs used to service, diagnose, and reprogram truck engine management systems are on-board diagnostic lights, DMMs, scanners, generic reader/programmers, proprietary reader/programmers, and PCs.
- Flash codes are an on-board method of accessing diagnostic codes. Most systems display active codes only, but some display active and historic (inactive) codes.
- ProLink 9000 with the appropriate OEM software cartridge or data card was for many years the industry standard, portable shop floor diagnostic and customer data programming EST.
- Most OEMs are currently using the PC and proprietary software as their primary diagnostic and programming EST.
- ESTs designed to connect with the vehicle ECM(s) do so via the SAE/ATA J1708/1939 connector. The J1708 uses a six-pin and J1939 a nine-pin Deutsch connector.

- Most electronic circuit testing requires the use of a DMM.
- A continuity test is a quick resistance test that distinguishes between an open and a closed circuit.
- A dark band identifies the cathode on a diode.
- Circuit resistance and voltage are measured with the test leads positioned in parallel with the circuit.
- Direct measurement of current flow is performed with the test leads located in series with the circuit.
- A Hall effect probe can be used to approximate high current flow through a DC circuit.
- The ProLink 9000 EST can be used to access active and historic codes, read system identification data, perform diagnostic testing of electronic subcomponents, reprogram customer data, and perform snapshot data analysis.
- A ProLink OEM software cartridge is updated by replacing the PROM chips or data card.
- A snapshot test is performed to analyze multiple data frames before and after a trigger, usually a fault code or manually keyed.
- SAE J1587 and J1939 protocols numerically code all on-board electronic subsystems, parameters, and failure modes.
- System parameter (PID) failures are identified by one of fourteen FMIs making circuit diagnosis easier.

REVIEW QUESTIONS

1. Which EST is used to read flash codes?
   a. Dash diagnostic lights
   b. Generic reader/programmer
   c. Diagnostic fork
   d. PC
2. The appropriate EST for performing a resistance test on a potentiometer-type TPS isolated from its circuit is a:
   a. DMM
   b. scanner
   c. Generic reader-programmer
   d. PC
3. Which of the following is the correct means of electronically coupling a diagnostic PC with a truck data bus?
   a. Modem
   b. ATA connector
   c. Jumper wires
   d. Parallel link connector
4. When resistance and continuity tests are made on electronic circuit components, the circuit should be:
   a. Open
   b. Closed
   c. Energized
5. DMM test lead resistance should never exceed:
   a. 0.2 ohm
   b. 1.0 ohm

c. 10 ohms per inch
d. 100 ohms per inch

6. The output (signal) of a pulse generator-type, shaft speed sensor is measured in:
   a. V DC
   b. Ohms
   c. V AC
   d. Amperes

7. Which of the following cannot be performed by a reader-programmer EST?
   a. Erase historic fault codes
   b. Customer data programming
   c. Snapshot testing
   d. Proprietary data programming

8. Which of the following procedures cannot be performed using a PC and the OEM software?
   a. Erasing historic fault codes
   b. Customer data programming
   c. Erasing active fault codes
   d. EUI cutout tests

9. Which of the following methods is the required method of upgrading ProLink software?
   a. Replacing the OEM software cartridge
   b. Upgrading the cartridge PROM chips or cards
   c. Reprogramming the software cartridge from a mainframe
   d. Replacing the RAM chip

10. When reading an open TPS circuit, which FMI would be broadcast to the data bus?
    a. 0
    b. 3
    c. 5
    d. 9

11. How many counts of resolution can be displayed by a 4½ digit DMM?
    a. 1,999
    b. 19,999
    c. 9,995
    d. 99,995

12. The specification that indicates how fine a measurement can be made by a DMM is known as:
    a. rms (root mean square)
    b. Percentage deviation
    c. Resolution
    d. Hysteresis

13. When a DMM is set to read AC, which is the term used to describe the averaging of potential difference to produce a reading?
    a. rms (root mean square)
    b. Percentage averaging
    c. Resolution
    d. Saw toothing

14. Which port on a generic reader-programmer is used to connect the printer?
    a. ATA data connector
    b. SAE J1939 connector
    c. Parallel output port
    d. RS–232 port

15. Which of the following components are inserted into terminal sockets to enable testing by DMM probes and minimize physical damage?
    a. Cycling breakers
    b. Noncycling breakers
    c. Breakout pins
    d. Diagnostic forks

16. Which electrical pressure value does a DMM output when in diode test mode?
    a. 0.3 V
    b. 0.6 V
    c. 0.9 V
    d. 1.1 V

17. *Technician A* states that the accuracy of an analog meter is usually no better than 3%. *Technician B* states that most DMMs have an accuracy factor that is within 0.1% of the reading. Who is right?
    a. A only
    b. B only
    c. Both A and B
    d. Neither A nor B

18. The component that is inserted into electronic circuits so that they can be tested without interrupting them is called a:
    a. Test light
    b. Diagnostic fork
    c. Breakout box
    d. DMM test probe

19. *Technician A* states that using the snapshot test mode when using an electronic service tool to troubleshoot an intermittently occurring fault code will help identify the conditions that produced the code. *Technician B* states that snapshot test mode can only analyze a historical code. Who is right?
    a. A only
    b. B only
    c. Both A and B
    d. Neither A nor B

20. When corroborating the signal produced by a potentiometer-type TPS, the DMM test mode selected should be:
    a. V AC
    b. V DC
    c. Diode test
    d. Resistance

Chapter

34

Electrical Wiring, Connector, and Terminal Repair

PREREQUISITE

Chapter 29 and 33

OBJECTIVES

After studying this chapter, you should be able to:

- Identify the Weather Pack- and Deutsch-type terminals and connectors.
- Assemble sealed connectors using the correct methods and crimping tools.
- Disassemble sealed connectors without damaging components.
- Splice wires where necessary in circuits where the practice is permitted.

KEY TERMS

butt splice
crimping pliers
Deutsch connector
locking tang
Metri-Pack connector
multiple splice
tang release tool
three-way splice

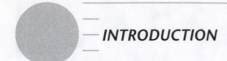

INTRODUCTION

One of the reasons today's truck electronically managed engines present relatively few wiring and terminal-related circuit problems is the overall quality of the weather-sealed components used. In a drive-by-wire chassis, it is crucial that wiring and connector repairs when required duplicate the original standards. Terminals may be platinum or gold coated to minimize resistance, so it is always important to observe OEM guidelines when undertaking any circuit repairs. The following instructions demonstrate general correct practices.

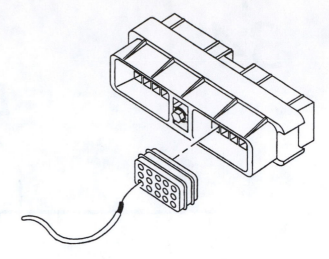

FIGURE 34–1   *Inserting wire in connector.*

CONNECTOR ASSEMBLY AND REPAIR

Most of the wiring and connectors used on electronically managed diesel engines in North America are manufactured by a couple of manufacturers, so although the procedures outlined in this section are specific to one OEM, they are representative of those required for all the major OEMs.

METRI-PACK CONNECTORS

The following sequences the disassembly, repair, and reassembly procedure required for the **Metri-Pack** 150 **connectors** used in DDEC electronic circuits (Courtesy of DDC). This procedure is similar to that required for the assembly and repair of most electronic wiring and connectors.

Installation of Metri-Pack 150 Connectors

Metri-Pack 150 connectors are the pull-to-seat design. The cable is pushed through the seal and correct cavity of the connector before crimping the terminal to the cable. It should be stripped of insulation after it is placed through the seal and connector body. Use the following instructions for terminal installation:

1. Position the cable through the seal and correct cavity of the connector (Figure 34–1).
2. Using wire strippers, strip the end of the cable to leave 5.0 ± 0.5 mm (.2 ± .02 in.) of bare conductor.
3. Squeeze the handles of the crimping tool together firmly to cause the jaws to automatically open.
4. Hold the "wire side" facing you.

5. Push the terminal holder to the open position and insert the terminal until the wire attaching portion of the terminal rests on the 20-22 anvil of the tool. Be sure the wire core wings and the insulation wings of the terminal are pointing toward the upper jaw of the crimping tool (Figure 34–2).
6. Insert the cable into the terminal until the stripped portion is positioned in the wire core wings and the insulation portion ends just forward of the insulation wings (Figure 34–3).
7. Compress the handles of the crimping tool until the ratchet automatically releases and the crimp is complete.

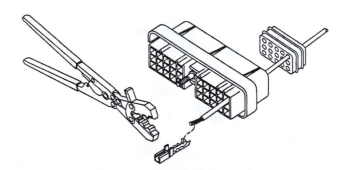

FIGURE 34–2   *Terminal and crimping tool position.*

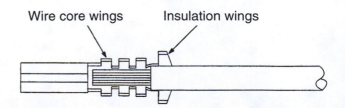

Wire core wings   Insulation wings

FIGURE 34–3   *Cable to terminal alignment.*

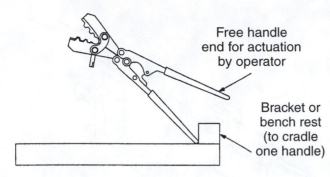

FIGURE 34–4   *Crimping operation.*

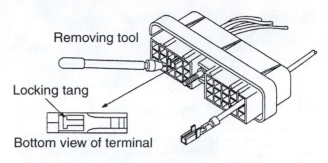

FIGURE 34–6   *Terminal removal.*

NOTE: For faster, more efficient crimping operation, a bracket or bench rest may be used to cradle one handle of the tool. The operator can apply the terminals by grasping and actuating only one handle of the tool (Figure 34–4).

---

8. Release the crimping tool with the lock lever located between the handles, in case of jamming.
9. Align the **locking tang** of the terminal with the lettered side of the connector.
10. Pull the cable back through the connector until a click is heard (Figure 34–5). Position the seal into the connector.

NOTE: For ECM 30-pin connectors, put locking tang opposite lettered side.

Removal and Repair

A tang on the terminal locks into a tab molded into the plastic connector to retain the cable assembly. Remove Metri-Pack 150 terminals using the following instructions.

1. Insert the **tang release tool** into the cavity of the connector, placing the tip of the tool between the locking tang of the terminal and the wall of the cavity (Figure 34–6).
2. Depress the tang of the terminal to release it from the connector.
3. Push the cable forward through the terminal until the complete crimp is exposed.
4. Cut the cable immediately behind the damaged terminal to repair it.
5. Follow the installation instructions for crimping the terminal and inserting it into the connector.

Installation of Metri-Pack 280 Connectors

Use the following instructions for terminal installation:

1. Insert the terminal into the locating hole of the crimping tool using the proper hole according to the gauge of the cable to be used (Figure 34–7).
2. Insert the cable into the terminal until the stripped portion is positioned in the cable core wings, and the seal and insulated portion of the cable are in the insulation wings (Figure 34–8).
3. Compress the handles of the crimping tool until the ratchet automatically releases and the crimp is complete. A properly crimped terminal is shown in Figure 34–8.

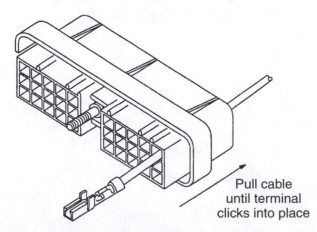

FIGURE 34–5   *Pulling the terminal to seat.*

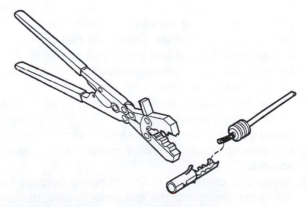

FIGURE 34–7   *Terminal position.*

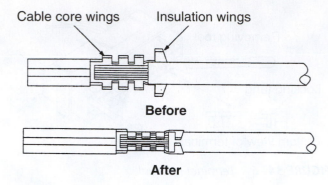

FIGURE 34–8   *Cable and terminal position before and after crimping.*

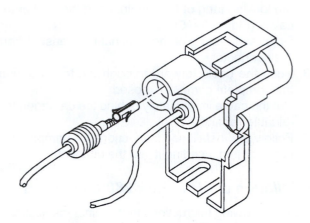

FIGURE 34–9   *Inserting terminal in connector.*

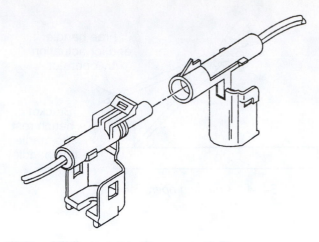

FIGURE 34–10   *Unlatched secondary lock.*

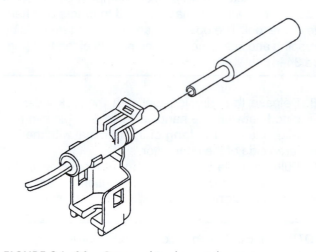

FIGURE 34–11   *Removal tool procedure.*

4. Release the crimping tool with the lock lever located between the handles, in case of jamming.
5. Push the crimped terminal into the connector until it clicks into place. Gently tug on the cable to make sure it is secure (Figure 34–9).

Removal and Repair

Two locking tangs are used on the terminals to secure them to the connector body. Use the following instructions for removing terminals from the connector body.

1. Disengage the locking tang, securing the connector bodies to each other. Grasp one-half of the connector in each hand and gently pull apart.
2. Unlatch and open the secondary lock on the connector (Figure 34–10).
3. Grasp the cable to be removed and push the terminal to the forward position.
4. Insert the tang release tool straight into the front of the connector cavity until it resists on the cavity shoulder.
5. Grasp the cable and push it forward through the connector cavity into the tool while holding the tool securely in place (Figure 34–11).

6. The tool will press the locking tangs of the terminal. Pull the cable rearward (back through the connector). Remove the tool from the connector cavity.
7. Cut the wire immediately behind the cable seat and slip the new cable seal onto the wire.
8. Strip the end of the cable strippers to leave 5.0 ± 0.5 mm (.2 ± .02 in.) of bare conductor. Position the cable seal as shown (Figure 34–12).

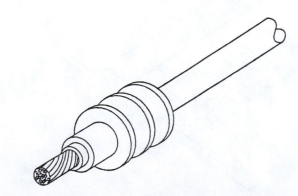

FIGURE 34–12   *Proper cable seal position.*

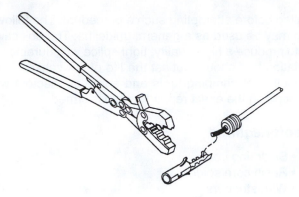

FIGURE 34–13   *Crimping procedure.*

9. Crimp the new terminal onto the wire using the crimp tool. (Figure 34–13).

DEUTSCH CONNECTORS

Deutsch connectors are used on the chassis data bus including the six- or nine-pin ATA connector. Deutsch connectors have cable seals that are integrally molded into the connector. They are push-to-seat connectors with cylindrical terminals. The diagnostic terminal connectors are gold plated.

Installation of Deutsch Connectors

Use the following instructions for installation:

1. Strip approximately ¼ inch (6 mm) of insulation from the cable.
2. Remove the lock clip, raise the wire gauge selector, and rotate the knob to the number matching the gauge wire that is being used.
3. Lower the selector and insert the lock clip.
4. Position the contact so that the crimp barrel is 1/32 of an inch above the four indenters (Figure 34–14). Crimp the cable.

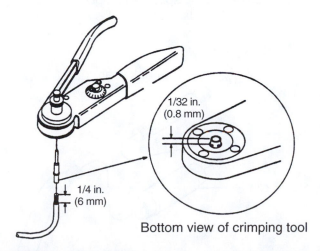

FIGURE 34–14   *Setting wire gauge selector and positioning the contact.*

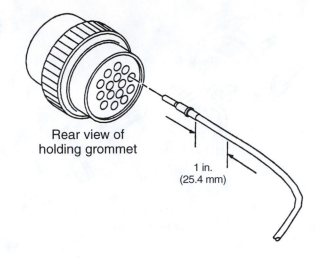

Rear view of holding grommet

1 in. (25.4 mm)

FIGURE 34–15   *Pushing contact into grommet.*

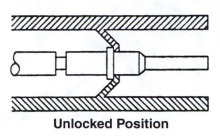

Unlocked Position

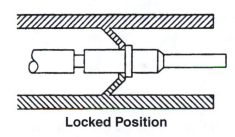

Locked Position

FIGURE 34–16   *Locking terminal into connector.*

5. Grasp the contact approximately one inch behind the contact crimp barrel.
6. Hold the connector with the rear grommet facing you (Figure 34–15).
7. Push the contact into the grommet until a positive stop is felt (Figure 34–16). A slight tug will confirm that it is properly locked into place.

Removal

The appropriate size removal tool should be used when removing cables from connectors.

1. With the rear insert toward you, snap the appropriate size remover tool over the cable of contact to be removed (Figure 34–17).

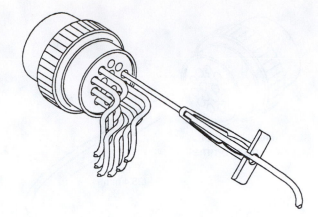

FIGURE 34–17   *Removal tool position.*

FIGURE 34–18   *Removal tool insertion.*

2. Slide the tool along the cable into the insert cavity until it engages and resistance is felt. Do not twist or insert tool at an angle (Figure 34–18).
3. Pull contact cable assembly out of the connector. Keep reverse tension on the cable and forward tension on the tool.

TECH TIP: When replacing connectors, ensure that each wire is labeled by cavity location to facilitate reassembly. Use strips of masking tape on wires to write each cavity code on if you have to: this is usually easier than reassembling a multiwire connector using the OEM wiring schematic.

SPLICING GUIDELINES

Not all wiring in electronic circuits can be spliced, so in all cases the OEM service literature should be con-

sulted before attempting such a procedure. The following may be used as a general guideline. The objective is to produce a high quality, tight splice with durable insulation that should outlast the life of the vehicle. The selection of crimping tools and splice connectors will depend on the exact repair being performed.

Tools Required

- Soldering iron
- Rosin core solder
- Wire strippers
- Heat shrink tubing
- Splice clips
- **Crimping pliers**

STRAIGHT LEADS

To splice straight leads:

1. Locate broken wire.
2. Remove insulation as required; be sure exposed wire is clean and not corroded.
3. Slide a sleeve of shrink wrap on the wire long enough to cover the splice and overlap the wire insulation, about ¼ inch on both sides.
4. Insert one wire into splice clip and crimp.
5. Insert the other wire into splice and crimp (Figure 34–19).

SOLDER

Soldering splice connectors is optional. To solder splice connectors:

1. You must use rosin core solder.
2. Check the exposed wire before the splice is crimped in its connector. The exposed wire must be clean before the splice is crimped.

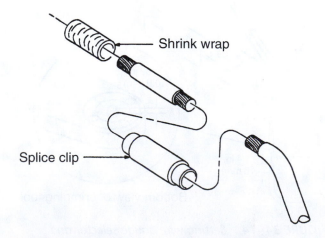

FIGURE 34–19   *Spliced wire. (Courtesy of DDC)*

3. Use a suitable electronic soldering iron to heat the wires. Apply the solder to the heated wire (not to the soldering iron), allowing sufficient solder flow into the splice joint.
4. Pull on connection to assure crimping and soldering integrity.

SHRINK WRAP

Shrink wrap is required. Alpha FIT-300, Raychem TAT-125 or any equivalent heat shrink dual wall epoxy encapsulating adhesive polyolefin is required.
    To heat shrink wrap a splice:

1. Select the correct diameter to allow a tight wrap when heated. The heat shrink wrap must be long enough to overlap the wire insulation about ¼ inch on both sides of the splice.
2. Heat the shrink wrap with a heat gun; do not concentrate the heat in one location, but play the heat over the entire length of the shrink wrap until the joint is complete.

MULTIPLE BROKEN WIRES

To **splice multiple** broken wires:

1. Stagger the position of each splice as illustrated in Figure 34–20.
2. You *must* stagger positions to prevent a large bulge in the harness and to prevent the wires from chafing against each other.

THREE-WIRE SPLICE

Three-way splice connectors are commercially available to accommodate three-wire splices. The technique is the same as a single **butt splice** connector (Figure 34–21).

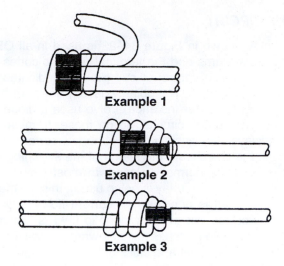

Example 1

Example 2

Example 3

FIGURE 34–21 *Three-way splice. (Courtesy of DDC)*

RELAYS

A relay is an electromechanical switch. Relays are used extensively on truck electrical circuits, and every technician should know exactly how one functions. A relay consists of two electrically isolated circuits. One of the circuits is used to change the switch status of the other. The other is a coil circuit used to electromagnetically switch the circuit. Relays are used so that a low-current circuit can control a high-current circuit. Look at Figure 34–22 showing an SAE standard relay when following this explanation.

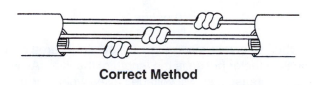

Correct Method

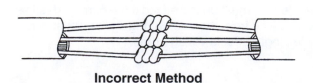

Incorrect Method

FIGURE 34–20 *Multiple splices. (Courtesy of DDC)*

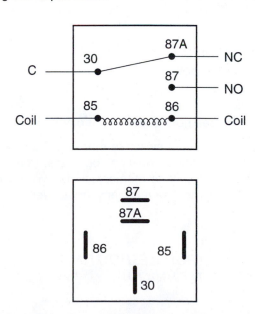

FIGURE 34–22 *Standard SAE relay terminal assignments.*

COIL CIRCUIT

The relay shown in Figure 34–22 is used in all OEM electrical systems and the terminal numeric codes do not change. The control or coil circuit is indicated by terminals 85 and 86. The polarity usually does not matter, but it might. For instance, Volvo uses a diode on some of its control circuits, so on these, terminal 85 must always be chassis ground. When no current flows through the coil circuit, the status of the switching circuit is in NC or normally closed. In most cases (depending on how the relay is being used), this will mean that the output is open. When current flows through the control coil, an electromagnetic field is created, and the movable switch is pulled toward the coil, opening the 30-87a circuit and closing the 30-87 circuit.

SWITCHED CIRCUIT

Referencing Figure 34–22 again, the switched circuit of the relay is represented by terminals 30, 87 and 87a. Terminal 30 is common. This means that a voltage reading could normally be measured at terminal 30 regardless of the status of the switch. When the coil circuit of the switch is not energized, 30 would close on the normally closed terminal 87a. In most cases, this would mean that the switching status of the relay is open, so no current flows through the switch circuit. When the control coil is energized, the switch is moved from the 87a normally closed pole, to the 87 normally open coil: this action permits current to flow from common terminal 30 to the NO terminal 87.

TERMINAL ASSIGNMENTS

Because this relay is standard and its use is widespread, it will pay you to remember the terminal designations:

30   common
87   normally open
87a normally closed
85   coil
86   coil (see note above about diodes in circuit).

Relays are simple switches that only cause confusion if you do not know how one operates. Commit the terminal assignments to memory, and it will pay off in troubleshooting time saved many times over.

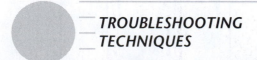

TROUBLESHOOTING TECHNIQUES

In an earlier chapter, we said that there were only three types of electrical circuit fault, namely, opens, shorts,

and high resistance. This said, locating a circuit fault can often be a tough task. The technician can make this task easier by understanding electrical basics and the actual circuit being diagnosed. Learn to rely on your DMM and make double-checking a habit.

WIRE GAUGE FACTORS

The larger the sectional area of wire, the more current-carrying capability it will have because there are more free electrons available to carry it. If too much current is forced down a wire with too small a sectional area, it will overheat due to the molecular friction created. Because some North American manufacturers are using metric wire gauge specifications, you should be able to convert them to American wire gauge (AWG) standards. Metric wire is identified by cross-sectional area expressed in square millimeters (mm²) and not by diameter. To calculate cross-sectional area:

wire sectional area = radius × radius X π (3.1416)*

Metric to AWG Conversion

The following list correlates metric wire size to its approximate AWG equivalent.

Metric Size	AWG
00,22	24 gauge
00,35	22 gauge
00,5	20 gauge
00,8	18 gauge
01,0	16 gauge
02,0	14 gauge
03,0	12 gauge
05,0	10 gauge
13,0	6 gauge
32,0	2 gauge
62,0	0 gauge

When selecting wire gauge, the sectional area is obviously a factor but so is the length of the wire. As wire length increases, so will voltage losses when current is pumped through the wire. This means that the wire gauge size must be increased if the current destination is far away from the source. Figure 34–23 shows how wire length affects voltage drop.

*π is the Greek letter pi that expresses the ratio of the diameter of a circle to its circumference.

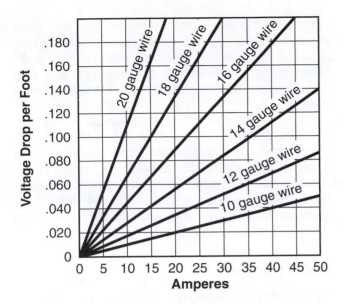

FIGURE 34–23 *Voltage drop and wire length chart.*

CIRCUIT VOLTAGE BEHAVIOR

When a discharged battery tested at 8 volts is connected to a battery charger with voltage measured at 16 volts, initially there will be high amperage flow from the charger to the battery. As the electrical pressure in the battery builds during charging, the rate of current transfer will slow due to the drop in charge differential, caused by battery voltage rising. The result is that as battery voltage rises to its fully charged level, the rate of amperage flow from the charger to the battery drops.

In Chapter 29, we said that the sum of voltage drops must always equal source voltage. In a perfect electrical circuit, voltage should be dropped only across the intended loads in that circuit. Should voltage be dropped elsewhere in the circuit due to high resistances, there will be insufficient voltage to allow the intended load in a circuit to operate properly.

TECH TIP: When the voltage reading does not stabilize but hunts, this is known as *ghost* voltage. Ghost voltage readings indicate an open circuit.

Voltage Drop versus Resistance

Every technician should understand the importance of voltage drop testing and that means understanding the limitations of the ohmmeter. If a copper starter motor cable has deteriorated to the extent that only 10% of its wire strands remain intact, an ohmmeter measurement will indicate that it checks out with a resistance reading similar to that of a cable in perfect condition because the ohmmeter is forcing a minute current through the test circuit. However, if a voltage drop test were to

be performed on the cable while cranking the engine, the voltage dropped would immediately indicate the problem.

TECH TIP: Voltage dropped across a high-resistance circuit is dissipated as heat. This accounts for the high temperatures felt across corroded terminals and wires.

TESLITE® CIRCUIT TESTING

TESlite voltmeter leads are a set of DMM leads that have circuitry added that helps locate corrosion in electrical circuits using digital voltmeter mode. The hardware consists of a set of normal DMM test leads connected through the TESlite module, a small plastic case with a trigger button shown in Figure 34–24. Until the button is depressed, the TESlite leads function exactly like regular DMM test leads. When the TESlite trigger button is depressed, a 510 Ω resistor is placed in parallel across the leads. This means that a component can be removed from its circuit, which can then be subjected to a dynamic load test. Because of the nearly universal adoption of sealed weatherproof connectors in truck electrical circuits making access to the wires difficult without damaging them, this permits voltage drop testing for corrosion in the wiring itself.

TESLITE® TEST PROCEDURE

The TESlite testing sequence is simple. The leads function as normal DMM test leads until the trigger button is depressed, so they are designed to replace the original set of leads.

1. Open the circuit at the component, for instance a relay.
2. Connect the voltmeter in the exact place where the load component was: the negative (black) lead

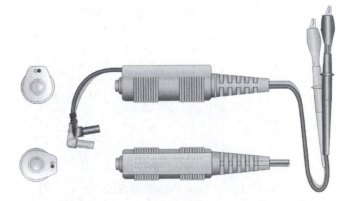

FIGURE 34–24 *TESlite® Enhanced Voltmeter Leads. (Courtesy of Sullivan Training Systems)*

should be placed on the ground terminal NOT to chassis ground.

3. Take a standard voltage reading with the circuit energized.
4. If nominal (system) voltage exists, depress the TESlite trigger button.
5. If voltage remains nominal, (does not drop more than 0.5 V) the circuit can be regarded as clean and the test can be concluded (Figure 34–25A).
6. If voltage drop exceeds 0.5 V, there is corrosion present. Proceed to the next steps in the test procedure.
7. Move the negative lead to a good chassis ground and depress the TESlite trigger.
8. If voltage remains low, the corrosion is in the positive wire (Figure 34–25C).
9. If voltage rises to nominal, the corrosion is in the negative wire (Figure 34–25D).

CIRCUIT ANALYSIS

Getting used to using your DMM on real world circuits should mean testing properly functioning circuits. Too often, a technician tests a circuit for a defect without knowing how the circuit would test if it were fully functional. Try to make a practice of testing good circuits as well as bad. It is important to remember that your DMM is capable of making exact measurements and the specific value is significant. When evaluating circuit performance, there is a difference between 12.6 and 12.5 volts: make a habit of writing down the results of circuit testing.

Figure 34–25 shows four schematics. In the first, Figure 34–25A, the circuit voltage values show what to expect when the circuit is energized and functioning properly. This is probably the most important of the four. The next three schematics show typical measurements made when the load is defective (34–25B), there is a defect on the positive side (34–25C), and a defect on the ground side of the circuit (34–25D).

CIRCUIT SYMBOLS

Most engine manufacturers like to use their own wiring schematic symbols. Despite some commonality, there are enough differences to make these symbols confusing if you are working with a variety of different manufacturers' product. However, you will almost always find these symbols are deciphered somewhere in the service literature or on-line service link. Figure 34–26 shows some examples of typical circuit symbols in current use.

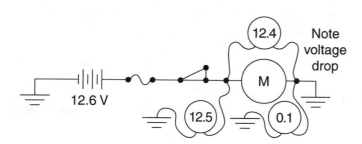

A. Circuit Functioning Properly

B. Defective Load

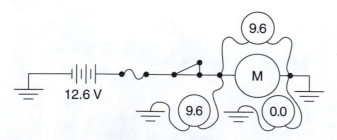

C. Defect on Positive Side

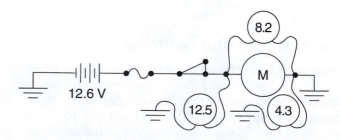

D. Defect on Ground Side

FIGURE 34–25   *Circuit analysis.*

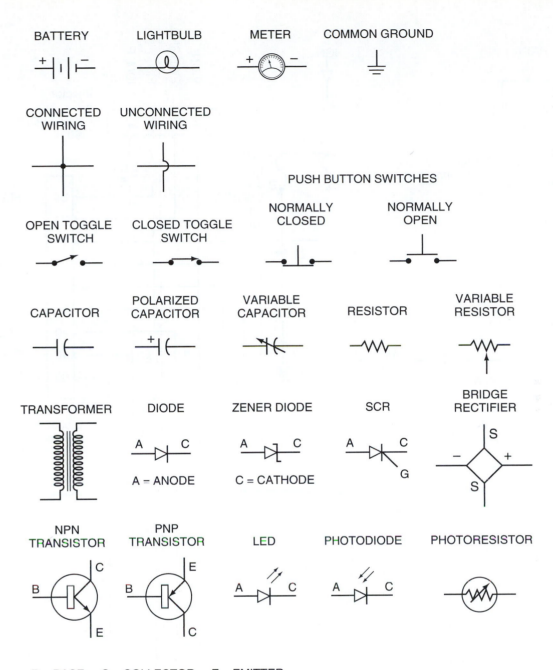

FIGURE 34–26 *Circuit symbols.*

CIRCUIT SCHEMATICS

There is no universal set of standards for wiring schematics in North America, which can make interpreting them challenging. Because of the recent European influence in truck manufacturing in this country (Freightliner/ Mack-Volvo/ DAF-Paccar), the same OEM often uses different wiring schematic protocols within their own operation. The best way to learn wiring schematics is simply to use them. Figure 34–27 shows a simple engine wiring schematic but it is probably better practice to use full chassis wiring schematics and trace circuits from load back to source.

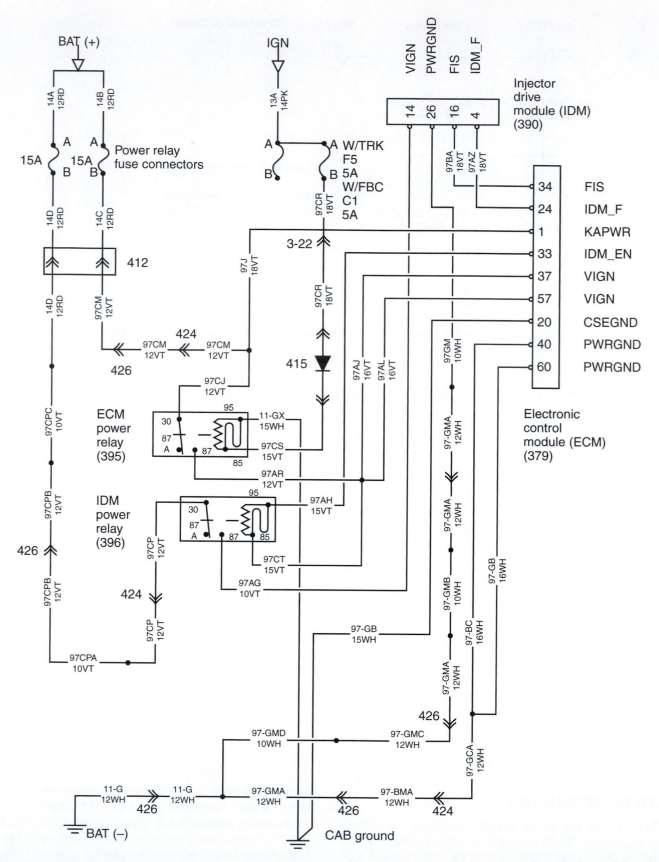

FIGURE 34–27 *Electrical circuit schematic. (Courtesy of International Navistar Corp., Designer and manufacturer of International Brand diesel engines)*

SUMMARY

- Always use the correct terminals and connector units.
- Avoid "temporary" repairs.
- As much as possible, duplicate the original standards.

REVIEW QUESTIONS

1. Metri-Pack terminals are retained in the connector unit by a:
   a. Clamp
   b. Jam bridge
   c. Locking tang
   d. Silicone
2. The correct tool for installing Metri-Pack terminals to the wiring is a(n):
   a. Crimping tool
   b. Electrical pliers
   c. Wire strippers
   d. Tang release tool
3. Which type of connector is the industry standard ATA data link connector?
   a. Weather-Pack
   b. Metri-Pack
   c. Siemens
   d. Deutsch
4. When using solder to splice repair wiring, which type should be used?
   a. 50/50
   b. Rosin core
   c. Silver solder
   d. Liquid set
5. When repairing a break through a section of a multiple wire cable, which of the following would be good practice?
   a. Use shrink cable on each individual wire.
   b. Stagger the splices.
   c. Twist the bared wires at each connection before soldering.
   d. Write "temporary repair" on the work order hardcopy.
6. What numeric code is used to represent the common terminal on a standard relay?
   a. 30
   b. 85
   c. 86
   d. 87
7. Which of the following describes the function of numeric code 87a on a standard relay?
   a. Common
   b. Coil
   c. Normally open
   d. Normally closed
8. Which metric wire gauge size is equivalent to 18 gauge AWG?
   a. 18 mm
   b. 00,22
   c. 00,8
   d. 05,0
9. What does ghost voltage indicate when testing a circuit?
   a. Short
   b. Open
   c. High resistance
   d. High current
10. When testing a 12.6 truck electrical circuit component, 12.5 volts are measured immediately on the positive side of the load, 9.2 volts are dropped across the load, and 3.3 volts are measured on the negative side of the load. Which of the following is true?
    a. Circuit is functioning properly.
    b. Load component is defective.
    c. Defect on positive side of circuit.
    d. Defect on negative side of circuit.

Chapter 35

Multiplexing

PREREQUISITES

Chapters 29, 31, and 32.

OBJECTIVES

After studying this chapter, you should be able to:

- Define the word multiplexing.
- Describe how multiplexing can make data exchange more efficient.
- Use some every day analogies to make multiplexing understandable.
- Define the role the data bus plays in a multiplexed electronic truck chassis.
- Explain how a "smart" ladder switch operates.
- List the seven essential fields that make up a data frame on a truck data bus transaction.
- Describe some of the features of the Freightliner M2 multiplex system.
- Explain how FETs are used as relays to effect data bus outcomes.

KEY TERMS

bandwidth

bus systems

client

controller area network (CAN)

data bus

data frame

electromagnetic interference (EMI)

J1850

J1939

ladder switch

multiplexing

packet

power line carrier

programmable logic controllers (PLCs)

protocols

servers

smart switch

twisted wire pair

INTRODUCTION

Multiplexing has been used in truck computer control systems since electronic control modules (ECMs) were first introduced in the late 1980s. Back then however, a truck technician did not need much of an understanding of what it was all about. Multiplexing means data sharing between multiple system control modules with processing capability. If you set up a chassis computer system so that all of the control modules "speak" the same language and provide a common, shared communication path between them, then you have a multiplexed system. In early systems, multiplexing meant communications between two interdependent modules, say a fuel injection control module and an engine controller. The way in which these communications took place did not much matter to the truck repair technician. For sure, these early systems might have been primitive, but it was nevertheless, multiplexing. But technology advances.

Although the term sounds complex, multiplexing was introduced to simplify truck electronics. It does this by giving electronic subsystems a common communication language, and, by using a **data bus** or information highway, it allows data signals to take the place of hard wire in the electronic input and output circuits. More recently, truck technicians have necessarily become more aware of the term multiplexing because truck manufacturers, notably Freightliner with its M2 chassis, are beginning to introduce more complex multiplex systems into the marketplace. This new generation of trucks has networked all of the electronic controllers in a chassis in a way that simplifies the hardware, eliminates miles of hard wiring, reduces the number of I/O (in-out) pins on modules, and optimizes vehicle operation by giving electronic subsystems an "accountability" that extends beyond the hardware they control. Many highway/city bus technicians already have some understanding of multiplexing as they have advanced to a higher level of sophistication in bus chassis.

So what do you really have to know about multiplexing today? Much less than you will have to know five years from now would be one answer. This chapter attempts to clearly define the building blocks of truck multiplexed electronics.

A MULTIPLEXING ANALOGY

Around the time of the release of the Freightliner M2 chassis, it became clear that truck technicians would have to come to terms with multiplexed chassis systems. This led to a lot of discussion about what analogies would work best to explain the concept in simple terms. I liked this one. Imagine a family of four is taking a shopping trip to the mall. The family consists of Mom and Dad and a couple of kids. Each member of the family has something different they want to accomplish at the mall. Each has some currency, yes money, with which they can achieve their individual objectives. In the world of the shopping mall these people are called customers. However, in the professional world, such as when consulting an accountant or lawyer, a customer is usually known as a **client**. In the computer world anyone, or anything, that wants something is referred to as a client. For a transaction to take place, there must be someone or something that fulfills that need. The fulfilling of a client need is provided by a **server**. So it makes sense that we refer to the stores in the shopping mall as servers.

To get to the mall, the family of four gets into an automobile driven by Dad. Each member is equipped with his or her own shopping list. The car with its four passengers, each with their own shopping list, is called a **packet** in the telecommunications world or for instance, in the world of courier services. Almost always there are rules about packets and how they are constructed — think of the difficulty and expense of sending an odd-shaped package by courier. So, Dad does the driving and Mom navigates, while the kids make lots of noise and argue. In this way, the packet travels from home to the shopping mall. It navigates a series of roads to get there. Some roads, like residential streets are small and slow. Others like an interstate are big and fast, capable of handling large volumes of traffic at high speed. In telecommunications language, the different types of roads that the package will navigate to get to its destination are known as transmission mediums or channels.

The roads used by telecommunications systems can be wires, fiber optic cables, light beams, or radio waves. The speed limit on the highway is the rated speed of a telecommunications channel measured in baud and K-baud. More K-baud is better and faster. An interstate with more lanes is generally considered to be better and the number of lanes on a road can be likened to the telecommunications term **bandwidth**. They both represent volumetric capacity of the road or transmission medium. One refers to the number of cars and trucks that can travel simultaneously, the other, the number of packets or data volume that can be pumped down a channel. On the highway, because there are other cars on the road each filled with other people with their own objectives, you can closely compare the dynamics of road travel with data multiplexing: multiple packets traveling a highway heading from one place to another, using common transmission mediums moving at pretty much the same rate.

So the vehicle arrives at the mall and Mom, Dad, and the two kids enter the mall with their individual shopping lists. The lists of items to be purchased are commands; they tell each individual what to purchase. Each will conduct a transaction with a server using some well-established **protocols**. In telecommunications, the term protocol can be defined as the rules and regulations of a communication transaction, for instance, communication only takes place if both parties are using a common language. At the stores, those shopping list items are purchased using money, they

are packaged, and transported back to the family vehicle. A command, such as purchase item X on the list, results in a transaction with a server and the effecting of a response, that is, the purchaser leaving the server with the item, transporting it back to the vehicle, which will route the item back home. Looking at the big picture, all of the family members meet back at the car, which then transports them home equipped with their individual shopping list items that will meet their individual needs. Figure 35–1 shows the shopping trip example in diagram form.

The Smith Family Visits the Mall

Mom's shopping list
Food
Book
Garden tools
Clothes

Dad's shopping list
New tools
Hose
Shoes

Daughter's shopping list
Cosmetics
Jewelry
Blouse
Pants

Son's shopping list
Shirt
Pants
Jacket
Sports gear

Clients = Customers

Channel = Road

Bandwidth = Capacity
— Speed: car is faster than a truck
— Four-lane highway has more capacity than two-lane highway
— Car with people and individual lists represent single package of information
— Multiple cars on highway = multiple packets all to their own destination

Servers = Stores
— Customers return home from mall with packages
— Shopping lists = commands/requests

FIGURE 35–1 *Example of multiplexing.*

Looking at the still bigger picture, a community of multiple families effecting similar journeys and transactions to achieve their distinct objectives but using the same transportation media (highways) and you begin to get an idea of what clients, servers, and multiplexing are all about. Do you need protocols? You bet! These rules and regulations are essential. Think what would happen in the absence of traffic lights, speed limits, and rules of the road such as driving on the right. Think about whether highway travel today would even be possible if each vehicle could only travel on its own dedicated road. Multiplexing modernizes electronic communications by making electronic circuits a lot more like our highway system. In doing so, miles of unnecessary hard wire and replicated components can be eliminated. Multiplexing uses electrons modulated by some exacting protocols to simplify subsystem-to-subsystem electronic transactions. Analog inputs are converted to digital signals by the primary multiplexing processor: these can then be sent digitally to slave processors, changed back to analog format to effect actions.

Other analogies can also be used. If you look at how any national courier company organizes a typical package delivery transaction, from pickup by a small van, sorting at a depot, trucking to a hub, air freighting to another hub, finishing with delivery to destination, you will see many similarities with how multiplexed data transactions are effected. A coast-to-coast telephone call would be another example.

MULTIPLEXING BASICS

If you have worked on trucks manufactured in the 1990s, you have probably worked with multiplexing, perhaps without knowing. Multiplexed truck chassis such as the Freightliner M2 herald a second generation of truck multiplexing. Many of the systems currently on bus chassis represent a third generation, which boasts more comprehensive interaction and control between the electronic subsystems. This third generation of multiplexing will find its way into trucks sooner rather than later.

Initially, we described multiplexing as chassis information sharing between the different onboard computer control systems to optimize vehicle performance. For instance, some of the key input sensor signals are required by more than one of the chassis control modules. The throttle position sensor (TPS) is a good example of such a signal. The TPS signal is a required input for the engine management, transmission management, dash display module, antilock braking system (ABS)/automatic traction control (ATC), and collision warning systems. Some electronic systems commonly multiplexed in first generation, multiplexed truck chassis included the following:

- Engine control module
- Fuel injection control module
- Transmission control module
- Dash display module
- Chassis control module
- Antilock braking system
- Automatic traction control system
- Electronic immobilizer
- Collision warning system

Exchange of information between computer-controlled systems reduces the total number of sensors required and optimizes vehicle performance by making each system perform according to the requirements of the vehicle system rather than each of its subsystems.

Conventional data transmission in a vehicle required that every input and output signal had to be allocated an individual conductor (dedicated wire and terminals) because binary signals can only be transmitted using the states "1" or "0" (binary code). On/Off ratios can be used to transmit continually changing parameters such as the status of an accelerator pedal-travel sensor. However, the increase in data exchange volume between vehicle electronic systems has reached dimensions that it does not make sense to attempt to handle it by conventional hard wire and plug-in connectors, (Figure 35–2). Today, the objective is to keep wiring-harness complexity down to a controllable level to reduce costs and downtime when there is a hard wire malfunction.

POWER LINE CARRIER

A more primitive method of multiplexing known as **power line carrier** has been in use since the introduction of trailer ABS. Power line carrier enables communications

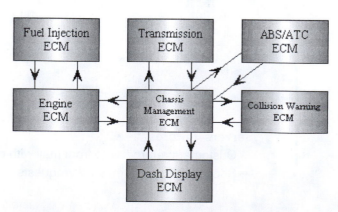

FIGURE 35–2 *Conventional data transmission: no common multiplex bus.*

transactions to take place on a nondedicated communication wire. Power line carrier communications were introduced when ABS was mandated on highway trailers, along with the requirement that a warning light signal the cab of the truck in the event of a malfunction. Because all the wires on a standard ATA seven-pin connector between truck and trailer were already dedicated, power line carrier technology was used to convert communication signals to radio frequency signals and broadcast them over the 12-volt auxiliary power wire. Translators on either end then convert the radio frequency signals back to signals that the ECMs can use. These signals used J1587/J1708 hardware and protocols and were diagnosed accordingly.

TYPES OF LOGIC

Most current engine management systems use largely open-loop (fuzzy) processing as opposed to closed-loop logic, so inputs from any onboard electronic circuit can influence how the engine is being managed at a given moment. This adds to volume of exchange data between the various control modules managing a chassis. To illustrate the difference between closed-loop and fuzzy logic, we'll use an example you should be familiar with if you have worked on or driven a truck—cruise control.

The older hard cruise, used in most autos, required that the driver set a road speed, say 60 mph. After the road speed has been selected, the road speed sensor drives the management of the powertrain to maintain the desired road speed regardless of conditions. This type of closed-loop operation puts the road speed sensor in command of the powertrain, and it can be a disadvantage. For instance, if the vehicle were traveling through mountainous terrain, attempting to maintain the input road speed at exactly 60 mph would waste fuel. Most trucks today use "soft" or "smart" cruise. When the driver inputs a road speed cruise control command into a soft cruise system, the circuit "thinks" for itself. It uses multiple inputs and programmed instructions to manage fueling. For instance, if the soft road speed were set at 60 mph and the vehicle was traveling through undulating hills, the processor algorithm "learns" from the terrain and uses a variety of inputs to plot an actual road speed. From a fuel economy point of view, it would make sense to allow the truck to slow slightly through each uphill climb and accelerate on each downgrade to prepare for the next hill. Hard cruise is an example of closed-loop operation, while soft or smart cruise is an example of fuzzy logic.

SERIAL DATA TRANSMISSION

You can see now that many of the problems of data transfer volume using conventional interfaces (hard wire) can be simplified by using **bus systems** or data

highways. A single wire delivers "instructions" rather than electrical signals to the controller modules. One example is **controller area network** or **CAN**, a data bus system developed by Robert Bosch and Intel for vehicle applications. CAN is a serial data transmission network used for the following applications in a vehicle:

- ECM networking
- Comfort and convenience electronics
- Mobile onboard and external communications

CAN 2.0 is the basis for SAE J1939, the high-speed network used on trucks and buses in North America. J1939 functions from 125 K bits per second (b/s) up to 1M b/s making it a class C bus: Class A and B buses use much slower speeds and no longer have applicability in trucks. For instance, the Mack V-MAC III module and EUP fuel injection control module communicate with each other at speeds up to 500 kBit/s. Transfer rates must be high to ensure the required real-time responses are met, so you can bet that transaction speeds will continue to increase in the coming years.

Multiplexing Clock Speeds

Microprocessor clock speeds of at least 16 MHz are required for J1939 transactions. A clock speed of 16 MHz translates to an ability to make up to 16 million decisions per second. Most current truck engine management processors have clock speeds of at least 16 MHz. Some specific examples are the Freightliner/ Mercedes-Benz 900 controller with a 24 MHz rating and a Caterpillar ADEM 2000 with a 32 MHz speed.

ECM Networking

CAN uses serial data transmission architecture. *Serial* means single track as opposed to multiple track which is known as parallel transmission. Several system management modules such as the engine, fuel system, ABS/ATC, electronic transmission, dash display, and collision warning modules are networked on the serial data bus. Each ECM is assigned equal priority and connected using a linear bus structure as shown in Figure 35–3. One advantage of this structure is that should one of the stations (subscribers) fail, the remaining stations continue to have full access to the network. The probability of total failure is therefore much lower than with other logical arrangements such as loop or star (hub) structures, like that shown in Figure 35–2. With loop or star structures, failure of a single stations or the central or command ECM necessarily results in total system failure.

Content-Based Addressing

CAN transactions are simplified. Instead of addressing individual stations in the network, an identifier label is

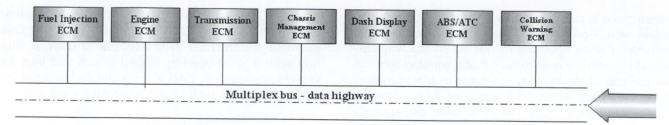

FIGURE 35–3   *Linear data bus structure.*

assigned to every message pumped out onto the data highway. Each message is coded with a unique 11- or 29-bit identifier that identifies the message contents. For instance, engine speed data is of significance to the engine, transmission, and collision warning systems but probably has no importance to the climate control module. Each station is designed to process only those messages whose identifiers are stored in its acceptance list. This means that all other messages are simply ignored.

CAN messages have no explicit addresses. Addressing is content-based. Content-based addressing means that a signal can be received by any of the multiple stations connected to the data bus. A sensor only needs to send its signal directly (or via one of the ECMs in the network) to the bus network from which it is broadcast. Also, because it is easy to add further stations to an existing CAN bus system, many equipment variations are possible.

Assigning Priorities

In a multiplex system, message identifiers can be labeled for data content and/or the priority of any message sent. In a vehicle management network, some signals are necessarily prioritized over others. An example of a high-priority signal would be the accelerator pedal angle or TPS signal. Any change in the TPS signal value must be responded to at high speed and is therefore allocated a higher priority than a signal that changes relatively slowly such as that from the coolant-temperature sensor. A brake request signal would have the highest priority status and a critical signal such as this can effectively shut down the data bus for a nanosecond or so.

Bus Arbitration

Handling traffic on the data highway requires rules and regulations. Just as every driver on an interstate is taught to cede to emergency vehicles, so must there be rules to handle data traffic. When the bus is free, any station can transmit a message. However, if several stations start to transmit simultaneously, bus arbitration awards first access to the message with the highest priority, with no loss of either time or data bits.

Lower-priority messages are shuffled by pecking order to automatically switch to receive and repeat their transmission attempt the moment the bus is freed up.

Message Tagging

In a busy airport with insufficient runways like Chicago's O'Hare, there is an orderly lineup at the beginning of each runway for takeoff. The protocol (rule) on the airport runway is usually to sequence takeoffs in lineup order. The protocol in the truck data bus is necessarily different. Each message is assigned a tag, a **data frame** of around 100–150 bits for transmission to the bus. The tag codes a message for sequencing transmission to the data bus and also limits the time it consumes on the bus. Tagging messages ensures a minimum queue time until the next (possibly urgent!) data transmission.

Each message or packet can be called a *data frame*. A data frame packet is made up of consecutive fields as indicated in Figure 35–4 that shows how a typical packet is constructed. Compare this with the slightly different, Freightliner version of a data packet shown in Figure 35–5.

Start of frame.   Start of frame announces the start of a message and synchronizes all stations on the data bus.

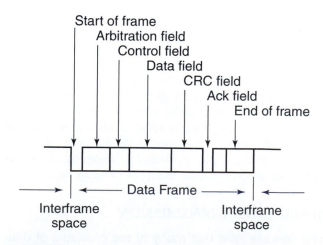

FIGURE 35–4   *CAN message format. (Courtesy of Robert Bosch Corporation)*

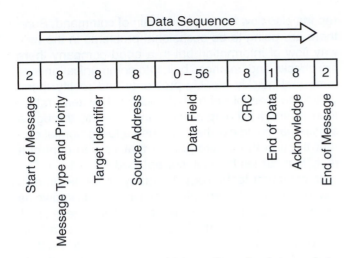

Data Sequence →

| 2 | 8 | 8 | 8 | 0 – 56 | 8 | 1 | 8 | 2 |

Start of Message — Message Type and Priority — Target Identifier — Source Address — Data Field — CRC — End of Data — Acknowledge — End of Message

FIGURE 35–5 *Message bit encoding of a data packet.*

Arbitration field. The arbitration field contains the message identifier and an additional control bit. When this field is broadcast, the transmitter tags the transmission of each bit with a check to ensure that no higher-priority station is also transmitting. The control bit classifies the message as a data frame or remote frame.

Control field. The control field simply indicates the number of data bytes in the data field.

Data field. The data field is sized between 0 and 8 bytes. A message with data length 0 can be used for synchronizing distributed processes.

Cyclic redundancy check (CRC) field. The CRC proofs a message to identify possible transmission interference.

Ack field. The Ack or acknowledgement field contains acknowledgement signals by which the receiver stations on the bus indicate they have received a non-corrupted message.

End of frame field. The end of frame (EOF) indicates the end of a message.

Message Bit Encoding

Figure 35–5 shows how a Freightliner data packet is sequenced and the number of bits dedicated to each segment. Remember that there are 8 bits to a byte.

Integrated Diagnostics

A vehicle multiplexing bus system must be provided with monitoring capability so that transmission, transaction, and reception errors can be detected. This would include the check signal in the data frame and monitoring components in a transmission. It is a feature of a multiplex bus that each transmitter receives its own transmitted message again, and, by file comparison, can detect any deviation or corruption. If a station identifies an error, it sends out an error flag that halts the current transmission. This prevents other stations from receiving a faulty message.

In the event of a major station defect, it is possible all messages, both good and bad, would be terminated with an error flag. To protect against this, a multiplex bus system has a built-in function that enables it to distinguish between intermittent and permanent errors. Furthermore, this feature can sometimes localize station failures and route the data network to an appropriate failure strategy. The process of localizing a station-type failure is based on statistical evaluation of error factors, so it is not entirely foolproof.

Multiplexing Standardization

Multiplexing standardization for vehicles has been orchestrated by ISO (International Organization for Standardization) internationally and by the SAE (Society of Automotive Engineers) in North America. In North America, SAE J standards, **J1850** (light-duty vehicles) and **J1939** (heavy-duty highway vehicles) define the hardware and software protocols of multiplexed components and data transfer. The international ISO CAN 2.0 is consistent with J1850 and J1939 standards used in the United States and Canada. The SAE J1939 combines and replaces the older J1587/1708 hardware and software protocols: current trucks tend to be both J1939 and J1587/1708 compatible. Hardware and software engineered in compliance with J1939 means that they can connect into the vehicle data bus without the translator/transducer modules that were required in older truck systems.

THE J1939 BACKBONE

A multiplexing data bus must be stable and be protected against unwanted radiation interference that could corrupt the data signals. The data bus or backbone used in a J1939 multiplexed chassis is a pair of twisted wires, color-coded yellow and green. The reason for using a **twisted wire pair** is to minimize the surface area on which outside low-level radiation or **electromagnetic interference (EMI)** can act on the bus. The idea is to prevent the data backbone from acting like a radio antenna. Reducing EMI (electromagnetism, radar, microwave) susceptibility is critical to protecting data from transmission corruption. The SAE standard for the J1939 data bus requires that the data bus wires twist through a full cycle once per centimeter, that is, 2.5 times per inch. At either end of the data bus, a terminating resistor is used. The function of the terminating resistors is to prevent the twisted pair from attracting signal interference by acting like a giant antenna. Also there cannot be any open T connector cavities. At points where the backbone has to be spliced, special T connectors with gold plated contacts are used. Gold plating minimizes resistance because the nominal voltage on the data bus is low, usually around 2.5 volts or lower.

In most trucks, if the J1939 connection is lost, the driver will be alerted by dash display and often a chassis signal such as flashing the hazard lights. As truck chassis management electronics become increasingly dependent on the data bus, the consequences resulting from a data bus malfunction increase in severity. Physical damage to any data bus described as high bus warrants replacement not repair on shielded twisted wire pairs. High bus lines are shielded. A typical J1939 data bus consists of the twisted wire pair (communication wires) already discussed, a shield wire, and often a couple of filler wires that function to keep the communication wires separate. The harness is usually wrapped with tin foil-like shielding.

Low bus wires are unshielded and may be repaired under some circumstances. Strictly observe the OEM instructions when repairing twisted wire pairs: this includes maintaining the existing twisting cycles, gauge size, and using the correct solder.

- High bus wires are shielded and coded J1939-11
- Low bus wires are unshielded and coded J1939-15

CAUTION: It is not good practice to attempt repairing physical damage to a data bus. Replace according to OEM guidelines.

TECH TIP: When repairing low bus, twisted wires, avoid twisting the wires together prior to soldering. Lay the wires you wish to solder contacting each other then apply tin solder. Twisting the wires together and applying a large blob of solder can create unacceptably high circuit resistance.

SUMMARY OF MULTIPLEXING BASICS

Although the definition of multiplexing is fairly specific when applied to vehicle electronics, if we broaden the definition somewhat, it is easy to see how it can be effective. In organizational theory, a company president tends not to do very much of the actual work himself. He delegates those responsibilities to those who report to him. Communications within the company flow through a chain of command. In an effective company,

reports also flow back up the chain of command. Sure, there are going to be occasional problems in the two-way flow of information but in a good company there are checks and balances built into communication systems that quickly identify problems and rectify them. Using the example of a truck electrical system, rather than control a 20 amp blower motor directly through a switch that could overheat, we use a low potential current through the control switch that acts to control a relay that carries the high current load.

In computer technology, for many years electronics have been used to simulate relays in devices known as **programmable logic controllers (PLCs)**. These small computers can simulate thousands of relays in a fraction of the space using elaborate stacking of transistors and SCRs. Now some of the principles used in PLCs are moving beyond the ECM housing. A key factor that has moved us into a second generation of multiplexing is this "smart" switching technology we look at more closely later in the chapter.

SECOND GENERATION MULTIPLEXING

In attempting to understand second generation truck multiplexing, it is probably best to use a real life example. We use a Freightliner M2 chassis as a model for our explanation. Most of the truck OEMs have introduced or are about to introduce second generation multiplexing electronics, and the decision to use the Freightliner as our example is simply based on the fact that Freightliner currently sells more trucks than any other manufacturer.

M2 ELECTRONICS ORGANIZATION

The key components in the M2 electronics network are the stalk switch, ICU3 (instrument cluster unit, generation 3)-M2, smart switches, bulkhead module, chassis module, engine control module (ECM), and under-hood power distribution module (PDM). These modules receive and/or broadcast data and control signals from the TPS, right headlamp, left headlamp, air manifold unit, rear lamps, fuel level circuit, and others. Many of these components seemingly have little connection with the engine and its management, but as we advance into the electronic era, ALL of the electronic apparatus in a vehicle becomes increasingly more co-dependent and integrated into the vehicle management logic. If you have studied this textbook more or less sequentially, you will have a pretty good understanding of most of the components we have listed here, so our focus is on how they are integrated into M2.

Instrument Cluster Unit

The instrument cluster unit used with M2 uses a seven-digit liquid crystal display (LCD) for displaying odometer and some other information. The ICU is equipped with twenty-six indicators and telltales to display engine warning alerts, turn signal, high beams: gauge needles are controlled by stepper motors actuated by the output circuit electronics. Additionally, the ICU is equipped with a mode switch that allows the operator to display trip information and chassis fault codes. Satellite download information can also be displayed on this latest version of the ICU.

The steering column stalk switch (multifunction switch) signals are routed directly to the ICU. Because the ICU receives signals from the stalk switch, it becomes the means of launching stalk switch signals onto the data bus. Also, because the ICU is integrated into the data bus, other chassis computers such as the engine management module can illuminate the ICU check engine light (CEL) without having to use a dedicated wire. Another advantage of the ICU used with M2 is that it can be updated with new software without having to replace the unit. Figure 35–6 shows the M2

ICU display used in U.S. chassis: chassis sold in Canada and Mexico use a slightly different display.

Bulkhead Module

The bulkhead module (BHM) is the first of two additional computers required by an M2 electronics circuit. This module can be looked on as the "brain" of the data bus. It is located in the lower left bulkhead and extends through the firewall into the engine compartment. It has three sealed connectors on the cab side and four on the engine side. Its inputs are too many to list here but include ignition-on, ignition-crank, V-Bat (five times), hazard warning-on, headlamp-on, A/C clutch request and five smart switches: we look at smart switches a little later. BHM outputs are also numerous and include ignition-on signal, wake-up ICU, left high and low beams, A/C clutch, and the J1939 data bus.

When the BHM is powered up, a chassis system self-check (or roll-call) sequence takes place. This uses the electronic processing capabilities of all the chassis ECMs that in turn report using the data bus. The results are then displayed on the ICU. Figure 35–7 shows the M2 self-check sequence.

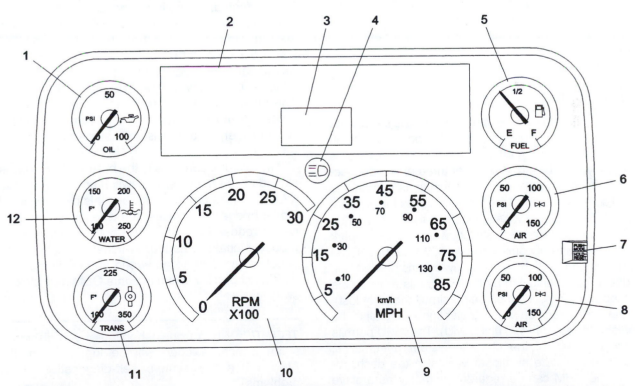

1. Engine Oil Pressure Gauge
2. Dash Message Center
3. Dash Driver Display Screen
4. Headlight High-Beam Indicator
5. Fuel Level Gauge
6. Primary Air Pressure Gauge
7. Mode/Reset Switch (optional)
8. Secondary Air Pressure Gauge
9. Speedometer (U.S. version)
10. Tachometer (optional)
11. Transmission Temperature Gauge (optional)
12. Coolant Temperature Gauge

FIGURE 35–6　*Instrument cluster unit (ICU display). (Courtesy of Freightliner LLC)*

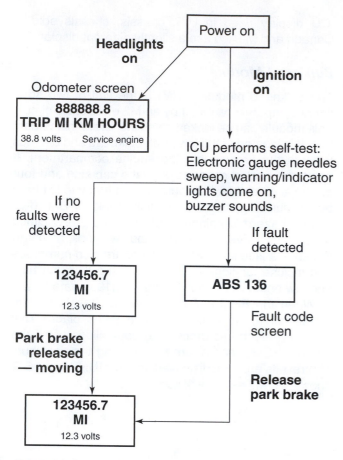

Power on

Headlights on

Odometer screen

| 888888.8 |
| **TRIP MI KM HOURS** |
| 38.8 volts   Service engine |

Ignition on

ICU performs self-test: Electronic gauge needles sweep, warning/indicator lights come on, buzzer sounds

If no faults were detected

If fault detected

| 123456.7 |
| MI |
| 12.3 volts |

| ABS 136 |

Park brake released — moving

Fault code screen

| 123456.7 |
| MI |
| 12.3 volts |

Release park brake

FIGURE 35–7   M2 self-check sequence. (Courtesy of Freightliner LLC)

Chassis Module

Located on a cross member behind the cab is the second of two computers required by the M2 electronics circuit, the chassis control module (CHM). This module is "slaved" by the BHM. It uses either three or five connectors, depending on the version. It functions primarily to control some front and all the rear lights, plus the air manifold's electric-over-air valves. Inputs to the CHM include three V-Bats, service brake switch, wake-up signal, park brake switch and of course, the J-1939 data bus. CHM outputs include the right high and low beams (the left side headlamps are controlled by the BHM: this reduces the chances of both headlamps being simultaneously lost), the rear lights, back-up alarm, trailer lights, and all of the signals to the air manifold unit (AMU). The AMU permits electrical control of the chassis system pneumatics, which can eliminate air flipper valves in the dash. Although the CHM can be regarded primarily as a slave module, it does require some limited processing ability.

Standard Switches

Every truck on the road today is equipped with dozens of switches, and multiplexed trucks will not reduce this

to any great extent. However, these standard switches are often backed up with "smart" switches so that their status can be tracked by the truck electronic systems. Some examples of typical standard switches would be check engine (MB engines), cruise control on/off, cruise control set/resume, hazard switch (in stalk assembly), and ignition switch.

Smart Switches

Smart switches are used to describe two distinct types of switches used in multiplex circuits. A smart switch may have some processing capability and uses that to broadcast switch status onto the data bus. Alternatively, the term smart switch can be used to describe a **ladder switch**, named because it contains a ladder of resistors, usually five per switch, known as a ladder bridge. The processor that receives data from the ladder switches on the data bus has a library of resistor values that enables it to identify switch status and its commands. The M2 truck uses a ladder bridge-type, smart switch in the dash whenever possible, and sometimes this is tagged to a standard switch so that switch status can be broadcast on the data bus. Each smart switch has a light-emitting diode (LED) designed to indicate (to the driver) that a switch request has actually been effected. Sometimes the LED will flash while a particular action is in the process of being effected and stay on when completed. Smart switches can be toggle (two position on/off), multiposition, or spring-loaded, momentary operation. At the present time on an M2 chassis, up to seventy-two smart switches are used. This means that the library of addresses in the receiver processor must contain at least 360 resistor values in order to interpret the resistance signals received off the data bus.

The system, specifically the bulkhead module, is designed to self-check smart switch operation and signal a fault if one is detected. It does this by analysis of the ladder bridge resistances. If a wire loosens or a terminal corrodes, the system will not only know it, it can probably locate it. Ladder switches are simple in construction with little to go wrong and eliminate the processing capability found in some OEM multiplex circuit switches.

TECH TIP: When troubleshooting, if you disconnect a smart switch, a code will be logged immediately. Always use the system self-diagnostics to locate problems.

Field Effect Transistors

Now it is time to see what happens at the other end of the processing circuit. When we have done that, we will study a simple multiplexing transaction from switch to outcome. In Chapter 29 we looked at field effect transformers (FETs). These are becoming commonplace in truck chassis electronics because they are cheap to manufacture and they function reliably. FETs are electronic relay switches. In their simplest formats, there are two types of FET, N channel and P channel. The channel behaves as a resistor that can conduct current from the source side to the drain side. The gate controls the resistance and therefore the operation of the semiconductor device by saying whether, or how much, current flows through the device. So, depending on the gate voltage, the FET can be used either as a straightforward switch (it is in most current truck applications) or as an amplifier. As a "relay," the FET has some great advantages in that gate-channel resistance is so high that first, there is almost no current flow in the gate-current circuit, and second, there is barely any effect on external components connected to the device.

Wake-Up

In the same way that the electronics in your automobile "wake-up" (usually on door open/dome light), the multiplexing electronics in a truck are designed to reactivate following a period of inactivity. Wake-up signals are sourced from the ignition switch, dome light, brake lights, and headlights.

Simple Multiplexing Transaction

To show how a simple multiplexing transaction takes place, we will use a simple toggle operation that actually has nothing to do with the engine or fuel system, but examines an operation from switch to signal to outcome. The suspension dump, smart switch is used to illustrate the execution of a data bus enabled command. The mechanical objective of this operation is to exhaust air from the chassis air suspension. The operator actuates the suspension dump, ladder switch. This causes the dash suspension dump to start flashing and alters the resistance status on the ladder bridge so a packet is generated on the data bus. The data on this packet means nothing to any of the modules on the data bus until it is delivered to the BHM. The BHM then broadcasts a command signal to the CHM, which in turn generates an output signal to the FET that acts as a relay for the suspension dump solenoid. The signal puts the FET in forward bias that completes the solenoid ground circuit, energizing the coil. Source power for the solenoid is the power distribution module. At the point the action is completed, that is, the suspensions dump has been effected, the dash LED ceases to flash and remains on. Simple enough? You bet! But get ready for a lot more use of this type of circuit, and increased use of FET potential in truck electrical circuits.

TECH TIP: When adding loads to a multiplexed truck chassis, *always* consult the OEM service literature. Splicing into circuits can cause electrical problems that become very difficult to troubleshoot.

CAUTION: *Never* splice into existing fuses in the chassis power distribution module (PDM) to source a V-Bat requirement. In dealerships today, it is not uncommon to hear horror stories that result when a truck driver splices into a "hot" wire to power up his CB radio. There are twelve nondedicated V-Bat terminals in the PDM that can be used for loads 30 amps or less. Use the open megafuse at the battery for a V-Bat requirement that exceeds 30 amps. Better still, consult the OEM literature: this will usually recommend several chassis locations for auxiliary electrical equipment.

BENEFITS OF NETWORKING

There are four primary benefits of networking a truck electronic system:

1. Greatly decreased hard wire requirement, reducing the size of the wiring harnesses. This impacts on cost, weight, reliability, and serviceability.
2. Sensor data such as vehicle speed, engine temperature, and throttle position is shared, eliminating the need for redundant sensors.
3. Networking allows greater vehicle content flexibility because functions can be added by making simple software changes. Existing systems would require additional modules and additional I/O pins for each function added.
4. Many additional features can be added at little or no additional cost. For instance, once installed in memory, driver preference data can be routed on the data bus to multiple processors to share such diverse information as seat preference, mirror positions, radio station presets and engine governor type (LS or VS).

SUMMARY

- Multiplexing means the ability of electronic components to exchange information by means of a common data bus.
- Multiplexing can eliminate miles of chassis harness wiring and duplication of hardware devices such as throttle position sensors by giving electronic subsystems a common communication language, and, by using a data bus or information highway, that

allows data signals to take the place of hard wire in the electronic input and output circuits.

- A good analogy that describes how data transactions take place on a data bus is that of a family of four leaving from home, each equipped with a separate shopping list, traveling on an interstate to a shopping mall, each making individual purchases and then traveling back on the highway to home.
- The data bus acts as the "information highway" in a multiplexed electronic truck chassis.
- A smart ladder switch contains a ladder of resistors: the processor that receives signal from the switch can interpret switch status data by comparing ladder resistances with a programmed library of resistance values that identify the switch, its status, and circuit integrity.
- The fields that usually make up a data frame on a truck data bus transaction are start of frame field, arbitration field, control field, data field, cyclic redundancy check field, ack field, and end of frame field.
- The Freightliner M2 chassis is a multiplex system that uses ladder switches and FETs to simplify the electrical system and make it easier to identify circuit problems.
- FETs are used in BHM and CHM modules to effect data bus processing outcomes.

REVIEW QUESTIONS

1. What is used to rate the speed of transmission on a data bus?
   a. Bandwidth
   b. K-baud
   c. Protocol
   d. CAN
2. Which of the following terms is used to rate the data volume that can be transmitted on a data bus?
   a. Bandwidth
   b. K-baud
   c. Frequency
   d. Bus arbitration
3. How many fields are used in a data packet in a J1939 data transaction?
   a. One
   b. One or two
   c. Five
   d. Seven
4. Which of the fields in a data packet contains the message identifier?
   a. Start of frame
   b. Arbitration

   c. Control
   d. Cyclic redundancy
5. Which field in a data packet is used by receiver stations to indicate that they have received an uncorrupted message?
   a. Arbitration
   b. Control
   c. Cyclic redundancy
   d. Ack
6. Which of the following is an accurate method of describing a smart switch as used on an M2 chassis?
   a. Standard switch
   b. Stalk switch
   c. Ladder switch
   d. Switch with processing capability
7. How many smart switches can be used on an M2 multiplexed chassis?
   a. 8
   b. 32
   c. 72
   d. 256
8. What is used to indicate the end of a data packet message?
   a. CRC field
   b. Ack field
   c. EOI
   d. EOF field
9. What current SAE J standard is used as the backbone of a multiplex data bus in current highway trucks?
   a. J1850
   b. J1939
   c. J1930
   d. J1667
10. Which of the following describes the role of an FET in a smart switch transaction?
    a. Transformer
    b. Electronic noise suppression
    c. Relay or amplifier
    d. Signal filter
11. What will result if a smart switch is disconnected at the connector?
    a. Vehicle will stop.
    b. Data bus will shut down.
    c. Default mode operation
    d. A fault code will be logged.
12. What current SAE J standard is used as the backbone of a multiplex data bus in current light duty trucks?
    a. J1850
    b. J1939
    c. J1930
    d. J1667

Chapter
36

Bosch Electronic Distributor and Common Rail Systems

PREREQUISITES

Chapters 18, 25, 29, and 32

OBJECTIVES

After studying this chapter, you should be able to:

- Identify Bosch electronically controlled rotary distributor pumps and common rail systems.
- Describe the Bosch fuel subsystems that supply electronically controlled, rotary distributor and common rail fuel systems.
- Trace fuel flow routing from tank to injector on both electronically controlled, rotary distributor and common rail, fueled engines.
- Identify Bosch electronic diesel control (EDC) system components.
- Recognize the components of a Bosch common rail fuel pump.
- Outline the low-pressure side components used in a Bosch CR system.
- Describe the operation of a VP 44, three-cylinder radial piston pump.
- Understand how rail pressures are managed in the Bosch electronic CR system.
- Outline the operation of an electrohydraulic injector.

KEY TERMS

accumulator

common rail (CR)

Cummins ISB

Duramax

Electronic Diesel Control (EDC)

electrohydraulic injector

Mack Trucks E-3

radial piston pump

rail

VP-44

INTRODUCTION

The subject matter covered by this chapter applies mainly to small-bore diesel engines. The truck technician who specializes in lighter duty highway engines will be familiar with Bosch Electronic Diesel Controls (EDC) because they have appeared on a number of popular fuel systems in recent years, sometimes as a stand-alone system, other times integrated by multiplexing into an existing engine management system such as with Cummins's popular 5.9 liter, ISB series engines.

The first part of this chapter deals with the electronic controls common to both Bosch rotary distributor, injection pumps and the electronically controlled common rail systems. Next, we look at the electronic controls specific to Bosch rotary distributor injection pumps. Because the hydromechanical version of these injection pumps was studied in Chapter 25, it is assumed that the operation of a Bosch VE pump is fully understood.

Most of this chapter studies the operation of the Bosch **common rail (CR)** diesel fuel injection system. This is a medium pressure, common rail system suitable for use on small-bore, direct-injected diesel engines with a wide range of applicability that goes well beyond light-duty, highway applications. In many of these applications, a Bosch **VP-44** high-pressure pump is used. A VP-44 pump is a three-cylinder, **radial piston pump** whose function is to charge the **rail**. In common with other recently introduced common rail, diesel fuel systems, Bosch CR permits injected fuel pressure to be controlled entirely independently from engine speed. This provides the advantage of enabling high fuel pressures at low engine rpm. The result is better transient response to a request for change in engine output, or put more simply, faster response to an acceleration request.

Although both Bosch and Mack Trucks use the acronym ECU (electronic control unit) in preference to ECM (electronic control module), Cummins and General Motors, two major users of Bosch fuel systems prefer the latter. In this textbook, the acronym ECM has been used to describe system controllers except when dealing with specific OEM technologies in which the term ECU is used. Because the OEMs using the technology described in this section use both acronyms in describing their systems, the term ECM is used throughout to avoid confusion.

ELECTRONIC DIESEL CONTROL

Bosch **Electronic Diesel Control (EDC)** was first introduced as a partial authority, engine management system that included drive-by-wire features but was otherwise fairly primitive in the way in which it adapted its popular rotary distributor injection pumps to control by computer. This has evolved into comprehensively controlled engine management systems used with more recently introduced systems such as its common rail (CR) and electronic unit injector (EUI) systems, which Bosch call its Unit Injector System (UIS). The operating principles of UIS are almost identical to the EUI systems used by DDC, Caterpillar, and Volvo in heavy-duty highway applications but are currently limited to use in Volkswagon light-duty diesel engines in North America.

In an EDC-controlled vehicle, the operator has no direct influence through the accelerator pedal on injected fuel quantity. Instead, injected fuel quantity is defined by the engine management computer based on a range of input variables and output requirements that would include accelerator pedal angle, operating temperatures, and emission requirements. EDC monitors ambient, chassis, and engine conditions and is capable of detecting faults, and depending on their severity, can strategize countermeasures such as limitation of torque and limp-home operating modes in addition to exchanging data with other chassis electronic systems in the vehicle.

INPUT CIRCUIT

The sensor circuit inputs data to the EDC microprocessor by means of protective circuitry and, where necessary, via signal transducers and amplifiers. The specific components used to input data to the ECM are no different from other engine management systems and are discussed in Chapter 32. Both analog and digital signals are input to the processing cycle. Manifold air quantity, engine fluid and intake-air temperatures, engine pressure values, and battery voltage are examples of analog signals that must be converted to digital values by an A/D converter in the electronic control module (ECM) microprocessor. Digital input signals include On/Off switching signals and digital sensor signals such as rotational speed/position pulses from Hall effect sensors, and these bypass the A/D unit to be processed directly by the microprocessor.

SIGNAL CONDITIONING

In order to suppress interference pulses and other electronic noise, pulse-shaped input signals from inductive sensors that carry information on shaft speed and position are conditioned by a special circuit in the ECM and converted to square-wave form.

Depending on the level of integration, some signal conditioning can take place either completely or partially in the sensor. Protective circuitry is used to limit the incoming signals to a maximum voltage level. Inputted signals are freed of superimposed interference signals by filtering and then amplified to match it to the ECM working voltage.

PROCESSING CYCLE

Because of the range of fuel system types and engine applications of EDC, the programming and switching apparatus within the ECM varies considerably: each specific ECM is provided with a variant code. Using this code, a selection of operational maps are stored in ECM memory (programmable by the manufacturer or in the workshop), with the objective of providing the specific functions required for the vehicle variant in question. Engine-specific curves and engine-management maps are also written to ECM memory and would include such data as immobilizer strategies, calibration, and manufacturing data. Current ECMs also have a write-to-self capability that will log fault codes and audit trails to EEPROM. The ECM is designed to interface with other chassis electronic systems. For a more detailed look at vehicle engine management computers and how a logic processing cycle functions, refer to Chapter 32.

OUTPUT CIRCUIT

With their output signals, Bosch microprocessors effect the results of the processing cycle into actions. When output commands are triggered within the ECM, the signals produced are usually powerful enough for direct connection to the actuators. While there is much in common in the input and processing hardware in both an electronically managed, rotary distributor pump and a CR fuel system, the output hardware is necessarily very different, so the triggering of individual actuators is dealt with later in the chapter. The output circuit is proofed for common electrical failures such as short circuit to ground, short to battery voltage, and electrical overload. Such faults are identified and reported to the microprocessor. Figure 36–1 shows a schematic of a typical Bosch EDC circuit.

MULTIPLEXING

Bosch EDC systems are capable of multiplexing, that is, sharing information with other chassis computer

control systems to optimize vehicle performance (see Chapter 35). Most current engine management systems use largely open-loop (fuzzy) processing as opposed to closed-loop logic so inputs from any on-board electronic circuit can influence how the engine is being managed. Bosch EDC is J1850 and J1939 compatible and uses the chassis data bus to interact with the following MIDs:

- Transmission control module
- Antilock braking system (ABS)
- Traction control system (TCS)
- Electronic immobilizer (EWS)
- Chassis control module
- Dash control module

Exchange of information between computer-controlled systems reduces the total number of sensors required and optimizes vehicle performance by making each system perform according to the requirements of the vehicle system rather than each of its subsystems. Conventional data transmission in a vehicle required that every signal be allocated an individual conductor (dedicated wire and terminals) because binary signals can only be transmitted using the states 1 or 0 (binary code). While On/Off ratios can be used to transmit continually changing parameters, such as the status of an accelerator pedal-sensor, miles of unnecessary wiring and component replication were required in a typical truck chassis. This, plus the increase in data exchange volume between vehicle electronic systems has reached dimensions that it no longer makes sense to handle it via conventional wiring and plug-in connectors. The hardware problems of data exchange by conventional interfaces can be solved by multiplexing, that is, using bus systems or data highways.

Perhaps most important to the technician troubleshooting vehicle electronic systems, multiplexing permits integrated diagnostics that monitor hardware functions and the data stream on the data bus because each transmitter receives its own transmitted message back again and can therefore detect any deviations. So, when a subsystem detects an error, it transmits an "error flag," which stops the current transmission and prevents other subsystems from receiving a possibly faulty message. If a whole subsystem were defective, it could happen that all messages, good and bad, would be terminated with an error flag. To prevent this happening, the multiplexing system incorporates a function that can distinguish between intermittent and permanent errors and can often localize station failures.

An introductory explanation of multiplexing is provided in Chapter 35. It is becoming increasingly important for truck technicians to have at least a fundamental understanding of how electronic information is managed and exchanged between the subsystems that are found on a typical chassis.

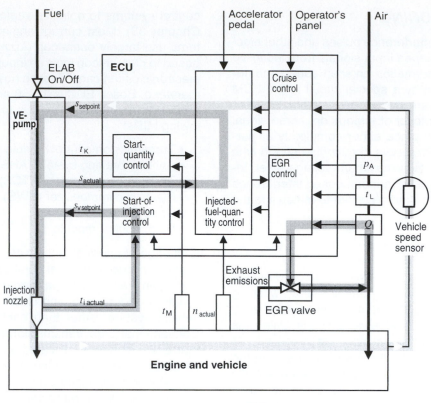

Q	Air-flow quantity	$s_{v\,set}$	Timing-device signal
n_{act}	Engine speed (actual)		(setpoint)
p_A	Atmospheric pressure	t_K	Fuel temperature
s_{set}	Control-collar signal	t_L	Intake-air temperature
	(setpoint)	t_M	Engine temperature
s_{act}	Control-collar position	$t_{i\,act}$	Start of injection (actual)
	(actual)		

FIGURE 36–1 *Closed control loop of the electronic diesel control (EDC). (Courtesy of Robert Bosch Corporation)*

ELECTRONIC CONTROL OF SLEEVE-METERING, SINGLE PLUNGER DISTRIBUTOR PUMPS

In order to understand this section, you must have a working knowledge of the Bosch sleeve-metering, single plunger distributor pumps introduced in the second part of Chapter 25. Because EDC adapts this hydromechanical injection pump for management by computer, it can be classified as a partial authority engine management system. As such, most of the hydromechanical components are retained in the electronic version and are simply adapted for computer control. The operating principles of the hydromechanical version of the pump are not repeated in this section. Mechanical diesel-engine speed control (mechanical governing) offers a limited variety of operating statuses and A/F mixture outcomes. Figure 36–2 shows a sectional view of an electronically controlled VE pump, highlighting

those components that differ from its hydromechanical predecessor.

DIFFERENCE BETWEEN ELECTRONIC AND HYDROMECHANICAL VERSIONS

Most EDC capabilities described in the first section of this chapter are integrated in the ECM used to control the electronic version of a sleeve-metering, single plunger distributor pump. Using an electronic input circuit, adopting flexible electronic data processing and governing, and replacing some of the mechanical controls with switching circuits enhances the versatility and performance of the Bosch rotary distributor injection pump. Most important, electronic controls enabled this diesel fuel injection pump to meet statutory emission requirements that extended its life in the on-highway marketplace. The EDC also permits data exchange with other chassis electronic systems such as elec-

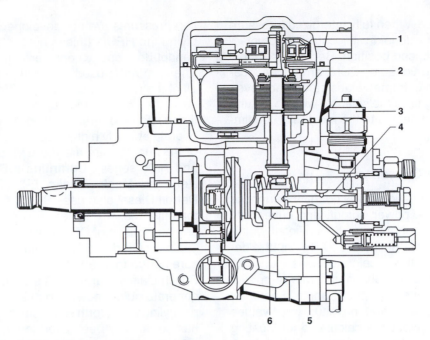

1. Control-collar position sensor
2. Solenoid actuator for the injected fuel quantity
3. Electromagnetic shutoff valve
4. Delivery plunger
5. Solenoid valve for start-of-injection timing
6. Control collar

FIGURE 36–2 *Distributor injection pump for electronic diesel control. (Courtesy of Robert Bosch Corporation)*

tronic transmission-shift control, allowing it to be integrated into the data bus, enabling an overall vehicle management system.

Because the input circuit and processing features are essentially similar to other partial authority systems, these are not examined in this chapter. The ECM output or actuator circuit makes the system distinct so we will take a closer look at this circuit. The ECMs used with sleeve-metering, single plunger injection pumps are usually located within the passenger compartment rather than under the hood.

OUTPUT CIRCUIT COMPONENTS

The ECM processes input circuit data and plots output commands that are sent to the output circuit. The hardware that converts the results of ECM processing into mechanical action are the actuator components. In a Bosch sleeve-metering, single plunger injection pump there are two critical actuators that manage injected fuel quantity and fuel injection timing.

Injected-Fuel Quantity Control

The solenoid actuator (rotary actuator) engages with the control collar or metering sleeve by means of a shaft as shown in Figure 36–2. In a manner similar to the mechanically governed version of the pump, the

spill/cutoff ports are opened or closed depending on control sleeve position. In this way, injected fuel quantity can be varied between zero and maximum because the control sleeve defines how much of the plunger stroke will be effective. By using a potentiometer-type sensor, the position of the control sleeve or collar is continually reported back to the ECM. In this way, the ECM has complete control over plunger effective stroke and therefore, engine output. When no voltage is applied to the sleeve actuator, return springs position it to reduce injected fuel quantity to zero. In other words, in the event of a system electrical malfunction or solenoid failure, the fuel management system is designed to default to no-fuel and shut the engine down.

Start-of-Injection Control

The pump is charged by a positive displacement type transfer pump unloaded into a defined flow area, so pump internal pressure is dependent on engine speed. Similar to the timing device used on the mechanical version of the pump, this pressure is applied to the timing-device piston as shown in Figure 36–2. In the EDC-controlled version, fuel pressure acting on the timing device, pressure-side, is modulated by a linear proportioning solenoid valve.

When the solenoid valve is electrically open (pressure reduction phase), the start of injection is at its most retarded location. When fully energized (pressure increase), the timing solenoid will locate start of injection to the most advanced position. Because solenoid position can be proportioned by ECM, any timing location between the two extreme values can be achieved by the ECM. Once again, if the solenoid were to fail, the system is designed to default to the most retarded timing value.

EDC Performance Outcomes

Improved control over injected fuel quantity influences vehicle starting, idling, power output, and driveability characteristics, but perhaps most significantly, improves engine particulate emissions. Driver input is delivered to the ECM by the accelerator sensor and the ECM will plot outcome responses in terms of speed, power, and torque based on pedal angle. Factoring programmed fuel map data and the actual input values from the sensors, a setpoint is calculated for locating the control sleeve actuator in the pump. Check-back signaling ensures that the control sleeve/collar is correctly positioned in any given moment of operation.

Injection timing (start of injection) has a decisive influence on startup, engine noise, fuel consumption, and exhaust emissions. Start-of-injection maps programmed to the ECM take these interdependencies into account to optimize engine operation. A needle-motion sensor (NBF, a German acronym) in the nozzle assembly signals the actual start of injection to the ECM and compares this with the desired or programmed start of injection. Any deviation in actual and desired timing values will result in a change to the on/off ratio of the timing solenoid valve.

COMMON RAIL ACCUMULATOR FUEL-INJECTION SYSTEM

Electronically controlled, common rail (CR) diesel fuel injection systems have been manufactured for automobile engines by Robert Bosch and Delphi Lucas for a number of years now. The Bosch and Lucas systems are very similar in operating principles but Lucas systems are used primarily offshore. Although their application is pretty much confined to light-duty applications, Bosch CR systems have found their way into North American built, light-duty trucks. For purposes of defining common rail diesel fuel injection, the term CR is used to refer to only those systems in which injection pressure values are held in the rail or **accumulator** feeding the injectors. For instance, some texts refer to the HEUI system as a CR system, but because injec-

tion pressures are not developed in the manifold supplying the HEUIs (this is held relatively constant), this cannot be properly considered a common rail system. The term *rail* is used to describe the gallery that feeds all of the fuel injectors: the principle is somewhat similar to the way in which automotive gasoline fuel injection systems operate, except that the diesel systems operate at vastly higher pressures. Current applications of this fuel system are the GM-Isuzu **Duramax** (6600 and 7800 series), **Cummins ISB** Series (3.9 and 5.9 liter) and **Mack Trucks E-3** (DCI 6-liter) engines. Detroit Diesel also uses Bosch CR in its light-duty automotive diesel engines. Figure 36–3 shows the on-engine location of CR components.

Bosch CR diesel fuel systems provide injection pressures of up to 1350 atms (± 20,000 psi) and although Bosch claims that the CR system has the potential to generate output power up to about 210 BHP (160 kW) per cylinder, in North American highway applications, it has so far only been used in small-bore engines with total output ratings up to 300 BHP. Advantages of the Bosch CR are as follows:

- Injection pressures of up to around 20,000 psi
- Computer-controlled (infinitely variable) start and end of injection
- Multipulse injection: pilot injection, main injection, and post injection shots
- Matching of injection pressure to the operating mode regardless of engine speed.

In common with most recent electronic diesel fuel management systems, pressure generation and fuel injection can be completely decoupled, meaning that injection pressure is generated independent of engine speed and the desired injected fuel quantity. Fuel is charged to, and stored at, injection pressures in the high-pressure accumulator or rail ready for injection. Injection events are triggered by ECM drivers that switch **electrohydraulic injectors**, which makes a diesel fuel injection system function similar to the gasoline injection system in your car, with the exceptions that the fuel is injected directly into the engine cylinder and at maximum pressures up to 20,000 psi instead of the 50 psi used on indirect gasoline fuel injection.

The Bosch CR system is a full authority system with an input circuit and processing capability similar to most other current diesel fuel injection systems. Because Bosch CR is used primarily in light-duty applications in North America, the input circuit and monitoring capability are a little more primitive than some of the systems used to manage contemporary medium- and large-bore engines. The ECM used may be Bosch engineered or in the case of the GM version, Delphi Lucas-engineered electronics are used. A typical input

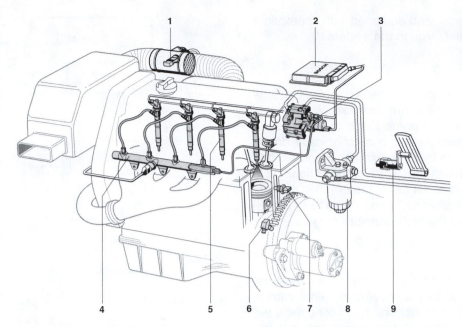

1. Air-mass meter
2. ECU
3. High-pressure pump
4. High-pressure accumulator (rail)
5. Injectors
6. Crankshaft-speed sensor
7. Coolant-temperature sensor
8. Fuel filter
9. Accelerator-pedal sensor

FIGURE 36–3 *Common rail accumulator injection system on a four-cylinder diesel engine. (Courtesy of Robert Bosch Corporation)*

circuit on drive-by-wire, light-duty truck system would comprise:

- Crankshaft-speed sensor
- Camshaft-speed sensor
- Accelerator pedal sensor
- Boost-pressure sensor
- Rail-pressure sensor
- Coolant sensor
- Air-mass meter

Using these input signals, the ECM registers driver request for output by the accelerator pedal angle and combines this with other input sensor monitoring data and programmed performance maps in memory to determine desired engine output. The ECM signals its output drivers to effect desired outcomes into actual outcomes. Effecting the results of the processing cycle in a Bosch fuel system is primarily about managing the fuel pressure in the common rail and switching the electrohydraulic injectors. Because there is nothing especially unusual about the input circuit and processing cycle, our approach is to address the operation of the fuel system components and how the injection cycle is managed, that is, the output circuit.

The common rail system is a modular system and for purposes of study, we divide it as follows:

- Fuel subsystem
- High-pressure pump
- Pressure accumulator or rail
- High-pressure distribution system
- Electrohydraulic injectors

FUEL SUBSYSTEM

The fuel subsystem in a Bosch CR system is made up of the low-pressure stage components. These are responsible for getting the fuel from the tank to the high-pressure stage components. The fuel subsystem components are:

- Fuel tank and prefilter
- Presupply pump
- Fuel filter
- Low-pressure fuel lines

Fuel Tank

The fuel tank is the responsibility of the chassis OEM and is not manufactured by Bosch. It should be made of noncorroding materials and designed to be free from leaks at double the normal operating pressure, which according to Bosch means 5 psi or 0.3 atms. The tank should be properly vented (currently, venting to atmosphere is allowed) and equipped with appropriate fittings for cycling fuel through the system.

Presupply Pump

The presupply pump can be located within the fuel tank or in-line. At present, there are two possible versions. An electric roller-cell fuel pump tends to be the standard option. An alternative is a mechanically driven gear-type pump driven by the engine or integrated into the high-pressure fuel pump, which tends to be used in truck applications. In either case, the pump is positive displacement and pushes fuel through the fuel subsystem and delivers it to the high-pressure stage.

Fuel Filter

The fuel filter is an OEM responsibility but it must meet Bosch specifications, that is, have a nominal entrapment capability of 8 microns. The fuel filter cleans the fuel upstream from the high-pressure pump, but because of the presupply pump, it is under charge pressure. In troubleshooting the system, check the OEM specifications for the rated charging pressure. Similar to other injection systems, Bosch CR requires a fuel filter with a water separator, from which collected water must be drained at regular intervals. Some are equipped with a water-in-fuel sensor that triggers a warning lamp indicating that water should be drained

Low Pressure Lines

Low pressure lines are used to route fuel through the fuel subsystem. Fuel is considered to be at low pressure until it arrives at the Bosch CR pump.

HIGH-PRESSURE STAGE COMPONENTS

Referencing Figure 36–4, the high-pressure stage components in a Bosch CR fuel system include:

- High-pressure pump (6) with pressure-control valve
- High-pressure fuel lines (7)
- The rail as the high-pressure accumulator (8) with rail-pressure sensor, pressure-limiting valve, and flow limiter
- Injectors (9)
- Fuel return lines (10)

High-Pressure Pump

The high-pressure pump is the heart of the Bosch CR fuel injection system and can pressurize fuel to maximum pressures often exceeding 20,000 psi (1,350 atms). The Bosch VP-44, one such high-pressure pump, is used on systems described in this section.

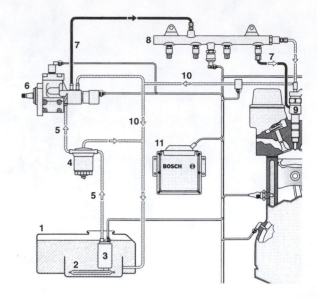

1. Fuel tank
2. Prefilter
3. Presupply pump
4. Fuel filter
5. Low-pressure fuel lines
6. High-pressure pump
7. High-pressure fuel lines
8. Rail
9. Injector
10. Fuel-return line
11. ECU

FIGURE 36–4 *Fuel system for a common rail fuel-injection system. (Courtesy of Robert Bosch Corporation)*

The VP-44 is a three-cylinder, radial piston pump responsible for developing rail pressure. Pressurized fuel from the high-pressure pump then passes through a high-pressure line and into the tubular high-pressure fuel accumulator (rail). The high-pressure pump continually generates the ECM desired pressure to the rail, with the result that in contrast to conventional systems, fuel does not have to be specially compressed for each individual injection pulse. The high-pressure pump is installed in the same location on diesel engine as a fuel injection pump and is driven by the engine at half engine speed by means of a coupling, gearwheel, chain, or toothed belt. The pump is internally lubricated by the diesel fuel it pumps. Figure 36–5 shows a schematic of the high-pressure pump.

Pump components. Pressure rise is created in the high-pressure pump by three radially arranged piston-type, pump elements, evenly offset at an angle of 120 degrees around the pump drive shaft. This means that each pump element is actuated once during a single pump rotation, with the result that the stress on the pump drive remains uniform. Each pump piston is actuated by eccentric cams machined to the drive shaft (Figure 36–6) and discharges fuel through an outlet valve.

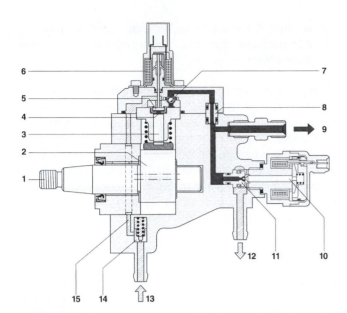

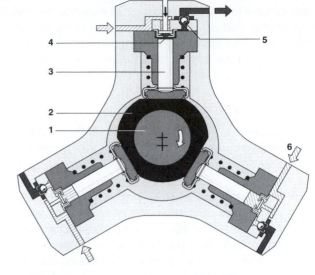

1. Driveshaft
2. Eccentric cam
3. Pumping element
   with pump plunger
4. Pumping element
   chamber
5. Suction valve
6. Element shutoff valve
7. Outlet valve
8. Seal
9. High-pressure
   connection to the rail
10. Pressure control valve
11. Ball valve
12. Fuel return
13. Fuel inlet from the
    presupply pump
14. Safety valve with
    throttle bore
15. Low-pressure passage
    to the pumping element

FIGURE 36–5 *High-pressure pump (schematic, longitudinal section). (Courtesy of Robert Bosch Corporation)*

1. Driveshaft
2. Eccentric cam
3. Pumping element
   with pump piston
4. Inlet valve
5. Outlet valve
6. Inlet

FIGURE 36–6 *High-pressure pump (schematic, cross section). (Courtesy of Robert Bosch Corporation)*

Operating principle. Fuel forced through the fuel subsystem is delivered to each pump element through the fuel inlet (Figure 36–5, Item 13) and the safety valve. Fuel passes through the safety valve throttle bore (14) and enters the high-pressure pump lubrication and cooling circuit, charging the pump elements. As the driveshaft (1) machined with its eccentric cams (2) rotates, it sequentially actuates the three pump plungers (3). These therefore reciprocate in their bores because they ride the cam profiles.

Any time delivery pressure (created in the fuel subsystem) exceeds the safety valve opening pressure, usually from 5 to 25 psi (0.5—1.5 bar), the pumping-element chamber (4) is charged with fuel as the pump piston moves downward on its suction stroke. The pump element inlet valve closes when the piston passes through BDC, and at this moment, fuel is trapped in the pump chamber. As the piston is driven into the pump chamber, pressure rise is created, and when pump chamber pressure exceeds the rail pressure, the outlet valve (7) opens and compressed fuel is charged to the high-pressure circuit.

The pump piston continues to unload fuel until it reaches TDC on the delivery stroke. At TDC, pressure collapses causing the outlet valve to close. As the piston moves downward during the fill stroke, the moment pressure in the pumping-element chamber drops below fuel subsystem pressure, the inlet valve opens and the pump element is charged with another slug of fuel, ready for the next effective stroke.

Fuel-delivery rate. Because the high-pressure pump is designed to comfortably deliver the required volume of fuel for rated speed and load performance, excess high-pressure fuel is delivered to the rail during idle and low-load operation. This excess fuel is returned to the tank by means of a pressure-control valve, which routes the fuel back to the tank. This action results in wasted energy because of the mechanical effort required to actuate the pumping elements. Some of this parasitic loss can be recovered by switching off one of the pumping elements.

Element switch-off. When one of the pumping elements (Figure 36–5, item 3) is switched off, the fuel volume unloaded into the rail is reduced. Element switch-off is effected by holding the suction valve (5) open. When the solenoid valve of the pumping-element switch-off is triggered, a pin attached to its armature holds the inlet valve open, meaning that fuel drawn into the pumping element cannot be compressed because the pump element does not seal. Instead fuel is pushed back into the low-pressure circuit: in essence, the result is similar to that of unloading an air compressor on a truck. With one pumping element switched off when less engine power is required, the high-pressure pump operates on two cylinders.

Pressure Control Valve

The pressure control valve is switched by the ECM. It functions to set the correct pressure in the rail and maintain the ECM plotted "desired" rail pressure. As such:

- If actual rail pressure is higher than desired rail pressure, the pressure control valve opens and fuel is spilled from the rail to the fuel tank return circuit.
- If actual rail pressure is lower than desired pressure, the pressure control valve closes and seals the rail permitting pressure rise.

A sectional view of the ECM controlled pressure control valve is shown in Figure 36–7.

The pressure control valve (Figure 36–5, item 10) is flange mounted either to the high-pressure pump or to the rail. To seal the high-pressure rail from the return circuit, the control valve armature forces a ball into a seat to create a seal. Two forces act on the armature. Mechanical force is provided by a spring, and opposing this, electromagnetic force is created by the solenoid coil when energized. Two control loops are used to manage pressure control valve operation:

1. A slow-response electrical control loop for setting (variable) mean pressure in the rail.

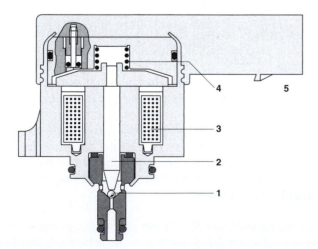

1. Valve ball
2. Armature
3. Electromagnet
4. Spring
5. Electrical connection

FIGURE 36–7 *Pressure control valve. (Courtesy of Robert Bosch Corporation)*

2. A fast-response mechanical control loop to compensate for the high-frequency pressure fluctuations.

Pressure control valve nonenergized. When desired rail pressure is higher than actual rail pressure, the pressure control valve must drop rail pressure. High pressure at the rail or at the high-pressure pump outlet acts on the pressure control valve via the high-pressure input. Because the nonenergized electromagnet in the control valve exerts no force, high-pressure fuel exceeds the spring force, opening the control valve and spilling rail fuel to the return circuit.

Pressure control valve energized. When desired rail pressure is lower than actual rail pressure, the pressure control valve must allow rail pressure to rise. If the pressure in the high-pressure circuit is to be increased, the force of the electromagnet must combine with the mechanical force of the spring. The ECM energizes the pressure control valve, causing it to close and remain closed until equilibrium is established between desired and actual rail pressures. This equilibrium means that a balance is reached between the high-pressure forces on the one side and the combined forces of the spring and the electromagnet on the other. The valve then remains open and maintains rail pressure constant. Any change in the pump delivery quantity/engine load is compensated for by the valve assuming a different setting. The electromagnetic forces in the pressure control valve are proportional to the energizing current control by the ECM driver circuit using a pulse-width modulated (PWM) signal. The 1 kHz pulsing frequency is stated by Bosch to be high enough to prevent unwanted electromagnet-armature motion and pressure wave fluctuations in the rail.

Common Rail

The common rail or accumulator receives fuel from the high-pressure pump and makes it available to the ECM-controlled, electrohydraulic injectors. The accumulator feature of the rail means that even after an injector has discharged a pulse of fuel into the engine, the fuel pressure in the rail remains almost constant. Rail volume therefore has a dampening effect on transient oscillations in pressure. This can be accounted for both by the accumulator effect and to a lesser extent, the small compressibility factor of the fuel. Rail pressure is monitored by the rail-pressure sensor and maintained at the desired value by the ECM pressure-control valve. A pressure-limiter valve acts to limit fuel pressure in the rail to maximum 22,000 psi (1,500 atms) in the event of a pressure control valve malfunction. Fuel in the rail is made available to the injectors by means of flow limiters, which prevent excess fuel from

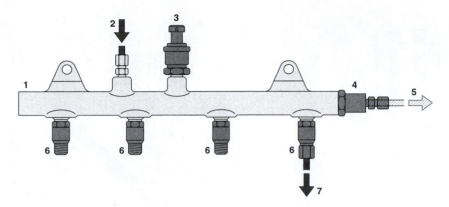

1. Rail
2. Inlet from the
   high-pressure pump
3. Rail pressure sensor
4. Pressure limiter valve

5. Return from the rail to
   the fuel tank
6. Flow limiter
7. Line to the injector

FIGURE 36–8 *High-pressure accumulator (rail). (Courtesy of Robert Bosch Corporation)*

being injected. The rail and its critical components are shown in Figure 36–8.

Rail pressure sensor.   The rail pressure sensor consists of a sensor housing, integral printed circuit, and sensor element. It uses a piezoresistive, Wheatstone bridge operating principle (see Chapter 32) and is supplied with a 5-volt V-Ref. Pressurized fuel acts on the sensor diaphragm through a blind hole, and the sensor element (semiconductor device) is mounted on this diaphragm. When the diaphragm shape changes as the result of pressure acting on it, the electrical resistance of the layers attached to the diaphragm also change. This deflection (approx. 1 mm at 1500 atms) of the sensing diaphragm resulting from fuel pressure alters its electrical resistance, causing a voltage change across the 5 V resistance bridge. The voltage change ranges between 0 and 70 mV (depending upon applied pressure) and is then amplified by the evaluation circuit to a signal value of 0.5 to 4.5 V that correlates with specific pressure values. A sectional view of a rail pressure sensor is shown in Figure 36–9.

Tight tolerances providing a high degree of accuracy apply to the rail pressure sensor during pressure measurement. Measuring accuracy is stated to be around ± 2%. In the event of a rail pressure sensor failure, the pressure control valve is triggered "blind" and defaults to a limp-home mode of operation.

Flow limiter.   The flow limiter (see Figure 36–10) prevents continuous injection in the event that one of the injectors sticks in the open position. Once the fuel quantity leaving the rail exceeds a predetermined volume, the flow limiter shuts off the line to the problem injector. The flow limiter consists of a metal housing with

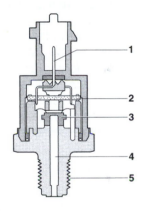

1. Electric connections
2. Evaluation circuit
3. Metal diaphragm with
   sensor element

4. High-pressure connection
5. Mounting thread

FIGURE 36–9 *Rail pressure sensor. (Courtesy of Robert Bosch Corporation)*

external threads for installation into the rail at one end and external threads at the other for connecting the injector lines. A plunger inside the flow limiter is forced toward the rail end by a spring. When fuel is injected, pressure drops at the injector end and causes the plunger to shift in the direction of the injector compensating for the fuel volume that exited the rail during injection. At the completion the injection pulse, the plunger moves to a midposition away from its seat without closing off the outlet completely, having a throttling effect. Next, the spring forces it back onto its at-rest position allowing fuel to flow freely through the throttle bore.

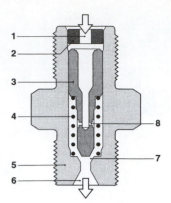

1. Connection to the rail
2. Sealing washer
3. Plunger
4. Spring
5. Housing
6. Connection to the injector
7. Seat
8. Throttle

FIGURE 36–10   *Flow limiter. (Courtesy of Robert Bosch Corporation)*

The spring and throttle bore are dimensioned so that even at maximum injection fuel quantity, the plunger returns to the stop at the rail end of the flow limiter, where it remains until the next injection. When a leak occurs, the flow limiter plunger is forced away from its at-rest position up against the seal seat at the outlet. It remains in this position up against the stop at the injector end of the flow limiter and prevents fuel from exiting to fuel the injector.

High-Pressure Fuel Lines

The fuel lines used in a Bosch CR system have similar requirements to the high-pressure pipes used in other high-pressure diesel fuel injection systems. They must be capable of sustaining the maximum system pressures with a wide margin of safety and the pressure wave reflections that result from high-speed hydraulic switching. The injection lines are required to be of identical length and normally have an outside diameter of 6 mm and an internal bore diameter of 2.4 mm.

ELECTROHYDRAULIC INJECTORS

The electrohydraulic injector (Figure 36–11) can be subdivided as follows:

• Nozzle assembly
• Hydraulic servo system
• Solenoid valve

Referencing Figure 36–11, fuel at rail pressure is supplied to the high-pressure connection (4), to the nozzle through the fuel duct (10), and to the control chamber (8) through the feed orifice (7). The control chamber is connected to the fuel return (1) via a bleed

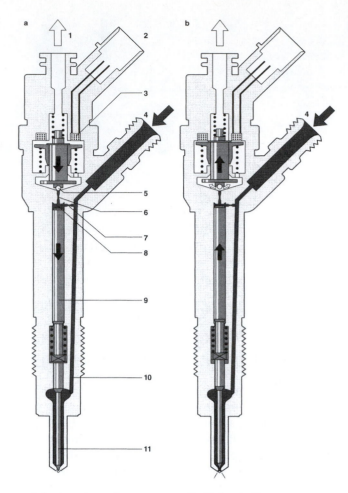

a. Injector closed
   (at-rest status)
b. Injector opened
   (injection)
1. Fuel return
2. Electrical connection
3. Triggering element
   (solenoid valve)
4. Fuel inlet (high
   pressure) from the rail
5. Valve ball
6. Bleed orifice
7. Feed orifice
8. Valve control chamber
9. Valve control plunger
10. Feed passage
    to the nozzle
11. Nozzle needle

FIGURE 36–11   *Injector. (Courtesy of Robert Bosch Corporation)*

orifice (6) that is opened by the solenoid valve. With the bleed orifice closed, hydraulic force acts on the valve control plunger (9) exceeding that at the nozzle-needle pressure chamber (located between the shank and the needle of the nozzle valve). Although the hydraulic pressure values acting on the top of the nozzle valve and that in the pressure chamber are identical, the sectional area at the top of the nozzle valve is greater. As a result, the nozzle needle is loaded into its seated position, meaning that the injector is closed.

When an ECM signal triggers the injector control solenoid valve, the bleed orifice opens. This immediately drops the control chamber pressure and, as a result, the hydraulic pressure acting on the top of the nozzle valve (11) also drops. When hydraulic force acting on top of the nozzle valve drops below the force on the nozzle-needle pressure shoulder, the nozzle valve retracts and allows fuel to pass around the seat to be injected through orifii into the combustion chamber. The hydraulic assist and amplification factor are required in this system because the forces necessary for rapid nozzle valve opening cannot be directly generated by an electrically actuated solenoid valve. Fuel used as hydraulic media to open the nozzle valve is in addition to the injected fuel quantity so this excess fuel is routed back to the tank. In addition to this fuel, some leak-by fuel losses occur at the nozzle valve to body clearance, and the valve plunger guides clearance. This leak-off fuel volume is also returned to the fuel tank via the fuel return circuit.

Injector Operating Phases

When the engine is stopped, all injector nozzles are closed, meaning that their nozzle valves are loaded onto their seats by spring pressure. In a running engine, injector operation takes place in three phases.

Injector closed. In the at-rest state, the solenoid valve is not energized and therefore the nozzle valve is loaded onto its seat by the injector spring combined with hydraulic pressure (from the rail) acting on the sectional area of the valve control plunger. With the bleed orifice closed, the solenoid valve spring forces the armature ball check onto the bleed orifice seat. Rail pressure builds in the injector control chamber, but identical pressure will be present in the nozzle pressure chamber. Given equal pressure acting on the larger sectional area of the nozzle control plunger (which is mechanically connected to the nozzle valve) and in the nozzle pressure chamber, this and the force of the nozzle spring combine to load the nozzle valve on its seat holding the injector closed

Nozzle opening. When the injector solenoid valve is energized by the ECM injector driver, it is actuated at high pressure, typically around 90 volts spike at a current draw of around 20 amps. Force exerted by the triggered solenoid now exceeds that of the valve spring and the armature opens the bleed orifice. Almost instantly, the high-level pickup current to the solenoid is reduced to a lower hold-in current flow required by the solenoid. Hold-in current is less than pickup current for the simple reason that the magnetic field air gap is reduced when the solenoid valve is fully actuated. As the bleed orifice opens, fuel flows from the valve control chamber into the cavity above it and out to the return circuit. This collapses the hydraulic pressure acting on

the valve control plunger that was helping to hold the nozzle valve closed. Now, pressure in the valve control chamber is much lower than that in the nozzle pressure chamber that is maintained at the rail pressure. The result is a collapsing of the force holding the nozzle valve closed and the nozzle valve opens, beginning the injection pulse.

The nozzle needle opening velocity is determined by the difference in the flow rate through the bleed and feed orifices. When the control plunger reaches its upper stop, it is cushioned by fuel generated by flow between the bleed and feed orifices. When the injector nozzle valve has fully opened, fuel is injected into the combustion chamber at a pressure very close to that in the fuel rail.

Nozzle closing. When the solenoid valve is de-energized by the ECM, its spring forces the armature downward, and the check ball closes the bleed orifice. The closing of the bleed orifice creates pressure buildup in the control chamber via the input from the feed orifice. This pressure should be the same as that in the rail, and now it exerts an increased force on the nozzle valve control plunger through its end face. This force, combined with that of the nozzle spring, exceeds the hydraulic force acting on the nozzle valve sectional area, and the nozzle valve closes, ending injection. Nozzle valve closing velocity is determined by the flow through the feed orifice. The injection pulse ceases the instant the nozzle valve seats.

CAUTION: Never attempt to locate an engine miss by cracking open the high-pressure lines that feed the electrohydraulic injectors. Use OEM technical literature to troubleshoot engine as well as fuel system malfunctions.

Nozzle Hole Geometry

Common rail injectors use both sac-chamber nozzles and what are generally referred to as valve-closes-orifice (VCO) nozzles. The latter tend to be required in engines that must meet North American emissions standards. Most CR injectors use a 4 mm nozzle valve diameter. Both sac-hole and seat-hole nozzles have the edges of each orifice rounded by hydroerosive (HE) machining. Hydroerosive machining helps prevent edge wear caused by the abrasive particles in the fuel and carbon coking that can reduce flow and disrupt the spray geometry.

PHASES OF INJECTION

Bosch CR fuel injection systems are capable of breaking up an injection pulse in two or three shots.

Multipulse injections require high-speed solenoid switching, which is achieved by using high voltages and currents. The solenoid valve triggering stage in the ECM is designed to be switched at the high speeds required for multiple shot injection. Start of injection is plotted by the ECM using the running conditions of the chassis and engine factored with programmed fuel maps. Typically, a full injection cycle can be broken into three parts, that is, three separate opening and closing events during a single combustion phase in the engine.

Pilot Injection

Pilot injection has been used by a number of diesel engine OEMs for several years now. First, a small shot of fuel is injected into the engine cylinder following which the injector closes. The ECM then calculates the point at which this initial shot of fuel is ignited using sensor data such as ambient temperature, boost pressure, and engine temperature. When the pilot charge has ignited (combustion begins), the ECM can then reopen the nozzle and inject the main pulse into the engine cylinder. The pilot injection volume used in the Bosch CR system is small, usually between 1 and 4 mm^3. Pilot injection has been long known to improve combustion efficiency and noise, making diesel knock a thing of the past. Fuel consumption and exhaust gas emissions are also improved. Because pilot injection reduces ignition lag by reducing the fuel in the cylinder at the point of ignition, engine torque is produced more evenly during combustion because the cylinder gas pressure curve is smoother. The ECM is responsible for the time duration between the pilot and main injection pulses.

Main Injection

Most of the fuel injected during a cylinder combustion event occurs during the main injection pulse. Although main injection is essentially similar to the main pulse in comparable engines equipped with a pilot injection feature, with CR injection systems, the injection pressure remains pretty much constant throughout injection regardless of duration or volume. This contrasts with other systems that trigger at a lower pressure value and increase injection pressure as the duration of the required pulse increases.

Secondary Injection

The Bosch use of the term secondary injection should not be confused with undesirable secondary injection caused by nozzle problems that follow the intended main injection pulse in other systems. In Bosch CR systems, secondary injections can be programmed into the management system. With certain types of reduction class, catalytic converters (NO_x reduction), secondary injection can help reduce emissions in a rather unusual manner. In contrast to the pilot and main injection processes, secondary injected fuel is not intended to be combusted but instead vaporizes due to the residual heat in the exhaust gas. When end gas is expelled from the cylinder, a mixture of exhaust gas and raw fuel exits through the exhaust valves and into the exhaust gas system. Some of this mixture is routed back into the engine cylinder via the EGR system where Bosch claims it has the same effect as a very advanced pilot injection pulse. When suitable NO_x catalytic converters are fitted, they utilize the fuel in the exhaust gas as a reduction agent to lower the NO_x content in the exhaust gas.

There can be significant disadvantages of secondary injection. While it might contribute to lowering NO_x emission, it necessarily increases HC emission. Also, it can dilute engine lube oil. Secondary injection is only used when approved by the engine manufacturer.

MULTIPLEXING

Communication between the common rail ECM, the engine controller, and the other chassis ECMs using a data bus uses either SAE J1850 or SAE J1939 communication protocols, depending on the weight classification of the vehicle. The data bus multiplexes shared sensor data, operating data, and status information needed for electronic subsystem operation and fault monitoring. Figure 36–12 shows a sequencing schematic of the data used to calculate fuel quantity for any given moment of operation. For more detailed information on multiplexing, see Chapter 35.

MALFUNCTION REPORTING

Malfunction detection, logging, and display are similar to other contemporary electronic control systems. Signal paths are classified as defective when out of range and present for longer than a predefined period. Such error events are stored in ECM memory together with an audit trail detailing the engine operating and environmental conditions that prevailed at the time of the error: this type of information would include such data as coolant temperature, engine speed, and ambient temperature. For many electronically monitored faults, it is possible for OK status to be re-established if the signal path is identified as intact for a defined period of time. In addition, in case of nonplausible signals from the accelerator pedal sensor, the engine can be run in default (limp-home) mode.

Figure 36–13 shows the critical component layout of a typical Bosch CR system.

Starting-switch position A: Start,
Starting-switch position B: Drive mode.

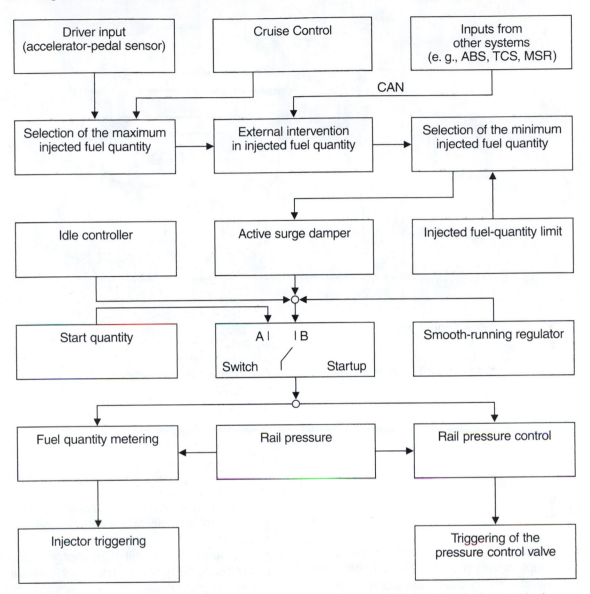

FIGURE 36–12   *ECU injected fuel quantity algorithm. (Courtesy of Robert Bosch Corporation)*

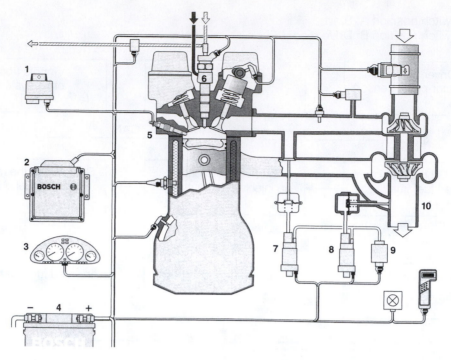

1. Glow control unit
2. ECU
3. Instrument panel with displays for fuel consumption, engine speed, etc.
4. Battery
5. Glow plug
6. Injector
7. EGR positioner
8. Charge-pressure actuator
9. Vacuum pump
10. Turbocharger

FIGURE 36–13   *Actuators and other components of the common rail system. (Courtesy of Robert Bosch Corporation)*

SUMMARY

- Bosch electronically controlled rotary distributor pumps adapt their popular hydromechanical sleeve-metering, rotary distributor pump to manage a partial authority, engine control system.
- The Bosch common rail (CR) system is a full authority, engine management system with multiplexing capabilities.
- The fuel subsystems that supply electronically controlled rotary distributor and common rail fuel systems have few functional differences from other types of low-pressure, fuel supply circuits.
- A Bosch electronic diesel control (EDC) system consists of an input circuit, processing hardware, and actuator circuit. The components in the input and data processing circuits are similar to other engine management systems. The output or actuator circuit components make this system distinct.
- The key actuator circuit components in a partial authority, VE pump are the control sleeve solenoid (controls effective pump stroke/fuel quantity) and the start of injection solenoid valve (controls timing).
- A Bosch common rail (CR) system consists of a fuel subsystem, radial piston high-pressure pump, a rail or accumulator, a high-pressure distribution circuit, and electrohydraulic injectors.
- The Bosch VP 44 pump is a three-cylinder radial piston pump responsible for charging the fuel rail with injection pressure values that can be varied by an ECM-controlled circuit.
- The ECM managing rail pressures in a Bosch CR system does so by sensing actual rail pressure using a sensor, plotting desired rail pressure based on input and fuel map data, and signaling a pressure control solenoid to attempt to match desired and actual pressures.
- Electrohydraulic injectors are used in a Bosch CR system because they can be switched at higher velocity and create fewer high-pressure circuit variables than hydraulically actuated injectors.
- Electrohydraulic injectors are PWM controlled by the ECM.

REVIEW QUESTIONS

1. *Technician A* says that Bosch electronic diesel control (EDC) is used on later versions of the VE sleeve-metering, rotary distributor pump. *Technician B* says that when a VE pump is managed electronically, the system would be classified as a partial authority management system. Who is right?
   a. A only
   b. B only
   c. Both A and B
   d. Neither A nor B

2. When a VE injection pump is managed by an ECM, which of the following is no longer required?
   a. Control sleeve/collar
   b. High-pressure pipes
   c. Timing advance
   d. Governor flyweights

3. Which of the following would best describe the injectors used in a Bosch CR fuel system?
   a. Hydraulic, multi-orifii
   b. Electronic unit injectors
   c. Electrohydraulic, multi-orifii
   d. Pintle type

4. In a Bosch CR fuel system, the rail could also be known as a(n):
   a. Charging gallery
   b. Accumulator
   c. Jumper pipe
   d. Injector tube

5. On a Bosch electronically controlled, sleeve-metering, rotary distributor injection pump (VE), what would happen if the ECM electrical signal to the control sleeve/collar failed?
   a. Default to hydromechanical operation
   b. Default to limp-home mode
   c. Engine runaway
   d. Engine shutdown

6. What is required to energize a Bosch CR fuel injector?
   a. 90 volts pressure and 20 amps of current
   b. V-Ref pressure and 12 amps of current
   c. V-Bat pressure and 5 amps of current
   d. 18 volts pressure and 93 amps of current

7. What does pilot injection tend to eliminate when starting a diesel engine with a Bosch CR fuel system?
   a. NO_x emissions
   b. Diesel knock
   c. Ignition lag
   d. Black smoke

8. Which of the following best describes the means by which the ECM drivers control the actuation of Bosch CR electrohydraulic injectors?
   a. V-Ref signal
   b. V-Bat signal
   c. Distributor spike
   d. Pulse width modulation

9. How many pump plungers are there on a VE pump used to fuel a six-cylinder engine?
   a. One
   b. Three
   c. Six
   d. Eight

10. How many pump plungers are there on a Bosch CR radial piston pump used to fuel a six-cylinder engine?
    a. One
    b. Three
    c. Six
    d. Eight

11. What device on a Bosch CR system would prevent constant fueling of a cylinder if one of the electrohydraulic injectors were stuck in the open position?
    a. Flow limiter
    b. Pressure limiter valve
    c. Pressure control valve
    d. Collapse of accumulator

12. *Technician A* says that when a CR injection pulse is broken into pilot, main, and secondary injection phases, the secondary injection phase is used to reduce NO_x emissions. *Technician B* says that secondary injection will increase HC emissions. Who is right?
    a. A only
    b. B only
    c. Both A and B
    d. Neither A nor B

13. *Technician A* says that a Bosch CR pressure limiter valve is a simple pressure relief valve designed to trip at a predetermined value. *Technician B* says that actual rail pressure values in any given moment of engine operation are managed by the pressure limiter valve. Who is right?
    a. A only
    b. B only
    c. Both A and B
    d. Neither A nor B

14. What force is used to hold the nozzle valve in its closed and seated position in an electrohydraulic injector used with a Bosch CR system?
    a. Spring force only
    b. Electrical force only
    c. Combined hydraulic and electrical force
    d. Combined hydraulic and spring force

15. When the control plunger in the electrohydraulic injector is energized, what should happen to the rail pressure charged to the injector?
    a. Significant decrease
    b. Significant increase
    c. Remains almost unchanged
    d. First increases, then drops-off

Chapter

37

Mack Trucks
V-MAC I and II

PREREQUISITES

Chapters 20, 21, 29, and 32

OBJECTIVES

After studying this chapter, you should be able to:

- Define the acronym V-MAC.
- Describe the layout used by Mack Trucks in their V-MAC chassis management system.
- Describe how Mack Trucks and Bosch adapted an in-line, port-helix metering injection pump for computerized management and control.
- Distinguish between the V-MAC I and V-MAC II systems.
- Define the function of the V-MAC ECU and the ECU software, and the input and output circuits.
- Identify the two Bosch injection pumps used on V-MAC I and II management systems.
- Outline the function of the Bosch rack actuator housing.
- Identify the subcomponents contained within the rack actuator housing.
- Describe the operating principles of the rack actuator housing subcomponents.
- Describe how Mack Econovance operates.
- Perform customer data programming of a V-MAC ECM.
- Outline the procedure for reprogramming a V-MAC ECM with proprietary (Mack Trucks) data.
- Describe the Co-Pilot Driver Data Display option on V-MAC II.
- Outline the procedure required to diagnose Mack Trucks V-MAC electronics.

KEY TERMS

Co-Pilot

Econovance

ELAB fuel shutoff solenoid

electronic control unit (ECU)

fuel injection control (FIC) module

fuel control actuator

linear magnet

proportional solenoid

rack actuator

rack actuator housing

rack position sensor

reference coil

road speed limit (RSL)

single speed control (SSC)

timing actuator

timing event marker (TEM)

variable speed control (VSC)

variable speed limit (VSL)

Vehicle Management and Control (V-MAC)

V-MAC module

INTRODUCTION

Mack Trucks **Vehicle Management and Control (V-MAC)** versions I and II are both classified as partial authority, electronic engine (chassis) management systems because the system is built around a hydromechanical, port-helix metering injection pump, adapted for computer control. V-MAC was introduced in the late 1980s, evolved into V-MAC II and the system was used until 1998, when it was replaced by V-MAC III. The fuel management system used in V-MAC I and II is distinct from the V-MAC III, E-Tech system, which is handled in Chapter 41 of this textbook. With the two early versions of V-MAC electronics, Mack Trucks chose to use a Bosch port-helix metering, high-pressure injection pump and manage it electronically. The actual metering and pumping apparatus differs little from that found on any previous Mack hydromechanical engine. The essential differences of V-MAC I and II focus on the means used to control injected fuel quantity and beginning of delivery timing. In addition, V-MAC electronics comprehensively manages vehicle and chassis systems, plots the fueling logic to meet a variety of chassis and running conditions, and permits a wide range of customer and proprietary (Mack) programming options. Mack Trucks was also the first of the truck OEMs to adopt the personal computer (PC) its primary diagnostic, system reader, and reprogramming instrument.

It is assumed that the reader has a sound knowledge of in-line, port-helix metering injection pumps before reading this chapter.

SYSTEM OVERVIEW

The V-MAC I system layout incorporated two modules bussed (electronically connected to each other) to manage engine fueling, diagnostics, and chassis functions (Figure 37–1). The **fuel injection control (FIC) module** was essentially the Bosch (EDC) (electronic diesel control) module used on other electronic applications of the PE 7000/8000 series of port-helix metering, injection pumps (used previously by John Deere/Volvo and other OEMs). The FIC module sensed and switched most of those functions directly associated with engine fueling. The FIC module was bussed to the **V-MAC module** by a dedicated CAN, which "managed" the FIC module from the broader perspective of the specific chassis application of the engine, performance, noxious emissions management, and command inputs

such as PTO and cruise controls. V-MAC II eliminated the separate FIC module, incorporating its functions in a single V-MAC II module, manufactured by Motorola until the system became obsolete in 1998. Both versions of V-MAC had the module(s) located in the cab dash compartment on the passenger side of the vehicle. In the case of V-MAC I, the FIC module was piggyback located on top of the V-MAC module.

V-MAC modules can be accessed for fault code display, reading system parameters, and customer and proprietary data programming via the standard six-pin ATA/SAE serial communications connector (SAE J1708) located under the dash on the driver's side of the vehicle. Active system faults can also be read by blinking them through the dash "electronic malfunction" light.

While a ProLink 9000 reader-programmer can be used to read system parameters, active and historical codes, plus perform most customer data programming, Mack Trucks prefers technicians to use a PC. Mack Trucks has developed a software package that simplifies sequential troubleshooting plotting, making it user friendly for those with moderate computer skills.

A PC is required to effect Mack data reprogramming, essentially a downloading of a mainframe-based file to diskette, reprogramming of the on-board modules from that diskette, and subsequent uploading of a verification file to the mainframe. This procedure has become more user friendly over time. It is backed up by technical support literature and helps lines and is unlikely to intimidate even the first time user. However, to effectively work with V-MAC as with any other current diesel electronic management system, OEM training is essential.

BOSCH PE7100/PE8500 INJECTION PUMPS

These are essentially "beefed-up" versions of the Bosch PE3000 series injection pump. Both can be managed with hydromechanical governors. The PE7100 was used with the first V-MAC I system (Figure 37–2). The injection pump itself was designed to produce peak injection pressures of 1,050 atms (106.4 MPa or 15,500 psi) using 10 mm plungers with lower helix geometry and a cam lift of 12 mm. The pumping elements were fitted with funnel spill deflectors and a reinforced housing. The more recent version, PE8500, is closely related to the PE7100 in most of its critical specifications but is capable of peak injection pressures of 1,150 atms (116.5 MPa or 16,900 psi). Both pumps are driven off the engine timing gear train at camshaft speed and use an electronically controlled, electric-over-hydraulic variable timing device, **Econo-**

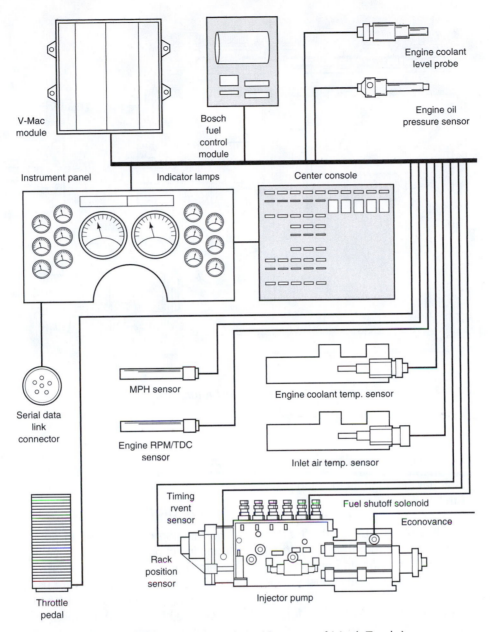

FIGURE 37–1 *V-MAC I system overview. (Courtesy of Mack Trucks)*

vance, located between the pump and the engine accessory drive gear. An injection pump with plungers of the lower helix design will have a constant beginning and variable ending of the fuel pulse, the length of which would be measured in crank angle degrees. Using a variable timing device "softens" the hard timing parameters represented by plunger and helix geometry by giving the ECM a window within which it can select effective stroke. The constant beginning, variable ending timing characteristic of the typical lower helix geometry is adapted into a variable beginning, variable ending of the delivery pulse, entirely managed by ECM processing logic. All V-MAC system injection pumps use Econovance, and the extent to which the port clo-

sure value can be moderated depends on the application; this is described in detail later in this section.

The function of the hydromechanical governor was to manage fueling on the basis of inputted fuel demand (accelerator/speed control lever), manifold boost (puff limiter), and engine rotational speed (sensed centrifugally). V-MAC entirely eliminates the hydromechanical governor, and the rack actuator housing that physically replaces it and the software programmed to the ECM provide fully computerized governing. The **rack actuator housing** incorporates sensors that input data to the ECM (V-MAC module) and a switched output from the ECM, capable of precisely controlling rack position and thus, injected fuel quantity.

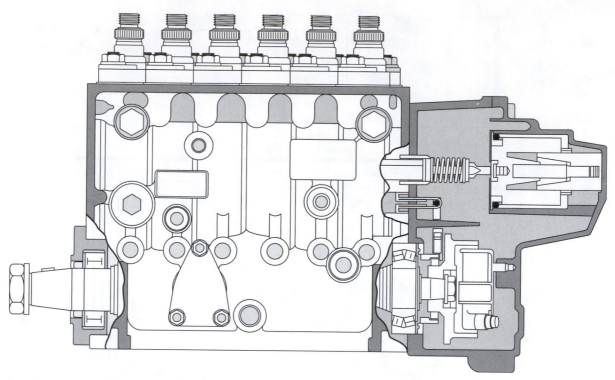

FIGURE 37–2   *Bosch PE7100 injection pump with RE30 rack actuator.*

RE24/30 RACK ACTUATOR HOUSINGS

It was stated previously that although its location positions it at the rear of the fuel injection pump, the rack actuator housing should not be confused with a governor. The governor is a self-contained unit that, after calibration by the technician, sets hard parameters (rigid values) for fueling the engine through its performance range. The V-MAC I and II rack actuator housing is a unit that contains mechanisms both for reporting data to the system module(s) (that is, sensors) and for effecting the fueling commands that are the result of ECM processing logic. These fueling commands are the result of the fuzzy logic computations (that is, based on multiple inputs and variables) and therefore the fueling parameters can always be described as "soft" or nonrigid. Truly, the programmed **ECU** (Mack Trucks uses ECU or **electronic control unit**, rather than the acronym ECM) software is the "governor" in a V-MAC management system. The rack actuator housing is not as a rule, field-repaired but its operation must be understood by the technician seeking to effectively troubleshoot V-MAC systems. Subcomponent testing is outlined in Mack technical support literature and detailed in Robert Bosch diagnostic literature.

The rack actuator housing consists of a **rack actuator** mechanism controlled electrically by a V-MAC II module or by an FIC module in V-MAC I. Within the housing there are also several critical sensors that

input engine data to the ECU(s). An RE30 rack actuator housing is shown in Figure 37–3.

Rack Actuator

The function of the "rack" in an in-line, port-helix metering injection pump is to rotate the plungers in unison within their barrels when it is moved linearly. When plungers with lower helices are used, port closure ("static" timing) determines the beginning of the fueling pulse at a constant timing location of the engine, while port opening sets the end of the pulse, which will vary according to the amount of fuel to be pumped.

The fueling window is therefore defined precisely by where the plunger helices register with the barrel spill ports. Plunger rotation is effected by linearly moving the rack, which is tooth meshed to plunger control sleeves. In Bosch rack actuator housings, the control rack extends into the rack actuator housing and is attached to the rack actuator mechanism (Figure 37–4); the rack actuator can be electrically defined as a **linear magnet** or **proportional solenoid**. The rack actuator is also known as the **fuel control actuator**. The rack assembly is spring loaded to the rack "no-fuel" position. This no-fuel rack position rotationally positions all the pump plungers so their vertical slots register with their barrel spill ports throughout their cam-actuated stroke. This plunger position permits them to do no more than displace fuel as the plungers

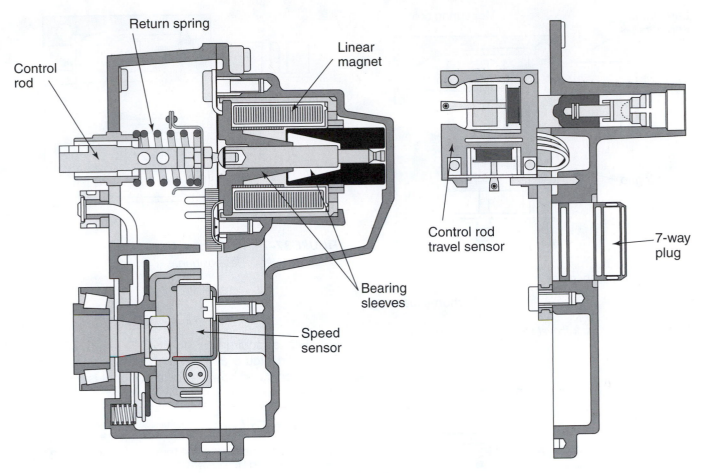

FIGURE 37–3 *Sectional view of Bosch RE30 rack actuator housing.*

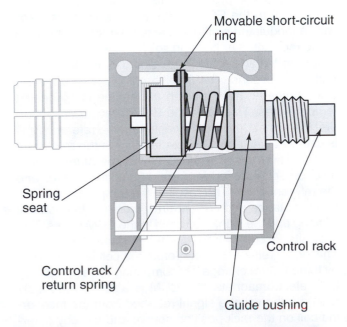

FIGURE 37–4 *Rack actuator assembly components.*

are actuated. When the rack actuator is energized by V-MAC module switching, the rack moves linearly in opposition to this spring pressure, thereby altering the rotational position of the plungers to increase plunger effective stroke. As the amount of current flowed through the rack actuator coil is increased, the further inboard the rack is forced. As the rack is forced inboard against the spring pressure, the length of the plunger effective stroke increases and therefore so does the quantity of fuel pumped per cycle. The rack retraction spring would therefore be maximumly compressed in the full fuel position when the rack is forced to its fully inboard position by the rack actuator proportional solenoid. The rack actuator is an ECU module switched output and therefore returns no data back to the module itself. It simply responds to command signals.

Rack Travel Sensor

The function of the rack travel sensor is to report exact rack position to the ECM. It does so 60 times per sec-

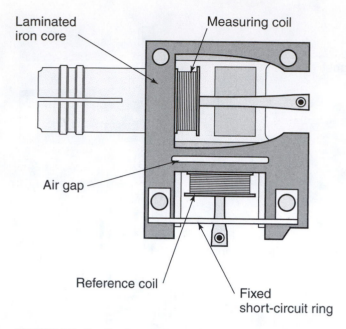

FIGURE 37–5   *Rack travel sensor components.*

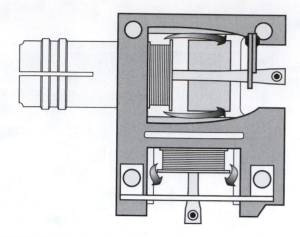

FIGURE 37–7   *Rack travel sensor at maximum rack (peak fuel per cycle) position.*

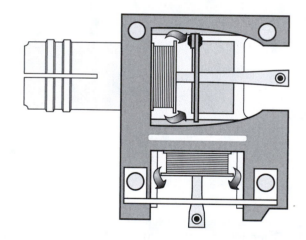

FIGURE 37–6   *Rack travel sensor at idle fuel rack position.*

ond. The assembly is mounted in a laminated iron core and consists of a measuring coil that is energized with reference voltage (5 V). Attached to the rack is a short-circuit ring and protruding from the fixed measuring coil is an iron bar (Figure 37–6).

The short-circuit ring attached to the rack is designed to slide over the iron bar without physically contacting it. As the fuel control rack is moved linearly by the rack actuator, the short-circuit ring will either move closer to or further away from the measuring coil thereby varying the electromagnetic field spread by the coil and therefore, the induced voltage signal returned to the V-MAC module. This component is necessarily sensitive and required to be precise; exact rack posi-

tion data is required by the ECM Figure 37–5, Figure 37–6, and Figure 37–7 show the rack travel sensor components and how the voltage signal is induced by the measuring coil.

Reference Coil

All electromagnet fields are temperature sensitive so in order to accurately interpret the voltage signal returned to the V-MAC module by the rack travel sensor, a precise "thermometer" is required to verify the signal. The **reference coil** is located close to the rack travel sensor coil; its location is shown in Figure 37–5. It is wound identically to the rack travel sensor coil. When it is energized, this coil provides a fixed magnetic field. The signal sent to the ECM by the **rack position sensor** will be modulated by the actual temperature conditions in the rack actuator housing so that when the rack is in a certain linear position, the signal outputted by the rack position sensor will be different when the engine is cold compared to that when the engine is at running temperatures. To enable the ECM to accurately interpret the signal from the rack actuator, the reference coil is used. For instance, as temperature in the rack actuator housing rises, so does the temperature in both electromagnets used in the rack position sensor and the reference coil. The result of the temperature rise is that resistance and current draw increase in each coil winding and the signal returned to the ECM varies. The opposite occurs when the temperature decreases. The signal returned by the reference coil is varied only by temperature change. By comparing the signals from both electromagnets, the ECM is able to accurately evaluate the voltage signal returned from the measuring coil on the rack position sensor and thereby determine exact rack position.

Pulse Wheel

The pulse wheel is the only rotating component in the rack actuator housing. The pulse wheel is a toothed impeller located at the rear of the injection pump camshaft and driven within the rack actuator housing. It is a sintered steel component whose total indicated runout (TIR) deviation must not exceed 0.03 mm. Straightening of impeller teeth should never be attempted and the impeller should be replaced if measured (or observed) to be defective. The impeller or pulse wheel (tone wheel) is responsible for two inputs to the V-MAC module. The first is engine speed (it is the primary reference for engine rpm) and the next is engine location reported by the **timing event marker (TEM)**. There are sixteen protruding teeth on the pulse wheel impeller that cut through the rpm sensor's magnetic field to produce an AC voltage signal to the ECM whose frequency increases proportionally with an increase in rotational speed. The TEM sensor consists of a single notch that cuts the magnetic field of the TEM sensor, producing a signal that works on the same principle as the rpm sensor, but whose input is used by the ECM to locate precise engine position. The TEM notch is also used for the electronic static timing of the pump to the engine (this procedure is covered later in this chapter). Both these sensors can be electrically classified as induction pulse generators that feed a small AC analog signal to the ECM.

ECONOVANCE

The PE7100 or PE8500 injection pump is driven off the engine crankshaft gearing at camshaft speed on Mack engines. The Econovance unit serves as an intermediary between the engine driven pump drive gear and the injection pump drive. Econovance provides an advance on the static timing, port closure location of 8-degree, 10-degree, or 20-degree crank angle depending on the engine model. Current versions of V-MAC tend to use a 20-degree Econovance, which is perhaps better described as a variable timing. Econovance establishes a port closure timing window within which the ECM can select port closure. For example, if static timing was set at 7 degrees BTDC on a V-MAC engine with 20-degree Econovance, V-MAC could select port closure occurrence at any value between 7 degrees BTDC and 27 degrees BTDC.

Econovance alters the position of the injection pump camshaft relative to that of the engine. The device is controlled electronically by the V-MAC module. The engine's injection pump drive gear rotates a hub within which is a sliding sleeve machined with a helical spline; this is spring loaded to a nonadvanced position. To ad-

vance timing, engine oil used as the hydraulic medium acts to move the sliding sleeve along the axis of rotation of the helical splines. The following components are critical in actuating Econovance:

- Proportional solenoid. The proportional solenoid used is known as the **timing actuator**. It is a linear magnet that can be smoothly moved to any position within its range of travel by varying current flow through the coil. The proportional solenoid is switched by the V-MAC module and its function is to control the Econovance spool valve.
- Hydraulic spool valve. This controls oil flow (engine lube) to the Econovance sliding sleeve. The hydraulic spool valve is directly controlled by the proportional solenoid, and it establishes the extent of advance by managing oil flow. Actual engine oil pressure values will not affect its operation (so long as the oil pressure is above the V-MAC "failure" value) because the operation of Econovance is flow dependent. The ECM monitors the advance location of the injection pump by reading the TEM signal.

ECONOVANCE OPERATION

Static timing values as retarded as 4 degrees BTDC seem to make little sense mechanically because the firing pulse and cylinder pressure rise would appear to be completely out of synchronization with the crank throw vector angles and act to "inefficiently" load the engine powertrain. However, it should be remembered that static timing the engine at 4 degrees BTDC with a 20-degree Econovance simply provides the ECU with the option of selecting a port closure value anywhere from 4 to 24 degrees BTDC, and an actual 4-degree BTDC port closure setting would seldom be used. Emission control in truck diesel engines is often a case of maintaining a tricky narrow cylinder temperature window of values in which the consequences of being slightly under optimum value would be excess HC emissions and the consequences of being slightly over produce higher NO_x emissions. When the V-MAC programming opts for port closure to occur at substantially retarded timing values, the reason is to reduce NO_x emission. When injection timing is this retarded, there is an adverse effect on performance and fuel efficiency.

INJECTION PUMP-TO-ENGINE TIMING

This procedure can be performed electronically with Mack V-MAC. Bosch PE7100 RE30 injection pumps supplied to Mack until April 1992 were fitted with **ELAB**, a **fuel** supply **shutoff solenoid**. This apparatus was located near the front of the pump, requiring that it be timed on the #6 cylinder. ELAB actually provided a secondary no-fuel method of shutting the engine down by gating off the supply fuel. On later PE7100 and the more recent PE8500, fuel shutdown is achieved by simply moving the rack to the no-fuel position. However, all V-MAC PE7100 injection pumps are timed to the engine referencing #6 cylinder. All V-MAC PE 8500 pumps are timed on the #1 cylinder.

Electronic timing is performed by electromechanically positioning the protruding TEM notch using a Mack timing device. Essentially, this consists of positioning the pump so a pair of lights on an EST illuminate simultaneously ensuring a timing accuracy within $\frac{1}{2}$ degree crank angle if the gear installation procedure has been performed accurately.

STATIC TIMING SETTING ON AN E-7 ENGINE WITH V-MAC ELECTRONICS

The following procedure is outlined according to Mack Trucks recommended procedure; the assumption here is made that the engine is timed to the #1 engine cylinder. The static timing procedure requires the use of Mack Trucks fixed timing tool; this consists of a display box with a pair of lights marked A and B, a ground wire and clamp, and a sensor. Attempting to static time a V-MAC engine without this tool will result in out-of-specification timing.

1. Using an engine barring device, preferably the toothed Mack barring ratchet adapter, turn the engine CCW as viewed from the front to a location ± 45 degrees BTDC of the #1 engine cylinder.
2. Next, rotate the engine CW until the timing pointer aligns with the calibration scale on the flywheel or vibration damper.
3. Remove the TEM sensor from the fixed timing port located on the side of the rack actuator housing, and install the fixed timing sensor tool. Take care to ensure that the sensor (fixed timing tool) is properly aligned with the locating groove in the TEM aperture.
4. Attach the ground clamp from the fixed timing tool sensor to the engine; ensure there is continuity between the two points. A poor ground will render the timing tool inoperative.
5. Using the hub rotation tool, rotate the engine timing gear hub CW until both lights on the fixed timing tool illuminate. It is important that the inner shaft nut is not used to rotate the hub or incorrect static timing will result.
6. Install the injection pump-driven gear on the Econovance outer shaft hub so that the screw

holes in the hub are properly centered in the gear slots.
7. Install the timing gear hub screws and install finger tight. This should permit the hub gear to be held in place while allowing relative movement between the Econovance hub and the pump-driven gear.
8. Using the hub rotation tool, rotate the timing gear hub CCW until the screws contact the stops at the end of the slots in the injection pump drive gear. At this point, both lights on the static timing tool should not be lit.
9. Rotate the timing gear hub CW until both lights on the fixed timing tool illuminate. If the hub is rotated too far CW so that the A light extinguishes, back up and repeat step 8.

WARNING: Both lights illuminate for a minute band of rotation (about $\frac{1}{2}$ degree crank angle) so the foregoing step must be undertaken carefully.

---

10. Torque the fasteners that clamp the injection pump driven-gear to the Econovance hub to the specified value. During the torquing procedure, it is normal for one of the two lights to extinguish, and this can be expected.
11. Next, bar the engine CCW to ± 45 degrees BTDC; this will cause both the A and B lights to extinguish.
12. Validate the timing by rotating the engine CW until both the A and B lights illuminate. Ensure that the engine is only rotated CW during this procedure; once again, if the engine is barred past the point where both lights are lit, back up and repeat from step 11. When both lights are illuminated, check the engine calibration scale and pointer. The specification should be within $\frac{1}{2}$ degree of the specified static timing (port closure) value; if out of specification the timing procedure will have to be repeated.

V-MAC ELECTRONICS

V-MAC incorporates comprehensive monitoring, engine management, customer and proprietary data programming, and fault/data retention consistent with a full option, partial authority management system. The V-MAC sensor system is as comprehensive as most of the full authority systems manufactured in the late 1990s. The primary sensors on which the V-MAC software relies to plot fuel ratio and timing logic are the two temperature thermistors, engine coolant, and ambient air. Figure 37–8 shows the V-MAC II ECU housing, Figure 37–9 is a schematic of inputs and outputs to the module.

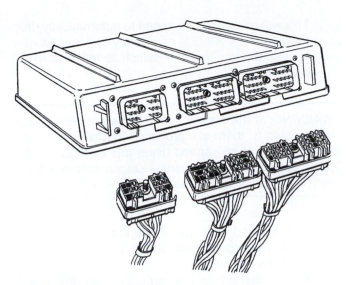

FIGURE 37–8   *V-MAC II consolidated ECU housing. (Courtesy of Mack Trucks)*

CUSTOMER DATA PROGRAMMING

All critical chassis data can be reprogrammed as with most electronically managed engines. Input data affecting ECM computation must be accurately reprogrammed and corrected when components are changed. Tire rolling radii, gearbox ratios, and final drive ratios would fall into this category and obviously, minor programming inaccuracies can make critical input data meaningless and thereby scramble V-MAC's computing logic. The ability to toggle the governor type from VS to LS is also a customer data option. Mack Trucks traditionally used variable speed (VS) governors in its hydromechanically managed engines, a feature that was not to every operator's taste. Other customer data categories would include **variable speed control (VSC)**, the setting of rpm using the dash cruise control switches, and **variable speed limit (VSL)**, which limits rpm while moving a ve-

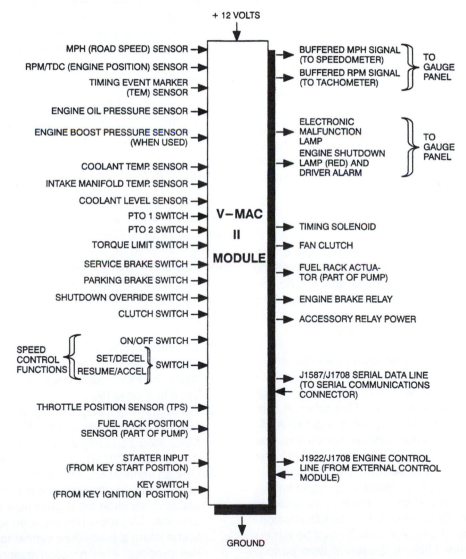

FIGURE 37–9   *V-MAC II ECU inputs and outputs. (Courtesy of Mack Trucks)*

hicle in power takeoff (PTO) mode. In early V-MAC software versions, fuel trip data could be logged into ECU files for subsequent readout; more recent versions of V-MAC offer Co-Pilot, a comprehensive driver data display tool described in detail later in this section. V-MAC customer data programming would also include a number of options designed to enhance the operating efficiency and response strategy to a critical monitored failure condition.

Vehicle speed control parameters including the **road speed limit (RSL)** cruise window and PTO management would fall into the first category. The RSL is a hard parameter that may be programmed to the ECM and require a password to alter. In common with most electronically governed engines, V-MAC in PTO mode provides **single speed control (SSC)** or isochronous governing; that is, as engine load changes, the engine rpm will be maintained with zero droop during the load changes. V-MAC may also be programmed with either alert/shutdown or alert-only failure strategy on the following sensor inputs:

CTS - coolant temperature sensor

OPS - oil pressure sensor

CLS - coolant level sensor

It is strongly recommended by Mack Trucks that the shutdown strategy (rather than alert-only) be programmed, especially in fleet applications. A V-MAC-managed engine shutdown can be overridden by an operator in 30-second increments any number of times, but it should be noted that override incidents will be logged to the ECM.

Resetting Low Idle Speed

This procedure is outlined as being an example of setting a typical customer data programming option. The V-MAC system permits the setting of the engine low idle speed between a range of 500 to 750 rpm. This procedure may be undertaken using the dash speed control switches using the following method:

1. V-MAC must have the Low Idle Adjust option enabled in customer data options; if not, a ProLink or PC (Figure 37–10) with the appropriate Mack Trucks software must be used. This function can, in other words, be disabled by the customer. Additionally the PTO minimum engine speed must be programmed to 475 rpm or once again, a ProLink or PC will be required to perform the procedure.
2. Ensure that no active fault codes are logged into V-MAC, the vehicle is stationary, the parking brake is applied, and the throttle pedal is in the idle position.
3. The Speed Control ON/OFF switch must be turned on/off three times within a 2-second time period.

This will cause the idle speed to automatically drop to 500 rpm. V-MAC is now ready to accept a new speed. Ensure that the switch is left in the ON position.

WARNING: If the engine speed does not drop to 500 rpm in step 3, there are problems that will not permit the resetting of idle speed using this method. At this point, consult the Switch circuit diagnostic procedures outlined in Mack Trucks technical service literature.

---

4. Depress and hold the accelerator pedal until the desired idle speed is produced. Then press and release the Set switch.
5. At the desired engine rpm, depress the clutch pedal and V-MAC will respond by locking the new speed as the idle speed.

TECH TIP: It is recommended that the low idle speed always be set at a value higher than 500 rpm so that a drop in rpm can be detected when resetting the idle speed.

ProLink or a PC may be used for customer data programming. However, only a PC may be used for proprietary (Mack Trucks) data programming.

MACK TRUCKS DATA PROGRAMMING

Proprietary data reprogramming is seldom required. However, when an ECU module is replaced for whatever reason, V-MAC module files are suspected of being corrupted or, exceptionally, the fuel map requires modification, the V-MAC module must be reprogrammed. (See Photo Sequence 5) The process required to perform this is undertaken in three distinct stages:

1. Download. This process electronically locates the appropriate file in the Mack Trucks Allentown mainframe via PC/modem using the VIN and downloading it to a diskette.
2. Reprogram the engine data file. The engine data files on the diskette are next transferred to the V-MAC module using a PC, serial link, and a standard SAE J1708/J1939 connector.
3. Upload verification. After the engine data files have been successfully (re)written to the V-MAC module, a verification file is automatically written to the diskette. This must be uploaded to the Mack mainframe using a procedure similar to the first stage in this process. This process essentially provides a

PHOTO SEQUENCE 5
Chapter 37 Mack Data Reprogramming

PS5-01 *On a computer with net access and loaded with Macknet software, open Mack data programming menu.*

PS5-02 *Access the Mack data hub using Macknet software, and the Mack engine serial number. Download the new engine files to PC memory and then record on disk.*

PS5-03 *Next, the downloaded engine files must be reprogrammed to the vehicle ECM. In this case, a simulator with functional electronics is used to demonstrate the process.*

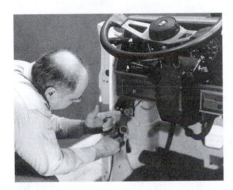

PS5-04 *First, a portable PC must be connected to the vehicle data bus by means of the ATA connector located to the left of the steering column.*

PS5-05 *Insert the disk with the downloaded engine files into the portable PC.*

PS5-06 *Using the Mack software, select the reprogram ECM option from the GUI menu.*

PS5-07 *The V-MAC II software GUI.*

PS5-08 *After reprogramming engine files is completed, the ignition key must be cycled before startup.*

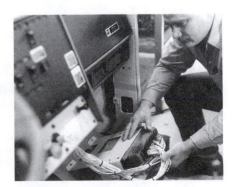

PS5-09 *Location of the V-MAC II module under the lower right side dash.*

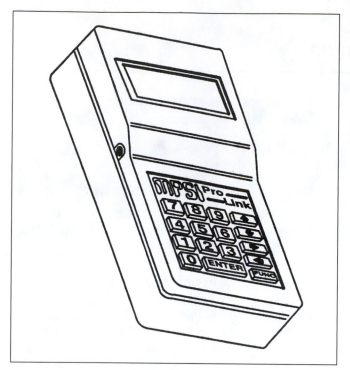

ProLink 9000

**Diagnostic Computer
(Any IBM PC Compatible Computer)**

FIGURE 37–10   V-MAC electronic service tools. (Courtesy of Mack Trucks)

confirmation to Mack Trucks data records that the reprogramming was successfully completed (Figure 37–11 and Figure 37–12).

Mack data programming can only be performed by Mack Trucks dealers. It requires access to one of the following:

- Macknet dealer communications network
- An auxiliary computer that has been equipped with MACKCOMM communications software

READING V-MAC ECUs

Mack ECUs may be read in three different ways, the most simple of which is using the on-board diagnostic capability of the system. The on-board diagnostics are necessary to read a condition when more appropriate tooling is not available, such as during an on-highway breakdown. ProLink and PCs loaded with the appropriated Mack Trucks software should be used when troubleshooting the electronics in the service facility.

Flash Codes

These will blink out active numeric codes using the "electronic malfunction" dash light. Multiple active codes may be blinked but only one is signalled per demand.

ProLink

ProLink can be connected to the truck chassis and the V-MAC module(s) by means of the standard six-pin ATA J1708 dash socket. A dedicated V-MAC cartridge is required to perform customer data programming or to use the unit as a PC serial link. This is upgradable as software versions change by replacing the two PROM chips in the cartridge circuit board. If an HD "general" ProLink cartridge is used, a portion of the system can be scrutinized on a read-only basis. A full description of the ProLink reader/programmer's capabilities is included elsewhere in this text, but it should be noted that while Mack's primary diagnostic and programming tool is the PC, ProLink is a versatile and durable shop diagnostic tool that can be mastered by technicians within a short period of time.

Personal Computer (PC)

Mack Trucks was the first of the OEMs to make the PC its recommended electronic service tool. It is required for proprietary data programming and access to the Mack Trucks data hub, but it also has many advantages over the use of flash codes and ProLink as a diagnostic instrument. Owners who do not wish to have their truck ECU programming altered by non-Mack dealers often program alpha-only passwords. Most

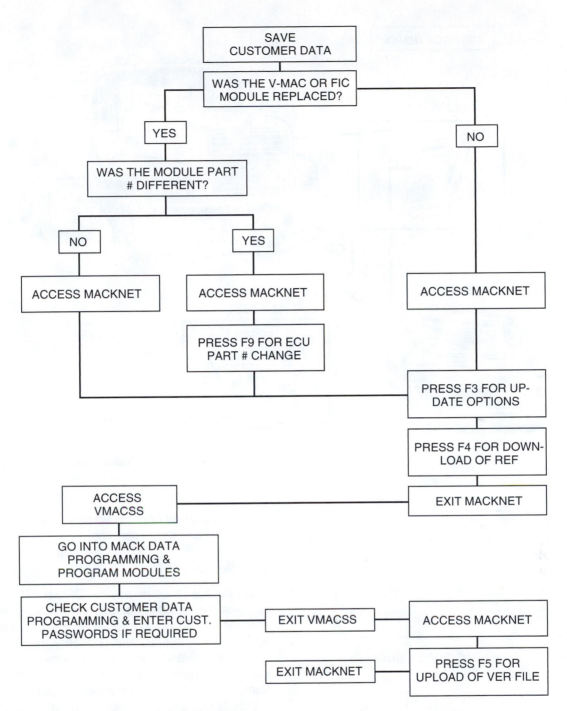

FIGURE 37–11 *Mack data programming flowchart. (Courtesy of Mack Trucks)*

Interface device

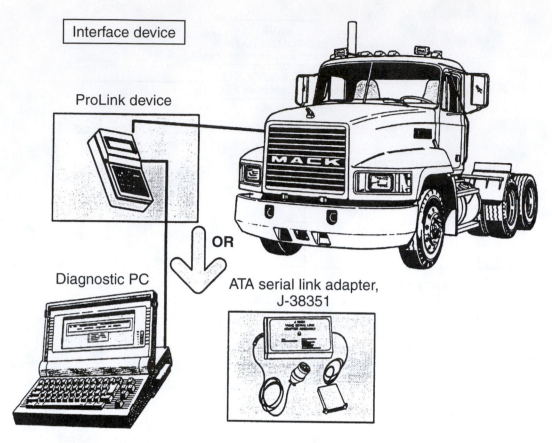

ProLink device

OR

Diagnostic PC

ATA serial link adapter,
J-38351

FIGURE 37–12   *The PC to on-vehicle ECU connection. (Courtesy of Mack Trucks)*

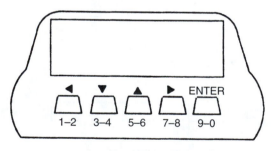

◄	▼	▲	►	ENTER
1–2	3–4	5–6	7–8	9–0

FIGURE 37–13   *The V-MAC Co-Pilot display unit. (Courtesy of Mack Trucks)*

general service facilities use ProLink, which (using a V-MAC cartridge) will not input anything but a numeric password.

Troubleshooting and customer data programming are simplified using a PC and the appropriate linkage hardware and software. Current V-MAC software is Windows-driven. Proprietary data programming must be downloaded from an Allentown, PA-based mainframe accessed by modem to a PC diskette using the three-stage process described previously. Mack programming manuals tend to be user friendly and the re-

programming procedure is simplified by clearly denoted input fields and menu screens.

Co-Pilot Display

Consistent with other engine OEMs, Mack Trucks's dash-located, digital displays known as **Co-Pilot** enable the driver to monitor specific ECU data relating to engine performance, fuel economy, trip information, time, and active and logged fault codes while the vehicle is in operation (Figure 37–13). It consists of a vacuum fluorescent screen located in the upper center of the dash display. The screen displays two rows of alphanumeric characters, and five buttons permit the operator to input display requests.

One of the real benefits of driver data displays (especially when used by a fleet) is the fuel economy readout, which has proved itself to be a cheap and highly effective driver training tool. Co-Pilot simply enables driver access to a portion of V-MAC running data that can enhance his/her ability to efficiently operate the vehicle (Figure 37–14). The system is user friendly to the extent that it can be mastered with a few minutes of instruction. Co-Pilot also provides coded access to the ignition circuit, a theft protection feature.

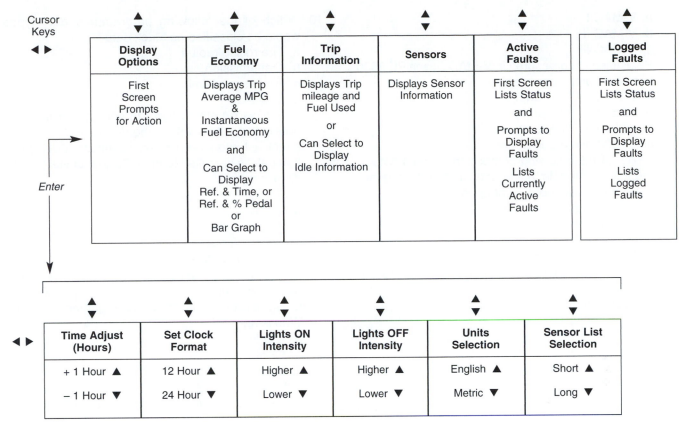

FIGURE 37–14 *V-MAC Co-Pilot quick reference. (Courtesy of Mack Trucks)*

SUMMARY

- V-MAC I and II are classified as partial authority engine management systems.
- V-MAC I and II use a Bosch hydromechanical, port-helix metering injection pump adapted for electronic control.
- V-MAC I featured a V-MAC module and a separate FIC module.
- The FIC module was specifically responsible for the Bosch fuel pump functions.
- V-MAC II may be identified by a single, dash-located module.
- V-MAC features comprehensive engine and chassis monitoring.
- The primary ECU outputs are the rack actuator and Econovance.
- The rack actuator is a linear magnet used by V-MAC to precisely control rack position.
- Econovance provides V-MAC with variable timing.
- V-MAC customer vehicle data programming includes such options as governor type, road speed governing, tire rolling radii, transmission and final drive ratios, failure mode strategies, cruise limits, and others.
- Customer data is V-MAC programmed data that an owner may wish to alter or rewrite as a vehicle's application or operating circumstances changed.
- Mack Trucks data programming must be downloaded from an Allentown, PA, located mainframe and is used to corroborate or reprogram system management files.
- Fuel map shaping would be classified as Mack Trucks data.
- Electronic troubleshooting is performed by following the diagnostic flow charts in the V- MAC service literature or software.
- Co-Pilot is V-MAC's driver data display option that may be used with V-MAC II.
- V-MAC III, introduced in 1998, is used on all current Mack Trucks E-Tech engines. This system uses EUP full authority engine management described in Chapter 41.

REVIEW QUESTIONS

1. Which version of V-MAC uses an ECU and a separate FIC module to manage an in-line injection pump?

a. V-MAC I
b. V-MAC II
c. V-MAC III

2. Peak injection pressures in V-MAC Bosch PE 8500 injection pumps can reach:
   a. 2,200 psi (150 atms)
   b. 7,350 psi (500 atms)
   c. 16,900 psi (1,150 atms)
   d. 29,400 psi (2,000 atms)

3. Which of the following components inputs data to the V-MAC module that defines metered fuel quantity in any given moment of operation?
   a. Rack actuator
   b. Rack position sensor
   c. Reference coil
   d. Timing event marker

4. Which of the following components could be described as a linear magnet?
   a. Rack actuator
   b. Rack position sensor
   c. Reference coil
   d. Pulse wheel

5. Which is the moving component on the Bosch rack position sensor assembly?
   a. Reference coil
   b. Short circuit ring
   c. Proportional solenoid
   d. Pulse wheel

6. Which of the following sensors is responsible for signalling engine position data to the V-MAC module?
   a. RPM/TDC sensor
   b. TEM sensor
   c. CTS
   d. OPS

7. What electrical operating principle is used by the V-MAC system TPS?
   a. Potentiometer
   b. Variable capacitance
   c. Pulse generator
   d. Wheatstone bridge

8. When performing the procedure required to reprogram a V-MAC module with Mack data, which of the following is the correct sequence?
   a. Upload, reprogram, download
   b. Download, reprogram, upload
   c. Reprogram, upload, download

9. Which of the following correctly describes Co-Pilot?
   a. A reader/programmer instrument
   b. A dash-located data display
   c. A customer data, programming instrument

10. Which of the following programmed data types would be classified as Mack data?
    a. Tire rolling radius
    b. Transmission ratio
    c. Road speed limit
    d. Fuel map

11. *Technician A* states that the component that signals rack position data to the V-MAC ECU is called a rack actuator. *Technician B* states that rack position data is signalled to the ECU in a digital format. Who is right?
    a. A only
    b. B only
    c. Both A and B
    d. Neither A nor B

12. *Technician A* states that the V-MAC II module is mounted on the right side of the E-7 engine. *Technician B* states that early versions of V-MAC used a two-module electronic management system. Who is right?
    a. A only
    b. B only
    c. Both A and B
    d. Neither A nor B

13. *Technician A* states that Mack data programming to a V-MAC ECU requires the use of a PC and modem connection. *Technician B* states that customer data programming can be undertaken with a ProLink and the appropriate Mack software cartridge. Who is right?
    a. A only
    b. B only
    c. Both A and B
    d. Neither A nor B

14. *Technician A* states that the fuel map would be classified as Mack data programming. *Technician B* states that tire rolling radii, transmission, and carrier ratios are all classified as Mack data programming. Who is right?
    a. A only
    b. B only
    c. Both A and B
    d. Neither A nor B

15. *Technician A* states that the V-MAC II rack actuator can be described as a linear proportioning solenoid. *Technician B* states that the rack position sensor is a piezoelectric device. Who is right?
    a. A only
    b. B only
    c. Both A and B
    d. Neither A nor B

Chapter

38

Detroit Diesel Electronic Controls (DDEC)

PREREQUISITES

Chapters 29, 30, 32, and 33

OBJECTIVES

After studying this chapter, you should be able to:

- Describe how DDEC electronics have evolved since their introduction.
- Outline the system layout of current DDEC EUI systems.
- Describe the main subsystems and components in DDEC electronic management system.
- Outline the components and function of the DDEC fuel subsystem.
- Understand the principles of operation of a DDEC ECM and its switching apparatus.
- List the primary inputs to the DDEC ECM and categorize them as command and monitoring sensors.
- Understand how a DDEC EUI is controlled and functions.
- Comprehend the principles of operation of the DDC EUI.
- Describe the importance of programming EUI calibration data to the ECM.
- Define the term pilot injection and its application.
- Explain injector response time.
- Describe the governing options that may be programmed to DDEC managed engines.
- Outline the basics of DDEC system diagnosis and troubleshooting.
- Explain the application of DDC ProDriver and ProManager.
- List the parameters that require calibration reprogramming to DDEC electronics.

KEY TERMS

audit trail

beginning of energizing (BOE)

beginning of injection (BOI)

breakout box

check engine light (CEL)

Detroit Diesel Corporation (DDC)

Detroit Diesel Electronic Controls (DDEC)

digital diagnostic reader (DDR)

Diagnostic Link Software

duty cycle

electronic distributor unit (EDU)

electronic foot pedal assembly (EFPA)

electronic unit injector (EUI)

ending of energizing (EOE)

ending of injection (EOI)

fuel pressure sensor (FPS)

injector drivers

injector response time (IRT)

pilot injection

pulse width (PW)

remote data interface (RDI)

split-shot injection

stop engine light (SEL)

stop engine override (SEO)

Series 50

Series 60

Series 92

synchronous reference sensor (SRS)

tattletale

throttle position sensor (TPS)

timing reference sensor (TRS)

turbo-boost sensor (TBS)

valve closing pressure (VCP)

vehicle speed sensor (VSS)

INTRODUCTION

Detroit Diesel Corporation (DDC) was the first diesel engine manufacturer to offer an electronically managed diesel engine in the North American truck and bus markets. DDC electronics are described by the acronym **DDEC**, which stands for **Detroit Diesel Electronic Controls**. DDEC I was introduced in 1985 and made generally available in 1987 on two-stroke cycle, **Series 92** engines (displacement of 92 cubic inches per cylinder). DDEC I was a full authority management system with mechanically actuated (camshaft) **electronic unit injectors (EUIs)** that were computer controlled by a dash-located ECM and an engine-mounted, **injector driver** unit called an **electronic distributor unit (EDU)**. The injector drivers are the ECM actuators responsible for effecting ECM-plotted fueling commands by switching the EUI duty cycle.

Shortly after the introduction of the full authority EUI system on the Series 92 engine, Detroit Diesel launched what was to become the most successful diesel engine of the 1990s, the Series 60. The **Series 60** engine is an in-line, six-cylinder, four-stroke cycle engine fueled by EUIs and managed by full authority electronics. The engine was the first truck diesel engine to be sold in North America that was designed specifically for computer controls. Electronic management was by DDEC II electronics. DDEC II combined the EDU and ECM assemblies into a single, engine-mounted module. Initially, the Series 60 engine was available in 11.1- and 12.7-liter displacements. DDC added a 14-liter displacement version in 2002. The **Series 50** engine is a four cylinder version of Series 60, which shares a high degree of component commonality with the 6-liter version and most of its operating principles.

DDEC electronics have been upgraded through the years and the current version DDEC IV was introduced in 1998. It is expected to be upgraded to DDEC V in the near future. The explanation of DDEC in this chapter is based on DDEC IV, but where significant differences exist between versions, these are indicated in text. DDEC electronics are designed to be SAE J-1939 compatible, that is, they will connect to current multiplexing backbones.

From 1992 to 2001, the DDC remained the biggest selling truck engine (12.7 liter version) in North America. Part of this success was the simplicity of the engine. While Series 60 competitors were continually exploring high-tech advances, Detroit Diesel made few real changes to its engine through the 1990s and relied on what seemed to work, good fuel economy, excellent reliability, and easy maintenance/overhaul. In 2001, the

ownership of DDC was transferred from Penske to DaimlerChrysler. DaimlerChrysler also owns Freightliner LLC and this alliance seems to have resulted in limitations in the availability of the engine in some truck chassis. Furthermore, it is no secret that Mercedes-Benz has been trying to get its MB-900 and MB-4000 engines accepted in the North American truck market and will presumably do this using its ownership of Freightliner and Detroit Diesel as vehicles.

Detroit Diesel opted to use cooled EGR to meet the reduce NO_x requirement on their 2004 emissions certified (by EPA agreement, introduced 10/02) engines. These engines have been widely tested since 2000 and DDC claims that fuel economy will be compromised by less than 3% at worst, and by as little as 1% in most applications. Engine hardware is little changed in 2004-compliant versions compared with earlier versions. Figure 38–1 shows a schematic representation of a typical DDEC system.

SYSTEM OVERVIEW

DDEC is a full authority computerized engine management system that uses cam-actuated, electronically controlled EUIs. Current versions of DDEC offer comprehensive engine monitoring, audit trails, self-diagnostics, other on-board electronic system interface capability, and customer and proprietary (DDC) data programming. The system can be divided as follows:

1. Fuel subsystem
2. ECM
3. Input circuit—sensors
4. Output circuit—EUIs and others

Each of DDEC's subsystems are dealt with sequentially but a knowledge of diesel fuel injection objectives and computer basics (covered elsewhere in this text) is assumed. DDEC IV is referenced in the context of a Series 60 engine in the description provided in this chapter.

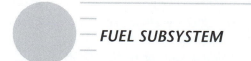

FUEL SUBSYSTEM

Movement of fuel from the fuel tank to the EUIs is accomplished in much the same manner as in a DDC hydromechanical engine. Fuel is drawn from the fuel tank

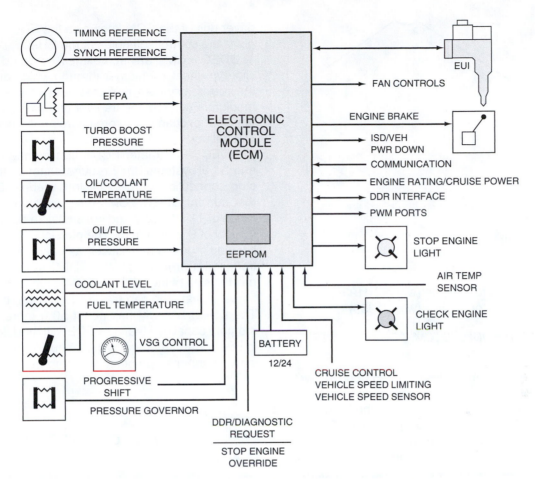

FIGURE 38–1 *DDEC management system layout showing input and output circuits. (Courtesy of DDC)*

by a gear pump; a primary filter is plumbed in series between the fuel tank and the transfer pump. Fuel discharged under pressure by the transfer pump is routed through a secondary filter, the ECM heat exchanger, and then to the fuel galleries located in the engine cylinder head. The Fuel Pro system, previously available as an option, is now standard on DDEC IV, replacing the two-filter system with one, extended service, filter separator system. The fuel galleries then charge the cylindrically shaped EUIs. Series 60 EUIs are manufactured with two external annular recesses in the cylindrical injector body, the lower of which is exposed to the charging gallery in the cylinder head. The upper annular recess in the EUI discharges fuel to the return gallery from which it is sent back to the fuel tank. In the older Series 92 engines, fuel is transferred to and from the EUIs by means of jumper pipes in much the same manner as in mechanical unit injectors. Fuel is constantly circulated through the EUIs whenever the engine is running. Charging pressures are typically around 50 psi (3.3 atms/334 KPa).

ELECTRONIC CONTROL MODULE (ECM)

The DDEC I computer was located within the vehicle cab, with the injector drivers (switching apparatus) or EDU (Electronic Distributor Unit) mounted on the engine close to the EUIs. DDEC I rapidly evolved into DDEC II, so there are few examples of the former in service. All current versions of DDEC house the computer and injector drivers in a single unit usually mounted on the engine. The physical size of the ECM housing has decreased in current DDEC systems despite having much greater computing capability and speed over the earlier versions (Figure 38–2).

The DDEC ECMs are microprocessor-driven management and switching modules. The single module therefore houses all of the microprocessing and output actuators. It is responsible for engine governing, fuel timing logic, self-diagnostics, comprehensive system

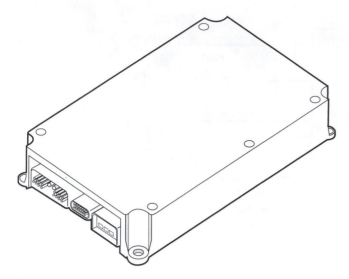

P/N 23513553 for 4 cycle (without J1939)
P/N 23516923 for 2 cycle (with J1939)

FIGURE 38–2   The DDEC III ECM. (Courtesy of DDC)

monitoring, **audit trails** (**tattletale**: open or covert writing of an electronically monitored event or condition to EEPROM), and customer and DDC data programming. Each DDEC ECM can manage up to eight EUIs; in off-highway engines that have more than eight cylinders, multiple ECM units are required. In multi-ECM-managed engines (off-highway applications), one of the ECMs masters the engine; the master ECM receives all inputs and processes data, while the other (or others) are referred to as receivers and are controlled by the master ECM. Receiver ECMs act as EUI actuators and basically effect EUI switching on command. However, in the event of master ECM communication failure, receiver ECMs can temporarily enable operation of each portion of the engine.

The DDEC ECM is programmed with engine management logic to fuel the engine through variable speeds and loads, while at the same time monitoring both engine and chassis operating conditions and continually performing self-diagnosis. Current DDEC versions feature electronically erasable programmable read-only memory (EEPROM), which qualifies the master engine management program in read-only memory (ROM) with application specific detailing such as engine governing, torque profile shaping, cold-start logic, and engine protection strategy. The results of ECM arithmetic, logic, and fetch-and-carry computations in the processing cycle are converted to actuator or control signals that switch the EUIs. Command pulses can be referred to as **duty cycle** or **pulse width (PW)** and are measured in milliseconds. Duty cycle or PW is the time period within the EUI cam-actuated downstroke in which fuel is actually being pumped. PW is calculated and controlled by the ECM; it is effected by the injector

driver unit of the ECM known as the EDU, which energizes the control solenoids of the EUIs.

DDEC ECMs are sometimes mounted on a heat sink-type heat exchanger through which diesel fuel in the fuel subsystem is routed and acts as a cooling medium. Fuel from the transfer pump is routed through this heat exchanger before charging the cylinder head fuel gallery.

DDEC III and later ECMs can manage both 12 V and 24 V systems; they require either a direct or bus strip connection to the vehicle batteries. DDC states that current DDEC ECMs have an operating voltage range of 11 V to 32 V and will sustain a transient cranking low of 9 V minimum. The physical size of the ECM housing tends to be reduced as each generation is introduced. For instance, the DDEC III ECM is roughly half the size of the DDEC II ECM, processes data eight times as fast, and can retain seven times the memory. The DDEC IV ECM has about 50% more speed and memory capability than the DDEC III module. Processing power and memory will continue to increase in the same manner as on personal computers.

INPUT CIRCUIT SENSORS

As DDEC has evolved, the sensor circuit has become more comprehensive. The following is a description of some current DDEC sensors without detailing the electrical operating principles, which are covered in detail in Chapter 32.

TIMING REFERENCE SENSOR

The **timing reference sensor** or **TRS** signals crank position to the ECM. An AC pulse generator is used. A six-tooth pulse wheel (reluctor) is used on DDEC II and thirty-six-tooth pulse wheel is used on current versions. In DDEC II, III, and IV, the pulse wheel is fitted to the front of the crankshaft (Figure 38–3).

SYNCHRONOUS REFERENCE SENSOR

The **synchronous reference sensor** or **SRS** signals the ECM to indicate a specific engine position and acts to confirm the TRS signal. An AC pulse generator is used. Both TRS and SRS must work for the engine to run; if either is not functional in DDEC II, a code is logged after 10 seconds cranking; in DDEC versions III, versions IV, and higher, this code is produced immediately. The SRS consists of a single dowel that acts as a reluctor tooth located on the cam gear and a stationary magnetic pickup in Series 60/50.

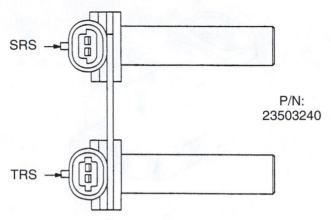

SRS →

TRS →

P/N:
23503240

**The SRS and TRS sensors —
6 V–92, 8 V–92, and 12 V–71 engines**

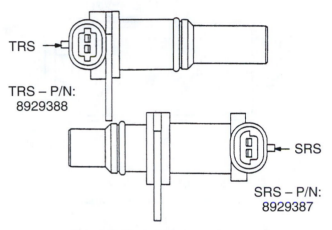

TRS →

TRS – P/N:
8929388

→ SRS

SRS – P/N:
8929387

**The SRS and TRS sensors —
series 50 and series 60 engines**

FIGURE 38–3 *DDEC SRS and TRS sensors on two-stroke cycle and four-stroke cycle engines. (Courtesy of DDC)*

NOTE: On Series 92, the SRS and TRS both pick up on the cam gear-located pulse wheel.

ELECTRONIC FOOT PEDAL ASSEMBLY

In common with most truck/bus electronically managed engines, DDEC is "drive by wire." The critical sensor in the **electronic foot pedal assembly (EFPA)** is the **throttle position sensor (TPS)**. The TPS receives reference voltage (V-Ref) of 5 V and returns a portion of it, proportional to pedal mechanical travel. The DDEC TPS is a three-terminal potentiometer or voltage divider; the resistance within the TPS that the V-Ref must overcome depends on wiper position on the TPS resistor. The actual voltage signal returned to the ECM is converted to counts, which can be read with an elec-

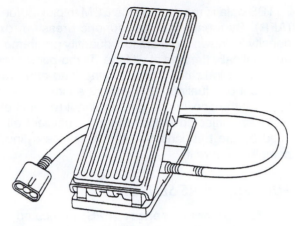

Electronic Foot Pedal Assembly (EFPA)

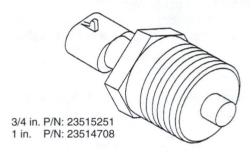

3/4 in. P/N: 23515251
1 in.   P/N: 23514708

**Coolant Temperature Sensor
(series 50 and series 60)**

FIGURE 38–4 *The EFPA and CTS. (Courtesy of DDC)*

tronic service tool (EST). TPS counts should be read as follows:

Zero travel, idle: 100–130 TPS counts

Full travel: 920–950 TPS counts

48 counts or less: low-voltage code generated

968 counts or higher: high-voltage code generated

The EFPA is supplied by the chassis manufacturer but must be consistent with DDC specifications (Figure 38–4).

TURBO-BOOST SENSOR

The DDEC **turbo-boost sensor (TBS)** is responsible for signalling the ECM with both manifold boost and barometric pressure data. Early versions of DDEC TBS used a variable capacitance principle. Pressures from 0 to 45 psi (3 atms/304 KPa) can be measured. Recent versions of DDEC use a piezoresistive (Wheatstone bridge) sensor to signal manifold boost to the ECM. These sensors are more accurate and less susceptible to tampering. Whichever type is used, the TBS is a DDC-provided sensor.

TBS data is required by the ECM to plot air/fuel ratio (AFR). By monitoring atmospheric pressure, oxygen density is known, so ECM fuel quantity programming is automatically derated at altitude. Turbo-boost pressure monitoring indicates actual engine load and prevents transient overfueling. Therefore, a sudden full fuel demand at the accelerator pedal will not result in excess fuel being injected before there is sufficient air (supplied by the turbocharger) in the engine cylinders to completely burn it.

FUEL PRESSURE SENSOR

The **fuel pressure sensor (FPS)** is located on the charge side of the transfer pump. It is a variable capacitance-type sensor capable of signalling pressure values from 0 to 75 psi (5 atms/505 KPa). The FPS is a DDC-supplied sensor that will warn the operator of a fuel subsystem problem such as a restricted filter.

OIL PRESSURE SENSOR

The oil pressure sensor (OPS) will trigger engine protection strategy if engine oil pressure drops below a specified value. It is a variable capacitance-type sensor capable of signalling values from 0 to 75 psi (5 atms/ 505 KPa). The OPS is a DDC-supplied sensor.

COOLANT PRESSURE SENSOR

The coolant pressure sensor (CPS) is a variable capacitance-type sensor used to warn the operator of low coolant pressure. It is a DDC-supplied sensor (Figure 38–5).

FUEL TEMPERATURE SENSOR

The fuel temperature sensor (FTS) is a thermistor supplied with V-Ref of 5 V whose resistance decreases as the temperature increases and therefore can be classified as a negative temperature coefficient (NTC)-type sensor; the voltage signal returned to the ECM increases with temperature rise. The FTS signal is required by the ECM to help factor air/fuel ratio and log fuel consumption data. It is a DDC-provided sensor.

COOLANT TEMPERATURE SENSOR (CTS)

The DDEC CTS is also a thermistor of the NTC type and will trigger engine protection strategy once a specific programmed limit is exceeded. It is a DDC-provided sensor.

AIR TEMPERATURE SENSOR (ATS)

The DDEC ATS is another thermistor of the NTC type whose signal input is critical to the ECM for calculating

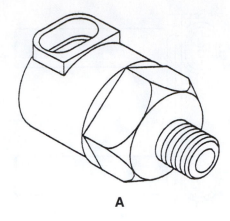

A

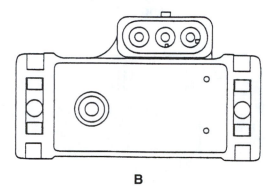

B

FIGURE 38–5   A. *Oil, fuel, and coolant pressure sensors;* B. *the turbo boost sensor. (Courtesy of DDC)*

cold-start fueling and timing, factoring AFR, and hot idle speed. It is a DDC-provided sensor.

OIL TEMPERATURE SENSOR (OTS)

The DDC OTS is a thermistor of the NTC type and is used to corroborate data from CTS and ATS sensors and can also trigger engine protection strategy once the specified limit is exceeded. It is a DDC-provided sensor.

FIRETRUCK PUMP PRESSURE SENSOR

The firetruck pump pressure sensor is used with DDEC III to master engine speed governing to produce a specific pump pressure value. It is of the variable capacitance type and is DDC supplied (Figure 38–6).

VEHICLE SPEED SENSOR

The **vehicle speed sensor (VSS)** is of the induction pulse generator type, usually a standard thirty-six-tooth reluctor located on the transmission tailshaft. Although the VSS is optional, it is required to enable cruise control and vehicle speed limiting ECM functions and it is

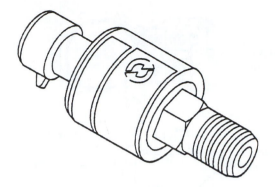

FIGURE 38–6   *Firetruck pressure sensor. (Courtesy of DDC)*

incorporated in most DDEC applications. The VSS is supplied by the chassis manufacturer (OEM).

COOLANT LEVEL SENSOR (CLS)

A switch-type sensor consisting of an integrated resistor that grounds through the engine coolant and is located in the top radiator tank. If the ground is interrupted, engine protection strategy is initiated after a prescribed time period if programmed. The CLS is OEM supplied.

MULTIPLEXING

Where DDEC electronics are required to interface with other chassis system ECMs, the J1587 and J1939 data communication buses are used. This synergizes separate electronic system operation and avoids the need to duplicate vehicle sensors such as the TPS and VSS. A full description of multiplexing is provided in Chapter 35.

OUTPUT CIRCUIT

Output circuit devices are components controlled by the ECM; they are the set of components that effect the results of the programmed software and the processing computations. Any component that is switched or can be controlled by the ECM can be called an output.

ELECTRONIC UNIT INJECTOR (EUI)

DDEC EUIs are integral pumping, metering, and atomizing devices. A DDEC EUI is shown in Figure C–3 in the color section. In common with the mechanical unit injectors (MUIs) used by DDC in the past, they are cam actuated but with that the similarity ceases. The EUI plunger has no cross and center drillings or helical recesses, but is a simple cylindrical plunger element that reciprocates within the pump stationary element or bushing (Figure 38–7). Actual plunger travel does not vary because this is dictated by cam profile. The EUI tappet spring holds the plunger in a retracted position for most of the cycle until the actuating cam on the engine camshaft ramps toward the cam nose and the EUI plunger is driven downward into the pump chamber.

Fuel is circulated through EUI internal ducting whenever the engine is run at charging or supply pressure (50 psi/3.3 atms/340 kPA). The EUI internal fuel circuit includes the pump chamber and circuitry routing fuel through the control solenoid; connected to the control solenoid armature is a poppet valve. Whenever the poppet valve is in the open position (that is, when the solenoid is not energized), fuel is permitted to pass through the solenoid spill ducting. The instant the ECM

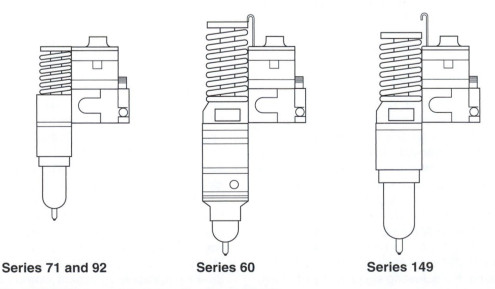

| Series 71 and 92 | Series 60 | Series 149 |

FIGURE 38–7   *The DDEC EUI by engine series. (Courtesy of DDC)*

energizes the EUI solenoid, the poppet control valve closes, blocking the passage of spill fuel through the solenoid ducting. This action effectively traps the fuel in the pump chamber under the EUI plunger. A duct in the base of the pump chamber connects it with a hydraulic injector nozzle. When the required nozzle opening pressure (NOP) is achieved, the nozzle valve opens, fuel passes around the nozzle valve seat, and exits the EUI through the nozzle orifii.

DDEC EUIs are designed for direct injection of a fuel charge to both two- and four-stroke cycle engines at NOPs typically around 5,000 psi (340 atms/34.4 MPa) and possible peak pressures of over 25,000 psi (1,700 atms/172 MPa). The fueling window within which the EUI can switch an effective pumping stroke is dictated by cam profile because the EUI plunger must be in the act of descending for this to occur. Plunger mechanical stroke is constant. Plunger *effective stroke* must occur within the mechanical stroke hard window, beginning when the EUI solenoid is energized and ending when it is switched open. DDEC EUIs are switched at an electrical pressure of 12 V (Figure 38–8).

DDEC III and IV EUIs are graded with a calibration code that ranges from 00–99 enabling the ECM to be programmed with data concerning how a specific EUI flows fuel. The **digital diagnostic reader (DDR)** is used to program this data to the ECM and enables highly accurate and balanced fueling of the engine. To program the calibration code the injector cylinder number would be selected first, followed by entering the digital calibration code value. Earlier DDEC II calibration codes are alpha, A, B, or C with A denoting the least fuel flow and C the most fuel flow. This critical procedure helps the ECM balance cylinder fueling.

TECH TIP: Whenever an EUI or set of EUIs is replaced, the fuel flow code(s) MUST be reprogrammed. Failure to perform this critical step will result in unbalanced engine fueling.

INJECTOR RESPONSE TIME

DDEC EUIs are switched at 12 V, unlike most other EUI systems that spike the application voltage to values around 100 V. This means that the time lag between **beginning of energizing (BOE)** or the instant that the EDU starts to flow current to the EUI and **beginning of injection (BOI)** solenoid is to some extent variable. DDEC electronics are programmed to measure the response time between BOE and BOI by studying the actuation voltage wave of the previous two actuations of each EUI on a continual basis. **Injector response time or IRT** is essentially the time lag between the output of the actuating signal at the EDU and the moment the EUI poppet control valve actually

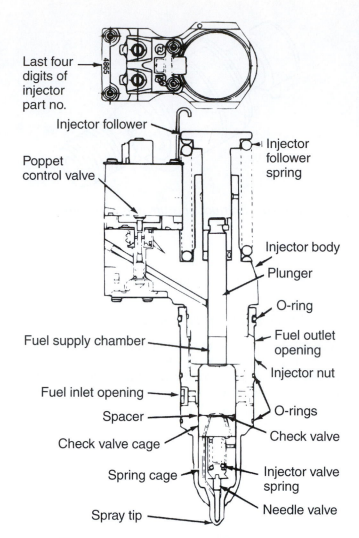

Last four digits of injector part no.

Injector follower

Poppet control valve

Injector follower spring

Injector body

Plunger

O-ring

Fuel outlet opening

Injector nut

Fuel supply chamber

Fuel inlet opening

Spacer

O-rings

Check valve cage

Check valve

Spring cage

Injector valve spring

Spray tip

Needle valve

Electronic Unit Injector Cross Section

FIGURE 38–8   Overhead and sectional views of a DDEC EUI. (Courtesy of DDC)

closes. It is measured in milliseconds and read on the DDR display. DDEC is capable of adjusting to minor electrical circuit resistance problems within a certain window to maintain balanced fueling and timing. It should be noted also that there is a fractional lag between EUI control valve closure (BOI) and the actual opening of the EUI hydraulic nozzle valve (NOP), which truly begins injection. Similarly, at the completion of the switched duty cycle or PW known as **ending of energizing (EOE)**, there is not an immediate cessation of injected fuel because some fraction of time is required to collapse the pressure in the EUI pump chamber to the **valve closing pressure (VCP)** value at which point the injection pulse truly ceases—**ending of injection (EOI)**. Figure 38–9, Figure 38–10, and Figure 38–11 graphically represent what is occurring electrically and hydraulically through the EUI duty cycle.

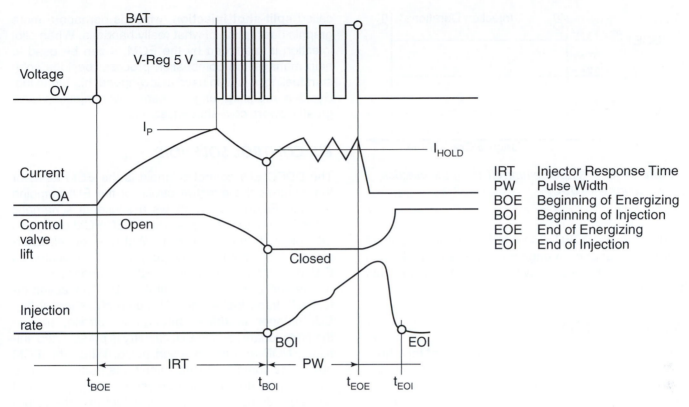

IRT   Injector Response Time
PW   Pulse Width
BOE   Beginning of Energizing
BOI   Beginning of Injection
EOE   End of Energizing
EOI   End of Injection

FIGURE 38–9   *DDEC injection cycle time graph. (Courtesy of DDC)*

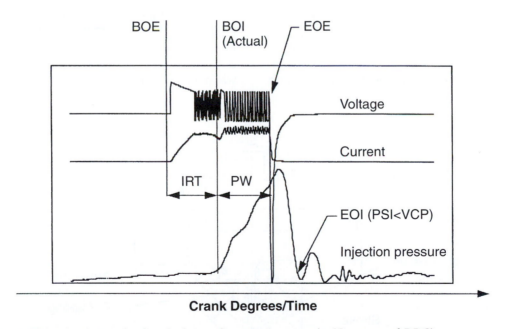

FIGURE 38–10   *DDEC injection cycle electrical waveform to time graph. (Courtesy of DDC)*

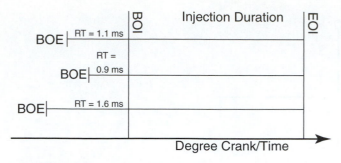

FIGURE 38–11 *Response time effect on the injection cycle. (Courtesy of DDC)*

The primary DDEC output is EUI management but DDEC will support an engine compression brake and bus into other chassis system ECMs via the data backbone.

PILOT INJECTION

Current DDEC electronics can manage **pilot injection**. This technology may be used as part of ECM cold-start strategy where it is used to eliminate diesel knock and minimize startup smoke emission. Pilot injection essentially breaks up the fueling pulse by switching the EUI at high speeds. Using this technology under cold-start conditions would mean delivering a short pulse of fuel to the engine cylinder, calculating the moment of ignition, and at that point resuming the EUI fueling pulse. Diesel knock is a cold-start detonation condition caused by delayed ignition and excess fuel in the cylinder at the point of ignition. Pilot injection can also be called **split-shot injection**, which is perhaps a more graphic description of what really happens. When pilot injection is managed by the ECM, it can be used to help manage the combustion process: besides cold-start fueling, it is also used at low-speed, high-load lug. Besides reducing engine wear, pilot injection also greatly lowers cold-start emissions.

EUI CONTROL SOLENOID

The DDEC EUI control solenoid is the ECM actuator that controls the effective stroke of the EUI pumping plunger. Figure 38–12 shows a sectional schematic of the EUI control cartridge. The DDEC ECM energizes the solenoid by grounding a PWM-type signal. Energizing the solenoid pulls the poppet valve (indicated as C.V. or control valve in Figure 38–12) inboard, closing off the spill or return circuit of the EUI. This action effectively traps fuel in the EUI pump chamber: as the EUI plunger is driven through its stroke, fuel in the pump chamber and EUI circuitry is pressurized, initiating NOP and the injection pulse. When the ECM opens the control solenoid circuit, the poppet valve is forced outboard by spring pressure, opening the spill passage, dumping high-pressure fuel into the return circuit. When pressure collapse is sufficient to allow the nozzle valve to close, the nozzle spring seats the nozzle valve ending the injection pulse.

The solenoid cartridge assembly may be removed and replaced from the EUI without pulling the assembly from the engine. The procedure for performing this is outlined later in this chapter, but it should only be done when indicated by DDEC troubleshooting.

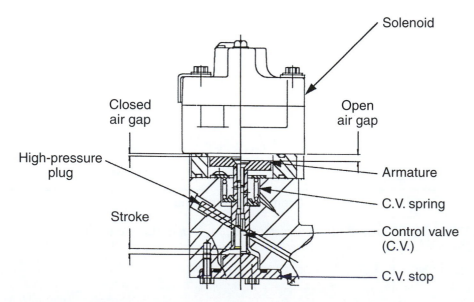

FIGURE 38–12 *Sectional view of EUI control solenoid and poppet valve. (Courtesy of DDC)*

GOVERNING

DDEC offers both limiting speed governing (LSG) and variable speed governing (VSG). Although managed electronically, both LSG and VSG are designed to function similarly to mechanical LSG/VSG. The LSG option fuels the engine based on percent throttle (pedal position) selected by the operator. The VSG option will hold engine speed at a specific requested value and attempt to maintain it as the load on the engine fluctuates; it is used for PTO operation and electronic cruise control.

The ECM is programmed with a detailed fuel map that manages engine fueling through both normal and transient abnormal conditions. Smoke control and cold-start fueling are managed by selecting optimum injection timing and injected fuel quantity based on such factors as ambient temperature, barometric pressure, manifold boost, and throttle demand. Cold-start strategy on DDEC-managed engines may produce an automatic increase in engine idle speed and advanced injection timing to minimize cold operation white smoke emission; this would return to normal as the engine approaches operating temperatures.

CRUISE CONTROL

Cruise control can be operated in any gear when engine speed is above 1,100 rpm and providing the road speed is above 20 mph (32 km/hr). Cruise control can be programmed to master engine brake operation so that if the vehicle accelerates above set cruise speed on a down grade, the engine brake will automatically actuate.

PROGRESSIVE SHIFTING

Progressive shifting is a feature designed to achieve better fuel economy by encouraging the driver to upshift before the engine reaches governed speed. When the speed of the engine is limited, the driver remains in the higher torque range of torque rise (that is, the lower rpm range) for longer where the engine operates at better fuel efficiency.

ENGINE PROTECTION

DDEC electronics monitors a range of vehicle and chassis functions and is programmed with engine protection strategy that may derate or shut down the engine when a running condition could result in catastrophic failure. The operator is alerted to system problems by illuminating dash lights, the **check engine light (CEL)**, and the **stop engine light (SEL)**. The CEL is illuminated when a system fault is logged to alert the operator that DDEC fault codes have been generated. The SEL is illuminated when the ECM detects a problem that could result in a more serious failure. DDEC may also be programmed to shutdown or ramp down to idle speed after such a problem has been logged. A shutdown override switch known as an **SEO** or **stop engine override** switch will permit a temporary override of the shutdown command.

ENGINE DIAGNOSTICS

Faults that occur are stored in ECM memory and may be recalled by any of the following means:

BLINK CODES

The CEL and SEL will flash codes when the diagnostic request switch is toggled. The CEL flashes inactive codes, and the SEL flashes active codes.

DIAGNOSTIC DATA READER (DDR)

The DDC (J38500-E)-recommended electronic service tool for DDEC is the ProLink 9000 with the appropriate level software, dedicated cartridge. The DDR connects to the DDEC ECM by means of an SAE/ATA J1708 six-pin/J1939 nine-pin Deutsch connector or as it is more commonly known, the ATA connector. This provides two-way communications for reading system parameters, calibration data, diagnostics, data programming, recalling faults, and audit trails. DDR data can be printed out using a printer (J38480) (Figure 38–13).

PERSONAL COMPUTER (PC)

Chassis OEMs can program engines on their assembly lines using a DDC software package called *Vehicle Engine Programming Station* and a PC with Windows software. A PC interface is required to translate the datalink signal of the ECM to the VEPS software. VEPS PC software can also be used to rewrite some of the parameters programmed to the DDEC III ECM.

A PC can also be connected to DDEC ECMs by using a serial link direct to the ATA connector. With DDEC IV and later versions, DDC recommends the PC as its primary electronic service tool (EST).

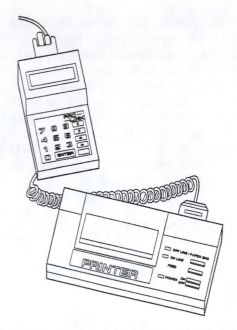

FIGURE 38–13   The DDEC DDR (ProLink 9000) and printer. (Courtesy of DDC)

FAULT CODES

DDEC logs two types of codes: active and inactive.

1. Active Codes—An active code is a fault that is present at the moment codes are checked. Active codes are flashed to the SEL when the diagnostic request switch is toggled.
2. Inactive Codes—Inactive codes are codes that have occurred sometime previously but are not active at the moment of request. Every code has an audit trail (tattletale) in DDEC and is tagged with the following data:
   a. First occurrence of each diagnostic code logged in engine hours
   b. Last occurrence of each diagnostic code logged in engine hours
   c. Total time in seconds that the engine code was active

INJECTOR RESPONSE TIME AND PULSE WIDTH

Injector response time (IRT) and pulse width (PW) were explained earlier in this chapter. Both are indicators of the overall electrical integrity and power balance of the engine. Both parameters can be displayed on any EST. When attempting to correlate PW, which is measured in milliseconds, with engine crank degrees, as an engine is accelerated from idle speed to high-idle speed, the number of crank angle degrees the engine passes through in one millisecond will triple. So do not be surprised if the PW time average decreases as an engine is accelerated and loaded down. Despite the reduced PW, EUI effective stroke, and therefore, injected fuel quantity, is increased.

INJECTOR CUTOUT TEST

There is no appropriate way to mechanically short out a DDEC EUI, so when a cylinder misfire is being diagnosed, the ECM electronics perform this task by electronically cutting out the EUIs in sequence and analyzing the performance effect. When an ECM commands an engine to run at a specific rpm and an injector cutout test is performed, first the average duty cycle of the EUI in milliseconds is displayed. If the engine rpm is to be maintained during the test (this test is normally run at idle or 1,000 rpm), as each EUI is cut out electronically by the ECM, the PW of the remainder of the injectors will have to be lengthened if engine rpm is held at the test value. This would be true until a defective EUI was cut out; in this case, there would be no increase in the average PW of the operating EUIs. If the example of a Series 60 engine is used with a dead EUI, the following would occur if an injector cutout test sequence were set with the engine running at 1,000 rpm. To run the engine at 1,000 rpm, the five functioning EUIs would produce fueling values read in PW milliseconds, which would be averaged on the display. When one of the five functioning EUIs is electronically cut out by the ECM, only four EUIs would be available to run the engine at the test speed of 1,000 rpm, causing the average PW to increase. As each functioning EUI is cut out in sequence, the average PW would have to increase to maintain the test rpm. However, when the defective EUI was cut out, there would be no change in the average PW because it is based on the engine running on five cylinders.

ENGINE PROTECTION STRATEGY

Engine protection strategy can be programmed to three levels of protection beginning at a low level of protection, essentially a driver alert, and progressing to a complete shutdown of the engine. The level of protection is selected by the vehicle owner and programmed to the DDEC ECM as a customer option. As a customer option, the failure strategy can be changed at any time by a technician equipped with the access password and an EST with the DDEC software. The parameters monitored are:

Low coolant level

Low coolant pressure

High coolant temperature

Low oil pressure

High oil temperature

Auxiliary digital input

When DDEC detects a fault that may result in catastrophic engine damage, it can be programmed to react to the problem in three ways:

1. *Warning only.* The CEL and SEL will illuminate but it is left up to the operator to take action to avoid potential engine damage. No power or speed reduction will occur. Engine protection is therefore left entirely to the discretion of the operator.

2. *Ramp down.* The CEL and SEL will illuminate when a fault is detected and the ECM will reduce power (fuel quantity injected) over a 30-second period after the SEL illuminates. The maximum power value (fuel quantity) maintained for this initial reduction is that operative just before the logging of the fault condition. A diagnostic request/stop engine override switch is required when this engine protection option is selected.

3. *Shutdown.* Both the CEL and SEL will illuminate and the engine will shut down 30 seconds afterwards. Once again, a diagnostic request/stop engine override switch is required if this option is selected.

STOP ENGINE OVERRIDE OPTIONS

Three types of stop engine overrides are available:

1. Momentary override—resets the 30-second shutdown timer, restoring engine power to the value operative at the moment the SEL was illuminated. To obtain subsequent overrides the switch must be recycled after 5 seconds. DDEC will record the number of overrides actuated after the triggering fault occurrence.

2. Continuous override, option 1—this stop engine override option is used when a vehicle needs full power during a shutdown sequence. Full power capability is maintained for as long as the override switch is pressed. It is intended for coach applications only.

3. Continuous override, option 2—this stop engine override option is used primarily for industrial and construction applications. It essentially disables the engine protection system until the ignition key is cycled.

TROUBLESHOOTING DDEC ELECTRONICS

The ESTs required are the DDR (ProLink 9000 and DDEC software cartridge) or a PC, depending on the DDEC version, and a DMM. DDEC troubleshooting also requires a basic knowledge of electrical circuits and schematics, DDEC operating fundamentals, and the DDC *DDEC Diagnostic and Troubleshooting Guide*. The latter is a large but user-friendly, sequential troubleshooting manual in which each stage progresses to a set of optional steps, the path routing being dependent on the test results of each. The key to using such a manual is to precisely execute the instruction and interpret the results of each step before proceeding to the next. With this type of electronic troubleshooting, relying on empirical knowledge (previous experience of troubleshooting the system) or making assumptions that short-cut a step, can result in wasted time. The *DDEC Diagnostic and Troubleshooting Guide* covers the following areas:

- Reading diagnostic codes
- Clearing codes
- Calibration reprogramming
- Connector checkout
- Using a DMM
- System testing
- Sequential troubleshooting
- Reprogramming engine parameters
- DDEC schematics

Troubleshooting DDEC requires the use of ESTs such as the DDR, PC-based hardware with DDC software, ProLink 9000 and DMMs, plus the appropriate **breakout boxes** and Ts, and the DDEC service literature.

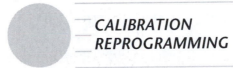

CALIBRATION REPROGRAMMING

A DDR or PC can be used to make the following calibration changes:

- Rewrite DDR calibration configuration password.
- Add/delete 5-minute idle shutdown.
- Change droop (governor speed deviation from set speed).
- Set PTO droop (isochronous if required).
- Enable/disable cruise control.
- Add/delete a VSS.
- Change vehicle speed limit.
- Change cruise control speed limit.
- Add/delete engine shutdown.
- Set PTO maximum speed.
- Switch between available engine ratings.
- Enter injector calibration values.
- Set progressive shift configuration.

ProDriver

DDEC III versions and later feature as an option, Pro-Driver, a dash-mounted data display that can provide the operator with immediate feedback on idle time, fuel economy, and other performance factors. ProDriver has a four-line display window and user-friendly, four-key controls. This type of instant data feedback to the operator has been proved to produce an almost immediate performance improvement; it can be regarded as low cost (compared to a human driver trainer) driver training.

ProManager

ProManager is a PC software program that permits data to be downloaded from the DDEC ECM for analysis. Fleets interested in improving vehicle and operator performance have found ProManager a useful tool. The latest DDEC versions are driven from MS Windows software.

REMOTE DATA INTERFACE

The **remote data interface** or **RDI** is a communications device that enables data exchange between the vehicle-based electronics, data hub products, and a fleet's computer network. System operation is automatic. The driver parks the vehicle adjacent to the RDI and connects the extraction cable to the ATA connector on the vehicle. RDI includes status lights to update the driver on the progress of the communications. The latest versions of RDI transfer are wireless.

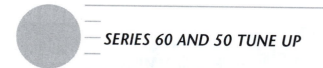

SERIES 60 AND 50 TUNE UP

Consistent with any other section of this textbook in which a hands-on procedure is outlined, the technician is reminded that the information presented in this text is no more than an outline and that the appropriate DDC service literature should be consulted before performing the tune up. The tune-up procedure is a much faster, simpler procedure on DDC Series 60 and 50 engines than on the DDC MUI engines because only two basic settings are being checked/adjusted. The basic settings are the valve lash dimension and the EUI follower height dimension. The procedure for setting DDC valves is consistent with that used to set valves on most other engines, and that used to adjust the EUI height dimension correlates directly with the injector height dimension setting of a DDC MUI. Both Series 60 and 50 engines use a four-valve, parallel port valve configuration. DDC has eliminated the need for yokes/ bridges to actuate the pairs of valves by using a distinct rocker arrangement in which each of the four valves are actuated by an individual rocker arm. The exhaust valves are set on one linear plane and the intake on another. This arrangement permits equal and uninterfered with cylinder gas flow through each valve. A fifth rocker located at the center of the cylinder in between the intake and exhaust valve planes is responsible for actuating the EUI. Each rocker is adjusted for lash/load by means of an Allen head screw and locknut. Table 38–1 may be used as a guide for correlating engine position with setting location on a 6-cylinder Series 60 engine.

A detailed account on adjusting valves is given elsewhere in this textbook. However, as with their MUI-fueled engines, DDC prefers the use of go-no-go gauges when adjusting valve lash. The procedure used to set the EUI follower height dimension is almost exactly the same as that used in DDC MUI engines. This procedure essentially synchronizes the actuating cam geometry with the EUI plunger. The correct height adjustment gauge must be selected by consulting the DDC specifications. A 78.8 mm (3.012″) is common. The engine must next be located in the correct position as shown in Table 38–1.

TABLE 38–1 *Series 60 Tune-Up Chart*

ENGINE POSITION	SET VALVES ON CYLINDER NUMBER	SET EUI HEIGHT DIMENSION ON CYLINDER NUMBER
#1 TDC compression stroke	1	6
#5 TDC compression stroke	5	2
#3 TDC compression stroke	3	4
#6 TDC compression stroke	6	1
#2 TDC compression stroke	2	5
#4 TDC compression stroke	4	3

1. Insert the dowel of the EUI height adjustment tool into the gauge hole in the EUI. Place a drop of the same engine oil used in the engine on top of the EUI follower; the lower ledge of the T on the height adjustment tool should just wipe the oil as it is rotated.
2. If the follower flange-to-EUI body height is incorrect, loosen the locknut on the injector rocker and, using a ³⁄₁₆ Allen socket, adjust the rocker train to obtain the exact specified height dimension. Torque the locknut to specification and recheck the EUI body-to-follower height. Because this procedure sets a dimension, in an emergency, that is, when the correct EUI height dimension tool is not available, an accurate electronic Vernier caliper can be used to check or adjust this setting.

EUI REMOVAL

When the troubleshooting sequencing indicates that an EUI must be removed, it is essential that fuel be drained from the cylinder head fuel supply gallery before attempting the removal using the following procedure:

1. First, clean off the exterior of the valve housing cover and remove it.
2. Drain the cylinder head fuel gallery by removing the supply hose from the inlet fitting at the rear of the cylinder head. Collect the fuel that drains from the cylinder head and next remove the outlet hose from the fitting at the cylinder head.
3. Use a shop air supply and an air nozzle to blow out any remaining fuel from the fuel supply gallery.

WARNING: Failure to observe the foregoing procedure will result in fuel exiting the gallery into the EUI bore the moment it is removed; this fuel will drain onto the piston crown and may cause a hydrostatic lock and other problems in the affected cylinder.

4. Remove the two rocker shaft through bolts and remove the rocker assembly from the engine.
5. Loosen the EUI terminal screws. The EUI connectors have keyhole slots that permit them to fit over the terminal screw caps. Avoid attempting to remove the terminal screws because this will damage the solenoid housing threads.
6. Use an injector pry bar to lift the EUI from its seated position.
7. Cap the EUI bore to prevent dirt from entering the engine. Handle the EUI with care.

EUI INSTALLATION

1. The O-rings should be lightly lubricated with engine oil and carefully installed on the EUI body.

2. Insert the EUI into the injector tube. Visually align the EUI body between the valve springs. The EUI is seated by pressing it firmly into position using the heel of one hand.
3. The EUI is retained by a hold-down crab; position the hold-down crab and install the hold-down fastener. Ensure that the hold-down crab does not interfere with either the EUI follower spring or the valve springs. Torque the hold-down fastener to the required specification.

WARNING: The hold-down washer must be installed with the hemispherical side located downward or the washer will be damaged.

4. Install the EUI wires by positioning the keyhole terminal over the terminal screw; next, position the terminal so that the narrow slot is immediately below the terminal screw and torque to the DDC specification.
5. Install the rocker shaft assemblies.

TECH TIP: The front and rear rocker shaft assemblies look similar but are not identical due to different bolt hole center distance. Identify the rocker shaft assemblies on removal and take care to return each to the proper location on reassembly.

6. Set the valve lash and EUI height dimension to specification.
7. Reinstall the fuel inlet and outlet lines to the cylinder head fittings and replace the valve housing cover.

EUI SOLENOID REPLACEMENT

DDC injector solenoids are serviceable without removing the EUI assembly from the cylinder head.

1. Loosen the EUI terminal screws. The EUI connectors have keyhole slots that permit them to fit over the terminal screw caps. Avoid attempting to remove the terminal screws because this will damage the solenoid housing threads.
2. Remove the four solenoid hex capscrews to remove the solenoid. The solenoid, load plate, follower retainer, and capscrews should be discarded. DDC stresses that the capscrews should always be replaced when undertaking this procedure.
3. Remove the spacer and seals from the EUI body. The seals may be discarded but the spacer *must* be reused. The spacer is matched to the EUI armature and cannot be interchanged.

4. Install a new seal in the spacer groove and position the spacer on the body with the seal facing downward. Lightly coating the seal with grease will help retain it during assembly.
5. Install a new seal in the solenoid groove and assemble the solenoid on the spacer.
6. Install the new solenoid capscrews through the load plate, follower retainer, solenoid, and spacer, threading the screws into the EUI body.

WARNING: Do not reuse the used load plate, follower retainer, capscrews, and washers or damage to the solenoids or screws may result.

---

7. Torque the capscrews in the correct sequence to the DDC specification.
8. Using an electric etching pencil, write the final four digits of the injector part number on the load plate.
9. Pressurize the fuel system and visually check for leaks.

DDEC IV FEATURES

Because vehicle management electronics are upgraded on a continual basis, it is difficult to identify the significant changes that mark a numeric graduation from one generation of electronics to the next. This section outlines some of the features introduced with DDEC IV.

FUEL ECONOMY DRIVER INCENTIVE

The vehicle owner can set a fuel economy goal and program a maximum vehicle speed. If the driver fails to meet the goal, nothing changes. However, for each 0.10 mpg increase over the fleet economy target, the maximum vehicle speed is increased by a speed value programmed by the owner. DDC claims that early testing of this feature can produce some dramatic results.

LOW GEAR TORQUE LIMITING

This permits the use of smaller transmissions by limiting torque in the lower ratios and allowing higher engine torque output in the higher gears.

JACOBS COMPRESSION BRAKE

DDEC IV permits three levels of braking, and DDC claims 29% more braking power with the DDEC IV Jake brake.

FUEL PRO FILTER

This has been available as an option but is now standard with DDEC IV versions and up. Fuel Pro eliminates the need for separate primary and secondary filters. It greatly extends filter service intervals and acts as a water separator. A Fuel Pro option is a thermostatically controlled fuel heater that prevents gelling during low temperature operation.

OPTIMIZED IDLE SYSTEM

DDEC IV can be programmed to an Optimized Idle mode that operates like a home heating thermostat when the vehicle is parked. The idea is to automatically stop and start the engine to:

- Maintain a comfortable cab/bunk temperature.
- Keep the vehicle batteries charged.
- Keep the engine warm.

ETHER START SYSTEM

This feature was available in later versions of DDEC III and all DDEC since. Ether Start provides automatic ether injection during startup. The ECM is programmed to use the Ether Start option only when the temperature conditions require its use; by eliminating driver-controlled ether injection, the chances of abusing an engine with ether are reduced. A low ether supply indicator is included with the package.

DATA ANALYSIS

As the computing power of vehicle ECMs increases, more parameters are monitored and the data recorded can be used for progressive analysis of operating conditions, as for instance, in the driver fuel economy incentive. The extent to which the data can be analyzed can be greatly increased by download to a PC-based system programmed with software designed to study every performance detail of the vehicle. Data analysis is useful to the service technician. First generation electronic systems identified what was malfunctioning in a system but offered few clues as to why the problem was occurring. Current troubleshooting packages can produce precise performance profiles that tell the diagnostic technician exactly how the vehicle has been driven and handled, and the conditions at the exact moment a trouble code was logged. ProDriver (described earlier in this section) data can be downloaded by a Fleet Manager for analysis and using DDC software produce a profile of exactly how a driver operates a piece of equipment. ProDriver data may be accessed with ProDriver Reports software (requires MS Windows software) or the RDI system. RDI is weatherproofed so it can be mounted in a convenient location such as a fuel island for hard wire downloads. ProDriver can now also be programmed for wireless extraction of data from the vehicle electronic system.

DIAGNOSTIC LINK SOFTWARE

Diagnostic Link Software is a MS Windows software troubleshooting package that includes a soft (electronic) service manual and can guide troubleshooting and view or modify vehicle speed settings, engine protection strategy and protection outcomes, cruise options, progressive shifting, fault codes, and just about any engine or trip data. Diagnostic Link software is designed with user-friendly architecture and is the preferred EST in all current DDEC systems.

- EUI calibration codes must be programmed to the DDEC ECM when EUIs are replaced to enable the EUI to balance engine fueling.
- DDEC electronic governing offers both LS and VS governing.
- DDEC systems may be read at different levels using flash codes, the DDR, and PCs.
- DDEC may be programmed to protect the engine when threshold values are exceeded at three different levels: driver alert, ramp down to default, and shutdown.
- DDEC supports ProDriver, a driver data display, and ProManager, a program that allows vehicle data to be downloaded from the ECM for analysis.

SUMMARY

- DDC was the first OEM to offer an electronically managed, truck diesel engine to the North American market.
- All versions of DDEC are full authority electronic diesel engine management systems that have most of their primary components in common: as the number value of the system increases, so does the processing speed, software detail, programming scope, and self-monitoring capability.
- Current DDEC versions used with EUIs are found on the following highway medium/heavy-duty trucks and buses: Series 92, Series 50, and Series 60.
- The fuel subsystem used on DDEC engines has many similarities to that used with DDC MUI engines: the transfer pump is of the gear type and the EUIs are supplied with charging pressure at around 50 psi.
- The DDEC ECM houses a microprocessor and the EDU is usually engine mounted. As each generation of DDEC is introduced, the ECM housings are reduced in physical size, but greatly increased in computing speed and power.
- The EDU is the DDEC injector driver unit responsible for switching the EUIs.
- DDEC ECMs have an electrical operating range of 11 V to 32 V and will sustain a transient cranking low of 9 V.
- DDEC may be programmed with customer and DDC data electronically to EEPROM. The removable PROM chip socketed to the ECM motherboard found in early DDEC versions has been replaced by the current system's EEPROM capability.
- DDEC electronics support pilot injection, which can eliminate cold-start diesel knock and significantly reduce warmup emissions.

REVIEW QUESTIONS

1. What force creates injection pressure values in DDC DDEC EUI?
   a. Electrical
   b. Hydraulic
   c. Spring
   d. Mechanical
2. What defines the hard limit window available to the DDEC ECM when it computes EUI effective stroke?
   a. Engine rpm
   b. Accelerator position
   c. Road speed
   d. Cam geometry
3. *Injection rate* or fuel injected per crank angle degree in the DDEC EUI is determined by:
   a. Cam profile
   b. Engine rpm
   c. Fuel demand
   d. Strength of EUI switch signal
4. What type of transfer pump is used to move fuel through the fuel subsystem to charge the DDEC EUIs?
   a. Diaphragm
   b. Plunger
   c. Swash plate
   d. Gear
5. DDEC EUI system charging pressure values typically approximate which of the following values?
   a. 15 psi (1 atm)
   b. 50 psi (3.3 atms)
   c. 735 psi (50 atms)
   d. 5,880 psi (400 atms)
6. In which version of DDEC was the EDU housed separately from the ECM?
   a. DDEC I
   b. DDEC II

c. DDEC III
d. DDEC IV

7. What is the primary function of the SRS?
   a. Signal engine position data to the ECM
   b. Signal engine rpm data to the ECM
   c. Signal vehicle road speed data to the ECM
   d. Signal ambient temperature data to the ECM

8. How many teeth does the TRS pulse wheel have on a current DDEC system?
   a. 6
   b. 12
   c. 24
   d. 36

9. What is the V-Ref value in DDEC highway engine management systems?
   a. 5 V
   b. 12 V
   c. 24 V
   d. 36 V

10. The DDEC firetruck pressure sensor is intended to master engine speed to:
    a. Govern isochronously
    b. Produce a constant pump speed
    c. Maintain a consistent pump pressure value

11. How is DDEC EUI effective stroke controlled?
    a. By varying plunger stroke
    b. By switching the EUI solenoid
    c. By altering the charging rail values

12. The hydraulic injector nozzles in DDEC EUIs usually have NOPs around:
    a. 2,200 psi.
    b. 3,300 psi.
    c. 5,000 psi.
    d. 25,000 psi.

13. Peak injection pressures in DDEC EUIs may exceed:

    a. 3,300 psi
    b. 5,000 psi
    c. 25,000 psi
    d. 40,000 psi

14. Which dash diagnostic light is used to flash active DDEC fault codes?
    a. Oil pressure warning
    b. SEL
    c. Electronic malfunction
    d. CEL

15. What is the primary objective when the progressive shifting option is selected?
    a. Extend the service life of the transmission.
    b. Improve fuel economy.
    c. Frustrate the driver.
    d. Ensure that engine oil temperature remains stable.

16. ESTs should be connected to the DDEC electronic circuit by which of the following means?
    a. Serial link
    b. ATA connector
    c. Interface module
    d. RS 232 serial port

17. When a DDEC engine programmed for *ramp down* as a failure strategy detects a low oil pressure condition, which dash light or lights will illuminate?
    a. CEL
    b. SEL
    c. CEL and SEL
    d. Ignition

18. The reason DDEC ECMs must be programmed with the EUI calibration code is to:
    a. Enable the ECM to balance cylinder fueling
    b. Maintain charging pressure values
    c. Synchronize EUI solenoid valve closure
    d. Maintain the same IRT for each injector solenoid

Chapter 39

Caterpillar ADEM and Volvo VECTRO EUI Systems

PREREQUISITES

Chapters 29, 30, 32, and 33

OBJECTIVES

After studying this chapter, you should be able to:

- *Describe the system layout and the primary components in a Caterpillar or Volvo, full authority, EUI, electronic management system.*
- *Define the acronyms ADEM, ACERT and VECTRO.*
- *Identify the Caterpillar and Volvo truck engines that use Delphi EUI fueling.*
- *Outline the role the four primary subsystems play in managing an EUI-fueled engine.*
- *Describe the operating principles of a Delphi EUI.*
- *Define the role of input circuit components.*
- *Describe how the ECM manages EUI duty cycle to control engine fueling.*
- *Outline some of the factors that govern the ECM fueling and engine management algorithm.*
- *Perform some basic troubleshooting on ADEM and VECTO EUI-fueled engines.*
- *Describe how <SMART> programming has replaced hard limits with soft parameters to optimize system operation.*
- *Identify the ESTs required to read ADEM and VECTRO electronic systems.*
- *Describe flash memory and programming.*
- *Access the data recording features used by ADEM and VECTRO.*

KEY TERMS

Advanced Combustion Emissions Reduction Technology (ACERT)

advanced diesel engine management system (ADEM)

algorithm

Fleet Information Software (FIS)

Caterpillar information display (Cat ID)

Cat Messenger

Electronic Technician (ET)

E-Trim

electronic unit injector (EUI)

Highway Master

injector drivers

personality module

Pocket Technician

SoftCruise

Volvo electronics (VECTRO)

INTRODUCTION

Caterpillar launched its first electronically controlled, mechanically actuated, unit-injected engine with the introduction of the 3176 in 1989. The term full authority is generally applied to all engines fueled by **electronic unit injectors (EUIs)**, although the fueling window is confined to the limitations of the cam geometry used to actuate the EUI. Both Caterpillar and Volvo use Delphi EUIs (Delphi-Lucas/CAV: referred to in this text as Delphi) to fuel their engines. The injectors used by each company are almost identical. Caterpillar also manufactures EUIs in its fuel systems manufacturing facility in Pontiac, Illinois, and the injectors used on the C-15 and C-16, although of similar appearance to the Delphi units are Cat manufactured. The EUIs discussed in this chapter appear on the following engines:

Caterpillar 3176	1989	10.3 liter displacement
Caterpillar 3406E	1994	14.6 liter displacement
Volvo VE D12	1995	12 liter displacement
Caterpillar C-10	1996	10 liter displacement
Caterpillar C-12	1996	12 liter displacement
Caterpillar C-15	1998	14.6 liter displacement
Caterpillar C-16	1998	15.8 liter displacement (Cat EUI)
Volvo V-Pulse D-12	2002	12 liter displacement (EGR)
Caterpillar C-11	2004	11 liter displacement ACERT engine
Caterpillar C-13	2004	13 liter displacement ACERT engine
Caterpillar C-15	2004	15 liter displacement ACERT engine

Both Caterpillar and Volvo build engines that are entirely metric engineered and all the engines in the foregoing list are six-cylinder engines designed for highway truck applications. The 3176 was engineered for EUI fueling and was Caterpillar's first EUI managed engine entry into the highway truck arena. The Caterpillar 3406E was a reengineered adaptation of the 3406 PEEC and hydromechanical 3406 engines using PLN fueling: this engine platform underwent some minor reengineering in the late 1990s and is currently marketed as a C-15. The Caterpillar C-10 and C-12 are essentially evolutions of the 3176 and 3196 engine. The Caterpillar C-16 engine was introduced in 1998 with power ratings up to 600 BHP to address the high horsepower niche of the truck engine market. This family of Caterpillar engines uses articulating pistons with a

FIGURE 39–1   *Caterpillar C-12 engine. (Courtesy of Caterpillar)*

cast steel crown and forged aluminum skirt. Three rings are used: the top compression ring being of a barrel faced, keystone design; the second, taper faced; and the third, the oil control ring is double railed, profile ground with a coil spring expander. All three rings are cladded, the top with a plasma face coating, the other two are faced with chrome. Figure 39–1 shows a C-12 engine that has become a popular fleet powerplant. The Caterpillar engine series use **ADEM (advanced diesel engine management system)** electronics: the ECM is currently a 32-bit processor that runs at 24 mHz.

Volvo entered the North American truck market with its purchase of White Trucks and the GM heavy trucks division. The Volvo VE D12 engine evolved from a similar 12-liter platform that used hydromechanical PLN fueling. The VE D12 underwent few changes until the October 2002-introduced, V-pulse version. This version was minimally reengineered to meet EPA 2004 emissions requirements and uses a constant geometry turbocharger and EGR. Current versions use two-piece articulating pistons in place of trunk pistons and an external exhaust compression brake is standard. The top power-rated version of this engine puts out 465 BHP and 1,650 lb.-ft. of torque. In the year 2000, Mack Trucks and Volvo merged their North American truck operations. No doubt this will have some future impact on the engine technology used by both companies. Because both manufacture a 12-liter engine with different fueling management, as the corporations consolidate, it is probable that only one of the engines will survive. The electronics used to manage the EUI fueling are known as **VECTRO**, a loose acronym for **Volvo electronic** control. Figure 39–2 shows an exploded view of a Volvo VE D12 engine.

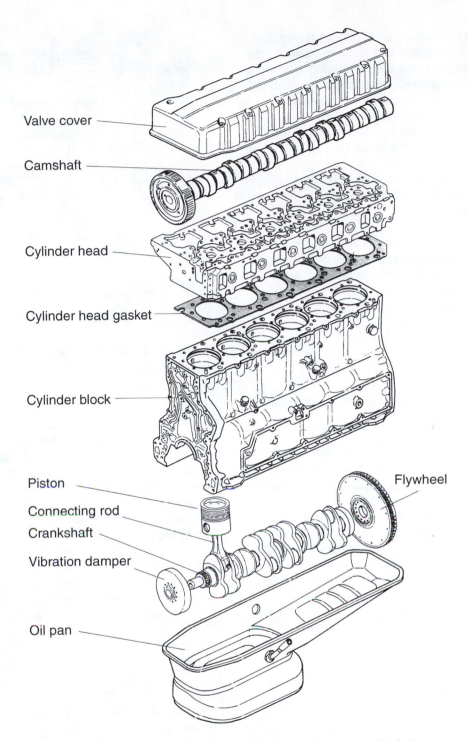

Valve cover

Camshaft

Cylinder head

Cylinder head gasket

Cylinder block

Piston

Connecting rod

Crankshaft

Vibration damper

Oil pan

Flywheel

FIGURE 39–2 *Exploded view of VE-D12 engine. (Courtesy of Volvo)*

ACERT

Caterpillar **Advanced Combustion Emissions Reduction Technology** or **ACERT** engines were introduced for the engine model year 2004. Three of these engines use the EUI fuel systems covered in this chapter. Unlike its competitors who have opted for C-EGR, Caterpillar has met the challenge of reducing NO_x emissions by improving in-cylinder combustion management. ACERT engines using EUIs include the C-11, C-13 and C-15 models. There are four primary features of ACERT in Caterpillar EUI fueled engines:

1. Series turbocharging. Twin, in-series, ECM-managed wastegated turbochargers provide ECM management of manifold boost over the widest possible engine load and rpm range.

2. Electronically-controlled variable valve timing. This provides ECM managed soft limits to valve timing with the result that cylinder breathing becomes engine-computer-managed rather being confined to the hard limits of camshaft management.
3. Pilot injection. The mechanically-actuated, electronically-controlled, unit injectors (MEUIs) are capable of electronically-managed pilot or split-shot injection.
4. Exhaust gas after-treatment. ACERT technology uses a single-stage oxidation catalyst to manage HC emissions.

Caterpillar indicates that the core of ACERT emissions management is the ability to manage in-cylinder combustion more effectively than its competitors, enabling it to meet tough emissions standards without compromising engine power, fuel economy, and engine longevity.

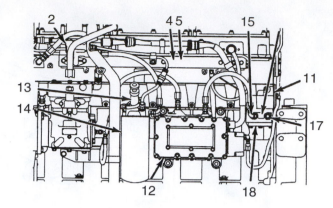

2. Adapter (siphon break)
4. Fuel return manifold
5. Fuel supply manifold
11. Fuel transfer pump
12. Electronic control module (ECM)
13. Fuel priming pump
14. Fuel filter (remote mounted fuel filter as an attachment)
15. fuel outlet (to ECM)
16. Spacer block
17. Fuel inlet (from tank)
18. Cover assembly

FIGURE 39–3   *C-10 fuel system components. (Courtesy of Caterpillar)*

SYSTEM OVERVIEW

Consistent with other full authority electronic engines, the mechanically actuated EUIs have an effective pumping stroke managed and switched by the ECM (electronic control module). However, because the actual EUI plunger stroke is cam actuated, the ECM is confined to the hard limit window represented by cam profile. This chapter addresses the features of Delphi EUI-fueled engines that make them distinct from other EUI, full authority diesel engines and describes the electronics used to manage them.

Both Caterpillar- and Volvo-built engines with full authority management systems may be divided into the following subsystems for purposes of studying fueling control:

• Fuel subsystem
• Electronic input circuit
• ECM
• Output circuit

FUEL SUBSYSTEM

The fuel subsystem incorporates those components that enable the transfer of fuel from the fuel tank to the EUIs. A fixed clearance gear pump is responsible for

fuel movement through the supply circuit to the ECM. This gear pump is flange mounted and is driven by the camshaft through a pair of helical gears. The circuit incorporates a check valve and pressure relief valve. Both Caterpillar and Volvo use a hand-actuated, plunger-type, priming pump located downstream from the gear pump. Fuel from the fuel subsystem is delivered to the fuel supply and return manifolds. An adaptor or siphon-break prevents fuel drain-back from the manifold when the engine is not running. While the engine is running, fuel is continually circulated through the system. In some Caterpillar applications, charge fuel is routed through a heat exchanger to which the ECM is mounted: this fuel acts as coolant. System charging pressure is maintained at values typically 60 psi or higher. Figure 39–3 shows the location of the fuel subsystem components on a Caterpillar C-10 engine while Figure 39–4 is a schematic showing the relationship of all the fuel system components on the same engine.

MANAGEMENT ELECTRONICS

Caterpillar uses ADEM to manage its EUI fuel system. This is one of the most comprehensive engine management packages found on highway truck engines, both from the programmability and diagnostics per-

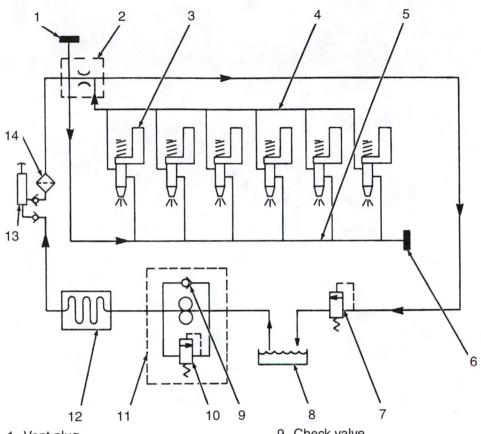

1. Vent plug
2. Adapter (siphon break)
3. Electronically controlled unit injectors
4. Fuel manifold (return path)
5. Fuel manifold (supply path)
6. Drain plug
7. Pressure regulating relief valve
8. Fuel tank
9. Check valve
10. Pressure relief valve
11. Fuel transfer pump
12. Electronic control module (ECM)
13. Fuel priming pump
14. Fuel filter (remote mounted fuel filter as an attachment)

FIGURE 39–4 *C-10 fuel system schematic. (Courtesy of Caterpillar)*

spective. It has rapidly evolved from its first introduction in the middle 1990s. Volvo's management electronics are known as VECTRO. Both ADEM and VECTRO connect to the J1587 and J1939 CAN buses enabling them to receive and broadcast on the data buses. EST access is by means of the ATA connector, either six- or nine-pin Deutsch socket.

CAT ADEM

The first generation of ADEM did not appear in any highway application Caterpillar engines. The second generation of ADEM was the electronics package used to manage 3176 and 3406E engines beginning in 1993, and the third generation of ADEM electronics is that used on its current engine lineup. In common with most truck engine OEM software, changes are continual and

do not necessarily result in a generation name change. The current ADEM IV ECMs use 32-bit processors running at 24 mHz and dual 70-pin connectors that enable up to 140 inputs and outputs. ADEM is a single module system. The processing hardware and all of the output switching apparatus are contained within the ECM housing.

VOLVO VECTRO

Volvo also uses a single module system, incorporating the processing hardware and injector drivers within a single housing. Like the Caterpillar ECM, this is mounted on the side of the engine. Volvo prefers to use the term ECU to describe its engine controller module, but to avoid confusion, the SAE-recommended term, ECM is used in this chapter.

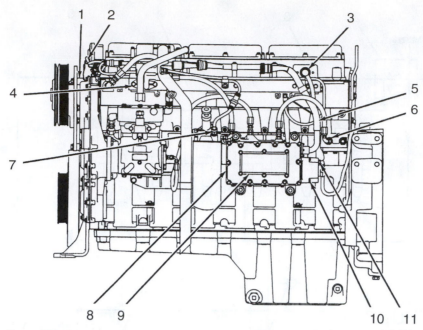

1. Speed/timing sensor
2. Coolant temperature sensor
3. Injector enable circuit
4. Fuel manifold
5. Engine wiring harness
6. Fuel transfer pump
7. Fuel pressure sensor
8. Electronic control module (ECM)
9. Personality module
10. Transducer module
11. Boost sensor

FIGURE 39–5 *C-12 electronic control system components. (Courtesy of Caterpillar)*

INPUT CIRCUIT

Command and monitoring sensors and switches in both the Caterpillar and Volvo applications are consistent with other comparable full authority management systems with the exception of the Caterpillar throttle position sensor (TPS). This subject is dealt with later in this section. The Volvo VECTRO-managed engine uses a standard potentiometer-type throttle position that is supplied with V-Ref and returns an analog signal representing a portion of the V-Ref that can be correlated with accelerator pedal travel. With this exception, the remaining input circuit components used by both Cat and Volvo are similar, and mostly generic. The operating principles of input circuit components are described in detail in Chapter 32. Figure 39–5 shows the location of some of the input circuit components on a C-12 engine and Figure 39–6 shows the sensor and connector locations on a C-15 engine.

CATERPILLAR TPS

Caterpillar ADEM ECMs require a pulse width modulated (PWM) signal to be delivered by the TPS used on

its engine. This PWM signal is generated in two ways depending on when the engine was manufactured.

Cat Can-Style TPS

The first type of TPS used by Caterpillar was a floor-mounted, variable capacitance device designed to output a constant frequency PWM signal. Caterpillar states that this digital PWM signal overcomes errors that can be produced by analog signals due to pin-to-pin leakage or contamination in either the connectors or harnesses, an important factor in any drive-by wire system. If the PWM signal is invalidated for whatever reason, the engine will default to run at the idle speed. Mechanically, the sensor is designed to have 30 degrees of active travel with an additional 5 degrees of undertravel and 10 degrees overtravel for linkage tolerance. The can-type TPS operates on a variable capacitance principle. The TPS is powered up by 8 volts (not the Caterpillar V-Ref, which is always 5 volts) and digital circuitry converts the capacitive signal to a PWM value, returned to the ECM. Figure 39–7 shows a Caterpillar floor-mounted TPS.

Cat Pedal-Mounted TPS

Caterpillar replaced the can-style TPS midway through the 1990s with a pedal-mounted unit. This unit functions

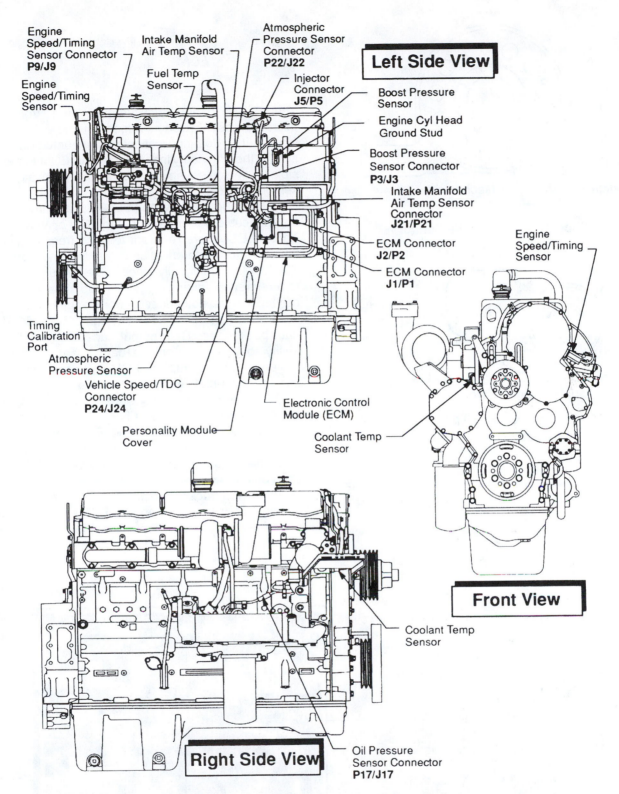

FIGURE 39–6   *C-15 sensor and connector locations. (Courtesy of Caterpillar)*

on a potentiometer electrical principle but that is all it has in common with other OEM TPS. The Cat pedal-mounted TPS must be powered up by 8 volts similarly to the can type. This TPS has built-in digital circuitry that converts the analog signal produced by the potentiometer to a PWM signal. The can-type TPS and the pedal-

type TPS cannot be regarded as interchangeable. Later version TPS units can automatically recalibrate themselves. They are a Caterpillar supplied component and must not be replaced with a generic TPS. The lower half of Figure 39–7 shows a pedal-mounted Caterpillar TPS.

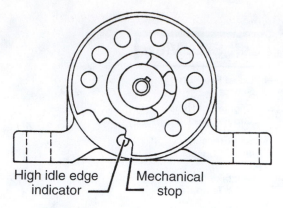

High idle edge   Mechanical
indicator   stop

Floor-Mounted Throttle Sensor

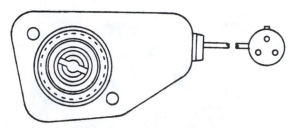

Pedal-Mounted Throttle Sensor

FIGURE 39–7   The floor- and pedal-mount throttle
position sensors. (Courtesy of Caterpillar)

ECM

The ECMs used by Caterpillar and Volvo are microprocessor-driven management and switching devices that, in both applications, are mounted on the left side of the engine cylinder block. The ECM is responsible for engine governing, fuel management algorithm, self-diagnostics, system monitoring, and creating data audit trails. Both ADEM and VECTRO are programmable with customer and proprietary (Cat or Volvo) data. Today's ECMs are resistant to radio frequency, electromagnetic interference, and other low level radiation. The ECM is additionally responsible for supplying system reference voltages, powering up sensors, and receiving inputs both directly from engine monitoring sensors and those broadcast on the J1587 and J1939 data backbones. Their output capability includes the ability to broadcast on the data buses and control the engine actuators. The ECMs also incorporate the injector driver units within the housing.

Despite the computing power and memory capability of ECMs, their physical size has generally decreased. Figure 39–8 shows a sectioned view of the VECTRO ECM used by Volvo VE D12 and its location on the engine while Figure 39–9 shows a Caterpillar ADEM ECM used on a C-15 engine.

Although a truck technician generally does not have to know how to repair engine ECMs, it can help if you understand a little about how they operate when

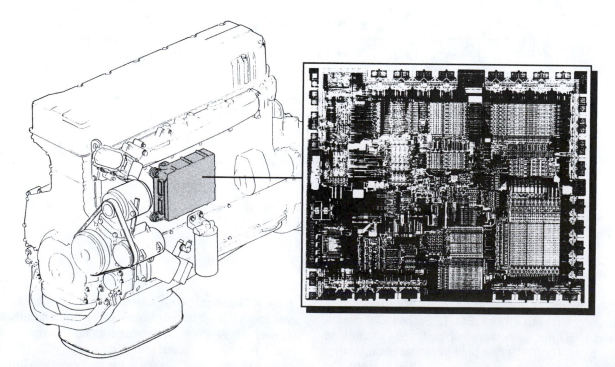

FIGURE 39–8   *VECTRO ECM. (Courtesy of Volvo)*

FIGURE 39–9 *ADEM ECM. (Courtesy of Caterpillar)*

it comes to troubleshooting malfunctions. The term **algorithm** means simply a set of rules and values that are designed to produce an outcome or result. If we say that 1 + 2 + 3 + 4 = 10, we can say that we have used the rules of addition, applied the values, and come up with an outcome, ten, the sum of the values. If any one of the values we used in that example changes, the outcome will change. So if the 1 is replaced by the figure 6, the outcome or sum of the values has to change. The control algorithm used by an engine management system uses the same means to produce outcomes, except that there are hundreds of rules and thousands of values. The values are received into the processing cycle by the input circuit, from memory, and off the chassis data bus. The rules are logged into the ECM memory components we know as ROM, PROM, and EEPROM. The computer manipulates this data in main memory or RAM and computes an outcome. Figure 39–10 is a schematic representation of a Caterpillar processing cycle.

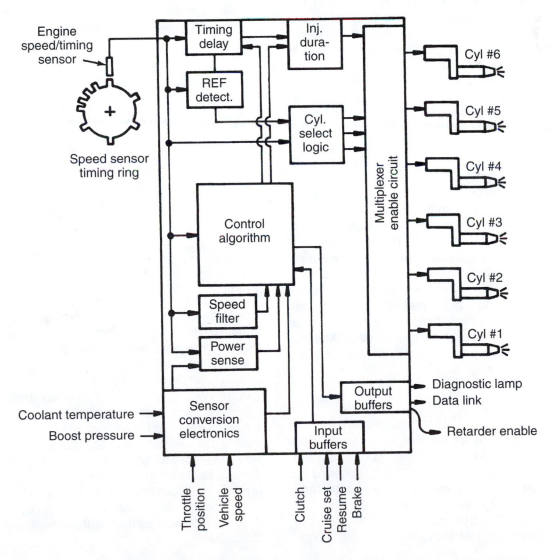

FIGURE 39–10 *Control algorithm used in a Caterpillar ECM processing cycle. (Courtesy of Caterpillar)*

The control algorithm contains a comprehensive fuel map. The ECM analyzes the data input from the command and monitoring sensors, and referencing the instructions programmed to its data retention media, plots a fueling profile. Startup fueling strategy, failure modes, and all the system default values are programmed into the ECM software. Some Caterpillar engines have software that enables the cycling of cylinders during certain conditions such as warm idling with the objective of saving fuel and minimizing engine wear. The ECM is also programmed to run diagnostic tests on input and output circuits and identify specific component and circuit faults using blink codes, reader programmers, or PCs. Most troubleshooting can be performed using the appropriate EST and a DMM.

CATERPILLAR PERSONALITY MODULE

The Caterpillar **personality module** combines the memory components we have called PROM and EEPROM in other systems. A full description of the PROM and EEPROM functions within an ECM is provided in Chapter 32 but in essence, this type of memory contains engine-specific data and the reprogram and write-to-self capability of the system. The personality module was bolted to older version ECMs so it could be removed and replaced if necessary. Also attached to the older version ECMs was a transducer module responsible for transducing all system pressure values to electrical signals to signal to the ECM.

Current version ECMs have eliminated the separate/attached personality and transducer modules and incorporated their functions in a single more compact unit with greatly improved data processing power and speed. When these are reprogrammed with data, the term used is flash programming. The flash memory chip in the personality module contains engine series specific control software. Caterpillar ECMs have a non-volatile RAM (NV-RAM) component for storing certain customer data and this is backed up by an internal ECM battery. If the vehicle batteries were disconnected or failed, it would not result in the loss of NV-RAM data. Figure 39–11 shows the complete ADEM electronic circuit used on a C-15 engine.

CAT ECM INJECTOR DRIVERS

Caterpillar **injector drivers** are integral with the ECM. They produce a spiked control signal to energize the EUIs to a voltage value of around 100 V, obtained by using induction coils. The voltage spike is initial and the circuit holds at 12 volts: the inductive kick produced by the coil when the EUI control solenoid circuit is opened is suppressed to prevent back-feed spikes.

The software control algorithm determines the desired duty cycle in any given moment of operation, which is converted to effective EUI pumping stroke by the duration the solenoid cartridge in the EUI is energized. This duration is measured in milliseconds.

VECTRO ECMs

Volvo VECTRO ECMs operate in similar fashion to the Caterpillar ADEM but there are some small differences. Most of these concern how the EUIs are managed so we look at the VECTRO injector driver unit.

VECTRO Injector Drivers

Volvo states that the trigger pulses delivered to the EUI control cartridges are driven at 90 V. The 90 V is induced by coils and the EUIs draw about 10 amps current during the initiation spike. Once the control solenoid has closed to put the EUI into effective delivery stroke, current draw by the cartridge drops to around 4 amps. The inductive kick produced on collapse is suppressed by a capacitor.

OUTPUT CIRCUIT

The Caterpillar and Volvo output circuit devices effect the results of the computer processing cycle into action. In an EUI fuel system, the primary outputs are the electronic unit injectors. You can also categorize as outputs the EC-conditioned V-Ref, dash display data, and any information broadcast to the data backbone.

EUIs

The Delphi EUIs used by Caterpillar and Volvo have changed little since their introduction in the late 1980s. EUIs are mechanically actuated by an injector train actuated in the same way cylinder valves are actuated. The EUI actuation train consists of a rocker, pushrod, roller tappet, and cam profile when the camshaft is cylinder block located: Figure 39–12 shows the injector actuation train used in a Caterpillar C-12 engine, which uses a cylinder block-located camshaft. In an engine with an overhead camshaft, a rocker is used to actuate the EUI as shown in the Volvo VE D12 example in Figure 39–13. Effective stroke of the EUI is electronically controlled by the ECM, specifically by the injector drivers discussed earlier in this chapter under ECM components.

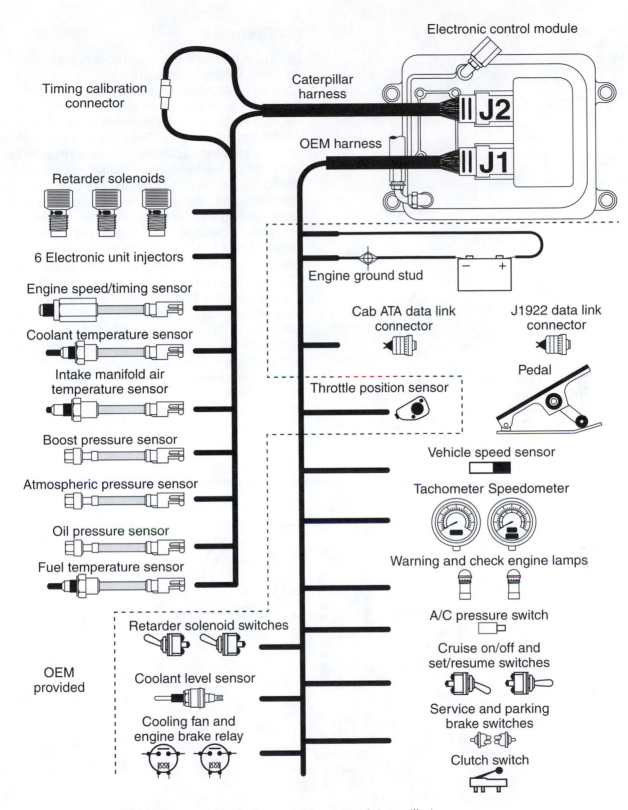

FIGURE 39–11 *C-15 electronic circuit block diagram. (Courtesy of Caterpillar)*

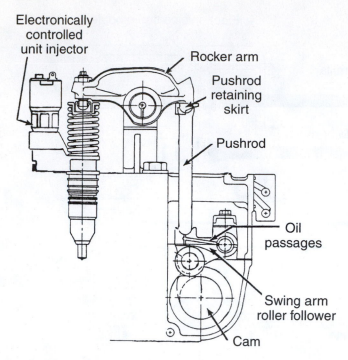

Electronically controlled unit injector

Rocker arm

Pushrod retaining skirt

Pushrod

Oil passages

Swing arm roller follower

Cam

FIGURE 39–12 *Cutaway showing C-12 EUI actuation by a cylinder block-located camshaft and injector train. (Courtesy of Caterpillar)*

EUI Operation

The EUI is fitted to a cylindrical bore in the engine cylinder head. Charging pressure fuel from the fuel supply manifold (drilled passages in the cylinder head) enters the EUI through the fill port and circulates through the EUI: this circuit feeds the pumping chamber in the barrel of the pump element. The EUI pumping chamber is formed by the cam-actuated plunger and the stationary barrel. When the injector train actuating cam is ramped off IBC toward the cam nose, the EUI plunger is driven downward and acts on whatever amount of fuel has been metered into the pump chamber. As the plunger descends, its leading edge first closes off the fill passage and next displaces fuel in the EUI pump chamber through the internal passages leading to the spill port. Effective stroke can only occur when the EUI solenoid is energized. This action moves the EUI control valve to close off the passage leading to the spill port, trapping fuel in the EUI pumping chamber and internal ducting. Located below the EUI pump chamber and connected by a passage is a hydraulic, multi-orifii injector nozzle. Pressure rise caused by trapping fuel in the EUI pump chamber acts on the nozzle valve pressure chamber and when the required NOP is attained, it unseats, be-

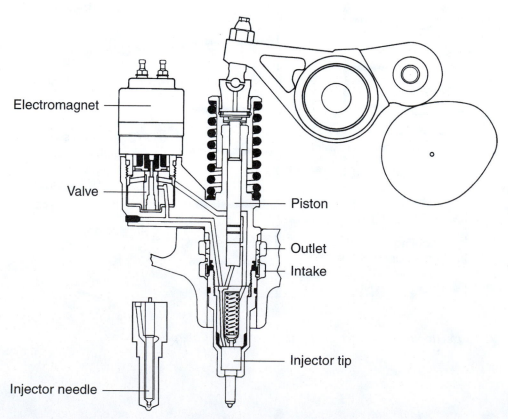

Electromagnet

Valve

Piston

Outlet

Intake

Injector tip

Injector needle

FIGURE 39–13 *Volvo VE-D12 EUI actuated by overhead cam. (Courtesy of Volvo)*

ginning the injection pulse. Pressure continues to rise after NOP, and the peak injection pressure achieved depends on the length of the duty cycle. Effective stroke ends when the ECM switches the EUI solenoid open, allowing fuel to spill from the EUI circuit and enabling pressure collapse.

EUI Subcomponents

Figure 39–14 shows a sectional view of the Delphi EUI with the subcomponents labeled. Reference the figure number codes in the description that follows.

1. Terminals. Connect to the injector drivers in the ECM.
2. Control cartridge. Solenoid consisting of a coil and armature with an integral poppet control valve. A spring loads the armature open. Energizing the solenoid closes the armature/ poppet control valve.
3. EUI tappet spring. Loads the EUI tappet upward. This enables the tappet/ plunger actuation train to ride the cam profile and retract the tappet after a mechanical stroke.
4. Poppet control valve. A valve integral with the solenoid armature. In Figure 39–14, the control valve is shown closed (solenoid energized): this prevents fuel from exiting through the spill port, trapping it in the EUI circuitry, enabling effective stroke.
5. Plunger. The reciprocating member of the pump element, the plunger, is lugged to the tappet, so it reciprocates with it. In Figure 39–14, the plunger is shown in its upward position.
6. Barrel. The stationary member of the EUI pumping element containing the fill port, pump chamber, and duct connecting the pump chamber with the injector nozzle.
7. Upper O-ring. Seals the upper fuel charging gallery.
8. Lower O-ring. Seals the lower fuel charging gallery.
9. Nozzle spring. Defines the NOP value typically, around 5,000 psi. NOP is initial and rebuild set by shims acting on the nozzle spring.
10. Spacer or shims. Define nozzle spring tension and therefore the specific NOP value.
11. Upper nozzle assembly body. Machined with the ducting that feeds fuel to the pressure chamber of the nozzle valve.
12. Nozzle valve. The nozzle assembly is a hydraulically actuated, multi-orifii, VCO nozzle. A full description of the operating principles of this type of nozzle is provided in Chapter 20.

EUI NOP values are typically around 5,500 psi and by factoring nozzle valve pressure differential ratio, closing pressure values are typically 3,700 psi. When the injector actuating cam has attained peak lift, the injector train is unloaded, and the tappet ramps down the

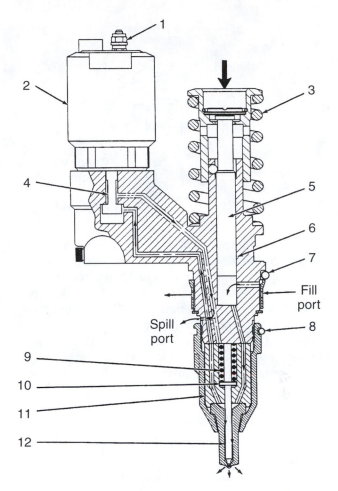

1. Solenoid connection
   (to the multiplex enable circuit)
2. Solenoid valve assembly
3. Spring
4. Valve (shown in the closed position)
5. Plunger
6. Barrel
7. Seal
8. Seal
9. Spring
10. Spacer
11. Body
12. Check

FIGURE 39–14 *Sectional view of a Delphi EUI and internal components. (Courtesy of Caterpillar)*

cam flank toward cam base circle. Simultaneously, the EUI spring lifts the plunger within the barrel, returning it to its retracted position and thereby exposing the fill port. Charging pressure fuel is then permitted to circulate throughout the EUI passages for purposes of cooling the assembly, exiting the spill port until the effective cycle is repeated. Figure 39–15 shows the four operating stages of an EUI.

Caterpillar states that peak injection pressures in C-10 and C-12 engines reach values of 25,500 psi. Peak injection pressures can achieve values exceeding 30,000 psi in C-15 and C-16 engines. Volvo states that peak injection pressures in the VECTRO VE D12 exceed 22,000 psi.

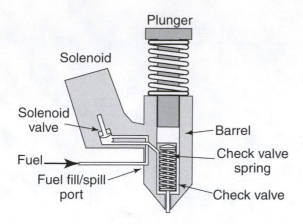

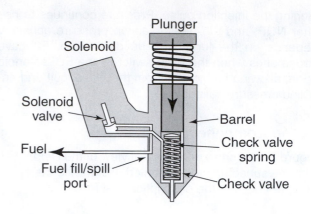

Without pressure applied to the plunger from the cam, a spring keeps the plunger retracted. Propelled by the new low-pressure fuel transfer pump, fuel flows into the injector through the fill/spill port. From there it flows past the solenoid valve, down through the internal injector passages to the spring-loaded check valve at the injector's tip and back up into the barrel. The pressure from the transfer pump is too low to unseat the spring loaded check valve at the injector's tip.

As the cam rotates, it starts to drive the plunger downward. Injection of the fuel may occur at any time after the plunger starts its downward travel. Until the ECM signals the start of injection, the displaced fuel is forced back out through the solenoid valve to the fill/spill port.

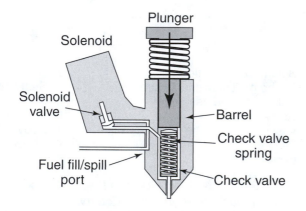

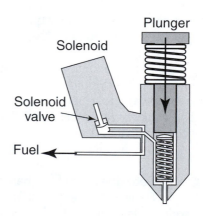

When the ECM signals the start of injection, the solenoid pulls the fuel valve closed, blocking the fuel's path to the fill/spill port. With this valve closed, pressure elevates at the injector tip to the 37 931 kPa (5,500 psi) needed to unseat the spring loaded check valve. Once this valve is overcome, fuel is injected into the cylinder.

Fuel will continue to be injected until the ECM signals the solenoid to open the valve, allowing fuel to exit through the open valve and out the fill/spill port. The pressure at the injector tip immediately drops and the check valve snaps shut ending the injection cycle. The plunger will continue on its downward path however, displacing fuel through the open valve to the fuel manifold and back to the tank. This flow of fuel helps to cool the injector.

FIGURE 39–15 *Operation of an EUI. (Courtesy of Caterpillar)*

E-TRIM

Caterpillar **E-Trim** is a four digit number printed on either the solenoid or the upper face of the EUI tappet flange. The E-Trim numeric value is established when an EUI is factory bench-tested and represents how each EUI flows fuel. Figure 39–16 is a graphic representation of the fuel flow differential that would be found in a set of injectors. E-Trim programming allows the ADEM ECM to adjust EUI duty cycle so that it can compensate for minor hydraulic differences that occur in different injectors and balance engine fueling. The E-Trim value for each injector must be programmed to the engine ECM using ET software.

TECH TIP: Whenever an injector has to be replaced, ensure that the new E-Trim value is programmed to ET. If this step is overlooked, the ADEM ECM will use the E-Trim values of the old injector, which can result in engine fueling balance problems.

VECTRO EUI

The VECTRO EUI is almost identical in operating principle to the version used on Caterpillar engines. The distinctions lie in how the unit is actuated and the spray emission characteristics that are obviously tailored to the VE D12 combustion requirements. The VECTRO EUI nozzle assembly has eight orifii measuring 0.2 mm (approx 0.008") apiece.

CAUTION: Always observe OEM instructions for draining the charging rail when removing EUIs from an engine cylinder head. When an EUI is removed, the contents of the fuel charging rail end up in the engine cylinder if the cylinder head fuel gallery is not first drained.

SYSTEM DIAGNOSTICS AND COMMUNICATIONS

Caterpillar electronic system self-diagnostics are consistent with its competitor OEM systems. The self-diagnostic capabilities of the system may be accessed by a range of means, beginning with simple flash codes and extending through a number of PC-based software, data management, and analyses programs. In the event of electronic system or component failure, the operator is alerted by the dash-mounted CEL (check engine light). The code may be read on the CEL using the cruise control switches on vehicles equipped with cruise control or with one of the ESTs listed previously.

Current Caterpillar electronic management systems have a built-in diagnostic recorder. This diagnostic recorder is a sort of constant snapshot mode that is triggered whenever a fault code is logged or by the driver toggling the cruise switches. The data recorded can

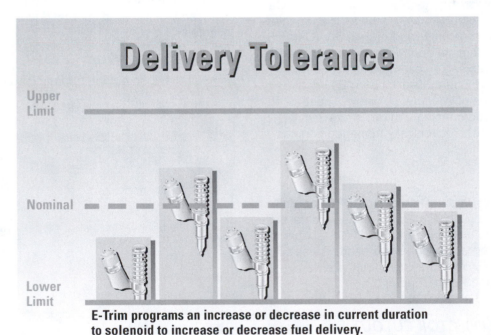

FIGURE 39–16 *E-Trim balancing of EUI fuel flow rates. (Courtesy of Caterpillar)*

be accessed using an EST and used to diagnose a set of conditions that may have contributed to the failure. A snapshot test is commonly used to diagnose intermittent problems in electronic management systems. The test is triggered by an event such as a trouble code and records data frames at the instant of, before, and after the event. These data frames display all the critical system parameters for analysis: the data immediately prior to the event can be critical in determining the cause of the code.

While the technician troubleshooting an electronically managed engine must not overlook the fact that an engine does have systems that are not electronic, in most of the current generation of electronically managed engines, the correct troubleshooting strategy is usually driven electronically. For this reason, whenever diagnosing any problem on an engine, the manufacturer's sequential troubleshooting sequence must be observed.

ESTS AND SOFTWARE

The recommended EST for accessing both Caterpillar and Volvo on-board electronics is a PC loaded with proprietary software. ProLink 9000 cartridges and MPC cards are made for both systems, but their usefulness is becoming limited as the systems become more advanced. Caterpillar had its own EST in the early days of its electronic systems known as the ECAP (Electronic Control Analyzer Programmer). For current systems, a PC with proprietary Windows-driven software called Caterpillar **ET (electronic technician)** is required. Cat ET is a user-friendly software package that enables system diagnosis, customer and proprietary (Caterpillar) data reprogramming, and fuel/mileage information accessing. You can see an example of the ET user interface (screen) in Figure 39–16. Using ET requires a minimal knowledge of Windows software. The read, diagnostic, and programming protocols are similar to those used on other electronically managed engines.

DATA LINK

All ESTs connect to the Caterpillar ADEM or Volvo VECTRO ECMs by accessing the chassis data bus using an ATA connector. Either a J1708 six-pin or J1939 nine-pin Deutsch connector is used. These connectors and the cavity pin assignments are shown in Figures 33–1 and 33–2 in Chapter 33.

CATERPILLAR INJECTOR CUTOUT

This ET driven test is described here because it differs somewhat from other OEM injector cutout tests. ET offers three cutout test options:

1. Single cylinder cutout
2. Four-cylinder cutout
3. Five-cylinder cutout

The four-cylinder cutout test is recommended by Caterpillar. The test protocol is based on running the engine at a specific rpm (idle), followed by the sequential cutting of EUIs while attempting to maintain the set rpm. For instance, if four EUIs are electronically shorted or cut, the running two are required to greatly increase their duty cycle if the engine is to maintain the set rpm. The idea is to get a pretty accurate determination of cylinder balance. Figure 39–17 shows the ET screen display following a cylinder cutout test procedure: the test indicates that performance on engine cylinder #5 is Not OK.

CAT POCKET TECHNICIAN

Cat **pocket technician** is a PalmPilot-based tool that plugs in to a dash adapter. The original version was designed for technicians to assist with diagnosis (great for road testing!) but a driver's version is also an option. The driver's version has Cat ID capability and can assist with analysis of fuel economy, engine operating conditions, and display fault codes.

CAT ID

Caterpillar information display (Cat ID) is a driver display option located on the truck dash that can either be integrated or added on to the engine electronics. Caterpillar ID accesses data from the ADEM 2000 ECM and displays it to the driver. Data displayed includes:

- Fuel usage
- Feedback on engine operating conditions

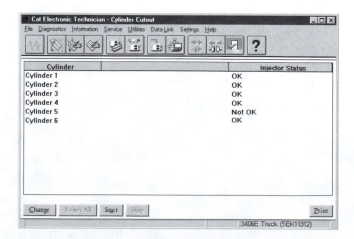

FIGURE 39–17   *ET screen display following a cylinder cutout test. (Courtesy of Caterpillar)*

- Fault codes
- First-level diagnostics

The driver is able to navigate through the displayed data using a four-button keypad. Cat ID displays critical fault codes as they occur with both the code number and an alpha description of the fault. Flash codes are provided for drivers who prefer operating the vehicle without a dash display. Cat ID has a built-in theft deterrent option: when activated, a four-digit password must be entered on engine shutdown and reentered before the engine will restart.

Cat Messenger

Cat Messenger is an enhancement of the Cat ID driver display unit. It provides real-time feedback on engine operating conditions (fuel usage, mpg, fluid temperatures, and so on), maintenance scheduling, diagnostic reporting, and theft deterrence. Drivers are required to enter a four-digit access code before the ignition circuit can be closed when the theft deterrence feature is enabled. A driver or owner can also set personal fuel economy goals and base oil change intervals on fuel consumed rather than mileage.

HIGHWAY MASTER

Remote programming technology can be used to reprogram engine parameters for a single truck or an entire fleet from an office location. This wireless technology is available through **Highway Master** working in conjunction with Qualcomm. The type of data that can be transferred to Highway Master would include fuel economy, mileage, and trip run/idle times. Highway Master can create fuel economy histograms (chart consisting of rectangular columns to show data on comparative scale), engine events, maintenance reports, and customer data programming.

CAT FIS

Caterpillar FIS (Fleet Information Software) permits data, tracked and stored in the ADEM 2000 software, to be downloaded to:

- ProLink 9000 generic reader-programmer
- Hand-held Argo mobile data terminal that allows a PC to extract data from up to 200 engine ECMs in less than 10 seconds per unit
- Caterpillar communications adapter

FIS software provides the ability to analyze driver and engine performance data by unit and can compare the data from any number of units in a fleet.

PROGRAMMING OPTIONS

Both Caterpillar and Volvo offer a wide range of programming options, some of which are listed in the following section. We have chosen to use the Caterpillar terminology here but equivalent options are mostly available in Volvo VECTRO.

<SMART> OPTIONS

Smart parameters or functions are those that will bend a hard programmed limit or value when it makes sense. Most current engine management electronics feature some <SMART> programming options. Some examples:

Vehicle Speed Limit (VSL)

The hard programmed VSL used in early versions of ADEM can be optionally reprogrammed with soft limit VSL technology. Soft limit programming permits a cushion on either side of a hard limit value. Soft VSL modulates engine fueling when the truck is running over rolling terrain permitting some latitude on either side of the programmed vehicle speed to maximize fuel economy. Vehicle maximum speed can be programmed at a value under peak cruise speed: this feature encourages the driver to use the cruise control feature and eliminate some of the irrational variables encountered when a driver's foot is managing the accelerator pedal.

VSL can be programmed as a soft parameter. This means that instead of an abrupt fuel cutoff when the vehicle achieves the programmed VSL, the engine is fueled to keep the manifold boost up and gain some speed advantage on the upcoming hill. Soft VSL improves fuel economy and driver comfort.

Programmable Extended Droop

Governor droop can be programmed to extend up to 150 rpm above the TEL (top engine limit) or rated speed value. This reduces driver shifting frequency.

ADEM MULTITORQUE

ADEM multitorque is an option that, when enabled, senses when the truck is in the top gear ratios (by analysis of engine rpm and vehicle speed ratio) and provides extra torque should running conditions require it. Extra torque would typically be provided when negotiating hills. The option results in less driver fatigue because the driver is shifting less often, and it has been proved to reduce driveline component wear, presumably as a result of less shifting.

SoftCruise

SoftCruise is a <SMART> cruise control option that essentially "learns" running and terrain conditions, then modulates engine fueling to eliminate abrupt cutoffs in fueling and engine braking actuations that are characteristic of hard cruise programming. Soft cruise can manage the engine retarder if programmed to do so: automatic brake activation reduces driver fatigue while also increasing the vehicle fuel economy.

Cooling Fan Control

The ECM constantly monitors the input from the temperature sensors monitoring the engine coolant, intake manifold, compression brake, and AC refrigerant system pressure to determine whether the cooling fan should be actuated. An ECM output switches and manages the engine compartment cooling fan, eliminating the requirement for such devices as fanstats. Additionally, the ECM may engage the fan when engine retarder high mode is selected to increase retarding effort.

PTO Ramp Rate

PTO ramp rate can reduce the PTO work cycle time by enabling programming rotational speed from 50 to 1,000 rpm/sec. Engine speed is therefore factored by PTO speed. The TPS is disabled when this option is in use to prevent PTO overspeed.

Gear Down Protection

This programmable feature prevents engine overspeed conditions by limiting engine rpm in the higher gear ratios with the objective of ensuring that cruising speed can only be attained in the top gear.

Progressive Shifting

Progressive shifting can assist less experienced drivers of trucks by providing engine speed prompts during different vehicle speeds. Progressive shifting can make a significant difference to vehicle fuel economy.

Idle Shutdown Timer

In common with most other truck electronic management systems, Caterpillar offers an idle shutdown timer programmable from 1 minute to 24 hours. The timer programming can be defeated by a driver activated override feature.

Trip Data Log

This standard feature permits monitoring and recording of the engine parameters by three methods:

1. Lifetime totals
2. Trip totals
3. Instantaneous readings

The following parameters are monitored:

Engine running hours/ECM hours

Machine miles

Idle mode miles

PTO mode hours

Total fuel consumed

Idle fuel consumed

PTO fuel consumed

Average engine load factor

DELPHI E3 EUIS

Delphi E3 injectors use the same general operating principles as the first generation of EUIs used in Caterpillar and Volvo applications but integrate the control solenoid and poppet control valve into the cylindrical injector body. This has reduced weight by about half and greatly reduces the obstructive effect of the solenoid cartridge on the injector/valve train. Perhaps of most significance is the addition of a second control valve within the nozzle assembly, which permits the ECM to change the NOP to adapt for different operating conditions.

SUMMARY

- Caterpillar full authority EUI engine management systems are found on the following pre-ACERT truck engines, in chronological order: 3176, 3406E, C-10, C-12, C-15 and C-16.
- Caterpillar full authority ACERT engines with EUI fueling introduced in 2004, include C-11, C-13, and C-15.
- The Volvo VECTRO EUI management system is used on its in-line, six-cylinder, twelve-liter engine known as VE D12.
- The fuel subsystems supplying the EUIs use either a gear or plunger type transfer pump to create charging pressures of around 60 psi.
- The Caterpillar ADEM and Volvo VECTRO ECMs used to manage EUI-fueled engines use a single

module incorporating both the processing capability and output switching apparatus.

- Caterpillar locates the injector drivers within the ECM housing and uses induction coils to spike the EUI actuation voltage to values around 100 V.
- Volvo also locates the injector drivers within the ECM housing and uses induction coils to spike the EUI actuation voltage to values around 90 V.
- Current version Caterpillar ADEM ECMs no longer use a bolt-on personality module but retain PROM type data and EEPROM data in flash memory. The personality module logs proprietary data, customer programmed options, and data audit trails.
- Some ECMs have a nonvolatile electronic memory component (NV-RAM) and an integral battery to sustain it, should battery power be interrupted for any reason.
- Both Caterpillar ADEM and Volvo VECTRO electronic management systems support an increasing number of <SMART> programming options that enhance formerly hard limits such as vehicle speed limit and programmed cruise settings with soft limits to optimize fuel economy and driver state of mind.
- Programming options in current systems include a data log that records critical engine and vehicle running parameters in three formats: lifetime totals, trip totals and instantaneous readout, and vehicle-to-base wireless communications capability.
- PC-based software is required to troubleshoot both Caterpillar and Volvo systems and more recently, hand-held ESTs such as PalmPilot have been added.
- The Cat ADEM cylinder cutout test can provide an accurate assessment of engine cylinder performance balance. The cylinder cutout test can be performed in three ways, single-, four- and five-cylinder cutout.

REVIEW QUESTIONS

1. What is a typical charging pressure used by Caterpillar and Volvo in their EUI fuel subsystem?
   a. 15 psi
   b. 60 psi
   c. 300 psi
   d. 1,200 psi
2. Which of the following injector cutout test options is recommended by Caterpillar to evaluate engine cylinder balance?
   a. One cylinder
   b. Two cylinder
   c. Four cylinder
   d. Five cylinder
3. Which of the following features makes the Volvo V-Pulse distinctive from earlier versions of the engine?
   a. Variable geometry turbocharger
   b. EGR
   c. Full authority electronic controls
   d. Single module engine control
4. What force creates injection pressure values in an electronic unit injector?
   a. Electrical
   b. Hydraulic
   c. Spring
   d. Mechanical
5. What defines the hard limit window available to the engine ECM managing an EUI fuel system, within which an effective stroke must occur?
   a. Accelerator position
   b. Road speed
   c. Cam geometry
   d. Engine rpm
6. Where are the Cat ADEM and Volvo VECTRO injector drivers located?
   a. Integral with the ECM
   b. On the firewall bulkhead
   c. Under the dash
   d. Integral with each EUI
7. Which of the following values would represent a typical EUI hydraulic nozzle NOP?
   a. 60 psi
   b. 120 psi
   c. 3,700 psi
   d. 5,500 psi
8. What type of input signal is produced by a Caterpillar can-type (pedal-mounted) TPS?
   a. Pulse width modulated
   b. Analog
   c. Spiked wave
   d. Sawtooth pattern
9. Peak injection pressures in a Caterpillar C-16 engine may reach:
   a. 5,000 psi
   b. 12,000 psi
   c. 22,000 psi
   d. 30,000 psi
10. Caterpillar ADEM EUIs are switched by the injector drivers at:
    a. 5 V
    b. 12 V
    c. 24 V
    d. ± 100 V
11. Caterpillar's electronic theft deterrent option is a feature of which of the following?
    a. Highway Master
    b. Qualcomm
    c. ET
    d. Cat ID

12. Which of the following systems uses wireless technology to download vehicle performance data such as fuel economy?
    a. Highway Master
    b. GPS unit
    c. ET
    d) Cat ID

13. Which of the following is the Caterpillar ADEM driver digital display?
    a. Highway Master
    b. Qualcomm
    c. ET
    d) Cat ID

14. How many spray holes are machined into the injector nozzle of a VECTRO EUI?
    a. One
    b. Four
    c. Eight
    d. Twelve

15. What electrical pressure is used to initiate EUI control valve opening on a Volvo VECTRO-managed engine?
    a. 5 volts
    b. 12 volts
    c. 36 volts
    d. 90 volts

16. *Technician A* states that Cat ADEM electronics permit the programming of the cruise control speed limit at a higher value than maximum vehicle speed limit. *Technician B* states that VSL is a hard-programmed parameter that cannot be exceeded. Who is correct?
    a. A only
    b. B only
    c. Both A and B
    d. Neither A nor B

17. A C-12 engine at operating temperature appears to have an audible, but inconsistent engine miss. Under load the engine appears to function normally. *Technician A* states that an EST should be used to identify the missing cylinder. *Technician B* states that the condition is normal, as management electronics are designed to cycle the engine on three cylinders under certain conditions. Who is correct?
    a. A only
    b. B only
    c. Both A and B
    d. Neither A nor B

18. Which of the following EUI components is electrically switched by the ECM to control the length of the EUI effective stroke?
    a. EUI solenoid
    b. Nozzle valve
    c. Pumping plunger
    d. EUI transistor

19. *Technician A* states that the force that creates pumping pressures in an EUI-fueled engine is electrical. *Technician B* states that the nozzle valve in an EUI is opened hydraulically. Who is correct?
    a. A only
    b. B only
    c. Both A and B
    d. Neither A nor B

20. While testing a Cat EUI-fueled engine on a chassis dynamometer, the fan clutch engages when the engine retarder high mode is selected. *Technician A* states that this occurs because of the additional heat the engine must dissipate during the engine retarding cycle. *Technician B* states that the ECM engages the fan to increase the retarding effort. Who is correct?
    a. A only
    b. B only
    c. Both A and B
    d. Neither A nor B

Chapter

40

Cummins CELECT Plus

PREREQUISITES

Chapters 29, 30, 32, and 33

OBJECTIVES

After studying this chapter, you should be able to:

- *Identify the Cummins engines that use the CELECT full authority electronic management system.*
- *Outline the system layout of a Cummins CELECT system.*
- *Describe the subcomponents used on a CELECT-managed engine.*
- *Outline the components and the purpose of the CELECT fuel subsystem.*
- *Understand the principles of operation of the CELECT System.*
- *Identify the primary inputs to the CELECT ECM.*
- *Describe the operating principles of a CELECT injector.*
- *Describe some of the governing options available in the CELECT ECM.*
- *Identify the ESTs and PC software used to diagnose and program the CELECT system.*
- *Identify customer and Cummins data programming options for CELECT.*
- *Outline the procedure required to troubleshoot and diagnose electronic problems on CELECT.*

KEY TERMS

CELECT injector

Cummins electronic engine
   control (CELECT)

charging pressure

Compulink

Echeck

electronic smart power (ESP)

electronic control module (ECM)

engine position sensor (EPS)

Jacobs C-brake

INSITE

road speed governing (RSG)

throttle position sensor (TPS)

vehicle speed sensor (VSS)

Windows (Microsoft)

INTRODUCTION

Cummins introduced **Cummins electronic engine control (CELECT),** its full authority electronic management system in 1991 on L10 and N14 engines, and later on its M11 engine when it was introduced in 1994. CELECT Plus electronics were released into the marketplace in 1996 and are used to manage current M11 Plus and N14 Plus engines (Figure 40–1).

It should be noted that L10 (10 liter) M11 (11 liter) and N14 (14 liter) engines could have PT, CELECT, or CELECT Plus fuel management but each engine incorporates different hardware. For instance, the M11 engine with PT fueling has been available until recently as an off-highway engine. CELECT Plus represents an enhancement of the CELECT engine management system, providing more comprehensive engine monitoring, faster processing speed, and greater memory retention and programmability. The CELECT Plus system is generally used as the basis for this description of Cummins electronic management systems.

SYSTEM OVERVIEW

CELECT Plus is a full authority, computerized engine management system that uses cam-actuated, electronically controlled **CELECT injectors** that can be compared to the EUIs on other full authority systems. In fact, CELECT Plus has much in common with its competitor's full authority management systems and to properly understand this section, the reader should have a thorough understanding of hydromechanical fuel systems and basic vehicle computer operation covered elsewhere in this text (Figure 40–2).

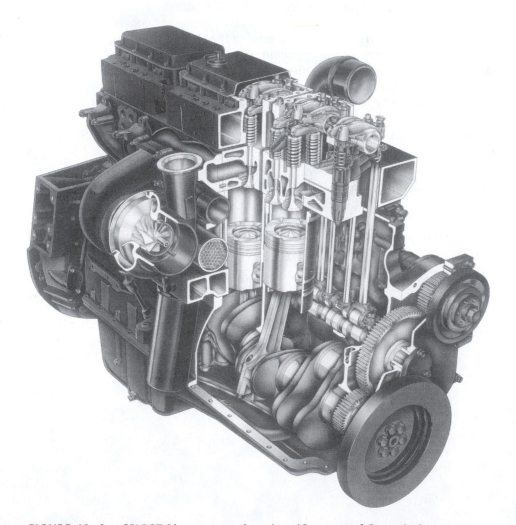

FIGURE 40–1   *CELECT Plus managed engine. (Courtesy of Cummins)*

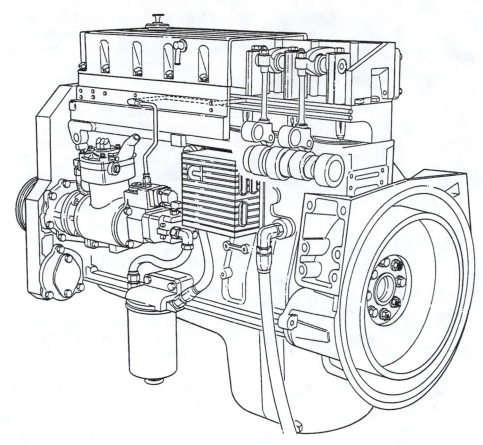

FIGURE 40–2   *Location of key components on a CELECT-managed engine. (Courtesy of Cummins)*

The CELECT Plus system can be divided as follows:

1. Fuel subsystem
2. Input circuit—sensors
3. ECM
4. Output circuit—CELECT injectors, C-brake, and so forth

FUEL SUBSYSTEM

Fuel movement in the fuel subsystem is provided by a gear pump driven by an accessory drive that rotates both the air compressor and gear pump at engine speed. The gear pump is flange mounted behind the compressor. It is fitted with a pressure regulator responsible for maintaining a 150 psi (10.2 atms/1.03 MPa) charge pressure to the CELECT injectors. An electrical shutdown valve is positioned on top of the fuel pump at the fuel outlet. It is energized to run. The fuel pump is also fitted with a small filter consisting of a mesh and magnet to prevent a cutting from the gear pump entering the charge pipe.

Fuel is drawn from the fuel tank by the gear pump. It is pulled through the cooling plate (heat sink) on which the CELECT **electronic control module (ECM)** is mounted; it circulates through this heat exchanger helping to dissipate heat from the ECM. From the ECM cooling plate, fuel passes through the fuel filter, which has a high-capacity 10-micron filtering medium. Both the ECM cooling plate and the filter are therefore under suction. From the filter head, the fuel is piped to the gear pump, which pressurizes then discharges fuel at a constant charge pressure of 150 psi (10.2 atm/1.03 MPa) to tubing that delivers fuel to the cylinder head charging galleries and made available to the CELECT injectors. After circulating through the CELECT injectors, fuel exits by means of a fuel drain and enters the cylinder head return gallery from which it is piped back to the fuel tank (Figure 40–3).

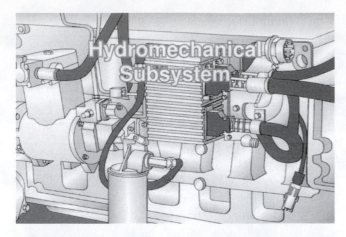

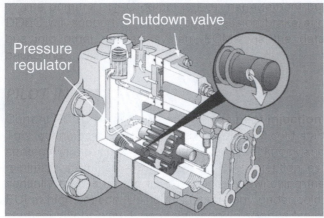

FIGURE 40–3 *The hydromechanical fuel subsystem on a CELECT engine. (Courtesy of Cummins)*

INPUT CIRCUIT

In the CELECT Plus electrical subsystem, components are connected to the ECM by a single wiring harness integrating the separate sensor and actuator harnesses of the CELECT system. CELECT Plus sensors are defined by type; the electrical principles on which they are based are explained in Chapter 32.

ENGINE MONITORING SENSORS

1. Ambient pressure sensor—piezoelectric type—for altitude fuel deration.
   M11—located in front of ECM
   N14—located behind ECM
2. Oil pressure sensor—piezoelectric type—located in the main oil rifle in the block.
3. Engine position sensor—a magnetic pulse generator-type sensor. The cam gear located pulse wheel has twenty-four evenly spaced recesses to feed

back rotational speed data and a small dowel at TDC for engine position data.

4. Oil temperature sensor—thermistor—located in the main oil rifle in the block.
5. Intake manifold temperature sensor—thermistor—located centrally on the intake manifold.
6. Boost pressure sensor—piezoresistive type—located in the intake manifold.
7. Coolant temperature sensor—thermistor—located in the thermostat housing.
8. Coolant level sensor—a switch-type sensor that grounds through the engine coolant with dual probes to indicate high or low coolant levels. This sensor is OEM supplied.

COMMAND SENSORS AND SWITCHES

See Figure 40–4.

1. Throttle pedal assembly—OEM supplied, accelerator pedal that incorporates a 5 V throttle position sensor (potentiometer) and an idle validation switch (open–close circuit). CELECT Plus is a drive-by wire system with no mechanical backup. An accelerator interlock is optional.
2. **Jacobs C-brake**—a Cummins designed, Jacobs manufactured, engine compression brake managed by the CELECT ECM. The current version permits engine braking down to 900 rpm (previously 1,500 rpm) and can be automatically actuated by the ECM in cruise control mode.
3. Cruise control—managed by the CELECT Plus ECM with some advanced features:
   a. Governor tailoring (see section on governor tailoring)
   b. Cruise auto resume after shift
   c. OEM cruise switch configuring
   d. Automatic engine braking in cruise mode
   The minimum cruise control speed limits the top vehicle speed while in cruise control mode, and this cannot exceed the maximum vehicle speed setting.
4. PTO control—this feature controls engine rpm at a constant speed selected by the operator. Either a cab or remote switch can be used, and PTO mode can be selected up to 6 mph (10 km/hr) vehicle speed.

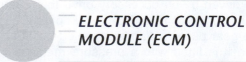

ELECTRONIC CONTROL MODULE (ECM)

The ECM receives and processes information from the various engine and chassis sensors and switches,

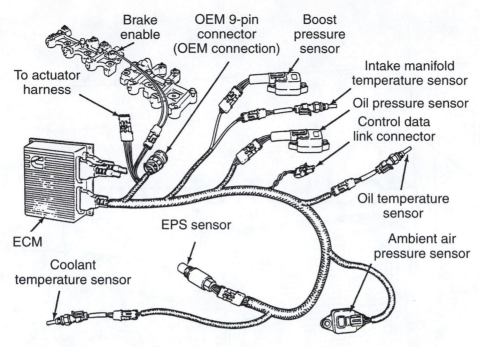

FIGURE 40–4 *The CELECT sensor harness. (Courtesy of Cummins)*

plots fueling and timing commands, and outputs signals to the injector driver circuitry (switching apparatus). A more detailed accounting of ECM functions is covered elsewhere in this text. The CELECT Plus injector driver unit uses coils to spike the EUI solenoid voltage to 78 V and is the primary output of the ECM.

Consistent with other manufacturers, the CELECT Plus ECM performs comprehensive self-diagnostics, which can be read by flash codes or electronic service tools (ESTs) and can be programmed to adopt a variety of failure mode strategies (Figure 40–5 and Figure 40–6). Cummins claims that CELECT Plus offers significant advantages over CELECT including:

- Faster processing speed
- Altitude fuel deration
- Cylinder pressure monitoring
- Faster, more accurate throttle response
- Low idle clutch engagement
- Multilevel programming security
- Engine warmup protection
- Accelerator interlock (i.e., bus door, vehicle brake, or PTO switched)

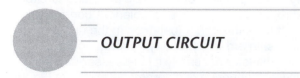

OUTPUT CIRCUIT

The primary ECM output function is control of the CELECT injectors whose operation is described in de-

tail. However, the Jacobs C-brake, PTO apparatus, and all ECM readouts can also be described as being part of the output circuit. The CELECT ECM drives output circuit functions through the actuator harness, shown in Figure 40–7. A brief description of C-brake and PTO operation is given in the input circuit description.

The primary diagnostic and programming tool for use with CELECT Plus systems is the PC with **Windows (Microsoft)**-driven, **INSITE** software, and the appropriate serial links. **Echeck** and **Compulink** (proprietary ESTs) software packages will also be made available.

THE CELECT PLUS INJECTOR

The CELECT injector is substantially different from the Cummins PT injector and probably the only real resemblance is that they both are cam-actuated injection devices (Figure 40–8). CELECT and CELECT Plus injectors use similar operating principles but the CELECT Plus injector is referenced in the following description. CELECT injectors are cam-actuated, electronically controlled injector assemblies that use multi-orifii, hydraulic injector nozzles and therefore have much in common with the EUIs used by their competitors; however, CELECT injectors differ in metering and timing principles. CELECT injectors are fitted to cylindrical bores in the engine cylinder head. They circulate fuel at

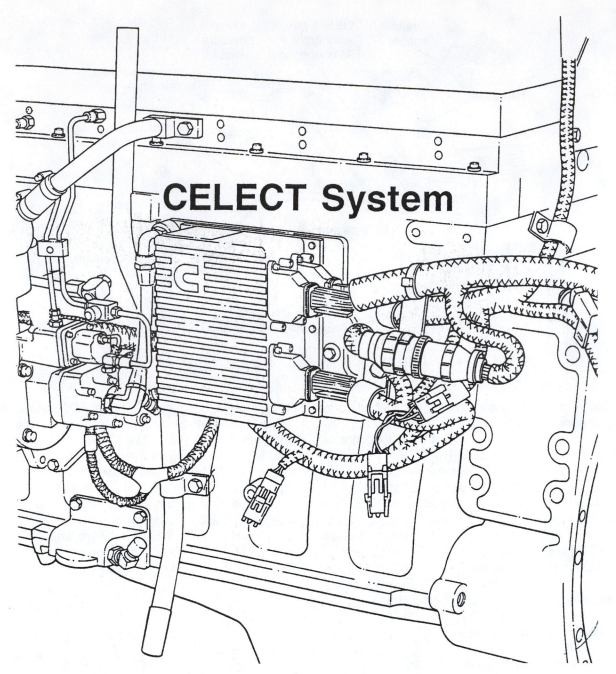

CELECT System

FIGURE 40–5 *The CELECT ECM showing the sensor and actuator harnesses. (Courtesy of Cummins)*

any time the engine is running and the gear supply pump is rotating. Fuel enters and exits the injector by means of annular recesses sealed by rubber O-rings. The lower annular recess is exposed to the supply gallery in the cylinder head and receives fuel at **charging pressure** (150 psi/10 atms/1 MPa). Fuel passes from the lower annular recess to the internal injector

circuitry, through the metering spill port, and is circulated through the injector and solenoid control valve ducting. It exits through a fuel drain in the upper external annulus. In order to properly understand the operation of this injector assembly it is necessary to refer to the following sequence of diagrams as shown in Figure 40–9.

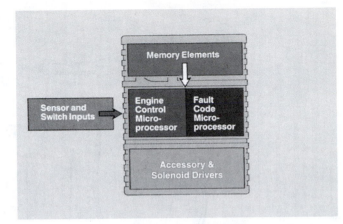

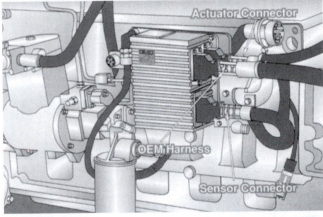

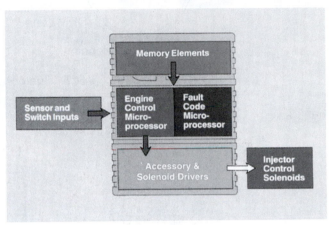

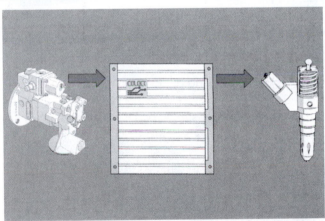

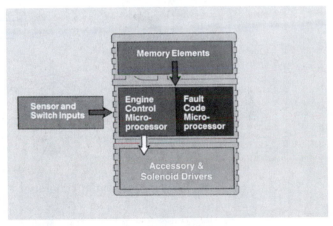

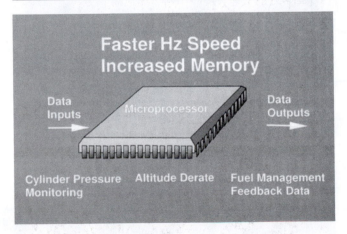

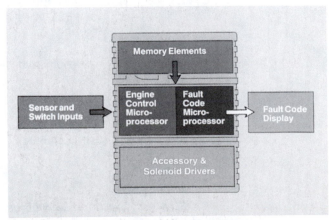

FIGURE 40–6 *CELECT Plus processing cycle graphics. (Courtesy of Cummins)*

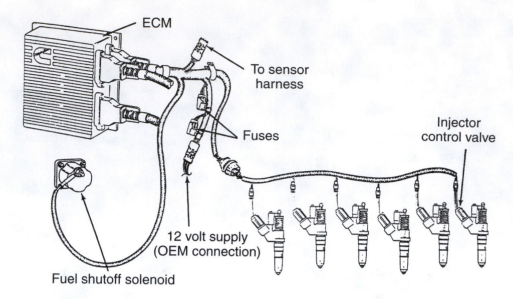

FIGURE 40–7   *The CELECT actuator harness. (Courtesy of Cummins)*

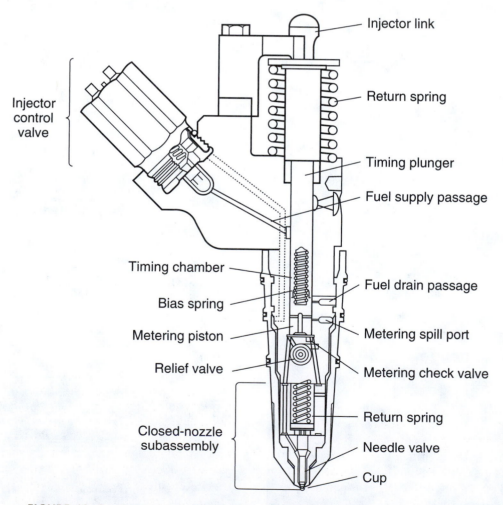

FIGURE 40–8   *Sectional view of the key CELECT injector components. (Courtesy of Cummins)*

Metering

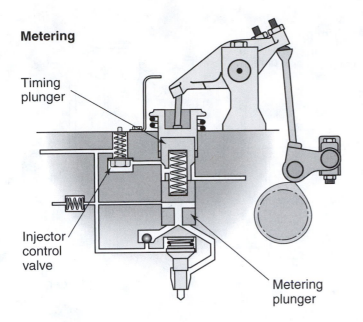

Metering

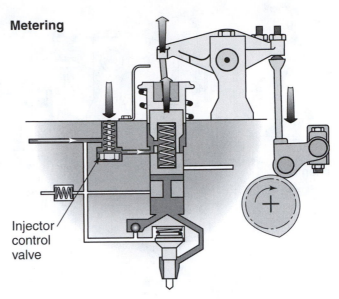

1. At the start of metering, both the metering plunger and the timing plunger are at their lower travel limit as their actuating cam is at peak lift. The injector control valve closes, actuated by a 78-V induction coil derived spike delivered from the CELECT injector driver unit.

3. The ECM will determine the end of metering by switching the injector control valve to its open position. This action causes the metering checkball to seat and permits fuel to pass around the injector control valve.

Metering

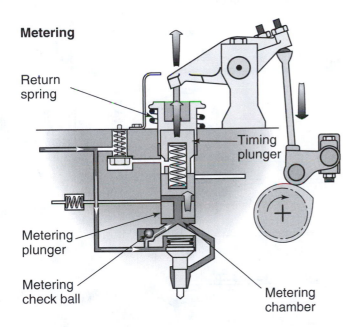

Metering

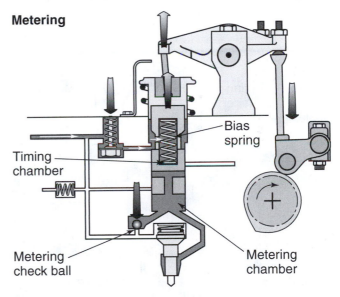

2. As the cycle continues, the cam ramps off the nose toward base circle, unloading the injector train and permitting the timing plunger return spring to lift the timing plunger. This enables fuel to flow past the metering checkball into the metering chamber. This flow continues as long as the timing plunger is moving upward and the injector control valve is closed; supply pressure acting on the bottom of the metering piston forces it to maintain contact with the timing plunger.

4. Fuel at supply pressure then flows into the timing chamber stopping metering piston travel. The bias spring ensures that the metering plunger remains stationary, preventing it from drifting upward as the timing plunger moves upward. This same force against the metering plunger results in sufficient fuel pressure below the metering piston to keep the metering checkball seated. The result is a precisely metered quantity of fuel in the metering chamber.

FIGURE 40–9 *The CELECT Plus injector operating cycle.*

Timing

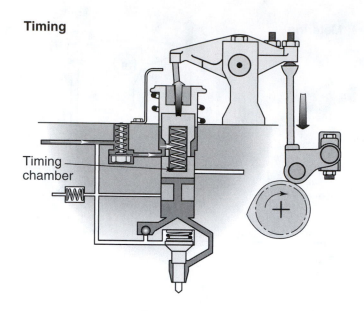

5. As the cycle continues, the injector train continues to ride toward cam inner base circle permitting the timing chamber to fill with fuel.

Timing

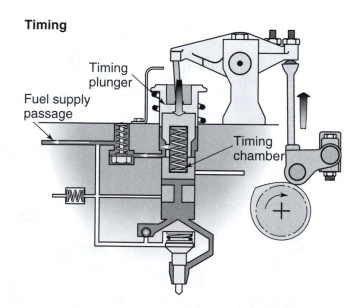

6. Next, the injector cam passes over the inner base circle location and begins ramping toward outer base circle. This action loads the injector train and consequently the timing plunger begins its downstroke. Initially the injector control valve remains open, allowing fuel to spill from the timing chamber reverse flowing it through the fuel supply passage.

Timing

7. The delivery sequence begins when the ECM switches the injector control valve to its closed position trapping fuel in the timing chamber. This trapped fuel acts as a hydraulic link between the timing plunger and the metering plunger; this forces the metering plunger downward with the timing plunger.

Injection

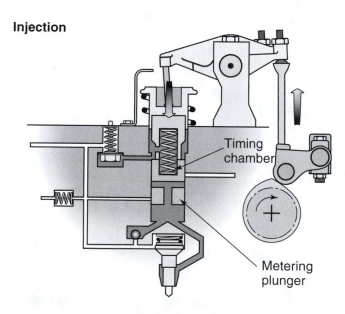

8. Because the metering plunger is being driven downward (hydraulically), rapid pressure rise begins in the metering chamber.

FIGURE 40–9 (Continued)

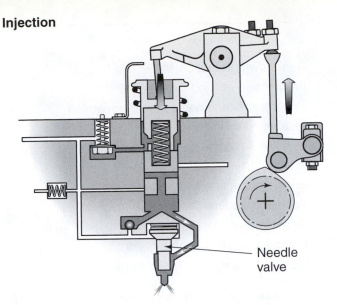

Needle
valve

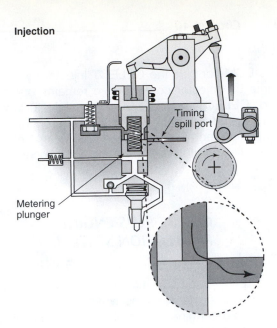

Timing
spill port

Metering
plunger

9. *Ducting connects the metering chamber with the pressure chamber of the hydraulic, multi-orifii injector nozzle located at the base of the CELECT injector. When the pressure in the metering chamber (and therefore in the nozzle pressure chamber) reaches the NOP value, approximately 5,000 psi (340 atms/34.442 MPa), the nozzle valve opens and injection begins. Fuel is forced through the nozzle orifii and atomized directly into the engine cylinder. The minute sizing of the nozzle orifii means that they are unable to relieve the pressure as fast as it is created and peak pressure is capable of rising well above the NOP value depending on the length of the effective stroke.*

11. *Immediately after the metering spill port is exposed by the downward travel of the metering plunger, its upper edge exposes the timing spill port.*

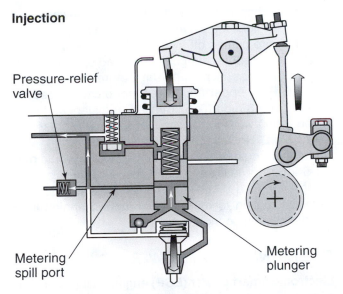

Pressure-relief
valve

Metering
spill port

Metering
plunger

10. *Injection continues until the metering plunger passes the spill passage. This action causes a collapse of metering chamber pressure permitting abrupt nozzle valve closure. At this moment, the pressure relief valve will relieve, minimizing the effect of the high pressure spike that occurs at metering spill—the relief valve passage connects to the fuel drain line.*

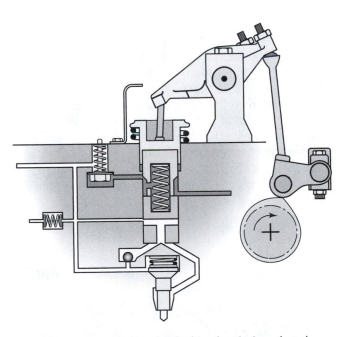

12. *This action permits the fuel in the timing chamber that was used as hydraulic medium to spill to the fuel drain. This completes the cycle.*

FIGURE 40–9 (Continued)

The ECM controls the CELECT injector by varying the time the injector control valve remains open and closed. CELECT and CELECT Plus injectors, while sharing operating principles, are *not* interchangeable, the latter incorporating increased orifii sizing, modified spray angles, and other performance enhancements.

CELECT PLUS ENGINE PROTECTION SYSTEM

The CELECT Plus ECM monitors critical engine sensors and will log fault codes when an out-of-normal condition is determined. If an out-of-range condition is detected in a critical sensor, an engine derate action may be initiated. The operator will be alerted by flashing of the appropriate dash light. The engine protection system monitors:

- Coolant temperature
- Oil temperature
- Intake manifold temperature
- Oil pressure
- Coolant level
- Engine overspeed

At engine derate, the operator is alerted by the flashing light, and engine power and speed deration will occur incrementally depending on the severity of the condition. The engine will not actually shut down unless the ECM is programmed with the engine shutdown option. The engine protection shutdown option automatically shuts down the engine when a fluid pressure or temperature out-of-normal range problem occurs that could lead to catastrophic damage. Should this occur, the dash engine protection lamp will flash for 30 seconds before the shutdown. Key off is required to restart the engine but if the shutdown trigger persists, the engine will be shut down again after 30 seconds.

CELECT PLUS PROGRAMMABLE FEATURES

CELECT Plus offers a full range of programmable features consistent with other full authority management systems including:

Cruise control

Automatic engine brake in cruise control

Power takeoff governing

Gear down protection

Progressive shifting

Limiting speed/variable speed governing

Disable air conditioner

Low idle adjustment

Idle shutdown option

Engine protection shutdown

Vehicle speed monitoring

Disable engine fan

Automatic transmission bussing

Maintenance monitor options

These features and their options are programmable as customer data using Compulink, Echeck, or INSITE ESTs.

ROAD SPEED GOVERNING TAILORING

Road speed governing or **RSG** adjusts or moderates the upper and lower droop values. Droop is governed speed deviation from a given set speed. Moderating droop can tailor the torque curve (by managing injected fuel quantity) to optimize engine fuel economy as road and speed conditions vary. Upper droop will tailor the torque curve upward (by providing extra fuel) before maximum vehicle speed is reached under conditions of high engine load.

Lower droop tailors the torque curve upward in a downhill or light load condition permitting increased speed before cutting off fueling. This allows a vehicle to accelerate above the governed road speed permitting an increase in momentum before going up the next hill. RSG upper and lower droop can be set between 0 and 3 mph (0–5 km/hr). For instance, if road speed droop were set at zero, this would represent a hard value. However, if it were programmed at 5 km/hr, this would be a *soft* value because it would permit 5 km/hr of latitude around the set cruise speed. Using increased droop settings enables the ECM to stategize fueling to road conditions and engine load to optimize fuel economy.

ELECTRONIC SMART POWER

Electronic smart power (ESP) engine ratings improve driveability and fuel economy in hilly terrain. In these conditions ESP results in more consistent road speed, faster trip times, and less shifting. It is essentially a cruise option that "learns" terrain and manages vehicle speed using soft limits. On level or near level terrain, ESP engines operate at a base power rating determined by the ESP calibration programmed into the ECM. The ECM continually monitors the average speed the operator is attempting to maintain and logs this as "learned speed." The engine then automatically switches to the high torque ESP mode if all ESP operating conditions have been met.

ESP Operating Conditions

1. Engine must be operating in cruise control or between 90% to 100% throttle (a momentary throttle position change such as required by a gear shift sequence will not disable ESP).
2. The transmission must be in a drive ratio of 2.62:1 or less.
3. Vehicle speed must not exceed the ECM programmed maximum cruise control speed.
4. Using the foot brake to reduce vehicle speed will not activate ESP mode, but momentary use will not disable it.
5. Any active fault codes that relate to the **throttle position sensor (TPS)**, **vehicle speed sensor (VSS)**, or **engine position sensor (EPS)** will prevent ESP operation.

The foregoing conditions met, the ECM will switch to the ESP high-torque mode any time the vehicle speed drops more than a calculated value (3 mph or 5 km/hr) below "learned speed." Once "learned speed" has been attained and the load reduced, the engine rating will return to the base power mode.

ROADRELAY

RoadRelay is a dash-mounted driver data display that keeps the operator informed of conditions on board. All RoadRelay data are recorded and may be downloaded to a PC database for analysis. The system will track the following data:

Trip length and average mpg

Trip running time and fuel consumed

Idle fuel consumed

Cruise control time

Top road speed

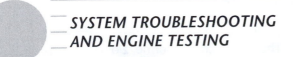

SYSTEM TROUBLESHOOTING AND ENGINE TESTING

Troubleshooting the CELECT Plus system requires the use of _Cummins Troubleshooting and Repair Manual 3666130–00 (9/95)_ volumes 1 and 2, a DMM, and the appropriate EST—preferably a PC with INSITE software. Troubleshooting is sequential and the stepped procedures must be undertaken precisely; skipping a step usually invalidates subsequent steps. Customer

data programming is fully covered in this manual. In common with other full authority, electronic management systems, CELECT Plus will perform self-diagnostic procedures. Some of these are briefly outlined in the remainder of this section with an emphasis on tests and procedures that have least in common with other OEM tests.

SETUP FOR DYNAMOMETER TEST

This test mode enables some advance diagnostic tests to run, structured by the CELECT ECM. The maximum engine speed without VSS, maximum vehicle speed in top gear ratio, and the maximum vehicle speed in lower gear are all set to their maximum values. At the same time, progressive shift programming and idle shutdown are disabled for the duration of the dynamometer test run. To properly enable the "setup for dynamometer test," some chassis electronic subsystems must be disabled for the duration of the test. Dynamometer testing is covered in detail in Chapter 15.

AUTOMATED CYLINDER PERFORMANCE TEST

The objective of the cylinder performance test is a thorough test of cylinder performance that is displayed by reporting cylinder load contribution in percentage terms. The driving software for the test is contained in the electronic service tool, which must be connected. The following requirements must be met to run this test:

1. The area around the engine and fan must be clear and the exhaust properly vented.
2. The engine oil temperature must be at a minimum of 170°F (77°C).
3. Shut down the engine.
4. Lock the fan clutch to the engaged position, disable the A/C system and any other auxiliary devices that could represent a parasitic load on the engine.
5. With the vehicle stationary, start the engine and allow it to idle.
6. Engage the PTO (the test cannot be performed unless the PTO feature is programmed to the ECM).

The test profile instructions should be run according to Cummins CELECT service literature. The test produces a PASS/FAIL message for each engine cylinder. In the event of cylinder performance outside the specification window, the percent contribution value will be displayed with the FAIL message. When the test is complete, the following sequence should be run:

1. Shut down the engine.
2. Return the fan clutch, A/C system, and all other disabled devices to normal operation.

CYLINDER CUTOUT TEST

The cylinder cutout test removes individual injectors from the cylinder firing cycle to monitor engine parameters while a selected cylinder is electronically disabled. The electronic service tool displays the percent load and engine rpm value while the cylinder cutout test is being run. Again, the test is performed while the vehicle is stationary and the electronic service tool is connected to the ECM. During the cylinder cutout test, only one cylinder can be electronically shorted out at a time in current versions of CELECT management.

SUMMARY

- CELECT and CELECT Plus are both full authority, electronic management systems that have almost no relationship with Cummins PT management other than sharing the same manufacturer.
- CELECT was introduced in 1991 on reengineered (from PT) L10 and N14 engines. It was also used to manage M11 when this series of engines was introduced in 1993.
- CELECT Plus is an evolution of CELECT with more comprehensive chassis monitoring, faster processing, greater programmability, and "smart" capability.
- CELECT Plus may be found on current versions of M11 and N14 engines.
- CELECT Plus management may be divided into four subsystems: hydromechanical fuel, input circuit, ECM, and output circuit.
- Charging pressures to CELECT injectors are the responsibility of a gear-type transfer pump. Charging pressures are typically 10 atms (150 psi).
- Charging pressures are converted to injection pressures by the CELECT injectors, which are actuated by cam profile. Effective stroke is ECM controlled by switching a solenoid on the CELECT injector.
- CELECT injectors are switched using a coil-spiked actuation signal of 78 V.
- CELECT systems have self-diagnostic capability and are programmable with customer and Cummins data.
- CELECT Plus has "smart" capability including RSG tailoring and ESP.
- Cummins RoadRelay is a dash-mounted, driver data display that keeps the driver updated with trip data such as mpg and service intervals. RoadRelay data can be downloaded to a PC for subsequent analysis.

REVIEW QUESTIONS

1. Which of the following CELECT Plus input sensors could be categorized as a "command" input?
   a. Ambient pressure
   b. Oil pressure
   c. Coolant level
   d. Throttle position
2. Cummins CELECT injectors are charged at a pressure of:
   a. 15 psi
   b. 50 psi
   c. 70 psi
   d. 150 psi
3. What type of fuel transfer pump is used in the Cummins CELECT system?
   a. Vane
   b. Diaphragm
   c. Gear
   d. Plunger
4. What operating principle does a Cummins CELECT intake manifold temperature sensor use?
   a. Piezoresistive
   b. Potentiometer
   c. Variable capacitance
   d. Thermistor
5. The CELECT boost pressure sensor would be described electrically as:
   a. Piezoresistive
   b. Potentiometer
   c. Variable capacitance
   d. Thermistor
6. At what electrical pressure are the CELECT injectors switched?
   a. 5 V
   b. 12 V
   c. 78 V
   d. 100 V
7. The Cummins Windows-driven software for accessing its data hub, programming CELECT ECMs, and performing system diagnosis is known as:
   a. Compulink
   b. Windows XP
   c. INSITE
   d. DOS
8. The hydraulic injector nozzle used in the CELECT injector is set to have an NOP of:
   a. 150 psi
   b. 5,000 psi
   c. 25,500 psi
   d. 30,000 psi
9. The "smart" road speed governing option on CELECT Plus is called:

a. Echeck
b. ESP
c. INSITE
d. RSG tailoring

10. Which of the following will prevent ESP operation?
    a. Any logged fault code
    b. Any active fault code
    c. TPS, VSS, or EPS active code
    d. 1:1 transmission drive ratio

11. _Technician A_ states that a CELECT Plus injector is energized with a 78 V spike. _Technician B_ states that the CELECT Plus injector voltage increases proportionally with the duty cycle. Who is right?
    a. A only
    b. B only
    c. Both A and B
    d. Neither A nor B

12. _Technician A_ states that CELECT engine position sensors use a magnetic pulse generator principle. _Technician B_ states that the engine position sensor is located on the engine crankshaft. Who is right?
    a. A only
    b. B only
    c. Both A and B
    d. Neither A nor B

13. _Technician A_ states that the CELECT engine position sensor is mounted on the engine camshaft. _Technician B_ states that the CELECT engine position sensor is fitted with 24 evenly spaced recesses with a dowel indicating the #1 piston, TDC position. Who is right?
    a. A only
    b. B only
    c. Both A and B
    d. Neither A nor B

14. _Technician A_ states that the ambient pressure sensor in a CELECT M-11 is located in front of the ECM housing. _Technician B_ states that the ambient temperature sensor in a CELECT N-14 is also located in front of the ECM housing. Who is right?
    a. A only
    b. B only
    c. Both A and B
    d. Neither A nor B

15. _Technician A_ states that the coolant and oil temperature sensors are both of the thermistor type. _Technician B_ states that both sensors must be supplied with a reference voltage value. Who is right?
    a. A only
    b. B only

c. Both A and B
d. Neither A nor B

16. _Technician A_ states that a CELECT injector has a single internal plunger that defines timing and metering. _Technician B_ states that injection pressures in the CELECT injector are created by varying the supply pressure. Who is right?
    a. A only
    b. B only
    c. Both A and B
    d. Neither A nor B

17. _Technician A_ states that the CELECT and CELECT Plus injectors use shared electronic operating principles. _Technician B_ states that CELECT and CELECT Plus injectors are interchangeable so long as the new fuel code is correctly programmed to the ECM. Who is right?
    a. A only
    b. B only
    c. Both A and B
    d. Neither A nor B

18. _Technician A_ states that the CELECT ECM is mounted on a heat sink and cooled by fuel. _Technician B_ states that both the ECM and the filter are charged by the gear pump at a pressure of 150 psi. Who is right?
    a. A only
    b. B only
    c. Both A and B
    d. Neither A nor B

19. _Technician A_ states that the function of the CELECT coolant level sensor is to shut the engine down should the sensor fail to ground through the coolant. _Technician B_ states that the CELECT coolant level sensor has a pair of probes that can read high and low coolant levels in the radiator. Who is right?
    a. A only
    b. B only
    c. Both A and B
    d. Neither A nor B

20. _Technician A_ states that the actuating force of a CELECT injector is mechanical and determined by cam geometry and engine rpm. _Technician B_ states that the CELECT injector metering plunger is actuated by a hydraulic column of fuel. Who is right?
    a. A only
    b. B only
    c. Both A and B
    d. Neither A nor B

Chapter

41

Bosch EUP on V-MAC III-E-Tech and Mercedes-Benz

PREREQUISITES

Chapters 29, 30, 32, 33, and 37

OBJECTIVES

- *Define the acronym EUP.*
- *Explain the differences between full authority EUP and EUI diesel engine management.*
- *Outline the system layouts on Mack Trucks E7-E-Tech and Mercedes-Benz engines.*
- *Describe the principles of operation of the EUP.*
- *Outline some critical differences between the Mack Trucks and Mercedes-Benz EUP systems.*
- *Describe how a EUP converts charging pressure fuel to injection pressure values.*
- *Identify the similarities between EUP fueled engines and those using EUI systems.*

KEY TERMS

beginning of energizing (BOE)
beginning of injection (BOI)
Co-Pilot
DataMax
electronic/engine control unit (ECU)
E-7 Series
E-Tech
InfoMax
network access point (NAP)
PLD module
RPM/TDC sensor

INTRODUCTION

Electronic unit pump (EUP) technology was introduced into highway diesel engines by Mack Trucks with its **E-Tech** engine that debuted in 1997 and the now-obsolete DDC Series 55. In 2000, Mercedes-Benz introduced the MB-900 series (4.2 to 6.4 liter) engine into light-duty Freightliner chassis, and this was joined by the MB-4000 (12 liter) in 2002. Because Daimler-Chrysler owns Detroit Diesel, Mercedes-Benz, and Freightliner, the MB engines are to be marketed and supported by Detroit Diesel, which may help them overcome a certain suspicion of off-shore engines held by the trucking industry. The MB-4000 was the platform used for the unsuccessful DDC Series 55 engine, and the fact that it has reappeared little changed is an indication of how much Mercedes-Benz wants to get its products accepted in the North American commercial truck markets.

Mack Trucks opted to use the Bosch EUP system on its V-MAC III-E-Tech engines because it proved to be good fit with a moderately re-engineered **E-7**, 12-liter engine that would come to be known as E-Tech. Those familiar with both E-7 and E-Tech quickly identify that the combustion dynamics and the components that determine this (cylinder head, injector type, piston geometry) are not so different between the two series. Mack Trucks manages E-Tech EUP using **V-MAC III** electronics, a two-module management system in which the chassis management "commands" an engine/fuel system controller. Figure 41–1 shows a cut-

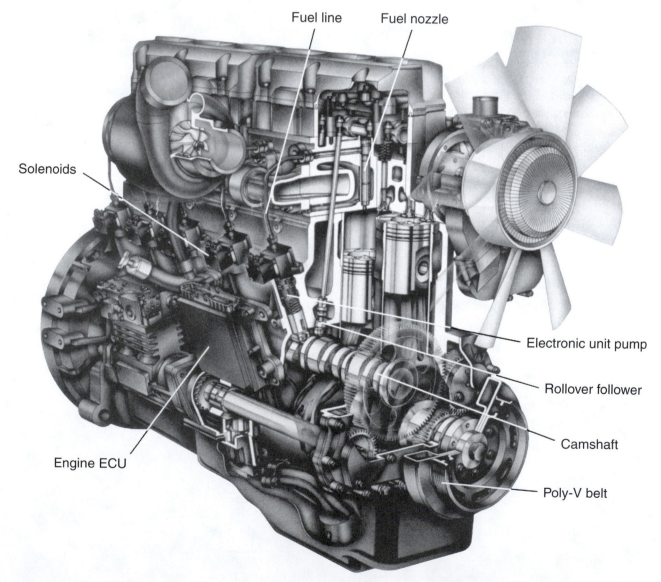

FIGURE 41–1 *Cutaway view of a Mack E-Tech engine with EUP fueling. (Courtesy of Mack Trucks)*

away view of a Mack Trucks E-Tech engine fueled by EUPs.

The Mercedes-Benz engines, while as yet unproven in North American conditions, boast some interesting features such as cracked connecting rods and a constant throttle, engine brake. The electronic controls used on the engine are similar to those used by Mack Trucks, with a vehicle control unit (VCU) mastering an engine-mounted control unit known as a **PLD module** (German acronym for ECM).

SYSTEM OVERVIEW

If you have already studied electronic unit injector systems, you will know much of what there is to know about EUP systems. In a EUP system, the functions of pumping and control are separated from that of the atomizing function of the nozzle. The pumping and control unit are hydraulically connected to a remote located, injector nozzle.

If we begin by outlining the functions of a diesel engine EUI, they can be divided into three subsystems:

1. Solenoid control
2. Cam-actuated pumping
3. Hydraulic injector nozzle

Bosch EUP technology incorporates the first two EUI functions in a single unit, flange-mounted directly over the engine camshaft that actuates the pumping stroke. One electronic unit pump manages fueling for a single engine cylinder. The EUP is connected by means of a high-pressure pipe to a hydraulic injector nozzle, much the same as would be found on any hydromechanically managed diesel engine with a port-helix metering injection pump. The hydraulic injector nozzle is of the multi-orifii type with the NOP trigger value adjusted by setting spring tension. The result is a system that provides full authority, pump-line-nozzle (PLN) fueling.

Figure 41-2 shows a broadside view of the Mack Trucks E-Tech engine. Compare this view with the fuel system schematic in Figure 41-3 that shows how fuel is routed through the system and note the location of the EUP units, high-pressure pipes, and the engine controller ECU. All the six high-pressure pipes are of equal length and shape, and compared to those found on a port-helix metering injection pump, of short length.

FUEL SUBSYSTEM

The fuel subsystem required to fuel the EUPs is fairly standard in that it can be divided into a suction and charge circuit in both the Mack and Mercedes-Benz

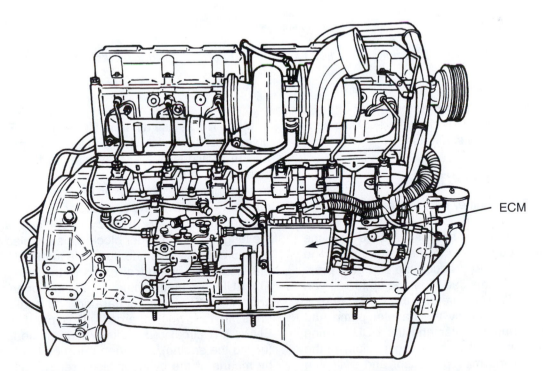

FIGURE 41-2 *Side view of the E-Tech engine showing the location of the ECM and EUPs. (Courtesy of Mack Trucks)*

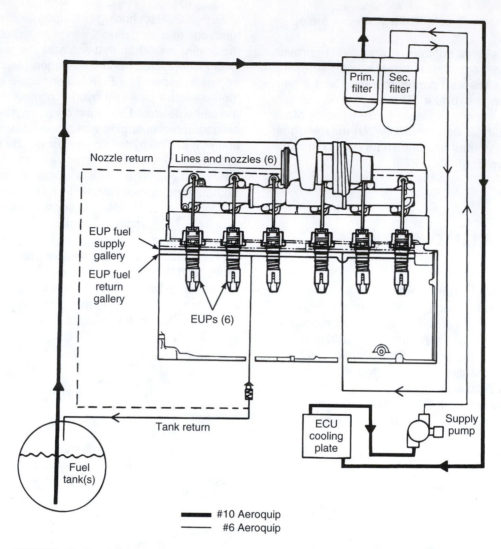

FIGURE 41-3   *Schematic of the EUP fuel system. (Courtesy of Mack Trucks)*

systems. Because there are minor differences in each system, we examine them separately.

MACK E-TECH FUEL ROUTING

In the Mack E-Tech engine, the plunger-type transfer pump used in V-MAC II is replaced by a gear-type pump driven off the accessory drive that the Econovance was coupled to in the E-7 engine. The gear pump is required to produce the charging pressures of 100 psi (6.8 atms/690 kPa) and up to 100 gph (gallons per hour) flow required by the EUPs. Fuel is drawn from the fuel tank by the transfer pump and pulled through a primary filter and the ECU cooling plate, within which it acts as the cooling medium for the heat generated by the microprocessor and switching units. Fuel exits the gear pump at the charging pressure and is next pumped through a secondary filter.

Mack Trucks use different filters on their E-Tech engines; despite the fact they are the same physical size as the previous filters, they filter at higher efficiencies and have metric mounting threads. The E-Tech filters are identified by a pair of circumferential black bands. Fuel exiting the secondary filter is then fed to the fuel supply gallery in the cylinder block, which runs the length of the cylinder block. This enables charging pressure fuel to be made available to the EUPs, which are located in cylindrical bores in the cylinder block. The EUPs are manufactured with exterior annular inlet and output recesses, separated and sealed with O-rings. In this way, charge fuel is pumped into the EUPs, it is circulated within (where some is used for fueling the engine), and then returned to the fuel tank by means of the cylinder block return gallery, which runs parallel with the supply gallery (Figure 41-3 and Figure 41-4).

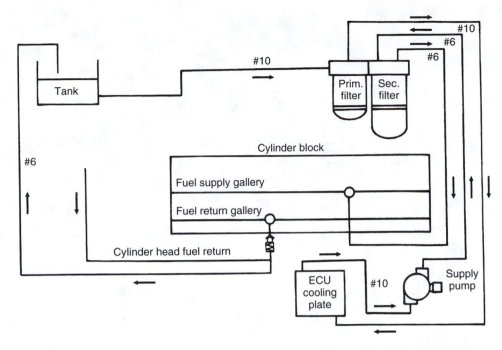

FIGURE 41–4   *Fuel routing in the E-Tech fuel circuit. (Courtesy of Mack Trucks)*

PRIMING THE FUEL SUBSYSTEM

This procedure should only be required when the fuel system has run completely dry. Mack Trucks stresses that when loss of prime occurs, using a hand primer pump is the only acceptable method. The following method should be used to prime the fuel subsystem.

1. Fill both the primary and secondary fuel filters with filtered fuel, pouring it into the inlet side of the filter.
2. Loosen the inlet hose fitting on the secondary filter mounting pad.
3. Actuate the hand primer pump until bubble-free fuel exits at the inlet fitting on the secondary filter mounting pad. No more than 50 strokes should by required. Tighten the hose line nut and crank the engine.

TECH TIP: Do not crank the engine for more than 30 continuous seconds. Allow 2 minutes between each 30-second cranking cycle.

WARNING: Applying air pressure to the fuel tank to prime the fuel system can blow out the transfer pump seals and pump fuel into the crankcase, contaminating the engine oil. Mack Trucks cautions that this proce-dure is prohibited.

MERCEDES-BENZ EUP FUEL ROUTING

The fuel subsystem layout used by Mercedes-Benz engines is similar to that used by Mack. The prime mover of the fuel subsystem is a gear-type pump. It pulls fuel from the tank, through a check valve and prefilter (primary filter), and charges a secondary circuit that consists of a secondary filter (known as "the" filter) and the manifold that supplies the EUPs. Fuel pressure downstream from the secondary filter should test at between 400 kPa (58 psi) and 650 kPa (94 psi), the actual values varying with engine rpm. Testing can be performed at no-load engine operation. Figure 41–5 is a schematic of the Mercedes-Benz fuel subsystem indicating the test points for verifying performance pressures.

INPUT CIRCUIT

Bosch EUP-fueled engines are full authority, electronic management systems with comprehensive engine/chassis monitoring and command circuits. The input circuit components used in the Mack Trucks and Mercedes-Benz systems are similar. However, both are dealt with separately here. Both Bosch and Mercedes-Benz often like to retain the original German

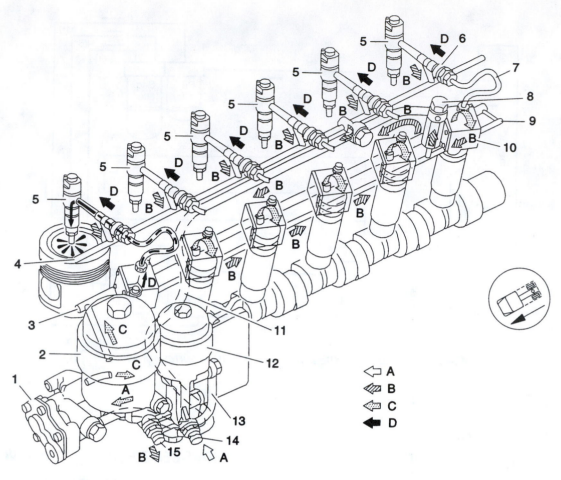

A. Fuel feed (suction side)
B. Fuel return (from leak line)
C. Fuel delivery (low pressure side—downstream of fuel pump and fuel filter)
D. Fuel delivery (high pressure side—downstream of unit pumps)

1. Fuel pump
2. Main fuel filter
3. Fuel delivery port (to unit pumps)
4. Fuel leak port
5. Nozzle holder
6. High-pressure connector
7. Injector line
8. Overflow valve
9. Fuel return port
10. Unit pump
11. Return line
12. Fuel prefilter
13. Fuel feed line (from prefilter to fuel pump)
14. Assembly valve (in fuel feed line)
15. Assembly valve (in fuel return line)

FIGURE 41–5   *EUP fuel circuit flow. (Courtesy of Mercedes-Benz)*

acronyms in translation and this can cause some confusion: for instance, the engine controller module is a PLD control unit rather than ECM.

V-MAC III INPUTS

Because V-MAC II was a partial authority system, there are some differences in the input circuit of V-MAC II compared with V-MAC III. In V-MAC versions I and II, many of the critical system sensors were located in the PE8500 injection pump and integral rack actuator housing. As this component is eliminated from V-MAC

III, an engine position sensor has been located in the front timing gear cover where the passage of holes in the front face of the camshaft gear is monitored. The function of the engine position sensor is similar to that of the timing event marker (TEM) on the E-7 RE30 rack actuator housing. The engine position sensor must be precisely installed using shims to ensure the correct gap. The V-MAC III engine speed sensor **(RPM-TDC)** is identical (the same part number) to the engine position sensor. It is located in the flywheel housing; the E-Tech-EUP flywheel has one more tooth than the E-7, two of which have half their width machined off. Again,

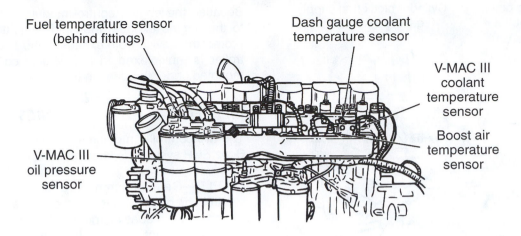

Fuel temperature sensor
(behind fittings)

Dash gauge coolant
temperature sensor

V-MAC III
coolant
temperature
sensor

Boost air
temperature
sensor

V-MAC III
oil pressure
sensor

Not shown
• V-MAC III engine speed sensor (on left side of flywheel housing)
• V-MAC III engine position sensor (on engine front cover)
• Dash gauge oil temperature sensor (on left side of oil pan)

FIGURE 41–6 *Location of V-MAC III input circuit components. (Courtesy of Mack Trucks)*

correct installation of this sensor is critical, and the gap must be calculated and then set by shims. Additional sensors used with V-MAC III are a boost air temperature sensor and a fuel temperature sensor on chassis fitted with the optional **Co-Pilot** Data display (displays V-MAC **ECU** parameters) (Figure 41–6).

MERCEDES-BENZ INPUTS

The locations of the sensors used on a six-cylinder version of a MB-900 engine are shown in Figure 41–7. The electrical principles on which these sensors operate are discussed in Chapter 32. Input circuit sensors feed their data to the PLD (engine controller module) that can then broadcast it on the J1587 or J1939 data backbones.

ENGINE CONTROLLERS AND MANAGEMENT ELECTRONICS

The way in which Mack Trucks and Mercedes-Benz electronically manage their EUP systems is similar. Both use a chassis management module with a CAN connection to an engine management module. In each case, the chassis and engine management modules can receive and broadcast data over the J1587 and J1939 data backbones.

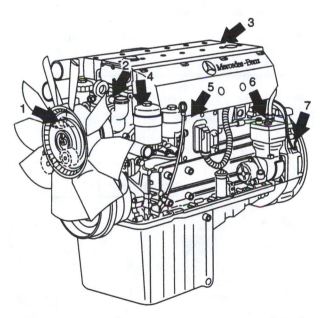

NOTE: The 6-cylinder engine is shown; sensor locations are similar on the 4-cylinder engine.

1. Oil pressure/temperature sensor
2. Coolant temperature sensor
3. Charge pressure/air temperature sensor
4. Fuel temperature sensor
5. Ambient air pressure sensor (integrated into PLD control unit
6. TDC sensor (on camshaft)
7. Crank angle position sensor (on timing case)

FIGURE 41–7 *Location of sensors on the MBE900 engine. (Courtesy of Mack Trucks)*

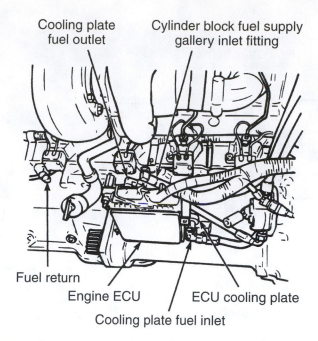

Cooling plate fuel outlet

Cylinder block fuel supply gallery inlet fitting

Fuel return

Engine ECU   ECU cooling plate

Cooling plate fuel inlet

FIGURE 41–8   *The E-Tech ECU. (Courtesy of Mack Trucks)*

V-MAC III ELECTRONICS

Mack Trucks V-MAC III on E-Tech returns to a two-module system, separating vehicle control functions from engine control functions. Mack Trucks uses the acronym ECU to describe its modules. The engine controller module is located on the right side of the engine, mounted on a fuel-cooled heat exchanger (Figure 41–8).

Because the engine control module is located close to the engine exhaust manifold, a heat shield helps protect the device from exhaust heat. The engine controller is multiplexed to the vehicle control module, mounted under the dash in the cab.

MERCEDES-BENZ ELECTRONICS

The two-controller module system used by Mercedes-Benz EUP systems is known as:

- PLD—German acronym for engine-mounted control unit or ECM. It is located on the left side of the engine and does not need to be mounted on a heat sink. Current versions have a clock speed of 24 mHz.
- VCU—the vehicle control unit that fulfills the same function as the V-MAC III module on the Mack Trucks system.

The two modules are multiplexed by a proprietary CAN bus. Input data signaled to the PLD module can be broadcast to the J1587 and J1939 data backbones by first routing it through the VCU module. Mercedes-Benz electronics are accessed for read, reprogram, and diagnostic operations through a standard six-pin or nine-pin diagnostic connector using Freightliner ServiceLink, PC-based software. For read-only purposes, the electronics can be interpreted through the standard SAE MIDs, SIDs, PIDs and FMIs. Figure 41–9 shows the multiplexing architecture used by Mercedes-Benz-powered chassis.

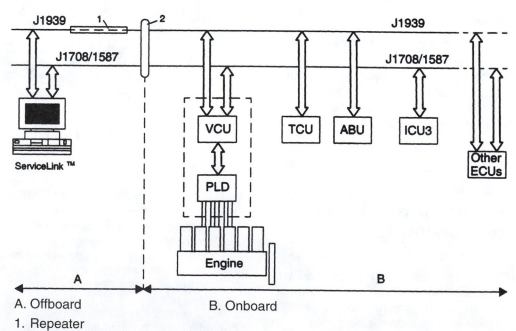

A. Offboard
B. Onboard
1. Repeater
2. 9-Pin diagnostic connector

FIGURE 41–9   *Mercedes-Benz multiplexing architecture. (Courtesy of Mercedes-Benz)*

ELECTRONIC UNIT PUMPS

The electronic unit pumps (EUPs) are flange mounted over a roller tappet that rides the center lobe of a set of cylinder cams. A separate EUP is required for each engine cylinder. Each is connected electrically to the ECM and hydraulically to an injector nozzle (Figure 41–10). Pumping action in the EUP is dictated by the actuating cam geometry. As the engine-driven cam rotates off its base circle ramping toward the nose, the roller lifter drives the EUP plunger into the pump chamber. Fuel is routed at charging pressure through the pump chamber at any time the engine is running, entering through an upper exterior annulus, circulating through the EUP internal ducting, and exiting through a lower exterior annulus. Control of the injection pulse is the responsibility of the EUP solenoid, itself switched by the engine-mounted V-MAC III module. When the EUP plunger is driven into the pump chamber by the actuating cam profile, fuel is merely displaced; however, the moment the EUP solenoid is energized, fuel is trapped in the pump chamber and pressure rise occurs. Connected to the EUP pump chamber by a high-pressure pipe is a multi-orifii, hydraulic injector nozzle. Actual injection to the engine cylinder begins when the preset NOP value is achieved (Figure 41–11).

The EUP plunger has a 10 mm diameter and the potential for an 18 mm stroke. Actual plunger stroke is determined by the cam profile. Peak injection pressures of up to 26,000 psi (1,770 atms/180 MPa) are possible. The EUPs are supplied with low pressure (charging pressure) fuel by a fuel supply gallery machined into the engine block. High-pressure fuel is routed from the EUP to the hydraulic injector nozzles in the cylinder head by a relatively short (compared with previous E-7 engines) high-pressure pipe. Each high-pressure pipe

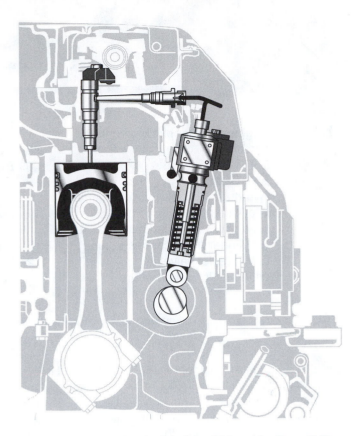

FIGURE 41–11 *Sectional view of Mercedes-Benz EUP fueled engine showing the hydraulic connection between the EUP and injector. (Courtesy of Mercedes-Benz)*

is of equal length and has an identical part number regardless of engine cylinder.

Controlling effective stroke is the responsibility of the V-MAC III engine ECU or the Mercedes-Benz PLD. Operation of the EUP almost identically parallels that of any typical diesel engine EUI system with one exception.

Because the EUP is connected to the hydraulic injector nozzle by means of a high-pressure pipe and in the EUI system the nozzle is integral, the lag time between the **beginning of energizing (BOE)** and **beginning of injection (BOI)** is obviously extended in the EUP-managed engine when compared to the typical EUI system. The nozzles, which are set to a specified NOP value (around 340 atms/5,000 psi/34 MPa), are serviced as units in the same manner as any other hydraulic injector nozzles; these procedures are covered elsewhere in this textbook.

Figure 41–12 is a sectional view showing how the EUP connects with the hydraulic injector in the engine cylinder head. Although the figure shows a Mercedes-Benz engine, this arrangement in almost identical in the Mack Trucks EUP-fueled engine.

Effective stroke can only take place within the ramp window between cam inner base circle and outer base circle. When the engine controller module (PLD or ECU)

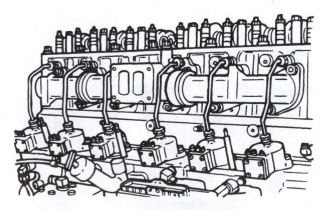

FIGURE 41–10 *The E-Tech engine showing the EUP line connection to the hydraulic injectors. (Courtesy of Mack Trucks)*

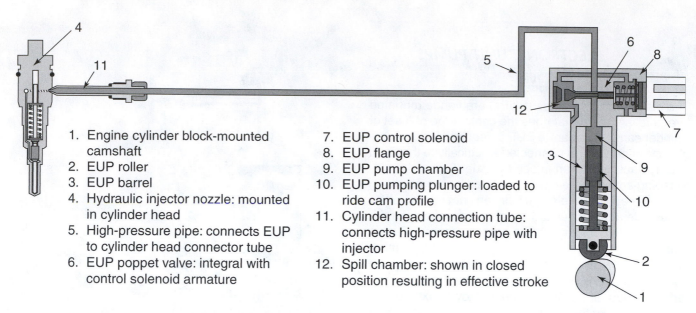

1. Engine cylinder block-mounted camshaft
2. EUP roller
3. EUP barrel
4. Hydraulic injector nozzle: mounted in cylinder head
5. High-pressure pipe: connects EUP to cylinder head connector tube
6. EUP poppet valve: integral with control solenoid armature
7. EUP control solenoid
8. EUP flange
9. EUP pump chamber
10. EUP pumping plunger: loaded to ride cam profile
11. Cylinder head connection tube: connects high-pressure pipe with injector
12. Spill chamber: shown in closed position resulting in effective stroke

FIGURE 41–12 *EUP operating principle. (Courtesy of Mercedes-Benz)*

switches the control solenoid EUP, fuel is trapped in the pump chamber as shown in Figure 41–12 creating pressure rise in the feed circuit to the injector. Effective stroke ends when the engine controller de-energizes the EUP solenoid, opening the spill circuit and creating pressure collapse: when there is no longer sufficient pressure to hold the nozzle valve open, it closes.

REMOVAL AND INSTALLATION OF EUPs

As with any engine and electronics system procedure, the appropriate OEM technical service literature is required before starting the work. The procedure is outlined here as a model so that the student has some awareness of what is required to remove and replace the distinct component on this fuel system. The following procedure is based on that required to perform it on a Mack Trucks E-Tech engine. Figure 41–13 and Figure 41–14 show external and internal views of an EUP.

REMOVAL

TECH TIP: When more than one EUP must be removed from an engine, paint mark each EUP with the engine cylinder number. This will eliminate the need to reprogram the EUP data to the ECU on reinstallation.

1. Remove the heat shields and thoroughly clean the exposed portion of the EUP and the cylinder block.

Any dirt or debris that drops into the EUP bores will enter the crankcase of the engine so this step is critical.

2. Drain fuel from the cylinder block fuel supply gallery. Consult the OEM service literature to determine how to do this most effectively on the specific engine the procedure is being performed on. In some cases, the filter pad assembly will have to be drained. It is important to remember that if the fuel supply gallery is not drained, all of the fuel in this gallery will drain into the crankcase when the first EUP is pulled. If this happens, the engine oil will be degraded to a level in which engine damage will result, so the oil must be changed.

3. Remove the high-pressure pipe from the EUP and seal it and the EUP inlet with a dust cap.

4. Remove the electrical terminals from the EUP solenoid.

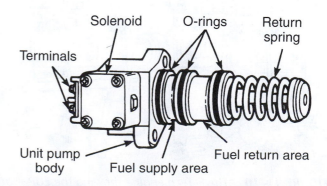

FIGURE 41–13 *EUP external subcomponents. (Courtesy of Mack Trucks)*

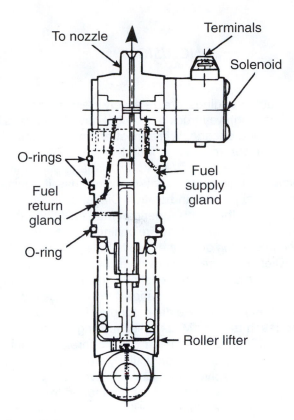

FIGURE 41–14 *Sectional view of an EUP. (Courtesy of Mack Trucks)*

Labels in figure: To nozzle; Terminals; Solenoid; O-rings; Fuel supply gland; Fuel return gland; O-ring; Roller lifter

5. Remove the EUP inboard screw and loosen and back out the outboard screw to about ½ inch.
6. Insert an injector heal bar under each of the EUP bolt bosses and lever on the EUP flange until the assembly stops on the outboard capscrew.

CAUTION: The EUP assembly may be loaded with considerable spring force depending on the position of the actuating cam profile under it. For this reason, it is essential that the outboard capscrew be backed out no more than ½ inch.

---

7. Back out the outboard EUP capscrew and remove the EUP assembly from the bore.
8. Remove the roller tappet by hand. It is important not to use any tool that could result in damage to the bore. Take precautions to ensure that dirt does not enter the engine through the exposed EUP bore. Protect the removed EUP and tappet assembly from contamination.

TAPPET AND EUP INSTALLATION

1. Clean the cylinder block at the EUP bore. Lubricate the tappet assembly using the same engine oil to be used in the engine.

2. Install the roller tappet assembly into the bore, being sure to orient it with the guide pin. Ensure that the tappet slides freely in the bore.

WARNING: Tappet guide pins are factory installed and should not be removed.

---

3. Install a set of new O-rings (3) on the EUP. The center O-ring is color coded to identify it; it is a different size from the upper O-ring.
4. Lubricate the O-rings with the same engine oil to be used in the engine. Press into position by hand.

TECH TIP: The cam profile under the EUP being installed must be on its base circle, that is, in its unactuated position. Manually bar the engine to ensure that the engine camshaft is correctly positioned. Using the starter may result in fuel being pushed through the supply gallery and ending up in the crankcase, so this should be avoided.

---

5. Install the EUP capscrews and torque to the required specification.
6. Install the high-pressure pipe and torque the line nut to specification.
7. Reinstall the EUP wire terminals and torque to specification.
8. Identify the calibration code usually recorded on the EUP nameplate. This is a four-digit code with a CAL prefix. Program the calibration code to the ECU using the appropriate EST, a ProLink, or PC with the appropriate software.
9. Install the heat shields. *Never* omit these as heat will affect the operation of the EUP solenoid coils.

TESTING INJECTORS

EUP engines use multi-orifii injector nozzles. These nozzles are fully hydromechanical devices that are tested and set up according to the procedures outlined in Chapter 20. Figure 41–15 shows a typical nozzle testing tool recommended for testing hydraulic injector nozzles.

INFORMATION MANAGEMENT SYSTEMS

Mack Trucks use information management software consistent with competing OEMs products. These are discussed briefly here.

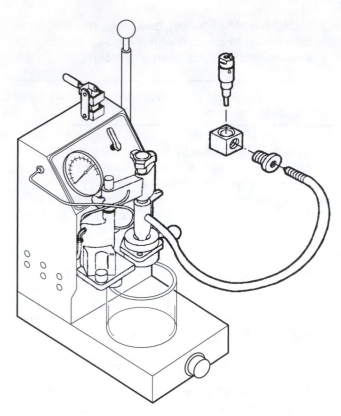

FIGURE 41–15   A typical nozzle test bench. (Courtesy of DDC)

DATAMAX

DataMax is an onboard data logger capable of recording and displaying engine and chassis running conditions. It is connected to the chassis data bus and can display information such as fuel economy, trip data, driver performance, and fault codes.

INFOMAX

InfoMax is Mack Trucks' vehicle to land station data transfer technology. It uses WiFi short range wireless data transfer with a direct line of sight range of up to 1,000 ft. InfoMax uses a truck dash-mounted broadcast unit that communicates with a terminal-based **network access point (NAP)**. Communications are two-way. InfoMax uses the NAP to download driver performance, electronic fault reporting, unauthorized reprogramming tamper detection, driver incentive fuel efficiency reports, and maintenance scheduling information. This information can then be analyzed by Info-Max software. When used in conjunction with satellite communications technology, the system can tabulate fuel mileage to journey coordinates.

The communication can be two-way. Fleets can reprogram every truck in their system at a single keystroke.

CONCLUSION

EUP fueled-engines are relatively new to the North American highway truck market but have been used in Europe for a decade. While most of the examples currently on our roads are found in the 12-liter Mack E-Tech engine, with V-MAC III management this is rapidly changing. Mercedes-Benz will use its ownership of Freightliner and Detroit Diesel to leverage its engines into the North American market. Although initially directly supported by Freightliner, current versions of the MB series of engines are to be supported by Detroit Diesel. At the present time, Mercedes-Benz engines are manufactured offshore, mainly in Brazil and Germany, but this could change if the engines are successful. Although there are a couple of isolated exceptions (such as the "Mack" Midliner engine, actually a Renault), it should be noted that no foreign manufactured engine has fared particularly well in America recently. Even Volvo, who has achieved over 10% in heavy-duty truck market share, has found that its customers more often than not will opt for an alternate engine. Volvo currently owns Mack Trucks, so the future of a specific Mack engine is not certain.

SUMMARY

- The Bosch EUP system may be classified as a full authority electronic management system.
- Three EUP-managed truck engine series are currently available in the North American market: Mack Trucks E-Tech and Mercedes-Benz 900 and 4000 Series.
- V-MAC III is a two-module system; the engine controller (ECU) is mounted on a fuel-cooled, heat exchanger located on the right side of the engine. It is multiplexed to the vehicle control module located in the cab under the dash.
- The Mercedes-Benz MBE-900 and MBE-4000 engines also use a two-module system. The PLD or ECM manages fuel and engine management functions while a multiplexed VCU takes care of command inputs and communications with other on-chassis systems on the J1587 and J1939 data backbones.
- The EUPs are supplied with fuel from the fuel subsystem at charging pressure. Fuel is constantly cycled through the system circuitry and the EUPs, following which it is returned to the tank.
- Actual plunger stroke is dictated by cam geometry. Plunger effective stroke is controlled by the engine ECM by switching the EUP solenoids.

- The EUP plunger is a simple cylindrical piston with no milled recesses and drillings. It is loaded to ride the cam profile and reciprocates within the EUP barrel.
- A high-pressure pipe connects each EUP unit with a multi-orifii hydraulic injector nozzle located over each cylinder for direct injection.
- It is important to observe OEM procedure when replacing EUPs. Fuel should be drained from the charging gallery in the cylinder block or it will spill into the engine oil. It should be remembered that EUPs can be under considerable tension if the actuating cam profile is at or close to outer base circle during removal.
- The hydraulic injector nozzles used with EUP-fueled engines are connected to the EUPs by a high-pressure pipe and cylinder head connection tube. They are tested in a similar manner to any other type of hydraulic injector nozzle.

REVIEW QUESTIONS

1. What force creates injection pressure values in an EUP-managed engine?
   a. Electrical
   b. Hydraulic
   c. Spring
   d. Mechanical
2. What defines the fueling window available to the ECM within which it can select an effective EUP stroke?
   a. Engine rpm
   b. Accelerator position
   c. Road speed
   d. Cam profile
3. "Injection rate" in an EUP-managed fueling system is determined by:
   a. Cam profile
   b. Engine rpm
   c. Fuel demand
   d. Strength of the EUI switch signal
4. How is EUP effective stroke controlled?
   a. By varying the pump plunger stroke
   b. By switching the EUP solenoid
   c. By varying the EUP charging pressure
   d. By controlling the rotational position of the plunger
5. Peak pressure values in an EUP injection pulse can reach:
   a. 4,700 psi
   b. 12,500 psi

c. 18,000 psi
d. 26,000 psi

6. Mack Trucks V-MAC III engine position and RPM/TDC sensors must be precisely gapped using:
   a. Shims
   b. An offset cam key
   c. Torque wrench
   d. Plastigage
7. Fuel is delivered to the EUP by the transfer pump at what pressure?
   a. 5,000 psi
   b. Charging pressure
   c. 26,000 psi
   d. Peak pressure
8. *Technician A* states that Bosch EUPs are mounted above and actuated directly by the engine camshaft. *Technician B* states that duty cycle in an EUP is controlled by the ECM. Who is right?
   a. A only
   b. B only
   c. Both A and B
   d. Neither A nor B
9. *Technician A* states that the EUP functions similarly to an EUI, with the exception that the injector nozzle function is separated. *Technician B* states that the EUP to nozzle lines on a multicylinder engine must be of identical length. Who is right?
   a. A only
   b. B only
   c. Both A and B
   d. Neither A nor B
10. When removing an EUP from an engine, which of the following should be performed first?
    a. Download customer parameters.
    b. Reprogram the fuel flow code.
    c. Ensure the actuating cam profile under the EUP is on inner base circle.
    d. Drain the fuel supply gallery.
11. *Technician A* says that the Mercedes-Benz PLD is connected to the VCU by means of a proprietary data bus. *Technician B* says that the Mercedes-Benz PLD can receive and broadcast data directly to the J1939 data bus. Who is right?
    a. A only
    b. B only
    c. Both A and B
    d. Neither A nor B
12. What software is required to read, program, and troubleshoot the Mercedes-Benz 900 and 4000 engines?
    a. ServiceLink
    b. Electronic Technician
    c. INSITE
    d. MackNet

Chapter

42

Caterpillar and International Trucks HEUI

PREREQUISITES

Chapters 29, 30, 32, and 33

OBJECTIVES

After studying this chapter, you should be able to:

- Define the acronym HEUI.
- Describe the system layout and the primary components in an HEUI-managed engine.
- Outline the role of the four primary HEUI subsystems.
- Describe the operating principles of injection pressure regulator (IPR).
- Describe the operating principles and subcomponents of an HEUI injector unit.
- Define the role played by the ECM input circuit and the factors that govern ECM processing logic.
- Describe how the ECM switches output devices to control engine fueling and manage combustion.
- Describe the role of the consolidated engine controller.
- Perform basic electronic troubleshooting on an HEUI system.
- Outline the procedure required to perform proprietary data reprogramming of an ECM.
- Perform customer data programming using an EST.

KEY TERMS

advanced combustion emissions reduction technology (ACERT)

amplifier piston

calibration parameters

camshaft position sensor (CPS)

consolidated engine controller (CEC)

engine family rating code (EFRC)

EZ-Tech

fuel demand command signal (FDCS)

hydraulically electronically controlled unit injector (HEUI)

injector driver module (IDM)

injector pressure regulator (IPR)

intensifier piston

keep alive memory (KAM)

pilot injection

preinjection metering (PRIME)

reference voltage (V-Ref)

self-test input (STI)

swash plate

VCO nozzle

vehicle personality module (VPM)

INTRODUCTION

When it was introduced in 1994, HEUI technology represented a new direction for the Caterpillar Engine Division because, for the first time, it launched the business of supplying fuel system engineering and components to competitor engine OEMs. The HEUI story began in 1987 when Caterpillar and International Trucks signed a joint development agreement. This resulted in the introduction of HEUI on the light-duty, International Trucks' 7.3-liter engine (444E) in 1994, an engine widely marketed as a Ford 7.3-liter engine. HEUI was developed in Caterpillar's Pontiac, Illinois, fuel systems production plant. All Caterpillar HEUI components are machined and assembled in a facility that uses extensive robotics, EDM (electrical discharge machining), and ECM (electrochemical machining) technology. EDM is used to produce the precision-sized orifii required in the HEUI hydraulic injector nozzles. This latest machining technology produces machine tolerances so precise that Caterpillar no longer has to match injector bodies to the nozzle valves as was the case when these components were lap finished. Product testing is fully automated and full function.

The application of HEUI technology continues to grow. During the 1990s, HEUI engine management was used in the International 466E and 530E engines, popular powerplants for medium- to light-duty trucks and school buses, and Caterpillar's own 3126 engine. More recently, International introduced a new Power-Stroke, 6-liter engine using the HEUI system, which replaced the 444E (7.3 liter) in 2003 and will be used in many Ford applications. This new 6-liter, V-8 engine uses a Bosch-built high-pressure oil pump and Siemens-built injectors in place of the Caterpillar units. Caterpillar has announced that HEUI technology will be used in its **advanced combustion emissions reduction technology (ACERT)** program for its range of highway diesel engines that will address EPA 2004 emissions standards. Initially, ACERT will be used on the C-7 and C-9 engines using HEUI fueling.

A major advantage of HEUI fueling is that it offers significant control over injection rate and duration compared with any diesel fuel system in which the pumping stroke is actuated by cam profile. HEUI injectors are actuated by oil pressure. The ECM that manages HEUI fueling controls the actuation oil pressure over a very wide range of pressure values resulting in precise control of injected fuel pressure entirely independent of engine speed. This rate-shaping ability provides the management electronics much greater control over fueling than any other system currently in the North American market. Figure 42–1, Figure 42–2, and Fig-

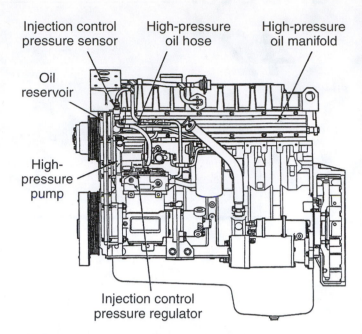

FIGURE 42–1   *Component locations on a International Trucks I-6 engine with the HEUI fuel system. (Courtesy of Navistar International Corp., Designer and manufacturer of International Brand diesel engines)*

ure 42–3 show some views of both International and Caterpillar engines using the HEUI fuel management system.

PRINCIPLES OF OPERATION

A disadvantage inherent in cam-actuated, EUI systems is that the fueling window available to the ECM is defined by the hard parameters of the injector train cam profile. The Caterpillar HEUI system uses engine lube oil as hydraulic medium to actuate the fuel delivery pulse in its HEUI assemblies. The delivery stroke is therefore actuated hydraulically, switched by the engine management ECM(s), and thus is not confined to any hard limits. HEUI is therefore truly a full authority system in that the ECM can select the fuel pulse (pump effective stroke) to occur at anytime it computes that it is required. HEUI signals a critical step toward eliminating the camshaft from a production diesel engine and achieving a fully regulated engine environment. HEUI technology is the first that permits injection rate shaping. This represents a significant departure from all other current systems that are confined to the hard parameters of actuating cam geometry. Figure C–5 in the color section compares EUI and HEUI fuel management systems.

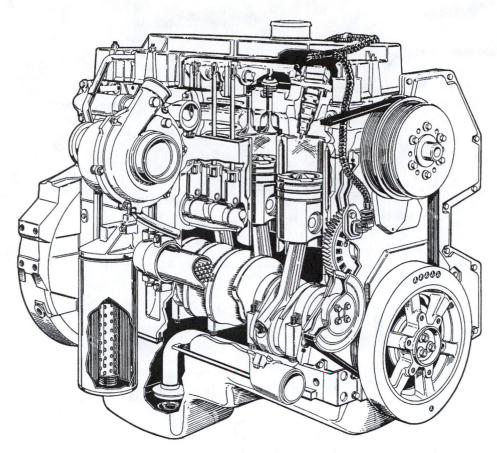

FIGURE 42–2 *Cutaway view of a International Trucks I-6 engine with HEUI: note the location of the HEUIs. (Courtesy of Navistar International Corp., Designer and manufacturer of International Brand diesel engines)*

HEUI management can be loosely divided into four subsystems as follows:

1. Fuel supply system
2. Hydromechanical injection system
3. HEUI assembly
4. Electronic management and switching

mately 30 to 60 psi (2–4 atms/206 kPa–412 kPa). Fuel is cycled through the fuel supply system and the HEUIs are mounted in parallel from the fuel manifold. A hand priming pump is located on the filter pad in the event of loss of prime on most Caterpillar and International engines using HEUI fueling. Figure 42–4 shows an HEUI fuel subsystem.

FUEL SUPPLY SYSTEM

The fuel supply system delivers fuel from the vehicle's tanks to the injector units. Fuel movement through the fuel supply system is the responsibility of either a cam-driven plunger pump or an external gear transfer pump. The transfer pump pulls fuel from the chassis fuel tank(s) through a fuel strainer. It then charges fuel through a disposable cartridge-type fuel filter and feeds it to the fuel gallery of fuel/oil supply manifold. A fuel pressure regulator at the fuel manifold outlet is responsible for maintaining a charging pressure of approxi-

INJECTION ACTUATION SYSTEM

The HEUI system uses hydraulically actuated, electronically controlled injector assemblies to deliver fuel to the engine's cylinders. The hydraulic medium used to actuate the pumping action required of the injector is engine oil. The engine lubrication circuit provides a continuous supply of engine lube to the HEUI high-pressure pump, a gear-driven, swash plate hydraulic pump used to boost the lube oil pressure up to values exceeding 3,000 psi (200 atms/20 MPa). The pump is actually capable of producing pressures of up to 4,000 psi (272

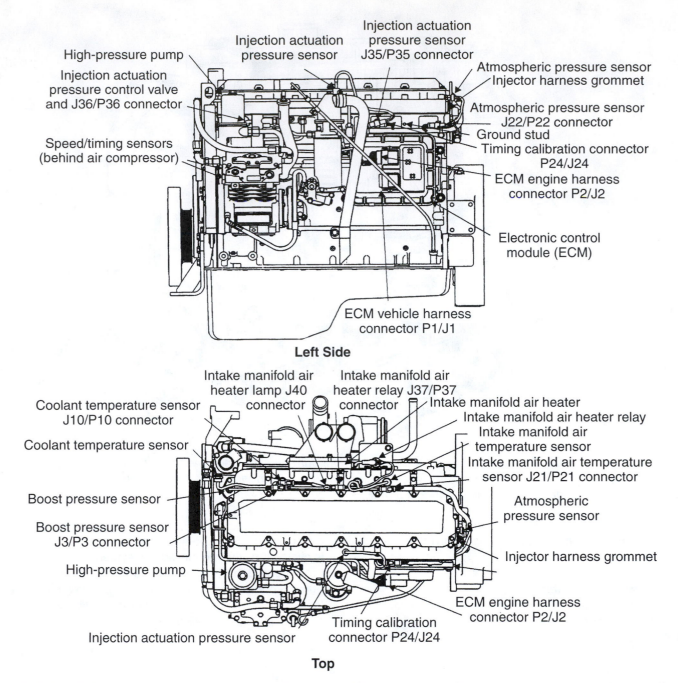

High-pressure pump

Injection actuation
pressure control valve
and J36/P36 connector

Speed/timing sensors
(behind air compressor)

Injection actuation
pressure sensor

Injection actuation
pressure sensor
J35/P35 connector

Atmospheric pressure sensor
Injector harness grommet

Atmospheric pressure sensor
J22/P22 connector
Ground stud
Timing calibration connector
P24/J24
ECM engine harness
connector P2/J2

Electronic control
module (ECM)

ECM vehicle harness
connector P1/J1

Left Side

Coolant temperature sensor
J10/P10 connector

Coolant temperature sensor

Boost pressure sensor

Boost pressure sensor
J3/P3 connector

High-pressure pump

Injection actuation pressure sensor

Intake manifold air
heater lamp J40
connector

Intake manifold air
heater relay J37/P37
connector

Intake manifold air heater

Intake manifold air heater relay

Intake manifold air
temperature sensor

Intake manifold air temperature
sensor J21/P21 connector

Atmospheric
pressure sensor

Injector harness grommet

ECM engine harness
connector P2/J2

Timing calibration
connector P24/J24

Top

FIGURE 42–3   *Side and overhead views of a Caterpillar 3126 engine with HEUI showing the component locations. (Courtesy of Caterpillar)*

atms/27.5 MPa), so future versions of HEUI may exploit this higher potential for peak pressures.

Actual high-pressure oil values are managed by the IPR or injection pressure regulator, which is controlled electronically and actuated electrically. **Swash plate** pumps use opposing cylinders and double-acting pistons actuated by a swash plate (circular plate, offset on a shaft to act as a actuating cam); they are similar in principle of operation to common automotive A/C compressors. The injection pressure regulator manages the

high-pressure oil pressure values between a low of 485 psi (33 atms) to highs up to 4,000 psi (270 atms) by receiving all the oil pressurized by the high-pressure pump and spilling the excess to the oil reservoir.

High-pressure oil is then piped to the oil gallery ducting within the fuel/oil supply manifold. From there, the oil is delivered to an exterior annulus in the upper portion of each HEUI. The HEUIs are mounted in parallel, fed by the high-pressure oil manifold. When the HEUI solenoid is energized, a poppet valve is opened by an

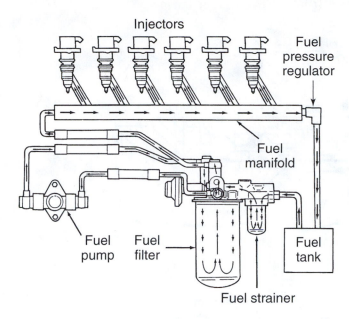

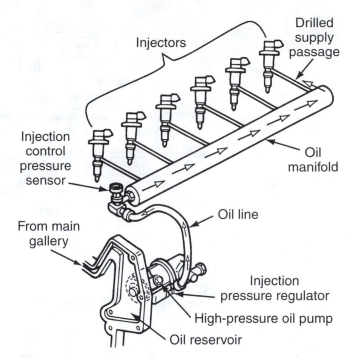

FIGURE 42–4 International Trucks HEUI fuel supply system. (Courtesy of Navistar International Corp., Designer and manufacturer of International Brand diesel engines)

FIGURE 42–5 International Trucks injection pressure oil flow schematic. (Courtesy of Navistar International Corp., Designer and manufacturer of International Brand diesel engines)

electric solenoid within the HEUI; this permits the high-pressure lube to flow into a chamber and act on the **amplifier piston** (International Trucks)/**intensifier piston** (Caterpillar), actuating the pumping stroke required to convert the HEUI charging pressure to injection pressure values. At the completion of the duty cycle, the HEUI is de-energized and the poppet valve retracts, spilling the oil acting on the amplifier piston to the rocker housing (Figure 42–5).

Injection Pressure Regulator (IPR)

The **injection pressure regulator (IPR)** is a spool valve, positionally controlled by a pulse width modulated signal to achieve a range of injection control pressures (ICP). It operates at 400 Hertz. The ECM pressure regulating signal determines the magnetic field strength in the solenoid coil and therefore, armature position. Integral with the armature is a poppet valve designed to control the extent to which high-pressure pump oil is spilled to the drain port and therefore the pressure of the oil that is used as the hydraulic medium for actuating the HEUIs (Figure 42–6).

HYDRAULICALLY ACTUATED ELECTRONICALLY CONTROLLED UNIT INJECTOR

The **HEUI** is an integral pumping, metering, and atomizing unit controlled by ECM switching apparatus. The

unit is essentially an EUI that is actuated hydraulically rather than by cam profile. While following this description of HEUI, reference Figures C-6 and C-7 in the color section. At the base of the HEUI is a hydraulically actuated, multi-orifii nozzle. When the HEUI pumping element achieves the required NOP value, the valve retracts, permitting fuel to pass around the nozzle seat and exit the nozzle orifii directly to the engine cylinder. A **VCO** (valve closes orifice) or sacless **nozzle** design is used.

The amplifier or intensifier piston is responsible for creating injection pressure values. This component is termed an amplifier piston in International Trucks technical literature and an intensifier piston in Caterpillar versions; amplifier piston is the term generally used in this text. When the HEUI is energized, high-pressure oil supplied by a stepper pump acts on the amplifier piston and drives its integral plunger downward into the fuel in the pumping chamber.

A duct connects the pump chamber with the pressure chamber of the injector nozzle valve. The moment the HEUI is de-energized, the oil pressure acting on the amplifier piston collapses, and the amplifier piston return spring plus the high-pressure fuel in the pump chamber retracts the amplifier piston, causing the almost immediate collapse of the pressure holding the nozzle valve open. This results in rapid ending of the injection pulse. In fact, the real time period between the

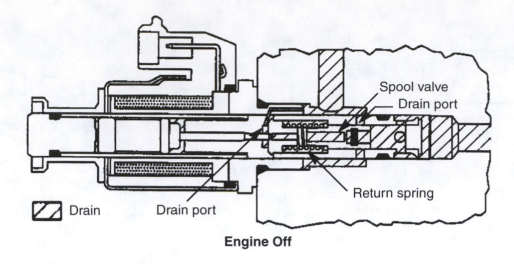

Engine Off

Drain
Drain port
Spool valve
Drain port
Return spring

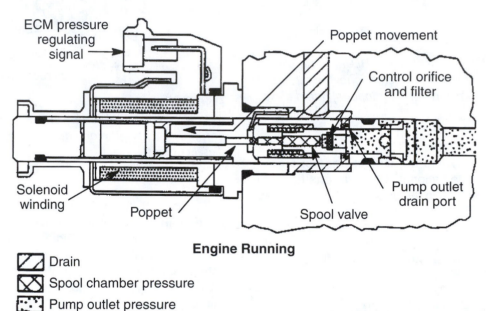

Engine Running

ECM pressure regulating signal
Poppet movement
Control orifice and filter
Solenoid winding
Poppet
Spool valve
Pump outlet drain port

Drain
Spool chamber pressure
Pump outlet pressure

FIGURE 42–6 *The International Trucks IPR valve operation. (Courtesy of Navistar International Corp., Designer and manufacturer of International Brand diesel engines)*

moment the HEUI solenoid is de-energized and the point that droplets cease to exit the injector nozzle orifii is claimed to be significantly less with HEUI than with equivalent EUI systems.

HEUIs typically have NOPs of 5,000 psi (340 atms) with a potential for peak pressures of up to 28,000 psi (1,900 atms/165 MPA) attainable depending on application. The oil pressure acting on the HEUI amplifier piston is "amplified" by seven times in the fuel pump chamber. This amplification is achieved because the sectional area of the amplifier piston is seven times that of the injection plunger, which means that the descent velocity of the injection plunger is variable and dependent on the specific actuation oil pressure value at a given moment of operation. Because the ECM directly controls the actuation oil pressure value, it can there-

fore control injection pressure. The injection pressure determines the emitted droplet size: the higher the injection pressure, the smaller the droplets, which is what is meant by rate-shaping ability of the HEUI. In short, rate shaping allows the ECM to optimize the extent of atomization to suit the combustion conditions at any given moment of operation.

Rapid pressure collapse enabled by HEUIs avoids the injection of larger-sized droplets toward the end of injection that would be difficult to completely oxidize in the "afterburn" phase of combustion. At the completion of the HEUI duty cycle or PW, the pressurized oil that actuated the pumping action is spilled to the rocker housing.

HEUI injectors are capable of being driven or switched at high speeds. The latest versions have plunger and barrel geometry that provide **pilot injection**. Pilot injection is a term used to describe an injection pulse that is broken into two separate phases. In a pilot injection fueling pulse, the initial phase injects a short duration pulse of fuel into the engine cylinder, ceases until the moment of ignition, and at that point, resumes injection, pumping the remainder of the fuel pulse into the engine cylinder. Pilot injection is used as cold-start and warmup strategy in EUI systems to avoid an excess of fuel in the engine cylinder at the point of ignition and as a result, minimize the tendency to cold-start detonation. HEUI systems with the pilot injection feature are designed to produce a pilot pulse for each injection. This has been proven to reduce both HC and NO_x emissions.

HEUI Oil and Fuel Manifold

Actuation oil and fuel at charging pressure are routed to the HEUI units by means of an oil/fuel manifold on the cylinder head. The HEUIs are inserted into cylindrical bores in the cylinder head, and they use a dedicated external annulus separated by O-rings to access the oil and fuel rifles. Figure 42–7 shows a cylinder head cross section showing the oil/fuel manifold and the ducts that connect them to the individual HEUIs.

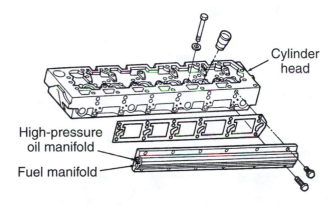

Cylinder head

High-pressure oil manifold

Fuel manifold

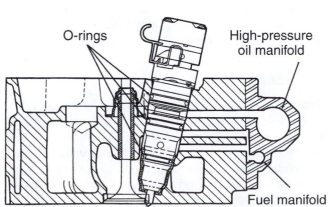

O-rings

High-pressure oil manifold

Fuel manifold

FIGURE 42–7 *International Trucks HEUI I-6 oil and fuel manifolds and HEUI cross section in cylinder head. (Courtesy of Navistar International Corp., Designer and manufacturer of International Brand diesel engines)*

HEUI SUBCOMPONENTS

The HEUI injector assembly can be subdivided as shown in Figure 42–8.

Solenoid

The solenoid is switched by the ECM using a 115 V coil induced voltage. The HEUI electrical terminals connect the solenoid coil with wiring to the ECM injector drivers.

Poppet Valve

The HEUI poppet valve is integral with the solenoid armature. It is machined with an upper and lower seat. For most of the cycle, the poppet valve seat loads the lower seat into a closed position, preventing high-pressure engine oil from entering the HEUI; the upper seat is open, venting the oil actuation spill ducting. When the HEUI solenoid is energized, the poppet valve is drawn into the solenoid, opening the lower seat and admitting high-pressure oil from the IPR. When the poppet valve is fully open, the upper seat seals, preventing the oil from exiting the HEUI through the spill passage.

Intensifier or Amplifier Piston

The intensifier/amplifier piston is designed to actuate the injection plunger, which is located below it. When the poppet control valve is switched by the ECM to admit high-pressure oil into the HEUI, the oil pressure acts on the sectional area of the amplifier piston. The actual oil pressure (managed by the ECM) determines the velocity at which the plunger located below the amplifier piston is driven into the injection pump chamber. The sectional area of the amplifier piston determines how much the actuating oil pressure is multiplied in the injection pump chamber. This value is specified as seven times in current HEUI systems. In other words, an actuating oil pressure of 3,000 psi (204 atms/21 MPa) would produce an injection pressure potential of 21,000 psi. The amplifier piston and injection plunger are loaded into their retracted position by a spring.

Plunger and Barrel

The plunger and barrel form the HEUI pump element. The first versions of the HEUI injectors did not offer the pilot injection feature. This description will use the recently introduced HEUI with the PRIME feature. The injection cycle shown in Figure C–6 shows a HEUI with the PRIME feature. The acronym **PRIME** is produced from the words **preinjection metering**. As the injection plunger is driven into the pump chamber, fuel is pressurized for a short portion of the stroke, actuating and opening the injector nozzle. The pressure rise is of

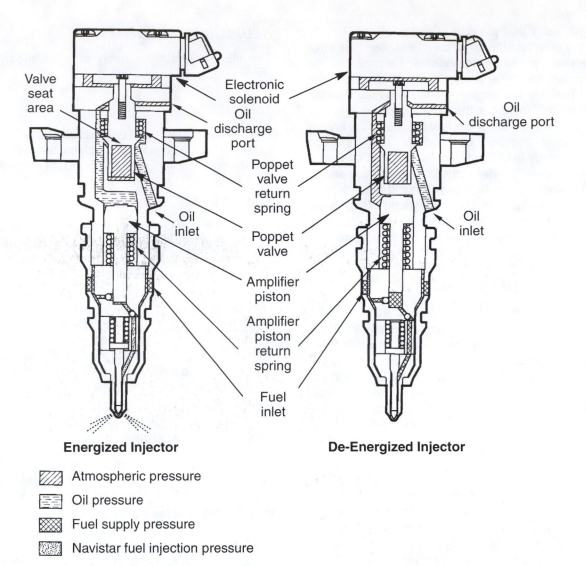

Energized Injector   **De-Energized Injector**

- ▨ Atmospheric pressure
- ▤ Oil pressure
- ▧ Fuel supply pressure
- ▨ Navistar fuel injection pressure

FIGURE 42–8   *International Trucks HEUI internal component identification and operating phases. (Courtesy of Navistar International Corp., Designer and manufacturer of International Brand diesel engines)*

short duration because when the PRIME recess in the plunger registers with the PRIME spill port in the barrel, the pressure in the pump chamber collapses as fuel spills through the PRIME spill port. This closes the injector nozzle and injection ceases. However, the moment the PRIME recess in the plunger passes beyond the spill port, fuel is once again trapped in the pump chamber and pressure rise resumes. This results in the injector nozzle being opened for the delivery of the main portion of the fuel pulse. The fueling pulse continues until the ECM ends the effective stroke by de-energizing the HEUI solenoid. At this point, the poppet control valve is driven onto its lower seat, opening the upper seat and permitting the actuating oil to be vented. With no force acting on the amplifier piston, the plunger is driven upward by the combined force of the high-pressure fuel in the pump chamber and the plunger return spring. This causes an almost immedi-

ate collapse of pump chamber pressure and results in almost instantaneous nozzle closure. A feature of the HEUI is its ability to almost instantly effect nozzle closure at the end of the plunger effective stroke.

Injector Nozzle

The HEUI injector nozzle is a multi-orifii injector nozzle of the VCO type that is a little different from any other injector nozzle used in a MUI or EUI assembly. A duct connects the nozzle pressure chamber with the HEUI pump chamber. A spring loads the injector nozzle valve onto its seat. The spring tension defines the nozzle opening pressure (NOP) value. When the hydraulic pressure acting on the sectional area of the nozzle valve is sufficient to overcome the spring pressure, the nozzle valve unseats, permitting fuel to pass around the nozzle seat and through the nozzle orifii. The nozzle valve functions as a simple hydraulic switch. Because of the nozzle differential ratio, the nozzle closure pressure is always lower than the NOP. For instance, a Caterpillar version of the HEUI with a NOP identified at 4,500 psi (306 atms/31 MPa) will not close until the pressure drops to 4,000 psi (272 atms/28 MPa). A full explanation of hydraulic injector nozzle performance is found in Chapter 20.

STAGES OF INJECTION

When the newer PRIME HEUIs are used, the injection pulse can be divided into five distinct stages. Use both the following sequence and Figure C–6 and Figure C–7 to follow the stages of injection.

Preinjection

The HEUI internal components are all located in their retracted positions as shown in Figure 42–9. In fact, they are in the preinjection position for most of the cycle. The poppet valve seat is spring loaded into the lower seat, preventing the high-pressure actuating oil from entering the HEUI, and the amplifier piston and plunger are both in their raised position. Fuel enters the HEUI to charge the pump chamber at the charging pressure value (Figure 42–10).

Pilot Injection

The pilot injection phase begins when the plunger is first moved into the HEUI pump chamber. The pressure rise created opens the injector nozzle to deliver a short pulse of fuel. The pilot injection phase ends when the PRIME recess in the HEUI plunger is driven downward enough to register with the PRIME spill port, causing the pump chamber pressure to collapse and the nozzle valve to close.

Delay

The delay phase occurs between the ending of the pilot injection phase and restart of the fuel pulse. The objective is to cease fueling the engine cylinder while the prime pulse of fuel is vaporized and heated to its ignition point. It is important to note that the plunger is still being driven through its stroke during this phase because the HEUI poppet control valve is in the open position, and oil pressure continues to drive the amplifier piston downward.

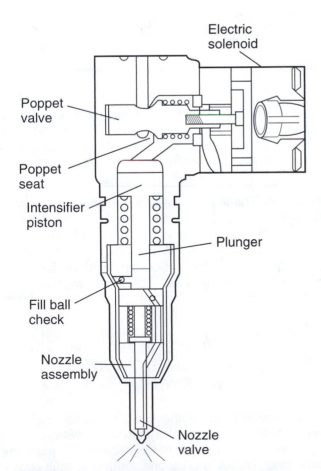

FIGURE 42–9 *Caterpillar HEUI preinjection cycle: subcomponents identification. (Courtesy of Caterpillar)*

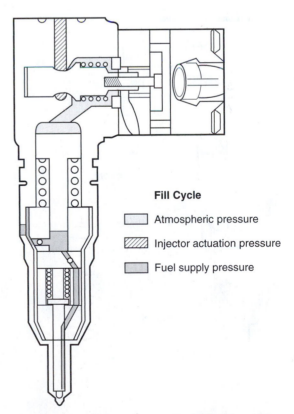

FIGURE 42–10 *Caterpillar HEUI fill cycle. (Courtesy of Caterpillar)*

Main Injection

When the PRIME recess in the plunger passes beyond the PRIME spill port, fuel is once again trapped in the HEUI pump chamber because it can no longer exit through the spill port. The resulting pressure rise opens the injector nozzle a second time to deliver the main volume of fuel to be delivered in the cycle. In an HEUI injector with no PRIME feature, the plunger has no cross and center drillings and PRIME recess, so main injection begins when the plunger leading edge passes the spill port on its downward stroke, as shown in Figure 42–11.

End of Injection

The end of injection begins with the de-energizing of the HEUI solenoid. The armature is released by the solenoid coil and a spring drives the poppet valve downward to seat on its lower seat. The instant the poppet valve starts to move downward, the upper seat is exposed, permitting the actuating oil inside the HEUI to spill. When the actuating oil pressure acting on the amplifier piston is relieved, the fuel pressure in the HEUI pump chamber combined with the plunger return spring collapses the fuel pressure almost instantly. Injection

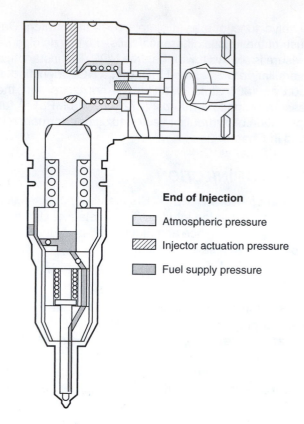

End of Injection

▢ Atmospheric pressure

▨ Injector actuation pressure

▥ Fuel supply pressure

FIGURE 42–12 *Caterpillar HEUI end of injection. (Courtesy of Caterpillar)*

ends when there is insufficient pressure to hold the nozzle valve in its open position, and the three moving assemblies (poppet valve, amplifier/plunger, and nozzle valve) in the HEUI are all in their return positions outlined in the preinjection phase (Figure 42–12).

TECH TIP: The importance of the engine oil type and condition in an HEUI fueled engine cannot be overemphasized. Because engine oil is the HEUI actuation media, it should meet OEM temperature and operating condition specifications. Contaminated and broken down engine oil can cause HEUI injector malfunctions.

HEUI ELECTRONIC MANAGEMENT AND SWITCHING

HEUI systems are full authority electronic management systems with comprehensive monitoring, vehicle management, and self-diagnostic capabilities. In early International Trucks applications, logging of fault codes was the responsibility of nonvolatile RAM (NV-RAM), which is described as **KAM** or **keep-alive memory**.

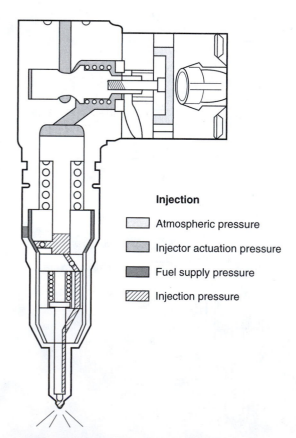

Injection

▢ Atmospheric pressure

▤ Injector actuation pressure

▥ Fuel supply pressure

▨ Injection pressure

FIGURE 42–11 *Caterpillar HEUI effective stroke. (Courtesy of Caterpillar)*

The circuit that provides continuous power to maintain KAM data is known as KAMPWR.

Additionally, the HEUI management ECM, vehicle personality module, and the switching apparatus are housed separately on International Trucks engines up to 1997; Caterpillar has always located the HEUI switching apparatus in a single ECM housing with no fuel cooling. In 1997, International Trucks introduced its **consolidated engine controller (CEC)**, which basically consolidated the microprocessing and switching functions of the ECM, the personality module, and **injector driver module (IDM)** into a single engine mounted unit. This housing is simply referred to as the ECM.

ECM FUNCTIONS

The ECM has four functions (see Figure 42–13).

1. Reference voltage regulator
2. Input conditioning
3. Microcomputer
4. Outputs

Reference Voltage

Reference voltage (V-Ref) is delivered to system sensors that divide this input and return a percentage of it as a signal to the ECM. Thermistors (temperature sensors) and potentiometers (TPS) are examples of sensors requiring reference voltage. Reference voltage values used are at 5 V pressure and the flow is limited by a current-limiting resistor to safeguard against a dead short to ground. Reference voltage is also used to power up the circuitry in Hall effect sensors used in the system such as the **camshaft position sensor (CPS)** (International Trucks). There is a full explanation of the electrical principles of input circuit components in Chapter 32.

Input Conditioning

Signal conditioning consists of converting analog signals to digital signals, squaring-up sine wave signals, and amplifying low intensity signals for processing.

Microcomputer

Both Caterpillar and International Trucks HEUI microprocessors function similarly to other vehicle system management computers. It stores operating instructions control strategies and tables of values, **calibration parameters**. It compares sensor monitoring and command inputs with the logged control strategies and calibration parameters and then computes the appropriate operating strategy for any given set of conditions. The current Caterpillar ECM, the ADEM IV, uses a 32-bit processor and a clock speed of 24 mHz. As with your home computing system, you can expect the computing power on truck engine management systems to increase as each year passes.

ECM computations occur at two different speeds, referred to as foreground and background calculations. Foreground calculations are more critical functions, and these are computed at a higher frequency than background calculations. Engine speed control and throttle position signals would be categorized as foreground calculations; in other words, an immediate response to a changing condition or command is required. Background calculations are processed at a lower frequency and include input signals such as ambient temperature and engine temperature. The difference in foreground and background computations is simply the speed at which the microprocessor is required to react to a change in operating characteristics. A change in throttle position requires an immediate adjustment in engine fueling, and therefore this command input requires almost instant response by the ECM.

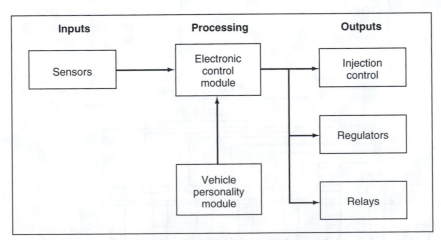

Electronic Control System Operation

FIGURE 42–13 *ECM processing cycle. (Courtesy of Navistar International Corp., Designer and manufacturer of International Brand diesel engines)*

However, while an increase in coolant temperature could have serious consequences if ignored, engine overheat conditions occur gradually, so an almost instant reaction by the ECM is not required.

Diagnostic strategies include monitoring input data on a continuous basis and flagging codes when an abnormal operating parameter is detected. Calibration tables and operating strategies are retained in ROM. This data is not lost by opening the ignition circuit or disconnecting the vehicle batteries as it is magnetically retained. RAM data is electronically retained and thus is only retained while a circuit is energized.

HEUI system ECM RAM stores information sourced from electronic monitoring and data processing/manipulation, which is volatile and as such, dumped each time the ignition circuit is opened. When KAM is used, it is nonvolatile RAM and functions to log fault codes. Out-of-normal parameters may also result in adaptive strategies being written into KAM; subsystem failure or component wear would be examples. KAM data is retained when the ignition circuit is opened but dumped when the vehicle batteries are disconnected. Both Caterpillar and International Trucks PROM is described as **vehicle personality module (VPM)** and is both customer and proprietary data programmable. The function of the VPM is to trim engine management to a specific chassis application and customer requirements. The **engine family rating code (EFRC)** is located in the VPM calibration list and can be read with an EST; this identifies the engine power and emission calibration of the engine.

Outputs

The switching apparatus within the ECM can be referred to as actuator control. The ECM controls the sys-tem actuators by delivering a signal to the base of the transistor output drivers. These drivers when switched, ground the various actuator circuits. The actuators may be controlled through a duty cycle (that is, percent time on/off), controlled by modulating pulse width or simply switched on and off, depending on the actuator type.

INJECTION DRIVER MODULE (IDM)

The injector driver module is responsible for switching the HEUIs. In early International Trucks' versions (then known as Navistar), the IDM was housed separately from the ECM housing. In current CEC versions, International integrates all the engine modules into a single housing as does Caterpillar.

The IDM has four functions:

1. Electronic distributor for the HEUIs
2. Powers the HEUIs
3. Output driver for the HEUIs
4. IDM and HEUI diagnostics

Electronic Distributor for the HEUIs

The ECM determines engine position from the CMP sensor located at the engine front cover. The ECM uses this signal to determine cylinder firing sequence and then delivers this command data to the IDM as a **fuel demand command signal (FDCS)**. FDCS contains injection timing and fuel quantity data. Figure 42–14 shows the relationship between the IDM and the ECM and the FDCS.

Powers the HEUIs

The IDM supplies a constant 115+ V DC supply to each HEUI. This 115 V DC supply is created in the IDM by

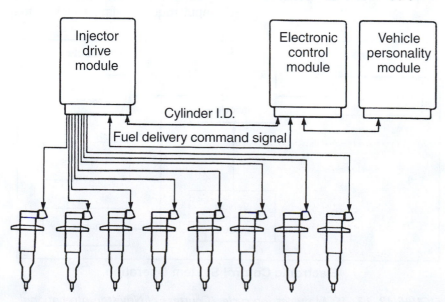

FIGURE 42–14 *IDM operation. (Courtesy of Navistar International Corp., Designer and manufacturer of International Brand diesel engines)*

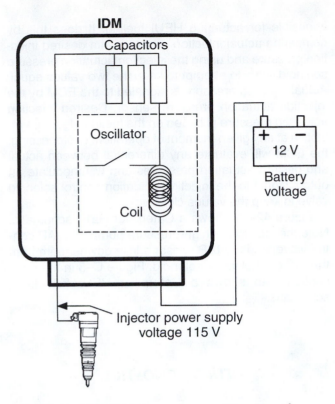

FIGURE 42–15 *IDM: power source. (Courtesy of Navistar International Corp., Designer and manufacturer of International Brand diesel engines)*

making and breaking a 12 V source across a coil using the same principles employed by the ignition coil in an SI engine. The resultant 115 V induced is stored in capacitors until discharged to the HEUIs (Figure 42–15).

Output Driver for the HEUIs

The IDM is responsible for switching the HEUIs. The unit controls the effective stroke of the HEUI by closing the circuit to ground. The direct control of the HEUI is managed by an output driver transistor in the IDM. When the FDCS signal is delivered from the ECM processing cycle, the beginning of injection (timing) and fuel quantity is determined. Figure 42–16 shows the role of the output drivers and Figure 42–17, the role of the timing sensor.

IDM and HEUI Diagnostics

The ECM software is capable of identifying faults within its electronic circuitry and can determine whether a HEUI solenoid or its wiring circuit is drawing too much (or too little) current. In the event of such an electronic malfunction, a fault code is logged. The self-diagnostics can also set an ECM code indicating that the module has failed and requires replacement (Figure 42–18).

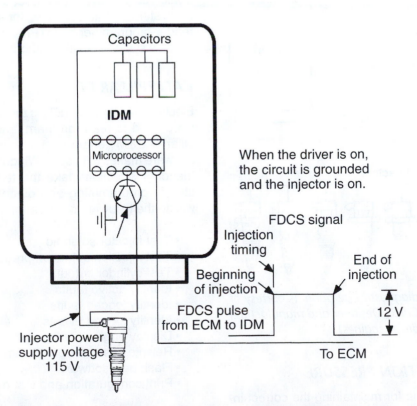

FIGURE 42–16 *Injector driver module operation: output driver operation. (Courtesy of Navistar International Corp., Designer and manufacturer of International Brand diesel engines)*

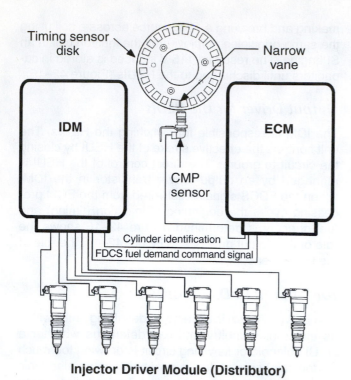

Injector Driver Module (Distributor)

FIGURE 42–17   IDM and ECM communications signals and relationship with CMP input. (Courtesy of Navistar International Corp., Designer and manufacturer of International Brand diesel engines)

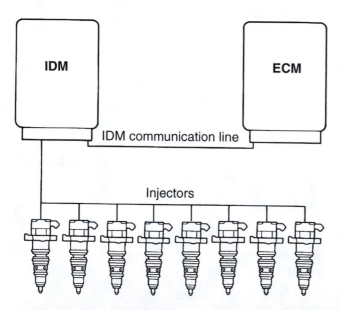

FIGURE 42–18   IDM diagnostic operation. (Courtesy of Navistar International Corp., Designer and manufacturer of International Brand diesel engines)

INJECTION ACTUATION PRESSURE

The ECM is responsible for maintaining the correct injection actuation pressure during operation. This means monitoring and adjusting the high-pressure oil circuit re-

sponsible for actuating HEUI fueling. It does this by comparing actual injection pressure with desired injection pressure and using the injection actuation pressure solenoid valve to attempt to keep the two values equal. Actual injection pressure is signaled to the ECM by the injection actuation pressure sensor. Desired injection actuation pressure is based on the fueling algorithm effective at any given moment of operation. In processing, the ECM will evaluate any differential between actual and desired actuation pressures and will modulate an output signal to the injection actuation control solenoid valve to keep the values close.

Figure 42–19 shows a Caterpillar HEUI schematic. Note the location of high-pressure pump, the IAP control valve, and the IAP sensor a little downstream from the IAP control valve solenoid. Figure C–5 in the color section also shows a Caterpillar HEUI full circuit schematic.

HEUI DIAGNOSTICS

Diagnostic procedures for Caterpillar and International versions of HEUI are distinct, as each has developed its own management software. Caterpillar uses Electronic Technician (ET) to read, diagnose, and program its systems. Until recently, International has used a reader-programmer base (ProLink) to connect to its on-chassis systems. Both methods are described here.

CATERPILLAR ET

Electronic Technician (ET) is the Windows platform software used to read, program, and troubleshoot Caterpillar ADEM engine management systems. The software makes full use of Windows graphics and user-friendly instructions take the technician from step to step in programming and diagnostic procedures. ET will do the following on a Caterpillar HEUI system:

- Test injector solenoid
- Test injection actuation pressure (IAP)
- Test cylinder cutout
- Identify active faults
- Identify logged faults
- Identify logged events
- Display engine configuration data
- Rewrite customer programmable parameters
- Flash new software
- Print configuration and test results

Two ET tests are distinct to HEUI systems so they are described here.

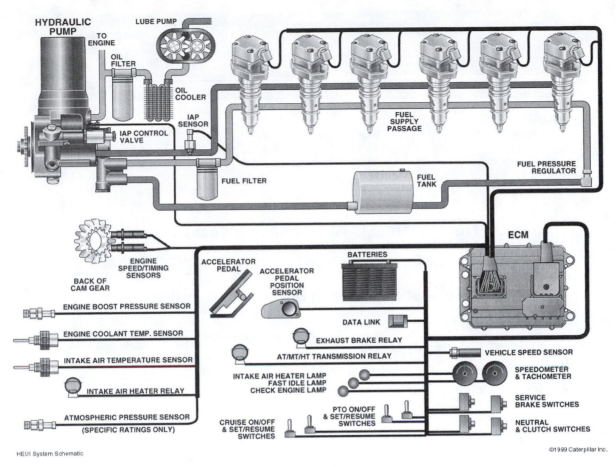

HYDRAULIC PUMP
TO ENGINE
OIL FILTER
LUBE PUMP
OIL COOLER
IAP SENSOR
IAP CONTROL VALVE
FUEL SUPPLY PASSAGE
FUEL FILTER
FUEL TANK
FUEL PRESSURE REGULATOR
ECM
ENGINE SPEED/TIMING SENSORS
BACK OF CAM GEAR
ACCELERATOR PEDAL
ACCELERATOR PEDAL POSITION SENSOR
BATTERIES
ENGINE BOOST PRESSURE SENSOR
ENGINE COOLANT TEMP. SENSOR
INTAKE AIR TEMPERATURE SENSOR
INTAKE AIR HEATER RELAY
ATMOSPHERIC PRESSURE SENSOR (SPECIFIC RATINGS ONLY)
DATA LINK
EXHAUST BRAKE RELAY
AT/MT/HT TRANSMISSION RELAY
INTAKE AIR HEATER LAMP
FAST IDLE LAMP
CHECK ENGINE LAMP
CRUISE ON/OFF & SET/RESUME SWITCHES
PTO ON/OFF & SET/RESUME SWITCHES
VEHICLE SPEED SENSOR
SPEEDOMETER & TACHOMETER
SERVICE BRAKE SWITCHES
NEUTRAL & CLUTCH SWITCHES

HEUI System Schematic   ©1999 Caterpillar Inc.

FIGURE 42–19   *Caterpillar HEUI schematic. (Courtesy of Caterpillar)*

Injector Cutout Test

ET will perform an injector cutout test on an HEUI managed engine. Unlike the cylinder cutout test on Caterpillar EUI systems in which the ability to cut between one and five cylinders during the test produces a comprehensive cylinder balance analysis, at the moment of writing, HEUI cylinder cutout is one cylinder at a time. The test sequence must be performed with the engine at operating temperature and with intermittent parasitic loads such as A/C disconnected. Test sequence:

1. Governor maintains programmed idle speed.
2. ET turns off one injector at a time: this means the functioning five HEUIs must increase their duty cycle if the specified engine rpm is to be maintained.
3. The ADEM ECM measures the average duty cycle of the five functioning HEUIs at each stage of the cutout test.
4. The ECM assigns a test value for each HEUI tested.
5. The cutout test cycle is then repeated.

Figure 42–20 shows an ET screen capture for an injector cutout test. The data in the cycle fields is fuel position in millimeters.

TECH TIP: Whenever making a warranty claim for a defective HEUI, a printout of the injector cutout test profile is required or the warranty claim will be rejected.

Injection Actuation Pressure Test

The injection actuation pressure (IAP) test checks high-pressure oil pump and IAP valve operation. The test is always performed at low idle. It functions by having ET read *desired* IAP pressure versus actual IAP pressure using ADEM ECM data. The IAP test sequence uses four desired pressure values:

1.   870 psi
2. 1450 psi
3. 2100 psi
4. 3300 psi

and compares these with actual IAP values read by the IAP pressure sensor.

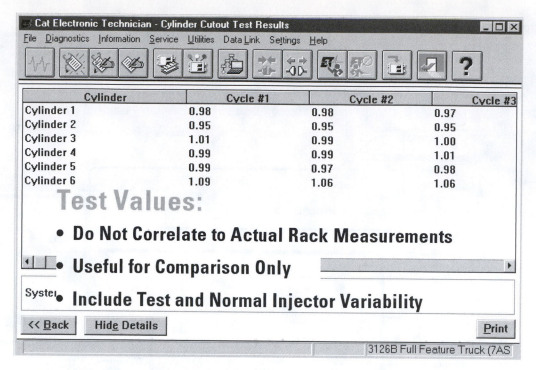

FIGURE 42–20 *ET graphical user interface (Windows) during a cylinder cutout test (Courtesy of Caterpillar)*

INTERNATIONAL HEUI DIAGNOSTICS

The International Trucks (Navistar) electronic engine, PC software known as **EZ-Tech** is used to read, diagnose, and program customer data to HEUI engines. Proprietary data programming can be effected by modem to the International Trucks data hub known as TechCentral and VEPS (vehicle electronic programming software). Until recently, International Trucks relied on ProLink 9000 to a greater extent than Caterpillar as its primary read, diagnostic, and programming instrument. However, the switch to PC-based software is consistent with the overall trends in the trucking industry. In general, EZ-Tech is functionally similar in its capability to Caterpillar's ET so it is not described again here.

International Trucks' Self-Test

International Trucks' HEUI electronics are capable of performing self-test procedures. When the dash-mounted **self-test input (STI)** button is depressed and the ignition circuit is closed (Figure 42–21), the ECM begins the self-test cycle. When complete, the oil/water and warn engine dash lights are used to signal fault codes. All International Trucks fault codes are three digits. Following is the sequence:

1. Oil/water light flashes once, indicating the beginning of the active faults display.

2. Warn engine light flashes each digit of the active fault code pausing between each. The oil/water light will flash once between each active code readout when multiple codes are logged. Code 111 indicates no active fault codes.

3. Oil/water light flashes twice to indicate that inactive fault codes will be displayed. Inactive fault codes are then blinked out in the same manner as active fault codes.

4. When all codes have been displayed the oil/water light will flash three times.

5. Test sequence can be repeated by retracing the preceding steps but all active faults should be repaired before progressing to later tests.

International Trucks' HEUI, like all electronic management systems, requires that troubleshooting strategies be sequentially and scrupulously followed using the correct instrumentation and tooling as follows:

MPSI ProLink 9000

International Trucks' Cartridge for ProLink (ZTSE 43667)

Fluke 88 DMM (Figure 42–21)

Hickok Breakout Box (ZTSE 4346) (Figure 42–21)

Breakout "T" (ZTSE 4347)

PC and proprietary software

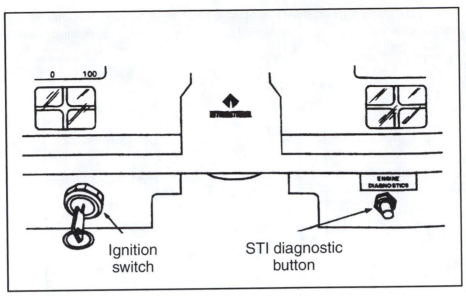

Self-Test Button Location

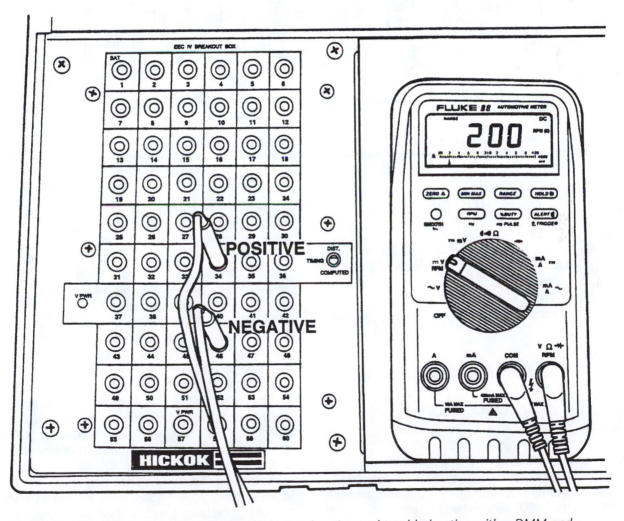

FIGURE 42–21 *International Trucks self-test button location and troubleshooting with a DMM and breakout box. (Courtesy of Navistar International Corp., Designer and manufacturer of International Brand diesel engines)*

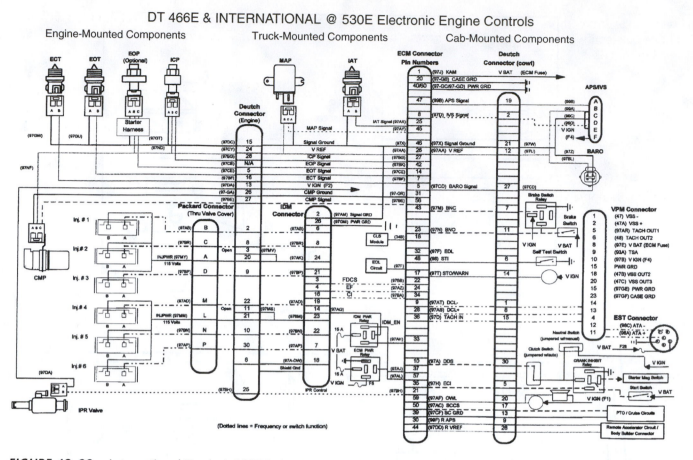

FIGURE 42–22   *International Trucks I-6 HEUI circuit schematics. (Courtesy of Navistar International Corp., Designer and manufacturer of International Brand diesel engines)*

International Trucks' HEUI Circuit Schematic

Figure 42–22 is a schematic showing the International HEUI system as it is used on its DT 466E and DT 530E engines. If you compare this schematic with the Caterpillar HEUI schematics, the only real differences are in some of the acronyms used to describe each system.

 **HEUI CONCLUSIONS**

With the EPA attack on diesel engine emissions likely to continue, Caterpillar's HEUI system is predicted to strengthen its already strong market position over the next few years. HEUI has already been successfully used in Caterpillar off-highway, large bore diesels and may yet be used in some large highway diesels.

The HEUI system is versatile and adapts well to small-bore diesel applications as evidenced by its use

in the PowerStroke diesels manufactured by International Trucks for the Ford Motor Company. Power-stroke engines include the original 444E (7.2 liter) and the more recent 6.0 liter engine that is replacing it. The hydromechanical diesel engines that found their way into automobiles and light-duty trucks a decade ago more often than not were badly engineered, low powered, and failure prone. The generation of diesels being introduced this millennium promises to change that image with well-engineered, high-tech powerplants boasting much higher longevity than gasoline-fueled equivalents. It is interesting to note how the original PowerStroke diesel has evolved. The original 444E was rated at 190 BHP, while the smaller displacement 6.0 boasts 325 BHP and peak torque of 560 lb.-ft., power and torque ratings equivalent to that on a class 8 truck a generation ago.

As the use of diesel engines becomes more popular in small-bore applications such as the powering of pickup trucks, truck diesel technicians should note that auto technicians often prefer not to work on diesel engines with the result that this type of work is often referred to a truck shop. These smaller diesels are produced in large volume compared to that of highway commercial truck engines, so the result will be that many truck technicians will have to adapt to working on small-bore diesels.

SUMMARY

- The HEUI engine management system is the first to use electronically controlled, hydraulically actuated injection pump units. Until the introduction of HEUI technology, pumping to diesel injection pressure values was always achieved mechanically.
- Effective injection pumping stroke in HEUI units is not confined by the hard limits of cam geometry as in other full authority electronic management systems.
- Fuel charging pressure to the HEUIs is at values between 30 psi (2 atms) and 60 psi (4 atms).
- Engine-lubricating oil is used as the hydraulic actuating medium for the HEUIs; it is boosted to pressure values of up to 4,000 psi by a swash plate-type, hydraulic pump.
- Actual HEUI-actuating oil pressure values are precisely controlled by the ECM, which switches the IPR to achieve oil pressure values of between 485 psi and 4,000 psi.
- The IPR is a spool valve, positionally controlled by a pulse width modulated ECM signal.
- The actuating oil pressure value determines the descent velocity of the HEUI pumping plunger and thereby manages injection rate and the emitted atomized droplet sizing.
- The HEUI solenoid controls a poppet valve that, when energized, admits high-pressure oil so that it acts on the amplifier piston; when the HEUI solenoid is de-energized, the high-pressure oil is spilled to the rocker housing.
- HEUIs use multi-orifii, hydraulic injector nozzles of the VCO type.
- HEUI NOP values are typically 5,000 psi (340 atms) with peak system values rising to between 18,000 and 28,000 psi.
- The HEUI ECM uses an IDM to switch the HEUI solenoids.
- The consolidated engine controller (CEC) introduced by International Trucks in 1997 uses a single ECM unit to drive all of the processing and switching functions previously performed by three separate modules.
- Caterpillar has always used an integral ECM incorporating processing and switching functions.
- The IDM steps the HEUI actuation voltage to 115 V DC using induction coils.
- Caterpillar ET (electronic technician) is used to diagnose and program HEUI engines.

- International uses EZ-Tech PC software to read, program, and troubleshoot HEUI.

REVIEW QUESTIONS

1. What force creates injection pressure values in an HEUI injector?
   a. Pneumatic
   b. Mechanical
   c. Electrical
   d. Hydraulic
2. What type of pump is used to increase engine lube oil pressure to the values required by HEUI?
   a. Gear
   b. Plunger
   c. Swash plate
   d. Diaphragm
3. Oil pressure values within the HEUI actuating circuit are controlled by:
   a. A pressure relief valve
   b. IPR
   c. An accumulator
   d. IDM
4. Which type of hydraulic injector nozzle is used in an HEUI?
   a. Pintle type
   b. Poppet type
   c. VCO
   d. Pinteaux
5. Which of the following components is used by HEUI electronics to sense the actuating oil pressure?
   a. Transducer
   b. IPR spool valve
   c. Relief valve
   d. IAP sensor
6. PROM data in Caterpillar International Trucks HEUI systems is retained in:
   a. Personality module
   b. IDM
   c. Identification module
   d. ICR
7. The HEUI amplifier piston increases actuation pressures by approximately how many times in the injection pump chamber of the unit?
   a. 2
   b. 3
   c. 7
   d. 10
8. HEUIs are switched at:
   a. 12 V DC
   b. 115 V AC
   c. 90 V AC
   d. 115+ V DC

9. International Trucks fault codes may be read from the vehicle dash when the STI button is depressed using which lights?
   a. Check engine and Stop engine
   b. Warn engine and Oil/water
   c. Electronic malfunction and Shutdown engine

10. *Technician A* states that the actual oil pressure acting on the HEUI amplifier piston will influence the sizing of droplets emitted from the injector nozzle. *Technician B* states that the HEUI effective stroke is not limited by any hard parameters such as those that limit effective stroke in EUIs. Who is right?
    a. A only
    b. B only
    c. Both A and B
    d. Neither A nor B

11. *Technician A* states that Caterpillar HEUI have always used a three-module controller system on its C-7 engine. *Technician B* states that current International Trucks HEUI uses a single engine controller module. Who is right?
    a. A only
    b. B only
    c. Both A and B
    d. Neither A nor B

12. *Technician A* states that HEUI foreground computations are those based on potential danger to the engine/vehicle in the event of a failure. *Technician B* states that HEUI foreground computations are those requiring almost immediate output reactions. Who is right?
    a. A only
    b. B only
    c. Both A and B
    d. Neither A nor B

13. *Technician A* states that ambient temperature input would be a good example of an ECM background computation. *Technician B* states that the input of the coolant temperature thermistor is another example of an ECM background computation. Who is right?
    a. A only
    b. B only
    c. Both A and B
    d. Neither A nor B

14. *Technician A* states that the Caterpillar *intensifier* piston and the International Trucks' *amplifier* piston both refer to the same HEUI component. *Technician B* states that an amplifier piston will produce fuel pressures seven times the value of the actuating oil pressure. Who is right?
    a. A only
    b. B only
    c. Both A and B
    d. Neither A nor B

15. *Technician A* states that a swash plate-type pump and ECM-controlled pressure regulator are used to supply the HEUIs with hydraulic medium. *Technician B* states that the hydraulic medium used to actuate the HEUI is high-pressure diesel fuel. Who is right?
    a. A only
    b. B only
    c. Both A and B
    d. Neither A nor B

Chapter

43

Cummins HPI-TP

PREREQUISITES

Chapters 18, 24, 29, and 32

OBJECTIVES

After studying this chapter, you should be able to:

- Identify the engine family using the HPI-TP fuel system.
- Describe the HPI-TP fuel subsystem.
- Trace fuel flow routing from tank to injector in an HPI-TP-fueled engine.
- Recognize the components of an IFSM module.
- Describe the operation of timing and metering actuators.
- Understand the operating principles of the HPI-TP system.
- Outline some of the features of an ISX engine.

KEY TERMS

High Pressure Injection-Time Pressure (HPI-TP)

INSITE

Integrated Fuel System Module (IFSM)

Interact System (IS)

Intebrake

ISX (Interact System, X-series)

metering actuator

Signature Series

timing actuator

trapped volume spill (TVS)

water in fuel (WIF) sensor

INTRODUCTION

The Cummins **High Pressure Injection-Time Pressure (HPI-TP)** fuel system is a common rail, open nozzle system with many similarities to the now obsolete (in truck applications) Pressure-Time or PT system. If you have previously worked with and understand the PT system, it might be a good idea for you to make the connection between the PT and TP system. However, in describing the TP system, this is not done, because the reality is that truck technicians today will seldom encounter the PT system.

The letters TP refer to two of the three factors of a hydraulic equation whose product is a volume of fuel flow. If you recall from earlier chapters, several times we have made the statement that the output of a diesel engine is determined by the volume of fuel metered into its cylinders. It follows that fuel quantity emitted per fueling pulse is critical. A hydraulic equation whose result is a volume of flow has 3 factors:

1. Flow area
2. Pressure
3. Time of flow

Change any of the values and the product, that is the volume of flow, will also change. In the Cummins TP hydraulic equation, both the critical flow areas and the pressure are designed to remain constant, so the variable used to control the "product," fuel metered, is time. Because time is the control variable it is the first letter in the acronym TP: the control variable in the older PT system was pressure, so P was the first letter in the acronym PT.

The time factor in the TP system is precisely controlled by the ECM. It accomplishes this by controlling a pair of actuators for each bank of the engine, front and rear, four in all. The actuators are PWM-controlled solenoid gate valves that either open or close. These gate valves are located on the fuel common rail that is charged at a more or less constant pressure. When energized, the actuator opens and permits fuel to be routed to a mechanical, open nozzle injector by means of riflings in the cylinder head. The timing actuator uses common rail fuel as hydraulic media to establish a timing value for the fuel pulse (this fuel is not injected), while the metering actuator feeds common rail fuel to the metering chamber in the TP injector.

TP injectors are open nozzle, mechanical injectors. While the fuel that is supplied to its timing and metering chambers is precisely controlled by the ECM, there are no electronic components integral with the TP injector. These injectors are actuated mechanically by an overhead camshaft.

The TP system was launched on the **Signature Series** engine introduced in 1998. This in-line six cylinder, 15-liter engine was initially rated at 600 horsepower with the capability of a little over 2,000 lb.-ft. of torque output. This became the platform for the current **ISX** engine series with BHP ratings that extend from 450 to 600 horsepower. The engine uses a pair of overhead camshafts mounted on a single slab cylinder head, gear driven by the engine timing geartrain. One camshaft is responsible for actuating the TP injectors and imparting drive to the fuel pump assembly, while the other is dedicated to the valve trains and engine brake. The internal engine compression brake known as **Intebrake** is capable of six stages of progressive braking by retarding on one through six cylinders. The engine uses a variable geometry turbocharger and for 2004 EPA emissions (effective October, 2002) will use cooled exhaust gas recirculation (EGR). In common with most other high-output, large bore, truck diesel engines, articulating pistons with steel crowns are used along with wet liners. Cummins introduced a fleet version of this engine, the ISX 385 ST (smart torque) in 2004. This engine has a fuel map algorithm trimmed for fuel economy.

ISX engine management electronics use Cummins common platform ECM architecture known as **Interact System** or **IS**. Engine performance and engine subsystems are precisely monitored and managed and the engine electronics connect to the J1939 data bus. Engine protection, extensively programmable customer and proprietary data programming options, and guided troubleshooting (in conjunction with **INSITE** software) are all features of the ISX Series. The engine uses a double overhead camshaft design: the camshafts are parallel mounted on a single slab cylinder head. The left side camshaft is responsible for actuating the HPI-TP injectors. The right side camshaft is responsible for actuating the intake and exhaust valves and the engine brake. The ISX engine has a built-in engine brake known as Intebrake. The ECM-actuated Intebrake uses three solenoids to permit progressive braking using from one to all six engine cylinders.

FUEL SYSTEM COMPONENTS

The Cummins HPI-TP common rail system is designed to manage the hydraulic fueling equation by holding the flow area constant, the pressure value constant, and using the time factor as the control variable. Figure 43–1 shows a black and white schematic of the Cummins HPI-TP system: this schematic is also shown in the color section as C8.

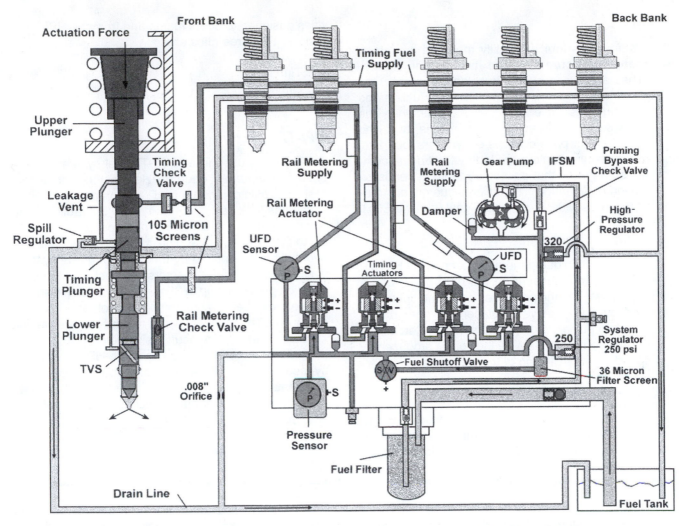

FIGURE 43–1   Heavy-duty HPI-TP fuel system schematic. (Courtesy of Cummins Engine Corporation)

The following components make up the HPI-TP system:

INTEGRATED FUEL SYSTEM MODULE

The **integrated fuel system module** or **IFSM** acts as the hydraulic command center of the fuel system and consists of the following subcomponents:

Check Valve

Fuel enters the IFSM through a check valve located behind the inlet. This prevents fuel from flowing out of the IFSM.

Fuel Filter/Water Separator

Fuel from the check valve is directed to a combination fuel filter and water separator with an integral **water-in-fuel (WIF)** sensor and drain. The filter medium uses a Stratapore™ 10-micron filtration medium. The combination fuel filter/water separator strips water out of the fuel using a centrifuge and drains the water to a sump. The WIF sensor is located in the sump, and it signals the dash maintenance lamp when it is covered with water.

Gear Pump

As with many other Cummins fuel systems, the entire fuel subsystem is under negative pressure whenever the engine is running. The gear pump acts as the prime mover. It uses an external gear pumping principle to move fuel downstream. The resulting low pressure upstream from the pump allows atmospheric pressure in the tank to force more fuel to the gear pump. The gear pump charges the IFSM. The gear pump is driven by a gear on the injector camshaft at 1.12 times crankshaft speed.

Actuators

The system uses four externally mounted actuators, two **metering actuators** and two **timing actuators**. They are electrohydraulic solenoid valves. One timing and one metering actuator manage fueling and timing parameters for each bank of the engine. The front bank consists of engine cylinders one, two, and three and the rear bank consists of cylinders four, five, and six. The four actuators are fed by the engine common rail so any fuel that passes by them depends on the duration of actuation. The timing actuators are controlled by a pulse width modulated signal from the ECM.

Pulsation Dampeners

Three fuel pulsation dampeners are used in the HPI-TP system. The first is located on the gear pump and it functions to dampen the pressure nodes and antinodes produced by any positive displacement gear pump. The other two pulsation dampeners are located on the common rail and act to stabilize the spikes produced by high-speed on/off switching of the actuators.

Fuel Shutoff Valve

The fuel shutoff valve is an energized to run (ETR), normally closed solenoid. It has a rapid restart feature and is opened any time the ignition circuit is turned on.

Fuel Pressure Sensor

The fuel pressure sensor is installed in the bottom of the IFSM close to the fuel filter pad. It signals fuel pressure values to the ECM. It is supplied with V-Ref and returns a signal to the ECM proportional with fuel pressure.

Unintended Fueling Sensors

Two unintended fueling sensors are used in the IFSM, one located on the front bank rail, the other on the rear. These are designed to detect a problem with the actuators, specifically that they are not sealing the rail. They are designed to signal the ECM in the event of actuator leakage.

IFSM Ports

There are two ports located on the IFSM used to test the fuel system. The front port is used to check the common rail pressure. The rear port is designed to test fuel subsystem inlet restriction. Both are equipped with quick-release nipples.

Fuel Manifold

A single slab cylinder head is used on ISX engines. The cylinder head is manufactured with three internal rifles designed to direct fuel to and from the HPI-TP injectors. The three rifles are:

1. Timing
2. Drain
3. Metering

HPI-TP Injectors

The injectors used in an HPI-TP fuel system are mechanically actuated, open nozzle injectors with no integral electronic components. The open nozzle principle eliminates leak paths found in many hydraulic injector nozzles and can be used to bring about an abrupt ending to the injection pulse. The collapse phase created in hydraulic nozzles used by most other current systems is eliminated. HPI-TP injectors are designed to deliver injection pressures up to 35,000 psi resulting in smaller droplet size, better combustion control, and improved emissions. The plungers within the injector are titanium nitride-coated for high scuff resistance.

Fuel is directed into and out of the HPI-TP injectors by means of three exterior annuli separated by four O-ring seals. The upper annulus feeds timing fuel to the injector, the middle annulus discharges drain/spill fuel, and the lower annulus receives the metering fuel. The operation and internal components of an HPI-TP injector are described in detail a little later in this chapter.

IFSM Covers

Three nylon covers snap on over the IFSM and other exterior fuel system components by means of clips.

FUEL FLOW

Figure 43–2 shows the fuel flow through the HPI-TP system on an ISX engine. Use this list and the system schematic to follow the fuel flow routing through the engine:

1. Fuel is pumped by the gear pump from the fuel tank, through the fuel filter/water separator.
2. The fuel filter is specified as 98% efficient and capable of 10 microns entrapment. Because it is upstream from the gear pump, the fuel filter/water separator is under negative pressure, so inlet restriction values are critical. Exceeding the maximum inlet restriction specification will result in fuel starvation.
3. Fuel is charged to the IFSM by the gear pump. A 320 psi high-pressure regulator is located upstream from the IFSM circuitry. If a fuel pressure

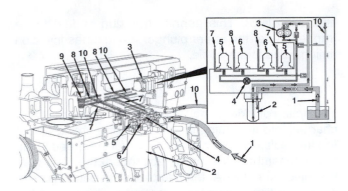

1. Fuel supply from tank
2. Fuel filter
3. Gear pump
4. Fuel shutoff valve
5. Rail metering actuator
6. Timing actuator
7. Rail metering supply to injector
8. timing fuel supply to injector
9. HPI-TP injector
10. Fuel drain to tank

FIGURE 43–2 *Fuel flow schematic. (Courtesy of Cummins Engine Corporation)*

exceeding 320 psi is achieved, the high-pressure regulator valve is tripped to route fuel back to the tank.

4. Fuel is next directed through a 36 micron screen within the IFSM. The screen is used to prevent a cutting from the teeth in the gear pump passing through the rest of the circuit.
5. Fuel is next flowed through a fuel shutoff valve. The shutoff valve is ETR and controlled by the ignition circuit.
6. Fuel is next charged to the common rail where it is made available to the two timing and two metering actuators. The four actuators used are shown in Figure 43–3. Two rail dampeners suppress spiking created by high-pressure, high-speed switching of the actuators.

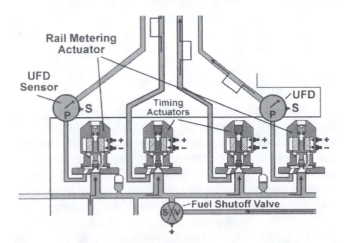

FIGURE 43–3 *Timing and metering actuators. (Courtesy of Cummins Engine Corporation)*

7. A 250 psi fuel system pressure regulator is located in the rail. When tripped, fuel is flowed back to the inlet side of the external gear, fuel pump.
8. The actuators are spring loaded to normally closed position. The ECM controls the actuators by means of PWM signals. This means that the four actuators are normally closed, on/off devices that only allow fuel to pass by them from the rail when energized by ECM signal.
9. Passages from the timing and metering rails connect to rifles in the cylinder head that in turn, connect to the exterior annuli in the injectors. The injectors are inserted into cylindrical bores in the cylinder head and each annulus is sealed by rubber O-rings.

TECH TIP: Make sure you understand how fuel is flowed through the HPI-TP fuel system and it will make the task of priming the fuel system much easier when using INSITE software guided instructions.

HPI-TP INJECTORS

HPI-TP injectors are mechanical with no integral electronic components. Figure 43–4 shows an internal view of the injector, and Figure 43–5 shows an external view. Each injector consists of the following subcomponents:

Upper Plunger/Tappet

The integral tappet and upper plunger assembly is loaded by the injector retraction spring into the injector actuation train to achieve zero lash. When actuation force drives the tappet downward, the upper plunger is driven into the timing chamber. A leakage vent directs plunger bleed back fuel into the return circuit following each downstroke of the upper plunger.

Injector Retraction Spring

The injector retraction spring serves two purposes. It loads the upper plunger tappet assembly into the injector actuation train, and it retracts the upper plunger to allow fuel to enter the timing chamber when the injection train ramps off outer base circle.

Inlet Screen

The inlet screen is designed to entrap any larger particulates that may have been created downstream from the gear pump responsible for moving fuel through the

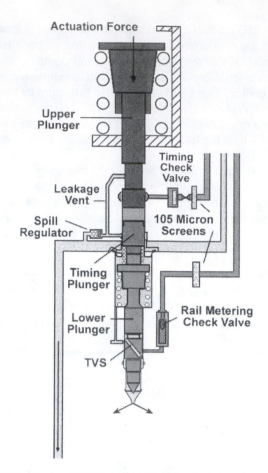

FIGURE 43–4 *HPI-TP internal components. (Courtesy of Cummins Engine Corporation)*

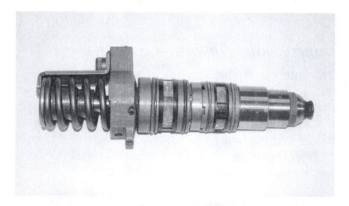

FIGURE 43–5 *External view of HPI-TP injector.*

fuel subsystem. It is a simple screen rated at 105 microns entrapment.

Timing Check Valve

The timing check valve prevents back pressurization of the fuel circuit from the injector timing chamber to the timing actuator. This permits the slug of fuel to be trapped under the upper plunger that defines injection timing value in any cycle.

Timing Plunger

The timing plunger is positioned by the amount of fuel charged into the timing chamber (by the timing actuator) immediately above it. The more fuel charged to the timing chamber, the more advanced the timing of the injection pulse. At the completion of each injection pulse, the timing plunger is mechanically trapped between the upper and lower plungers of the HPI-TP injector.

Timing Plunger Spring

When the injector actuation cam ramps off outer base circle, the main injector spring lifts the upper plunger, and the timing plunger spring does likewise for the timing plunger. This means that the spring permits the timing plunger to be loaded against the slug volume of fuel metered into the timing chamber. Because the hydraulic pressure of the fuel metered into the timing chamber is much higher, timing plunger spring tension plays no role in determining the volume of fuel charged to the timing chamber.

Timing Chamber

The timing chamber is charged by fuel that passes through the timing actuator. The actual volume of fuel charged to the timing chamber depends on:

- Time duration the timing actuator is energized, which is controlled by the ECM
- Rail fuel pressure: designed to be held relatively constant in the HPI-TP system

The amount of fuel in the timing chamber positions the timing plunger. This slug of fuel acts as a hydraulic lock when the HPI-TP injector is mechanically actuated by the injector train.

Spill Regulator

The spill regulator is responsible for directing timing chamber fuel back into the return circuit at the completion of the injection pulse. As the injector train drives the upper plunger downward into the timing chamber following the injection pulse (that is, after the metering plunger has bottomed into the cup), the fuel trapped in the timing chamber has no place to go other than through the spill regulator. Pressure rise causes the spill regulator to trip (open) and it spills fuel until the upper plunger mechanically contacts the timing plunger.

Lower Plunger/Annular Groove

The lower plunger is machined with an annular groove. This routes leak-by fuel upward to hydraulically center the lower plunger. It also prevents side-loading of the lower plunger and lubricates it. The lower plunger is mechanically actuated by the timing plunger located directly above it. Remember, this is positioned by the slug of fuel charged to the timing chamber. Fuel is directed by the metering actuator into the cup of the HPI-TP injector. Once again, the amount of fuel metered to the cup depends on:

- Time duration the metering actuator is energized, which is controlled by the ECM
- Rail fuel pressure: designed to be held relatively constant in the HPI-TP system

Lower Plunger Spring

The lower plunger spring serves to load the lower plunger against the timing plunger. When the injector train is ramped off outer base circle, the lower plunger spring lifts the lower plunger to make it possible for metering to take place.

Metering Chamber

The metering chamber is located under the lower plunger in the cup. HPI-TP injectors use an open nozzle principle, so fuel directed into the cup is subject to whatever is happening in the engine cylinder. At high engine speeds, this is unlikely to impact on the metering process because there is little time for cylinder pressure bleed through, but at lower engine speeds there is more time for cylinder gas leakage into the cup, which can influence metering.

Rail Metering Check Valve

The metering check valve isolates the cup whenever cup pressures exceed the actual metering pressure. Because metering fuel pressures are designed to be a little less than 300 psi and cylinder pressures tend to be more than twice that value, it is possible at lower engine speeds to raise cup pressure above metering pressure. In this event, the metering check valve locks off to isolate the cup. Also, when the lower plunger is actuated to deliver a fuel pulse, fuel in the cup is effectively trapped in the metering cup when the metering check valve locks off, preventing back pressurization of the metering circuit. For those students familiar with the older Cummins PT system, the metering check valve fulfills the same function as the lock-off ball in the PT injector.

Trapped Volume Spill Port

The **trapped volume spill** port or **TVS** is designed to provide an escape route for fuel just as the lower plunger bottoms into the cup at the completion of an injection pulse. This helps provide a more abrupt closure to the injection pulse. When the TVS passage aligns with the TVS port, fuel is optioned to the return circuit rather than to the cup orifii.

Cup Orifii

The cup orifii define the flow area through which all metered fuel has to pass. Their actual sizes combined with the pressure developed in the metering chamber determines the exact sizing of the metered fuel droplets.

INJECTION CYCLE

The injection cycle begins with all three plungers loaded into the cup by mechanical crush created by the injector camshaft which is on outer base circle (OBC). When the injector cam ramps off OBC toward inner base circle (IBC), springs lift all three plungers as the injection train is unloaded. As the lower plunger moves upward, it exposes the metering port. However no fuel enters the metering chamber at this point. The lower plunger rises until it contacts the lower stop. The upper and timing plungers continue to move upward by spring pressure until the timing plunger exposes the timing feed port. The upper plunger continues to move upward until the tappet contacts the top stop. With the plungers in this position, both timing and metering ports are exposed, but no fuel feeds into the timing and metering chambers until the actuators are switched by the ECM.

The timing and metering actuators are opened at the appropriate time and duration to charge the injector timing and metering chambers. By varying the open time, the ECM controls the amount of fuel charged to the timing and metering chambers. The longer the metering actuator is open, the more fuel charged to the cup, also known as the *metering chamber*.

Fuel is charged to the timing chamber between the upper plunger and timing plunger, forcing the latter downward. Next, when the injector cam ramps off base circle, the upper plunger is driven downward. The slug of fuel in the timing chamber now acts as a hydraulic lock because the timing check valve prevents fuel from flowing back to the rail; as a result, the upper plunger and timing plunger move downward in unison. The more fuel charged to the timing chamber, the lower

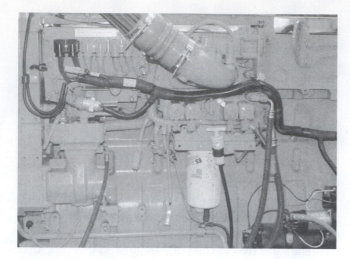

FIGURE 43–6   *Left side view of an ISX engine.*

the position of the timing plunger at the beginning of travel, resulting in more advanced timing. The less fuel metered to the timing chamber, the higher the position of the timing plunger at the beginning of travel, resulting in more retarded timing.

When the timing plunger mechanically contacts the lower plunger, all three now move downward in unison. The metering check valve prevents fuel backflow from the metering chamber to the rail. When the lower plunger contacts fuel in the cup, pressure rise is created in the metering chamber, forcing fuel through the cup orifii. Injection continues until the annular groove in the lower plunger aligns with the TVS port. This permits the small amount of fuel remaining in the cup to flow out of the injector into the drain rifle, providing an almost immediate cessation of injection. At the same time, the timing spill port is open to the drain allowing fuel in the timing chamber to spill. As the lower plunger seats, a ring valve (exterior drain annulus) flexes to regulate the timing spill drain, preventing pressure from dropping too rapidly in the timing chamber; this holds the lower plunger seated until the injector cam is fully ramped to OBC.

The cycle is completed when all three plungers are loaded into the cup with mechanical crush created by the injector train. The fuel injector drain flows fuel back into the IFSM and from there is returned to the tank. Figure 43–6 shows a left side view of an ISX engine featuring the IFSM and electrical harnesses.

 *OVERHEAD ADJUSTMENTS*

For truck technicians familiar with setting valves and injectors, the procedure on ISX is straightforward. When performing overhead adjustments, use the OEM litera-

ture and always reference specifications and check for technical service bulletins. Bar engine clockwise as seen from front during the procedure.

Engine timing indices are marked on the vibration damper as A, B, and C. Use the following table as a reference for location. Note that in a set position on the timing index, both the injector and valves are adjusted over the SAME cylinder. The firing order is 1-5-3-6-2-4.

PULLEY POSITION	SET INJECTOR	SET VALVES
A	1	1
B	5	5
C	3	3
A	6	6
B	2	2
C	4	4

SETTING VALVES AND INJECTORS

Bar engine to align A index with the pointer on the front timing gear cover. Check the valve rockers over #1 engine cylinder: both should move easily indicating lash. That is, the valves on #1 should be fully closed.

Adjust HPI-TP Injectors

Back off the #1 injector adjusting screw locknut and back out the adjusting screw a couple of turns. Next, turn the adjusting screw clockwise and torque to 70 in.-lbs. Continue to hold the adjusting screw in position and torque the locknut. Next, bar the engine over to the B setting and adjust the injector over #5 cylinder using the same procedure. Continue in engine firing sequence until all the injectors have been adjusted.

Adjust ISX Valves

The valves and injectors may be adjusted simultaneously on ISX engines. Position the engine with timing index A and aligned with the pointer on the timing gear cover. In this position, both the intake and exhaust valves should show lash. To verify intake valve lash, insert a 0.014″ thickness gauge between the rocker pallet and the valve crosshead. It should indicate minimal drag. If not, back off the rocker adjusting screw locknut, and reset the valve lash. Use the same procedure to set exhaust valve lash, which is specified at 0.027″. Progress through the firing order in sequence checking the valve lash: adjust only where necessary.

CAUTION: When setting injectors and valves on an ISX engine, both are checked/adjusted over the same cylinder over one timing location. Always consult Cummins specifications and check for technical service bulletins (TSBs) before attempting the procedure.

ISX ELECTRONICS AND ENGINE FEATURES

The electronic input circuit used on the ISX engine is consistent with those on a late generation, large-bore truck diesel engine. Sensors requiring reference voltage are fed a 5-volt DC V-Ref signal, precisely modulated by the ECM. The ECM mounting is integrated into the cylinder head. Rather than use fuel as cooling media, the ECM is vented to atmosphere and air-cooled. A unique method of sealing ambient moisture from the circuit board and driver hardware is incorporated. A Gore-Tex™ seal is used in a vent port, so the ECM housing is aspirated to atmosphere (as the unit heats, warm air is discharged; as it cools, moisture-free air is pulled in), eliminating a vacuum seal that would fail in time.

The ISX engine package is trim and uncluttered. It is also notably light, at up to 600 BHP and capable of outputting up to 2,000 lb.-ft. of torque, it weighs less than many of its competitors that put out 30% less power. Engine electronics are simple to diagnose using InSite computer guided, sequential electronic troubleshooting. The acronym IS, or Interact System, means that the electronic architecture has plenty in common with other Cummins electronic management systems. The system is designed to connect to the J1939 data bus. Engine servicing tends to be relatively straightforward with the possible exception of timing the double overhead camshafts: the Cummins procedure for timing the camshafts must be rigidly followed. This is accomplished working through the engine timing geartrain, working from the crankshaft drive gear up to the camshaft gears, which are physically wedged to position for the procedure. All ISX engines use articulating piston assemblies (Figure 43–7) and engines manu-

FIGURE 43–7 *ISX articulating piston assembly.*

factured after October 2002 will use cooled EGR as a means of cutting back on NO_x emissions.

CAUTION: Do not attempt to time ISX overhead camshafts without the Cummins special tool timing wedges. The consequence of out-of-time camshafts can be a lost engine!

TROUBLESHOOTING ISX

Troubleshooting problems in all electronically managed engines should always be performed by technicians with OEM training using guided software and service literature, specifically Insite when working on ISX engines. The following data is intended to give technicians some broad guidelines for diagnosing performance complaints.

Gear pump pressure:	Cranking: 50 psi min.
	Idle: 245 to 275 psi min.
	High idle: 245 to 320 psi min.
Maximum fuel inlet temp.:	160°F
Maximum air intake restriction:	10 in. H_2O at rated/new filter.
	20 in. H_2O at rated/used filter.
Maximum exhaust restriction:	3 in. Hg at rated

INTEBRAKE

ISX engines use six-stage engine braking achieved by using three engine brake solenoids that function as follows:

Solenoid 1: Actuates braking on #1 cylinder only.

Solenoid 2: Actuates braking on cylinders #2 and #3.

Solenoid 3: Actuates braking on cylinders #4, #5, and #6.

In this way engine braking is possible on one, all six, or any number of cylinders in between depending on which solenoids are energized by the ECM. For instance, to achieve maximum engine braking, all three solenoids would be energized. Engine braking is not enabled when cruise control is active, when an engine fault code is active, or when engine rpm drops below 850. The engine brake should immediately deactivate when the accelerator pedal is depressed.

SUMMARY

- The ISX engine family uses the HPI-TP fuel system. These engines are currently available in the 385 to 600 BHP power range with peak torque values slightly exceeding 2,000 lb.-ft. of torque.
- The HPI-TP fuel subsystem is consistent with most other Cummins fuel subsystems in that it is held under suction with a gear pump acting as prime mover.
- Fuel is pulled through the HPI-TP fueled engine by a gear pump, charged to a common rail and from there, metered by timing and metering actuators to mechanically actuated injectors.
- The IFSM module houses the electronics required to manage the fuel system and routes fuel to the common rail used to make fuel available to the timing and metering actuators.
- The timing and metering actuators are electrohydraulic solenoids controlled by PWM actuation by the ECM. One timing actuator and one metering actuator are used to manage three engine cylinders: a six-cylinder ISX engine uses a total of four actuators, one timing and one metering actuator to manage the front engine cylinders, and another pair to manage the rear cylinders. When energized, each actuator opens to permit rail pressure fuel to pass into the circuit it controls.
- The HPI-TP system uses computer-controlled actuators to feed fuel to the timing and metering chambers of an entirely mechanically actuated, open nozzle type injector.
- Two overhead camshafts are used in an ISX engine with HPI-TP fuel system. The left side camshaft is responsible for actuating the HPI-TP injectors while the right side camshaft is responsible for actuating the valves and the six-stage engine brake.
- The ISX engine series that uses HPI-TP injectors is a 15-liter displacement engine that can produce up to 600 BHP and more than 2,000 lb.-ft. of torque in its current trim. The engine weighs less than many of its competitors that output 30% less power. It uses a single slab cylinder head equipped with double overhead camshafts, dedicated to actuating the HPI-TP injectors and the actuation and operation of a six-stage engine brake and the cylinder head valves. Engines manufactured after October 2002 use cooled EGR to reduce NO$_x$ emissions.

REVIEW QUESTIONS

1. How is a Cummins HPI-TP injector actuated?
   a. Hydraulically
   b. Electrically
   c. Electronically
   d. Mechanically
2. Which of the following describes how an ISX ECM prevents moisture from entering the housing?
   a. Vacuum sealed
   b. WeatherPac sealed
   c. Vented with a Gore-Tex seal
   d. WeatherProof II technology
3. Which of the following would approximate rail pressure in an HPI-TP fuel system?
   a. 50 psi
   b. 150 psi
   c. 285 psi
   d. 5000 psi
4. How many timing actuators are used on the HPI-TP system used on an ISX engine?
   a. One
   b. Two
   c. Four
   d. Six
5. How many metering actuators are used on the HPI-TP system used on an ISX engine?
   a. One
   b. Two
   c. Four
   d. Six
6. What happens to fuel charged to the HPI-TP injector timing chamber following its pumping stroke?
   a. It is spilled to the rocker housing.
   b. It is directed to the drain circuit.
   c. It is flowed into the injector cup.
   d. It is routed back to the transfer pump.
7. Which of the following is true of the fuel subsystem used to supply an HPI-TP system?
   a. It is divided into suction and charge circuits.
   b. It is divided into primary and secondary circuits.
   c. It is held at charging pressure.
   d. It is entirely under suction.
8. Which of the following is true of the moving components of an HPI-TP injector after the actuating cam profile has been ramped to OBC?
   a. They are retracted by the injector retraction spring.
   b. They are loaded into the cup with mechanical crush.
   c. Timing and pumping plungers are separated by fuel in the timing chamber.
   d. The injector push tube is under compression.
9. *Technician A* says that the timing actuator on an HPI-TP fuel system determines fuel quantity by the

amount of time it is energized and held open. *Technician B* says that each metering actuator used on the same engine is responsible for fueling of three engine cylinders. Who is right?

   a. A only
   b. B only
   c. Both A and B
   d. Neither A nor B

10. Fuel that passes through the metering actuator is routed to what component on the HPI-TP injector?
   a. Hydraulic nozzle assembly
   b. Timing chamber
   c. Cup
   d. Accumulator

11. Which of the following best describes an HPI-TP injector?
   a. Open nozzle, mechanically actuated
   b. Closed nozzle, electronically actuated
   c. Mechanically actuated, hydraulic nozzle
   d. PWM controlled and actuated

12. What type of fuel transfer pump is used on the HPI-TP system on ISX engines?
   a. External gear
   b. Double acting plunger
   c. Vane
   d. Single acting plunger

13. What software is used by Cummins to troubleshoot the ISX engine?
   a. Electronic Technician
   b. InSite
   c. DeNet
   d. ServiceLink

14. What is the function of the pulsation damper integral with the transfer pump assembly on an ISX engine?
   a. Dampens pressure waves produced by the transfer pump.
   b. Insulates the ECM from vibration.
   c. Dampens drive torque oscillation produced by the gear pump.
   d. Insulates the transfer pump from shock loading.

15. What defines the window, measured in crank angle degrees, within which the ISX ECM can select an effective stroke of an HPI-TP injector?

   a. Infinitely variable
   b. Cam profile
   c. Engine rpm
   d. Injector timing height

16. How is the mechanical stroke of a HPI-TP injector actuated?
   a. Lobes on left side camshaft
   b. Lobes on right side camshaft
   c. Front bank by left side camshaft, rear bank by right side
   d. Front bank by right side camshaft, rear bank by left side

17. What type of pistons does an ISX engine use?
   a. Trunk type
   b. Anodized
   c. Crosshead
   d. Two-piece, articulating

18. Which of the following best explains why ISX engines use cooled EGR?
   a. Improve fuel economy
   b. Improve power
   c. Reduce HC emissions
   d. Reduce NO_x emissions

19. *Technician A* says that the lower plunger seals the cup orifii at the completion of each injection stroke in an HPI-TP injector. *Technician B* says that at the completion of the injection cycle, the lower plunger is loaded into the cup with mechanical crush created by the injection train. Who is right?
   a. A only
   b. B only
   c. Both A and B
   d. Neither A nor B

20. *Technician A* says that the ISX Intebrake is actuated by the injector cam profile on the right side camshaft. *Technician B* says that intake and exhaust valves are both actuated by the right side camshaft. Who is right?
   a. A only
   b. B only
   c. Both A and B
   d. Neither A nor B

Chapter

44

Cummins Accumulator Pump System

PREREQUISITES

Chapters 18, 20, 29, and 32

OBJECTIVES

After studying this chapter, you should be able to:

- Identify the engines that use the CAPS fuel system.
- Describe the CAPS fuel subsystem.
- Trace fuel flow routing from tank to injector in a CAPS-fueled engine.
- Recognize the components of a CAPS fuel pump.
- Understand the operating principles of the CAPS system.
- Perform some simple repairs on the CAPS modular pump.

KEY TERMS

accumulator

Cummins Accumulator Pump System (CAPS)

injection control valve (ICV)

Interact System (IS)

pressure control valve (PCV)

transorb diode

INTRODUCTION

Introduced in 1998, the **Cummins Accumulator Pump System**, or **CAPS**, was initially used on the ISC engine and more recently on the ISL engine. It is a medium-pressure, **accumulator** system suitable for use on direct-injected diesel engines classified as small- to medium-bore in truck applications. Accumulator pumps are sometimes described as being common rail pumps and certainly, in terms of their general operating principles they are not much different. In common with other recently developed diesel engine fuel management systems, CAPS fuel pressure can be controlled entirely independently from engine speed permitting high fuel pressures at low engine rpm. This permits better transient response to a request for change in engine output, or put more simply, much faster response to a sudden acceleration input request.

CAPS shares many of the ECM software and hardware advantages of Cummins larger engines because it is **Interact System (IS)** managed. IS uses common software protocols and hardware architecture for the full family of Cummins engines and, although the output circuit for a CAPS-fueled engine is necessarily distinct, it has many electronic features associated with large-bore truck diesel in a small bore package.

The CAPS pump uses a modular design and therefore is easy to field repair. The Cummins procedure used to repair CAPS pumps is referenced in this text, but technicians should understand that Cummins training is required prior to undertaking repairs. Figure 44–1 shows an external view of a CAPS fuel pump.

FUEL SYSTEM COMPONENTS

We begin with an overview of the CAPS fuel system components. The components are simply introduced in this chapter and because most of them will be familiar to readers, they are not re-examined in great detail. The distinct component in this system is the CAPS accumulator injection pump and its operation is presented in more detail later in the chapter. Fuel flow in the ISC is nearly double that on the earlier C-series engine (600 lb./hr. versus 375 lb./hr.), so much larger supply and drain plumbing is required. You can follow the fuel routing using Figure 44–2.

ELECTRIC LIFT PUMP

A 12 VDC electric lift pump is used to initiate the movement of fuel through the fuel subsystem. The pump operates for 30 seconds when the ignition circuit is first switched on, whether the engine is running or not. The lift pump produces a minimum pressure of 10 psi. If additional priming is required, cycling the ignition key will trigger another 30 second priming sequence.

The lift pump is used only for starting the engine. After start-up, the lift pump is shut off and bypassed, leaving the gear pump to pull fuel from the fuel tank.

GEAR PUMP

After the CAPS fuel subsystem has been primed by the lift pump, a gear-type transfer pump is used to move

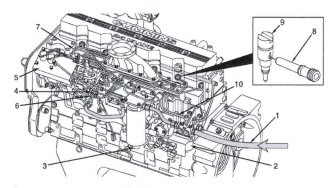

1. Fuel suction line
2. Electric lift pump
3. Fuel filter WIF sensor
4. Regulator valve
5. Accumulator module
6. Distributor snubber valve
7. High-pressure pipe
8. Fuel injector connector and edge filter
9. Hydraulic injector
10. Return circuit

FIGURE 44–2   *CAPS fuel flow routing. (Courtesy of Cummins)*

FIGURE 44–1   *CAPS pump. (Courtesy of Cummins)*

fuel through the system. The priming pump uses an external gear pumping principle.

TECH TIP: CAPS fuel flow requirements are nearly double that of its C-series predecessor. Do not use C-series components on ISC engines and consult ISC fuel inlet restriction specifications when troubleshooting. Maximum inlet restriction to the CAPS pump is 4 inches Hg.

COMBINATION FILTER/WATER SEPARATOR

A combination filter/water separator is used, located upstream from the gear pump. Cummins claims that the filter functions at 97% efficiency using Stratopore™ filtering media effective to 10μ particulate. The assembly is fitted with a water-in-fuel sensor (WIF) that alerts the operator when it is immersed in water.

CAPS PUMP

The heart of the CAPS fuel system is the accumulator injection pump, which is driven by a camshaft drive gear to which it is timed. This pump is presented in some detail later in the chapter. Its function can be briefly described by saying that it uses a pair of camshaft-actuated plungers to charge fuel to an accumulator: the pressure in the accumulator is precisely managed by the ECM, which can therefore control the exact pressure delivered to the injectors during each cycle. Injection pressures are consistent with accumulator pressures. The accumulator delivers fuel to the distributor head, which responds to ECM output switching and routes fuel to the injectors in engine firing sequence.

HIGH-PRESSURE CIRCUIT

Steel high-pressure lines connect the CAPS distributor head with fuel connectors located in the cylinder head. The internal dimensions and elasticity of the high-pressure pipes are critical requirements for maintaining engine fueling balance. The fuel connectors are transversely mounted in the cylinder head, and a flared nipple seats each one to a recessed angled seat in the injector. The fuel connectors are fitted with an integral edge filter that screens anything that might have somehow escaped the extensive upstream filtering processes.

TECH TIP: Torque on the transverse fuel connector nut is critical because this defines the linear force that seats the connector nipple into the injector seat.

Always check the OEM torque specification during installation.

HYDRAULIC ORIFICE-TYPE INJECTORS

Bosch cylindrical style, hydraulic injectors are used in the CAPS fuel system. They use a multi-orifice nozzle, and shims are used to define nozzle-opening pressure (NOP) at 4,500 psi, resulting in a nozzle closing pressure of around 2,700 psi. The injector nozzle is a seven orifice, concentric geometry type. A leak-off circuit routes nozzle valve leakage back to the fuel tank. NOP is bench set in the usual way with shims used to trim the exact value. The injectors are field serviceable, but replacement of nozzle assemblies is recommended over reconditioning. Hydraulic injector servicing is covered in Chapter 20.

It is important to remove the transverse fuel connector in the cylinder head before attempting to remove an injector. A 1.5 mm copper alloy washer seals the nozzle from the engine cylinder and precisely positions nozzle protrusion into the cylinder. The center location of the injector and design of the cylinder head permit removal and reinstallation of the injectors without disturbing the rockers.

CAUTION: When removing a CAPS injector for service or testing, the transverse fuel connector in the cylinder head must first be removed. If you forget, the result will be costly fuel connector and nozzle body damage.

TECH TIP: The axial positioning of these injectors into their cylindrical bores in the cylinder head is critical so that the angled seat in the injector inlet properly aligns with the fuel connector nipple. The hold-down clamp is offset so that the injector is properly positioned. Always reference Cummins service literature when removing and replacing CAPS injectors.

CAPS PUMP

The CAPS accumulator pump uses a modular design that facilitates field diagnosis and repairs. It is the heart of this fuel system and technicians should have some understanding of how it operates so that they can perform accurate first-level diagnoses in conjunction with Cummins INSITE software. Field repairs are not difficult but should only be attempted by Cummins trained technicians.

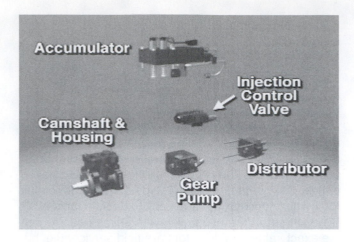

FIGURE 44–3   CAPS modules. (Courtesy of Cummins)

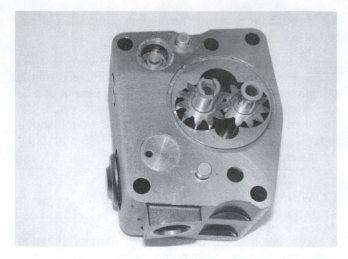

FIGURE 44–4   CAPS gear pump assembly.

The five modules that comprise a CAPS unit are:

1. Gear pump module
2. Camshaft and housing module
3. Accumulator housing module
4. Injection control valve module
5. Distributor module

The operation of each of these modules will now be studied in detail. An exploded view of the modules is shown in Figure 44–3.

GEAR PUMP MODULE

The subcomponents of the CAPS gear pump module are a drive shaft, a pair of external pumping gears, front and rear couplings, and a piston-type pressure regulator. The gear pump module is located behind the CAPS camshaft. The gears are driven by the CAPS camshaft. Fuel flow through the gear pump module begins when fuel enters the inlet: as the gears rotate, fuel passes between the teeth and liner to the outlet. Once the system is initially primed, the gear pump is responsible for movement of fuel through the subsystem. Pump pressures maximize at 1,100 rpm engine speed and stay consistent as engine rpm rises above this value. A piston-type regulator manages fuel pressure at a maximum of 165 psi with excess fuel spill, rerouted back to the inlet side of the gear pump. Fuel is discharged from the gear pump to the **pressure control valves (PCVs)** located in the accumulator module. Figure 44–4 shows the CAPS gear pump module separated from the main pump housing.

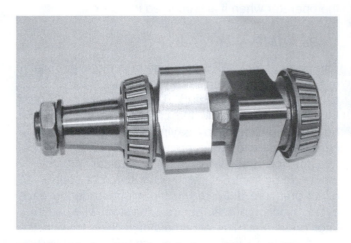

FIGURE 44–5   CAPS Camshaft: Each cam profile actuates one of the two pumping elements.

CAMSHAFT AND HOUSING MODULE

The subcomponents of the CAPS camshaft (Figure 44–5) and housing module (Figure 44–6) are the main

FIGURE 44–6   CAPS main housing module.

FIGURE 44–7 *CAPS: main housing, camshaft, and piston.*

FIGURE 44–8 *Accumulator housing module*

housing, a camshaft, camshaft main bearings, and two tappet assemblies.

The camshaft is supported by taper roller bearings at either end and is driven by the main CAPS pump drive at half engine speed. The camshaft has a pair of three lobe cam profiles. Each cam profile is identically dimensioned. A pair of roller tappets is located above each cam profile in bores so that they ride the cam profiles. As the cam turns through one revolution, each tappet is linearly actuated three times. The rear of the camshaft imparts drive to the gear pump module. By actuating the pair of tappets, the camshaft provides the reciprocating force required by the accumulator module to develop injection pressures. Figure 44–7 shows the disassembled camshaft and housing module components.

ACCUMULATOR HOUSING MODULE

The accumulator housing module contains the high-pressure pumping elements required to develop injection pressures. The subcomponents include a pair of plungers and barrels, barrel retainers, retraction springs, a pressure control valve (PCV), rate shape tube and snubber, pressure and temperature sensors, and an air bleed valve. The accumulator housing module is shown in Figure 44–8.

High-Pressure Pumping

Internal drillings in the CAPS housing direct fuel from the gear pump to the PCVs located above each of the two plunger and barrel pumping elements. The plungers are ceramic with no recess geometry: each is lugged to a tappet and the tappet retraction springs en-

sure they ride the cam profiles. As the CAPS camshaft is rotated, each plunger reciprocates within its stationary barrel. When the PCVs are de-energized, fuel is allowed to flow into the pump chambers formed by the plunger and barrels at the gear pump pressure of 165 psi. When a PCV is energized, it closes, trapping fuel in the pump chamber and subjecting it to pressure rise as the plunger rises through its stroke. As the pump plunger is driven into the slug of fuel trapped in the pump chamber, when the pressure exceeds that in the accumulator, a check valve opens, forcing pressurized fuel into the accumulator. The effective pump stroke continues until the plunger reaches the top of its travel when the accumulator check valve closes.

Managing Accumulator Pressure

The timing of PCV valve opening and closing is managed by the ECM that senses accumulator pressure. Accumulator pressure is therefore precisely controlled by the ECM prior to each injection event. A bleed valve located at the top of the accumulator permits any air in the system to bleed off. When fuel enters the accumulator, it is stored in six cylindrical accumulator passages awaiting injection: accumulator volume is designed to store a sufficient volume of fuel so that a maximum pressure drop-off of 10% follows each injection event.

Fuel leaves the accumulator by means of the rate shape snubber valve. The rate shape snubber limits fuel injected in the initial portion of the fuel injection pulse, resulting in a pressure dip as fuel is unloaded.

INJECTION CONTROL VALVE MODULE

The **injection control valve (ICV)** module is an ECM-controlled electrohydraulic solenoid. The subcomponents are a housing and pressure regulator assembly made up of a stator, stator housing, and outer (armature) and inner pins. When the ICV stator is de-energized, spring pressure forces the outer pin onto its seat, creating a flow path between the distributor port and drain port. Energizing the ICV causes the outer tube to be pulled toward the stator, shutting off the drain port and permitting pressure to build on the distributor port. Fuel now flows through one of the six drillings to the injection snubber valves in the distributor module. De-energizing the stator allows spring pressure to force the outer pin off its seat with the drain port, creating a flow path between the distributor port and drain, collapsing the injection pressure, and permitting the injector nozzle to close rapidly. The regulator allows fuel to spill to the drain until the pressure drops to 330 psi, then closes, retaining a residual line pressure of about that value until the next injection cycle. At the completion of each injection pulse, the outer pin seats, closing off source fuel.

By controlling the energizing and de-energizing of the stator, the ECM manages both injected fuel quantity and timing. Both mechanical and electronic variables and lags are built into the ECM software algorithms (processing rules/logic). A **transorb diode** in the harness protects the ECM from the inductive kick created by the ICV when its magnetic field collapses. If the transorb diode fails either by shorting or opening, an ECM fault code will log.

CAUTION: Transorb diodes produce heat. Do not touch after the engine has been operated before allowing time to cool!

DISTRIBUTOR MODULE

The distributor module functions a little like the distributor in your car except that its medium is high-pressure fuel. Its components are the distributor housing, rate shape inlet, rotor coupling, rotor and six (one for each engine cylinder) distributor snubber valves. The rotor is rotated at camshaft speed, that is, through one complete rotation per full effective cycle of the engine (two complete rotations). The distributor module is shown in Figure 44–9.

There are six injection ports in the distributor housing, each connected to an injection line. The port in the rotor (Figure 44–10) is of sufficient size so that it does not play any role in controlling either the beginning or ending of injection. It merely identifies the window within which the ECM can select an injection pulse for any engine cylinder. The beginning, duration, and ending of injection is managed by the ECM by energizing and de-energizing the ICV. The six injection high-pressure pipes are connected to the CAPS distributor module to provide the 1-5-3-6-2-4 firing sequence of the

FIGURE 44–9   *CAPS distributor: note the location of the distributor rotor.*

FIGURE 44–10   *CAPS distributor rotor.*

engine cylinders. The injection pumping cycle described in this section is repeated for each cylinder.

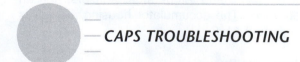

CAPS TROUBLESHOOTING

Troubleshooting problems in all electronically managed engines should always be performed by technicians with OEM training using guided software and service literature, specifically INSITE when working on ISC and ISL engines. The following data is intended to give technicians some guidelines for diagnosing performance complaints.

Minimum lift pump pressure: 10 psi (25 secs/key-on)

Fuel filter restriction: max. 6 in. Hg new filter
max. 10 in. Hg used filter

Gear pump pressure: Idle: 20 psi min.
High idle: 120 psi min.

Max. fuel drain restriction: 10 in. Hg

Max. fuel inlet temperature: 160°F

NOP: 4,500 psi (300 atms)

CAPS accumulator pressure: 4,800 psi min at idle
15,000 psi min at peak torque

Maximum air intake restriction: 25 in. H_2O at rated
Maximum exhaust restriction: 3 in. Hg at rated

SUMMARY

- The Cummins Accumulator Pump System (CAPS) fuel system is a Cummins-patented system used on ISC and ISL engine series.
- The CAPS fuel subsystem consists of an electric lift pump, combination filter/water separator, and external gear pump assembly located within the CAPS pump assembly.
- CAPS is a common rail system using a gear pump to charge a pair of cam-actuated piston pumps that unload to an accumulator. The accumulator supplies the fuel injectors by means of dedicated high-pressure pipes.
- The CAPS fuel pump is a modular assembly consisting of five modules: gear pump, camshaft and housing, accumulator housing, injection control valve, and distributor modules.
- The CAPS fuel system is a full authority, electronically controlled common rail fuel system.
- The CAPS pump should be diagnosed in accordance with the guided procedures in Cummins IN-SITE software.

REVIEW QUESTIONS

1. For how long does the CAPS electric lift pump run following startup?
   a. During cranking only
   b. Until the engine starts
   c. 30 seconds or until the engine starts
   d. 30 seconds whether the engine is running or not
2. What types of injectors are used in a CAPS fuel system?
   a. Hydraulic, multi-orifice
   b. Electronic
   c. Mechanically actuated
   d. Pintle
3. How many lobes are machined onto each cam profile in a CAPS fuel pump camshaft?
   a. One
   b. Two
   c. Three
   d. Four
4. Which of the following would best describe the high-pressure pumping elements used in a CAPS fuel pump?
   a. Port helix
   b. Plunger and barrel
   c. External gear
   d. Vane type
5. What results when a CAPS WIF sensor is immersed in water?
   a. Alarm buzzer sounds
   b. Engine shutdown
   c. Driver alert
   d. Power ramp-down
6. *Technician A* says that a WIF sensor is used to alert the driver that the fuel filter needs to be replaced. *Technician B* says that the WIF sensor triggers an alert when it detects water in the combination filter/separator assembly. Who is right?
   a. A only
   b. B only
   c. Both A and B
   d. Neither A nor B
7. *Technician A* says that a CAPS pump is divided into five modules, any one of which can be replaced if diagnosed defective. *Technician B* says CAPS pumps are simple enough that they can be field repaired without any Cummins technical training? Who is right?
   a. A only
   b. B only
   c. Both A and B
   d. Neither A nor B
8. *Technician A* says that to pull a CAPS injector, the rocker assembly must be backed off. *Technician B* says that transverse fuel connector nut torque is responsible for defining the sealing force of the connector nipple to injector seat. Who is right?
   a. A only
   b. B only
   c. Both A and B
   d. Neither A nor B

9. After a CAPS engine has been run for some time, the engine is shut down and the transorb diode in the harness is too hot to touch. *Technician A* says that the diode is shorted and should be replaced. *Technician B* says that a heated transorb diode indicates that the ICV is about to fail. Who is right?
   a. A only
   b. B only
   c. Both A and B
   d. Neither A nor B

10. What is the function of a transorb diode in the harness of a CAPS fuel system?
   a. Protects the ECM from inductive voltage spikes caused by the ICV
   b. Protects the ICV from inductive spikes caused by ECM output drivers
   c. Isolates the CAPS electronics from the chassis electrical system
   d. Isolates CAPS ECM from the CAPS sensor and actuator circuit

11. What is the maximum inlet restriction permitted measure downstream from the filter and upstream from the CAPS gear pump?
   a. 25 in. H_2O
   b. 10 in. H_2O
   c. 10 in. Hg
   d. 6 in. Hg

12. What is the minimum permissible fuel pressure measured at a CAPS accumulator when the engine is run at idle speed?
   a. 4,500 psi
   b. 4,800 psi
   c. 15,000 psi
   d. 24,000 psi

Chapter 45

Emissions

OBJECTIVES

After studying this chapter, you should be able to:

- Define the origin of the word smog.
- Define photochemical smog and describe the conditions required to create it.
- Describe the role vehicle emissions play in the formation of smog.
- Describe how ozone is formed at ground level.
- Identify the compounds exhausted in engine end gases and identify those that are classified as noxious.
- Identify some EPA and CARB emissions tests required for diesel engine certification.
- Outline some of the means used to control emissions in current diesel engines.
- Explain the effect fuel injection timing can have on diesel engine end gas.
- Outline the principles of operation of an opacity meter.
- Describe the SAE J1667 test procedure.
- Perform a J1667 opacity smoke test.
- Correlate opacity test failures to an engine or engine management malfunction.

KEY TERMS

aneroid

California Air Resources Board (CARB)

carbon dioxide (CO_2)

carbon monoxide (CO)

cooled exhaust gas recirculation (C-EGR)

constant volume sampling (CVS)

end gas

Environmental Protection Agency (EPA)

fueling algorithm

gas analyzer

hydrocarbons (HC)

inert

infrared

internal exhaust gas recirculation (I-EGR)

J1667

nitrogen dioxide (NO_2)

noxious emissions

opacity meter

oxides of nitrogen (NO_x)

ozone

photochemical reaction

photochemical smog

smog

sulfur dioxide (SO_2)

ultraviolet (UV)

volatile organic compounds (VOCs)

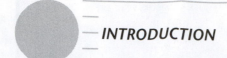

INTRODUCTION

Until the late 1990s, the **Environmental Protection Agency (EPA)** focus on the **noxious emissions** from diesel engines was primarily on the engine manufacturer and testing of new engines prior to certification. The state of California has traditionally led the way for the rest of the jurisdictions in North America when it comes to identifying vehicle emissions standards and enforcing them. The California agency responsible for emissions legislation and enforcement is the **California Air Resources Board**, or **CARB**. Any new engine introduced to the marketplace requires testing to EPA and CARB standards prior to certification. However until recently, once an engine series had received certification, it was seldom tested in service. This has changed dramatically in the past few years. Field testing of diesel engines for noxious emission output has become commonplace with most states and provinces having some kind of emissions testing program in place. Jurisdictions vary on how they test and enforce compliance to emissions standards: vehicles may have to be tested on a schedule (yearly, every two years or five years), each time a vehicle changes ownership, and/or when observed by enforcement agencies to be smoking.

Although precertification exhaust emission standards on diesel engines are increasingly demanding, those standards that field-tested diesel engines currently are required to meet are much less stringent than those required of gasoline-fueled, spark-ignited engines, because most field testing currently used in North America is based on a rather dated SAE standard known as **J1667**. Published in 1991, SAE J1667 was based on a trucking world that used mostly hydromechanical diesel engines. Very few of these engines are around today. J1667 test requirements only address visible smoke emissions, and the test procedures are entirely based on highway diesel engines commonly used in the 1980s. You might ask why this is so? While the EPA plays hardball with the diesel engine manufacturers prior to certification, field enforcement of highway diesel engine emissions is left up to state and municipal jurisdictions. Their enforcement agencies are caught in a squeeze between the voting public who are requesting tougher standards and the hard fact that aggressive enforcement and tough penalties can hurt them economically. By enforcing the J1667 standard, enforcement agencies can respond to environmentalists by saying that they are adhering to an accredited national standard. The truth is that the test does not begin to examine the type of emission system shortcomings of modern, computer-controlled truck engines.

Because J1667 tests visible smoke emission, the primary system being tested was that used to sense

manifold boost and limit maximum fueling until turbo-boost pressure reached a specified pressure value. Turbo-boost devices are known as **aneroids**. An aneroid is a device used to measure low-pressure values. The aneroid devices used in hydromechanical truck engines were known by names such as turbo-boost sensor, manifold boost sensor, puff limiter, or air-fuel control, and most of them could be shorted out by a person with minimal mechanical knowledge. They were shorted out because drivers believed that they were being cheated of low end power by aneroid devices. A shorted-out aneroid produced a tell-tale puff of smoke from the exhaust stack at every upshift, and it was often difficult to explain to a driver that if fuel exited the exhaust pipe in the form of smoke, then it was probably being wasted. The good news is that a J1667 opacity will identify any of these engines of nearly a generation ago that have been tampered with.

Today, truck highway diesel engines are managed by a computer to meet the stringent emissions standards required for EPA certification. Today, truck technicians have to be aware of the emission control devices on the vehicle. Tampering with any emission control devices is illegal. Sooner or later, the widely used SAE J1667 testing will be replaced by a testing that more accurately assesses the emission control equipment on the vehicle and the original EPA certification standards the engine had to meet. As every year passes, truck technicians will have to know more about the nature of exhaust gas emissions, how to test vehicles for noncompliance, and how to repair engines that fail to meet whatever standards are in place.

This chapter begins by providing the technician with a basic understanding of the causes of smog, how diesel engine exhaust gases contribute to fouling the environment, the components used to help reduce noxious emissions, and the methods used to test vehicle exhaust. We then briefly look at some methods the EPA uses for testing engines for certification and finally, a detailed look at how to perform opacity testing.

PHOTOCHEMICAL SMOG

A **photochemical reaction** is any type of chemical process initiated by exposure to visible **ultraviolet (UV)** or **infrared** radiation. For example, photosynthesis, the process by which green plants transform light energy into chemical energy and gaseous oxygen is a photochemical reaction. To say that a photochemical reaction has occurred, light energy has to be absorbed by a substance. Another natural example of a photochemical process is the production of **ozone** in the Earth's upper atmosphere. Ozone (O_3) is responsible for filtering out most of the sun's harmful ultraviolet radiation. It is produced when sunlight breaks the bonds of some oxygen molecules (O_2) to form atomic oxygen (O), which then combines with O_2 to form O_3.

The word **smog** comes from the words fog and smoke. Throughout the world there are two main types of smog. The first type is **sulfur dioxide (SO_2)** smog produced by the burning of sulfurous fuels such as coal and heavy oils. Sulfurous smog conditions are aggravated by dampness and therefore fog. It is produced mostly by industry especially in those countries that have lax standards for combusting industrial coals and heavy oils. The extent to which the burning of modern day vehicle fuels contributes to this problem is minimal due to the use of low sulfur fuels.

Photochemical smog (also known as photosynthetic and photoelectric smog) requires neither smoke nor fog in its formative stage. It is produced mainly by the gases emitted from vehicle exhaust systems. Photochemical smog is characterized by a yellowish/light brownish haze. The effects of photochemical smog are as serious as those produced by sulfurous smog: plant damage, eye irritation, and respiratory failure in animals (including humans) are characteristic. Photochemical smog is produced in two stages. First, **hydrocarbons (HC)** and nitrogen oxides (emitted mostly by vehicles) react with exposure to sunlight to produce ozone. Ozone then reacts with gaseous hydrocarbons to produce smog. Ozone by itself is highly toxic in low concentrations, and the smog that results from exposure to light is a major problem in America today.

Most large cities have experienced problems with photochemical smog especially during the summer months. The location most noted for the problem is Los Angeles, California. Conditions in Los Angeles are ideal for producing photochemical smog. It is the second largest population center in the United States, per capita automobile ownership is the highest in the world, the area experiences a high ratio of sunny days, and the San Gabriel mountains act as a wall that helps entrap air masses that tend to move from the west to the east. Many years of exercising the toughest emissions standards in the world have contained, but by no means eliminated, the photochemical smog problem in Southern California. In fact, throughout North America, photochemical smog problems have worsened slightly since the first emission control legislation several decades ago due to increased per capita ownership of vehicles and greatly increased yearly distances traveled. Doing something about this problem is not simple. Money is the problem. There is a cost attached to developing low emissions engines and fueling them with friendlier fuels can be more costly and less fuel efficient. Environmentalists counter this argument by saying that the "hidden" costs of polluting the environment are borne by health care and the generations that follow ours.

The next major issue, currently being addressed by CARB, is **carbon dioxide** emission. Currently, carbon dioxide **(CO_2)** is classified as a greenhouse gas (responsible for global warming) and *not* as a noxious emission. Reducing carbon dioxide emissions will mean taking a serious look at the chemical composition of the fuels we burn, specifically reducing or eliminating the carbon content. As both gasoline and diesel fuel are composed of approximately 85% carbon, some major changes in fuel formulations are probable in the coming years.

INGREDIENTS OF SMOG

Understanding a little basic chemistry helps in understanding the nature of smog. Combustion is a chemical reaction. Under optimum conditions, when an HC fuel is combusted using the oxygen available in ground-level air as the reactant, the HC fuel is oxidized to form H_2O (water) and CO_2 (carbon dioxide). Neither H_2O nor CO_2 are classified as noxious products of the combustion process, so if these two gases were the only products of the reaction, the term *perfect* combustion could be applied to what has happened. The problems arise when the combustion of the HC is not "perfect" or the nitrogen that makes up more than 80% of air at ground level becomes involved in the combustion process. When a hydrocarbon fuel is not completely oxidized, gaseous HC, particulate HC, and **carbon monoxide (CO)** result. When emitted from the tailpipe of a vehicle these chemicals are all classified as noxious emissions. Nitrogen can become involved in the combustion process under certain conditions. Within the combustion temperature window of a typical internal combustion engine, this tends to be when temperatures are higher. When nitrogen is oxidized, it forms a number of different compounds known collectively as NO_x, but **nitrogen dioxide (NO_2)** presents the most problems. Nitrogen dioxide is a key to the formation of ozone and acid rain. All NO_x emissions are classified as noxious emissions by the EPA.

Sulfur

In recent years, the maximum sulfur content of diesel fuels has been dramatically reduced. At the time of writing, the maximum permitted sulfur content in an ASTM 1D or 2D fuel is 0.05%, and the adoption of ultralow sulfur fuels containing a maximum of 0.0015% sulfur is being considered by some jurisdictions. When sulfur is oxidized in the combustion process, it forms sulfur dioxide (SO_2), an ingredient of sulfurous but not photochemical smog. Sulfur is acidic, toxic, and an ingredient of acid rain. While sulfur compound emissions from the combustion process are serious, highway trucks are no longer a major contributor. Sulfur emissions produce an unpleasant, acidic odor. Because of the acidity, high sulfur emissions from a diesel engine will usually produce tattletales such as rust-out of exhaust piping.

Hydrocarbons

The HC emitted in the combustion process of an internal combustion engine burning liquid or gaseous petroleum product may be gaseous, liquid, or solid (particulate) state. They are only observable in liquid or solid states when they may be seen as white smoke (liquid emission) or black smoke (particulate emission). For a more detailed analysis of diesel engine smoke emission see Chapter 27. A truck diesel engine emits hydrocarbons in liquid state (that is, condensing in the exhaust gas) at cold startup and in particulate state under conditions of overfueling or air starvation. HC emission can be detected only if it is the solid or liquid states by smoke density-measuring equipment such as an opacity meter.

Particulate Matter

Anything in the exhaust gas that is in the solid state can be classified as particulate matter (PM). Carbon soot and dust form particulate matter. Incomplete combustion is the main culprit in the formation of PM. These minute particulate compounds can cause respiratory problems in human and animal life. The EPA classifies particulates into two general categories: (1) inhalable particulates that range between 2.5 and 10 microns in diameter, and (2) respirable particulates that tend to be 2.5 microns or less in diameter. Respirable particulates are not entrapped by nasal filters and are small enough to get deep into respiratory systems, causing wheezing, coughing, and shortness of breath. PM emission is always visible from a diesel engine and is easily detected with smoke density-measuring equipment such as opacity meters.

Volatile Organic Compounds

Volatile organic compounds or **VOCs** are hydrocarbons in the gaseous state that boil off fuels during production, distribution, and pumping. The transportation sector accounts for approximately 30% of the VOCs found around population centers. VOCs are more likely to be an atmospheric problem in hot weather conditions when the most volatile fractions of fuels more readily boil off. VOCs react with sunlight to produce ground level ozone.

Carbon Monoxide

Carbon monoxide is a colorless, odorless, tasteless, and highly poisonous gas. It is a result of the incomplete combustion of a hydrocarbon fuel. Carbon monoxide can be combusted to form harmless CO_2.

Approximately two-thirds of the atmospheric levels of CO in North America can be attributed to vehicle emissions. Exposure to CO can impair brain function, cause fatigue, and may be fatal.

Ozone

Atmospheric oxygen is diatomic, that is, into combines to form molecules consisting of two oxygen atoms (O_2). Ozone is triatomic, that is, three oxygen atoms covalently bond to form a molecule of ozone (O_3). It occurs naturally in small quantities in the Earth's stratosphere where it absorbs solar ultraviolet radiation. Ozone can be produced by passing a high-voltage electrical discharge through air. It is manufactured commercially in this manner and is produced during a thunderstorm. As a pollutant, ozone is produced in the photochemical reaction between NO_x and various hydrocarbons (especially those categorized as VOCs). It is known to irritate the eyes and mucous membranes.

Ozone is the subject of some recent controversy. Although most experts agree that ozone is a threat at concentrations of 80 parts per billion or higher, some medical scientists believe that exposure at any level can be a health threat.

Oxides of Nitrogen

Despite the fact that nitrogen under usual atmospheric conditions and temperatures is **inert** (unlikely to participate in chemical reactions), when subjected to certain conditions in an engine cylinder it reacts. That is, it can be oxidized. It forms nitrous oxide (N_2O), nitric oxide (NO), and nitrogen dioxide (NO_2) when oxidized. **Oxides of nitrogen** produced in the combustion process of internal combustion engines are known collectively simply as NO_x. Nitrogen dioxide (NO_2) is the most reactive of the **NO_x** compounds emitted from vehicle tailpipes. In North America, on- and off-highway vehicle engines are the source of about 60% of the NO_x our atmosphere. NO_x directly affects persons with respiratory problems, and medical opinion states that children's lungs are susceptible to concentrations of 0.1 ppm (parts per million). This accounts for the reason that the EPA has so aggressively attacked NO_x emissions from diesel engines in its certification requirements.

AIR QUALITY INDEX

Most jurisdictions today monitor air quality and use the monitoring results to alert the public on those days that it is above acceptable standards. If you suffer from asthma or other respiratory problems, you probably notice that each year that passes usually has more smog alert days when you are advised to avoid the outdoors. Table 45–1 is an air quality index (AQI) used by one jurisdiction.

TABLE 45–1   Air Quality Index and Environmental Effects

AIR QUALITY INDEX AND ENVIRONMENTAL EFFECTS	CATEGORY	CARBON MONOXIDE (CO)	NITROGEN DIOXIDE (NO_2)	OZONE (O_3)	SULPHUR DIOXIDE (SO_2)	SUSPENDED PARTICLES (SP)	SO_2 + SP	TOTAL REDUCED SULPHUR (TRS)
100 and over	Very poor	Increasing cardiovascular symptoms in nonsmokers with heart disease. Some visual impairment.	Increasing sensitivity of patients with asthma and bronchitis.	Light exercise produces respiratory effects in patients with chronic pulmonary disease.	Increasing sensitivity in patients with asthma and bronchitis.	Increasing sensitivity in patients with asthma and bronchitis.	Significant respiratory effects in patients with asthma and bronchitis.	Sensitive individuals may suffer nausea and headache due to severe odor.
50–99	Poor	Increased cardiovascular symptoms in smokers with heart disease.	Odor and discoloration. Some increase in bronchial reactivity and asthma attacks.	Decreasing performance by athletes exercising heavily.	Odorous. Increasing vegetation damage.	Visibility decreased. Soiling evident.	Increased symptoms in patients with chronic respiratory disease.	Extremely odorous.
32–49	Moderate	Blood chemistry changes but no detectable impairment.	Odorous.	Injurious to many vegetation species (e.g., white beans, tomatoes)	Injurious to some species of vegetation.	Some decrease in visibility.	Injurious to vegetation due to sulphur dioxide.	Odorous.
16–31	Good	No effects.	Slight odor.	Injurious to some vegetation species in combination with ozone.	Injurious to some vegetation species in combination with SO_2.	No effects.	No effects.	Slight odors.
0–15	Very good	No effects	No effects	No effects	No effects	No effects	No effects	No effects

SMOG SUMMARY

We can summarize the consequences of vehicle emissions as follows:

NO$_x$ + VOCs + Sunlight = Ozone

Ozone + Particulate matter = Smog

Because the costs of doing something about vehicle emissions are upfront and levied on big business, they tend to be resisted by powerful political lobbying. However, the cost of the effects of pollution are delayed and impact on the health of individuals who then have to resort to the big business of our health system. Whether you are an environmentalist or antienvironmentalist probably has much to do with where you live. Residents of geographical areas that are susceptible to smog tend to advocate stronger emissions legislation whereas those in relatively unaffected areas dismiss such controls as unwanted government interference. While the implementation of tougher emission controls and their enforcement is going to be politicized and controversial, expect them nevertheless to become more aggressive by the year, simply because there are more voters in big cities than rural areas who are demanding such controls.

EPA CERTIFICATION TESTING

It is not really necessary for a truck technician to know much about the precertification testing used by the EPA, although you should recognize that it is highly exacting and goes well beyond the opacity or smoke density field testing used by most jurisdictions. Gaseous emission of toxic combustion by-products cannot be measured using an **opacity meter** because gaseous emissions do not show as smoke when discharged from the vehicle tailpipe. Before EPA certification, diesel engine OEMs must conduct gaseous exhaust analysis and what is known as **constant volume sampling (CVS)** test standards. CVS is a process that involves diluting exhaust gas with purified ambient air and subsequently routing the air through three tests:

1. Filtration. The gas is forced through filters that are then weighed. The increase in mass indicates particulate emission.
2. Flame ionization detection (FID). This test indicates HC concentration.
3. Exhaust sample bags. Measures CO, CO$_2$, and NO$_x$.

TABLE 45–2 *EPA Emissions Requirements by Year*

YEAR	HC (GRAMS BH/HR)*	PM (GRAMS BH/HR)	NO$_x$ (GRAMS BH/HR)
1990	1.3	0.60	6.0
1991	1.3	0.25	5.0
1994	1.3	0.10	5.0
1996	1.3	0.10	5.0
1998	1.3	0.10	4.0
2004	0.5	0.05	0.5
2007	0.28	0.05	0.4
2010	0.14	0.01	0.2

*bh/hr = brake horsepower hour

Noxious gas emissions can also be tested by sampling and analyzing raw exhaust gas using **gas analyzers** not unlike those used for SI engine testing and described later in this chapter.

ACTUAL EPA EMISSIONS REQUIREMENTS BY YEAR

Given the fact that emissions from highway truck diesel engines were unregulated until 1970, what has been accomplished since is nothing short of astounding. In 1970, a typical diesel engine emitted 16 grams per hp/hr of NO$_x$ and 10 grams of PM per hp/hr. The NO$_x$ emission standard required in 2007 is 0.2 grams per hp/hr or an 8,000% reduction. Similarly, the PM standard required for 2007 has been reduced to 0.01 grams per hp/hr, a 10,000% reduction. Table 45–2 outlines the progress of heavy duty highway truck EPA emission standards from 1990 to 2010. Figure 45–1 shows the EPA diesel emissions story over the 40-year period extending from 1970 to 2010.

TRUCK ENGINE EMISSION CONTROLS

The primary emission control device in the current truck/bus diesel engine is the computer that manages the fueling. The requirement to meet tough emission standards is the reason that truck engines were managed by full-authority computer systems well ahead of

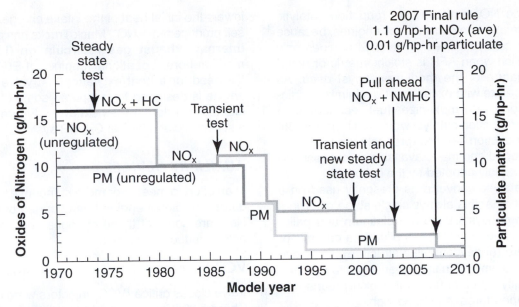

FIGURE 45–1 *EPA heavy-duty engine emission standards.*

their automobile counterparts. The result is that more of the emission control devices used on the SI gasoline engine are external. External devices include catalytic converters, air pump units, EGR systems, and PCV systems. Recently, the role of the engine management computer in automobile engines has increased, and the number of external emission control devices on truck diesels has increased. An example would be the **cooled exhaust gas recirculation (C-EGR)** that many truck engine manufacturers have opted to use to meet the NO_x requirement for 2004 emissions standards. We now look at some of the devices used on today's trucks to limit noxious emissions.

ENGINE CONTROL MODULE

Those of you who have studied automobile technology will be familiar with the term closed loop fueling. Closed-loop fueling is the term used to describe the running an SI, gasoline-fueled engine using the exhaust gas sensor signal to manage a stoichiometric air/fuel mixture. In the lean burn technology of the highway diesel engine, things are a little more complex in that the input of no single sensor is ever used to drive the **fueling algorithm** written to the ECM software. The fueling algorithm is the set of ECM programmed rules and procedures designed to produce the desired performance and emissions from the engine at any given moment of operation.

Managing fuel injection quantity and timing in a diesel engine requires managing combustion temperature within an often narrow temperature window. Generally, within the typical operating temperature window of a diesel engine, HC emission will tend to increase when

combustion temperatures are in the lower range and the air-fuel-ratio (AFR) is richer, and NO_x emission will increase when AFR leans out and the combustion temperatures are in the higher range. It is also important to note that injection timing or more specifically, ignition timing, affects noxious emission characteristics. Retarded timing tends to reduce NO_x emission but overretarded timing can greatly increase HC emission. Timing deviations of as little as 1° crank angle can increase NO_x or HC content in the **end gas** by as much as 10%. This timing sensitivity has made computer management of CI engines essential in attempting to meet emission requirements. In general terms, advancing the engine timing will increase engine temperatures and therefore tend to increase NO_x emission, while retarding timing can lead to incomplete combustion of the fuel and therefore an increase in HC emission.

PARTICULATE TRAPS

Particulate traps have been used by some OEMs especially when attempting to meet unusual emissions criterion such as indoor operation. An example would be garbage packer units operating in the garbage decks of large buildings. Particulate traps are essentially soot filters that use a ceramic filtration medium. In other words, they filter particulate that can be subsequently combusted during a electrically managed burn cycle.

CATALYTIC CONVERTERS

Despite the fact that a major challenge facing truck diesel engine manufacturers over the coming years is how to reduce NO_x emission, no reduction catalytic converters are used on current truck engines, because the reduction catalysts used on auto SI engines only function as such when AFR is stoichiometric or richer due to the nature of the rhodium catalyst used. As CI engines operate with excess air (to minimize particulate emission and maximize fuel economy), a rhodium-type reduction catalyst would not function. But if a suitable reduction catalyst (probably urea injection type) is developed and made available at low cost, expect to see it rapidly adopted by the industry.

When a catalytic converter is presently used on a truck diesel engine, it is always single-stage oxidizing. The oxidation catalysts used are platinum and palladium. These noble metals are applied to a ceramic or metal substrate (honeycomb-like structure). They oxidize HC and CO emitted in the exhaust gas to H_2O and CO_2, essentially by combusting it, which means that operating temperatures tend to be high.

COOLED-EGR

Until recently, diesel engine manufacturers had been able to meet NO_x emissions requirements without the use of exhaust gas recirculation (EGR) systems. This changed with EPA 2004 emission standards! The objective of EGR is to dilute the intake charge with "dead" gas, that is, the spent gas that will make a percentage of the cylinder volume unreactive. The idea is to reduce NO_x emission by lowering combustion heat but you do this at the expense of engine power and fuel efficiency, so it has never been popular with engine OEMs. It is even less popular with the end users who have to pay fuel bills.

Most diesel engine OEMs have opted to use C-EGR in an effort to meet the 2004 EPA requirement. Only Caterpillar with ACERT has decided against the use of EGR systems on its engines. If an EGR system must be used on an engine, C-EGR makes sense because it

lowers the initial heat of the intake charge that can in itself produce more NO_x. Mack Trucks has introduced an **internal exhaust gas recirculation (I-EGR)** on its medium-bore, vocational engines. I-EGR eliminates the need for a heat exchanger. At valve overlap, the engine is designed to reaspirate some of the end gas into the cylinder for the next cycle. Figure 45–2 shows a schematic of a typical C-EGR system.

AUXILIARY DEVICES

In an effort to meet ever more stringent emissions standards, the diesel engine has changed over the years. Here are some of these changes and how they have helped reduced emissions.

VCO Injectors

Valve closes orifice (VCO) injectors were introduced in Chapter 20. Elimination or reduction of nozzle sac volume in hydraulic injection injectors reduces the cylinder boil/dribble of sac fuel at the completion of injection.

Charge Air Cooling

Effective cooling of intake air lowers combustion temperatures, making it less likely that the nitrogen in the air mixture is oxidized to form NO_x. Air-to-air charge air coolers cool air more effectively than those that relied on engine coolant. Note that anything that compromises the charge air cooler's ability to cool will result in higher NO_x emissions.

Variable Geometry Turbochargers

Variable geometry turbochargers that perform effectively over a much wider load and rpm range can make a significant difference to both HC and NO_x emissions, especially when ECM controlled, by providing the ability to manage boost on the basis of the fueling and emissions algorithms.

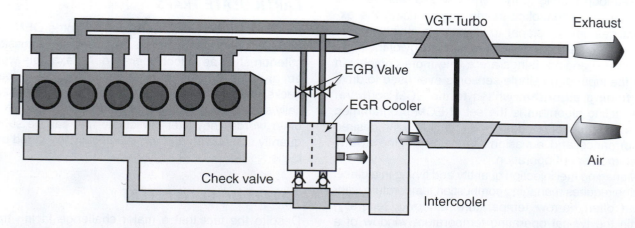

FIGURE 45–2 *Typical C-EGR system.*

Low Headland Pistons

Low headland volume pistons raise the upper compression ring close to the leading edge of the piston crown. This keeps the headland gas volume close to minimum. Headland gas volume tends to be unreactive and can increase HC emissions. The requirement for low headland volume pistons by engine designers resulted in some movement away from aluminium alloy truck pistons toward two-piece, articulating pistons.

Cerium Dioxide

Cerium dioxide is used by some major fleets as an in-cylinder combustion catalyst. It works as an oxidizing catalyst. Cerium dioxide has the approval of at least one major engine OEM, which claims increased fuel efficiency and improved power performance. A more detailed account of cerium dioxide appears in Chapter 12.

FIELD TESTING OF SMOKE DENSITY

Field testing for emissions refers to any testing performed after an engine series has been EPA certified. In the past couple of years, most U.S. states and Canadian provinces have launched highway diesel testing programs that conform to the SAE J1667 standard. The implementation of J1667 test programs can be regarded as a first step in the development of a comprehensive field testing program that examines real world emissions of computer-controlled diesel engines. At present, few jurisdictions require exhaust gas analysis.

OPACITY METERS

Opacity meters are simple to use and usually require that the unit's sensor head probe be fitted to the outlet of the exhaust pipe. Two different types are used: partial flow and full flow. In the partial flow-type, during the test procedure, a portion of the exhaust gas flow is diverted to a sensing chamber in which the opacity of the smoke can be read by the light sensor in the head assembly. A partial flow opacity meter is shown in Figure 45–3. The full flow type is more commonly used and makes the opacity measurement directly at the stack outlet. Figure 45–4 shows the operating principle of a full flow opacity test meter. Both types are classified as light extinction test instruments: a beam of light from a light-emitting diode is targeted at a photo diode sensor through the exhaust gas stream. The amount of light blocked from the photo diode sensor is determined by

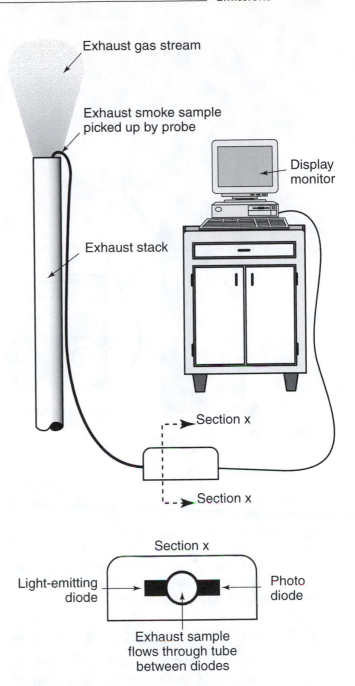

FIGURE 45–3 _Partial flow opacity meter._

the particulate density in the engine exhaust. The higher the particulate density, the less light is capable of penetrating it to be read by the sensor. Smoke density is expressed as a percentage reading by the opacity meter. In most cases, the opacity meter is equipped with an extension handle for the sensor head, so the device can be placed at the exhaust stack outlet by the technician working at ground level.

Opacity meter readings are displayed in percentage readings and the actual readings must always be factored with engine power output and the exhaust system outlet pipe diameter. Most should produce accurate results regardless of weather conditions and many will

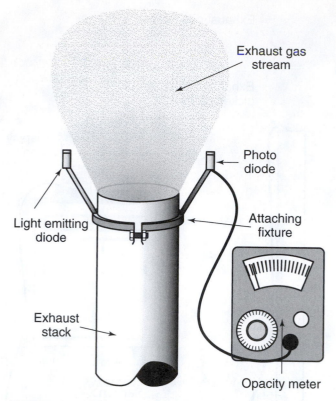

FIGURE 45–4 *Full flow opacity meter.*

log the data electronically. The data logged into the opacity meter can be hard copy printed or transferred to PC-based systems for analysis. Manufacturers of opacity meters claim accuracy factors within 2% and programmed response times within one second. Any opacity meter can be used to evaluate smoke density in diesel smoke. However, those used for official testing and to enforce compliance must be approved and have PC-managed and logged test sequences.

PERFORMING J1667 TESTING

In 1997, the EPA recommended that all jurisdictions adopt the SAE J1667 procedure for the testing of heavy-duty diesel vehicles perhaps because it would promote some consistency in test standards between different jurisdictions. The test routine outlined here adheres closely to that performed by compliance enforcement agencies. While the emission standards do vary by jurisdiction and year of manufacture of the truck, there is some consistency and the specifications used in this text are those generally accepted. The actual test is described a little later in this chapter but the data in the following list reflects the maximum puff cutpoint on a throttle snap used in many jurisdictions:

Model Year	Applicable Opacity Limit
1990 and earlier	55%
1991 and later	40%

Ambient environmental conditions can have some impact on smoke density, and changes in air density influence the operation of any internal combustion engine. Air density is influenced by ambient temperature, and generally, high air density will result in higher air/fuel ratios, which will tend to produce lower smoke opacity readings. For this reason, SAE J1667 imposes restrictions on the environmental conditions during the administration of an "official" snap acceleration test.

Environmental Conditions

The following environmental conditions are required when performing a J1667 test:

- Temperature in the test area must be between 35°F and 86°F.
- Testing must not take place outdoors when there is visible precipitation such rain, snow, or fog.
- Testing must not take place when the temperature is at or below the dew point, such as during fog or high humidity conditions.
- Vehicles using downward-directed exhaust systems should be tested over ground surface conditions that are not dusty so that dirt particulates are not combined with the exhaust gas being measured.

Snap Test

Each J1667 throttle snap consists of a three-phase cycle that must be precisely executed by the test administrator:

1. Accelerator is snapped to high idle and held for 1 to 4 seconds or until prompted by the opacity meter to release.
2. Engine rpm must be allowed to drop to the specified low idle speed.
3. Engine must run at the specified low idle for a minimum of 5 seconds and a maximum of 45 seconds before initiating the next snap as prompted by the opacity meter.

SAE J1667 Test Cycle

The SAE J1667 test cycle consists of four phases and again, these must be precisely executed by the test administrator:

1. Preliminary snaps. Three full snaps are required to clear the exhaust system of loose particles and precondition the vehicle.
2. Official snaps. Three official snaps should be undertaken as prompted by test instrument software.
3. Validation. The difference between the highest and the lowest maximum opacity readings of the three official test snaps should be within five opacity per-

centage points. If the difference is less than 5%, the meter software will compute the average. If the difference is greater than 5%, additional official snaps must be undertaken up to a maximum of nine.

4. Drift factor. Validation is required that the drift between three official test cycles does not exceed 2% opacity. A variation greater than ± 2% will result in an invalid test.

Stack Dimensions

Exhaust pipe diameter has an effect on the opacity reading of exhaust gas. Before undertaking a snap test, the diameter of the exhaust stack has to be known. For instance, an engine that tests at 30% opacity when tested with a 3-inch exhaust stack would only measure 20% opacity when tested with a 5-inch tailpipe. This is why you have to input the stack diameter into the opacity meter.

Although there are exceptions, because horsepower rating usually determines the stack diameter, it is recommended that the horsepower rating be inputted into the opacity test instrument before the exhaust pipe diameter. Horsepower rating is often specified on the engine ID plate or in the EST accessible read-only data. Table 45–3 correlates power with stack diameter.

EVALUATING J1667 TEST RESULTS

If a truck fails a J1667 snap test, a quick glance at the emitted exhaust smoke will suggest there is something seriously wrong with engine, especially if it is an engine manufactured after 1991. If the engine is hydromechanically managed and manufactured before 1991, the cause is frequently in the boost management system, a circuit that is often managed by an **aneroid** device.

Aneroids

Although most types of manifold boost measuring devices have been covered elsewhere in different chapters of this book, some of the more common methods of smoke control used on hydromechanical engines are briefly examined here for ease of reference. The way in which each operates has dictated how the actual test compliance procedure has been determined.

- Caterpillar AFRC (air fuel ratio control). An aneroid device that is not effective until after the first boost application. This feature permits the engine to be full-fueled for cold engine startup but not after the AFRC has been activated.
- Cummins AFC (air fuel control). An aneroid built into the PTG pump. Manifold boost is ported directly to the AFC diaphragm, which can then modulate fueling based on actual boost values.
- Detroit Diesel throttle delay cylinder. Used on the two-stroke cycle, MUI engines, this creates a hydraulic lag between a full fuel request and the act of moving the rack control tube to full fuel. Consists of a throttle delay cylinder and piston that reciprocates within it: the piston is attached by means of a lever actuated by the control rack. Engine oil is drizzled into the throttle delay cylinder so that when the control tube attempts to move from a low or idle fuel position, oil has to be displaced from the throttle delay cylinder first.
- Mack Trucks puff limiter. Used on many Mack engines of the 1980s, the puff limiter used an aneroid device located on the intake manifold to sense manifold boost and system air pressure to signal a plunger-type bimba valve located on the fuel injection pump and designed to block rack travel to full fuel. It functioned by cutting system air pressure to the bimba valve when boost pressure was sufficient: this allowed devices such as air brake status and transmission gear ratio to be used to limit unnecessary full fueling.
- Robert Bosch LED. LED is a German acronym meaning aneroid. The device is a direct-acting aneroid that requires that manifold boost be ported to it. Like the puff limiter, this type of aneroid is on/off and rated with a trigger value.

OPACITY READINGS

Because J1667 is used for compliance testing diesel engine emissions, the compliance test profiles have to

TABLE 45–3 *Horsepower and Recommended Exhaust Pipe Diameter*

RATED ENGINE POWER (BHP)	RATED ENGINE POWER (kW)	EXHAUST PIPE DIAMETER INCHES	EXHAUST PIPE DIAMETER mm
Less than 101	Less than 75	2	51
101 to 200	75 to 149	3	76
201 to 300	150 to 224	4	102
301 or more	225 or more	5	127

be simple and not vulnerable in court if offenders decide to fight fines. That is why the snap test is used. However, if you have an opacity meter, you can also look at some of the other criteria that were components of J1667. Smoke opacity readings of 5% or less are not easily observed. When using an opacity meter, it is important to meticulously observe the test procedures if any credence is to be given to the results. Remember that all of the data relates to engines manufactured before 1991 (despite being used on current engines) and note that J1667 suggested that maximum puff on throttle snap should be 50% opacity and not the 55% generally used as the limit by compliance enforcers.

Idle speed	10%
High idle speed	15%
Maximum puff	50% (transient response to throttle snap)
Rated speed/load	20% (requires chassis dynamometer)
Engine lug	25% (requires chassis dynamometer)
Motoring	25% (requires chassis dynamometer)
Full acceleration	30% (requires chassis dynamometer)

SMOKE ANALYSIS

Any truck diesel engine that fails an opacity snap test has a *severe* smoking problem, one that could not pass unnoticed under any circumstances. Many sick engines emitting large amounts of visible smoke can still pass the opacity standards defined in J1667 testing. Chapter 28 explains in some detail how to evaluate smoke emission from diesel engines, but the following section underlines some examples. Smoke appearance can be an indication of engine condition, fuel injection timing, fuel quality, and fuel delivery problems.

BLACK SMOKE OR GRAYISH HAZE

Black smoke or a grayish haze indicates that the engine is producing particulate emission. The cause usually has something to do with overfueling or air starvation. Some possible causes:

- Restricted leak-off circuit
- Restricted air intake
- Defective aneroid device
- Restricted exhaust
- Low NOP or enlarged nozzle orifii
- Overfueling
- Defective injector nozzles
- Inappropriate or degraded fuel

- Low cetane fuel
- Retarded injection pump timing
- Defective timing advance mechanisms
- Low cylinder compression
- Valve adjustment or timing problems

WHITE SMOKE

White smoke emission is caused by liquid condensing in the exhaust gas stream. This could be the result of coolant getting into the combustion chamber or low combustion temperatures resulting in fuel not properly vaporizing in the engine cylinders. Some possible causes are:

- Coolant leakage into the combustion chamber
- Injection timing problems
- Low combustion chamber temperatures
- Low cetane rating or degraded fuel
- Poor atomization
- Uneven fuel spray dispersal
- Dead engine cylinder
- Low compression pressures

BLUE SMOKE

Blue smoke is almost always an indication of an engine that is burning lubricating oil. Some possible causes are:

- Worn or broken piston rings
- Leaking valve seals or guides
- High engine oil crankcase level
- Defective crankcase ventilation system
- Leaking turbocharger seals

SUMMARY

- The word smog is derived from the words smoke and fog.
- Photochemical smog is formed by gaseous and particulate hydrocarbons (HC) combining with ozone and oxides of nitrogen (NO_x) followed by a period of exposure to sunlight.
- Because creating photochemical smog requires a period of exposure to sunlight, relatively still air mass is required for its formation.
- Vehicle emissions are the largest single contributor to the photochemical smog in most geographic areas of North America. Exhaust emissions for NO_x will have been reduced by 8,000% in a period extending from 1970 to 2010.
- Conditions in Southern California are ideal for the formation of photochemical smog. These include a

high percentage of per capita vehicle ownership, plenty of sun, and a natural wall formed by the San Gabriel mountains to the east that acts to trap air mass movement.

- VOCs are hydrocarbon fractions boiled off fuels during their production, transportation, storage, and production. VOCs are a more severe problem in high-temperature climatic conditions.
- Ozone is known to be toxic in concentrations of 80 parts per billion but some experts believe that any exposure to ozone can cause health problems.
- Opacity meters use a light extinction method of measuring exhaust gas density that consists of directing a light beam through the exhaust gas and using a sensor that measures the amount of light that succeeds in penetrating it.
- Opacity meters produce readings indicated as percentage density: the higher the percentage reading, the higher the smoke density.
- Some opacity meter tests require the use of a dynamometer, but SAE J1667 throttle snap testing is a field test that can be performed with no load on the engine.
- Smoke density testing based on the SAE J1667 standards is used by many jurisdictions in North America as a compliance standard.
- Field testing of diesel engine exhaust emissions currently requires engines manufactured before 1991 to meet smoke density, throttle snap standard of better than 55%. Engines manufactured after 1991 are required to meet a smoke opacity, throttle snap standard better than 40%.

REVIEW QUESTIONS

1. Which of the following compounds is not classified by the EPA as a noxious emission?
   a. Oxides of nitrogen
   b. Hydrocarbons
   c. Sulfur dioxide
   d. Carbon dioxide
2. By mass, which of the following is present in the largest quantity in the diesel engine cylinder during combustion?
   a. Oxygen
   b. Nitrogen
   c. Fuel
   d. Carbon monoxide
3. Which of the following best describes the catalytic converter used on a diesel engine?
   a. Three-way, two-stage
   b. Three-way, three-stage
   c. Single stage, oxidizing
   d. Single stage, reduction

4. Which of the following exhaust gas test instruments uses a light extinction principle to measure smoke density?
   a. CVA sampler
   b. Seven-gas analyzer
   c. Two-gas analyzer
   d. Opacity meter
5. Which of the following would properly describe atmospheric oxygen at ground level?
   a. O
   b. O_2
   c. O_3
   d. O_4
6. Which two compounds would result from theoretically "perfect" combustion of a HC fuel reacted with oxygen?
   a. Carbon monoxide and carbon dioxide
   b. Carbon dioxide and water
   c. Nitrogen dioxide and water
   d. Carbon dioxide and nitrogen dioxide
7. When a diesel engine is operated at lower-than-normal combustion temperatures, which of the following categories of noxious exhaust emissions is likely to increase?
   a. Oxides of nitrogen
   b. Carbon dioxide
   c. Ozone
   d. Hydrocarbons
8. When a diesel engine is operated at higher-than-normal combustion temperatures, which of the following categories of noxious exhaust emissions is likely to increase?
   a. Oxides of nitrogen
   b. Carbon dioxide
   c. Ozone
   d. Hydrocarbons
9. Which SAE standard defines the guidelines for testing diesel engine exhaust opacity emissions?
   a. J1587
   b. J1667
   c. J1708
   d. J1939
10. Which of the following compounds is classified as a greenhouse gas responsible for contributing to global warming?
    a. Oxides of nitrogen
    b. Hydrocarbons
    c. Sulfur dioxide
    d. Carbon dioxide
11. _Technician A_ states that advancing the static timing of a fuel injection pump by one crank degree can increase NO_x emission by up to 10%. _Technician B_ states that retarding static timing could produce higher HC emissions. Who is correct?
    a. A only
    b. B only

c. Both A and B
d. Neither A nor B

12. *Technician A* states that the maximum puff test when using an opacity meter can be performed with the vehicle stationary and zero load on the engine. *Technician B* states that the full acceleration test when using an opacity meter is a throttle snap test that can be performed with the vehicle stationary. Who is correct?
    a. A only
    b. B only
    c. Both A and B
    d. Neither A nor B

13. *Technician A* states that the combustion of diesel fuel is the major cause of ambient sulfur dioxide in North America. *Technician B* states that sulfur dioxide is the main cause of sulfurous smog. Who is correct?
    a. A only
    b. B only
    c. Both A and B
    d. Neither A nor B

14. Which of the following would more readily produce ambient VOCs (volatile organic compounds)?
    a. Diesel fuel in a sealed reservoir
    b. Diesel fuel in an open vessel
    c. Gasoline in a sealed reservoir
    d. Gasoline in an open vessel

15. *Technician A* states that the transportation industry accounts for 30% of ambient VOCs. *Technician B* states that the transportation industry accounts for 100% of the photochemical smog problem in North America. Who is correct?
    a. A only
    b. B only
    c. Both A and B
    d. Neither A nor B

16. *Technician A* states that photochemical smog requires both HC and NO_x plus a period of exposure to sunlight. *Technician B* states that the presence of ozone is a requirement for creating photochemical smog. Who is correct?
    a. A only
    b. B only
    c. Both A and B
    d. Neither A nor B

17. *Technician A* states that maximum puff on a throttle snap of a pre-1991 diesel engine is 55% in most jurisdictions. *Technician B* states that a seven-gas analyzer is required to verify whether an engine complies with state smoke density specifications. Who is correct?
    a. A only
    b. B only
    c. Both A and B
    d. Neither A nor B

18. How many official throttle snaps are required when performing a J1667 opacity test?
    a. One
    b. Two
    c. Three
    d. Four

19. A hydromechanically managed, diesel-powered truck is road tested and produces a puff of smoke at each upshift. Which of the following devices should be checked out first?
    a. Shift tower air signal
    b. Aneroid
    c. Turbocharger wastegate
    d. Governor torque trim spring

20. Which of the following is *not* an aneroid device?
    a. DDC throttle delay cylinder
    b. Caterpillar AFRC
    c. Mack Trucks puff limiter
    d. Cummins AFC

Appendix A
Acronyms

A ampere
ABP air box pressure (DDC)
ABS antilock brake system
AC alternating current
A/C air conditioning
ACERT Advanced Combustion Emissions Reduction Technology
ADEM advanced diesel engine management (system)
ADS Association of Diesel Specialists
AFC air/fuel control
AFR air/fuel ratio
ALCL assembly line communications link
ALDL assembly line diagnostic link
ALU arithmetic and logic unit
AMU air manifold unit
ANSI American National Standards Institute
API American Petroleum Institute
APT American pipe thread
AQI air quality index
ASCII American Standard Code for Information Interchange
ASE (National Institute for) Automotive Service Excellence
ASME American Society of Mechanical Engineers
ASTM American Society for Testing Materials
At ampere turns
ATA American Trucking Association
ATAAC air-to-air after-cooling
ATC automatic traction control
ATDC after top dead center
ATF automatic transmission fluid
atm unit of pressure equivalent to one unit of atmospheric pressure
ATM asynchronous transfer mode
AWG American wire gauge

BARO barometric pressure sensor
BDC bottom dead center
BHM bulkhead module
BHP brake horsepower
BIOS basic input/output system
BMEP brake mean effective pressure
BOE beginning of energizing
BOI beginning of injection
BP brake power
BPS bits per second
bsfc brake specific fuel consumption
BTDC before top dead center
BTM brushless torque motor
Btu British thermal unit

C carbon
C-7 Caterpillar 7 liter ACERT version of 3126
C-9 Caterpillar 9 liter, HEUI fueled, ACERT engine
C-10 Caterpillar 10-liter engine
C-11 Caterpillar 11 liter, MEUI fueled, ACERT engine
C-12 Caterpillar 12-liter engine
C-13 Caterpillar 13 liter, MEUI fueled, ACERT engine
C-15 Caterpillar 14.6 liter engine
C-16 Caterpillar 15.8 liter engine
CAC charge air cooling
CAD computer assisted design
CAFE corporate average fuel economy
CAM computer assisted machining/manufacturing
CAN Controller Area Network
CAPS Cummins Accumulator Pump System
CARB California Air Resources Board
CAT-ID Caterpillar information display
cc cubic centimeter
CCA cold cranking amps
CCV closed circuit voltage
CCW counterclockwise or left hand rotation
CD compact disk
CDC Consolidated Diesel Corporation (Case/ Cummins)
CD-ROM compact disk–read-only memory
C-EGR cooled exhaust gas recirculation
CEL check engine light
CELECT Cummins electronic engine control
C-EGR cooled exhaust gas recirulation
CEO chief executive officer
cfm cubic feet (per) minute
CFR ceramic fiber reinforced
CHM chassis module
CI compression ignition
cid cubic inch displacement
CMAC chassis mounted charge air cooling
CMOS complementary metal oxide semiconductor
CMP camshaft position
CMVSS Canadian Motor Vehicle Safety Standard
CN cetane number
CNG compressed natural gas
CO carbon monoxide
Codec coder/decoder
COE cab-over-engine truck chassis
CPL control parts list (Cummins parts #)
CPS camshaft position sensor
CPS characters per second
CPU central processing unit
CRC cyclic redundancy check
CRT cathode ray tube
CSA Canadian Safety Association

CTS coolant temperature sensor
CTV continuously open throttle valve
CVS constant volume sampling
CVSA Commercial Vehicle Safety Association
CW clockwise

DAT digital audio tape
DC direct current
DCA diesel coolant additives
DCL data communication link
DDC Detroit Diesel Corporation
DDEC Detroit Diesel electronic controls
DDL diagnostic data link
DDR digital diagnostic reader
DDT digital diagnostic tool
DFF direct fuel feed (Cummins STC injector)
DI direct injection
DIMM double in-line memory module
DIP dual in-line package (chip)
DMM digital multimeter
DOS disk operating system
DOT Department of Transportation
DRAM dynamic RAM
DVD-ROM digital video disk–read-only memory
DVOM digital volt ohmmeter
DW double weight (DDC governor)

ECAP electronic control analyzer programmer (Caterpillar EST)
ECB electronic circuit breaker
ECI electronically controlled injection
ECM electronic/engine control module
ECS evaporative (emission) control system
ECT engine coolant temperature
ECU electronic/engine control unit
EDU electronic distributor unit (DDEC)
EEC electronic engine control
EEPROM electronically erasable programmable read-only memory
EFPA electronic foot pedal assembly
EFRC engine family rating code
EG ethylene glycol
EGR exhaust gas recirculation
EIA Electronics Industries Association
ELC extended life coolant
EMF electromotive force
EMI electromagnetic interference
EMM extended memory manager (ECMs)
ENIAC electronic numeric integrator and calculator
EOE ending of energizing
EOF end of frame (multiplexing packet)
EOI ending of injection
EOL end of line (programming)
E-9 Mack Trucks V8 engine
EPA Environmental Protection Agency
EPS engine position sensor
E-7 Mack Trucks in-line, 6-cylinder, 12-liter engine
E7-EUP Mack Trucks V-MAC III managed, in-line, 6-cylinder, 12-liter engine
ESP electronic smart power (Cummins)
ESS engine speed sensor

EST electronic service tool
ET Electronic Technician
E-Tech Mack Trucks V-MAC III, EUP 12-liter engine
ETR energized to run
EUI electronic unit injector
EUP electronic unit pump

FARC fuel air ratio control-Caterpillar aneroid
FE iron
FET field effect transistor
FDCS fuel demand command signal (Navistar)
FIC fuel I module injection pump controller
FIS fleet information software (Caterpillar)
FM frequency modulation
FMI failure mode indicator (SAE)
FMVSS Federal Motor Vehicle Safety Standard
FPS fuel pressure sensor
FRC fuel ratio control

GCM governor control module
GM General Motors (Corporation)
gnd ground
gph gallons per hour
GPS global positioning satellite
GUI graphical user interface

H hydrogen
HC hydrocarbon
HDEO heavy-duty engine oil
HE hydro-erosive (machining technology)
HEUI hydraulically actuated electronic unit injector
Hg mercury
HPI high-pressure injection (Cummins)
HPI-PT high-pressure injection-pressure time (Cummins off-highway system)
HPI-TP high-pressure injection-time pressure (Cummins)
HVAC heating, ventilating and air-conditioning
H_2O water

IAP injection actuation pressure
IBC inner base circle
I/C integrated circuit
ICP injection control pressure (Navistar)
ICU instrument control unit
id inside diameter
ID identify
IDI indirect injection
IDM injector driver module (Navistar)
I-EGR internal exhaust gas recirculation
IFSM integrated fuel system module (Cummins)
IMEP indicated mean effective pressure
IHP indicated horsepower
INFORM Cummins data analysis software
INSITE Cummins PC software
INSPEC Cummins data analysis software
IPR injection pressure regulator (Navistar)
IRT injector response time
ISB Interact System B series (Cummins)
ISC Interact System C series (Cummins)
ISM Interact System M series (Cummins)
ISO International Standards Organization

ISX   Interact System X series-Cummins 15 liter Signature Series

KAM   keep-alive memory
KAMPWR   electrical circuit that powers KAM (Navistar)
km   kilometers
K-19   Cummins in-line, 6-cylinder, 19-liter engine
KPa   kilopascals

LAN   local area network
LED   light-emitting diode
LCD   liquid crystal display
LNG   liquefied natural gas
LPG   liquid petroleum gas
LS   limiting speed
LS   low sulfur
LSG   limiting speed governor
L-10   Cummins in-line, 6-cylinder, 10-liter engine

m   meter
Mack E-Tech   Mack 12-liter, V-MAC III engine
MAP   manifold actual pressure
MB-906   Mercedes-Benz 6.4 liter engine
MB-4000   Mercedes-Benz 12 liter engine
M-11   Cummins in-line, 6-cylinder, 11-liter engine
MEP   mean effective pressure
MEUI   mechanically-actuated, electronically-controlled, electronic unit injector.
MHz   megahertz
MID   message identifier (SAE)
MIL   malfunction indicator lamp
MIPS   millions of instructions per second
mm   millimeter
mmf   magnetomotive force
MON   motor octane number
MOSFET   metal oxide semiconductor field effect transformer
MPa   megapascals
MPC   multi-protocol cartridge/card (ProLink)
MPP   massively parallel processors
MSDS   material safety data sheets
MUI   mechanical unit injector

N   nitrogen
NAP   network access point (wireless data transfer module)
NBF   German acronym for needle motion sensor (Bosch)
N-14   Cummins I-6, 14-liter engine
Nm   Newton-meter
NO₂   nitrogen dioxide
NOP   nozzle opening pressure
NOₓ   oxides of nitrogen
NPN   negative-positive-negative (semiconductor)
NTC   negative temperature coefficient
NV-RAM   nonvolatile random-access memory

O   oxygen
OBC   outer base circle
OBD   on-board diagnostics
OC   occurence count
OCR   optical character recognition
OCV   open circuit voltage

od   outside diameter
OEM   original equipment manufacturer
OOS   out-of-service
OPEC   Organization of Petroleum Exporting Countries
OS   operating system
OSHA   Occupational Safety and Health Administration

Pa   Pascal
PACE/PACER   Cummins partial authority, electronic PT fuel system management
Pb   lead
PC   personal computer
PC   port closure (spill timing)
PCM   pulse code modulation
PCM   powertrain control module
PCMCIA   personal computer card International Association
PCU   powertrain control module
PCV   positive crankcase ventilation
PDI   predelivery inspection
PE   pump (injection) enclosed (actuation—integral camshaft)
PEEC   programmable electronic engine control (Caterpillar)
PDM   power distribution module
PF   pump (injection) foreign (actuation — external camshaft)
PG   propylene glycol
pH   power hydrogen (measure of acidity/alkalinity)
PHC   partially-burned hydrocarbons
PID   parameter identifier (SAE)
PLC   power line carrier (multiplexing)
PLC   programmable logic controller (smart relay)
PLD   German acronym meaning ECM (MB engines)
PLD   engine control unit (Mercedes Benz)
PLN   pump-line-nozzle (diesel fuel injection)
PM   particulate matter
PM   preventative maintenance
PN   positive-negative (junction semiconductor)
PNP   positive-negative-positive (semiconductor)
POST   power-on self-test
POTS   plain old telephone service
ppb   parts per billion
ppm   parts per million
PRIME   preinjection metering (Caterpillar HEUI)
PT   pressure-time (Cummins)
PTCM   PT (pump) control module
PTG   pressure-time governor controlled (Cummins)
PTO   power takeoff
PW   pulse width
PWM   pulse width modulation

R   resistance
RAM   random-access memory
R&D   research and development
RDI   remote data interface
RCRA   Resource Conservation and Recovery Act
RE   Bosch rack actuator
rms   root mean square
ROM   read-only memory
RON   research octane number

RQV   Bosch VS governor
RSG   road speed governing
RSL   road speed limit
RS-232   port standard telephone jack
RSV   Bosch VS governor

S   sulfur
SAE   Society of Automotive Engineers
SAE J1587   data bus software protocols
SAE J1667   emission testing standards
SAE J1708   data bus hardware protocols
SAE J1930   recommended acronyms/terminology
SAE J1939   data bus hardware/software protocols
SCA   supplemental cooling (system) additive
SCFR   squeeze cast, fiber-reinforced
SCR   selective catalytic reduction
SCR   silicone-controlled rectifier
SCSI   small computer system interface
SEL   stop engine light
SEO   stop engine override
Series 50   DDC I-4, DDEC engine
Series 55   DDC I-6 EUP, DDEC engine
Series 92   DDC two-stroke cycle engine
Series 60   DDC I-6, DDEC engine
SFC   specific fuel consumption
s.i.   Système international (metric system)
SI   spark ignited
SID   subsystem identifier (SAE)
SIMM   single in-line memory module
SOI   start of injection
SPL   smoke puff limiter
SRAM   static random access memory
SRS   synchronous reference sensor (DDEC)
SSC   single speed control
SSU   seconds Saybolt universal (fluidity rating)
STC   step timing control (Cummins)
STEO   stop engine overide
STI   self-test input
STOP   stop engine light
STS   Service Technicians Society
SW   single weight (DDC governor)
SVGA   super video graphics array

TBS   turbo boost sensor
TCP   transmission control protocol
TCU   transmission control unit
TDC   top dead center
TDS   total dissolved solids
TEL   tetraethyl lead (gasoline)
TEL   top engine limit (Cat: high idle)
TEM   timing event marker
TIG   tungsten inert gas (welding)

TIR   total indicated runout
TMC   Truck Maintenance Council
TML   tetramethyl lead (gasoline)
TQM   total quality management
TP   throttle position
TP   time pressure (Cummins)
TPS   throttle position sensor
TRS   timing reference sensor (DDEC)
TSB   technical service bulletin
TT   tailored torque (DDC)
TTS   transmission tailshaft speed

UHC   unburned hydrocarbons
UIS   unit injection system (Bosch EUI fuel system)
ULEV   ultra low emissions vehicle
ULS   ultra-low sulfur (fuel)
UNC   unified (thread) coarse
UNF   unified (thread) fine
UPC   universal product code
USB   universal serial bus

V   volt
V-Bat   battery system voltage
VCO   valve closes orifice (nozzle)
VCP   valve closing pressure
VCU   vehicle control unit
VDT   video display terminal
VEPS   vehicle electronics programming system
VGA   video graphics array
VI   viscosity index
VIA   visible image area
VIN   vehicle identification number
VIP   vehicle interface program
V-MAC   vehicle management and control
VOC   volatile organic compound
VPM   vehicle personality module (Caterpillar/ Navistar)
VR   voltage regulator
V-Ref   reference voltage (almost always ±5 VDC)
VS   variable speed
VSC   variable speed control
VSG   variable speed governor
VSL   variable speed limit
VSL   vehicle speed limit
VSS   vehicle speed sensor

W   watt
WHMIS   Workplace Hazardous Materials Information System
WIF   water in fuel (sensor)
WORM   write once, read many
WOT   wide open throttle
WT   World Transmission (Allison)

Glossary

absolute maximum power the highest power an engine can develop at sea level with no limitations on speed, fuel/air ratio or fuel quantity.

accumulator device for storing energy. Often used to describe the high-pressure chamber or common rail used in some electronically controlled, fuel injection systems.

ACERT Advanced Combustion Emissions Reduction Technology Caterpillar technology that uses a four-phase emissions reduction strategy to meet 2004 EPA standards. Critical ACERT components include series turbocharging, electronically managed variable valve timing, MEUI and HEUI injectors, and exhaust gas after-treatment.

acronym a word formed by the initial letters of other words.

active codes an electronically monitored system circuit, condition, or component that is malfunctioning and logs an ECM code, which may be displayed or read with an EST.

actuators hardware that effects the results of computer processing into action. Examples of actuators would be the injector drivers in a diesel EUI system.

advanced diesel engine management system (ADEMS) Caterpillar acronym used to describe its management electronics. The current version uses a 32 bit processor and has dual 70 pin connectors.

aeolipile a reaction turbine. The first heat engine and fore-runner of the modern jet turbine, invented and built by Hero, a Greek, around AD 60.

AFC circuit the turbo boost sensing and fuel management components on a Cummins PTC-AFC pump.

afterburn term that can be used to describe the normal combustion of fuel in a diesel engine cylinder after injector nozzle closure or random ignitions of fuel pockets after primary flame quench in an engine cylinder.

after top dead center (ATDC) any engine position during piston downstroke.

air box the term used to describe the chamber charged by a Roots blower in a DDC two-stroke cycle engine; the air box supplies the engine cylinders with the scavenging air charge.

air conditioning (A/C) the cooling circuit in an HVAC climate control system.

air/fuel control (AFC) usually refers to turbo boost and/or altitude compensation fuel management in diesel engines.

air/fuel ratio (AFR) the mass ratio of an air-to-fuel mixture.

air-to-air after cooling (ATAAC)

alcohol any of a group of distillate hydrocarbon liquids containing at least one hydroxyl group; sometimes referred to as oxygenates.

algorithm software term that describes a programmed sequence of operating events.

alloy the mixing of a molten base metal with metallic or nonmetallic elements to alter the metallurgical characteristics.

all speed governor another term for variable speed governor.

alpha data represented by letters of the alphabet.

alternating current (AC) current flow that cyclically alternates in direction, usually produced by rotating a coil within a magnetic field.

altitude compensator device used on older diesel fuel injection systems containing a barometric capsule: used to measure atmospheric pressure and derate fueling at altitude to prevent smoking.

altitude deration the engine fuel delivery cutback that is managed to occur on the basis of increase in altitude, to prevent engine overfueling as the air charge becomes less oxygen dense. Power deration is typically 4% per 1,000 feet of altitude in boosted engines.

amber the Greek word for electron. A translucent, yellowish substance evolved from fossilized trees.

American Petroleum Institute (API) classifies lubricants and sets standards in the petroleum-refining industry.

American Society for Testing Materials (ASTM) agency that sets industry standards and regulations including those for fuel.

American Standard Code for Information Interchange (ASCII) widely used data coding system found on PCs.

American Trucking Association (ATA) organization with a broad spectrum of representation responsible for setting standards in the U.S. trucking industry.

ampere unit of electrical current flow equivalent to 6.28×10^{18} electrons passing a given point in a circuit per second.

ampere turns (At) the basic unit of measurement of magnetomotive force.

amplification term used in electronic circuits to describe what happens when very small currents are used to switch much larger ones using transistors.

amplifier piston hydraulically actuated piston that pumps fuel to injection pressure values in a Cat/International HEUI; also known as *intensifier piston*.

anaerobic sealant pastelike sealant that cures (hardens) without exposure to air.

analog the use of physical variables, such as voltage or length, to represent values.

analog signal a communication line signal consisting of a continuous electrical wave.

AND gate an electronic switch in which all the inputs must be in the ON state before the output is in the ON state.

aneroid a device used to sense light pressure conditions. The term is used to describe manifold boost sensors that limit fueling until there is sufficient boost air to combust it and usually consists of a diaphragm, spring, and fuel-limiting mechanism.

annular ring shaped.

annuli plural of *annulus*.

annulus a ring.

737

anode positive electrode; the electrode toward which electrons flow.

anodizing oxide coating on the surface of a metal formed by an electrolytic process.

antifreeze a liquid solution added to water to blend the engine coolant solution that raises the boil point and lowers the freeze point. Ethylene glycol (EG), propylene glycol (PG), and extended life coolants (ELC) are currently used.

antinodes portion of a wave below the zero, mean, or neutral point in a waveband.

antithrust face used to describe the minor thrust face of a piston; the outboard side of the piston as its throw rotates off the crankshaft centerline through the power stroke.

antithrust side piston term meaning *minor thrust side*.

application software programs that direct computer processing operations.

arcing bearing or gear failure caused by electric arcing.

articulating piston a two-piece piston with separate crown and skirt assemblies, linked by the piston wrist pin and afforded a degree of independent movement. The wrist pin is usually full floating or bolted directly to the conn rod, in which case it is known as a *crosshead piston*.

ash 1. the powdery/particulate residues of a combustion reaction. 2. solid residues found in crude oils. Present in trace quantities in engine lubricating oils and diesel fuels.

Association of Diesel Specialists (ADS) Organization to which fuel injection specialty shops belong and which monitors industry standards of practice and education.

ASTM #1D fuel fuel recommended for use in high speed, on-highway diesel engines required to operate under variable load and variable speeds. Minimum CN must be above 40. In theory, the ideal fuel for highway truck and bus diesel engines but in practice, it is not used as often as #2D fuel because it has less heat energy by weight, making it less economical.

ASTM #2D fuel fuel recommended for use in high speed, on-highway diesel engines required to operate under constant loads and speeds. Like #1D fuel, the minimum CN is required to be above 40. Widely used in highway truck operations because it produces better fuel economy than #1D fuel due to its higher calorific value, albeit at the expense of slightly inferior performance.

asynchronous transfer mode (ATM) a method of transmission and switching that can handle vast amounts of data at high speed.

ATA connector the common method of describing the *ATA data link*.

ATA data link an SAE/ATA standard J1587/J1708/J1939, six-pin Deutsch connector currently used by all truck and truck engine OEMs to access the on-board ECMs.

atms a unit of atmospheric pressure equivalent to 14.7 psi (101.3 kPa). Used as a unit of measurement in the US and UK, especially on fuel calibration instruments. Close but not exactly equivalent to a European unit of bar.

atom the smallest part of a chemical element that can take part in a chemical reaction; composed of electrons, protons, and neutrons.

atomization the process of breaking liquid fuel into small droplets by pumping it at a high pressure through a minute flow area.

atomized droplets the liquid droplets emitted from an injector nozzle.

audit trail a means of electronically tracking electronically monitored problems in an engine management system. May be discreet, that is, not read by some diagnostic ESTs and programs; also known as *tattletale*.

AUTOEXEC.BAT a batch file loaded into the DOS kernel that governs boot-up protocol.

Automotive Service Excellence National Institute (ASE) organization dedicated to setting test standards for auto and truck technicians.

automotive governor a term sometimes used to describe a limiting speed, mechanical governor.

auxiliary power unit (APU) power supply unit used on many trucks to provide electrical power when the engine is not running. Gasoline, diesel, or fuel cell sourced.

axis the point about which a body rotates; the center point of a circle. Plural: *axes*.

B20 standard petroleum-based diesel fuel cut with 20% biodiesel.

B100 term used to described pure biodiesel fuel meeting the ASTM standard D6751.

backbone data bus consisting of a twisted wire pair.

backfire ignition/combustion of the fuel in an oxyacetylene torch in the torch tip causing a popping and squealing noise.

background computations computer operating responses of lower priority than *foreground* operations, that while important do not require immediate response; monitoring of engine fluid temperatures would be classified as a background computation.

back leakage test an injector bench fixture test in which nozzle valve to nozzle body leakage is measured.

balance orifice the inlet orifice of a Cummins PT injector. Collectively, the balance orifii (one in each PT injector) define the flow area the rail unloads to.

balanced atom an atom in which the number of electrons and protons are equal.

balanced rack setting infers that the rack adjustment of an MUI-fueled engine has been set so the fuel delivery in each injector is identical through the travel of the rack.

bandwidth volume or capacity of a data transmission medium; the number of packets that can be pumped down a channel or multiplex backbone.

bar a metric unit of pressure 10^5 Newtons per square meter; approximately, but not exactly one unit of atmosphere or 1 atm.

barometric capsule a barometer device used on some hydromechanical injection pumps to limit high-altitude fueling

barometric pressure sensor (BARO) an electronic barometric pressure sensing device.

base circle the smallest radial dimension of an eccentric. Used to describe cam geometry, the train that the cam is responsible for actuating would be unloaded on the cam base circle; also known as *inner base circle* or IBC.

basic input/output system (BIOS) when a computer is booted, the CPU looks to the BIOS chip for instructions on how to interface between the disk-operating system and the system hardware.

baud   times per second that a data communications signal changes and permits one bit of data to be transmitted.

baud rate   the speed of a data transmission.

bay   a vacant location in the computer housing/system designed to accommodate system upgrades.

bearing shell   a half segment of a friction bearing such as would be used as a crankshaft main bearing.

before top dead center (BTDC)   a piston location in the cycle before full piston travel usually abbreviated BTDC.

beginning of energizing (BOE)   moment that an EUI or EUP is electrically energized.

beginning of injection (BOI)   in engine management, a specific point at which injection begins.

bell crank   a single arm lever with its fulcrum at the apex of a shaft, often used as a mechanical relay. The word originates from medieval church bell ringing mechanisms.

benzene   hydrocarbon fuel fraction obtainable from coal or petroleum; known as a carcinogen.

big end   the crankshaft throw end of a connecting rod.

biodiesel   fuel derived from farm products with a vegetable and alcohol base; when used in current diesel engines should meet ASTM standard D6751.

binary system   a two-digit arithmetic, numeric system commonly used in computer electronics.

bipolar transistor   a three-terminal transistor that functions as a sort of switched diode.

bit   a binary digit that can represent one of two values, on or off; presence of voltage or no voltage; the smallest piece of data a computer can manipulate. There are 8 bits to a byte.

bits per second (BPS)   a measure of the speed at which data can be transferred.

black smoke   smoke that appears black to the observer is caused by particulate (solids) emission in the exhaust gas stream; light is blocked by the particulate, making it appear black.

blink codes   fault codes blinked out using diagnostic lights; also known as *flash codes*.

blotter test   an inaccurate and generally obsolete method of testing used engine oil for viscosity and contamination.

blue smoke   usually associated with engine oil combusted in the engine cylinder; caused by the mixture of condensing droplets and particulate emitted when oil is burned in an engine.

boil point   the temperature at which a liquid vaporizes.

bomb calorimeter test   a test used to calculate the heating value of a fuel: a known quantity of the substance is combusted and the heat released calculated.

boosted engine   any turbocharged engine; turbo-boosted.

boot   the process of loading an operating system into RAM or main memory.

boot-up   to load an operating system into RAM, electronically reload a system program, or reset a computer.

bore   an aperture. The internal diameter of a pump or engine cylinder or the act of machining a cylindrical aperture.

bottom dead center   lowest point of travel of piston in an engine cylinder during its cycle. Usually abbreviated BDC.

boundary lubrication   thin film lubrication characteristics of an oil.

Boyle's Law   states that for a given, confined quantity of gas, pressure is inversely related to the volume, so as one value goes up, the other goes down. In compressing gas in an engine cylinder during piston upstroke, cylinder volume is reduced so cylinder pressure accordingly is increased.

brake fade   a vehicle braking characteristic caused by excessive heat that can expand brake foundation components, such as drums, and lower the coefficient of friction of the critical friction surfaces.

brake horsepower   standard expression for *brake power* commonly used in the truck industry. See definition for **brake power**.

brake power   power developed by an engine measured at the flywheel measured by a dynamometer or *brake*. Factored by *torque* and rpm.

BrakeSaver   a Caterpillar engine-mounted, hydraulic retarder.

breakout box   a diagnostic device fitted with coded sockets that accesses an electrical or electronic circuit by teeing into it; used in conjunction with a DMM.

breakout T   term used to describe a breakout box or in some cases a diagnostic device that tees into two- or three-wire circuits to enable diagnoses by DMM of a single component such as a sensor.

brake specific fuel consumption (bsfc)   a measure of the fuel required to perform a unit of work; used in graphs of engine data designed to show fuel efficiency at specific engine loads and rpm.

bridge   the software and/or hardware used to make electronic connections such as that used to connect nodes in a network.

broach   a boring bit used for final, accurate bore sizing.

British thermal unit (Btu)   The amount of heat required to raise the temperature of 1 pound of water, 1°F at 60°F. The standard unit of heat energy measurement.

brushless torque motor (BTM)   Caterpillar rotary proportional solenoid used for PEEC timing and rack position control.

bubble collapse   the condition caused by wet liner combustion pressure impulses acting on the coolant resulting in vapor bubbles that collapse and cause *cavitation*.

buffers   memory locations used to store processed data before it is sent to output devices.

buffer screw   a DDC governor adjustment that lightly loads the governor differential lever to reduce engine rpm hunting.

bundle   multiple arrangement of cooling tubes that form the core of a heat exchanger.

buret   see *vial*.

bus   1. a transit vehicle. 2. an electronic connection; transit lines that connect the CPU, memory, and input/output devices; increasingly used to mean "connected."

bushing   any of a number of types of friction bearings designed to support shafts.

bus systems   term used to describe data highways.

buttress   an additional/auxiliary support device such as a gusset.

butt splice   the joining of two pieces in a series connection.

bypass filter   a filter assembly plumbed in parallel with the lubrication circuit, usually capable of high filtering efficiencies.

bypass valve   a diverter valve fitted to full flow filter (series) mounting pads, designed to reroute lubricant around a plugged filter element to prevent a major engine failure.

byte   unit of measure of computer data, comprised of 8 bits; used to quantify computer data memory.

cab-over-engine (COE)   truck chassis in which the engine compartment is located directly underneath the driver cab, eliminating the hood. Usually abbreviated to *COE*.

cache   high-speed RAM located between the CPU and main memory used to increase processing efficiency.

cage   a computer system housing location accommodating two or more bays.

calibration   adjusting performance specifications to a standard: fuel trimming of diesel fuel injection components is known as calibration.

calibrating orifice   see *balance orifice*.

calibration correlate   a set of readings with a standard or the process of adjusting to a standard.

calibration parameters   the specific values required when setting performance to specification.

California Air Resouces Board (CARB)   The state of California agency responsible for driving emissions legislation and enforcement. By establishing standards that exceed federal standards and effecting them earlier, CARB has led the emission control initiative throughout North America.

calipers   comparative measuring instrument used for measuring od or id.

calorific value   the heating value of a fuel measured in Btu, calories, or joules.

cam   an eccentric. An eccentric portion of a shaft, often used to convert rotary motion into reciprocating motion.

cambox   the lower portion of a port-helix metering injection pump in which the actuating camshaft is mounted and the lubricating oil sump is located.

cam follower housing timing   the method used by Cummins on its PT N series engines to synchronize the actuation of the PT injector pumping stroke by the cam profile with engine position.

cam geometry   the shaping of a cam profile and the effect it produces on the train it actuates.

cam ground   trunk-type pistons that are machined slightly eccentrically. Because of the greater mass of material required at the wrist pin boss, this area will expand proportionally more when heated. Cam ground pistons are designed to assume a true circular shape at operating temperatures.

cam heel   the point on a cam profile that is exactly opposite the *toe* or center point of the highest point on the cam.

cam nose   the portion of the cam profile with the largest radial dimension; its center point would be the *cam toe*. That portion of the cam profile that is *OBC*.

cam plate   the input shaft driven, rotating-reciprocating member used to actuate the distributor plunger in a Bosch type, sleeve-metering, rotary distributor injection pump such as the VE.

cam profile   the cam geometry; simply, the shape of the cam.

camshaft   a crankshaft-driven shaft, machined with eccentrics (cams) designed to actuate trains positioned to ride the cam profiles; the engine feedback assembly actuator responsible for timing/actuating cylinder valves and fuel injection apparatus. Driven at half engine speed on four-stroke cycle engines and at engine speed on two-stroke cycles.

camshaft position sensor (CPS)   any of a number of types of engine position sensors using either an inductive pulse generator or Hall effect electrical principle.

canister   a cylindrical container.

capacitance   measure of how much electrical charge can be stored for a given voltage potential: measured in *farads*.

capacitor   an electrical device that can store an electrical charge or block AC and pass DC. Also known as *condenser*.

carbon (C)   an element found in various forms including diamonds, charcoal, and coal. It is the primary constituent element in hydrocarbon fuels. Atomic number 6.

carbon dioxide (CO_2)   the product of combusting carbon in the oxidation reaction of a HC fuel. An odorless, tasteless gas that is nontoxic and not classified as a noxious engine emission, but that contributes to greenhouse gases that concern environmentalists.

carbon monoxide (CO)   a colorless, odorless, and poisonous gas that is produced when carbon is not completely oxidized in combustion.

carcinogen   a cancer-causing agent.

Carnot cycle   relates to ratio of work output to heat input that should equal the difference between the temperatures of the heat source and rejected heat combined, divided by the temperature of the heat source.

cartridge   a removable container; used to describe the housing that encloses a filter.

cartridge tape   data storage medium of the sequential type, currently used for PC data backup.

Cat Messenger   an advancement on the Cat ID driver display unit providing feedback on engine operating conditions, maintenance tracking, theft deterrence and chassis performance.

catalyst   a substance that stimulates, accelerates, or enables a chemical reaction without itself undergoing any change.

catalytic converter   an exhaust system device that enables oxidation and reduction reactions; in lean burn truck diesel engines, only oxidation catalytic converters are currently used.

Caterpillar Engine Company   a major diesel engine manufacturer. Corporate center is Peoria, Illinois.

Caterpillar Fleet Information Software (Cat FIS)   program that permits data, tracked and stored in the ADEM 2000 software, to be downloaded to a PC for analysis.

Caterpillar information display (CAT-ID)   the Caterpillar digital dash display that provides the driver with ECM feedback data such as fuel economy and engine parameters.

cathode   negative electrode; the electrode from which electrons flow.

cathode ray tube (CRT)   device used as a computer monitor.

cavitation   describes metal erosion caused by the formation and subsequent collapse of vapor pockets (bubbles) produced by physical pulsing into a liquid such as that of a wet liner against the wall of coolant that surrounds it. Bubble collapse causes high unit pressures and can quickly erode wet liners when the protective properties of the coolant diminish.

CD-ROM   an optically encoded data disk that is read by a laser in the same way an audio CD is read and is designed for read-only data.

CENTINEL Cummins on-board, engine oil management system that can extend engine oil change intervals up to 300,000 miles (483,000 km) by using monitoring electronics and a makeup oil tank.

central processing unit (CPU) computer subcomponent that executes program instructions and performs arithmetic and logic computations.

centrifugal filter a filter that uses a centrifuge consisting of a rotating cylinder charged with pressurized fluid and canted jets to drive it; centrifugal filters often have high efficiencies and are often of the *bypass* type.

centrifugal force the force acting outward on a rotating body.

centrifuge a device that uses centrifugal propulsion or a centrifugal force principle of operation.

CELECT injector a Cummins EUI.

cetane improvers see *ignition accelerators*.

cetane number (CN) the standard rating of a diesel fuel's ignition quality. It is a comparative rating method that measures the ignition quality of a diesel fuel versus that of a mixture of cetane (good ignition characteristics) and heptamethylnonane (poor ignition characteristics). A mixture of 45% cetane and 55% would have a CN of 45. Diesel fuels refined for use on North American highways are classified by the ASTM as 1D and 2D and must have a minimum CN of 40.

chain hoist a mechanical or power-operated ratcheting lifting device consisting of an actuating block, lift chains, and hook.

characters per second (CPS) speed rating of a printer device.

charge air cooling the cooling of turbo boost air by means of ram air or coolant medium heat exchangers.

charge differential electrical pressure usually described as *potential difference* and measured in *voltage*.

charging circuit the portion of the fuel subsystem that begins with the charging or transfer pump and is responsible for delivering fuel to the injection pumping/ metering apparatus. In a port-helix metering pump, this extends through the charging gallery of the injection pump.

charging pressure a term used to describe the pressure on the charge side of the transfer pump in a fuel subsystem. Charging pressure parameters are defined by the cycle speed of the charging pump, the flow area it unloads to, and regulating valve.

charging pump the pump responsible for moving fuel through the fuel subsystem. Plunger, gear, and less commonly, vane-type pumps are used.

Charles Law states that the volume occupied by a fixed quantity of gas is directly proportional to its temperature if the pressure remains constant.

chassis dynamometer a test bed that measures brake power delivered to the vehicle wheels by having them drive roller(s) to which torque resistance is applied and accurately measured.

chassis-mounted charge air cooling (CMAC) method of cooling turbo boost air using a ram air heat exchanger: effective in highway applications, less so in off-road service.

chatter a nozzle bench test characteristic in which a nozzle valve rapidly opens and closes; caused by the slow rate of pressure rise when testing nozzle valves.

check engine light (CEL) a dash warning light that is often used as a first level alert to the driver.

chemical bonding the force holding atoms together in a molecule or a crystal.

chief executive officer (CEO) the head of an organization.

chip a complete electronic circuit that has been photo-infused to a semiconductor material such as silicon; also known as I/C (integrated circuit), microchip.

chopper wheel the rotating disc that cuts a magnetic field to produce rotational speed or rotational position data to an ECM either by producing an AC voltage value or by pulse width modulation.

clearance volume the volume in an engine cylinder when the piston is at top dead center.

clevis a yoke that is often used in conjunction with a clevis pin and a lever to convert rotary motion to linear or vice versa.

client anything in a computer processing cycle or multiplex data transaction that can be described as having a need.

clipboard temporary storage location for data during cut, paste, and program transfer operations.

clock speed the measure of how fast a CPU can process data measured in MHz (megahertz) or millions of cycles per second.

clockwise (CW) right-hand rotation.

closed circuit an electrical circuit through which current is flowing.

closed circuit voltage (CCV) voltage measured in an energized circuit.

cloud point the temperature at which wax crystals present in all diesel fuels become large enough to make the fuel appear hazy. It is also the point at which plugging of fuel filters becomes a possibility. The cloud point is usually 5°F (3°C) above the fuel's pour point.

cluster the smallest data storage unit on a diskette.

CMP sensor camshaft position sensor.

coalesce to combine to form a single whole.

coaxial cable type of wiring used to transmit signals with almost unlimited bandwidth but unable to carry two-way signals.

coder/decoder (Codec) device that converts analog voice signals to digital signals and vice versa.

coefficient of friction a means of rating the aggressiveness of friction materials; alters with temperature and the presence of any kind of lubricant.

coefficient of thermal expansion the manner in which a material behaves as it is heated and cooled. For instance, aluminum has a higher coefficient of thermal expansion than steel meaning that when a similar mass of each material is subjected to an identical amount of heat, the aluminum will expand more.

coils electromagnetic devices used as the basis of solenoids, transformers, and motors, and in electronics, to shape voltage waves.

cold-start strategy a programmed startup sequence in an electronic management system in which the timing, fuel quantity, and engine-operating parameters are managed on the basis of ambient and engine fluid temperatures. During this process, other inputs such as throttle position may be ignored by the ECM.

combustion   the act of burning a substance. An oxidation reaction.

combustion pressure   usually refers to peak cylinder pressure during the power stroke.

command circuit   used to describe input sensors such as the TPS (throttle position sensor) that commands (requests) an output from the ECM.

common rail (CR) system   fuel injection system in which injection pressures are created by a pump that then supplies fuel to an accumulator or common rail connected to fuel injectors. The fuel injectors are then electrically or electrohydraulically actuated by the ECM.

compact disk (CD)   optically encoded digital data storage.

companion cylinders   term used to describe pistons paired by their respective crank throws to rotate together through the engine cycle such as #1 and #6 in an in-line, six-cylinder engine.

comparative measuring   measuring instruments that gauge a dimension but require another instrument to produce an actual value. Dividers would require a tape measure to convert the dimension measured to a value.

comparitor   anything used to compare one value to another.

comparitor bench   a fuel injection pump test fixture used to compare the performance and output values of an injection pump with a set of master specifications. Usually consists of a means of driving the injection pump (as if it were being driven by the engine it is designed to fuel), a drive-turret equipped with a protractor (for phasing), and graduated vials (means of measuring fuel quantity injected). The term can also be used to describe the test fixtures used to set up mechanical and electronic unit injectors.

compound   1. a substance consisting of two or more elements held together by chemical force and not necessarily retaining any characteristics of the composite elements. 2. the process of increasing the force acting on a plunger or piston by using both mechanical and fluid forces.

compressed air   the means of powering many tools and truck chassis equipment in many applications. It is usually pressurized to between 90 and 150 psi, plumbed throughout the shop, and accessed by quick couplers.

compressed natural gas (CNG)   pressurized natural gas used for commercial and automotive vehicles; consists largely of methane.

compression ignition (CI)   an engine in which the fuel/air mixture is ignited by the heat of compression.

compression ratio   the ration of piston swept volume to total cylinder volume with the piston at bottom dead center: a volumetric ratio not a pressure ratio.

compression ring   the ring(s) designed to seal cylinder gas pressure located in the upper ring belt.

compressional load   a force that attempts to compress or squeeze from diametrically opposite directions to a common point in the component under load.

compressor housing   the section of a turbocharger responsible for compressing the intake air and feeding it into the intake circuit; also known as *impeller housing*.

Compulink   Cummins CELECT EST.

computer-assisted design (CAD)   the commonly used industrial component design tool.

computer-assisted machining (CAM)   programmable computer-managed machining.

concentric   circles having a common center.

concept gear   found in some diesel engine timing gear trains, a concept gear is a two-piece assembly that uses coaxial springs between the hub and outer toothed ring to maintain zero lash tooth contact with the gears it is in contact with.

condensation   the changing of a vapor to a liquid by cooling.

condenser   see *capacitor*.

conductance   the ability of a material to carry an electrical current.

conduction   heat transmission through solid matter.

conductors   materials that readily permit the flow of electrons from atom to atom; usually metallic elements that have less than four electrons in their outer shells.

connecting rod   the rigid mechanical link between the piston wrist pin and the crankshaft throw.

consolidated engine controller   an ECM that houses the microcomputer and output switching such as injector drivers.

constant horsepower   sometimes used to describe a *high torque rise* engine.

constant volume sampling (CVS)   an exhaust gas measurement procedure used before certification.

continuity   an unbroken circuit; used to describe a continuous electrical circuit. A continuity test would determine if a circuit or circuit component was capable of current flow.

continuously open throttle valve (CTV) brake   Mercedes-Benz variation on the internal engine compression brake using small valves (the CTVs) are fitted into the engine cylinder head that allow some cylinder leakage to the exhaust during both the compression and exhaust strokes under braking.

control rack   the fuel control mechanism on an MUI or multicylinder port-helix metering pump that when moved linearly, rotates the pumping plunger(s) in unison.

control rod   in a DDC two-stroke cycle MUI engine, it links the governor with the control tube, which when rotated moves the MUI racks.

control sleeve   the component that is tooth meshed to the control rack and connects to the plungers by means of slots, used to rotate the plungers in the barrels.

control strategy   the manner in which an ECM has been programmed to manage the engine especially in the event of an electronically monitored problem.

control unit   the part of a computer CPU responsible for fetching, decoding, executing, and storing.

controller area network (CAN)   a data bus system developed by Robert Bosch and Intel for vehicle applications. CAN is a serial data transmission network used as the basis for SAE J1850 (automotive) and SAE J1939 (truck) data backbones.

convection   heat transfer by currents of gas or liquids.

conventional memory   the first data logged into RAM on boot-up used primarily to retain the operating system.

conventional theory   (of current flow) asserts that current flows from a positive source to a negative source. Despite the fact that it is fundamentally incorrect, it is nevertheless widely accepted and used. See *electron theory*.

cooled exhaust gas recirculation (C-EGR)   introduced in 2002 to address EPA 2004 diesel emission standards. Dead end gas is first cooled, then rerouted into the intake system to dilute the intake charge and lower temperatures, reducing NO_x emission.

CoPilot   Mack Trucks V-MAC digital monitor and driver display unit.

coprocessor   a chip or CPU enhancement designed for specific tasks such as mathematical calculation.

corrosive   alkaline or acidic substances that dissolve metals and skin tissue.

coulomb   one coulomb is equal to 6.28×10^{18} electrons.

counterclockwise (CCW)   left-hand rotation.

counterflow radiator   a double pass radiator in which coolant is cycled through U column tubes from usually a bottom-located intake tank to a bottom located output tank; they have higher cooling efficiencies than other radiator designs.

covalent bonding   the atomic condition that occurs when electrons are shared by two atoms.

covert   term that means *undercover,* but is commonly used in vehicle electronics to describe the logging of data that cannot be read using the commonly available diagnostic software. Events such as engine overspeed conditions that could impact on the system warranty are often written *covertly* to an electronic system.

cracked rod   connecting rod manufactured and machined in one piece following which the big end is separated by a precisely defined fracture. This ensures a cap-to-rod fit of the highest precision.

crank angle   a location in an engine cycle noted by rotational degrees through the cycle.

crank axis   center point about which a crankshaft rotates.

crankcase   the lower portion of the engine cylinder block in which the crankshaft is mounted and under which is the lubrication oil sump.

crankshaft   a shaft with offset throws designed to convert the reciprocating movement of pistons into torque.

crank throw   the offset journal on a crankshaft to which a connecting rod is connected.

creep   describes the independent movement of two components clamped by fasteners when they have different coefficients of thermal expansion or have different mass, which means their expansion and contraction rates do not concur.

crimping pliers   pliers designed to crimp a terminal to a wire without crushing or damaging the terminal.

crossflow radiator   a usually low profile design of radiator (used with aerodynamic hood/nose), in which the entry and output tanks are located at either end and coolant flow is horizontal.

crossflow valve configuration   a cylinder head valve configuration in which the intake and exhaust valves are located in series in the cylinder head, meaning that gas flow from the inboard valve differs from (and may interfere with) that of the outboard valve.

crosshead piston   an articulating piston with separate crown and skirt assemblies in which the connecting rod is bolted directly to the wrist pin.

crossover   a pipe that connects a pair of fuel tanks mounted on either side of a truck frame at the sump level

enabling fuel to be drawn from one tank while enabling equal fuel load in each tank.

crown   the leading edge face of a piston or in articulating pistons, the upper section of the piston assembly. Crown geometry (shape) plays a large role in defining the cylinder gas dynamic.

crown valve   a now obsolete DDC MUI nozzle valve.

crown valve nozzle   an obsolete DDC hydraulic injector nozzle integral with early version MUIs.

crude oil   the organic fossil fuel pumped from the ground from which diesel fuel, gasoline, and many other petroleum products are refined; raw petroleum.

Cummins Accumulator Pump System (CAPS)   a Cummins innovative variation of common rail fueling that uses full authority electronic management and is used on the ISC and ISL engines.

Cummins electronic engine control (CELECT)   computer subcomponent that executes program instructions and performs arithmetic and logic computations.

Cummins Engine Company   a major manufacturer of truck diesel engines. Corporate center is Columbus, Indiana.

Cummins ISB   Interact System B series, in-line 3.9 (4 cylinder) and 5.9 liter (6 cylinder) engines.

Cummins ISC   Interact System C series, in-line 8.3 liter, 6 cylinder engine.

Cummins ISL   Interact System L series, in-line 8.9 liter, 6 cylinder engine.

Cummins ISM   Interact System M series, in-line 11 liter, 6 cylinder engine.

Cummins ISX   Interact System X series, in-line 15 liter, 6 cylinder engine.

current   the flow of electrons in a closed electrical circuit.

current transformer   a DMM accessory that permits high electrical current flow values to be transduced and read.

cursor   the underline character or arrow that indicates the working location on a computer screen display.

cybernetics   the science of automated (computer) control of machines, systems, and nature. Diesel engine controllers or ECMs are *cybernetic* devices.

cycle   1. a sequence of events that recurs such as those of the *diesel cycle.* 2. one complete reversal of an alternating current from positive to negative.

cylinder block   the main frame of any engine to which all the other components are attatched.

cylinder gas dynamic   engine cylinder gas movement during the cycle: high turbulence was an objective in many older diesel engines while lower turbulence or quiescent dynamics are used in many newer diesels with high injection pressures.

cylinder head   the components clamped to a cylinder block containing the engine breathing and fueling control mechanisms.

cylinder leakage tester   device used to test cylinder leakage by applying regulated air to the cylinder at a controlled volume and pressure and producing a percentage of leakage specification.

cylinder volume   total volume in an engine cylinder with the piston at BDC: the sum of swept volume and clearance volume.

Darlington pair   two transistors arranged to form an amplifier that permits a very small current to switch a large one.

data   raw (unprocessed) information.

database   a data storage location or program.

data bus   multiplex backbone consisting of a twisted wire pair.

data compression   a means of reducing the physical storage space for data by coding it.

data frame   a data tag consisting of 100–150 bits for transmission to the bus. Each tag codes a message for sequencing transmission to the data bus and also serves to limit the time it consumes on the bus.

data hub   the hub of a network system. Used by most truck engine OEMs to log data such as warranty status, repair history, and proprietary programming of on-board ECMs.

data link   the connection point or path for data transmission in networked devices.

DataMax   Mack Trucks onboard data logger and driver display system. Used in conjunction with InfoMax can enable downloading of vehicle performance data for analysis.

data processing   the production and manipulation of data by a computer.

dead volume fuel   fuel that is statically retained for a portion of the cycle; usually refers to the fuel retained at residual line pressure in a high-pressure injection pipe that connects injection pump elements with hydraulic injectors in PLN system.

decoding   a CPU control unit operation that translates program instructions.

default   pre-selected option in computer processing outcomes that kicks in when a failure occurs outside the programmed algorithm. Failure strategy that permits limited functionality when a critical input is lost. Revert to basics. Limp-home mode.

delivery valve   a combination check and pressure management valve that is used on many hydromechanical diesel fuel injection systems.

desktop   a computer term that either describes a nonportable, desk-based PC system or the screen display at any given moment of PC operation.

detonation   combustion in an engine cylinder occurring at an explosive rate, accelerated by more than one flame front; caused by a number of different conditions but in diesel engines often by prolonged ignition lag when ambient temperatures are low when it is known as *diesel knock*.

Detroit Diesel Corporation (DDC)   a major diesel engine manufacturer, part of the DaimlerChrysler/Freightliner Corporation. Corporate center is Dearborn, Michigan.

Detroit Diesel electronic controls (DDEC)   DDEC I was introduced in 1985 and marketed in 1987. It was the first full authority engine management system available on a North American engine. DDEC has evolved through a number of versions.

Deutsch connector   a widely used, weatherproof, proprietary electrical and electronic connector.

device drivers   software used to control input and output devices.

Diagnostic Link   a DDEC PC-based troubleshooting software package driven from MS Windows, designed to guide the technician through troubleshooting sequences, customer data programming, DDEC programming, and data analysis.

diagnostic pressure sensor   Cummins HPI-TP sensor located downstream from the rail actuator in each bank.

dial bore gauge   an instrument designed to facilitate rapid bore comparative measurements, much used by the diesel engine rebuilder.

dial indicator   an instrument designed to measure movement, travel, or precise relative dimensions. They consist of a dial face, needle, and spring-loaded plunger. They can measure values down to one hundred thousandth of an inch or thousandths of a millimeter.

diamond dowels   diamond-shaped alignment dowels used on flywheel housings that are less inclined to deformation than cylindrical dowels.

diatomic   a molecule consisting of two atoms of the same element.

dielectric   insulator substance such as the separation plates used between the conductor plates in a typical capacitor.

diesel coolant additives (DCA)   proprietary supplemental coolant additives.

diesel cycle   the four stroke, compression ignition cycle patented by Rudolf Diesel in 1892. Though the term *diesel* can be used to describe some 2-stroke cycle CI engines, the diesel cycle is necessarily a 4-stroke cycle.

diesel fuel   a simple hydrocarbon fuel obtained from crude petroleum by means of fractioning and usually containing both residual and distillate fractions.

diesel knock   a detonation condition caused by prolonged ignition lag.

diffuser   the device in a turbocharger compressor housing that converts air velocity into air pressure.

digital audiotape (DAT)   high-density data storage tape written to by a helical scan head.

digital calipers   a precise id and od measuring instrument with the appearance of Vernier calipers, the accuracy of a micrometer, and the ability to convert from the standard to metric system at the push of a button.

digital computer   a calculating and computing device capable of processing data using coded digital formats.

digital diagnostic reader (DDR)   DDC term for an EST.

digital diagnostic tool (DDT)   term used to describe a reader/programmer EST.

digital micrometer   a micrometer that displays dimensional readings digitally.

digital multimeter (DMM)   A voltage, resistance (ohms), and current (amperes) reading instrument.

digital signals   data interchange/retention signals limited to two discernable states; combinations of ones and zeros into which data, video, or human voice must be coded for transmission/storage and subsequently reconstructed.

digital video disk—read-only memory (DVD-ROM)   An optically encoded data storage medium that can retain 25 times the data of a CD-ROM.

digitizing   the process used to convert data to digital format.

direct current (DC)   current flow through a circuit in one direction only.

direct fuel feed (DFF) A Cummins PT injector type that stops fueling under motoring conditions.

direct injection (DI) describes any engine in which fuel is injected directly into the engine cylinder and not to any kind of external prechamber. Most current diesel engines are direct injected.

diskettes the primary portable data storage media for PCs.

disk-operating system (DOS) The set of software commands that govern computer operations and enable functional software programs to be run.

distillate any of a wide range of distilled fractions of crude petroleum, some of which would be constituents of a diesel fuel. Refers to the more volatile fractions in a fuel. Sometimes used to refer to diesel fuels.

distributor head section of a Bosch-type, sleeve-metering, rotary distributor pump in which the plunger moves; contains delivery passages, an electric fuel shutdown, screw plug with vent screw, and delivery valves.

distributor plunger center and cross-drilled plunger used in rotary distributor pumps to feed injection fuel to the hydraulic head or distributor head supply passages.

distributor rotor drive shaft-driven rotor on an inlet-metering, opposed plunger distributor pump that connects the pump chamber with the supply passages in the hydraulic head for fuel delivery to the injectors.

dividers a comparative type measuring compass usually with an adjusting screw for setting precise dimensions.

doping the process of adding small quantities of *impurities* to semiconductor crystals to provide them with either P or N electrical characteristic.

double helix a port-helix plunger design with both upper and lower helix characteristics that results in a variable beginning and ending of the pump effective stroke.

double pass radiator a counterflow radiator in which the coolant is routed to make two passes, therefore, entering and exiting from separate tanks both located either at the top or the bottom of the radiator. A high-efficiency radiator.

downflow radiator a typical radiator in which hot coolant from the engine enters at the top tank, flows downward, and exits through a bottom tank.

downlink the transmission signal from a communications satellite to an Earth receiver or the receiver itself.

download data transfer from one computer system to another; often used to describe proprietary data transfer when reprogramming vehicle ECMs.

driver another name for a *power transistor*. A transistor capable of switching high-current loads.

droop an engine governor term denoting a transient speed variation that occurs when engine loading suddenly changes.

droop curve a required hydromechanical governor characteristic in which fueling drops off in an even curve as engine speed increases from the rated power value to high idle.

dry liners liners that are fitted either with fractional looseness or fractional interference that dissipate cylinder heat to the cylinder block bore and have no direct contact to the water jacket.

dry sump an engine that uses a remotely located oil sump; not often seen on highway diesel applications but used in some bus engines to reduce the profile of the engine.

dual helices a plunger geometric design with identical helices machined diametrically opposite each other on the plunger. Commonly used, it helps prevent side loading of the plunger at high-pressure spill.

dumb node a network node with no independent processing or data retention capability.

Duramax Isuzu-built, GM diesel engines available in 6600 and 7800 versions.

dynamic RAM (DRAM) RAM with high-access speed.

dynamometer a testing device that loads an engine by applying a resistance to turning effort (torque) and factors this against time to produce brake power values. Often used to performance test or break in engines after reconditioning.

Dynatard Mack internal engine compression brake; not used in Mack engines after 1996.

dyno short form for *dynamometer*.

EEPROM electronically erasable, programmable read-only memory. Vehicle computer memory category that can be rewritten or flashed with customer or proprietary reprogramming. Includes an ECMs write-to-self capability.

E-9 Mack Trucks V-8 engine.

E 7 Mack Trucks in-line, 6-cylinder, 12-liter engine.

E7-EUP Mack Trucks V-MAC III managed, in-line, 6-cylinder 12-liter engine.

eccentric not circular; axes that are not common.

Echeck Cummins EST.

Econovance Mack Trucks mechanical or electronic variable timing device for port-helix metering injection pumps.

effective stroke describes that portion of a constant travel plunger or piston stroke used to actually pump fluid.

ELAB Bosch fuel shutoff solenoid used on early V-MAC I PE7100 injection pumps.

electricity a form of energy that results from charged particles, specifically electrons and protons, either statically (accumulated charge) or dynamically such as current flow in a circuit.

electrohydraulic injector an ECM-switched injector used on common rail injection systems.

electromagnetic interference (EMI) Low level radiation (such as emitted from electrical power lines, vehicle radar, etc.) that can interfere with signals on data buses unless suppressed.

electrolysis a chemical change produced in electrolyte by an electrical current often resulting in decomposition.

electrolyte a solution capable of conducting electrical current.

electromagnetism describes any magnetic field created by current flow through a conductor.

electromotive force (EMF) voltage or charge differential.

electron a negatively charged component of an atom.

electronically erasable, programmable read-only memory see EEPROM.

electron theory the theory that asserts that current flow through a circuit is by electron movement from a negatively charged point to a positively charged one. See *conventional theory*.

electronic control analyzer programmer (ECAP) Caterpillar PC-based reader/programmer instrument.

Electronic Diesel Controls (EDC) Bosch term for its engine ECMs.

electronic/engine control unit (ECU) refers to the computer and integral switching apparatus in an electronically controlled system. Some engine OEMs use this term rather than the more commonly used ECM.

electronic/engine control module (ECM) refers to the computer and integral switching apparatus in an electronically controlled vehicle system. Most engine OEMs prefer this term to describe their engine controllers.

electronic engine management computerized engine control.

electronic distributor unit (EDU) DDC term for injector drivers.

electronic foot pedal assembly (EFPA) pedal mechanical travel managed by the TPS.

electronic governor any kind of governing using computer controls.

electronic management system management by computer or computers.

electronics branch of electricity concerned with the study of the movement of electrons through hard wire, semiconductor, gas, and vacuum circuits.

electronic service tool (EST) covers a range of instruments including DMMs, diagnostic lights, generic and proprietary reader-programmers, and PCs.

electronic smart power (ESP) (Cummins)

electronic technician (ET) Caterpillar PC-based software that enables the technician to diagnose system problems, reprogram ECMs, and access system data for analysis to produce fuel mileage figures and driver performance profiles.

electronic unit injector (EUI) The cam-actuated, electronically controlled pumping mechanism used to fuel most full authority, electronically controlled truck diesel engines.

electronic unit pump (EUP) a cam-actuated, ECM-controlled pumping and metering unit that supplies a hydraulic injector by means of a high-pressure line.

electrostatics the force field that surrounds an object with an electrical charge.

element 1. any of more than one hundred substances (most naturally occurring, some man-made) that cannot be chemically resolved into simpler substances. 2. a component part of something such as a pump *element*.

e-mail network messaging software that enables communication between PCs using modems and the telephone system.

emulsify the dispersion of one liquid to another or the suspension of a fine particulate in a solution.

emulsion the dispersion of one liquid into another such as water in the form of fine droplets into diesel fuel.

end gas the gas that results from combusting fuel in engine cylinders; usually means the gases present at flame quench, that is, before any exhaust gas treatment so a mixture of CO_2, H_2O, and whatever noxious gases are present.

ending of energizing (EOE) denotes the end of the switched duty cycle of an EUI.

ending of injection (EOI) instant that fuel injection ceases.

energized-to-run (ETR) any of a group of solenoids that must be electrically energized to remain in ON status, a non-latching type solenoid.

end of line (EOL) usually in reference to terminating a programming procedure.

engine a machine that converts one form of energy to another.

engine brake any type of engine retarder. The term usually describes an internal engine compression brake but may also refer to an exhaust compression brake or an engine-mounted hydraulic retarder.

engine displacement the sum of the swept volume of all the engine cylinders.

engine dynamometer a dynamometer used for testing the engine on a test bed outside of the chassis.

engine family rating code (EFRC) engine series ratings (Caterpillar).

engine hours a means of comparing engine service hours to highway mileage. Most engine OEMs equate 1 engine hour to 50 highway linehaul miles (80 km), so a service interval of 10,000 miles (16,000 km) would equal 200 engine hours. The term *service hours* is also used.

engine longevity the engine life span. In highway diesel engines it is usually reckoned in miles on highway engines and hours in off-highway applications.

engine position sensor (EPS) shaft position sensor using a reluctor pulse generator or Hall effect principle.

engine silencer a *muffler* that uses sound absorption and resonation principles to change the frequency of engine noise.

Environmental Protection Agency (EPA) Federal regulating body that sets and monitors noxious emission standards among other functions.

EPS engine position sensor.

etching bearing or other component failure caused by chemical action.

E-Tech Mack Trucks V-MAC III, electronic unit pump (EUP) fueled, E-7 engine.

ethylene glycol (EG) an antifreeze of higher toxicity that the EPA hopes to phase out.

E-Trim Caterpillar EUI fuel flow specification required to be programmed to the ADEM ECM whenever an EUI is removed and replaced. E-Trim data is important because it enables the ECM to balance fueling to each cylinder.

execute effect an operation or procedure.

executive the resident portion of a computer or program operating system.

exhaust blowdown the first part of the cylinder exhaust process that occurs at the moment the exhaust valves open.

exhaust brake an external engine compression brake that operates by choking down the exhaust gas flow area; sometimes used in conjunction with an internal engine compression brake, meaning that the piston is contributing to retarding effort on both its upward strokes.

exhaust gas recirculation (EGR) A means of routing "dead" end gas back into the intake to "dilute" the intake charge of oxygen, reducing combustion heat and therefore NO_x. In most truck diesel engines EGR, exhaust gas is cooled by heat exchanger before rerouting to the intake.

exhaust manifold the cast-iron or steel component bolted to the cylinder exhaust tracts responsible for delivering the end gases to the turbocharger and the exhaust system.

expansion board a circuit board added to a computer system to increase its capability.

explosion an oxidation reaction that takes place rapidly; high-speed combustion.

extended life coolant (ELC) Coolant premix that claims a service life of up to 6 years with almost no maintenance.

external compression brake refers to an engine exhaust brake.

EZ-Tech International engines PC software.

failure analysis diagnosis of a failed component usually out of engine.

fanstat a combination temperature sensor and switch (usually pneumatic) used to control the engine fan cycle.

farad a measure of *capacitance*. One farad is the ability to store 6.28×10^{18} electrons at a 1 V charge differential.

FARC valve Caterpillar turbo aneroid.

fault mode indicator (FMI) Defines a component or circuit failure numerically to an EST by ascribing to it one of twelve possible failure modes (SAE).

fax short form of *facsimile*; a method of reproducing an image or text digitally and transmitting it using the phone system.

feedback assembly the engine's mechanical, self management components consisting of a gear train, camshaft, valve trains, MUI and EUI actuating trains, fuel injection pumping apparatus and valves.

fetching CPU function that involves obtaining data from memory.

fiber optics the transmission of laser light waves through thin strands of fiber used to digitally pulse data more cheaply and at much higher speeds than copper wire.

FIC module fuel injection control module. Usually a slave module connected by a proprietary CAN bus to the engine ECM, responsible for controlling an injection pump (such as on V-MAC I) or electronic unit pumps (V-MAC III and MB 900/4000).

field effect transistor (FET) group of transistors used to switch or amplify within a circuit.

fields specific items of (electronic) information.

file a collection of related data.

fire point the temperature at which a combustible produces enough flammable vapor for a continuous burn; always a higher temperature than *flash point*.

fire ring normally used to refer to the fixed ring that may be integral with the cylinder head gasket responsible for sealing the cylinder. Sometimes used to refer to the top compression ring but this usage is uncommon.

fixed disk a data storage device used in PCs and mainframe computers consisting of a spindle and multiple stacked data retention platters.

flame front during flame propagation, the leading edge of the flame in an engine cylinder.

flame propagation the flame pattern from ignition to quench during a power stroke in an engine cylinder.

flame quench the moment that the flame ceases to propagate or extinguishes in an internal combustion engine.

flammable any substance that can be combusted.

flash term used to describe the downloading of new software to EEPROM.

flashback a highly dangerous condition that can occur in operating oxy-acetylene equipment in which the flame may travel behind the mixing chamber in the torch and explode the acetylene tank using the system oxygen. Most current oxy-acetylene torches are equipped with flashback arresters.

flash codes the ECM-generated fault codes that are usually displayed by means of diagnostic lights and alert the driver or technician as to the nature of an electronically monitored malfunction; also known as *fault codes, blink codes*.

flash memory term used to describe write/ overwrite memory in a computer or ECM such as EEPROM.

flash point the temperature at which a combustible produces enough flammable vapor for momentary ignition.

flash programming term that has come to mean any reprogramming procedure.

flash RAM nonvolatile RAM; NV-RAM.

floppy disks see *diskettes*.

floptical a diskette that combines magnetic and optical technology to achieve high density storage capacity.

flow area the most restricted portion of a fluid circuit; for instance, a water tap sets a flow area and as the tap is opened, the flow area increases, thereby increasing the volume flow of water.

flow control refers to any device that can proportionally control flow through a circuit. A thumb over the end of a hose is a flow control device.

fluid any substance that has fluidity. Both liquids and gases are fluids. Fluid power incorporates both hydraulics and pneumatics.

fluid friction the friction of dynamic fluids, always less than solid friction.

fluidity a substance in state that permits it to conform to the shape of the vessel in which it is contained. Both liquids and gases possess *fluidity*.

fluid power term used to describe both *hydraulics* and *pneumatics*.

flutes protruding lands with grooves in between.

flywheel an energy and momentum storage device usually bolted directly to the crankshaft.

flywheel housing concentricity a critical specification that ensures that the relationship of the flywheel and anything connected to it is concentric.

follower used to describe a variety of devices that ride a cam profile and transmit the effects of the cam geometry to the train to be actuated; also known as *tappet*.

font typeface size and appearance.

force the action of one body attempting to change the state of motion of another. The application of force does not necessarily result in any work accomplished.

foreground computations computer-operating responses that are prioritized, such as the response to a critical command input such as the TPS (throttle position sensor) whose signal must be acted on immediately to generate the appropriate outcome.

format 1. To alter the appearance or character of a program or document. 2. To prepare data retention media, such as diskettes, to receive data by defining tracks, cylinders, and sectors, a process that removes any previously logged data.

forward leakage an injector bench fixture test that tests the nozzle seat sealing integrity.

fossil fuel unrenewable, organically derived fuels such as petroleum and coal.

fractions refers to separate compounds of crude petroleums separated by distillation and other fractioning methods such as catalytic and hydrocracking and classified by their volatility.

Freightliner Corporation   currently the truck chassis manufacturer with the largest market share. Corporate center is Portland, Oregon.

friction   the resistance an object or fluid encounters in moving over or through another.

friction bearing   a shaft-supporting bearing in which the rotating member can directly contact the bearing face or race.

fuel   substance that can be used as source for heat energy.

fuel conditioner   usually unknown quantities of cetane improver and pour point depressants suspended in an alcohol base.

fuel control actuator   any of a number of electronically controlled devices used as fuel control mechanisms.

fuel demand command signal (FDCS)   International HEUI output driver signal.

fuel filter   device for filtering sediment from fuel rated by entrapment capability.

fuel heater   a heat exchanger device used in extreme cold to prevent diesel fuel from waxing in the fuel subsystem.

fuel map   a diagram or graph used to indicate fueling through the entire performance range of an engine; also used to describe the ECM fuel algorithm.

fuel pressure sensor (FPS)   a pressure sensing mechanism usually of the variable capacitance type that measures the charging pressure in the fuel subsystem and signals its value to the ECM.

fuel rate   actual rate of fuel pumped through an injector to an engine cylinder; factored by cam geometry and engine rpm.

fuel ratio control (FRC)   Caterpillar aneroid mechanism for limiting fueling in low boost conditions.

fuel subsystem   the fuel circuit used to pump fuel from the vehicle fuel tank and deliver it to the fuel metering/injection apparatus. The fuel subsystem typically comprises a fuel tank, water separator, primary filter, transfer or charge pump, secondary filter, and the interconnecting plumbing.

fuel tank   the fuel storage reservoir on a vehicle.

fueling actuator   Cummins HPI-TP metering control solenoid; also known as the rail actuator.

fueling algorithm   the set of ECM programmed rules and procedures designed to produce the desired performance and emissions from an engine at any given moment of operation

fueling chamber   the lower chamber in the Cummins TP injector that forms the injector pumping chamber.

full flow filter   a filter plumbed in series on the charge side of the pump that feeds a circuit.

function keys   numerical keys prefixed by the letter F that act as program commands and shortcuts.

fuzzy logic   an ECM or computer processing outcome that depends on multiple inputs, operating conditions, and operating system commands: opposite to *closed loop.*

gallons per hour (gph)   the means of rating liquid flow in a hydraulic circuit.

galvanometer   a meter used to measure small electrical currents.

gas analyzer   a test instrument for measuring and identifying the exhaust gas content.

gas dynamics   the manner in which gases behave during the compression and combustion strokes and the processes of engine breathing.

gasket yield point   the moment that a malleable gasket is crushed to its desired shape to conform to the required shape between two clamped components to provide optimum sealing.

gasoline   a hydrocarbon fuel composed of the volatile petroleum fractions from the aromatic and paraffin ranges.

gates   routing switches with either digital or mechanical actuation.

gear pump   a positive displacement pump consisting of intermeshing gears that uses the spaces between the teeth to move fluid through a circuit.

genset   a complete electricity generating unit consisting of an internal combustion engine and an electricity generator.

geosynchronous orbit   the park orbit of communications satellites 35,400 km (22,300 miles) from the Earth's equator.

gerotor   a type of gear pump that uses an internal crescent gear pumping principle.

gigabyte   a billion bytes; a measurement of digital memory capacity.

glazing friction   wearing of a component to a mirror finish.

global positioning satellite (GPS)   refers to communications using telecommunications satellites. Currently used for vehicle tracking, navigation, and data exchange.

governor   a component that manages engine fueling on the basis of fuel demand (accelerator) and engine rpm; may be hydromechanical or electronic.

governor barrel   the stationary governor component in a PT pump, within which the governor plunger rotates and into which the fuel passages are machined.

governor button   the component in a Cummins PT pump that helps define pressure within the governor barrel and therefore greatly influences fuel flow to the rail.

governor differential lever   a double-bell crank-type, lever device that pivots on a fulcrum.

governor gap   a DDC tune-up setting.

governor plunger   the rotating member of the Cummins PT governor, driven by the governor weight carrier.

governor spring   the force, usually variable, that opposes centrifugal force in mechanical governors; often amplified by accelerator pedal travel.

governor weight forks   a means of jamming governor centrifugal weights in their outermost position for purposes of tuning an engine.

graduates   see *vial.*

graphical user interface (GUI)   software such as MS Windows that is icon and menu driven.

gray scaling   used by monochrome monitors and scanners to code color to black, white, and shades of gray.

ground   describes the point or region of lowest voltage potential in a circuit; the portion of a vehicle electrical circuit serving multiple system loads by providing a return path for the current drawn by the load. Used in vehicle systems using 48 V or less and ideal for the commonly used 12 V vehicle systems.

ground strap a conductive strap, usually braided wire, that extends a common ground electrical system.

groupware software that allows multiple users to work together by sharing information.

gumming a term used to describe unburned fuel and lubrication oil residues when they sludge in piston ring grooves and other areas of the engine.

gusset a triangular bracket used to strengthen two perpendicularly joined beams.

Hall effect a method of accurately sensing rotational position speed and digitally signaling it. A rotating metallic shutter alternately blocks and opens a magnetic field from a semiconductor sensor.

handshake establishing a communications connection especially where two electronic systems are concerned. The communication protocols must be compatible.

hard copy computer-generated data that is printed rather than retained on disk.

hard disk see *fixed disk; hard drive.*

hard drive a data storage device used in PCs and mainframe computers consisting of a spindle and multiple stacked data retention platters.

hard parameter a fixed value that usually cannot be altered or rewritten. Maximum rpm would be an example of a hard parameter.

hardware computer equipment excluding software.

Hazard Communications Legislation federal legislation that incorporates the Right to Know clauses that pertain to workplace hazards. Administered by OSHA.

head crash occurs when a computer disk head collides with the hard disk surface causing loss of memory.

headers the manifold deck to which the coolant tubes are attached in a heat exchanger bundle or the term used to describe low gas restriction, individual cylinder exhaust pipes that converge at a point calculated to maximize pulse effect.

headland the area above the uppermost compression ring and below the leading edge of the piston.

headland piston term used to describe a piston design that has minimized the *headland volume.*

headland volume the headland gas volume in a cylinder.

heat energy an expression of the energy potential a substance possesses. It is actually the amount of kinetic energy at the molecular level in an element or compound.

heat engine a mechanism that converts thermal energy into mechanical work.

heat exchanger any of a number of devices used to transfer heat from one fluid to another where there is a temperature difference using the principles of conduction and radiation.

heating value the potential heat energy of a fuel; also known as *calorific value.*

heating, ventilating and air-conditioning (HVAC) Acronym used to describe the climate control system, usually integrated on current trucks.

helical gear a gear with spiral cut teeth.

helical scan technology used to write data at high density on tape helically as opposed to longitudinally.

helices plural of *helix.*

helix a spiral groove or scroll. The helical cut recesses in some injection pumping plungers that are used to meter fuel delivery. Plural: *helices.*

Hg manometer a mercury (Hg) filled manometer.

high-idle speed the highest no-load speed of an engine.

high-pressure injection-time pressure (HPI-TP) Cummins' electronically controlled open nozzle common rail system using ECM-controlled metering and timing actuators to manage fuel injection.

high-pressure pipes the pipes or lines that deliver fuel from an injection pump element to the injector nozzle.

high-pressure washer a high-pressure water pump used to clean equipment and components before repair and inspection that has generally replaced steam cleaners.

high spring injector a type of hydraulic injector nozzle that locates the injector spring high in the injector/nozzle holder body. NOP is usually adjusted by an adjusting screw that acts directly on the spring.

Highway Master Caterpillar remote communications technology that enables remote programming of vehicles. Used in conjunction with Caterpillar Fleet Information software.

histogram a graphic display in which data is represented by rectangular columns used for comparative analysis.

historic codes fault codes that are no longer active but are retained in ECM memory (and displayed) for purposes of diagnosis until they are erased; also known as *inactive codes.*

hone any of a number of types of abrasive stones used for finishing metals. Rotary hones are electrically or pneumatically driven and are used for sizing and surface finishing cylinder liner bores.

horsepower the standard unit of power measurement used in North America defined as a work rate of 33,000 lb.-ft. per minute; equal to 0.746 kW.

host computer a main computer that is networked to other computers or nodes.

HPI-PT electronically managed, common rail, open nozzle fuel system used to fuel the Cummins K-19 engine in which *pressure* is the control variable.

H_2O manometer a water-filled manometer.

hunting rhythmic fluctuation of engine rpm usually caused by unbalanced cylinder fueling.

hunting gears an intermeshing gear relationship in which after timing, the gears may have to be turned through a large number of rotations before the timing indices realign.

hydraulically actuated electronic unit injector (HEUI) Caterpillar oil pressure actuated, high-pressure fuel pumping/injecting element.

hydraulic governor, nonservo a hydraulic governor that uses fuel pressure unloaded into a defined flow area by a positive displacement pump, as the basis for determining rpm; not used on any current truck engines.

hydraulic governor, servo type a hydraulic governor that uses centrifugal weights to sense rpm but the force responsible for actually moving the fuel control mechanism is hydraulic either engine oil or fuel pressure.

hydraulic head stationary member of an inlet metering, opposed plunger rotary distributor pump within which the distributor rotor rotates: as the distributor rotor turns it is brought in and out of register with each discharge port connecting the pump with the fuel injectors.

hydraulic injectors   any of a group of injectors that are opened and closed hydraulically as opposed to electronic: this would include the nozzle assemblies used in many EUI and HEUI units. One OEM uses the term *mechanical injector* in place of *hydraulic injector.*

hydraulics   the science and practice of confining and pressurizing liquids in circuits to provide motive power.

hydrocarbon (HC)   describes substances primarily composed of elemental carbon and hydrogen. Fossil fuels and alcohols are both hydrocarbon fuels.

hydrodynamic suspension   the principle used to float a rotating shaft on a bed of constantly changing, pressurized lubricant.

hydrogen (H)   the simplest, most abundant element in the known universe, occurring in water and all organic matter. Colorless, odorless, and tasteless, explosive. Atomic number 1.

hydromechanical engine management   all engines managed without computers.

hydromechanical governing   engine governing without the use of a computer: requires a means of sensing engine speed (centrifugal force exacted by flyweights/ fuel pressure) and a means of limiting fuel.

hydrometer   an instrument designed to measure the specific gravity of liquids, usually battery electrolyte and coolant mixtures. Not recommended for measuring either in truck engine applications where a *refractometer* is the appropriate instrument due to greater accuracy.

hypermedia   a multimedia presentation tool that permits rapid movement between screens to display graphics, video, and sound.

hypertext link   the highlighting and bolding of a text word/ phrase to enable Web page or program selection by mouse/click. Increases user-friendliness of Internet and many other programs.

hypothesis   a reasoned supposition, not necessarily required to be true.

hysteresis   1. in hydromechanical governor terminology, a response lag. 2. molecular friction caused by the lag between the formation of magnetic flux behind the magnetomotive force that creates it.

icons   pictorial/graphical representations of program menu options displayed on-screen.

idle speed   the lowest speed that an engine is run at usually managed by the governor.

ignition accelerators   volatile fuel fractions that are added to a fuel to decrease ignition delay. They increase CN.

ignition lag   the time period between the entry of the first droplets of fuel to the engine cylinder and the moment of ignition based on the fuel chemistry and the actual temperatures of the engine components and the air charge.

impeller   1. the driven member of a turbocharger, responsible for compressing the air charge. 2. the power input member of a pump such as on a torque converter or hydraulic retarder.

inactive codes   fault codes that are no longer active but are retained in ECM memory (and displayed) for purposes of diagnosis until they are erased; also known as historic codes.

indicated power   expression of gross engine power usually determined by calculation and in the United States, expressed as *indicated horsepower.*

indirect injection (IDI)   describes any of a number of methods of injecting fuel to an engine outside of the cylinder. This may be to an intake tract in the intake manifold or to a cell adjacent to the cylinder such as a pre-combustion chamber.

induction circuit   refers to the engine air intake circuit but more appropriately describes air intake on naturally aspirated engines than on boosted engines.

inert   chemically unreactive. Any substance that is unlikely to participate in a chemical reaction.

inertia   in physics, it describes the tendency of a body at rest or in motion to want to continue in that state unless influenced by an external force.

InfoMax   vehicle to land station data transfer system that uses WiFi short range wireless technology. Permits download of vehicle and driver performance data and upload of customer data programming.

INFORM   Cummins PC-based data management system.

infrared   the wavelength just greater than the red end of the visible light spectrum, but below the radio wave frequency.

infrared thermometer   accurate heat measuring instrument that can be used for checking cylinder fueling balance.

ink jet printer   nonimpact printer that uses a nozzle to shoot droplets onto a page.

injection actuation pressure (IAP)   Caterpillar's HEUI actuation oil pressure.

injection control pressure (ICP)   actuation oil pressure in HEUI fuel systems.

injection control valve   an ECM-controlled, electrohydraulic valve used in the Cummins CAPS fuel system.

injection lag   a diesel fuel injection term describing the time lag between port closure in a pumping element and the actual opening of the injector.

injection pressure regulator (IPR)   ECM-controlled device that manages HEUI-actuating oil pressure.

injection rate   a diesel fuel injection term that is defined as the fuel quantity pumped into an engine cylinder per crank angle degree. In systems except for the HEUI, and common rail (CR) injection rate is determined by the pump actuating cam profile geometry.

injector   a term broadly used to describe the holder of a hydraulic nozzle assembly. May also be used to describe PT, TP, MUI, EUI, and HEUI assemblies.

injector driver module (IDM)   separate injector driver unit used in International's versions of HEUI up to 1997. IDM functions are integrated in a single engine controller ECM in current applications.

injector drivers   the ECM-controlled components that electrically switch EUI and HEUI assemblies. Injector drivers may be integral with the main ECM housing or contained in a separate module or housing.

injector response time (IRT)   time in ms between injector driven signal and EUI control valve closure.

inlet metering   any injection pump that meters fuel quantity admitted to the pump chamber. For instance, in an inlet

metering, opposed plunger injection pump all of the fuel admitted to the high pressure pump chamber is injected each time the plungers are actuated.

inlet restriction a measure of the pressure value below atmospheric, developed on the pull side of a pumping mechanism. Air inlet restriction and fuel inlet restriction are common specifications used by the diesel technician.

inlet restriction gauge instrument that measures (usually air) inlet restriction often on-chassis.

inner base circle (IBC) in cam geometry, the portion of the cam profile with the smallest radial dimension; also known as *base circle*/IBC. When the train riding the cam profile is on IBC, it is unloaded.

input the process of entering data into a computer system.

input devices the hardware, such as a keyboard on a PC, or *sensors* on a vehicle system responsible for signaling/switching data to a computer system.

inside diameter (id) diametrical measurement across a bore.

inside micrometer standard or metric micrometer consisting of a spindle and thimble but no anvil: used for making internal and bore measurements.

INSITE Cummins PC software.

INSPEC Cummins Windows-driven PC vehicle ECM diagnostics and programming software.

insulators materials that either prevent or inhibit the flow of electrons; usually nonmetallic substances that contain more than four electrons in their outer shell.

intake circuit the series of components used to route ambient air into engine cylinders. In a diesel engine, includes filter(s), piping, turbo compressor housing, charge air cooler and intake manifold.

intake manifold the piping that is clamped to the intake tract flange faces responsible for directing intake air into the engine cylinders.

Intebrake internal engine compression brake used on the Cummins ISX Series engine, capable of six-stage progressive braking.

integrated circuit (IC) an electronic circuit constructed on a semiconductor chip, such as silicon, that can replace many separate electrical components and circuits.

Integrated Fuel System Module (IFSM) module containing the fuel management hardware for Cummins HPI-TP managed engines such as ISX.

intensifier piston Caterpillar HEUI hydraulically actuated piston that pumps fuel to injection pressure values; also known as *amplifier piston*.

Interact System (IS) Cummins' term used to describe its integrated electronic engine management systems with the ability to connect with fleet management and analysis software. Acronym IS used ahead of the engine series letter.

Interact System B (ISB) series Cummins' in-line six-cylinder 5.9-liter engine.

Interact System C (ISC) series Cummins' in-line six-cylinder 8.3-liter engine.

Interact System L (ISL) series Cummin's in-line six-cylinder 8.9-liter engine.

Interact System M (ISM) series Cummin's in-line six-cylinder 11-liter engine.

Interact System X (ISX) Cummins in-line, six cylinder, 1.5 liter engine.

interface the point or device where an electronic interaction occurs. Separate vehicle system ECMs will sometimes require interface hardware.

interference fit the fitting of two components so that the od of the inner component fractionally exceeds the id of the outer component. Liners are sometimes interference fit to cylinder bores. Interference fitting requires the use of a press, chilling, heating, or other forceful means.

internal cam ring the means of actuating the pumping plungers in an opposed plunger, inlet-metering rotary injection pump. Cam profiles are machined inside the cam ring and as the plunger rollers rotate within it, they are forced inboard to effect a pump stroke. See Chapter 25 for a full explanation.

internal compression brake any of a number of engine brakes that use the principle of making the piston perform its usual work through the compression stroke and then negate the power stroke by releasing the compression air to the exhaust system at TDC on the completion of the compression stroke.

internal exhaust gas recirculation (I-EGR) introduced by Mack Trucks in smaller vocational diesels in 2002 to achieve EGR within the cylinder head using cylinder head valve timing.

International Truck & Engine see *Navistar*.

Internet the global computer multimedia network communications system accessed through the phone system. A network of networks.

Intranet an internal computer network designed for organizational communications using Internet protocols; also known as *LAN*.

ion an atom with either an excess or deficiency of electrons, that is, an unbalanced atom.

iron (Fe) the primary constituent of steel.

isochronous governor a zero droop governor or one that accommodates no change in rpm on the engine it manages as engine load varies. In electronically managed truck engines, the term is sometimes used to describe engine operation in PTO mode.

J 1667 SAE standards for emission testing of highway diesel engines manufactured before 1991 EPA standards. Currently used by many jurisdictions.

J-1850 data backbone hardware and protocols used in light duty CAN multiplexing systems.

J-1939 data backbone hardware and protocols used in heavy duty CAN multiplexing systems.

Jacobs retarders Jacobs is known mainly for its internal engine compression brakes but also manufactures driveline retarders.

Jacobs brake see *Jacobs retarders*.

Jacobs C-Brake a Jacobs internal engine compression brake designed for certain Cummins engines.

joule unit of energy that describes the work done when an electrical current of 1 ampere flows through a resistance of 1 ohm in 1 second or in mechanical terms.

jumper pipes a term used to describe the pipes that connect the charge and return galleries with DDC MUIs or with each other in multicylinder heads.

kaizan Japanese word meaning "continuous improvement." It has become a catchword in industry and is often linked with the practice of TQM.

KAMPWR electrical circuit that powers KAM (Navistar).

keep alive memory (KAM) nonvolatile RAM.

Kenworth truck OEM owned by Paccar DAF.

kernel the resident portion of a disk or program operating system.

kerosene a petroleum derived fuel with a lower volatility than gasoline and fewer residual oils than diesel fuel.

keyboard the data entry device used on PC systems enabling alpha, numeric, and command switching.

keystone the trapezoidal shape that gets its name from the trapezoidal stones used in a classic Roman arch bridge.

keystone ring a trapezoidally shaped piston ring commonly used in diesel engine compression ring design.

keystone rod a connecting rod with a trapezoidal eye (small end) to increase the loaded sectional area.

kilobyte a quantitive unit of data consisting of 1,024 bytes.

kilowatt a unit of power measurement equivalent to 1,000 watts. Equal to 1.34 BHP.

kinetic energy the energy of motion.

kinetic molecular theory states that all matter consists of molecules that are constantly in motion and that the extent of motion will increase at higher temperatures.

Kirchhoff's first law states that the current flowing into a point or component in an electrical circuit must equal the current flowing out of it.

Kirchhoff's second law states that the voltage will drop in exact proportion to the resistance in a circuit component and that the sum of the voltage drops must equal the voltage applied to the circuit; also known as Kirchhoff's law of voltage drops.

kPa kilopascal. Metric unit of pressure measurement. Atmospheric pressure is equivalent to 101.3 kPa.

L-10 Cummins in-line, six-cylinder, 10-liter engine produced in versions for PT and CELECT management.

lacquering the process of baking a hard skin on engine components usually caused by high-sulfur fuels or engine oil contamination.

ladder switch a 'smart' switch named because it contains a ladder of resistors, usually five per switch, known as a ladder bridge. The processor that receives data from the ladder switches on the data bus has a library of resistor values that enables it to identify switch status and its commands.

lambda the Greek letter λ used as a symbol to indicate stoichiometric combustion. See definition of *stoichiometric*.

lambda sensor an exhaust gas sensor used on electronically managed, SI gasoline-fueled engines to signal the ECM the oxygen content in the exhaust gas.

lamina a thin layer, plate, or film.

lands the raised areas between grooves especially on the ring belt of a piston.

laptop computer a portable PC, larger than a notebook but smaller than a desktop PC.

laser any of many devices that generate an intense light beam by emitting photons from a stimulated source. Used in computer technology to read and write optically.

laser printer a common PC printer device that aims a laser beam at a photosensitive drum to produce text or images on paper.

latching solenoid a solenoid that locks to a position when actuated and usually remains in that position until the system is shut down.

latent heat thermal energy absorbed by a substance undergoing a change of state (such as melting or vaporization) at a constant temperature.

leakoff pipes/lines the low-pressure return circuit used in most current diesel fuel injection systems.

lever a rigid bar that pivots on a fulcrum and can be used to provide a mechanical advantage.

lifters components that ride a cam profile and convert rotary motion of the camshaft into linear motion or *lift*. Lifters used in truck diesel engines are generally solid or roller types.

light-emitting diode (LED) diode that converts electrical current directly into light (photons) and therefore is highly efficient as there are no heat losses.

limiting speed governor (LSG) a standard automotive governor that defines the idle and high idle fuel quantities and leaves the intermediate fueling to be managed by the operator within the limitations of the fuel system.

limp-home see *default*.

linear magnet a proportional solenoid. This term is used by Mack Trucks to describe its rack actuator mechanism.

linehaul terminal-to-terminal operation of a truck, meaning that most of its mileage is highway mileage.

liners the normally replaceable inserts into the cylinder block bores of most diesel engines that permit easy engine overhaul service and greatly extended cylinder block longevity.

link and lever assembly Caterpillar intermediary between the governor and MUIs.

liquefied petroleum gas (LPG) another term used to describe propane, a petroleum-derived gas consisting mostly of methane.

liquid crystal display (LCD) flat panel display consisting of liquid crystal sandwiched between two layers of polarizing material. When a wire circuit below is energized, the liquid crystal media is aligned to block light transmission from a light source producing a low-quality screen image.

load ratio of power developed versus rated peak power at the same rpm.

local area network (LAN) also known as Intranet, a usually private computer network used for communications and data tracking within a company or institution.

local bus an expansion bus that connects directly to the CPU.

locking tang a tab on a component, such as a bearing, that may help position and lock it.

logical processing data comparison and mapping operations by the CPU.

log-on an access code or procedure used in network systems (such as truck OEM data hubs) used for security and identification.

longevity long life or lifespan.

lower helix the standard helix milled into most port-helix plungers used in truck diesel applications. These produce a constant beginning, variable delivery characteristic when no external variable timing mechanism is used.

low spring injector an injector design that locates the spring directly over the nozzle valve, thereby reducing the mass of moving components compared to a high spring model. Injector spring tension is usually defined by shims.

LS fuel low sulfur fuel required for use on highway diesel engines: contains a maximum of 0.05% sulfur.

lubricity literally, the *oiliness* of a substance.

lugging term used to describe an engine that is run at speeds lower than the base of the torque rise profile (peak torque) under high loads, that is, high cylinder pressures.

M-11 Cummins in-line, six-cylinder, 11-liter engine managed by CELECT Plus, or Interact System (IS) electronics.

machine cycle the four steps that make up the CPU processing cycle: fetch, decode, execute, and store.

Mack Trucks E-3 In-line, 6-cylinder light-duty diesel engine. Current versions use Bosch CR fuel management.

Mack Trucks Inc. a major manufacturer of medium- and heavy-duty trucks and engines. Corporate center is Allentown, PA.

magnetic flux test magnetic flux crack detection used to identify defects in crankshafts, connecting rods, cylinder heads, and other parts. An electric current is flowed through the component being tested and iron particles suspended in liquid are then sprayed over the surface. The particles will concentrate where the magnetic flux lines are broken up by cracks.

magnetism the phenomenon that includes the physical attraction for iron observed in lodestone and associated with electric current flow. It is characterized by fields of force, which can exert a mechanical and electrical influence on anything within the boundaries of that field.

magneto an electric generator using permanent magnets and capable of producing high voltages.

magnetomotive force (mmf) the magnetizing force created by flowing current through a coil.

mainframe large computers that can process and file vast quantities of data. In the transportation industry, the data hubs to which dealerships and depots are networked are usually mainframe computers.

main memory RAM; electronically retained data pipelined to the CPU. Data must be loaded to RAM to be processed by a computer.

major thrust side when cylinder gas pressure acts on a piston, it tends to pivot off a vertical centerline: the major thrust side is the inboard side of the piston as its throw rotates through the cycle.

malleable possessing the ability to be deformed without breaking or cracking.

manifold boost turbo-boost.

manometer a tubular, U-shaped column mounted on a calibration scale. The tube is water or mercury filled to balance at 0 on the scale and the instrument is used to measure light pressure or vacuum conditions in fluid circuits.

mass the quantity of matter a body contains; weight.

master bar a test bar used to *check* the align bore in engine cylinder blocks.

master gauge a diagnostic gauge of higher quality used to corroborate readings from an in-vehicle gauge.

master program the resident portion of an operating system. In a vehicle ECM, the master program for system management would be retained in ROM.

master pyrometer an accurate thermocouple pyrometer used when performing dynamometer testing.

material safety data sheets (MSDS) a data information sheet that must be displayed on any known hazardous substance: mandated by WHMIS.

matter physical substance; anything that has mass and occupies space.

MB-900 Mercedes-Benz 6.4 liter, I-6 engine.

MB-4000 Mercedes-Benz 12.0 liter, I-6 engine.

mean average.

mean effective pressure average pressure acting on a piston through its complete cycle, the net gain of which, converts to work potential. Usually calculated by disregarding the intake and exhaust strokes, and subtracting mean compression pressure from mean combustion pressure.

mechanical advantage the ratio of applied force to the resultant work in any machine or arrangement of levers.

mechanical efficiency a measure of how effectively *indicated horsepower* is converted into *brake power:* factors in pumping and friction losses.

mechanical governor a governor in which the centrifugal force developed by the rotating flyweights used to sense rpm is the force used to move the fuel control mechanism.

mechanical injectors one OEM's term for *hydraulic injectors.*

mechanical unit injector (MUI) cam-actuated, governor-controlled unit injectors used by DDC and Caterpillar.

megabytes one million bytes. A quantitive measure of data. Often abbreviated to "meg."

megahertz a measure of frequency. One million cycles per second. The system clock is speed rated in megahertz.

megapascal (MPa) One million pascals. Metric pressure measurement unit.

memory address the location of a byte in memory.

menu a screen display of program or processing options.

message identifier (MID) identifies an on-vehicle electronic circuit by numeric code when read by an EST (SAE).

metals any of a group of chemical elements such as iron, aluminum, gold, silver, tin, and copper that are usually good conductors of heat and electricity and can usually form basic oxides.

metallurgy the science of the production, properties, and application of metals and their alloys.

metering the process of precisely controlling fuel quantity.

metering actuator one of two ECM-controlled electrohydraulic solenoids used to manage injected fuel quantity on Cummins common rail, HPI-TP fueled engines. Duration of energization determines injected fuel quantity delivered to the engine cylinder by an HPI-TP injector.

metering chamber the lower chamber in a Cummins TP injector located under the pump plunger.

metering orifice the component within a Cummins PT injector that defines the flow area to the cup pump chamber.

metering recesses the milled recesses in MUI and port-helix plungers that are used to vary the fuel quantity and timing during pumping.

meter resolution a measure of the power and accuracy of a DMM.

Metri-Pack connector a type of commonly used, sealed electrical/electronic connector.

Mexican hat piston a piston design in which the center of the crown peaks in the fashion of a sombrero. Commonly used in DI diesel engines.

micron one millionth of a meter equivalent to 0.000039 inch. The Greek letter mu is used to represent micron and is written as μ.

microorganism growth a condition that may result from water contamination in fuel storage tanks.

microprocessor a small processor. Sometimes used to describe a complete computer unit.

microwaves radio waves used to transmit voice, data, and video. Limited to line of sight transmission to distances not exceeding 30 km.

millions of instructions per second (MIPS) rating of processing speed.

minor thrust face the outboard side of the piston as its throw rotates away from the crankshaft centerline on the powerstroke. See *thrust faces*.

minor thrust side when cylinder gas pressure acts on a piston, it tends to pivot off a vertical centerline: the minor thrust side is the outboard side of the piston as its throw rotates through the cycle.

mixture the random distribution of one substance with another without any chemical reaction or bonding taking place. Air is a mixture of nitrogen and oxygen.

modem a communications device that converts digital output from a computer to the analog signal required by the phone system.

modulation in electronics, the altering of amplitude or frequency of a wave for purposes of signaling data.

module a housing that contains a microprocessor and switching apparatus or either of each.

monatomic a molecule consisting of a single atom.

monitor the common output screen display used by a computer system; a CRT.

motherboard the primary circuit board in the computer housing to which the other components are connected.

motive power automotive, transportation, marine, and aircraft.

motoring running an engine at 0 throttle, with chassis momentum driving engine.

mouse input device that controls the curser location on the screen and switched program options.

muffler an *engine silencer* that uses sound absorption and resonation principles to alter the frequency of engine noise.

multimedia the combining of sound, graphics, and video in computer programs.

multi-orifii nozzle a typical hydraulic injector nozzle whose function it is to switch and atomize the fuel injected to an engine cylinder. Consists of a nozzle body machined with the orifii, a nozzle valve, and a spring. Used in most DI diesel engines using port helix injection pumps, MUIs, EUIs, EUPs, and HEUIs.

multiple splice an electrical connection that joins a number of wires at a single junction.

multiplexing used to describe the connecting of two or more electronic system controllers on a data backbone to synergize system operation and reduce the number of common components and hard wiring.

multitasking the ability of a computer to simultaneously process multiple data streams.

N 14 Cummins evolution of the 855 cu. in. engine produced for PT, CELECT, and CELECT Plus management—an in-line, six-cylinder engine.

Nalcool a brand of antifreeze solution.

nanosecond one billionth of a second.

natural gas naturally occurring subterranean organic gas (gaseous crude oil) composed largely of methane.

naturally aspirated (NA) describes any engine in which intake air is induced into the cylinder by the lower-than-atmospheric pressure created on the downstroke of the piston and receives no assist from boost devices such as turbochargers.

Navistar a major manufacturer of truck chassis and engines. Previously *International Harvester* and often referred to by the slang term *binder* sourced from the strong agricultural heritage of the company. Corporate center is in Chicago, Illinois.

needle valve nozzle another way of describing a multiorifii, hydraulic injector nozzle (DDC).

network a series of connected computers designed to share data, programs, and resources.

network access point (NAP) wireless data transfer module used for data transfer from vehicle to land station.

networking the act of communicating using computers.

neutron a component part of an atom with the same mass as a proton, but with no electrical charge. Present in all atoms except the simplest form of hydrogen.

new scroll pump A Caterpillar port-helix metering injection pump.

newton unit of mechanical force defined as the force required to accelerate a mass of 1 kilogram through 1 meter in 1 second.

nibble four bits of data or half a byte.

Ni-Resist insert a high strength, nickel alloy piston ring support insert in an aluminum trunk-type piston with a similar coefficient of heat expansion as aluminum.

nitrogen (N) a colorless, tasteless, and odorless gas found elementally in air at a proportion of 76% by mass and 79% by volume. Atomic number 7.

nitrogen dioxide (NO$_2$) one of the oxides of nitrogen produced in vehicle engines and a significant contributor in the formation of photochemical smog.

no-air set screw a PT pump adjustment that defines the maximum flow area to the rail until the AFC circuit is activated by manifold boost.

nodes 1. dumb terminals (no processing capability) and PCs connected to a network. 2. portion of a wave signal above a zero, mean, or neutral point in the band.

noise in electronics, unwanted pulse or wave form interference that can scramble signals.

normal rated power the highest power specified for continuous operation of an engine.

NOT gate   any circuit whose outcome is in the *on* or *one* state until the gate switch is in the *on* state at which point the outcome is in the *off* state.

notebook computer   briefcase-sized PC designed for portability.

noxious emissions   engine end gases that are classified as harmful. Includes NO_x and HC but does not include CO_2 (a greenhouse gas) and H_2O.

nozzle   the component of most hydraulic and electronic injector assemblies responsible for switching and atomizing fuel.

nozzle closing pressure (NCP)   the specific pressure at which hydraulic injector nozzle closes, always lower than NOP due to nozzle differential ratio. Also known as valve closing pressure.

nozzle differential ratio   the ratio of nozzle valve seat to nozzle valve shank sectional areas. This ratio defines the pressure difference between NOP and nozzle closure values.

nozzle opening pressure (NOP)   the trigger pressure value of a hydraulic injector nozzle.

nozzle seat   the seat in an injector nozzle body sealed by the nozzle valve in its closed position.

nucleus   the center of an atom incorporating most of its mass and usually made up of neutrons and protons.

numeric   data represented by number digits.

numeric keypad   microprocessor-based instrument with numeric-only input keys such as on a ProLink EST.

Occupational Safety and Health Administration (OSHA)   U.S. federal agency responsible for administering safety in the workplace.

octane rating   denotes the ignition and combustion behavior/rate of a fuel, usually gasoline. As the octane number increases, the fuel's antiknock characteristics increase and the burn rate slows.

offset camshaft key timing   a Cummins method of timing engine position with cam-actuated, injector pump mechanisms.

ohm   a unit for quantifying electrical resistance in a circuit.

Ohm's law   the formula used to calculate electrical circuit performance. It asserts that it requires 1 V of potential to pump 1 A of current through a circuit resistance of 1 Ω. Named for Georg Ohm (1787–1854).

oil cooler   a heat exchanger designed to cool oil usually using engine coolant as its medium.

oil pan   the oil sump normally flange mounted directly under the engine cylinder block.

oil window   the portion of the upper strata of the Earth's crust in which crude petroleum is formed.

opacimeter   see *opacity meter*.

opacity meter   a light extinction means of testing exhaust gas particulate and liquid emission that rates density of exhaust smoke based on the percentage of emitted light that does not reach the sensor, so the higher the percentage reading, the more dense the exhaust smoke.

open circuit   any electrical circuit through which no current is flowing whether intentional or not.

open circuit voltage (OCV)   voltage measured in a device or circuit through which there is no current flow.

open nozzle   refers to an injector that is sealed by its pump plunger such as Cummins PT and HPI-TP used in the ISX series.

opens   an electrical term referring to open circuits/no continuity in a circuit, portion of the circuit, or a component.

operand   machine language directive that channels data and its location.

operating environment   defines the monitor display character and the GFI (graphical user interface) consisting of icons and other symbols to increase user friendliness.

operating system (OS)   core software programs that manage the operation of computer hardware and make it capable of running functional programs.

opposed plungers   reciprocating members of an inlet-metering, opposed plunger rotary distributor pump. Usually a pair of opposed plungers are used but some pumps use two pairs. The plungers are forced outward as fuel is metered into the pump chamber: when the plunger actuating rollers contact the internal cam profiles, they are driven inboard (toward each other) simultaneously pressurizing the fuel in the pump chamber.

optical character recognition (OCR)   scanners that read type by shape and convert it to a corresponding computer code.

optical codes   graphic codes that represent data for purposes of scanning such as bar codes.

optical disks   digital data storage media consisting of rigid plastic disks on which lasers have burned microscopic holes. The disk can then be optically scanned (read) by a low power laser.

optical memory cards   digital data storage media the size of a credit card capable of retaining the equivalent of 1,600 pages of text.

OR gate   a multiple input circuit whose output is in the *on* or *one* state when any of the inputs is in the *on* state.

Organization of Petroleum Producing Countries (OPEC)   a cartel of oil-producing countries that regulates oil supplies to maintain pricing.

orifice   a hole or aperture

orifii   plural of orifice

orifice nozzle   a hydraulic injector nozzle that uses a single orifice (unusual) or a number of orifii through which high-pressure fuel is pumped and atomized during injection.

original equipment manufacturer (OEM)   term used to describe the manufacturer of original product, distinct from after-market manufacturer (replacement product).

oscilloscope   an instrument designed to graphically display electrical waveforms on a CRT or other display medium.

Otto cycle   the 4-stroke, spark ignited engine cycle patented by Nicolas Otto in 1876. The four strokes of the cycle are induction, compression, power, and exhaust.

outer base circle (OBC)   the portion of a cam profile with the largest radial diameter.

output   the result of any processing operation.

output devices   components controlled by a computer that effect the results of processing. The CRT and printer on a PC system and the injector drivers on a diesel engine are all classified as output devices.

outside diameter (od) outside measurement of a shaft or cylindrical component, but can also be used to mean any outside dimension.

outside micrometer a standard micrometer designed to precisely measure od or thickness. Consists of an anvil, spindle, thimble, barrel, and calibration scales.

overspeed a governor condition in which the engine speed, for whatever reasons, exceeds the set high-idle speed or top engine limit.

oxyacetylene a commonly used cutting, heating, and welding process that uses pure compressed oxygen in conjunction with acetylene fuel.

oxygen colorless, tasteless, odorless gas; the most abundant element on the Earth; occurs elementally in air and in many compounds including water.

oxidation the act of oxidizing a material; can mean combusting or burning a substance.

oxidation catalyst a catalyst that enables an oxidation reaction. In the oxidation stage of a catalytic converter, the catalysts, platinum and palladium, are used.

oxidation stability describes the resistance of a substance to be oxidized. It is a desirable characteristic of an engine lubrication oil to resist oxidation so one of its specifications would be its oxidation stability.

oxides of nitrogen (NO$_x$) any of a number of oxides of nitrogen that may result from the combustion process: they are referred to collectively as NO$_x$. When combined with HC and sunlight, reacts to form photoelectric smog.

ozone an oxygen molecule consisting of three oxygen atoms (triatomic). Exists naturally in the Earth's ozonosphere (6 to 30 miles altitude) where it absorbs ultraviolet light, but can be produced by lightning and by photochemical reactions between NO$_x$ and HC. Explosive and toxic.

PACE/PACER Cummins partial authority, electronic PT fuel system management.

packet a data message delivered to the data bus when a ladder switch resistance changes indicating a change in switch status.

palladium an oxidation catalyst often used in catalytic converters.

Palm Pilot hand sized, microprocessor units with PC compatibility, increasingly used loaded with proprietary software to troubleshoot truck electronic systems.

parallel circuits electrical circuits that permit more than a single path for current flow.

parallel hybrid term used to describe a commercial vehicle powertrain in which a diesel engine drives a genset producing electricity to drive electric motors that provide torque to the wheels.

parallel ports peripheral connection ports for computer devices that require large volume data transmission such as printers and scanners.

parallel port valve configuration engine cylinder valve arrangement that locates multiple valves parallel to crank centerline permitting equal gas flow through each (assuming identical lift).

parameter a value, specification, or limit.

parameter identifier (PID) codes components within a MID system.

parity the even or odd quality of the number of 0s (zeros) and 1s (ones), a value that may have to be set to handshake two pieces of electronic equipment.

partial authority term widely used in truck technology to describe a hydromechanical system that has been adapted for management by computer: a good example is the port-helix metering injection pump that was adapted for electronic management in early V-MAC engines.

particulate matter (PM) solid matter. Often refers to minute solids formed by incompletely combusted fuel and emitted in the exhaust gas.

passive matrix an LCD screen display used in older notebook-type PCs. They use a single transistor for each row and column producing a low quality visual image.

password an alpha, numeric, or alpha-numeric value that either identifies a user to a system or enables access to data fields for purposes of download or reprogramming.

peak pressure the highest pressure attained in a hydraulic system.

peak torque maximum torque. In an internal combustion engine, peak torque always occurs at peak cylinder pressure and in most cases this will be achieved at a lower speed than rated power rpm.

Peltier effect a heat exchanger principle used in some exhaust gas analyzers to drop a test sample below the dew point to remove water.

pencil injector nozzle a slim pencil-shaped hydraulic injector that often uses an internal accumulator that eliminates the need for a leakoff circuit. Nearly obsolete.

peripherals input and output devices that support the basic computer system such as the CRT and the printer.

periphery in cam geometry, the entire outer boundary of the cam; cam profile.

peristaltic pump a pump used in some exhaust gas analyzers to remove water from a test sample while allowing the gas to have minimal contact with the water.

personal computer (PC) any of a variety of small computers designed for full function in isolation from other units but which may be used to network with other systems.

personal computer card international association (PCMCIA) credit-card size data storage cards that can be inserted into PC expansion slots.

personality module Caterpillar and Navistar PROM/ EEPROM component.

personality ratings term used by Caterpillar and Navistar to describe PROM and EEPROM functions.

Peterbilt truck OEM owned by Paccar-DAF.

petroleum any of a number of organic fossilized fuels found in the upper strata of the Earth's crust that can be refined into diesel fuel and gasoline among other fuels.

pH used to evaluate the acidity or alkalinity of a substance. From a logarithm of the reciprocal of the hydrogen ion concentration in a solution in moles per liter: p = power, H = hydrogen.

phasing the precise sequencing of events; often used in the context of *phasing* the pumping activity of individual elements in a multicylinder injection pump.

photochemical reaction a chemical reaction caused by radiant light energy acting on a substance.

photochemical smog smog formed from airborne HC and NO$_x$ exposed to sunlight; also known as photoelectric smog or photosynthetic smog.

photoelectric smog see *photochemical smog*.

photonic semiconductors semiconductors that emit or detect *photons* or light.

photons a quantum of electromagnetic radiation energy; when visible, known as *light*.

photovoltaic the characteristic of producing a voltage from light energy.

pickup tube a suction tube or pipe in a fuel tank or oil sump.

pilot ignition a means of igniting a fuel charge that might normally require a spark, by injecting a short pulse of diesel fuel into a cylinder to ignite a premixed charge of gaseous fuel and air.

pilot injection the injection of a short duration pulse of diesel fuel, followed by a pause to await ignition, followed by the resumption of the fuel pulse. Used as a cold start strategy in some systems to prevent *diesel knock* and throughout the fueling profile by others, notably, later versions of HEUI. Can also be used as the ignition means in applications using an alternative fuel that does not readily compression ignite. In such instances, a short pulse of diesel fuel is injected to act as the ignition means for the primary fuel.

pin boss the wrist pin support bore in a piston assembly.

pintle nozzle a type of hydraulic injector nozzle used in some IDI automobile, small-bore diesel engines until recently.

pipelining rapid sequencing of functions by the CPU to enable high-speed processing.

piston the reciprocating plug in an engine cylinder bore that seals and transmits the effects of cylinder gas pressure to the crankshaft.

piston pin a wrist pin that links the piston assembly to the connecting rod eye.

piston speed the distance traveled by one piston in an engine per unit of time.

pitting a wear pattern that results in small pock marks or holes.

pixels picture elements. A measure of screen display resolution—each dot that can be illuminated is called a pixel.

plain old telephone service (POTS) telecommunications using the telephone system for part or all of the transaction, still the backbone of most telecommunications.

plasma a gas of positive ions and free electrons with an approximately equal positive and negative charge.

Plastigage a shaft-to-friction bearing clearance measuring system consisting of nylon cord that deforms to conform with the clearance dimension so it can be measured against a coded scale on the packaging envelope.

platinum an oxidation catalyst often used in catalytic converters.

PLD term used to describe the engine controller module on Mercedes-Benz diesel engines: a German acronym meaning ECM.

plunger the reciprocating member of a plunger pump element.

plunger and cup the PT injector components that form the high-pressure pump element.

plunger geometry term used to describe the shape of the metering recesses/helices in a pumping plunger and therefore the pump timing and delivery characteristics.

plunger leading edge the point on a pumping plunger closest to the pump chamber.

plunger link on a Cummins PT injector, the rod located in the plunger flange that prevents side loading of the plunger by the rocker arm.

plunger pump any pump that uses a reciprocating piston or plunger and, in most cases, is hydraulically classified a *positive displacement*.

pneumatics the science of the mechanical properties of gases, especially in confined circuits designed to provide motive power.

pocket technician PalmPilot-based tool that plugs in to a dash adapter used to assist technicians with diagnosis but a driver's version is also an option. Capable of analyzing fuel economy, engine operating conditions, and display fault codes.

polymer a compound composed of one or more large molecules, formed by chains of smaller molecules.

poppet nozzle a forward opening nozzle valve used in older IDI diesel engines.

popping pressure see *nozzle opening pessure*.

pop test the testing of NOP on a hydraulic injector nozzle using a bench or pop tester.

port 1. an aperture or opening. 2. a computer connection socket used to link a computer with input and output devices.

port closure the beginning of effective stroke in a plunger and barrel pumping element, occurring when the plunger leading edge closes off the spill/fill port(s).

port opening the ending of effective stroke in a plunger and barrel pumping element, occurring when the fill/spill ports are exposed to the chamber.

positive displacement describes a pumping principle in which the quantity of fuel pumped (displaced) per cycle does not vary so the volume pumped depends on the rate of cycles per minute. When a positive displacement pump unloads to a defined flow area, pressure rise will increase in proportion to rpm or cycles per minute.

positive filtration a filter in which all of the fluid (gas or liquid) to be filtered is forced through the filtering medium. Most air, fuel, coolant, and oil filters used today employ a positive filtration principle.

POST power-on self-test. A BIOS test run at boot-up to ensure that all components are operational.

potential difference electrical *charge differential* measured in *voltage*.

potentiometer a three-terminal variable resistor or voltage divider used to vary the voltage potential of a circuit. Commonly used as a throttle position sensor.

pour point a means of evaluating a fuel or lubricants low temperature flow characteristics. The pour point of a fuel is slightly higher in temperature than its gel point.

power the rate of accomplishing *work*, it is necessarily factored by time.

power line carrier term used to describe communication transactions delivered through a power line. Signals are converted to radio frequencies for the transaction and subsequently decoded by the receiver ECM. An example is the power line carrier use of the auxiliary (blue) wire in a standard seven-pin trailer connector for trailer to tractor ABS communications.

PowerStroke   International-built, HEUI-fueled V-8 engine used by Ford. Available in 7.3 (obsolete) and 6.0 liter displacement versions.

power takeoff (PTO)   an engine or transmission located device used to provide auxiliary power. Can also mean the primary coupling between the engine and powertrain, so in a truck engine, this would be the flywheel.

powertrain   the components of a system directly responsible for transmitting power to the output mechanisms. In an engine, the powertrain components are piston assemblies, connecting rods, crankshaft, and flywheel.

power transistor   a transistor used as the final switch in an electronic circuit to control a solenoid or other output; sometimes known as a *driver*.

prefix   addition of a syllable or letter(s) or numbers at the beginning of a word or acronym.

preinjection metering (PRIME)   a Caterpillar term for the pilot injection concept used in its HEUI injectors.

prelubricator   a pump used to charge the lubrication circuit on a rebuilt engine before startup.

pressure control valve (PCV)   ECM-controlled, rail pressure management valve used in the accumulator of a Cummins CAPS pump assembly.

pressure-time (PT)   Cummins common rail hydromechanical fuel system used in highway applications until 1994. An electronic version known as HPI-PT is used on K-19 engines.

pressurizing   the process of raising the pressure in a circuit.

preventative maintenance (PM)   routine scheduled maintenance on vehicles.

primary filter   usually describes a filter on the suction side of a fuel subsystem whereas the term secondary filter describes the filter on the charge side of the transfer pump.

prime mover   an initial source of power: for instance, the prime mover of a diesel fuel subsystem is the transfer pump.

processing   the procedure required to compute information in a computer system. Input data is processed according to program instructions and outputs are plotted.

ProDriver   DDC-DDEC driver digital display.

program   set of detailed instructions that organize processing activity.

programmable electronic engine control (PEEC)   describes an electronically managed Caterpillar 3406B engine.

programmable logic controllers   small computers that perform switching functions in much the same way as a relay.

programmable read only memory (PROM)   a chip or chips used to qualify ROM data to a specific chassis application. In early vehicle computers, this was usually the only method of reprogramming data to an ECM: this PROM function has now been superseded by the EEPROM capability found in most ECMs.

ProLink 9000   a generic reader/programmer capable of scanning all MIDs connected to a J1850 or J1939 data bus and performing diagnostic and limited programming

operations on some systems. Software cartridges or multi-protocol data cards are used to read each OEM system.

ProManager   DDC-DDEC program that permits ECM data to be downloaded to a PC for analysis.

propagate   to breed, transmit, or multiply. The word is often used to describe the combustion process in an engine cylinder such as in *flame propagation*.

propane   a petroleum-derived gas consisting mostly of methane often known as liquefied petroleum gas or LPG.

proportional solenoid   a solenoid whose armature will be positioned according to how much current is flowed through its coil. Often an ECM-actuated output. Proportional solenoids may be linear such as the V-MAC rack actuator or rotary such as the Caterpillar BTM.

proprietary OS   OS that are privately owned and are specific to a manufacturer or operator.

propylene glycol (PG)   A less toxic glycol base antifreeze solution than EG. PG mixture strength must be tested with a refractometer with a PG scale and not mixed with EG.

protocols   sets of rules and regulations. Often used to define communication language.

proton   positively charged component of an atom located within its nucleus.

proton exchange membrane fuel cells (PEM)   a fuel cell using solid polymer membranes (a thin plastic film) as the electrolyte and used in most current motive power applications of fuel cell technology.

psi   pounds per square inch. Standard unit of pressure measurement.

pulsation damper   on a Cummins gear type supply pumps, pulsation dampers are used to smooth the pressure waves caused by a gear type pump as it loads fuel into its outlet.

pulse exhaust   a tuned exhaust system used to optimize the gas dynamic of exhaust gas delivered to the turbocharger.

pulse wheel   the rotating disc used to produce rpm or rotational position data to an ECM. The term is most often applied to the rotating member of a Hall effect sensor, but at least one manufacturer uses the term to describe an AC reluctor wheel.

pulse width (PW)   usually refers to EUI duty cycle measured in milliseconds.

pulse width modulation (PWM)   constant frequency, digital signal in which on/off time can be varied to modulate duty cycle.

PT (pump) control module (PTCM)   the ECM used by Cummins in their PACE/PACER partial authority, electronic management system.

pump drive gear   the gear responsible for imparting drive force to a pump.

pump-line-nozzle (PLN)   the hydromechanical or electronically managed injection pump to line to nozzle fuel injection principle used in most diesel fuel systems until the introduction of EUI engines. The term can, but is not usually, applied to some of the more recent systems such as the Mack E-Tech and Mercedes-Benz EUP systems.

pushrods   cylindrical solid rods located between a follower and a rocker assembly that transmit the effects of cam profile to action at the rocker arm.

push tubes   hollow, cylindrical tubes located between a follower and a rocker assembly that transmit the effects of cam profile to action at the rocker arm.

pyrometer   a thermocouple type, high temperature sensing device used to signal exhaust temperature. Consists of two dissimilar wires (pure iron and constantan) joined at the hot end with a millivoltmeter at the read end. Increase in temperature will cause a small current to flow, which is read at the voltmeter as a temperature value.

Qualcomm transfer   PC base, remote communications technology that enables wireless data downloads for analysis for data programming uploads. Used in conjunction with proprietary Fleet Information Software, Highway Master, and other data management packages.

quantum   a defined quantity of energy, proportional to the frequency of radiation it emits.

quiescent   a term used to describe any low turbulence engine cylinder dynamic. Its root is from the word *quiet*.

rack actuator   a proportional solenoid (Bosch) or hydraulic servo (Cat) responsible for moving the rack in a computer-controlled, port-helix metering injection pump. An ECM output.

rack actuator housing   the housing at the rear of a computer-controlled port-helix metering injection pump that contains sensors and the rack actuating mechanism. It is located in place of the governor on a hydromechanical engine.

rack clevis   the link on a DDC MUI rack to which the rack control lever is fitted.

rack lever   the actuating mechanism on a DDC MUI system that converts the rotary movement of the control tube into linear movement of the rack.

radial piston pump   means of creating injection pressures in some current common rail diesel fuel injection systems, notably Cummins CAPS and Bosch CR. Multicam profiles actuate reciprocating plungers that unload to an accumulator or rail.

rack position sensor   an electromagnetic sensor used to signal rack position data to the ECM on electronically controlled, port-helix metering injection pumps.

radial vector   the radial angle off a reference point, say TDC, in a crankshaft that indicates the mechanical advantage of a throw in its relationship with the crankshaft centerline.

radiation   the transfer of heat or energy by rays not requiring matter such as a liquid or a gas.

radiator   a heat exchanger used in liquid-cooled engines designed to dissipate some of the engine's rejected heat to atmosphere.

radioactive   any substance or set of physical conditions capable of emitting radioactivity. Exposure to high-level radioactivity can be life threatening while low-level radioactivity (such as electrical or radar waves) represents debatable hazards.

rail   a manifold.

rail actuator   see *fueling actuator*.

rail pressure sensor   Cummins HPI-TP rail fuel pressure sensor.

ram air   air fed into engine cooling and intake circuits by the velocity of a moving vehicle; increases proportionally with vehicle speed.

ramps   in cam geometry, the shaping of the cam profile between the IBC and the OBC. The ramp geometry defines the actuation/unload characteristics of the train that rides its profile.

random access memory (RAM)   electronically retained "main memory" of a computer system.

rapid start shutoff valve   Cummins common rail fuel system electric (solenoid) shutoff valve that traps fuel in the rail on shutdown to enable an almost instant restart.

rate shaping   a fuel injection term that describes the ability of a fuel system to control fuel delivery to the cylinder independent of the hard limitations of cam geometry and engine rpm. Because HEUI injectors are actuated hydraulically and the hydraulic actuation pressure can be controlled by the ECM, this system is capable of rate shaping. The Cummins CAPS is also capable of rate shaping.

rated power   the highest power specified for continuous operation.

rated speed   the rpm at which an engine produces peak power.

ratio   quantitative relationship between two values expressed by the number of times one contains the other .

reaction turbine   an aeolipile, the first heat engine.

reactive   substances that can become chemically reactive if they come into contact with other materials resulting in toxic fumes, combustion or explosion.

read-only memory (ROM)   data that is retained either magnetically or by optical coding and designed to be both permanent and read-only.

reader-programmer   a generic or OEM electronic service tool (EST) designed to both scan, reprogram, and perform some diagnostics on an electronic system.

ream   the machining process of accurately enlarging an orifice using a steel boring bit with straight or spiral fluted cutting edges.

recording density   number of bits that can be written to an inch of track on a disk; measured in bpi or bits per inch.

rectifier   device used to convert AC into DC.

reference coil   Mack Trucks rack position sensor magnetic field temperature reference; validates input from the rack position sensor.

reference voltage (V-Ref)   the ECM-controlled output to on-board sensors.

refraction   the extent to which a light ray is deflected (bent) when it passes through mediums such as water, coolant or fog.

refractive index   truly the ratio of the speed of light in a vacuum versus the speed of light through a specified medium, but in practice, is used to express natural light, refractometer readings in coolant or battery electrolyte.

regenerative braking   vehicle retarding effort achieved in a parallel hybrid drive unit when the drive electric motor magnetic field is reversed, both applying retarding torque and generating electricity that can be used to charge the batteries.

register   alignment or track point of two components.

registers   temporary storage locations in a CPU.

rejected heat that portion of the potential heat energy of a fuel not converted into useful kinetic energy.

relief valve a commonly used valve in hydraulic circuits (such as fuel subsystem and lubrication circuits) that defines maximum circuit pressure. The simplest type would consist of a ball check, loaded by a spring to seal a return line. When circuit pressure was sufficient to unseat the ball check, circuit fluid would be diverted from the main circuit to the return.

reluctance resistance to the movement of magnetic lines of force.

reluctor a term used to describe a number of devices that use magnetism and motion to produce an AC voltage.

remote data interface (RDI) DDEC communications link between the vehicle electronics and a fleet's PC or PC network.

reprogram general term used to cover a range of rewrite and over-write procedures in computer technology.

residual line pressure the pressure that *dead volume fuel* is retained at in a high-pressure pipe in a PLN fuel system that uses delivery valves at the injection pump.

resistance opposition to electrical current flow in a circuit.

resolution the smallest interval measurable by an instrument. In computer terminology, it usually describes the image clarity of a CRT display in pixels. It also defines range in a DMM.

resonation principle a noise-reducing principle used in engine silencers that scrambles sonic nodes and antinodes by reflecting sound back toward its source, thereby altering the frequency.

Resource Conservation and Recovery Act (RCRA) U.S. federal legislation that regulates the disposal of hazardous materials.

retarder generally refers to braking action, that is, the retarding of vehicle movement.

retraction collar/piston the component on a delivery valve core that is designed to seal before it seats and therefore helps define the *residual line pressure* value.

retraction spring any spring in any component that causes an assembly to mechanically withdraw or retract.

rhodium a hard white metal occurring naturally in platinum ores and used as a NO_x reduction catalyst in gasoline fueled engines.

Right to Know legislation a provision of the U.S. federal Hazard Communications legislation that imposes on employers the duty of fully revealing the potential dangers of hazardous materials to which their employees may be exposed.

ring belt the area of the piston in which the piston ring grooves are machined.

road speed governing (RSG) the managing of engine output on the basis of a specific road speed. Can also be used to mean the maximum programmed road speed limit.

road speed limit (RSL) usually the maximum programmed road speed value programmed to a vehicle management system meaning that the vehicle should not travel faster than this speed. However some OEMs permit maximum cruise speeds to be programmed above RSL to encourage drivers to use cruise control so this is not always true.

road speed sensor a sensor usually of the pulse generator type located at the transmission tailshaft or a wheel assembly that signals the ECM road speed data.

rocker arm see *rockers*.

rocker assemblies the entire rocker assembly consisting of rockers, rocker shaft, and pedestals.

rocker claw/lever PT or mechanical unit injector hold-down or short-out tool.

rocker pallet the end of a rocker that contacts the injection pumping tappet, the valve stem, or the valve bridge.

rockers shaft-mounted, pivoting levers that transmit the effects of cam profile to valves and injection pumping apparatus.

rod eye the upper portion of a connecting rod that connects to the piston wrist pin; also known as *small end*.

root mean square (RMS) a term used to describe averaging when performing machining operations and also used describe AC voltage measurement by ascribing it an equivalent DC value.

Roots blower a positive displacement air pump consisting of two gear-driven, intermeshing spiral fluted rotors in a housing; used to scavenge DDC two-stroke cycle engines.

router connection between two networked computers or two vehicle ECMs.

RS-232 port a serial communications port (often Com 1) in a computer system that accepts a standard phone jack. ProLink is equipped with an RS-232 port, which can drive a printer or PC data display.

run-in usually describes the engine break-in procedure following a rebuild outlined by the OEM.

sac a spherical cavity. Refers to the chamber in some multi-orifii injector nozzles beyond the seat and from which the exit orifii extend.

SAE horsepower a structured formula used to calculate brake power data that can be used for comparison purposes.

SAE J standards standards developed by SAE industry committees and generally agreed to, usually without any statutory obligation.

SAE J1587 electronic data exchange protocols used in data exchange between heavy-duty, electronically managed systems.

SAE J1667 standards for emissions testing.

SAE J1708 serial communications and hardware compatibility protocols between microcomputer systems on a J1587 data bus. Its data link is a 6-pin Deutsch connector.

SAE J1939 the set of multiplexing standards that incorporate both J1587 and J1708. Both software and hardware protocols and compatibilities are covered by J1939, which is updated by simply adding a suffix. The standard SAE/ATA, 9-pin Deutsch connector is referred to as a J1939 diagnostic/data link. A CAN 2.0 data bus.

SAE viscosity grades the industry standard for grading lubricating oil viscosity.

sampling the process a computer system uses to monitor noncommand type input data from the sensors in a system. Inputs from oil pressure, ambient temperature, coolant temperature, and so on, would be monitored by the ECM by sampling.

saturation condition of an electromagnet in which a current increase results in no increase in the magnetic flux field.

scavenge term used generally to describe the process used to expel end gases from an engine cylinder and specifically to describe: 1. the final stage of the exhaust process in a four-stroke cycle engine that occurs at valve overlap. 2. cylinder breathing on a two-stroke cycle diesel engine.

scissor jack an air-actuated floor jack with a pair of clevises that lock to the truck frame rails and can lift one end of the chassis well clear of the floor.

screen any computer output display from LCDs through CRTs.

scrolling the moving of lines of data up or down on a computer display screen.

SCSI controller controller that can support up to seven disc drives and enable high data transfer rates.

SCSI port a type of parallel port that can support up to seven different devices to a single port.

scuffing a superficial scraping of metal against metal damage mode.

search engine software that when provided with a key word or phrase scans and retrieves data from memory.

secondary filter usually refers to a filter downstream from the transfer or charge pump in a typical fuel subsystem. It is in most cases under pressure and capable of much finer filtration than a primary filter, which is usually under suction.

section modulus relates the shape of a beam, cylinder, or sphere to section and stiffness; the greater the section modulus, the higher the rigidity and resistance to deflection. A factor of RBM or resist bending moment.

sector in computer terminology, a pie-shaped section of a disk or section of track.

selective catalytic reduction (SCR) term used to describe NO_x reduction converters for truck applications using urea and water injection. On injection, the urea converts to gaseous ammonia that reacts with NO_x compounds to produce elemental nitrogen and water.

self-test input (STI) International HEUI diagnostic scan initiated by depressing a dash STI button.

semiconductor materials that neither conduct well nor insulate; they have four electrons in their outermost shell.

sending unit a variable resistor and float assembly that signals a gauge and/or ECM the liquid level in a tank.

sensor a term that covers a wide range of command and monitoring input (ECM) signal devices.

sequential storage storage of data on media such as magnetic tape where data is read and written sequentially.

sequential troubleshooting chart commonly used by engine OEMs to structure electronic troubleshooting. The technician routes the troubleshooting path through the chart on the basis of test results in each step.

serial port port connection that transfers data one bit at a time and therefore more slowly than a parallel port. A mouse is connected to a serial port.

series circuit a circuit with a single path for electrical current flow.

Series 50 DDEC controlled DDC in-line, four-cylinder engine.

Series 55 obsolete DDC 12-liter, in-line, six-cylinder managed by DDEC III - EUP.

Series 92 DDC 6/8 cylinder, V configuration engine available with DDEC; common transit vehicle engine.

Series 60 DDEC-controlled DDC in-line, six-cylinder engine available in 11.1-, 12.7- and 14-liter displacements.

server in the computer processing cycle or multiplex transaction, the fulfilling of a *client* need is provided by a server.

service hours a means of comparing engine service hours to highway mileage. Most engine OEMs equate 1 engine hour to 50 highway linehaul miles (80 km), so a service interval of 10,000 miles (16,000 km) would equal 200 engine hours. The term *engine hours* is also used.

Service Technicians Society (STS) a branch of the SAE dedicated to the information and training needs of motive power technicians.

shear the stress produced in a substance or fluid when its layers are laterally shifted in relation to one another. Viscosity describes a fluid's resistance to shear.

shell term used to describe the concentric orbital paths of electrons in atomic structure.

shutdown solenoid an ETR or latching solenoid that functions to no-fuel the engine to shut it down.

shutterstat a temperature sensing, pneumatic switch used to manage air shutter operation.

signals codes, signs, or symbols used to convey information.

Signature Series in-line, six-cylinder engine introduced by Cummins as 600 BHP with 2000 lb.-ft. of torque. A 15-liter engine using HPI-TP common rail fueling with open nozzle injectors, part of the ISX family.

silicon a nonmetallic element found naturally in *silica*, silicone dioxide in the form of quartz.

silicone any of a number of polymeric organic compounds of the element silicon associated with good insulating and sealing characteristics.

silicone-controlled rectifier (SCR) similar to a bipolar transistor with a fourth semiconductor layer added. See full explanation in Chapter 29. Extensively used to switch DC in vehicle electronic and ignition systems.

single pass radiator any radiator through which flow is unidirectional.

single speed control (SSC) Mack Trucks isochronous PTO governing.

sinter a means of alloying metals in which the constituent materials are mixed in powdered form and then coalesced by subjecting them to heat and pressure. Produces more uniform metallurgical characteristics than alloying.

sintered steel a steel produced by a sintering process; used in certain engine and fuel system components to produce especially tough and durable material characteristics.

sleeve metering a means of varying the effective stroke in injection pumps by using a movable (by the governor) control sleeve. Used in a number of older injection pumps such as the American Bosch M-100 but more recently, only in the Bosch sleeve-metering, rotary injection pumps such as the VE.

sleeves see *liners*.

small end the connecting rod eye.

smart   1. used to describe computed outcomes that use a broad range of input and memory factors to produce "soft" outcomes rather than adhere to hard values. *Smart cruise control* will learn road terrain patterns and permit some latitude around the programmed road speed value to improve fuel economy. 2. may describe a computer peripheral that possesses some processing capability.

smart switch   so named because it contains a ladder of resistors, usually five per switch, known as a ladder bridge; switch status can be determined by the processor mastering the multiplex using a programmed resistance library that identifies the switch and its status.

smog   a word formed by combining the words fog and smoke. A haze produced by suspended airborne particulates. Two major types exist: sulfurous smog produced by combusting sulfur-laden fuels such as coal and heavy oils and photochemical smog, a primary cause of which is vehicle emissions.

snap rail test   Cummins rail pressure spike test.

snapshot test   a diagnostic test performed on a PC or Pro-Link that captures frames of running data before and after an event that can be identified by either an automatic (such as a fault code) or manual trigger.

soak tank   a usually heated tank that is filled with a detergent or alkaline solution; used to clean engine components.

Society of Automotive Engineers (SAE)   organization responsible for setting many of the manufacturing standards and protocols of the motive power industries and dedicated to educating and informing its members.

soft copy   data that is retained electronically or on disk as opposed to being on paper.

soft cruise   a cruise control mode programmed into some vehicle/engine management electronics in which the road speed is managed within a window extending both above and below the set speed. Soft cruise can increase fuel economy and is often used in conjunction with vehicle maximum speed programming *below* maximum cruise speed.

soft parameter   a value that varies and depends on input and processing variables (see *fuzzy logic*). The term is often used to describe current cruise control systems that permit a cushion both above and below the set value. See *soft cruise.*

software   the programmed instructions that a computer requires to organize its activity to produce outcomes.

solar cells   PN or NP silicon semiconductor junctions capable of producing up to 0.5 V when exposed to direct sunlight.

solenoid   an electromagnet with a movable armature.

solid state   components that use the electronic properties of solids such as semiconductors to replace the electrical functions of valves.

solid state storage   volatile storage of data in RAM chips.

sound absorption principle   a means of converting sound waves to friction then heat, which is then dissipated to atmosphere, used in engine silencers/mufflers.

sound card   multimedia card capable of capturing and reproducing sound.

spark ignited (SI)   any gasoline-fueled, spark-ignited engine usually using an Otto cycle principle.

specific fuel consumption (SFC)   fuel consumed per unit of work performed.

specific gravity   the weight of a liquid or solid versus that of the same volume of water.

spectrographic analysis   a low-level radiation test that can accurately identify trace quantities of matter in a fluid; used to analyze engine oils.

spike   an electrical (voltage) or hydraulic pressure surge.

spill deflector   a cylindrical sleeve that fits outside the bushing on a DDC MUI, and protects the MUI body from high-pressure spill at port opening.

spill timing   term used to describe port closure timing of a port-helix metering injection pump to the engine it fuels, and when establishing injection pump bench phasing.

spindle   an intermediary, responsible for transmitting force. In a hydraulic injector, it relays spring force to the nozzle valve.

splice   to join.

split shot injection   term used to describe *pilot injection.* Diesel fuel injection intentionally broken into more than one pulse during a single injection cycle to one engine cylinder.

spreader bar   a rigid lifting aid permitting an engine to be raised on a single point chain hoist using two, three, or four lift points on the engine.

spreadsheet   software that enables numeric data organization and calculation.

spur gear   a gear with radial teeth.

square engine   an engine in which the bore and stroke dimensions are the same or nearly the same.

star network   network set up to operate from a central hub computer.

starting aid screw   a device used on DDC two-stroke cycle to limit fuel during cranking on turbocharged engines when the turbocharger acts as a restriction.

start of injection (SOI)   describes the moment atomized fuel exits nozzle orifii, an event that always occurs

static electricity   accumulated electrical charge not flowing in a circuit.

static friction   the characteristic of a body at rest to attempt to stay that way: see the definition of inertia.

static RAM   a RAM chip with a medium to large volume memory retention and high-access speed.

static timing   term used to describe *port closure* timing of a fuel injection pump to an engine.

STC control valve   the Cummins STC component that controls the supply of engine oil to the STC tappets and determines whether the system operates in advanced or normal modes.

STC-PT injector   a Cummins PT injector with an STC tappet used on STC timing advance systems.

steel   an alloy of iron produced by the addition of a small percentage of carbon.

step timing control (STC)   Cummins PT injector, hydraulic timing advance tappet, integral with a PT injector.

Sterling   truck OEM owned by Freightliner. Previously owned by Ford.

stoichiometric ratio   an air-fuel ratio (AFR) term meaning that at ignition, the engine cylinder has the exact quantity of air (oxygen) present to combust the fuel charge: if more air is present, the AFR mixture is rich, if less air is present, it is lean.

stoichiometry   the science of determining the ratio of reactants required to complete a chemical or physical reaction.

stop engine light (SEL)   a level III driver alert indicating the driver should shut down the engine or if programmed to do so, the management system will shut down the engine. Usually illuminated 30 to 60 seconds before shut down.

strategy   action plan. In computer technology it relates to how a series of processing outcomes is put together: for instance, start-up strategy deals with how the sequence of events are switched by the ECM that results in the engine being cranked and started.

stroke   linear travel of a piston or plunger from BDC to TDC. In an engine, piston stroke is defined by the crank throw dimension.

storage media   any nonvolatile data retention device: floppy disks, data chips, CD-ROM, and PCMCIA cards are examples.

sublimation   the process of converting a solid directly to a vapor by heating it; may also mean refine.

subsystem identifier (SID).   branch circuit within a MID off the data bus used for diagnostic reporting.

substrate   1. the supporting material on which an electric/electronic circuit is constructed/infused. 2. thermally stable, inert material on which active catalysts are embedded on a vehicle catalytic converter.

suction circuit   the portion of a lubrication or fuel subsystem that is on the pull side of the transfer pump.

suffix   addition of a syllable or letter(s) or numbers at the end of a word or acronym.

sulfur   an element present in most crude petroleums, but refined out of most current highway fuels. During combustion, it is oxidized to sulfur dioxide. Classified as a noxious emission.

sulfur dioxide   the compound formed when sulfur is oxidized that is the primary contributor to sulfurous type smog. Vehicles contribute little to sulfurous smog problems due to the use of low-sulfur fuels.

sump   the lubricating oil storage device on an engine more commonly referred to as an *oil pan*.

supercharger   technically any device capable of providing manifold boost, but in practice used to refer to gear-driven blowers such as the Roots blower.

supplemental cooling (system) additive (SCA)   conditioning chemicals added to antifreeze mixtures.

SVGA   a super version of VGA computer graphics.

swash plate pump   a pump that uses a rotating, circular plate (the *swash* plate) set obliquely on a shaft to act as a cam and convert rotary motion into reciprocating movement to actuate pistons or plungers; also known as *wobble* plate.

swept volume   volume displaced by a piston as it travels from BDC to TDC.

synchronizing position   the Caterpillar MUI fuel-balancing procedure.

synchronous reference sensor (SRS)   DDEC engine location sensor consisting of a single dowel located on the cam gear on Series 60/50 engines that inputs an analog voltage signal to the ECM.

synergize   the act of creating synergy.

synergy   any process in which the sum of a group of subcomponents combine to exceed the sum of their individual roles. Used to describe somewhat independent vehicle systems that share hardware and software to cut down on duplicated components and increase overall effectiveness or processing speeds.

synthetic oil   petroleum-based and other elemental oils that have been chemically compounded by polymerization and other laboratory processes.

system board   see *motherboard*.

system clock   device that generates pulses at a fixed rate to time/synchronize computer operations.

system pressure regulator   a usually hydromechanical device responsible for maintaining a consistent line pressure; located downstream from a pump.

system unit   the main computer housing and its internal components.

tailored torque (TT)   highway fuel trim used on some DDC engines: Belleville springs are used in the governor spring pack.

tang release tool   a lock release tool required to service the sealed connector blocks on electronic engines.

tappets   used to describe a variety of devices that ride a cam profile and transmit the effects of the cam geometry to the train to be actuated; also known as followers.

tattle tale   an audit trail that may be discreetly written to ECM data retention; the recording of electronic events for subsequent analysis.

Tech Central   International Trucks central data hub.

telecommunications   any distance communication regardless of transmission medium.

teleconferencing   audio/video communication using computers, a camera, and modem linkages.

teleprocessing   networked station-to-station data exchange and processing.

Tempilstick   a heat-sensing crayon used for precise determination of high temperatures.

template torquing   a torquing procedure that usually involves torquing to a specified value with a torque wrench followed by turning the fastener through an arc of a specified number of degrees measured by a protractor or template. Produces more consistent clamping pressures than torque-only methods.

tensile strength   unit force required to physically separate a material: in steels, tensile strength exceeds yield strength by around 10%.

terabyte   a trillion bytes.

terminal   1. a computer station or network node. 2. an electrical connection point.

thermal efficiency   measure of how efficiently an engine converts the potential heat energy of a fuel into mechanical energy, usually expressed as a percentage.

thermatic fan   a fan with an integral temperature-sensing mechanism that controls its effective cycle.

thermistor   a commonly used temperature sensor that is supplied with a reference voltage and by using a temperature sensitive variable resistor, signals back to the ECM a portion of it.

thermocouple   a device made of two dissimilar metals, joined at the "hot" end and capable of producing a small voltage when heated.

thermostat   a self-contained, temperature-sensing/coolant flow modulating device used to manage coolant flow within the engine cooling system.

thick film lubrication   lubrication of components where clearance factors tend to be large and unit pressures low.

three-way splice   the uniting of three wires at junction.

threshold values   outside limits or parameters.

throttle   air flow to the intake manifold control mechanism used in SI gasoline and diesel engines with pneumatic governors. The term is commonly used to describe the speed control/accelerator/fuel control mechanism in a diesel engine.

throttle delay   a mechanical device used to create a lag between accelerator demand and fuel delivered, usually to cut down on smoke emission.

throttle position sensor (TPS)   device for signaling accelerator pedal angle to the TPS. Most receive a V-Ref signal and use a potentiometer to return a portion of that voltage as a signal that can be correlated with pedal angle.

thrust   linear force.

thrust bearing   a bearing that defines the longitudinal or end play of a shaft.

thrust collar   in a mechanical governor, the intermediary between the centrifugal force exacted by the flyweights and the spring forces that oppose it. The thrust collar in a governor is usually connected to the fuel control mechanism.

thrust faces   a term used to describe loading of surface area generally but most often of pistons. When the piston is subject to cylinder gas pressure there is a tendency for it to cock (pivot off a vertical centerline) and load the contact faces off its axis on the pin.

thrust washer   see *thrust bearing*.

thyristor   three terminal solid state switches.

timing   the manner in which events or actions are sequenced. The term is used in many applications in engines relating to valves, injection, ignition, and others.

timing actuator   one of two ECM-controlled electrohydraulic solenoids used to manage timing on Cummins common rail, HPI-TP fueled engines. Duration of energization determines fuel quantity delivered into the timing chamber of an HPI-TP injector defining start of injection.

timing advance unit   a hydromechanical or electronically controlled timing advance mechanism used with port-helix metering pumps.

timing chamber   term used by Cummins to describe the upper chamber in its two-stage injector units such as STC, CELECT, and TP injectors.

timing check valve   device found in Cummins two-stage injector units that holds fuel in the timing chamber during downstroke.

timing bolt   means of locking a component or engine to position for purposes of timing.

timing dimension tool   a tune-up tool used to set the tappet height on MUIs and EUIs.

timing event marker (TEM)   Mack Trucks Bosch engine position sensor located in the rack actuator housing.

timing reference sensor (TRS)   crankshaft position sensor. The acronym was originally used by DDC and current versions of DDEC use a 36-tooth tone wheel.

tip turbine   a charge air cooler fan device driven by turbo-boosted air designed to blow filtered air through the heat exchanger and increase boost air cooling efficiencies.

tone wheel   the rotating disc used to produce rpm and rotational position data to an ECM. The term is most often applied to the rotating member of a Hall effect sensor, but at least one manufacturer uses the term to describe an AC reluctor wheel.

top dead center   outboard extremity of travel of a piston or plunger in a cylinder. Usually abbreviated to TDC.

top engine limit   a Caterpillar term for *high idle*.

top stop injector   a Cummins PTD injector in which plunger stroke is set internally and not as part of the injector train adjustment.

topology   the configuration of a computer network; a network schematic.

torched piston   a piston that has been overheated to the extent that meaningful analysis of cause is not possible.

torque   twisting effort or force. Torque does not necessarily result in accomplishing *work*.

torque rise   the increase in torque potential designed to occur in a diesel engine as it is lugged down from the rated power rpm to the peak torque rpm, during which the power curve remains relatively flat. High torque rise engines are sometimes described as *constant horsepower* engines.

torque rise profile   diagrammatic representation of torque rise on a graph or fuel map.

torsion   twisting force.

torsional stress   twisting stresses. A crankshaft is subject to torsional stress because a throw through its compression stroke will travel at a speed fractionally lower than mean crank speed, whereas a throw through its power stroke will accelerate to a speed fractionally higher than mean crank speed. These occur at high frequencies.

total dissolved solids (TDS)   dissolved minerals measured in a coolant by testing the conductivity with a current probe (TDS tester). High TDS counts can damage moving components in the cooling system such as water pumps.

total indicated runout (TIR)   a measure of eccentricity of a shaft or bore usually measured by a *dial indicator*.

total quality management (TQM)   a customer service philosophy that maintains that customer service is the core of all business activity and that every employee in an organization should be made to feel part of a team with a common objective.

tower computer   a PC housed in an upright case, usually capable of greater system expansion than the horizontal desktop style.

toxic   materials that may cause death or illness if consumed, inhaled, or absorbed through the skin.

trackball   the cursor control device often used in some notebook computers consisting of a rotating ball integral with the keypad.

train   a sequence of components with a common actuator. See *valve train*.

transducer module   (Caterpillar) component responsible for transducing all engine pressure values (such as oil/turbo boost, etc.) to electrical signals for ECM input.

transfer pump describes the fuel subsystem pump used to pull fuel from the fuel tank and deliver it to the injection pumping/metering apparatus.

transformer an electrical device consisting of electromagnetic coils used to increase/decrease voltage/current values or isolate subcircuits.

transient short lived, temporary; often refers to an electrical spike or hydraulic pressure surge.

transistors any of a large group of semiconductor devices capable of amplifying or switching circuits.

transmission control protocol/Internet protocol (TCP/IP) a standard for data communications that permits seamless connections between dissimilar networks.

transorb diode used to protect sensitive electronic circuits (such as an ECM) from the inductive kick that can be created by solenoids when their magnetic field collapses.

transponder a ground-based satellite uplink—can be either mobile or stationary.

transposition error a data entry error.

trapezoid a quadrilateral with one pair of parallel sides.

trapezoidal ring see *keystone ring*.

trapezoidal rod a connecting rod with a trapezoidal small end or rod eye designed to maximize the sectional area of the rod subject to compresive pressures; more commonly referred to as a *keystone rod*.

trapped volume spill (TVS) used on HPI-TP injectors, the TVS provides an escape route for fuel just as the lower plunger bottoms into the cup bringing an abrupt closure to the injection pulse.

triatomic a molecule consisting of three atoms of the same element.

tribology the study of friction, wear, and lubrication.

trilateration the locating of a common intersection by using three circles each with different centers: the mathematical basis of GPS technology.

troubleshooting the procedures involved in diagnosing problems.

Truck Maintenance Council (TMC) division of the ATA that sets and recommends safety and operating standards.

turbo boost sensor (TBS) a device used to signal manifold boost values to the ECM. An aneroid is used in hydromechanical engines and either a piezo-resistive or variable capacitance device is used in electronically controlled engines.

trunk-type piston a single piece piston assembly usually machined from aluminum alloys in diesel engine applications.

truth table a table constructed to represent the output of a multiswitch circuit, based on the switch status within the circuit.

turbine a rotary motor driven by fluid flow such as water, oil, or gas.

turbocharger an exhaust gas driven, centrifugal air pump used on most truck diesel engines to provide manifold boost. Consists of a turbine housing within which a turbine is driven by exhaust gas, and a compressor housing within which an impeller charges the air supply to the intake manifold.

twisted wire pair used as the data backbone in multiplex systems. The wires are twisted to minimize EMI. In a J1939 data backbone, the positive wire is color coded yellow and the negative wire color coded green.

two stage filtering any filtering process that takes place in separate stages.

typeface the design appearance of alpha characters.

ultra-low sulfur fuel (ULS) required in some jurisdictions. At this moment in time, ULS is required to have a maximum sulfur content of 0.0015%: some are advocating its adoption as the national standard for highway diesel fuel.

ultraviolet radiation having a wavelength just beyond the violet end of the visible light spectrum. Emitted by the Sun but much of it is filtered out by the ozone shield in the Earth's stratosphere.

ultrawide SCSI an SCSI unit that can support up to fifteen devices in any combination of internal and external devices. The last device in the chain must be terminated both internally and externally on shutdown.

undersquare engine an engine in which the bore dimension is smaller than its stroke dimension. Refers to high compression engines. Most diesel engines are undersquare.

universal product code (UPC) the commonly used commercial bar code designed for optical scanners.

universal serial bus (USB) a means of connecting peripherals (external devices) to a PC or network system.

uplink signal transmission from a stationary or mobile ground station to a telecommunications satellite.

upper helix a helix milled into the upper portion of a pumping plunger and giving the characteristic of a variable beginning of pump effective stroke.

upload the act of transferring data from one medium or computer system to another.

unit injector a combined pumping, metering, and atomizing device.

urea crystallized nitrogen base compound injected with water into the reduction stage of late technology exhaust catalytic converters.

user ID a password used for security and identification on multiuser computer systems.

vacuum restriction used to describe a restriction on the suction side of a fluid circuit; for instance, a plugged filter.

valence number of shared electron bonds an element can make when it combines chemically to form compounds. A valence electron is one in the outer shell of an atom.

validation 1. data confirmation/corroboration. 2. the third and final stage of a vehicle ECM reprogramming sequence in which a successful data transfer is confirmed to a central data hub.

valve any device that controls fluid flow through a circuit.

valve bridges a means of actuating a pair of cylinder valves with a single rocker; also known as valve yoke.

valve closes orifice (VCO) (nozzle) a sacless hydraulic injector nozzle.

valve closing pressure (VCP) the specific pressure at which hydraulic injector nozzle closes, always lower than NOP due to nozzle differential ratio. Also known as nozzle closing pressure.

valve float a condition caused by running an engine at higher-than-specified rpms in which valve spring tension becomes insufficient, causing asynchronous (out of time) valve closing.

valve margin   dimension between the valve seat and the flat face of the valve mushroom: critical valve machining specification.

valve pockets   recesses machined into the crown of a piston designed to accommodate cylinder valve protrusion when the piston is at TDC.

valve train   all the components between the cam and the valve and typically would include followers/tappets, push tubes/rods, rocker assemblies and valve bridges/yokes.

valve yoke   see *valve bridge*.

vaporization   changing the state of a liquid to a gas.

variable speed control (VSC)   Mack Trucks' electronic setting of rpm using cruise switches.

variable speed governor (VSG)   a governor in which the speed control/throttle mechanism inputs an engine speed value and the governor attempts to maintain that speed as the engine load changes.

variable speed limit (VSL)   rpm limiting on a moving vehicle for PTO operation.

vector   a straight line between two points in space; a line that extends from the axis of a circle to a point in its periphery.

VECTRO   Volvo electronic engine management.

VED-12   Volvo VECTRO managed, in-line 12 liter, 6-cylinder engine.

vehicle electronic programming system (VEPS)   Caterpillar and International initial ECM proprietary programming, usually performed on the assembly line.

vehicle management and control   see V-MAC. Mack Trucks chassis management electronics currently in version III.

vehicle personality module (VPM)   (Caterpillar/Navistar) term used to describe the PROM and EEPROM memory components in a chassis management module.

vehicle speed sensor (VSS)

venting   the act of breathing an enclosed vessel or circuit to atmosphere to moderate or equalize pressure.

vial   calibrated cylindrical glass vessel (test tube) used for precise measuring of fuel delivery on a fuel injection pump, calibration bench. Also known as *graduate* or *buret*.

video display terminal (VDT)   a monitor or CRT.

videographics array (VGA)   the quality category of a video display monitor.

viscosity   often used to describe the fluidity of lubricant but correctly defined, it is a fluid's resistance to sheer.

visible image area (VIA)   the means of assessing the actual size of a video display monitor by measuring diagonally across the screen. A nominal 15″ monitor may actually measure only 13.5″.

volatile memory   RAM data that is only retained when a circuit is switched on.

volatile organic compounds   the boiled-off, more volatile fractions of hydrocarbon fuels. The evaporation to atmosphere occurs during production, pumping, and refueling procedures; also known as VOCs.

volatility   the ability of a liquid to evaporate. Gasoline has greater volatility than diesel fuel.

volt   a unit of electrical potential; named after Alessandro Volta (1745–1827).

voltage   electrical pressure. A measure of charge differential.

voltage drop   voltage drops in exact proportion to the resistance in a component or circuit. The voltage drop calculation is made to analyze component and circuit conditions.

volute   a snail-shaped diminishing sectional area such as used in turbocharger geometry.

Volvo Trucks   manufacturer of trucks and engines. Corporate center is in Sweden.

VP-44   one type of radial piston type pump manufactured by Bosch for common rail fuel injection systems.

VSS   vehicle speed sensor.

water in fuel (WIF) sensor   usually located in the filter/water separator sump, it signals an alert when covered with water.

water separator   a canister located in a fuel subsystem used to separate water from fuel and prevent it from being pumped through the injection circuit.

watt   a unit of power commonly used to measure mechanical and electrical power. Named after James Watt (1736–1819).

wavelength   frequency or the distance between the *nodes* and *antinodes* of a radiated or otherwise transmitted wave.

Weather Pack connector   a commonly used proprietary, sealed electric and electronic circuit connector system used in many electronically managed circuits.

Western Star   manufacturer of trucks owned by Freightliner. Corporate center is in Portland, Oregon.

wet liners   cylinder block liners that have direct contact with the water jacket and therefore must support cylinder combustion pressures and seal the coolant to which they are exposed.

white smoke   caused by liquid condensing into droplets in the exhaust gas stream. Light reflects or refracts from the droplets making them appear white to the observer.

wide open throttle (WOT)   a term usually used in the context of SI gasoline-fueled engines to mean full fuel request. Used by one diesel OEM to describe *high-idle speed*.

Windows   the Microsoft Corporation graphical user interface program manager widely used in PC systems.

wireless   the use of radio frequencies to transmit analog or digital signals.

wobble plate pump   slang term for a swash plate pump.

word processing   using a computer system to produce mainly text in documents and files.

word size   the number of bits that a CPU can process simultaneously, a measure of processing speed. The larger the word size, the faster the CPU. A 286 CPM is a 16-bit processor while a Pentium is a 64-bit processor.

work   when force produces a measurable result, work is accomplished.

Workplace Hazardous Materials Information System (WHMIS)   the section of the Hazard Communications legislation that deals with tracking and labeling of hazardous workplace materials.

wrist pin   the pin that links the connecting rod eye to the piston pin boss; also known as piston pin.

write once, read many (WORM)   optical disk that can be written to once (permanently) and read many times.

yield strength   unit force required to permanently deform a material: in steels, yield strength is approximately 10% less than tensile strength.

zener diode   a diode that will block reverse bias current until a specific breakdown voltage is achieved.